**International Geography
1972
La géographie internationale**

International Geography 1972
La géographie internationale

Papers submitted to the 22nd International Geographical Congress, Canada
Communications présentées au 22e Congrès international de géographie, Canada

Edited by / sous la direction de
W. Peter Adams and Frederick M. Helleiner
Trent University

Published for the 22nd International Geographical Congress
Publié à l'occasion du 22e Congrès international de géographie
Montréal 1972

University of Toronto Press

© University of Toronto Press 1972
Toronto and Buffalo

Printed in Canada

ISBN 0-8020-3298-2 (Cloth)
Microfiche ISBN 0-8020-0271-4

Preface
Préface

This two-volume collection of papers contains most of the written submissions to the various events of the Centennial International Geographical Congress held in Canada in 1972. This Congress is the quadrennial conference of the International Geographical Union, the major world-wide association of geographers. The first Congress was held in Antwerp, Belgium in 1871, and the most recent ones in Stockholm (1960), London (1964), and New Delhi (1968).

The volumes have been produced to provide a broad basis for the discussion-format proposed for this particular Congress. In order to strengthen this basis, authors were requested to submit 'short papers,' of no more than 1200 words including references, rather than the usual abstracts of their work. These papers provide delegates with a very substantial briefing on the principal topics of the meetings.

In addition to having an important function in connection with the Congress, these volumes, unlike the usual collections of conference abstracts, provide a remarkable overview of work in progress in geography around the world. The work of geographers in over 60 countries is represented, and the topics encompass the entire range of the spectrum of modern geography. The articles published here include those submitted to the central programme of papers in Montreal and substantial groups of papers submitted to special Congress events, Commission meetings and Symposia, on specialized subjects such as 'Water Resources' and 'Geography in Education.' We feel that the inclusion of references with most of the papers greatly enhances the value of this collection as an 'overview' of geography.

Preparing such volumes for use at a large international conference is an exceedingly complex operation. The editorial function must be performed within a wide range of constraints not normally encountered in collecting and publishing the works of others. Users of these volumes should therefore recognize the limitations and biases of the sample of geographical research represented by these papers.

Even without the emphasis on discussion

Cet ouvrage en deux volumes renferme la plupart des communications écrites soumises dans le cadre des différentes activités du Congrès international de géographie qui se tient au Canada en 1972. Ce congrès est la conférence quadrienniale de l'Union géographique internationale, la principale association de géographes du monde. Le premier congrès s'est tenu à Anvers, Belgique, en 1871, et les plus récents congrès à Stockholm (1960), à Londres (1964) et à New Delhi (1968).

On a préparé ces volumes dans le but de fournir un document de travail pour les réunions des diverses sections, commissions et symposiums du Congrès. Les auteurs avaient été invités à soumettre de cours textes, n'excédant pas 1200 mots, y compris les références, plutôt que les classiques résumés, afin que les délégués aient en main un important dossier sur les principaux sujets de discussion à leurs réunions.

En plus d'être un outil de travail important lors du congrès, et contrairement aux compilations habituelles de résumés des communications, cet ouvrage présente une vue d'ensemble exceptionnelle de la recherche géographique actuelle à travers le monde. On trouvera dans ces volumes les textes soumis en vue des réunions des sections, ainsi que plusieurs textes qui seront présentés ou discutés lors des activités spéciales du congrès, de réunions des Commissions, et des symposiums; ces travaux traitent de sujets spécialisés tels que 'Les ressources en eau' et 'La géographie dans l'éducation.' Nous croyons que cet ouvrage offre une vue d'ensemble d'autant plus importante que, dans la plupart des cas, une liste de références est incluse à la suite du texte.

Il est extrêmement compliqué de préparer un ouvrage en vue d'un congrès d'une telle envergure. On doit en effet tenir compte de multiples contraintes qui n'existeraient pas en temps normal dans la compilation et la publication de travaux collectifs. Les lecteurs de ces volumes doivent être conscients des limites de l'échantillon de la recherche géographique que représentent ces textes.

Au cours d'un tel congrès, où l'accent sera mis sur la discussion, il est impossible que

which is a special feature of this Congress, it would be impossible to have all worthwhile submissions presented orally. Instead, organizers of the various Congress events were encouraged to use this volume as a means of distributing acceptable papers submitted to them while building their actual Congress programmes around *a selection of those papers published*. Thus many excellent papers published here will not be presented, in the formal sense, at the Congress.

We hope that all authors who submitted papers to this Congress will be sympathetically aware of the difficulties faced by ourselves and by organizers of the programmes of papers involved. The event-organizers are dispersed widely over this rather large country and the sheer mechanics of correspondence between them and us, in addition to the complexities of communicating, in two languages, with over seven hundred persons scattered around the world, have inevitably resulted in mistakes and omissions. The absolute imperative of publication before the Congress has not simplified the task. Under other circumstances we would have consulted more closely with more authors before exercising our editorial prerogatives.

Given the problems presented by the scale and complexity of the Congress, we could not attempt to produce here a collection of papers which is homogeneous in style and quality but, rather, we have assembled a readable collection which will provide a useful working basis for the Congress and, even more, one which will form a valuable reference work after the Congress.

In preparing the volumes, our more important editorial objectives were: to publish 'short papers,' of about 1200 words including references (the majority of papers do not conform to this ideal but our editing has been undertaken with it clearly in mind); to publish in either French or English, as submitted; to include the name, initials, professional affiliation, and country of all contributors; to eliminate all except the most essential subheadings (this includes 'Acknowledgments' and 'References,' which appear, in this order, immediately following the text of each paper, as discussed below); to ensure that the text is 'readable' (no effort has been made to develop uniformity of style).

All papers were read for content and tech-

les auteurs puissent tous présenter oralement leur travail, aussi important qu'il soit. On a donc encouragé les organisateurs des différentes activités du Congrès à utiliser cet ouvrage pour diffuser les communications qui leur ont été soumises et pour élaborer leur programme autour d'un choix de textes publiés. Ainsi, plusieurs textes d'excellente qualité qui font partie de cet ouvrage ne seront pas présentés de façon conventionnelle au congrès.

Nous espérons que tous les auteurs qui ont soumis des communications écrites à ce congrès se rendront compte des difficultés auxquelles les organisateurs et les soussignés ont dû faire face. La complexité qu'implique la correspondance avec les organisateurs disséminés dans ce grand pays, ainsi que la nécessité d'entrer en communication dans deux langues avec plus de sept cent personnes à travers le monde, ont occasionné des omissions et des erreurs inévitables. La tâche n'a pas été simplifiée par le fait qu'il était absolument nécessaire de publier cet ouvrage avant le Congrès. En d'autres circonstances, nous serions entrés en contact avec un plus grand nombre d'auteurs avant de nous prévaloir de notre rôle de rédacteurs.

Compte tenu des problèmes occasionnés par un congrès de cette envergure et de cette complexité, il nous était impossible de publier un recueil de travaux dont le style et la qualité soient homogènes; nous avons plutôt compilé un ouvrage qui se lit bien, du moins nous l'espérons, et qui servira de document de travail utile pour le Congrès; par dessus tout, il constituera une importante source de références après le Congrès.

Les principales normes dont nous avons tenu compte dans l'édition de ces volumes sont les suivantes: la publication de courts textes d'environ 1200 mots, y compris les références (la plupart des études ne se conforment pas à cet idéal mais c'est à partir de cet objectif que nous avons dirigé notre travail d'édition); la publication en français ou en anglais, selon la langue utilisée; l'inclusion du nom, des initiales, de l'appartenance professionnelle et du pays de tous les collaborateurs; l'élimination de tout sous-titre à l'exception de ceux qui sont absolument nécessaires (y compris 'Remerciements' et 'Références' qu'on trouve dans cet ordre à la suite de chaque communication); le désir

nical language by the event organizers named at the beginning of each section. These persons, their selected referees, and Professor J. Warkentin, Chairman of the Programme Committee, were responsible for screening the papers.

Further details of the organization of the volume, numbering of articles, pagination, etc. are given below, and additional notes are included at the beginning of each major division of the books.

We gratefully acknowledge the assistance of: The event organizers named at the beginning of each section; Jill Adams; Douglas R. Barr; J.B. Bird, Chairman, Congress Organizing Committee, for consistently firm encouragement; R.I.K. Davidson, Lorraine Ourom, and other members of the staff of the University of Toronto Press; Neil Durford; J.K. Fraser, Congress Executive Secretary, and his staff, especially Janet McDonald and Louise Moore for devotion to duty; Nicole Gaudet; Lois Helleiner; Cathy Hewton; D.P. Kerr; Louise Lettelier; Department of Geography, Laurentian University, Canada; Judy Pack, for almost two years of careful work; Marion Robertson, for assistance with translation from the Russian; Louis Trotier; J. Warkentin.

W.P.A., F.M.H.
Trent University, Canada
June 1972 Juin

de publier des textes 'lisibles' (on n'a pas essayé d'établir un style uniforme).

Toutes les communications ont été examinées quant au contenu et au vocabulaire technique par les responsables dont le nom apparaît au début de chaque section de ces volumes. Ces personnes et les arbitres choisis par eux, de même que le professeur J. Warkentin, président du Comité du programme, se sont chargés d'examiner et de sélectionner les textes soumis.

On trouvera d'autres renseignements concernant la disposition de l'ouvrage, le numérotage des articles, la pagination, etc., sur une page supplémentaire au début de chaque chapitre.

Nous remercions sincèrement: les responsables dont le nom paraît au début de chaque section; Jill Adams; Douglas R. Barr; J. Brian Bird (président du Comité d'organisation, pour son encouragement sincère et constant); R.I.K. Davidson, Lorraine Ourom et autre personnel de l'University of Toronto Press; Neil Dunford; J. Keith Fraser (secrétaire exécutif du Congrès, et son personnel, tout spécialement Janet McDonald et Louise Moore pour leur dévouement; Nicole Gaudet; Lois Helleiner; Cathy Hewton; Donald Kerr; Louise Letellier; Département de géographie, Université Laurentienne, Canada; Judy Pack, pour presque deux ans de travail consciencieux; Marion Robertson, pour son aide dans la traduction de travaux en russe; Louis Trotier; J. Warkentin.

Users' Guide
Guide de l'usager

There are fifteen major divisions of these volumes, one for each of the thirteen 'Sections' of the International Geographical Congress Programme in Montreal, a fourteenth for *all* Commission meetings held in conjunction with the Congress, and a fifteenth for *all* Symposia. These major divisions are listed in the Contents. The Symposia and Commission sections are subdivided by individual symposium and commission.

Papers are designated by the division, or subdivision, and the number of the paper within it. Because of the exigencies of the processing and printing schedules, the papers are *not* in alphabetical order within divisions, although they may, in places, appear so.

The first thirteen divisions have the prefix P (for *Programme*) so that, for example, P01 is the first division (Geomorphology) and P13 is the thirteenth division (Remote Sensing). Thus paper P0127 is the 27th paper in Programme Section 1.

Commission papers bear the prefix C, and Symposium papers the prefix S. Thus, C0113 and C0801 are respectively the 13th paper in Commission 1 and the first paper in Commission 8.

The titles of the thirteen Programme divisions are self-explanatory and provide the simplest means of access to the papers in the collection. At the beginning of each of these divisions is a list of the papers which it contains, in numerical order, by short title and first author. Contributors were asked to include key words in their titles so that a scan of these lists should provide a useful indication of the content of the papers.

In addition, at the beginning of each major division, we have noted groups of papers located elsewhere in the volumes, of interest to readers of that division. For example, at the beginning of P01 (Geomorphology), we draw attention to S05 (Karst Geomorphology), a collection of papers devoted to this one geomorphological topic.

To complement the list of titles and authors at the beginning of each division, we have included two indexes at the end of each volume. One is an index to paper number by author and co-author, and the other an index

Ces volumes sont divisés en quinze grands chapitres. A chacune des treize 'Sections' du Programme de Montréal correspond un *chapitre*; *tous* les textes soumis en vue des Commissions se trouvent dans le quatorzième chapitre; le quinzième chapitre comprend les textes soumis à tous les Symposiums dans le cadre du Congrès. Une liste de ces chapitres apparaît dans la Table des matières.

On a assigné un sigle à chaque texte selon le chapitre du livre où il se trouve et selon son numéro d'ordre à l'intérieur de ce chapitre. A cause de l'énormité du travail et des échéances, les textes ne sont pas en ordre alphabétique dans chaque chapitre de l'ouvrage.

Les treize premiers chapitres sont identifiés par le sigle P, pour *Programme*: P01, par exemple, réfère au premier chapitre, qui est réservé à la géomorphologie, et P13 réfère au treizième chapitre, qui porte sur la télédétection. La communication P0127 est donc la vingt-septième dans la Section 1 du Programme.

Les communications relevant des Commissions sont identifiées par la lettre C, et celles des Symposiums par S. Ainsi, C0113 et C0801 sont respectivement la treizième communication de la Commission 1 et la première de la Commission 8.

Les titres des treize premiers chapitres ne requièrent aucune explication; ils permettent de trouver sans difficulté les textes qui se trouvent dans l'ouvrage. Au début de chaque chapitre, il y a une liste des textes par numéro d'ordre d'après le titre abrégé et l'auteur. On a demandé aux collaborateurs d'inclure des mots clés dans le titre, ce qui permettra de déterminer plus facilement, en examinant la liste, de quel sujet traite chaque étude.

De plus, au début de chaque chapitre, on mentionne les communications qui se trouvent dans une autre section de l'ouvrage, mais qui peuvent intéresser le lecteur de ce chapitre. Au début de P01 de géomorphologie, par exemple, nous attirons l'attention sur S05 (Géomorphologie karstique), où l'on trouve une série d'études traitant spécifiquement de ce sujet de géomorphologie.

of the geographical locations to which papers relate.

To save space, we have reduced all acknowledgments to a simple list immediately following the text of an article. Individuals' titles and explanations of the contributions of those acknowledged have been deleted.

It should be noted that in some cases the 'references,' which follow immediately after the 'acknowledgments,' may be references *sensu stricto*, whereas in other cases they might be better described as 'general bibliography' in that they may contain more than the references actually cited.

Afin de compléter la liste des titres et des auteurs au début de chaque chapitre, nous avons inclus deux index à la fin de chaque volume. Le premier est un index par auteur et coauteur; l'autre index énumère les lieux géographiques dont traitent les communications.

Pour économiser l'espace, tous les remerciements se réduisent à une seule liste qui paraît immédiatement à la suite du texte. On a retranché les titres des personnes et les détails concernant la contribution de ceux qu'on remercie.

Dans certains cas, il est fort possible que les références apparaissant à la suite des 'remerciements' soient des références *sensu stricto*; dans d'autres cas, on devrait plutôt les qualifier de 'bibliographie générale,' car on y trouve d'autres références que celles qui avaient été citées par l'auteur.

To find a topic or paper of interest to you
1. Scan the Contents page for the most appropriate major division of the volumes.
2. Scan the list of titles and authors at the beginning of that division.
3. Find the paper by its number (e.g. P0101, P0102 are the first and second papers in the first division).

To find an author use the author index at the end of each volume, where the papers are cited by number as in item 3 above.

To find papers related to a particular geographical location use the locational index at the end of each volume, where papers are cited by number as in item 3 above.

Note that the bibliography following a paper may often contain more than references actually cited.

Pour trouver un sujet ou une communication qui vous intéresse
1. Examinez la Table des matières afin de trouver le chapitre approprié.
2. Examinez la liste des titres et des auteurs au début de ce chapitre.
3. Trouvez le texte que vous cherchez en utilisant le sigle (p.e. P0101, P0102 sont les première et deuxième communications dans le premier chapitre).

Pour trouver un auteur, consultez l'index par auteur à la fin de chaque volume où les textes sont énumérés comme dans la section 3 susmentionnée.

Pour trouver des communications traitant d'un lieu géographique spécifique, consultez l'index des lieux à la fin de chaque volume où les textes sont énumérés comme dans la section 3 susmentionnée.

A noter: *il est fort possible que la bibliographie à la fin d'un texte contienne d'autres références que celles qui avaient été citées par l'auteur.*

Contents
Table des matières

VOLUME 1

Preface / *Préface* v
Users' Guide / *Guide de l'usager* viii
1 Geomorphology / *Géomorphologie* 1
2 Climatology, Hydrology, Glaciology / *Climatologie, Hydrologie, Glaciologie* 127
3 Biogeography and Pedology / *Biogéographie et Pédologie* 245
4 Regional Geography / *Géographie régionale* 321
5 Historical Geography / *Géographie historique* 393
6 Cultural Geography / *Géographie culturelle* 473
7 Political Geography / *Géographie politique* 499
8 Economic Geography / *Géographie économique* 525
9 Quality of the Environment / *Qualité du milieu* 629
   Index of Authors and Co-authors / *Index des auteurs et coauteurs* xiii
   Selected Index of Locations / *Index des noms de lieux* xxi

VOLUME 2

Preface / *Préface* v
Users' Guide / *Guide de l'usager* viii
10 Agricultural Geography and Rural Settlement / *Géographie agraire et Peuplement rural* 695
11 Urban Geography / *Géographie urbaine* 789
12 Geographic Theory and Model Building / *Théorie géographique et Elaboration des modèles* 891
13 Remote Sensing, Data Processing, and Cartographic Presentation / *Télédétection, Traitement des données et Représentation cartographique* 963
14 Commissions / *Commissions* 1001
15 Symposia / *Symposiums* 1231
   Addenda 1351
   Index of Authors and Co-authors / *Index des auteurs et coauteurs* xiii
   Selected Index of Locations / *Index des noms de lieux* xxi

P10
Agricultural Geography and Rural Settlement
Géographie agraire et Peuplement rural

CONVOCATEUR/CONVENOR: John McClellan, *Brock University, St Catharines*

This listing of short titles and first authors' surnames will assist in identifying articles, topics, and places of interest. A complete author and co-author index is located at the end of this volume, as well as a selected index relating papers to geographical locations. Note that the papers are *not* listed in alphabetical order. The organization of the volumes is described in full in the Preface.

Other papers of particular interest to readers of this section are to be found under c06 (Humid Tropics), c07 (Man and Environment), c09 (Agricultural Typology), c11 (Geography of Arid Lands), c12 (Population Geography), c15 (World Land Use Survey), c17 (Medical Geography), and s04 (Frontier Settlement on the Forest/Grassland Fringe).

Cette liste des titres abrégés et des noms des auteurs principaux permettra d'identifier les communications, les sujets et les lieux qui présentent un intérêt quelconque. Un index complet d'auteurs et de coauteurs se trouve à la fin de ce volume, ainsi qu'un index des lieux géographiques. Prière de noter que les textes ne sont pas classés par ordre alphabétique. On explique en détail le plan de ces volumes dans la Préface.

D'autres études qui intéresseront peut-être les lecteurs de cette section se trouvent dans c06 (Les tropiques humides), c07 (L'homme et son milieu), c09 (Typologie de l'agriculture), c11 (Géographie des pays arides), c12 (Géographie de la population), c15 (L'utilisation du sol dans le monde), c17 (Géographie médicale) et s04 (L'aménagement du territoire en régions frontalières prairies/forêts).

p1001 Non-urban settlements of southern Canada and Europe north of the Alps and Danube ZABORSKI 697
p1002 Physical and cultural ecology of rural settlements, western-central Himalaya KAUSHIC 699
p1003 Changes in rural settlement structure AMIRAN 700
p1004 Possibilities for typological studies of American agriculture ANDERSON 702
p1005 Behavioural aspects of rural settlement (Georgia, USA) BOHLAND 704
p1006 Transformation des paysages ruraux en paysages touristiques, Canton de Valais, Suisse BUGMANN 706
p1007 Agricultural land-use planning, village Banwaripur district, Meerut, U.P., India CHAUHAN 707
p1008 Multivariate approach to analysis of changing crop-livestock regions (South Carolina, USA) COWEN 709
p1009 Rural resettlement in Singapore DE KONINCK 711
p1010 High diversity alternatives in tropical landscape management DICKINSON 713
p1011 The green revolution in India DUTT 714
p1012 New agrarian settlements on the Polar Circle in the marshwoods near Rovaniemi, Finland ENDRISS 717
p1013 Changing farming patterns in southern prairies of Canada FLOWER 718
p1014 Features of rural population in Mexico AGUILAR 720
p1015 The quasi-plantation, a conceptual model (USA) GREGOR 722
p1016 Feasibility of an agro-industrial complex on the Gaza-Sinai sea coast HAUPERT 723
p1017 Development of services for rural population in a peripheral region (Finland) HELLE 725

696 / Agricultural Geography

P1018 The tea culture system of East Asia HUNG 727
P1019 Efficiency of tropical, small farm agricultural practices INNIS 729
P1020 Chinese mining settlement: features of an intrusive pattern in Southeast Asia JACKSON 731
P1021 Changes in distribution of agricultural land, Denmark KAMPP 733
P1022 Hotel food supply systems (Jamaica) LUNDGREN 734
P1023 Perceptional change in attitudes toward rural land use, Southern Ontario MICHIE 737
P1024 Agricultural regionalization: origins and diffusion in the Upper South before 1860 (USA) MITCHELL 740
P1025 Types of agricultural land (India) NITZ 742
P1026 Recent rural Georgian abandonment as an adjustment (USA) PAYNE 744
P1027 Types of agriculture in Ecuador PECORA 746
P1028 Multiple land use, Mälar valley (Sweden) PORENIUS 748
P1029 Critères et indices de la typologie de l'agriculture mondiale RAKITNIKOV 750
P1030 Crop combination regions of Chhattisgarh Basin, India PANDA 752
P1031 Pattern of rural land use, Malehra village (India) SAXENA 754
P1032 Delimitation of the food productivity regions of India SHAFI 756
P1033 Tobacco cultivation, an alternative to chena cultivation, Walapane Division, Ceylon SILVA 758
P1034 Indigenous settlement planning in Rajasthan (India) SINGH 759
P1035 Intra-village space: Varanasi District, India SINGH 761
P1036 Canal irrigation and agriculture, Gangapar plain, UP, India SINGH 763
P1037 Recent changes in land-utilization, Saryupar plain, India PANDEY 764
P1038 Rural settlement regions at the ecumene's edge: Europe and North America STONE 767
P1039 Effect of productivity and environment on rural population density STOUSE 769
P1040 Land policy, farm migrations, and rural planning, Nigeria UDO 771
P1041 Two settlement simulations WALKER 773
P1042 Rural settlements in ancient central India NIGAM 774
P1043 From aeroplanes to agriculture (Canada and UK) FRASER 777
P1044 Irrigation agriculture in Sind, Pakistan RAHMAN 778
P1045 Land utilization maps in planning economic complexes SHOTSKIY 780
P1046 Calculation of gross margin in agriculture and the productivity of arable farming, Finland VARJO 783
P1047 An isolation index for Queensland cattle stations (Australia) WEIR 786

P1001
Comparison of non-urban settlements of southern Canada and of the European mainland north of the Alps and the Danube
BOGDAN ZABORSKI *Sir George Williams University, Canada*

Physiographic conditions in southern Canada roughly resemble those of northern Europe, but their human history is quite different. There are several similarities and many differences in the settlement patterns which we shall discuss.

Some of the agricultural villages in Europe are supposed to have existed in the neolithic epoch. A large majority of the others were created before the era of industrialization.

The pre-17th century settlement in Canada plays an insignificant role in the present-day human cultural landscape. The European farm settlements started in the 17th century in the Maritime provinces, Newfoundland, and Quebec, towards the end of the 18th century in Ontario, in British Columbia in the 19th century, and towards the end of the 19th century in the Prairies. The pioneers settled mainly along the Atlantic coast, lakeshores, the banks of the lower St Lawrence River and its major tributaries. At the beginning of the 18th century they numbered about 25,000, approximately 85,000 by 1763. Towards the end of the 18th century about 40,000 of the refugees from the south, known as United Empire Loyalists, reinforced Canada's English-speaking population. The bulk of the European population, however, settled in Canada in the 19th and 20th centuries. The farming ecumene adopted its present day extension in the 1930s. Settlements created after this date were mainly mining and recreational.

There is an essential difference between the settlements which were established and grew spontaneously and those of a planned community. The first ones grew uncontrolled by the authorities in an irregular way. Small settlements of this type are called hamlets. They are common in Europe, especially on a morainic landscape. In Canada they occur mainly in Newfoundland and the Maritime provinces. In Europe large-sized irregular multi-street villages are indicators of the most ancient settled areas in open lands. They cover extensive areas in France, Germany, Hungary, Rumania, and the Ukraine. After the withdrawal of the Turks, in the 18th century, villages on the lower Tisa plain were reconstructed in a geometrically regular form, with a grid pattern. There are no comparable farm villages in Canada. Villages composed of a couple of streets, situated on both sides of a stream in the form of a fork or of a ladder, are common in eastern Europe, but scarce in Canada. Round and oval (or spindle-like) villages, which are found in central Europe east of the Elbe River, are absent in Canada.

Common sense and the sound instinct of settlers guided them often to locate the village on river terraces along the slope. The resulting form of the settlement is an elongated village called a street village. In Europe small size of scattered fields, scarcity of wells and social habits are often responsible for the compactness of the settlement. Several Mennonite villages in the Canadian Prairies adopted for social reasons the form of compact street villages.

Medieval colonization of the forests in the Carpathians, Sudets, and Ore Mountains in the 13th and 14th centuries was planned. Elongated villages followed the axis of the valleys and thus repeated their sinuosities. This colonization is the fruit of the epoch and has no counterpart in Canada.

The land survey in mid-16th century Poland and Lithuania resulted in the consolidation of fields, construction of grids of roads and regular street villages in the form of elongated rectangles. It is the only early example of rural road grids found in central Europe which compares in part to the large-scale Canadian surveys of land.

Farming on the wet floors of the river valleys was started in the Low Countries. Dams, parallel to rivers, were built. Numerous drainage creeks, transverse to the river and dam took away the surplus of water. Creeks also represented property boundaries. Lines of farmsteads followed dams with roads on the crest. This type of village, known as Marschhufendorf or row village, expanded in the 16th century to northern France, but also across northern Germany to Poland. This scheme was brought from northern France to Canada and settlers who colonized the moist St Lawrence valley followed it. Thus within the framework of the seigneuries the 'rang' village originated and greatly resembled European row villages.

British administrators liked the plan of the row village with its farmsteads arranged along one straight road, but instead of imitating the seigneuries introduced their own system. Towards the end of the 18th century a great land survey started in Ontario. The terrain was covered by grids of mutually perpendicular roads, which also represented limits of particular townships – small units of land, which in turn were subdivided into Plots, corresponding to future farms. One century later the Canadian Prairies were covered by a dense network of roads which divided the land into square miles. Scattered farmsteads were established on quarter mile sections. The colonization of the Canadian prairies by farmers, mostly newcomers from Europe, was done during the first three decades of the present century.

This pattern has an equivalent in Europe in Westphalia, central Poland, and Lithuania. However, in none of these regions does the road scheme resemble that of the Canadian Prairies.

In most of Poland and the Baltic countries farmsteads have been built of timber, often in the form of log cabins which are easy to dismantle and transport. This was an important factor encouraging agrarian reforms and the reconstruction of villages. The density of population was great and still growing, and in its early stage industrialization could not absorb the surplus farm population. The demand for land on the part of small farmers was great. Many big estates were divided into small plots to be sold to small-holders.

To make farming more efficient in the existing villages the holdings have been consolidated: each owner received one piece of land instead of dozens of long, narrow scattered fields. Both reforms encouraged settlement in the form of scattered farms or in the form of row villages. This system has been considered more progressive, as the farmstead was approximately in the centre of the farmland and safer in the case of a neighbourhood fire.

Those regions of Europe covered with loosely built settlements resemble the farm patterns of Ontario, parts of Quebec, and the Canadian Prairies, but differ greatly as to the road grid. The Canadian system, as created by the big land survey, has only a few regions in Europe for comparison.

During the last quarter century the Soviet Union's administration started the transformation of dispersed villages in the newly incorporated Baltic States into compact villages. The agglomerated settlements are considered, by the Soviet Union, more appropriate for collective farming and more easily controlled by the administration.

Blomkvist, E.E., 1956 *Peasant settlements of the Russians, Ukrainians and Beloruthenians*, East Slavic Ethnographical Series (in Russian) (Moscow), 3–458, esp. 27–59.

Demangeon, A., 1927 La géographie de l'habitat rural, *Annales de Géographie*, 36.

Jackson, D., 1948 A Geographic Study of the Early Settlement of Southern Ontario, unpublished MA thesis, U. Toronto.

Kielczewska-Zaleska, M., 1969 *The Geography of Settlement* (Warsaw) (in Polish).

Klute, F., ed., 1933 *Ländliche Siedlungen in Verschiedene Klimazonen* (Wroclaw).

Krenzlin, A., 1952 Dorf, Feld und Wirtschaft im Gebiet der Grossen Täler und Platten Östlich der Elbe, *Forschungen zur Deutschen Landeskunde* (Remagen).

Mackintosh, A., and W.L.G. Joerg, eds., 1936 *Canadian Frontiers of Settlement* (Toronto).

Trewartha, G.T., 1946 Types of rural settlements in colonial America, *Geog. Rev.*, 568–96.

Zaborski, B., 1926 *Sur les formes de villages en Pologne et leur répartition*, Comité ethnographique de l'Academie polonaise des sciences (Cracow). (Polish with a French summary; German trans., Osteuropa Institut, 1930.)

Zaborski, B., 1972 *Atlas of Landscapes and Settlements of Eastern Canada* (Montreal), 63–73 and 160–1.

P1002
Physical and cultural ecology of rural settlements in western-central Himalaya
S.D. KAUSHIC *Meerut University, India*

The physical and cultural ecology of rural settlements in the Himalaya has five aspects, which make up one complex whole. The morphology and processes of rural settlements, their location, size, pattern, layout, functions, socio-economic tempo, house types, building materials, and architectural style are governed by the synthetic-ecologic setting of the people-region-resources-adjustments-and-values. (i) The physical milieu (spatial distances, relief, configuration, climate, water-supply, soils, flora, and fauna) governs the origin, distribution, and density of settlements. (ii) The population of the region is the 'seed' and the 'region' the 'plot' of habitation, in which people adjust to the milieu according to their socio-cultural values and the stage of succession. (iii) Regional resources govern the density and stage of progress. (iv) The tempo of group-economy works as the 'urge' of growth, and the cultural milieu and stage of economy and technology serve as the 'nurture' for further growth and development. (v) Periodic changes in history and successive socio-economic trends mould the 'age-cycle' of the habitations.

Himalayan settlements are, thus, related to both physical and socio-cultural processes, yet the predominant governing factors of the settlement-ecology are (i) relief, (ii) gradient of the slope, (iii) climate, (iv) sunny or shady aspect of the slope, and (v) soil.

The topography of the region is highly complex. The Zaskar Range, the Great Himalayan Range, the Dhaula Dhar Range and Pir Panjal, and other Lesser Himalayan ranges cross the region diagonally, in a NW-SE direction. A criss-cross of hundreds of spurs has created a network of valleys over the whole area. Thousands of streamlets pour their waters into the main rivers, which have cut deep gorges.

Concave and gentle slopes (below 30°), alluvial fans, tillite fans, and river-flat terraces have comparatively deep soil deposits, and they provide the most suitable sites for human settlement and agricultural field-terraces. Convex slopes are either made gentle by back-breaking human labour, or avoided. Scarp slopes are avoided or used for pasture.

Climate determines the zones of permanent and itinerant settlements. (i) The subtropical warm temperate zone of the Lesser Himalaya, below 1800 metres altitude, is the zone of maximum permanent settlement. (ii) The cool temperate zone between 1800 and 2400 metres is rather sparsely populated, with rural settlements lying along the river valleys only. (iii) The cold zone of the Main Himalaya has an alpine climate with seven or eight months of winter snowfall. This zone is inhabited by the Bhotia pastoral people, who practise agriculture as a subsidiary occupation. The climate forces the people to migrate with their herds of cattle, sheep, and goats to warmer valleys below 1800 metres in the winter, but they live in the alpine zone during the summer. The people have two or three sets of permanent seasonal habitations.

Villages and hamlets are situated on mid-slopes of spurs running from ridges, or along the valleys of rivers and streamlets. The settlements on slopes have a stair-pattern, whereas the houses built on river-flat terraces are of a compact amorphous pattern although some houses can extend above the flat terrace on to the slope, assuming again a staircase pattern. For every habitation, a sunny aspect is preferred to a shady one, the latter being left for forests and pasture. Gentler slopes provide easier travel, particularly because all loads of corn, fodder, grass, fuel, dung, and litter are carried on the human back.

Villages are situated below the forests, neither too far for transporting fuel and fodder, nor too near to discourage visits of wild beasts. Cold and snowy ridge-tops are avoided. The forests supply timber, fuel, fodder, as well as wild fruits, and pasture, but they do not provide land for cultivation unless they are burned and cleared. The strip of newly cleared forest is called *Katil* and is used for intermittent cultivation; it is unterraced. Below the *Katil*, there are field-terraces given to hilly millets depending on rainfall. These are called *Upraon* field-terraces. Crops of *koda* (*Elucine coracana*), *jhangora* (*Panicum frumentaceum*), *kauni* (*Panicum italicum*), *ogala* (*Fagopyrum esculentum*), and *phaphra* are grown on these dry-cultivation terraces.

Below the *Upraon* field-terraces lies the residential part of the villages, near any source of water-supply, such as a streamlet or spring. Below the habitational part lie the irrigated field-terraces, which are given over to paddy, potato, beans, *gahath* (*Dolicus biflorus*), and a little mustard in some villages. Irrigated paddy terraces lie in the lowest zone of the villages.

Most houses are of two storeys. The lower storey serves as cattle-stall and store of fodder and fuel. In the front part of a house is a courtyard for drying cereals or threshing out grain. It is also used for basking in the sun, spinning wool, weaving woollen cloth, and gossiping.

Five architectural styles are used for houses: (1) plain two-storeyed house; (2) three-storeyed house of *Dandayala* type; (3) *Evan* type two-storeyed house; (4) *Tibari* type house; and (5) poor man's hut type. The types will be illustrated by diagrams and colour-slides. The roofs are sloping. Building materials are rubble, stone, slate, and wood.

The people of this region are very superstitious. The overwhelming environment overpowers the human mind. The stupendously high ranges cut man off from the outside world; the dense forests, roaring streams, horrifying landslides, and terrific blizzards create an atmosphere of vague consternation.

The eerie sounds indicate the presence of some spirit world. Every peak and streamlet has been dedicated to some deity. Deities are invoked to bring rainfall or cause bumper harvests. They are invoked at the time of sowing, transplanting, harvesting, and storing. There are family-deities, village-deities, and tract-deities for a group of several villages.

Certain magical rituals are performed when a new house is built, or the bridegroom goes to the house of the bride, or the bride comes to her husband's house. Magico-religious flags of family-deities are put up on every house, in the zone of summer residences, to scare away evil spirits and ghosts and bring fortune to the family.

In this region, the social organization of community life has been very sound for many centuries. In spite of social castes (of farmers and artisans) the solidarity of the whole village as one community has been absolutely perfect in all villages. People of the entire village cooperate in schemes of maintaining irrigation *kuls*, footpaths, and mule-tracks.

To mitigate the drudgery of Himalayan life, there are periodic social gatherings and festivals. In religious fairs, all the people of a village (or group of villages) participate in deity-worship, which is accompanied by feasting, music, and dancing. Music and dance are the life-breath of the Himalayans.

P1003
Changes in rural settlement structure
DAVID H.K. AMIRAN   *Hebrew University, Israel*

This paper deals with rural settlement in its wider context, i.e., both villages and associated service towns. It, therefore, refers to the whole settlement fabric based on agriculture, as distinct from settlement based on non-agricultural functions.

From the point of view of settlement geography, the twentieth century is distinguished by the transformation from agricultural to non-agricultural functions. This process develops at different degrees of intensity in various areas. In the countries which have progressed furthest along this trend, less than 10 per cent, and in recent years even less than 5 per cent, of the working population are engaged in agriculture. At the same time the whole range of modern agricultural technology permits this vastly reduced labour force to produce an increasing volume of agricultural goods.

Throughout history and until the turn of the last century agriculture had been man's primary activity. Consequently, the vast majority of settlements in all countries were villages and farmsteads, housing the agricultural population and a range of service towns of different density and hierarchical structure which as *central places* provided the necessary services for the agricultural population. A specific and well-developed type of such a hierarchical pattern was analysed by Christaller, and served as take-off for central place theory (Christaller 1933).

The change from agricultural to non-agricultural employment makes an increasing part of the existing settlement structure obsolescent. It brings about a whole range of adaptations, listed as follows.

Migration of rural population, including that from towns in rural areas, to the large cities. In many countries the largest metropolitan cities with their strong attractions are absorbing a high percentage of migrants from rural areas (Amiran 1971; Jefferson 1939). In some of the countries where the rural to urban migration is the most intensive and the gap between most rural and some urban standards of living is the most glaring, the majority of these migrants end up in central or peripheral slums where living conditions are barely better than the rural misery from which they wanted to escape – and sometimes worse.

As far as the rural area is concerned, this brings about the *depopulation and regression* of some villages and towns. For obvious reasons this process is selective. It will affect marginal, distant, and isolated areas more strongly than those well integrated into the settlement fabric. In extreme cases it leads to abandonment of individual settlements or of settlement areas. Generally it brings about regression of villages and towns so affected. They lose some of their manpower, usually those with the greatest initiative, and some of their economic force. Consequently, the towns lose some of the volume and at certain stages even range of the demand for services. The result of this process of regression is a deterioration in services and in quality of the population, as well as a deterioration in opportunities for the remaining population. All these factors in turn serve to aggravate and accelerate the process of depopulation and regression in those rural areas affected by them.

On the positive side de-agriculturization leads to an increasing *occupational polarity of village populations*. Part of them become highly specialized farmers employing all the paraphernalia of modern mechanized science-based agriculture, very different from the simple peasants of former generations, who still form the majority of the agricultural population in many less advanced countries and thus the majority of people in the world at the present time.

But a growing majority of people in the villages become part of the non-agricultural labour force. This takes a variety of forms, partly dependent on the location of the rural area in question and its proximity to major urban and industrial centres of the nation.

Increasing numbers of people formerly employed in agriculture find employment in industry, manufacture, and services in nearby towns or in the area in general. They remain residents of the village for socio-economic reasons but become *commuting workers*. The pressure on housing in the fast-growing towns and its consequent high cost are a frequent reason for this choice. The environmental and social disadvantages of urban life versus the ease of accessibility are an additional potent factor. In extreme cases, erstwhile villages turn into dormitory settlements, especially those located near cities.

Introduction of non-agricultural branches of the economy into villages is common. First and foremost is industry. This has long-time antecedents in Central Europe, e.g., the Swiss watch industry, or village (home) industries elsewhere. In recent decades this trend has intensified and expanded considerably, with sizable industrial plants being located in villages and transforming them. The availability of land for plants and of a resident-labour pool are powerful incentives, as are the absence of industrial congestion and environmental deterioration, and incentives in taxation offered by the local authorities.

Of lesser importance is the development of the *resort industry* in villages. Here the village is making use of its rural and scenic environment as a resource base. Often the mountain village has a particular resort potential, offsetting in part its locational and technological disadvantages in agriculture. In a very general way, as the urbanization rate grows in a population so the demand increases for recreational facilities, and therefore the opportunity of villages to utilize this demand. It is important to note that recreation makes considerable demands on land use, both for areas directly used for recreational purposes, and in prohibiting deleterious uses in adjacent areas.

All these changes, eroding traditional village functions or transforming them, bring about basic changes in the structure of rural settlement. An increasing number of villagers adopt non-agricultural occupations. Many of these people are in regular working contact with larger towns, the more so as modern transportation makes the towns much more easily accessible than in the horse-and-buggy days of which Christaller wrote. The result is a decreasing importance of the service functions of the traditional central place in the rural settlement fabric, i.e., the obsolescence and consequent regression of many a small town.

These trends of development which spread to an ever larger number of nations should by the end of the century, for the first time in human history, lead to an occupational structure in which agriculture is no longer the first-ranking branch of occupation in the world's population.

As this situation is approached and will most probably intensify during the 21st century, our settlement structure based on villages as its main element is becoming obsolete. Agricultural villages will in future be accessories to the settlement fabric only, whereas towns, the population of which engages in non-agricultural occupations, will form its basic element.

This fundamental change in settlement structure which was initiated by the industrial revolution requires a thorough re-assessment of priorities in planning for land-use. A rational allocation of reasonably large continuous tracts of land for urban and for agricultural use, as well as reserves of recreational land adequate for the projected size of future populations should guide future planning. Agriculture can no longer have automatic first-call on land; but as far as possible the best soils should be set aside for agricultural use. Villages will have to yield to the needs of urban expansion, including such space-consuming installations as the expansion of airports. The high economic and social cost of these adaptations is part of the price man has to pay for basically changing his settlement structure.

Amiran, D.H.K., 1971 The structure of settlement: needed adaptations to change, *Geog. Helvetica* 26, 2–4.
Christaller, W., 1933 *Die zentralen Orte in Süddeutschland* (Jena).
Jefferson, Mark, 1939 The law of the primate city, *Geog. Rev.* 29, 226–32.

P1004
Possibilities for typological studies of American agriculture
JAMES R. ANDERSON   *University of Florida, USA*

The publication by the Commission on Agricultural Typology of a carefully organized and well-defined table of characteristics to be considered in agricultural typology presents an opportunity to analyse the possibilities for typological studies of American agriculture within the frame of reference being proposed (Kostrowicki and Tyszkiewicz 1970, 48). The main objective of this paper is to present: (1) a brief commentary on those characteristics which are most significant for understanding the agriculture of the United States and (2) to stress the difficulties at the *regional or national scale* in using some of the criteria recommended by the Commission. Some of the tentatively selected criteria are being tested by the author in a study of agriculture in the American South.

A country as large as the United States should have a strong interest in and a definitive need for typological studies of its agriculture. Over the past 50 years several studies dealing with the typology and regionalization of agriculture have been published. Baker published a series of articles and a map on the Agricultural Regions of North America in the late 1920s (Baker 1926). Although Baker's work was more of a regionalization than a typing of agriculture, he was very much concerned with the comprehensive analysis of the agricultural characteristics that must be studied in order to determine types of agriculture. Elliott and others, who prepared a detailed map on 'Type of Farming Areas in the United States,' refined and added to the prior efforts of Baker (Elliott 1933). Percentage of gross sales of major agricultural products was the main criterion used in those studies. Later, in the early 1950s, Weaver undertook a study of agriculture in the midwestern United States in which he used a more refined methodological approach in the analysis of the agriculture of that region (Weaver 1954).

More explicit methodological approaches in typological studies are needed and are now possible in the United States. Careful selection of those characteristics which have the most significance for American agriculture should be made. Furthermore, the data constraints for regional and national scale studies will need to be clearly recognized. The county is the basic areal unit for gathering statistical information on agriculture in the United States. Even though case studies and other selected approaches for the study of some characteristics can be used to supplement county-unit data, statistics from the US Census of Agriculture, which is made every five years, and data gathered every year on a sampling basis by the Statistical Reporting Service of the US Department of Agriculture must be the main sources of information.

Social characteristics recommended for consideration in typological studies at the regional or national levels do not give a really good notion of the control of agricultural land in the United States, which is mainly in private ownership except for areas of federal land used for grazing. The tenure status of the operators of such privately-owned land must be examined for meaningful typological analysis. Yet in many parts of the country there is a great mixing within counties of tenure types such as full owners, part owners, tenants, and managers, thus making meaningful typing difficult. The growing importance of the corporate farm and the incorporation of some family farms have made the family farm concept much less important and meaningful in recent years. The part-owner type has become increasingly important because of the need of operators to expand the size of their operating units by renting land. Size of farm contrasts markedly from place to place. However, the format of census information on size of farm presents problems in obtaining results for typological study that can be effectively interpreted.

Organizational and technical characteristics can be placed into two categories based mainly on the availability of statistical information. Only the sub-characteristics relating to the mechanization of agriculture and the density and degree of agricultural overpopulation lack significance as typological criteria in the United States. Agriculture is now so completely dependent on the tractor that only in the harvesting of some crops and in the care of livestock does this dimension offer an appreciable opportunity for differentiation with available county data on the use of mechanical power and family and hired labour. Agricultural overpopulation is no longer of great significance other than in a few areas where older persons remain underemployed on small farms.

Among the sub-categories of organizational and technical characteristics which present problems in typological studies in the United States because of lack or inadequacy of statistical data at the county level are: degree of fragmentation of agricultural holdings, permanent grassland types, percentage of land cultivated with particular implements, percentage share of particular land and/or crop rotation systems, permanent grassland use systems, percentage dependence upon permanent grasslands, fodder crops or forages acquired from other farms or counties, composition and age structure of productive livestock according to species and productive types of animals, and systems of livestock breeding.

In spite of these characteristics for which data are lacking or inadequate, there remains an impressive list of organizational and technical characteristics for which data are available for typological studies. Some of these data will lack comparability with data being collected in other countries; however, by careful definition and explanation it will be possible to use a significant amount of information on American agriculture for studies undertaken at a world scale.

Production characteristics of American agriculture can be studied quite effectively with information collected by the Bureau of the Census and the Department of Agriculture. It is not possible to obtain net returns from the production of specific crops and livestock products. Information on land productivity with the exception of the productivity in some animal categories can be analysed in a fairly refined way, although the analysis of such data does not of course approach the level of accuracy and sophistication possible in individual case studies of farm management. Labour and capital productivity are not so easily evaluated at the regional or national scale. Since nearly all agricultural production in the United States can be classed as commercial, the orientation or emphasis of production and the degree of specialization offer much more pertinent opportunities for study than studies dealing with the degree and level of commercialization.

Baker, O.E., 1926 Agricultural regions of North America: Part I – The basis of classification, *Econ. Geog.* 2, 459–93.
Elliott, F.F., 1933 *Type of Farming Areas in the United States* (US Department of Commerce, Washington).
Kostrowicki, J., and W. Tyszkiewicz, 1970 *Agricultural Typology: Selected Methodological Materials* (Instytut Geografii, Warszawa).
Weaver, J., 1954 Changing patterns of cropland use in the Middle West, *Econ. Geog.* 30, 1–47.

P1005
Behavioural aspects of rural settlement
JAMES R. BOHLAND *University of Oklahoma, USA*

An analysis of dispersed settlement in a portion of the United States suggests that the concept of spatial behaviouralism has applicability to the field of rural settlement. The purpose of this paper is to report the methodology and results of an analysis in which rural residential behaviour is associated with the formation of contemporary rural settlement patterns. The region in question is the northeastern corner of Georgia in the United States, an area where 'dispersed' settlement predominates.

Spatial behaviouralism is defined as 'any sequence of consciously or subconsciously directed life processes which results in changes in location through time' (Golledge 1969, 101). This concept provides geography with a research paradigm that has significant ramifications. Associating behaviour *with* rather than inferring *from* spatial patterns provides the geographer with the necessary methodology for developing a sound theoretical framework. Preliminary steps have been taken in settlement studies in relating residential behaviour to the formation of urban neighbourhoods; however, little has been accomplished in the rural sector.

From a methodological viewpoint a severe limitation on a behavioural approach in rural settlement is the absence of a systematic taxonomy for rural dwellings at something less than an aggregate area level. To associate behaviour with settlement patterns requires that both behaviour and dwelling types be identifiable at a singular level. Two popular techniques, nearest neighbour and quadrat analysis, illustrate the weakness of aggregate measures. While both enable the researcher to measure and test hypotheses concerning the random or non-random quality of dwelling dispersion, they deal with arrangements in relation to a specific areal level. Consequently, individual dwellings within the area cannot be differentiated utilizing either technique. Thus, we speak of a random pattern of dwellings within an area and not of a randomly located house.

Since aggregate pattern classification does not adequately differentiate specific dwelling types, it is suggested that Stone's designations of *single dwellings* and *dwelling clusters* be utilized (Stone 1968, 311). Single dwellings are defined as dwellings which are separated from their nearest neighbouring dwelling such that they appear to stand as spatial entities in themselves. Thus, they represent dwellings that are spatially isolated. Stress is placed on the concept of spatial isolation, for social or economic isolation is not implicit with single dwellings.

Clusters, in turn, are dwellings that are located in close proximity to one another so that two, three, or more dwellings are considered as one settlement unit. Subdivision of clusters by size, i.e., number of dwellings, further enhances the specificity of the classification.

While derived differently, the proposed taxonomy can be related to the dispersion levels of the aggregate methodologies. A *uniform* or *even* distribution as indicated by aggregate techniques would consist of a pattern composed entirely of singles; a *random* one includes an admixture of singles and clusters while *agglomeration* indicates a proliferation of clusters.

Identification of singles and clusters is dependent on determining a distance parameter that can differentiate the various dwelling types. Once the criterion is established, 240 metres in this case, the distance from one dwelling to its nearest neighbouring unit is

TABLE 1

Dwelling type	Average percentage in the study area	Maximum percentage within a single county	Minimum percentage within a single county
Singles	25.0	35.5	12.2
Clusters of two	15.4	19.3	8.4
Clusters of three	10.0	12.4	7.6
Clusters of four	6.8	7.1	5.4
Clusters of five	4.9	5.5	4.8
Clusters of six	3.9	3.7	3.7
Clusters of seven or greater	34.7	16.0	57.9

measured. Singles are those separated by a distance in excess of the stated value while clusters consist of several dwellings separated by a distance less than the critical value. Thus all dwellings within a cluster are at least within the specified distance but are spatially isolated from dwellings outside the group.

In northeastern Georgia a significant variation in the occurrence and distribution of rural dwelling types exists. By focusing on just single dwellings it was possible to isolate several important factors that were influential in creating the significant differences, both in frequency and distribution, that existed in the percentage of dwellings that were spatially isolated. To test the hypothesis that this spatial variation was a function of individual behaviour, a 5 per cent random sample of rural residents in each settlement type was selected. Interviews and tests were utilized to determine how the individual selected his specific site and situation, what were his perceptions and preferences of rural sites, and how he scaled these factors in reaching a decision regarding his dwelling location.

In combining interview results with an analysis of the settling process, two distinctive sets of factors that contributed to the spatial variance in singles were identified. The first were those over which the individual had no control (exogenous factors), i.e. what Golledge calls nonsensate behaviour (Golledge 1970, 5). The second consisted of goal-oriented behaviour (sensate) or instances in which the individual made conscious judgments of alternatives in his selection of dwelling sites.

The two significant exogenous factors were the density of dwellings and the forced movement and resultant abandonment of rural dwellings. While density affects the number of singles, the relationship being inverse, a significant variation existed between the frequency of singles predicted by a random density function and the actual number within an area. This lack of correspondence was due in part to the wholesale abandonment of dwellings, abandonment resulting from governmental expansion of recreational areas, and the expansion of corporate timber ownership. Both processes restricted the number of alternative sites available to the resident and, consequently, fostered the clustering of rural units.

While exogenous forces were important, the variance in singles was found to be primarily a function of variations in the behavioural patterns of residents regarding their selection of a rural site. Four different preference dimensions were recognized among the rural residents, and the individual's hierarchy in ranking these dimensions resulted in the differing occurrences and spatial patterns of singles. These four were: (1) an accessibility dimension; (2) a site aesthetic dimension; (3) a topographic perception dimension; and (4) a social interaction dimension.

The single dwelling resident ranked low in his desire for an accessible location, had a high preference for an aesthetic site, perceived topography in an aesthetic frame-of-reference, and expressed a low desire for a site that facilitated social exchange. Because of these preferences, the individual tended to select rural sites that were spatially removed from neighbouring dwellings. Residents in the larger clusters, on the whole, displayed the reverse pattern of preferences, a fact which increased the potential for the creation of large dwelling clusters in certain rural sectors.

Based on the evidence for northeastern Georgia, it is suggested that a residential behavioural approach has potential as a research methodology in rural areas. Additional studies of the rural resident's behaviour regarding the selection of his dwelling will lead to a greater understanding of the dynamic processes operative within rural environments.

Golledge, R.G., 1969 The geographical relevance of some learning theories, in *Behavioral Problems in Geography: A Symposium*, ed. by Kevin Cox and Reginald Golledge (Northwestern U. Dept. Geog., no. 17), 101–46.
– 1970 Process approaches to the analysis of human spatial behavior (Ohio State U., Dept. Geog., Discussion Paper no. 16).
Stone, K., 1968 Multiple scale classification for rural settlement geography, *Acta Geog*. 20, 307–28.

P1006
La transformation des paysages ruraux en paysages touristiques au Canton de Valais (vallée alpestre de la Suisse)
E. BUGMANN   *Société Suisse de Géographie*

Pendant les cent dernières années l'industrialisation croissante du plateau suisse effectuait une dislocation permanente de la population des alpes suisses dans les grands centres de la plaine.

Dans les alpes valaisannes, certaines stations touristiques classiques, comme par exemple Zermatt, Saas Fee et Montana, jouissaient pendant la même période d'un développement et d'une consolidation continus. A part les structures touristiques ces stations ont largement conservé un caractère rural.

Par suite d'une mentalité traditionnelle et conservatrice, des vallons et des régions entières ne s'étaient qu'à peine ouverts à l'établissement du tourisme. Pour cette raison on trouvait, il y a 30 ans, au Val d'Anniviers des villages montagnards intacts avec une économie rurale intensive : à St Luc par exemple on arrosait les prés jusqu'à la région alpine, on plantait du seigle à l'altitude de 1600m s.m. et la culture du vignoble au hameau de Muraz près Sierre formait une partie essentielle de l'exploitation traditionnelle. On trouvait ce nomadisme alpin extrême dans tous les villages du Val d'Anniviers, mais il s'est presque complètement perdu il y a une dizaine d'années environ. En même temps les villages ont perdu les jeunes qui préféraient travailler dans la plaine au lieu de continuer le dur travail de leurs parents.

Dès 1945 le gouvernement du Canton de Valais essayait d'arrêter l'émigration du peuple montagnard par l'établissement de l'industrie dans la vallée principale du Rhône. Les usines de la région de Sierre provoquaient une évacuation accélérée du Val d'Anniviers. La diminution de la population de St Luc entre 1941 et 1965 était énorme, c'est à dire que la plupart des émigrants changeait du domicile du village montagnard à la résidence vigneronne d'été à Muraz près Sierre. Un grand nombre d'entre eux restaient des bourgeois actifs et gardaient les devoirs et les droits à l'exploitation des forêts et des vignobles du village natal.

En 1965 St Luc forçait le développement du tourisme en construisant plusieurs remontées mécaniques, deux centaines de chalets et en ouvrant un office de tourisme. Environ 1800 lits ont été mis à disposition du tourisme familial. L'aménagement du territoire est achevé en vue de la création d'une station calme et récréative.

Au début le développement touristique ne pouvait pas retenir la population au village. De 1965 à 1971 le nombre d'exploitations rurales diminuait de 22 à 3 fermes. Cette situation de crise retenait la création des services centraux nécessaires pour une station de tourisme. La station se plaint d'un manque grave d'employés pour les besoins touristiques; par exemple on ne compte qu'un seul moniteur de ski à St Luc. Les autorités communales espèrent qu'une partie des habitants de l'ancienne résidence d'été à Muraz reviendra dans le cours des années prochaines à St Luc.

A l'été 1971 le gouvernement du Canton de Valais s'est décidé à soutenir l'installation de petits complexes industriels aux centres des vallons latéraux. Dès 1965 un atelier d'horlogerie à Vissoie (centre du Val d'Anniviers) occupe 45 ouvriers du vallon dont un seul de St Luc. Ce type d'industrie offre des places à la population montagnarde, mais il est favorable à un déclin forcé de l'agriculture alpine.

Des genres de maisons et de villages traditionnels, des formes d'exploitation ou même des réseaux de canaux d'arrosement sont sans doute une attraction spéciale pour le touriste. Une station avec l'atmosphère rurale est beaucoup plus attrayante que les stations artificielles comme on les a créées à Verbier, Anzère et Thyon et qui n'ont aucune relation à un noyau historique. Transformer des villages montagnards en villages touristiques exige de conserver l'élément rural. Le paysan attaché à la terre et au village garde les traditions. Il peut d'ailleurs être intéressé à un emploi touristique saisonnier. L'élevage du bétail garantit l'existence des pâturages ouverts qui sont indispensables pour les pistes de ski et les promenades ouvertes.

La transformation d'un village montagnard en une station touristique ne doit pas s'effectuer intégralement. A part les services spécifiquement touristiques l'entretien de l'élément rural est à projeter. L'élément rural forme une partie de l'infrastructure touristique. Il est nécessaire de soutenir et de garantir ce facteur important du développement touristique par des investissements financiers. Dans un village transformé en station récréative, le paysan a la possibilité de réduire l'intensité de l'exploitation en utilisant de grandes parties de terrain premièrement pour le pâturage et pour faire le foin pour l'hiver sans jamais l'arroser ni faire du fourrage vert.

Bär, O., 1969   Chandolin 1968, *Geog. Helv. Bern* 1.
Bugmann, E., 1972   St-Luc, ein Walliser Bergdorf, wird Touristen station, *Geog. Helv. Bern* 2.
Schwendener, Peter, 1965   Commune de St-Luc, aménagement local, Sion.

P1007
Benefits of agricultural land-use planning in village Banwaripur district, Meerut (U.P., India)
V.S. CHAUHAN and SURENDRA SINGH   *Meerut University, India*

Since land-use planning is the right and balanced allocation of land between rival claimants (Stamp 1960, 65), it is being strongly advocated in countries like India where the term 'over-population' is often applied. In this paper the benefits of agricultural land-use planning have been adjudged in a micro-areal study. Optimum land-use involves consolidation of scattered holdings (which, in India, are typical, and which are being consolidated as an integral part of the agricultural production program) and changes in cropping pattern: benefits occurring from them may be assessed with reference to two significant years, 1960–1 (in the pre-consolidation period) and 1968–9 (in the post-consolidation period).

The Village of Banwaripur, as a sample, depicts the typical agricultural environs found in the northwestern plain of India. It is situated (approximately 20°5′N and 77° 32′E) about 70 km northeast of Delhi. The village area is small, having an area of about 126 km². The usual monsoon climate prevails. *Kharif* and *Rabi* are the two main cropping periods, while *Zaid*, a third crop, is very limited. (These terms locally mean summer, winter, and rainy season crops, respectively.) Most of the rain (about 430mm) falls during *Kharif* and therefore irrigation is needed less. During *Rabi*, with lower rainfall (54 mm), agriculture mostly depends upon irrigation.

The soils of the village are broadly categorized into four types, namely *Gohra* (richly manured soil), heavy loam, light loam, and sandy loam. The groundwater table is usually at a depth of 4 to 5 metres. Of course, it rises by a metre during the rainy season.

The statistics showing over-all changes with regard to land-use and inputs in 1960–1 and 1968–9 point out the following facts. The irrigated area has increased by 17 per cent and now only 6 per cent of the NAS (Net Area Sown) is unirrigated. There is little change in the use of agricultural implements. The village does not possess any tractors. Total population has considerably increased (26 per cent) from 397 to 502. On the other hand agricultural population has decreased, which shows that new occupations other than agriculture have been introduced. Because of an increase in population, the area under settlement has also increased. NAS and ASMO (Area Sown More than Once) have increased by about 3 and 15 per cent respectively. The area under food grains has decreased in both *Kharif* and *Rabi* seasons, while that under sugarcane has increased enormously (42 per cent).

To estimate total benefits accruing out of this planning the land may be classified on the basis of returns available. Classification of land on the basis of crops grown, or per unit production, has been considered inappropriate in agronomic studies (Stamp 1958, 1–15; 1960a, 108; Shafi 1960, 296; 1967, 24–7) because on two equal pieces of land the one devoted to millet would produce more in weight than the one devoted to wheat, but wheat has more food value than millet, when this is conveniently calculated in calories. It is proper, therefore, to classify land on the basis of calories produced, and calorie production in the two relevant years (1960–1 and 1968–9) may be compared. The net quantity of farm production in a year is made available in kilograms and divided by 365 to obtain daily production. For the calorie values, the conversion factors from Aykroyd have been used (Aykroyd).

Since plotwise classification would not be convenient because of the large number, the land under cultivation had been divided into certain agricultural areas on the basis of soils, irrigation, and crops grown (Buchman 1953, 303), using 1960–1 as the base year. As the village is more or less wholly irrigated, soils become the main guiding factor, and therefore considering different crops grown in the various soil zones, as many as 15 agricultural areas (Chauhan and Singh 1971, 27) have been demarcated.

Daily calorie production in percentages from all the agricultural areas was calculated. Now a comparative study of these areas in 1960–1 showed that the units situated near the settlement had a fairly high percentage (over 6) of calorie production and the units toward the village periphery had a very low percentage (less than 4). Calorie production near the settlement was naturally high, on account of greater use of natural manures and the same was not possible in the distant areas.

By the year 1968–9 the work of consolidation in the village was over. Since the size of holdings increased, a different cropping pattern emerged and therefore the pattern of calorie production also changed. Daily calorie production was now higher (over 6) in areas nos. 12, 13, 14, and 15 situated towards the village periphery than in areas situated near the settlement. This was possible mainly on account of consolidation of holdings, i.e., better land use planning. The approach to distant fields through suitable roads was provided and therefore improved manures, fertilizers, seeds, and agricultural implements were made available.

On bigger holdings in 1968–9 the farmers sometimes produced two or three and even more crops for a balanced economy. The percentage increase in calorie production over 1960–1 in the village as a whole was 130. The increase was 50 per cent or below in areas 1, 2, 3, 4, and 5 near the settlement. Perhaps the saturation point for higher production had been reached here. On the other hand the increase was as high as 150 to 220 per cent in areas 11, 12, 13, 14, and 15 situated towards the village boundary. Obviously, the gap between actual production and the optimum capacity was considerable in those areas in the year 1960–1, and this was gradually overcome by 1968–9. Another reason was the preference given to crops such as sugarcane, which had a high selling price in the market and also a high caloric value.

Aykroyd, W.R. *The Nutritive Value of Indian Food and Planning of Satisfactory Diets*, ICMR, New Delhi.

Buchman, K.M., 1953 *Delimitation of Land-Use Regions in a Tropical Environment*.

Chauhan, V.S., and Surendra Singh, 1971 Changing agricultural land-use and calorie production in V. Banwaripur (Meerut) – use of some working formulae, *Geog. Observer* 7, 26–38.

Shafi, Mohd., 1960 Measurement of agricultural efficiency in U.P., *Econ. Geog.* 296.

– Food production efficiency and nutrition in India, *The Geog.* 14, 24–7.

Stamp, L.D., 1958 The measurement of land resources, *Geog. Rev.* 48, 1–15.

– 1960a *Our Developing World*.

– 1960b *Appl. Geog.* 65.

P1008
A multivariate approach to the analysis of changing crop-livestock regions
DAVID J. COWEN and PAUL E. LOVINGOOD, JR.   *University of South Carolina, USA*

Traditionally, research in agricultural regionalization has emphasized the determination of homogeneous regions of crop land use, thereby largely neglecting the role of livestock production (Kendall 1939; Weaver 1954). In this paper, we demonstrate the way in which cash receipts from farm marketings, an increasingly available statistic, can be used to form regions which include the important role of livestock. We believe that regions of agricultural specialization formed by the use of this type of data are more important from an economic viewpoint than other regionalization schemes which solely investigate crop acreages.

To demonstrate the applicability of this form of data, in conjunction with the use of multivariate grouping procedures, we will investigate the changing structure of the agricultural economy of the state of South Carolina between 1959 and 1969. Beginning with the matrix of cash receipts from ten crop and six livestock categories, expressed as percentages of total cash receipts (South Carolina Crop and Livestock Reporting Service 1960) for each of the forty-six counties of South Carolina, the problem is one of forming a smaller number of homogeneous groups of counties. The basic procedure followed in this paper is similar to Berry's method of determining multi-factor uniform regions in which the original data matrix is orthogonalized by computing principal components scores and applying a hierarchical grouping procedure (Berry 1961; Ward 1963). In an earlier paper we demonstrated, by the use of a simple analysis of variance test, the superiority of the use of a grouping algorithm over the Weaver deviation approach (Cowen and Lovingood 1970). In the paper we apply a step-wise discriminant analysis to the original data matrix in order to evaluate the groups formed as well as to determine the relevant discriminating crop or livestock categories.

In the first year of this study (1959) South Carolina's agricultural income was dominated by the importance of the two traditional crops, cotton and tobacco. In 1969, tobacco continued to be an important cash crop while income from the sale of cotton had dropped from 20 per cent of the total income in 1959 to only 6 per cent. This decline in cotton receipts represents an absolute drop of $50,000,000 during the decade. During the same period, receipts from the sale of the six livestock categories increased by $62,000,000. Thus, the state underwent a rapid transition from traditional southern row crop specialization to a much more diversified agricultural structure. We will describe in detail only the internal patterns of change within the state, not the causes for these changes.

Principal component analysis of the 1959 matrix of percentage of cash receipts for the 46 counties and 16 crop-livestock categories reveals six components that could account for 76.28 per cent of the total variance. While this represents a sizable reduction, it also suggests a more complex agricultural structure than one might have expected in South Carolina. Hierarchical grouping of the unrotated component scores indicated four basic patterns of agricultural specialization. By far the most prevalent pattern was found in the northwestern part of the state. The 27 counties in this first region had a structure consisting of relatively similar proportions of cotton, forest products, dairy, cattle, and eggs. A second region of five counties was found along the lower coastal plain. These counties basically formed an area of crop specialization with large percentages of cash receipts from cotton, vegetables, and soybeans. The third region of the six intercoastal plain counties formed an area of great cotton specialization (40.79%). The remaining eight counties, all occupying the northern coastal plain, were closely linked to tobacco (59%) and cotton (17.5%).

Step-wise discriminant analysis of the four regions, based on the original data, revealed that no members of the groups formed were more similar to the membership of any other group. Additionally, the analysis indicated that tobacco was the overwhelming discriminatory variable between the groups. The lack of importance of cotton as a discriminating variable further indicates that cotton at that time was an important cash crop throughout the state. On the other hand, soybeans and vegetables, as well as tobacco, proved to be highly localized in 1959.

Following the same procedure for 1969 the data matrix collapsed to seven components, accounting for 80.4 per cent of the original variance. The component scores grouped into five subgroups. The primary difference between the two study years was the division of the twenty-seven piedmont counties into two regions. Specialization within the regions changed markedly, with cotton all but disappearing in importance except in the inner coastal plain. The seventeen upper piedmont counties in 1969 derived close to a fourth of their agricultural income from the sale of cattle, which in conjunction with eggs and dairy products, accounted for more than sixty-four per cent of the income. The sandhills region of the centre of the state, and the lower coastal plain around Charleston, had developed into areas of specialized vegetable and peach production, with eggs, hogs, and cattle being of secondary importance. A third group of counties in the inner coastal plain had evolved into an area of specialized soybean production, with other large proportions of income from dairy products, cattle, and hogs. Cotton remained important only in three disjoined counties of the inner coastal plain. Probably because of governmental controls on tobacco production the eight tobacco counties of the northern coastal plain remained intact during the ten-year period.

Again discriminant analysis revealed the homogeneity of the groups formed and indicated the continued importance of tobacco as a discriminatory variable. However, cotton was a pronounced secondary variable, indicating the development of a specialized cotton region within the state since 1959. The next three important discriminatory variables were cattle, eggs, and dairy products.

Examination of the absolute change in cash receipt percentages for the ten-year period was undertaken to form regions of agricultural change as well as to identify the nature of the changes which had occurred. The data matrix of the change variables reduced to eight components, from which scores were used to form five regions of agricultural change. The first region, corresponding to the piedmont, exemplifies the change from cotton (-14.4%) and the increased importance of eggs and cattle in the south.

Throughout the centre of the state, cotton declined by 22.47 per cent per county and was replaced largely by soybean production. In a third area, along the inner coastal plain, cotton was replaced by a combination of soybeans and eggs. Along the coast smaller declines in cotton have been supplemented by increases in soybeans and vegetables. As previously noted, the northern coastal plain demonstrates the stability of tobacco as a cash receipt, while a shift from cotton to soybeans as a cash crop has also taken place.

Discriminant analysis revealed that a county along the southern border with Georgia was misclassified by the grouping procedure. Since cotton had declined in importance in almost every county since 1959, it was only of minor importance as a discriminant variable. Similarly, since soybeans had made large increases throughout the state it was likewise of only secondary importance as a discriminatory variable compared to the other crops.

This paper demonstrates, through the use of cash receipt data, how both crops and livestock can be combined in forming agricultural regions. The multivariate procedures successfully identified distinct regions of agriculture specialization for two time periods, as well as identifying the pertinent changes in agriculture which are occurring in South Carolina.

Berry, Brian J.L., 1961 A method for deriving multi-factor uniform regions, *Przeglad Geog.* XXXIII, 263–82.

Cowen, David J., and Paul E. Lovingood, Jr., 1970 A hierarchical development of crop-livestock regions in South Carolina, *Bull. South Carolina Acad. Sci.* XXXIII.

Kendall, M.G., 1939 The geographical distribution of crop productivity in England, *J. Roy. Statist. Soc.* 102, 21–48.

South Carolina Crop and Livestock Reporting Service, 1960, 1970 *Cash Receipts from Farm Marketings* (US Department of Agriculture, Clemson, SC).

Ward, J.H., 1963 Hierarchical grouping to optimize an objective function, *J. Am. Statist. Assoc.* 58, 236–44.

Weaver, J.C., 1954 Crop combination regions in the Middle West, *Geog. Rev.* 44, 175–200.

P1009
Aspects of rural resettlement in Singapore
R. DE KONINCK   *Université Laval, Canada*

In Singapore, rural resettlement areas represent an increasingly common type of settlement. Of recent occurrence, they are an integral part of modern Singapore, having been planned by the same people responsible for the republic's gigantic urban renewal and housing programs.

Resettlement is no peculiarity to Singapore. Many countries carrying out planned urban and industrial development are faced with the problem of resettlement of farmers. Large cities of the world either absorb or force out and preserve their own food-producing hinterland. Singapore has chosen to preserve at least a large part of its food-farming land and over the last two decades resettlement has become inevitable in the dynamic development of the city state. Not only is the resulting pattern of particular interest but the workings of the agricultural resettlement program and the broader socio-economic consequences are also of great significance to a large number of Singapore residents. This paper will attempt a brief description of this program and an analysis of some of its impact on resettled Chinese farmers.

Although large-scale resettlement of farmers began in 1949, it was only in 1957 that a specific government agency was put in charge of the difficult task of co-ordinating eviction and relocation of squatters. This Resettlement Department has been under the jurisdiction of the powerful Housing and Development Board (HDB) since 1962.

Clearance of squatters or occupiers of land earmarked for public development has reached enormous proportions in Singapore over the last years. From 1957 to 1969 nearly 35,000 households, representing at least 10 per cent of Singapore's population have been cleared (De Koninck 1970, 245). Of these, approximately 1500 have been resettled as farming cases; perhaps as many farming households have resettled themselves without the help of government agencies. The larger proportion of these cases was found in fringe areas now occupied by housing and industrial estates. In most instances, the land on which farmers are being resettled was previously under plantation crops and belonged to various private companies. Concurrently with the urban and suburban areas clearance schemes, the HDB acquires these old estates (mostly rubber) which are then cleared of trees, surveyed, and divided into lots ready to be taken over by the new settlers who have just received a moving allowance and compensation for their immovable properties.

In 1968, a survey was carried out among 1063 of these resettled households. (Some of the results of this survey, in which the author participated actively, have been published [Yeh 1970].) Theoretically, the resettlement areas are for farmers and should therefore be occupied virtually only by Chinese as opposed to Malay settlements where all occupants are Malays. However, for reasons too complex to be outlined here, 242 of the households included in this survey were actually Malay. All others are Chinese and all farming cases encountered belong to this community. The great majority of these Chinese farmers are engaged either in market gardening or in livestock rearing (pig or poultry) or in a combination of both. Very few have been resettled on lots of more than 0.8 hectare.

Following resettlement, the rhythm of social, economic, and cultural change undergone by the farming community has been greatly accelerated (De Koninck 1970, 257–300). Here we shall be able to examine only some examples of this impact, particularly on the occupational structure of the resettled population.

Major changes have indeed occurred in terms of employment and household occupational structure but not so much between the non-farming and farming groups as within the latter itself. A large number of households have seen the size of their farm labour reduced. In 310 cases, the number of full-time farmers has been reduced and, in an additional 150 instances, no full-time farmers are left. In other words 64 per cent of the displaced farming households have seen their full-time farm labour either reduced or eliminated. Among those 460 households, at least 584 farmers have become redundant.

The consequence of such drastic curtailments in the total farm labour is not so much the reduction in the number of farming families, as the multiplication of the households depending both on farm income and on off-farm wages. It would appear that, following resettlement, few households readily give up farm income; many do it partially and, one might suspect, in phases.

While the percentage of households depending solely on income from the farm has dropped from 49 to 28 per cent, there is only a small corresponding rise in the percentage of non-farming households, from 33 to 39 per cent. In reality, the bulk of the affected households has filled the ranks of the part-time farming community whose representation nearly doubled, jumping from 18 to 32 per cent of all units.

How much of this evolution can be attributed to resettlement itself and how much to more general factors of industrialization, modernization, rising expectations, etc., is difficult to evaluate with precision. However, an assessment of the more tangible aspects and policies of resettlement is possible.

'It was also considered to be to the economic benefit of the Colony that occupiers of Crown Land producing foodstuffs should be given the opportunity of maintaining or increasing food production by the allocation of food agricultural land in areas where they would not again be disturbed' (Damu 1963, 21). This policy, advocated in the 1950s, seems to have remained the official one under the People's Action Party government. It is self-evident that farmers are being resettled *because* they are farmers. However, there is a distinction between being resettled on a farm plot and being given the opportunity of maintaining or increasing food production; for to maintain or increase production may depend to a large extent not only on where, of what size, even on what soil or slope the lot is, but on numerous other factors. In other words, by providing a displaced household with a farm plot, it does not necessarily follow that its members will resume their agricultural activities to the previous level of intensity.

For many households, particularly if the conditions are to them unsatisfactory, resettlement will promote the trend to seek off-farm employment. Resettlement appears for many the crisis that forces them off the farm; nevertheless, the farm, meantime, will continue to function at a rhythm sufficient to support the household throughout the transitional period. The need for the farm as a temporary source of supplementary income is the first reason why even those who have decided to move off the land will generally accept the resettlement plot even if they can occupy it only on a Temporary Occupation Licence (TOL). The second is that on this piece of land they can invest the compensation money in a house of their own, an attractive alternative to the prospect of living in an HDB flat. Also, although the land is on a TOL, the house belongs to them, and resettled squatters know by experience the value of investment in a house.

That farmers in resettlement areas be given a 'special' or 'preferential' treatment is not necessarily desirable. What might be beneficial is for the authorities to take advantage of the 'special' opportunities they have to fill the vacuum created around the farmers when they have to start anew. However, no special efforts are made to 'sell' more government services to the new settlers. There are no particular policies to organize nor even to supervise more closely investment, production, and distribution.

It implies that the authorities are leaving – by choice or necessity – economic recovery and modernization of the farms mostly to the ingenuity and spontaneity of the farmers. Planning seems to focus on the farmers' way of living (location, housing, social amenities) rather than on their means of earning that living.

'Because the powerful forces of spontaneity, if rightly guided, are an ally to national economic policy, thus is saved the superhuman task of planning everything down to the very last detail' (Lösch 1954, 508). This has evidently been the prevailing philosophy in farming Singapore during the prosperous past decade.

Damu, A., 1963 Problems of resettlement in Singapore, academic exercise, U. Singapore.

De Koninck, R., 1970 Chinese farmers of Singapore: a study in social geography (PH D thesis, U. Singapore).

Lösch, A., 1954 *The Economics of Location* (New Haven).

Yeh, S.H.K., 1970 *Report on the Census of Resettlement Areas Singapore, 1968* (Singapore).

P1010
High diversity alternatives in tropical landscape management
JOSHUA C. DICKINSON   *University of Florida, USA*

The thesis of this paper is that productive landscape management can be achieved by simulating the structure and diversity of tropical ecosystems as an alternative to the monocultural systems being transplanted to the tropics. These systems attain high productivity only by subduing natural processes with massive fossil fuel subsidies at considerable cost to society and the environment.

The system of monocultural agriculture that has been developed to its highest level of output in the United States involves: (1) essential independence from natural mineral cycling processes through heavy annual fertilization and irrigation; (2) suppression of plant and animal competition by use of chemicals and cultivation; (3) replacement of human labour by machines; and (4) the breeding of specialized plants and animals to reflect the above subsidies in high yields of food and fibre. A system that causes major environmental stress and a high rate of non-renewable resource exhaustion is not viable as a means of sustained life support. The limitations are, if possible, more serious in the tropics than in the mid-latitudes (Dickinson 1971, 16). Why? Higher rates of oxidation and leaching compounded by breakdown of soil structure make fertilization more costly and potentially damaging. The greater diversity of plants and animals makes the subsidy cost of maintaining a monoculture higher, both economically and environmentally. In most tropical countries characterized by high population density and low levels of industrialization, few alternatives to agricultural employment exist. Any mechanization or other form of subsidy which supplants people from gainful employment or even subsistence may be socially disfunctional. Agriculture on the US model has been implanted in the tropics through the assumption of a satellite role by the country; a role in which technical and material inputs are imported to produce a high-value product largely for export at a price determined elsewhere. Some examples are cotton, bananas, sugar, cacao, and cattle. Generally it is not profitable to produce foodstuffs in this fashion for the mass of people because they lack purchasing power.

The imbalance between tropical peoples and their life support system is attributable in part to population pressure triggered by imported medical technology but also caused by imported systems of land and water use which may be ecologically inappropriate and socio-economically prejudicial to a segment of the population. In many areas the best-quality land is dedicated to export crops and cattle. Expansion of export systems often includes the displacement of food-producing traditional agriculturists. These displaced people crowd the marginal land at destructive levels or they move to cities ill-prepared to absorb them. Profits from this modern satellite economy accrue to a small socio-economic elite with a low multiplier effect elsewhere in the economy.

The alternative is what might be called agricultural judo – tapping the natural system's own strengths to yield food, rather than forcing output through massive application of power. In systems language, one would say that small inputs are strategically applied in the control circuit which then amplify and redirect major power flows to man's use. This philosophical alternative may find many expressions in the landscape, each dictated by environmental fitness, cultural preference, and economic reality. Rough models have existed in the tropics for many centuries. In shifting agriculture, forest succession is manipulated by burning, to switch energy stored in the forest to man's early successional crop plants for relatively brief periods. The structure and diversity of the tropical forest is simulated in seemingly chaotic multistoreyed dooryard gardens. Where peoples have persisted in a place with a relatively stable culture and population level, the resultant agricultural systems often have been quite sophisticated. The Hanunoo in the Philippines, the Kekchi in Guatemala, and the Polynesians provide instructive examples (Conklin 1957; Barrau 1968; Carter 1969). Heretofore we have studied these agricultural systems more as anthropological anachronisms rather than as models and gene pools for a new approach to tropical agriculture.

The failure of these systems in modern times has often been inextricably tied to the disturbance or competitive exclusion of the parent culture by elements of energy-rich industrial civilization. In light of the finite nature of fossil fuel resources and their unavailability to many tropical countries, it would appear worthwhile to emulate aspects of primitive and natural systems in designing alternatives to monocultures.

Sophisticated research in modern agriculture has been devoted primarily to improving the production of specific plants and animals. Systems analysis has been concerned with economic inputs and outputs. In contrast, research on alternative approaches treats each plant and animal as an integral part of an ecological and social system with outputs tapped by man for food and economic return.

Actual systems have been, and are, managed with negligible inputs from the fossil fuel economy. However, a realistic program to hold a rural population on the land will require that demands be met for desired consumer goods available to city dwellers. The income requirement would almost undoubtedly be in excess of that possible from the output of a system running solely on solar energy. The fossil fuel inputs would differ quantitatively and qualitatively from inputs applied to monocultures in the sense that the subsidy would augment and redirect energy flows in a functional system rather than obliterating natural processes.

Research will initially require a thorough gleaning of existing work of geographers, anthropologists, and agricultural scientists. The data required will include names of plants and animals, their metabolic requirements and net yield, light requirements and shade tolerance, transmission of solar energy through plant canopies, rate of nutrient production in the detritus cycle of plants and animals, and symbiotic, synergetic, and parasitic relations among food plants and animals, pests, predators, and the environment. Available hard data and estimates will be incorporated in an analog simulation model developed by a systems ecologist, H.T. Odum (Odum 1971). Energy flow coefficients can be manipulated to simulate biological processes, management inputs, and resultant useful energy outputs. Data gaps identified in the model can be filled through field research. The final step in research will be field testing of prototype systems. This step will require considerable interdisciplinary cooperation. Instrumented field plots will reveal energy flow, actual effects of management inputs, net yield, labour requirements, and costs.

This research project is in its initial stages although considerable data on actual systems have been gathered by the author and others in Guatemala. The alternative approach offers no solution to the population problem. Its major virtue is the potential for sustained moderate output of quality food with few or even no fossil fuel inputs. If seriously considered the approach would require drastic changes in thinking by research and technical assistance agencies with no less drastic social change in many tropical countries. For this reason a high diversity alternative may be considered more realistically as part of the recovery mechanism for the residual population following a catastrophe induced by overpopulation, environmental mismanagement, or social upheaval.

Barrau, J.L., 1968 Humide et le sec: an essay on ethnobiological adaptation to contrastive environments in the Indo-Pacific area, in *People and Cultures of the Pacific*, ed. A.P. Vadya (Garden City, NY).

Carter, W.E., 1969 *New Lands and Old Traditions: Kekchi Cultivators in the Guatemalan Lowlands* (Gainesville, Fla.).

Conklin, H.C., 1957 *Hanunoo Agriculture: A Report of an Integral System of Shifting Cultivation in the Philippines*, FAO Forestry Development Paper 12 (Rome).

Dickinson, J.C., 1971 Letters, *Science*, 171 (8 Jan.).

Odum, H.T., 1971 *Environment, Power, and Society* (New York).

P1011
The green revolution in India
G.K. DUTT *National Atlas Organisation, India*

The two successive years of widespread drought in 1965–6 and 1966–7 caused a severe set-back to food grain production in India. As compared to the previous five years' average annual production of 82.3 million tons and the record production of 89 million tons in 1964–5, the food grain production in 1965–6 was 72 million tons and the production in 1966–7 was only a shade better, 74.2 million tons.

The need for food grains in the country was much higher than the quantity produced, and the gap had to be filled by frantic imports. Imports of food grains during the drought years were obviously the highest ever made. Thus a record import of 10.3 million tons was made in 1966, while in 1965 and 1967 the amount imported was also high, being in the range of 7.5 million tons and 8.7 million tons respectively.

These two bad years brought to the surface the unstable and non-dependable agriculture production that we are confronted with because of lapses in our agricultural methods and practices. This also indicated the need for making an all-out effort for a radical improvement in all aspects of agriculture.

The Ministry of Food and Agriculture in the government of India, with the active collaboration of organizations like the Rockefeller Foundation and the State Agricultural Directorates planned and adopted in 1966–7 'the new strategy' of agricultural development. This has the following important elements: (i) cultivation of high-yielding varieties of seed, (ii) development of multiple cropping, i.e. bringing additional areas under crop production in the irrigated and assured rainfall areas, (iii) development of irrigation for intensive cultivation, (iv) soil and water management measures, (v) using a package of practices including high-yielding seeds, optimum quantities of fertilizers, and pest control measures, (vi) emphasis on research and its application, (vii) farmers' training and education, (viii) development of the infra-structure of credit, marketing, and distribution system for supply of inputs, etc.

Implementation of the programs set up in the scheme was carried out mainly by the Community Development Program. The Indian Agricultural Research Institute in collaboration with agricultural universities and colleges in the country and organizations like the International Rice Research Institute introduced exotic high-yielding varieties in the Indian farming system quite successfully. Thus, Taichung Native-1, a variety of rice adopted from Taiwan, recorded a yield of 6000–8000kg/ha as compared to 3000–4000kg/ha for the local variety. In the case of wheat, a beginning was made with dwarf Mexican wheat varieties of Lerma Rojo and Sonara 64. These exotic varieties were subsequently crossed with Indian strains with a view to evolving new strains to suit a wide variety of climatic and water availability conditions existing in different parts of the vast subcontinent. Intensive research has thus yielded results in developing new 'tailor-made' strains of all types of food grains, oilseeds, cotton, and pulses. In 1966–7 the high-yielding varieties were sown in an area of 1.89 million hectares, and this had reached 6.4 million hectares in 1967–8. Under the Fourth Plan it is expected that by 1973–4 the coverage of high-yielding varieties will go up to 24.1 million hectares. For implementing successfully the High Yielding Varieties Program, provision of sufficient quantities of irrigation facilities and fertilizers is necessary. To meet the increasing demand for water, canal irrigation is being supplemented by minor irrigation projects which are directed towards tapping underground water by shallow and deep tube wells. Thus the total area of 10.9 million hectares under canal irrigation is to be supplemented by 7.2 million hectares irrigated by minor irrigation projects by the end of the Fourth Plan. Production and consumption of fertilizers have vastly increased. Consumption of about 780 thousand tons of fertilizers in 1965–6 has rapidly increased, which, by the end of the Fourth Plan, is expected to be in the order of 6.6 million tons. Supply of quality seeds is being organized by the National Seed Corporation. Various agricultural research institutes, central seed farms, and state seed farms are producing quality seeds to meet the demand. Cooperative seed stores are being organized in all the rural areas. Although farming in India is far behind in the process of full mechanization, considerable progress has been made in improving the implements used and introduction of simple, manually-operated or power-operated modern tools in all stages of farming. The use of tractors, for example, has gained considerable popularity and has increased from 54 thousand in 1966 to about 90 thousand in 1970. The use of modern threshers, harvesters, mould-board ploughs, sprayers, dusters, and other improved implements is also gaining in popularity. The gross area benefitted by plant protection measures increased from 16.5 million hectares in 1965–6 to the level of about 54 million hectares in 1968–9. Adequate supplies of pesticides being available, the farmers have gradually adopted the use of pesticides which are so very essential for growing high-yielding varieties of crops. Facilities for availability of credit for agricultural needs have been considerably extended. Cooperative institutions and individual farmers are being assisted by the government and nationalized banks in this regard. The level of credit advanced by the cooperative credit societies had increased from Rs.396 crores in 1965–6 to about Rs.545 crores in 1967–8. What is most encouraging is the fact that the general consciousness and awareness by the common farmer for growing more by adopting new seeds and improved techniques is rapidly increasing. The inherent apathy towards any change for the better has been shaken off.

After the scarcity years of 1965–6 and 1966–7 the total food production has been considerably stepped up, each year bettering the records of the previous year. The food grain production in 1969–70 broke all records, amounting to 99.5 million tons. This represented a 5.8 per cent increase over 94 million tons of 1968–9. In 1970–71, production was 107.8 million tons, an increase of 8.4 per cent over the previous year. In some parts of the country, particularly in Haryana, the target of producing two tons of grain per acre has already been achieved. In wheat production the country has achieved a 'revolution' when compared to an average annual production of 11 million tons in the first half of the sixties; wheat production had successfully moved up to 23.2 million tons in 1970–1. As a result of two successive years of record food-grain production and a consequent increase in market availability, there had been a significant increase in the procurement of food grains. Thus in 1970–1 a record amount of 8.7 million tons was procured as against 6 to 6.5 million tons during the previous three years. This made it possible for the government to build up a sizable buffer stock which was over 7.6 million tons in 1970–1. Imports of food grains were rapidly declining. During 1970–1 the imports were expected to be not more than 1.5 million tons as against 8.7 million tons in 1967–8. The government has decided not to undertake any concessional imports under PL480 after 1971.

In spite of all-out efforts by various government and non-government agencies, the New Strategy could not cover more than 20 per cent of the arable land of the country. Unirrigated farms occupy nearly 80 per cent of our total cropped area of 138 million hectares; special strains are to be developed and farming techniques evolved for bettering output on those areas.

The 'green revolution' so far has been extremely restricted in its scope. It is still largely confined to wheat although new high-yielding strains of rice have been developed and are being grown in several states. Since rice covers 31 per cent of the arable land compared to 15 per cent of such land covered by wheat, high-yielding varieties of rice would affect the total food output to a greater degree. The same is the case with other food grains like millets, pulses, and oilseeds.

Implementation of the New Strategy had not been uniform in all states because of extraneous factors like smallness of holdings, poverty of cultivators, their inability to obtain inputs in time, or financial and credit difficulties. Even in states like Punjab and Haryana where the program has been most successful, it is confined, because of the high expenses involved in its operation, to the comparatively rich farmer with the small farmer still remaining largely where he was. The rich farmer is getting richer while the poorer farmer with all his aspirations and consciousness for modernization of farming is denied any betterment. This creates potentially disturbing social and political issues.

Another economic problem that the 'green revolution' may create is that of rural unemployment. The old squeeze whereby tenants are reduced to share croppers and eventually to landless labourers is being accelerated as more of the bigger owners become involved with modern technology. The increasing landless labourers pose problems social, economic, and political.

For the first time after independence, the Ministry of Food and Agriculture has to tackle a problem of plenty rather than scarcity; perhaps the former needs more attention since we have now a continuous, expanding, and self-regenerating dynamism in agriculture. The problems are of storage, marketing, movement, and provision of price support. The Agricultural Prices Commission, Food Corporation, Directorates of Marketing and Inspection have pooled their efforts to tackle the problems. Procurement targets in the governmental level in food grains have already been increased. The long-term storage plans are also being reviewed by an expert committee on storage set up by the Planning Commission. The price support now in operation for bumper crops in Punjab and Haryana may have to be extended to other states as well, in future.

P1012
New agrarian settlements on the Polar Circle in the marshwoods near Rovaniemi, Finland
GERHARD ENDRISS   *Badisches Generallandesarchiv, West Germany*

In recent years, someone travelling from the Black Forest (southwest Germany) to northern Scandinavia might have been surprised to see that here, at the edge of Europe, forests were still being cleared for farmland. The Black Forest was opened to farming in the period between 800 and 1300 AD. The open land served for pasture-farming ('Feld-Graswirtschaft'), and the woods for burn-beating ('Rutte'). In addition, there were extensive pastures in the forest, since deciduous trees are very common there.

Many parallels present themselves between forest clearing in Black Forest settlements in the Middle Ages and new settlements in Finnish Lapland. We will treat only those Finnish settlements which were created after the Second World War; the settlements of the Laplanders, which are of a different sort, cannot be considered because of space limitations.

The settlement of the area on the northern border of Finland did not proceed steadily. The interests of the settlers and those of the foresters or wood buyers have always been opposed. The brisk pace of forest clearing in Lapland after the Second World War has little to do with refugee settlements. The latter were very quickly built; they lie primarily in central and southern Finland, where the old farmers were dispossessed of a part of their land.

The extension of farmland in northern Finland came in the form of 'Kalte Betriebe.' The settler had at first neither house nor arable land, but this extension of farmland comprises more than what we think of today in central Europe as forest clearing. Most of the land to be settled is marshland. Forest clearing is only a small part of making the land fit for cultivation; it gives the farmer little trouble. But draining this land, which is mostly flat, takes much care. Earlier, many a farmer failed because drainage was poor. Today the new areas of settlement are drained according to the latest methods and with the most modern equipment. Even in central Europe we must consider that much more of the land to be cleared was wet and swampy than we might realize today.

The job of cultivating takes time. Here are three examples from Välijoki, observed in the year 1965. Farm A has ten children, of whom most go to school. It has been in operation for five years. There are only two cows. Pasture comprises 5.5ha; it is to be extended to 10ha. Potatoes and barley are cultivated only to a small extent; therefore it is not possible to keep pigs and chickens. The father does forestry part-time.

Farm B has been in operation for ten years. The many flowers in the living room are conspicuous; they are for the long winter. Radio and television are not lacking. 6.5ha of land have been cleared. Eight cows and four calves are in the stable; they are of a native, hornless race. The cows are milked by hand. The farm has had electricity for three years; before that the family used oil lamps. The milk goes to the dairy. The road is kept open all winter. There is snow from October until May; sometimes there is two metres of it. The farmer has no horse and no tractor. He wants to clear more trees for pasture land. He told me that the forest is good, and that means money. He speaks of pines and birches, and even of spruce.

Farm C has been occupied for 4½ years. The farmhouse is a new type of ranch house. The stable receives more light than older types. The floor-heating uses birch wood, which does not have to be completely dry.

The cows are of an English breed (Ayrshire). During the year of our visit, potatoes were planted for the first time. As yet there is no grain, and therefore no chickens. 'Perhaps we'll have them later sometime,' said the farmer's wife. There was a death in the family: a son drowned in a nearby river. Another son is still living.

Two farms have their land divided as follows:

	Farm 15	Farm 44
Grassland (and cropland)	6.13ha	6.00ha
Land suitable for clearing	27.57ha	24.18ha
Forest	165.40ha	129.01ha
Land unfit for cultivation	17.62ha	33.92ha
	216.72ha	193.11ha

All these farms were deliberately built farther away from the city (in this case Rovaniemi) than the refugee farms, one of which we visited near Kajaani. This family of 12 persons cannot live on its 8ha of cultivated land. The men now earn more from factory work than they do from the farm.

At the Jormua research station near Kajaani, about 250km to the south of the Polar Circle, we heard that the potatoes had frozen ten years in succession; not until this year was there a harvest, because frost damage was small.

Meanwhile, the flight from the land, which is widespread in Sweden, has taken hold of Southern and Central Finland and is now infecting Lapland too, as is particularly evident in Varjo. The causes are many; among them are a slight worsening of the climate, and the mechanization and rationalization of the state's forestry operations, which now hire only full-time, year-round foresters. Thus farmers have no second source of income any more. Other reasons for the flight from the land are the paucity of jobs in Lapland and the fact that Finland, being a sparsely populated land, provides only a limited market for food products.

Endriss, G., 1968 Die innere Kolonisation Finnlands im 19. Jahrhundert nach der Darstellung in Alexis Kivis 'Die sieben Brüder' (1870), Regio Basiliensis 9, 193–8.

Häkkilä, M., 1971 Einige naturgeographische Grundzüge Nordfinnlands, Geog. Helvetica 26, 63–71.

Helmfrid, St., 1970 Der Norden heute, Deutscher Geographentag Kiel 1969 (Wiesbaden), 39–49.

Soosten, H.P.v., 1970 Finnlands Agrarkolonisation in Lappland nach dem zweiten Weltkrieg (Geog. Inst., Marburg/Lahn).

Varjo, U., 1971 Development of human ecology in Lapland, Finland, after World War II, Geoforum 5, 47–74.

P1013
Changing farming patterns in the southern prairies of Canada
D.J. FLOWER Medicine Hat College, Canada

In 1859 Palliser designated a large area of the southern prairies as semi-arid, and unsuitable for farming. By 1879 this assessment had been reversed by Macoun's report claiming the land was extremely fertile. This latter view led to the construction of the CPR line through the south, followed later by homesteading.

From the outset, wheat was the base of the arable farm with the size of the harvest resting on unpredictable precipitation. Successful farming depended upon either irrigation, where water was available, or dry farming methods, many of which were developed internally to overcome local problems. In both cases the small family farm was the universal unit. After the droughts of the 1930s, brief variations in the traditional pattern occurred, but only in the mid-1950s, when huge wheat surpluses weakened the national and international markets, did large scale changes appear. The traditional response to the land is no longer practical and changes must be made by those who wish to survive. This paper examines the problems of adjustment to the difficult natural and economic circumstances faced by farmers around Medicine Hat, Alberta.

An examination of statistics, a review of literature, and first-hand knowledge suggest

the following farm categories:
1. Farms entirely supporting the operator:
   A. RESIDENT
      (i) large farm: (*a*) ranch, (*b*) wheat farm
      (ii) traditional farm
      (iii) small irrigated farm
   B. NON-RESIDENT
2. Farms partially supporting the operator:
   A. MAJORITY OF INCOME
      (i) resident small farmer
      (ii) non-resident small farmer
   B. MINORITY OF INCOME
      (i) resident part-time farmer
      (ii) non-resident part-time farmer

1A(i) The large farm is typified by two features: size and use of hired labour. The size varies from 1215 to 40,486ha. These holdings need not be contiguous and the close road network prevents this being an economic handicap. Because of its size, the large farm needs hired labour to operate successfully. Two divisions come within this category:

1A(i)(*a*) The ranches are located principally in the south where the land is either too dry or too rough to produce grain. These regions were leased to ranchers by the federal government in the 1890s and, despite inroads by homesteaders in the wet years, ranching continues today. Many ranch lands are still leased, some are owned, but in any case company control is likely because of the size and complexity of the operation. Because of the dryness of the land and the poor quality grass, cattle need 8 to 12ha each for grazing, necessitating large tracts of land for a reasonably sized herd.

1A(i)(*b*) In the north, with its flatter topography, large wheat farms are increasing in number. The farming of vast areas provides a high return for capital investment, and unlike the smaller traditional wheat farms, these units, often company owned, are not as vulnerable to fluctuating markets. Stocks of wheat, held on the farm from year to year, act as an insurance against poorer yields or drought.

1A(ii) The traditional farm differs in two respects. It is farmed by the operator and his family without recourse to hired labour, and is much smaller. The 1957 Royal Commission (Drummond and Mackenzie 1957, 268) suggested that a unit of 5.12km² was the smallest size appropriate for future successful farming and this appears to be true in 1971. However, its survival rests on the owner's acceptance of changing needs and economic conditions. Wheat is invariably the preferred crop, though the farmer's desire to provide sufficient income for his family makes him aware of current lucrative crops. As a secondary insurance against crop failure, many farmers now maintain a small beef herd.

1A(iii) The small irrigated farm emerges as a separate unit in the west. Its size, often less than 200ha is compensated for by intensive cropping. Irrigation agriculture is an acceptable replacement for dry farming methods wherever it is possible. Ever since the CPR established its first irrigation area in 1914, proposals for large scale extensions have never reached fulfilment.

1B The non-resident farmer is typical of the American 'suitcase' and 'sidewalk' farmer (Kollmorgen and Jenks 1951, 450). He and his family are eventually attracted to the village or town by many facilities. Such moves are not always voluntary but may be enforced by school closures, curtailment of services and other factors, making farm living unacceptable. Fortunately, wheat growing, apparently providing high returns for the least work, permits such movements to take place. The farmer need only return to his land for seeding, spraying, harvesting, and summer fallowing. With such a small proportion of his time spent on the farm, he can live up to 160km away.

2A(i) The resident small farmer often exists because of his own determination to stay on his land. He owns insufficient acreage to provide an adequate income, and though he may rent more land from absentee landowners, he still needs to supplement his income. Part-time jobs, such as 'custom' spraying or combining, driving a school bus, electrical work, etc., provide this supplement. He may have one or several such jobs. In all cases the farm provides the bulk of the income. Additionally, hens, pigs, and cattle are kept to augment the family's food supply.

2A(ii) The non-resident small farmer, living and working part-time in the town or city, is the typical town farmer (Williams 1970, 181). His land is too small to provide an adequate income and so, attracted by the amenities of the town, he adds to his income with seasonal or commission paid work, where he is often his own boss. Jobs in real estate and insurance provide sufficient time and opportunity to work the farm.

2B(i) The part-time resident farmer is not a very common phenomenon. His land is often inherited and his full-time job takes precedence over whatever farming he does. Often the farm provides a convenient residence for his work. The land may even be rented out.

2B(ii) The non-resident part-time farmer often farms as a 'hobby.' Having purchased some land near his town or village, he may farm the land himself or rent it to another; ownership then being on a speculative basis rather than for its agricultural worth.

Such are the simplified farming categories in the semi-arid prairies. The smaller farms often do not exist as single economic units, yet they still find a place in the farming economy. The persistence of such units is shown by the fact that in 1966 farms of less than 300ha still made up 49 per cent of the total.

The changing emphasis consequent on the continuing decline in world demand for wheat, and the increasing North American market for beef, must lead to substantial adjustments in the farming patterns of the drier areas of the prairies. Such changes will be most severely felt by the small farmer, whose whole future must be very uncertain.

Drummond, W.M., and W. Mackenzie, 1957 *Progress and Prospects in Canadian Agriculture* (Ottawa).

Kollmorgen, W.M., and G.F. Jenks, 1951 A geographic study of population and settlement changes in Sherman County, Kansas, Part I: Rural, *Trans. Kansas Acad. Sci.* 54, 449–94.

Williams, M., 1970 *Town farming in the Mallee Lands of South Australia and Victoria* (Australian Geog. Studies, 8), 173–91.

P1014
Features of the rural population in Mexico
LUIS FUENTES AGUILAR   *Universidad Nacional Autonoma de Mexico*

Before the Mexican Revolution was finalized, rural property fell under the power of a few people, its consequence being the creation of *large properties*, characterized by big estates or 'haciendas' from 10 thousand to 100 thousand hectares in area. In order to increase the estate's productivity, the owner hired day-labourers, who could be task-labourers or year-labourers. Task-labourers were those who rendered service during seed time or harvest time, and the year-labourers, also called 'acasillados' (semi-slaves), were those who had contracts for the year, being urged to move with their family to the estate.

Within the estates, *credit stores* functioned where the labourer could buy on credit the products he needed to support his family; thus the labourer mortgaged his children's future and his own. If he ran away from the estate, special guards employed by the owner made him return on the grounds that he had left when indebted to the owner. In addition, there were in this organization estate jails, punishment systems, and penalties ordered by the owner.

Trusting the established organization in his property, the estate owner moved to the big cities where he could show off his wealth; it was then that a new wage earner appeared, *the manager*, who in various cases caused the agricultural production to decrease and the tyranny over the labourers to increase.

Because of government prerogatives, in the form of tax exemptions, the big haciendas annulled competition, in production and prices, with small property-holders. The semi-slavery conditions imposed on the labourers and the salary system which recouped the owners' wages by means of the credit stores were sufficient by themselves to put the hacienda in an extremely advantageous condition ahead of the small property, in terms of market competition.

These were the prevailing circumstances when the revolution started. We can now observe the social and economic consequences of the revolution in the great variety of socio-economic strata of which the rural population in Mexico is composed. These strata are the following:

1. *The old estate owners.* Those who were not affected by the revolution or who under its protection acquired large properties,

mainly consisting of unirrigated land for agricultural and cattle raising exploitation; on such estates, human labour is exploited, and backward techniques are general.

2. *The new estate owners.* To this category belong persons of varied social background: public officials, former public officials, merchants, old estate owners who have adopted new technology, and agriculturists who were small contractors. All of them, evading the agrarian laws, obtained large landed properties in the irrigation zones, with the addition of hundreds of hectares they rent to public landers or 'ejidatarios' and to small landowners in need of the necessary means to cultivate them. Modern production techniques are employed and high profits obtained.

3. *Mid-rural landowners.* They seed in general between 25 and 100 irrigated hectares. They work under precarious conditions with the problems caused by credit scarcity, and lack of commercial organization, but they form part of a group whose situation has gradually improved, particularly in those modern holdings with more than 50 hectares, where they resort to using salaried workmanship, and belong to the bourgeoisie devoted at the same time to trade, money-lending at high interest, and financing the farming operations of weaker landowners.

4. *Small rural landowners.* 60 per cent of all rural landowners possess less than 10 hectares. The majority of peasants possessing cultivable land are small private landowners. These persons are distinguished by the fact that they are unemployed or underemployed during most of the year. The labourers employed in wage labour work 55 days a year. They constitute the middle-class peasant, having an individualistic mentality, not only because of their education, but also because of their lack of technical preparation to cultivate land in common and to give and obtain cooperative social services.

5. *There are a million and a half public land peasants whose land measures from half an hectare to four furrows.* They constitute a vast sector in which we can clearly distinguish the miserable conditions of the peasants who lack water, credit, technical assistance, and education, and who are exploited by community land functionaries, by private intermediaries, and who obtain starvation salaries which oblige them frequently to abandon their lands and move to the United States as day labourers or to cities in search of whatever job they can find.

6. *The Indians.* Land scarcity or the exploitation of the Indian by the half-breed causes the inferior condition, poverty, and the lack of culture which they suffer. These Indians tend to emigrate to other agricultural zones to sell their labour incredibly cheaply, or they go to the capital to become lost inhabitants in the city or continue being discriminated against, when they don't find work in small businesses or as artisan labourers in provincial cities.

7. *Skilled workmen of commercial agricultural zones.* They develop their activities in the irrigation districts as tractor operators, mechanics, managers, chauffeurs, assistants, etc. Although disorganized, still they live in less precarious conditions than the rest of the peasants.

8. *Day-labourers.* They perform unskilled tasks; though they receive a minimal salary, they earn it only from three to six months a year.

9. *Migratory or ambulant workingmen.* Nomadic, these peasants remain two months in an irrigation district, to leave later to go to another one; from tomato picking, they go to cotton picking, and so on. They normally live in cardboard and tin huts, made also with straw; they have neither electric lights, water, nor sanitary services. Employers do not build lasting housing because they fear agrarian reform: settling in a certain place gives the agricultural workmen the right to apply for these lands. Wild plant and forest product collectors like the 'chicleros,' the palm collectors, the ixtle collectors, the palm oil collectors, etc., can be included within the latter group.

One of the major problems which Mexican agriculture has to face is unemployment or underemployment of rural labour. In order to resolve this situation, the displacement of one part of the rural population from an agricultural sector to other sectors with better conditions has been suggested.

Another solution consists in extending the cultivable area, clearing virgin lands, and intensifying irrigation works. This has been, in fact, the prevailing tendency during the recent past and the government has invested large sums. In the new irrigated zones, land has been preferably delivered to small landowners; nevertheless, the characteristics of the Mexican hydraulic resources fix a material limit to this process; and the rural population growth index keeps on increasing as new lands are equipped.

A partial solution would consist of extending the industrialization process to agricultural zones, in order to offer the rural population employment possibilities not far from the countryside. Up to now, little has been done in this area. Industry and capital are still under the power of a few important urban zones, and private enterprise does not want to invest in less developed regions of the country. The formulation of an adequate planning policy will be the only way to rectify the present imbalance.

P1015
The quasi-plantation as a conceptual model
HOWARD F. GREGOR   *University of California, Davis, USA*

Agricultural specialization brings with it new farming types, many of which, through borrowings and eliminations, become intermediate to other types. Such is the 'quasi-plantation' version of livestock farming, one that lies between livestock ranching with its generally lesser intensity and plantation farming with its greater intensity and emphasis on crops. Models are a natural tool for locating and characterizing such intermediate types, and the simple conceptual model that was used to delineate plantations in the United States in 1936 offers one such means (Woofter et al. 1936).

The basic assumption of this model was that the operation would be large enough to support at least five resident tenants and their families. Wage and transient labour have continued to increase in importance, however, until today they are much more important than sharing arrangements and tenancy in farming operations. For this study of the American quasi-plantation (QP), therefore, the model was modified to include all units that spend annually for labour the equivalent of support for five resident tenant families. Expenditure minima were computed from wage levels obtained from Department of Agriculture statistics (*Farm Labor* 1960) and applied, through linear interpolation, to expenditure classes for livestock farms in the published and unpublished 1959 agricultural census data (US *Census of Agriculture* 1959). Unfortunately, wage statistics were obtainable only for states, and referred only to farms in general. The resulting distribution pattern of minimum wage expenditures needed for support of the QP nevertheless appears logical. Expenditures are lower in the south and higher in the north, particularly the agricultural-intensive northeast. Increasing distance from labour supplies accounts in part for the unusually high wages of the west, although the difference between west and northeast would probably not be quite so great if labour-intensive irrigation agriculture were eliminated. The resulting QP pattern then would show a somewhat better ranking for western states than what was actually derived.

QPs appear in every state, but are particularly prominent in California and Texas, followed by a significant showing in the southeastern states and Pennsylvania and New York. Slightly less important are the QPs of the western plains states. When compared with all livestock farms, QPs are a minute number, with just over one per cent of the total 1.2 million farms. No less striking is their mild representation in the midwest, the area of greatest concentration of the total livestock farm group. So great is this disparity that QPs show practically no positive areal association with all livestock farms (coefficient of association, $A = 0.539$). With far more livestock farms than QPs, it appears that repelling forces in the midwest, such as many small farms, rather than attracting forces elsewhere are primarily responsible for the overall distribution pattern of QPs. Secondary attractions are nevertheless important, particularly when the number of QPs in each state is compared with the total number of livestock farms in the same state. QPs are especially important locally in the extreme southwest and southeast, reflecting at least partly the strong influence of large landholdings. But if large holdings seem to favour these areas, their sparsity in the extreme northeast, where QPs are also important locally, indicate that intensity can compensate for smaller properties.

Each of the three principal QP types, dairy, poultry, and livestock (other than dairy and poultry), has its own characteristic distribution pattern. Coefficients of localization (L), computed on the basis of percentage-point differences reflecting the deviation of each QP type from a theoretical uniform distribution, reveal livestock QPS as the most evenly dispersed among the states ($L = 0.327$). Because they are also most numerous, 48 per cent of all QPS, livestock QPS also best characterize the over-all QP pattern ($A = 0.761$). Texas has the most livestock QPS, with California, Virginia, and the states of the extreme southeast and the northern and central plains also having above-average numbers. Dairy QPS, with 38 per cent of all QPS, are the most irregularly distributed among the states ($L = 0.490$). They are also the only QP type to associate negatively with the distribution of all farms of its specialization ($A = 0.379$). California leads overwhelmingly in dairy QPS, although Texas and the majority of states bordering the Atlantic also stand out. Poultry QPS, on the contrary, associate closely with over-all poultry farm distribution ($A = 0.833$) and are intermediate to the other two types in localization ($L = 0.392$). Georgia and California lead the states, the other important ones concentrating in an almost continuous coastal belt from Texas to New York and in scattered areas of the Midwest and Pacific Northwest.

To determine more than numbers and distributions of QPS required still other assumptions in the model. One was that mean labour expenditures completely correlate with farm size, so as to allow linear interpolation of farm size classes in order to obtain QP acreages. The resulting pattern of mean QP size was, unsurprisingly, one of largest farms in the western half of the country and smallest farms in the extreme northeast. However, the moderate product-moment coefficient of correlation ($r = 0.601$) between mean labour expenditures for all livestock farms and mean livestock farm size warrants caution. The distribution of residuals from regression also suggests that sizes of livestock QPS were seriously underestimated for the extreme southwest, Florida, and some of the New England states; whereas areas were overestimated for the continental interior.

Another assumption was necessary in order to interpolate classes based on value of products sold so as to estimate QP productivity. This was that perfect correlation exists between mean labour expenditures and mean value of products sold. An r of 0.724 obtained between these two variables as they apply to all livestock farms again cautions against complete acceptance of the derived patterns. Thus the distribution of residuals points to an exaggerated margin for the southwest over the east coast in total QP sales value and an insufficient margin for leaders on the east coast in QP importance relative to state livestock farm sales (Florida) and sales value per QP acre (Delaware, New Jersey, Connecticut, Massachusetts).

Simple conceptual models then provide important information springboards, but even with rich statistical sources cannot guarantee a large range of information. The problem in agricultural typology therefore is to produce models that are complex enough to account for additional variables and yet simple enough to prove workable in areas where statistics are sparse, a situation unfortunately still common for much of the world.

Farm Labor, Crop Reporting Board and Agricultural Marketing Service, US Department of Agriculture, January 11, 1960, 15.

U.S. Census of Agriculture 1959, I, *Counties*, Parts 1–50, Table 18; and unpublished worksheets at the Bureau of the Census, Washington, DC.

Woofter, T.J., Jr., and others, 1936 *Landlord and tenant on the plantation*, Works Progress Administration Research Monograph no. 5, Washington, DC.

P1016
Feasibility of an agro-industrial complex on the Gaza-Sinai sea coast
JOHN S. HAUPERT   *State University of New York at Buffalo, USA*

With two exceptions, the Nile Delta and the Gaza Strip, the desert littorals of North Africa and the Mediterranean Sinai are sparsely inhabited, punctuated only by a few oases with limited settlement concerned primarily with intensive subsistence agriculture. Unlike the Nile cultivators, who are well endowed with water for perennial irrigation, the excessive populations of the Gaza–Northern Sinai regions are beyond the range of water diversion necessary for effective agricultural development.

A large segment of the total population, estimated at 380,000 in 1971, has led an artificial and hopeless existence. Approximately 205,000 are classified as refugees, of whom 179,000 exist in camps under the auspices of the United Nations Relief and Works Agency (*Statistical Abstract of Israel* 1970, 624–39). Citriculture, supported by limited amounts of groundwater, is the main economic activity and source of income (50 per cent) in the Gaza–Khan Yunis areas, and comprises 90 per cent of all exports. This alone is insufficient to absorb a registered labour force of 62,300. Since 1967, however, about 20,000 labourers have found work in Israel and Judea-Samaria (Kashiv 1971, 14). Industrial development lags behind agriculture and consists of food processing and small-crafts workshops employing all but 10,000 of the labour force (*Israel Economist*, Oct.–Nov. 1970, 213–15). Although development is encouraged by Israeli investments and marketing privileges, population pressures tend to keep living standards generally at a low level particularly among those in the refugee camps.

To these coastal oasis dwellers beyond the range of water diversion projects from perennial streams in Egypt and Israel, limitless supplies of sea water form a tantalizing picture. About 45 per cent of the arable land in the Gaza Strip is currently irrigated, thus utilizing the absolute maximum of available water. The alternative is the existence of a number of highly sophisticated and efficient desalinization methods that require sizable amounts of power, particularly where irrigation is involved. In recent years, nuclear power has been in vogue, but the initial costs of a nuclear facility would be so great that few developing countries can afford such a venture without massive assistance from countries with large amounts of capital investment. In the Middle East, locally available oil and gas have proven to be a cheaper source of energy and require a more simple technology to construct and implement than larger complex nuclear centres. In order to absorb huge amounts of electric power generated by petroleum or nuclear sources, current proposals suggest the construction of large agro-industrial complexes that would utilize energy in associated chemical and metal industries, and produce sufficient fresh water for agriculture and related industries (*Nuclear Energy Centers, Industrial and Agro-Industrial Complexes*, 1968).

The massive Aswan Dam project in the United Arab Republic undoubtedly generates more power than it can use efficiently at the present time. Given the present political situation, however, it is not available for transmission to the northern Sinai and Gaza population centres. Israel, on the other hand, is now emphasizing industrial development on a scale that is taxing available power to the limit. Either a large nuclear plant or one powered by conventional means could provide a great array of agro-industrial products for both Israel and the Gaza-Sinai coasts within the framework of a possible Arab-Israeli accord that would promote interregional cooperation. Israeli expertise has evolved from experience with the nuclear reactors at Dimona, Nahal Sorek, and the desalinization plant at Elat.

The siting of a similar plant on the eastern Mediterranean coast has certain advantages. In the first place, the Gaza-Sinai coastal plain extends five to fifteen miles into Israel's northwestern Negev Desert. There favourable soils and low relief exist which could be extensively irrigated if more water were available on a year round basis at reasonable cost. Comparable physical features are also found on the northeastern Sinai littoral westward to the Wadi El Arish (Haupert 1971). Secondly, chemical fertilizers, a product of an agro-industrial complex, could be made available at low cost. Another advantage is the frost-free climate favouring year-round crop production. Improved farm to market roads, cooperatives, modern food processing plants, and port facilities for export are required in an integrated design.

Obvious disadvantages, of course, are the costs of desalting water that would have to be produced in excess of 100 million cubic metres annually. Although the initial capital investment is higher for a nuclear plant, fuel costs over the estimated thirty-year life span of a reactor should be considerably lower.

Large agro-industrial projects require several years to construct if funds are available; however, during the interim period, food requirements stemming from continuing population pressures in these coastal communities entail more practical solutions. In the case of the Gaza-Sinai coast, population dispersals and transfers to nearby employment opportunities are an inadequate solution to solve the exigencies of the moment. Potential investment capital from abroad is at present withheld pending some sort of political détente. In the short run, a smaller desalting plant operated by conventional fossil fuel generators coupled with careful manipulation of water and soils, and enlightened farm management, are the necessary measures which are the precondition of successful agricultural development anywhere.

Haupert, J.S., 1971 The Lachish and Besor regional settlement schemes as models for agricultural development in Israel, *Proc. Assoc. Am. Geog.* 3, 73–5.

Kashiv, I., ed., 1971 Trade with the administered areas, *Econ. Rev.* (Tel Aviv, Israel: Bank Leumi), 69.

Nuclear Energy Centers, Industrial and Agro-Industrial Complexes, 1968 Oak Ridge Nat. Lab. Rept. no. 4290.

Statistical Abstract of Israel, 1970 Jerusalem, Israel: Central Bureau of Statistics, no. 21.

P1017
Development of services for rural population in a peripheral region
REIJO K. HELLE   *Helsinki School of Economics and Business Administration, Finland*

In northern Finland, which is the northern frontier of human existence, the population is decreasing and its average age is rising. The peripheral areas of communes and other localities are also becoming relatively more sparsely settled than the central agglomerations and their immediate surroundings. These developments are associated with changes in the service functions available to people in rural areas.

In the study region of northern Finland the phenomenon involves a rather young pioneer settlement. The major part of the area has been settled within the last 200 or 300 years. An important phase, especially in the northern half of the area, was the colonization activity organized after World War I by the government in the form of several different colonization programs. Of course, the process of change in the service functions, which is examined here, is not only a result of the time of settlement and the sex and age structure of the population, but is also influenced by changes in the standard of living and the economic situation as well as the amount of information received by the people.

The background of the actual developments has been the trend of economic evolution as influenced by the exploitation of natural resources in the whole of northern Finland. During the last 25 years, a great program of harnessing hydroelectric power has been completed. Simultaneously, forest resources have been used effectively. Toward the end of the 1960s, the power plants were completed and the expanded woodworking and paper industries rapidly almost exhausted the forest resources at the same time as employment in forestry underwent rationalization and less human labour was needed. Economic growth ceased and unemployment became steadily worse. The situation became serious because the colonization programs were based on logging operations as well as on agriculture. Thus the economic circumstances of the population do not look promising now. This has led to heavy emigration. The younger part, especially, of the population emigrates. Therefore, the population is becoming older and decreasing in number, besides which the regional distribution is changing.

Along with the economic developments described, the economic growth of the nation as a whole has been usually 3–4 per cent yearly. Thus the growth of the GNP per capita has been rather rapid. This also means that the standard of living has risen, which is revealed by, among other things, the increased mobility of people. Of course, the rise in the standard of living is not distributed evenly over the whole country. However, the standard of living in the

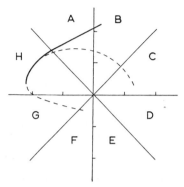
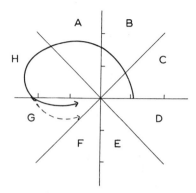

Fig. 1. Types of population change according to Webb (1963) and Naukkarinen (1969). Dashed line: theoretical population development. Solid line: actual post-war population development. A: natural gain exceeds net out-migration. B: natural gain exceeds net in-migration. C: net in-migration exceeds natural gain. D: net in-migration exceeds natural loss. E: natural loss exceeds net in-migration. F: natural loss exceeds net out-migration. G: net out-migration exceeds natural loss. H: net out-migration exceeds natural gain.

Fig. 2. Development of retailing. C: Lower part of the sector period 1900–20, period of self-support, first groceries, upper part of the sector 1920–45, beginning of commercial agriculture, several groceries. B: period 1945–50, several groceries. A: period 1950–60, groceries are common, occasional mobile shops. H: period 1960–5, mobile shops compete with fixed shops. G: fixed shops disappear (solid line) and become replaced by mobile shops (dashed line), period from 1965 on.

northern peripheral regions also is much higher in comparison to what it was earlier. In connection with these facts, it is perhaps noteworthy that, together with the rising standard of living, new demands arise. This is the consequence of increasing information and mobility. Thus younger people, in particular, demand more payment for their work and more of the comforts of life.

The population trend referred to is accompanied by a change of available services. The change in these service functions is due to the fact that a decreasing number of people use them. Private services, retailing in particular, are very sensitive to the circumstance of a diminishing population. On the other hand, public services are influenced by the decreasing population only very slowly. Only the elementary school system is sensitive to it. These differences in reaction stem from the basic reason that private services are run for economic profit, but, in the case of public services, profit is of secondary importance. They exist primarily because people need them.

These principles determine how the service posts are operated. It has been found that

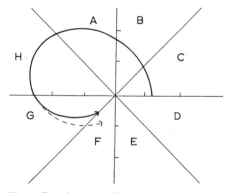

Fig. 3. Development of elementary education and other public services. C: lower part of the sector period before 1920, local elementary schools appear; upper part of the sector period 1920–45, local elementary schools are common, some other public institutions appear. B: public institutions become more effective, period 1945–50. A: expansion of public services, period 1950–60. H: expansion of social services, period 1960–5. G: closing of local elementary schools (solid line), appearing of distant services in public sector (dashed line), period from 1965 on.

the quantity and quality of the different services available to people in a locality closely relate to the types of development of the population (Fig. 1) in any given locality. Figure 2 illustrates retailing by grocers' shops as it usually occurs in the countryside. It is also intended to represent the private sector of services in general. The years indicated are rough estimates and as such tentative. Figure 3 illustrates the sector of public services with special reference to elementary education, which is the most common in the countryside. The quantities of services are measured by annual sales, people employed or pupils educated.

The trend is approaching the origin. This raises the question of future development.

It is improbable that settlement activity would disappear with services. But there might begin a new cycle of expansion and contraction. However, there is a higher probability that the redistribution of population and activities will remain permanent. Thus the developments discussed here are in process of adjustment to a stable situation with regard to the new degree of exploitation of natural resources and employment.

Helle, Reijo, 1964 Retailing in rural northern Finland: particularly by mobile shops, *Fennia* 91 (3).

Naukkarinen, Arvo, 1969 Population development in northern Finland 1950–1965, *Nordia* 8.

P1018
The tea culture system of East Asia
FREDERICK HUNG *University of Guelph, Canada*

The tea industry might best be studied as an open-ended system consisting of the growing, manufacturing, marketing, and distribution of tea. This is how the planters, brokers, auction houses, agency houses, estate owners, tea companies, and millions of tea farmers view the industry.

Six different systems of tea production, consumption, and trade could be identified. These are: (i) the millennium-old system of family tea industry in East Asia; (ii) the revolutionary *tea-commune* system in the People's Republic of China, concerning which little is known; (iii) the well-known tea estate or plantation in the humid tropics, which in fact is a modern agricultural factory; (iv) the small-holding tea cultivation in former European colonies of Afro-Asia, which is a native enterprise distinguished from the former colonial estate; (v) the highly centralized Soviet system of tea *sovkolz* (state tea farm), the *kolkhoz* (collective farm) with tea gardens, the regional tea factories, and the state tea distribution system, a system which is unique in world tea industry; (vi) lastly, the new nucleus-estate-factory scheme which is being promoted in some parts of tropical Afro-Asia.

The object of this study is to examine the traditional family system of tea growing and processing, as distinguished from the modern tea estate system developed in former European colonies. The traditional tea farmer in China and Japan also differs radically from the small holder in tropical Afro-Asia. The latter sells his fresh leaves to an estate (which generally has a factory) or to an independent tea factory, and is thus a satellite of the estate system. The tea estate does not exist in China and Japan. The tea farmer often has his own crude-tea processing unit, that is a small and simple factory in his own farm, usually contiguous to his dwelling. The product is called crude tea, which is sold in the local market or to a 'tea refinery' for further processing before it is shipped to distant domestic markets or exported overseas.

The typical tea farm in East Asia is small. In Taiwan, of the 4878 tea households (*ch'a-hu*) in seven *hsien* which received technical assistance from the provincial government in 1967, the average size of a family farm is only 1.17 hectares. This area is again divided into tea plantings, paddy land, woodland, orchards, dry fields (*han-tien*), and non-farm land-use. Nearly all tea farms practice inter-cropping which is a characteristic of tea farming in Taiwan. In a normal year, tea growing is more profitable than rice culture.

In Japan, the average size of tea farms is

even smaller. In 1966, there were 1,269,000 households (*to*) growing tea on 48,400ha, averaging 2.5 ares per household. Ninety per cent of the tea farms have an area of less than 3 ares (*Shizuokaken Chagyo no Genjo to Mondai Ten* 1967, 5). Gardening seems a more appropriate term than tea farming. It is often a side-line in East Asian farming. Despite the small size of his farm the Japanese tea farmer prefers to grow tea in his own garden and to process it himself (*jien jisei*). It provides him with a cash income and a satisfying way of life with a millennium-old tea tradition. As early as 801 AD the Japanese monk Saicho brought back from China some seeds of tea and planted them in Yeishan (Okakura 1935, 19–20). Tea is Japan's national drink (*O-Cha* = 'the honourable tea'), and the 'way of tea' (*Chado* or *Ochayu*) is universally practised in Japan as a mark of education and national culture.

The small tea garden is cultivated intensively in Japan, giving the highest yield in the world, 1755kg (1967), of made-tea per hectare. Corresponding figures from other countries for the same year were as follows: India, 1100kg; Ceylon, 914kg; Taiwan, 657kg (*Cha* 1968, 170–1). However, labour costs in Japan are much higher than in India and Ceylon, while by comparison with other occupations in Japan today, the profit in tea is low. The future of the Japanese tea industry depends on the cost of labour and the rationalization of the industry, such as integration of small holdings into larger units and the streamlining of its complex marketing system.

The Japanese tea farmer leads the world in the modern development of small tea gardens. Research and experimentation has successfully developed superior breeds of the tea plant giving high yields, and which are adapted to the humid subtropical climate with cool winters. Japan is carrying out a systematic replanting program to rejuvenate old tea gardens which have degenerate bushes. The modern transformation of crude-tea processing is in full swing. In all these endeavours, the tea gardener sets the pace of the renovation of the ancient system of family tea industry. Recently, Japan's status as a major tea exporter has shifted to that of a net importer, despite a 73.5 per cent increase in output from 1936 to 1966. This trend is explained by the increasing domestic consumption of tea, reflecting a higher living standard. Taiwan has become a major supplier of tea to Japan, as the Taiwanese practice the same system of tea culture, but Taiwan's labour cost is lower. Recently, the Taiwan tea industry has started to produce Japanese green tea (quite different from Chinese green tea) specifically for the Japanese market.

China has the longest history of tea culture in the world (Hung 1932*a*, *b*). The Communist Revolution has brought radical changes, the latest being the commune system. Unfortunately, little is known of the tea communes in China. Considerable progress has been reported in the tea industry and in the tea trade in the 1950s, before the agricultural calamities of 1959, 1960, and 1961. China has 50 million peasants who work on tiny tea gardens, with an average earning of 1.0 yen (= 42½ cents) a day. An ambitious expansion plan was in progress during the 'Great Leap Forward' (1958–9) which aimed at quadrupling the area under tea by 1962 (China: Streiflichter der Teewirtschaft 1959), but the target does not appear to have been reached. No further data are available. An appraisal of the tea commune system must await some future time.

Canada Council.

CHA ('Tea' monthly in Japanese), 21 (10), Oct. 1968.
China: Streiflichter der Teewirtschaft, 1959 *Kaffee-u. Teemarkt* 9 (13), 9–10.
Hung, F., 1932*a* The early history of tea in China, *The Chinese Social and Political Science Review*, XVI (2), 160–9.
– 1932*b* *La géographie du thé*, Bibliothèque de l'Institut de Géographie de l'Université de Lyon et des Etudes Rhodaniennes, 1, 54–8.
Okakura, K., 1935 *The Book of Tea.*
Shizuokaken Chagyo no Genjo to Mondai Ten, 1967 (in Japanese).

P1019
The efficiency of tropical, small farm agricultural practices
DONALD QUAYLE INNIS   *State University College, Geneseo, USA*

Research on the efficiency of small-farm tropical agriculture presents several problems. An almost universal technique, for example, is to mix crops in each field; intercropped fields are hard to describe, and the total yield per acre is hard to determine. Intercropping, planting long-term and short-term crops together, and relay cropping are often dismissed in geographical and agricultural literature as 'primitive' or 'traditional.'

In central India small farmers grow half a dozen short-term crops like chickpeas, corn, and radishes with young sugar cane. Apparently this does not lower the yield of the mature cane, and it substantially increases the food supply. In India and Jamaica, where most of the author's research has been done, there is a large labour force, and since the traditional intensive agricultural techniques produce more food per acre, it seems most unwise to advocate the greater use of machines and large farms, which can benefit only a few farmers, while increasing landlessness and unemployment.

Leguminous crops are often grown with potatoes or grain. In Jamaica kidney beans and Irish potatoes are usually grown together; in the Deccan, pigeon peas, cowpeas, hyacinth bean, and moth beans are usually grown with bullrush millet. In both examples small farm experimenters have discovered that legumes will help a non-leguminous crop. One of the few pieces of controlled research which has been carried out on this practice in India shows that chickpeas and wheat grown together always give a greater total yield than either crop grown by itself. Over a five-year period the average increase was 41 per cent. Critics of small farm agriculture point out that wheat doesn't do as well with chickpeas as it does by itself. Intercropped wheat or chickpeas cannot be planted as thickly as a pure crop because some room must be left for the other crop.

The question of yield per hectare in mixed versus unmixed fields is a crucial one in evaluating peasant agriculture. In the above example the question would take the form: would the spaces in a wheat-chickpea field which are devoted to chickpeas provide a greater yield if wheat were planted there instead? The answer is 'no,' because by giving up some wheat plants to provide space for chickpeas we introduce a crop with partially complementary water, light, and nutrient requirements, instead of the completely competitive requirements which other wheat plants would have. These are some basic reasons why intercropping produces more than single cropping, when all other factors are equal.

In the tropics many long-term crops (8 months to a year or more) such as bananas, cassava, yams, taro, and evergreen fruit trees are grown. Intercropping, or relay cropping, in which one crop is planted before the previous one is harvested, therefore makes much more sense in the tropics than in temperate lands. Often two or three short-term crops, such as beans or Irish potatoes, can be grown while one or two long-term crops are maturing. Long-term crops leave the soil more exposed to erosion than fast-growing short-term crops. Intercropping not only protects the soil from erosion better than single cropping; it utilizes water and nutrients at several levels in the soil, retards the spread of disease, and provides crops to eat and to sell at the same time. In tropical countries where a large labour force already exists, most of which cannot at present be otherwise employed, the use of intensive labour to produce intensive crop yields is the wisest use of land resources. Many aid programs, and much advice from agricultural experts and extension workers, are useless to peasant farmers because they are already using better techniques than those which are being suggested to them.

To discover the most widely used crop combinations it is useful to observe as many fields as possible, and make notes on the crops which occur in each field. Questioning the farmers produces much valuable information, but direct observation is an invaluable corrective. In Maharashtra, India, for example, most farmers will say that there is no intercropping with sugar, and for most of the sugar plants' lifespan this tends to be true. But field observation shows that chickpeas, coriander, corn, gourds, onions, okra, wheat, sorghum, radishes, and a dozen other crops are often planted along with sugar and are harvested before the sugar is much more than a metre high. Shevri, a small tree, is usually grown with sugar here to provide wind protection throughout the lifespan of the cane plants.

From a thousand lists of the crop combinations, one can determine, using computer cards and a card sorter, which plant combinations are most popular. The results indicate clear patterns of crop combination, which have evidently been worked out over long periods of empirical experimentation. Ingenious, well-integrated plant combinations and rotations have been developed, and the tradition is one which includes continuous experimentation. The invention of intercropping techniques, like the domestication of plants and animals, probably took several millennia. Pliny noted complicated intercropping patterns in a Saharan oasis. The fact that experimenting continues is proved by the Indian development of plant associations which include New World crops such as corn, tomatoes, sweet potatoes, pumpkins, and above all red peppers. Farmers of African descent in Jamaica grow combinations such as corn and yams, cacao and bananas, kidney beans and ginger. Analysis of crop lists shows that certain combinations have proved to be the best. These combinations are extremely valuable cultural artifacts.

Assessment of plant yields in order to test the superiority of peasant techniques is more difficult than the identification of crop combinations. Yields vary greatly from year to year in the tropics, especially in areas with no irrigation water, or when irrigation water fails. Farmers tend to cite the yields they hope for instead of average yields. Cross-checking must be attempted by asking for poor yields, average yields, and high yields, by asking farmers for yields for different areas of land (acre, square-1/10 acre, and gunta-1/40 acre), by asking for yield per plant or per seed planted, and then measuring average numbers of plants per acre, and by obtaining official yield figures from government agencies or other publications.

Printed or mimeographed figures are not necessarily the most reliable ones because yields are reported as if each field held only a single crop. The traveller will see intercropping in most fields, but publications seldom mention this fact. It is of course more difficult to give yields per acre when a field holds several crops. In Jamaica statisticians publish figures for a 10-acre mixed field of coffee and bananas, as if it were 4 acres of coffee and 6 acres of bananas. Farm plans and aid schemes for small farms employ the same fiction, which would inflate yield figures while hiding the contribution intercropping makes to higher yields. Factors which tend to lower published figures, compared to actual figures, are that some crops, such as taro and trees in a banana-coffee field, are omitted altogether; and the fact that what a family eats is usually subtracted from the total yield before the crop is marketed so that reported figures are often only the crop which is sold, and don't include bruised, damaged, or small-sized produce. Short of camping beside a field in order to weigh everything removed from it over a given period of time, the best way to improve the accuracy of yield figures would be to have more controlled experiments on experimental farms.

A great deal more attention should be given to tropical farming techniques such as those discussed here, partly because they are very efficient and it seems foolish for scientists to try and rediscover everything that tropical farmers have already worked out, and partly because if scientists would work on improving peasant agriculture, instead of trying to replace it, they would probably find that the small farmer could be helped. Farmers in the central highlands of Jamaica, for example, are currently having trouble adapting fertilizer to their intercropped fields. The right amount of chemical fertilizer for Irish potatoes seems to be too much for kidney beans and taro. Some farmers put 2.5cm of earth over the seed material before sprinkling on the fertilizer, other farmers don't know of this technique, while still others know of it, but feel it is too much work. Research scientists in the tropics must stop avoiding this sort of research problem merely because it is not valuable to the big land holder.

Growing populations in the tropics will mean that contributions to knowledge by small farmers will have to be recognized, and every effort made to help solve problems which modern small farmers face.

Edwards, David, 1961 An economic study of small farming in Jamaica (Inst. Social and Econ. Res., University College of the West Indies, Mord, Jamaica).

Innis, D.Q., 1959 Human ecology in Jamaica, PhD thesis, U. California, Berkeley.

– 1961 The efficiency of Jamaican peasant land use, *Can. Geog.*, 5 (2), 19–23.
– 1964 Problems of the Jamaican peasant economy, *Spotlight* magazine, Kingston, Canada, 25–31.

Kumar, Aggarwala, Arakeri, Kamath, Moore, and Donahue, 1963 *Agriculture in India*, vol. II: *Crops* (Bombay).

Subbiah Mudaliar, V.T., 1960 *South Indian Field Crops* (Madras).

P1020
Chinese mining settlement: features of an intrusive pattern in Southeast Asia
JAMES C. JACKSON *University of Hull, England*

The genesis and characteristics of settlement patterns in southeast Asia have received little attention from geographers. The present paper seeks to identify common features in the system of settlements generated by the intrusion of Chinese miners during the eighteenth and nineteenth centuries. In the areas affected – principally the tinfields of western Malaya and Bangka and the goldfields of west Borneo – this was a significant phase in the process of settlement evolution. Each area experienced a period of feverish activity when a form of settlement largely devoid of integration with indigenous patterns spread rapidly into hitherto almost unoccupied lands. With these explosive frontier thrusts extending the settlement network, it was the Chinese who imposed the initial lattice of settlement nodes in the mining fields and, since their goals, heritage, and organizations were peculiarly distinctive, the resulting patterns differed from those in previously peopled zones; as one manifestation of an emergent pluralism, a discontinuity developed between indigenous settlement patterns and those in the enclaves of Chinese mineral exploitation. Most studies of this kind of pioneering focus on the processes and directions of frontier movement; here attention is directed to the nature of the settlements and to their interrelationships.

Since mining was essentially a group activity for the Chinese, isolated homesteads were unusual. The smallest unit of settlement was the mining camp and, on the basis of size and potential life-span, these may be divided into three types. The simplest and most ephemeral elements in the settlement structure were the small camps associated with the ground-sluicing of hillside deposits. Somewhat more complex and scattered in considerable numbers throughout all the mining fields were the medium-sized camps with anything from a dozen to eighty or more workers. Least numerous but most permanent were the largest camps, those with populations of several hundred; frequently employing the most sophisticated techniques, these mines entailed the heaviest capital investment and had a relatively long productive life. Each camp comprised a cluster of buildings constructed of local materials. Resembling their present-day counterparts and with no greater air of impermanence than contemporary *kampongs*, their most distinctive features were the huge thatched sheds built of rough-hewn timber serving as the communal dwelling places for the miners. Additionally, to cater to persistent Chinese food preferences, there were vegetable gardens and pig-sties at the larger camps, many of which also had their own smelting furnaces. Indeed, Chinese villages frequently sprang up beside these larger mines with individual accommodation for community leaders and the aged, together with a few Chinese shops supplying imported goods and provisions and locally produced fruits and vegetables. In an evolutionary sense, these signify a transition towards the next level on the hierarchy: the local mining town.

As mining operations were extended, there developed within each field a network of small local service centres, most with populations of one to two thousand at the time of peak mining activity. These invariably consisted of one or two long, narrow streets of single-storey thatched wooden shophouses; functionally simple, they served merely as the local base of operations for the mining camps in their immediate vicinity, distributing supplies directly to them and providing facilities both for the onward shipment of produce and remittances and for the occasional relaxation of the miners. In all senses, however, these small Chinese townships were subsidiary to the major collecting-distributing centres such as Ipoh or Kuala Lumpur – each with several thousand inhabitants, a larger range of services, and a more varied population – which came to dominate the mining regions and which provided the indispensable 'head-link' between the mines and the markets for their produce.

Predominantly males in the working age groups, the Chinese miners were temporary sojourners intent on returning to the homeland when they had amassed a sufficient 'competency.' Focussing on a single economic activity, their object was maximum short-term profit, and consequently they adopted a technique of 'picking the eyes' of those deposits currently offering the most remunerative returns before moving on to fresh locations. Thus, in addition to peculiarities of form resulting from an irregular age and sex structure, an outstanding feature of the settlement pattern was its fluidity: the population of each settlement changed continuously; within specific fields the location of the workings and therefore of the camps altered as the miners worked across the deposits; local service centres mushroomed overnight and occasionally disappeared almost as quickly; and, as the mining frontier moved on, so the settlement network was extended. Intriguing unanswered questions concern both the manner in which settlement differentiation began in this phase of instability and the processes whereby this flexible pioneer pattern was transformed into that existing today.

The object of Chinese mining enterprise was of course production for export; moreover, working as they did on the frontier, the Chinese relied heavily on the inward movement of supplies, labour, and capital. In consequence, the settlement network in the mining fields was tied, through a series of intermediate stages, to the acknowledged entrepots of the archipelago. Put simply, after about 1830 Singapore was the region's commercial hub; it was linked by sea to more localized entrepots such as Malacca and also had maritime connections with river-mouth ports such as Klang at which cargoes were trans-shipped on their journeys to or from the interior. Rivers served as the arteries of trade between these coastal settlements and the inland distributing centres; the linkages between the latter and subsidiary service towns and mining camps were maintained by bullock carts, human porters, and elephants using rough tracks through the forest.

Despite the importance of transport, however, the spatial arrangement of settlements was not simply the product of the free play of general marketing or traffic principles. The trade links between settlements in this system were paralleled by the hierarchy of control within the Chinese community. In the absence of any other effective form of administration, the individual settlements in the mining fields were part of a broader Chinese framework which, with secret society connotations, performed those functions necessary for the successful conduct of mining on the frontier. Best exemplified by the West Borneo *kongsis*, the territorial basis of this framework usually reflected the outcome of inter-group conflict for the control of resources; whatever their form, particular Chinese groupings became dominant in each area, with the larger settlements serving as local headquarters, subordinate officials in the smaller centres and the general mass of Chinese owing financial and other allegiance to the structure.

These facts go far to explain the location and spacing of settlements in the mining areas. Siting was closely conditioned by the precise distribution and quality of deposits and by the need for access to an exportation routeway, be it directly to a river or – as often happened when the mining frontier penetrated virgin territory – via new tracks leading into the existing transport system. But the relationships between settlements were largely determined by the specifically Chinese network of socio-economic control and this had powerful influences on the evolving patterns. The full implications of this interaction of settlement instability, transport requirements, and the Chinese organizational framework in terms of the nature and spatial arrangement of settlements in the mining fields and the ways in which they differ from the patterns in areas of indigenous settlement clearly merit further, more refined analysis.

P1021
Changes in distribution of agricultural land in Denmark
AAGE H. KAMPP   *The Royal Danish School of Educational Studies*

The whole of Western Europe has experienced a noticeable migration from agriculture in recent years as part of a general social development. In this connection there has occurred a structural change, which is partly to be seen in an amalgamation of farms. Quite regardless of all attempts at artificially counteracting this development, it is continuing at an increasing rate.

An increase in the number of farms has occurred during several time periods, just as their number has decreased at other times, both tendencies arising for different reasons on each occasion.

From the Middle Ages the Danish farming community has undergone a social structuralization. After the Reformation, the number and influence of the squires grew rapidly as a result of favourable export conditions for corn and oxen, conditions which were particularly rewarding for big producers.

In the early 1500s it was established that all land which was not laid claim to by any other person belonged to the king. When the Crown confiscated the episcopal and monastic estates after the Reformation, almost half the land in Denmark became state property. In the seventeenth century there were almost no farmers owning their own land; the majority were tenants under the Crown or on noblemen's estates.

In the period following 1660, more and more state land was sold off; around 1760 the nobility owned 60 per cent of the land, the middle-class landowners 15–20 per cent, and the farmers only 4 per cent. The Land Act of 1766 aimed at increasing the number of freeholders, and after the abolishment of the open field system in 1781 an exchange of strips made it possible for the individual farmers to own undivided properties. In the course of the following generations, Denmark achieved the best redistribution of land in Europe, which offered good conditions for labour, transport, and crop rotation, thereby furthering efficiency and competitiveness. In the same period, many farmers were able to buy off their tenancy, thanks to the high corn prices and, in addition, thousands of smallholders' lots were set up in areas inconvenient for farmers and squires, and on the outlying commons.

From 1900 to 1960, when the parcelling out was given legislative and economic assistance, an average of 500 farms per year were set up.

Already around 1909 there was a general desire for larger units than the state farms that had so far been set up. This desire was still evidenced by remaining state parcelholders after 1940.

European agriculture has reached its present stage by means of three phases: (1) the agricultural population increases at a slower rate than the population as a whole, (2) the agricultural population rate decreases, (3) not only does the number of people decrease, but the number of properties also falls. Denmark entered the third phase around 1945 with 208,000 agricultural properties; in 1970 there were 140,000. Farms are being closed down at an increasing rate.

Throughout the first half of the twentieth century the government set up small state holdings by means of parcelling out land, and until 1960, legislation attempted to prevent farms from being amalgamated, but even then practice had long been working counter to the Land Parcelling Acts, so that there were fewer farms in 1960 than at the turn of the century, despite 30,000 new state holdings.

The decision to discontinue parcelling out of land was thus a consequence of the development, and the movement in the opposite direction is now proceeding at an increasing rate.

Freehold ownership has dominated Danish

agriculture for more than 150 years, and today over 66 per cent of Danish farmers are owner-farmers. However, many have begun to understand that the right of use is more important than the right of ownership, and it has become common practice for the owner of a property to lease his neighbour's land, so that the latter uses the buildings only for habitation. The joint operating of farms is becoming more and more common; for the whole country the total has reached 25 per cent of all properties of over 60 hectares (for the Islands up to 40 per cent). It is the holdings of between 5 and 10ha that are disappearing.

In recent years the demand for larger units has been growing faster than the actual increase in size, and as in many cases there is only one middle-aged or elderly couple to run the farm, a structural rationalization by means of amalgamation will take place at a faster rate in the future.

Three localities have been traced through four, and a fourth through five different phases of division and amalgamation:

I. In the Middle Ages, *Spanager* was a village. Spanagergaard, established on the basis of several farms, became a home farm through a royal concession of 1680. At the conversion of entailed estates into fee simple in 1923 it was parcelled out into 44 ground rent farms. Today joint operation has become common in the same area, 26 original farms being now run by 13 farmers by purchase or lease, in addition to which 3 other farms have been leased by farmers outside the area.

II. *Lindersvold* was set up on the basis of two villages in 1596. The parcelling out at the conversion of entailed estates into fee simple took place in 1922–6, when 40 ground rent farms were established. Nine of the farmers have now purchased or leased land outside the area, and in addition 36 holdings have now been merged into 15 farms.

III. In 1689 the village of *Nislev* was with royal assent placed under the home farm tariff as Nislevgaard. At the conversion of entailed estates into fee simple, it became state property in 1922, and the whole of Nislevgaard was parcelled out in 1925; the old village reappeared on its original land, though admittedly in the form of a state agricultural colony, consisting of 45 ground rent farms. The remaining Nislevgaard land was sold as building land to the borough. Of the 45 holdings there were 7 market gardens, 1 orchard, and 37 farms, 9 being run jointly with one or several others, with the result that in all there are now 11 farm units fewer than in 1925.

IV. Out of the sale of the Mön Crown estates in 1769 *Aalebaekgaard* arose on land from the 14 farms of Aalebaek village. In the same area a home farm existed already in the Middle Ages. The parcelling into 16 state holdings took place in 1922. There are now three properties fewer than in 1922, and one property is run jointly with a farm outside the area. The number of properties has thus varied as follows: 1–14–1–16–12½.

During the village phase the land was cultivated by copyholders; during the home farm phase first by copyholder villeinage, later by paid workers. The state-subsidized tenant farms are family holdings.

Kampp, Aa.H., 1971 Changes in distribution of land in Denmark. *Geog. Tidsskrift*.

P1022
Hotel food supply systems
JAN O.J. LUNDGREN *McGill University, Canada*

In countries with advanced economic systems agricultural products usually travel longer distances to reach ultimate markets than in developing countries. This greater geographic commodity mobility is a function of closer producer/consumer co-operation, i.e. of a greater market awareness among potential producers tied with the existence of an effectively functioning marketing mechanism, and of a better knowledge among the marketing operators as to requirements in the ultimate markets. Agricultural-geographic research has, since von Thünen, mostly focused upon the basic relationships between the market place and agricultural land use.

On interregional and international scales, trade flows in tropical agricultural export commodities have been analysed by numerous scholars. However, on the subnational scale, the economic link-up between a newly established (hotel) enterprise, generating a variety of demands in areas and regions with few dynamic economic activities, and the functional arrangements of the food supply lines deserves more attention. The reasons are obvious: for developing countries international commodity agreements, well-established marketing systems, and smoothly integrated financial arrangements guarantee markets for a number of the export staples, giving the exporting country precious export revenues; but on a wider domestic socio-economic scale, the role(s) of the export staple is minor; a large, economically backward, agricultural sector tries, often unsuccessfully, to meet domestic needs and the needs of convenient markets which have developed domestically, caused by the expanding international tourist trade and the needs of resort hotels in low-latitude countries of the world. The hotel demands constitute often the only dynamic market within convenient distance for large groups of farmers. However, such markets often set specific requirements, which perhaps need not be met in the larger metropolitan market place alternative, which usually exists in most Caribbean islands. The metropolitan market can nevertheless be less accessible, transportwise, less lucrative, because of tougher competition, and more difficult to supply under tropical conditions, where agricultural products are quite perishable and thus can be marketed only within limited geographic areas.

The von Thünen model with subsequent modifications has always viewed the relationship between the supply area and the market place as consisting of two elements with instant mutual responses to changes in the economic situation. Seldom has the fact that the major entrepreneurial components in an agricultural system perceive economic situations and economic opportunities in different lights constituted an integral part of the over-all model. Von Thünen's pre-industrial agrarian economic system could ignore this aspect. More recent modifications of the von Thünen model could likewise leave such considerations aside, because they were usually operating under assumptions basic to economic behaviour in a developed industrial system. However, these assumptions lose a great deal of their validity when we consider such disparate entrepreneurial phenomena as represented in the food supply/demand structures of the modern international resort hotel and the response of the surrounding, from a metropolitan point of view, economically less articulate agricultural system to these demands.

Four basic components in the agricultural marketing systems have been analysed in order to assess the interaction in the food supply system of a sample of six Caribbean resort hotels in Jamaica. The result is a conceptual model which, although containing certain von Thünen characteristics, emphasizes more the importance of the organization of transfer mechanisms and the differences in entrepreneurial abilities among producers and distributors, influencing the geographic patterns of product flows. The following constitute the main findings of the investigations.

1. The crops purchased by the hotel originate, with few exceptions, from the small-scale farm sector; some 30 products were recorded, including a few plantation crops as well as small growers' crops. The majority of crops, each one accounting for over 0.1 per cent in the aggregate food purchases, are usually grown by small farmers. Six product items accounted for over 50 per cent of all purchases in the July 1970 survey, which reduces the market potential for small-scale farm products. The hotels showed no consistent behaviour regarding shipment size/unit price, which was anticipated, thus discounting bulk purchase as a feature in the purchase pattern. The hotel mark-up price over the national farm gate price levels tended to increase with upward changes in the farm gate price quotations, indicating the attractiveness of the hotel as a market place. The anticipated characteristics in the relationships between frequency of delivery and unit price levels, also indicative of bulk purchase policies, materialized only partially and had to be modified so the degree of perishability and ubiquity of crops were accounted for. The frequency tended to rise with increasing unit price on products in lower price categories, and then decline. (Only by analysing the perishability and occurrence of individual products could an explanation be given.) The regression analysis of the distance-decay characteristics of purchase gave a low correlation, although purchases decline con-

stantly by each consecutive distance zone. This trend is broken only at extreme distance where major metropolitan centres are deployed as suppliers. Over 50 per cent of all purchases are made from suppliers within the 15-mile zone. Variations from one hotel to another are considerable.

2. Modern management principles are generally the rule among the resort hotels. Thus the food supply system would have to adapt to requirements set by the hotel as regards business confidence, regularity in deliveries, packaging, quality specifications, etc. With the general requirements set by the hotel the tendency would be to enter stable business contacts with a limited number of supply agents and to ignore irregular, minor potential suppliers. The empirical material could only partially support this hypothesis. After grouping the suppliers into five distinct categories based on their characteristics of operation, their relative supply role and location were analysed. Although there are a few, large wholesale supply agents, conveniently located, carrying a substantial trade with the hotel, the trade with other supply agents was diffused and varied, though the scale of operations often was smaller. There was no significant concentration on specific crops among any special category, although when single crops were supplied, in minor volumes, they were often delivered from single crop suppliers. Large wholesaling agricultural marketing companies, private or government-operated, played a small role. The Jamaican Agricultural Marketing Corporation (JAMC) supplied less than 5 per cent of all registered hotel purchases. However, for specialized bulk crops (Irish potatoes) over 20 per cent was supplied by the JAMC for which its distribution logistics are admirably suited. The geographic distribution of the different supply groups shows a concentration of large suppliers to convenient urban locations, with unevenly distributed short-distance and long-distance specialized suppliers.

3. In the original and subsequent von Thünen elaborations, the land use pattern was a function of production economies and market place. It was always assumed that these factors were easily perceived by the farmer. If this is true, one would expect a completely vertically integrated supply system in the hotel model, where profits are returning to a greater extent to the original producers. However, the farmers that initially put products into the hotel markets operate mostly through available supply agents, and on their assessment of the market. In only a few cases do we find that, even in the small Caribbean geographic territory, the specialized farmer-producer-marketer is the same person; where he occurs his market awareness is high, but usually he shares a small market with many competitors, and therefore his lot is hard, and his returns are usually meagre. Indirectly, he is the supplier to all the supply lines ultimately leading to the resort hotel, but as an entrepreneur he is a marginal feature on the supply scene. When he exists, however, he is an excellent example of economic man. Crop dominance/distance to market was determined by measuring the relative importance of the first-second-third dollar value crop combinations. The ensuing regression analysis gave a correlation of 0.673, using as market centres the sample hotels and metropolitan Kingston. Geographically one finds crop specialization well established at both short and extreme distance zones from markets.

4. The term 'crop availability' is used to indicate regional surplus of a particular crop. The supply lines are basically a function of the degree of geographic dispersion of grown crops and entrepreneurial perceptions of market potential for different crops: thus geographically ubiquitous crops should have shorter supply lines and less complex supply structures than geographically more concentrated crops. The mechanics of the supply lines and the transport requirements for a crop are mutually dependent, but the resulting shipment patterns can be understood only by considering regional crop availability. The degree of crop concentration was measured for administrative regions on Jamaica and schematically mapped: the anticipated fit for high crop availability/low unit price/bulk shipment character/short supply lines, and the reverse for crops having low regional crop availability, occurred partially. The reason for deviation from the expected pattern was primarily the variation in entrepreneurial behaviour among individual farmer-producers, who would overcome market distance, or high costs of production either by extreme crop specialization on a less ubiquitous crop, in areas where land was

at a premium, or by careful analysis of supply-demand changes in the market place and ultimately selecting crops with anticipated high unit price and strong demand, if peripherally located.

The geographic analysis of the food demand structures of these hotels has involved a consideration of the four components constituting the integral over-all supply system. A major finding is the distinct market awareness among some suppliers, reflected in the length of supply lines, the location of intermediaries, the crop specializations in different supply zones, and the inconsistencies in the geographic distribution of suppliers of selected crops. Thus the assumption introduced – the importance of entrepreneurial perception in agricultural marketing in developing countries – is basically correct. Entrepreneurial perception is an important factor in forming the geographic characteristics in the system. The urban/agrarian dichotomy is less pronounced; however, shorter distances to the market place tended to diminish the dichotomy: there exists a 'demonstration effect,' clearly visible among farmer-producers located near the market; there exists a remarkable marketing perception also among some peripherally located intermediaries and farmer-producers. Both features indicate the difficulty in applying theoretical designs originating in industrialized countries to situations in developing countries.

Baker, O.E., 1925   The potential supply of wheat. *Econ. Geog.* 2, 15–52.
Chisholm, M., 1962   *Rural Settlement and Land Use* (London).
Dunn, E.S., 1954   *The Location of Agricultural Production* (Gainesville).
Grotewald, A., 1959   Von Thünen in retrospect, *Econ. Geog.* 35, 346–55.
Henshall, J.D., 1967   Models of agricultural activity, chap. 11 in *Socio-economic Models in Geography*, ed. by R.J. Chorley and P. Haggett (London).
Isard, W., 1956   *Location and Space-Economy* (Cambridge, Mass.).
Jonasson, O., 1925   The agricultural regions of Europe, *Econ. Geog.* 1, 277–314.
Moyer, R., 1965   *Marketing and Economic Development*.
Schultz, T., 1964   *Transforming Traditional Agriculture*.
Symons, L., 1966   *Agricultural Geography* (London).
Von Thünen, J., 1826   *Der isolierte Staat in Beziehung auf Landwirtschaft und Nationaloekonomie* (Rostock).
Wolpert, J., 1964   The decision process in a spatial context, *Ann. Assoc. Am. Geog.* 54, 537–58.

P1023
Perceptional change in attitudes toward rural land use in Southern Ontario
GEORGE H. MICHIE   *York University, Canada*

Rural land use to date has been typified by unplanned development which occurred more or less spontaneously to meet economic and related demands of the moment. Ontario is entering a new stage in the human drama on land use. Disturbing evidence is before us that urban problems can be severed no longer from rural problems. If land use is not circumscribed very soon by a long-term plan which provides an ordered layout and distribution of uses, the opportunity to do so will have been lost. Even now tentative plans are the object of scorn by those who wish land use developments to be the prerogative of economic interest groups, developers, and speculators. Happily, the beginnings of perceptual change in attitudes toward rural land use may be noted in the public utterances of government spokesmen, private planners, environment interest groups, and others among the concerned public.

The rigid land survey and land registry system which confronted most of the early settlers resulted frequently in the rather too generous allocation of swamp, hill, and poor soil in the lots allocated to some settlers, while other land ticket holders found themselves with properties with excellent prospects (Thomson 1966). As a result the present rural areas of central southern Ontario support a variety of buildings reflecting the quality of the land and the energy and vision of the operator.

There was rapid city growth in the first

half of the twentieth century but urban pressure on rural land intensified sharply after 1950. Formerly the chief supplier of perishable farm commodities, the adjacent rural area was called upon now to supply new urban space. Without updating its form, private enterprise extended the city into the country. Speculative spoliation and an acceleration of land prices resulted. As the new population advanced, rural roads were widened and paved and new traffic arteries constructed. Improved accessibility and increased leisure time encouraged urbanites to make new forays into the country seeking recreation and open space. The exploration process often led to a desire to own country property. To separate, convey, and sell became the feverish activity of the later 1950s and 1960s. Even the freezing of subdivision of rural land which became general in Ontario after 1964 did not greatly deter the land separators and developers. Many of the smaller parcels of land posted for sale in the area are subject to local and provincial government approval for separation.

The appetite for rural land by developers and urban buyers has become exceptionally voracious in the 1970s. Any land well built upon by a settler and his descendants is quickly converted to a rural estate. Smaller patches of land with trees, a view, and either water or potential for storing it, become rural estates, and roadsides convenient to place of work are turned into strips of rural-urbanization. The new urban demands on the countryside threaten to annihilate it.

The effect on agricultural land use in the areas within 50km of the larger urban areas has been catastrophic. Experts on agriculture in the colleges and government departments have called for an end to small farm units and their conversion to large industrialized 'agri-businesses.' In order to expand in these times of high land values and assessment many farmers have abandoned traditional and proven land husbandry for expansionist methods. Their operations have been automated because of the scarcity of manpower available to work in agriculture. The change-over has been effected at a time of high interest rates so that repeated expansion of plant has been necessary to increase output ever higher. The result has been that enough agri-businesses have been created that prices of the commodity that they have specialized to produce have fallen. The quality of the product has had to be averaged downward and the number of producing units increased within a smaller space.

Concentrated specialized farming endeavours have created new problems. Large amounts of animal waste must be disposed of on land not capable of using it in the effective ways of the past. The use of increasing quantities of synthetic fertilizers, herbicides, and pesticides contributes to air, soil, and water pollution as well. In an attempt to dispose of waste and to increase output the agri-business has come into a head-on collision course with the interests of the new rural-estate owners and the new roadside rural-urbanites. The farmers' expanded plant with its cluster of silos and storage towers is an even more visible feature of the landscape. In the process of expansion, the farmers have very seriously overextended their economic base on borrowed money. Equipment prices soar while produce prices are falling, waste disposal is difficult, and growing public interest in food quality makes the consumer question the methods used to produce the product. The unhappy results are fairly new to Ontario but have been heralded by similar developments in Great Britain and the United States of America years ago.

Rural townships in southern Ontario are beginning to learn the economic frailties of releasing small lots on country sideroads for low-cost subdivision-type homes (Coleman 1969). Although the demand for such properties is flattering, the resulting homes frequently do not provide an assessment sufficient to cover the education taxes necessary to teach the basics to the offspring of the home, let alone contribute enough to municipal coffers to cover the cost of road repairs and the other services the home-owners demand.

Esquesing Township, between Toronto and Hamilton, has undertaken a survey into land use in that area. The report raised a furor among those wishing for perpetual expansion of every aspect of progress in the township's economy. It was reported that the township was in an enviable position with a reasonable tax base and very little indebtedness. Further scattered urban settlement in the country is to be discouraged (Dahms and Pearson 1969; Hills et al. 1970). The major conclusion was that there is no gain in being big, that urban development should take place only in the existing towns where the services are available, and that all valley lands and the numerous ravines should be free of urban encroachment (cf. Little 1968). This township is just one of several in the area already very badly scarred by quarries operating in the limestone of the Niagara Escarpment. Any land use plan for the area comes a little late.

Rural municipalities now question the right of individuals to locate where they like and the right of cities to use the country as a source of sand, gravel, and rock where the resulting excavations may be filled with city wastes. Among the public there is a new perception of a unique historic countryside which is to be valued and preserved. It is this countryside which in 1970 drew 1.6 million visitations to the conservation parks within a 48km radius of Toronto. Such use is incompatible with plundering and despoiling. Yet the very amenities sought for in the country are being destroyed as wider and wider highway networks improve accessibility.

Strong leadership in planning has been provided for a variety of discussions led mainly by departments of the government (Ontario Department of Municipal Affairs 1967). Whereas government agencies are concerned with rationalizing growth, the public is gradually becoming more and more interested in totalities of the environment and concerned about balances and imbalances in land use. Rather than allow private initiative to organize itself and use up rural space as it sees fit, government planners speak of aligning urban development along transportation corridors. Decentralization is to be planned and certain key areas developed in zones well removed from present urban congestion. Open space is to be conserved for recreation and maintenance of the natural life of the region. An attempt is to be made to preserve the unique attributes of the regional landscape, to cut to a minimum urban uses of productive agricultural land, to minimize air and water pollution, to preserve identifiable communities, to improve accessibility, services, and developments consistent with the new technology, and in so doing to provide for a flexible approach to future needs.

These are ambitious plans and impossible to implement without public co-operation. Private planners and consultants have other points of view but outside of government cannot implement them on a wide scale. 'A utopian formulation is the statement of a purpose so broad, so lofty, that if it means anything at all it denotes an unattainable goal' (Peterson 1967, 182).

If there is to be rational response of private economic interests to competent government-led planning, there must be some change in its concepts of highest and best use. Until now, highest and best use has referred to uses which led to the highest economic return. To get such returns urban developments were replicated out into the country on the pattern of the city. Progress might have been defined as skyscrapers, eight-lane expressways, canyon streets, and vast acreages of new buildings. The problem with such progress is the problem of irreversibility when the landscape is used in this way. Open space is still multi-purpose space and a variety of alternatives for its use are possible.

Attitudes toward rural land use in central southern Ontario have altered since the days of the first settlers but not as yet on a wide scale. While certain uses are dominated by circumstances of the economy, it is notable that the public now suggests alternative solutions to some of those put forward by private enterprise or government. Possibly an ideal of progress concerned with bigness and flashiness still stands in the way of impartial considerations of alternatives. But a new interest in historical chains of development is arising along with a new vision of the environment as an interrelated coherent system under tremendous stress at key points in the cycle.

Rational, well thought out, government planning must decide the growth of Ontario's cities and towns, the preservation of unique rich agricultural land, the setting aside of conservation and recreation areas, the means of moving population, and other matters for preservation of open space. The engineering of rational planning has been made at least possible by the recent perceptional change in attitude toward land use on the part of some part of the public. It is to be hoped that the lessons learned from the experiences of other countries which are at a more advanced stage of dealing with similar problems will shorten the task of effecting change in this area.

There is an opportunity to create a dazzling new pattern of land use which will be exciting to live in.

Coleman, A., 1969 *The planning challenge of the Ottawa area*, Geog. Paper no. 42, Dept. of Energy, Mines and Resources, Ottawa.

Dahms, F.A., and N. Pearson, 1969 *A study of the fringe area north of Metropolitan Toronto* (Central Ontario Regional Development Council, Toronto).

Hills, A., D. Love, and D. Lacate, 1970 *Developing a better environment* (Ontario Economic Council, Toronto).

Little, C.E., 1968 *Challenge of the Land: Open Space Preservation at the Local Level* (Toronto).

Ontario Dept. of Municipal Affairs, 1967 *Metropolitan Toronto and Region Transportation Study: choices for a growing region* (Toronto).

Peterson, W., 1967 On some meanings of planning, in S.V. Ciriacy-Wantrup and J.J. Parsons, eds., *Natural Resources, Quality and Quantity* (Berkeley).

Thomson, D.W., 1966 *Men and Meridians* (Ottawa).

P1024
Agricultural regionalization: origins and diffusion in the Upper South before 1860
ROBERT D. MITCHELL *University of Maryland, USA*

The literature of regional geography in the United States is replete with discussions of the typology and identification of geographic regions. Recognition of the processes of regionalization has taken two forms: (1) definition of contemporary regions as the result of the analysis undertaken as part of the study; (2) examination of the processes involved in the origins and development of regions through time. The historical approach has been much less exploited. For example, only recently has there been any serious attempt to explain the origins of the former Corn Belt. This paper examines the pre-1860 origins of an adjacent agricultural complex in the Upper South, a region defined as the Corn and Winter Wheat Belt in the twentieth century and now reduced to the status of a 'general farming' area. Its nineteenth-century character was quite different.

The agricultural distinctiveness of an Upper South was clear by 1860. In contrast to a Lower South dominated by the production of cotton (or in some areas rice or sugar) for export, together with corn, cattle, and mules, large proportions of slaves, and plantation forms of agriculture, the Upper South had a more balanced system of crops (tobacco, hemp, flax, wheat, and corn) and livestock (beef cattle and hogs), and more emphasis on mixed farming than on plantation agriculture. Consequently, the proportion of slaves in the region was only half that of the Lower South by 1860. In that year, as a section, Maryland, Virginia, North Carolina, Kentucky, Tennessee, and Missouri dominated US production of tobacco, hemp, and flax, were in the second level of producers in corn, wheat, beef cattle, and hogs, and produced more than 80 per cent of the South's output of hay, oats, rye, barley, and buckwheat. If one had defined the area in which the above crop and livestock combinations were most typical by, say, 1830 it would have included all of the above-mentioned states (except southeastern North Carolina, western Tennessee, and northern Missouri), as well as the southern third of Ohio, Indiana, and Illinois. How was this set of agricultural specialties achieved and what relationship did it have to the origins of the Corn Belt?

In their brief examination of the origins of the Corn Belt, Spencer and Horvath identified four eighteenth-century source areas for the region's agricultural characteristics (Spencer and Horvath 1963, 74–6). Three of these, southeastern Pennsylvania, the Kentucky bluegrass, and the Nashville basin, were intimately associated with Upper Southern agriculture. The authors do not indicate precisely how the farming features of these areas were diffused to other areas. Southeastern Pennsylvania may well have been a pivotal area in the evolution of trans-Appalachian agricultural patterns, but it cannot account by itself for the distinctive features of the Upper South or the early Midwest. As Spencer and Horvath point out, corn and hogs were not the dominant products of this hearth. Tidewater Virginia would appear to have been a second primary source area. Viewed from this standpoint, the Kentucky bluegrass and the Nashville basin were really western expressions of the fusion of agricultural elements contributed from the general areas of southeastern Pennsylvania and tidewater Virginia.

A fusion of components from these two agricultural subcultures had occurred earlier in the Shenandoah valley of Virginia (and also to some degree in the eastern North Carolina piedmont) between 1760 and 1790. Here yeomen-farmers from Pennsylvania and adjacent areas mingled with small and large planters from eastern Virginia to produce an essentially mixed-farming region based on corn, wheat, hemp, tobacco, flax, rye, barley, oats, cattle, hogs, and horses (Mitchell 1969). Because most of these elements were to be found elsewhere along the Eastern Seaboard it is often difficult to identify their primary source area of specialization. Therefore the remarks which follow are necessarily tentative.

It seems fairly clear that the origins of the plantation component of Upper Southern agriculture, together with the cultivation of tobacco and the institution of slavery, were to be found in tidewater Virginia. They were diffused across the Virginia piedmont after 1720, had appeared in more limited form in the Shenandoah valley by 1775, and in the Kentucky bluegrass and Nashville basin by the 1790s. Flax, rye, barley, oats, and buckwheat in the Shenandoah valley seem to have been introduced mainly from Pennsylvania. Corn was ubiquitous as a livestock feed but it was the large Southern gourdseed varieties which were the basis of the early livestock activities in the Ohio valley (Power 1953, 49, 151–4).

Wheat was not well adapted to physical conditions along the Atlantic coastal plain, but increasing demands for flour and bread during the late colonial period encouraged the development of wheat specializations in southeastern Pennsylvania and in the piedmont areas of Maryland, northern Virginia, and North Carolina. By the 1780s, the Shenandoah valley was a major wheat producer, and by 1800 this specialization had spread to Kentucky and Tennessee. Hemp followed a similar diffusion pattern except that its commercial hearth area seems to have been the Virginia piedmont in the late 1750s. From there it diffused rapidly because of Revolutionary War demands to the Shenandoah valley and later, in the 1790s, to Kentucky, which remained the leading producing area until 1860, followed closely by Missouri.

With respect to livestock, steer-hog-horse complexes appear to have been well developed in western Virginia and North Carolina by 1775. Similar combinations emerged in Kentucky and Tennessee before 1790. Corn and oats, as livestock feeds, also diffused in this manner. The use of rye and barley as livestock feeds declined as they diffused westward. Rather, their importance in the production of whiskey, which had been a regional specialization of the settled Appalachians from Pennsylvania to North Carolina, was the main reason for their appearance in western agricultural systems at all, particularly in eastern Kentucky and Tennessee.

A major clue to the diffusion of these agricultural characteristics west of the Appalachians lies in the nature of the source areas from which westward migrants came. Most of the early settlers in Kentucky, Tennessee, and the Ohio Valley came from Virginia and North Carolina, and to a lesser extent from Pennsylvania and Maryland. By 1850, of the 400,000 Virginians who had emigrated since 1800, 37 per cent were to be found in Kentucky, Tennessee, and Missouri, and 40 per cent in the Old Northwest; of 265,000 migrants from North Carolina 38 per cent had relocated in Kentucky, Tennessee, and Missouri, and 20 per cent in the Old Northwest (Lynch 1943).

However, the agricultural diffusion patterns westward across the Mississippi and northward across the Ohio differed significantly after 1800. Although all of the cash crops cultivated in the Upper South could be grown in the Old Northwest, hemp and flax were much more important in Missouri; for physical and economic reasons corn and wheat became more significant in the Old Northwest. Cattle and hogs were present in both areas, but they were much more thoroughly integrated into the grain economy north of the Ohio. Equally important, some essential features of Upper Southern agriculture were legislated out of the Old Northwest. Congress, by passing the (Northwest) Ordinance of 1787, prohibited the expansion of slavery north of the Ohio, thus affecting the viability of such labour-demanding crops as tobacco and hemp, and discouraged the perpetuation of large landed plantations. The (Land) Ordinance of 1785 established a national land survey system which was applied first in the Old Northwest. Thus, the 1785 act circumscribed a new cadastral framework and the 1787 act a more flexible social framework for the agricultural exploitation of the Old Northwest which distinguished that region from the area immediately to the south.

Lynch, W.O., 1943 The westward flow of southern colonists before 1861, *J. Southern Hist.* 9, 303–27.
Mitchell, R.D., 1969 The commercial nature of frontier settlement in the Shenandoah Valley of Virginia, *Proc. Assoc. Am. Geog.* 1, 109–13.
Power, R.L., 1953 *Planting Corn Belt Culture* (Indiana Hist. Soc., Indianapolis).
Spencer, J.E., and R.J. Horvath, 1963 How does an agricultural region originate?, *Ann. Assoc. Am. Geog.* 53, 74–92.

P1025
Types of agricultural land
H.J. NITZ   *University of Göttingen, West Germany*

In agricultural typology the farm is regarded as a basic unit. We try to identify the various regional types by a set of specific characteristics which Kostrowicki summarized in Delhi in 1968. As Olmstead has shown, this basic unit can be explained as a functioning system. Most of the characteristics, respectively, of the subsystems and their elements are related to the operational unit as a whole. But, as land use studies have shown, quite a number of characteristics are not properties of the whole utilized area of a farm but primarily refer to different spatial sub-units of it, such as arable land, pasture land, garden land, or tree crop land. A combination of such 'types of agricultural land' is the rule for most mixed farms, which combine different branches of production such as animal husbandry, growing of different field crops, vegetable gardening, and production of tree crops. For each of these branches there may be quite specific types of agricultural land, as for instance wet-rice land, vineyards, tea gardens, or rough pasture for sheep. Each of these kinds of land is characterized by specific phenomena and functions, a specific crop combination and rotation, or in the case of pasture land, pasture rotation and kind and density of livestock, specific equipment and techniques of cultivation, of irrigation, a specific sequence of working operations during the agricultural year, a specific intensity and productivity of land use, etc.

Such types of agricultural land are the spatial sub-units or links of the functioning agricultural system or type of agriculture. In the case of the wheat farming system of the Prairie Provinces of Canada there may be only one type of agricultural land, but generally we will find several types combined within a farming system.

I shall demonstrate the utility of the concept of agricultural land types by the example of the type of agriculture in the Lesser Himalaya region of Kumaon in north India. In the villages at altitudes of 1000–2500 metres the utilized area of the peasant holdings with their fragmented and dispersed field plots consists of five types of agricultural land:

1. The first type is situated on the alluvial terraces of the valley bottoms. It is irrigable all the year by small canals constructed by the village community. During the monsoon the embanked and terraced fields are cropped with transplanted rice, with all the well-known operations connected with this crop. In autumn the fields are manured and sown with winter wheat, mixed with some mustard and lentils. This type of irrigated land on the valley bottoms is thus characterized by a one-year crop rotation of wet rice and wheat.

2. The lower slopes are covered with flights of artificial terraces and form a second type of agricultural land. Irrigation is not possible. Rain-fed crops on this type of land are dry rice and finger millet during the monsoon, wheat and barley in winter. Pulses are mixed with all grains. A two-year crop rotation is applied. First year: monsoon – finger millet; winter – pasture on fallow,

followed by fallow plowing during late winter rains. Second year: monsoon – dry rice, manured; winter – wheat and barley. Winter fallow is necessary because dry rice has to be sown in April before the harvest of the winter crops; furthermore, dry rice needs the soil moisture accumulated in the fallow fields during the winter rains for germination. The fallow fields are pastured commonly by all the villagers. To protect the winter crops from cattle, a common field system is applied; all the fallow fields of a village form one or a few contiguous blocks, separated from the wheat and barley fields. Intensity and productivity of the upper land are noticeably lower than on the irrigated land, where 3.5–4.0 tons of rice per hectare are harvested compared to 0.8–1.2 tons on the slopes.

3. Above the terraced slopes we find a third type of land: common grass and bush land, where the farmers, after a slash and burn cultivation, may grow two unpretentious monsoon crops such as small millets, amaranth, and buckwheat. Only the hoe is used. During the 3–9 year period of grass and bush fallow the fields return to common pasture. There are no artificial terraces. Intensity and productivity are very low. Year by year, plots of good grass are selected by the village community for haymaking and protected from the cattle by fencing. From these unimproved meadows each villager gets his share. Thus this type of agricultural land is characterized by a land-use pattern consisting of an irregular rotation of pasture, haymaking, and shifting cultivation.

4. The higher slopes are covered by forest which, though state forest, by traditional right is used by the farmers as common pasture for the cattle all the year round. In the forest the females from the villages gather loads of pine needles and oak leaves for litter and manure. These functions characterize the forest as another type of agricultural land within the farming system.

5. The highest hills of the lesser Himalayas with altitudes of 2400–2700 metres are covered by oak forests. They are used for mountain summer range. Neighbours join in taking their cattle, with some young family members, to the alpine huts. The mountain summer range is quite a distinct type of agricultural land and an important element of the agricultural system of the lesser Himalayas of Kumaon.

As has been shown, each of these five types of agricultural land has its specific characteristics and operates as a spatial as well as a functional sub-unit in the farming system, each different in importance, in extent, and location.

For a systematic approach to agricultural land types the following formal and functional characteristics should be considered:

I. The natural potentialities of the ecotype on which the type of agricultural land is located (soil fertility, natural water supply, relief).

II. Social characteristics, in particular the form of land tenure, which may be private, communal, tribal, or alternate from communal to private, which again may result in different field patterns.

III. Organizational and technical characteristics of land type: (1) Leading agricultural land use (such as forest pasture land, improved grassland pasture, field crop land, garden land). (2) Land rotation system (such as field-grass systems on improved land or shifting systems). (3) Crops or livestock on the land type (such as one-crop or crop-combination systems). (4) Crop rotation system (which may be a common field rotation) on grassland; system of pasture rotation or rotation of meadow and pasture (hay and grazing). (5) Techniques of preparation for land use such as slash and burn, terracing, drainage. (6) Soil water accumulation and irrigation, dry farming techniques on rainfed land; irrigation systems and techniques, as well as water rights, on irrigated land. (7) Forms and sequence of work on the land type during the agricultural year, and the tools and draught power used.

IV. Intensity characteristics: land intensity measured by the number of crops per year or the rate of stocking of livestock; input of labour and capital per unit area (e.g. the input of dung may differ depending on the distance of the fields from the farmstead, resulting in the formation of types of dungland and unmanured land with long fallow, e.g. infield and outfield systems).

V. Production characteristics of the land type measured by the output, such as land productivity and labour productivity per unit area; the function of the agricultural land type for the production system of the farm as expressed by the destinations of the land types' production which may be used for stable feeding, for home consumption, or for sale. Assessing these components will be easier for a particular type of agricultural land than for the operational unit as a whole.

Kostrowicki, J., 1970 Agricultural typology, summary of the IGU Commission for the years 1964–1968, in J. Kostrowicki and W. Tyszkiewicz, eds., *Essays on Agricultural Typology and Land Utilization*

(Geographia Polonica 19; Warszawa), 11–29.

Olmstead, C.W., 1970 The phenomena, functioning units and systems of agriculture, *ibid.*, 31–41.

Nitz, H.-J., 1970 Agrarlandschaft und Landwirtschaftsformation (Agricultural landscape and agricultural formation), *Mod. Geog. in Forsch. und Unterricht*, Hanover, 70–93.

P1026
Recent rural Georgian abandonment as an adjustment
G. FREDERICK PAYNE *Georgia Southern College, USA*

What is the nature of rural settlement change? Neglect of the study of process in rural settlement leaves many unanswered questions. What are the most appropriate measures of process (Stone 1968, 321)? Can rural settlement patterns change rapidly (Clawson 1966, 283)? How important is the magnitude of change in settlement numbers in the delineation of an area of settlement or abandonment? Are rural settlement distributions becoming more clustered or more uniform? Obviously, answers are dependent upon the circumstances of location and the time period in question.

This paper reports on the nature of rural settlement change in the southeastern USA, specifically in Oglethorpe County, Georgia, between 1940 and 1967 – a location and period in which abandonment was the dominant process. In this respect, Oglethorpe County is representative of a number of counties in eastern Georgia which have experienced severe depopulation and abandonment since about 1940 (Payne 1971; Zelinsky 1962, 510–11). The research indicates that rural settlement patterns can adjust rapidly under the influence of a proper combination of dynamic economic and social conditions, since, in less than three decades in Oglethorpe County, (1) the number of occupied rural dwellings declined by more than one-third and (2) the distribution of occupied rural dwellings became more clustered.

1940–67, a period of adjustment. Some variables related to rural settlement, such as agriculture, population, and roads, have recorded dramatic changes since 1940. Oglethorpe County's population was stabilized above 12,000 between 1930 and 1940 when the Great Depression temporarily slowed the rural exodus, and then declined by about two-fifths between 1940 and 1970. Losses were most dramatic in the farm and Negro components of the population.

Income-producing agricultural activity shifted from one dominated in 1939 by crops, especially cotton, to one dominated by poultry products in 1964. Between 1939 and 1964 neither forest products nor livestock and dairy products changed significantly in their proportion of the value of all farm products sold. However, pine forests became the dominant land use as land management was transferred from many individual farmers to timber companies and the federal government.

The paving of selected dirt roads was the major change in the transportation network between 1940 and 1970. More than 291km of road was paved by 1970, compared with only about 32km – the length of the primary east-west route through the county – in 1943.

Under such conditions of change, one could expect the rural settlement pattern to exhibit characteristics of adjustment. The number of occupied dwellings should decline in response to agricultural and population change. The loss of labour stimulated the search for alternative crops and for more efficient machinery. The adoption of poultry as the predominant economic activity led to a decline in the demand for a large number of workers and permitted the conversion of cropland to timber production. Settlement distribution should become more concentrated as improved accessibility along paved roads increased their desirability as residential locations.

Changes in rural dwelling numbers, 1940–67. Using a grid of 1.6km square quadrants which covered the county (but excluded incorporated places and some marginal areas outside the grid) rural dwellings shown on the earliest and most recently published Oglethorpe County highway maps were counted (Georgia State Highway Department 1940, 1950, and 1967). The data revealed a marked decline in numbers between 1940 and 1967. In 1940 there were 2604 occupied and unoccupied dwellings within the study area of the county. The total number of dwellings dropped drastically to 1978 in 1950, and to 1897 by 1967. In a parallel fashion, the number of occupied dwellings fell from 2265 in 1940, to 1710 in 1950, and finally to 1568 in 1967. Thus, in less than thirty years, the number of occupied dwellings decreased by more than thirty per cent.

Changes in the patterns of rural dwelling densities, 1940–67. Between 1940 and 1967, adjustments in the distribution of rural dwellings tended to produce a greater degree of clustering. In 1940 dwellings were scattered over the countryside with some clustering near the scattered service centres and along the major, but still unpaved, roads. Quadrants with high densities (greater than 10) of occupied rural dwellings were usually identified as service centres with a community name. Generally quadrants with higher densities are connected by roads which have been paved since 1940. Several high-density cells were located adjacent to incorporated places along major highways. Low-density cells were scattered all over, but several areas of contiguous quadrants of unoccupied land were present in the southern and eastern portions of the county.

In 1940 there were distinctive differences in the pattern of dwelling densities in the northern and southern halves of the county. The latter had a large number of quadrants with low dwelling densities and only a few with high densities, but in the northern segment the reverse was true. The differential, which persisted in 1967, is unexplained at present.

In 1950 high-density cells of occupied rural dwellings were less numerous in both halves of the county and there was an increase in the number of low-density cells. Dwellings were abandoned on large tracts of land which passed from agricultural into timber production. The maintenance of high-quadrant dwelling densities in service centres and along paved roads reflected the higher degrees of accessibility.

In 1967 the trend toward further polarization of the occupied-unoccupied quadrants continued. In the eastern and southern portions of the county the area of unoccupied land expanded, and quadrants which formerly had the highest rural densities had indices that were as high or higher. The increased densities of rural settlement on the county's western border, adjacent to Clarke County, probably were directly related to the rapid growth of the Athens urban area.

The correspondence between quadrants with high dwelling density and the occurrence of paved roads is striking. The chronological relationship between these two variables, with its implications for settlement redistribution, is clear. In Oglethorpe County, rural settlement tended to cluster around rural service centres; when roads joining the clusters were paved, the density of occupied dwellings along them tended to increase. Conversely, such densities tended to decrease along unpaved roads joining the smaller service centres.

A comparison of Lorenz curves showing the proportion of the area of the county covered by quadrants in which occupied dwellings were present provides further evidence that the occupied dwellings are being concentrated onto smaller and smaller segments of the county's area (Cole and King 1968, 243). Slopes of the curves are relatively parallel, but the proportion of land in quadrants over which the occupied dwellings are scattered decreases significantly from the 1940 level of 90 per cent, to 81 per cent in 1950, and to 65 per cent in 1967.

The measurement of point patterns by quadrant frequencies also demonstrates increased clustering between 1940 and 1967 (Birch 1967). The ratio (R) of the arithmetic mean of the points per quadrant to the variance would yield a value of 1.00 if the point distribution were random. Grouping is indicated by a ratio below 1, and a value above 1 for the actual pattern reflects a tendency towards a dispersal of points more regular than in a random situation. R values were computed for both total and occupied dwellings in Oglethorpe County for 1940,

1950, and 1967 and for the occupied dwellings in the northern and southern halves of the county. In every case the ratio indicated the pattern of change was from a clustered toward a more clustered distribution. The distribution which tended most toward a random distribution was the 1940 ratio ($R = 0.3818$) for the northern part of the county. The 1967 county distribution of occupied dwellings was the most clustered ($R = 0.1480$). In 1967 the occupied dwelling pattern changed so that there was virtually no difference between the ratios for the northern and southern portions ($R = 0.1552$ and 0.1559).

In summary, recent changes in agriculture, population, and transportation in Oglethorpe County, Georgia, have been accompanied by adjustments in rural settlement which are characteristic of abandonment. The adjustments in this representative county of eastern Georgia include a decrease in dwelling numbers and an increasingly clustered distribution. Additional research planned is necessary to identify additional relationships and to establish the generality of these findings for other locations and time periods.

Kirk H. Stone.

Birch, B.P., 1967 The measurement of dispersed patterns of settlement, *Tijdschrift voor Econ. en Soc. Geografie*, 58, 73–4.
Clawson, Marion, 1966 Factors and forces affecting the optimum future rural settlement pattern in the United States, *Econ. Geog.* 42, 283–93.
Cole, J.P., and C.A.M. King, 1968 *Quantitative Geography* (New York).
Georgia State Highway Department, 1940, 1950, 1967 General Highway Map of Oglethorpe County.
Payne, G. Frederick, 1971 A geographic analysis of the relationship between population change and rurality in Georgia, 1940–1970, paper read at the Georgia Acad. Sci., April 24.
Stone, Kirk H., 1968 Multiple-scale classifications for rural settlement geography, *Acta Geog.* 20, 307–28.
Zelinsky, Wilbur, 1962 Changes in the geographic patterns of rural population in the United States, 1790–1960, *Geog. Rev.* 52, 492–524.

P1027
Types of agriculture in Ecuador
ALDO PECORA   *Facolta di Magistero Universita, Italy*

Many inherent characteristics, or properties, representing three principal groups of criteria (social, technical, and economic) to determine the type of agriculture of a region (Kostrowicki and Helburn 1970) cannot be investigated in many developing countries for lack of suitable statistical data. Therefore, although it is possible to give a synthetic picture of the agricultural typology of small areas through local surveys, any study on a national scale can have a considerable significance in obtaining a deeper knowledge of the agriculture of the country considered, but runs the risk of being of very little value for a comparative analysis on a world scale.

In spite of this, I intend to make a preliminary attempt at a typological classification of agriculture in a developing country – Ecuador – in order to stimulate this type of research so as to achieve reasonable results, good either for these countries or for a more ambitious comparative analysis on a world scale.

For some important characteristics of Ecuadorian agriculture – such as size of holdings, land ownership, mechanical power, organic manuring, mineral fertilizing, value of agricultural products, level and degree of commercialization – only national data exist, while for other properties statistical data are available on a provincial basis (Secretariá General 1969), but the administrative areas are so large as to conceal a great variety of agricultural conditions.

For this paper, the graphic method of star-diagrams has been used, with the following indices:
A. GENERAL LAND USE:
1. Percentage of land in farms (L)
2. Percentage of arable land (A)
3. Percentage of pastures, natural and artificial (P)

4. Percentage of forests (F)
B. TYPES OF CROPS:
5. Subsistence crops (S)
6. Speculative food crops (C)
7. Industrial crops (I)
C. LIVESTOCK RAISING:
8. Number of cattle per 100ha of land in farms (c)
9. Number of sheep per 100ha of land in farms (s)
D. SOCIAL CHARACTERISTICS:
10. Density of agricultural population (D)
11. Average size of farms (FS)
12. Percentage of owners, tenants, and their families (CT)
13. Percentage of permanent labour (PL)
14. Percentage of occasional labour (OL)

In order to determine land use, I adopted the Polish Land Utilization Survey's method (Kostrowicki 1970); but with a view to emphasizing the importance of crops in the pioneer fringes of many provinces, crops were considered even if they cover less than 10 per cent of the agricultural land. Other indices were calculated in the same way.

On the basis of the above structural characteristics, nine types of agriculture were distinguished. They differ mainly in the amount of land used (a very important feature of all countries with a scanty population), in the types of crops (especially in the relations between subsistence and speculative crops), and in the social composition of agricultural labour.

The main characteristics of Ecuadorian agriculture are as follows:

1. *Northern Altiplano provinces* (Carchi-Imbabura). Land in farms consists of about 40 per cent of the total area. Farms are usually of small or medium size (7–12ha on the average), and are operated mostly by peasants and occasional labour. Production is orientated towards maize and potatoes (with a little sugar-cane in the lowlands), with cattle and sheep. This type of agriculture can be presented in a formalized way as follows:
Carchi: L_4/A_3 (S_4zm st + C_1so) + P_3 ($c_4 + s_1$) + $F_1/D_5 - FS_3 - OT_4 + OL_4$

2. *Southern Altiplano provinces*. Land in farms consists of 40–75 per cent of the territory, with very small farms (2–6ha) inhabited by a dense peasant population. Maize, or maize and barley and potatoes, and sometimes barley and wheat, with a small sector of speculative crops, and good cattle breeding, form the most common production orientations. It is possible to distinguish at least three subtypes:
Cotopaxi: L_5/A_3 (S_3zm hv st + C_2tc ms) + P_3 ($c_4 + s_2$) + $F_1/D_4 - FS_2 - OT_5 + OL_2$
Bolívar: L_4/A_4 (S_4zm hv ts + C_3ca) + P_3 ($c_3 + s_3$) + $F_1/D_5 - FS_2 - OT_5 + OL_3$
Azuay: L_3/A_3 (S_6zm + C_1so) + P_2 ($c_5 + s_3$) + $F_1/D_6 - FS_1 - OT_5 + OL_2$

3. *Pichincha*. Its type of agriculture is peculiar owing to the presence of medium-sized and large farms (24ha); therefore occasional and permanent labour play a more important role. The production orientation is similar to that of other Altiplano provinces, but cattle is more productive and forests are more widespread.
Pichincha: L_4/A_3 (S_3hv zm st + C_2ms) + P_3 ($c_3 + s_1$) + $F_3/D_3 - FS_4 - OT_4 + PL_3 + OL_3$

4. *Esmeraldas*. This Costa province has less than 30 per cent of its territory in farms, with a large sector occupied by forests. Farms are medium or large in size (35ha); 20 per cent of the labour force is made up of occasional workers. The production orientation is maize and especially speculative crops, with a modest amount of livestock.
Esmeraldas: L_3/A_3 (S_1zm + C_3ms mp + I_1gh) + P_3 (c_3) + $F_3/D_2 - FS_4 - OT_5 + OL_3$

5. *Manabí*. A much greater percentage of land in farms, a higher population density, and a smaller size of farms (14ha) are the striking features of this province, but production orientation is similar to that of Esmeralda.
Manabí: L_4/A_3 (S_2zm + C_3ca + I_1gh) + P_3 (c_3) + $F_2/D_3 - FS_3 - OT_5 + OL_2$

6. *Southern Costa*. In these regions forests occupy a very low proportion of land in farms, which occupy about 60–80 per cent of their territory. There is little maize, but a large amount of speculative agriculture with rice, or coffee, cocoa, and bananas as leading crops. Farms are medium or large in size (16–26ha), and occasional workers form three-fifths of the agricultural labour force. Livestock is poor.
Guayas: L_4/A_4 (S_1zm + C_4os tc + I_1gh) + P_3 (c_3) + $F_1/D_3 - FS_3 - OT_5 + IL_2 + OL_3$

7. *Pastaza*. Oriente is a reserve for future

development: pioneer fringes are still hampered by serious environmental problems and a lack of roads. Everywhere land in farms consists of less than 5 per cent of the territory. In Pastaza forests prevail with about 60 per cent of the land in agriculture. Farms are large (64ha), and hired labour represents more than 50 per cent of the agricultural population. Production orientation is yuca and maize, used by settlers, and sugar-cane and citrus fruit for export, with few cattle.
Pastaza: L_1/A_1 (S_1mu zm + C_2so ci) + P_3 (c_2) + F_5/D_2 − FS_6 − OT_4 + PL_2 + OL_3

8. *Other Oriente provinces.* These differ from Pastaza in that they have a lower percentage of forested land, although it is still prevalent both in the landscape and in the economy, a smaller size of farms (35–45ha), a greater incidence of owners and tenants, and more intensive cattle breeding. Crop orientation is similar to Pastaza, maize or maize and yuca prevailing as subsistence crops, and platanos or platanos and sugar-cane as speculative crops.

Morona Santiago: L_1/A_1 (S_1zm mu + C_1mp so) + P_3 (c_3) + F_4/D_2 − FS_4 − OT_5 + PL_2

9. *Colón.* In these islands, secluded in the Pacific Ocean, land in farms is as small a proportion as in Oriente. Farms are large (more than 50ha), but the labour force is three-fourths owners and tenants. The subsistence crop is maize, but coffee is much more important, and cattle breeding is good.
Colón: L_1/A_2 (S_1zm + C_3ca) + P_3 (c_4) + F_2/D_2 − FS_5 − OT_5

Kostrowicki, J., 1970 Some methods of determining land use and agricultural 'orientations' as used in Polish land utilization and typological studies, *Geog. Polon.* 18, 93–120.

Kostrowicki, J., and N. Helburn, 1970 Agricultural typology. Principles and methods. Preliminary conclusions, *Documentacja Geog.* 1, 20–48.

Secretaría General de Planeación Económica, 1969 *Encuesta agropecuaria nacional 1968* (Quito).

P1028
Multiple land use in the Mälar valley
PER PORENIUS *The Agricultural Board, Sweden*

The length of the Mälar valley is about 180km and the width at the broadest part is about 100km.

The population is expected to increase from today's 2 million to 3 million in the year 2000. Most of the increase will be due to in-migration. People will settle in urbanized areas with a preference for Stockholm, its suburbs and satellites. At the same time there will be a population shift from rural to urban areas within the valley as a result of rationalization in agriculture and forestry. Today about 70,000 persons depend on those industries. At the end of this century only 20,000 may be left. The Mälar valley is, nevertheless, a region with qualities for expanding agricultural production. The output of products, according to recent estimates, should rise about 100 per cent within 30 years. Today the products of the valley represent about 15 per cent of the total for Sweden, but before the year 2000 this proportion should reach 25 per cent.

Since the early 1960s the need for a master plan for development in the valley has been recognized. Much basic data has been assembled and many suggestions have been made. Discussions have been lively for seven or eight years, and the disagreement on principles has grown stronger and stronger. One reason is that the valley contains several administrative units with independent planning authorities and their own planning departments. Another and perhaps stronger reason is the lack of scientifically established principles for comprehensive planning of a big region. However, that is not a fully recognized lack in the background material. Many politicians and planners would formulate the reasons for planning difficulties differently, probably, referring to traditional problems in urban planning. One first step towards securing a grip on the situation is therefore to formulate and present the actual planning problems in a clear manner.

Most of the actual planning is done by the

local planning units where they have resources and experienced staffs. Such a staff is of course dominated by technicians specialized in urban problems. Until now other experts have not had sufficient reason to take part in planning and thus they now have difficulties in gaining proper consideration for the various functions in modern society which they represent. Within the agricultural sector we are indeed worried. Agricultural land use and many other forms of land use must be considered along with urban land use. Development in agriculture and forestry must be programmed as part of the regional planning. We must not draw a line between rural and urban planning when the object is a region like the Mälar valley.

A few of the problems from the viewpoint of agriculture and forestry are as follows. Food will remain a limiting resource for mankind. One tends to forget this fact during today's surplus situation in industrialized countries. Good agricultural land cannot be replaced and the reserves of cultivable land are limited. Bringing new land under cultivation cannot compete with the improvement and conservation of already cultivated land.

Regarding factors limiting production we have two types of agriculture in the world: *irrigation agriculture* and *fertilization agriculture*. The former is old, and the latter has existed for only 150 years. Modern industrial society has developed in regions of fertilization agriculture, that is in humid zones. Cultural soils of good quality in such zones have a very high potential productivity and can be cultivated indefinitely under clever management. They are the best food production resources in the world. That fact must be observed when it comes to consider competition for the use of land. However, it is not always a question of either-or. Many forms of multiple use of land must be introduced. A rationalization program carried out without regard to society as a whole should not be acceptable within any industry.

Over the rural environment already looms a heavy threat of highly mechanized industrialization within agriculture and forestry. The result will be undue enlargement of enterprises, very big fields with removal of shaws, copses, single trees, and irregularities in the layout, growth of industrial types of farm buildings, problems of large-scale intensive livestock units with attendant problems of effluent disposal, heavy machines that cause soil packing, specialized cropping that will increase the vulnerability to plant diseases and attacks from insects and other pests that will make abundant use of chemical pesticides necessary. The decrease in the need for labour might cause a regional depopulation that would make the cultural landscape poor and create administrative problems of many kinds. Within forestry, mechanization and other forms of rationalization are best applied on stands consisting of one single species of trees of the same age. Even here the production is furthered by the use of fertilizers and pesticides, and thus we will soon have a forest landscape with fewer and different qualities than today.

Similar results will follow from the unhampered development of other industries. At the same time we can see signs of a more realistic conception of human demands on living standards and environment. Man is physically and mentally equipped for a life much different from that of modern city dwellers. Even if human adaptability is great the adjustment to unnatural circumstances will cause pressure on the individual and an urge to alter the situation. With rising economic standards we can already note a change in people's leisure habits as well as in their dietary habits. It is a switch from passive indoor to active outdoor activities, and from high consumption of food based on bulk vegetables to high-quality foods with more animal proteins and with fewer calories. The demand for outdoor activities is expected to rise as a function not only of better economy and shorter working weeks and more holidays, but also of a higher general level of education.

It is true that rather small recreational areas with adequate facilities will be attractive to many people. This might diminish the danger of a damaging use of land. However, the demand will keep on rising and at the same time become more and more varied. We want a nice living environment, landscape beauty, scenic views, a rich fauna and flora, special nature reservations, a right to access to land for walks and hikes, possibilities for fishing, hunting, and riding. At the same time we have to restore biological ecosystems and develop operational techniques for recycling wastes. It will be necessary to accept, encourage, or enforce many forms and degrees of multiple land use upon farming and forestry as well as on other industries. We may need directions and new lines for the management of agricultural land and forests within the Mälar valley, but they have to come about as a result of good planning for development as a whole within the valley.

Will it be necessary to launch a new planning philosophy to get such a plan? Do we have the scientific background material we need?

Skiss 1966 till regionplan för stockholmstrakten.
Förslag till regionplan 1970 för stockholmstrakten.
Regionplan för västeråsbygden 1965.

P1029
Critères et indices de la typologie de l'agriculture mondiale
ANDRÉ RAKITNIKOV *Université de Moscou, URSS*

Pour établir une classification des formes de l'agriculture dans le monde correspondant aux buts pratiques qu'elle doit atteindre, il est nécessaire de se servir des indices caractérisant: (1) les types sociaux des exploitations; (2) la place occupée par les exploitations dans le système de la division du travail (spécialisation dans la production de certains produits et degré de commercialisation); (3) les modes de gestion des exploitations caractérisés par: (*a*) les procédés de l'utilisation des facteurs du milieu naturel dans la culture des plantes et dans l'élevage et les différents niveaux d'intensité de l'agriculture qui correspondent à ces procédés; (*b*) les niveaux de l'équipement technique et ceux de la productivité du travail agricole qui en dépendent. Une question se présente: dans quelle mesure chacun de ces aspects importants ou certains d'entre eux peuvent-ils être exprimés par un seul ou quelques indices quantitatifs?

Pour caractériser le premier des aspects susmentionnés (la structure sociale de l'agriculture) on peut en principe appliquer le pourcentage des exploitations de divers types sociaux (dont la classification a été préalablement établie) dans la production totale de l'agriculture.

Pour caractériser le second aspect on peut utiliser le pourcentage des diverses branches de l'agriculture et de divers produits dans l'ensemble de la production agricole vendue et aussi le rapport de la production vendue à la production globale.

Pour le troisième aspect – en ce qui concerne les procédés de l'exploitation du milieu naturel et conséquemment le niveau d'intensité de l'agriculture (3*a*) on peut, semble-t-il, utiliser les indices quantitatifs du niveau de l'intensité. Cependant ces indices synthétiques ne peuvent servir de base à une classification typologique sans qu'on la complète par les classifications qualificatives des systèmes d'agriculture et de l'élevage.

Enfin, le niveau technique atteint dans l'agriculture (3*b*) peut dans une certaine mesure être évalué par les indices du rendement du travail généralisés. Cependant ces indices doivent être complétés par les caractéristiques qualitatives de l'équipment technique.

Il est certainement admissible d'employer les moyens de classification des formes de l'agriculture fondées sur un nombre restreint d'indices qui peuvent être tirés des sources statistiques existantes et qui ne présentent qu'indirectement et insuffisamment les aspects de l'agriculture cités ci-dessus. Mais dans ce cas nous serons encore loin du but final adopté.

Les considérations exposées indiquent la nécessité d'études typologiques plus spéciales comme base de départ pour établir une classification synthétique des formes de l'agriculture: celles des types sociaux des exploitations, de la classification des exploitations suivant leur spécialisation, de la classification des systèmes agricoles, de ceux de l'élevage, des types de l'équipement technique. Ainsi la classification doit se faire inévitablement à plusieurs degrés.

Dans cet exposé nous nous bornerons à citer certains schémas de ces classifications spéciales qui nous semblent les plus réalisables (simplifiés au maximum).

SYSTÈMES D'AGRICULTURE
Les systèmes d'agriculture peuvent être classifiés suivant les combinaisons de deux catégories d'indices: (1) suivant les moyens d'améliorer le milieu où se développent

TABLEAU 1

Moyens de l'amenée des eaux sur les terres cultivées	Origine des eaux d'irrigation					
	Prises d'eau en rivières		Mobilisation des eaux de ruissellement		Eaux souterraines	
	Débit contrôlé	Débit non contrôlé	Débit contrôlé	Débit non contrôlé	Débit contrôlé	Débit non contrôlé
Irrigation sans élévation de l'eau						
Irrigation avec élévation de l'eau						

les végétaux et (2) suivant les plantes cultivées et leur combinaisons.

Moyens d'améliorer le milieu:
(a) Les engrais minéraux et organiques ne sont pas employés en quantité déterminant, dans une grande mesure, les rendements.
(b) Les engrais minéraux et organiques sont employés en quantité déterminant, dans une grande mesure, les rendements.

On considère sous (a) et (b) cinq types de culture:
  I. Culture sans irrigation ni contrôle du niveau des eaux de sous-sol.
  II. Culture irriguée.
  III. Culture fondée sur le contrôle du niveau des eaux de sous-sol.
  IV. Culture sur terrasses.
  V. Culture sur sol radicalement modifié.

Plantes cultivées et leurs combinaisons:
Agriculture sans assolement:
  1. Culture d'arbres et d'arbustes.
  2. Prairies permanentes améliorées.
  3. Culture continue d'une plante herbacée (ou de plusieurs plantes biologiquement semblables) sans jachère.
  4. Culture épisodique des herbacées après laquelle le terrain est abandonné (culture itinérante).

Assolements où pendant une année on cultive une seule plante et par conséquent la rotation est effectuée au cours de plusieurs années:
  5. Alternances des cultures et de la jachère ou de la friche herbacée.
  6. Alternances des cultures et de la friche à brousse.
  7. Alternances des cultures sarclées et non-sarclées.
  8. Alternances des cultures non-sarclées et des herbes fourragères pérennes.
  9. Alternances de cultures sarclées, non-sarclées et des herbes fourragères pérennes.
  10. Culture de plantes surtout ou exclusivement sarclées.

Assolement où la rotation est effectuée au cours d'une année:
  11. Au cours d'une année on obtient une récolte de la culture principale et, en plus, une récolte des herbes fourragères (culture dérobée après la moisson).
  12. Au cours d'une année se succèdent plusieurs cultures principales qui forment la rotation.
  13. Au cours d'une année on cultive plusieurs fois la même plante.

Culture de plantes associées (cultures intercalaires):
  14. Culture de plantes ligneuses associées à des plantes herbacées.
  15. Culture de plantes herbacées associées.

Dans les régions irriguées il est nécessaire d'indiquer les systèmes d'irrigation. Leur classification la plus simple serait présentée dans le tableau 1.

SYSTÈMES D'ÉLEVAGE

Les systèmes d'élevage se distinguent: (1) d'après le type de ravitaillement du bétail, (2) d'après les espèces et le genre de leur productivité exploitée. La subdivision par espèces et par l'orientation de la productivité étant évidente, nous nous limiterons à une classification des types de ravitaillement qui nous paraît la plus pratique en l'occurence:
  A. L'élevage à base des herbages naturels (pâturages et prairies de fauche non améliorés).
  B. L'élevage à base d'herbages naturels (pâturages et prairies de fauche non améliorés) et de sous-produits de l'agriculture.
  C. L'élevage à base de sous-produits de l'agri-

culture et de la culture des plantes fourragères.
D. L'élevage à base des plantes fourragères occupant les terres cultivées.
E. L'élevage à base de prairies améliorées,
suppléées par la culture des plantes fourragères.
F. L'élevage à base de fourrage transporté des autres régions.

P1030
Crop combination regions of Chhattisgarh Basin in India
B.P. PANDA *Jiwazi University, India*
J.P. SAXENA *M.L.B. Arts & Commerce College, India*

Crops are rarely grown in absolute isolation. The combinational associations in which crops of a region are raised not only reflect the influence of various physical factors but mark the economic orientation of the use of arable land as well. Individual crops hold variable positions among those crops grown within a region or a unit. A more synthetic study of agricultural land use should, therefore, include not only the rank of individual crops but their character and differences in extent also. Crop combination analysis provides such a synthetic study, based on measurable and comparable methods. It gives an integrated understanding of the geographic pattern of cropland use and provides an essential basis for establishing valid agricultural regions. Crop combinations are now recognized as important typological characteristics of agriculture.

The grouping of crops into crop-combination regions was first attempted by Weaver (1954), who used a standard deviation formula for delimiting the crop combination regions of the American Middle West. He devised a standard curve against which variations could be measured. All the crops which showed minimum deviations from the standard value were included in the crop combination region.

Values for each tahsil of Chhattisgarh Basin were computed according to Weaver's formula (Map 1; maps not shown here). The region shows a wide range of crop combinations. The pattern changes from a more or less monocultural pattern in the north and east to multiple associations of five or six crops in the west. In the southern region minimum deviations in Raipur, Dhamtari, and Durg tahsils could not be obtained for fewer than 10 crops, which included all crops occupying up to 0.75 per cent of the gross cropped area. Nine out of 21 tahsils of the region gave such marginal statistical results that some sort of subjective adjustment became necessary.

Weaver's method of forming crop combination regions, though an ingenious and valuable device, was not found suitable for the region mainly because of the following drawbacks:

(i) The method is purely statistical and is based on relative land occupancy strength of various crops, without taking into account the agro-technical properties of the crops.

(ii) An element of subjectivity is involved in determining the crop combination of a unit where the percentage grouping is found to produce marginal statistical results. Thus, the number of crops in the crop combinations of the west and south becomes so large that even unimportant crops of small areal extent have also been included.

(iii) Some speciality crops important for their value or yield but low in acreage do not find a place in the crop combination region. For example, the groundnut is an important cash crop of Mahasamund and Sarangarh tahsils, occupying 3 to 6 per cent of the gross cropped area, but it is excluded from the crop combination region because of the predominant role of paddy.

Professor K. Doi (1959) made some improvement over Weaver's method, by modifying Weaver's formula slightly by substituting Σd^2 for $\Sigma d^2/n$. The group of crops having the smallest Σd^2 will form the combination for the region, and others are dropped out. With this modification only major crops enter the combination. The number of crops in the crop combination region is reduced to a desirable and more sensible level. The labour involved is also minimized to a great extent, and no subjective adjustment is required since none of the units produce marginal statistical results.

TABLE 1

Area under crop	Role	Rank	Orientation
Over 80%	Dominant	5	Highly paddy
60–80%	Preponderant	4	Paddy with teora
40–60%	Equal	3	Paddy and linseed
20–40%	Accompanying	2	Paddy with blackgram
10–20%	Secondary	1	Paddy with teora and linseed

Map 2 provides the distribution pattern of the crop combination regions using Doi's method. As the map shows, the monocultural region of the north and east is adjoined by two-crop combinations in the south and three- to four-crop combinations in the more diversified west. The north and east is predominantly a paddy area with more than 70 per cent of the gross cropped area devoted to this crop. In the south rice is grown with kodo-kutki, teora, or linseed.

Though Doi's method improves significantly Weaver's method, she fails to integrate the agrotechnical properties with the rank and areal strength of the crops. The approach is still statistical and the basis the same as that of Weaver. This gives the groups or crop combination regions a strength, but is not concerned with the qualitative characteristics of the crops.

Credit goes to Kostrowicki for blending the qualitative and quantitative aspects of crop combination in an altogether new method devised on a different basis. In his crop combinations Kostrowicki has tried to synthesize the two essential features of cropland use: (i) the agronomic properties of the crops grown in a region; (ii) the areal strength, rank, and differences in extent of crops raised.

In small-scale village agriculture of subsistence or semisubsistence type numerous crops are grown within the area of a village. Crops with similar or complementary physical requirements, methods of cultivation, places in the crop rotation, and amount of input have been grouped to form one category. Based on these factors, Kostrowicki has distinguished three groups of crops: (i) intensifying crops, (ii) structure-forming crops, (iii) extractive or exhaustive crops.

After classifying the various crops into their respective groups the proportion in each is examined. A group occupying below 20 per cent of the total area under crops is excluded in Polish land use surveys, but this can be reduced to 10 per cent for the Chhattisgarh Basin in view of the fact that the bulk of the area is devoted to cereals and millets (extractive crops) with only a small margin left for other groups. Kostrowicki allows such additional thresholds to be adopted in suitable cases.

A dominant crop is then selected in each group. A crop which occupies more land than others within a given group is designated as 'dominant' while a crop covering 80 per cent of the area of the dominant crop and 60 per cent of the area of the group is considered as co-dominant. Their role, rank, and orientation have been determined as shown in Table 1.

A symbolic nomenclature is given to each crop after its Latin or botanical names. The orientation is shown in a formula in which (i) capital letters represent the groups, (ii) dominant and co-dominant crops are shown by small characters, (iii) figures represent the rank of the particular group. The whole orientation is named after the dominant crop and the role of other crops is shown according to their position.

Map 3 gives the crop combination regions determined using Kostrowicki's technique. Only broad groups with the same orientation have been shown and micro-differences of ranks, if any, have been noted in the description. It may be remarked that the resulting orientations reflect the crop rotation cycle actually prevalent in the region. The following crop combination regions are identified in Chhattisgarh Basin:
1. Rice–Teora–Linseed Region (E_4os+S_2pt Iilu)
2. Rice–Blackgram–Linseed Region ($E_4os+S_2pt+Iilu$)
3. Rice–Teora Region (E_4os+S_2pt)
4. Rice–Teora–Groundnut Region (E_5os+S_1Pt & ah)
5. Rice–Blackgram Region (E_5os+S_1pr)

6. Rice–Kodokutki–Gram–Linseed Region (E_4os, ps, pm + S_2ca + Iilu)
7. Rice–Horsegram–Blackgram Region (E_4os+S_2 & d & pr)
8. Rice–Horsegram–Blackgram–Maize and Mesta Region (E_4os + S_2pr & dd + 1_1zm & hc)
9. Rice–Horsegram Region (E_5os+S_1dd)

The basin as a whole gives a combination of E_4os + S_2pt with an orientation which can be described as preponderantly rice with accompanying teora. A critical study of the crop combinations will disclose that extractive crops hold a dominant position in the cropland use. Soil deterioration is a normal phenomenon inherent to the cropping and unless the soil is liberally fertilized to the extent required it is not possible to increase the growth rate of agricultural production. The contribution of S and I crops to total cropland use is small and is unlikely to make good the loss by a natural process. It may further be noted that cultivation of linseed becomes unimportant in the eastern and northern parts of the basin. It is restricted only to the central and western parts, where soils for the most part are silty clay. With the exception of linseed, intensifying crops are quite unimportant and need greater attention by the extension services for their quantitative and qualitative increase. Among structure-forming crops, teora holds a predominant position, mainly because of its drought-resistant qualities and its fodder value.

Kostrowicki's crop combination technique was found to be eminently suitable for the Chhattisgarh Basin of India. It was found not only to provide a good foundation for delimiting the types of agriculture of the basin but also to give an integrated understanding of the cropland use. It answers some of the basic problems of agriculture and is helpful in planning a more rational use of land.

Doi, K., 1959   The industrial structure of Japanese prefectures, *Proc. IGU Regional Conf., Japan, 1957*, 310–16.
Kostrowicki, J., 1968   Some methods and techniques to determine crops and other landuse combinations as used in the Polish land use studies, *Proc. IGU, India*, 1–11.
Weaver, J.C., 1954   Crop combination regions in the Middle West, *Geog. Rev.* 44, 179–80.

P1031
Pattern of rural land use in Malehra village
J.P. SAXENA   *Jiwaji University, India*

Malehra is a large village situated in the granite country of Bundelkhand region in India. With a population of over 7000, this medieval settlement was recognized as a town in the 1961 census. It is situated on the Chhatarpur-Mahoba road at a distance of about 13km from Chhatarpur. Its geodetic location is lat. 25°02′N and long. 79°05′E.

Considered at the regional level, there is no cash crop of any major significance in Bundelkhand. This is, however, not true at the village level. Malehra (popularly known as Garhi-Malehra) engages in agricultural specialization of considerable local importance, viz. the cultivation of betle-vine or 'pan,' which it has monopolized since medieval times.

The scenery around Malehra is typical of Bundelkhand granites criss-crossed by numerous dolerite dikes and quartz reefs. The physical setting of the village is between two parallel quartz reefs running in a SW and NE direction. The depression between the two is occupied by an elongated water tank, which is the scene of intense agricultural activity, betle farming. Hills around the village are about 120 to 150 metres above the village surface, and are strewn with granite boulders. The tank, which has been embanked on the lower side, receives most of the local rain in the rainy season.

Temperature records of Malehra do not exist, but rainfall data are available for about 20 years. The mean annual precipitation of the village is 115cm, which comes mostly from July to September. Temperatures vary from 15.5°C in December-January to 34.0°C in May-June. May is the hottest month, when the mean maximum temperature frequently rises above 40.5°C.

The hills are covered with bushy jungles and stunted trees and contain little firewood for the village. Some of the hillocks are devoid of vegetation, except seasonal grasses, which spring up during the wet season. These are, however, soon overgrazed to give the landscape a barren and dreary look in the dry months.

Land use statistics of Malehra village reveal that out of the total village area (2000ha) about 10 per cent is forests, 18 per cent is 'not available for cultivation,' 15 per cent is rough pasture and grazing lands, 16 per cent is fallow land, 1.6 per cent is cultivable wastes, and the rest, 40 per cent, is net sown area. The double-cropped area constitutes only 2 per cent of the village area. These percentages clearly suggest the general poverty of the local soils, which restricts not only the area under commercial forests and permanent pastures but also the area under effective cultivation.

Roughly 60 per cent of the agricultural activity in Malehra is centred around the cultivation of betles as a cash crop of great commercial value. Normally betles worth rupees one crore are produced annually and sold in the towns of Uttar Pradesh. Strange as it may appear, these do not find markets in the state of Madhya Pradesh to which the village belongs. To be sure, the betles, like tobacco, have local varieties and, therefore, find markets in particular regions only.

Betle-vine flourishes best on well-drained and well-aerated 'rankar' (lateric) soils in a tropical climate, but the tender leaves of the creeper are very susceptible to direct insolation, especially in summer. The entire field, therefore, has to be thatched to provide proper shade to the creeper. It needs plenty of water, which should come in regular showers; otherwise, watering, thrice a day, is necessary.

It may be noted that the physical conditions under which the cultivation of betles is carried on in Malehra are repeated at many places in Bundelkhand, but the centres of its cultivation are only a few – Maharajpur, Kusma, Bari, Nivari, Gaurari, and Mahoba. The reason for this centralization is to be found in socio-cultural, rather than physical factors. These villages are mostly settled by the Chaurasias – a community which traditionally monopolizes the cultivation of and trade in betle leaves throughout India. Betle cultivation is a specialized branch of farming, the techniques of which have come down through generations and are lacking entirely in other communities.

Betle leaves are planted in neatly drawn rows about 75cm apart. Bamboo poles are fixed in these rows at an interval of about 25 or 30cm and the entire field is thatched with grasses. The initial cost of a plantation is high, about Rs. 250/ per row; deducting all the input expenditures, each row of betle-vine yields, in net returns, about Rs. 150/ to Rs. 200/ in favourable seasons. The plantation is established in October–November and plucking of leaves starts in May and continues up to July. Fresh leaves are sold at Rs. 4/ per kg, but fully matured leaves fetch Rs. 7/ or Rs. 8/ per kg.

All the betle gardens seek water frontage and are located either on the gently sloping banks of the lake or at places which could be approached by water channels from the tank or the irrigation wells. At least 400 to 500 gallons of water is needed for each row of betle-vine. Sprinkling of water on the creeper is in itself an art, and especially trained and skilled labourers are hired for this purpose (they are usually paid Rs. 60/ per month). They use an earthen pot (capacity 19 to 23 gallons) to fetch water from the holes facing the garden; they also apply manure (oil cakes) in the rows.

It has been reported to the author during the village survey that huge losses are being suffered by the 'pan' cultivators at present because of the infestation by some pest, which the agricultural scientists have not as yet diagnosed and for which no remedy has been suggested.

The pattern of crop land use of Malehra referred to above covers only a small portion of the total arable land. One might note the following:

1. There is a definite plan and pattern in the distribution of betle-vine gardens, which occupy the area around the tank. There are two distinct advantages, first the proximity to the source of irrigation and secondly the soils on the banks are well drained – a condition which is as essential to the creeper as the frequency of watering.

2. Most of the gardens on the map are shown as 'Bareja' or fallow lands of betle-vine fields. In fact many of the fields have been permanently withdrawn from the culti-

vation of this vine because of the outbreak of the pest since the agricultural season of 1965-6.

3. Much of the village area including arable land is infertile and barren and therefore a large portion of the map is shaded to show wastelands, including cultivable wastes. This wasteland has been largely used for village settlement.

4. The map further shows a great variety of field crops in the village. Variations in the local relief and soils together with cultural practices of the farmers alone can explain this feature, which is not peculiar to this village alone.

5. A further feature of rural land use pattern of the village is that the grazing belt lies at the outskirts of the village in the west and between this grazing belt and the betle gardens (in the centre) lies the area earmarked for raising cereals, pulses, and oil seeds. Irregular distribution of a variety of these crops again reflects the dependence on the suitability or otherwise of the total soil characteristics for these crops (wheat, gram, barley, jowar, arhar, small millets, oilseed, and rice).

Kharif (autumn) and Rabi (winter) are the major agricultural seasons of this village. Out of the total cultivated area (843.3ha) roughly 40 per cent is devoted to Kharif and the other 60 per cent to Rabi crops. The area under the former is mostly rain-fed, but 30 per cent of the Rabi area is irrigated by wells and tanks, the only sources of irrigation.

Yields of various crops are low and their total production falls short of village requirements, being sufficient for five months only. For the rest of the period all food comes from Chhatarpur. This is, however, possible because people of Malehra (thanks to betle-vine cultivation) are comparatively well off. They provide an example for the rest of the people in Bundelkhand of a way to economic improvement in a rather infertile tract.

University of Saugar.

P1032
A new approach to the delimitation of food productivity regions in India
MOHAMMAD SHAFI   *University of Aligarh, India*

Indian agriculture is crop-oriented. About nine-tenths of the newly sown area of the country is devoted to food crops and one-tenth to commercial crops. In 1971, the Indian population was estimated at 550 million people; by the turn of the century, there will be about 1000 million – two for every one Indian living today. This means that if the future need is considered in terms of the minimum nutrition for a 10 million annual increase in population, India will need a minimum of 3.7 million tons of cereals, including starchy roots and sugar, to feed the extra mouths being born every year. The objective is not only to feed the additional population but also to remove the immediate deficit, which falls somewhere near 14 million tons in terms of food grains, starchy roots, and sugar. If 1961 is taken as the base year, the country should increase its food production (food grains, starchy roots, and sugar excluding animal products, fruits, and vegetables) to the extent of 32.5 million tons by the end of the Fourth Five Year Plan (Shafi 1969, 21). If India's food grain production continues to grow at no more than the trend shown for the years since 1961 (2.36 per cent per year), the per capita local supply will decrease because of an even greater growth rate in population, and the gap between food grain supply and demand in 1986 could be 50kg per person. Thus careful examination of the areas where production can be increased is needed. This in turn requires in the first instance the precise delimitation of food productivity regions.

Kendall tried to measure crop productivity per unit area by taking the per acre yield of ten leading crops in each of the 48 counties of England for selected years and applying the 'ranking coefficient method.' Similarly L.D. Stamp attempted to measure the crop productivity of 20 countries on the basis of the per acre yield of nine selected food crops applying the same method. The countries were placed in the order of output per acre for each crop. The places occupied by each country relative to the selected crops were then averaged and, from these averages, the ranking coefficient of crop productivity of each country was obtained (Stamp 1960, 108). This technique was also employed by the author to measure the crop productivity of Uttar Pradesh in terms of the per acre yield of eight selected crops (Shafi 1960). One of the drawbacks of this method is that insignificant acreages under certain crops which show a high adaptation relative to physical factors in the same or different regions may have higher yields per acre than those crops which occupy substantial acreage but have poor adaptability to physical conditions. The ranking coefficient on the basis of average yields will therefore be biased.

Professor Enyedi (1964, 61), in a discussion of geographical types of agriculture, refers to a formula for determining an index of productivity coefficient:

$(y/y_n) : (t/T_n)$,

where y = the total yield of the respective crop in the unit area, y_n = the total yield of the crop at the national scale, t = the total crop area of the district, T_n = the total crop area at the national scale.

The writer adopted this formula to determine the productivity coefficient index for twelve food crops of India. From the productivity index of each crop of a district the percentage of the productivity level in relation to the national scale for that crop was obtained. The percentages of all the twelve crops thus obtained were added to indicate the food–crop productivity level of that district relative to the national level. The plus figures of productivity percentages of all the districts were arranged in descending order, and medians, quartiles, and octiles were worked out to give eight ranks (I to VIII). The minus figures of the productivity percentages were arranged separately in descending order, and the median was worked out to give two ranks, IX and X (Shafi 1970, 9).

Despite the value of the formula in determining the productivity index of an area relative to the national scale, there are certain cases where the results are influenced by the size of the area under a particular crop when the yield of that crop in the district is the same as or less than the national yield. When the yield of the district is the same as the national yield, the district, by the computation of the formula, would show a higher productivity coefficient than that of the national scale for that crop. Similarly, when the yield of a crop in a district is even less than the yield at the national level, the productivity coefficient index for that crop may be higher than at the national level (Shafi 1971, 6).

The writer has made an attempt to modify the formula so that the productivity coefficient of a particular crop may conform with higher or lower yields per hectare of that crop in the district than at the national level.

In the modified formula the summation of the total yield of all the crops in the district is divided by the total area under the crops considered in the district and the position thus obtained is examined in relation to the total yield of all the crops considered at the national level divided by the total area at the national level under those crops. The formula would read thus:

$(y_w/t + y_r/t + y_{mi}/t ... n)$:

$(y_w/T + y_r/T + y_{mi}/T ... n)$

or $\Sigma_n(y/t) : \Sigma_n(y/T)$.

The writer has made an attempt to determine the productivity index of India on the basis of the above formula and has delimited the productivity regions of India. The paper draws attention to those specific areas which have productivity far below the national level and which should receive first attention by planners in improving the productivity of the area. It also indicates the areas where productivity varies from low to medium and the areas where productivity ranges between high and very high. This may help to calculate precisely the total increase in food production which may accrue if the areas having low and medium productivity are brought to the level of the existing high-productivity areas.

Enyedi, G.Y., 1964 *Applied Geography in Hungary* (Budapest).
Shafi, M., 1960 Measurement of agricultural efficiency in Uttar Pradesh, *Econ. Geog.* 36 (4), 296–305.
– 1969 Can India support five times her population?, *Science Today*, May.
– 1971 Measurement of agricultural productivity of the Great Plains of India, Paper presented at the European Regional Conference of the IGU, Budapest.
– Measurement of crop productivity in India, *Indian Council of Geographers Trans.* 1 (no. 3). (In Press)
Stamp, L.D., 1960 *Our Developing World* (London).

P1033
Tobacco cultivation as an alternative to chena cultivation in the Walapane Division of Ceylon
W.P.T. SILVA *University of Ceylon*

The Central Highlands of Ceylon, occupying about one-third of the surface area of the island, are characterized by the presence of two distinct land classes, the irrigable valley bottoms and the unirrigable slopes. Based on these two land classes there has evolved a threefold system of land use: irrigated paddy cultivation, village gardening, and highland cultivation.

Prior to 1850, the major form of agriculture practised on the forested unirrigable slopes was shifting cultivation. This type of cultivation, locally referred to as the 'chena' system, is generally confined to the forested unirrigable slopes. It consists of the burning of the forest cover in a selected area of land once in about 10–15 years, cultivating it for 2–3 years with a variety of crops, mostly sown broadcast, and abandoning it until such time as the forest has regenerated itself.

With the spread of plantation agriculture in the latter half of the nineteenth century, the forests in the wetter half of the highlands were gradually cleared, first for coffee, and subsequently for tea and rubber. In the eastern half, on the other hand, the low rainfall prevented the cultivation of permanent tree crops and as a result chena cultivation persisted as the dominant form of agriculture on the unirrigable slopes.

Chena cultivation, as practised by the local villager, is essentially a subsistence type of agriculture. Although it has continued to support him by providing a narrow range of food crops and at times a supplementary source of income, it has generally been condemned as being wasteful of timber and land. Even those who have defended it as a wise concession to the Dry Zone environment have supported the argument that chena cultivation becomes pernicious once the population in an area has reached a certain limit. Therefore, the consensus of opinion has been that this type of cultivation should be replaced by some sort of permanent commercial arable farming.

A change in this direction has taken place in the Walapane Division, where, within the last one and a half decades, traditional chena crops have given way to commercial tobacco cultivation. This division, which forms one of the four revenue divisions in the Nuwara Eliya District, covers an area of approximately 260km^2 and in 1963 had a population of a little over 62,000 persons.

From the point of view of settlement and land use, the Walapane Division may be divided into two broad regions, the plantation area in the south and the area of peasant agriculture in the north. In the former area the traditional threefold system is absent, partly because the lower temperatures inhibit the cultivation of paddy, and partly because the higher rainfall has permitted the cultivation of permanent tree crops, especially tea. In the northern part of the region, on the other hand, the pattern of land use conforms to that of the Dry Zone in general, with irrigated paddy fields in the valley bottoms and village gardens and chena cultivation on the unirrigable slopes.

In contrast to the southern part of the region, which is characterized by a prosperous plantation economy, the northern part has for long remained one of the more backward areas within the Central Highlands. Some have tried to explain this backwardness through relative isolation while others have stressed the stubbornness and apathy of the local inhabitants. Although these may be treated as contributory factors, the more important reason has been the relatively low rainfall which has prevented the effective utilization of not only the irrigable lowlands but also the uplands. Most of the people have been dependent on chena cultivation and the inevitable result has been poverty and stagnation.

The situation, however, has changed within the last one and a half decades with the gradual replacement of traditional chena crops by tobacco. Although tobacco has become the dominant upland crop, nevertheless the method of cultivation follows the same lines as chena cultivation. A selected forest area is burned and cleared, after which tobacco is grown continuously for a period of two to three years. The land is then abandoned until the forest has had enough time to regenerate itself. Thus, unlike in the past, the villagers in the area have been able to fit into the chena system a crop of genuine commercial importance. This has not only brought about a change in the pattern of utilization of the unirrigable uplands, but has also rescued the region from its earlier economic backwardness. This paper makes an attempt to examine the spread of commercial tobacco cultivation in the Walapane Division and to assess its impact on the economic development of the region.

Abeyratne, E.F.L., 1956 Dryland farming in Ceylon, *Trop. Agr.* 112 (3).
De Rosayro, R.A., 1949 Some aspects of shifting cultivation in Ceylon. *Trop. Agr.* 105 (2).
Farmer, B.H., 1957 *Pioneer Peasant Colonization in Ceylon* (London).
Paul, W.R.C., 1949 Roving agriculture and the problem of dry farming, *Trop. Agr.* 105 (1).
Udagama, P.A., 1947 Some observations on shifting cultivation in Ceylon, *Ceylon Geog.* 2 (2).
Wikkramatillake, R., 1957 Whither chena? The problem of an alternative to shifting cultivation in the Dry Zone of Ceylon, *Geog. Studies* 4 (2).

P1034
Indigenous settlement planning in Rajasthan (India)
BASANT SINGH *University of Rajasthan, India*

In popular imagination today the word 'planning' seems to evoke some kind of pious feelings. Much can be brought within the term because it deals with the progress of human beings in the background of an area and space with equal emphasis. The geo-historical trends and political and economic upheavals and crises have led to a great territorial demarcation and redemarcation of the villages and cities of the world. The genesis, concept, and the objectives of village planning in India have their origins in the ancient past, when planning was considered to be a sacred art, as can be seen in the oldest literatures, the Purānās, the Vēdic literature, and Kautalya Arthasastra. But crystallization has come gradually and haphazardly through the efforts of various individual planners. The grids of the streets in indigenously planned villages are made up of parallel lines cut at right angles. The villages enjoyed planned and measured roads, houses, efficient drainage systems, swimming pools, and other civic amenities. They definitely represent the supreme advancement of the Indian technological achievement in the field of settlement planning.

The concept of settlement starts with the cave dwellings of ancient India. Siva, the Hindu's God, lived in caves and is therefore called *Guhavasin* and *Guha-priva*. There were climatic crises, and thus climatic extremes were experienced, bringing multiple hardships to human beings in *Treta Yuga*. People previously of nomadic habits must have constructed some artificial shelters amidst the hills, inside the earth, and along the riverbanks, just to enjoy some relief from the heat and cold. Fixed and permanent dwellings may have begun with the construction of huts using branches of trees. At this stage people must have adopted hunting and food-gathering as their main occupation. These hut sites must have been converted into strategic points. The rulers of these settlements definitely planned them in conformity with geo-strategy, and the social, economic, and cultural needs of the stronghold. However, it appears that these indigenously planned villages had no over-all socio-economic development as an aim. Mention has been made in the Purānā (1946) about the site of the village and town. The tops of hills, bases of mountains, and riverbanks have been suggested as the best sites for these early dwellings. Measurements of village and town boundaries, roads, lanes, and paths have also been given. The roads have been categorized as Rajpath (10 Dhanus wide; 1 Dhanu = 1.8m), directional roads (20 Dhanus), streets (4 Dhanus), and other streets between residential houses and the main road (also 4 Dhanus wide). Planners of those days did not visualize the growth potentialities of planning at different levels in a long perspective. These villages primarily collected revenue, fed the bigger capitals, embraced the maximum possible civic amenities, and strictly maintained and safeguarded the exclusive culture of the ruling class. In terms of the eradication of illiteracy the planning movement was centred in only the capitals of the respective chiefs, resulting in the rapid and untimely ruin and quick disappearance of the planning technique from the land.

Two distinct types of indigenously planned settlements are clearly visible in the area under study: (i) developed and (ii) decayed. In the absence of law and order, a civic code of conduct, or other measures, the villages which were numerically strong formed a group of strong people and continued to preserve their identity even during times of aggression and otherwise adverse conditions and therefore emerged as stronghold centres in the later part of the medieval period. Now they are identified as marketing, educational, cultural, medical, and administrative centres. Along with others, Jaipur furnishes one of the best examples in this respect. The facilities and privileges, appropriate to the chief's residence, administrative headquarters, and economic and social centre, were thrust upon Jaipur from the time of its selection as a site. Subsequently it provided the proper incentive, security, and environment for industrial and economic activities among the inhabitants, which led to the rapid and regular growth of the modern city of Jaipur. Money dominated the modes of transaction in these cities.

In due time such settlements emerged as towns and cities, and the functions of these settlements grew to include the processing, transportation, and manufacturing of agricultural products and implements. According to Dickinson's (1971) classification, these settlements may be grouped into four categories, viz., small rural villages, urban villages, towns, and cities. In contrast to the larger and geographically more suitable settlements, capitals of smaller chiefs remained stagnant or grew increasingly weaker to the extent that some failed to survive. The village Garh Himmat Singh is the best example of this sort; along with other capitals, though originally planned and privileged in a similar way to Jaipur, it stopped progressing and now remains as a very small village. These stagnant settlements remained backward, ignorant, and traditional; a barter system called 'Zazmani' prevailed in them, reflecting a unique and planned assimilation of service and castes. Almost all the service-class people were persuaded to live together to form a 'mohalla,' named after the quality of service. The population must have shared certain common attitudes in planning the villages, to take into account the values, needs, and desires of the different classes of people serving in different capacities. Such villages with common community interests formed the best possible combination of morphological structure, location, organization, and function.

From a study of sample villages, it may be concluded that in Western Rajasthan streets are randomly arranged and do not enjoy any geometric regularities, that there is great variation in morphology and size of villages, and that buildings receive greater emphasis than the streets (unlike in Eastern Rajasthan, where both receive equal emphasis).

Similar to other capital villages, locally called *Thikana*, Garh Himmat Singh was very prosperous, self-sufficient, and abounding in farm products. Agro-based industries, such as oil-pressing, thread-making, jaggery-making, and cotton-spinning, were very well developed here. Farmers from the nearby villages used to come to exchange their products. In 1897 a railway line with a gauge of one metre passed through Mandawar, then a tiny hamlet or *dhani* lying 4km north of Garh Himmat Singh. Gradually Mandawar experienced a very rapid and regular growth in population. In search of a better livelihood the villagers deserted their homes in Garh Himmat Singh, moved to the railway site, established their businesses and settled there permanently. This railway line provided quick accessibility to cities like Agra, Bharatpur, and Jaipur. Garh Himmat Singh has recently been connected with the railway station by a metalled road. On both sides of this road one can see new houses, orchards, small manufacturing establishments, stone-carving centres, stores of building materials, oil machines, and a variety of shops. Since the nature of the administration changed, the number of houses and population have decreased, and Garh Himmat Singh enjoys relatively fewer civic amenities. To meet the daily needs of vegetables, cloth, and medicines, the settlement is completely dependent on Mandawar.

Founded by Sawai Jai Singh II in 1727, Jaipur, a well-planned city, has emerged as the largest city serving Rajasthan; it is the capital at present. The city within the wall is laid out along the checker-board plan of the ancient Mansar Shilpa Shastra. Situated on a focal point in the central Arvallis, the city from its very inception served as a politico-cultural centre which now has become a commercial, industrial, educational, and medical centre as well. The outer walls of Jaipur have many gates, e.g. the Ajmeri gate, Sanganeri gate, named after the city lying in the particular direction. The ruler planned and constructed a very beautiful market, palaces, houses, streets. Ram Singh gave a pink colour to every structure, most of which are planned and have either square or rectangular shapes. The main business areas include Johari Bazar, Tripolia, and Chandpole. Here also the side streets have traditionally been named for the type of commercial commodity, e.g. Ghee ki Gali, Khajane-Walon-ka-Rasta. Residential areas (30 per cent of the total land) are common within the wall. Commercial land makes up about 5 per cent of the total land. After Indian independence was declared in 1947, development outside the city wall occurred, with forest clearing and well-planned residential colonies. Almost all the modern educational, medical, recreational, industrial, administrative, and civic amenities are found in the new colonies.

Dickinson, R.E., 1931 The distribution and function of smaller urban settlements of East Anglia, *Geography* 7.
Purana, A.P. Agri, 1946 Anandasram Sanskrit Series, 19–21.
Purana, V.P. Vayu, 1946 Anandasram Sanskrit Series, 51–5.

P1035
Intra-village space: a case study of Varanasi District, India
RAM BALI SINGH *University of Gorakhpur, India*
and R.L. SINGH *Banaras Hindu University, India*

Geographers have been accustomed to use the term 'village-farm distance' to denote the space between the place of residence and the cultivated fields of a farmer. Conditions in India, however, are different, in general, the term 'village' representing a parcel of land of defined boundaries and settled by a revenue survey or by a cadastral survey; it may be, but need not always necessarily be, a single house-cluster with a local name, marking its distinctiveness as a residential locality. In Varanasi District, in particular, no fewer than 706 villages are uninhabited and most of the remaining 3524 villages have been defined as semi-compact inasmuch as they include one or more secondary and subordinate aggregates of residential units named variously as *Pura, Purwa*, or *Toli* and usually physically detached from the main central group. Thus (*a*) the dwelling site forms the core of a rural settlement, which may consist of more than one site; (*b*) the location of the core (cores) is functionally related to the areas of primary production (arable fields, pastures, gardens, ponds, etc.); and (*c*) activities other than primary production, for example shops, cottage industries, artisan activities, are found in varying degrees within a rural community. Appreciating the trilogy of intra-village space – sociocultural, economic, and physical – and the complex spatial distribution of settlement units within the village the term 'intra-village space' has been used to denote the distance, in the physical sense, separating the farm from the residential foci. Here the main concern is the internal space relationships between areas of primary production and the residential foci.

A nucleated settlement pattern precludes any spatial contiguity between the individual farmer's farm and his residence, whereas in a dispersed pattern both are contiguous, and a semi-compact pattern exhibits a transitional character. The space between the dwelling and the production area is interposed by peasant dwellings and fields belonging to other farmers. Moreover, the farm holding of an individual farmer itself is usually divided into several detached plots.

Nucleation of farm dwellings and fragmentation of holdings are the two concurrent causes of excessive intra-village space. The latter also varies widely according to the over-all size and shape of the village area, degree of fragmentation, and scattering of holdings as well as the pattern of ownership and operation. Twisted and circuitous pathways add to the travel distance to the farm. The laws of inheritance, soil fertility, and physiography have been the causes of a mosaic of fragmented agricultural areas which forms a dominant structural characteristic of the long-settled and fertile Ganga Plain. The average size of agricultural holdings is about 1.62ha in the Ganga Plain and 4.05ha over the Vindhyan Plateau. The partition of fertile lands along the total length of the plot leads to narrow, strip-type parallelogram field patterns, small uneconomic holdings, and large intra-village spaces. The fields further afield are less disturbed and less subdivided. The large rectangular blocks of the Vindhyan Plateau are of very recent origin.

Under the assumption of circular or hexagonal village lands with the settlement foci situated at the centre, the average intra-village space comes to 675 metres in the study area. Table 1 gives an idea of this 'intervening space' for the 22 community development blocks of the study area, calculated under the following assumptions: (1) all the villages in a development block are of equal size; (2) the entire village population is organized into one village-settlement; (3) the settlement is situated in the centre of the village area; (4) the entire village-land forms one continuous block; (5) the village-land has a circular or hexagonal shape; (6) the villagers do not cultivate any plot outside their village-boundary; and (7) every plot of a village is accessible in a straight-line fashion.

The intra-village space increases to a maximum of 2.1km in Kesar village of Chakia Tahsil, provided we leave aside the first two assumptions. In a few compact large villages of the study area it varies from 1138 metres to 2101 metres.

These figures do not provide an accurate picture of the intra-village space as there exist great variations in shapes and sizes of the village lands and the sites of farm residences within them do not follow the least-cost-location principle. Now retaining only the last assumption of straight-line accessibility, a few villages have been taken as case studies. In Kanwar village the main hamlet is more or less centrally situated with reference to the village lands. The latter consist of a compact rectangular block. The intervening space does not exceed 900 metres from the central hamlet, but it increases to 1200 metres from the secondary hamlet. Babauri has all the village houses huddled together in a corner of the village lands at one site, and intra-village space reaches a maximum of 2.2km. Formerly the village land was subdivided into large rectangular blocks on the basis of soil fertility and the lie of the land.

TABLE 1. Intra-village space in Varanasi District

Block	Distance in metres
1. Chahania	694
2. Dhanapur	779
3. Barahani	773
4. Sakaldiha	673
5. Niamatabad	645
6. Chandauli	635
7. Suriawan	561
8. Bhadohi	582
9. Gyanpur	551
10. Deegh	690
11. Aurai	515
12. Baragaon	673
13. Pindra	641
14. Cholapur	656
15. Chiraigaon	667
16. Harhua	521
17. Sewapuri	555
18. Arazi Lines	574
19. Kashi Vidyapeeth	523
20. Chakia	537
21. Shahabganj	607
22. Naugarh	703

TABLE 2. Intra-village space in a few large villages

Village	Distance in metres
Kesar	2101
Khempur	1664
Talsamdha	1508
Mahuar Kalan	1484
Barahani	1427
Gorari	1265
Chhitauna	1253
Jai Mohani	1244
Tekaria	1241
Dandpur	1228
Eonti	1218
Jorari	1210
Gurehun	1197
Bishunpura	1188
Katesar	1183
Uruwa	1174
Pharsand Mohanpur	1167
Chamerbandh	1159
Ganeshpur	1157
Semar Sandhupur	1143
Diya	1142
Khore	1138

The present field pattern of narrow strip dispersal is of recent origin. Fonti has an elongated shape; 460ha of village area measures about 3.42km in length. The village population is organized into three detached hamlets situated almost halfway in length. The maximum space distance is 2.4km. The entire population of Torwa is organized into one central hamlet and three appended hamlets spread over an area of 261ha. The village would have represented an ideal situation with a maximum space distance of about 1510 metres, but the projection of the village land in the southwest corner has enhanced the intra-village space to a maximum of 2.61km.

P1036
Canal irrigation and agriculture in the Gangapar plain, UP (India)
UJAGIR SINGH *University of Gorakhpur, India*
VIJAYA RAM SINGH and B.S. TYAGI *Banaras Hindu University, India*

The present case study deals with one administrative area (Pargana Bhuili of district Mirzapur, U.P., $82°56'E-83°11'E$ and $25°3'N-25°14'N$), comprising 245 villages, stretching over an area of 224.5km^2 in the Gangapar plain. The nearest boundary lies about 12km from Varanasi and 50km from Mirzapur, the district headquarters.

The area is well served by roads and railways which link it with the industrial complex of Mirzapur. A road passes through the heart of this pargana in the north-south direction. The northern section is traversed by the road and main railway line, which run parallel to one another and connect Mughalsarai in the east and Mirzapur in the west. Regular bus service plies the road, providing accessibility to the major portion of the area, though a small part of the eastern section remains unserved by any modern transport system.

The *Gangapar* plain is a contact zone between the Ganga plain in the north and the Vindhyan plateau in the south. This plain has a typical topography of flat-topped hillocks, isolated knolls, spurs, and eroded scarps of the peninsular upland scattered all over in small numbers. The pargana has a few knolls notably around Bhuili and in the southeast corner. Excepting these two features, the whole region is a plain having less than 1° slope towards the north.

Soils vary from sandy in the west to clayey in the east. The sandy soils occupy a narrow strip on the southern bank of the Ganga. Brownish and yellowish sandy loam soils border the sandy soils. The rest of the area is under clayey soils which are grey coloured and heavy textured. These soils are agriculturally very fertile and when properly irrigated pay heavy dividends.

Canal irrigation in Bhuili pargana plays a very important role as 66 per cent of the total arable land is under irrigation. Before independence (1947) about 5 per cent of the area was irrigated, by various irrigation systems, i.e. tanks and wells. After Independence a network of canals was constructed by damming the Garai and the Jirgo. These systems have played a major role in transforming the agriculture of this area. Most of the villages have more than 60 per cent of their net cultivated area under irrigation, while a few scattered villages and a narrow belt along the Ganga receive nil or very little irrigation as they are far off from irrigation channels. These canals not only check the floods but also provide irrigation to about 13,359 hectares (33,000 acres). The introduction of canal irrigation has caused tremendous changes in agriculture, and similarly the patterns of land use have also changed.

Agriculture is mostly dependent on efficient irrigation systems for better returns, which increase the prosperity of the region. This paper presents some of the most prominent effects of canal irrigation on the farming structure as well as the agriculture.

About 86 per cent of the total area is under agriculture, of which 66 per cent is under canal irrigation. Fallow lands occupy 4 per cent, and water bodies and barren land cover about 9 per cent of the land. The very fertile area is mostly devoted to field crops, and only 1 per cent of the land is under gardens.

The double-cropped area occupies as much as 74.5 per cent of the net cropped area and presents a checker-board pattern in the pargana. There is considerable regularity in the distribution pattern; most of the villages have more than 50 per cent of their area under double cropping. In 59 villages double cropping is done on 70 to 80 per cent, in 48 villages to 80 to 90 per cent, and in 57 villages to more than 90 per cent.

There is not much difference in *Kharif* and *Rabi* acreage as they cover 17,425 and 16,783 hectares (43,065 and 41,471 acres) respectively. The general pattern of cultivation indicates that in 144 villages *Kharif* cultivation exceeds 90 per cent of the cultivable area. *Rabi* crops (sown in winter) occupy 49 per cent of the gross arable land. In most of the villages *Rabi* cultivation exceeds 70 per cent of the cultivable area, though in 74 villages it ranges from 80 to 90 per cent and in 117 villages it even exceeds 90 per cent. The general practice in the area is that the lands put under the early *Kharif* crops are widely cultivated in *Rabi*.

Rice, wheat, barley, and pulses are the major crops and occupy 43.8, 10.3, 8.8, and 6.6 per cent of the gross cultivated area respectively. There are other minor crops, but they occupy very little acreage. Paddy alone occupies 86 per cent of the total *Kharif* land, while the remaining crops share 14 per cent. Thus it is rice which is most suitable and profitable in this intensively irrigated tract. In *Rabi* a variety of crops is grown. Wheat occupies only 30 per cent of the *Rabi* area, leaving behind much scope for other crops such as barley, peas, gram, *Masur,* and oilseeds.

Cultivation of paddy starts with the start of monsoon rains and continues up to the last week of August. It is divided into two groups – early paddy and late paddy. Early paddy is not so important as it occupies only 4 per cent of the *Kharif* acreage and is an inferior variety with a low yield. Some fields used for the early crop are also used for transplanting late paddy, provided they are in a fertile area. But usually early paddy is sown on comparatively poor soils and in areas where irrigational facilities are inadequate.

The late variety occupies 82 per cent of the *Kharif* acreage which is in rice. The area is a level and fertile tract having mostly loamy and clayey soils which are famous for their suitability for rice cultivation. Canal irrigation provides an added advantage and is a boon for its cultivation. In the whole region a minimum of 60 per cent of the village's cultivated acreage is devoted to this crop. There are 134 villages where rice cultivation exceeds 80 per cent of the *Kharif* land.

It is sown either alone or mixed with barley, gram, peas, and oilseeds. Traditionally, the sowing of wheat alone is not very common in the area under reference. It occupies only 6 per cent of the *Rabi* acreage and only in 24 villages does it exceed more than 10 per cent. Wheat mixed with other cereals, as mentioned above, covers 24.7 per cent of the area under *Rabi* crops. It is a tender crop so the clayey texture of soil does not favour its cultivation, and hence the other hardy crops are combined with it so that they make up the loss in the event of its failure. In 70 villages its acreage exceeds 25 per cent. It is usually sown in areas where a minimum of three irrigations are available.

Barley is a hardy crop and is never sown alone in this area. It is sown mixed with wheat, pulses, and oilseeds. It occupies the highest acreage among the *Rabi* crops claiming 28 per cent of the total. It is widely sown, and in 64 villages it occupies more than 30 per cent of the *Rabi* acreage. As the soil moisture remains sufficient because of the previous irrigated crops it never requires irrigation.

Singh, U., 1955 A sample study in land utilization near Sarnath, Varanasi, *Nat. J. of India,* 3, 51–64.

Singh, V.R., 1970 *Land Use Patterns in Mirzapur and Environs* (Banaras Hindu University, Varanasi).

Velde, E.J.V., 1970 Irrigation and spatial change in Haryana, India, paper presented at the 22nd Annual Meeting of the Association for Asian Studies, San Francisco.

P1037
Recent changes in land-utilization in Saryupar plain, India
J.N. PANDEY and UJAGIR SINGH *University of Gorakhpur, India*

Saryupar plain, including the five districts of Uttar Pradesh, India, in the north of the Ghaghra river, comprising an area of 33,472.4km^2, displays the physical and climatic characteristics of the Middle Ganga Valley, excluding the forest-clad northern terai. The region has a simple geological history; it forms an almost flat alluvial land, with an elevation of less than 140m above sea level. The land slopes gradually but imperceptibly from northwest to southeast with a gradient of one metre per five kilometres.

The land use pattern of an area is an outcome of its physical environment and human endeavours. The quality and configuration of the soil are main determinants of the use of land. The major part of the region has fertile alluvial soil deposited by the Ghaghra and its tributaries, but it varies from place to place in texture and colour.

The average rainfall of the region is 100 to 120cm, and it is generally concentrated in the months of July, August, and September. A long break in rain causes desiccation, whereas a long, sudden burst results in erosion and floods. The climate of the region is subhumid tropical. The temperature varies from 17.5°C (normal) in the coldest month, January, to 30.5°C in the hottest month, May. The average annual and monthly distribution of major climatic elements, i.e. rainfall, temperature, pressure, and relative humidity, in the area correspond with the monsoonal characteristics prevailing in the Middle Ganga Plain.

Land use data for the years 1965-6 and 1969-70 for 21 tahsils of the region have been collected. The variation of the various land-use classes for the year 1969-70 over the year 1965-6 has been calculated. The area of positive and negative changes has been categorized. Statistical methods such as mean and standard deviation have been used for the purpose. Finally the data have been plotted and mapped.

The general land-use of the region shows that the percentage of the net sown area to the total area of the region is the highest (77.74%) among all other land-use categories. The very ancient history of the land occupance and an almost overwhelming dependence on agriculture, to the almost complete exclusion of all other economic activities, have practically forced the proportion of the net sown area to be persistently very high.

Non-agricultural land, including barren and uncultivable land and land put to non-agricultural uses, is a sizable proportion (9.93%) closely followed by cultivable waste (3.36%), forest (3.15%), land under miscellaneous tree crops and groves (2.26%), current fallow (2.23%), old fallow (1.15%), and permanent pastures and other grazing lands (0.18%) for the year 1969-70.

The percentage changes in the proportion of various land-use classes in the period show both positive and negative changes. 'Net sown area,' 'non-agricultural land,' and 'area sown more than once' have changed positively, while current fallow, cultivable waste, land under miscellaneous tree crops and groves, and forests have changed negatively.

The general land-use variation of the region is given in Table 1, which reveals the general patterns of the region. There are great variations among the tahsils. It is interesting to note that changes are both negative as well as positive in every land-use class. Although some of the land-use classes shows negative or positive changes as a whole, yet the spatial distribution of the same classes reveals positive as well as negative changes during the period for smaller parts. Table 2 depicts the positive and negative changes of land-use.

The negative and positive changes in the various land-use classes among the tahsils have been classified statistically in five groups for each of the land-use classes: (i) mean + SD2, (ii) mean + SD, (iii) mean, (iv) mean − SD, and (v) mean − SD2. Thus a vivid picture of the changes in land-use may be easily gained.

Cultivation being the main occupation and agriculture the base of the economy, a large portion of the total area is put under plough. It was 77.17 per cent of the total area in 1965-6 and 77.74 per cent in 1969-70. The changes range from −4.34 per cent to +2.39 per cent in net sown area during 1969-70 as compared to the period 1965-6. Sixty-six per cent of the area shows positive changes. Gonda and Balrampur tahsils have positive changes of mean + SD2, whereas Tarabganj tahsil shows a negative change of mean + SD2. This is caused by increases in current fallow and non-agricultural land in Tarabganj tahsil. The developing irrigation facilities, fertilizers, and advancement in technology are responsible for the increase in the percentage of the net sown area. Decrease in cultivable wastelands was also responsible for the positive changes in net sown area.

TABLE 1. Land-use variation

Classification of area	1965–6 percentage of total area of the region	1969–70 percentage of total area of the region	% variation of 1969–70 over 1965–6
Forest	3.40	3.15	−7.63
Non-agricultural land	9.84	9.93	+0.91
Cultivable waste	3.61	3.36	−6.91
Permanent pastures and other grazing lands	0.18	0.18	—
Land under miscellaneous tree crops and groves	2.36	2.26	−4.24
Fallow lands	3.44	2.23	−35.18
Net area sown	77.17	77.74	+0.73
Total	100.00	100.00	

TABLE 2. Changes in land-use

Classification of area	No change		Positive change		Negative change	
	No. of tahsils	% of total	No. of tahsils	% of total	No. of tahsils	% of total
Forest	1	4.7	4	19.0	15	71.3
Non-agricultural land	5		9	42.8	7	33.3
Cultivable waste	2	9.5	4	19.0	15	71.3
Permanent pastures and other grazing lands	3	13.3	14	66.6	4	19.0
Land under miscellaneous tree crops and groves	3	13.3	14	66.6	4	19.0
Fallow lands	0	—	3	13.3	18	87.6
Net area sown	1	4.7	14	66.6	1	4.7

Six tahsils experienced negative changes in the percentage of net sown area to total area of the tahsil. These negative changes occurred as a result of increases in fallow land and land under miscellaneous tree crops and groves.

The amount of non-agricultural land in an area is mostly controlled by physical factors and soil characteristics. This is unproductive land under settlement, roads, and barren lands. It was 9.84 per cent of the total area in 1965–6 and 9.93 per cent in 1969–70. 42.8 per cent of the tahsils showed a positive change in which Tarabganj and Kaisarganj tahsils have positive changes up to mean + SD2.

The higher percentage of land in non-agricultural use is due mainly to frequent flooding of areas adjacent to the Ghaghra, Rapti, Gandak, and other tributaries on the one hand, and to the marshy character of the soil on the other.

With the increasing pressure of the population on the cultivated land, most of the cultivable wasteland had been reclaimed, which was estimated at 3.61 per cent of the total area in 1965–6 and 3.36 per cent in 1969–70. Thus there is 6.91 per cent negative change in the cultivable waste during the period. Fifteen tahsils experienced negative changes, whereas four tahsils experienced positive changes. Domariaganj tahsil has a positive change due to a reduction of forest land, which was changed into cultivable waste.

Permanent pastures and other grazing lands, in general, have not changed greatly. There is a slight increase in pasture lands recorded in Kaisarganj, Tarabganj, and Gorakhpur tahsils. A large number of tahsils experienced no changes during the period 1965–6 to 1969–70.

Land under miscellaneous tree crops and groves has experienced negative changes in general (−4.24 per cent). This is due to the increase in net sown area.

Fallow land experienced considerable negative changes (−35.18 per cent) in the region. 85.1 per cent of the tahsils have experienced negative changes, whereas only three tahsils of the region, i.e. Tarabganj, Bansgaon, and Nanpara, have experienced positive changes. The negative changes are high (mean + SD2) in the tahsils Kaisarganj (−57.13%) and Bahraich (−51.85%), because of the increase in net sown and non-agricultural land.

Double cropping of the region experienced positive changes. It was 34.14 per cent of the net sown area in 1965–6 and 35.38 per cent in 1969–70. Thus there would appear to be great possibilities for the development of the area under double cropping.

The above discussion concludes that there are some possibilities of reclaiming more land for cultivation purposes. A large number of tahsils experienced only slight changes in net sown area during the period 1965–6 to 1969–70, due to the fact that cultivation in these tahsils has reached such a saturation level that no further extension is possible. Cultivable waste lands are also lacking in most of the tahsils. The area under double cropping may be increased, which can provide sustenance for a large increasing population and can be an asset in improving the standard of living of the area after the area is reclaimed through the application of soil conservation methods and irrigation facilities. Intensive cultivation is essential to improve the methods of cultivation. Besides irrigation facilities, scientific methods of cultivation should also be introduced.

P1038
Rural settlement regions at the ecumene's edge: Europe and North America
KIRK H. STONE   *University of Georgia, USA*

Since the beginning of geography the edge of the inhabited world has usually been mapped as a line. Perhaps long ago the boundary was sharp. At present, however, it is not. In fact, the world's fringes of settlement are a region of areally discontinuous settlement, divisible into four distinct and complexly distributed zones, and because much new settling, and some abandoning, is quite normal in the region its distribution and character should be known to planners in many disciplines.

Originally a classification of northern Europe's ecumene was undertaken. The basis of the technique is the distribution of permanently occupied residences; the assumption is that people are dependent upon each other and if they are separated by more than an hour's foot travel (10km maximum) they are too far apart for effective short-time aid. Thus, permanently occupied dwellings plotted on maps of scale 1/500,000 or smaller as black circles of 5-km (3-mile) diameter provide the data for regionalizing. Where areas several tens of kilometres in at least one dimension are discernible, a region is delineated. If mapping shows it solid black, it is continuous settlement; if mixed black and white, it is a discontinuous settlement region; and if all white, it is unpopulated. When this system was applied to the northern edges of settlement in Europe and North America, regions were easily recognizable and, also, comparable (Stone 1967).

However, at the southern edges of Europe and North America the mapping technique needed change. There, as in Spain, the distribution of residences is only in a concentrated form (e.g. clusters, hamlets). But by doubling the previous basic distance the concept of effective short-time aid is maintained; the assumption is that each settlement concentration has at least one telephone so one hour's one-way travel is the same maximum of 10km assumed previously. Thus, settlements less than 10km apart in all directions form a region of continuous settlement (CS) and those farther in one or more of the central-place hexagonal directions form the discontinuous settlement region. Thus, intercontinental comparability is attained.

The European fringes of the *ecumene* are in three corners of the continent. The largest is the northern one, which extends from coastal Iceland eastward to include most of Norway, the northern three-quarters of Sweden and Finland, and the northern Soviet Union poleward of a line from Lake Onega southeastward but north of Kirov and Sverdlovsk. Included in the region are uninhabited parcels smaller than the minimum mapping unit of 38,000 square kilometres (15,000 sq. mi.), but there are five occurrences of the

Uninhabited Region, two of which are on the Soviet mainland. Otherwise the settlement pattern is one of scattered dwellings.

A second region is the southeastern corner of the European Soviet Union. There the boundary goes southwest from south of Sverdlovsk and Kuybyshev to the lower Don River and thence southeast to the northwestern Caspian Sea. The result is an area about 240 by 1450km but it does, of course, continue southeastward into Asia. In the region, as in Spain, settlement is largely concentrated.

In Spain is the third region. It has two parts separated by a cs region. At the northern Spanish coast is one section about 125 by 475km in size. The other is the southern third of Spain, an area of about 325 by 525 km; at a larger scale this is seen to extend northward on the eastern side so as to cut the cs in two (Stone 1971b).

Unlike its European counterpart, the North American discontinuous settlement region is uninterrupted. It is larger and its geographic range from the southern tip to the northern shore and the Pacific Ocean to the Atlantic is much greater. Still, the minimum mapping unit of about 38,000 square kilometres is maintained.

At the north almost all of the Alaskan-Canadian part is in the region. Exceptions are five unpopulated regions across the mainland edge, a sixth in the archipelago, and the seventh in Greenland. To the south are three more in the cs regions in British Columbia, Saskatchewan-Manitoba, and Ontario-Quebec, extensions from the USA.

Within the United States the discontinuous settlement region is most of the western third. Its eastern edge is roughly based at the 100th meridian but three westward extensions to about 105° are found in North Dakota, Nebraska-Colorado, and Texas; in most parts the change to discontinuity is gradual, even as seen from high-altitude jet aircraft. It is notable that the region breaks the two western occurrences of cs by going between them to the Pacific coast and that it is interrupted by an oval Unpopulated Region in the Nevada area.

In the southeastern USA is an exception, a strip of discontinuous settlement region through central Florida. Clearly the state's eastern coast has a long, very narrow strip of continuous settlement and so does the Tampa area on the west coast. Here the minimum mapping unit is relaxed somewhat to show the complexity of distribution; in fact, a larger-scale mapping discloses a number of cs interruptions along the coasts of west Florida, Georgia, and South Carolina.

The western US discontinuous settlement region continues southward to cover nearly all of Middle America. Two exceptions are the continuous settlement region of the Mexico City area and that in western Cuba, the latter being just the minimum unit mapped herein. Both the mainland and Antilles have combinations of scattered and concentrated settlement.

Now that two continents have been mapped provisionally four more immediate needs are clear. First, the remaining continents need to be mapped. In general, the mapping to date has been done from 1/250,000 to 1/125,000 base maps supplemented by larger-scale air photos and study. Preliminary mapping of other continents has been somewhat unsatisfactory at such scales. Perhaps satellite photography will be helpful; until then much time is necessary to use existing air photos and maps. Certainly international co-operation is desirable.

Second, the present minimum mapping unit needs to be smaller. Reduction of the 38,000-square-kilometre minimum is desirable in order to recognize enclaves within the region for these are quite important to planners in all disciplines.

Third, all discontinuous settlement regions need to be subdivided to show their complexity. This has been done for northern Europe and northern North America. In them four fringe zones have been mapped by varying degrees of isolation; these are defined by the locations of residences relative to one another and relative to interregional and local lines of transportation. The results provide primary data for settlement planning and, equally important, are comparable areally (Stone 1971a, 1971c, 1971d, 1972).

Last and most important, regional and zonal boundary locations must be explained. At present it is certain that most are founded on multiple qualities that are variously and intricately related in each area. The explanations will add to geography's theoretical and applied values in planning for the continuing changes in distribution and character of the edges of the inhabited world.

Stone, K.H., 1967 Geographic aspects of planning for new rural settling in the Free World's northern lands, in S.B. Cohen, ed., *Problems and Trends in American Geography* (New York), 221–38.
- 1971*a* Isolations and retreat of settlement in Iceland, *Scot. Geog. Mag.* 87, 3–13.
- 1971*b* Spanish fringe of settlement zones, *Proc. Assoc. Am. Geog.* 3, 195.
- 1971*c* *Norway's Internal Migration to New Farms since 1920* (The Hague).
- 1971*d* Regional abandoning of rural settlement in northern Sweden, *Erdkunde* xxv, 36–51.
- 1972 Geographical results of post-war northern Finnish colonizing and emigration (in press).

P1039
Determining the effect of productivity and environment on rural population density
PIERRE A.D. STOUSE, JR. (deceased), and BETTY HOLTZMAN
University of Kansas, USA

Population densities of the rural regions of the world vary significantly from one area to another. The magnitude of the variation has been recorded in a number of sources and is said to range from several hundred to less than one person per square kilometre. In the Western Hemisphere alone the range is from 514 in Barbados to less than 0.2 in the Brazilian states of Para and Amazonas (Ruddle and Mukhtar 1970, 55, 75; James 1959, 539). Variability in human densities is not easily explained by climate or other environmental variables, or by the economy, technology, or other cultural variables. Thus the student of man-environment relations is faced with the problem of determining how differential densities can be explained for each separate set of multi-variate situations (Zelinsky 1966; Zelinsky, Kosinski, and Prothero 1970; Ackerman 1959; Clarke 1965). Once this determination has been initiated he may begin to answer the vexing question: When has a specific region reached the limits of the population it can support?

The study of this question concerning population densities may be begun by selecting one or more factors for analysis while attempting to hold others constant. We shall look at those factors that we consider particularly important and see how much about the population density we can explain with them, specifically the physical environment and what the inhabitants are able to extract from it. This paper describes a method of analysing the environment and suggests ways of measuring the productivity of forms of livelihood with the objective of presenting a technique for determining when an area is over- or under-populated (for a review see Butler 1966; for criticisms and problems see Beshers 1967, 166–7; Taeuber 1970, 56–7).

Measuring the productivity and analysing the environment of an area is a complex task. The technique employed must be flexible in order to allow for the many different levels of technology and widely varying cultural choices of crops, livestock, and natural products. To assess population densities for any given region the information needed includes the products grown and gathered, the animals kept and hunted, and the level of living of the inhabitants of the area. Additional information needed includes yield figures for the agricultural systems employed and knowledge about what parts of the physical environment are used. Standard climatic data such as temperature and rainfall are also required. With the information available we can proceed with our analysis.

The first step is to create a series of physical-geographic regions that have equal potential for the people living in the area being analysed. This may be done by taking small unit areas, such as one square kilometre or some multiple thereof, and measuring the important physical variables for that unit area. The key word is 'important.' Examples of this type of variable include the length of the frost-free period in areas of high altitude, the amount of sunshine in cloudy areas, soil characteristics such as compactness or availability of a particular nutrient, the presence of annual flooding, etc. For any given area, and for any given culture or level of technology, there is a single set of important physical variables. Each of these variables and the segments into which they are separated must relate significantly to some aspect of the subsistence activities carried on in the area. Once these variables are determined and measured for each small unit area, clustering techniques may be employed to provide uniform regions based on physical criteria.

The result of this first step will be a map of the area being analysed that shows a set of physical regions. They may be contiguous, but not necessarily. There may be two regions, five, or a dozen; the number will vary from case to case and will be dependent on the level of generalization or the degree of refinement being employed. The most important feature of these regions is that they will be homogeneous in terms of potential productivity for the inhabitants of the area. They need not and probably will not have the same productivity when compared to each other.

The second step is to calculate a productivity potential for each ecologically different region. This potential is best determined by measuring average agricultural output under the best operant conditions currently existing in each region.

The use of existing productivity norms to determine productivity potential will keep the population-carrying capacity we ultimately determine within the cultural framework and reality of the region chosen for analysis. We think it is of little value to hypothesize a productivity level using unrealistic assumptions such as the presence of crops or animals that are not actually in the area or the use of advanced technology that is beyond the means, economically and conceptually, of the inhabitants. Neither are our purposes well served by trying to derive a productivity potential from some measurable quantity such as solar radiation or potential photosynthesis. These indices are currently still too crude.

A common measurement factor which will equalize different types of agricultural production, or a common denominator, is needed. Two come readily to mind: cash equivalency or caloric output. The choice of the common denominator will depend on its appropriateness for the area being studied (Clark 1967; Bennett 1954; Keys 1958).

The result of our first and second steps will be a map of physical regions each of which has a known productivity potential in either cash or calories.

The third step is to determine the cash or calories needed to supply the life style to which the people are accustomed – in the area being analysed. With this figure in hand the number of people that can be supported in any of our several regions can be calculated. We call this figure the maximum population-carrying capacity. We would like to reiterate that it has been derived by using existing productivity norms for the area being analysed and existing levels of living. Our figures represent a rough measure of the maximum population density that can be achieved by the people living in the area using their interpretations and analyses of the environment and their choice of life styles. One caveat here is that this index is a reflection of variation in two factors only and if others were added to the analysis the result might not be equivalent.

The fourth step is to measure the actual population densities in the different regions. Obtaining these data may require enumeration in the field to get the desired population-area correspondence or it may be available in published or unpublished censuses. Needed is the population size of the community and the area it encompasses. If the area is entirely within one of the previously established physical regions a maximum carrying capacity figure and an actual density are available and ready for comparison. If the area used by the community is in more than one physical region an appropriate and simple adjustment must be made prior to comparison.

The result of this fourth and last step will be a map that reflects the difference between the hypothesized carrying capacity and actual population of each unit area giving us an indication of the areas of overpopulation and areas of underpopulation.

The usefulness of the technique we have just described should be self-evident. Grossly overpopulated areas are the source regions for migrations, frequently the centre of political unrest, and in less-developed countries the place where starvation is most likely to occur. They are the areas that demand attention because the condition represented there will not persist – in some way it will be resolved (Tricart 1970; Kosinski and Prothero 1970). Underpopulation, on the other hand, indicates areas where population can grow, areas suitable for immigration, areas where resources are underutilized. This is most important information to have available, and the technique outlined will provide us with these data.

Ackerman, E.A., 1959 Population and natural resources, in P.M. Hauser and O.D. Duncan, eds., *The Study of Population* (Chicago), 621–48.
Bennett, M.K., 1954 *The World's Food: A Study of the Interrelations of World Populations, National Diets, and Food Potentials* (New York).
Beshers, J.M., 1967 *Population Processes in Social Systems* (New York).
Butler, C., 1966 Theories of optimum population, *Cornell J. Social Relations*, 1 (no. 2), 119–34.
Clark, C., 1967 *Population Growth and Land Use* (New York).
Clarke, J.I., 1965 *Population Geography* (Oxford).
James, P.E., 1959 *Latin America* (New York).
Keys, A., 1958 Minimum subsistence, in R.G. Francis, ed., *The Population Ahead* (Minneapolis), 27–29.
Kosinski, L.A., and R.M. Prothero, 1970 Migrations and population pressures on resources, in Zelinsky, Kosinski, and Prothero, ed., *Geography and a Crowding World* (New York and London), 251–8.
Ruddle, K., and H. Mukhtar, ed., 1970 *Statistical Abstract of Latin America*, 1969 (Latin American Center, U. California, Los Angeles).
Taeuber, I.B., 1970 Population dynamics and population pressures: geographic-demographic approaches, in Zelinsky, Kosinski, and Prothero, eds., *Geography and a Crowding World* (New York, London), 55–70.
Tricart, J., 1970 Physical environment and population pressure, in *ibid.*, 157–71.
Zelinsky, W., 1966 *A Prologue to Population Geography* (Englewood Cliffs, NJ).
Zelinsky, W., L.A. Kosinski, and R.M. Prothero, 1970 *Geography and a Crowding World: A Symposium on Population Pressures upon Physical and Social Resources in the Developing Lands* (New York and London).

P1040
Land policy, farm migrations, and rural planning in Nigeria
R.K. UDO   *University of Ibadan, Nigeria*

The persistence of traditional land tenure systems and the maldistribution of land and population are two basic and related problems facing agricultural development in Nigeria. The need for a land policy that will lead to a planned development of rural areas has long been recognized and action on the issue is considered to be overdue. At present traditional ideas of communal land ownership persist and an alien (this includes any Nigerian who is not a member of the particular village, clan, or subtribal group) cannot acquire outright title to land within the territory of the group. The result is that people who originate from very densely settled districts where farmland is in short supply find it difficult to acquire outright titles to land in those sparsely settled areas where there is an abundant supply of farmland. Rather a system of farm migrations has developed whereby those in need of farmland migrate to settle and cultivate land leased to them for short periods on payment of an annual rent which varies from 20 shillings to over 200 shillings depending on the size. The planning problems posed by these migrations are discussed in this short paper, which emphasizes the urgency of a land policy that will make for a planned redistribution of rural population in Nigeria.

The volume of rural-rural migrations in Nigeria has increased considerably since the end of the Second World War (Udo 1970) and out of 1200 migrants interviewed in 1966, 586 (48.9%) were tenant farmers cultivating food crops, while 276 (23.3%) were tenants who exploited wild palm trees on payment of an agreed rental per annum. Our discussion is focused on migrant tenant food farmers who have spread unscientific farming methods to other parts of the country, thereby increasing the incidence of soil impoverishment and soil erosion.

Admittedly, these migrant tenant farmers

have made a considerable contribution to the growth of the rural economy of the districts where they settle. One such district is the cocoa belt of southwestern Nigeria, where the indigenous Yoruba people concentrate on cashcrop production to the extent of neglecting to produce enough food for their families, thereby converting much of the cocoa belt into one of the food deficit areas of Nigeria (Udo 1971). The activities of Igbira and Ibo migrant tenant farmers have, however, transformed certain areas in the cocoa belt such as parts of Ekiti and Ondo Divisions into food surplus areas exporting local staples to urban centres as well as to rural food deficit areas.

Our contention is that the present system of uncontrolled movement of people who squat about the territory of the host community does not make for the most efficient utilization of either the land or the human resources of the country. Indeed the present system of land occupation by migrant farmers simply results in the spread of traditional farming methods which are destructive of soils and do not yield much income to the migrant farmers. Since the migrants have no security of tenure, their interest in the land which they cultivate is transitory. It is also significant that in spite of the increase in the number of migrants into the cocoa belt during the last twenty years, indigenous cocoa farmers still find it difficult to hire adequate wage labour to work on their cocoa as well as food farms. This is because an increasing number of migrants who originally provided wage labour in the cocoa belt have since found it more profitable to become independent foodcrop farmers in this food deficit area.

As a rule, these migrant tenant farmers plan to return to their villages of origin since they cannot acquire land outright or become citizens of the area in which they settle to farm. Such migrants are therefore obliged to retain the small portion of farmland which they inherited in their village of origin. The result is that farm integration or the redistribution of farmland to those who stay behind in the congested districts is hindered and, in consequence, a planned development of the source regions of the migrants is made impossible.

An early attempt in 1948 to redistribute the rural population of what was then Eastern Nigeria failed largely because of the lack of a land policy and, so far, it is migrant tenant farming that has proved the most effective way of redressing the imbalance between rural population density and land resources. But in view of the planning problems posed by an uncontrolled colonization of the sparsely settled parts of the country, a firm and progressive land policy is required on the lines suggested by the FAO report on land tenure (FAO 1966). Although the state governments are responsible for matters dealing with land tenure, the idea of setting up a National Commission for Land Policy has much to recommend it in view of the manpower problems of the less-developed states in the country. The Commission will advise the state governments on suitable land policies taking into consideration existing institutions and availability of land in each state.

Government acquisition of land for development is not a new thing in the country as is evidenced by the establishment of many farm settlements and Federal Agricultural Research projects in many parts of the country. The problem with the capital-intensive farm settlements which have proved a huge failure is that for political reasons the state governments failed to adopt a firm policy in selecting settlers. Rather the settlers consisted largely of people from the clan which owned the land, and since the farm settlements came to be regarded as government amenities to various groups, they were not necessarily located in areas which have abundant farmland. In urban areas, on the other hand, government has acquired vast areas of crownland for urban development. What is required is the extension of the crownland concept to rural areas so that vast areas of uninhabited land can be acquired and made available in economic units to people who are in dire need of farmland, irrespective of their ethnic group. Security of tenure will be guaranteed and fragmentation prohibited.

Apart from opening up new areas for planned development and satisfying the legitimate aspirations of those who are desperately in need of farmland, this process of resettlement can be seen as a first step to transforming the agricultural economy of the congested districts which are characterized by fragmented and dispersed farm holdings which rarely exceed 0.4ha in area. Once some local people are resettled elsewhere on the understanding that they forego title to the land they inherited at their village of origin, preferably by selling the land or transferring it to a relative, farm integration would become feasible and the process of rural planning will be extended to the congested districts.

FAO, 1966 *Agricultural Development in Nigeria, 1965–1980* (Rome), 331–8.

Udo, R.K., 1970 Rural-land migrations in Nigeria, *Nigeria Mag.* 103, 616–24.

– 1971 Food deficit areas of Nigeria, *Geog. Rev.* 61 (no. 3), 415–30.

P1041
Two settlement simulations
GERALD WALKER   *York University, Canada*

Two models which generate point patterns, similar to settlement distributions, are presented. Both operate on simple rules which can produce a variety of distributions comparable to those found in the real world.

Simulation I, LIFE, is based on a mathematical game, developed by John Horton of Cambridge University and presented in an article by Martin Gardiner in *Scientific America* (Gardiner 1970). The computer program, in Fortran IV, was written by Douglas Marsh, an undergraduate in physics and mathematics at the University of California, Santa Barbara.

The simulation, LIFE, sets rules for behaviour. Distributions are generated which could not be predicted otherwise. The rules are extremely simple. A series of points are initially placed on a grid of any size. The number of generations to be computed is specified. Mr Marsh's program specifies a 200 × 200 matrix, but this is readily changeable. In each generation all births and deaths occur simultaneously. The eight adjacent cells surrounding each occupied cell are evaluated. If two or three neighbours are occupied, the counter survives. Death may occur in two ways. If four or more adjacent cells are occupied, or if only one or fewer adjacent cells are occupied, the counter dies. Finally, each empty cell adjacent to three neighbours produces a birth on the next generation.

The operating set of initial cell occupants is read into the memory matrix and centred on the larger field. Then evaluation of the points for births, deaths, and survivals is performed and stored in memory. Finally the new distribution of cell occupants is made. This constitutes a generation. The print-out is in an alpha-numeric matrix of blank and occupied cells.

Horton maintains, and preliminary running of the program supports his contention, that several classes of outcomes will result. In a few cases the society created will disappear. In the majority of cases either of two other outcomes will occur. First, after some number of generations, a stable pattern, 'a still life,' will emerge. Second, a pattern which oscillates over two or more generations will develop. In general, a non-symmetrical initial pattern will generate and maintain patterns with symmetry. Several of the symmetrical, stable patterns are variants of hexagonal (Christaller's hexagons?) structures. Further, Horton maintains that rules will not permit the generation of a continuously growing distribution.

Though the rules established by Horton can have some common-sense relationships to behavioural rules in human societies, the potential isomorphisms are not developed. For all of that, the rules are robust, and the distributions are patterns of interest to geographers.

The second simulation, CNPROB, combines a conditional probability model with a Monte Carlo process to generate a distribution. Values are initially distributed over a number of locations in memory. Values may be any positive integers and the number of locations may be up to 999, arrayed in a vector. Arbitrary low values are added to all empty locations to provide a basis for some settlement probability. The number of births and deaths is specified. This program is also in Fortran IV and was designed and carried out by the author. From this initial distribution a probability surface is generated. A vector of expected final values, against which the results of the simulation will be tested, is read into memory.

For a birth, a random number is generated. A new settler is assigned to the cell containing the probability number matching his random number. For a death the same procedure is used. After each birth or death the probability surface is recast. Thus, a new probability surface is used for each iteration of the program. The last operation of the program is testing the final simulated distribution against the expected values. The Kolmogorov-Smirnov goodness of fit test (at 1 per cent and 5 per cent significance levels) is used. Output includes the values and probabilities in each location at each iteration, and the results of the Kolmogorov-Smirnov tests.

Though this simulation was designed for individual settlers moving into a region, it could have other applications. The model seems appropriate to testing development related to cumulative and circular models of the sort initially suggested by Myrdal and expanded upon by Pred (Myrdal 1957; Pred 1969).

In both simulations simple, arbitrary, and not especially theoretically satisfying rules are used. A priori theory does not suggest any particularly strong reason to use a conditional probability model over the Hägerstrand varieties of Monte Carlo simulation. But both simulations could be used as a base to develop more interesting and more realistic models. The conditional probability model has the serious procedural shortcoming of only generating a vector and not a matrix. However, this may be rectified shortly.

Gardiner, Martin, 1970 Mathematical games, *Scientific American*, 223 (no. 4), 120–3.

Myrdal, Gunner, 1957 *Rich Nations and Poor* (New York).

Pred, Alan, 1969 *Behaviour and Location*, parts I and II (Lund Studies in Geography, ser. B, nos. 27 and 28).

P1042
Rural settlements in ancient central India
RAJ KUMAR NIGAM   *University of Kanpur, India*

Rural settlements are characteristic of the agricultural landscape of central India. They are living geographical entities for they represent a synthetic complex of its physical and cultural environments. In advanced countries which carry a high degree of social freedom, each natural region tends to develop a characteristic form of arrangement of its rural population, while in backward regions like the one in hand, social organization is no less a powerful factor than their physical milieu. 'The historic tendency is, in general, for an initial social arrangement to break down with advancing civilization and to be replaced by a more appropriate response to physiographic conditions' (Aurousseau 1920, 224). As such, the study of settlements involves not only a knowledge of the facts of land but also of the physical qualities, the psychological attitudes, inherited traditions, and taboos of the people, and all the countless trivial accidents that play so important a part in the irrational course of human affairs.

The prehistoric stage is shrouded in mystery. However, historians postulate early human settlements in the Ganga Valley, long before the Aryan occupation. Before the advent of prehistoric peoples whose cultures had flourished in the valley of the Euphrates, Tigris, and Indus in the period 3000 to 2500 BC (Majumdar 1952, 28), the Ganga valley was still the home of the Proto-Indies or Proto-Australoids, whom the Sanskrit text designates as Nishads and Savanas. The surviving memories and prevailing traditions of these ancient peoples remind us of their blending with the later human currents, viz., the mongoloids and the Mediterranean-Armenoid. The Mediterranean peoples, in popular usage referred to as Dravidians, 'introduced a city culture as opposed to the village culture of their predecessors (Majumdar 1952, 16).

Many authorities hold that the Aryan settlements were first established in the Punjab about 2500 BC (Prasad 1947, 18) and that the early vedic culture was *not* located in the

Ganga valley, as the Rig veda makes but occasional mention of the Ganga and the Yamuna. From the Punjab the early Aryan colonists probably trickled southeast and eastward in two branches, another branch moving east also but crossing the Ganga and entering the Ghagra valley, where flourished the famous Koshal kingdom extending to the foot of Nepal hills (Majumdar 1924, 467). The aborigines were slowly assimilated and where assimilation was resisted forced further into their mountain fastnesses (Panikkar 1950). Another branch purely of Indo-Aryans moved south to central India and the Narbada valley. Here 'the Aboriginal tribes, such as the Bhils, believed to have been the earliest inhabitants of India, pressed by the advancing Indo Aryans retreated to the forest recesses' (Ministry of Transport and Communication 1958, 8). Before the fourth and fifth century BC, the period extending from the era of the composition of the vedas to the construction of the law code which we know as the Institute of Manu, the present Hindu polity was established.

These Indo-Aryan settlements bespeak our glorious efforts towards rural planning. To quote Ram Raj: 'The extent of villages or towns is declared to admit of forty varieties, consisting of from five hundred to twenty thousands dandas square, each sort exceeding to one immediately below it by five hundred dandas. The whole area of a village, with the lands there into belonging, being divided into twenty parts, one is assigned for the occupation of the Brahmans, six or more for that of the three other classes, and the remainder for agriculture. Each village was surrounded by a street called "*Mangalaritihi*," generally one to five dandas in width. The street that ran from east to west was called "*Rajpath*"; that which had gates at both ends, "*rajavithi*"; that which had "*andhis*" or angles, "*Sandhi-vitihi*"; and that which was in a southerly direction, "*mahakala*" or "*vaman*" ' (Ram Raj 1823, 41).

Prior to the building of a village, the 'Sthapt' traced upon the ground selected for the purpose any of the mystical figures, particularly that which was called '*param-sayika*.' Thereupon he offered the prescribed sacrifices to the deities presiding over its various sectors, laid out the street patterns, and marked out sites for building temples and private houses, etc., according to rules of shastras.

According to their function the Aryan settlements would appear mainly to have been of six types: (1) Goshala Vraja (cattle ranch); (2) Palli (a small barbarian settlement); (3) durga (fort); (4) grama (village with the durga as its focus); (5) kharvata or pattan (town); and (6) nagar (city). (This classification is based on the Mahabharat. Manu distinguishes among three kinds of settlements: villages [grama], town [pura], and city [nagar].) According to the Mahabharat, the village was the fundamental unit of administration and had its own head, the gramini, who was its leader and administrator (Dube 1955, 2; Radha Kumud 1950, 140–5). One of the major responsibilities of this headman was to protect the village and its boundaries in all directions within a radius of 3.2 km (Dube 1955, 2). The inter-village organization was based on a decimal system.

The village communities formed groups of 10, 20, 100, and 1000 villages (Belvalkav 1950), and the ruler of each group was called dasgramini, vimsatipa, satgramini, or gram satedhyaksha and adhipati respectively. According to Manu, the village was the fundamental unit and its head was gramini. The villages were under a dasi; twenty villages were under a vimsi; one hundred villages were under a satesa; and one thousand villages under a sahasresa.

According to the plan there were eight types of villages, defined according to their shapes: (1) dandaka (that which resembled a staff); (2) sarvatob hadra (in every respect happy); (3) nandy-varta (the abode of happiness); (4) padmaka (that which has the forms of a lotus); (5) svastika (that which resembled the mystical figure so named); (6) prastava (that which had the shape of a couch); (7) karmuka (that which resembled a bow); and (8) chaturmukha (that which had four walls).

The village called 'dandak' (Acharya 1933, 68–9) was quadrangular in plan and was surrounded by a wall. It had from one to five streets and also one crossing them in the middle. The width of the streets varied from one to five dandas, the innermost wider than the rest. The streets at the extremities or near the walls had a single row of houses, the inner ones having a double row or one on each side of the enclosing wall of the villages and as many smaller ones at the several angles. On the part presided over by the varma or maytra was erected a temple Vishnu, and in that presided over by Adita, at the northeast angle, one dedicated to Shiva; a shrine for Chamunda Bhawani was built nearer the north gate outside the wall. There used to be two tanks or reservoirs, one towards the southwest and the other towards the northwest. This village was particularly intended for the residence of Brahmins. It might contain 12, 24, 50, 180, 300, or more houses. The smallest, or that which contained 12 houses, was called 'ashrama' (hermitage) and was located near mountains or forests for the habitation of the hermits. The village containing 24 houses was to be situated, on the banks of a river and inhabited by 'yatis' or holy mendicants, and was called 'puram.' That which contained 50 houses was occupied by those who had performed holy sacrifices or by householders in general. In the former case it enjoyed the appellation of 'puran' and in the latter that of 'mangalam.' The village containing 180 houses was commonly known as 'kost ham.'

The village called sarvata bhodra (Acharya 1933, 69–71) had a quadrangular (square) pattern, containing at the centre a temple dedicated to any one of the triad, Bramha, Vishnu, or Mahesh, treated in Indian mythology – the creator, the preserver, and the destroyer. It had four streets of equal length on the four sides within the wall, meeting one another at right angles, and two more crossing each other in the middle. Between these might be laid out, three, four, five, or as many as the dimensions of the village would admit on each side parallel to the middlemost street. Outside the walls were erected the shrines of the deities who presided over and defended the several quarters of the village. At the angular points were built halls porticoes, schools, and other public buildings, and towards the quarter of 'Agni' (southeast), a rainshed for the accommodation of travellers and passengers. The whole village was surrounded by a wall with an accompanying moat, with four large and as many small gates in the middle of the sides and at the angular points. Outside the northern gate was erected a temple for the goddess Mahakali. The huts of the Chandals or outcastes lay a 'krosa' distant from the village. A tank or reservoir was built either on the south or north side, or near either of these two points, for ablutionary or culinary purposes.

Thus we have considered the anatomic features of the first two types of villages. In the same manner we can describe the forms and arrangements of the remaining types as well (Acharya 1933, 71–90), but as the accompanying ground plans will amply suffice, their description is avoided. In conclusion it may be just pointed out that most of the villages carried quadrangular plans and that they had some common features, although differing in meticulous details. Generally they had a surrounding wall and a moat for defence purposes and the centre was often occupied by a temple, tank, or public pavilion. The villages were thoroughly planned. Street patterns were well laid out, drainage properly controlled, communal compartments separately marked out, and a residential atmosphere created.

Aclarya, P.K., 1933   *Architecture of Mansar* (London).
Auroussean, M., 1920   'The arrangement of rural populations,' *Geogr. Rev.* 10.
Belvalkav, et al., 1950   *Mahalharat*, Shanti Parva, 12.88.3 Poona, Bhandarkar Oriental Research Institute.
Dube, S.C., 1955   *Indian Villages* (London).
Majumdar, R.C., 1952   *Ancient India* (Benares).
Majumdar, S.N., 1924   *Cunningham's Ancient Geography of India* (Calcutta).
Ministry of Transport and Communication (tourist dept.), 1958   *Madhya Pradesh*, Publications Division, Ministry of Information and Broadcasting (New Delhi).
Panikkar, K.M., 1950   *A Survey of Indian History* (Bombay).
Prasad, Ishwari, 1947   *History of India*, The Indian Press.
Radhakumud, Mukerjee, 1950   *Hindu Civilization* (Bombay).
Ram Raj, 1823   *Essay on the Architecture of the Hindus* (London).

P1043
From aeroplanes to agriculture
D.A. FRASER   *Sir George Williams University, Canada*

Man's 'development' of the environment is not a straight line. Thus we find abandoned farms, where farming on the marginal land was unprofitable; abandoned strips of highway where new technology helped to straighten the route. While a few papers (Booth 1941; Rice et al. 1960; Tomanek et al. 1955) concern themselves with revegetation of abandoned fields, the bits of abandoned hardtop have not been investigated. These will certainly be distinctly influenced by climatic conditions which will result in primary disruption of the surface. Plants might be the original invaders or occupy minute niches. Yet, as will soon become evident, the large areas surveyed did not depend on natural plant invasion for their disintegration.

During World War II there was a great proliferation of aerodromes both in Canada and particularly in the United Kingdom. There, construction produced a severe modification of relatively large land areas.

It is the purpose of this paper to survey the fate of these areas 25 years later, when the original need for them has been lost. It is unnecessary to explain that the diversity of the surrounding population has a direct effect on the acceptance of a well hardtopped surface and a complex of adjoining buildings. We are, of course, excluding from our survey those aerodromes that are still actively used by the military (e.g., Newton, Nottinghamshire, England; Trenton, Ontario, Canada).

Cities that have 'graduated' to a municipal airport have made use of some of the facilities (e.g. the airport at Wick, Scotland, and Mount Hope outside Hamilton, Ontario). Some of the facilities have also been adapted for use. Primarily, available buildings have been used for human residences and places of work, as in the establishment of a community college at Centralia, Ontario, and the establishment of an industrial park at Wick, Scotland. At Wick we notice that a division between the facilities has made two important contributions to the economic development of this northernmost part of Scotland. A comparable split exists at Silloth, Cumberland, England, where both an industrial park and an agricultural purpose is being served. Here the great hangars have been modified to house about half a million chickens. Enough of the runway is maintained in usable condition to allow the executive planes of resident companies to land. In addition, part of the area previously occupied by the aerodrome is developed to accommodate a large number of campers, for Silloth is trying to establish its reputation as a seaside resort. The dual type of utilization is also evident at Snaith, Yorkshire, though here we understand that official use is slowly being abandoned. Most of the buildings were, as late as the summer of 1971, still used for storage of government equipment, both military and non-military (National Fire Service) in origin. The heavy equipment of the National Fire Service, which was tested for efficiency here, was instrumental in dislodging most of the runways. The runway building material was then buried at a depth which would not interfere with cultivation, for the aerodrome was located on agricultural land which is once again seeded with barley. A less happy incorporation of an aerodrome exists at Skitten, Scotland, where the runways, while invaded by weeds which are getting a foothold in the hardtop, can still carry car traffic. No large hangars were located at this aerodrome and the cement buildings used for administration and residences are now converted into a home and several animal shelters.

Of the whole group, the most desolate reminder was that at Penfield Ridge, NB, Canada. Here all the buildings have been removed, and, although we understand that one part of the runway is kept up for drag racing, most of them have begun to disintegrate. Only seagulls, cinquefoil, and blueberries make permanent homes, where 25 years ago the hum of aeroplanes prevailed.

While the trend today is towards larger and larger areas being converted into blacktopped runways (Dansereau 1971), it should not be taken for granted that such trends are irreversible. New aeronautical developments forecast by eminent engineers such as Sir Barnes Wallis (of R-100, Wellington bomber, and dam-buster fame) already visualize a decreased need for exceedingly long runways in the near future. It is, therefore, not a farfetched idea that, while new airports which will envelop tens of thousands of hectares are considered or under construction, the potential of future reconversion into original land use should be planned.

Booth, W.E., 1941 Revegetation of abandoned fields in Kansas and Oklahoma, *Amer. Jour. Bot.* 28, 415–522.
Dansereau, P., 1971 EZAIM – An interdisciplinary adventure, *National Research Council of Canada Newsletter* 3, no. 3, 1–2.
Rice, E.L., W.T. Penfound, and L.M. Rohrbaugh, 1960 Seed dispersal and mineral nutrition in succession in abandoned fields in central Oklahoma, *Ecology* 41, 224–8.
Tomanek, G.W., F.W. Albertson, and A. Riegel, 1955 Natural revegetation on a field abandoned for thirty-three years in central Kansas, *Ecology* 36, 407–12.

P1044
Irrigation agriculture in Sind, Pakistan
MUSHTAQ-UR RAHMAN *University of Karachi, Pakistan*

In the province of Sind, Pakistan, irrigation appears to be as old as agriculture. Archaeological sites dating back about 6000 years indicate the presence of irrigation and agriculture in Iran, Baluchistan, and the alluvial plains of Sind. The locations of these sites, relatively younger in age towards Sind, indicate a possible migration of agriculture and irrigation from Southwest Asia. Following Sauer (1952), Southwest Asia is accepted as a cradle of seed agriculture and also a place of possible origin of irrigation (Rahman 1960).

Irrigation agriculture overwhelmingly dominates the man-made landscape, especially in the valley section of Sind. One cultivated field lies next to another in a seemingly endless succession. Canals and related structures delimit the fields and largely influence their layout, shape, size, and settlements, together with other aspects of land occupance in Sind. The twin problems of salinity and waterlogging, unconsciously induced by excessive canal irrigation, have been threatening the regimen in the recent years.

The physical geography of Sind, in simplest terms, consists of the Indus Valley, the oecumene of Sind, the flanking barren Kirthar/Kohistan mountains, and the Thar Desert. Southward in the string of ever changing distributaries is the delta land. The Indus furnishes huge volumes of silt-laden water, which has made life possible in Sind. The river brings water and new soil during the annual flood, which starts in late April, reaches its maximum in July, and ends in September. Since the river flows on a ridge of its own silt which averages 6.1m above the lateral valleys bordering it on both sides, and stands 15m to 21m above them in the 386km stretch between Sukkur and Hyderabad (Kingsbury 1961, 204), flood irrigation has been easy and widespread. Today the Indus is largely an artificial river, diked on both sides with openings for irrigation at only three places.

Climatically the entire area is a desert, with high temperatures and little rainfall. Summer temperatures ranging from 40° to 45°C are not uncommon; temperatures of 50° or above are sometimes recorded (Naqvi and Rahmatullah 1962, 13). During the winter the temperature sinks down to 20°C near the coast, and to 15°C in the north. The average rainfall amounts to hardly 178mm; 127mm in the summer and about 51mm in the winter.

In such a setting irrigation is done by canals, wells, and other means, which include tube wells, pumps, and springs. Wells seem to be indigenous in Sind, and well irrigation is represented by two systems: *Chahi* and *Charkhi*. The term *Chahi* refers to simple well irrigation, or boka-well irrigation. In boka-wells water is lifted in a leather bag pulled up by a pair of bullocks. Similar wells are reported in Yemen, Iran, the Persian Guf, and also from the southern parts of India.

The second system is known as *Charkhi* or *Nar*-well irrigation. The word *Nar* seems to have been derived from *Noria*, the original name of the device, sometimes wrongly called the *Persian wheel* in the literature. According to the present beliefs, these *Nars* were invented by the Arabs and introduced by them into the area around 712 AD (Rahman 1966, 326). They were quite an effective device until the 1930s, but are now being abandoned because of canal irrigation or an abnormal rise in the water table.

Tube wells and pumps were introduced in 1953 by the government of Pakistan. In the beginning there was general resistance against the installation of tube wells, and in some cases they are still resisted. Pumps are installed to raise water from the lakes or *dhands*.

Canal irrigation is the one which dominates the landscape today. Much has been done to improve the system after British administration in Sind began in 1843. Before and during the early phases of the British occupation all canals were excavations away from the river in an oblique direction (Fife 1855, 4). The agricultural economy was exposed to many risks, arising from the behaviour of the river. Too little or too much water, the supply coming too soon or too late, and the problems arising from sowing the crops at the wrong time combined to make yields uncertain.

The first efforts of the British were directed to deepening the canals, and keeping them clear of silt. By 1900 the total length of the canals was about 11,973km, without any order or arrangement. The major change in canals came in 1932, with the introduction of perennial canals from a barrage at Sukkur. The approximate length of the canal network associated with Sukkur Barrage is 76,910km; it is designed to irrigate 3.02 million hectares. Since the establishment of Pakistan two more barrages have been added in Sind at Kotri (1955) and Guddu (1962). The canals coming from these barrages irrigate most of the alluvial plains and present an interesting case study with regard to the availability of water and its withdrawals from barrages for irrigation.

Of the 104 MAF (million acre feet; 1 acre foot = $1.23km^3$) of water which flowed into Sind in 1963–4, 36 MAF were retained by the barrages; 61 MAF flowed out to the sea and 7 MAF were lost by seepage and evaporation between the Guddu and Kotri Barrages. The barrages use the full river discharge in December, February, and March, and in the remaining months water passes to the sea (Huntings-MacDonald 1966, 90). By far the greatest quantities escape in July, August, and September.

The immediate impact of these barrages was to increase substantially irrigation and cultivation, but they have disturbed the dynamic equilibrium between ground water recharge and discharge. The deep percolation of seepage water from canals and from water applied to irrigation lands has provided a new source of ground water recharge. As a result, the water table has risen ever since the barrage-controlled irrigation systems were placed in operation. Between the years 1930 and 1940 the water table has risen over much of the irrigated area to within 2.7m to 3.6m of the ground surface, and has been continually rising ever since. At some places the ground water table is right at the surface or has reached the root zone of the crops.

As the water table continues to rise, more and more lands become adversely affected by waterlogging and salinity. Statistics relating to the extent of damage caused by these twin problems vary widely from one source to another. Based on the first aerial survey, the Hunting Survey Corporation of Canada gave detailed statistics on the area, according to which about 91 per cent of the agricultural land has been affected to some degree by waterlogging and salinity (Colombo Plan Cooperative Project 1958, 53). The Water and Power Authority of the Government of West Pakistan estimated that more than 1.214 million hectares of agricultural land was damaged by 1962 (WAPDA 1961, 36). The White House–Interior Panel, specially constituted to study the problem, reduced the figures of the Hunting Survey Corporation by more than half.

In spite of such an alarming growth of waterlogging and salinity and the contradictory statistics, agriculture is carried on by about 75 per cent of Sind's population. The agricultural seasons are divided into *Kharif* (summer) and *Rabi* (winter), largely producing rice, cotton, millet, wheat, oil seeds, and sugar cane.

In non-perennial areas rice is by far the most important *Kharif* crop. Where water is insufficient or the supply is irregular, rice is replaced by sorghum, or millet, or the land is left fallow. The rice crop is usually followed by a winter crop of gram, wheat, or oil seeds, grown on residual moisture (*dubari* cropping). In the perennial areas, where water is available all year, cotton dominates the *Kharif* cropping. Sorghum and millets are also important. Summer pulses are the only other crops of significance. Wheat, oil seeds, and *berseem* are the chief winter crops.

This pattern of irrigation agriculture is barely satisfactory. Per hectare yield is decreasing every year. While waterlogging and salinity are the most important problems, there are many other factors. In terms of farm management, these include inefficient watering, poor seed bed preparations, use of seeds of low viability, inadequate use of fertilisers, and, above all, the satisfied contented outlook of the Sindhi farmer, who is happy with a good harvest and equally reconciled to a bad one. Any calamity which affects agriculture is looked upon as a stroke of bad luck.

Under such conditions all-out efforts are being made to improve the conditions by national and international agencies. Sixteen reclamation projects, which embody construction of drainage channels, tube wells, pump houses, and other related structures, have been planned for Sind. Changes are being brought about in the field patterns and field sizes. This is being done to economize water and save the land which goes waste in irregular water courses, field boundaries, and others. Other major changes in irrigation agriculture are coming with the introduction of new crops. Predominant among the new crops are bananas in lower Sind, jute in the Indus Delta, and emphasis on sugar cane cultivation. The banana was brought in from India, and has been widely adopted in the area. Jute was introduced in 1962 from East Pakistan, after the introduction of perennial irrigation in the Indus Delta. Sugar cane is being emphasized to counter the waterlogging and boost the sugar industry in Sind, recently with the assistance of the People's Republic of China.

The present position of agriculture and irrigation in Sind is that of continued changes and challenges to Sind. In addition to physical treatment, efforts are being made to break the persistent age-old traditions and outlook of the peasants. If all the plans are successful, the results for agriculture and the economy will be encouraging, with a complete change in the landscape and in old entrenched practices and methods.

Colombo Plan Cooperative Project, 1958 *Report on a Reconnaissance Survey of the Land Forms, Soils, and Present Land Use of the Indus Plains, West Pakistan* (Government of Canada for the Government of Pakistan).

Fife, J.G., 1855 *A Sketch of the Irrigation in Sind with Proposals for its Improvements* (Karachi).

Huntings-Technical Services Ltd., and Sir MacDonald Partners, 1966 *Lower Indus Report*, Part I: *Present Situation*, West Pakistan Power and Water Development Authority (Lahore).

Kingsbury, R.C., et al., 1961 *Pakistan: A Compendium* (New York).

Naqvi, S.N., and M. Rahmatullah, 1962 Weather and climate of Pakistan, *Pakistan Geog. Rev.* XVII (1), 1–18.

Rahman, Mushtaq-ur, 1960 Irrigation and field patterns in the Indus Delta, PH D thesis, Louisiana State U.

– 1966 Kunstvading Med Øseværket Nar I Vest Pakistan, *Kulturgeografi* 99, 325–31.

Sauer, Carl O., 1952 *Agriculture Origins and Dispersals* (New York).

WADPA, Government of West Pakistan, 1961 *Master Plan for Waterlogging and Salinity Control*.

White House and Department of Interior Panel, 1964 *Report on the Land and Water Development in the Indus Plain* (Washington).

P1045
Maps of land utilization in the system of cartographic provision for planning economic complexes
V.P. SHOTSKIY *Institute of Geography of Siberia and the Far East, USSR*

In distributing and in perspective planning of territorial economic complexes, the inventory of land acquires more and more significance. Growing production requires an ever increasing volume of natural resources to be drawn into its sphere, which in turn calls for a detailed working out of the spatial localization of all forms of man's economic activity. Considering that the land fund is an indispensable means of production in agriculture, the demand that all qualities of land and the spatial localization of all the branches of its economic utilization should be taken into account becomes more and more important. This problem acquires a particularly great practical interest when planning (or, to be more exact, forecasting) production, development, and the utilization of resources for a more distant perspective period. Historical experience of economic development in many countries shows that even during times of lower rates of production growth in the past an insufficiently grounded solution of problems for the utilization of land resources later caused quite perceptible economic damage. All this explains the growing interest in problems of rational land utilization. This is reflected correspondingly in geographical, economic, and even demographic scientific literature.

As the value of various lands depends on a number of natural factors varying in space, and also the territorial distribution of land, the most effective means of studying land resources is the compilation of various maps. The most important ones are those which give a more universal idea about the land resources of territories, that is, maps of land utilization.

Working out maps of land utilization is not a new problem. In west European countries where there is a shortage of land and the necessity for its more effective utilization appeared long ago, some experience in compiling such maps has been acquired. Especially significant changes in this respect have taken place during the past 10–15 years. At this time in various countries considerable efforts were made to prepare maps of land utilization in separate publications, and also in national and regional atlases. Such maps, generally of medium and large scale, were issued in Great Britain (Stamp 1964; *Atlas of Britain and Northern Ireland* 1963), Czechoslovakia (*Atlas CSSR* 1966), Poland (Polska przegladowa ... 1969), Spain (*Atlas National de España* 1965), West Germany (*Atlas der deutschen Agrarlandschaft* 1969).

The principles for mapping land utilization in various countries are very diverse, but it is possible to trace two major directions. The first is the illustration of land utilization in agricultural production on the map, and the second land utilization by all branches of the economy.

In Great Britain and France most maps are compiled on the materials of land utilization in agricultural production. From the published materials it may be seen that, in compiling maps, seven categories of agricultural land utilization (including forest plots) may be defined. Some time later, in mapping land utilization in the north of France (Flatres 1966), the system of crop rotation was taken as a basis to distinguish land types. In Yugoslavia cadastral maps showing all types of agricultural lands (Crkvencie 1962) were used as a basis. At the beginning of the 1960s in the FRG data about the structure of arable lands, or more precisely, the structure of sowing areas, were used as a basis for characterizing soil types (Telbis 1966). Polish cartographers included indices characterizing the types of agricultural economy in the legend. Thus, according to Kostrowicki's data (Kostrowicki 1964), land classification, worked out for compiling maps, included the following indices: character of land property, size of farms, trends in agricultural utilization of lands, specific weight of technical crops, character and ways of land utilization, degree of agricultural development of territory, intensity of economy.

During the last decade the tendency to compile maps that characterize more completely the utilization of the land fund by all branches of the economy intensified. This is proved by a number of works carried out in Great Britain, FRG, Norway, and other countries. Thus, while compiling maps in arid zones, data not only on agricultural lands, but also on territories of urban settlements, mining, forest, marshy, and non-productive lands (Stamp 1964) were used.

The experience of cartographers and geographers in the USSR in compiling maps of land utilization demonstrates the necessity for more complete and thorough cartographic characterization of the land fund of the country and separate regions. If earlier (for instance, while compiling maps for the atlas of the virgin lands (Pertzeva 1966)) only the characteristics of lands with agricultural utilization (arable lands, natural fodder lands) were given, now programs are investigated which reflect the utilizing of lands by all branches of the economy. So, for example, in 1965 M.I. Nikishev (1965) wrote that 'a map is expected to show: lands of towns, industry, transport, arable lands, perennial plantations, hay lands, pastures, deer pastures, forests, bogs, peat fields, and non-utilized lands.' Besides, when compiling maps of land utilization for the cartographic provision for district economic complexes, it is necessary to take into consideration the peculiarities of geographical conditions, perspective trends of economic development, and the working out of forecasts for the rational utilization and restoration of the resources in these territories.

Definite methods for making such maps, taking into account natural and economic specifications of the mapped regions, are being worked out.

The mapping of land utilization in the Upper Yenesei territory was based on maps made according to collective and state farm plans, with the boundaries of utilized farm lands marked, and divisions made according to the main land categories: arable lands, natural fodder lands, forests (with used lands), and inconvenient lands. After that the farmsteads were marked on the maps. Then field reconnaissance was carried out and the map made more accurate: all the changes in the location of the used land contours were marked. Reconnaissance may be carried out along key routes or by farms having obsolete plans of land utilization. Then the following are marked on the map, according to corresponding materials of forests, with subdivisions into categories of types of land utilization, areas of future construction work, irrigation systems, and other data characterizing the peculiarities of land utilization in different parts of the territory being mapped. As a special addition, data on geobotanic research with a simplified legend are marked on the map.

The representation of relief is removed from the map itself in the process of compilation, but, as the character of the surface is very important in solving a number of concrete questions of land utilization, the relief is given on a special base, on a transparent plate made to the scale of the main map. This makes it possible, when necessary, to combine the land utilization map with the map of the relief.

In such a way the general map legend for a given region may be presented as follows:
LANDS OF AGRICULTURAL IMPORTANCE
Arable lands (with characteristics of perennial crop capacity, as distinguished from irrigated lands)
  Real steppes
Natural fodder lands (with characteristics of degrees of flooding)
  Meadow steppes
  Stony, sandy, saline steppes
  Bottom lands, valleys, hollows, and forest meadows
  Forests and shrubs within limits of lands used by agricultural enterprises
  State land fund
LANDS NOT UTILIZED FOR AGRICULTURE
  Eroded soils
  Rocks, stony screes of steep slopes, sand-pebble drifts
  Territories of inhabited localities
  Areas for industrial and urban construction
LANDS OF STATE FOREST FUND
  Forests of local importance (productive-protected-raw material)
  Industrial forests
  Forests used for hunting
  Mountainous forestless territories
  Other conventional signs (boundaries of the Krasnoyarsk Region, Khakas Autonomous Region, districts, collective farm land-using enterprises, reservoirs, irrigation networks, systems for supplying fodder lands with water)

If such a method and legend were used for the compilation of a map, it would be an important cartographic contribution to any region. With some elements omitted, such a map may be used as a basis in compiling state maps of land utilization at the scale of 1:2,500,000.

The Atlas of Britain and Northern Ireland (Oxford), 1963.
Atlas Ceskoslovenske Socialisticke Republiky (Praga), 1966.
Atlas der deutschen Agrarlandschaft, 1969.
Atlas National de España Institute Geografico y Catastral (Madrid), 1965.
Crkvencie, I., 1962   Praze geogr. PAN, *Inst. geogr.*, no. 31.
Flatres, P., 1966   *Hommes et terres*, no. 2.
Kostrowicki, J., 1964   *Poznai swiat.*, no, 5.
Nikishev, M.I., 1965   Working out maps of land utilization for large economic regions, *Information, Higher Education Institutions; Geodesy and aerophotography*, no. 1.
Perpillon, A., 1963   *Acta geogr.* (France), nos. 46–7.
Pertzeva, A.A., 1966   Methods of mapping land utilization, *Vestnik of Moscow University. Geography*, no. 3.
Polska przegladowa mapa nzytkowania ziemli (Lublin), 1969.
Stamp, D., 1964   *Arid Zone Res.*, no. 26.
Telbis, H., 1966   *Ber. dtsch. Landeskunde*, no. 2.

P1046
The calculation of gross margin in agriculture and the productivity of arable farming in Finland
UUNO VARJO *Oulu, Finland*

There are many factors involved in agricultural productivity. The principal commercial factors and their mutual dependencies are presented in Fig. 1. Perhaps the most com-

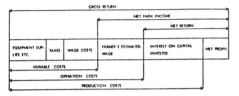

Fig. 1.

mon expression of agricultural productivity is net return. Since the wages of the farmer and his family constitute part of the operational costs, the financial success of the farm may be seen in the variations in net return. The farmer is considered to receive a fixed wage, while his net profit and interest on capital vary with his economic fortunes. Net return itself may be a suitable starting-point for the expression of agricultural productivity on farms whose profit and interest on capital are considerably more significant than the wages of the farmer and his family, as is usually the case on larger farms, but in Finland the farms are small and the estimated wages of the farmer's family comprise the majority of the net farm income. Here wages may be taken as a better indicator of agricultural productivity than the sum of interest on capital and net profit, and since such small farms are generally managed without resort to outside labour, the earnings of the farmer's family represent the entire labour costs of the farm.

The use of labour costs as an index of agricultural productivity presupposes both that production is not achieved at a loss and that the harvest yield does not vary very greatly. If these conditions hold, it may serve to depict regional differentiation in productivity resulting from variations in the intensity of field cultivation (Varjo 1958, 57–61). On the other hand, in an area where harvests vary considerably labour costs alone would provide an inadequate or even misleading impression of productivity, for this figure does not show whether production succeeds in covering labour costs, let alone what clear profit the farmer obtains in addition to his wages.

More recently the concept of 'gross margin' has been introduced, chiefly for agricultural planning. This is obtained by deducting variable expenditure from gross return. Thus it closely resembles net farm income, though differing in that it may be calculated separately for each branch of agricultural production. Gross margin thus enables comparisons to be made between different branches of agriculture, different regions, and also the yields in different years. As gross margin is an absolute figure, the gross margins for various branches of agriculture may be combined, so that the gross margin for arable farming, for instance, is the sum of the gross margins for all the field crops cultivated.

The effect of harvest yield on gross margin is illustrated in Fig. 2, which presents gross margins for different harvests of wheat (MKJ 518). Gross margin increases with harvest yield, but its proportional increase is distinctly greater than that of the yield itself, for barely a threefold increase in the harvest gives a fivefold increase in gross margin. In this sense gross margin indicates the productivity of the crop better than does gross or net return. Similar results are obtainable for other crops. In addition, it can be seen from Fig. 2 that the sector lying below curve 1, variable expenditure, does not vary appreciably with increased yield. Thus, without introducing any serious error it may be treated as a con-

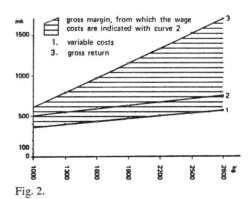

Fig. 2.

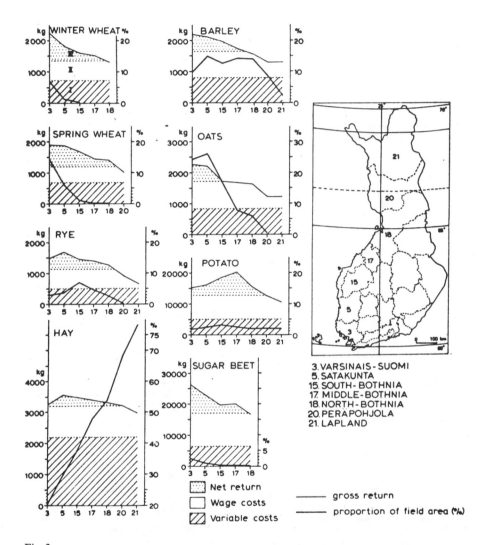

Fig. 3.

stant for each crop, directly subtractable from the value of the harvest to give the gross margin for that crop.

Finland provides very suitable material for the evaluation of gross margin as an indicator of productivity differences in arable farming. It is an extensive country in a north-south direction, stretching from 60°N lat. to 70°N lat., and thus shows large, chiefly climate-determined productivity differences. Weather conditions are characteristically variable, and early frosts are a frequent hazard, especially in the north (Valmari 1966, 194–7). Although specialized strains which to some degree tolerate the climate and ripen within the short growing season have been developed for cultivation in northern Finland (Paatela 1953; 1960, maps 21/7–9), early frost may destroy the crop or reduce it to an uneconomic level. For this reason arable farming in Finland is zoned in such a way that the more demanding and productive crops are confined to southern Finland, while in northern Lapland, for example, only timothy and potatoes are generally cultivated (Alestalo 1965, 23; Varjo 1971, 59).

The influence of harvest yield on the productivity of arable farming, expressed by gross margin, is described for northern and western Finland in Fig. 3. The harvest yields are the averages of the 1965–6 values (SVT III 1967–70, 62, 63); the labour costs, variable expenditures, and gross margins are after Westermarck (1964, 49; see also MTTJ 16 1968, and the gross margin tables in MKJ 518). The value of the winter wheat harvest in Varsinais-Suomi (Fig. 3, area no. 5) exceeded labour costs, but in Ostrobothnia it no longer succeeded in covering these. The spring wheat harvest similarly covered its labour costs and variable expenditure in Varsinais-Suomi, but was grown at a loss from northern Ostrobothnia (18) onwards. Rye still yielded a profit in northern Ostrobothnia, but in the province of Lapland (20, 21) it no longer covered labour costs entirely. Barley was uneconomic in northern Ostrobothnia and oats in southern Ostrobothnia.

Fig. 3 also shows the proportion of field area devoted to each crop. It is interesting to note that the cultivation of most crops decreases as gross margin diminishes, and ceases before becoming uneconomic. There are some exceptions to this: for example, the proportion of field area devoted to hay increases even where its gross margin is decreasing, and in the northernmost parts of the country where it becomes uneconomic. The same is true of barley and potatoes, which are grown right into Lapland, even though at a loss.

By combining the gross margin per hectare for each crop, which may be calculated once the proportions of field area, harvest yield, and variable expenditure for each crop are known, the productivity of arable farming may be determined. This is presented for Finland in Fig. 4; the harvest yields are the averages for the years 1967–9, the proportions of field area are for 1969 (SVT III 1967–70, 63–5), and the variable expenditures for the harvest of 1965–6 (MKJ 518). The highest gross margin for arable farming was achieved in Varsinais-Suomi. The figure then decreases until in the extreme north it is scarcely a sixth of this value. Here the farmers are obliged to sacrifice part of their earnings in order to maintain production. It is essential to do this, however, as it is not worthwhile to buy animal fodder from elsewhere because of the great distances involved.

Alestalo, Jouko, 1965 Die Anbaugebiete von Ackerpflanzen in Finnland, *Fennia* 92, 4.

Paatela, Juhani, 1953 Täkeimmät viljalajikkeemme ja niiden viljelysalueet, *Acta Agralia Fennica* 80, 1.

– 1960 Cultivation of most common varieties of spring wheat, barley and oats, 1955, *Atlas of Finland*, maps 21/7–9.

MKJ 518 Maatalousseurojen keskusliiton julkaisuja no. 518.

MTTJ 16, 1968 *Tutkimuksia Suomen maatalouden kannattavuudesta, 1966*, publications of the Agricultural Economics Research Institute (Finland) 16.

SVT III, 1967–70 *The official statistics of Finland*, III, *Agriculture*, 62–5.

Valmari, Arvi, 1966 On night frost research in Finland, *Acta Agralia Fennica*, 107.

Varjo, Uuno, 1958 Zonengliederung in südwestfinnischer Landschaft und Landwirtschaft, *Fennia*, 82, 4.

– 1971 Development of human ecology in Lapland, Finland, after World War II, *Geoforum* 5.

Westermarck, Nils, 1964 Viljelijän suunnitelmaopas. *Maatalousseurojen keskusliiton julkaisuja* 513.

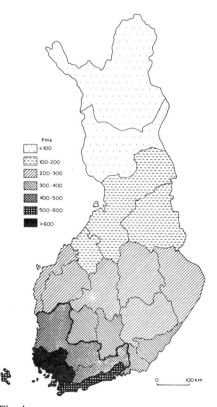

Fig. 4.

P1047
An isolation index for Queensland cattle stations
T.R. WEIR   University of Manitoba, Canada

Queensland, with 6 million cattle located on more than 9000 properties (stations) averaging 145km² in area, provides an excellent framework within which to study physical isolation. We may define physical isolation as location in respect to two basic components: (1) transportation and (2) communication.

Regardless of the carrier used, transportation by road and rail may be expressed in the distance-time relationships of a station to service centres and market points. Communications on the other hand may be expressed as the weighted sum of the facilities by which outside contact is maintained. A set of weighted indices for both components was derived and finally expressed as a single 'isolation index' for each of 97 stations. The index is objective and physical rather than psychological and perceptive.

Distance-time relationships involved in transportation were measured in respect to centres ranging from hamlets to coastal cities, providing vital services, namely, post-office, major source of provisions, banking, marketing.

1 *Post office*. Although most stations have mail delivery once or twice a week, the location of the post office usually decides the location of such related services as telephone and telegraph, the ubiquitous hotel with its 'pub,' social centre of the outback, and clearinghouse of local information.

2 *Provisions*. Food, clothing, and miscellaneous articles may in part be procured at small centres near at hand, but many stations go far afield for the bulk of their needs. Where the source is local, isolation is regarded as reduced.

3 *Banking*. Again, this service is usually at a point different from the first, and more often corresponds to the source of provisions. Because it is less significant it has been assigned a lower rating.

4 *Marketing*. Most have two options for marketing their cattle. Some, especially large stations, prefer shipping, by rail or road, to a packing plant at a coastal port such as Rockhampton, Townsville, or Mackay. Others prefer selling by auction at a local town. In most stations both means were used, one being preferred over the other. Almost no cattle are sent to feed-lots for fattening, but in companies which operate a chain of stations specialization into breeding stations and fattening stations occurs, causing some intermediate shipping of young stock before slaughtering. Large mobs are usually moved by rail; small lots are trucked, especially where shorter distances are involved.

Because it was considered that certain of the factors were more decisive than others, each was given a proportional weighting as follows: post office 2.5; provisions 1; banking .05; principal market .15; secondary market .05. A summation of the weighted five factors for each station comprises the *distance index* (DI) thus:

$$\text{DI} = \sum_{i=1}^{5} (W_i D_i T_i),$$

where $W_i = i$th weighting factor; $D_i = i$th distance factor; $T_i = i$th time factor. This was followed by a calculation of the standard normal deviate.

The second component of the isolation index is *communications*, which consists of four elements: (1) communication facilities, (2) floods, (3) educational facilities, and (4) mail deliveries. Each in turn was weighted by assigning a simple numerical, non-quantitative index to suggest its relative importance in affecting physical isolation.

Included in this group (with weighted indices included in parentheses) were the wireless transceiver (2); telephone (5); television (2); air strip (3); aeroplane (3). The transceiver is used to communicate locally between groups of stations over a radius of 80 to 160km. Items of concern are discussed by individuals but the entire group may listen in. It is indicative of great isolation. Telephones are common in eastern Queensland, but are non-existent in the outback of northern and western Queensland. The service is frequently poor and the range unpredictable. The presence of an air strip permits emergency plane landings such as visits from the Flying Doctor or occasionally from government and commercial agents. Very few possess aeroplanes, the exception being large company-operated stations using 'pastoral managers.'

Most stations are cut off by road during the monsoon period for one to four or even five weeks. This factor is regarded as *negative* and rated according to the number of weeks of isolation (0 to 5).

Some of the greatest limitations arising from remote location were problems related to the education of children. Most coped with the problem in one of three ways: use of government-provided correspondence-by-mail lessons; if near enough to a town, commuting in a school bus; or supplying a teacher and operating a school. In many cases the first and third alternatives were combined. A fourth alternative was sending a child to a boarding school, usually for his secondary education. The latter was not considered in the index as it applied to older children, but the first three were rated in order as: correspondence courses (1), commuting (2), providing a teacher (5).

The delivery of mail was provided regularly by government carriers either once (1) or twice (2) a week.

A tally of the weighted indices was made for each of 97 stations. A large tally indicated a plethora of communication devices, optimum mail delivery, little flooding, and good educational facilities. The untreated communications index (CI) is represented as $CI = Y_i$. The standard deviation, as for the transportation index, was then calculated. Since minimum isolation in the case of *transportation* was indicated by a low index, and since minimum isolation in the case of *communication* was indicated by a high index, the latter was subtracted from the former after first weighting each (expressed as the standard normal deviate) by a ratio of 2/3 to 1/3. In this way an overall isolation index (F_i) was achieved as indicated by the formula:

$$F_i = \tfrac{2}{3} Z x_i - \tfrac{1}{3} Z y_i, \qquad i = 1, 97$$

$$= \frac{2}{3}\left(\frac{x_i - \bar{x}}{sx}\right) - \frac{1}{3}\left(\frac{y_i - \bar{y}}{sy}\right), \qquad i = 1, 97$$

$$= \frac{2}{3}\left(\frac{x_i - \sum_{i=1}^{n} x_i/n}{\sqrt{\frac{n\sum_{i=1}^{n}(x_i^2) - \left(\sum_{i=1}^{n} x_i\right)^2}{n(n-1)}}}\right) - \frac{1}{3}\left(\frac{y_i - \sum_{i=1}^{n} y_i/n}{\sqrt{\frac{n\sum_{i=1}^{n} y_i^2 - \left(\sum_{i=1}^{n} y_i\right)^2}{n(n-1)}}}\right),$$

where Zx_i and Zy_i are the standard normal deviates for the transportation and communication indices respectively.

The 97 station indices were plotted on a map of Queensland to reveal a pattern of distribution as follows:

1 Indices within 80km of the coast were low and defined a north-south strip (Zone I) parallel to the sea.

2 Indices extending inland from the coast along four principal routes of travel were nearly as low and could be grouped with the coastal strip (Zone II–III). Each focused on a principal port city: Rockhampton, Mackay, Townsville, and Ingham.

3 Indices tributary to these east-west routes were higher both to the north and south of these routes, thereby indicating a third group (Zones IV–V–VI–VII–VIII). These might be compared to the interfluves of a drainage system with the principal routes of travel corresponding to main streams.

4 Indices in the far west declined in value and the direction of decline was from north to south. These combined to form Zone IX. One group focuses on the Carpentaria coast, a second on Conclurry and Mt Isa.

5 Finally, those stations in the southwestern part of Queensland known as the Channel Country had indices much higher than elsewhere; they form Zone X. Here remoteness reaches its extreme with the average station being 2300km^2 in area.

6 Anomalies occurred in all zones, often the result of the unusual choice of an operator in respect to several of the factors going into the index.

7 The isolation index brings out the fact that remoteness is not only a matter of distance-time relationships, but the latter may be reduced by taking advantage of modern techniques in communication and improved systems of marketing. However, costs of reducing isolation are correspondingly high.

Keating, C., 1967 The Queensland beef cattle industry, *Quart. Rev. Agri. Economics* 20 (4), 192–203.

Stone, K.H., 1971 Regional abandoning of rural settlement in northern Sweden, *Erdkunde* (Archiv für wissenschaftliche Geographie), Band 25, Lfg. 1 (Bonn), 36–51.

P11
Urban Geography
Géographie urbaine

CONVOCATEURS/CONVENORS: Frederick A. Dahms, *University of Guelph*; John N. Jackson, *Brock University, St Catharines*

This listing of short titles and first authors' surnames will assist in identifying articles, topics, and places of interest. A complete author and co-author index is located at the end of this volume, as well as a selected index relating papers to geographical locations. Note that the papers are *not* listed in alphabetical order. The organization of the volumes is described in full in the Preface.

Other papers of interest to readers of this section are to be found under c01 (Geographical Data Sensing and Processing), c07 (Man and Environment), c12 (Population Geography), c13 (Applied Geography), c14 (Quantitative Methods), c16 (Geography of Transport), s03 (Role of the City in the Modernization of Developing Countries), and s04 (Frontier Settlement on the Forest/Grassland Fringe).

Cette liste des titres abrégés et des noms des auteurs principaux permettra d'identifier les communications, les sujets et les lieux qui présentent un intérêt quelconque. Un index complet d'auteurs et de coauteurs se trouve à la fin de ce volume, ainsi qu'un index des lieux géographiques. Prière de noter que les textes ne sont pas classés par ordre alphabétique. On explique en détail le plan de ces volumes dans la Préface.

D'autres études qui intéresseront peut-être les lecteurs de cette section se trouvent dans c01 (Sélection et traitement de l'information géographique), c07 (L'homme et son milieu), c12 (Géographie de la population), c13 (Géographie appliquée), c14 (Méthodes quantitatives), c16 (Géographie des transports), s03 (Les villes dans les pays en voie de développement) et s04 (L'aménagement du territoire en régions frontalières prairies/forêts).

P1101 The intra-urban migration decision: a place utility formulation AUSTIN 791
P1102 Budapest among the cities of Europe BERNÁT 792
P1103 Concepts for urban studies at school level (UK) CLARK 795
P1104 Some vector representations of intra-urban residential mobility (Milwaukee, USA) CLARK 796
P1105 Political ecology of a town in Trinidad CLARKE 798
P1106 Changing residential structures, South African cities, 1950–70 DAVIES 801
P1107 Factor ecology of British cities DAVIES 805
P1108 Interaction of two scales within ecological structure of Haifa, Israel GRADUS 808
P1109 Urbanization in the arid zones of Mexico GUTIERREZ DE MacGREGOR 810
P1110 Stability of urban dimensions (Southeast Asia) HOFFMAN 812
P1111 Cultural convergence and divergence and changes of city structure – urban typology HOFMEISTER 813
P1112 Pathology of urban environment in developing countries KAR 816
P1113 Different forms of semi-urbanization processes, Poland KIELCZEWSKA-ZALESKA 819
P1114 Lorry or motor park: land use pattern element in West African cities (Nigeria) KIRCHHERR 819
P1115 Residential morphology and urban social structure (Edmonton, Canada) McCANN 820
P1116 Change in retail structure of cities in developing nations (Lagos, Nigeria) McNULTY 823
P1117 Changes in localization of manufacturing in megalopolis: 1954–67 (USA) MILLER 824

P1118 Urban dimensions of selected Indian cities and towns MOOKHERJEE 827
P1119 Pattern of settlement hierarchy in mountainous plantation landscape: Darjeeling, Himalaya MUKHERJI 829
P1120 Urban geography or urban studies? NADER 832
P1121 Impact of Jamshedpur industrial complex (India) PAL 834
P1122 Les migrations journalières de travailleurs à Brasilia (Brazil) COSTA BARBOSA 837
P1123 Toward a dynamic theory of urban plat development (Buenos Aires, Argentina) SARGENT 837
P1124 Features of urbanization in densely populated developing countries (India) SDASYUK 839
P1125 Character of urbanization in Uttar Pradesh (India) SINGH 842
P1126 Friendship and urban inter-residential social trip length STUTZ 843
P1127 Urban ethnic acculturation: a functional approach (North America) VILLENEUVE 846
P1128 Quantification of city-hinterland relationships through input-output (Peru) WALLER 848
P1129 Urbanization in Manchuria, 1907–40 WANG 850
P1130 Modelling the social geography of urban areas: a Canadian example YEATES 852
P1131 Structure of common urban constructs (Ontario, Canada) DEMKO 854
P1132 Qualité du commerce et valeur du paysage urbain (W. Europe) SPORCK 856
P1133 Application de l'analyse morphologique dans la planification des villes WITKOWSKI 858
P1134 Territorial aspects of local recreation in agglomerated areas of the GDR (Berlin) KOHL 861
P1135 Inner functional zoning and rationalization of use of town territories (USSR) LITOVKA 863
P1136 Développement urbain en économie socialiste, République démocratique allemande SMOTKINE 865
P1137 L'insertion et le rôle des villes touristiques dans le système urbain régional (France) VEYRET-VERNER 866
P1138 L'effet de synergie et le développement des villes en pays d'économie développée (France) VIGARIE 868
P1139 Semi-urban centres in Thailand WITAYARUT 870
P1140 Recent trends of urbanization, Madhya Pradesh (India) AGARWAL 872
P1141 Proposition d'analyse comparée du commerce de détail (Iran) ALBERTI 875
P1142 Les tendances de l'urbanisation dans l'Europe de l'est BLAZEK 876
P1143 Le développement des réseaux urbains en France BEAUJEU-GARNIER 879
P1144 Pattern of urbanization in developing countries of Asia: political significance KULARATNAM 880
P1145 La structure et l'évolution des systèmes urbains régionaux de l'URSS KONSTANTINOV 884
P1146 Micro-scale spatial behaviour: distance and interaction in Christchurch cul-de-sacs (New Zealand) GREER-WOOTTEN 886

P1101
The intra-urban migration decision: a place utility formulation
C. MURRAY AUSTIN   *University of Kansas, USA*

Many writers have dealt with the intra-urban migration process. Wolpert (1965) and Brown and Longbrake (1970) are interesting because of their use of the concept of place utility. However, the concept, first explicitly introduced by Wolpert (1965), had not been developed formally until recently (Austin 1971). The purpose of this paper is to present briefly a more rigorous formulation of place utility and to demonstrate how utility analysis can lead to important conclusions about the intra-urban migratory process.

The general result of this analysis may seem trite; a person will move if he feels that he is 'better off' after such a move. However, it is both useful and important to understand the various conditions and meanings which underlie such a statement.

An individual can be assumed to have a utility or preference relation which relates the satisfaction he receives from his existence and consumption at a given time. This total level of satisfaction can be separated into the utility he receives from his expenditure on consumption and the satisfaction he receives from his residential location:

(1) $T_i(y) = C_i(y - r_i) + U_i(r_i)$,

when T_i is total satisfaction of the ith person, C_i is consumption utility of the ith person, U_i is the place utility of the ith person, (y) is annual income of the ith person, r_i is the annual cost of the residence of person i.

The utility the individual receives from consumption is discussed in the *Theory of Consumer Behavior*. The satisfaction the individual receives from his residential location is here termed 'place utility.' I (Austin 1971, chap. 4) have formally defined this concept as being a function of the generalized environment (the set of attributes which affect the individual at his location, but over which he has little control). Examples of such environmental attributes are: the level of air and noise pollution; the crime rate in his neighbourhood; the availability of parking, the level of traffic congestion and the employment level; the accessibility of goods and services; etc. These, and other, attributes compose the generalized environment. An individual's place utility depends upon the values of these attributes at his neighbourhood and upon his household bundle (his house, lot size, furnishings, etc.):

(2) $U_i(r_i) = f_i(E_i H_i)$,

when U_i is the ith person's place utility, E_i is the generalized environment at his location, H_i is his household bundle, $f_i(\cdot)$ is the place utility function, and r_i is the 'rent' he pays for the residential location and household bundle.

For an individual choosing his first residence, it is assumed he chooses a site (for a specified rent) for which the allocation of his income between rent and consumption expenditure yields a maximum personal total satisfaction. Using this as the initial condition and assuming that the individual is a utility maximizer, the decision process of intra-urban migration can be analysed.

At a given point of time the household bundle and rent can be considered fixed. Thus, for a given income and fixed prices of consumption goods, the individual's satisfaction will depend upon the generalized environment. Similarly, other variables can be allowed to vary. This section attempts to show how such changes are related to the intra-urban migration process. There is one important difference between this analysis and the more traditional utility analysis in consumption theory. The latter usually ignores any friction, while this analysis specifically includes friction in its analysis. Friction in an intra-urban migration decision is largely due to the fixed cost of moving, and (for home owners) the fixed investment in the current property. Another element of friction is the cost of real estate transfer (for home owners), or the cost (in time or money) caused by fixed time leases (for renters). There are several cases to be analysed. Because of space limitations, it is assumed that the household bundle is constant for each individual. Thus rent, income, and the generalized environment are the only variables that are considered. The more complex situation where the household bundle is allowed to vary can be handled by the same analytic procedure, but the added complexity requires more space than is available here. Three general cases are examined, each of which may lead to a positive decision to change residential location. The first involves an exogenous change in income, the second is an exogenous change in the rent (relative to other locations), and the third case is where the generalized environment changes. In each case all other variables are held constant. The analytic procedure is demonstrated only for case one; for the other cases the procedure is similar, and only the results are presented. In all cases the individual is assumed to have been at the maximum level of possible satisfaction.

CASE 1 INCOME CHANGES

A change in income must be allocated between consumption expenditure and rent in such a way that satisfaction is maximized. Expenditure on rent can be changed only by a move. Such a move will occur only if the person is better off after he moves than he would be if he did not move.

(3) $T_i(y + \Delta y) = C_i(y + \Delta y - r_i)$
$+ U_i(E_i, r_i),$

(4) $T_i'(y + \Delta y) = C_i(y + \Delta y - r_{i'} - M_{ii'})$
$+ U_i(E_{i'}, r_{i'}).$

For equations (3) and (4) above and, if there exists a location i' such that

$T_i' > T_i,$

then the individual will move to i' from i. This simply shows that the utility gained from an expenditure of the new income ($y + \Delta y$) minus the new rent ($r_{i'}$) minus the cost of moving (including all related costs discussed above) ($M_{ii'}$) plus the new place utility must be greater than the gain in utility if the entire new income were spent on consumption goods (Austin 1971, chap. 4, sec. 6).

CASE 2 A CHANGE IN RENT (r_i to r_i')

A change in the cost of the residential site (through rent, property taxes, unexpected repairs, etc.) is seen to lead to a move if there is a new location for which the total satisfaction after subtracting moving costs is greater than at the present site. This is likely to occur only for increases in rent since, if rent (or costs) are lowered at a site (but not elsewhere), what was previously the maximum possible utility is simply increased.

CASE 3 A CHANGE IN THE GENERALIZED ENVIRONMENT

Such a change can be important if the environment changes either at the current location or at a potential alternative.

In the first situation, a change in the environment of the current site, a necessary condition for a move is that the change is such that place utility is reduced, a sufficient condition is that the added utility gained at an alternative site compensates for both the loss of consumption utility due to the moving cost plus the change in rent.

In the latter situation, a change in residential location will be made only if the change at other site increases place utility enough to compensate for the loss of consumption utility caused by the cost of moving plus the change in rent.

Recall that in each of these cases the household bundle was considered fixed. If this assumption is relaxed, then the conclusion, although remaining similar, becomes more complex, with several alternative decisions becoming possible as opposed to the simple choice of moving or not moving.

It is possible to use the analytic technique above to generate information about the aggregate results of migration behaviour on the land market and on residential location patterns of the urban population. I have elaborated on this elsewhere (Austin 1971, chap. 6).

Austin, C.M., 1971   Place utility, the generalized environment, and urban public facilities: impacts and locational effects (U. Pennsylvania dissertation).
Brown, L., and D. Longbrake, 1970   Migration flows in intraurban space: place utility consideration, *Ann. Assoc. Am. Geog.* 60 (2).
Brown, L., and E. Moore, 1970   The intraurban migration process: a perspective, *Geografiska Annaler*, Ser. B, 52.
Wolpert, J., 1965   Behavioral aspects of the decision to migrate, *Papers*, R.S.A., 15.

P1102
Budapest among the cities of Europe
TIVADAR BERNÁT   *Karl Marx University of Economics, Hungary*

According to population size, Budapest is the eighth city in Europe, forming the largest agglomeration of east-central Europe. The Hungarian capital owes its distinguished rank among the European cities to the population increase which has taken place during the past century. When the three towns along the Danube (Pest, Buda, Óbuda) were integrated into Budapest, the new capital, with more than 280,000 inhabitants, occupied the seventeenth place in Europe. (Population data refer to the prevailing area of Budapest.) The political sphere of action gained by integration served as the starting point of the development of Budapest. However, the chief motive force of the city's development was the industrial revolution of the last third of the nineteenth century; this took place in Hungary later than in western Europe and displayed specific features. It started the structural change of the national economy by enhancing industrialization and the establishment of modern infrastructures. Urbanization was concentrated primarily in the capital. As a result of a backward rural market, subsistence farming, low purchasing power, and low-level internal market, the provincial towns were deprived of the dynamism of capitalistic urbanization, and, apart from Budapest, practically no other city evolved in Hungary. This specific development, which is characteristic of east-central Europe, is responsible for the fact that the capitals of its countries fulfil the central role within the urban network.

The urban development of Budapest unfolded during the last decade of the nineteenth century. At the same time, the Hungarian capital experienced a rapid population increase (by about 250,000) by which it leapt ahead to become the eighth city of Europe. After World War I, the centrality of the capital became more accentuated in the diminished territory of Hungary, despite the temporary decrease of its population. The population of Budapest attained a million only by 1930. Despite the doubling of population during the subsequent 40 years, the city did not get ahead among the European cities. According to the 1965 International Statistical Yearbook of Large Towns, Budapest is a commercial and servicing city with a developed industry. This functional classification of a city is defined by the International Statistical Institute as a centre in which one-fifth to a half of all active earners are employed in industry; one-fifth to one-third in commerce and communication; and one-fifth to one-third in services or public administration.

According to the data of 1961, 48 per cent of the active population of Budapest was employed in industry, 20 per cent in commerce and communication, and 23 per cent in services and public administration. Comparing the occupational structure of Budapest to ten other European capitals of similar size (Vienna, Brussels, Copenhagen, Madrid, Athens, Rome, Warsaw, Lisbon, Stockholm, and Prague), we come to the following conclusions: (*a*) The ratio of *industrial* employees within the active population of Budapest indicates a nationally significant industrial concentration employing more than 600,000 workers, including the daily commuters. With regard to the ratio of industrial employees, the Hungarian capital comes after London, Paris, and Moscow, and it forms the largest industrial concentration of east-central Europe. For example, taking the absolute amount of industrial employment, Budapest offers about three times as much employment as Prague and Vienna, and about half as much as Paris. (*b*) During the early sixties high industrial employment and an underdeveloped retail network were responsible for the slightly lower than average ratio of *commercial and communicational* employees within the active population of Budapest. Among the comparable cities, Madrid shows a similar ratio. The ratio of employees in the non-productive branches grows parallel with the development of the cities. Concerning the ratio in service employment, Budapest takes last place among comparable urban centres, partly because of the long-held concept that the development of services deprives the productive branches of resources, and so restrains the rapid-pace development of the national economy.

The decrease in the birth rate, its stagnation, and the prolonging of the average life span led to considerable senescence and an unfavourable age-split in the Budapest population in the sixties. Though this tendency also prevailed formerly, it was tolerable compared to the relatively older west European capitals. Only in Warsaw and Lisbon was the ratio of inhabitants over 65 years in age lower than in Budapest. The Hungarian capital shows a relatively advantageous demographic structure, as both juveniles and the so-called working-age population represent a higher ratio than in the majority of the European capitals. Owing to senescence, the tendency is for the working-age population of Budapest to decrease. (c) As a consequence of a relatively low birth rate and the increasing death rate (resulting from the unfavourable age split), the *natural population increase* of Budapest is 0.1 per cent. Though this ratio does not suffice for the normal reproduction of the population, it is not unfavourable compared to the cities under analysis. That is, the natural population increase necessary for normal reproduction is lacking in Prague, Stockholm, and Vienna, where decrease is higher than it was in Budapest formerly. Those cities develop by natural reproduction most rapidly where high birth rate goes parallel with a relatively low or normal death rate, as in Rome and Madrid (12 to 18 per cent). (d) Budapest represents an average *population density* compared to other European capitals. The highest residential density is found in Brussels and Prague (60 and 54 tenants per ha) which have relatively narrow administrative boundaries. Cities with an average residential density are Vienna (39) and Budapest (34), while those with low densities are Madrid (28), Copenhagen (24), and Warsaw (24).

Budapest suffered from a grave housing shortage even before World War II. After the war the stock of dwellings could not keep pace with the sudden population increase. Improvements can be achieved only by the use of the large-scale methods of housebuilding, which began to spread only at the time of our analysis in Hungary. Comparing the structure of the stock of dwellings of Budapest with that of nine European capitals, it emerges that Budapest is the only capital where the ratio of one-room dwellings is more than half of the total. (Comparison was hindered by the fact that in the case of some cities the kitchen was also included with the number of rooms.) This ratio is lower in other capitals: 34 per cent in Vienna, 26 in Warsaw, 19 in Prague, and 5 to 15 per cent in the rest of the cities. Among the nine capitals Budapest represents the lowest ratio in four-room dwellings or more: 16 per cent (it is 6 to 8 per cent in Vienna and Warsaw, 19 to 26 per cent in Copenhagen, Brussels,

and Stockholm). With regard to the *equipment of dwellings*, Budapest enjoys a favourable position among the ten capitals. As a result of the great differences of level between the former Little Budapest and the suburban districts, the dwellings of Budapest are still moderately equipped compared to other cities.

Currently, the stock of cars represents a small ratio in Budapest and therefore public transit plays a prominent role in the communication system, accounting for 88 per cent of all journeys. Among the cities compared to Budapest, only Warsaw has a similarly significant public transit system. In Budapest in 1967 more than 1400 million passengers were transported by 1790 tramcars, 226 trolleybuses, and 1265 buses on 800km of lines. A similar passenger traffic load was realized in Warsaw with a slightly smaller stock of vehicles. Mass transport in Vienna, with a similar tramway network and stock of cars but with a smaller stock of buses, carried only 30 per cent of the passenger traffic of Budapest; in Prague and Rome public transit took care of over 500 million passengers, while only 3 to 25 per cent of the Budapest volume of public transit was handled in other capitals. This comparison suggests the crowded nature of public transport in Budapest, and urges the necessity of establishing an up-to-date underground network.

Bernát, T., ed., 1969 *Magyarország gazdasági földrajza / Economic geography of Hungary* (Budapest).
Budapest Statisztikai Évkönyve, 1971 / Statistical Yearbook of Budapest, 1971 (Budapest).
Fodor, L., and I. Illés, 1968 Some problems of metropolitan industrial agglomeration, VIII, European Congress of the Regional Science Association, 27–30 August 1968, Budapest.
Information Hungary, 1968 (Budapest).
Josse, R., 1969 Budapest: Croissance et structure urbaine, *Information Géographique*, no. 4, 174–9.
Preisich, G., 1969 *Budapest városépitésének története, 1919–1969 / Urbanistic history of Budapest between 1919 and 1969* (Budapest).

P1103
Concepts for urban studies at school level
RICHARD CLARK   New University of Ulster, UK

How to treat the nature and implications of cities is clearly a vital present consideration in the planning of education in all parts of the world. How geographers may best contribute to this aspect of the general education of young citizens is dependent not only upon the various approaches geographers bring to their studies of cities but also upon the norms and trends of their particular societies and the influence these have upon interpretations of urban phenomena.

In many countries the major systematic studies of cities which school pupils undertake is organized by geography teachers; such studies have been, until quite recently, based largely upon concepts of original and subsequent advantages of location, of major occupation, of functional zones, and morphology. Lately, developments in geographical theory and technology have led to refinements of description and analysis of these. There has been a relative neglect in schools of the study in an urban context of dynamic spatial relationships of sociological and ecological types. There is also a lack of explicit comprehensive conceptual structures which facilitate the interpretation of cities at school level. An effective contribution to this educational objective is more important than any concern over whether or not the identity and purity of a discipline are sustained.

The formulation of major organizing concepts does not in itself lead to effective teaching and learning but is necessary for the conjunction of the nature of a discipline to the aims of educational courses. Thus it is necessary to pursue such questions as 'understand what about cities?' and 'to what ends?' Answers to such questions depend not only on intrinsic characteristics but also on judgments informed by an image of society upon what individuals should be able and willing to do, upon what and how opinions and attitudes might be formed, in consequence of learning about cities.

In educating the young, and especially the youth of cities, about cities, it is evident that a degree of sensitivity to urban heterogeneity is called for; the pupils of town schools represent such a variety of perceptions and aspirations, of opportunities and achievements, that the propagation of one set of cultural characteristics can be divisive and offensive. It is useful therefore to consider the idea of 'neutral' concepts; these avoid transmitting assumptions and value judgments, but form a basis upon which pupils may come to their own conclusions, e.g. upon the implications of urbanization.

It is probable that in the contemporary situation much study of urbanization will be focused on the present and the future, but perspective may be enhanced by an historical review which can employ a similar conceptual structure. Particularly the view that civilization and urbanity are nurtured by cities can be shown never to have been more than partially true, and, even in its limited truth to have been based upon the creation and perpetuation of inequalities.

It is important that abstract ideas about cities be exemplified through case studies of accessible areas and as far as possible reinforced by comparative studies. Any development of generalizations or theories of city development should be derived from the first-hand study of specific situations, and tested against other instances.

If dynamic, process-and-effect, ideas are to be given a large niche in the study of cities there is a need to restore the focus from the material characteristics of towns to a 'people-centred' approach to location and distribution. Thus, for example, it is not wholly adequate to differentiate the residential areas of town solely in terms of age and type of housing when the more important attributes which vary across a residential area are space and privacy, aspiration, opportunity, mobility, income, ethic, etc. Thus there may be more-or-less clearly recognized sub-cultural areas within the town, and each such area, besides having its own internal relationships, also has particular relationships with other parts of the town, e.g., industrial, retail, etc.

The relationships between city and rural areas also hold a number of significant concepts both simple and sophisticated. It is not difficult, using mean food consumption data for a country, to obtain a rough approximation of the area of farmland upon which any city depends, even though that area is divided and rural areas perhaps not singularly associated with any one city. The consequences of increasing town populations for size of dependent area and intensity of food production and also for intra- and international movements of food may be inferred. It seems surprising that this type of relationship is usually left implicit and unexamined, or considered only in qualitative terms. The subtler relationships between city growth and commercial-farming areas rarely enter the considerations of school pupils. These include the expenditure of wealth produced in the cities to divert resources and energy to farmland where it is used to sustain simple, short-chain, immature ecosystems with high-energy throughputs (Margalef 1968, 47). Moreover as cities grow, and technologies develop, the increased production of food per unit area leads towards increased energy and material consumption per unit area tending to increase 'resource cost' per food unit; the 'price' of urban man is progressively raised. It is important that the food which enters the city is not costed in currency terms only. These concepts may also be used as the basis of historical-ecological studies of change in urban-rural relationships.

The general alienation of urban man from an understanding of food production is curiously matched by an increasing dependence of some town dwellers upon rural areas as a relief from urban stress; the unequal opportunity to do this is perhaps a further form of sub-cultural diversity noted above as demanding attention in town study.

The metabolic characteristics of a city can be treated in both general and locational terms. As well as the external relationships, e.g. in sizes of and implications for food and water supplying areas, growth in route capacity, increased consumption of energy and materials in transport to and from the city, there are internal characteristics. Study may determine the relative intensity within the city of emission of particulate and gaseous waste, where solid waste is most in evidence, where is most noise, physical danger, etc. The concentrated effects of a town in decycling or attenuating the cycling of nutrients, in dispersing energy as residual heat, may readily be demonstrated.

The study of economic activities or functions of cities has long been a part of geography and to some extent that study has included the recognition of consequences of industry in terms of waste, dirt, and variations in earning ability. A comprehensive conceptual basis for relating the results of town growth at school level does not appear to be widely used. This may, in part, be due to a general acceptance, as desirable and normal, of a view of economics rooted on a separation of costs and prices (Mishan 1967).

It would seem important, and not only in its relevance to cities, to match the concepts of 'service' and 'commodity' with those of 'disservice' and 'discommodity' to show that these greatly extend the idea of economic transactions, and to note their relevance to the internal inequalities in city environments (Coddington 1970, 595).

It is claimed that the ideas in this short selection around which learning about cities might be organized are 'neutral' in the sense that they neither rely on nor form attitudes and judgments. They are manageable and realistic at school level and local scales. They facilitate an understanding and experience upon which opinions and attitudes may grow, and they form a humane approach to the study of cities.

Furthermore, they form a continuum in which social, economic, and ecological systems are seen to be mutually related and which can be sympathetically associated with the spatial and locational approaches which the geographer more usually brings to urban study.

Coddington, A., 1970 The economics of ecology, *New Society* 393, 595–7.
Margalef, R., 1968 *Perspectives in Ecological Theory* (Chicago).
Mishan, E.J., 1967 *The Cost of Economic Growth* (London).

P1104
Some vector representations of intra-urban residential mobility
W.A.V. CLARK *University of California, USA*

In recent published and unpublished research there has been an increasing concern with the interrelationships of behaviour and spatial structure, and in particular with the interrelationships as they apply to the processes of change within cities (Adams 1969; Brown and Holmes 1971; Clark 1971a, b; King 1970). A paper by King (1970) suggested that a concern with process-form relationships, or what are termed in this paper behavioural-structural relationships, is most important for the further development of theory. Two papers (Clark 1971a, b) have suggested ways in which these links between behaviour and spatial structure may be investigated, and have gone some way towards testing the extent and nature of the relationships for some aspects of intra-urban residential mobility. Residential mobility within the city is an appropriate substantive area of concern for investigations of behaviour and structure, as elements of behaviour (actual moves), structure (neighbourhoods), and changing urban structures (declining and changing neighbourhoods) are all involved in the residential relocation process. It is important to establish that in using the terminology of behaviour and structure the focus is on behaviour in space, as distinguished from spatial behaviour, which has been defined as the rules by which residents make decisions within the spatial framework (Rushton 1969). At the present time it seems more likely that progress will be made in the direction of understanding the rules by which urban residents make decisions affecting their actual spatial choices within the city, by first focusing on the more easily identified actual behaviour.

The specific element of spatial structure which has received some attention from urban geographers interested in residential mobility and the influence of structure is the nature of the sector structure of the city. This interest stems directly from Hoyt's early suggestions of the influence of a high rent sector on the moving behaviour of high status families. Adams (1969) argued from Hoyt's work that households move sectorally to maintain certain socio-economic characteristics. In further explanation Adams emphasized a wedge or sector shaped image of the city, which is derived by urban residents from a radial commuting and travelling pattern. Whereas Adams found some evidence supporting a directional bias in the movements of households, Brown (1970) was not able to find any directional bias in residential movements. Clark (1971) found only limited evidence in a general survey of an urban area, but some evidence of directional bias related to some specifically identified sectors within the same city. One explanation for the contradictory results is a possible dichotomy of long and short distance moves. Long distance moves out of the neighbourhood may be directionally biased, while short distance moves are random in direction, and do not reflect a directional bias. This paper is directed to the general problem of further measuring and identifying directional bias, and specifically to the problem raised by a division of long and short distance moves.

Directional data have long been of interest to structural geologists, who have developed a number of techniques for measuring the directional nature of their data. It has already been established that the arithmetic mean of the angles of direction does not adequately represent the nature of directional data, and that vector means are the only thorough representation of such data (Pincus 1956). The vector mean is defined as

(1) $\quad V\bar{x} = \arctan \sum_{i=1}^{n} \sin x_i \Big/ \sum_{i=1}^{n} \cos x_i.$

Although this measure gives an adequate representation of the mean of the angles of movement for sets of households which relocated over some specific period of time, it is necessary also to calculate some measure of dispersion for the data. Several measures have been suggested including the consistency ratio:

(2) $\quad \text{CR} = \sqrt{(\Sigma n \sin \theta)^2 + (\Sigma n \cos \theta)^2} / \Sigma n,$

which is a measure of the consistency of the angles of direction and the measure ranges from 0 to 1, or 0 to 100 per cent. A random distribution of orientations will give a measure of 0, and a set of moves which are perfectly aligned with one another will yield a measure of 100 per cent. By calculating a modification of Rayleigh's measure (Potter and Pettijohn 1963).

(3) $\quad \text{RM} = 1 - e^{-R^2/n},$

it is possible to test whether or not the data are distributed uniformly in the interval 0° to 360°.

The final test statistic to be used in the present analysis is directly related to the dichotomy of long and short distance moves. The following test of the extent of bias in long distance moves is suggested. Calculate a vector mean in which the length of the move is held constant. This is identical with (1). Define the vector of the moves, including the distance moved, as:

(4) $V\bar{x}_1 = \arctan \sum_{i=1}^{n} (X_F - X_S) \Big/ \sum_{i=1}^{n} (Y_F - Y_S)$

where X_S, Y_S indicate the origin of a vector and X_F, Y_F the end of a vector in cartesian coordinates. The difference between these measures (1) and (4) is a measure of the extent to which long distance moves are different from short distance moves.

The test statistics were applied to a set of moves collected for the Milwaukee urban region for the interval 1960–2. Milwaukee was subdivided into a set of 128 subregions so that the test statistics would be applied to relatively homogeneous areas of the city. Each of the subregions was approximately 2.4km square. Any subregion with fewer than ten moves was eliminated from further analysis as it would be unlikely to reflect a general pattern of movements.

The calculations were computed for thirty subregions. Although no general directional bias exists for the city as a whole, in six specific subregions of the city the consistency ratio yields results which indicate the presence of directional bias. This of course emphasizes that only where there is the possibility of a strong structural component will it be possible to identify a well-developed directional bias. On the other hand, the results of the tests using the difference measure do not support the hypothesis of a significant difference between long and short distance moves. While five regions have difference measures greater than 45 degrees, and one region has a difference in excess of 90 degrees, in general the difference between the vector calculations holding length constant and allowing length to influence the calculations was less than 20 degrees. Thus, only in specific well-identified sectors within the city are long distance moves likely to be significantly different from moves in general. However, at the present time the tests suggested here have not been broadly applied, and only further tests using the consistency and the difference ratios will enable a complete report to be made on the nature of directional bias in residential mobility. For specific well-identified sectors within the city the possibility of directional bias in residential mobility appears strong.

The investigation of behaviour and structure is an important area of endeavour within urban geography. The development of models and statistical tests which will yield results about specific elements of the behavioural-structural relationship is important and productive. The paper offers evidence in support of a limited directional bias in residential mobility and offers support for King's suggestion that emphasizing process or behavioural formulations in which process and structure are handled simultaneously will accomplish more than further attempting to analyze spatial structure alone.

Adams, J., 1969 Directional bias in intra-urban migration, *Econ. Geog.* 45, 302–3.
Brown, L., and J. Holmes, 1971 Intra-urban migrant lifelines: a spatial view, *Demography* 8, 103–22.
Clark, W.A.V., 1971a A test of directional bias in residential mobility, in H. McConnell and D. Yaseen, eds., *Perspectives in Geography – Models of Spatial Variation* (DeKalb, Ill.).
– 1971b Behavior and the constraints of spatial structure. Unpublished paper, Dept. Geog. U. Cal., Los Angeles.
King, L.J., 1970 The analysis of spatial form: some theoretical and applied shortcomings. Unpublished paper, Conf. on Process and Form, Ann Arbor, Mich.
Pincus, H.J., 1956 Some vector and arithmetic operations on two-dimensional orientation variates, with applications to geological data, *J. Geol.* 64, 533–57.
Potter, P.E., and F.J. Pettijohn, 1963 *Paleocurrents and Basin Analysis* (Academic Press).
Rushton, G., 1969 Analysis of spatial behavior by revealed preference, *Ann. Assoc. Am. Geog.*, 59, 391–400.

P1105
The political ecology of a town in Trinidad
COLIN G. CLARKE *University of Liverpool, UK*

This paper has two objectives: to examine political attitudes and voting patterns in a segmented urban community in Trinidad, and to compare the political behaviour of individuals with aggregate results derived from areal data (Linz 1969, 93).

San Fernando is the second largest settlement in Trinidad and in 1960 recorded nearly 40,000 inhabitants. The principal feature of its social structure is the dichotomy between Creole and East Indian. Census data on race, religion, and occupation depict a Creole majority among whom whites (3 per cent of the town's population), coloureds (21 per cent), and Negroes (47 per cent) rank in decreasing order of socio-economic status, and an East Indian minority of approximately equal standing that comprises a relatively high-ranking stratum of Christians (12.3 per cent), and two low-ranking groups of Hindus (7.6 per cent) and Moslems (5.8 per cent). Creoles and East Indians are residentially polarized and highly endogamous; and little intermarriage occurs between Hindus, Moslems, and Christian East Indians (Clarke 1971, 208–13). Nevertheless, the town possesses a small number of Creole-East Indian mixtures known locally as *douglas*. At the national level the division between Creoles and East Indians is reinforced by politics (Rubin 1962, 449–50; Oxaal 1968, 172). Voting has followed racial lines, East Indians supporting the Democratic Labour Party (DLP) and the majority of Creoles, the People's National Movement (PNM) (Lewis 1962, 10; Bahadoorsingh 1968, 16–19). The PNM has won every general election since it was founded in 1956, including the 1971 poll which was boycotted by the opposition and precipitated a constitutional crisis.

During fieldwork in San Fernando in 1964, samples of respondents of both sexes were drawn from the electoral roll, and the political alignment of the major groups was measured by questionnaire survey. Information about the predominantly Hindu village of Debe was included for comparison with the urban East Indian samples. The questionnaire measured reactions to the PNM government, assessed party preference, and solicited responses to three key issues: the use of voting machines, approval of the involvement of race with politics, and the desirability of the Hindu leader of the opposition becoming prime minister (Table 1). The tabulation demonstrates a variable but strong polarization of political opinion between Creoles and East Indians, though East Indian discontent at that time was somewhat muted. The *douglas* emerged as an intermediate element significantly different from the Creoles and Christian East Indians (according to the chi-square tests). There was a good deal of East Indian solidarity in San Fernando, but the Christians were significantly different from the Moslems and Hindus in certain political attitudes. The material reveals minor shades of distinction between urban Hindus and Moslems, but many clear differences between them and the Debe sample. The segregated rural Hindus clearly comprised the hard-core opposition to the PNM, and constituted what the prime minister once called a 'recalcitrant and hostile minority' (Williams 1969, 275). Taking the Creoles as the starting point, it is possible to rank the groups in approximate order of social, cultural, *and* political proximity: Creoles, *douglas*, Christian East Indians, Hindus, Moslems, and Debe.

By collapsing the enumeration and polling districts to twenty-one common units, statistical comparisons may be made between the 1960 census for San Fernando and the 1961 election. Only the major parties contested the town, which was divided between two political constituencies: each was a safe PNM seat. The percentage of those voting who favoured the DLP has been calculated, and Kendall rank correlations computed with the various social categories. The DLP vote was positively but not strongly correlated with Christian East Indians (0.20), Hindus (0.18), and Moslems (0.12). Only the coefficient linking the DLP and the entire East Indian population was as high as 0.28. A larger correlation with the spatial pattern of the DLP votes was achieved by the Creoles but the sign was negative (-0.33). This material, too, shows that Creoles and East Indians are politically opposed, and confirms that Creole support for the PNM is probably stronger than East Indian support for the DLP. The positive but weak Spearman rank correlation between the spatial pattern of abstentions and the East Indians (0.17), among whom only the coefficient with the Moslems (0.48) is significant, together with the small but negative correlation with the Creoles (-0.21), confirms the slightly equivocal position of some East Indians. In 1961 many Moslems in San Fernando West were embarrassed by a PNM candidate of their own faith.

The ecological analysis for San Fernando reveals approximately the same political patterns as the questionnaire, despite the fact that census data for small areas cannot be disaggregated by age and the political corre-

TABLE 1. Political opinions in the San Fernando and Debe samples

		Social group					
Statements or questions	Answers	Creole (%)	Dougla (%)	Christian East Indian (%)	Hindu (%)	Moslem (%)	Debe (%)
Has the PNM done a good job for people like you? (a)	good	46.9	23.0	9.8	8.7	7.1	11.0
	fair	48.3	74.4	66.9	61.7	66.7	29.3
	not good	2.4	2.6	18.7	20.8	19.1	37.6
	d.k.	2.4	0.0	4.6	8.8	7.1	22.0
Which party can best handle the problem of education? (b)	PNM	83.9	66.6	25.4	21.5	23.8	22.0
	DLP	1.4	10.3	26.6	40.9	35.7	40.3
	d.k.	14.7	23.1	48.0	37.6	40.5	37.6
Which party can best handle the problem of unemployment? (c)	PNM	75.4	48.7	18.8	9.4	10.3	8.2
	DLP	2.4	10.3	36.3	44.9	35.7	44.9
	d.k.	22.2	41.0	44.9	45.6	54.0	46.8
Voting machines have improved the elections (d)	agree	85.3	59.0	14.9	14.1	11.1	7.3
	disagree	7.6	33.3	65.7	68.4	77.0	80.7
	d.k.	7.1	7.7	19.6	17.4	11.9	11.9
I should vote for a good candidate even if I don't support his party (e)	agree	14.2	41.0	39.5	43.6	55.6	28.4
	disagree	83.9	56.4	57.0	53.0	42.1	67.8
	d.k.	1.9	2.6	3.5	3.4	2.4	3.8
Dr Capildeo would make a good prime minister	agree	14.7	33.3	68.8	79.2	77.8	81.6
	disagree	68.7	48.7	16.8	8.7	5.6	6.5
	d.k.	16.6	18.0	14.4	12.1	16.6	11.9
Size of samples		211	39	256	149	126	109

(a) *Dougla*-Creole $\chi^2 = 9.5$; d.f. = 3; $p < 0.05$; *Dougla*-Christian East Indian $\chi^2 = 12.65$; d.f. = 3; $p < 0.01$; Hindu-Debe $\chi^2 = 28.21$; d.f. = 3; $p < 0.001$.
(b) *Dougla*-Creole $\chi^2 = 11.82$; d.f. = 2; $p < 0.01$; *Dougla*-Christian East Indian $\chi^2 = 27.09$; d.f. = 2; $p < 0.001$; Hindu-Debe $\chi^2 = 0.013$; d.f. = 2; $p > 0.99$.
(c) *Dougla*-Creole $\chi^2 = 13.55$; d.f. = 2; $p < 0.01$; *Dougla*-Christian East Indian $\chi^2 = 20.49$; d.f. = 2; $p < 0.001$; Hindu-Christian East Indian $\chi^2 = 7.17$; d.f. = 2; $p < 0.05$.
(d) *Dougla*-Creole $\chi^2 = 21.64$; d.f. = 2; $p < 0.001$; *Dougla*-Christian East Indian $\chi^2 = 40.26$; d.f. = 2; $p < 0.001$.
(e) *Dougla*-Creole $\chi^2 = 16.05$; d.f. = 2; $p < 0.001$; *Dougla*-Christian $\chi^2 = 0.11$; d.f. = 2; $p > 0.99$; Creole-Debe $\chi^2 = 10.87$; d.f. = 2; $p < 0.01$; Christian-Debe $\chi^2 = 4.06$; d.f. = 2; $p > 0.10$.

lations are made with the entire social group in question. In general, political attitudes and voting are similar. The ecological material is less satisfactory than the questionnaire for differentiating among Hindus, Moslems, and Christians. The various East Indian groups occupy similar neighbourhoods (Clarke 1971, 207); moreover, voting is a simple matter compared with the opinions which have been analysed. The ecological results also underestimate the true extent of political opposition between Creoles and East Indians, and this is due to geographical patterns in the town (Robinson 1950, 351). Creoles and East Indians are spatially polarized but not highly segregated (Clarke 1971, 207). If the more substantial part of the DLP vote appears to come from the Christian East Indians, this is primarily because of the group's large size. Racial and cultural factors influence political affiliation; but socio-economic factors are responsible for the fairly low level of residential segregation, which, in turn, generates small ecological correlations. These correlations supply excellent information about the social geography of San Fernando which complements, but does not provide a substitute for, survey analysis in this particular town.

Research Institute for the Study of Man.

Bahadoorsingh, Krishna, 1968 *Trinidad Electoral Politics* (London).
Clarke, Colin G., 1971 Residential segregation and intermarriage in San Fernando, Trinidad, *Geog. Rev.* 61, 198–218.
Lewis, Gordon K., 1962 The Trinidad and Tobago general election of 1961, *Caribbean Studies* 2, 2–30.
Linz, Juan J., 1969 Ecological analysis and survey research, in Mattei Dogan and Stein Rokkan, eds., *Quantitative Ecological Analysis in the Social Sciences* (Cambridge, Mass.), 91–131.
Oxaal, Ivar, 1968 *Black Intellectuals Come to Power* (Cambridge, Mass.).
Robinson, W.S., 1950 Ecological correlations and the behavior of individuals, *American Sociological Review*, 15, 351–7.
Rubin, Vera, 1962 Culture, politics and race relations, *Social and Economic Studies* 11, 433–55.
Williams, Eric, 1969 *Inward Hunger* (London).

P1106
Changing residential structures in South African cities, 1950–70
R.J. DAVIES   *University of Natal, South Africa*

The consequences of contact between ethnic groups in cities may be polarized about two theories. One theory maintains that common interests develop from contact and shared experiences and provide the basis for ethnic co-operation. The other postulates that physical, social, cultural, and economic differences between peoples are incompatible and that contact leads to friction.

Neither theory can be shown to be adequate under all circumstances and the problem of race relations rests with the identification of conditions which produce one or other result. Both theories, however, are action oriented and administrators proceed on the assumption that the theory they adopt is, in fact, valid. Thus, in the United States and Britain for example, it is upon the co-operation theory that social action is being taken in the search for solutions to the minority ethnic ghetto. In South Africa, on the other hand, apartheid policy is based on the friction theory – the concept that harmonious relations between ethnic groups can be secured only by reducing points of contact to a minimum. From a history of pragmatic segregation has grown an attempt to create vertically separate ethnic societies in which horizontal contact will (in theory) be reduced to a minimum and each society identified within its own territory.

This paper is focused upon changes in residential structure in South African cities brought about by 'apartheid planning' between 1950 and 1970. Attention is centred chiefly, though not exclusively, upon the city of Durban. Change will be measured by consideration of levels of segregation and elements of spatial organization of ethnic residential districts.

Development processes which underlie residential patterns in the contemporary 'apartheid city' differ radically from those which underlay the earlier 'segregation city.' Differences are complex in detail, but in essence are expressed in a contrast of pragmatism and compromise between natural social processes, voluntary and legally imposed segregation, on the one hand, and a rigid, uncompromising framework of controls over interethnic changes in ownership and occupation of property under Group Areas legislation (1950) on the other. Legal controls in the segregation city lacked an overall framework of spatial and functional organization, were incomplete, differentially applied, and permissive in practice. In the apartheid city, comprehensively structured controls are uniformly imposed on all cities and include the radical element of retrospective planning which subordinates conventional planning practice and rights of ownership to racial zoning.

Despite differences in development processes, residential segregation in both cities is basic to their structure. Segregation Indexes for Durban in 1951, for example, were high and comparable with extreme cases of segregation in American cities of the 1950s (Duncan and Davis 1953) as shown in Table 1. Only Indians and Coloureds occupied broadly similar residential space. Segregation was not complete, however, and transition zones of ethnic heterogeneity were characteristic of the segregation city. Nevertheless,

TABLE 1

	Segregation indexes (Durban)				Centralization indexes (Durban)			
					CBD focus		Industrial focus	
	Whites	Col.	As.	Black	(1951)	(1970)	(1951)	(1970)
		(1970)						
White		0.91	0.98	0.91	0.19	0.54		
Col.	0.84		0.79	0.96	−0.13	−0.44	0.06	0.16
As.	0.91	0.46		0.98	−0.28	−0.61	0.05	0.09
Black	0.81	0.81	0.81		−0.27	−0.82	0.15	−0.08
	1951							

s.i.: Ranges from 0 = even distribution to 1 = absolute segregation.
c.i.: Ranges from −1 = absolute decentralization through 0 = even distribution to +1 representing absolute centralization. Calculations based on centralization in relation to the White group.

it would have required only a small proportional movement of population to achieve absolute segregation within the existing residential structure (defined on a census tract basis). In Durban, a minimum out-movement from 18 tracts of 14.71 per cent of Whites and a corresponding in-movement of 6.25 per cent of Indians, for example, would have achieved absolute segregation between these groups.

By 1970 segregation had reached nearly absolute levels. On the basis of the 1951 population it is estimated, however, that planned group areas (proclaimed from 1958) required the movement of over 50 per cent of the Coloured, 50 per cent of the Indian, and 67 per cent of the Black population, and less than 20 per cent of the Whites to achieve absolute segregation at that time.

It is clear that, if segregation alone were at issue, no great stress need have been placed upon existing residential structures in the segregation city. Change was imposed because the spatial form and ethnic functional organization of the segregation city did not conform to criteria set for urban apartheid.

Elements of two schematic models of the segregation city appear in Fig. 1a (Pretoria: simple ethnic structure; 49 per cent White, 47 per cent Black, Coloured and Asiatic minorities) and 1b (Durban: complex ethnic composition: 31 per cent White, 37 per cent Asiatic, 32 per cent Black, Coloured minority). Predictable structural elements in both cities included:

(a) a relatively high degree of centralization of Whites with peripheral Non-White residence. Despite their poverty, development processes did not permit the centralization of Non-Whites about either the CBD or the major industrial focus to any appreciable degree.

(b) In theory Black residence was controlled, but in practice, because of official inertia, lack of compulsion, and inadequacies in housing programs, it developed randomly with no specific, planned relationship to other ethnic groups. A complex situation arose and by 1951 most Blacks were housed in peripheral shack slums, in working quarter barracks, and in a limited number of overcrowded official townships (locations).

(c) Under permissive regulations Indians and Coloureds competed with Whites for residential space. Competition was governed by a high level of selectivity exercised by Whites (through wealth and political power), while Non-White residence grew on remaining available land, rejected by Whites as undesirable, and through invasion of older White residential areas. In both cities, central Indian trading-residential communities were typical, and in Durban outward growth of Indian residence gave rise to an inner transition zone of ethnic mixture. The majority of Indians and Coloureds, however, were peripherally located and in Durban an outer Non-White zone developed, hemming in the predominantly White inner city.

These structures may be set against the criteria drawn for an apartheid city model. The model includes: the creation of consolidated group areas for each ethnic group with growth hinterlands or, in the case of minorities, areas large enough to permit

Géographie urbaine / 803

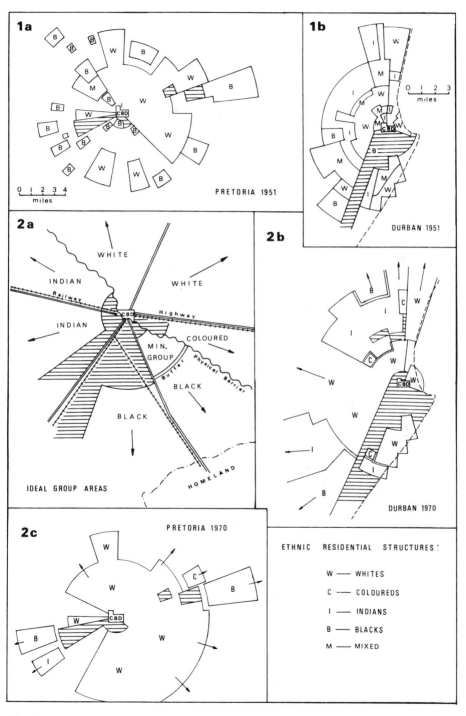

Fig. 1.

TABLE 2. Proportion of each ethnic group resident in proclaimed group areas (Durban 1970)

Group areas	White	Coloured	Asiatic	Black	Percentage pop. *resident* in work-zones of city
White	*99.65*	14.87	10.78	12.25	3.37
Col.	0.05	*72.52*	0.38	0.29	13.40
As.	0.29	12.61	*88.83*	0.48	13.78
Black	0.01	0.00	0.00	*86.98*	10.05

growth over 30–50 years; the use of strong physical or man-made barriers and buffer strips of sterilized vacant land to achieve spatial separation; direct access by each group to work zones to avoid traversing of areas set aside for other groups and the location of Non-White areas close to major industrial foci; the orientation of Black areas towards Reserve Lands (Homelands) where these are present, and the avoidance of ethnic 'islands' within large group areas. Ideally, the division of space in the city is structured within a sectoral framework.

Functionally each group area is conceived as an independent community, ultimately to have its own local authority and services. Carrying apartheid to a logical conclusion, each community should also ultimately be economically self-sufficient and vertically separate in all respects. Vertical economic separation is unrealistic, however, and the model continues to envisage integrated work zones. It should be noted, however, that proposals have been made for the complete functional and spatial separation of the Black population of Pretoria by developing an independent Black city within the Tswana 'Homeland' 20 kilometres north of Pretoria (Moolman 1969).

The application of apartheid criteria to existing cities has inevitably led to compromise (common work zones have already been cited), to avoid undue disturbance of existing conditions and particularly existing White residential areas and excessive costs. Contemporary models of Durban and Pretoria illustrate the nature of spatial changes accomplished. In both cities ethnic patterns have become strongly sectoral and intergroup area movements show the effect of population controls. In Durban the level of segregation achieved is high and a high proportion of each ethnic group is resident in proclaimed group areas. In achieving these patterns, however, the Non-White population has been subjected to the greatest changes and is now more strongly decentralized in relation to Whites, while only marginal overall improvements have been achieved in locating Non-Whites near to major work centres (Table 1). It has been necessary also to create several group areas for individual ethnic groups largely vitiating any convenient functional organization of local authorities and communities.

Although the restructuring of South African cities has been substantial, urban apartheid has in essence achieved only a more tidy arrangement of ethnic groups and higher levels of segregation – attained at very considerable financial and social costs. Substantial housing programs for Non-Whites which preceded population resettlement, while commendable, should not be attributed to apartheid. They are improvements which would have been necessary under any circumstances. Greater levels of vertical social segregation have been achieved, but cities remain economically integrated and in general the goal of complete, vertically separate, urban ethnic communities remains unrealistic.

Dept. of Statistics, 1951–1970 Special tabulations of the population by enumerators' districts for Durban.
Duncan, O.D., and B. Davis, 1953 *Chicago Urban Analysis Project* (Chicago).
– 1955 A methodological analysis of segregation indexes, *Am. Soc. Rev. 20*.
Kuper, L., H. Watts, and R. J. Davies, 1958 *Durban: A Study in Race Ecology* (Cape and Columbia).
Moolman, J.H., 1969 *Die vestiging van die Bantoe in Pretoria*, Paper presented to the Congress of the Society for the Teaching of Geography, Potchefstroom.

P1107
The factor ecology of British cities: a preliminary report
WAYNE K.D. DAVIES   *University of Calgary, Canada*

Factor ecological studies of British cities became possible only after 1961, when the UK Census first provided data on an enumeration district basis, and since then a series of studies has revealed the structure of individual British cities (Robson 1969; Giggs and Herbert 1970). However, it has proved difficult to make rigid comparisons between the case studies, in order to produce a general typology of the social structure of British cities, despite the influence of an informal working party on the use of small area statistics (Census Research Group 1971). Of major significance in this respect are the variations in data input and areal basis of the existing studies, as well as the limitations in the types of factor model chosen by the various investigators. These problems have even greater impact when cross-cultural comparisons are attempted, and it is hard to avoid the conclusion that the substantive results of the majority of British studies are of an ideographic rather than nomothetic value, despite their technical sophistication.

This short paper represents a preliminary report on a comparative study of the ecological structure of some of the larger British cities outside the conurbations. As comparability is the primary objective, both the data input and the techniques were standardized for the three primary study areas (Swansea, Leicester, Southampton), and similar investigations of smaller regional centres such as Colchester, Llanelli, and Pontypridd provided an opportunity to introduce the effect of size differentials. It is worth emphasizing that the study adopted the use of oblique factor rotations and higher-order factor solutions, thereby making it possible to remove one of the major restrictions of previous factor ecological studies, namely the assumption of orthogonality between the factors.

Limitations of space make it possible to discuss only the results of the Leicester study, although parallel results have been produced for the other centres. A Principal Component Analysis of the 139 area by 55 variable matrix for the built-up area of Leicester produced eight factors accounting for 69.6 per cent of the original variance, and these results were subjected to a series of orthogonal and oblique rotations (Dixon 1969, 90). This was designed to search for invariant factor axes (Davies 1971). A biquartimin solution using the reference axis technique provided the most satisfactory results, and the distribution of factor loadings over ± 0.3, together with a short title of the variables, is shown in Table 1.

Contrary to the findings of previous British studies, the three dimensions associated with American cities (Berry and Horton 1970) – socio-economic status, stage in life cycle, ethnic-migrant status – proved to be the primary factor dimensions of Leicester, accounting for 46.9 per cent of the original variance. Table 1 shows that the first two of these factors form a typical factor scale with high negative and high positive loadings, and that comparability with the North American example is again indicated by the pattern of factor scores. Thus, the 'stage in life cycle' dimension shows the broadly concentric zonation of the Burgess hypothesis, whereas the pattern of 'socio-economic status' is sectorial, with the highest positive scores attained by the southwestern sector of the built-up area, in contrast to the usual westerly orientation of the highest status area. The major discrepancies with the patterns established for North American cities come from the 'ethnic-migrant' dimension. Here, only high positive loadings are recorded, with the migration variables attaining the highest values, probably testifying to the comparative recency of this dimension, and the recently built public housing estates on the edge of the city scoring almost as high as the inner city area.

Five other oblique factors individually accounting for between 6.0 and 3.0 per cent of the original variance were also identified, and are named in Table 1. Although each of these dimensions represents a distinct aspect of modern British society, their dependence upon the primary factors was revealed by a higher-order factor analysis. The correlations between the oblique factors were used as input to an initial component analysis in

TABLE 1. Distribution of loadings: eight oblique factors (Only abbreviated titles of variables used in the analysis are shown; high = >0.70%; medium = 0.50–0.69%; low = 0.30–0.49%)

FACTOR I. SOCIO-ECONOMIC STATUS

Positive
High — Two-car households; car ratio; employers and managers; high social class; professional workers
Medium — Non-manual workers (intermediate); car-commuter; distribution and service workers; owner-occupiers
Low — Non-residents; government workers

Negative
Low — Irish; commonwealth-colonial; born overseas
Medium — Industrial workers; unskilled manual workers; households without car; pedestrian commuters; foremen and skilled workers; tenants of unfurnished property
High — Personal service–agricultural workers; low social class

FACTOR II. STAGE IN LIFE CYCLE

Positive
High — Old age
Medium — Middle aged; rooms per person (area)
Low — Households with pensioners; owner-occupiers; tenants of unfurnished property; pedestrian commuters

Negative
Low — None
Medium — Council tenants; mature adults
High — Children; single adults; large households

FACTOR III. ETHNIC AND MIGRANT

Positive
High — Movers within 5 years; female movers; single person movers; recent movers; new residents
Medium — Sharing dwellings; born overseas; foreign born; Irish; commonwealth-colonial
Low — Mature adults; tenants of unfurnished property; one-person households; local movers

Negative
Low — Council tenants
Medium — None
High — None

FACTOR IV. RESIDENTIAL MOBILITY

Positive
High — None
Medium — None
Low — Cross-commuters; new residents; agricultural workers; tenants of unfurnished property

Negative
Low — Transport workers; medium social class; council tenants
Medium — None
High — Local movers

FACTOR V. AGE-SUBSTANDARD

Positive
High — Substandard houses; dense occupancy households
Medium — One-person households; households with pensioners; households without car
Low — Pedestrian commuters; large households

Negative
Low — Rooms per person (area)
Medium — None
High — None

TABLE 1 (Continued)

FACTOR VI. MOBILE YOUNG ADULT

Positive
High	Young adults
Medium	Non-residents
Low	Employed females; sharing dwellings

Negative
Low	Industrial workers; transport workers; medium social class; works transport commuters; owner-occupiers
Medium	Married females
High	None

FACTOR VII: ECONOMIC PARTICIPATION

Positive
High	Employed persons; employed females
Medium	Economically active
Low	Medium social class; government workers; young adults

Negative
Low	Old age; females; households with pensioners
Medium	None
High	None

FACTOR VIII. PERIPHERAL-URBAN

Positive
High	Agriculture workers
Medium	Crossing-commuters; industrial workers
Low	Fertile female ratio; car-commuters

Negative
Low	None
Medium	None
High	None

which three components accounted for 50.8 per cent of the variance. These three higher-order factors were rotated to a simple structure using the biquartimin oblique solution. Associated primarily with the first of these oblique higher-order factors were the life cycle (negative), residential mobility, and economic participation dimensions; associated with the second were the socio-economic scale, the age substandard (negative), and peripheral urban dimensions, and the third picked out the ethnic-migrant and mobile young adult dimensions. Hence, the eight dimensions accounting for almost 70 per cent of the original variance collapsed to three higher-order dimensions accounting for half of this variability.

The preliminary results of this comparative study suggest that the existing ideas of the factor ecology of British provincial centres need to be revised in the light of the limitations of the original studies. Both the social patterns and their spatial distributions seem to be closer to the American experience than has hitherto been revealed. Research is continuing on the effect of the size differential upon social dimensions, and upon the technical issues of factor ecology, in particular the utility of Q mode analysis and the invariance of rotation solutions.

Gareth J. Lewis.

Berry, B.J.L., and F. Horton, 1970 *Geographic Perspectives on Urban Systems* (Englewood Cliffs, NJ).

Census Research Group, 1971 The latest meeting of the Study Group took place on 7 July 1971, Centre for Urban Studies and Regional Studies, University of Birmingham. See also *Area*, 3 (2), 95.

Davies, W.K.D., 1971 Invariance in factor rotations. Unpublished paper, U. Calgary, Geog. Dept.

Dixon, W.J., 1969 *Biomedical Computer Programmes: X Series Supplement* (no

77) (Los Angeles). The calculations were carried out on the IBM 360-350 system of the University of Calgary, using the above package program.

Giggs, J.A., and D.T. Herbert, 1970   Chaps. 6 and 5 in H. Carter and W.K.D. Davies, *Urban Essays: Studies in the Geography of Wales* (London).

Robson, B.T., 1969   *Urban Analysis* (Cambridge).

P1108
The interaction of two scales within the ecological structure of metropolitan Haifa, Israel
Y. GRADUS   *University of Negev, Israel*

The impact of government and public policies on the urban ecological structure is a matter which has received relatively little attention in geographic literature. Yet it is a vital concern in certain economic and political systems. Recent empirical studies have linked the classical spatial ecological models with social area analysis by the use of factor analysis techniques, resulting in the emergence of the factorial ecological approach. None of these models have considered the effect of public or governmental policies. Rather, they assume that residential patterns are a result of the free play of economic forces. Berry, when he initiated the systematic cross-cultural factorial-ecological studies, encouraged case studies along a scale from pre- to post-industrial form, though he, too, ignored the degree of government control as a component shaping the urban ecological structure (Berry and Rees 1969, 445). Only recently has Berry suggested that we limit our overly general focus and compare cities within three categories: (*a*) western industrial societies with some spontaneous freedom of choice; (*b*) 'planned' or directed societies; and (*c*) traditional societies (Berry 1971, 216).

It seems that the urban ecological structure can be viewed both as a reflection of the degree of a society's development, and as a manifestation of its political and economic system and its specific sociocultural characteristics. We may see, therefore, that there are actually two scales interacting within a given culture. One scale concerns a society's degree of development, while the other refers to the level of direct governmental involvement in the processes of development.

Applying the two scales to an urban ecological analysis reveals that they suffer from oversimplification and can be viewed only as a guideline for systematic cross-cultural ecological analyses. It is necessary to identify urban ecological forces in each case study regardless of any preconceived biases, considering the two scales which have been suggested only as guidelines.

Metropolitan Haifa, Israel's second largest population concentration, was chosen as a case study of urban ecology in order to demonstrate a mixture of spontaneous and planned system as well as a mixture of industrial and pre-industrial (traditional) society interacting in urban space. Keeping in mind the two scales as guidelines, an attempt was made to identify forces which have affected and shaped Haifa's urban ecological structure. For this purpose 39 selected variables on 53 observations (statistical zones) were obtained from data collected in 1965 by the Israel Central Bureau of Statistics and Haifa Area Transportation Planning. The heterogeneity and the multi-cultural character of the Israeli society is reflected in variables such as continent and country of birth. The differences in the length of stay in the country are presented in variables expressing periods of immigration. The analysis also employed the conventional variables expressing economic, demographic, educational, occupational, and housing characteristics. Haifa's topography is expressed in the variable 'elevation above sea level.'

Control of the land and the housing market have been identified as the most significant governmental control variables in metropolitan Haifa. About 70 per cent of the total residential land is owned and leased by the government, with the remainder privately owned. The public sector controls a high portion of the housing market. Approximately 60 per cent of the total housing stock in Haifa was planned and constructed by the public sector mainly for new immigrants, but some for veterans, young couples, and slum evacuees as well. In a country of continuous immigration and with a rate of growth unmatched anywhere else in the world, the government had to assume responsibility for the absorption and integration of the newcomers. Government housing projects have been erected mainly on publicly owned land, while the private sector has tended to build on privately owned land. Thus two variables were added to the analysis, 'percentage of residential land owned privately' and 'percentage of residential land owned publicly,' which are equally representative of spatial distribution of construction activity by sectors.

Factor analysis and correlation methods were used in the present analysis to account for the interrelationship between the variables: the first technique for the broad descriptive structure, and the second for more specific relationships. Four major factors were extracted, accounting for 70 per cent of the total variance. The first two dimensions were identified as socio-economic status and ethnicity of the Jewish population. The third was an Arab minority dimension, and the fourth was a density and land value gradient.

The differences between the pre-independence European origin groups and the post-independence immigrants of Middle Eastern origin are reflected in the loadings on the first two factors. Generally, Middle Eastern groups are associated with lower income, lower level of education, large family size, and a higher degree of overcrowding per dwelling unit. Analyzing the correlation matrix for the specific relationships revealed that the North African immigrants are at the bottom of the socio-economic spectrum, while immigrants from central and eastern Europe, with the exception of Rumania, are at the top of the scale. The group 'born in Turkey & Iran' were higher than the rest of the Middle Eastern groups on this scale. The overlap and link between socio-economic status and ethnic factors is basically an aggravating situation in any society. No factorial separations were found between indicators of socio-economic status and indicators of family status dimension or family cycle state, in contrast with the normal separation of these two sets in factor analysis of most western cities. Mapping the factor scores to account for the spatial pattern of the socio-economic/ethnicity dimensions, and correlating them to distance from the centre and elevation above sea level revealed clearly the impact of public control of land and housing. The lower socio-economic groups, mainly of Middle Eastern origin, were distributed as a U shape around Mount Carmel on publicly owned land and lived mostly in new government-built housing, but some also in substandard dwelling units abandoned by their former Arab residents. The higher socio-economic status group of European origin can be found mainly on the upper part of the city on privately owned land. Given the elevation and land ownership of a section in the city almost enables one to predict its residents' socio-economic and ethnic background. Generally no association was found between socio-economic status and distance from the centre in contrast with the filtering process which characterizes cities within a system of extreme spontaneous freedom of choice. The absence of distinct filtering and family cycle processes indicates the relative immobility of residents within the metropolitan area, reflecting, perhaps, the predominance of the public sector in the Israeli economy, while the higher mobility of residents in an extremely free enterprise economy is a function of groups or individuals competing for residential urban space.

The factor 'density and land value gradient' was strongly associated with distance from the centre. Land values and density appeared to drop with distance from the centre as in the case of most free-enterprise systems. The well known paradox of rich living on cheap land and poor on expensive is not very applicable in our case, since, as already indicated, socio-economic status is not rising with distance. Those in the higher socio-economic status live in Haifa on expensive privately owned land on the upper part of the mountain, while those of the lower socio-economic status live mostly on the periphery in dwelling units built by the government on public land and purchased on relatively easy terms by its inhabitants. Some poor can be found also near the centre as in most free-enterprise western cities (transitional zone), but in the case of Haifa this is due mainly to historical events and less to the filtering process. After Israel's War of Independence (1948) most of Haifa's Arab inhabitants preferred not to stay in the newly established state and left the town. Their abandoned dwellings were reoccupied mainly by new North African immigrants.

The Arab minority group was identified as a separate dimension, highly correlated to the variables 'construction workers' and 'proportion male,' indicating rural-urban male migration. A large portion of Arab males have been attracted to the big city, especially by the construction industry, which enables them to support the family left behind in the traditional village. The other portion is the indigenous of the city who preferred not to leave in 1948. Mapping the factor scores revealed that spatially they are highly concentrated in the old section of the city which has always been predominantly Arab.

It is obvious that the attempt being made by the government to integrate all ethnic groups was not implemented spatially in Haifa's case mainly because of the distribution of land ownership. Apparently the solution of the problem of bridging cultural gaps seems to require more than a balanced distribution and spatial planning of ethnic groups in urban space.

The present factorial ecological structure of metropolitan Haifa is in transition. With the country moving even closer toward a free-enterprise system and the integration process proceeding quite rapidly, a different factorial structure may emerge in which ethnic background and socio-economic status will be independent dimensions not associated with land ownership. A family status dimension may emerge following this process.

Berry, B.J.L., 1971   Introduction: the logic and limitations of comparative factorial ecology, *Econ. Geog.* 47, 209–19.
– and P.H. Rees, 1969   The factorial ecology of Calcutta, *Am. J. Soc.* 74, 445–91.

P1109
Urbanization in the arid zones of Mexico
M.T. GUTIERREZ DE MACGREGOR and CARMEN VALVERDE   *National University of Mexico*

This paper examines the evolution of the urban population which is required as a basis of a program for regional development of the arid zones in Mexico. Four main facts show us the importance of the arid zones in Mexico: the arid zones occupy a large surface, about 1,140,000km^2 equivalent to 57% of the surface of the country; of the 21,196,416 rural inhabitants registered in 1960, 6,111,500 (29%) were living in the arid zones; of the 23,526,586 urban inhabitants registered in 1970, 7,025,143 inhabitants (30%) were living in the arid zones; of the 286 cities registered in 1970, 101 (35.0%) were located in the arid zones.

There are different criteria which can be used to draw the limits of the arid zones. For the present study we make use, primarily, of Köppen's climatic classification as modified by García (1964). The main reason for the drought of these areas is that a great part of Mexico lies in the desert belt zone of the northern hemisphere. The influence of the topography is also important because the general orientation along a NW–SE axis of the two main chains of mountains, the Sierra Madre Oriental and the Sierra Madre Occidental, rising to an average elevation of about 2500m, acts as a barrier. These mountains impede the passage of humid air masses that come from the oceans to the interior high plateau, named Altiplanicie Mexicana. It is not uncommon to find some areas receiving an annual precipitation of only 200mm. The arid lands in Mexico are characterized by high temperature and low precipitation.

On the basis of general climatic characteristics, the region can be divided into two major areas: a very arid zone corresponding to the BW type of climate, and a semi-arid zone corresponding to the BS type of climate. These arid zones extend between the parallels 20° to 32° north latitude; to the south of these limits we find only a very few small arid areas.

In Mexico a town of less than 2500 inhabitants is considered rural; over that limit it is considered urban. This quantitative criterion does not take into account the functions of the town, or if it has public services such as running water, electricity, or medical services. It would be better to take a limit of 10,000 inhabitants, as that could give us a more realistic idea of the urban population of Mexico while some intensive studies permit us to give a more accurate definition.

An analysis of the urban population growth of the arid zones during the last 70 years shows a slow growth from 1900 to 1940 and a very rapid growth from 1940 to 1970. The increase of the last 30 years has occurred as a consequence of an increase in the natural growth rates, owing to a very high birth rate (for the period 1946–67 the average birth rate was 47.3 per thousand inhabitants) and a rapidly decreasing death rate, from 18.6 in 1946 to 8.4 in 1967, due to improvements in health and living conditions.

It is necessary to point out the importance of the rapid increase of the urban population in the arid zones. If we take into account the number of inhabitants between 1900 and 1970, an increase of almost twelve times is registered, from 591,151 inhabitants in 1900 to 7,025,143 in 1970. This rapid urban population growth is also due to technological advances in drought-resistant seeds, in pest control, and in a mechanized cultivation that has increased the yield of crops; to industrial development from 1940 until now, which has been observed in almost all big cities; to migration movements from rural areas to urban ones, due to the attraction of Monterrey (the main industrial centre of the region) and of the cities of the northern border of Mexico that offer more opportunities for employment and better salaries. We must also remember that in urban nuclei there are more cultural and amusement facilities than in rural ones, which again attracts people. Equally fundamental was the development of improved means of transportation and roads.

The distribution of urban population in the arid zones is very uneven. It is obvious that the past population distribution influences present distribution. This is particularly true in the northern part of the country since 1900, where the influence of a population pattern survives from colonial times.

If we relate the distribution by latitude in 1900, 1940, and 1970, we notice that the percentage of total urban population north of the Tropic of Cancer varies from 37% in 1900, to 62% in 1940, to 76% in 1970. That shows a tendency of the urban population of the arid zone to concentrate in the northern part of the country.

Analyzing the vertical distribution of the urban population in the arid zones between 1900 and 1970, we notice a change of pattern: in 1900 more than 84% of the whole urban population was living in altitudes between 1000 to 3000m but by 1970 only 48% were living there.

The reason for this change is the establishment and increase of irrigation farming in altitudes between 0 and 1000m, mainly in the NW and NE parts of the country in the valleys of Mexicali, Yaqui, Mayo, and Fuerte, and the lower valley of Rio Bravo, respectively; demographic development of the northern border cities of Tijuana, Mexicali, and Ciudad Juárez; and finally, the development of the agricultural and oil region of Reynosa in the lower valley of Río Bravo in the NE of the country.

The map of distribution of urban population in 1900 shows that the influence of physical factors such as climate, natural resources, and historical facts was more accentuated than in the following periods. The historical factor also intervenes in that the origin of many cities located in the arid zones comes from colonial times in which the mining cities were very important.

The map of 1970 showing the urban population distribution indicates the influence of the United States frontier region upon the population distribution of Mexico's border cities by absorbing surplus population and labour because of the higher wages attainable in the cities of the United States and its area of agricultural influence. The cities possessing the greatest dependence are, in order: Ciudad Juárez, Tijuana, Neuvo Laredo, and Matamoros in each of which 30 per cent or more of the wage income derives from the United States (Dillman 1970).

The government policies of establishing irrigation systems in the arid zones and introducing modern techniques in agriculture and industry, and oil exploitation, have permitted rapid and very recent economic and demographic development. The rapid demographic increase is also a consequence of the high birth rate together with a reduction in mortality rate and the internal migration movements from small urban towns to the big cities, changing the structure of their population.

Dillman, C.D., 1970 Recent developments in Mexico's national border program, *Prof. Geog.* 23 (5), 243–7.

García, E., 1964 *Modificaciones al sistema de classicación climática de Köppen* (México).

P1110
A note on the stability of urban dimensions
WAYNE L. HOFFMAN *Western Kentucky University, USA*
and GERALD ROMSA *University of Windsor, Canada*

The recent multiplicity of research articles dealing with inter- and intra-urban spatial structure is a recognition of the need to understand more fully urban processes and their association with the social, political, and economic fabric of society. A common thread of these studies has been their technique orientation. With large masses of data generally available multivariate statistical procedures have been employed. The most popular of these methods has been factor or component analysis. This popularity is due to the ability of the various forms of the model to handle large amounts of data and, theoretically, to achieve order out of chaos. Furthermore, it is often implied that the factor analysis method arrives at basic dimensions of urban structure which are stable and thus comparable through time and space. It is with this stability of derived urban dimensions that this paper is concerned. Derived components of urban structure must have recognizable consistency if comparisons of observed values with theoretical ones are to be valid. The purpose of this paper is to test the invariance of urban dimensions through both time and space.

The literature concerning the stability of urban dimensions is quite inconclusive. Hadden and Borgatta (1965), and Ahmad (1965) have demonstrated that dimensions do remain stable under varying conditions. King (1966) in his study of Canadian cities draws the conclusion that, although the Canadian urban systems do appear to shift through time, the city groupings based on these dimensions are relatively stable. Hoffman and Romsa (1970), on the other hand, found that urban dimensions and city groupings of southeastern American cities were not stable through time. Finally, Sweetser (1965) in his work on the ecology of Helsinki and Boston surmised that stability among dimensions varies, with certain characteristic dimensions tending to be more stable than others.

To test the stability of urban dimensions through time and space, basic components of urban structure for selected cities were derived and tested for consistency through the employment of the coefficient of congruence. This test has been outlined by Harmon (1960, 257–259) and applied by Sweetser (1965, 215) and Murdie (1969, 78). For the temporal analysis 86 cities located in the southeastern portion of the United States were examined for two time periods, 1950 and 1960. Cross-regional comparisons were made for three Canadian cities – Halifax, London, and Regina. Each of these Canadian cities has a different historical, economic, and social setting.

Data were obtained from the American and Canadian censuses. Forty-one variables reflecting socio-economic and demographic characteristics were used for the American cities, while a similar but smaller set of twenty-four were used for the Canadian examples. These data were factor analysed by the principal-components method and normal varimax rotations were used. The first six factors derived for the American cities were then compared. More than six groupings were calculated for the Canadian cities to indicate some of the problems in deciding which pairs of dimensions were identical.

The results of the factor analysis and tests of congruence found in Tables 1 and 2 indicate that, although basic urban dimensions may appear to be similar, they are not necessarily so. Among the American cities only one pair had a fairly high coefficient of congruency – dimension 1 identified as representing social economic status. The other factors revealed weaker and in one instance negative associations. The Canadian findings are similar. In general, it appears that the first dimensions retain some sort of stability, while the other components appear to be less invariant. Essentially what the findings of the study indicate is a tendency for urban dimensions to be unstable both through time and space. These findings raise a serious question as to the general comparability of basic urban dimensions that have not been tested in some manner for consistency.

The London-Regina case is of added interest because it points out that one basic dimension may reveal a structure which is represented by two dimensions in another city – dimension 1 of London and 1 and 6 of Regina. Furthermore, as represented by dimensions 1 and 2 of London and 1 and 6 of Regina, the findings illustrate the problem experienced in comparing certain dimensions that seem similar and/or the difficulty in selecting the number of components to be used in a study. The problem of using factor scores in a linear regression causal model especially for comparative purposes also becomes readily apparent.

The use of urban structure components derived by means of factor analysis does have limitations. These limitations are due in part to the problem of standardizing the dimensions so that accurate temporal and spatial comparisons can be made. Unless some standardization exists, comparative hypothesis testing becomes difficult, if indeed not impossible. The authors of this study suggest some test of consistency be applied to generated components to ensure stability, better identification of dimensions, and more meaningful conclusions.

TABLE 1. Similarity among American urban dimensions (temporal analysis)

1950 dimension	1960 dimension	Coefficient of congruence
1	1	0.79
2	6	0.21
3	5	0.11
4	4	0.37
5	3	0.50
6	2	−0.38

TABLE 2. Similarity among Canadian urban dimensions (spatial analysis)

Dimension	Dimension	Coefficient of congruence
LONDON	HALIFAX	
1	1	0.84
3	2	0.64
6	6	0.34
4	4	−0.70
5	5	−0.11
2	3	−0.30
LONDON	REGINA	
1	1	0.66
1	6	0.56
6	3	0.20
4	5	−0.15
5	3	0.14
4	4	0.18
3	2	−0.48
2	1	0.50
HALIFAX	REGINA	
1	1	0.69
2	2	−0.78
3	3	−0.19
5	6	0.35
5	5	−0.39
4	4	0.01
6	3	−0.62

Ahmad, Q., 1965 *Indian Cities: Characteristics and Correlates* (Chicago).

Hadden, J.K., and E.F. Borgatta, 1965 *American Cities: Their Social Characteristics* (Chicago).

Harmon, H., 1960 *Modern Factor Analysis* (Chicago).

Hoffman, W., and G. Romsa, 1970 A multivariate grouping of southeast cities, 1950 and 1960, Annual Meeting, Southeast Division of Assoc. Am. Geog., Columbia, SC.

King, L.J., 1966 Cross-sectional analysis of Canadian urban dimensions, *Can. Geog.* 10, 205–22.

Murdie, R.A., 1969 *Factorial Ecology of Metropolitan Toronto 1951–61: An Essay on the Social Geography of the City* (Chicago).

Sweetser, F.L., 1965 Factor structure as ecological structure in Helsinki and Boston, *Acta Sociol.* 8, 205–25.

P1111
Cultural convergence and divergence and the changes of city structure: a study in urban typology
B. HOFMEISTER   *Technical University of Berlin, West Germany*

This paper is an attempt to set up a model of structural city types as they have developed through time and space. It is an attempt to link the cross-cultural theory of urbanization as it has been promoted by Sjoberg, Horvath, and others with the concept of regional or cultural-genetic city types elaborated by Holzner.

The degree and kind of inner differentiation of an urban place vary with (*a*) its size,

(b) the time necessary for the differentiating processes, which is mainly a function of changing technology and land evaluation, and (c) the direction of such processes, which is dependent upon both the culture realm of which any particular city is a part and the influences from other culture realms. Here we shall concentrate on points (b) and (c). The working hypothesis is set forth that the basic criterion for the urban mosaic is the spatial relations of city folks as they depend upon land evaluation and land use patterns.

The preindustrial West European city showed a certain inner differentiation according to people's occupational and social rank. Craftsmen and merchants not only exercised control on the numbers of masters, journeymen, and apprentices but also assumed political power in city councils. A master used to live and work under the same roof with his journeymen and apprentices. The family and the occupation group formed the house community. The members of any particular guild lived next to one another in the same street 'within the precinct of the guildhall' (Vance 1971). While in river towns fishermen occupied a section close to the waterfront around a church devoted to St Peter, the Jewish group often lived in a ghetto-like quarter, and the Hansa merchants used an extraterritorial area reserved for them. The land was often owned by the community; it was used, but not possessed, by the residents.

The preindustrial New World city does not appear to have been profoundly different from its Old World counterpart so that we need not go into further details.

In the Islamic and South Asian realms, here referred to as the Third World, the pattern of terrestrial space was based on ethnic relationships. Various city sections could be identified as the residential quarters of people of similar ethnic background, creed, and language. In so far as particular occupations were reserved for certain ethnic groups, they performed guild-like functions. Each quarter's residents were more or less self-sufficient, each group having religious and cultural centres and commercial outlets of their own. 'A particular family, or an extended kinship unit, may control a given street' (Sjoberg 1960), hence the cul-de-sac street pattern which was often superimposed upon an earlier Hellenistic grid. Only the wall and the centripetal forces of the main mosque and bazaar made for some unity of such a heterogeneous place. In contrast to the European city, the bazaar as a working place was distinct from the living quarters surrounding it and the mosque.

From such backgrounds the *industrial city* of the nineteenth and early twentieth centuries developed with its differentiation of socio-economic or functional sections such as the CBD, a wholesale area, manufacturing areas, residential quarters for lower, middle, and upper class residents, and secondary commercial and administrative nuclei. Since the sixteenth century the land-rent gradient brought about an increasing segregation of land uses. Production was separated from the household. Accessibility became a decisive factor for land evaluation. The biological family became the basic unit of urban life. 'The emphasis on the individual and the biological family worked toward the extinction of ethnic identity while enlarging economic-class identity' (Vance 1971). Land was converted to real estate and evaluated as to its functional use and financial returns. The industrial city required a great deal of social contact, and a multitude of interrelations developed between various city sections.

During the nineteenth century some European nations penetrating into Africa and Asia created the colonial city by founding administrative and residential sections adjacent to indigenous towns such as the British cantonments which were later encroached upon and finally annexed by the growing cities of India and Pakistan. The essential feature of the colonial city is domination, 'where one group controls part of the affairs of another group' (Horvath 1969) and where there is an élite of exogenous origin differing from the other group in language, religion, and economic orientation. Often intervening groups developed from interracial mixing and the migration of people from a country other than the colonial country. We refer to the role of the Levantine and Indian merchants in the cities of West Africa or of East and South Africa respectively. The colonial city has, however, been a foreign body beside the Afro-Asiatic city rather than a part of it.

At the same time distinct settlements in the older sections of North America's cities developed by the influx of increasing numbers of immigrants from south and east European countries many of whom had but little command of the English language and a low level of education. This process was later reinforced by the development of the Negro ghetto. The existence of such ethnic neighbourhoods beside socio-economic sections such as the CBD or manufacturing district added an element to America's cities that had earlier characterized the cities of the Third World and that has, until recently, been unknown to the European city. After World War II this trend toward self-sufficient quarters has been reinforced by the mushrooming of residential neighbourhoods for people of nearly the same income level around large modern shopping centres, which are, indeed, the cultural and administrative nuclei of great portions of the outskirts of the American metropolis. The neighbourhoods of minority groups as well as those residential sections around the great shopping centres brought about a structural type of city somewhere between the Old World and Third World types.

In the postwar era the West European city showed a similar tendency, although to a lesser degree, with the concentration of foreign labourers and – as far as they were accompanied – of their families in the older and partly run-down sections around the CBD, this tendency making for a recent convergence toward the New World type.

In the era of trusteeships and, at an accelerated rate, since World War II many cities in the emerging nations of the Third World adopted 'western' traits of architecture, forms of organization of the economy, and features of the western way of life. Beside indigenous traditions such as the bazaar with its internal pattern of higher and lower ranking crafts, modern commercial outlets and modern residential neighbourhoods began to characterize the westernized Third World city.

In the Old World and New World cities slowly transforming toward a post-capitalist stage, the economic-class identity of the residents will no longer be predominant. 'With the assertion of socialist doctrines during the last century, the situation has been sharply reversed: class has been played down whereas the identity with a community, often ethnic in the United States and Canada, has been encouraged' (Vance 1971).

In the countries with a true command economy, i.e. the bloc of the Council of Mutual Economic Aid, certain demands of the socialist society have brought about new morphological features such as broad thoroughfares and large central places for military parades, culture parks and culture palaces for educational and recreational purposes, and public rooms for societal functions in rental housing units. Such elements make for a physiognomic uniformity within the cities of this region of the world.

Bobek, H., 1959   The main stages in the socio-economic evolution from a geographical point of view, *Die Erde,* 259–98 (German with English summary).
Friedmann, J., 1961   Cities in social transformation, *Comp. Studies in Soc. and Hist.* IV (Nov.), 86–103.
Hauser, P.H., and L.F. Schnore, ed., 1965 *The Study of Urbanization* (New York).
Hofmeister, B., 1970   Anglo-America's great cities: major characteristics, recent trends, regional variations, *Geoforum* 3, 17–29.
Holzner, L., 1967   World regions in urban geography, *Ann. Assoc. Am. Geog.* 704–12.
Horvath, R.J., 1969   In search of a theory of urbanization: notes on the colonial city, *East Lakes Geog.* 5.
Manshard, W., 1970   Some urban developments in tropical Africa, *Geoforum* 4, 63–74.
Masai, Y., 1970   The contemporary Japanese townscape, *Assoc. Jap. Geog. Spec. Publ.* 2, 97–108.
Mumford, L., 1961   *The City in History* (New York).
Redfield, R., and M.B. Singer, 1954–55   The cultural role of cities, *Econ. Devel. and Cult. Change.*
Sjoberg, G., 1960   *The Preindustrial City* (New York).
Soja, E.W., 1971   The political organization of space, *Comm. on College Geog. Resource Pap. No. 8* (Washington, DC).
Vance, J.E., Jr., 1971   Land assignment in the precapitalist, capitalist, and post-capitalist city, Festschr. f. R.E. Murphy; *Econ. Geog.* 101–20.
Wirth, L., 1938   Urbanism as a way of life, *Am. J. Sociol.* (July), 1–24.

P1112
Towards a pathology of urban environment in the developing countries
NISITH R. KAR *Government of West Bengal, India*

In recent times, much attention has been given to the question of urban growth and urbanization in developing countries of the world. Hoselitz was one of the pioneers to emphasize the role of cities in the economic growth of undeveloped countries and the process of urbanization and economic growth in Asia in general (Hoselitz 1953, 1955, 1957). The trend of urbanization in Asia and the Far East, together with all its demographic, economic, and social implications, was the central theme of a seminar sponsored by the UN Economic Commission for Asia and Far East (Hauser 1957). The role of urbanization in economic development at the international perspective was also dealt with by Hoselitz in another publication (Turner 1962). Davis has exhaustively dealt with the relations of urbanization and industrialization with particular reference to the preindustrial and underdeveloped countries of the world (Davis 1962). On a regional basis, several authors have focused their attention on urbanization problems of southeast Asia (Chabot 1964; McGee 1967), and some have dealt with urban growth and urban explosion in Latin America and consequent social and human problems. There have also been various studies on patterns of city growth and trends of urbanization in individual countries of the underdeveloped world, such as on the Philippines (Spencer 1958; Ullman 1960), Malaya (Hamzah 1962; Caldwell 1963), India (Davis 1962; Hoselitz 1962), Indonesia (Wertheim 1964; Milone 1966), tropical Africa (Steel 1960), and Nigeria (Mobegunje 1969). On the question of the changing role of large-sized and primate cities in the process of the modernization-industrialization of developing countries, numerous studies exist such as the pioneering work on primate cities of SE Asia (Fryer 1953; Ginsburg 1955), cities of China (Murphey 1955, 1957), Indonesian cities (Heeren 1955; Keyfitz 1961; Withington 1962), Malaysian cities (McGee 1963; Wikramtelike 1965), Nigerian cities (Mobegunje 1968). More recently systematic studies have been carried out towards the formulation of typology of urbanization and city growth together with their attendant urban problems in the underdeveloped countries (Ginsburg 1965; Breese 1966, 1969; Beaujeu-Garnier 1970).

In spite of such a wide understanding of the historical process of urbanization and the characteristic pattern and form of cities and their role in the economic development of the developing countries, no systematic study of the character and quality of the urban environment has yet been undertaken. Though there are admirable accounts of density of population, housing congestion, traffic concentration, and problems of civic services and amenities in cities of Asia, Africa, and Latin America in the various reports of tcwn planning and social surveys, there is ample scope for undertaking a study of the quality of urban environment in these countries, in the context of the economic processes and social behaviour for the conditioning of such typical urban habitat. While in the industrially developed countries much attention has recently been focused on the qualitative aspects of urban environment as a resource for man (Perloff 1969), no such serious attention has so far been given by geographers towards study of urban environmental quality in the developing countries of the world. Cities in the developing world have often been christened as deplorable dens for human beings to which a large number of country folk rush in search of shelter and sustenance. In countries of low level of urbanization-industrialization, these cities manifest signs of overurbanization (Breese 1966). The qualities of the urban environment in these countries are thus marked by inhuman congestion, substandard housing, widespread pollution, lack of sanitation and effective sewage, ill-planned physical layout, and the hopeless inadequacy of the civic authorities in the functions they are called upon to discharge. Absence of any marked physical expansion and lack of healthy suburban growth are an added feature of these urban areas of underdeveloped countries. In fact such conditions are so much the characteristic symptoms of urbanization in developing countries that they may be

TABLE 1. Urban environment, ethology

	Physical conditions	Living conditions	Service conditions	Health, sanitation, and pollution
SYNDROME	Highly congested cities and suburbs. Sprawling urban areas. Narrow streets and lanes. Heavily congested transport system (intracity and intercity). Lack of open spaces, recreation grounds, etc.	Extreme physical congestion in housing. Abnormally high population density. Large number of persons living per room. Dilapidated housing and poor sanitation. Paucity of sun and air. Mixture of living and work places.	Inadequacy of good water supply, electricity, sewage, etc. Deficient transport facilities. Paucity of marketing, school, and recreational facilities. Low standard of municipal services.	Overall polluted working conditions, congestion, insanitation and polluted water supply leading to epidemics, infection, endemic diseases. High mortality in slums due to airborne dust, soot, gases, leading to respiratory and eye diseases.
TECHNOLOGICAL FACTOR	Lack of technology to effect urban decentralization and physical expansion of cities. Poor technology for transport, sewage, water supply, utilities, etc. inhibiting suburban growth.	Lack of technology to produce mass housing at cheapest possible cost. Problem of producing low-cost housing materials.	Poor technological equipment for civic services, transport, communication, recreation, etc.	Location of workshop, factories, etc. within residential areas. Absence of modern fume-, gas-, soot-control techniques. Uncontrolled industrial waste. Lack of scientific investigation.
ECONOMIC FACTOR	Lack of financial resource to develop rapid mass transport system or provide utilities in suburban areas. Low tax resources inadequate for cost of land acquisition, city planning, urban renewal, suburban expansion, etc.	Lack of financial resource to create mass housing, replace slums, and undertake urban renewal. Inability of slum-dwellers and low income class to pay for the minimum cost of rehousing. Middle class unable to pay high rentals and undertake renovation and maintain housing standard.	Inadequate tax resource to cope with soaring demands of city service. Poor financial resources to meet the heavy costs of municipal services, repairs, and equipments.	Lack of funds for changeover to non-polluting industrial techniques. Limitation of resources for disposal and utilization of industrial wastes.
CULTURAL AND BEHAVIORAL FACTOR	Tendency to huddle together and move by walking or by animal-drawn or human-drawn carriages. No respect for distance and speed in life.	Tendency to live together or in big extended families. Habit of living in congestion without much space or privacy. No liking for separation of living and work places for low income group. Rural mode of life in urban areas by migrants.	Laxity in tax collection. Delinquency in tax payment. Migrants, slum dwellers, floating population not contributing to city coffers but using them. Tendency to live under inadequate utilities, poor sanitation, and without recreational facilities.	Apathy towards harmfulness of polluted air and water. Use of coal, kerosene, and cow dung as domestic fuel producing soots, fumes, coal dust. Lack of concern and awareness of health hazards due to pollution.

termed as 'syndrome.' On the other hand, there are inherent physical constraints, conditioning economic forces and limiting social-cultural behaviour which have given rise to such unfortunate symptoms in urban environment of these countries and these may be called the etiology of such environment (Table 1).

In many such Asian and African cities, the physical site and situation have been very disadvantageous, discouraging physical expansion but impossible to change because of poor technology. These cities are also the vortex of economic activities in a country of predominant agricultural society, but under the spell of vigorous industrialization. There is always a lag between the pace of industrial growth and the expansion of agricultural base and population explosion, thereby leading to the tragic process of overurbanization, hyperurbanization, or pseudourbanization. Over and above, the predominance of the rural way of life of the majority of the country's population, rampant migration from rural habitat to urban areas for economic reasons, the stupendous unemployment problem facing the urban proletariat and white collar class, not to speak of a large floating population in metropolitan cities, create conditions of poverty, privation, and misery in every sphere of urban life in these parts of the world. The lack of financial resources and capital for meeting the cost of providing adequate civic services, slum clearance, urban renewal, etc. is another feature of the overurbanized cities. The standard of living and demands, the life-style and cultural behaviour of a large section of uprooted rural folk and of unacculturated urban squatters, slum and pavement dwellers thus create an environment equivalent to that of an 'expanded village' rather than true urban habitats. All these physical, technological, economic, and sociocultural factors interplay symbiotically to give rise to the poor quality of urban environments in these countries. The cause of such an unwholesome urban environment and its effects on the quality of life of urban dwellers, which themselves contribute towards the aggravation of urban maladies, are intertwined in a kind of vicious cycle; this cycle may be called the pathology of urban environment in the underdeveloped world (Table 1).

Beaujeu-Garnier, J., 1970 Large overpopulated cities in the underdeveloped world, in *Geography and a Crowding World* (London).
Breese, G., 1966 *Urbanization in Newly Developing Countries* (New Jersey).
– 1969 *Cities in the Newly Developing Countries* (New Jersey).
Bulsara, J.F., 1964 *Problems of Rapid Urbanization in India* (Bombay).
Chabot, H.T., 1964 Urbanization problems in south-east Asia, *Trans. V World Cong. Sociol*. 3.
CMPO, 1966 *Basic Development Plan, Calcutta Metropolitan District 1966–1986* (Calcutta).
Davis, Kingsley, 1965 The urbanization of human population, *Sci. Am*. 213 (3), 41–53.
Ginsburg, N.S., 1965 Urban Geography and 'non-western' areas, in the study of Urbanization.
Hauser, R., ed., 1957 *Urbanization in Asia and Far East*, Economic Commission for Asia & Far East, 1957 (Calcutta).
Herbert, J.D., and A.P. Van Huyck, 1968 *Urban Planning in the Developing Countries* (New York).
Hoselitz, B.F., 1953 The role of cities in the economic growth of underdeveloped countries, *J. Pol. Econ*. 61 (3), 195–208.
– 1955 Generative and parasitic cities, *Econ. Dev. & Cult. Change* 3 (3), 278–94.
– 1957 Urbanization and economic growth in Asia, *Econ. Dev. & Cult. Change* 6 (1), 42–54.
IIPA, 1960 *Problems of Urban Housing* (Bombay).
McGee, T.G., 1967 *The South East Asian City* (London).
Murphey, R., 1954 The city as a center of change: W. Europe and China, *Ann. Assoc. Am. Geog*. 44, 349–62.
– 1957 New capitals of Asia, *Econ. Dev. & Cult. Change* 5 (3), 216–43.
Perloff, H.S., ed., 1969 *The Quality of the Urban Environment* (Baltimore).
Sen, S.N., 1960 *The City of Calcutta* (Calcutta).
Stokes, Ch. J., 1962 A theory of slums, *World Econ*., Aug., 187–97.
Turner, Roy, 1962 *India's Urban Future* (Berkeley).

P1113
Different forms of semi-urbanization processes in Poland
MARIA KIELCZEWSKA-ZALESKA   *Polish Academy of Sciences*

Not published because of length.

P1114
The lorry or motor park: an element of the land use pattern in West African cities
E.C. KIRCHHERR   *Western Michigan University, USA*

Approaches to the study of urban land use rely on familiarity with Western cities, and the geographer frequently overlooks or fails to appreciate the functional role of special land uses in other cultural realms. One such land use which has been given scant attention in the literature is the 'lorry park' or 'motor park' of West African urban places. This land use evolved in connection with the developments in road construction and the use of motor transport in developing countries, and the emergence of privately controlled, non-scheduled lorry (truck) services. The lorries perform as all-purpose carriers for the transport of people, agricultural produce, manufactured goods, and, not uncommonly, even for live animals. For a few decades, the lorries have provided relatively cheap and readily available transport throughout large parts of the West African countries. The colourful 'mammy wagon' has been supplemented, if not superseded, by modern vans and buses, still operating as unscheduled carriers, but at least offering a more comfortable ride to passengers.

As the lorries do not follow fixed schedules and are not restricted as to the number of stops that can be made between major cities, passengers and goods are picked up at almost any point en route. Even small villages along principal roads provide a roadside parking area. In the larger cities and towns, lorries park at a number of convenient places where space is available, e.g. along major streets, at petrol stations, or at small markets. Although these unofficial stopping points are not designated on maps, their locations are known to local residents and to those who visit a town frequently. This study, however, is concerned with the larger units of land that have been officially designated as off-street parking areas, where lorries normally pick up and discharge passengers and goods. Seeking to derive a clearer understanding of the nature and functional role of this land use, a survey of lorry parks was undertaken in thirteen Nigerian cities, all of which, with one exception, were located in the Western state. What soon became apparent was that students of land use, seeking to explain the lorry park by analogy, could be misled if they compared this land use with terminal facilities associated with highly structured systems of scheduled transport in advanced countries.

The survey data revealed some general characteristics of the Nigerian lorry parks. First, it was found that all were publicly owned sites, with fee collection and maintenance the responsibility of a local government body. Secondly, despite variations in the size and activity of lorry parks, a standard fee of two shillings was charged to lorry drivers, but the parks were supervised and fees collected only during daylight hours. Thirdly, the lorry park proved to be a point towards which other functions and services gravitate. Food hawkers are usually found in the vicinity of the park, selling food and drinks to tired and hungry passengers who may have been travelling for many hours. In the larger lorry parks, stalls have been constructed and are rented to small traders. Lastly, the large open lots used as lorry parks are poorly maintained.

In some towns, lorry parks may be a plot of less than one-third of an acre in area where there are seldom more than three lorries at any one time. In contrast are other lorry parks distinguished not only by their size – the largest measured in this study was two acres – but also as foci of activity. A good part of this activity can be attributed to the concentration of lorries, drivers, and passengers. The lorry passenger may have arrived realizing that he will have to stop off until he can find another lorry proceeding to his final destination. This means inquiries will have to be made among drivers, and bargaining about the fare when a lorry is found. If none of the parked lorries is making a trip to his destination, the traveller simply waits until one arrives; in small towns, this may entail a lengthy layover.

In the developing nations of West Africa, there is a constant flow of commodities between port cities and urban centres of the interior. Not surprisingly, the lorry park often functions as a collection centre and market, where goods and produce are being bought, sold, and transferred. This is evidenced in the stacks of goods awaiting sale and transshipment, which often give the park the appearance of a freight depot. The lorry park has become an integral component in the complex system of commodity flows involving numerous buyers, middlemen, and petty traders.

Although data were compiled on lorry parks in only thirteen cities, it can be suggested that there does not appear to be a strong correlation in the size and activity of a lorry park with the population size of the urban centre. Some small towns located at or near road junctions have large lorry parks with a considerable amount of traffic, whereas some large cities with 50,000 or more population but a limited economic base generate little lorry traffic.

The investigation also attempted to ascertain the possible association of the lorry park location with the principal foci of the towns of southwestern Nigeria, namely, the oba's (chief) palace and the traditional market. Almost all of the lorry parks were found contiguous, or in close proximity, to the palaces and markets. This appeared especially true in urban centres with major markets where large numbers of visitors want to terminate their journey as close as possible to the market.

This interim report has attempted to identify some salient features and the functional role of the lorry park. The findings should be regarded as tentative and suggestive, being based on a survey of this special land use in thirteen urban places in southwestern Nigeria; substantiation or modification will necessitate additional study of lorry parks not only in other regions of Nigeria but in other countries of West Africa. Moreover, as more data is compiled, quantitative methods should be employed to secure more precise statements with regard to features and associations noted in this study. The results of continued research on the lorry park should prove valuable to both local planners and national transportation administrators in the developing states of West Africa.

Students, University of Ife.

African Urban Notes (Quarterly), African Studies Center, Michigan State University.
Buchanan, K.M., and J.C. Pugh, 1958 *Land and People in Nigeria* (London).
Hance, William A., 1970 *Population, Migration, and Urbanization in Africa* (New York).
Hodder, B.W., and U.I. Ukwu, 1969 *Markets in West Africa* (Ibadan).
Lystad, Robert A., ed., 1965, *The African World: A Survey of Social Research*, Frederick A. Praeger, New York.
Mabogunje, Akin L., 1968 *Urbanization in Nigeria* (London).
Morgan, W.B., and J.C. Pugh, 1969 *West Africa* (London).
Nigeria: Haring after greyhound, 1971 *Newsweek* 6 Sept., 58.
Ojo, G.J.A., 1966 *Yoruba Culture* (London).
Udo, Reuben K., 1970 *Geographical Regions of Nigeria* (London).

P1115
Residential morphology and urban social structure
L.D. MCCANN and PETER J. SMITH   *University of Alberta, Canada*

Recent factorial ecologies of American and Canadian cities have postulated three summary constructs of urban social structure: namely, socio-economic, family, and ethnic status. These constructs are implicit in the classical models of urban growth and structure. Also implicit is the relationship between social structure and morphology, the physical fabric of the city. But this relationship has seldom been tested in a spatial context, let alone proved (Herbert 1967).

The relationship between residential morphology and social structure has been analysed in the 1951 built-up area of South Edmonton. This area was chosen because of the need to set reasonable limits on data collection time and also because the area had established its character by the 1961 analysis date, showing both change and stability in its subareas. The population in 1961 totalled over 60,000.

Social data were derived from the 1961 census, while morphological data were re-created and drawn for the same year from property assessment files, using a 20 per cent random sample of residential properties, stratified by census enumeration areas. Data sets of 80 social and 50 morphological variables for 100 enumeration areas were each subjected to a principal components analysis with varimax rotation.

Five components, resolving nearly 60 per cent of the total variance in the data, were identified. In order of importance, they are socio-economic status (23 per cent), family status (22 per cent), clerical occupations (5 per cent), transient population group (5 per cent), and occupancy status (4 per cent). The two major components correspond with the findings of other studies (Rees 1971). No distinct ethnic status component was revealed but, as in Winnipeg (Nicholson and Yeates 1970), variables relating to recent immigrants and non-British persons loaded highly (positively) with low socio-economic status, whereas variables characterizing persons of British background loaded highly (negatively) with high socio-economic status. The isolation of the minor components reflects the unique nature of the input variables and the study area.

The housing variables were grouped under residential land use; property and house size; age, life expectancy, and period of house construction; house types and architectural styles; original construction quality, maintenance, and present condition; basement characteristics; sanitary facilities; depreciation factors; and building value. These variables were reduced to five principal components which together accounted for 61 per cent of the data variance. They have been termed housing age (24 per cent), housing quality (17 per cent), housing size (9 per cent), housing condition (7 per cent), and land use transition (5 per cent). Because no other factor analyses of detailed morphological data are known to the authors, these components will be briefly summarized.

Under housing age, high positive loadings were recorded for such variables as average physical depreciation, average age of houses, older architecturally styled homes, houses built before 1920, one-and-three-quarter and two-storied house types, and houses that have been converted to multiple-family occupancy. High negative loadings were obtained for modern architecturally styled homes, houses built after 1949, bungalows, and single-family residences.

Housing quality distinguishes areas on the basis of original construction. Positive loadings are characteristic of houses of fair construction quality, with full and high ceilinged basements, and valued between $12,000 and $23,000. In contrast, negative loadings are identified with houses which are small, typically low-valued cottages, and of poor construction and surface foundations.

Housing size can be interpreted mainly from high negative loadings which include the average size of houses, properties, and lot frontages; properties over 7500 square feet; houses with ground floor areas greater than 1000 square feet; and homes valued over $23,000. Positive loadings reflect smaller-sized housing.

Housing condition distinguishes areas with housing in need of minor and major repairs (high positive loadings) from areas with housing in good condition (high negative loadings).

Land use transition is characterized by dwellings with shared bath and toilet facilities, apartment units, and different types of converted houses with basement, upstairs, or multiple suites (high positive loadings). Only single-family houses achieved a moderately high positive loading.

To what extent, as implicitly recognized in theories of urban structure, is there a relationship between the physical fabric and the social mosaic of the city? Comparison of the mapped component scores for the two data sets revealed that there were definite spatial relationships. For example, family status and housing age showed similar concentric patterns, while socio-economic status and housing quality displayed corresponding sectoral distributions. For these pairs of components, then, a high degree of areal correlation might be expected. In contrast, the spatial patterns of the minor components were not always so clear. In a number of cases, components showed tendencies towards both concentric and sectoral pattern-

TABLE 1. Contingency coefficients: residential morphology and urban social structure

	Socio-economic status	Family status	Clerical occupations	Transient population	Occupancy status
Housing age	0.20*	0.47*	0.24*	0.01	0.10
Housing quality	−0.37*	0.16	−0.15	−0.11	0.13
Housing size	0.19*	−0.08	−0.10	−0.02	0.17
Housing condition	0.01	0.12	0.01	−0.02	0.04
Land use transition	0.28*	0.33*	−0.03	0.16	0.29*

*Significant at 0.05 level.

ing (housing size and land use transition) or towards clustering (housing condition and occupancy status). In such cases, only low or moderate correlations might be expected between these and other components.

To correlate the spatial relationships between the social and morphological components, the enumeration areas were first dichotomized according to whether their scores for a particular component were positive or negative. Then for each possible pair of components between the two data sets, areas of positive and negative scores were compared and similarities and differences recorded. That is, twenty-five 2 × 2 frequency tables were set up and for each the contingency coefficient (C) was calculated. The maximum value C can attain in a 2 × 2 table is ±0.707; thus, resulting correlations must be interpreted with this upper limit in mind (Table 1).

The strongest relationship is between family status and housing age ($C = +0.47$). Areas of older housing in which many houses have been converted to multiple-occupancy coexist in concentric fashion with areas of non-family households, families with few or no children, and an older population base. Conversely, newer housing districts of single-family and bungalow development are associated with areas comprised of larger families and young children.

The correlation between socio-economic status and housing quality ($C = −0.37$) suggests how the quality of building construction in a neighbourhood influences socio-economic patterns. The negative value is the result of people of higher socio-economic status shunning a sector of poorer-quality wartime housing. Further, lower socio-economic status coincides spatially with land use transition ($C = +0.28$), older housing ($C = +0.20$), and smaller-sized housing ($C = +0.19$). Of course, converse relationships between higher socio-economic rank and land use stability and newer and larger housing can be stated.

Areas of land use transition are clearly related to areas of non-family status ($C = +0.33$) and to areas of rented dwellings with short-term occupancy rates ($C = +0.29$). Significantly, the sectoral patterning of housing size relates more closely with socio-economic status ($C = +0.19$) than with family status ($C = −0.08$). Note, too, that housing condition, which displays a clustered pattern, correlates only marginally with the social dimensions.

This study has shown that the social and housing structure of an urban area can be summarized by several basic constructs. The social structure components which are revealed are consistent with previous findings. Of more importance, what are apparently major constructs of residential morphology have been identified. That these constructs show meaningful spatial relationships with the social dimensions provides additional evidence for building a general theory of urban structure.

Real Estate Research Committee, University of Alberta; Dianne Dodd.

Herbert, D.T., 1967 The use of diagnostic variables in the analysis of urban structure, Tijd. voor Econ. en Soc. Geog. 58, 5–10.
Nicholson, T.G., and M.H. Yeates, 1970 The ecological and spatial structure of the socio-economic characteristics of Winnipeg, 1961, Can. Rev. Sociol. Anthropol. 7, 163–78.
Rees, P.H., 1971 Factorial ecology: an extended definition, survey, and critique of the field, Econ. Geog. 47 (Suppl.), 220–33.

P1116
Aspects of change in the traditional retail structure of cities in developing nations
M.L. MCNULTY *University of Iowa, USA* and
PIUS O. SADA *University of Lagos, Nigeria*

Cities in developing nations are undergoing considerable transformations both functionally and structurally. A major aspect of such transformations has been the growth of the retail structure to accommodate the efficient distribution of goods and services to the rapidly expanding urban populations. The traditional market place has been a conspicuous feature of urban retail structure in most developing nations. These markets represent modes of exchange which existed in the industrial cities and rural areas and which persisted in the growing urban centres. Whether these traditional elements of retail structure can efficiently function within a rapidly urbanizing environment is a question of considerable practical importance to administrators and planners in developing nations; more so, since they are faced with increasing competition from established 'modern' retail shops and shopping centres. While a number of useful studies have examined the impact of development on rural marketing systems, few have attempted to assess the functions of such markets within an urban context. The discussion below is intended to outline certain elements of a model designed to explain the likely consequences of urbanization for these traditional components of retail structure.

Urbanization amounts to a rapid intensification of the marketing landscape resulting from (i) an increase in the density of households and (ii) an increase in the degree of household participation in the marketing process (Skinner 1965, 208). These conditions result from the processes of rural-urban migration and the transformation of labour from self-sufficiency to wage earning and consuming. The impact of this intensification upon the *functional, locational,* and *temporal* features of traditional markets will be assessed.

Markets serve a variety of economic and social functions. Herein we are primarily concerned with the economic functions performed where markets are viewed as points of spatial articulation in the collection and distribution of goods and services. A number of important questions concerning the changing functions of these markets must be considered. What are the original functions performed and how are they expected to change? What types of changes may be expected in the number and variety of goods being offered? With the intensification of the marketing landscape, which commodities are likely to receive strong competition from the more fixed retail establishments?

Initial functions of the markets generally are confined to the provision of low order goods and services, especially foodstuffs, to a limited trade area. A wider variety of goods and services begins to be offered once demand density, to use Stine's term, increases to a point where the minimum range (threshold) is achieved (Stine 1962, 77). Demand density, defined as a function of population density and level of disposable income, increases as urbanization proceeds. The initial effect of this increase is that lower order goods are the first to be offered in the markets. Gradually, higher order goods are added. However, owing to their generally higher threshold and capital requirements, higher order goods are the first to seek locations outside the market in fixed shopping districts. Thus one may posit that higher order goods are the last to enter the traditional markets and the first to disappear as intensification of urban demand proceeds, while lower order goods are both first to enter and last to be displaced. It is also likely that such functional change will be spatially differentiated with markets located near the centre of the city being the first to be affected.

A second major consideration deals with the location and spacing of the markets. These, of course, are not unrelated to the total number and size of the markets. Initially, each section of the city may have a small local market aimed at the convenience of residents and benefitting from the generally low mobility of consumers. However, these markets will increasingly begin to compete with one another as the transportation system, and therefore the mobility of the consumer, increases. This provides the consumer with the opportunity of becoming more discriminating in his selection of goods and of possible sites for satisfying his wants. This means that markets which have some type of competitive advantage, locationally or otherwise, will gain customers at the expense of other, less advantageously located markets. Thus, intensification of the marketing landscape together with an improved transportation system means that the number of markets is gradually reduced and their size increased.

Moreover, this reorganization of the pattern of markets must be viewed in relation to the location and growth of established shopping centres within the urban area. The increase in demand density is greatest at the centre of the city and declines toward the periphery. This means that initial changes in retail structure take place in the central parts of the city while the periphery is affected much later. It is therefore possible that while certain markets are declining or even becoming defunct near the centre of the city, new markets may still be established at the periphery. Such changes in the pattern of markets are also conditioned by changes in the population distribution of the urban area. The rapid expansion of the urban area may outrun the ability of the marketing system to adjust resulting in a disequilibrium between the distribution of population and that of markets. The result could be that older areas have several markets, generally underused, while newer areas are ill-served because new markets have not developed. Here, of course, the role of the local urban administration is crucial. If the administration views the existence of markets as an anachronism it may actively discourage their development and improvement.

Urban markets in many developing countries initially have been periodic owing partly to the generally low levels of demand together with the imposition of traditionally rural marketing schedules upon the urban areas. One consequence of urbanization is that the timing of the markets is altered. As demand density increases, and as occupational specialization results in full-time traders, periodic markets change their schedules in order to meet more frequently. Thus, frequency of meeting increases and after some time periodicity ceases. It is unlikely that such a change in market schedules will come about abruptly, except in those cases where some administrative decision regulates the days of meeting. Rather, the volume of trading on days preceding and following market day gradually increases making it difficult to distinguish the actual market day from non-market days. Intensification of the marketing landscape gradually leads to the development of daily markets.

It must be noted that these changes in traditional markets are highly interrelated, so that the functional, locational, and temporal changes envisaged are seen to be part of the general dynamic process of change in the urban retail structure of developing nations. Empirical investigations currently being conducted in Lagos, Nigeria, will provide the basis for testing certain of the hypotheses suggested by the model outlined in this paper (Sada and McNulty 1970).

Rockefeller Foundation.

Sada, P.O., and M.L. McNulty, 1970   Market structure in Metropolitan Lagos, 15th Ann. Conf. Nigerian Geog. Assoc., U. Ife, Nigeria, Dec. 18–23 (Mimeo).
Skinner, G.W., 1965   Marketing and social structure in rural China, Part II, *J. Asian Studies* 24(2), 195–228.
Stine, J., 1962   Temporal aspects of tertiary production elements in Korea, in F.R. Pitts, ed., *Urban Systems and Economic Development* (School of Business Admin., U. Oregon).

P1117

Comparative changes in the localization of manufacturing in megalopolis: 1954–67
E. WILLARD MILLER   *Pennsylvania State University, USA*

Scholarly interest in the trends of industrial location has gained considerable attention in geographic and economic literature in recent years. The increase in locational studies reflects a growing recognition that the solution of many industrial problems requires an understanding of spatial changes of industrial activities. The purpose of this study is to investigate the concentration and comparative changes in the localization of manufacturing

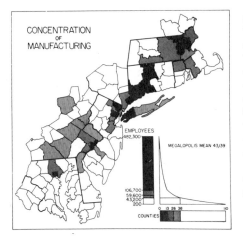

Fig. 1.

in the east coast megalopolis of the United States. The eventual goal of such an inquiry is to achieve a framework of analysis in order to raise pertinent questions, and, if possible, arrive at certain decisions for the organization of industrial spatial patterns.

Manufacturing is highly concentrated in megalopolis. If manufacturing were evenly distributed in 1967, each of the 110 counties would have had 43,139 employees. In contrast, county manufacturing employment varied from 200 to 482,300 employees. There were 38 counties above the mean employment and 72 counties below the mean (Fig. 1). The 38 counties above the mean had 3,904,600 employees, or 82.2 per cent of the total. Further, the 13 counties in the top one-third above the mean had 2,200,700 employees, or 46.8 per cent of the megalopolis total.

The New York metropolitan area, consisting of 14 counties that lie above the mean, had a manufacturing employment of 2,299,800, or about 48 per cent of the regional total. The Philadelphia, Boston, and Baltimore metropolitan areas were the next largest.

These four metropolitan areas had about 72 per cent of the manufacturing employment of megalopolis. Smaller concentrations were located in Providence, Rhode Island; New Haven, and Hartford, Connecticut; and a concentration of manufacturing extended from Allentown to Lancaster and York in southeastern Pennsylvania.

COMPARATIVE CHANGE

The technique of comparative change is not concerned with absolute change over time, but with the relative change in one area as compared with others. The analysis is based on the actual manufacturing data for each county for 1954 compared with a theoretical figure showing the amount of manufacturing the county would have had if it had grown at the same rate as megalopolis between 1954 and 1967. If the county grew more rapidly than megalopolis, it is said to have experienced a 'comparative gain'; if it grew less rapidly, it is said to have experienced a 'comparative loss.' A comparison of the actual 1967 figures with the theoretical 1967 figures for the counties gives the magnitude of the growth or decline.

Between 1954 and 1967 employment in manufacturing in megalopolis increased from 4,343,000 to 4,745,000 or an increase of 9.25 per cent. Thus if each county was to maintain its position relative to megalopolis the 1967 manufacturing employment in each county had to be 109.25 per cent of that of 1954. Of the 110 counties, 74 experienced a comparative gain and 36 experienced a comparative loss. The comparative change ranged from a gain of 34,021 to a loss of 89,770.

The counties that experienced a comparative decline are essentially the largest and oldest manufacturing areas of megalopolis (Fig. 2). The New York metropolitan area, with six core counties with a comparative decline, had a loss of 208,800 employees, or

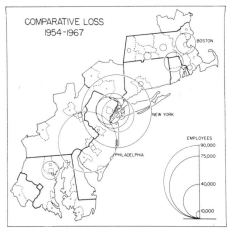

Fig. 2.

Fig. 3.

about 48 per cent of the region's total. The core counties of the Philadelphia, Boston, and Baltimore metropolitan areas also experienced large comparative declines. These four regions had a comparative loss of 344,195 employees, or about 81 per cent of the total comparative decline in megalopolis.

The counties with a comparative gain are predominantly grouped around the counties that experienced the greatest comparative decline (Fig. 3). In the New York metropolitan area 11 counties of comparative gain surrounded the region of comparative loss. These 11 counties had a comparative gain of 159,186 manufacturing workers, or about 38 per cent of the total regional comparative growth. The Philadelphia, Boston, and Baltimore metropolitan areas exhibited the same characteristic pattern of decline in the core and growth in the surrounding areas. Of the comparative growth regions Connecticut, southeastern Pennsylvania, and eastern Maryland have a number of counties that lie outside the great metropolitan areas.

The spatial patterns of industry are undergoing marked changes in megalopolis. The attraction of industry to the largest metropolitan regions of megalopolis remains evident, but the industrial growth is not being concentrated in a relatively small core area. Rather, as the metropolitan areas increase spatially, industrial growth is concentrating in the developing areas.

As a result of this study a number of questions arose which are worthy of study: (1) what are the factors influencing both comparative growth and decline within metropolitan complexes? (2) Has the industrial structure of the metropolitan areas changed when both comparative growth and decline are experienced? (3) Is there a relationship between the magnitude of the comparative growth or decline and the infrastructure of specific subregions? (4) What accounts for the larger area of comparative decline in the northern portion of megalopolis and the greater areas of comparative growth in the central and southern portion of megalopolis? (5) How has the change in the spatial patterns of manufacturing affected other economic and cultural activities?

These are only a few of the questions which require study, not only in megalopolis but in other industrial regions of the nation. If we are to formulate developmental industrial policies they should be based on careful research that recognizes the dynamics of the spatial patterns of manufacturing.

Bucks County (Pa.), 1965 Industrialization with order, *Industrial Development and Manufacturers Record* 134, 41–56.

Estall, R.C., 1964 Planning for industry in the distressed areas of the us (examples in New England), *Town Planning Inst. J.* 50, 390–6.

Estle, Edwin F., 1967 The region's [New England] roving industries: industrial dispersion within the region since World War II, *New England Business Rev.* 2–12.

Fellmann, Jerome D., 1967 Industrial problems and prospects of the central city, *AIDC (American Industrial Development Council) J.* 2, 1–16.

Gottman, Jean, 1961 *Megalopolis* (New York).

Herrera, Philip, 1 June 1967 That Manhattan exodus, *Fortune* 75, 106–9.

Newman, Dorothy K., 1967 The decentralization of jobs: job opportunities multiply in the suburbs, *Monthly Labor Review* 90, 7–13.

Zschock, Dieter, ed., 1969 *Economic Aspects of Suburban Growth: Studies of the Nassau-Suffolk (NY) Planning Region*, Economic Research Bureau, State U. New York at Stony Brook.

P1118
Urban dimensions of selected Indian cities and towns
DEBNATH MOOKHERJEE Western Washington State College, USA

In view of the massive increase in the urban population in India over the 1951–61 decade, because of heavy urban-industrial concentration in the big cities, a great deal of research interest has been focused on the problems and prospects of urbanization in India. The class I centres (cities of 100,000 population and over) have shown by far the highest growth rate, and have received the most attention in a number of studies in recent years (e.g. Ahmed 1965; Berry and Spodek 1971).

An examination of the urban growth pattern in India reveals that along with an intergroup differential, the urban centres also exhibit a wide range of variations in their rates of growth within groups of the same size, and it is believed that a look at the basic similarities or differences in the structural patterns of these cities and towns with different growth rates will be of interest to social scientists and urban planners.

In this paper a modest attempt is made to identify the structural patterns of two groups of urban centres selected on the basis of availability of data, (a) fast growing and (b) moderately growing, classified according to their individual rates of growth over 1951–61, as compared to the national average growth rate for the same class size. The major purpose of the research is two-fold: to examine the dimensional characteristics of the two groups and to analyse the similarities and diversities between the groups with the underlying assumption that growth is associated with structure and that, therefore, the patterns in the structural characteristics between the groups are also likely to differ.

The analysis is based on 44 urban places of 50,000 population and over, using a set of 17 occupational, demographic, and land-use variables (*Census* 1961a; Town and Country Planning Organisation 1968, 188–90). The occupational variables are calculated for male workers in the total labour force and in primary, secondary, and tertiary sectors as a percentage of the total population. Literacy refers to male literates as a percentage of the total population, and the sex ratio indicates the female-to-male ratio. The land-use variables are calculated as a percentage to total land; and the total population divided by the total, the developed, and the residential land yields the density variables. A principal components analysis was performed to discover the fundamental dimensions of urban places in the two groups, using varimax rotation procedure to satisfy a simple structure solution. The data were assumed to be linear and no transformation was made.

The preliminary data prepared for factor analyses on the two groups indicate that, on an average, group (b) is lower in population size, literacy rate, total labour force, and employment in the tertiary sector with a higher proportion of developed and residential land and a lower density, particularly in the residential area.

The dimensional structures between the groups do not reveal any substantial difference (Table 1). Dimension one in group (b) is a 'mirror image' of the first dimension in group (a), both indicating a polarization between industrialization and congestion variables on the one hand and primary occupations on the other. These dimensions may be identified as industrial congestion. The highest negative and positive factor scores for the densely populated, predominantly industrial cities in group (a) (e.g. Howrah, Kamarhati, Bangalore, Coimbatore) and in group (b) (e.g. Bhatpara, Kanchrapara, Kanchipuram), respectively, are consistent with the findings in these dimensions. (To save space, the factor scores are not presented in this paper.)

The second dimension in group (a) indexing a high literacy level coupled with high positive loadings in labour force and tertiary occupations and a low female-male ratio is generally comparable to the third dimension in group (b); these may be termed as commercial dimensions. Noticeably, in accordance with the lower proportion of workers in the tertiary sector in group (b), the commercial dimension in this group accounts for a lower proportion of total variance than in group (a). It is to be noted that along with the cities characterized by a high level of employment in service sectors such as Gauhati and Muzaffarpur, also the cities

TABLE 1. Component structures: Indian cities and towns

	I		II		III		IV		Total
(a) Fast growing									
Primary	0.73		Literacy	0.84	Commercial area	0.87	Residential area	0.93	
Secondary	−0.89		Labour Force	0.72	Developed area	0.75	Residential density	−0.75	
Industrial area	−0.85		Tertiary	0.72	Roads and streets	0.67			
Developed area density	−0.62		Sex ratio	−0.79					
Density	−0.56								
Per cent variance	32			18		15		11	76
(b) Moderately growing									
Density	0.95		Developed area	0.95	Tertiary	0.85	Sex ratio	0.74	
Developed density	0.75		Residential area	0.94	Literacy	0.69	Commercial area	0.58	
Industrial area	0.74		Commercial area	0.63			Roads and streets	−0.61	
Residential density	0.69		Residential density	−0.61			Population	−0.51	
Secondary	0.68		Developed density	−0.51					
Labour force	0.50								
Primary	−0.76								
Per cent variance	34			16		12		10	72

that have been classified in the Census as monofunctional with predominant activities in the industrial sector (Census 1961b, maps 54, 55) such as Howrah, Kamarhati, and Coimbatore, have high positive factor scores in group (a). The existence of a large external economy generated by the industrial population, as well as a high level of literacy and total labour force, may be partially accountable for these high scores. Cities high on the scale in group (b), such as Puri, Kottayam, Ranchi, Udaipur, are also the ones with service as the 'predominant form of activity.' Consistent with the findings of the TCPO, although an association between secondary occupation and industrial land use is clearly discernible in the industrial-congestion dimensions, no such pattern is revealed between commercial land use and employment in the tertiary sector (TCPO 1968, 168, 172).

The third and fourth land-use dimensions in group (a), indicating a high proportion of commercial and developed land along with an extensive network of roads and streets, and a residential land-use-density pattern respectively, are somewhat comparable to the second dimension in group (b). Interestingly, cities with the highest positive scores in the second dimension in the latter group, and the two cities with highest scores in dimension 4 in group (a), indicating a high proportion of residential area in their land-use patterns, are located in Kerala. In the light of the rather unique garden-enclosed residential settlement pattern of Kerala, this is not unexpected (TCPO 1968, 166).

Since the present analysis is based on only 44 observations and 17 variables, it serves as a limited introduction to a much-needed detailed study of the structural characteristics of India's urban centres with differential growth patterns. However, despite the basic limitations, some preliminary conclusions can be drawn.

The dimensional structures of the two groups analysed have revealed a highly similar pattern. Considering the fact that many other components could have come forth in the study, the three that have emerged, namely, (1) industrial-congestion, (2) commercial, and (3) land use, provide an empirical basis for further research on aspects of urban growth phenomena.

Secondly, the high loadings of the land-use variables in the component structures indicate their importance in urban analyses in India. These have been largely unexplored in prior research, primarily because of lack of data.

Although the two groups did not show any substantial difference as expected, the delineation of a number of comparable components in urban centres with different growth rates presented in this study does indicate the need for further analyses with a more comprehensive set of socio-economic and land-use variables.

Ahmad, Q., 1965 *Indian Cities: Characteristics and Correlates*, Research Paper No. 102, Dept. Geog., U. Chicago.

Berry, B.J.L., and H. Spodek, 1971 Comparative ecologies of large Indian cities, *Econ. Geog.* 47 (Suppl.), 266–85.

Census of India 1961a Paper No. 1 of 1962, New Delhi, for variables 1–7. The Census refers to class I centres of 100,000 and over as cities, the rest as towns. In this paper, these terms are used interchangeably.

Census of India 1961b 1970, Vol. 1, Part IX, Census Atlas.

Town and Country Planning Organisation, 1968 Land use pattern of India's cities and towns, *Urban and Rural Planning Thought*, 11, 188–90, for variables 8–17.

P1119

The pattern of settlement hierarchy in a rugged mountainous plantation landscape: Darjeeling Himalaya

S. MUKHERJI   *University of Hawaii, USA*

Nestling under the shadow of ethereal Kanchenjungha, Darjeeling Himalaya is a land of deep dissected valleys and rugged mountains. Still, it shows definite signs of embryonic central places of a local nature nested in a hierarchy. It is surmised that a study of the functions of settlements in this area may answer the query whether or not a hierarchical pattern of central places may develop even in a mountainous plantation country.

This region displays a unique landscape in which the set criteria for delineating a hierarchy are not applicable. Tea gardens have neither shops nor bus services, but they offer one special kind of industrial-occupational services. Each tea garden has its own local finance and administration, and all the tea gardens are incorporated, financed, and controlled by the 'Tea Planters' Association,' located in the highest ranking centre, the town of Darjeeling. The existence of a nested order of areal functional organization is evidenced by the appearance of an increasing number of tea garden-servicing enterprises in central places of ascending order. Similar phenomena are also present in the cinchona plantation region. Again in the forest region certain types of forest-based services with increasing complexity are offered successively from the lowest to the highest-order centres. Further, in agricultural regions the smallest farm villages are served and administered through the 'Village Panchayat Office' (smallest self-government) which provides seeds and fertilizers to the peasants, settles local disputes, and often organizes cultural-religious festivities. In a progression from smaller to larger centres, the agricultural settlements offer similar services with an increasing level of functional specialization, for instance, through Anchal office (local office), Anchal Pradhan's office (local chief's office), Block Development and Agricultural Extension office, and lastly the District Agricultural Department.

In all three cases the specialization in the respective services and organizations gives a keynote to each higher-order centre, and the highest-order centre embraces all the functions of each of the lower-order centres. Such being the case, three new fields of central activity are considered here – 'tea or cinchona plantation services,' 'forestry,' and 'agricultural services.' These groups of central functions might not be of general application to developed countries but are found to be more telling and useful in an accurate appraisal of status of the lower order centres, especially in developing countries.

Keeping in mind the speciality of central functions in Darjeeling Himalaya, about 115 central functions have been considered in the present study. Each of the central functions found in about 190 settlements has been grouped under various fields of central activity on the basis of its relation to the field of central activity. These twelve fields are administration, health services, agricultural services, finance, public utilities, retail and wholesale business, industrial services, tea or cinchona services, forest organization, traffic, and educational and recreational institutions. All the central functions are weighted according to the order of goods sold and the quality of services offered at each centre. Then, all the aforementioned fields of central activity are combined and brought into a meaningful balance and integration in ascertaining the hierarchy. The total score in each centre is computed and each settlement is ranked on the basis of its score. It has been found that the central functions tend to occur in groups at different levels, so that different categories of settlement can be recognized. Eventually, a seven-tier hierarchy is arrived at.

The accompanying map (Fig. 1) shows the spatial arrangement of this hierarchy and all the functional characteristics of central places. The twelve arrow-like phenomena in the star diagram show the intensity and the quality of central functions. In each case the length of an arrow indicates the score obtained and represents the total number and quality of central functions occurring in each centre, thus showing the degree of centrality. The twelve arrows point out the twelve groups of central functions with their kinds and intensity which, when considered together, form the twelve 'components' or fields of central activity shown within the circle. The components are also graded according to their intensity, as portrayed within the circle in the diagram on the left-centre of the map. The circle in each centre is split into 12 compartments which are all shaded according to the grades of the respective component to which the centre belongs. Ultimately, the components are grouped to form a 'complex' which is shown in the innermost circle. The symbol used inside the innermost circle represents the rank of a settlement in the hierarchical pattern.

The lowest-ranking central places are termed 'hill farm villages and forest bastees.' In functional characteristics such villages may roughly be compared with the 'neighbourhood' (Kolb 1933, 20), or with the

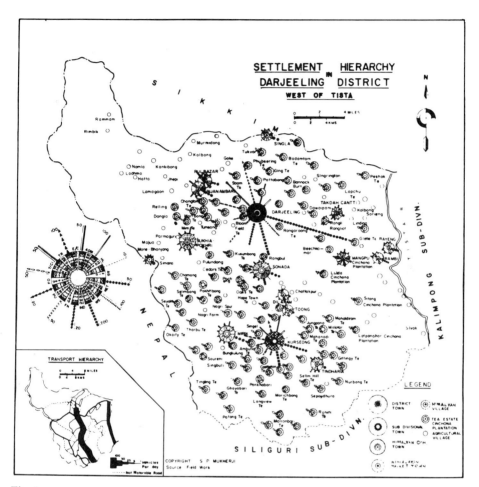

Fig. 1.

'hamlet' in the USA (Philbrick 1957, 301).

The next higher-ranking central places are denoted 'tea or cinchona plantation settlements.' These can be compared with the second-order 'focal' places of Philbrick's hierarchical system (Philbrick 1957, 302). The functional characteristics of the two are not fully comparable, but the inherent nature of their functional organization is, more or less, the same.

The next higher-ranking central places are called the 'Himalayan village.' From functional point of view this unit may be comparable to the 'village' of lower West Bengal (Kar 1960, 17), or the 'market village' of Coorg, South India (Folke 1968, 68); and it is transitional between the 'hamlet' and 'village' of Wisconsin (Brush 1953, 390).

'Himalayan market town' represents the fourth-order central place in the system. Functionally, these are similar to the 'market' of Germany (Christaller 1933, 55), or 'urban village' of England and Wales (Smailes 1944, 48). These are also comparable to 'agricultural collecting centre' of lower West Bengal (Kar 1960, 18).

The fifth-ranking central places, named 'Himalayan commercial centre' or 'tea-garden service centre,' are worthy of comparison with 'amtsort' of Germany as postulated by Christaller, and with 'sub-towns' in England and Wales after Smailes. Their suitable parallels in the USA would be the 'urban community centre' (Kolb 1933, 122), or 'towns' in southern Wisconsin (Brush 1953, 391). In Europe its counterparts would be 'lower-order gemeinden' in Bavaria (Boustedt 1960, 223). In lower West Bengal these types are called 'semi-towns.'

The 'sub-divisional town' forms the sixth-ranking central place in the hierarchy. Such central places are similar to the 'full-fledged town' of England and Wales, or the 'second-order gemeinden' of Bavaria, 'kreistadt' of Germany, or 'sub-divisional town' in lower West Bengal.

The highest rank of urbanism can be seen in the district town, Darjeeling. It had a population of about 45,000 in 1971, but from the functional perspective this central place is similar to 'bezirkstadt' of Germany, 'resort town' of England and Wales, or 'fourth-order centre' in the hierarchical system of Philbrick (Philbrick 1957, 305).

The findings establish that a pattern of settlement hierarchy, though some central places may be weak or embryonic, does really exist even in a mountainous terrain having a 'subsistence agriculture cum forestry cum plantation economy.' The whole pattern may be regarded as a 'miniature pattern' of the central place system, which may be best appreciated if viewed in its Himalayan background. This pattern is characterized by a diachronic and uneven spatial distribution of service centres over the territory, gaps in the service areas, hiatus in functional association of central places, and a stepped nature of the hierarchy. And, all these are mainly caused by inaccessibility and a rugged topography.

Boustedt, O., 1960   Die Zentralen Orte und ihre Einflussberiche, IGU symposium, Lund, 201–26.
Brush, J.E., 1953   The hierarchy of central places in southwestern Wisconsin, *Geog. Rev.* 33, 380–402.
Christaller, W., 1933   Die Zentralen Orte in Suddeutschland, Jena.
Folke, S., 1968   An analytic hierarchy in comparative regional study, collected papers, Denmark, 21 IGU New Delhi, 55–70.
Kar, N.R., 1960   Urban hierarchy and central functions around Calcutta in lower West Bengal, India and their significance, IGU Lund Studies in Urban Geog., Ser. B, Lund, 6–20.
Kolb, J.H., 1933   Trends of country neighbourhood, *U. Wisconsin Agr. Expt. Station Res. Bull.* 28, 20–40.
Philbrick, A.K., 1957   Areal functional organization in regional geography, *Reg. Sci. Assoc.* 33, 299–326.
Smailes, A.E., 1944   The urban hierarchy in England and Wales, *Geog.* 29, 41–51.

P1120
Urban geography or urban studies?
G.A. NADER   *Trent University, Canada*

This paper examines the role of geography in urban studies. It is postulated that geography's contribution to urban studies must become more specialized if the discipline is to maintain its importance in this interdisciplinary field. The main obstacle to the development of a specialist contribution seems to be the lack of a subject matter focus, and it is therefore contended that geography needs to identify the distinct elements of urban phenomena which it is best able to describe and explain. The importance of a discipline approach to the teaching of urban geography in schools and universities is also examined.

Before considering the distinctive contribution of urban geography, it is necessary to define the role of a discipline within the interdisciplinary field of urban studies. According to Popenoe (1963): 'The primary task of a discipline is to analyse a variety of concrete phenomena by abstracting certain aspects from them ... The primary task of a field, on the other hand, is to focus a variety of such analytical disciplines on one set of concrete phenomena.'

In illustration of the interdisciplinary approach to the analysis of urban phenomena it may be useful to adopt the analogy of a set of overlays which represents the total array of phenomena to be analysed, embodying different elements in each layer. The analysis of each horizontal layer may be the responsibility of one specialist, but the total evaluation of the system requires vertical integration. For instance, in the analysis of the internal system a *horizontal* layer of urban phenomena is abstracted by each discipline and subjected to specialist analysis; that is, each discipline studies the total spatial extent of the city, but only some aspects of the available phenomena (for instance, land use or socio-economic characteristics). A *vertical* integration of the specialist contributions of each discipline is the objective of urban studies. The horizontal/vertical analogy appears to be a meaningful way of regarding the discipline/field relationship.

Briefly, there seem to be two schools of opinion on the distinctiveness of the geographical approach: geography is viewed either as a predictive space science, or as a synthesis of the specialist contributions of other subjects. It is clearly possible for geographers to function either as specialists or synthesists within the field of urban studies, but external forces may compel the discipline to move in one direction or another since geography must, like all other disciplines, 'continually re-examine its position vis-à-vis the larger structure of knowledge' (Simmons 1967). In the modern era of professional specialization it is doubtful whether urban geography can function as a synthetic discipline.

If the role of urban geography is to be a specialist one, it is necessary to identify those concepts and methods which are distinctively or uniquely geographical. According to Abler, Adams, and Gould (1971, 56) the distinctively geographical question is 'why are spatial distributions structured the way they are?' This is in accordance with the growing emphasis on the evolution of the discipline to the status of a predictive space science; Morrill (1970, 3), for instance, has stated that 'space, spatial relations, and change in space ... are the core elements of geography.' Regardless of a consensus among geographers as to the definition of the discipline's primary emphasis, there is undoubtedly no other discipline which is as strongly oriented to spatial analysis, and this must be accepted as the geographical method. It is, however, important to note that the method is not uniquely characteristic of geography since the social, economic, technological, and other processes of urbanization cannot be divorced from their spatial manifestations.

There remains, however, the question of the phenomena to be studied by geographers.

Although it is quite possible for a science to consist of distinctive ways of looking at things without a unique content or subject matter, nearly all sciences have a definable core area of interest which overlaps little with other sciences – given the many urban disciplines, it is also inevitable that distinctive foci will emerge in each. The most distinctive characteristics of the cultural phenomena described by geographers seem to be their *physical* properties; that is, geographers are primarily interested in how physical phenomena are spatially distributed. Typically, urban geographers study areal variations in function and the spatial interactions that make such territorial specializations possible *in order to describe and explain regularities in the physical structure of urban systems.*

Even if one accepts that there is a distinctively geographical contribution to urban studies, there is the further question of how to teach urban geography. There are really two sets of decisions to be made in teaching an urban geography course. The first is whether to maintain a sharp distinction between the urban and societal components of phenomena that occur in cities; for instance, consideration only of those aspects of poverty in cities that are urban in nature. The second question is whether to teach urban geography or urban studies, that is, whether to adopt a disciplinary or multi-disciplinary approach. This is an important question in both schools and universities since there is undoubtedly some pressure on the designers of urban courses to adopt a multi-disciplinary or 'comprehensive' approach: the demand for 'relevant' courses, concern over contemporary urban problems, and so on. This question is especially pertinent at the university level, where geography students generally have the opportunity of taking a variety of urban courses in economics, sociology, political science, and anthropology, to mention some.

Above all it seems necessary to accept that urban geography must be distinguished not only on the basis of its analytical methods but also by its objectives – the description and explanation of certain elements of urban phenomena. Those aspects of urban phenomena that are best explained by geographers are the 'physical' elements. Hitherto, the lack of a generally accepted objective has led to a diminishing role for geography within urban studies. For instance, the current concern with urban problems and the demand for improved urban systems have left geographers in an embarrassing position, since they have little or no specialist contributions to offer. Geography's only significant contribution to urban studies has been central place theory, which essentially seeks to explain certain *physical* properties of urban systems, the size, number, and spacing of settlements. In terms of the internal system, geography has as yet made no major contribution, but the discipline's tradition of land use analysis, with its strong emphasis on physical structure, offers the most promising potential.

Abler, R., J.S. Adams, and P. Gould, 1971 *Spatial Organization: The Geographer's view of the World* (Englewood Cliffs, NJ).
Morrill, R.L., 1970 *The Spatial Organization of Society* (Belmont, Cal.).
Popenoe, D., 1963 On the meaning of 'urban' in urban studies, *Urban Affairs Quart.* 6 (Feb.).
Simmons, J.W., 1967 Urban geography in Canada, *Can. Geog.* 11, 341–56.

P1121
Impact of Jamshedpur industrial complex
INDRA PAL *University of Rajasthan, India*
and Y.B. VISHWAKARMA *Xavier Institute, India*

A city does not exist in a vacuum; rather it is intimately related to areas larger than the mere site it occupies. As such it leaves a realistic impact stamped on its environs in respect of habitat, economy, and society. This paper proposes to investigate and interpret the extent of geographical interrelationship between Jamshedpur and its environs. It is also to study and analyse the elements of impact and to assess its degree of intensity and finally to determine the limits of the urban influence.

Jamshedpur (22°47′N and 86°12′E) in the Singhbhum district of Bihar, has a picturesque site in an undulating micaschist plain surrounded by rugged hills and hillocks of the Chota Nagpur plateau. It is situated at the confluence of the Subarnarekha and the Kharkai Rivers. It is very well connected by roads with its district headquarters, Chaibasa, 68 kilometres to the west, and with the state capital, Patna, 469 kilometres to the north. Proximity to the port of Calcutta, market, mineral belts of iron-ore, coal, limestone, and manganese and, above all, abundant supply of water and power, are responsible for the development of a great industrial complex at this site. In respect of modern layout, cleanliness, and the absence of congestion, it is the most prominent industrial city of India, heading soon to become the country's industrial metropolis.

While conclusively delimiting the sphere of influence of Jamshedpur, the respective impact zones, prepared on the basis of each of the following eighteen indices have been taken into account: (*a*) five indices relating to rural supplies to the city; (*b*) five related to centralized services available to the rural areas; (*c*) seven elements dealing with demographic changes; and (*d*) the last relating to the changes in land use and habitation.

It includes under impact all those villages which fall within a radius of 45 kilometres (28 miles) of the city and which have at least a density of 247.11 persons per square kilometre, over 27.5 per cent of population decade variation, and over 39 per cent of non-agricultural workers, i.e. more than the regional average.

A superimposition of the zonal boundaries of the distribution of these 18 indices reveals the formation of a boundary girdle, though not very regular and perfect. Some protuberances extend radially and axially along the transport routes, railway lines, and major roads. The negative force of physical handicaps acts as barrier to the expansion of the impact zone; the development of transport facilities has been instrumental in extending the sphere of city influence as a positive factor.

From a synthesis of these 18 indices of impact (both conventional and demographic) it is obvious that the sphere of influence is directly associated with proximity to the steel city and oriented along the principal lines of communication. As such, the sphere of city influence is broadly divided into three ranges: high (intensive), medium, and low (extensive) based on four different methods: (1) according to the number of indices the village qualifies for; (2) according to the degree of intensity of impact exerted on the villages; (3) according to the rank of those eighteen indices in conformity with their respective importance, utility, and accessibility, and finally, (4) according to the composite and integrated impact brought about by the combination of all the preceding three methods. The results also are checked by other empirical methods.

Three maps (not published here) based on the first three methods reveal that the impact zone of high, medium, and low intensity do not form regular and circular shapes due to lack of development of roads and bridges on the northwest and northeast corner of the steel city. The integration of the aforesaid three maps into one gives a composite picture of the impact zone.

The final results of the city impact have also been checked by the gravity potential method (Olsson 1965). The following formula is used to test the influence on ten towns selected from within and outside the impact zone:

$A_i = P_i \, a.b.c/d_{ij} \, X.Y,$

where A_i = association index (propensity to move), P_i = population of town i, $a.b.c$ = attributes of or weights on population, d_{ij} = distance between town i and desired place, XY = attributes of our weights on distance (cost and time).

It is found that there is a close fit between the theoretically derived sphere of influence and the empirically derived sphere of influence of Jamshedpur except in a small area in the direction of the Tata-Chaibasa road.

The industrial complex has brought about a change in both the physical and cultural landscape in an area of about 4300km^2 covering 1790 villages. One of the earliest and most tangible impacts of the growth of Jamshedpur has been the denudation of forests which had been the source of earning to the indigenous population and are now being consumed by factories for furnaces or packing material. Secondly, a few of the aboriginals who had primarily lived by scratching the soil and extracting a few hundredweights of iron from their tiny indigenous furnaces which they had to close down later, got employment in the factories or took to marketing the products of the surrounding area in the newly developed bazaars of Jamshedpur.

The pressure of urbanization resulting in shortage of accommodation in Jamshedpur acted as a centrifugal force driving people to cheap residences in the nearby bustees. Multistoried flats belonging to the varied industries not only grew within the city, but they also scraped away the old huts and rural settlements in the surrounding forest lands. Jugslai, south of the railway station, has a concentration of large residences of the rich Marwaris, capitalists, traders, and the industrialists. Ancillary industries born of the Tata Iron and Steel Company have developed on the east, southeast, and the west of the city.

The immigrants mostly in the age group of 15 to 44 years and from Singhbhum district have been drawn for employment. Adivasi women in considerable number find employment in mines, factories, and other places, adopting a new outlook and new adjustments.

A high standard of services and amenities maintained by the city constitutes an unconscious, soothing, and healthy impact on the environs. Medical facilities, training of nurses and midwives, many of them *adivasi* women, a higher standard of liberal education, both general and technical, a network of well maintained roads, development of a national highway no. 33, and almost all sorts of communications, have brought about a great change in the life of the people and their activities. The dairies and the poultries and the co-operative societies, all have unconscious and deep influence on the society.

Milk from a distance of 15 kilometres, vegetables from a radius of 25 kilometres, rice from 25 to 35 kilometres, and poultry from 45 kilometres from the city centre find their way into it mostly by roads. Among the centralized services medical facilities extend to a radius from 25 to 35 kilometres, education up to 15 kilometres, while only 2 per cent of the villages in the city region receive benefit of electricity and 72 per cent of the villages get benefit of the traffic flow which forms the largest zone of influence.

Before the Tatas entered the scene, the rugged terrain of the region had been entirely rural, primitive, and undeveloped, sparsely peopled by aboriginals, and covered with forest interspersed with a few small patches of poor cultivation. With the establishment of the Tisco in 1907 the impact was first noticeable on the land, then on the people, and finally on the society and the culture.

The population of 800 (1907) had risen to 5672 in 1917 and soared to 57,360 in 1921. It had increased from 0.15 million in 1941 to over 0.3 million in 1961. Simultaneously its area had also risen from 14.19 square kilometres in 1912 to 64.75 square kilometres in 1961. As a result of this revolutionary change, land values have risen fantastically from Rs 5.02 to Rs 1931.25 per hectare. Like the European and American cities, there is a commuter zone, a few urbanized villages and residential neighbourhoods of the city.

The developments started toward the north in the 1940s with the construction of the Sakchi bridge and with the construction of the Kharkai bridge in 1962. New avenues of life and industrial expansion started towards the west where Adityapur may develop into an industrial complex double the size of Jamshedpur. The ribbon-type settlements are developing along the roads and the neighbouring villages are developing as urban settlements or satellite towns. Some of them have already reached a saturation point in population. Jugsalai, the nearest suburb, has attained the highest rate of literacy (48 per cent) in the region.

The shift of an economy from agricultural pursuits to non-agricultural activities is decidedly a reliable determinant of progressive industrialization in the region. Patches of high concentration of non-agricultural population (over 50 per cent) are discernible on the south, west, and north of the steel city.

The village economy has also changed from nomadic or fire-field culture to settled agriculture. They also appreciate the use of developed methods while earlier they used to pass their leisure in cock-fighting, hunting, and dissipation. Dairy and poultry farming has also developed around the steel city.

Only a small cultural impact has been exerted among the poor and backward people who are psychologically impervious to such a change. Among the change in food habits, some have given up eating pork and beef while some others have given up chickens too. Change has also occurred in their dresses.

Three divisions of the sphere of city influence may be distinguished: the area of city dominance, the area of city prominence, and the area of city association according to the degree of the intensity of impact. These are irregular in shape on the map. The major and minor axes of the first division are 20 kilometres and 16 kilometres in length, of the second 40 and 35 kilometres, and the third division is peripheral up to a radius of 45 kilometres. Thus the sphere of influence of Jamshedpur has a gradient pattern, extending, but in spheres of decreasing influence, whose character and range are largely controlled by accessibility and functional character of the service centres and the villages. Their boundaries are, however, of fluctuating nature due to the stages of growth and decline in size and function of the city. These are reflections of the normal behaviour of the dynamic character of Jamshedpur.

Alam, S.M., 1965 Hyderabad and Secundrabad (Twin Cities), A Study in Urban Geography (Bombay).
Ellefsen, R.A., 1962 City Hinterland Relationship in India, in Roy Turner, ed., India's Urban Future (London).
Elwin, V., 1958 The Story of Tata Steel (Bombay).
Fraser, Lovate, 1919 Iron and Steel in India (Bombay).
Mahadeva, P.D., and D.C. Jaysankar, 1970 Concept of a City Region: An approach with a case study, Umland of Mysore, Indian Science Congress Association, Abstract.
Olsson, G., 1965 Distance and Human Interaction. A review and bibliography, series no. 2, Regional Science Association Institute (Philadelphia).
Percival, F.G., 1933 India's indigenous Smelters, Tisco Review (Jamshedpur).
Prasad, S.D., 1967 Census of India, 1961, Bihar, District Census Handbook, no. 17, Singhbhum (Patna).
Roy, Chowdhary, 1958 Bihar District Gazetteers (Singhbhum).
Sinha, J.N., 1955–56, 1964–65 Working Plan For Forests of Manbhum Division, Bihar Govt. (Patna).
Smailes, A.E., 1966 The Geography of Towns (London).
Vishwakarma, Y.B., 1970 Impact of Jamshedpur on Neighbouring Villages, thesis, U. Rajasthan, vs. I and II.

P1122
Les migrations journalières de travailleurs à Brasilia
IGNEZ COSTA BARBOSA and ALDO PAVIANI *Universidade de Brasilia, Brazil*

ENGLISH ABSTRACT
As a working hypothesis, it is assumed that the majority of the workers living in the vicinity of Brasilia work away from their domiciles, particularly in the Plano Pilôto (the city of Brasilia). The intensity, quality, and direction of commuting in the Brazilian Federal District were investigated. Direct research was conducted in the analysis with a random stratified sampling procedure. Such a procedure yielded information which indicated the spatial distribution of the population and workers in the Federal District and the trend of their daily commuting. The following elements were then characterized: employment/residence ratio and the resulting capacity of retention of the active population, that is, the respective migratory residues. The lack of similarity observed among different localities was subsequently characterized. The studied localities are not fully equipped to function as satellite towns as yet, and are largely dependent upon Brasilia. This fact results in a flow of a large number of workers to Brasilia.

This study suggested that the forced commuting to which the workers of the peripherial localities are submitted will persist in the direction of Brasilia. New policies have to be formulated in the sphere of spatial organization and the availability of jobs: once the construction work in Brasilia is concluded there will be a surplus of available labour.

New policies will relieve Brasilia of the 'daily turbulence of population' and will provide the conditions for new job openings in the residential localities.

Lysia Bernardes, Celso Chiarini, Gilséa Malvar, Henrique Malvar, Maurício Gama, Azize Drumond.

Corona, Eduardo, 1970 Gostar, conhecer e respeitar Brasília, *Revista Acrópole*, 79–80 (julho/agôsto), 375–6.
Fundação, I.B.G.E./I.B.E., 1971 Sinopse preliminar do censo demográfico, VIII Recenseamento Geral, 1970, Distrito Federal, Departamento de Censos.
Grupo de Áreas Metropolitanas, FIBGE/IBG, 1969 Áreas de Pesquisa para determinação de áreas metropolitanas, *Revista Brasileira de Geografia*, 31(4), 53–127.
HMSO, 1969 Sample Census 1966 – England and Wales, workplan and transport tables, General Register Office, *apud* Jean Rossano, Notes et Comptes Rendus, *Annales de Géographie*, 78 (425), 103–8.
Juillard, Etienne, 1961 Europa industrial e Brasil: Dois tipos de Organização do Espaço peri-urbano, *Boletim Baiano de Geografia*, 1 (4), 3–10.
Snyder, David E., 1964 Alternate perspectives on Brasilia, *Ec. Geog.* 40 (1), 34–45.

P1123
Toward a dynamic theory of urban plat development
CHARLES S. SARGENT JR. *Arizona State University, USA*

Central place theory explains urban networks, and other theories have been formulated and refined to deal with the internal structure of the city (Burgess, Hoyt, Harris, and Ullman), but there has not yet evolved a dynamic theory relating to the spatial development of the urban plat itself, i.e. the forces that influence urban morphology have received less attention than has the nature of land occupance within the city by discrete social or economic groups (Vance 1960; Simmons 1965). Hence the need for the formulation of a dynamic theory that will further an understanding of (1) the forces behind the outward spread of the urban area, (2) the pace of areal changes in urban growth, as well as (3) the intensity of occupance.

In a study of the spatial development of Greater Buenos Aires from 1870 to 1930 (Sargent 1971), the location, physical extent, and occupance of residential areas were inductively found to be a function of the interaction of many individual *areal, occupational,* and *temporal* factors.

The *areal* factors that determined the

direction, shape, and extent of the urbanized area include inter- and intra-urban transportation, the location of land relative to the city centre, the pattern of land ownership, land speculation and development schemes, terrain, public services, local politics, and municipal controls over urban growth. The *occupational* factors that influenced the settlement of the residential districts by different socio-economic and ethnic groups include the level of personal income, spatial differentials in land costs and rents, the location of employment and its expression in the journey-to-work, social and ethnic values, housing policy, and perception of the environment. The *temporal* influences simultaneously affected the operation of both the areal and the occupational factors. These include the application of new developments in transportation and sanitation technology, population change, changes in real personal income, the availability of credit, shifts in the location of industry, price levels, the perception of speculative gain to be derived from real estate transactions, and the perception by land-buyers of improved and less expensive living conditions in new residential plats being promoted.

In the construction of a dynamic model of urban spatial growth emphasis is more properly placed upon the areal and temporal factors that influence the location and timing of development (process) rather than upon the factors that explain the specific nature of the occupance of residential space. The intent is not to deny the importance of these occupational factors, but instead to give primary attention to the forces that created the residential areas rather than, as is the case with most urban models and studies in factorial ecology (Berry 1971), to examine what made them socially or culturally distinct.

When the areal and temporal dynamic factors are considered together, it is hypothesized that a framework is created within which there is 'nesting' of the forces that influence urban morphology and land use (Fig. 1). In such a nesting arrangement, the areal and technological evolution (or *stage*) of the various transportation modes is most responsible for the extent of the modern metropolitan area and can be viewed as the primary force behind the creation of a transportation frame that determines the outer limits of urban growth. Historically, the primary and traditional role of the transportation system, i.e. the diurnal movement of workers into and out of the central city, with both the trolley and the railroad focusing on this major zone of conflux, created de facto linear fields of potential settlement, fields which define the physical extent of the urbanized area. The earlier horse tram operated within a more highly restricted nuclear time-distance matrix.

Within this transportation frame, it is hypothesized that the operation of urban lot

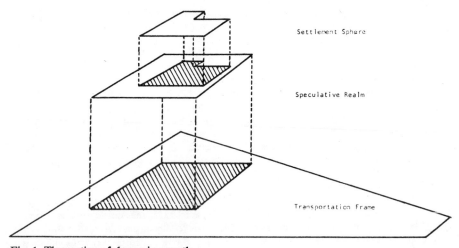

Fig. 1. The nesting of dynamic growth factors.

speculation by real estate interests largely determines in what sequence and where the transportation frame is subdivided into lots, thus creating speculative realms within which ultimate residential development takes place. Over time, the role of transportation technology and its expression in trolleys, faster trains, and increased schedulings tends to become more permissive than determinative in influencing residential development while urban lot speculation becomes relatively more significant as a force in the spatial build-up of the urban area. By changing the time and cost isolines of passenger movement, transportation technology becomes a tool of urban lot speculation and the actual direction, shape, and extent of urban growth on the periphery is increasingly a function of successful land development endeavours. Other spatial forces such as terrain, municipal controls, and even the strong influence of the journey-to-work prove increasingly malleable under the pressure of land speculation.

Nested within the realms delineated by speculative decisions and actions are *settlement spheres* related in both extent and stage of development to the occupational factors that determine the location of individuals within the speculative realm. In a very real sense, the speculative realm and the settlement sphere can be viewed as 'nets' which may be cast within the transportation frame in a number of random ways, giving an almost infinite variety of occupational combinations within the urban frame.

It is further argued that the hypothesis derived from a study of Buenos Aires can be applied to the American and West European city, as the operation and significance of the dynamic elements is less restricted by national or cultural boundaries than by temporal and economic boundaries that limit the application of new transportation innovations and the development of urban land speculation.

Berry, B.J.L., ed., 1971 Comparative factorial ecology, *Econ. Geog.* 47 (2; Suppl.).
Sargent, C.S., Jr., 1971 *Urban Dynamics and the Changing Pattern of Residential Development: Buenos Aires, Argentina, 1870–1930*, PH D dissertation, U. California, Berkeley.
Simmons, J.S., 1965 Descriptive models of urban land use, *Can. Geog.* 9, 170–4.
Vance, J.E., Jr., 1960 Labor-shed, employment field, and dynamic analysis in urban geography, *Econ. Geog.* 36, 189–220.

P1124
Some features of urbanization in densely populated developing countries
GALINA V. SDASYUK *Academy of Sciences, USSR*

The worldwide process of urbanization taking place under present conditions of the scientific and technical revolution extends to the developing countries too.

In addition to a number of global features ('explosive' growth of major cities, agglomeration as a basic form of modern urban life, emergence of giant conurbations transforming into vast urbanized zones, megalopolises, etc.), the urbanization in the developing countries is distinguished by some specific trends which make impossible the mechanical application of the notions and ideas derived from the experience of urbanization in the West to their specific conditions.

The experience of India as the largest and a highly typical representative of the 'third world' shows that the mechanism of interaction between her urban centres and rural areas is an extremely specific one; associated with this are some important features in the process of formation of India's economic regions that are typical of many developing countries. One of the principal features of urbanization in the developing countries is that their towns grow not only because of attraction on their part (pull factor), but in greater measure because of the push of migrants from heavily populated rural areas (push factor). 'Urbanization is ahead of industrialization' – this has been an important feature (and complication) of the present stage of development in almost all countries of the third world. In India the action of still another factor (push-back) was found to be the urban population itself in cities which were originally large, and had a high natural increase and pronounced unemployment (Bose 1966). All this creates a 'damper' on the inflow of rural migrants, the result being the deformation and weakening of territorial ties.

The functional structure of towns in the developing countries has been highly specific (Breese, ed. 1969, 555), accounting for the specific nature of their economic-territorial and social links and for their particular role in the processes of region-formation. Service activities are widespread here, with 'poverty-oriented' services (disguised forms of unemployment) dominating. The situation of a greater part of numerous urban residents, who are among those engaged in petty trade, is similar. In 1961, of all urban workers in India, 30.6 per cent were engaged in service activities with 16.3 per cent more in trade (*Census of India* 1961, 1965). Thus, almost half of the total urban workers of India are outside productive occupations.

An extensive spread of agricultural activities is also characteristic of the developing countries. In this way agriculture is a basic occupation for 10 per cent of the workers in Indian towns. The towns with such a functional structure have great internal reserves for developing a 'true' urbanization and for strengthening economic functions (first and foremost, industrial ones).

Indeed, in India throughout all of the twentieth century, the towns in which industrial functions are predominating have been steadily growing at a faster rate than the towns of other functional types (Mitra 1967b). Although the total number of industrial towns is still far less than the number of service ones (39.7 per cent against 47.2 per cent in 1961), in the towns and cities of higher classes (population above 50,000) the industrial functions have already become dominant. Since independence, further consolidation of an industrial (active) economic basis of urbanization has been taking place in India and other industrializing countries of the 'third world.'

However, in densely populated developing countries industrialization is not always accompanied by an accelerated urbanization. This is proved, for example, by the low rates of urban growth in India in 1951–61 (26.4 per cent, as compared to 21.5 per cent total population increase). Evidently India and other densely populated countries of the 'third world,' above all South and South-East Asia, existing at a low level of urbanization (in 1961 India's urban population was just 18 per cent of the total), have limited potentials to accelerate substantially their urban growth even under the present conditions of industrialization now making headway. Thus, according to the competent opinion of Asok Mitra, a former Registrar General, now the Secretary of the Planning Commission of India, the Indian towns are so oversaturated by labour force that 'urban population alone ... is sufficient to meet foreseeable demands of industrial expansion in the current decade particularly when we remember that the gross capital investment per capita in the public sector is apt to be quite high' (Mitra 1967a).

Relatively weak production ties between urban industries and agriculture due to the general economic backwardness, low level of capital accumulation in the village and the fact that rural communities still retain their more or less isolated status are characteristic of the developing countries. For example, according to the FAO estimates, the share of industrial commodities, construction, and services (repairs, etc.) was less than 2 per cent of the total cost of India's agricultural output whereas in the advanced countries it exceeds one-quarter or one-third (Chattopadhya 1969).

Inadequate development of ties between industry and agriculture in the developing countries underlies the production-territorial duality of their economy, and is also the cause of their generally poor economic integration.

Thus, cities plus the transport network, by forming 'a framework, a skeleton that shapes up a territory and gives it a specific configuration' (Baransky 1956), determine the macroscheme of the spatial structure of the economy in the developing countries, acting at the same time mainly as a relatively closed system.

The spatial structure of the economy features greater stability and is markedly slow in keeping pace with the changes in a branch structure of the economy. The study of the spatial structure of India's economy, as it has reflected in her urban geography, reveals a clearly defined territorial division of specific, relatively independent regional systems and town groups (Qazi Ahmad 1965; Datt 1967) that are closely correlated to the regional levels of economic development (Mitra 1965; Schwartzberg 1962).

The four largest conurbations of Calcutta, Bombay, Madras, and Delhi exert a decisive impact on the spatial structure of the Indian economy. The three major port-cum-industrial cities and the capital extend their influence all over the country. These oppositely located centres with the highest urban population potentials of India (Berry 1966) are the region-forming cores of the most mature economic regions in the country. They appear to form the apexes of a huge quadrangle, a framework for the spatial pattern of India's economy. The sides and the latitudinal diagonal of the quadrangle stand out as the 'growth corridors' covering the country's main railways and highways on which are strung out the cities featuring higher growth rates and particularly favourable prospects for future development. The Bombay-Calcutta corridor stands out here as one of ever increasing national importance, due to the combined effect of such potent factors as the growing interdependence of the Indian primate cities (which rank among the largest conurbations of the world) and the accelerated development of the country's major coal and metallugical base (the Damodar basin–Chhota Nagpur plateau, Central India) that has been taking place since independence. It is natural, therefore, that a number of new large public-sector projects have emerged in this corridor to give rise to such towns as Bhilainagar, Rourkela, and others.

Thus, the decentralization of industries in India (an objective process supported and controlled by the government) takes place alongside the strengthening of the role of major metropolitan centres and hinges on these to a great extent. The decentralization of industries and first of all of large public-sector undertakings (representing a most important town-forming factor in modern India) is not evenly dispersed throughout the country but follows a specific pattern gravitating first and foremost to the 'growth corridors.' Evidently it is one of the most significant trends in the shaping up of the spatial structure of the Indian economy. Trends of this kind can be traced in the 'geometry' of the development in other developing countries as well. Consequently it is essential for the vast agrarian areas outside the zone of urban influence, outside the growth corridors, that a number of developmental measures based on local resources be elaborated and implemented.

By focusing industrial, trade and transport, administrative and cultural, and other activities of society the towns in the developing countries play a particularly important role at a present stage of the transformation of the sectoral and spatial structure of their economy. A paramount consideration is given to the growth of towns (recognized as essential in many of the developing countries) in regional planning of the various taxonomic levels: from the highest macro-regional to the lowest local ones. At the same time these plans must be complemented by extensive studies to be carried out with a view to comprehensively evaluating natural and economic potentials of the vast rural areas remaining outside the zone of the urbanization impact.

Ahmad, Qasi, 1965 *Indian Cities: Characteristics and Correlates* (Chicago), 188.
Baransky, N.N., 1956 Economic geography, *Econ. Cartography*, 368.
Berry, Brian J.L., 1966 *Essays on Commodity Flows and the Spatial Structure of the Indian Economy* (Chicago), 377.
Bose, Ashish, 1966 *Studies in India's Urbanization*, Institute of Economic Growth (Delhi), 377.
Breese, G., ed., 1969 The city in newly developing countries, in *Readings on Urbanism and Urbanization* (Princeton), 555.
Census of India 1961, 1965 Vol. I, part II–B(i), General Economic Tables, 702 (Delhi).
Chattopadhaya, A., 1969 Recession and India's Agrarian Sector, Alternative Policies for the Fourth Five Year Plan, Trivandrum, 397.
Datt, Kusum Lata, 1967 Urban zones of India, *Nat. Geog. J. India* 13, part 2, 97–109.
Mitra, Asok, 1965 Levels of regional development in India, in *Census of India 1961*, vol. I, part I-A(i), text, p. 357; part I-A(ii), 892 (Delhi).
– 1967a *Internal migration and urbanization in India*, part I, 183 (Delhi).
– 1967b *Internal Migration and Urbanization in India*, part II, 632 (Delhi).
Schwartzberg, J.E., 1962 Three approaches to the mapping of economic development in India, *J. Annals Assoc. Am. Geog.* 52, (4), 455–68.

P1125
Character of urbanization in Uttar Pradesh (India)
UJAGIR SINGH *University of Gorakhpur, India*

The state of Uttar Pradesh is located almost in the centre of northern India, extending from 20°52′N to 30°18′N and from 77°30′E to 85°39′E. A part of the Siwalik hills and Himalayan mountains falls within the state boundary in the north while the northern part of the Vindhyan uplands extends into the south. But the greater part of the state consists of the upper and middle Ganga plain built by the Ganga and its large tributaries like the Yamuna and Ghaghara.

The region as a whole is the mirror of Indian civilization in which almost all events of the country's history are reflected through its cities. Varanasi, Ayodhya, Allahabad, Mathura, and Haridwar are the ancient seats of Indian culture and civilization, while Lucknow and Agra are the relics of long Muslim rule in India. Kanpur and some other industrial centres represent the modern trend.

In 1951, the state consisted of 463 towns divided into 6 categories – class I with over 100,000, class II: 50,000–99,999, class III: 20,000–49,999, class IV: 10,000–19,999, class V: 5000–9999; and VI less than 5000 persons. But their number decreased to 244 in 1961 due to change in definition of urban centre. This change reduced the number of class V towns to 74 from 169, while the category VI towns declined to 10 from 153. But the first four classes of towns have not been affected. Class I towns shared 54.4 per cent of the total urban population in 1961, while the figure for 1901 was only 23.8. On the other hand, the share of other classes of towns has declined considerably. Thus the large urban centres are highly progressive while the lower order towns are either growing slowly or showing the signs of decay.

The upper Doab and the Ram Ganga plain are highly urbanized with 130 towns, comprising 54.3 per cent of the total urban population of the state and divided into three large urban clusters: zone 1 containing 52 towns, zone 2 comprising 32 towns, zone 3 containing 42 towns; Varanasi-Mirzapur-cum-Vindhyachal forms zone 4 in the southeast. Some large urban centres like Kanpur, Lucknow, Allahabad, Gorakhpur, and Jhansi stand in isolation. A greater concentration of towns in the west is thus obvious.

The first seven cities of the state (Kanpur, Lucknow, Agra, Varanasi, Allahabad, Meerut, and Bareilly) have continued to remain class I towns, ever since 1901. Except for Allahabad all changed their rank once or more during the last seven decades. Little change in the rank has been noted in the case of other towns.

By using a population and centrality index the towns may be divided into four hierarchical orders: (i) regional cities (6), namely Kanpur, Agra, Varanasi, Lucknow, Meerut, and Allahabad with high population and very high centrality indices; (ii) the cities (10), including Bareilly, Gorakhpur, and Jhansi etc.; (iii) major towns (48), including some district and tahsil headquarters; (iv) minor towns (remainder of the 189 towns) with low centrality indices and low population.

Further, the study reveals that a positive correlation exists only in higher ranks (above 1 lakh population) of towns – the regional cities in general are located one and a half times farther apart from each other (133km) than the cities (88km). Similarly the case between the major and minor towns.

Only 34.8 per cent of the state's urban population is considered to be outborn; most of them are local, only 5.2 per cent coming from outside the state. The percentage of outborn in the industrial city of Kanpur is 46.3 per cent. Most of the towns have a very low percentage of immigrants, except some hilly towns with over 60 per cent.

A large number of cities have 850 or less females per 1000 males, but in smaller centres the number of females is a little higher (about 950). Generally, the towns in the western sector have a comparatively high sex ratio (about 800 females). In general, the sex ratio is related inversely to the percentage of migration. The higher the percentage of migration the lower is the ratio of females to males.

Trade and commerce show the least differential mean percentage and mean percentage index of the state as they are the most sustaining function in these towns. Most of the class V towns are administrative and commercial centres. Towns belonging to class VI are mostly administrative, being tahsil headquarters. Thus there is a predominance of administrative and service towns in all size classes. Class I towns are mostly industrial and manufacturing. But the majority of towns of the state are multifunctional.

Most of the cities are riverine and the rivers have played a great role in their origin and development in the beginning. But during the last one hundred years, the growth of towns has been greatly influenced by the advantage of railroad transport. At present, the role of automobiles in the establishment of industrial concerns along the main thoroughfares on the periphery is becoming quite significant. Further, with the establishment of the civil lines and cantonments on one hand and several cultural centres like the hospitals and educational centres etc. on the other, a considerable growth in the built-up area of the cities has taken place within the last one hundred years.

The cities in general have four distinct urban zones: (i) the crowded and haphazardly grown old city with its associated business core, (ii) semi-planned and more open civil lines and cantonments adjacent to the old city, (iii) the housing schemes and new residential areas developed in different parts as a result of the filling-up process, (iv) the developing urban fringe along the main thoroughfares.

Though the principal business districts have retained their old aspect with about 70 to 80 per cent of commercial establishments, a part of the business (20 to 30 per cent) has been taken up by the submarketing centres and shopping ribbons. Similarly, though the manufacture of handicrafts is still carried on in different parts of the main town, manufacturing and large scale industries have recently been established on the periphery. The tremendous increase in population and the resultant migration of people to urban centres has resulted in overcrowding and a high percentage of slum dwellers. In the large towns and cities, the percentage of persons living in one room varies from 38.1 to 62.2 per cent.

As no details of planned urbanization are possible here, the following broad principles for planned urban growth in the state may be presented. (i) In the highly urbanized western part, the growth of urbanization may be restricted, while planned and controlled development of new centres may be encouraged in the central and eastern parts of the state. (ii) The existing growth points and nodal centres should be encouraged to grow by establishing industrial estates and marketing centres. (iii) With a view to accommodating the growing population of towns vertical expansion of the towns is highly desirable. (iv) The entire area comprising the city, the suburbs, and the fringe should be considered as one unit for planning purposes.

Census of India, 1961 Uttar Pradesh, part I-A(i), *General Report on the Census*, 168–325.

Singh, K.N., 1959 Functions and Functional Classification of Towns in U.P., *Nat. Geog. J. Ind.* 5 (3), 121–48.

Singh, O., 1967 The trend of urbanization in Uttar Pradesh, *ibid.*, 13(3), 141–57.

– 1969 Towns of Uttar Pradesh, PhD thesis, Banaras Hindu U.

Singh, U., 1966 Changes in the built-up area of Kaval towns in the Ganga Plain (India), *Nat. Geog. J. Ind.* 12 (4), 203–17.

– 1967 Distribution and character of cities of the Ganga Plain, *ibid.*, 13(3), 1–12.

P1126
Friendship and urban inter-residential social trip length
F.P. STUTZ   *San Diego State University, USA*

Urban residents generally establish and maintain friendship relationships with other residents. These friendships are both neighbourhood-centred and spatially dispersed over the city. There is little doubt that intimate social relationships are important components of urban integration and that they bind the city functionally through social travel linkages. This paper deals with the explanation of the spatial distribution of both neighbourhood-centred and spatially diffuse friendships. Given this inherently spatial theme, it is surprising that only a small amount of research in geography has been done on friendship interaction at the neighbourhood or city-wide level (e.g. Cox 1969; Wheeler and Stutz 1971; Stutz 1971).

Accepted ecological theory suggests that spatial distances among clusters of people of the same occupational class are closely related to their social distances (Duncan and Duncan 1955, 493). Mate selection, as well as other types of social interaction, has been shown to correlate significantly with the degree of similarity in occupational class (Laumann 1966, 141; Ramsøy 1966, 773). Social interaction has also been shown to increase with increased residential tenure (e.g. Smith et al. 1954, 282; Festinger et al. 1950, 31).

One approach to the question of the spatial distribution of friendships is to ask city dwellers to give the locations of their closest friends. This was done for a small sample of households in the Lansing, Mich. study area. An alternative approach is to use distances of actual social trips made in the city and obtained from transportation surveys locally available. The latter data source was used for the quantitative analysis in this paper. The data were obtained from the Tri-county Regional Planning Commission of Lansing, Mich., and are based on a 1965 Home Interview Survey of over 4500 households representing a 5 per cent sample of the population. The 44,860 trip records that were supplied to the researcher on magnetic tape were reduced to approximately 2000 interresidential social trips.

Travel inputs have been shown to vary positively with measures of socio-economic status. This fact, along with the assertion that individuals of higher socio-economic status, as measured by occupation, are generally more sparsely settled, should indicate that their social trips will be longer in length than those of lower socio-economic status. Therefore, it is hypothesized that the mean social trip length is greater for individuals of high occupational class than it is for those of low occupational class.

When examining social trip length for different occupational classes, one must consider the density of spatial opportunities for each occupational class. Spatial opportunities for social interaction vary from one part of the urban area to another, depending on the associated urban spatial distribution of individuals. It has been shown for the study area that there is a strong social stratification of interaction (Wheeler and Stutz 1971, 381). People living in areas with a large number of individuals of similar status nearby are likely to have shorter social trips than those living in a sparsely populated area. If the residential structure of the study area displays the common inverse relationship of occupational status and population density with distance from the city centre, then the mean length of social trips should increase with an increase in occupational status, *ceteris paribus*.

In order to measure the density of social opportunities, interresidential distance by occupational class is measured. It is expected that the minimum interresidential separation among individuals of an occupational class will rise as the status of the class rises. Social trip length by occupational class will then be compared to the interresidential separation of groups.

The linear programming model used to solve the transportation problem of optimum allocation of transported goods from m supply places to n demand points is used to give a measure of minimum interresidential separation of a given occupation. The output of the model is the shortest possible aggregate distance for all individuals of occupational class k, X^k, from their residences in zone i, S_i^k (supply places), to residences of k in zones j, D_j^k (demand points), as computed between zones. A unique feature of the application of the linear programming model of transportation to this problem is that every household of a particular occupational class acts both as a supply and a demand unit, so that the total supply of a zone equals the total demand of the same zone. The supply of one zone cannot be used to fill the demand of the same zone, however, or the shipments would be effectively zero. This means that the results will be aggregated on the zone level, which is a small enough unit of analysis to give meaningful results, but, more important, it is the same data unit on which social trip length is computed. Associated with the movement of individuals between zones is a cost, C_{ij}, of travel from i to j, which is considered to be a linear function of distance in this analysis.

In the objective function of the linear programming model, the total cost, Z^k, of moving all individuals of occupational class k is minimized. The objective function is expressed as

$$Z^k = \sum_{i=1}^{m} \sum_{j=1}^{n} X_{ij}^k C_{ij},$$

where $m = n$, subject to the constraints that total supply from residences and total demand at residences equal the number of individuals shipped. This is written

$$\sum_{j=1}^{n} X_{ij}^k = S_i^k \quad (i = 1, 2, \ldots, m),$$

and

$$\sum_{i=1}^{m} X_{ij}^k = D_j^k \quad (j = 1, 2, \ldots, n),$$

respectively. Or, the total number of individuals moving from residences is the same as the total amount demanded at residences,

$$\sum_{i=1}^{m} S_i^k = \sum_{j=1}^{n} D_j^k.$$

Finally, movements cannot be zero or negative, or the solution would be mathematically trivial: $X_{ij} \geq 0$. The average minimum interresidential separation, MIS, for occupational class k is $\text{MIS}^k = Z^k / \Sigma X$.

MIS values for each occupational class are shown in Table 1. High values for the highly ranked occupations indicate that these groups have greater mean separation of residences among individuals of that class. Low values for low occupational classes suggest residential concentration of these groups. MIS values in Table 1 are rank correlated with occupational class at the level $r = 0.90$, $p = 0.01$. The assumption of residential stratification by class with distance from the city centre appeared to be justified after mapping occupational classes by traffic zone. In any case, the density of social opportunities for lower occupational classes is greater than that for higher occupational classes.

Mean social trip length by occupational class is shown in Table 2. An examination of the distances for each occupational class shows a general decrease in distance with an increase in occupational class. These values are rank correlated with occupational class at the level $r = -0.74$, $p = 0.05$. The two extreme classes are the only exceptions to the general relationship. The hypothesis that individuals of lower status make shorter average social trips must be rejected. Lower status individuals are apparently not more distance sensitive for social travel than other individuals.

TABLE 1. Minimum interresidential separation by occupational class

Occupational class	Distance (km)
1. Labourers	1.66
2. Service workers	2.18
3. Operatives	2.63
4. Craftsmen and foremen	2.26
5. Clerical and kindred workers	2.47
6. Sales workers	2.71
7. Managers	3.40
8. Professionals and technical workers	3.24

TABLE 2. Mean social trip distance by occupational class

Occupational class	Distance (km)
1. Labourers	2.74
2. Service workers	4.91
3. Operatives	4.72
4. Craftsmen and foremen	4.60
5. Clerical and kindred workers	4.21
6. Sales workers	3.92
7. Managers	3.69
8. Professionals and technical workers	4.02

Because social interaction is so personal, individuals prefer to travel the necessary distance, whatever it might be, to interact with a friend, rather than going to the closest person available. Individuals need to receive a certain level of satisfaction from a person or group with whom they interact and are thus willing to travel to the individual's residence location. Within the city the interresidential separation of the two parties seems to have little relevance.

In conclusion, interviews show that neighbourhood-centred friendship is found among high occupational classes. The social trips are shorter for them, on the average, than social trips for lower occupational classes. Yet, residents of low occupational classes state that present and former neighbours comprise the majority of their friendships. This apparent contradiction is at least partially due to the fact that, although interresidential social travel is carried on over a larger social space among lower occupational classes, friendship within these groups was originally formed through neighbourhood associations. Because of their relatively high residential mobility, these groups display social interaction which is not neighbourhood-centred. It is spatially dispersed over the portion of the city in which they had former residential locations. Due to relatively long residential tenure, higher occupational classes have smaller social spaces which are based on neighbourhood-centred friendships.

Cox, K.R., 1969 The genesis of acquaintance field spatial structures: a conceptual model and empirical tests, in K.R. Cox and R.G. Golledge, eds., *Behavioral Problems in Geography: A Symposium*. Evanston, Ill., Northwestern U., Dept. Geog., Studies in Geog. no. 17, 146–68.

Duncan, O.D., and B. Duncan, 1955 Residential segregation and occupational stratification, *Am. J. Sociol.* 60, 493–503.

Festinger, L., S. Schachter, and K. Back, 1950 *Social Pressures in Informal Groups* (New York).

Laumann, O.E., 1966 *Prestige and Association in an Urban Community: an Analysis of an Urban Stratification System* (Indianapolis).

Ramsøy, N.R., 1966 Assortative mating and the structure of cities, *Am. Sociol. Rev.* 31, 773–86.

Smith, J., W.H. Form, and G.P. Stone, 1954 Local intimacy in a middle-sized city, *Am. J. Sociol.* 60, 276–84.

Stutz, F.P., 1971 *Research on intra-urban social travel: introduction and bibliography*, Council of Planning Librarians, Exchange Bibliography no. 173.

Wheeler, J.O., and F.P. Stutz, 1971 Spatial dimensions of urban social travel, *Ann. Assoc. Am. Geog.* 61(2), 371–86.

P1127
Urban ethnic acculturation: a functional approach
PAUL Y. VILLENEUVE   *Université Laval, Canada*

Ethnicity is an important basis for the formation of large groups in urban areas. Not unlike social class, it differentiates urban populations and often becomes a source of divisiveness and conflict. Ethnicity, however, also plays an integrative role by providing the individual with a reference group through which he can maintain his identity while learning the 'urban language' (Hyland 1970, 69).

The impact of ethnicity, as both a differentiating and an integrating factor, is compounded by the degree to which ethnic communities are self-contained in urban space. Residential segregation is perhaps the most easily detectable manifestation of the relation between ethnicity and territory. Specific neighbourhoods come to be identified with certain ethnic groups.

A considerable sociological literature has described the role of ethnic colonies as one of 'port of entry' to the urban environment. Too often, however, the integrative aspect of this role has been overemphasized at the expense of the differentiating aspect (Michelson 1970, 63). Moreover, the study of ethnic colonies as aggregates has often masked important differences within groups with regard to indicators of integration such as residential and occupational mobility.

A theoretical framework is therefore needed which will be general enough to provide an understanding of urban acculturation for most ethnic groups in most cities, and be specific enough to be applicable at the level of individual behaviour.

Five conceptual variables considered as necessary and sufficient components of urban ethnic acculturation are related with the aid of functional analysis. These five variables are: ethnic identity, socioeconomic status, psychological stress, spatial structure, and time. Each household in a given ethnic community can be placed in the multidimensional space formed by these five variables.

Ethnic identity can be thought of as the result of a shared perspective held by the group members and derived primarily from intensive collective transactions. In isolated societies, group identification has a territorial basis because communication networks are spatially limited. This also holds, to a lesser degree, for residentially segregated ethnic groups in large North American cities (Suttles 1968).

Socioeconomic status refers to the class position of individuals and groups in urban industrial society. It has an objective and a subjective reality: objectively, status is primarily based on education, occupation, and income, while subjectively, it is based on the 'opinions and judgments of some members of a community about the class position or class reputation of other members of the community' (Porter 1965, 9).

Psychological stress designates the state of disequilibrium in which ethnic members find themselves during the transitional stages typical of acculturation processes. It is related to a condition of relative deprivation which affects the household during the process and which is likely to result from a perceived lack of ethnic identity and/or socioeconomic status.

In its simplest form, spatial structure can be equated to the geometrical properties of territorial occupancy. It results from the residential mobility of individuals and can be deduced additively from it. Time underscores the dynamic nature of urban ethnic acculturation; from a process viewpoint, ethnic identity can be thought of as ethnic identification, socioeconomic status as socioeconomic integration, and spatial structure as spatial dispersal (or concentration).

A functional explanation can be defined as one in which 'the consequences of some behaviour or social arrangement are essential elements of the causes of that behavior' (Stinchcombe 1968, 80). Such an explanation is formed from a homeostatic variable, a structure or behaviour which has a causal impact on the homeostatic variable, and tensions which tend to upset the homeostatic variable. In this system, the stronger the tensions, the more the behaviours which tend to maintain the homeostatic variable will be selected.

Using this functionalist terminology, each of the five conceptual variables can be characterized as follows: (1) the homeostatic variable is the psychological stress to be maintained at a minimum for each household; (2) the tensions increasing the level of stress are ethnic identity and socioeconomic status; (3) the behaviour selected to minimize the stress is residential mobility at the household level and spatial structure at the group level; and (4) a certain time period is needed during which a feedback effect takes place between psychological stress and spatial structure (Golant 1971).

As either one or both tensions increase the level of stress of a given household, the propensity for this household to move increases accordingly. Ethnic identification tensions are likely to induce moves away from the community core while socioeconomic tensions will stimulate moves toward the core of the community. The collected propensities for all households in the community yield a respective change in spatial structure. This change can in turn generate a new set of tensions, thus bringing about a repetition of the process.

It is likely that, during earlier time periods, more households will experience stress due to competition for status in the larger society, while, at later time periods, ethnic identity may be a source of stress for a greater number of households. However, if immigration to the community is an ongoing process, a continuous distribution of households should be found with respect to both the ethnic identity and the socioeconomic status dimensions, for any given point in time.

In this sense, the hypothesis can be phrased in terms of the notion of spatial equilibrium (Claval 1970, 110–24). Individual households, if they locate farther and farther away from the historical core of the community, substitute socioeconomic status rewards for ethnic identity rewards. The substitution effect taking place over distance between status and identity also takes place over time, with distance held constant, mainly because of the impossibility for ethnic households to totally isolate themselves from the various media diffusing urban values. The net structural result is a spatial deconcentration of the community.

Theoretically, it is possible for the whole mechanism to operate in the other direction, i.e. identity being substituted for status, thus yielding spatial concentration. It could be argued that this was the case during the formative years of the Black ghettos – and is now the case where the Black Power movement is strong.

The hypothesis developed above represents a holistic approach to urban ethnic acculturation. The variables, or dimensions, included in the framework either comprehend other more specific variables or act as mediating variables.

A partial test of the hypothesis has been performed for the French Canadian community of Maillardville, British Columbia, and further empirical validation in other ethnic colonies throughout Canada is under way.

Winnie Frohn-Villeneuve, Stephen M. Golant, Canada Council.

Claval, P., 1970 L'espace en géographie humaine, *Can. Geog.* 14, 110–24.
Golant, S.M., 1971 Adjustment process in a system: a behavioral model of human movement, *Geog. Anal.*
Hyland, G.A., 1970 Social interaction and urban opportunity: the Appalachian inmigrant in the Cincinnati central city,

Antipode 2, 68–83.
Michelson, W., 1970 *Man and His Urban Environment* (Reading, Mass.).
Porter, J., 1965 *The Vertical Mosaic: An Analysis of Social Class and Power in Canada* (Toronto).
Stinchcombe, A.L., 1968 *Constructing Social Theories* (New York).
Suttles, G.D., 1968 *The Social Order of the Slum: Ethnicity and Territory in the Inner City* (Chicago).

P1128
The quantification of city-hinterland relationships through input-output
PETER P. WALLER *German Development Institute, Federal Republic of Germany*

Geographers traditionally have studied many aspects of city-hinterland relationships (CHR), but have not been able to quantify these relationships in their totality (cf. Spelt 1958). This is indeed a pity, because sociologists and economists have established some theories about the role of CHR for economic development that have been generally accepted, but never proved empirically. Hoselitz has developed the notion of parasitic and generative cities, which are distinguished by their different interrelationships with their hinterlands (Hoselitz 1954/5, 279). Hirschman and Myrdal explain the regional growth process by means of spread and backwash effects that radiate from cities (Hirschman 1958; Myrdal 1957).

A first step towards the quantification of CHR is the economic base approach that quantifies all export activities of a city. However, no distinction is made between such exports that go into the hinterland and those that go out into the rest of the world. Also, imports into the city are not measured (Boesler 1962, 145).

Both shortcomings can be overcome through the use of the input-output method, which has to be specially adjusted for the particular problem. Such a special city-hinterland input-output table was developed by this author and empirically tested in a study on Arequipa in Southern Peru (Waller 1970).

The table is of the interregional type, with the city of Arequipa as Region I and it hinterland as Region II (see Table 1). Government consumption and households are included in the matrix, because these sectors contribute heavily to the CHR, i.e. hospital services commuters' earnings, etc. Conceptual and statistical problems within primary and secondary sectors are well known in the literature, but little has been written on the tertiary sectors which however are typical for their central functions or CHR. One of the various problems will be described briefly here.

If one includes government within the matrix, one has to measure government outputs as well as inputs. Whereas inputs (wages, material, etc.) are similar to those of other sectors, outputs are not usually paid by those who benefit from government services, but rather by quite different taxpaying sectors. If a university in the city has 50% of its students coming from the city and 50% from the hinterland, in reality no payment from the hinterland for these services occurs; rather, the costs of the university are paid out of the general funds of the government. In order to make CHR appear in the input-output table, an indirect accounting was introduced. It was assumed that the central government (outside the matrix!) paid the costs of the university to the students (households) and that those in turn paid the university. Thereby an output of the university to the households of the hinterland, i.e. a typical CHR, appeared in the table.

The main problem with input-output is the collection of data. Here a combination of indirect and direct methods was used. Certain basic data were derived from general statistics and from an input-output table of an earlier period (Isard 1953, 116). The bulk of the data, especially those on interregional flows, however, was obtained through ques-

TABLE 1. Aggregated input-output table of the city-hinterland relationships, Arequipa, Southern Peru (1968, in millions of Soles)*

	Arequipa						Hinterland							Exp.	Oth.	Outp.
	1	2	3	4	5	Total	6	7	8	9	10	11	Total	12	13	14
1. Agriculture	36	37		8	51	132	42		144		23	244	453	27	28	187
2. Industry	14	338	34	60	578	1024	8	6	20		3	199	236	1203	213	2893
3. Commerce	23	61		26	603	713		19	45		15	222	383	107	19	1075
4. Services	11	152	102	60	960	1285	41	2	28	41			30	200		1368
5. Households	64	411	787	994		2256									720	3006
Total (Arequipa)	148	999	923	1148	2192	5410	91	27	237	41	41	665	1102	1537	980	9029
6. Agriculture	19	303		1	159	482	1131	5	436		2	2724	4298	1004	48	5832
7. Mining		6				7			11		1		12	3875		3894
8. Industry		97		3	81	181	9	1	91		70	507	677	640	310	1808
9. Commerce				1	3	5	33		23		10	959	1026	37	1	1069
10. Services	3	4	3	4	8	22	109	179	106	79	67	1162	1702	67		1791
11. Households		7	2	98		107	4342	426	289	800	1023		6880		789	7776
Total (Hinterland)	22	418	5	108	251	804	5624	611	956	879	1173	5352	14595	5623	1148	22170
Total 1–11 (Southern Peru)	170	1417	928	1256	2443	6214	5715	638	1193	920	1214	6017	15697	7160	2128	31199
12. Imports	10	888	38	236	563	1172	117	474	338	29	247	1759	1088			
13. Others	7	588	109	376		1643		2782	277	120	330		5385			
Total Inputs	187	2893	1075	1868	3006	9029	5832	3894	1808	1069	1791	7776	22170			

*The original table consists of 38 × 38 sectors.

tionnaires handed out to firms and institutions of the various sectors (Waller 1970, 26). The preliminary results for Arequipa (Table 1) indicate a rather minor position of CHR within total outputs and inputs of the city. Only 12% of all outputs go into the hinterland but 17% into exports. Only 9% of all inputs come from the hinterland but 12% are imports. These figures, however, become really meaningful only if one is able to compare them with those of other cities for which similar tables have been established. Then a sound classification of cities according to their predominant interrelations with their hinterlands or with the national market will be possible.

A comparison of flows between city and hinterland shows that outputs of the city by far exceed inputs from the hinterland. This might classify Arequipa as a 'parasitic' city in Hoselitz' terms, since, like a colonial power, it draws more money from its hinterland than is returned. However, here again, only after comparison with the structures of other cities can general conclusions be drawn.

The Arequipa table also allows a first comparison between the importance of central functions of the city and other flows into the hinterland. At a first glance, the industrial sectors (452 million Soles) range before services and trade. These, however, are gross flows. Important for the income of the city are net figures, i.e. the fact that of every dollar output earned in industry only 14 cents stay in the city, whereas in services it is 53 cents. Therefore, on a net basis, services range before trade and industry. Broken down into subsectors, no. 1 is education followed by trade in food and beverages, the railroads, health services, textile trade, banks, and textile industry. A similar analysis can be made for the sectors of the hinterland.

Besides giving some insights into the structure of CHR, an input-output table is of great importance for regional planning. Through the use of an inverted matrix, alternative programs and projects can be tested as to their repercussions on the various sectors of the city and the hinterland. Especially if a policy of establishing growth poles and maximizing spread effects for the development of the respective hinterland is adopted, an interregional input-output table is an excellent quantitative basis for sound decisions.

Boesler, K.A., 1962 Zum Problem der quantitativen Erfassung städtischer Funktionen, in *Proc. IGU Symp. Lund 1960* (Lund), 145–55.
Hirschman, Albert O., 1958 *The Strategy of Economic Development* (New Haven),
Hoselitz, Bert, 1954/5 Generative and parasitic cities, *Econ. Dev. & Cult. Change*, 279.
Isard, Walter, 1953 Some empirical results and problems of regional input-output analysis, in Leontieff, W., ed., *Studies in the Structure of the American Economy* (New York), 116ff.
Myrdal, Gunnar, 1957 *Economic Theory and Underdeveloped Regions* (London).
Spelt, J., 1958 Towns and umlands, in *Econ. Geog.* 362–9.
Waller, Peter P., et al., 1970 *La Cuantificación de las Relaciones Ciudad-Area de Influencia mediante el Metodo de Insumo-Producto* (informe provisional), German Development Institute (Berlin). The final report will be available early in 1972.

P1129
Urbanization in Manchuria, 1907–40
I-SHOU WANG   *San Fernando Valley State College, USA*

Cities have existed in China for centuries, but Chinese urbanization has received little attention from geographers. Lack of such studies apparently relates to a shortage of readily available and reliable data. For Manchuria, however, population and household enumeration reports for 1907, compiled by the Ch'ing government, a preliminary population census for 1935, and a population census for 1940, both taken by the Manchukuo, provide data on cities and towns. Reliability of the 1907 data is unclear, but these statistics provide the best available view of Manchurian urbanization at the turn of the century. The 1935 and 1940 tabulations were taken by the Japanese during a stable period, and represent the finest data thus far available for studying Manchurian urbanization.

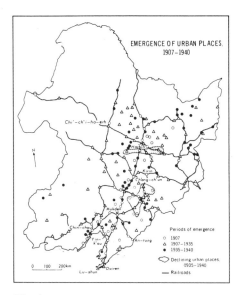

Fig. 1.

The purpose of this paper is to describe and analyse briefly Manchurian urbanization during the period covered by the data identified above, 1907–40.

No official definition of 'urban' existed in China. Thus, the definition of the term 'urban' used here required adoption of two arbitrary rules. An urban place must have (1) at least 5000 inhabitants and (2) one of the following administrative designations, shih (municipality or city), chieh (town), or ch'en (walled city). All populations within the administrative boundaries of such places are considered urban.

There were sixty-two urban places in Manchuria in 1907, and they housed 1.4 million persons, a figure less than ten per cent of the total Manchurian population. Most centres were small, having fewer than 20,000 inhabitants, and most were concentrated in South Manchuria (Fig. 1). Only Mukden and Ch'ang-ch'un had populations greater than 100,000. Between 1907 and 1935, an expanding railway system brought millions of Chinese farmers to Manchuria. Accordingly, the Manchurian agricultural frontier pushed outward and numerous service centres emerged. New and intensified exploitation of mineral resources also stimulated urban growth. New centres with rich resource endowments grew rapidly, and some like Harbin, Dairen, and Fu-shun outgrew the established centres (e.g. Kirin and Ch'i-ch'i-ha-erh) and quickly became major cities. Established towns also experienced urban growth, however.

The number of urban places increased from 62 in 1907 to 145 in 1935. Total urban population tripled from 1.4 million to 5.1 million over the same period. However, the majority of urban places were still small in size (Table 1). Mukden doubled in population from about 250,000 to 500,000 and the number of urban places with over 100,000 people increased from two to eight.

Efforts by Manchukuo and Japan to develop Manchuria's industrial potential accelerated urbanization in the late 1930s. Establishment of manufacturing in large cities, construction of factories and residential building created jobs. Concomitantly, demand for services grew, more urban jobs were created, and still more people from both rural Manchuria and North China were attracted to the growing cities.

By 1940, Manchuria's urban population reached 9.7 million, an increase of over 90 per cent in five years. The number of urban places grew from 145 to 206 over the same period. Urban population growth was concentrated in large cities which accounted for

TABLE 1. Number and size of urban places, 1935–40 (Compiled by author from *Second Annual Report of Statistics* and *Reports of 1940 Census of Population*)

Size (in 1000s people)	Number			Population (in 1000s)			
						Change	
	1935	1940	Change	1935	1940	Net	% of total
Small (5–20)	77	90	13	930	1307	377	8
Medium (20–100)	60	101	41	1984	3456	1472	32
Large (100 and over)	8	15	7	2193	4983	2790	60
Total	145	206	61	5107	9746	4639	100

over 60 per cent of Manchuria's urban growth (Table 1). Over half of Manchuria's urban dwellers were in large cities by 1940, and the leading city Mukden, had 1.1 million. Harbin and Dairen maintained second and third rank with populations of 661,000 and 656,000. Ch'ang-ch'un stood fourth with 555,000. Rapid urbanization between 1935 and 1940 was also reflected in an increase in average urban place size. In direct contrast with 1935, most urban places had over 20,000 inhabitants in 1940. Moreover, the number of places with over 100,000 people increased from eight to fifteen (Table 1). During the same period only 20 had lost population.

In terms of growth rate between 1935 and 1940, large cities, Mukden, Fushun, and Ch'ang-ch'un, more than doubled their populations. Mining towns like Fu-hsin, An-shan, and Pen-ch'i in South Manchuria, as well as service towns in the northeast, also experienced high growth rates of more than 200 per cent. By contrast, most small towns, except newly emergent ones and those in the northeast, experienced either a low growth rate or an actual decrease in population. Declining towns were concentrated in the foothills immediately to the east of Mukden and the area south of Harbin (Fig. 1). New urban places emerged as service centres in rural territory. For example, some thirty places emerged in the core area of Manchuria, where the spatial pattern of new centres suggests a filling-in process complementing the continuing expansion of the agricultural frontier in the north and northeast (Fig. 1).

Japan, Kwan-tung Bureau, 1911 *Gazetteer of Manchuria* (in Japanese), 7 vols. (Dairen).
Manchukuo, Department of Civil Affairs, 1936 *The Second Annual Report of Statistics* (in Japanese) (Hsinking).
Manchukuo, Bureau of Population Census, 1941–1943 *Reports of 1940 Census of Population* (Hsinking).

P1130
Modelling the social geography of urban areas: a Canadian example
MAURICE H. YEATES   *Queen's University, Canada*

This is a report of a project that has been designed to test the efficacy of a series of models and research procedures that can be used to determine the general structure of the social geography of North American cities. The conceptual basis of the structure has been proposed by Rees (1968, 16–18) and is that a family's position in social space (S_k) should have a direct housing space (H_i) counterpart, and that these two should be linked via the social area constructs of economic and family status (Shevky and Bell 1955) to community space (C_j), the units of which have particular locational characteristics (Berry 1965).

With respect to Canadian cities, there is some empirical evidence to indicate the existence of the basic dimensions of community space for Montreal (Greer-Wootten 1968), Toronto (Murdie 1969), and Winnipeg (Nicholson and Yeates 1969). The evidence with respect to the social space and housing space models is, however, rather more inferential.

The Rees model hypothesizes two basic dimensions with respect to a family's position in social space, the axes of which are socio-economic status and the life-cycle. Porter (1965) has demonstrated the existence of a hierarchy of families in urban Canada according to socio-economic status, and Simmons (1968) has emphasized its importance as a determinant of residential location. The life-cycle is a major factor causing people to change their residence (Rossi 1965), and can therefore be considered a second independent dimension of a family's social space. Krieger (1969) in fact suggests that the life-cycle should be the single most important *social indicator* used for developing and evaluating social policy in urban space.

The housing space model, referring to the physical characteristics of a dwelling unit, comprises two independent dimensions (or axes). Each dwelling unit can be described with respect to type, that is whether the unit is an apartment, a row house, or a detached single family dwelling. The inference, of course, is that the type of dwelling unit that a family occupies is very much related to its stage in the life-cycle (Lansing et al. 1969). The second axis refers to the quality of the home, and this is determined by such factors as spaciousness, degree of luxury, and price (or rent).

THE HYPOTHESIS

The model can be expressed symbolically as:
(1) $C_j = f [H_i; S_k]$,
where C_j is the position of the jth census tract in community space; H_i is the position of the ith house in housing space; and S_k is the position of the kth family in social space. In situations of this kind where the symbols refer to vectors, a common procedure is to evaluate the interrelationship using univariate multiple regression techniques. In this case, however, the symbols refer to matrices, not vectors, and so, therefore, multivariate regression techniques are required (Anderson 1958).

Thus, if there are n observations and m variables (axes or dimensions) for each matrix
(2) $C_{n,m} = [H; S]_{n,2m} B_{2m,m}$,
where $B_{2m,m}$ is a matrix of regression coefficients. The hypothesis can therefore be phrased as whether $B_{2m,m} \neq 0_{2m,m}$; in other words, we are seeking to reject the null hypothesis that the matrix B is a zero matrix. Anderson (1958, 196) indicates that this hypothesis can be tested using the F-distribution with $2m$ and $2(n^*-1)$ degrees of freedom, where $n^* = n-2m$.

CALIBRATION OF THE MATRICES: WINNIPEG

In situations where there are a number of characteristics that together define a scale along which units (families, houses, census tracts, etc.) are to be calibrated, factor analysis is a useful procedure (Rummell 1970, 30). It is useful because it not only divides the characteristics into independent sources, or axes, of variation, it also derives factor scores, which are the desired calibrations of each unit along each axis. The scores of the census tracts in community space (C_j) have been calibrated using 1961 census data for Winnipeg, and the method, procedure, and results of this study have been reported elsewhere (Nicholson and Yeates 1969).

HOUSING SPACE AND SOCIAL SPACE

There are many individual characteristics of houses and families that are summarized by the axes of social space and housing space. A random sample of 200 direct interviews using a single page questionnaire-observation sheet yielded 101 usable returns. From this questionnaire-observation sheet fifteen variables (the maximum number that can be derived from the particular questionnaire used) are analysed (Table 1), some of which involve rank-order measurement and others an interval scale.

TABLE 1. Winnipeg sample: varimax rotation loadings*

Variable	Wealth and quality of home	Type of home	Life cycle	Social status
Total income	0.8936			
Income, male	0.8484			
Living condition	0.6531			
Price of home	0.6300			
Lot size		0.8998		
Type of home		0.8587		
Home ownership		0.7501		
Family size			0.7701	
Space ratio			0.6521	
Age of family heads			0.5988	
Family age			0.5314	
Education, male				0.7859
Education, female				0.7113
Occupation, female				0.6232
Occupation, male				0.5844
Proportion of variance (per cent)	20.24	16.39	12.16	13.98
Factor concentration (per cent)	77.44	85.88	91.00	88.32

*All loadings less than 0.5 are excluded.

If the housing and social space models and the axes of each represent separate sets of characteristics, the data should reveal this objectively through factor analysis. Accordingly, the 101 × 15 data matrix was subjected to common factor analysis, using the principal axes solution with unities on the diagonal (Rummell 1970, 104–12), and the first four eigen-values accounted for 63 per cent of the variation in the original data matrix. A varimax rotation of these four factors yields a very clear structure which can be interpreted with some ease (Table 1). The clarity of the structure suggests that the models and axes do in fact represent separate sets of highly interrelated characteristics. The position of each family and each dwelling unit along each of these four dimensions can therefore be calibrated by calculating the factor score matrices.

TESTING THE HYPOTHESIS

The hypothesis to be tested is that the matrix $B_{2m,m} \neq 0_{2m,m}$ (equation 2). In this case $m = 2$ as there are two axes for each model, and $n = 90$. This latter figure represents a reduction from 101, for eleven houses (and therefore families) lay in census tracts not included in the Nicholson-Yeates (1969) study. Using the derived standardized factor score matrices, the F-statistic is calculated to be 4.64, and this is to be compared with the significance point of F_{8170}. This is significant at both the 5 per cent and 1 per cent level and the null hypothesis is rejected. The research hypothesis, which is that the models are linked into a single framework, can therefore be inferred.

Thus the models and research procedures outlined above provide a promising approach to the construction of models of the social geography of urban areas. Further studies with larger and more carefully designed samples are required to establish the structure with greater clarity. The multivariate approach is useful because it permits the impact of change along any of the axes to be analysed in a planning context.

Anderson, T.W., 1958 *Introduction to Multivariate Statistical Analysis* (New York).
Berry, B.J.L., 1965 Internal structure of the city, *Law and Contemporary Problems* 30, 111–9.
Greer-Wootten, B., 1968 Cross-sectional social area analysis: Montreal, 1951–1961, Dept. Geog., McGill U., mimeo.
Krieger, M.H., 1969 *The Life Cycle as a Basis for Social Policy and Social Indicators*, working paper no. 106, Center for Planning and Development Research, U. California, Berkeley.
Lansing, J.B., et al., 1969 *New Homes and Poor People: A Study of Chains and Moves*, Institute for Social Research, U. Michigan (Ann Arbor, Mich.).
Murdie, R.A., 1967 *Factorial ecology of Metropolitan Toronto 1951–1961*, research paper no. 116, Dept. Geog., U. Chicago.
Nicholson, G.T., and M.H. Yeates, 1969 The ecological and spatial structure of the socio-economic characteristics of Winnipeg, 1961, *Can. Rev. Soc. Anth.* 6, 162–78.
Porter, J., 1965 *The Vertical Mosaic* (Toronto).
Rees, P.H., 1968 *The Factorial Ecology of Metropolitan Chicago, 1960*, MA thesis, Dept. Geog., U. Chicago.
Rossi, P.H., 1965 *Why Families Move* (Glencoe, Ill.).
Rummell, R.J., 1970 *Applied Factor Analysis* (Evanston, Ill.).
Shevky, E., and W. Bell, 1955 *Social Area Analysis: Theory, and Illustrative Applications* (Menlo Park, Cal.).
Simmons, J., 1968 Changing residence in the city: a review of intra-urban mobility, *Geog. Rev.* 58, 622–51.

P1131
The structure of common urban constructs
DONALD DEMKO *Queen's University, Canada*

Strong arguments have been presented by Downs and Horsfall (1971) and Winkel (1971) for re-assessing the directions that the behavioural approach is taking in geography, and to evaluate if there are indeed rewards to be gained from pursuing these routes. The former statement concludes that the behavioural approach can be fruitful if a new methodology is adopted.

One of the shortcomings of researchers

who have adopted the behavioural approach in geography is that little concern has been given to the most basic questions. As Downs and Horsfall suggested, the stimuli, cognitive categories, and mediating influences in environmental behaviour have not been identified. Instead, concepts which have geographic implications or meaning for the researcher are assumed to be meaningful for all spatial decision-making. An interesting alternative to this approach has been presented by Kasmar (1970), who argues for the necessity of developing a scale for describing and distinguishing among environments and architectural spaces. Her study is devoted to the first step in this direction, that is, to the development of a relevant, meaningful, and unambiguous lexicon of spatial descriptors which can be used to identify dimensions which are 'central' and 'peripheral' determinants of the perception of architectural space. The same argument can be applied to geographic studies. If we are to develop an understanding of spatial decision-making, it would undoubtedly be efficient to begin by identifying the basic constructs which people utilize in organizing their conception of space.

One approach which has been utilized attempts to identify some process which is conceptually satisfying to describe spatial decision-making (Golledge 1967; Harvey 1968). Another tactic has been to measure the mental images or preferences that groups of individuals have for geographic entities (Demko and Briggs 1970, 1971; Gould 1967; Rushton 1969). Both of these methods suffer from the same shortcoming. Neither necessarily takes account of the basic stimuli and mediating influences which are behaviourally relevant. If we wish to develop testable models and an understanding of spatial behaviour, it would seem that it is imperative to consider these basics. It is with regard to these notions that personal construct theory becomes useful in geographic research.

Personal construct theory (Kelly 1955) is concerned with how man organizes and interrelates with his environment. The theory assumes that man – all men – are scientists who anticipate events rather than simply reacting to them and who understand the universe only insofar as they can make interpretations of it. Further, integration in the universe can be appreciated only by means of a system of ideas applied by an individual. This suggests that the system of interrelated interpretations or meanings of the world should be examined in order to uncover this integrated viewpoint and its composition.

Man-the-scientist is thought to construe relationships constantly in an attempt to relate what was hitherto seen as diverse into an integrated and meaningful system. Each man erects for himself a representational model or construct system of the world which enables him to make sense of it and which makes it possible for him to act in relation to it. Such a construct system may constitute a crude facsimile of the world and be tested and altered in an attempt to make predictions in the future. Thus, man moves in the direction of increased meaning in his own individual terms.

The fundamental postulate of Kelly's theory is that a person's processes are psychologically channelled by the ways in which he anticipates events. This means that man is a form of motion with the direction of the motion controlled by the ways events are anticipated, with these anticipations being defined by each person's personal constructs.

The reality of a construct is in its use by a person as a device for making sense of the world and so anticipating it more fully. It is the construct system which sets limits beyond which it is impossible for a person to perceive, and in this way constructs are seen as controls on a person's outlook and also, in an ultimate sense, as controls on his behaviour. The corollaries summarizing the use, organization, and modification of constructs can be found in Bannister and Mair (1968), on which much of the above summary is based.

Given a group of familiar urban places in southern Ontario which vary in size, functional role, and relative location with respect to large and small places within the system, the objective in this study was to extract those constructs which were commonly utilized by a group of persons for their interpretation and functioning within such urban areas. Then, by supplying these and some additional constructs to a second group of respondents, it was desired to reveal the structure of the constructs as they were utilized to organize and interrelate with the various city types.

The constructs were derived by presenting 60 students with a list of 20 urban areas ranging from village to metropolitan size. Each person was then asked for adjectives or phrases which would differentiate two centres or which characterized more than one place. The opposite word or phrase was then asked for in order to provide bi-polar constructs. The responses were then tabulated and the 26 constructs or complements which were used by more than 75 per cent of the respondents were retained. To this list were added 11 constructs which might be of interest to geographers, resulting in 37 bi-polar constructs in all.

The 37 constructs were then presented as seven-point rating scales (Bannister and Mair 1968) to a group of respondents selected from cities in southern Ontario. Each person was to indicate the amount of the characteristic which could be attributed to each element (city) by rating the cities on each of the 37 scales. The data are then in a format which can be readily factor analysed for the structural characteristics which are contained in the element ratings for each construct. The resulting scales which can be extracted utilizing this framework can shed insight on the structure and content of the city concept as seen by the respondents. That is, it is possible to inspect various characteristics of urban areas by the use of constructs and uncover those which are meaningful to people in organizing their behaviour in cities.

The application of a factorial procedure to the battery of constructs provided by the subject is assumed to yield the structure which characterizes features of the person's view of cities. The results, which will be forthcoming, will present the structure summarized by the dimensions along which individuals organize their interrelated system of constructs referring to the urban environment. In operational terms this procedure results in a geometric or mathematical structure of a person's psychological space.

Canadian Council on Urban and Regional Research; Arden Brummell.

Bannister, D., and J.M.M. Mair, 1968 *The Evaluation of Personal Constructs* (New York).

Demko, D., and R. Briggs, 1970 An initial conceptualization and operationalization of spatial choice behaviour: a migration example using multidimensional unfolding, *Proc. Can. Assoc. Geog.* 1, 79–86.

Demko, D., and R. Briggs, 1971 A model of spatial choice and related operational considerations, *Proc. Am. Assoc. Geog.* 3, 49–52.

Downs, R., and R. Horsfall, 1971 Methodological approaches to urban cognition, paper presented to Am. Assoc. Geog. annual meeting, Boston.

Golledge, R.G., 1967 Conceptualizing the market decision process, *J. Reg. Sci.* 7 (2) (suppl.), 239–58.

Gould, P.R., 1967 Structuring information on spatio-temporal preferences, *ibid.* 7 (2) (suppl.), 259.

Harvey, D.W., 1968 Some methodological problems in the use of the Neyman type A and negative binomial probability distributions in the analysis of spatial series, *Trans. Inst. Brit. Geog.* 43, 85–95.

Kasmar, J., 1970 The development of a usable lexicon of environmental descriptors, *Environment and Behaviour* 2 (2), 153–69.

Kelly, G., 1955 *The Psychology of Personal Constructs* (New York).

Rushton, G., 1969 The analysis of spatial behaviour by revealed space preferences, *Ann. Am. Assoc. Geog.* 59 (2), 391–400.

Winkel, G.H., 1971 Theory and method in behavioural geography, paper presented to Can. Assoc. Geog. annual meeting, Waterloo, Ontario.

P1132
Qualité du commerce et valeur du paysage urbain
JOSÉ A.L. SPORCK   *L'Université de Liège, Belgique*

Le commerce de détail joue un rôle croissant dans notre civilisation de 'consommation.' Certes, la contestation s'en prend volontiers à cet aspect de notre genre de vie. Mais nous lui devons le passage à une durée moyenne de vie double de celle des hommes des pays sous-développés.

Certains excès de publicité sont regrettables. Le 'conditionnement' des individus, organisé pour faire naître artificiellement des besoins pour des produits futiles, se situe au niveau d'une atteinte à la liberté individuelle. Mais, mis à part ces abus, moins graves en Europe qu'aux Etats-Unis, le commerce de détail apparaît aujourd'hui comme un des services des plus utiles et des plus agréables. Le commerce joue aussi un rôle en tant que créateur d'un aspect essentiel du paysage urbain.

L'animation des villes d'Europe occidentale et l'attrait exercé par leurs centres commerciaux contrastent violemment avec l'aspect beaucoup plus monotone des centres urbains des pays d'économie socialiste. Le commerce de détail, entreprise d'Etat, y est considéré comme un mal nécessaire. L'aspect des magasins y est généralement beaucoup plus terne, tant par le cadre de présentation que par l'étroitesse de la gamme des produits vendus et par le manque de recherche ('industrial design') des produits eux-mêmes. Ce contraste saisissant permet de se rendre compte de l'importance du rôle joué par le commerce de détail dans la valeur du paysage urbain.

La dégénérescence du commerce des centres de la plupart des villes américaines offre un exemple, à contrario, de l'importance du standing et du modernisme des magasins pour l'obtention d'un centre urbain vivant, agréable, prospère et de qualité. A ce point de vue, les villes d'Europe occidentale, constituent un exemple significatif qui devrait servir de modèle. Or, trop souvent, dans le domaine de la distribution, on ne propose comme idéal que ce qui vient en ligne droite des Etats-Unis.

L'avenir du commerce du Centre-ville apparaît pour certains comme assez sombre. Une telle évolution compromettrait aussi la valeur économique, sociale et récréative de la cité. Cette vue pessimiste provient du risque d'asphyxie du centre urbain par suite du développement de la circulation automobile et de l'apparition des centres commerciaux, dits périphériques, basés sur la facilité d'accès en voiture. Ces deux arguments doivent être interprétés dans une perspective à la fois dynamique et hiérarchisée des problèmes.

L'asphyxie du centre urbain des villes européennes n'est pas inévitable. Des travaux imposants sont exécutés pour y faciliter l'accès de ceux qui doivent pouvoir y venir en voiture tandis que des mesures sont prises pour dissuader ceux qui feraient mieux d'utiliser les transports en commun sérieusement améliorés. Des parties de plus en plus amples des centres historiques et très commerciaux sont en outre progressivement rendues aux piétons. De plus, les centres urbains s'équipent en parkings qui permettent de se débarrasser de la voiture.

Les rues très étroites, souvent aussi très commerçantes, ont comme fonction logique de redevenir piétonnières, étant particulièrement agréables pour le 'lèche-vitrine' et le coude-à-coude si nécessaires pour la vie urbaine.

Les centres des villes rénovés, réorganisés et restaurés dans leurs parties valables au point de vue historique, architectural et archéologique ont devant eux un avenir absolument remarquable. Leur 'qualité' et la multitude de leurs fonctions complémentaires et très personnalisées les placent très au-dessus des nouveaux shopping centers, même le plus élaborés.

Venons-en à la concurrence, considérée comme dangereuse pour le commerce du centre-ville et provenant des équipements périphériques. En fait, il importe de ne pas confondre deux problèmes. Tout comme le réseau urbain d'un pays doit être hiérarchisé, les grandes agglomérations urbaines doivent être structurées de façon polynucléaire, avec un centre principal ('urban core'), des centres secondaires et, suivant leur importance, une hiérarchie de centres à 3, 4 ou 5 niveaux, descendant jusqu'aux unités de voisinages dont la population se situe vers 10,000 ou 12,000 habitants. Dans une telle organisation de l'espace urbain, les commerces dits périphériques ne sont que les équipements indispensables pour la desserte des populations des diverses parties d'agglomération.

Or, presque toutes les conurbations se caractérisent par des banlieues sous-équipées sur le plan commercial et des services. Compte tenu de cette réorganisation des banlieues, le centre urbain principal doit évoluer vers un type de commerce qui ne se rencontre pas dans ces centres secondaires soit parce que plus spécialisé; plus artistique; de plus haut standing; de plus large assortiment; de type rare; de type spécifiquement métropolitain: à savoir, par exemple, des 'show rooms' que les firmes réalisent à destination de leurs revendeurs installés dans tout l'hinterland d'une métropole, ou des salles d'expositions louées par les firmes de construction d'automobiles, par des maisons de couture (défilés), pour les expositions de service de table, objets de décoration, etc.

Notre méthode d'étude du commerce de détail permet de faire apparaître les caractéristiques qualitatives sans lesquelles on ne peut obtenir une analyse correcte du commerce et de son rôle dans l'urbanisme commercial.

Quelques exemples montreront que, pour connaître la valeur d'un centre commercial et son niveau, il est indispensable de tenir compte des aspects qualitatifs tels que le standing, le modernisme et le rayonnement probable des magasins.

La bijouterie-joaillerie m'a conduit à énoncer presqu'une loi significative. On ne peut totaliser purement et simplement les bijouteries d'un centre-ville ou d'une agglomération pour obtenir une indication de la valeur ou de la position hiérarchique de ce centre. Une bijouterie où l'on vend des bagues et médailles pour communions, fiançailles et des anneaux de mariage n'a rien de commun avec la bijouterie-joaillerie où l'on trouve des bijoux de grande classe et de très grand prix. Il m'est apparu que le prix du bijou le plus cher du centre principal d'une ville est à peu près égal, en francs belges, au nombre d'habitants de l'hinterland commercial de cette ville. Ce rapport, a été vérifié dans plusieurs villes belges, françaises et allemandes. Il est valable jusqu'au niveau des villes rayonnant sur des populations atteignant le 1/2 million d'habitants.

Au niveau des métropoles, le rapport s'établit plutôt avec le chiffre de population de l'agglomération elle-même et dans les toutes grandes capitales mondiales comme Paris, New York et Londres, on retrouve des prix correspondant, en francs belges, aux chiffres de la population du pays tout entier, voire davantage.

Des gradations analogues s'observent dans tous les types de commerces. On passe successivement des fourrures quelconques à l'astrakan, puis au vison mais en cravates et chapeaux seulement, puis au manteau de vison en 2 ou 3 exemplaires et ensuite au magasin avec un réel choix de visons de diverses teintes et avec manteaux en ocelot ou en panthère, voire en Chinchilla.

Ces divers niveaux font l'objet dans notre méthode qualitative d'une cotation allant de 1 à 8: (1) commerce de quartier dégénéré; (2) magasin insatisfaisant à l'heure actuelle; (3) magasin satisfaisant et correspondant à ce que l'on trouve généralement dans les toutes petites villes; (4) magasins les meilleurs dans des villes rayonnant sur 20 à 50 mille habitants; (5) magasins les meilleurs dans des villes rayonnant sur 150 à 250 mille habitants; (6) magasins du standing le plus élevé au niveau des métropoles régionales; (7) magasins du standing le plus élevé dans des capitales telles que Bruxelles ou des très grandes métropoles telles que Milan et Francfort; (8) magasins les plus 'chics' des capitales mondiales: Paris, Londres, New York (certains de Washington), Tokyo.

La cote huit peut se rencontrer même au niveau de magasins d'alimentation, il suffit de songer aux magasins 'Vachon' près de la Madeleine à Paris.

Ces exemples ont pour but de faire saisir la gradation qui permet de mettre en évidence la qualité des centres commerciaux. Cette technique permet de traduire la qualité du paysage urbain produite par les commerces des centres-villes. Ce système a de plus l'avantage de mesurer l'amélioration ou le déclin de certaines parties des centres commerciaux lorsque de telles études sont réalisées à quelques années de distance.

L'évolution dans le commerce est tellement rapide qu'une période de 5 ou 10 ans révèle des évolutions très significatives. Dans l'agglomération liégeoise, nous entreprenons maintenant une telle étude pour la deuxième fois, huit ans après la première. Nous sommes convaincus que des enseignements très importants pourront en être tirés.

Sporck, J.A., 1964 Etude de la localisation du commerce de détail – aspects méthodologiques, *Bull. Soc. d'Etudes Géographiques* XXXIII (1), 63–70, et *Travaux Géographiques de Liège*, n° 149, 63–70.
– 1966 Le commerce de détail à Liège, aujourd'hui et demain: implantation, structure et qualité, *Revue Habiter* 36 (Bruxelles), 20–37, plus 2 planches hors texte.
– 1967 Problèmes industriels, rénovation urbaine et centres commerciaux. Livres-guides I et II des excursions des 8 et 9

septembre du Colloque International de Géographie Appliquée, Séminaire de Géographie de l'Université, Liège.
– 1968, 1969 Le réseau urbain hiérarchisé, base de l'aménagement du territoire et du développement économique, *Bull. Soc. Géographique de Liège* 4 et 5.

– 1969 Problèmes du commerce de détail en Wallonie, *Bulletin d'Information de l'Institut Economique et Social des Classes Moyennes* (Bruxelles), 21e année, n° spécial, mai: L'aménagement du territoire et les petites et moyennes entreprises.

P1133
Application de l'analyse morphologique dans la planification des villes
STEFAN WITKOWSKI   *L'Ecole Polytechnique, Radom, Pologne*

Une ville est un organisme composé de beaucoup d'unités de colonisation qui s'étaient formées dans les différentes époques historiques. L'organisation urbaine et l'occupation de l'espace de ces unités se sont conformées au modèle de l'occupation du temps de leur origine. Ce modèle comprend des éléments du réseau de rues, de la localisation de l'occupation de l'espace, des sortes de matériaux de construction, de la forme technique des objets. Ces éléments forment des ensembles fixes et constituent une base pour distinguer dans le plan d'une ville des unités morphogénétiques qui peuvent être considérées comme héritage de la conception urbaine dans le passé. Dans les étapes successives de l'extension des villes, ces unités passent par une évolution de forme et de fonction se conformant aux nouvelles demandes de production et de consommation de nouvelles générations de la société urbaine. Dans l'extension des villes il y a donc une unité de processus de la prise de l'héritage et de la formation de nouvelles valeurs matérielles dans des unités particulières de l'espace possédant des formes de construction bien différentes et appelées: unités morphologiques.

La base de l'analyse morphologique est constituée par les éléments de la physiographie, du plan urbain et des aménagements fixes résultant de l'activité de l'investissement de la société urbaine, en tenant compte de leur forme actuelle et de la manière de leur utilisation.

L'analyse morphologique permet de distinguer les unités de l'espace typiques, formant des modèles et qui possèdent des traits caractéristiques et particuliers de la forme et de l'utilisation.

L'organisation urbaine d'une ville est aussi bien l'expression de la structure sociale et économique que de la structure fonctionnelle et de la forme technique qui se manifestent dans les bâtiments, dans les constructions et dans les aménagements techniques.

L'analyse de la structure bien compliquée d'une ville permet, en la décomposant en éléments simples et homogènes, d'approfondir les recherches et de déterminer plus minutieusement le caractère des unités morphologiques.

Les unités morphologiques, les unités particulières de l'occupation de l'espace dans une ville, formées dans les différentes époques du développement d'une ville et destinées aux différents programmes fonctionnels, présentent des traits distinctifs aussi bien dans le plan d'une ville que dans les rapports écologiques formés dans ces unités sous l'influence de l'état de l'investissement et de l'utilisation des installations fixes.

Dans la structure urbaine d'une ville il y a toujours des unités morphologiques uniques, par exemple les unités provenant du Moyen Age, les unités qui s'étaient développées hors des murs d'une ville, les unités de la construction de l'époque du baroque ou du classicisme.

D'autres unités du volume bâti et en particulier celles qui proviennent de l'époque de l'agrandissement intensif d'une ville se répètent dans certaines constructions comme par exemple dans celles de la fin du XIX et du début du XX siècle, dans les ensembles de construction des maisons pour une seule famille, dans les ensembles de la construction contemporaine des cités, bien que ces constructions portent aussi l'empreinte de la période de leur réalisation. Les aménagements fixes, liés inséparablement avec l'activité humaine et l'adaptation du milieu géographique à la construction des cités, constituent l'élément fondamental de la structure urbaine d'une ville.

Elles rendent possible les fonctions productives, elles déterminent les conditions d'existence de la population et elles constituent la base matérielle des villes.

Selon la théorie de la reproduction socialiste on peut considérer le développement d'une base matérielle comme des dépendances quantitatives agissant dans le processus du développement de l'économie nationale.

L'objet de la reproduction sur le territoire des villes est constitué par la sphère de la production matérielle où sont formés les biens et les services matériels et par la sphère de la consommation et des services non-matériels. Cette dernière sphère se développe dans les limites déterminées par les fonctions productives dont le fondement est constitué par la base matérielle et technique de la production et de la consommation dans les villes.

L'analyse de la base économique de la production des biens et des services dont dépend en général l'agrandissement (le développement économique) des villes, constitue le problème fondamental des recherches économiques et urbaines des villes.

La base économique, dans les cadres de la production des biens et des services, détermine le degré de l'indépendance économique d'une ville.

Les aménagements fixes dans les cadres de la production participent dans la production des biens et des services directement ou indirectement en rendant possibles les fonctions productives dont résultent les effets économiques. La base économique de la production est la condition indispensable du développement économique des villes, c'est à dire de l'accroissement dans le temps de la production des biens matériels sur le terrain des villes et, per capita cet accroissement peut être réalisé soit par la construction de nouveaux établissements, soit par le remplacement des installations usées et surannées de la production par des installations plus effectives, soit en complétant les installations déjà existantes par des installations supplémentaires. Ce dernier moyen de l'accroissement de la production entraîne en général des investissements supplémentaires des aménagements de la production. Il en résulte que la base matérielle et technique d'une ville, tout en grandissant, est soumise aux transformations et aux perfectionnements continuels. Ces processus obéissent aux droits de la reproduction sociale (Marx), qui concernent le renouvellement, entre autres, des produits sous la forme des moyens de la production et aussi des aménagements fixes de la production. La totalité de ressources des forces productives d'une ville est donc composée d'aménagements fixes provenant de diverses périodes de l'investissement et qui sont dans des états différents de vieillissement technique et économique et aussi de rénovation.

L'efficacité économique des installations productives dépend aussi du choix de la variante optimum de la localisation de nouveaux investissements. Cette variante dépend aussi de l'état des installations fixes qui rendent possible la production (des installations de l'infrastructure économique et sociale) et du coût de l'adaptation du milieu géographique. La tendance à arriver à l'optimum économique qui dépend des conditions économiques existantes décide du choix entre la technique de la production exigeant un apport considérable de finances et de ressources fixes et la technique exigeant un apport plus grand de travail.

L'agrandissement de la base productive des villes offre des possibilités d'embauche supplémentaire (des fonctions exogéniques) qui entraînent aussi l'accroissement de la population urbaine.

Le niveau de vie des habitants d'une ville dépend des revenus de la population qui exercent une grande influence sur l'état de la consommation.

Dans les unités urbaines réprésentant une économie spécialisée (non-fermée), le processus de la production se poursuit indépendamment de la consommation aussi bien du point de vue de la quantité que de la valeur. Dans de telles villes donc les indices relatifs à la production ne caractérisent point le niveau de vie des habitants puisque les biens produits ne sont pas destinés à la consommation de la population. Les conditions de vie des habitants d'une ville sont influencées uniquement par les éléments de la production qui influencent les revenus de la population et peuvent être destinés à la consommation. A ces conditions répondent les revenus pécuniaires de la population ainsi que la consommation naturelle et une partie du revenu national destinée à la distribution sur le terrain de l'unité urbaine en question. Le niveau de la consommation dans les villes dépend des aménagements qui rendent possible la satisfaction des besoins d'existence des habitants des villes et aussi des aménagements fixes de la consommation de longue durée qui satisfont aux besoins des habitants par les services non-matériels (habitat, moyens de communication, établissements municipaux, sociaux, culturels, sanitaires et de l'instruction publique). Il en résulte les effets utilitaires des aménagements fixes.

L'accroissement de la population et de ses revenus ainsi que le relèvement du standard de vie entraînent la demande de nouveaux aménagements fixes pour la consommation de longue durée. L'augmentation de la demande des ressources fixes des aménagements municipaux pour la consommation a donc un caractère qualitatif et quantitatif. Les besoins des habitants d'une ville sont satisfaits par les installations fixes les plus modernes et au plus haut niveau technique.

Le niveau général de vie dans une ville dépend de l'étendue des investissements municipaux réalisés dans des périodes de développement les moins reculées. Ces investissements doivent concerner aussi bien la construction de nouveaux objets que la modernisation des objets construits dans des périodes précédentes. De même que, dans les cadres de la production, les aménagements fixes de la consommation de longue durée, réalisés dans différentes époques d'existence d'une ville, se trouvent dans des états divers de vieillissement technique et économique, ce qui influence aussi l'état et le niveau du satisfaction des besoins des habitants qui profitent de ces aménagements. Puisque les unités morphologiques d'une ville proviennent en général d'une seule période bien déterminée de réalisation, les effets utilitaires des aménagements fixes dans chaque unité se trouvent pour la plupart au même niveau exerçant une grande influence sur les rapports écologiques dans cette unité.

P1134
Some territorial aspects of local recreation in agglomerated areas of the GDR (demonstrated with the example of the capital, Berlin)
H. KOHL and ALFRED ZIMM   *Humboldt University, East Germany*

The monocentric agglomeration found on the territory of the GDR capital, Berlin, is one of the most important agglomerated areas in the country. Though Berlin covers only 0.4 per cent of the country's total territory, it has 6.4 per cent of its population and 5.4 per cent of the total industrial production.

As elsewhere in the country, the administrative bodies of Berlin, too, are making efforts to develop the human environment as well as possible with the help of socialist planning. The systematic satisfaction of the demand for local recreational facilities is one of the problems within the framework of improving the working and living conditions of the people. An analysis prepared by the geography section of Berlin's Humboldt University at the request of the city government has resulted in a number of new findings with regard to the territorial aspects of this problem, some of which we are going to present in this paper.

Local (open-air) recreation is an integral part of the reproductive complex of relations between the fundamental spheres within the life of the population: working, living, and local recreation. In territorial terms these relations manifest themselves in the functional connections between the place of work, the residential area, and the local recreational facilities.

In this relationship, the place of work and the residential area are the places where the demand for local recreation arises; both of them, taken together, are called 'recreation demand area.' In functional terms the area where the place of work is situated can be characterised as the place where a demand for local recreation is, so to speak, 'produced' through the process of work. In general, however, the demand for local recreation arising at a person's place of work is not directly satisfied from there, but is 'transferred,' in most cases, to the residential areas to be satisfied from there. By contrast, the residential areas have a threefold function: they 'produce' a demand for recreation, 'accumulate' both this demand arising in the residential areas themselves and the 'transferred' demand arising at the place of work,

TABLE 1

Day of the week	Season	Predominant motivation	Predominant activities
Monday to Friday (i.e., working week)	all seasons	local recreation area must be within easy distance	going for walks
Saturday and Sunday (i.e., weekend)	warm season only	local recreation area may be somewhat distant, but should be quiet	bathing, going for walks, hiking
Saturday and Sunday (i.e., weekend)	cold season only	local recreation area must be within easy distance	going for walks, winter sports

and constitute the point of departure from which the demand is satisfied. Thus, the relationship between the residential area and the local recreation area is the basic territorial relation between areas where the demand for recreation is produced and areas where it is satisfied. All efforts to ensure optimum planning of local recreation must, therefore, proceed from this fundamental relationship.

Scientific and technological progress generally entails an increase in the number of jobs requiring a minimum of movement on the part of the worker. At the same time, observation has shown that a considerable proportion of people's leisure time is also spent on activities requiring a minimum of movement. This makes it imperative for planners to see to it that active forms of recreation are offered and actually made use of. For this reason, territorial planning in the GDR has been oriented toward ensuring that, in accordance with the demands of the population, recreational facilities are developed in such a way that recreation, education, adventure, entertainment, and compensating physical activities are sensibly combined with each other. This combined form of recreation can to a large extent be achieved in local recreation areas (and, of course, during people's holidays).

Local recreation areas are mass recreational facilities which, thanks to their natural resources and additional man-made equipment, can be developed in such a way that people's desire for recreative activity, rest and relaxation, culture, and education is satisfied. In the agglomerated area of the capital of Berlin there are four basic types of local recreation areas, viz. (1) large green plots (parks, etc.) with unspecific equipment, of emphatically local significance, (2) large green plots with special equipment, of emphatically supra-local significance, (3) wooded areas of restricted supra-local significance, and (4) wooded areas with lakes, of national significance. These four basic types are the basis and the constituent elements of planning in the field of local recreation areas.

Fully aware of the individual differences in motivation prompting people to prefer one or the other local recreation area, we can, nevertheless, define two main motives, depending on the day of the week and the season, and two corresponding groups of activities (see Table 1). The predominance of the two motivations is explained by the amount of leisure time people are prepared to spend on local recreation and the special conditions of life in a metropolitan agglomeration. The amount of time spent by the inhabitants of the capital on recreation in the local recreation areas of Berlin is as follows:

Day of the week	Warm season (minutes)	Cold season (minutes)
working week	120	70
Saturday	330	140
Sunday	360	150

The current typical time limit results in the fact that, during the working week in the warm and cold seasons and on Saturdays and Sundays in the cold season, the demand for minimum distances between the demand areas and the local recreation areas outweighs all other motivations. In the circumstances obtaining in Berlin, it was found that fifteen minutes' walking is generally considered as the longest acceptable distance to local recreation areas, as most people want to reach the local recreation area on foot. During the working week the average ratio between the time a person spends walking to the recreation area and the duration of his stay there is 1:3 in the warm season and approximately 1:1 in the cold season, assuming that he visits the recreation area once only. Taking these facts into account, town planners must provide for optimum communications between residential areas and local recreation areas. Only thus can the utilisation of the reproductive effects of local recreation areas be fully guaranteed, i.e., the utilisation of specific social and natural facilities for the individuals in a given catchment area. The distribution of housing construction and green plots must be planned accordingly. The socialist production relations have created the conditions for doing so in the interest of the entire population.

An entirely different mobility scheme, in terms of time and territory, obtains for weekends during the warm season. A considerable increase in the amount of leisure time available for local recreation on Saturdays and Sundays is accompanied by increasing readiness on the part of the population to use the bigger part of their leisure time spent on local recreation for covering the distance between the residential area and the local recreation area. The way of life typical of agglomerated areas causes people to seek recreation in quiet areas with plenty of woods and lakes, so as to balance the manifold environmental stimuli to which they are exposed during the working week. This motive is stronger than any other motive – so strong, in fact, that people accept rides of two and more hours (there and back). With regard to the choice of destinations, there is a tendency towards spending as little time on transport within the agglomeration as possible. On weekends, however, the range of deviations from this tendency is so wide that it is extremely difficult to establish any firm territorial relationship between a given residential area and a local recreation area.

Whereas long distances are the main obstacles to satisfying the demand for local recreation during the working week, it is overcrowding that constitutes the principal nuisance on weekends. It appears, however, that this factor is significant mainly for the population of such residential areas that have unfavourable communications to the local recreation areas. As a result, people have to spend more time on going to, and returning from, the local recreation areas, which reduces the radius of their efforts to seek recreation on the outskirts of the city. Consequently, those local recreation areas are particularly popular which are both comparatively near to the demand areas and marked by an especially high number of visitors. Thus, the time necessary to reach a recreation area plays a regulative role even on weekends.

Being part of a more comprehensive complex of effects, the functional and territorial relationships explained in this paper are of fundamental significance for reaching the planned goal, viz. ensuring optimum conditions for local recreation and optimum utilisation of existing local recreation areas. They are important for reaching the goal set forth in the directive of the five-year plan for the development of the GDR's national economy: continued development of local recreation, improvement of natural environmental conditions, provision of more possibilities for sports and cultural activities.

P1135
Inner functional zoning and rationalization of use of town territories
OLEG LITOVKA *Leningrad Financial-Economic Institute, USSR*

In spite of the constantly increasing role of towns in the life of community and great attention being paid to the study of urbanization processes by geographers, the problems of inner territorial zoning of towns are not yet fully studied.

The urban population of the Soviet Union is approximately 140 million and its growth is more extensive than that of the state's population. The areas of urban territories are also growing from year to year. According to rough calculations of the author, 5500 urban settlements of the USSR now occupy about 80,000km^2. Thanks to the vast expanses of the Soviet Union the problem of territorial growth of towns in our state is not so urgent as to compare with many other countries, but urban expansion causes the exclusion of land areas which are valuable in agricultural, sanitary, and other respects and also complicates the normal functioning of the town's organism itself.

The inner functional zoning makes it possible to use urban territories more rationally and optimize urban population in the whole, so as to form the most favourable conditions for urban life and productive activity of the town's organizations and enterprises. Functional zoning is sure to take into consideration the perspectives for development of the urban place which is in no degree a static formation, but one that is constantly transforming itself in time.

Within the borders of a modern town it is possible as a rule to distinguish the following functional zones: dwelling zone, which includes regions of inhabited and public building; a zone of public use and town's communications; a zone for external transportation; and some other zones. For historical reasons, these zones do not always form pronounced functional zones, but proper organization of such zones can restrict territorial growth of the town and find inner resources for its future development.

Dwelling zones are the most important parts of urban territories. In proportion with the town's population growth the proportion of residential land as a rule declines in all parts of the town, caused largely by the rise of average number of storeys in inhabited and public buildings of cities. Nevertheless it should be mentioned that even in many large cities the proportion of regions occupied extensively by one-storeyed buildings is often very high, but the degree of territorial utilization is ten and more times as much as, for example, in the regions with nine-storeyed buildings. That means that one-storeyed buildings in large cities are economically and functionally unsuitable and should be gradually replaced by higher buildings. The existing regions with one-storeyed buildings are an important inner territorial resource.

In many towns, especially in central regions, the right-angled layout of streets is still preserved. That results in a very high proportion of streets, whose traffic capacity doesn't meet the requirements of constantly rising urban transport streams. The creation of a more rational street network will make it possible to solve transport problems, with a simultaneous reduction of the proportion of streets, and will also improve the communication between residential districts and districts of application of labour and of siting of public-service institutions; in other words it will enable the labour and cultural and daily outings of inhabitants to be more convenient and fast.

Public planting of trees and shrubs are obviously the single part of dwelling territories one must not aim to reduce, but in their turn the planting of trees and shrubs must be placed in such a way as to be the perfect answer of their purpose, in other words to improve microclimatic conditions of the town and its air space. For this purpose the planting of trees and shrubs must chiefly represent compact formations connected if possible with a green belt outside the town.

In accordance with the fact that the basis of the urbanization process in the USSR lies in industrialization (the emergence of three-quarters of soviet towns created in the last fifty years is due to industry), industrial territories in the towns of our country are very large and the area of industrial territories per thousand of population is bigger than in other countries. The proportion of that category of urban land may be considerably reduced by blocking of the industrial buildings and facilities, distribution of new industrial enterprises in clusters, or creation of urban industrial regions, that permit the creation of common spur-tracks satisfactory for several industrial enterprises, the systems of engineering equipment, and store houses, and also lead to absolute reduction of size of sanitary inspection protective zone from enterprises with unhealthy trade; otherwise such zones overlap. The work of designing the urban industrial clusters and regions that has been conducted in the USSR shows that group distribution of enterprises brings great economy of capital investments, apart from cutting down of needed territories.

Technical progress made possible the development of the territories which were considered to be useless in the past. The use of territories on the coast of Finnish Bay in Leningrad and in the floodland of the Dnieper in Kiev thanks to hydroalluvium illustrates this progress. Possibilities of assimilating land with steep slopes are also revealed. Many Soviet specialists on town-building suppose that building on land with an inclination of 30 per cent and more is profitable; that fact favours the drawing of rather great areas into active urban life. An important reserve lies in a gradual change within the limits of urban building from surface lines of electrotransmission to underground cable lines that will sharply improve the town-planning structure. Very often in the existing town large territories are used for residential and public-building purposes in proximity to industrial enterprises with unhealthy trades. Sanitary inspection protective zones from such enterprises, especially when situated in central regions, eliminate from rational use very large areas of urban territories.

It should be noted that forms of urban settlement have recently undergone many changes and in many cases the objective realities are systems (or agglomerations) of urban settlements with very complex social and economic connections between town and surrounding territories. The number of inhabitants of many of such systems achieves a considerable value. At present in the USSR there are approximately thirty agglomerations with a population that exceeds or approaches one million. In the territories which are parts of the composite of agglomeration there must be reserve areas for further development of settlements, industrial enterprises, and transport, territories of agricultural use, territories of mass recreation for the population. In that case special care should be given to rapid conversion of the neighbouring large tracts of forest into forest-parks; otherwise the big flow of visitors will lead to destruction of these large tracts of forest. Outside the limits of main urban cores there should be placed several higher educational establishments and scientific-research institutes.

In conclusion, it may be said that nowadays there is an urgent need of transition from functional zoning of separate urban settlements to zoning of whole agglomerations and urbanized areas.

P1136
Aspects du développement urbain en économie socialiste d'après l'exemple de la République démocratique allemande
HENRI SMOTKINE *Université de Paris, France*

Si l'on considère le taux d'emploi de la population active dans l'industrie et la valeur de la production industrielle par tête d'habitant, la RDA est le plus industrialisé de tous les Etats socialistes. Son développement urbain, dans le cadre d'une économie très évoluée, est influencé par l'organisation socialiste des rapports de production.

Les agglomérations de plus de 2,000 habitants groupent 73.2 pour cent de la population en 1968 contre 72.2 pour cent sur le même territoire en 1939. La relative stabilité du pourcentage global ne rend pas compte d'une double évolution: la population des villes de moins de 50,000 habitants s'est élevée de 40.1 pour cent de la population totale en 1939 à 44.9 pour cent en 1968; par contre celle des villes de plus de 50,000 habitants est passée de 32.1 pour cent de la population totale en 1939 à 28.3 pour cent en 1968.

La croissance des villes petites et moyennes est liée à la décentralisation industrielle dans le cadre d'une économie socialiste; la planification impérative permet de donner aux ouvriers et aux employés la sécurité de l'emploi et soustrait les entreprises dans lesquelles ils travaillent aux aléas de la libre concurrence. Par ailleurs, l'attrait de ces villes petites ou moyennes est accru par la décentralisation du commerce et des activités culturelles.

Les grandes villes (à l'exception de Halle, ville non détruite pendant la guerre) n'ont pas retrouvé leur chiffre de population de 1939. Leur destruction partielle au cours des hostilités a entraîné le départ de nombreux habitants qui ont conservé leur résidence dans les villes ou les villages avoisinants. En outre, la reconstruction, faisant disparaître les taudis hérités du XIXe siècle, a mis fin à l'entassement des maisons ouvrières dans les noyaux urbains. Cependant, on n'observe pas en RDA un étalement des zones suburbaines et périurbaines comparable à celui des pays d'économie libérale ayant le même niveau de développement technique.

Ces modifications dans l'habitat urbain ancien sont allées de pair avec la création de villes nouvelles qui s'inspirent des expériences soviétiques, britanniques et scandinaves: autour d'un centre urbain, commercial, culturel et récréatif, qui donne une âme à la ville, des quartiers d'habitation aux architectures plus ou moins variées possèdent les magasins pour la satisfaction des besoins quotidiens (boulangeries, charcuteries, épiceries, débits de journaux ...) et les établissements assurant la prestation des services sociaux (crèches, jardins d'enfants, écoles polycliniques, terrains de jeux ...). Eisenhüttenstadt (qui s'est appelée jusqu'en 1961 Stalinstadt) née en 1950, comptait, après avoir englobé la petite ville voisine de Füstenberg-sur-l'Oder, 38,138 habitants en 1965, et 44,605 habitants en 1969; Hoyerswerda, où logent les travailleurs du nouveau combinat chimique de Schwarze Pumpe, a vu sa population croître de 7,365 habitants en 1950 à 56,200 en 1969; Schwedt-sur-Oder, où aboutit l'oléoduc 'Amitié' venu d'URSS et où se construit, à côté d'une nouvelle entreprise fabriquant le papier et le carton, un puissant combinat pétrochimique, est devenu une véritable ville, sa population passant de 6,606 habitants en 1950 à 31,390 en 1969.

Halle-Neustadt, en construction depuis le 15 juillet 1964 et dont la population s'élevait au 31 décembre 1969 à 27,898 habitants (la ville achevée en comptera 70,000) n'est pas une ville nouvelle au sens britannique du terme. Son urbanisme est assez semblable à celui des villes nouvelles, mais sa population ouvrière travaille dans les grandes usines chimiques du complexe industriel de Merse- bourg, distantes de 10 à 25km. Pour cette ville en construction qui se trouve à la lisière de la grosse agglomération de Halle, sur la rive gauche de la Saale, la vieille cité exercera les fonctions de centre de services sociaux et culturels, fonctions favorables à la création d'un sentiment d'appartenance à une grande collectivité urbaine.

Dans les pays socialistes, le développement urbain planifié se fait par une adjonction à l'agglomération, née du développement historique, de quartiers périphériques constitués par de grands ensembles dont la localisation est en rapport avec les implantations industrielles. Chacun de ces ensembles a ses centres commerciaux et assure à ses habitants les prestations de services sociaux et culturels. Cependant, le vieux centre urbain conserve ses fonctions d'animation sociale et culturelle afin de donner une âme collective à la ville. Ce souci nous apparaît comme un des aspects fondamentaux de l'urbanisme socialiste qui, par suite de l'effacement de certaines fonctions tertiaires, n'a pas à tenir compte du phénomène de 'citysation,' tout au moins sous sa forme traditionnelle de centre des affaires.

Par ailleurs, la décentralisation du commerce et de la vie culturelle, liée à la planification socialiste, favorise les villes petites et moyennes: en Saxe, en Bohême, en Silésie, le travailleur n'hésite pas à s'installer dans une agglomération d'importance restreinte où se trouve assurée sa sécurité d'emploi.

Enfin, dans les régions industrielles densément peuplées, il se produit une interpénétration des habitats urbains et ruraux, mais cette suburbanisation n'est pas spécifique de l'économie socialiste; le même phénomène peut s'observer en Europe occidentale. Toutefois, le développement de l'agriculture collective crée des conditions favorables à l'implantation ouvrière dans les campagnes: les villages s'agrandissent à cause de la concentration de la population rurale autour des bâtiments co-opératifs, et, agrandis, ces villages acquièrent progressivement, par suite des changements dans les structures socio-professionnelles régionales, une population ouvrière industrielle dont les modes de vie et les comportements sociaux exercent leur influence sur les travailleurs agricoles, favorisant ainsi la formation d'une nouvelle société industrielle.

Smotkine, Henri, 1967 'La République démocratique allemande' in A. Blanc, P. George, H. Smotkine, *Les Républiques socialistes d'Europe centrale,* Collection Magellan (Paris).
— 1970 'L'individualité géographique de la RDA,' *Revue d'Allemagne,* T11 n° 3.

P1137
L'insertion et le rôle des villes touristiques dans le système urbain régional
GERMAINE VEYRET-VERNER   *Université de Grenoble, France*

Dans l'étude des systèmes urbains régionaux il est assez rare que l'attention ait été attirée sur les villes touristiques; les méthodes d'analyse prospective ont jusqu'à présent laissé de côté ce problème, mettant l'accent sur le rôle de la métropole régionale, sur la hiérarchie, la dépendance et les zones d'influence des autres villes, sur l'opportunité des villes nouvelles ou de certaines zones industrielles en fonction des grands axes de communication et des flux économiques. Les problèmes de services et d'emplois industriels, de migrations de travail sont les préoccupations dominantes des schémas directeurs pour l'organisation d'un système urbain régional. Notre but est d'essayer de déterminer au contraire quels peuvent être la place et le rôle des villes touristiques dans ce domaine en nous référant plus spécialement aux exemples alpins. Pour celà il nous a paru intéressant de montrer la spécificité de la ville touristique, son intégration dans le système urbain régional, de déterminer dans quelle mesure le tourisme peut être un facteur d'évolution des systèmes urbains régionaux.

On peut définir une ville touristique comme un organisme dont 75 pour cent au moins de la population active vit du tourisme et de ce fait occupe des emplois tertiaires, qu'il s'agisse des commerces, des transports, des services, de l'hôtellerie, des activités d'animation au sens large etc. Ainsi définie, cette ville diffère des autres types de ville pour plusieurs raisons évidentes:

– C'est un type de ville entièrement tributaire de la société de consommation et qui n'apporte aucune contribution à la production proprement dite. Si certaines villes touristiques sont devenues aussi des villes en partie industrielles (Nice par exemple), il s'agit en général de grandes villes; la fonction touristique peut encore y être prépondérante mais a perdu son caractère absolu, se doublant d'activités administratives, judiciaires, politiques ou industrielles. Au contraire, la véritable ville touristique consomme mais ne produit pas; les transports de marchandises se font à sens unique car elle achète mais ne vend pas (sauf des loisirs). Ce simple fait implique des transports plus coûteux puisqu'il n'y a pas de fret de retour.

– C'est une ville dont le rôle principal est le loisir et non le travail puisqu'en saison la population active (commerces, services) ne représente environ que 10 pour cent de la population résidente, celle-ci pouvant quintupler la population permanente. L'organisation du repos, de la détente, des sports, des distractions ne développe guère les facultés créatrices hors du domaine très spécialisé du tourisme. Les jeunes de la population permanente seront donc en grande partie obligés de chercher dans d'autres villes, régionales ou hors de la région, les possibilités d'enseignement, de formation ou d'emplois autres que celles du tourisme.

Dans le domaine de l'urbanisme proprement dit, la ville touristique est également très spécifique. L'urbanisme fonctionnel est soumis à des impératifs particuliers aussi bien en ce qui concerne les équipements collectifs que les services. Les infrastructures y sont hors de proportion avec la population permanente (accès, voirie, parking, équipements sanitaires, sportifs, de remontées mécaniques etc.). Les services et les commerces eux-mêmes sont très supérieurs en nombre et en qualité à ceux d'une ville de même population (banques, agences de voyages, galeries marchandes). Ainsi les équipements et les services nécessitent une conception différente de celle des villes actives de même population aussi bien catégoriquement que spatialement.

L'urbanisme résidentiel est plus différent encore. Si la concentration peut s'adapter aux commerces et aux services, les quartiers résidentiels exigent en général (sauf dans le cas des stations entièrement axées sur les sports d'hiver) un coefficient beaucoup plus faible d'occupation des sols et plus d'espaces verts. Le site urbain a donc des limites et des besoins différents de ceux d'un organisme urbain classique.

Autre différence fondamentale, *la faiblesse du poids démographique* dans la région. Aucune ville typiquement touristique n'atteint 100,000 habitants permanents, ce qui est peu à l'échelle des villes actuelles. En montagne, la plupart d'entre elles ne dépassent pas 10,000 habitants permanents. Cette spécificité étant très nette, comment la ville touristique s'insère-t-elle dans le réseau urbain régional?

L'insertion de la ville touristique dans le système urbain régional et son évolution: Nous envisagerons deux cas, celui d'une région à prédominance touristique, celui d'une région à fonctions multiples et complexes.

(1) Il n'existe que très peu de régions dominées par la fonction touristique et, dans ce cas, il s'agit de petites régions ne correspondant pas forcément à une entité administrative. Nous pourrions citer la Côte d'Azur française et son hinterland immédiat ou les lacs Tessinois. Dans ce cas, la ville touristique

est un maillon plus ou moins satellisé de la chaîne. En effet, il peut alors exister une ville touristique majeure et une quantité de petites villes satellites avec une disposition en étoile ou linéaire suivant les conditions géographiques. Dans ce cas, la ville majeure peut jouer le rôle de petite capitale régionale (Lugano en Tessin). Dans d'autres cas, la ville qui avait primitivement la fonction régionale peut être influencée par la fonction touristique prédominante, s'agrandir et évoluer en fonction du tourisme (la ville de Coire dans le canton suisse des Grisons). Allant plus loin, le tourisme peut être à l'origine d'un développement urbain complexe d'une ville qui, primitivement touristique, peut aspirer à devenir une véritable capitale régionale (Nice en France).

(2) Dans le cas d'une grande région à activités multiples (agricole, industrielle et commerçante), les villes touristiques plus petites que la moyenne s'insèrent plus difficilement dans le système urbain, surtout si la région est dominée par une métropole économique (Milan, Marseille, Lyon par exemple). Alors plusieurs cas se présentent. La ville touristique devient un lieu de détente à proximité des autres villes actives qui lui fournissent l'essentiel de sa clientèle, mais reste indépendante pour les principaux services et ne participe à l'activité régionale que par sa fonction de loisirs. Elle a ses propres problèmes de croissance et d'urbanisme très différents des problèmes urbains régionaux, mais elle contribue à la vie urbaine régionale par la spécificité de sa fonction. Cette contribution est d'autant plus sensible que la ville touristique est plus proche des grandes villes. Par voie de conséquences, les investissements peuvent être financés par des capitaux d'origine régionale et les travaux d'équipements confiés à des entreprises régionales ayant leur siège dans les principales villes de la région.

Un autre cas peut se présenter lorsque la ville touristique atteint un niveau national ou international. Alors elle reste en dehors du réseau régional, surimposée à la région dont elle dépasse le cadre par son influence et son pouvoir d'attraction. Chamonix par exemple ne participe qu'indirectement au système urbain régional Rhône-Alpes.

En conclusion, il apparaît que le tourisme peut être un grand facteur d'évolution des systèmes urbains régionaux.

(1) Les possibilités du tourisme dans une région (ski, alpinisme, lacs, montagne) peut avoir un pouvoir attractif et contribuer au développement de villes non touristiques en favorisant une immigration définitive et des implantations industrielles. Il peut donc par ce biais renforcer le système urbain régional dans une région active susceptible de bénéficier des possibilités touristiques (développement de Grenoble, d'Annecy, Nice etc.).

(2) Le tourisme peut revigorer des moyennes et petites villes en voie de déclin en leur adjoignant une fonction de relais touristique, ou en permettant le développement de petites industries du tourisme (articles de sport).

(3) Enfin, il peut créer des stations nouvelles, très urbanisées et qui contribuent à renforcer le semis urbain, donc à maintenir ou à créer des services dans de petites régions où ils étaient en voie de disparition.

Dans cette évolution, la grande victime est le plus souvent l'activité agricole mais c'est un autre problème!

Veyret, G., 1969 *Essai de classification des moyennes et petites villes et leur insertion dans un réseau urbain* (Grenoble).
Veyret, Paul, G. Veyret-Verner, & Gilbert Armand, 1967 'L'organisation de l'espace urbain dans les Alpes françaises du Nord,' *Revue de Géographie alpine*.
Veyret, P. et G., 1967 *Au cœur de l'Europe, les Alpes* (Paris).
– 1964 'Petites et moyennes villes,' *Revue de Géographie alpine*.

P1138
L'effet de synergie et le développement des villes en pays d'économie développée
ANDRÉ C. VIGARIE   *Université de Nantes, France*

L'urbanisation est un facteur de développement et de croissance économique et culturelle; et cependant l'observation révèle l'existence de villes en état de stagnation ou de recul, à côté d'autres marquées de dynamisme. Il est parfois difficile pour le géographe d'expliquer ce contraste. Les causes en sont nombreuses; certaines seront rappelées ci-dessous; parmi elles, il en est une, encore mal connue dans ses caractères: *l'effet de synergie*. La présente communication a pour but de le distinguer des autres facteurs d'évolution des économies urbaines, et ainsi d'en faciliter l'étude.

De nombreux auteurs, économistes ou géographes, ont étudié l'importance de cet environnement pour l'essor des fonctions industrielles et tertiaires. L'on sait d'abord qu'il doit présenter un certain nombre de *cohérences* internes: cohérence verticale résultant du processus d'induction technique (cycle de production avec industries lourdes de tête et industries induites), dont on a tiré la théorie des pôles d'entraînement étudiée par J. Boudeville (1968); cohérence horizontale, c'est-à-dire intéressant un même plan de la production et dont un exemple est fourni par la nécessité des industries' auxiliaires (entretien, sous-traitance...) étudiées par E.S. Tosco et Ital Consult pour le compte de l'ONU (1966).

L'on sait aussi qu'il existe pour chaque ville une masse critique, un seuil quantitatif à partir duquel apparaissent de façon optimale les *économies d'échelle* par branches de production, et les *économies externes* résultant de l'utilisation en commun des équipements et avantages issus du contexte économique et urbain, et permettant d'accroître productivité et rentabilité.

Aux effets dûs à ces aspects quantitatifs de l'environnement, il faut joindre ceux dûs aux caractères qualitatifs. Jean Gottmann (1961, 1966, 1970) a montré dans des études très connues que la concentration provoque la *valorisation mutuelle* des différents services, de sorte que la valeur et la signification économiques d'un centre tertiaire – un centre-ville par exemple – sont supérieures à la somme des constituants; et le phénomène est transposable dans le domaine industriel. S'appuyant sur cet auteur, J. Remy (1966) montre que la ville est un milieu spécifique, un tout qu'on ne peut pas totalement appréhender par la seule étude de ses éléments, qu'elle crée un contexte humain et économique différent en qualité et quantité de la somme de ses entreprises. L'on s'accorde pour reconnaître que les ensembles urbains possèdent, à des degrés divers, une dynamique interne qui transforme l'homme, la production, la consommation.

Ces divers phénomènes: cohérences, économies d'échelle ou d'agglomération, transformations qualitatives, reposent sur des processus d'activités secondaires ou tertiaires, dont certains ont été bien étudiés; et selon qu'ils se réalisent avec plus ou moins d'intensité, ils fournissent une part d'explication quant au dynamisme des villes, en relation avec d'autres faits que l'on ne méconnaît pas (démographie, politique, planification, etc.).

L'étude de l'effet de synergie apporte un complément utile aux processus ci-dessus rappelés, dont il se différencie, et qui vont aider à le caractériser.

Il est un phénomène de création d'activités nouvelles, d'autre nature que celles préexistantes, mais rendu possible par la présence de ces dernières; c'est un produit de l'environnement urbain. Ainsi, dans l'estuaire de la Loire (France) préexistait une industrie des produits azotés fabriqués à partir de l'air: l'oxygène était un sous-produit dont la commercialisation était très difficile. Se crée une usine d'engrais azotés; à proximité est une raffinerie. Les conditions d'environnement sont réunies pour que, grâce à la production massive et peu onéreuse d'oxygène et à un écoulement local assuré (constructions navales, etc.), la création d'une petite aciérie à l'oxygène, très spécialisée, soit mise à l'étude et ait de fortes chances de réalisation dans l'avenir. Autre exemple dans le domaine des services: à la suite d'une reconversion industrielle et commerciale, entre 1950 et 1965 environ, l'agglomération de Nantes (France) a vu se créer des établissements de marketing, de développement de la recherche sous contrat, d'études techniques diverses, de traitement à façon de l'information. Des faits semblables pourraient être analysés à travers la récente croissance d'Anvers, de Rotterdam, etc.

Ces exemples permettent de caractériser le processus de synergie.

Il exige la *présence préalable d'un environnement propice* né de la juxtaposition des formes de productions favorables à la naissance de nouvelles entreprises; cette juxtaposition sous-entend une concentration spatiale suffisante, même si elle est atténuée par les facilités de transports et de télécommunications: elle doit permettre des contacts aisés fréquents. Dès lors se posent un certain nombre de problèmes: celui de la masse critique de cette concentration, et de la combinaison qualitative, à partir desquelles le processus de création est possible; celui des contraintes de localisation auxquelles doivent obéir, ou desquelles sont affranchies les entreprises nouvelles; celui de l'extension spatiale des aires d'intervention des activités d'origine synergétique: un bureau d'engineering s'appuie sur des firmes nationales, voire étrangères, mais sa localisation primitive est motivée par un certain contexte économique local ou régional. Localisation, extension spatiale, environnement: faut-il souligner que ce sont bien là des domaines d'études privilégiés pour le géographe?

Autres séries de caractères: la *nature* des activités économiques nouvellement créées est différente de celles préexistantes. Il y a nécessairement une complémentarité fonctionnelle entre elles, mais sans dépendance étroite, absolue, comme celle que l'on trouve dans les faits d'induction technique. L'effet de synergie apparaît le plus souvent dans les secteurs de pointe: technologie avancée ou services de niveau supérieur. A cause de cela, il renforce beaucoup, en qualité autant qu'en quantité, le potentiel urbain; c'est pourquoi il est à la fois signe et cause de dynamisme.

Enfin, il faut le différencier des autres processus ci-dessus évoqués: de l'induction technique, car il n'entre pas dans un cycle de production intégrée; des faits de cohérence horizontale, car il n'est pas au service exclusif d'un secteur d'activité comme les industries d'entretien et de sous-traitance; des économies externes ou d'échelle et de la valorisation mutuelle des entreprises, car il ne tire pas de l'environnement urbain une meilleure productivité: il en tire son origine et sa raison d'être. Il est une réponse à un besoin général ou à une virtualité offerte.

Maîtriser la connaissance du processus de synergie est de grande utilité pour le géographe. Cela peut être un instrument d'analyse et d'explication de l'évolution des villes et de l'appréciation du haut niveau des fonctions urbaines. Cela peut devenir un outil essentiel en géographie appliquée à l'aménagement de l'espace: le choix des activités à implanter dans les zones industrielles, dans les centres-villes, dans les groupements de services doit être déterminé par la possibilité de faire naître l'effet de synergie; l'analyse de l'évolution de Mezzogiorno ou de l'Ouest de la France, par exemple, en révèle la nécessité. D'où l'intérêt qu'il y avait à en étudier les mécanismes souvent subtils et encore mal précisés.

Boudeville, J., 1968   *L'espace et les pôles de croissance*, Cahiers ISEA (Paris), 232p.
Gottmann, J., 1961   *Megalopolis, the Twentieth Century Fund* (New York).
— 1966   'Why the skyscraper?' *Geogr. Rev.*, 16, no. 2, 190–212.
— 1970   'Urban centrality and the interwearing of quaternary activities,' *Ekistics Rev.*, 29, no. 174 (mai).
Remy, J., 1966   *La ville, phénomène économique* (Bruxelles).
Tosco, E.S., 1966   *Importance des industries auxiliaires et des services industriels pour une bonne planification des implantations*, ONU, Division des Politiques et des Programmes (novembre).

P1139
Semi-urban centres in Thailand
PRASERT WITAYARUT   *College of Education, Thailand*

Since the Second World War the economy in Thailand has changed its form impressively. Subsistence farming has died out and commercial farming has replaced it. Crops have been diversified from traditional food crops to several cash crops. Within this new type of economy, small centres are needed for changing cash crops into living essentials. Therefore, during the past twenty years, many semi-urban centres have sprung up throughout the country. These centres serve as trading posts, where agricultural products are converted into money and consumer goods. The introduction of highways and small vehicles is also an important contribution to the development of the centres which, in many cases, rapidly grow to large urban complexes. In 1955 there were only 70 such centres serving half a million people. By 1970 there were more than 600 centres housing more than three million people.

Three distinctive characteristics differentiate semi-urban centres from rural villages. Although both are similar in the sense of population concentration, the semi-urban centres are officially recognized, have good physical arrangements, and organized economic structures. They are better organized and established, with observable patterns.

Semi-urban centres in Thailand are known officially as Sanitation Districts. Any population agglomerated area will be officially recognized as a Sanitation District if: (1) the area has at least 1500 people confined to a space between one to 4km^2; (2) the centre of the area has at least 100 active commercial establishments; (3) public opinion in the community consents to the government official declaring their community to be a Sanitation District. The advantage of being a Sanitation District is that it becomes well organized in public utilities, such as sanitation works, garbage disposal, water supply, and electricity. A Sanitation District may also ask for a loan from the central government to be used in improving the community. At the same time, people in the Sanitation District have to pay some extra taxes for use towards the well-being of the community.

The physical structure of the semi-urban centres can be considered in terms of population, location, arrangement of buildings, and distribution patterns.

The average population concentration at each centre is 4000 people. The range of population is between 1000 and 10,000 people. These people are engaged mostly in trading and processing agricultural products for shipment. However, almost 30 per cent of the population still work either full-time or part-time in agriculture. This is the main reason why the area is called a semi-urban centre. The agricultural population usually live on the peripheries of the centre. Economic opportunities in the semi-urban centres are therefore higher and the agricultural people will eventually, as the centres grow up, change their occupations from agriculture to commerce or other occupations. People in the surrounding rural areas prefer to migrate to these semi-urban centres because life in such places is not greatly different from their previous environment.

Rows of shop houses along the main streets or waterways are the dominant characteristic of the semi-urban centres in Thailand. Location of the rows could be classified into two types. The first type is located on an intersection of the main streets or waterways. The second type is located along a main highway or waterway where there are many villages in the vicinity. The location of the semi-urban centre serves primarily as an outlet of the hinterland; therefore a distinctive physical structure at each centre is in all cases a market place.

The market place is the centre of buildings arrangement in a semi-urban centre. A market place consists of a rectangular or square wall-less building with corrugated roof and platforms for displaying goods inside. The activities at the market place may be held every morning, or every other day, or every other two days. One of these systems is selected according to the needs of the locality. Shop rows extend from both sides and on the street opposite to the market place. In proximity to the market place, there is also a traffic station where minibuses or motor boats wait for passengers. These vehicles are important as a means of communication of the centres and perform a dual purpose, i.e., to carry passengers and to transport goods. Besides the mentioned structures, there are various other community buildings in each centre, for instance, a health centre, a children's playground, a police station, a fire station, a slaughter house, an incinerator, and a water works. These facilities may differ from centre to centre.

Semi-urban centres in Thailand are distributed on a regular pattern along main transportation routes. The main highways throughout the country are dotted with semi-urban centres. These centres are situated at an average distance of twenty kilometres in the densely populated areas and forty kilometres in the sparsely populated areas. In the central part of the country where river transportation is highly developed, semi-urban centres are well estabished along the river banks. The distribution pattern of the centres along the rivers is similar to those along the highways.

Economic structure: semi-urban centres serve as outlets for villages in the hinterland. As the economy of the country changes from subsistence farming to a commercial type economy there is a need in each locality to set up a centre to exchange agricultural products for other consumer goods. The semi-urban centres grow up according to this economic requirement. Therefore, in economic terms the centres perform the functions of buying, supplying, repairing, and processing. For buying, at each centre there are stores buying agricultural products from the surrounding areas. These stores act as sources of information to farmers about the kinds of crops which are currently demanded in the world market. They also control the quality of the products by regulating prices. Some of these stores, in many cases separate stores, supply the goods needed by farmers. The farmers normally buy fertilizer, agricultural tools, clothing, food, and some luxury goods from the centres. The buying and selling activities are performed at the market place where a fair market price for a certain commodity is fixed. Normally, the activities at the market place, particularly in the early hours of the day, are enormous, but will decline in the afternoon. Repair facilities are essential in a semi-urban centre, and all small motors, namely, minibus, motor cycle, motor boat, and farm tools, can be repaired at the centre. At least one small workshop is operated at each centre. In many cases where mechanized farming is popular, workshops for repairing machines are prosperous. Many farm products require primary processing; for instance rice mills for rice, flour or processing mills for cassava, pressing mills for copra, etc. Thus in many centres, a factory or factories are established to cater for the demand. These factories create jobs at the centres and in turn make the centres prosperous.

The semi-urban centres have many implications for a developing country such as Thailand, where the economy of the country is traditionally based on a few crops. A diversification of agricultural products encourages the development of these centres. At the same time well established centres will help the expansion and improvement of the farms in the surrounding rural areas. Semi-urban centres can be used as relay stations to the remote areas of the country for new innovations. So far the government sector has not put much effort into utilizing these centres to diversify the economy of the country toward a form of industry which supports agriculture. However, many private enterprises have successfully established many small industries which support local agriculture, in several centres. If industry is to be introduced to the economy of the country, these semi-urban centres are ideal places to begin with. They have raw materials, i.e., agricultural products, cheap labour, and cheap land. What they lack are technology, management, and capital. In the case of Thailand the development of these semi-urban centres may yield better results for the economy of the country than the development of large rural areas with a sparse population.

P1140
Recent trends of urbanization in Madhya Pradesh
P.C. AGARWAL *Ravishankar University, India*

Urbanization refers to the fact that 'a sizable proportion of the population lives in cities' (Davis 1955, 429–37). The growth of urbanization in the world was greatly accelerated by the Industrial Revolution and took place mostly in the industrial regions (ibid.).

The Madhya Pradesh State, in the middle of India, is essentially an agricultural area. Hence, urbanization is still quite low here. It had only 16.26 per cent urban population in 1971 (Pandya 1971, 14).

This paper analyses the trends of urbanization in Madhya Pradesh during 1951–71, i.e. in the Planning Era of India. It examines the districtwise degree and rate of urbanization and the population of individual large towns and employs the Spearman's Rank Order Correlation and Student's t test of significance to arrive at its conclusions.

It may be noted that the urban population of Madhya Pradesh State (as reorganized on 1 November 1956) decreased 10.91 per cent from 1,458,045 in 1901 to 1,299,007 in 1911 due to epidemics and subsequently grew to 6,770,323 in 1971 (Jagathpathi 1963, passim). The growth was unprecedented during the Planning Era as shown in Table 1. This accelerating urban growth is generally attri-

TABLE 1. Urban population and its variation in Madhya Pradesh 1911–71

Year	Urban population	Per cent variation in past 20 years	Growth (no. of times in past 20 years)
1911	1,299,007	—	—
1931	1,771,871	+36.40	1.36
1951	3,132,937	+76.82	1.77
1971	6,770,323	+116.10	2.16

TABLE 2. Degree and rate of urbanization in Madhya Pradesh, 1951–71

Group	No. in group	Degree of urbanization (urban per cent of population) in 1951	Rate of urbanization, 1951–71		
			Mean (no. of times)	SD	SE
A	5	25 and above	1.11	0.06	0.03
B	6	15–25	1.14	0.21	0.09
C	7	10–15	1.12	0.02	0.01
D	15	5–10	1.40	1.29	0.33
E	9	below 5	2.36	1.18	0.39

buted to the developing economy, influx of refugees, and the white collar jobs and amenities of living which attract rural people to the towns.

During 1951, the degree of urbanization, i.e. the urban percentage of population, in the present 43 districts of Madhya Pradesh varied from 0 per cent in Sidhi to 59.99 per cent in Indore (Kerawalla 1952, passim). Similarly, in 1971, the urban percentage ranged between 3.73 per cent in Bastar and 63.17 per cent in Indore.

The Rank Order Correlation Coefficient (rho) of the 1951 and 1971 Degree of urbanization in the 43 districts of Madhya Pradesh, calculated by its formula,

$\rho = 1 - 6 \Sigma D^2/n(n^2-1)$

comes to 0.808 (Haber and Runyan 1969, 116). This shows a high positive correlation which is significant beyond .001 level of confidence.

The important cotton textile industrial Ditricts nearly *retained* their high ranks during 1951–71: Indore (1, 1), Gwalior (2, 2), Ujjain (3, 5), and Ratlam (4, 6). Durg (32, 11) and Sehore (6, 3) *improved* considerably because of the establishment of Bhilainagar (steel town) and development of Bhopal (capital) respectively. Mostly ex-princely state areas declined: Datia (8, 14), Chhatarpur (18, 23), Rewa (21, 35), Panna (22, 34), and Tikamgarh (33, 42), etc. Least urbanized districts generally retained their low ranks: Mandla (38, 41), Surguja (39, 37), and Bastar (41, 43).

The 42 urbanized districts of 1951 showed negligible correlation between their 1951–61 and 1961–71 rates of urbanization, as determined by the Rank Order Correlation Coefficient, whose value came to 0.22 only. The rate of urbanization here means the 'number of times growth in the urban percentage of the District population.'

High rank changes characterized, e.g. Bastar (22, 1), Shahdol (38.5, 3), Jhabua (1, 32), and Raisen (2, 20). On the other hand consistent growth was noticed in Durg (3, 6), Sehore (12, 13.5), and Raigarh (31.5, 35), etc.

The Student's *t* test was applied to find if the rate of urbanization in 1951–71 differed significantly between district-groups of different degrees of urbanization in 1951. The calculations shown in Table 2 were made. The values shown in Table 3 were obtained

TABLE 3. *t* values of means of rate of urbanization in district-groups in Madhya Pradesh, 1951–71

Group	A	B	C	D	E
A	—	0.30	1.00	0.88	3.21*
B	—	—	0.22	0.76	3.05*
C	—	—	—	0.85	3.18*
D	—	—	—	—	1.88

*P is less than .01.

TABLE 4. Size and growth-rate of large towns in Madhya Pradesh, 1951-71

Town group	Town population in 1951	Growth rate of town size 1951-71		
		Mean (no. of times)	SD	SE
A	100,000 and above	2.21	1.795	0.80
B	50,000–99,999	1.85	0.18	0.08
C	30,000–49,999	1.97	1.245	0.56
D	25,000–29,999	2.26	1.25	0.42

(Garret 1965, 214). These values indicate that group E differs significantly (*P* less than .01 in each case) from each of the groups A, B, and C in its rate of urbanization in 1951–71. Group E consisting of Shivapuri, Bhind, Bilaspur, Mandla, Surguja, Raisen, Jhabua, Balaghat, and Bastar districts showed the highest (2.36 times) mean rate of urbanization in 1951–71. Bilaspur, Surguja, and Bastar developed mining. Forest produce extraction, agriculture, transport, and social services developed widely in this group. The Chambal Project helped Shivapuri and Bhind, etc. In short, regional inequalities in development and urbanization were reduced in this Planning Era.

The present area of Madhya Pradesh had 32 large towns, i.e. with 20,000 and more population in each during 1951. They ranged from Satna (20,183) to Indore (310,859). In 1971, the same towns ranged from Sehore (36,141) to Indore (572,622). Ranking them by population in 1951 and 1971 and comparing, the Rank Order Correlation Coefficient comes to 0.81. This shows that the relative position of large town-sizes in Madhya Pradesh during 1951 and 1971 had a high positive correlation which is significant at .001 level of confidence.

Indore, Jabalpur, and Gwalior maintained their respective *high ranks* 1, 2, and 3 in 1971 as in 1951. These are important industrial and administrative centres. Bhopal, developing new capital improved in rank (5, 4). Raipur and Sagar, diversified towns, maintained their respective ranks 6 and 7. Satna, cement industry centre, shot up (32, 16), as did Durg (31, 12) because of the proximity of the Bhilainagar steel centre. Unimportant places receded, such as Raigarh (16, 24) and Jaora (18, 29). However, in 23 out of 32 towns the differences between the 1951 and 1971 size-ranks remained less than 5 and in 15 towns, less than 3.

Classifying the 32 large towns of 1951 into size-groups and comparing the group-means of growth-rate during 1951–71, the *t* test has been applied to find if the size-groups differ significantly in their growth rate. The calculations in Table 4 show this. The *t* values

TABLE 5

Group	A	B	C	D	E
A	—	0.45	0.25	0.62	0.06
B	—	—	0.21	1.36	0.96
C	—	—	—	0.46	0.41
D	—	—	—	—	1.27

shown in Table 5 are obtained from the data in Table 4. Table 5 shows that the size-groups of large towns of 1951 had too little difference in their growth rate from 1959 to 1971 to be significant at any acceptable level of significance.

M.C. Joshi; S.N. Upadhyay; M.P. Gupta.

Agarwal, P.C., 1967 Some aspects of urbanization in Madhya Pradesh (unpub.), Summer School in Geography, Patna.
Davis, Kingsley, 1955 The origin and growth of urbanization in the world, *Am. J. Soc.* 60, 429–37.
Garrett, H.E., 1965 *Statistics in Psychology and Education*, 3rd Indian ed. (Vakils, Feffer and Simons), 214.
Haber, Audrey, and R.P. Runyon, 1969 *General Statistics* (Reading, Mass.), 116.
Jagathpathi, G., 1963 *Census of India, 1961*, 8, Madhya Pradesh Part II-A General Population Tables, Manager of Publications, Delhi, passim.
Kerawalla, J.D., 1952 *Census of India, 1951*, 7, Madhya Pradesh Part II-A, General Population Tables, Nagpur, passim.
Pandya, A.K., 1971 *Census of India, 1971*, Ser. 10, Madhya Pradesh, Paper 1 of 1971 (supplement) Provisional Population Tables, Bhopal, 14.

P1141
Proposition d'analyse comparée du commerce de détail
MARIA PAOLA PAGNINI ALBERTI *Université de Trieste, Italie*

Les nombreuses études existantes sur le commerce de détail visent à se centrer sur une seule ville ou sur quelques villes comparables; il n'y a pas d'études organiques qui permettent la comparaison du commerce de détail parmi plusieurs villes de pays différents.

Notre programme de comparaison s'est servi de l'aide de l'ordinateur électronique, auquel on a confié la partie la plus lourde du travail.

Comme exemple de relevé on a choisi la ville de Mashhad (Iran, Khorassan), ville ayant une vaste zone d'attraction, qui la qualifie comme localité centrale d'ordre supérieur. Les données ont été ramassées sur les lieux, à cause du caractère incomplet des relevés officiels: cela a permis de relever tous les éléments nécessaires pour des recherches géographiques, en évitant des adjonctions compliquées et des corrections sur des données déjà relevées avec des buts différents. On a relevé toutes les activités de détail situées au rez-de-chaussée, l'artisanat et quelques services comme bureaux, établissements publics, bains publics.

Les données dans l'entrée (input), codifiées comme il en découle d'après les annexes, ont été perforées sur des cartes. Les données sont constituées par des lettres et des chiffres. Après la date d'élaboration, pour chaque activité on a perforé: un code à trois caractères qui indique le continent et la nation (A = Asie, IR = Iran); la région (KO = Khorassan) et enfin la localité à laquelle on se réfère (MAS = Mashhad). L'organisation a été conçue en tenant compte du fait qu'on aurait employé le code postal où cela était possible.

La ville a été divisée en trois parties, plutôt générales, selon l'âge: vieille, nouvelle et mixte. Puisque dans une position de la carte on peut insérer 10 symboles numériques (0–9), dans des relevés ultérieurs on pourrait distinguer avec une plus grande précision les âges relatifs aux différentes parties de la ville.

On a ensuite divisé la ville en une partie centrale, semi-périphérique et périphérique, pour souligner la localisation préférentielle de l'activité dans le cadre urbain.

Les rues on été classées en: principales, latérales, bazars, places pour souligner des diversités ou des similitudes; en outre elles ont été classées en base au trafic (des piétons, véhiculaire, mixte) et à son intensité (intense, moyen, rare), à leur longueur et largeur. A chaque rue on a donné un numéro qui la qualifie.

En ce qui concerne les activités on a étudié un code à cinq caractères dont le dernier caractère est = 0 si l'activité se trouve sur la rue principale, tandis qu'il a un nombre si l'activité se trouve dans une rue transversale: les numéros de 1 à 3 indiquent les activités sur le premier côté de la rue transversale; les numéros de 6 à 9 celles de l'autre côté. La numération est conçue de façon à donner les numéros pairs au côté sud et au côté est, outre aux places, tandis que les numéros impairs concernent les activités du côté nord et ouest. Le côté nord et le côté sud commencent la numération de ouest vers est, le côté ouest et est de nord vers sud. Il arrive souvent que dans le même milieu se développent deux ou plusieurs activités; ce fait est révélé, mais on les considère comme un seul point de vente.

Il y a ensuite un code d'orientation qui spécifie, approximativement, la direction cardinale de la façade et qui tend à souligner des localisations préférentielles de quelques activités étant données les conditions climatiques particulières de la ville.

Le relevé de l'activité a été fait sur les lieux par un code prédisposé. Il prévoit des groupements avec 16 lettres alphabétiques qui permettent de spécifier en détail l'activité.

On a prévu aussi un code pour les dimensions, surtout en fonction de l'étude d'autres villes, étant donné la dimension standard de Mashhad.

Les données dans l'entrée (input) sont lues par l'ordinateur, enregistrées sur un support magnétique, classées par: pays, région, localité, position et âge relatif de la partie de la ville où se localise le magasin, genre de rue ou de place, trafic, rue, numéro.

Les données ainsi classées sont contrôlées, en écartant les données incomplètes, doubles ou sur lesquelles on remarque quelques erreurs: ces données sont imprimées sur une liste qui reproduit aussi le genre de faute relevée. Aux cartes contrôlées et classées on ajoute une 'clef de classement' qui sera successivement utilisée pour classer les données selon des critères différents. La 'clef' résulte de l'union d'une 'clef de 1er ordre' avec une 'clef de 2ème ordre.' En particulier: clef de 1er ordre: pays, région, localité ... rue; clef de 2ème ordre: A1, A2, A3, A4 qui sont des caractéristiques particulières selon lesquelles on classera les données dans le cadre de la clef de 1er ordre. Au début on pose: A1 = activité; A2 = les deux activités qui précèdent dans le cadre de la même rue: A4 = les deux activités qui suivent dans le cadre de la même rue.

A ce point on fait la 'presse topographique'; pour chaque rue on met en liste les données relatives à toutes les activités, en soulignant la position dans laquelle elles se trouvent (côté nord ou sud, rues transversales, etc.).

A partir de ce point on fait une série d'élaborations qui produisent plusieurs listes. Le schéma est le suivant: de l'ensemble des données enregistrées sur un support magnétique et ajoutant la clef de classement on tire des records où la première clef est constante, tandis que la deuxième est compilée chaque fois selon des exigences diverses. Par exemple, la première élaboration construit ainsi la clef de classement: 1ère clef comme plus haut; 2ème clef: A1 = activité, A2 = 0, A3 = 0.

On crée ainsi des records qui reproduisent le nombre total d'activités pour chaque rue. Les données sont classées selon l'entière clef de classement et imprimées. On obtient ainsi une liste des activités classées pour activité dans le cadre de chaque rue. Les records avec le nombre d'activités totales par rue qui, après le classement, précèdent tous les records relatifs à la même rue, permettent de calculer les pourcentages. Successivement on obtient des listes où les activités sont groupées par genre de rue, par âge et position de la zone prise en considération, par localité, c'est-à-dire considérées comme totales.

Tous les programmes d'élaboration ont été écrits en COBOL et on leur a donné la dimension d'après un petit ordinateur (Honeywell 115) aux caractéristiques suivantes; mémoire centrale (12 kilobyte), lecteur de cartes, presse rapide, 4 unités à bande magnétique.

Outre les problèmes techniques ce programme présente plusieurs problèmes d'interprétation. Le premier est donné par la nécessité d'étudier d'abord à fond l'activité de détail dans chaque ville et de ne développer que dans un deuxième temps la comparaison entre villes différentes. La comparaison parmi ces villes naturellement présente plusieurs limites: même si nous essayons de les niveler en posant des nombres index comme facteurs de réduction, relatifs au revenu de la ville ou au nombre des habitants ou à leur condition sociale, nous ne réussirions qu'à avoir des modèles très approximatifs. Ce qui est fondamental est le facteur dimensionnel, donné par le nombre d'habitants de la ville, qui devrait être le point de départ pour les comparaisons. Des activités de détail en excès ou en défaut mettraient en lumière des facteurs comme le revenu moyen des habitants, les différents genres de vie, la différente organisation du territoire, l'umland et donc les fonctions de la ville, les communications, les systèmes politiques, économiques, etc. Puisque ces éléments ne peuvent avoir qu'une évaluation objective, il est nécessaire que l'analyse comparée, rendue possible à l'aide de l'ordinateur, soit soutenue par une connaissance personnelle de chaque situation ambiante.

P1142
Les tendances de l'urbanisation dans l'Europe de l'est
MIROSLAV BLAZEK   *Académie Tchécoslovaque des Sciences, Tchécoslovaquie*

Les progrès de l'urbanisation font l'objet d'études de la Commission des problèmes et processus de l'urbanisation auprès de l'Union géographique internationale. Le développement de l'urbanisation dans les pays d'Europe orientale/centrale présente certains traits communs auxquels se rattachent aussi les problèmes relatifs à l'évolution future de leur urbanisation.

Par les pays d'Europe orientale, et avec moins de précision par certains d'Europe centrale, nous entendons ceux qui ont traversé, à l'issue de la Seconde Guerre mondiale, la même évolution sociale et économique, à l'exception de l'URSS qui représente un ensemble à part.

L'évolution économique de l'Europe orientale, et en partie de l'Europe centrale, accuse, à travers l'histoire, un certain retard, d'où aussi un niveau relativement faible de l'urbanisation. Cela est également vrai pour les pays plus avancés, tels que le territoire actuel de la République démocratique allemande ou la partie occidentale de la Tchécoslovaquie, avec une urbanisation inférieure à celle que l'on trouve sur le territoire de l'actuelle République fédérale d'Allemagne, en Suisse, etc.

L'urbanisation et son niveau étaient aussi sous l'influence de la considérable dispersion des industries et des traditions peu développées de l'édification de villes. A cela s'ajoutaient en outre les importantes destructions infligées aux villes durant la Seconde Guerre mondiale.

Par leur taux de population urbaine, les pays de l'Europe orientale – désignation du territoire soumis à l'étude employée par la suite à titre de simplification – se rangent parmi les derniers pays de l'Europe. En 1946, ils ne possédaient pas une seule ville d'une population supérieure à un million d'habitants, à l'exception de Budapest.

Après 1946, les pays de l'Europe orientale se sont mis à réaliser de vastes programmes d'industrialisation. Parallèlement à l'effort général de croissance économique, ces programmes se proposaient de mettre à profit les réserves de main d'œuvre, disponibles notamment à la campagne. Les migrations internes qui en résultaient se trouvaient encore accélérées et multipliées par la collectivisation de l'agriculture. A défaut de celle-ci, la main d'œuvre excédentaire cherchait des emplois en dehors du pays; c'est, par exemple, le cas de la Yougoslavie avec les déplacements partiels de la main-d'œuvre vers les pays de l'Europe occidentale et septentrionale. Une certaine partie des réserves disponibles a aussi été absorbée par les mouvements vers les nouvelles zones d'habitation (Tchécoslovaquie, Pologne).

Une évolution pareille s'accusait aussi ailleurs, mais l'intensité qui la marquait dans l'Europe de l'Est tenait au fait qu'elle devait s'accomplir dans une période relativement courte. Bien entendu, elle n'était pas exempte de l'influence des différences d'ordre démographique, que l'on peut toutefois négliger dans ce contexte.

Les programmes d'industrialisation adoptés par les pays d'Europe orientale se basaient sur des principes idéologiques et des considérations socio-politiques, ayant pour but d'atteindre une répartition plus régulière des industries et, par conséquent, des agglomérations urbaines. Ces programmes imposaient l'investissement de fonds importants, mais en même temps les possibilités d'accumulation étaient fort restreintes et il ne pouvait pas en être autrement surtout au départ. L'édification de l'industrie ne s'accompagnait qu'en partie de la construction de logements et d'ensembles d'habitation qui devait de plus rattraper, par rénovations, les dégats causés par la guerre. En 1968 encore, les pays d'Europe orientale n'achevaient en moyenne que 5 logements par an et 1000 habitants, donc un taux incapable de tenir le pas avec les besoins toujours croissants.

Au bout d'un quart de siècle, entre 1946 et 1970, la population de l'Europe orientale a augmenté de 15.6 pour cent, mais celle des villes a enregistré un rythme de croissance trois fois plus rapide (de 37.1 à 61.6 millions d'habitants). Le tableau 1 montre les chiffres qui caractérisent cette évolution, par pays et par rapport à l'accroissement de la population totale. Le tableau 1 n'indique que les chiffres essentiels, se rapportant à la structure géo-

TABLEAU 1

Pays	Population en 1970 (millions)		Accroissement de la population (1946 = 100)			
	Totale	Urbaine	Villes au total	Métropoles	Villes plus de 100,000	Autres villes
Bulgarie	8.5	4.3	247	217	383	228
Albanie	2.1	0.7	368	175	—	510
Roumanie	20.3	8.2	216	189	913	164
Pologne	32.6	16.7	223	170	228	229
Yougoslavie	20.4	8.1	198	147	253	193
Hongrie	10.3	5.6	164	143	575	154
Tchécoslovaquie	14.3	6.9	152	118	166	159
RDA	17.1	11.1	94	67	111	95
TOTAL	121.5	61.6	166	134	218	160

graphique, mais ne répondant pas toujours aux définitions statistiques. Parfois, il fallait se contenter d'approximations.

Au bout d'un quart de siècle, les pays d'Europe orientale ont nettement élevé le taux de leur urbanisation. Comparée à l'année de départ 1946, l'année 1970 marque des accroissements importants de la population urbaine, comme le montre, par pays, le tableau 2. On remarque qu'un moitié de la population de l'Europe de l'est vit dans les villes.

TABLEAU 2

RDA	de 62.0%(?)	à 65.0%
Hongrie	37.0	54.0
Pologne	30.7	51.2
Bulgarie	24.8	50.5
Tchécoslovaquie	37.4	48.3
Roumanie	23.0	40.2
Yougoslavie	25.0(?)	40.0
Albanie	16.0(?)	33.0(?)

Dans le cadre de l'évolution générale, il convient d'attirer l'attention sur les phénomènes suivants: on trouve que les progrès ont été les plus rapides et d'une envergure remarquable dans les anciennes grandes villes (plus de 100,000 habitants) ou les villes de moyenne taille, ayant franchi la limite de 100,000 habitants. La population des grandes villes a augmenté plus que deux fois et représente actuellement un cinquième de la population urbaine de tous les pays d'Europe orientale. Dans ce domaine, la République démocratique allemande et la Pologne se trouvent en tête. A la périphérie de grandes villes, on voit se développer rapidement la semi-urbanisation. Les grandes villes, situées dans les zones bénéficiant d'une longue tradition de petites agglomérations urbaines (Tchécoslovaquie), aussi bien que les villes de moyenne taille (entre 50,000 et 100,000 habitants) se transforment rapidement en régions urbaines. Le processus de semi-urbanisation est cependant difficile à chiffrer. Les allers et retours entre le domicile et les lieux de travail, attribuables aux difficultés du bâtiment dans les centres, sont un trait caractérisant tous les pays d'Europe orientale.

Quant au volume, les métropoles qui représentaient dans le passé les concentrations économiques décisives ont plus ou moins conservé leur importance, mais elles sont en même temps un des groupes accusant un rythme d'urbanisation relativement le plus lent (les métropoles ne sont pas comprises dans le groupe des grandes villes). A la différence de l'Europe occidentale, la croissance ralentie des métropoles et l'expansion rapide de centres régionaux répartis d'une manière assez régulière sont des traits caractéristiques de l'Europe orientale. Somme toute, les métropoles ont augmenté de quelque 2.5 millions d'habitants, les autres grandes villes d'environ 8.5 millions. Ainsi, plus de deux cinquièmes de l'accroissement de la population urbaine reviennent aux grandes villes, y compris les métropoles et capitales.

Au point de vue quantitatif, les accroissements de la population dans les villes de moyenne et petite tailles jouent toujours un rôle décisif. Au sein de ce groupe hétérogène et étendu, on remarque des allures différenciées. Dans leur ensemble, ces villes accusent une croissance plus rapide que les métropoles et les indices respectifs sont voisins du taux d'expansion moyen de l'urbanisation. On peut dire que les villes de moyenne taille enregistrent les progrès les plus rapides que l'on trouve dans le groupe entier. L'impossibilité d'une démarcation géographique rigoureuse n'a cependant pas permis de chiffrer cette croissance avec précision. Il n'est pas rare que précisément les villes de moyenne taille représentent d'importants nouveaux centres d'industrialisation, assumant en même temps nombre de fonctions administratives et culturelles.

On souligne souvent, par voie de publicité, la naissance de villes entièrement nouvelles, notamment dans les zones d'industrie métallurgique, d'extraction de combustibles, etc. C'est là, sans conteste, un trait fort remarquable de l'évolution moderne, mais un trait qui n'est pas décisif au point de vue quantitatif. Les nouvelles villes n'ont rien à faire avec les villages de départ et ne dépassent qu'exceptionnellement le niveau de 50,000 habitants.

Les efforts de décentralisation ont facilité les progrès d'une série de petites villes qui étaient autrefois d'un intérêt industriel très faible, sinon nul. Le nombre de petites villes industrialisées avec une croissance rapide de leur population est un multiple de celui de villes entièrement nouvelles, issues de la transformation d'anciens villages.

Au pôle opposé de cette évolution on trouve une dégradation souvent importante de certaines petites villes, centres marchands traditionnels d'intérêt local. La cause en est, le plus souvent, la perte d'attributions administratives, résultant de la réduction du nombre de districts, etc. Parfois cette dégradation est aussi due aux migrations internes de la population (zones limitrophes de la Bohême, ruine de la minorité juive en Pologne, etc.). Une certaine partie des petites villes se caractérise par un niveau sans changements de la population, mais ces villes gardent toujours le rôle de centres locaux. Dans ces conditions, une sélection étudiée est fort difficile dans ce groupe. Toutefois, les fonctions de centre local perdent en importance à la suite des progrès rapides des transports, de l'expansion des villes plus grandes et en particulier du rôle des agglomérations rurales en général.

Ce qui caractérise les anciennes zones non-urbaines (bassin de la rivière Tisza en Hongrie, Multenie en Roumanie, etc.), c'est surtout la transformation d'anciens villages en centres de services, d'industrie locale et de fonctions agraires spécialisées (stations de tracteurs, ateliers de réparation, ferme d'élevage de grande envergure, etc.). De cette façon, le réseau de petites villes s'agrandit dans les régions autrefois non urbaines. Cela et l'évolution générale ont pour conséquence une répartition plus régulière des agglomérations urbaines.

L'évolution qui a été sommairement décrite se poursuit et obéit aux mêmes tendances. Les pays d'Europe orientale ont ainsi la possibilité de sauter la période de concentration excessive de la population urbaine dans un nombre restreint d'agglomérations, pour le remplacer par une évolution proportionnelle des métropoles régionales et par le développement de régions urbaines de 500,00 habitants au plus, pouvant être avantageusement complétées d'un réseau régulier de menues agglomérations urbaines.

Les études se proposant d'orienter l'évolution vers ces buts désirables doivent être organisées avec le concours des géographes. L'étape de départ, pouvant constituer une base sûre des programmes respectifs, consiste à examiner minutieusement l'évolution précédente. Le présent rapport en est une récapitulation sommaire.

Blažek, M., 1971 Les perspectives de l'urbanisation en Tchécoslovaquie, *Studia geographica* 21 (Brno).
Cucu, V., 1967 Probleme de geografie urbana in Romania, seria *Geografie* 6 (Bucarest).
Dziewonski, K., 1967 Les notions de réseau urbain et d'armature urbaine, *Geographia Polonica* 12 (Varsovie).
Lettrich, E., 1965 *Urbanization in Hungary – Urbanizácič Magaroszágon Földrejzi Tanulnáyek* (Budapest).
Lichtenberger, E., 1970 The Nature of European Urbanism, *Geoforum* no. 4 (Braunschweig).
Velčev, I., and P. Orechkova, 1971 Ossobenosti v promenite na broia i teritorialnoto razpolozjenie na gradskoto naselenie – NRB, Problemi na geografiata na naselenieto (Sofia).

p1143
Le développement des réseaux urbains en France
JACQUELINE BEAUJEU-GARNIER   *Université de Paris, France*
YVES BABONAUX, ROGER BRUNET, PIERRE BRUYELLE, RAYMOND DUGRAND,
HENRI ENJALBERT, PIERRE GABERT, BERNARD KAYSER, ETIENNE JUILLIARD,
HENRI NONN, GERMAINE VEYRET, MAURICE VOLKOWITSCH   *France*

Etude collective réalisée à travers différentes parties du territoire français (notamment les Alpes du Nord et du Sud, la Provence, la Côte d'Azur, le Bassin d'Aquitaine, les pays de la Loire et la région du Nord). L'expérience d'autres études, notamment de celles faites à propos de la région parisienne, a été utilisée pour les conclusions générales.

Les différents thèmes abordés ont été le rôle du milieu naturel (surtout le relief) à l'échelle de la situation générale et à celle du site (relief et micro-climats); l'époque de création des villes et leur 'réussite' actuelle; l'influence passée des villes traduite par les limites des circonscriptions administratives ou religieuses dont elles étaient le chef lieu au temps de Rome, à la fin du Moyen Age et à la Révolution; l'existence du réseau urbain préindustriel et son évolution jusqu'en 1968; les phénomènes d'accessibilité liés aux transports et aux infrastructures vers 1840 et vers 1970; l'organisation actuelle des réseaux urbains; l'analyse détaillée des espaces urbains au niveau de l'îlot au point de vue de l'utilisation des sols (résidence et de quel type, bâtiments industriels, administratifs, espaces verts, infrastructure de transport ...) et au point de vue des composantes socio-professionnelles de la population.

Les principales conclusions qui ressortent de ces études menées par neuf équipes de géographes appartenant aux Universités d'Aix, Bordeaux, Grenoble, Lille, Montpellier, Paris, Toulouse, Tours et Strasbourg sont les suivantes:
- la valeur traditionnelle des sites est actuellement dépassée par les phénomènes d'expansion spatiale urbaine; les géographes doivent participer à l'étude de la géomorphologie et de la microclimatologie des espaces où peuvent détendre les nouveaux quartiers (exemple de Tours et de Grenoble).
- la persistance du rôle des villes à travers l'histoire française est remarquable: la plupart des villes créées dans l'Antiquité et spécialement par Rome n'ont jamais cessé d'avoir un rôle important et sont devenues souvent les centres fondamentaux des réseaux actuels. On constate un véritable mécanisme 'boule de neige' qui a favorisé les complexes de l'accumulation de fonctions sur les mêmes villes. Au contraire bien des créations du Moyen Age, faites souvent pour des motifs militaires ou politiques locaux (exemple des Bastides de l'Aquitaine) n'ont rien donné à l'époque actuelle.
- le réseau urbain français apparaît encore très lié au milieu naturel (morcellement et compartimentage dans les pays de montagne, quadrillage peu propice à l'éclosion de métropoles régionales à large obédience dans les pays de la Loire). Pourtant l'apparition des nouveaux moyens de transport et en particulier des chemins de fer a provoqué des perturbations dans le réseau préindustriel (par exemple les villes de la vallée de la Loire, où la batellerie a disparu en quelques années devant la concurrence ferroviaire, ont connu depuis cette époque une régression sensible dans le classement statistique des grandes villes françaises). D'autre part, certaines activités industrielles très liées à une implantation particulière (bassins miniers) ont déterminé l'éclosion d'un nouveau type d'urbanisation: la nébuleuse. Enfin actuellement le tourisme favorise la création ou le développement d'organismes urbains dans les lieux restés jusque-là vides de villes (hautes montagnes, littoraux et surtout rivages méditerranéens).
- la question de l'accessibilité semble jouer un rôle très important. Une étude de l'accessibilité des villes d'Aquitaine en 1840 montre la coïncidence entre les fonctions administratives et l'existence d'un carrefour routier important à cette époque. Actuellement le développement des transports modernes crée autour des villes des aires de dessertes plus ou moins vastes et plus ou moins denses qui permettent d'exprimer visuellement la puissance d'attraction et de rayonnement des principales villes françaises (enquête portant sur toutes les villes de plus de 100,000 habitants; comparaison de l'aire de desserte par les transports publics et privés avec les aires d'attraction commerciale, les densités de population ...).
- on a donc tenté de classer les villes d'après leur fonction et d'esquisser la structure du réseau urbain dans différentes parties de la France telle qu'elle existe à l'époque actuelle.
- enfin, les géographes participants ont également procédé à des analyses de détail au niveau de l'îlot permettant de découvrir la trame de la structure urbaine à la fois au point de vue de l'utilisation du sol et des catégories de la population résidente. Cette approche actuellement en cours sera très précieuse pour esquisser un modèle du développement urbain à l'échelle de la France et probablement de l'Europe occidentale.

P1144
The pattern of urbanization in the developing countries of Asia and its political significance
K. KULARATNAM   *Colombo, Ceylon*

This paper attempts to place on record some impressions gathered primarily through on-the-spot studies of several urban places in developed and developing countries of the world spread over a number of years. It corroborates the studies in the list of references, which portray with deep insight and examples many highlights of the urbanization phenomenon in the region.

According to United Nations Population Projections, 70 to 80 per cent of the population of Asia is rural. This is expected to diminish to 70 per cent by 1985, though the absolute increase of rural population will be 400 million. Excepting Mongolia and Afghanistan, most of the Asian countries have less than one hectare of land per person. In 1950, only 15 per cent of Asia's population was urban. This figure was expected to rise by 1970 to 29 per cent for east Asia and 25 per cent for south Asia, and by 2000 AD to 50 per cent for east Asia and 33 per cent for south Asia. By 2000 AD, east Asia is expected to have 722 million persons in urban places and south Asia 793 million. The total urban population will number seven times that in 1950, which will be greater than the total population of south and east Asia in 1950.

However, it must be stated at the outset that the available statistics are inadequate for most areas, not comparable and largely 'guesstimates,' and the criteria adopted for defining urban places also vary from country to country.

Historically, most of the urban places in the region started as trading posts where tribute-receiving protectors and leaders took control as rulers and later 'divine kings.' Some became 'cult centres,' with a convenient identification of king with God. Morphologically, there were sharply demarcated foreign quarters right from very early times, as witness the hydrological civilization of Mantota and Anuradhapura (Ceylon) and later Nakorn Pathom (Thailand), Pagan (Burma), and Angkor (Khmer). Their pattern was the microcosm of extant cosmological beliefs, e.g. Madurai (south India). Angkor was a replica of the same Hindu cosmology. There were sometimes transfers of locations according to the whims and fancies of rulers or due to military, strategic, and political circumstances or natural calamities, e.g. Ayuthiya and Bangkok. Their historical evolution and the vicissitudes they have passed through provide an interesting study.

With the advent of the Europeans, the colonial type of city came into being after the 15th century. The missionaries who followed the sword with cross did much to accentuate the process of establishing 'western transplants.' With the advent of the foreigner grew also some mining and collecting centres. Saigon-Cholon, Singapore, Batavia, Manila, Rangoon, Bangkok, etc. are good examples of 'western transplants.' Being nerve-centres of colonial exploitation and administration, they developed downtown cores, near the ports where possible, with banking, trading, transport, insurance, and shipping. In order to staff the institutions, an imported tertiary population, comprising Indians and Chinese, arrived and settled down to do clerical and other middle-level work, while the executives came from the foreign ruling class and the indigenes provided the minor labour; for example, in 1931, fifty per cent of the population of Rangoon comprised Indians.

With the colonial powers came also cheap imported manufactured goods and thus began the decline of rural centres which were becoming semi-urban as centres of industry. Thus was set into operation the growth of the 'primate' urban place.

After the end of the Second World War, many of the colonial territories gained independence, characterized by various types of pseudo-nationalism. The 'primates' began to burgeon, borrowing heavily on the future and thus urban sprawl, squatter-settlements, slums and blighted areas, congestion, etc. as well as irreversible environmental degradation began to characterize these 'monstrocities.' Where former single colonial units broke up into more than one as in the case of Malaysia and Singapore, French Indo-China, etc. previous second urban places and ports became new capitals and repeated the primate process with renewed vigour as illustrated by Kuala Lumpur, Georgetown, and Phnom Penh. Unlike the core and ring morphology of the gradually evolved western cities, these do not exhibit any clear morphological pattern. Kuala Lumpur, which covered 4662 hectares in 1947, spread out to include 7770 hectares in 1957. This rapid expansion was accompanied by burgeoning unemployment and problems of housing and sanitation.

Government is centred in these primates; the urban-based elite have stepped into the shoes of the foreign masters and are ruling in the same way that they did. They are getting more and more alienated from the rural masses. Unlike the old religious, ceremonial, administrative centres, these pseudo-cities are

TABLE 1. Some examples to illustrate urban pattern characteristics

Country		Largest city/town pop. (approx.)		Next two towns pop. (approx.)	
Afghanistan		Kabul	435,000	Kandahev	115,000
				Herat	62,000
				Sardez	46,000
Burma		Rangoon	1,617,000	Mandalay	317,000
				Moulmein	157,000
Khmer (Cambodia)		Phnom Penh	600,000	Battambang	40,000
				Kompeng Cham	30,000
Ceylon		Colombo	512,000	Jaffna	100,000
				Kandy	70,000
China (Taiwan)		Taipei	1,221,000	Kaoshiung	663,000
				Tainan	430,000
India		Bombay	4,152,000	Calcutta	2,928,000
				Delhi	2,061,000
Indonesia		Djakarta	4,500,000		
Iran		Teheran	2,800,000	Abadan	270,000
				Ahwaz	206,000
Laos (no census)		Vientiane	132,000	Luang Prabang	22,000
Malaysia	West	Kuala Lumpur	400,000	Georgetown	235,000
	Sabah	Sandakan	29,000	Jesselton	21,000
	Sarawak	Kuching	70,000		(Kota Kinabalu)
				Sibu	40,000
				Miri	19,500
Mongolia		Ulan Batar	250,000	Darkhan	25,000
Nepal		Kathmandu	195,000	Patan	135,000
				Bhatgdon	84,000
Pakistan		Karachi	1,900,000	Lahore	1,296,000
				Dacca	557,000
The Philippines		Manila	1,138,000	Quezon City	398,000
				Cabu	251,000
Republic of Korea		Seoul	3,800,000	Pusan	1,430,000
				Taegu	847,000
Republic of Vietnam		Saigon	2,000,000	Da Nung	144,300
				Hue	103,000
Singapore					
Thailand		Bangkok	2,500,000		
		Thonburi	540,000	Chiengmai	66,000

mere vestiges of western-dominated, now native, rule. The situation is tending to become explosive where, as in China, a rural-based revolution can overthrow the city-based structure, at present somehow or other maintained through national slogans, emergency or army control, etc.

Fortuitous circumstances like disruption of normal life in Saigon are giving temporary artificial respiration to cities like Bangkok through burgeoning tourism, or helping to develop urban places like the American naval-base town of Olongopo in the Philippines, which has grown mushroom-like into a city of 80,000; likewise rest and recreation centres for American troops too are helping maintain certain urban places, but these are transitory phenomena in the over-all urbanization picture.

The sets of figures in Table 1 illustrate clearly the development of the abnormal hierarchy of urban places in the region and speak for themselves.

What is clear is that (1) there is an irresistible trend of pseudo-urbanisation; (2) the tendency for the urban hierarchy to be disproportionately dominated by one giant 'monstro-city' with all its attendant shortcomings, is accentuating at an alarming rate, e.g., Bangkok is 30 times bigger than Chiengmai; and (3) These urban places are characterized by the legacies of colonial rule in the form of heterogeneous and cosmopolitan plural societies (with, in many cases, no grassroots in the country), and are becoming powder kegs. In 1960, only 9 per cent of the population of southeast Asia lived in centres of over 100,000, and the highest proportion of such urban dwellers lived in just one centre. Of the urban population resident in centres of over 20,000, when we consider Thailand, 70 per cent lived in Bangkok. Likewise in the Philippines, 70 per cent lived in Manila-Quezon city.

Indonesia has to be taken as an exception because of its vast size and stretch as well as its composite character, but even there the same phenomena can be seen.

The factors accentuating such pseudo-urbanisation are: (1) rising over-all population, especially high rural densities; (2) falling death rates accompanied by increasing unemployment and under-employment among the rural population, exerting a 'push out' on the people, rather than urban industrialisation 'pulling' them to the urban centres; (3) natural resources as well as their rates of utilisation being inadequate to cope with the needs of the growing population; (4) the existing types of education geared to the needs of colonial exploitation and administration have outlived their purpose in catering to the needs of urban employment. Unless the type of education given is changed, there will continue to be an influx of people in search of jobs to the urban places; (5) influx of refugees due to war situations, natural calamities, etc., and (6) most jobs are still associated with international trade and therefore people tend to gravitate towards urban places, especially harbour towns.

Though Thailand has not been under western colonial rule, its economy has been patterned on the same model and therefore urbanisation shows the same characteristics as in the neighbouring countries. All these cities are continuing still to run on the geographical momentum generated in the colonial era. They are not true urbanized cities. They continue to be mere clearing houses for the supply of raw materials to ex-imperial nations and receivers of manufactured goods from them, for internal distribution. They are tied more to the economies of the cities of the industrialised world and are their 'outposts,' and hence centres of urban implosion. Unless rational economic development, fostering industrialisation in urban places concurrent with improvements in the agricultural sector, takes place fast enough to keep pace with population growth, and also unless environmental degradation resulting from rapid population growth is arrested through population engineering, revolution and political explosion cannot be ruled out.

The pattern of urban evolution is the same, whether it be in south Asia or southeast Asia; the differences are only in degree. The purpose of this paper is to add weight to the notes of warning already given by several writers regarding the gravity, magnitude, and urgency of the situation and to call for timely action-oriented planning. It has been reported that 60 per cent of all Thailand's qualified physicians and surgeons in 1960 were in Bangkok which has only 8 per cent of the total population of the country. Such situations are fraught with grave danger unless corrected in good time.

Burnley, I.H., 1971 ECAFE Pop/Sem.ERUP/WP/5.
Davis, Kingsley, 1969 *World urbanisation, 1950–1970*, 1, *Basic data for cities, countries and regions* (Institute of International Studies University of California, Berkeley).
ECAFE, 1971 Pop/Sem.ERUP/BP/2.
Fryer, D.W., 1953 The million city in southeast Asia, *Geog. Rev.* (Oct.).
Ginsburg, N.S., 1955 The great city in southeast Asia, Am. Jour. of Soc., 60, no. 5.
Goldstein, Sidney, 1971 Interrelations between internal migration and the environment in the ECAFE region. ECAFE Pop/Sem. ERUP/BP4.
McGee, T.G., 1967 *The southeast Asian city* (London).
United Nations World population prospects, 1965–2000, as assessed in 1968.

P1145
La structure et l'évolution des systèmes urbains régionaux de l'URSS
OLEG KONSTANTINOV *Leningrad, URSS*

Les rythmes rapides de l'industrialisation et de la réorganisation socialiste de toute la vie en URSS mènent aux changements dans le réseau des centres urbains: les uns disparaissent, les autres apparaissent, la situation géographique de ces centres change, ainsi que leurs fonctions, le chiffre de la population, les liens et les influences réciproques. Un dynamisme extrême est propre au réseau des localités en URSS. Ceci concerne, surtout, les centres urbains dont le réseau grandit continuellement et se caractérise par une grande mobilité. Pendant les années 1926–70 ont été fondées en URSS 947 villes nouvelles, ainsi que 2216 localités de type urbain. Le réseau des localités à type urbain devient de plus en plus dense.

Pendant les années d'avant la dernière guerre, l'accroissement de la population urbaine devançait celui de la population rurale; après la guerre, le total absolu de la population rurale en URSS diminue. De sorte que, pendant les années 1926–70, le chiffre de la population urbaine est multiplié presque par cinq, le chiffre de la population rurale diminue d'un cinquième, environ. Pendant la période indiquée, la part de la population urbaine passe de 18 à 56 pour cent. C'est pourquoi, aujourd'hui, ce sont les centres urbains qui déterminent le réseau des localités du pays et les systèmes régionaux de l'établissement de la population. Le rôle des centres urbains augmente et ces centres forment un squelette pour les systèmes régionaux de l'établissement de la population.

En URSS, on assiste à un processus continuel d'agrandissement des centres urbains. Le nombre d'habitants dépasse 100,000 en 1926 dans 31 villes, en 1970 dans 221 villes. Le nombre de villes à plus de 500,000 habitants augmente encore plus rapidement: de 3, ce nombre passe à 33. Il y a 25 ans, nous n'avions que 2 villes dont la population atteignait un million; aujourd'hui nous en avons 9. Le nombre des petites villes et des villes moyennes (20,000–100,000 habitants) augmente également, ainsi que celui des localités de type urbain. Pendant les années 1959–70, le nombre des petites villes augmente de 17 pour cent, celui des villes moyennes – de 25 pour cent, celui des grandes villes (100,000–500,000 habitants) – de 53 pour cent et celui des très grandes villes – de 32 pour cent. Ce dynamisme renforce le rôle des grandes villes dans les systèmes régionaux. Leur nombre va continuer d'augmenter plus vite que celui des villes à 500,000 habitants.

Le processus d'agrandissement des villes mène à une concentration de la population dans les villes les plus populeuses. Pendant les années 1959–70, le chiffre de la population augmente dans les très grandes villes (sans considérer Moscou et Léningrad) de 76 pour cent, dans les grandes villes (100,000–500,000 habitants) – de 57 pour cent, dans les villes moyennes – de 22 pour cent et dans les petites villes – de 13 pour cent. Ces différences dans le dynamisme des centres urbains, au chiffre de population divers, ont mené à des changements profonds dans la structure des systèmes urbains régionaux. En 1970, plus de 75,000,000 d'habitants – 30 pour cent de l'ensemble de la population de l'URSS – étaient répartis dans 221 villes, à la population dépassant 100,000 habitants. Ces villes forment la base des systèmes urbains régionaux de l'URSS.

Les vastes étendues du territoire en URSS et une grande variété des conditions de l'établissement (expansion territoriale) de la population ont conduit à une grande diversité structurelle du réseau des localités et de la formation des systèmes urbains régionaux.

La diversité dans l'établissement de la population et entre les systèmes urbains régionaux, diversité que l'on constate dans les différentes régions, est due aux traits particuliers de la structure économique des régions du pays; cette diversité est liée aux conditions et aux ressources naturelles, à la marche historique des choses et à la structure de la population suivant les nationalités.

Ces différences régionales se manifestent dans différents indices: entre autres:

(*a*) dans la corrélation entre la population urbaine et rurale (par ex.: dans la région de Mourmansk, la part de la population urbaine est de 89 pour cent, dans la région Sourkhandarienne 16 pour cent et la moyenne pour l'URSS est de 56 pour cent).

(*b*) dans la corrélation entre les centres urbains de taille diverse (la moyenne pour l'URSS, pour les centres urbains de 20,000 habitants, 21 pour cent de la population urbaine totale, pour les villes de 20,000 à 100,000 habitants, 23 pour cent, pour les villes dont le nombre d'habitants dépasse 100,000, 56 pour cent).

(c) dans la corrélation entre le nombre de villes et de localités de type urbain (pour la région de Tarnopol, par exemple, cette corrélation se chiffre par 1:1, pour la région de Kamtchatka, par 1:16, la moyenne pour l'URSS marque 1:1.8).

La situation géographique des centres urbains, formant les systèmes régionaux, les fonctions des centres urbains, les liens et les interactions, y compris les liens avec la population rurale entourant ces centres, exercent une grande influence sur la structure et l'évolution de ces systèmes; c'est pourquoi les systèmes urbains régionaux se forment en tenant compte des traits particuliers propres à l'établissement de la population rurale. En Estonie, en Lettonie et en Lituanie, où prédominent les fermes (khoutor) ou de très petits villages, apparaît nécessairement un très dense réseau de très petites villes, tandis que dans la zône nord du Caucase, où une grande partie de la population rurale habite des villages très populeux, le réseau des centres urbains est bien moins dense que dans les Républiques Baltes, et les villes moyennes prédominent.

Le dynamisme de la population exerce également une influence considérable sur les différences géographiques qui existent dans la structure et l'évolution des systèmes urbains régionaux. La population quitte certaines régions, ce qui mène à une diminution de la population dans ces régions. Dans d'autres régions, le chiffre de la population augmente, ce qui est dû, d'une part, à une migration venant des autres contrées, et, d'autre part, à un grand accroissement naturel; ainsi, par exemple, en Lettonie, l'accroissement naturel a été de 2.9 pour cent, pour 1969, mais il a été de 28.6 pour cent au Tadjikistan.

Ces traits, caractérisant le dynamisme de la population, exercent une influence directe sur la structure et l'évolution des systèmes urbains régionaux et déterminent l'extrême variété de ces systèmes.

Les systèmes régionaux présentent une hiérarchie des localités; les villes occupent les degrés supérieurs de cette hiérarchie. A la tête de chaque système, on trouve, habituellement, le chef-lieu de la région, du pays, de la république. Au-dessus de ces chefs-lieux sont les centres des grandes régions économiques (Gorki, Rostov-sur-Don, Sverdlovsk, Novossibirsk, Kharkov, etc.). Presque tous ces centres figurent parmi les très grandes villes de l'URSS.

Voici les changements les plus importants qui caractérisent les systèmes urbains régionaux: (a) Sous l'influence de l'afflux mécanique de la population et aussi de son accroissement naturel, la plupart des villes passent graduellement, d'après l'indice du nombre d'habitants, dans une catégorie plus élevée; (b) Un accroissement rapide de la population dans quantité de centres urbains qui ne sont pas chefs-lieux des régions, pays, républiques, ainsi que la formation de nouveaux centres urbains, ont eu pour résultat la diminution de l'importance des chefs-lieux de beaucoup de régions, pays, républiques (ainsi, pendant les années 1959–70, la part qui revient aux chefs-lieux des régions, pays, republiques dans l'ensemble de la population urbaine diminua pour Karaganda de 25 pour cent, de 19 pour cent pour Volgograd, de 15 pour cent pour Kouïbychev, de 14 pour cent pour Bakou, de 15 pour cent pour Kazan, de 11 pour cent pour Tblissi, etc.); (c) à l'intérieur des systèmes régionaux apparurent des grandes villes nouvelles: en 1970, 90 villes, de plus de 100,000 habitants, n'étaient pas chefs-lieux des régions, pays, républiques; parmi ces villes, beaucoup sont devenues (ou deviennent) centres des sous-systèmes nouveaux (Krivoi-Rog et Nikopol, dans la région de Dniepropetrovsk, Magnitogorsk, dans la région de Tcheliabinsk, Nijni-Tagil et Serov, dans la région de Sverdlovsk, Novo-Kouznetsk, dans la région de Kemerovo, et Kirovobad dans l'Azerbaidjan, etc.); (d) Beaucoup de localités de type urbain se transforment en villes; (e) De nouveaux centres urbains apparaissent là où avant se trouvaient des villages (surtout, là où se trouvaient des localités rurales, non agricoles ou à la population mélangée) ou, encore, là où aucune localité ne se trouvait; (f) Un accroissement des liens multi-latéraux, unissant la ville à son entourage rural, un développement du mouvement de pendule, s'établissent alors, et on voit croître le nombre de 'citadins cachés,' c'est à dire de ceux qui vivent (ou, parfois, ne font que passer la nuit) à la campagne, mais travaillent en ville et y trouvent à satisfaire leurs différents besoins, d'ordre matériel ou culturel; (g) Un accroissement de l'établissement par groupes et la formation de nouvelles agglomérations, plus ou moins considérables.

La migration de la population rurale vers les villes, la présence, dans les villages, des 'citadins cachés,' et la transformation des villages en centres urbains, tout ceci crée des liens étroits entre les localités rurales et urbaines et crée une interaction entre ces deux formations. Ceci nous oblige à considérer chaque système urbain régional en tant qu'une partie et un élément-chef dans l'ensemble, formé par le système des localités de cette région.

P1146
Micro-scale spatial behaviour: distance and interaction in Christchurch cul-de-sacs
B. GREER-WOOTTEN   *York University, Canada* and
R.J. JOHNSTON   *University of Canterbury, NZ*

There has been a great surge of geographical interest during recent years in the influence of distance on spatial organization. Most of this work has been concentrated at certain scales of enquiry: spatial interaction at a small scale (such as individual street block interaction) has been relatively ignored. Indeed, a recent review text (Michelson 1970) includes only two geographical works (Gould 1966; Wolpert 1966) in its bibliography.

The classic study of micro-scale spatial behaviour is by Festinger, Schachter, and Bach (1950) on two MIT post-World War II housing projects. Their results showed strong relationships between distance – both physical and functional – and friendships formed during the projects' early months: 'a striking relationship between these ecological factors and sociometric choice' (Festinger, Schachter, and Bach 1950, 58). But by working in these project areas the investigators were able to hold constant other factors which may have influenced friendship choices, such as age, occupation, and length of residence in the subdivision. Indeed, few studies have referred to 'normal' residential situations (for exceptions, see Whyte 1956; Gans 1961, 1967).

The present study used a selection of cul-de-sacs within Christchurch, New Zealand, which were picked (a) to give a good coverage of various parts – and therefore social areas – of the city, and (b) because they contained about a dozen houses. It was hypothesized that in each street strong correlations would exist between the distance separating homes and the amount of contact between them. Further, it was anticipated that deviations from these relationships could be accounted for by reference to variables such as socio-economic status differences, age differences, family structures, whether or not wife goes to work, and attitudes towards the street. People who deviated from the street's norm on the variables would be less strongly linked to the local network of friendships.

Data were collected by questionnaires which obtained basic information on household characteristics. Sociometric information was obtained in the same way as in the Festinger et al. study, as answers to the question 'which three homes *in this street* do you have most contact with?' For direct comparability, wherever possible the wife of the household head was interviewed. The aim was to get a complete coverage of respondents in each street, but this was usually thwarted by occasional refusals and inability to make contact with some households.

Analysis of the data, street by street, was based on aspects of digraph theory (Harary, Norman and Cartwright 1965). Each set of sociometric choices was treated as an adjacency matrix, with the nominations represented by 1s, and the non-nominations by 0s. These matrices were raised to the fifth power, five being the diameter of several matrices (the maximum distance between any two households), the others having a diameter of four. From these powered matrices two others were produced: a minimum link matrix, which showed the shortest 'path' through the adjacency matrix linking two households; and a total links matrix, indicating the total number of 'paths' from A to

TABLE 1. Correlations (*gamma*) between distance and friendship links

Independent variable	Street										
	1	2	3	4	5	6	7	8	9	10	11
Adjacency matrix	−0.75	−0.38	−0.19	−0.68	−0.19	−0.43	−0.36	−0.73	−0.58	−0.46	−0.35
Minimum link matrix	0.74	0.22	0.07	0.60	0.03	0.33	0.24	0.59	0.41	0.26	0.24
Total links matrix	−0.03	0.21	−0.05	−0.15	0.20	−0.26	0.12	−0.20	−0.05	−0.05	−0.00
No. of residents	14	13	13	16	13	12	12	10	13	10	14
Entries in adjacency matrix	17	27	28	31	26	32	38	14	31	26	28

TABLE 2. Group means for total accessibility

	Street										
	1	2	3	4	5	6	7	8	9	10	11
Working wife											
Yes	19(5)	62(4)	110(6)	7(1)	203(6)	214(4)	149(1)	50(7)	243(2)	216(3)	133(3)
No	18(6)	52(5)	73(6)	175(11)	260(3)	239(8)	240(9)		213(9)	221(6)	151(8)
Length of residence in street											
1–3 years	20(2)	46(4)	81(4)	192(2)	224(5)	228(1)	240(9)	47(2)	260(1)	121(1)	142(2)
4–7 years	12(1)	56(5)	97(8)	196(4)	230(3)	251(2)	251(1)	51(5)	221(5)	225(1)	76(2)
8– years	18(8)			128(6)	242(4)	226(9)			207(5)	232(7)	167(7)
Attitude to street											
Like it very much	22(2)	67(2)	89(5)	154(11)	265(8)	281(5)	243(3)	59(4)	227(6)	231(8)	135(8)
It's ok	20(7)	53(7)	108(6)	243(1)	154(3)	202(6)	227(6)	38(3)	209(3)	121(1)	174(3)
Don't like it much	10(2)		5(1)			149(1)	217(1)		260(1)		
Don't like it at all					196(1)				151(1)		

NOTE: Figures in parentheses indicate number of respondents in that group.

TABLE 3. Group means for total accessibility

	Street										
	1	2	3	4	5	6	7	8	9	10	11
Age group											
Under 30	18(3)	21(1)			196(3)	228(1)	229(9)		260(2)	208(2)	135(1)
30–49	16(7)	70(6)	85(9)	165(4)	267(5)	215(9)	251(1)	50(7)	232(5)	200(6)	147(9)
50–69	32(10)	31(2)	111(3)	211(5)	239(4)	302(2)			180(4)	251(1)	150(1)
70–				73(3)							
Children under 10 living at home											
Yes	19(8)	21(1)	80(5)	233(3)	256(5)	240(7)	248(8)	46(4)	260(3)	219(9)	135(1)
No	16(3)	60(8)	100(7)	137(9)	214(7)	257(5)	165(2)	55(3)	203(8)		150(10)
Children over 10 living at home											
Yes	17(3)	74(2)	85(9)	199(6)	260(2)	227(6)	251(1)	46(5)	213(7)	211(2)	
No	19(8)	51(7)	111(3)	123(6)	226(10)	234(6)	229(9)	58(2)	228(4)	222(7)	146(11)
Occupation of household head											
Professional	19(1)				283(5)		236(3)		214(2)	205(3)	178(1)
Managerial	24(3)		96(3)	226(3)		175(3)	281(3)		151(1)	225(1)	
Sales			121(1)		192(2)	180(2)	213(1)		260(1)		
Clerical	12(1)		51(2)						234(2)		
Crafts/skilled	9(2)	69(1)	94(4)	138(2)	231(3)	222(3)	165(2)	32(2)	259(2)		133(3)
Semi-/unskilled	20(4)	59(3)		125(2)	216(1)	300(3)	217(1)	58(4)			
Other					76(1)			54(1)	188(1)	78(1)	
Retired		52(5)	106(2)	146(5)		310(1)			196(2)	264(4)	147(7)

NOTE: Figures in parentheses refer to number of respondents in that group.

B. A total accessibility vector was computed from this last matrix: comprising the sums of its row vectors, it represented the total links from each person.

A distance matrix was computed separately for each street. Physical distance on the ground between households was replaced by an approximate measure (as in Festinger's work), based on a simple scheme of one distance unit between adjacent dwellings, two units between dwellings separated by one another, and so on. Crossing the street was counted as one unit.

Correlation of the elements of the distance matrix with the corresponding elements of the others used Goodman and Kruskal's (1955) *gamma*, which can be interpreted in the same way as a squared correlation coefficient. The roles of the hypothesized variables on sociometric choices were assessed by comparing group means (age groups, etc.) within the total accessibility vectors.

Though each respondent was asked to make three nominations, many were unable to. Together with the non-responding households this led to considerable variation in the number of contacts analysed (Table 1).

In the correlations of the distance and adjacency matrices, the former was able to predict at least one-third of the variation in the latter for 9 of the 11 cases. The negative signs (Table 1) indicate that the number of nominations decreased as distance increased. This general finding is partly confirmed by correlations involving the distance and minimum links matrices. The positive signs indicate that the longer the distance, the longer the minimum link (the number of intermediates connecting two households), but the correlations were never as strong as in the first analysis. The final set of correlations, between the distance and total links matrices, were generally negligible. One would have expected that the shorter the distance, the greater the number of links, producing a negative *gamma*: in fact, however, several of the correlations indicated a slight tendency in the opposite direction.

It is not possible to compute residuals from the relationships outlined above: in any case, since the relationships were virtually zero in the final analysis, residuals would have been largely meaningless. As a first test of the expectation of important social group as well as distance influences of sociometric choices, group means on the total accessibility measures were investigated.

The general expectation of this part of the analysis was that the largest mean total accessibility values would be recorded for the majority group in the street: social group deviates were expected to be relatively isolated from the local network of friendships, receiving few nominations. This expectation was particularly related to the age, children at home, and occupational variables (Table 3). The evidence, however was not conclusive. Among the four age groups there was some indication that the majority age group were the most linked within the street's social network, but there was no direct relationship: the same was true for the presence or absence of children in the households. This pattern was clearer in a larger number of streets with regard to occupational status: people of the majority group tended to be the most linked.

For the other three variables a direct relationship was expected. Working wives, for example, should have been less well linked because of their daily absence from the street: in several cases the evidence suggested the opposite, however (Table 2). There was, however, clear evidence that both length of residence and attitude to the street were directly related to degree of membership in the friendship network: in most streets, the longer people had lived there and the more they liked it, the more linked they were.

This limited initial analysis of eleven sociometric networks has suggested that distance, social group membership, and attitudes all influence participation rates for friendships within small, 'closed' streets. The distance effect, it would seem at this stage, is most influential on the actual choice of friends. More sophisticated analyses of the data are needed, however, before the role of the spatial factor in micro-scale interactions can be specified.

My graduate students.

Festinger, L., S. Schachter, and K. Back,
    1950 *Social Pressures in Informal Groups* (New York).
Gans, H.J., 1961 Planning and social life: friendship and neighbor relations in suburban communities, *J. Am. Inst. Plann.* 27, 134–40.

Gans, H.J., 1967 *The Levittowners* (New York).

Goodman, L.A., and W.H. Kruskal, 1955 Measures of association for cross-classifications, *J. Am. Stat. Ass.* 49, 732–64.

Gould, P.R., 1966 *On Mental Maps*, Michigan Inter-University Community of Mathematical Geographers, Discussion Paper 9.

Harary, F., R.Z. Norman, and D. Cartwright, 1965 *Structural Models* (New York).

Michelson, W., 1970 *Man and His Urban Environment* (Reading, Mass.).

Whyte, W.W., 1956 *The Organization Man* (New York).

Wolpert, J., 1966 Migration as an adjustment to environmental stress, *J. Social Issues* 22, 92–102.

P12
Geographic Theory and Model Building
Théorie géographique et Elaboration des modèles

CONVOCATEUR/CONVENOR: Ross D. MacKinnon, *University of Toronto*

This listing of short titles and first authors' surnames will assist in identifying articles, topics, and places of interest. A complete author and co-author index is located at the end of this volume, as well as a selected index relating papers to geographical locations. Note that the papers are *not* listed in alphabetical order. The organization of the volumes is described in full in the Preface.

Other papers of particular interest to readers of this section are to be found in most major divisions of these volumes.

Cette liste des titres abrégés et des noms des auteurs principaux permettra d'identifier les communications, les sujets et les lieux qui présentent un intérêt quelconque. Un index complet d'auteurs et de coauteurs se trouve à la fin de ce volume, ainsi qu'un index des lieux géographiques. Prière de noter que les textes ne sont pas classés par ordre alphabétique. On explique en détail le plan de ces volumes dans la Préface.

D'autres études qui intéresseront peut-être les lecteurs de cette section se trouvent dans toutes les autres sections des volumes.

P1201 Maps of oceanographical fields from an electronic computer (Black Sea) ANDRYUSHCHENKO 893
P1202 Application of graphs and net analysis to meteorology and oceanography BELYAEV 895
P1203 Spatial spread of economic development CASETTI 897
P1204 Spatial entropy CURRY 899
P1205 Stochastic stationarity and analysis of geographic mobility GALE 901
P1206 The Poisson and the simple random distribution in nearest neighbour analysis HSU 904
P1207 Computing factor scores (Kathmandu, Nepal) JOSHI 906
P1208 La systématique typologique et la division en régions des paysages de montagne KAVRICHVILI 908
P1209 Spatial association of qualitative maps MAXFIELD 909
P1210 Problems of an international multilingual geographic dictionary MEYNEN 911
P1211 Estimation of a spatially-autoregressive model MIRON 913
P1212 Renaming the field MOOKERJEE 915
P1213 Spatial distribution computer for geographical analysis NUNLEY 917
P1214 Ocean-tundra-glacier interaction model (Axel Heiberg Is., Canada) OHMURA 919
P1215 Population density and quality of environment PAPAGEORGIOU 921
P1216 Continuity and discreteness of the geographical shell PREOBRAGENSKY 922
P1217 L'analyse discriminatoire des correspondances typologiques de l'espace géographique (Montréal, Canada) RACINE 924
P1218 The spatial viewpoint: essence and prospect of geography ROGLIC 926
P1219 Analysis of spatial distributions SCARLETT 928
P1220 Theoretical foundations of geographical sciences SEMEVSKIY 931
P1221 Applications of dummy variable analysis (London, England) SILK 933
P1222 Mathematical analysis of mobility matrices STONE 935
P1223 Spectral analysis and planning of observations of oceanographical fields BELYAEV 937
P1224 Griffith Taylor and the beginnings of academic geography in Canada TOMKINS 939
P1225 The gamma function in geographic research (USA) WELLAR 941

P1226 Graph-theoretical analysis of ridge patterns WERNER 943
P1227 On metageography ZABORSKI 945
P1228 Simple spatial diffusion model ANDERSON 947
P1229 A Geographica Generalis before Varenius BUTTNER 948
P1230 Importance de la théorie géographique dans l'orientation de la recherche IANCO 950
P1231 Growth of geography in India CHATTERJI 951
P1232 Modèles mathématiques en géographie PREOBRAGENSKY 954
P1233 Regional forecasting and school-leaving patterns in United Kingdom CLIFF 956
P1234 Toward some dynamic concepts of spatial decision-making (North America) HOMENUCK 958
P1235 Factorial-dynamic rows of elementary geosystems, a basis of modelling natural regions (USSR) KRAUKLIS 960

P1201
Construction of maps of oceanographical fields on an electronic computer
A.A. ANDRYUSHCHENKO, V.I. BELYAEV, A.I. ERMOLENKO, and I.E. TIMCHENKO
Ukrainian Academy of Sciences, USSR

Construction of maps of physical fields of the ocean from measurements is the main aim of the majority of modern oceanographical investigations.

At the present time this problem is being solved by the complete automation of the processes of collection, transmission, and processing of oceanographical information. Since measurements of physical fields of the ocean are performed at individual geographical points, the mathematical aspect of the problem of map construction is to determine the values of the physical field throughout the area for which the chart is prepared.

In the development of methods of construction of maps of oceanographical fields, one should consider the fact that in addition to a deterministic component, they have a substantial random component. The latter arises from oceanic turbulence and from those random components which are present in external forces both of terrestrial and cosmic origin affecting the ocean. In this connection dividing of them into a deterministic component and a random component is important in map construction of physical fields. Finding the deterministic component, in principle, is possible by three methods: theoretical, semi-empirical, and empirical. Theoretically, the deterministic component of hydrophysical characteristics is estimated using equations of the theory of currents and equations of the theory of other physical processes in the ocean. When the theory offers expressions for characteristics requiring accurate empirically determined constants, the semi-empirical method is used. With the empirical approach to the problem, a low-frequency component is considered to be deterministic, the characteristic period of which is comparable to the period of observation of the process. This component is determined by filtering high frequency fluctuations from experimental data using averaging of various kinds or by applying other smoothing filters.

In fact the random component of hydrophysical characteristics is obtained as a difference between initial and deterministic components found by this or other methods.

The field value at some point of space is reconstructed as a result of combining the predicted values of deterministic and random components of the field.

Analysis and reconstruction of the random component in the investigated area are naturally to be performed on the basis of principles of the theory of probability and mathematical statistics.

Let us discuss basic properties of the mathematical probability model of the field. We shall accept the studied random field $f(\mathbf{x})$ as random and isotropic, having zero average value $M\{f(\mathbf{x})\} = 0$ and normal distribution of probabilities. We also assume that the correlation function and dispersion of the field $K(\mathbf{x}), K(0)$ are known. Each measurement of the field in some system of points disturbs the *a priori* indeterminacy of its values. Samples induce conditional laws of distribution of probabilities with dispersions $E(\mathbf{x})$ throughout the field, changing from point to point. Values of these dispersions will characterize the *a posteriori* indeterminacy of field values at each point.

The aim of the calculation of the random field is to determine conditional average values $f'(\mathbf{x})$ induced by measurements $\{f(\mathbf{x}_k)\}$. It is evident that root-mean-square (rms) error $\varepsilon(\mathbf{x})$ of the calculation of the conditional average cannot be less than the actual existing indeterminacy $E(\mathbf{x})$, i.e., dispersion of the conditional law of distribution of probabilities corresponding to point $\mathbf{x}$. Therefore the problem of calculation of the field consists in finding a method of estimation of $f'(\mathbf{x})$ at which error $\varepsilon(\mathbf{x})$ on the average differs least of all from $E(\mathbf{x})$.

(1) $\varepsilon(\mathbf{x}) = E(\mathbf{x}) = M\{[f'(\mathbf{x}) - \hat{f}(\mathbf{x})]^2\}$,

where $\hat{f}(\mathbf{x})$ is a result of the calculation of the field. Following the terminology generally accepted, we shall call the algorithm for reconstruction of values of the field satisfying this condition the 'optimal interpolation.'

It is shown in the theory of stationary random functions that for a Gaussian distribution of probabilities linear approximation of field values according to formula (2) is optimal interpolation.

(2) $\hat{f}(\mathbf{x}) = \sum_K g(\mathbf{x}, \mathbf{x}_K) f(\mathbf{x}_K)$.

Function $g(\mathbf{x},\mathbf{x}_k)$ is a weight function for samples of the field $\{\mathbf{x}_k\}$. According to Kholmogorov's fundamental theory its values are determined from the system of equations to which minimizing of rms error $\varepsilon(\mathbf{x})$ leads (Kholmogorov 1941, 3–14). Substituting (2) into (1) and equating to zero the derivative with respect to $g(\mathbf{x},\mathbf{x}_k)$ from (1), we obtain

(3) $K(\mathbf{x}, \mathbf{x}_m) = \sum_n K(\mathbf{x}_m, \mathbf{x}_n)g(\mathbf{x}, \mathbf{x}_n)$.

The number of equations in system (3) is equal to the number of field samples, involved by interpolation, and, consequently, coincides with the number of unknown $g(\mathbf{x},\mathbf{x}_n)$ to be determined.

Substituting expression for $K(\mathbf{x},\mathbf{x}_k)$ from (3) into (1), we obtain a formula for the rms error of the optimal estimation of the field

(4) $\varepsilon(\mathbf{x}) = K(\mathbf{x}, \mathbf{x}) - \sum_{K=1}^{n} K(\mathbf{x}, \mathbf{x}_K)g(\mathbf{x}, \mathbf{x}_K)$.

It follows from the latter formula that the less the error of field reconstruction the stronger are correlation connections between point x and points $\{\mathbf{x}_k\}$. To underline values of the correlation function in reconstructing the field from formulae (2)–(4) let us agree to call the latter the correlation algorithm of **optimal interpolation**.

Optimal interpolation from given formulae is successfully used in meteorology in objective analysis (Gandin 1963, 151–67).

Let us discuss the application of this method to oceanography. As an example we shall choose the problem of constructing a map of the Black Sea surface temperature. Data of synchronous surveys performed by ships, as well as data of airborne surveys of surface temperature carried out by the radiation method have been used by us to investigate the characteristics of statistical structure of this field. In dividing the initial field into deterministic and random components, it is assumed that the deterministic component of the temperature field involves the component of the diurnal change dictated by regular causes, a similar component of seasonal change, and a low frequency component resulting from the change in physical-geographical conditions along co-ordinates. The completed analysis of the diurnal change of sea temperature indicates that its influence on the statistical structure is insignificant, i.e., the radius of correlations of random components is not in fact changed after subtraction of the diurnal change from it (Belyaev et al. 1969, 291–300).

Owing to this, in the objective analysis of realizations of the temperature field one can neglect the diurnal and the seasonal run of temperature variations.

The low frequency component of the field dictated by the change of physical-geographical conditions along co-ordinates, was approximated by a linear function of the form $\bar{t}(\lambda,\varphi) = a + b\lambda + c\varphi$ (λ and φ are geographical co-ordinates) and the field of deviation was accepted as the random component $t'(\lambda,\varphi)$

(5) $t'(\lambda,\varphi) = t('\lambda,\varphi) - \bar{t}(\lambda,\varphi)$.

The random component thus obtained offered the property both of stationarity and isotropy within the accuracy of its determination, and the spatial correlation function of its $K(\zeta)$ was well approximated by the curve $K(\zeta) = e^{-0.016r}$, where

(6) $r = |\mathbf{r}|$ — distance expressed as kilometres.

Accepting the distance in which $K(R)$ drops to $+0.1$ as the radius of correlation, we obtain from (6) that $R = 150$ km.

The temperature value at some point of the Black Sea surface is obtained by the addition of the values of a deterministic component $\bar{t}(\lambda,\varphi)$ at this point and random component $t'(\lambda,\varphi)$ calculated by the method of optimal interpolation. The mentioned algorithm was realized on an electronic computer. Values of parameters a,b,c of the computer input, co-ordinates of the required network of stations of observations, values of the random component on them, as well as co-ordinates of interpolation nodes were prescribed. Fifty measurements of the field, uniformly spaced throughout the water area of the Black Sea were used as a required network of stations. This number of measurements was used because of the condition that 5 or 6 stations on the average should fall within the correlation circle. Results of the calculation of the temperature field and the field of errors in nodes of a regular net were given by an automatic digit-printing device.

The rms error of optimal interpolation throughout the field was $0.53°$.

Calculations made by us indicate that the map of values of the surface temperature field can be constructed on an electronic computer by the objective method with a relatively small quantity of empirical data. The possibility of automation of data processing and calculation of maps by the method of optimal interpolation makes this method highly promising not only in meteorology and oceanology but in geography too which uses charts as a means of generalization of information.

Belyaev, V.I., A.I. Zhilina, L.V. Timofeeva, and E.A. Shevchenko, 1969 The objective analysis of data observations used to calculate sea temperature, in *Automation of Research in Seas and Oceans*, part I (Sevastopol), 291–300.

Gandin, L.S., 1963 Objective analysis of meteorological fields, L., *Hydrometeoizdat*.

Kholmogorov, A.N., 1941 Interpolating and extropolating of stationary random sequences, *Izvestiya AN SSSR*, ser, math. 5 (1), 3–14.

P1202
Application of graphs and net analysis to meteorology and oceanography
VALERY I. BELYAEV *Ukrainian Academy of Sciences, USSR*

Keeping conditions on the earth's surface from deterioration as a result of man's activity is associated with the necessity of developing methods of mathematical description of environment for carrying out appropriate prognoses. At the present time only detailed calculations can be the basis for development of measures directed both toward preservation of favourable properties of environment and toward its remaking in the required direction.

The problem of mathematical description of environment in all its diversity is highly complicated. Physical, chemical, biological, and geological processes take place in close interaction in the air and water cover of the earth. Methods of mathematical description are far from being developed for all of them. In those cases where corresponding methods of calculation are devised they are based on simultaneous consideration of a rather limited number of phenomena.

At the same time, much information on quantitative interrelations of individual phenomena obtained both empirically and theoretically has already been accumulated. In our opinion, the most effective way to generalize this information in common mathematical expressions susceptible to analysis on electronic computers is its representation as graphs.

Since methods of analysis of graphs using electronic computers present no problem now, then the stated problem reduces to the development of a method of transition from all the models of phenomena in meteorology and oceanology to their appropriate graph models. Such a method was proposed by us as applied to the description of interaction between phenomena of various kinds in the ocean-atmosphere system (Belyaev 1969, 137–53; Belyaev and Mikhailova 1969, 115–25).

It consists of the following: let us assume that we have information on interaction of two phenomena as a functional relation between their parameters found theoretically or empirically. If we divide the fields of the observed values of parameters into intervals satisfying the qualitative estimation of their intensity (intervals of weak, average, strong, etc., values) then the model of interaction of corresponding phenomena will be depicted as a graph. Nodes will correspond to intervals of values of parameters, and the graph's edges to relations between them.

Having 'sewn together' particular graphs concerned with individual groups of phenomena from their full aggregate, we obtain a complete graph describing the interrelationship between all the phenomena in the field under study.

In the ocean-atmosphere system we mark out the area G_{XT} with spatial scale **X** and time T which is of interest to us.

'Natural' phenomena A_i will be considered by us as objects to be determined in the discussed area, putting a finer meaning to this idea than is usually customary. We shall submit the choice of objects A_i to the following conditions: (1) Multiplicity of objects A_i chosen in the field G_{XT} contains submultiplicity of such objects, association of states of which provides the notion about hydrometeorological situation in the area G_{XT} required for practical purposes. (2) The state of object A_i is characterized only by one positive parameter a_i taking a limited number of discrete values with a degree of intensity of α_{ki}. (3) Cause and effect relations are fixed between states of objects A_i:

(1) $A_i \to A_j$, $\alpha_{ki} \to \alpha_{lj}$,

meaning that A_i causes the effect A_j at corresponding relationships between their parameters. According to the first condition, the choice of objects in the area G_{XT} can be carried out as follows. First, those phenomena A^*_i are separated, states of which provide the required practical notion about the hydrometeorological situation in the area G_{XT}. Then, those objects are separated which are directly or indirectly connected with objects A^*_i according to relation (1). Among the objects thus separated, phenomena A^0_i are naturally found also which are outside the area G_{XT}. The aggregate of objects A^0_i forms external conditions under the influence of which the state of the area G_{XT} is formed. Time-spatial scales of external phenomena $X^0 T^0$ must satisfy the condition

(2) $X^0 T^0 \gg XT$.

The second condition affecting the character of objects A_i relates to the procedure of their determination. So phenomena determined by vector values are dismembered into several phenomena described by one positive parameter. For example, instead of wind, described by the vector parameter velocity, several independent phenomena are introduced: the north wind, the northeast, the east, etc. In the same way, two different phenomena are put in accord by a parameter taking positive and negative values.

Methods for construction of a network model of hydrometeorological phenomena are developed by us as a deterministic graph, which for many reasons seems to us preferable to a stochastic graph. This means that relations between degrees of parameters of phenomena in expression (1) must always be realized with probability equal to unity. The choice in system G_{XT} both of phenomena themselves and parameters characterizing them is subject to this basic condition.

If A_i causes A_j with a probability which differs from unity, then it means that some association of phenomena exists, which, acting simultaneously with A_i, causes phenomena A_j with probability equal to unity.

In addition, there is another possibility: phenomenon A_j is complicated and A_i causes with probability equal to unity only some part of phenomenon A_j. Thus, by widening this phenomenon for A_i or narrowing it for A_j (or both at the same time) one can, in principle, obtain the condition when probability of transition $A_i \to A_j$ is equal to unity.

It is worth noting that it is not necessary for the usual hydrometeorological elements to be parameters of phenomena chosen by us. These can be their combinations, new values not used before. A network model of interrelation between phenomena in field G_{XT} will be presented as an oriented (directed) graph, which is basically a branching system in which edges of inverse direction as well as edges connecting nodes of one and the same rank are possible. Intensities of parameters α_{ki} correspond to nodes of the graph, and relation between them correspond to edges. Each edge is put in accord with the degree of rapidity of realization of relation reflected by it. In construction of a graph infinite cycles and non-simplicity of passing its nodes are excluded. The network model is not a formal means of solving the stated problem. It depicts only that fact which exists in nature and is a convenient form of storage of information obtained as a result of investigations on interrelation of phenomena.

Analysis of a graph is performed within limits of time T beginning from the prescribed state of external phenomena. Prediction of situation in G_{XT} is given for the end of the period T. Relations are analysed with due regard to the degree of rapidity of their realization.

In field G_{XT} a situation is thus established on one hand under the influence of external large-scale phenomena, and on the other hand due to local factors.

On the basis of dividing of all the phenomena into scales satisfying the conditions

$X_1 T_1 \ll X_2 T_2 \ll ... X_n T_n$,

where microscales are on the left-hand side and scales of cosmic phenomena on the right-hand side, a scheme of construction of a network model has been developed by us involving hydrometeorological phenomena on the listed scales.

To consider the inverse effect of phenomena of smaller scales on large-scale phenomena, smoothing of degrees of intensity of parameters over the large-scale is applied.

Belyaev, V.I., 1969 On the application of the method of graphs to studies of the

P1203
Concerning the spatial spread of economic development
E. CASETTI, L. KING, and F. WILLIAMS   *Ohio State University, USA*

Modern economic development originated in England at the end of the eighteenth century and spread from there into Europe, into the countries inhabited by north European settlers, into Japan, and into South America. At present we are witnessing the early stages of its inroads into most of the remaining countries of the world. In the course of this process, the initial origin of economic growth was overtaken by other centres which supplemented or replaced it in its role of generator and propagator of economic growth.

The diffusion of economic development has a decided, although somewhat unclear, spatial connotation. It involves a spreading of economic development among spatially distinct and separated countries. The extent and speed of this spread is conditioned by the intensity and nature of the interaction among the countries involved and by their cultural similarities, but both of these are related, in turn, to spatial factors such as physical distance.

This paper is concerned with the problem of identifying and measuring modalities and relevant parameters of the spatial spread of economic growth, particularly its speed. The paper contributes to articulating propositions such as the ones discussed above into testable hypotheses and measures. Since the complications involving multiple origins, and physical or cultural barriers are not appropriate for this initial discussion, the analysis is here restricted to the case of Europe for the period 1850–1913. The hypothesis that during this period economic development propagated from England throughout Europe will be tested.

A suitable analytical model was generated using the expansion method (Casetti 1972). Specifically, a spatial dimension was added to the following aspatial model of income growth over time:

(1) $y(t) = \exp(a + bt) + v, \quad b, v > 0,$

where $y(t)$ is income per capita at time t, and a, b, v are parameters. According to equation (1) the income per capita in excess of a base level v, increases exponentially with time; y tends to an asymptotic low level of v as $t \to -\infty$. The parameter b measures the rate of increase of y over time: the larger is b, the faster the growth of y over time. The parameter a positions the y curve on the time axis; increasing a is equivalent to shifting the curve to the left without altering its shape.

Assume that at any point in a region a value of y is defined throughout a time interval. This is equivalent to assuming that y is a function of its spatial-temporal co-ordinates. Furthermore, let y grow according to equation (1) with v spatially invariant over the region and a and b being some well-behaved function of the spatial co-ordinates. Then at any point in time, the income per capita in the region may be thought of as a surface above the region, the height of which corresponds to the local level of y. At $t = -\infty$ the y surface is a plane at height v everywhere over the region. Subsequently, as incomes per capita grow, this plane would be uplifted, warped and stretched (without tearing) in a manner determined by the spatial variation of the coefficients a and b.

The pattern of concern here involves a spread of economic growth from an origin into the region. This means that any given income per capita larger than v should be attained earlier at the origin and elsewhere in the region after a time lag related to distance from the origin. Consider one arbitrary level of income per capita, h, greater than v. The locus of the points in the region with an income per capita of h, will include at first a point in correspondence with the origin. Subsequently, it forms a closed curve, an h-isoline, that divides the region into two subregions, having an income per capita respectively higher and lower than h. Over time this isoline expands, making larger and larger the subregion where $y > h$.

If the spreading out of economic growth is unaffected by orientation, the h-isoline is a circle centred at the origin, that grows larger through time.

Indicate distance from the origin by s. In order to generate a model capable of representing a spatial spread of economic growth, equation (1) may be 'expanded' by redefining its coefficients a and b as linear functions of s:

(2) $\quad a = a_0 + a_1 s$,
(3) $\quad b = b_0 + b_1 s$.

Substituting (2) and (3) into (1) yields a model in which income per capita is a function of its spatial temporal coordinates:

(4) $\quad y(s,t) = v + \exp(a_0 + a_1 s + b_0 t + b_1 st)$.

The expansion of a and b serves two purposes, corresponding to the roles that the two parameters play in determining the shape of the y curve. The expansion of a makes the positioning of y curves on the time axis a function of distance from the origin, thus allowing for the possibility of delaying the attainment of any given income level by a time lag related to distance. The expansion of b makes it possible for the time gradient of $(y-v)$ to become greater or smaller (if b is greater or smaller than zero, respectively) with distance from the origin.

For appropriate values of its parameters equation (4) does represent the outward spatial spread from an origin. For example, set y equal to a given constant level h. Replacing h for y in (4), and after a few manipulations, the following equation is obtained

(5) $\quad H = a_0 + a_1 s_h + b_0 t + b_1 s_h t$,

where $H = \ln(h - v)$.

Equation (5) specifies the distance s from the origin where at time t the income per capita equals h. For a given t, if $s_h < 0$, an income of h is nowhere attained yet in the region; if $s_h = 0$, an income of h is attained at the origin only; if $s_h > 0$, it is found in a circle at distance s from the origin that divides an interior subregion including the origin, where $y > h$, and an exterior subregion where $y < h$. In equation 5 s_h is implicitly defined as a function of t, and for appropriate values of the parameters s_h increases as t increases, producing the widening of the subregion where $y > h$. The speed of spread of economic growth, ds_h/dt, may be obtained by taking the implicit derivative of s_h in equation (5) with respect to t and solving. Thus,

(6) $\quad ds_h/dt = (-b_0 - b_1 s_h)/(a_1 + b_1 t)$.

Solving equation (5) for s_h and substituting in (6) yields

(7) $\quad ds_h/dt = (a_0 b_1 - b_0 a_1 - H b_1) / (a_1 + b_1 t)^2$.

Equations (6) and (7) show that if b_1 equals zero, that is, if the percentage rate of change of $(y-v)$ over time is spatially invariant, the speed of spatial spread equals $-b_0/a_1$, and is a constant value unaffected by either t, s, or h. If instead, $b_1 \neq 0$, then ds_h/dt is a function of t and h.

THE SPREAD OF ECONOMIC GROWTH IN EUROPE

The analytical framework outlined above was applied to test whether economic development in Europe between 1860 and 1913 spread outward from England, and if so, to determine the modalities and speed of such a spread. Income per capita in constant 1953 dollars for European countries for the years 1860, 1880, 1900, and 1913 (Zimmerman 1962), and the distances between London and the capital cities of the different European countries, were used. The resulting 58 observations, each consisting of a value of income per capita (y), of time (t), and of distance from London (s), were used in estimating the parameters of equation (4). The value of v was arbitrarily set at 60 1953 dollars. A stepwise multiple regression program was used. In the analysis the step retained was that having the highest correlation coefficient and all the regression coefficients significantly different from zero at the 5 per cent significance level. The resulting equation was:

$y = 60 +$
$\exp(4.40979 - 0.00127s + 0.01642t)$
$\quad\quad (17.97) \quad\quad (10.59) \quad\quad (6.44)$
$R = 0.848; R^2 = 0.719$ (the t values are in parentheses).

The equation indicates (1) exponential growth over time, since $b_0 = 0.01642$ is >0, and (2) spatial spread outward from England, because $a_1 = -0.00127$ is <0.

Since b_1 was not significantly different from zero, there is no evidence of increase or decline of the percentage rate of change in income per capita with distance from London. The income per capita in the countries considered appeared to conform to a time trend identical in shape but shifted to the right on the time axis; the more so the greater the distance from London. Finally, the velocity of spatial spread of economic growth from England into continental Europe, given by equation (6), is approximately 21km per year. This result implies that any one iso-income line moved away from London into continental Europe during the period 1860–1913 at a velocity of 21km per year.

us National Science Foundation.

Casetti, E., 1972   Generating models by the expansion method, *Geog. Anal.* 4 (1).

Zimmerman, L.J., 1962   The distribution of world income: 1860–1960. In E. de Vries, ed., *Essays on Unbalanced Growth* (Mouton).

P1204
Spatial entropy
LESLIE CURRY   *University of Toronto, Canada*

There appears to be a real need for a map measure which provides a common yardstick between maps: this should be a measure of structure independent of possible generating mechanisms. We want to talk about any map however it was generated, but we still wish to ascribe probability measures to it. To do so, we invoke an outside agent observing the map or acting within the real world distribution represented by the map. Consider Figure 1(a) representing a map formed from Gaussian random numbers and the values obtained by twofold cascading averaging and differencing. If the squared deviations from the mean are used in this procedure and if the process generating them is stationary Gaussian, the marginal totals provide an estimate of the spectrum. In a similar manner, and without invoking normality, we may use the marginal totals of the absolute values to describe the apportionment of the map deviations by scale. It is in order, then, to take the total of the absolute values of a complementary pair and regard it as the proportion of the total value contributed at that scale and in that position (Figure 1b).

Since we now have a 'probability set' let us use Shannon's measure of entropy, $-p_i \log p_i$, to obtain a summary measure of the map.

(1) $$H = - \sum_{n=1}^{N} \sum_{q=1}^{Q} p_{nq} \log p_{nq},$$

where summation is over scales from $q = 1$ to Q and over 'cells' at the same scale from 1 to N (Figure 1c). If we repeat this analysis for a completely concentrated distribution of equal mean, the marginal 'probabilities' turn out to be equal at all scales, as in 1b, but the entropy measures are:

Scale	$-\Sigma p_n \log p_n$
2	0.15923
3	0.25934
4	0.35951
$-\Sigma\Sigma p_{nq} \log p_{nq}$	0.77808

On the other hand a uniform distribution will clearly have an entropy of zero at all scales.

It is interesting that the concentrated and the uniform distributions no longer find themselves at either end of the range of density functions where the mean/variance ratio has habituated us to expect them. Yet structurally, a totally concentrated distribution has $n - 1$ of its unit areas spatially undifferentiated while the uniform has n. The

(a)

	2		9		1		0		4		6		8		5		
	−3.5		3.5		0.5		−0.5		−1.0		1.0		1.5		−1.5		13
		5.5				0.5				5.0				6.5			
	2.5		2.5		−2.5		−2.5		−0.75		−0.75		0.75		0.75		13
			3.0								5.75						
	−1.375		−1.375		−1.375		−1.375		1.375		1.375		1.375		1.375		11
								4.375									

(b)

	0.2010	0.0280	0.0580	0.0850	0.3720
	0.1435	0.1435	0.0425	0.0425	0.3720
	0.0785	0.0785	0.0785	0.0785	0.3140

(c)

	0.1401	0.0435	0.0718	0.0911	0.3465
	0.1203	0.1203	0.0583	0.0583	0.3572
	0.0868	0.0868	0.0868	0.0868	0.3472

Figure 1

notion of negentropy, the complement of entropy, must thus be treated circumspectly. In that it is usually read as the degree of 'structure' in a system, this implies that a uniform distribution with its zero entropy would exhibit maximum structure. This could be accepted as a convention, but then a totally concentrated distribution which surely has at least as much 'structure' would fall somewhere between the independently random and the uniform.

For temporal stochastic processes, it has long been known that the stationary Gaussian process produces a distribution having maximum entropy for a given variance, so that the same result must apply in two dimensions. However, we wish to discuss all maps. It will be recalled that one major characteristic of a Gaussian process is that its derivatives are independent of each other. If we regard the cascade procedure as describing gradients, and instead of speaking of derivatives use gradients, we can require that they exist at all scale levels and that they be equal everywhere to obtain maximum entropy. Clearly, no random process can match this score since only expectations of deviations are equal. There will always be a 'deterministic' map which has greater entropy than a stochastically generated map.

Consider the simplest map of all, the uniform distribution, so commonly used as the initial condition of location theories. The reason for its popularity in this context is that it provides no constraints on the operation of the distance dependent forces introduced. As a forcing function it must be zero; it cannot produce areal differentiation no matter what the forces involved. The only differentiation which could occur on a uniform map would be dependent on the boundaries of the map so that we assume the map to be infinite, or at least very large relative to the scale of the range of the forces in operation. We can put this in a slightly more useful form: no spatial convolution operation will change the form of a uniform map. The entropy of a uniform map is zero. We may thus use entropy as a measure of the degree to which a map distribution will be changed by performing convolution operations on it. Since it has a scale component, it will reflect the frequency response characteristics of the operator and the degree of uniformity component provided by using logarithms will show the extent to which averaging will affect it. It is a measure of the degree of differentiation of the map.

Spatial entropy has application to Horton's law of stream numbers, which is obtained from combinatorial arguments as a maximum entropy state for a given number of tributaries where only two may form a junction and all join before leaving the region. It can be shown that the maximum entropy configuration of channels is associated with low spatial entropies of elevations and less probable states with greater elevational entropies.

The joint information of the two maps may be defined as
$$\bar{I} = H(x) + H(y) - H(x,y),$$
where $H(x,y)$ refers to the joint entropy.
$$H(x,y) = H(y) + H(x|y),$$
$$\bar{I} = H(x) - H(x|y)$$
$$= H(x) - H(y)(1 - \rho_{x,y}),$$
where the correlation coefficient $\rho_{x,y}$ is obtained for each hierarchical level and appropriately weighted by the number of 'cells.' In similar manner to channel networks, the joint spatial information of two maps measures the degree of constraint on ascribing one map as the output from the input of the other map to a convolution operation. Put another way, flows (in the gravity model) will assume their maximum entropy state only when spatial joint information is zero. When the latter is maximum, uncorrelated Gaussian distributions or their deterministic analogues, considerable constraints are put on travel configurations so that travel entropy is low. The precise numerical relationship has not been sought.

Because entropy is something which is added or subtracted from maps we have the basis of the common yardstick to which we referred earlier. This allows several possible interpretations of spatial entropy and corresponding applications, following the ways it is currently used in non-spatial situations. It can be used as a measure of the information content of a map: any general numerical statement about the map such as convolution will reduce its content, transferring it to the statement itself. Similarly, the map can be built up by statements and their information incorporated in the map. Entropy can also be used as a measure of uncertainty in the sense that general statements about the map will leave residual structures waiting to be explained: we are uncertain as to their cause. Notice that the 'noise' term of a regression analysis represents a causal statement in this context so that when it is removed, the map is uniform and has zero entropy. We can thus use entropy in a statistical framework similar to sums of squares or residual variance.

From equation (1),

$p_{qn} \log p_{qn} = p_{qn}{}^{p_{qn}}$ in antilogs,

and

$$\sum_{n=1}^{N} p_{qn} \log p_{qn} = \prod_{n} p_{qn}{}^{p_{qn}} \text{ in antilogs.}$$

Call this value λ_q and the following value λ.

$$\sum_{n=1}^{N} \sum_{q=1}^{Q} p_{qn} \log p_{qn} = \prod_{n} \prod_{q} p_{qn}{}^{p_{qn}}$$

$$= \prod_{q} \lambda_q \text{ in antilogs.}$$

λ may be recognized as the likelihood ratio. The statistic $-2 \log \lambda$ has an asymptotic χ^2 distribution under the null hypothesis. Since $p_Q \log p_Q = $ antilog $p_Q{}^{p_Q} = \lambda$, and the statistic $= 2 \log \lambda = $ antilog λ^2, then $\lambda^2 = p_Q{}^{p_Q}$ and $\lambda = p_Q{}^{0.5 p_Q}$. It would be possible, for example, to test whether two or more apparently different spatial regression relationships were in fact 'real' under the null hypothesis and thus whether separate operators were necessary. The 'minimum discrimination information statistic' could be used to test the homogeneity of the relationship.

$$\text{mdis} = -2 \log \lambda = 2 \sum_{n}^{N} \sum_{q}^{Q} p_{qn} \log p_{qn}.$$

In general, spatial entropy appears as a more relevant measure for spatial statistics than sum of squares.

Anderson, T.W., and L.O. Goodman, 1957 Statistical inference about Markov chains, *Ann. Math. Stat.* 28, 89–109.

Curry, L., 1972a A bivariate spatial regression operator, *Can. Geog.*

– 1972b A spatial analysis of gravity flows (submitted for publication).

Shreve, R.L., 1965 Statistical law of stream numbers, *J. Geol.* 74, 17–37.

Wilson, A.G., 1967 A statistical theory of spatial distribution models, *Transportation Res.* 1 (Pergamon), 253–69.

P1205
Stochastic stationarity and the analysis of geographic mobility
STEPHEN GALE *Northwestern University, USA*

In the past few years there has been an increased interest in the application of finite space stochastic models (e.g. Markov chains) to the analysis of geographic mobility (Beshers 1967, 838–42; Gale 1970; Nordbotten 1970, 193–201; Rogers 1968). Although models of this kind often incorporate the pertinent features of human locational behaviour, there are nevertheless a number of problems which limit their applicability. Several of these issues have been treated in earlier papers (Gale 1970, 1971, 1972a; Olsson and Gale 1968); in particular, the notion of stochastic stationarity was shown to be an important property of law-like processes. As there has been some confusion over the use of the term, in the present paper we will define and give operational characterizations of three general types of stochastic stationarity; all will be shown to have a direct bearing on the analysis of geographic mobility.

Let $\{X(t_r); r = 1, 2, ..., T\}$ be a discrete parameter, finite space, stochastic process in which the conditional probability that the state of the system, X, takes on some particular value, μ_r, at time t_r is expressed by

(1) $\Pr[X(t_r) = \mu_r | X(t_1) = \mu_1, ...,$

$X(t_{r-1}) = \mu_{r-1}]$.

Most commonly, we also assume that the Markov property holds:

(2) $\Pr[X(t_r) = \mu_r | X(t_1) = \mu_1, ...,$

$X(t_{r-1}) = \mu_{r-1}]$

$= \Pr[X(t_r) = \mu_r | X(t_{r-1}) = \mu_{r-1}]$.

Now, if we denote μ_r and μ_{r-1} by the representative variables j and i respectively, we may rewrite expression (2) in matrix form as

(3) $[P(t_r)] = \{p_{ij}(t_r)\}$

$= \{\Pr[X(t_r) = j | X(t_{r-1}) = i]\}.$

By assumption, i and j are discrete, finite variables – i.e., $i, j = 1, ..., m$ – and as usual

(4a) $0 \leq p_{ij}(t_r) \leq 1$,

(4b) $\sum_{j=1}^{m} p_{ij}(t_r) = 1 \quad (i = 1, ..., m)$.

Extensions of this basic model to the case where X and/or t are themselves matrices involving several conditional predicates have also been outlined; the model is then expressed in the so-called N-way matrix format (Gale 1970; Oldenburger 1934; Olsson and Gale 1968). A further extension which includes growth and attrition has also been described (Gale 1972b).

Of interest here are the conditions under which the $\{p_{ij}(t_r); i, j = 1, ..., m\}$ are constant over time, i.e., where the probability process is stationary (or time-homogeneous). In this regard, expressions will be given for three general types of stochastic stationarity: local stationarity, functional stationarity, and differential stationarity. Each of these will also be given substantive interpretation in terms of geographic mobility.

Consider expression (3). If the conditional probability that the system is in the jth state at time t_r given that it was in the ith state at time t_{r-1} is independent or r, then the process will be said to exhibit *local stationarity*, which is the usual definition of stationarity (Feller 1968, 419–21). More specifically, if for each $p_{ij}(t_r)$, $(i, j = 1, ..., m)$, the following condition holds:

$$p_{ij}(t_r) = p_{ij}(t), \text{ for every } r \ (r = 1, ..., T)$$

and each pair (i, j),

the process is locally stationary. At least two statistical tests for local stationarity have been discussed: Anderson and Goodman's chi-square test (Anderson and Goodman 1957) and Kullback's information statistic (Kullback 1968, 168–9). Both give essentially the same results.

Local stationarity in the case of expression (3) thus implies that the process of change (e.g., residential mobility, shopping behaviour, etc.) depends solely on the conditional probability of the present state given the state in the immediate past; it is independent of the specific time interval over which the probabilities are measured. Furthermore, since each $p_{ij}(t_r)$ is treated as if it were independent of all factors except the immediately preceding state, local stationarity can also be regarded as a form of general parameter (i.e., functional) stationarity; the probabilities are not only independent of the specific points in time over which they are measured, but remain constant under all other internal and external influences. The relaxation of the Markov assumption (expression (2)) changes the specification of the number of time intervals over which the condition of temporal independence must hold, but does not affect

the related conditions of functional independence.

The notion of *functional stationarity* can be viewed as an extension of the concept of local stationarity as it is applied to the basic conditional probability model given by expression (3). This extension can, in turn, be stated in two different model-forms: the N-way matrix model (NMM) and the regression-matrix model (RM). The NMM is a general case of expression (1) in which either X or t (or both) are themselves matrices; as an analogue to expressions (3) and (4) we thus obtain higher dimensional transition matrices of the form

(6) $\{p_{n_1}, ..., n_N\} = \{\Pr[X_{q+1} = \mu_{q+1} \wedge ... \wedge X_N = \mu_N | X_1 = \mu_1 \wedge ... \wedge X_q = \mu_q]\}$,

where

(7a) $0 \leqslant p_{n_1, ..., n_N} \leqslant 1$ for every $n_\alpha (\alpha = 1, ..., N)$,

and

(7b) $\sum_{n_\alpha \varepsilon Q_2} \cdots \sum_{n_\alpha \varepsilon Q_2} p_{n_1}, ..., n_N = 1$

for each $n_\alpha \varepsilon Q_2$ ($\alpha = q + 1, ..., N$). Here $Q = \{Q_1, Q_2\}$ is a partition of conditional predicates (e.g., location, time, social class, economic status, etc.), with $Q_1 = \{n_1, ..., n_q\}$ and $Q_2 = \{n_{q+1}, ..., n_N\}$ (Gale 1970). Stationarity is treated in the same manner as in the preceding section except that here it can also be defined with respect to conditional predicates other than time.

The RM model is also an extension of expression (1); in this case, however, the 2-way matrix form of expression (3) is preserved by treating the functional properties of the system in terms of the method of computing the $\{p_{ij}(t_r)\}$s. In the usual case, estimates of the transition probabilities are based on maximum likelihood estimates of the movements of individuals between states (Lee et al. 1965); thus, if $s_{ij}(t_r)$ is defined as the *number* of individuals who move from state i to state j in the interval $(t_r - t_{r-1})$, then the maximum likelihood estimate of the transition probabilities is given by

(8) $\hat{p}_{ij}(t_r) = s_{ij}(t_r) \Big/ \sum_{j=1}^{m} s_{ij}(t_r) \quad (i = 1, ..., m)$

(Anderson and Goodman 1957). Now, where we desire to include information concerning several related variables (e.g., as in the NMM model), an alternative estimation procedure can be employed. Let $Z = \{z_1, ..., z_k\}$ be the set of such factors; it has been shown then that functionally dependent estimates of the transition probabilities can often be obtained by fitting a least squares regression equation of the form

(9) $\hat{p}_{ij}(t_r) = \hat{\alpha}_{ij} + \sum \hat{\beta}_{ijk} z_k(t_r)$

for each of the m^2 cells of the transition matrix (Hallberg 1969). Stationarity in this model is analogous to the use of the concept in regression analysis; that is, if the regression coefficients, the $\{\beta_{ijk}\}$s, are constant over time and for different sample populations, and the specification of the model is also consistent, then the model is taken to be stationary.

In both the NMM and RM models, the notion of functional stationarity is an extension of the notion of local stationarity; the difference is in the number of factors over which the stationarity is defined. This is, of course, of great value in the explanatory analysis of a complex empirical phenomenon such as geographic mobility. It has been shown, in fact, that in the initial stages of research, this type of functionally defined model is a prerequisite for valid explanations.

Differential stationarity may be regarded as a means for discerning stationarity in particular kinds of non-stationarity processes. As before, consider the basic matrix model given by expression (3) and let it be a non-stationary stochastic process. This system may be said to exhibit differential stationarity if, for example, for each pair (i, j) there exists a k_{ij} such that the transition probabilities at t_r and t_{r-1} differ only by this constant. In effect, this is a means of decomposing non-stationary processes into two components, one stationary, the other non-stationary. Clearly, higher-order and non-linear measures of differential stationarity can also be employed in this framework.

A related method of treating non-stationary stochastic processes has been discussed by Harary, Lipstein, and Styan (1970, 1168–81). Their approach is a 'form of calculus for stochastic matrices [in which] the derivative of conventional calculus is replaced by ... [a] ... causative matrix, which represents the rate of change from one transition matrix to another' (Harary et al. 1970, 1177–8). More specifically, given a non-stationary Markov process (where each transition matrix is non-singular), for $\{P(t_r); r = 1, ..., T\}$ let $\{C(t_r); r = 1, ..., T\}$ be the associated *causative matrices*; the $\{C(t_r)\}$ are defined by the following equations:

$P(t_1)C(t_1) = P(t_2); ... ; P(t_{r-1})C(t_{r-1}) = P(t_r); ... ;$
$P(t_{T-1})C(t_{T-1}) = P(t_T).$

Causative matrices may thus be regarded as describing the changes between transition matrices. For example, where the process is locally stationary, then $C(t_r) = I$ (the identity matrix) for all r; where the $\{C(t_r)\}$ are not identity matrices but are all equal, the chain is said to be a *constant chain* (Harary et al. 1970, 1171–2).

The reduction of non-stationary to stationary processes via differential stationarity is potentially of considerable importance in the analysis of social and geographic mobility. Very often, where the rules which govern a process are found to be changing, they are nevertheless changing in a consistent way. By employing some form of differential calculus, it thus may be possible to isolate certain regularities and transform the process into a stationary model.

The discussion indicates that, rather than being a very limited model (in the sense that stationarity occurs only rarely), the application of finite space stochastic processes to the analysis of geographic mobility could be extended by employing a broader perspective on the notion of stationarity itself. In addition, since any empirical research program which uses such models is concerned with the *explanation* of the process of change, this broader view is clearly of importance in that it also permits more significant insights into the concomitants and structural properties of the mechanism of change.

Anderson, T.W., and L.A. Goodman, 1957 Statistical inference about Markov Chains, *Ann. Math. Stat.* 28, 89–110.
Beshers, J.M., 1967 Computer models of social process: the case of migration, *Demography* 4, 838–42.
Feller, W., 1968 *An Introduction to Probability Theory and Its Applications* 1, 3rd ed. (New York).
Gale, S., 1970 Explanation theory and models of migration, paper presented to the third conference on the mathematics of population, U. Chicago.
– 1971 Evolutionary laws in the social

sciences, paper presented to the fourth international congress on logic, methodology and philosophy of science (Bucharest).
– 1972a Inexactness, fuzzy sets, and the foundations of behavioral geography, *Geog. Anal.*, forthcoming.
– 1972b Some formal properties of Hägerstrand's model of spatial interaction, *J. Reg. Sci.*
Hallberg, M.C., 1969 Projecting the size distribution of agricultural firms – an application of a Markov process with nonstationarity transition probabilities, *Am. J. Agri. Econ.* 51, 289–302.
Harary, F., B. Lipstein, and G.P.H. Styan, 1970 A matrix approach to nonstationary chains, *Op. Res.* 18, 1168–81.
Kullback, S., 1968 Information Theory and Statistics (New York), 168–9.
Lee, T.C., G.G. Judge, and T. Takayama, 1965 On estimating the probabilities of a Markov chain, *J. Farm Econ.* 47, 747–62.
Nordbotten, S., 1970 Individual data files and their utilization in socio-demographic model building in the Norwegian Central Bureau of Statistics, *Rev. Int. Stat. Inst.* 38, 193–201.
Oldenburger, R., 1934 Composition and rank of n-way matrices and multilinear forms, *Ann. Math.* 25, 622–57.
Olsson, G., and S. Gale, 1968 Spatial theory and human behavior, *Papers, Reg. Sci. Assoc.* 21, 229–42.
Rogers, A., 1968 *Matrix Analysis of Interregional Population Growth and Distribution* (Berkeley).

P1206
A comparison between the Poisson and the simple random distribution with respect to the nearest neighbour technique in map analysis
SHIN-YI HSU *State University of New York, USA*

The nearest neighbour technique was originally developed based upon the Poisson distribution of points in space by astronomers in the late nineteenth and early twentieth centuries (Hertz 1907). In the late 1940s and early 1950s this technique was reviewed in the same mathematical framework by ecologists (Skellam 1952; Moore 1954; Clark and Evans 1954). Since the middle 1950s geographers have employed it in their research. It has been recognized that the Poisson model concerning the distribution of points in space is not usually applicable to geographic phenomena because the model neglects the existence of boundaries in map analysis. The purposes of this paper are (1) to examine the effect of boundaries in map analysis in terms of theoretical consideration and empirical check of computer simulation, (2) to develop a simple random model that fits a bounded point set to solve this problem, and (3) to compare these two models in terms of their practicability in geographic research.

There are two basic assumptions in the Poisson type of distribution of points in space (Parzen 1962, 31–2). Consider a group of points distributed in a space S, where S is a Euclidean space of dimension ≥ 1. For each region R, let $N(R)$ denote the number of points (finite or infinite) contained in the region R. The array of points is said to be distributed according to a stochastic process if for every region R in S, $N(R)$ is a random variable. The array of points is said to be a Poisson distribution with density ρ if the following assumptions are fulfilled: (i) The number of points in non-overlapping regions is an independent random variable; more precisely, for any integer n and n non-overlapping (disjoint) regions $R_1, R_2, ..., R_n$, the random variables $N(R_1), ..., N(R_n)$ are independent; and (ii) for any region R of finite volume, $N(R)$ is Poisson distribution with mean $\rho V(R)$, where $V(R)$ denotes the volume of the region R.

The derivation of the equation of expected distance has been given by Skellam (1952), Moore (1954), and Clark and Evans (1954), summarized by Miller and Kahn (1962) and Dacey (1964), and reviewed by Kendall (1963). Consider a collection of points, fixed in a randomly chosen area of specific size. A point is selected at random and the distance between it and its nearest neighbour is recorded. This process is repeated for all points (sampling without replacement) yielding a resulting distribution of distances, which are representative of the spatial relationships of the point pattern in that region.

Let ρ represent the mean density of an observed distribution, which equals the total number of points (N) divided by total area (A), i.e., $\rho = N/A$. Let r represent distance in any specified units from a given point to its nearest neighbour.

For a random distribution of points on a plane, the probability that a randomly chosen unit area will contain exactly x points is the Poisson function (1) where the parameter λ is the mean density:

(1) $P(x) = e^{-\lambda}(\lambda)^x/x$.

Let the specified area be a circle of radius r. If ρ is the number of points per unit area, then $\rho\pi r^2$ is the mean density. From equation (1) it can be derived that the probability of distances of nearest neighbour less than or equal to the distance r is

(2) $P(r) = 1 - e^{-\rho\pi r^2}$.

The expected value of r is

(3) $E(r) = \int_0^\infty 2\rho\pi r^2 e^{-\rho\pi r^2} dr = 1/(2\sqrt{\rho})$

since equation (1) is a countable but infinite mathematical function; the nearest neighbour statistic of equation (3) assumes no boundary. This is not the usual case in geographic map analyses.

The proposed model is intended for fitting distributions having small population size with a definite boundary. This theoretical distribution is a simple random distribution: each point is placed randomly with respect to its x and y co-ordinates. The expected nearest neighbour distance is to be used as a reference to check the degree of departure of a given distribution from the theoretical one.

Assume that points are randomly distributed in a circle with respect to their x and y co-ordinates. Given that points P_i and P_j are independent, P_i is selected randomly as the centre of origin for measuring the distance to its nearest neighbour P_j. The question then becomes, what is the probability that only P_j falls inside the circle with radius $\leq r_1$ (Fig. 1)? This probability function is constructed in the following manner. Assume the area of the big circle equals 1, and consider the geometric probability of points with respect to the small circle with radius r_1 in Fig. 1.

Then P (one point falls inside the circle) $\leq \pi r^2, 0 \leq r \leq 1/\sqrt{\pi}$. If the equality case, $P(P_i) = \pi r^2$, is used, the expected nearest neighbour distance will be slightly underestimated because of the boundary problem as indicated in Fig. 1. Hence, P (any one point falls outside the circle) $= 1 - \pi r^2$, and P (only P_j falls inside the circle, given that P_1 is the centre of origin) is given by

(4) $P(r) = 1 - (1 - \pi r^2)^{n-2}$

since there are $(n - 2)$ points outside the small circle with radius r_1. The mean or expected value of r is seen in equation (5):

(5) $E(r) = \int_0^{1/\sqrt{\pi}} (n - 2)(1 - \pi r^2)^{n-3} 2\pi r^2 dr$.

If we take the inequality case, the probability function of r is conditional on the distance from the centre to the given point (v), and the form of $P(r, v)$ depends on whether or not $r + v \leq 1/\sqrt{\pi}$ (see Fig. 1). We have

(6) $P(r/v) = \begin{cases} 1 - (1 - P(\sqrt{\pi}r, \sqrt{\pi}v))^{n-2}, \\ \quad \text{if } 1/\sqrt{\pi} - v \leq r \leq 1/\sqrt{\pi} + v, \\ 1 - (1 - \pi r^2)^{n-2}, \\ \quad \text{if } 0 \leq v \leq 1/\sqrt{\pi} - v, \end{cases}$

where

$1 - P(x, y) = (1 - x^2)/2 - (x^2 - 1 - y^2)$
$\times (1 - (x^2 - 1 - y^2)^2/4y^2/4y^2)^{\frac{1}{2}}/2y\pi$
$- (1/\pi) \sin^{-1}((x^2 - 1 - y^2)/2y)$
$+ (x^2 - 1 + y^2)(x^2 - (x^2 - 1 + y^2)^2/4y^2)^{\frac{1}{2}}/2y\pi$
$+ (x^2/\pi) \sin^{-1}((x^2 - 1 + y^2)/2xy)$,

$f(v) = \begin{cases} \int_0^{1/\sqrt{\pi}} Q_1(r, v) dv + \int_{1/\sqrt{\pi}-r}^{1/\sqrt{\pi}} Q_2(r, v) dv, \\ \quad \text{if } 0 \leq r \leq 1/\sqrt{\pi}, \\ \int_{r-1/\sqrt{\pi}}^{1/\sqrt{\pi}} Q_2(r, v) dv, \\ \quad \text{if } 1/\sqrt{\pi} \leq r \leq 2\sqrt{\pi}, \text{ model I,} \end{cases}$

where

$Q_1(r, v) = 4\pi^2 v(n - 2)r(1 - \pi r^2)^{n-3}$,
$Q_2(r, v) = 2\pi v(n - 2)(1 - P(\sqrt{\pi}r, \sqrt{\pi}v))^{n-3}$
$\times (\partial p(\sqrt{\pi}r, \sqrt{\pi}v)/\partial r)$,

and

$f(r) = (n - 2)(1 - \pi r^2)^{n-3} 2\pi r$,

if $0 \leq r \leq 1/\sqrt{\pi} - v$, model II.

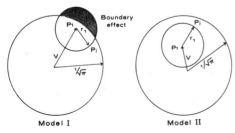

Fig. 1

The relation $1 - P(x, y)$ is derived by Graney (1969, 25). Equation (6) will be called model I, which takes into consideration the boundary effect, and equation (5), which reflects the boundary effect, will be called model II.

To check whether the above mathematical models are correct, a Monte Carlo simulation of a simple random distribution was performed. The population mean nearest neighbour distance and its standard error were calculated by simulating each n 10 times for $N = 100, 250, 500$, and 750, and 5 times for $N = 1000$. The result confirms that our models are much better than the Poisson model in terms of solving the boundary effect.

To achieve statistical inference or hypothesis-testing by the use of model I, it is necessary to determine the sampling distribution of the population mean. By applying the first four central moments of the population mean to Pearson's *K-criteria* (Kendall and Stuart 1969, 151), it is found that beta-distribution is the sampling distribution. However, it is not realistic to employ the beta-functions because there is essentially no difference between the obtained beta-distribution and a truncated normal distribution. Normal distribution is thus determined to be the sampling distribution of population mean.

It is clear that model I is superior to model II because the former considers the boundary effect and the latter neglects it. However, it will still be of great interest to compare the Poisson model with model I and model II with respect to the critical regions. The formula of sample mean and sample standard deviation of the Poisson model are $1/(2\sqrt{n})$ and $0.26136/n$, respectively (Clark and Evans 1954). The Poisson model and model II determine essentially the same critical regions. However, model I has significantly different critical regions. The difference is attributable to the boundary effect, which is significant even for population size as large as 500 points. Therefore, a given point pattern as determined by the Poisson model to be random may not be random at all.

J.D. Mason.

Clark, P.J., and F.C. Evans, 1954 Distance to nearest neighbor as a measure of spatial relationship in population, *Ecology* 35, 445–53.

Dacey, M.F., 1964 Two dimensional random point patterns: a review and an interpretation, *Reg. Sci. Assoc. Papers and Proc.* 13, 41–5.

Graney, R.W., 1969 Tests of concentration and identification of mixed samples, unpublished dissertation in the Dept. Statistics, U. Georgia.

Hertz, P., 1907 Über den gegenseitigen durchnittlichen Abstand von Punkten, die mit bekannter mittleren Dichte im Raume angeordnet sind, *Mathematische Annalen* 67, 387–98.

Kendall, M.G., 1963 *Geometrical Probability* (New York).

– and A. Stuart, 1969 *The Advanced Theory of Statistics,* v. 1, 3rd ed. (New York).

King, L.J., 1969 *Statistical Analysis in Geography* (Englewood Cliffs, NJ).

Miller, R.L., and J.S. Kahn, 1962 *Statistical Analysis in the Geological Science* (New York).

Moore, P.J., 1954 Spacing in plant population, *Ecology* 35, 222–7.

Parzen, E., 1962 *Stochastic Process* (San Francisco).

Skellam, J.G., 1952 Studies in statistical ecology. I. Spatial pattern, *Biometrika* 39 (3, 4), 346–462.

P1207
Toward computing factor scores
TULASI R. JOSHI   *University of Pittsburgh, USA*

Factor analysis has been widely used in urban studies to identify basic dimensions and patterns of socio-economic aspects of urban areas. Orthogonal varimax rotation is used by almost all geographers in their factorial studies of cities to simplify the factor loading matrix. Recently, some researchers have raised questions as to whether such orthogonally rotated factors are representative of the real world (Berry 1971, 213; Johnston 1971, 316–23; Haynes 1971, 326; King 1969, 187; Cattell 1965, 209). The notion that factors may not be independent is substantiated by several empirical studies (Johnston 1971, 316–9). Hence more detailed spatial models may be developed via alternative factor models if it can be shown that factors are not independent (Johnston 1971, 323; King 1969, 187).

A very interesting insight of orthogonal factors was gained from a factorial study on metropolitan Kathmandu, Nepal, using 80 intercorrelated socio-economic variables on 52 'statistical tracts' (equivalent to census tracts). A computer program (Dixon 1970, 90–103) was used to perform factor analysis with orthogonal varimax rotation on seven factors which accounted for 70 per cent of the total variance and had eigen-value of more than 2.0. By using a significance test of the correlation coefficient, it was found that a factor loading of $|0.3|$ was not significant at 0.01 level but $|0.4|$ was significant, for this research. Therefore, $|0.4|$ loading was considered as a dividing line between important variables and unimportant variables of a factor. Some researchers have considered only factor loadings greater than $|0.6|$ to avoid spurious loadings in their factorial studies (Bourne and Barber 1971, 265). In short, the logic is that variables which do not load over a specified minimum value should be eliminated in a factor analysis problem to get a picture representative of reality (Haynes 1971, 326). Important variables were distributed among the factors as follows: 31 variables each on factor 1 (ethnic-immobility) and factor 2 (socio-economic status); 13 variables each on factor 3 (stage in life cycle) and factor 4 (educational status); 6 variables on factor 5 (salary worker status); 8 variables on factor 6 (self-employed status); and 3 variables on factor 7 (foreign origin status). The factors were identified on the basis of their important variables and named accordingly. It was possible to interpret the rotated factor loading matrix satisfactorily, but when factor scores were mapped, it was difficult to interpret their spatial patterns in a meaningful way, especially for the minor factors with small eigen-values. The question arose as to why these contradictory conclusions were derived from the same analysis.

At this point it was conjectured that the inclusion of unimportant variables in computing factor scores might not be meaningful since their loadings could be spurious, and their cumulative effect would overshadow the net contribution of important variables since absolute factor loadings for many of them were greater than zero. Especially, such factor scores may extremely distort the picture of minor factors in a factor analysis in which there is a large set of variables and the basic factors number more than three. For example in this study, the cumulative total of squared loadings of unimportant variables for factor 7 is 1.548 which is 50 per cent more than the cumulative total (1.035) of squared loadings of important variables. This finding reveals that the inclusion of unimportant variables in computing factor scores, which is done quite uncritically in computer programs, certainly distorts spatial patterns of factors. Therefore, unimportant variables were eliminated assigning zero value for their loadings, and only important variables were used to compute factor scores employing the formula $F = AB$, where $F =$ standardized factor score matrix, $A =$ standardized original data matrix, and $B =$ rotated factor loading matrix. This is a straightforward method of computing factor scores if unities are used for communality estimates in diagonal elements of the correlation matrix (Cooley and Lohnes 1962, 164).

Factor scores computed by eliminating unimportant variables as mentioned before, turned out to be very meaningful in exhibiting their spatial patterns, and factors in this case were in fact found to be correlated, as surmised by the several persons mentioned earlier. Two major factors which apparently were independent of each other were found to be significantly correlated with other factors: ethnic-immobility factor was correlated with stage in life cycle factor ($r = -0.571$), salary worker status factor ($r = -0.406$), and self-employed status factor ($r = 0.464$); and socio-economic status factor was correlated with educational status factor ($r = 0.574$), and foreign origin status factor ($r = 0.792$). These correlated factors were conceptually meaningful. Thus for metropolitan Kathmandu, a correlated factorial model was designed in which the ethnic-immobility dimension with its three subdimensions, and the socio-economic status dimension with its two subdimensions, were considered as independent dimensions.

As an alternative to orthogonal rotation, some individuals tend to favour the use of oblique rotations in which factors may be correlated (Berry 1971, 213; King 1969, 197;

Cattell 1965, 209), whereas others have indicated that oblique rotations have produced no better results (Anderson and Bean 1961, 119–24), and it is doubtful whether they would always meet the requirements (Johnston 1971, 320). In this research, the factor loading matrix resulting from the orthogonal varimax rotation was quite satisfactory and meaningful. Here the main problem seems to concern an effective method of computing factor scores rather than the choice of rotations, since the issue of cumulative effect of unimportant variables, unless they are eliminated, still exists in factor scores even in oblique rotations. Therefore, the method presented in this paper of computing factor scores by eliminating unimportant variables may be a meaningful step toward making factor analysis a more useful tool.

Anderson, T.R., and L.L. Bean, 1961 The Shevky-Bell social areas: confirmation of results and reinterpretation, *Soc. Forces* 40, 119–24.

Berry, B.J.L., 1971 Introduction: the logic and limitations of comparative factorial ecology, *Econ. Geog.* 47, 209–19.

Bourne, L.S., and G.M. Barber, 1971 Ecological patterns of small urban centers in Canada, *ibid.*, 258–65.

Cattell, E., 1965 Factor analysis: an introduction to essentials I and II, *Biometrics* 21, 190–215, 405–35.

Cooley, W.W., and P.R. Lohnes, 1962 *Multivariate Procedures for the Behavioral Sciences* (New York).

Dixon, W.J., ed., 1970 *BMD: Biomedical Computer Programs, X-series Supplement* (Los Angeles).

Haynes, K.E., 1971 Spatial changes in urban structure: alternative approaches to ecological dynamics, *Econ. Geog.* 47, 324–35.

Johnston, R.J., 1971 Some limitations of factorial ecologies and social area analysis, *ibid.*, 314–23.

King, L.J., 1969 *Statistical Analysis in Geography* (Englewood Cliffs, NJ).

P1208
Principes fondamentaux de l'élaboration de la systématique typologique et de la division en régions physico-géographiques des paysages de montagne
K. KAVRICHVILI *L'Académie des sciences, RSS de Géorgie, URSS*

Cet exposé est trop long pour être reproduit ici.

Armand, D., 1964 Logique des classifications géographiques et schéma de division en régions (en langue russe), dans *Evolutions et transformation des conditions géographiques*, éd. Naouka (Moscou), 39.
– 1968a La réalité du paysage (en langue russe). *Les problèmes de la méthode de recherche de la science du paysage* (Moscou), 34.
– 1968b Géographie physique contemporaine (en langue russe), éd. Znanié (Moscou), 15.
– 1970 Objectif et subjectif dans la division en régions physico-géographiques (en langue russe). *Izvestia de l'Académie des Sciences de l'Union*, série géog., N1, 129.

Birot, P., et J. Dresch, 1953 La Méditerranée et le Moyen-Orient, dans *La Méditerranée Occidentale*, tome 1 (Paris), 57.

Davitaya, F., 1970 Regionalizacion climatica a base de los paisajes geograficos, dans *Atlas nacional de Cuba* (La Habana).

Issatchenko, A., 1968 Theoretical and practical landscape science – its purpose (en langue russe), *Przeglad geograficzny*, tome XL, 4 (Warszawa), 69.
– 1971 Les systèmes et les rythmes de la zonalité (en langue russe), *Izvestia de la Société de géographie de l'Union Soviétique*, tome 103, publication 1, 17–20.

Kavrichvili, Kh., 1968 Systématisation des paysages de montagne et de certains aspects de leur analyse par région (en langue russe), Rapports des géographes soviétiques au XXI-ème congrès international de géographie, éd. Naouka, 94.
– 1970a Principles of landscape classification for practical purposes (en langue géorgienne), *Bull. Acad. Sci. Georgian SSR* 57, N1, 108.
– 1970b On systematics of mountain landscapes (en langue géorgienne), *ibid.*, N3, 601.

Kalesnik, St., 1964 Rapport fait au VI-ème congrès de la science du paysage de l'URSS

(en langue russe), dans *Problèmes des recherches dans le domaine du paysage pour les pays de montagne* (Alma-Ata), 32.
- 1971 Some misunderstandings in the theoretical views of soviet geographers (en langue russe), *Izvestia de la Société de géographie de l'Union soviétique*, tome 103, publication 1, 28.

Preobragensky, V., 1969 Critical analysis of authors' essays of theses on landscape science, *Vestnik de l'Université de Moscou*, série v, géographie, N4, 69.

Sheffer, E., 1970 Landscape investigations for planning in agriculture (en langue russe), *Vestnik de l'Université de Leningrad*, N12, série II géog.-géol., 96.

P1209
Spatial association of qualitative maps
DONALD W. MAXFIELD   *Northern Illinois University, USA*

The concept of spatial association has been with geography a long time. Geographers make frequent references to the degree of similarity between maps in their professional lecturing and writing. In the past decade and a half there has been considerable research and writing concerning the spatial association that exists between two or more *quantitative* distributions. This research centres around a combination of statistical and cartographic analysis involving isopleth and choropleth mapping, simple and multiple correlation and regression analysis, factor analysis, and residual mapping. However, little attention has been given to a methodology for analysing a spatial association between two or more *qualitative* distributions. Such questions as the spatial association between a map showing climatic types and a map showing soil types, have not been answered except in vague or intuitive terms, such as, 'The general outlines of the major soil groups of the world coincide roughly with climates in their distribution, as can be seen by comparing the two world maps' (Murphey 1966, 101). It is the intent of this paper to present a statistical method to analyse objectively a spatial association between qualitative maps.

The statistical technique of contingency analysis refers to the idea of correlation applied to non-quantitative data (Blalock 1960, chap. 15). It provides a method of studying the relationship between two nominal scaled variables that have been cross-classified. Contingency analysis permits a test of the 'existence' of a relationship as well as a measurement indicating the 'strength' or degree of the relationship.

Contingency analysis can be applied to spatial problems when the following two conditions are met. First, the two nominal scales are the categories that represent the grouping of information on two qualitative maps, for example, a climatic map with the categories of tropical rainforest climate, humid microthermal climate, etc. and a soil map with the categories of arctic forest soils, podzolic soils, chernozem soils, etc. Second, each location where climate and soil types are identified, is selected by an areally stratified random sample method.

To illustrate the use of contingency analysis the world distribution maps of climate and soils were selected from *Goode's World Atlas*, 12th ed. The two sets of data from the world climate map and the world soils map are cross-classified in a contingency table as shown in Table 1.

Table 1 shows that there were a total of 1823 sample locations. Each entry in the table shows the number of locations of the various combinations of soil and climate. Also shown is the total number of cases for each climate type and each soil type.

This table represents the raw data from which an analysis of the association can be made.

(1) Does a significant relationship exist between these two sets of data? χ^2 statistic is used to determine if there is a significant relationship. From this table the computed $\chi^2 = 52.98$. Therefore, the null hypothesis of no difference among soil types with respect to climatic types is rejected at the 0.001 level (25 degrees of freedom). We have thus shown that there exists a dependence (or relationship) between the climate data and the soils data.

(2) What is the strength of the relationship? Pierson's contingency coefficient, $C =$

TABLE 1. Raw data

Climate		No soils A	Alluvial B	Tundra C	Soil Arctic forest podzolic, red-yellow podzolic-latosolic D	Chernozem, prairie, chestnut, sierozem, red desert E	Latosolic of humid tropics, wet & dry tropics F	Total
Tropical rain forest	1	0	19	0	5	78	358	460
Dry	2	0	6	1	8	496	29	540
Humid mesothermal	3	0	12	4	93	49	55	213
Humid microthermal	4	0	13	32	297	60	2	404
Polar	5	21	0	78	5	0	0	104
Undifferentiated highlands	6	0	1	20	21	33	27	102
Total		21	51	135	429	716	471	1823

TABLE 2. Percentage of total

Climate		No soils A	Alluvial B	Tundra C	Soil Arctic forest podzolic, red-yellow podzolic-latosolic D	Chernozem, prairie, chestnut, sierozem, red desert E	Latosolic of humid tropics, wet & dry tropics F	Total
Tropical rain forest	1	0	1	0	0	4	20*	25
Dry	2	0	0	0	1	27*	2	30
Humid mesothermal	3	0	1	0	5*	3*	3*	12
Humid microthermal	4	0	1	2	16*	3	0	22
Polar	5	1	0	4*	0	0	0	5
Undifferentiated highlands	6	0	0	1	1	2	2	6
Total		1	3	7	23	39	27	100

0.79. This coefficient needs to be carefully interpreted. It does not have the range of ±1.00. C has a shifting upper limit and depends upon the number of rows and columns. C can be correct or made a comparable test statistic by dividing C by the maximum possible value of C for the particular number of rows and columns. The upper limit of C for a six by six table is 0.91. Therefore, the corrected C for this table = 0.87. (Tschuprow's T, and Cramer's V, are other measurements of association.)

Table 2 represents table percentages based on the raw data of Table 1. It can be observed that all the entries total 100 per cent. The last column identifies the percentage land area of each climatic type. For example, 12 per cent of the land area was mapped as a humid mesothermal climatic type. The bottom row identifies the percentage of land area of each soil type, for example, 39 per cent of the land area was mapped in a soil category that included chernozem, prairie, chestnut, sierozem, and red desert. The individual table entries indicate the percentages of land area that were mapped in the various combinations of climate and soils. These individual entries are important in terms of the concept of spatial association that exists between soil and climate. It is claimed that there is a direct causal relationship between soil formation and the prevailing climatic conditions under which it was formed. Therefore, Table 2 should demonstrate this claim. Data from Table 2 support the following observations (see asterisks in Table 2):

(a) Latosolic soils tend to form under tropical conditions (col. F, row 1 = 20 per cent). (b) Chernozem, prairie, *et al.*, soils tend to form under arid and semi-arid conditions (col. E, row 2 = 27 per cent). (c) Podzolic soils tend to form under microthermal conditions (col. D, row 4 = 16 per cent). (d) Tundra soils form under tundra climate (col. C, row 5 = 4 per cent). (e) Humid mesothermal climates created a wide range of soils depending upon the relative amounts of temperature and moisture. It occupies somewhat of a large transition zone between dry, humid, warm, and cold conditions. Therefore, there are podzolic, lateritic, semi-arid, and arid soils (row 3, col. D, E, F = 11 per cent).

If we consider only the combinations just identified they represent 78 per cent of the total area. Recall that the non-corrected contingency coefficient $C = 0.79$. If we removed the columns of no soils (A) and alluvial soils (B) and the row of undifferentiated highlands climate (6), since these all represent exceptions to the spatial association concept of climate and soil, we will increase our percentage to 88. By analysing Table 2 in this manner we have shown that 88 per cent of the spatial distribution of soil has the same areal representation as controlling climates.

Two additional tables can be compiled to complete the analysis of the raw data of Table 1. One table would show the percentage of each soil type found within the various categories of climate and the other table would show the percentage of each climatic type within the various categories of soils. For one example – the tundra soils were found to be mapped within the following climatic zones: tundra 58 per cent, humid microthermal 23 per cent, undifferentiated highlands 15 per cent, humid mesothermal 3 per cent, and dry 1 per cent.

Richard A. Stephenson.

Blalock, Hubert M., 1960 *Social Statistics* (New York).
Murphey, Rhoads, 1966 *An Introduction to Geography*, 2nd ed. (Chicago).

P1210
Problems of an international multilingual geographic dictionary – the selection of terms and the layout of the planned dictionary
E. MEYNEN   *Federal Republic of Germany*

The IGU Commission on International Geographical Terminology, established at New Delhi in 1968, has concentrated on the development of a plan for an IGU international dictionary of technical terms used in geography.

When we speak of a 'Dictionary of Geographical Terms,' the aims of the enterprise are not entirely clear. To begin with, one has to distinguish between three kinds of terms. It was felt that the planned dictionary should include more or less all terms of strictly geographical pertinence or origin (e.g. 'spatial relationship' or 'transhumance'). Interdisciplinary terms (e.g. 'diluvial,' 'adret,' or 'acculturation') should only be included, if they have a certain significance in geographical research. General ones (like 'mountain,' 'anomaly,' or 'quotient') should in principle not be taken in.

Furthermore, we have to distinguish between topical terms, the designations of topographical features (e.g. 'canal,' 'glacier,' 'cotton field,' 'farmhouse') and conceptual terms, those relating to notions, processes, and properties (e.g., 'zonal soils,' 'aeolian erosion,' 'ecology,' 'conurbation,' 'central places').

What is needed and what we do not find in general dictionaries are the conceptual terms used particularly in science and their definitions. A defining dictionary of those terms is the task justifying the efforts of the IGU commission. The dictionary, limited to the conceptual terms used particularly by geographers, would comprise about 1000 to 1200 terms. This is the approximate amount which the ISO considers appropriate for the extent of a defining dictionary.

If we request an international dictionary of technical terms of the IGU, it should serve present demands. Also from this point of view, a clear limitation of the thematic range of the dictionary is necessary. The compilation of the dictionary should be carried out within four to six years.

With regard to the presentation of the selected terms, one could think of a logical arrangement as well as of an alphabetical sequence. Considering that most of the users will want a quick answer to their question without wasting time looking over many pages, the Commission proposes an alphabetical ordering; the Commission, however, is aware that the selection of the terms requires a systematic approach.

A definitive dictionary requires a uniform system of terms. Cross-references from the main term to subterms and vice-versa should form a uniform conception. A term should be defined only once. Where it reoccurs in the definition of another, a reference should be made to this fact; its definition ought not to be repeated.

The following three types of definitive texts are to be distinguished: (1) The definitive text of a main term where the definition is followed by quotations of subterms and other related methodological or process describing terms. (2) The more or less short definitive text of a single term. (3) The reference note, mostly without definition, or if so with a very short one (e.g., in the case of an adjective referring to the term it modifies: 'polar' referring to 'polar region,' 'polar front,' 'polar climate').

Regarding the layout of the text the following arrangement is proposed: (1) The term (e.g., 'erosion') with the mention of language origin (Latin: *erodere*, taking out). (2) Information about the field of application (e.g. 'geomorphology'). (3) Definition according to present-day usage. In many cases the definition will be a present-day definition given by the author of the text; in some others it will be a definition quoted from literature. (4) A historic note (e.g. at what time, in which connection the term occurred for the first time as a geographical term or by whom it was introduced). (5) Comments relating to the present application of the term (e.g. whether the term is used in different meanings, maybe within geography's own range, or varying from country to country, or from language area to language area, or maybe in a given case; whether the term has a different meaning outside geography, a peculiar meaning, as it often occurs that statistics or other state agencies or planning authorities define a term differently). (6) Listing of all cross-references. (7) Bibliographical notes of quoted or selected literature of relevance (of course only in selection). Naturally, this subdivision of the definitive text is only a guide.

The Commission was charged to prepare a sample publication of the dictionary as it will later appear in English, French, German, Russian, and Spanish. The ICA *Multilingual Dictionary of Technical Terms in Cartography* is regarded as a model. However, this is a different kind of dictionary as it relates to terms of techniques, working processes, and their documents. The enterprise of a dictionary of conceptual terms used in geography is more difficult. It requires the explanation of many variants and in many cases also the definition of related subterms, the articles thus being much longer. Recognizing this and considering the advantages and disadvantages, the Commission proposes to publish the international dictionary of technical terms in geography not as one volume of five languages, but as a multilingual dictionary in five editions, one edition for each of the five languages. Each one will give the equivalents in the other four languages in the margin.

In this way not only the five indicated languages could participate in the enterprise, but all other language communities also. Another important reason for the Commission's proposal is that a very voluminous publication would be found only on the shelves of libraries, whereas a book of about 300 to 350 pages could be bought also by the single research worker and student.

The Commission's proposal, however, is based on the understanding that the selected terms as well as the defining texts will be included uniformly in all five editions, that the equivalent of the other four languages is given in the margin in the same manner, and last, but not least, that the appearance of the books (title, format, etc.) will be uniform.

Regarding the equivalents to be placed in the margin there are three different marks:

(1) An equivalent which corresponds exactly to the term is marked '='. (2) One that does not entirely correspond is marked '±'. (3) Where an equivalent does not exist and a paraphrase or a translation is given, this is marked '≠'. Of course, this is only a guide for the user. If he wants more precise information he will refer to the edition of the language in question.

A uniform definitive text does not mean that four editions have to make literal translations from the leading language. In each case the definitive text will follow the approach of the language concerned. Where one language uses a noun another may use a verb, e.g., for the French nouns 'glaciaire' and 'périglaciaire' the English has only the adjectives 'glacial' and 'periglacial.' The text, however, must be analogous in all editions.

The compilation of a multilingual dictionary requires openminded co-operation. It is a big task, but it is worth the effort. The preparation of the planned international dictionary will best realize the general aims of the IGU. It will serve as a means of promoting the exchange of information at an international level and at the same time be a step in the direction of future national and international standardization.

p1211
A note on the estimation of a spatially-autoregressive model
JOHN ROBERT MIRON *University of Toronto, Canada*

In a recent paper, Cliff and Ord (1971) have discussed the use of multiple regression in regard to the problem of spatial forecasting. In that paper, it is acknowledged that the use of ordinary least squares (OLS) in the estimation of a spatially-autoregressive model may be invalid. In this paper, an estimate of the size of the bias in OLS estimators is presented.

Let a set of variables have defined values for each of N zones partitioning a map. A nonspatial hypothesis concerning these variables might take the form of

(1) $z_j = \alpha_1 x_{j1} + \alpha_2 x_{j2} + \ldots + \alpha_k x_{jk}$

$$j = 1, 2, \ldots, N,$$

where z_j is the dependent variable, x_{ji} is the value of the ith independent variable, α_i is the ith coefficient to be estimated, u_j is a random error term, and j refers to zone j.

We make the usual assumptions about the variables as required by the general linear model. These are, in review, (i) x-variables are mutually linearly independent; (ii) $E(u_j) = 0$, $j = 1, 2, \ldots, N$, (iii) $E(u_i u_j) = 0$, if $i \neq j = \sigma^2$, if $i = j$, and (iv) $E(u_i x_{jk}) = 0$ for any i, j, k, where E is the expected-value operator.

If we define β, an autoregressiveness measure, to be the effect of the size of all other z-values on the z-value in zone j, then a simple spatial hypothesis might take the form of

(2) $z_j = \alpha_1 x_{j1} + \alpha_2 x_{j2} + \ldots + \alpha_k x_{jk}$
$\qquad + \beta a_j + u_j, \quad j = 1, 2, \ldots, N$

where a_j is defined as

(3) $a_j = \sum_k m_{jk} z_k \quad j = 1, 2, \ldots, N,$

and such that

(4) $\sum_k m_{jk} = 1, \quad m_{jj} = 0, \text{ and } m_{ij} \geqslant 0$

$$j = 1, 2, \ldots, N.$$

We can redefine (2), in matrix notation, by

(5) $Z = XA + \beta MZ + U$, where

$$Z = \begin{bmatrix} Z_1 \\ Z_2 \\ \vdots \\ Z_N \end{bmatrix} \quad X = \begin{bmatrix} X_{11} X_{12} \cdots X_{1k} \\ X_{21} \\ \vdots \\ X_{N1} \cdots X_{Nk} \end{bmatrix} \quad A = \begin{bmatrix} \alpha_1 \\ \alpha_2 \\ \vdots \\ \alpha_k \end{bmatrix}$$

$$M = \begin{bmatrix} M_{11} M_{12} \cdots M_{1N} \\ M_{21} \\ \vdots \\ M_{N1} \cdots M_{NN} \end{bmatrix} \quad U = \begin{bmatrix} U_1 \\ U_2 \\ \vdots \\ U_N \end{bmatrix}.$$

Letting $X^* = (X \; MZ)$ and $A^* = \begin{bmatrix} A \\ \beta \end{bmatrix}$,

(5) can be rewritten as

$Z = X^* A^* + U$.

The OLS estimator of A^* is given by

(6) $\hat{A}^* = (X^{*\prime} X^*)^{-1} X^{*\prime} Z$.

Using the calculus of mathematical expectations, we can derive that

(7) $E(\hat{A}^*) = A^* + (X^{*\prime} X^*)^{-1} V$, where

$$V = \begin{bmatrix} 0 \\ 0 \\ \vdots \\ E(Z'M'U) \end{bmatrix}.$$

The typical element in the summation making up $Z'M'U$ is

(8) $a_j u_j = u_j \sum_s m_{js} z_s$.

TABLE 1. Bias

β	$\Sigma(\Sigma(m_{js}w_{sJ}))$	$\sigma^2 = 0.50$		$\sigma^2 = 0.75$	
		Bias α	Bias β	Bias α	Bias β
0.5	3.89	−0.90	0.14	−1.35	0.21
0.6	5.27	−1.22	0.19	−1.83	0.28
0.7	7.27	−1.68	0.26	−2.52	0.39
0.8	10.45	−2.42	0.39	−3.61	0.59
0.9	17.39	−4.00	0.63	−6.00	0.94

Note that we can re-express (5) as

(9) $Z = (I - \beta M)^{-1}XA + (I - \beta M)^{-1}U$,

where I is an $N \times N$ identity matrix. Let w_{st} be the typical element of $(I - \beta M)^{-1}$ and reformulate (8) as

(10) $a_j u_j = u_j \sum_s \left[m_{js} \sum_t \left[w_{st} \sum_r [x_{tr}\alpha_r] + w_{st} u_t \right] \right]$,

for which

(11) $E(a_j u_j) = \sigma^2 \sum_s (m_{js}w_{sj})$,

and

(12) $E(Z'M'U) = \sigma^2 \sum_j \left[\sum_s (m_{js}w_{sj}) \right]$.

Unfortunately, w_{sJ} is a function of β and, therefore, no simple adjustment to the OLS estimator, to make it unbiased, is possible.

However, suppose that $|\beta| < 1$. Then,

(13) $(I - \beta M)^{-1} = I + \beta M + \beta^2 M^2 + \beta^3 M^3 + \ldots$

and

(14) $w_{sJ} = \beta(M)_{sJ} + \beta^2(M^2)_{sJ} + \beta^3(M^3)_{sJ} + \ldots$

Note that M is a Markov Transition Matrix and that

(15) $\sum_j (M^k)_{sj} = 1$ for any k and s.

Substituting (14) into (12) yields

(16) $E(Z'M'U) = \sigma^2 \sum_j \left[\sum_s (m_{js}(\beta(M)_{sj} + \beta^2(M^2)_{sj} + \ldots)) \right]$

(17) $= \sigma^2 \sum_j \left[\sum_k (\beta^k(M^{k+1})_{jj}) \right]$.

If M satisfies the Ergodic Theorem, then

(18) $\lim_{k \to \infty} (M^k)_{sj} = P_j$ for any s, j,

where P_J is the ergodic probability of zone j. In the limiting case where β approaches unity, we can approximate $E(Z'M'U)$ by

(19) $\lim_{\beta \to i} E(Z'M'U)$

$\approx \lim_{\beta \to i} \sigma^2 \sum_j \left[\sum_s (m_{js} P_j \beta/(1 - \beta)) \right]$,

(20) $\approx \lim_{\beta \to i} \sigma^2 \beta/(1 - \beta)$,

(21) $\approx \infty$.

From (17), we can see that when $\beta = 0$, $\hat{A}^*$ will be unbiased. From (21), we can see that as β approaches unity, $\hat{A}^*$ becomes totally unreliable. From (17), we can also conclude that, for values of β between zero and unity, the size of the bias is partly dependent on M.

A different estimator of A^* might be given, after (7), by

(22) $\tilde{A}^* = \hat{A}^* - (X^{*\prime}X^*)^{-1}V$,

where the parameter values in the non-zero element of V are estimated by $\tilde{A}^*$. Then (22) can be solved iteratively for $\tilde{A}^*$. It is not clear, however, that $\tilde{A}^*$ will, itself, be an unbiased estimator of A^* and the intractability of this problem has prompted consideration of the use of Monte Carlo simulation to test the properties of this estimator.

Cliff and Ord have estimated, using OLS, a model of the form

(23) $z_t = \alpha + \beta a_t \quad t = 1, 2, \ldots, 25$,

where z_t is automobiles per 100 population, a_t is the average value for z_t in contiguous counties (binary case), and the map consists of 25 Irish counties. Their estimates are

(24) $z_t = -1.165 + 1.1649 a_t$.

Using their published data, it can also be derived that

(25) $(X^{*\prime}X^*)^{-1} = \begin{bmatrix} 3.0208 & -0.4644 \\ -0.4644 & 0.0723 \end{bmatrix}$

$\sigma^2 = 0.585$.

Using (17), we can derive estimates for

(26) $\sum_j \left[\sum_s (m_{js}w_{sj}) \right]$

for different levels of β. Since σ^2 might be expected to lie between 0.50 and 0.75 in this case, let us calculate estimates of the biases in α and β at these levels (see Table 1). It is concluded, from this table, that when β lies between 0.7 and 0.8, the bias produced is large enough to account for the $\hat{\beta}$ value of 1.1649.

Using (9), it can be seen that, having estimates for β and A and forecasts for X, we can predict values for Z. Suppose, however, that β, A, and X are to be changed for the prediction of Z. We would like to be able to interpret both β and A so that we could make some meaningful statements as to how and why these are changing.

Let us consider the simplest spatially autoregressive model; that given by (23). We have already interpreted β as the effect of z-values in neighbouring zones on zone j. We can also show that

(27) $\hat{\alpha} = \bar{z} - \beta\bar{a}$

where bars indicate variable means. Thus, $\hat{\alpha}$ is directly related to the average value of Z. This suggests that, in order to predict a new map in this case, one needs only an old map, an estimate for β, and an estimate of the new $\bar{z}$. The value of α which produces this $\bar{z}$ can be determined iteratively. Further, it might be suggested that, if one wanted to use (9) for forecasting solely, an accurate and unbiased estimate for α is not critical.

Cliff, A.D., and K. Ord, 1971 A regression approach to univariate spatial forecasting, in M. Chisholm, et al., eds., *Regional Forecasting* (London).

p1212
Renaming the field
S. MOOKERJEE   *NMV College, India*

It is my apprehension that geography could be reduced to a mere skill or point of view, unless its thought-structure is lifted to a philosophical level from time to time. Not that all geographers should be doing this intellectual exercise, but at every phase of its development there must be a few 'arm-chair' geographers who would do some deep-thinking in the discovery and narration of a geographical philosophy. One great tradition (Pattison 1964) of geography has been in some kind of equation between man and environment, but of late a serious challenge seems to have been posed to convert geography into an analysis of spatial diffusion. The intellectual content, or the lack of it more often, in the field of geography has bothered me all along, and I have even been dissatisfied with the continued use of the very name of our field. Who does not agree that the name 'geography' does not signify today even a substantial part of the knowledge which has come to be associated with this discipline? The name is definitely inadequate to denote either the core or the periphery of the subject.

But is geography integrated enough to hold its own or likely to go into specializations? In the Soviet Union Academician I.P. Gerasimov has been accused of stressing research on physical components such as climates, soils, land forms, and vegetation and of supporting views which will lead to the liquidation of geography. On the other hand, the decline of physical geography in the land of W.M. Davis is well known and much regretted, as will be evident from the following statement of an American geographer: 'In some departments, physical geography has been dropped, perhaps in the expectation that it will be taken up by meteorology, geology or forestry' (Miller 1965, 3).

There appears to be, then, a 'cold war' between the Soviet and the American views on the nature and scope of geography. There is no doubt that a wide discrepancy exists between the Soviet and American points of view regarding the subject matter of geography. While the Soviet geographers have stuck to the 'mother of Science' concept of geography, post-world war II American geography has drifted towards social sciences. Both countries have, however, contributed valuable research to their respective areas of geography and also made considerable success of practical applications of the geographical method. Yet in both countries criticism has been put forward suggesting that the subject has been pushed ahead too much in one direction, to the neglect of the other. The Russians tend to utilize geographical research in the socialistic reconstruction of their regions: the Americans prefer to undertake resource-inventories and improvement of individual regions, through free enterprise. Over 10 years ago, while analysing this growing divergence of viewpoints between Soviet and American geography (Mookerjee 1957) I had spelled out a new statement on geography, 'as the study of man in the ordinary existence of life' (following Marshall's well-known definition of economics). In this regard geography breaks some common ground with the controversial philosophy of existentialism, whose famed exponent Jean-Paul Sartre observed: 'The first effect of existentialism is that it puts every man in possession of himself as he is, and places the entire responsibility for his existence squarely upon his own shoulders.'

The divergence became evident in the presentation of 'view points' on geography in the plenary session of the 21st International Geographical Congress held in New Delhi in December 1968, the principal participants therein being, Academician I.P. Gerasimov (USSR), Professor N.S. Ginsburg (USA), Professor C.G. Sautter (France), and Professor M.J. Wise (UK). While Soviet geography is going into depths of physical phenomena as related to human life and economic development, the Americans, after having gone through a glorious era of cultural and urban geography in the broad framework of social sciences, are now on their mark to set forth new goals in quantification. In the latter area of development, measurement of terrestrial phenomena, their circulation, and evolving patterns have become important. If this development goes uninterrupted, 'geo-graphy' may very well become another form of 'geometry' which will measure earth forms, not necessarily the distances and directions between astronomical lines and points, but those between nuclei of human enterprise, or in other words, between such 'central places' or poles of growth, where the action is.

There are, fortunately, the British and West European schools of geography where physical geography continues to enjoy the same status as cultural geography, particularly in Britain and Germany. Or, to look at it in the other way, here human aspects of geography have not yielded position to the demands of physical geography and its numerous subfields. In many Asian countries where university education is still largely fashioned after the British system this West European pattern continues to hold ground.

In American geography, already there is a demand for a trek back to physical geography. 'There is no alternative to physical geography being done by a physical geographer' (Miller 1965). Further, complained Garnier (1963): 'Advanced physical geography boldly proclaiming the geographic point of view, is difficult to find. Yet there is plenty of scope for it' but 'many geographers are content to allow the study of earth, air, and water to be left almost entirely to the systematic sciences.' Happily, however, quite a few US universities have opened up, during the past decade only, earth science faculties where physical elements are being emphasized, including laboratory work. Also, in the new regional science programs of the University of Pennsylvania and elsewhere, mathematical analysis is becoming increasingly important. In the East European countries again, the study of physical geography is not without its economic bias and practical content. The ultimate aim of each school of geography is to attempt a balanced and all-round economic growth of the regions it serves. On the American side, regional analysis is gathering momentum with computerization and quantification of an astonishing array of available data. On the Soviet side, economic regionalization is pledged to pursue the same objectives – with the help of comprehensive and also extraordinarily minute physical studies. In the Central European countries, of course, regionalism has always been a strong unifying force. In France, for example, it has long been the predominant concept of geography. And there are, in each country, a few old-fashioned but rather competent geographers who are still fond of a vivid description of a region which will bring before the eye a lively picture of an area, anywhere in the world, not visited personally. And this knowledge is intended to come more through 'perception' and without the rigours of 'systems analysis,' as in the USA, or through the technicalities of the geosciences as in the USSR.

While on this subject of diversity of viewpoints in geography I again want to emphasize the need for a change in the nomenclature of the subject itself. Geography seems to be the only discipline in which there is constant introspection as to its philosophy and point of view. Almost every issue of a journal contains an article on the nature or purpose of geography and wherever professional geographers meet, this subject inevitably crops up and in it most present seem to take part with great conviction. However geographers may differ in the definition of regions and even if one might agree with the view that 'there is no such a thing as region except for a purpose,' space still forms the hardcore of geography. In Hartshornian language, after all, what one studies in geography is the areal, regional differentiation of patterns, all over the globe. Ritter called it 'une impression ensemble' brought about by a complexity of physical, biological, and cultural factors. In the diversity of viewpoints, therefore, the unity of approach is provided by the areal or regional factor in geography. This is borne out by the findings of a survey recently undertaken by the National Academy of Sciences and the Social Science Research Council of the USA. In a study of problems which were of concern to and investigated by American geographers (Taaffe 1970) in the last decade there emerged the following six categories: (1) spatial distribution and interrelationships, (2) circulation, (3) regionalization, (4) central-place systems, (5) diffusion, and (6) environmental perception. It is striking to note that except for the last one, all of these study-groups veer round the dynamics of the 'region,' par excellence.

Why not then call the science of geography itself, the science of regions? To the layman, the good old term geography may continue to denote a variety of concepts as it has done hitherto, but in its technical and scientific aspect the term 'Regionics' seems to be an acceptable alternative, out of the two I had proposed (regionics and regionology) in a communication to *Professional Geographer* in July 1960. If this is accepted, the various branches of geography will then be re-named as physical regionics, cultural regionics, political regionics, and economical or economic regionics (or regional economics?), and so on. The name areal regionics, in lieu of regional geography, will hold good for the study of regionics of a specified area. In a recent article in the same *Professional Geographer*, the names 'geography' and 'geographical science' were suggested for two distinct purposes of the content and the subject respectively. This does not appear to be satisfactory, if one would only compare a similar situation prevailing in such group-names as physics and physical science, biology and biological science, geology and geological science. In the case of geography the problem becomes very special indeed, even urgent, for the reason that the time-honoured name 'geography' no longer refers to much that is recognized as the core, or even might mislead the non-geographical scientist, not to speak of the layman, as to what is done in and expected of the field of geography. If, on the other hand, the name of our discipline is changed to 'regionics,' it would at least meet this dilemma squarely and fairly. Political economy and political philosophy changed into economics and politics not many decades ago. It is my fear that if we do not adopt regionics, our field might fast become a too specialized technique, *sans* its traditional philosophy, and be called regionometry. Let us then, in the centenary year of the international geographical congress in 1972, vote for a new name to our field: regionics.

Garnier, B.J., 1963   *Prof. Geog.*
Miller, 1965   *Prof. Geog.*
Mookerjee, S., 1957   *Geography as a Philosophy of Existence: Geographical Outlook* (Ranchi).
Pattison, W.D., 1964   The four traditions of geography, *J. Geog.*, 63.
Taaffe, E.J., ed., 1970   *Geography* (Prentice-Hall).

P1213
A spatial distribution computer for geographical analysis
ROBERT E. NUNLEY and GEOFFREY W. ROPER   *University of Kansas, USA*

A spatial distribution computer system has been designed to simulate geographic problems. The programming, analysis, and interpretation of results can be done efficiently and in a manner intuitively obvious to persons trained in geography. Construction of the system will be completed by June 1972 at a cost of $200,000.

The system is a much extended, more flexible version of the analogue field plotter (Commission on College Geography 1971). It will permit experiments that are too complicated for complete field plotter solutions and experiments designed to develop further the field plotter as a research and educational tool.

In geography, as in other areas of study, the modelling of physical processes is assuming an increasingly important role. The digital computer and associated mathematical techniques have played an important role in the development of modelling and simulation in recent years. However, this tool has not been employed as effectively in geographical analysis as in other fields. Numerous attempts have been made, and some positive results have been achieved, but it is becoming increasingly apparent that digital computers are not well suited for addressing many, if not most, geographical problems. Three of the principal reasons are: (1) Language: existing programming languages were not designed to express geographic relationships. Moreover, they are not based on the systems and structures of the language of geographers. As a result, the best work done using digital computers for geographic analysis has primarily consisted of statistical analyses of geographical phenomena. It has not allowed the simulation or investigation of direct geographical relationships. (2) Organization: digital computers are organized to receive and process data in a one-dimensional sequence. There are several systems using so-called parallel processes that attempt to overcome this one dimensional limitation. However, these systems are essentially parallel duplicates of linear devices. No present digital computer effectively uses parallel arithmetic units capable of simultaneously processing stored data. This linear organization presents a severe limitation to geographers who base their understanding of geographical problems on maps having a minimum of two spatial dimensions. (3) Processing time: once the factors describing a real world problem are reduced and simplified to a point where the problem can be addressed on a digital computer, the results frequently do not apply to the problem. If several thousand alternatives to a model are to be considered, then the necessary iteration process requires considerable time. This results in prohibitive costs and the loss of insight into the process through tedium. This is often the case with geographical analysis. It would be desirable to have an 'instant' model iteration capability, which would allow the experimenter to observe trends by simply 'dialling in' or programming model variations.

The above points suggest what we feel are the most important criteria for a geographic analysis system: (1) an appropriate language; (2) a map-like organizational basis; and (3) minimum time delay (i.e., seconds) between program modifications and the presentation of modified results. The spatial distribution computer system satisfies these three criteria.

The basis of the system is an assumed analogy between (ideal) geographic processes and processes that occur throughout the physical sciences. This analogy, and the conservation and continuity principles that give it credibility, have been discussed briefly elsewhere (Commission on College Geography 1971). Both the validity and the proper application of the assumed analogy are not well understood today. It is only through the use of this system that the extent of validity can be tested.

In order to meet the criteria specified previously, the spatial distribution computer is a specially designed hybrid system (interconnected analog and digital devices). It is characterized principally by the way in which it interacts with the operator. The machine elements involved in this interaction are: (1) Programming mapboard. The programming mapboard is a map-like device on which the operator can specify topography and symbology. While creating or modifying a map, the operator is simultaneously providing input data for a computer capable of solving various distribution equations. (2) Display. The display device portrays the results of the computer calculation. Although the results displayed may consist of a single number, a table, or a graph, the most common output is a distribution map. The distribution may be represented by contours, by patterns of colour or brightness, or by texture patterns. One of the possible representations is an isometric projection which gives a three-dimensional effect to the results. (3) Point source controls. The point source controls allow the operator to select several points on the map as sources or centres of attraction, and to experiment with the values of these sources while observing their effect on the displays. The strength of these source points can be constant (with arbitrary value), or they may be time varying. They may also be functionally dependent on values measured elsewhere on the map. (4) Area source controls. The area source controls allow source points of various value and functional dependence to be selected for entire regions, or for the entire map. These are set up by the computer using empirical data that have been stored in local memory or using a program that generates a desirable pattern of values.

The spatial distribution computer is capable of several modes of operation. It should be emphasized that these modes are quite different from those of conventional computer systems. This system uses previously prepared and stored programs that are suitable for a class of distribution problems. Once a particular class is selected, the computer configures the system so that any manipulations performed by the experimenter are consistent with the selected distribution class. Some of the modes of operation are described: (1) Display modes. In either the intensity, contour, or isometric projection modes, the system is capable of representing potential patterns for solutions or for the resistance network. Other modes will permit the display of time variations for potential patterns over an area or at specific points. (2) Iterative solution mode. This mode of operation permits convergence to a desired solution through an iterative process which constantly improves upon the set of input values. (3) Model simplification mode. This mode permits the synthesis of spatial phenomena by simulation. It is characterized by rapid response and visual feedback. It allows the operator to determine the location and magnitude of sources, barriers, and expeditors required to reproduce an observed distribution. Unlike a determination of statistical relationships, this model then makes a definite statement about the mechanisms that created or affect the observed distribution.

In summary, the system described has the speed, capacity, and versatility to allow the geographer to conduct effective and relevant investigations of spatial distributions.

National Science Foundation; Floyd Preston; Robert M. Haralick; Gerry L. Kelly; George Dalke; Michael Buchannan.

Commission on College Geography, 1971
Living Maps of the Field Plotter: A Guide to Analog Simulation of Geographic Processes, technical paper no. 4, Assoc. Am. Geog. (Washington, DC).

P1214
Ocean-tundra-glacier interaction model
ATSUMU OHMURA *Swiss Federal Institute of Technology*

The precise understanding of the relationship between mass-balance and the meteorological elements is of fundamental importance not only for glacial hydrology but also for reconstructing the climate of the past. One of the reasons why this is a difficult problem is the fact that a glacier-surface is three dimensional with a complicated geometry and is of vast dimension.

The practical problem of how to identify the most important heat source for melt, short-wave incoming radiation, on the actual surface arises. This important problem has already been solved and there is little possibility for further development of the theory (Garnier and Ohmura 1969, 21). The sensible heat flux is usually the second most important heat source for melt. The present paper describes the structure of the model of the atmosphere over the meso-scale region, by means of which the regional distribution of sensible heat flux on the glacier can be computed and the source region of the heat identified.

The sensible heat flux on glaciers varies significantly from place to place and micrometeorological point measurement is not satisfactory to describe the entire glacier-surface.

As the highest temperature which the glacier-surface will attain is $0°C$, at which melt takes place, the heat must be brought from outside of the glacier by advection. This means that the melting process of the glacier is significantly influenced by the conditions that prevail in some remote areas.

To understand the distribution of the sensible heat flux on the glacier and the influence of the external region, the two-dimensional steady state equation of diffusion was solved numerically with the boundary conditions obtained during the Axel Heiberg Island Expedition of 1969.

The model simulates, as a first approximation, a glacier near the centre of the island extending westwards to within 10km of the Arctic Ocean. In a vertical direction, the model represents the bottom 500m of the atmosphere. The logarithmic wind profile and the linearly increasing eddy diffusivity

TABLE 1. Comparison of measured and computed sensible heat fluxes for 1969 (in cal cm^{-2} hr^{-1})

	Jun. 16–30	Jul. 1–15	Jul. 16–31	Aug. 1–15	Aug. 16–23
Measured	10.0	9.0	6.0	3.0	3.4
Computed	11.1	8.1	4.5	2.6	3.3

in the bottom 40m were accounted for in the model. The upper 40–500m was assumed to have the wind profile given by Taylor's solution of the Ekman spiral and the constant eddy diffusivity. The east-west cross-section for setting-up the two-dimensional model was decided by the fact that prevailing wind in this region is westerly in summer.

The diffusion equation to be solved is

$\bar{u}(\partial I/\partial z) = \partial[K_H\{(\partial I/\partial z) + r_d\}]/\partial z$

together with

$K_H \simeq K_M = u_* k z$, for 0–40m,

$K_H = $ constant, for 40–500m,

$\bar{u} = (u_*/k)\ln(z/z_0)$, for 0–40m,

$\bar{u} = $ Taylor's solution of the Ekman spiral, for 40–500m. The cross-isobar angle at the bottom of the spiral layer was assumed to be 10°. All the symbols have the usual meanings. u_* is calculated by Sutton's empirical formula with the observed wind speed at 2m level. z_0 was assumed to be 1cm.

The computation of the temperature field was performed by the forward difference method based on the differential equation of diffusion. The horizontal grid space is 1m. The vertical grid space is 1m for the bottom 0–40m and 10m for 40–500m. The temperature of the advecting air was obtained from the climatological station located on the west coast. The lower boundary condition was given by the ground surface temperature measured by Cu-Co thermocouples at the base camp. The temperature of the surface of the glacier was assumed to be 0°c throughout the period. The upper boundary was assumed to keep its original temperature. The temperature field solved by this method enables one to compute the sensible heat flux on the surface (H_0). By applying the conservation law of thermal energy for a finite column of air (height z) on the surface, we have

$$H_0 = H_z + \int_0^z \rho c_p \bar{u}(\partial I/\partial x)dz.$$

The symbols have the usual meanings. In the numerical solution, $\partial I/\partial x$ was replaced by the temperature gradient computed by the already established temperature field. Similarly dz was replaced by the vertical grid size. A value of z equal to 10m was chosen.

The computation was performed for five periods, each of which was two weeks long. Table 1 shows the comparison of the computed sensible heat flux on tundra at 10km east of the shore, with the measured values at the base camp.

After testing the validity of the model, the model was altered by removing the tundra surface between the sea and the glacier. By eliminating tundra, the contribution of the heat produced on tundra can be estimated, in comparison with the first model. The computation with the second model gave only 40 per cent of the first model for the marginal 10km of the glacier. It is interesting that some 60 per cent of the sensible heat flux for this region is derived from the adjacent tundra surface, while the more interior region does not experience much change due to the conditions of the tundra. This interior region of the ice cap may be influenced by a much more distant region, such as mid-latitude, in summer.

Fritz Müller.

Douglas, J., Jr., 1961 A survey of numerical methods for parabolic differential equations, in *Advances in Computers*, 2 (New York).

Dyer, A.J., 1967 The turbulent transport of heat and water vapour in an unstable atmosphere, *Quart. J. Royal Meteor. Soc.* 93, 501–8.

Garnier, B.J., and Atsumu Ohmura, 1969 The evaluation of surface variations in solar radiation income, *Solar Energy* 13, 21–34.

Sutton, O.G., 1953 *Micrometeorology* (New York).

Webb, E.K., 1970 Profile relationships: the log-linear range, and extension to strong stability, *Quart. J. Royal Meteor. Soc.* 96, 67–90.

P1215
Population density and quality of the environment
GEORGE J. PAPAGEORGIOU   McMaster University, Canada

Phenomena related to the distribution of population around a single centre have been given one type of explanation, based upon the existence of fundamental locational preferences at the level of the household (Papageorgiou 1971). Given that these preferences are expressed throughout as a desire for space abundance and for accessibility, it was shown that a possible form of the utility attained by a residential location is

(1) $u(D, \delta) = hD^{-\alpha} \exp(-w\delta)$,

where D is the population density and δ is the commuting costs associated with this location. If $u = \bar{u}$, a constant throughout, there is no motivation to relocate. This condition reflects a state of spatial equilibrium. Given $\bar{u}$, the corresponding $\bar{D}$ can be expressed in terms of δ as

(2) $\bar{D}(\delta) = (\bar{u}^{-1}h)^{1/\alpha} \exp(-\alpha^{-1}w\delta)$.

The spatial equilibrium distribution of population was obtained from (2) by defining δ as a function of the distance s from the centre. It was seen that certain forms of $\delta(s)$ imply certain empirical density gradient functions. None of these was of the 'density crater' type.

Clearly, this conclusion depends upon the fact that space abundance and accessibility were postulated as the only significant determinants of residential location. It may be argued, however, that environmental quality is also significant. In this paper the existing framework is extended to incorporate the quality of the environment as a factor affecting the locational decisions of the household. As a result, both strictly decreasing and 'density crater' empirical forms emerge quite naturally. It is seen that the new framework is a generalization of the existing one: whenever the quality of the environment is held constant, the two frameworks become identical.

Consider a centre within an unbounded region and a number of identical households distributed around it. Suppose that the locational decisions of the households are completely regulated by a desire for space abundance, for quality of the environment, and for accessibility. Under these conditions, a possible modification of (1) is

(3) $u(D, E, \delta) = D^{-\alpha}E^{\beta} \exp(-w\delta)$,

where E denotes quality of the environment.

If a state of spatial equilibrium does exist,

(4) $\bar{D}(E, \delta) = (\bar{u}^{-1}E^{\beta})^{1/\alpha} \exp(-\alpha^{-1}w\delta)$.

It is natural to assume that $\delta = \delta(s)$. The quantity E, however, may depend upon several factors, including D. On the other hand, it could be argued that high density–high quality planned developments abound: thus, E may not depend upon D. The strategy adopted here is to examine certain aspects of the alternatives $E = E(D)$ and $E = E(s)$. Given $\delta(s)$ and E, $\bar{D} = \bar{D}(s)$ and (4) defines an equilibrium distribution of population over space.

If $E = h$, a constant, $u(D, E, \delta)$ reduces in effect to (1). If the relationship between E and D is assumed to be of the form

(5) $E(D) = hD^{-v}$,

then

(6) $\bar{D}(\delta) = (\bar{u}^{-1}h^{\beta})^{1/(\alpha+\beta v)} \exp(-(\alpha + \beta v)^{-1} w\delta)$,

which is identical in structure with (2). Thus, in both cases, the analysis in Papageorgiou (1971) holds. It follows that, if E is a significant determinant of location, the strictly decreasing empirical forms of the density gradient may imply an undifferentiated environment, or that $E \propto D^{-v}$.

Suppose now that transportation costs are proportional to distance: $\delta(s) = kfs$, where k is the transportation rate and f is the frequency of interaction with the centre. Moreover, suppose that the quality of the environment is increasing away from the centre. If

(7) $E(s) = h(1 + vs)$,

then

(8) $\bar{D}(s) = (1 + vs)^{\beta/\alpha} \exp(a_0 - as)$,

which represents a generalization of the 'density crater' form proposed by Reinhardt (1959). If

(9) $E(s) = hs^v$,

then

(10) $\bar{D}(s) = s^{\beta v/\alpha} \exp(a_0 - as)$,

which represents another 'density crater' form proposed by Ajo (1965). It should be stressed, however, that (8) and (10) refer to net density. Thus, since empirical 'density crater' forms refer to gross densities, (8) and (10) should be treated as new hypotheses rather than explanations of established empirical functions. It should also be noted that since E in (5), (7), and (9) is unbounded, the corresponding gradients seem to be applicable only within a limited distance from the centre: all of them may be labelled consistently as 'urban' density gradients.

Certain spatial equilibrium population distributions were justified in terms of three fundamental locational preferences, expressed as a desire of the household for space abundance, for a better environment and for accessibility. This construct is an extension of an older one that recognized space abundance and accessibility as the only determinants of location. The two constructs are formally related: the old construct evolves as a special case of the new one, whenever the environment is held 'constant.' Given alternative assumptions in respect to the structure of transportation costs and to the environmental quality, both strictly decreasing and 'density crater' empirical gradients can be generated. It is shown that the strictly decreasing forms correspond to either an undifferentiated environment or to an environment that is affected by congestion. On the other hand, certain 'density crater' forms correspond to situations where the quality of the environment depends upon distance from the centre.

Emilio Casetti.

Ajo, R., 1965 On the structure of population density in London's field, *Acta Geog.* 18, 1–17.

Papageorgiou, G.J., 1971 A theoretical evaluation of the existing population density gradient functions, *Econ. Geog.* 47 (1), 21–6.

Reinhardt, F.R., 1959 A test of hypotheses specifically related to the cross-sectional distribution of population densities of cities, MA thesis, U. Minnesota.

P1216
Continuity and discreteness of the geographical shell
V.S. PREOBRAGENSKY *Academy of Sciences, USSR*

The introduction of a systematic approach and the mathematization of geography are accompanied by an increase in demand for clearness of initial non-formalized concepts. In physical geography (landscape science and earth sciences) the concepts 'geographical shell' (biogeosphere, biosphere, etc.), 'spatial natural complex' (landscape, geochora), and 'component of nature' are fundamental ones. Very important are both definitions of the objects themselves and characteristics of the links between them.

Real objects, represented by the above-mentioned concepts, are perceived by us as natural geosystems. While studying systems as basic ones in physical geography, as well as in other principal geographical disciplines, two models are regularly being used. In the first model, elements of the system are the components of nature, and in respect to the components of nature – their separate properties (a monosystem model); in the second model, spatial complexes of lower orders (polysystem model). It is quite natural that elements of the first and the second models may be considered as objects for an independent study.

In combination, these two models reflect more completely the complexity of structure of objects under consideration, than each of them do separately.

For example, together they explain better as complicated a phenomenon as dialectic unity of discreteness and continuity of geographical space.

The problem of continuity and discreteness of systems has created rather hot discussions in many sciences (beginning with physics and including biology).

In physical geography there has been no such discussion in a direct form. However, the presence of opposite points of view in certain investigations is quite obvious. The contradictions are especially noticeable when we consider the problem of the nature of borders of natural geosystems, and when we work out principles and methodological questions of regionalization and classification.

The adherents of the concepts of geographical space as a *continuum*, deny the reality of the borders of natural objects, the possibility of existence of clear definite frontiers, and the qualitative definiteness of geographical objects, and in their investigations they use only the monosystem model. As a cartographic model they use only a map of isolines and build mathematical models mainly on the basis of the mechanism of a mathematical analysis.

Geographers who are adherents of the concept of discreteness doubt the existence of transitional strips and try to single them out as independent objects. In their research they use mainly the polysystem model, a cartographic expression of which is a typological map. They make efforts to build mathematical models mainly on the basis of the concepts of finite mathematics. Both these extreme points of view probably reflect only two sides of the real existing unity of the discontinuity and continuity of the geographical shell. In reality, these two properties interact and coincide. For instance, it is well known that anthropogenic effects spread in the geographical space either gradually die out, or stop rather abruptly on some comparatively sharp boundaries, corresponding to the borders of natural complexes.

To dismiss the argument about discreteness or continuity of the geographical shell means to dismiss the question of 'either-or'?. But the statement 'yes or no' does not answer the question of the relation between these two phenomena. As this question in a pure form was put forward by geographers only several years ago the available data have been insufficiently analyzed.

Nevertheless, even a brief consideration of the existing material shows that: (1) continuity is very likely expressed in the geographical sphere more clearly than discontinuity, (2) relations between continuity and discontinuity are different in different parts of the three-dimensional geographical sphere.

In particular, it has been noticed that discreteness of the geographical shell weakens as it moves upwards from the border between lithosphere and atmosphere; in atmosphere discreteness weakens. This feature is being used by meteorologists and geophysicists when they apply quantitative methods of forecasting, taking the upper layers of atmosphere for a continuum, a continuous field. At considerable altitudes these forecasts are well justified. But such a model turns out to be inadequate for forecasting the weather of the layer of atmosphere near the earth's surface, where one should take into account films of cold in relief depressions, and the influence of water reservoirs. In these cases the concept of atmosphere as a continuum contradicts reality. Here it is necessary to introduce concepts of air masses as relatively discrete objects or fronts of the stability of circulation mechanisms, of local air masses, and local weather.

The same circumstance is reflected in the character of the biocomplexes of vertical belts. A comparative homogeneity of conditions of the higher layers of atmosphere, their slow transformation, and the insignificant influence of the land surface, discretized (graded) by relief and vegetation, lead to the state when even in different mountain systems the upper vertical belts – nival and alpine – are characterized by a high degree of monotypeness, homogeneity of biological and geomorphological processes and phenomena, as compared to other vertical belts. By way of contrast, the discretization of territorial geomorphological, biological, climatic, and hydrological objects is especially well manifested in deep intermontane hollows, which very often have natural features differing sharply from those of the surrounding ridges and plains.

Analogous is the influence of oceans, where the presence of some discreteness is combined with a more pronounced phenomenon of continuity (compared with dry land). It has long ago been noticed that the systems of biocomplexes of the ocean shores are more homogeneous than the biocomplexes of the central parts of continents.

Observations have also been accumulating (in particular, in geobotany) which show that discreteness of biocomplexes is more pronounced in high latitudes and weaker in the equatorial part of the landscape sphere.

Continuity and discreteness of the geographical sphere indicate more general processes of integration and differentiation of matter. We may speak about two trends in these processes, which may be illustrated by the vertical and horizontal examination of the geographical sphere or natural complexes, having on the whole as a minimum a three-dimensional nature.

In a vertical section they have a 'storey' structure, as they consist of separate spheres (components): lithosphere, biosphere, hydrosphere, and atmosphere that overlap one another. These spheres (components) are characterized by interpenetration and interconnection with the predominance of vertical constituents.

Considering the geographical shell 'on a

plane' one first of all notices that it consists of spatial natural complexes. In the links among the complexes of all orders there predominates a horizontal constituent (for instance, the transfer of heat and moisture by air masses, surface and underground runoff). A landscape typological map, on which separate natural complexes are plotted, is practically an indication of the results of the process of horizontal differentiation of the landscape shell.

There is no doubt that in nature vertical and horizontal processes of differentiation and integration are united. By dividing the geographical shell into components, we 'tear apart' the vertical links, while by dividing it into natural complexes, we tear 'horizontal links.' The detection of the unity of horizontal and vertical links makes it possible to raise the question of going from models of landscape maps to three-dimension models, which will be as simple as the economic map that is so customary for many hundreds of thousands of naturalists, but which we geographers will consider a great achievement. We will consider it a great achievement even though we know that the three-dimension model is also only a simplified and incomplete concept of our four-dimension world, which besides the measures of length, width, and height, has a fourth most complicated coordinate – time, which would inevitably be considered from the point of view of continuity and discreteness.

P1217
L'analyse discriminatoire des correspondances typologiques de l'espace géographique
J.-B. RACINE   *Université d'Ottawa, Canada*

Cet essai est né de la confrontation d'un géographe formé à la problématique et aux outils traditionnels de la géographie d'expression française avec les exigences mais aussi les riches possibilités aujourd'hui largement reconnues en milieu anglo-saxon et scandinave de l'analyse quantitative et de la géographie moderne. De cette rencontre est née chez l'auteur la conviction que, s'il n'est pas de connaissance géographique sans différenciation et comparaison des paysages, il n'est pas non plus de connaissance scientifique sans expérience et sans mesure. De même, en admettant qu'il n'est pas de connaissance géographique sans raisonnement par *analogie typologique*, encore faut-il demander à la mathématique de *soutenir la typologie* en créant des références qui ne seront plus décrites par un ensemble de qualitatifs, mais qui seront enserrées entre des extrêmes quantitatifs (Libault 1967). Cet essai montre justement comment l'analyse multivariée mathématiquement conduite peut considérablement féconder l'analyse typologique et surtout la dépasser en mesurant le *degré de correspondance existant entre les différentes structures typologiques* selon lesquelles s'organisent les paysages géographiques. Le *modèle de recherche* que nous proposons devrait en d'autres termes permettre à la géographie de type idiographique d'aller de pair avec une géographie de type nomothétique.

Combinée à la réflexion théorique sur les formes actuelles de la prise en compte de l'espace par les géographes (Pinchemel 1968), l'induction quantitative appuyée sur l'écologie factorielle comparée (Berry 1971) peut conduire à identifier *toute une série de trames différentes* qui peuvent être ordonnées dans autant de matrices partielles d'information spatiale (ou chrono-spatiales au besoin) $E_1, E_2, ..., E_p$ associant m attributs spatiaux (sur l'axe des X) à n unités d'observations (sur l'axe des Y). C'est l'étape fondamentale et indispensable de la transcription des données définissant l'espace géographique dans un langage et sous une forme susceptibles de recevoir un traitement informatique.

Le premier problème qui se pose alors au géographe désireux de définir en termes de résultats optimalisés la 'structure typique' qui sous-tend chacune de ces matrices est de trouver la clé des correspondances statistiques qu'elles contiennent, c'est-à-dire, sinon la 'loi' du système, du moins et plus exactement sa structure latente puisqu'aussi bien c'est la structure d'un système qui détermine ses comportements possibles. La technique consiste à *réduire* ces matrices d'information sur leurs deux axes, en en dégageant les principales articulations structurelles pour s'élever à une généralisation typologique. Nous avons montré (Racine 1971 et 1972), en nous fondant sur les résultats du test discriminatoire de classification optimale proposé par Casetti (1964), comment l'utilisation successive de l'analyse factorielle pour réduire l'axe des X et de l'analyse des chaînes d'association statistique maximale – 'linkage analysis' de McQuitty (1957) – pour réduire l'axe des Y (*après inversion de la matrice originale pour obtenir d'abord des corrélations entre unités d'observation (n × n) et non plus entre attributs spatiaux*) permettait de dégager pour chacune des matrices d'information spatiale $E_1, E_2, ..., E_p$ une structure typologique optimale.

Le deuxième problème qui *peut* se poser au géographe (et c'est en cela que cette communication est particulièrement neuve) est de trouver les 'corrélations' ou, en termes plus adéquats, les *correspondances* qui peuvent exister entre les *structures typologiques* acquises à partir des séries d'attributs spatiaux ('variates') définissant successivement dans les matrices partielles $E_1, E_2, ..., E_p$ les différentes trames physiques, sociales, économiques, démographiques, financières, relationnelles, saisies en termes statiques comme en termes dynamiques, dont la superposition définit l'espace géographique. Nous proposons ici qu'en utilisant comme référentiels les poids locaux obtenus sur les différents facteurs de la différenciation physique et humaine de l'espace analysé, on affecte successivement à la typologie issue de E_1 les scores issus de E_2 puis issus de E_3 et ainsi de suite jusqu'à E_p. Si on affecte ainsi aux différents membres des groupes définis par la structure typologique issue de E_1 les valeurs obtenues par l'analyse factorielle de la matrice E_2, l'analyse discriminatoire de ces groupes (ou de ces types) affectés de valeurs empruntées à une trame différente montrera *le degré de stabilité de la structure typologique* de E_1 en regard de celle qui serait définie à partir des dimensions latentes de E_2.

Admettons que la zone métropolitaine de Montréal soit découpée en 200 secteurs de recensement que l'on aura regroupés en dix groupes, grâce à la généralisation typologique définie à partir des articulations structurelles des attributs spatiaux de la trame physique. Si, affectés des poids locaux des facteurs issus de l'analyse des attributs spatiaux de la trame d'utilisation du sol, ces dix groupes restent stables sans qu'aucun de leurs membres ne permute avec d'autres ou aille se localiser dans un autre groupe, on devrait pouvoir en conclure qu'il existe une correspondance parfaite (100 pour cent) entre les deux structures typologiques. Mais si le test discriminatoire montre qu'un groupement optimal des unités d'observations telles que définies par leurs attributs spatiaux physiques se traduit, une fois nourri des valeurs décrivant l'utilisation du sol, par 100 permutations, la correspondance tombe évidemment à 50 pour cent. Si par exemple on découvre que l'analyse discriminatoire, à sa *première itération*, provoque 64 permutations sur les 200 possibles (soit 136 observations qui restent stables), les coefficients de non correspondance ou de permutation (P) et de correspondance Ct (correspondance typologique) s'établissent ainsi:

$P = 64 \times 100/200 = 32$ pour cent,
$Ct = 100 - 32 = 68$ pour cent, ou encore:
$Ct = 136 \times 100/200 = 68$ pour cent.

L'extraction de la racine carrée de Ct donne évidemment un coefficient analogue au coefficient de corrélation (ici 0.825) mais il n'a guère de sens puisque de toutes façons il ne peut être négatif. Le coefficient Ct correspond plutôt au coefficient de détermination R^2.

Nous avons utilisé cette méthode pour évaluer les correspondances typologiques pouvant exister entre les structures typologiques des différentes trames (visibles ou invisibles) qui composent le paysage géographique de la couronne métropolitaine de Montréal au sud du Saint-Laurent: trame physique (ou naturelle), trame d'occupation du sol, trame du logement et des conditions d'habitation, trame socio-économique, trame de l'activité économique, trame des finances municipales. L'intérêt des résultats (Racine et Lemay 1972) est tel (coefficients de correspondance oscillant entre 38 et 95 pour cent) qu'il nous a conduit à proposer ici cette méthode. Elle comporte cependant des limites.

Les coefficients obtenus ne sont pas automatiquement réversibles. C'est pourquoi il faut renoncer à parler de 'corrélation' et s'en tenir à la notion de 'correspondance.' On doit se souvenir en outre que l'analyse discriminatoire suppose à priori que soit déterminé *au départ* un certain nombre de groupes ou de types et que les observations qui vont être classées appartiennent à l'un ou l'autre de ces groupes (King 1969, 204). Nous avons d'abord pensé que la première de ces deux conditions pouvait expliquer en partie la non réversibilité des coefficients puisque la structure typologique issue de la matrice E_1

pouvait avoir un nombre de types supérieur, égal ou inférieur à celle issue de la matrice E_2. L'expérience nous a démontré cependant que la fixation à priori du nombre de types pour toutes les matrices ne permettait pas d'obtenir automatiquement cette réversibilité que nous cherchions.

Ainsi pour la Rive-Sud métropolitaine de Montréal, la structure typologique de la trame du logement 'correspond' à 81.00 pour cent à la structure typologique de l'utilisation du sol. Mais si à l'inverse on choisit comme découpage de base les groupes issus de l'analyse de la trame du logement, cette dernière ne reste stable que dans la proportion de 76.19 pour cent, alors que dans les deux cas six groupes avaient été définis.

La seconde de ces conditions oblige le géographe à choisir avec le plus grand soin non seulement les trames à zéro, une, deux ou n dimensions dont la superposition sous-tend l'organisation de l'espace géographique, mais aussi les limites spatiales au sein desquelles il va chercher à associer à n observations m attributs spatiaux. Il reste que le jour où nous pourrons disposer, peut-être avec la généralisation du 'géocodage,' d'unités d'observations homogènes en taille et assez nombreuses pour que l'on puisse se dispenser des habituelles triturations artificielles devant assurer la normalité des distributions, peut-être pourrons-nous repenser le problème des rapports entre la géographie inductive et la géographie déductive.

L'induction par réduction permettra de décrire de façon optimale un système spatial grâce à une matrice typologique T, la déduction permettant de son côté, après la formalisation des hypothèses explicatives en modèles théoriques, de reconstruire une matrice typologique T' *déduite*, que l'on pourra comparer, en suivant notre méthode, avec la matrice T de façon à tester la pertinence des hypothèses et du modèle. Nous osons croire que ce jour là notre discipline, science de la différenciation et de l'organisation des éco-systèmes spatiaux, aura fait un grand pas en avant.

Le Conseil des Arts du Canada; l'Université d'Ottawa; Bryn Greer-Wootten; Guy Lemay.

Berry, Brian J.L., guest ed., 1971 'Comparative factorial ecology,' *Econ. Geog.* 47 (2) (suppl.), 209–367.
Casetti, Emilio, 1964 *Classificatory and Regional Analysis by Discriminant Iterations*, Technical report 12, Computer Applications in the Earth Sciences Project, Dept. Geog., Northwestern U.
King, Leslie, 1969 *Statistical Analysis in Geography* (Englewood Cliffs, NJ).
Libault, André, 1967 La mathématique, méthode ou auxiliaire de la géographie?, dans J. Duculot, éd., *Mélanges Tulippe* II (Gembloux), 523–34.
McQuitty, Louis L., 1957 Elementary linkage analysis for isolating orthogonal and oblique types and typal relevancies, *Educational and Psychological Measurement* 17, 207–29.
Pinchemel, Philippe, 1968 Redécouvrir la géographie, Ann. U. *Paris*, no. 3, 350–60.
Racine, Jean-Bernard, 1971 et 1972 Modèles graphiques et mathématiques en géographie humaine, *Revue Géog. Montréal* XXV (4), et XXVI (1).
Racine, Jean-Bernard et Guy Lemay, 1972 L'analyse discriminatoire des correspondances typologiques dans l'espace géographique, *L'Espace Géographique* (3) (à paraître).

P1218
The spatial viewpoint is the essence and prospect of geography
J. ROGLIC   *University of Zagreb, Yugoslavia*

The geography of our time can be defined as the science studying the surface of the earth on the basis of the aspects, contents, and significance of its components. The spatial viewpoint is the most essential characteristic of geography.

In our time the surface of the earth may be studied in a variety of means of observation and experience. It has become the object of everyday interest. The contemporary inhabitant of the earth needs to know the natural qualities of his global surroundings and the life of his neighbours. This is not an encyclopaedic interest, but the need of the present human generation which, by the fruits of its common toil, lives a global life.

In neighbourly co-operation, achievements are reached which effect a Promethean revolution, while mutual ignorance and misunderstandings may lead to unthinkable catastrophes. Spatial geographical knowledge is important now as never in the past, and everything points to its becoming still more important in the future.

Unjustified are the views that with the onset of global relationships, regional realities are losing their significance. The reverse is true: by integration into the larger wholes, the components do not become lost but by interaction they gain new qualities.

The processes of integration are complex and differ in space and change with time, and mutual influences lead to new polarizations. It is not a question of accumulation but of mutual influences and the appearance of new elements both in the whole and in its component parts. New relationships make geographical study more complex, and it therefore needs to be continually adapted and perfected.

Geography studies the present-day processes and relationships. It is important for us how much our geographical outlook harmonizes with the present-day possibilities and needs. It seems to me that we look into the past needlessly, that we discuss the geographies of past times and project them inorganically into the present. The starting point needs to be the fact that the object of the geographical scientific interest is changing in connection with the processes that condition these changes. The problem of geographical scientific adaptation and modernization is continually present. Static schools do not correspond to the essence of our discipline, and this is true for science in general. Theoretical discussions are a specific need of geography and will always be present. Discussions ought to be connected with our experiences and needs, and not provoked by uncritical acceptance or introduction of what is alien to geography.

The concepts of the geographical outlook have to be modified as time passes. A new, diverse and dynamic global society is appearing. The blood circulation of this process is communication in the widest sense of the word: exchange of goods, travel, and especially the exchange and merging of ideas.

The social processes have won global dimensions through the means of communication. A harmony is being established between global dimensions of natural phenomena and social processes, and that should be reflected in geography.

Social activities are reflected in global dimensions and are evaluated according to what they signify in human society. In this transitional period the spatial differences are great, but the process of association and adaptation is in progress. The basic task of geography – a history of the present – is to define the actual relationships because our time needs this and it will be useful for the future.

The spatial viewpoint also includes the ecological viewpoint because the spatial reality reflects its natural basis and the influences of social interactions. Here the genetic viewpoint becomes apparent because the present situation reflects the inheritance from the past, and comparison makes the comprehension of the future developments possible.

Objections that the spatial viewpoint is not scientific are unjustified, since the entire geographical experience confirms the opposite. The most durable – we might say an eternal – value belongs to works that elaborate the spatial reality at a definite time.

It is unrealistic to attribute past subjective descriptions to present-day spatial studies. The former correspond to the facilities of their times whereas reliable ways of objective presentation are now available.

Much more realistic is the objection that the spatial viewpoint is difficult. This viewpoint is different in space and changes with the time, and demands extensive knowledge. That is true and has to be faced. On the other hand, it ought to be admitted that spatial knowledge, i.e. finding the way in man's global homeland, is a basic social need. If geography does not meet this need, a new discipline will develop; certainly none of us wishes such a suicide. In the program of this Congress the suspicion is justifiably expressed that regional studies are 'a responsibility neglected by geographers.'

The fear of systematic specialities – should the spatial viewpoint prevail – is unjustified. The knowledge of spatial relationships and processes feeds the corresponding systematic specialities, harmonized and organically integrated into the scientific geographical system. The knowledge of actual regional climates, for example, would correct schematic divisions and promote climatology. Dangerous urban and industrial pollution offers the basis for a complex study of location. A new era of geo-human processes and their study has begun. The new processes of urban dispersion and of the human valuation of space stress the need for systematic study, more real than that which made the urbanists foresee apocalyptic agglomerations. The evolution of markets, the differentiation of production and the adaptation to technological processes are changing the agrarian landscape and are geographically more important than the inherited parcelling, which is necessarily changing and being abandoned.

The organization of the 22nd International Geographical Congress means a turning point and spreads optimism. Thirteen sections and several symposia with their themes have abandoned those static and tragic patterns of earlier Congresses, which have jeopardized not only the reputation of geography but also its independence. The Canadian organizers of the Congress have linked us more closely than ever before to the time in which we live and to the country where we are meeting.

As a consequence, let us remain within the framework of our geographical opinion and logic. The essential part of the impulse that has made us, foreigners, come to this Congress, was our desire for the greatest acquaintance possible with Canada; and it is also a considerable stimulus for the hosts to present us their interesting country and its society. We are thus meeting in space, i.e. in essence, in geography. An important part of this Congress is thus a meeting of real and potential 'Canadologists.'

We ought to remain on the specific path of geography and to devote our attention above all to spatial specialization. This will strengthen our cohesion, will lead to a renaissance of scientific geographical work and increase the reputation of geographers in society. This is the need of our time, which is characterized by the strengthening of global co-operation and reconciliation, both of them depending on mutual acquaintance and understanding, i.e. on geographical knowledge.

Spatial orientation secures the co-operation of the geographers. Systematic specializations will send deep roots into the knowledge of spatial reality, and geography will become organically integrated. Progress will lead from details to the integral, which corresponds to the nature of geography and to scientific work in general.

Congrès national du 24 mai, 1958 *Bull. de l'Assoc. de géogr. français* 275 (Paris), 20–63.

Roglic, J., 1961 Gegenwärtige Probleme der Geographie, *Geogr. Rundschau.* 13, 11 (Braunschweig), 425–31.

Schmithusen, J., 1966 Rundgespräch 'Theorie der Geographie,' *Geographica Helvetica* 21, 1 (Bern), 36–7.

Uhlig, H., 1970 Organisationsplan und System der Geographie, *Geoforum* 1 (Braunschweig), 19–52.

P1219
Problems of analysis of spatial distributions
M.J. SCARLETT   *University of Calgary, Canada*

At the outset we distinguish three forms of spatial distribution: those without any focus; those of multiple foci; and those of single focus. In the present paper attention is given to the last, since they present interpretive simplicities and allow discernment, with maximum clarity, of the mode of analysis used. Examples might be the distribution of users of a single facility: such as the town's one library, the zoo, university, or major league hockey arena. We can conceive of the arena, for example, generating a demand surface which is manifested in the spatial distribution of patrons of major league hockey in the city. However, unless patrons are required to record their address it may be impossible to map the population of patrons; and sampling, apart from being difficult, introduces familiar problems of inferential statistics at the analysis stage. In the approach suggested here variables examined should satisfy the criterion of offering data of the total population. Season ticket holders of the Calgary Philharmonic are a population which is assumed to have empirical meaning, and distinguishable from a larger population of all concert-goers, or all music lovers. It is certainly impossible to infer anything about these larger populations from the phenomena of the season ticket populations.

Season ticket holders are not only recorded, but also must move to the focus, the auditorium, to satisfy their desire to listen to the symphony orchestra. We may examine distribution distinguished by frequency of movement required, with a view to revealing the importance of this on a distance-decay element in patronage but this has not been attempted here. We may also suppose that there should be a fundamental difference between all these variables and those having a nominal focus without necessitating physical movement. A case might be patronage of a local radio station; but (ignoring the impossibility of measuring the population except via a sample) this distribution would only differ from the forms actually used in the absence of a comparable distance-decay element in explanation.

Unique-focus distributions rarely have identical foci; and therefore can seldom be compared by simple means. Geographers have hardly begun to tackle this most basic of our problems, but an important approach to it can be derived from the works of geologists and, more powerfully, from Sneath's anthropological approach (Sneath 1967, 65–122). An alternative to this is suggested in the present study, as part of a sequential approach to analysis.

Trend surface analysis is by now a familiar phenomenon in the literature of geography, even though it is much less commonly used there than by geologists and geophysicists. The latter, in particular, have explored most of the important questions concerning its concepts and applications, so that in their hands it has become an often used and powerful tool, both for economic geology and for more academic approaches to chronology and genesis via mineral and other physical characteristic distributions. Despite some good reviews by geographers of the literature, important substantive studies using trend surface analysis in geography are rare (Chorley and Haggett 1965; Norcliffe 1969, 338–48). Briefly, for those unacquainted with the technique, a statistical surface measured in values for a set of areal units satisfying certain criteria, can be approximated in form

and height with varying degrees of success by fitting mathematically derived surfaces of increasing complexity (Chayes 1970, 1273–78). Each surface can be regarded as 'explaining' some part of the variance in the original surface. The simplest surface is a plane, with equation
$Z = k + \alpha x + \beta y$,
where k gives an intercept value in the Z plane, and x and y are cartesian co-ordinates. Second degree and higher surfaces add, successively, 3, 4, ..., n additional terms to the predicting equation for Z in terms of x and y. Chayes has demonstrated conclusively the importance of evaluating the merits of increasingly higher degree surfaces by an F-ratio which tests the mean increase in explanation from each added set of terms rather than of *all* terms of lower degree surfaces together with the new terms (Chayes 1970).

However, clarity of interpretations seems often to have escaped workers using trend surface analyses. One reason derives from an apparent desire to justify and to use a *single* criterion for accepting any given surface as meaningful. So we find the inevitable dilemma: one surface either explains a 'large' part of the variance or gives a 'large' increase in explanation (in Chayes' terms, or others) but is empirically meaningless; another, less competent to 'explain,' still has empirical or theoretical meaning (Howarth 1967, 619–25). Certainly, many studies have been virtually valueless from a more or less simplistic acceptance of the notion that more (variance explained) is better. A Chayesian approach may in contrast indicate 'that a handsome map contains no new information and should not be published' (Chayes 1970). Often workers seem to have forgotten, or been ignorant of, three characteristics of surface fitting. It is always possible to fit a mathematical surface, or set of superimposed surfaces, to *any* statistical distribution; so that progression towards one hundred per cent 'explanation' may not imply any true explanation at all unless accompanied by other conditions (for example, theoretical sense). The second neglected characteristic is that polynomial fitting is not the only form available. Despite occasional warnings, most workers have operated as though, for example, double-Fourier series did not exist, or at least had no relevance, without suggesting why (Norcliffe 1969). It is perfectly permissible to fit polynomials without having tried an alternative, so long as we are suitably reticent about our conclusions; and are willing to admit any failures we may have could in part be due to a faulty choice of technique.

The third neglected matter concerns orthogonality. Much work has been done as though there were no danger in departing from orthogonality. The development of efficient computer programs capable of producing trend surfaces from data not evenly spaced has lulled us into a sense of complacency on this question. Yet it has hardly been explored by geographers, despite the significance for geographic inquiry. In the present study, two, possibly inadequate, precautions are suggested: (1) spatial units should be broadly rectangular, evenly spaced and equal sized, and thus yielding a roughly uniform density of 'points.' (2) The aggregate of spatial units used has to be made convex, on the grounds that lack of convexity can be easily demonstrated to yield meaningless results, and that no acceptable measure of error has been devised to handle departures from convexity.

Some statistical surfaces are analysable in purely mathematical terms. Examples are found frequently in geological literature. In these cases, any well-fitting surface can be thought of as having meaning capable of explanation in terms of geologic process or form. By contrast, many surfaces derivable from human geography may only be so analysed if: (1) theory predicts some such form or, (2) the form correlates with some other distribution having empirical validity. For example, theory might predict a cone-shaped distribution around a focus of activity and thus allow testing of an hypothesis using a cubic surface. On the other hand, the distribution in greater New York of the homes of the members of the New York Stock Exchange may be empirically explicable by reference to the socio-economic status of districts which may, nevertheless, reveal an extremely complex pattern of distribution. The search for a simple explanatory surface can become like a quest for the philosopher's stone. Mathematical clarity and elegance may need tempering with empirical good sense, if we are to make progress.

Thus stage one of the analysis requires the extraction of the elements which are quantifiable and are known or thought to be partial explanants. This in turn requires operation in terms of normalized values, so that throughout we should use z-scores. We subtract z-scores of these elements to leave a residual surface which then becomes the subject of further analysis. If trend surfaces are fitted to the residual surface and each tested for significance using Chayes' method, three possible outcomes may prevail: (1) one surface only is significant; (2) more than one is significant; (3) no surface is significant. In the first case we may assume that although we cannot immediately attach empirical meaning to the surface it *may* reveal such meaning under close examination. Therefore, we remove it as a further (unknown) explanatory variable and proceed to treat its residual surface in the same way. In the second and third cases the analysis might stop, since multiple surfaces which satisfy a significance test suggest that there exists a complex of possibly single simple variables, or that one or more inherently complex surfaces exist. There seems to be no ready means without new empirical or theoretical input of determining which conditions occur. However, suppose we have two surfaces, one of first degree, significant at the 1 per cent level, and one of second degree, significant at the 5 per cent level (i.e., the new terms of the second degree surface add a significant increase to the explained variance), what should we then do? The literature has no answer. We can certainly extract a first degree surface, assuming empirical validity, and then a second degree surface with slightly less confidence as to its statistical significance. But is our belief in its empirical significance in any way affected? All that can be suggested here is that it *may* deserve serious consideration, but that the decision to look for empirical validity for any surface depends on subjective evaluation by the researcher. Certainly a look at the absolute increase in explanation gives no sure guide; and neither do the weak rules for acceptance based on the computer randomization work of Howarth (1967), or the multiple regression view of Norcliffe (1969). In any case, using Chayes' approach, the choice is *not* between one surface or the other, even if both are significant at, say, the 1 per cent level. The x, y, k terms of the second degree surface are not identical with those of the previously extracted first degree surface since the second degree surface is being fitted to a new, residual surface.

It is clearly impossible to generalize about the form of the residual surface when we finally cease the subtraction process in the first outcome; whatever that surface may be, it is conceptually identical with those abandoned in outcomes (2) and (3). We arrive finally at a set of surface variables expressible in the form:

$$Z = X_e^1 + X_e^2 + \ldots + X_e^n + X_m^1 + X_m^2 + \ldots + X_m^p + \varepsilon$$

where the subscripts e imply empirical elements, subscripts m mathematically derived elements, and ε is an 'error' or 'noise residual variant element.' No attempt is made to extract elements strictly according to their ability to explain variance, both because some empirical variables can be assumed to have relevance and because unknown, mathematically approximable variables can only be extracted on the ground of their significance as shown by a Chayes F-test.

Chayes, F., 1970 On deciding whether trend surfaces of progressively higher order are meaningful, *Bull. Geol. Soc. Am.* 81, 1273–8.

Chorley, R.J., and P. Haggett, 1965 Trend-surface mapping in geographical research, *Inst. Brit. Geog. Pub.* 37.

Howarth, R.J., 1967 Trend-surface fitting to random data – an experimental test, *Amer. J. Sci.* 265, 619–25.

Norcliffe, G.B., 1969 On the use and limitations of trend-surface models, *Can. Geog.* XIII (4), 338–48.

Sneath, P.H.A., 1967 Trend-surface analysis of transformation grids, *J. Zoo.* 151, 65–122.

P1220
Theoretical foundations of the geographical sciences
B.N. SEMEVSKIY   *University of Leningrad, USSR*

The world is material by its nature: everything that exists is a kind of matter and has a certain form. Matter never is, and cannot be, in a state of complete rest or immobility; there is a constant movement in it, which goes on in space and time. Movement, space, and time are the main forms of matter's existence. Geographical sciences pay most attention to the distribution of matter in space, but space cannot be separated from time and movement; space cannot exist without time and movement. Therefore, Hettner's one-sided interpretation of geography as purely spatial and distributive is wrong. All geographical investigations made without taking motion into account will be of a scholastic, artificial character. The movement of matter is far from being its distribution in space; it first of all means constant inner changes of matter – its development.

Continuous movement of matter is realized in many forms and ways. The main kinds of movement of matter are: mechanical, physical, chemical, biological, and social.

The highest form of the movement of matter is the social form; the process of development of human society. This form of the movement of matter differs qualitatively from all the others and arises with the appearance of human society, which comes into being in the process of the generation of *material values*, the process that defines all the other sides of social life.

Geographical sciences deal, to some extent, with all forms of the movement of matter, and therefore, considered as a whole, form a complex of interconnected sciences.

The base of unity and interconnection of all forms of the movement of matter is the material unity of the world. Under certain conditions every form of movement can be transformed into another one. But compared with the other forms of the movement of matter, the social movement occupies a different, peculiar place. All the other forms of movement are natural and behave according to the laws of nature. Nature created man as a complicated biological organism, but human society arose not under the effect of the laws of nature, but as a result of the productive relations among people, which united them into a human society. Therefore, human society has been developing not under the action of the laws of nature, but on the base of peculiar laws – laws of social development. This principal qualitative difference of the development of society from all the other forms of movement of matter is based on the qualitative difference between the natural-geographical and economic-geographical sciences. Nevertheless, the general laws of philosophical materialistic dialectics act in all the geographical sciences and form their methodological base.

The dialectical-materialistic philosophy stated the most common laws for the development of nature and society, that is, the laws which, due to their universal character, act in all forms of the movement of matter. The geographical sciences, just as all the others, are based on these laws.

The essence, the very core, of dialectics is the law of unity and the struggle of contradictions, the law that shows the sources and real reasons of continuous movement and development of the material world. The importance of this law for the geographical sciences is enormous and universal. Here is the simplest example: two poles of the earth – the south and the north poles – are simultaneously united and in opposition, because they oppose one another (opposition), yet one pole cannot exist without the opposite one (unity).

A striking example of the struggle of contradictions is seen in the ever-existing contradictions between the old or dying, and the young or growing. The importance of this feature of the struggle of opposites is clearly demonstrated in social life, and this law greatly influences the problems dealt with by economic geography. Contradictions form the main source of the development of matter and consciousness.

Another law of the greatest importance for the geographical sciences is the law of the transformation of quantitative changes into qualitative ones, which is the universal law of the development of the material world. Numerous transformations of matter from one aggregate state into another, each of them having quite different qualities (the simplest example: ice-water-vapour), the change of one 'facia' of landscape by another one that qualitatively differs from it, the change of one economic region by another one, all these, and many other examples, are striking demonstrations of this law.

All objects and phenomena possess both quantitative and qualitative definity. The quality and quantity of objects and phenomena are interconnected: in the process of development slight and gradual quantitative changes are transformed, by means of a jump, into principal qualitative ones.

The law of the denial of denial is also of great importance for geographical sciences. This law shows the general direction and tendencies of development of the material world. The overcoming of the old by the new rising on the base of the old is what the term 'denial' means in philosophy. Nothing can develop without 'denying' its previous forms of existence. For example, the development of the earth's crust has gone through a number of geological epochs and periods, and each new epoch, each new period, though it arises on the base of the previous one, is the denial of old, former epochs and periods. Thus, the quaternary period arose from the tertiary period, on its base, but once it had come into existence it became the denial of the tertiary period. Once the quaternary period appeared, the tertiary one died, went into the past. Thus, the quaternary period is the denial of the tertiary one. The denial is the result of their own inner development of all objects and phenomena. Objects and phenomena being developed on the base of their inner contradictions create conditions for their own liquidation, for a transition into a new, higher quality.

Any 'denial' is the denial of the previous 'denial.' The quaternary period is the denial of the tertiary period, but the tertiary period in turn is the denial of the cretaceous period which existed earlier still. Therefore, the quaternary period is the 'denial of the denial.' However the cretaceous period is the denial of the previous Jurassic period. In this sense, the tertiary period is also the 'denial of the denial,' and so on, and so forth.

Geographical sciences, while studying all the objects and phenomena, are concerned with the process of development, of stating the genesis and evolution of every object and phenomenon. They are based on the law of the denial of the denial, one of the principal laws of materialistic dialectics.

The theoretical foundations of the natural geographical sciences and the general geographical regularity they deal with are discussed in detail in academician S.V. Kalesnik's work, 'General geographical regularities of the earth' (*M 'Mysl.'* (Thought), 1970), which also gives the classification of the geographical sciences. But the laws dealt with by economic geography have not received such clear definitions as yet.

Economic geography separated itself from natural-geographical sciences and has become an independent social science closely connected with physical geography and with the problems of opening lands for development by human society. Economic geography is being formed as a science that does not deal with the landscapes of the earth's surface, but is concerned with the geographical distribution of production (by this we mean the unity of productive forces and production relations, with conditions and peculiarities of its development in different countries and regions).

What is especially important for the foundation of the theory of economic geography is that law of historical materialism which affects all stages of the development of the human society; namely, the law stating the decisive role of the means of the generation of material values. It is this general law that allows economic geography to find out peculiar economic-geographical laws of the distribution of social production and economic regions, and enables it to overcome the erroneous influence of geographical fatality.

Another law of historical materialism which acts in all stages of the development of society is the law that defines the principal role of the social being wth respect to social consciousness. As far as economic geography is concerned, this law stresses the importance of studying different categories of social production. This law allows us to define the proper place to be occupied by political geography in the system of economic-geographical sciences, and to state the role of political and ideological superstructure in economical-geographical investigation. The development of production is a law of social life and at the same time the objective necessity. It is the development of production applied to its distribution in space that is the main object investigated by economic geography.

The laws of political economy are also very important for economic geography, because they define specific economic-geographical laws of the distribution of production and economic regions.

The spheres of action of the laws of political economy are not equal. Some of them act in all stages of social development. Examples of this are the laws of the necessary accord between the production relations and the level of the development of productive forces. According to this law, economic geography pays the greatest attention to the processes of production of something or other, and also to the implements used, and only from these points of view does it study peculiarities of the foundation of economic regions and characteristic features of the distribution of social production.

In spite of their undeniable importance, quantitative data and their computer manipulation show the consequences or results of production, but fail to expose its essence, its base, its inner moving forces. They show what is being produced, how much, what quantitative changes and combinations they may bring to a given model, but they cannot answer the most important question: how is a product produced, what methods, means, and manipulations are to be used, what interconnections appear between different branches of production or inside those branches, or inside a region and between regions, what main problem forms the complex? Nor do they say what roles separate parts of the whole play, nor how to overcome the changes that are bad for man and stimulate those that are good for him.

That is why it is not enough to rely on the analysis of only quantitative data and models based on them, but it is necessary to learn the inner essence of every economic-geographical phenomenon and process, the tendency of its development. In other words, it is necessary to have a problem approach to economic geography.

P1221
Applications of dummy variable analysis
JOHN A. SILK   *University of Reading, England*

Dummy variable analysis (DVA) is a form of multiple regression analysis permitting estimation of parameters of relationships between a binary 'response' and sets of categorical independent variables (Suits 1957; Goldberger 1964, 218–31). For applications see Orcutt et al. (1961), Yeats (1965), Starkie (1970), and Bayliss and Edwards (1970).

A dummy variable is a dichotomous variable taking on the value 1 if a particular contingency occurs, and 0 otherwise (the regression coefficients are easier to interpret if the values are 0 and 1, or −1 and +1). Variables considered inherently categorical, or strongly discontinuous in their influence on a response, are suitable for such treatment. DVA need involve relatively few or no assumptions about the form of the response surface, and nominal, ordinal, and continuous data are assimilable within a common framework. In practice, the procedure is more efficient if an additive model representing the 'main effects' is employed initially, and this assumption relaxed to the extent necessary to obtain the desired 'fit' with respect to the problem at hand. To this end, categories may be combined and interaction terms introduced, using standard variable transformation options, and estimates of the relative importance of different factors, and of different levels making up each factor, in explaining variation in the response obtained. The technique also permits more straightforward treatment of cross-classifications for which the number of observations varies from cell to cell (Goldberger 1964, 230–1). (Formal equivalence between DVA and the analysis of variance has been pointed out by Draper and Smith (1966, chap. 9).) Computationally efficient solutions are obtained, provided the dummy variables take on values of 0 and 1 or −1 and +1, rounding errors being reduced due to the small numbers involved and the fact that the dummy variables are mutually orthogonal (Draper and Smith 1966, 143–5, 156).

The first example uses data on movements of goods and service vehicles from 256 factories in Harringay, London, England, obtained from a one-day survey carried out by the Harringay Department of Planning (Starkie 1970). Little substantive significance is attached to the results at this stage, as data on factors, such as the relative location of industrial premises, are lacking. Factories were classifiable into 12 types according to the Standard Industrial Classification (SIC), but 4 categories, each containing less than 10 observations, were combined to yield 9 categories over all. A dummy variable represented each SIC class, and the continuous variable, employment, taken as a surrogate for size. The employment values were multiplied by each dummy variable to obtain sets of values – 'continuous dummies' – specific to each SIC category (Gujarati 1970).

One each of the dummy and continuous dummy variables was not explicitly included, so that a computer routine employing a matrix inversion procedure to obtain estimates of the regression coefficients could be used. Stepwise regression analysis produced the following equation: number of trips = 8.409 − 0.1639 (employment) + 0.2913 (SIC3 × EMPL) − 0.0766 (SIC6 × EMPL). The constant term represents the 'effect' of *all* SIC categories included in the analysis. None of the regression coefficients associated with the individual dummy variables proved to be significantly different from this over-all value. However, differences in trend shown by values in categories SIC3 (food, drink, and tobacco) and SIC6 (engineering and electrical goods) from that estimated for all other categories are shown to be significant. The trend for SIC3 is greater than 'average,' 0.4552 (0.1639 + 0.2913), that for SIC6 less, 0.0873 (0.1639 − 0.0766). Inclusion of differences in trend increases R^2 from 54.76 per cent ($R = 0.74$) to 81 per cent ($R = 0.904$). A 'better explanation' might be obtained by introducing further terms, but the temptation to approximate all the vagaries that may occur in a large data set should be resisted if the technique, and its results, are to be of wider significance.

If the *response* is a dichotomous variable taking on only values of 0 or 1, its estimated value is interpreted as the conditional probability of occurrence, $P(Y_i = 1)$, of the event for which it takes on the value 1 (Cox 1970, 16–8; Bayliss and Edwards 1970, app. 2). Difficulties may arise because the dependent variable is not normally distributed and the error term is heteroscedastic, but a crude test of the adequacy of such models is suggested by Cox (1970). The constraint $0 \leq P(Y_i = 1) \leq 1$ may not be satisfied, but we accept Bayliss and Edwards' argument (1970, 144) that an additional variable taking $P(Y_i = 1)$ 'out of range' is to be interpreted as having no effect on an already 'certain' situation.

Our second example is from a study of a sample of 155 18-year-olds interviewed in

Inverness in March 1970 (Blaikie et al. 1970). Independent variables were dichotomous (sex, born in Inverness, student/worker, part-time education, continuance of full-time education), continuous (income), and ordered categorical data assumed to be measured on an interval scale, these being obtained by asking respondents to evaluate their own career prospects in Inverness in terms of five-point rating scales labelled narrow-wide, dull-exciting, and badly-paid-well-paid. The 'response' was interpreted as the conditional probability of intending to leave Inverness, 'students' being asked if they intended to return, and 'workers' if they intended to leave 'in the near future.'

It was found that workers *not* pursuing part-time education, students, and those regarding career prospects in Inverness as 'dull' were most likely to leave (or not return). A crude test of the adequacy of the model, suggested by Cox (1970), proved positive, although only 18.5 per cent of the total variance was explained. It appears that young people in Inverness are attracted by interesting, challenging work, rather than income.

Regression of narrow-wide scores on all variables except the other rating scales produced an interesting interaction effect: NARROW-WIDE SCORE $= 2.15 - 0.27X_1 + 0.06X_3$, where $X_1 = 1$ for males, and -1 otherwise, and X_3 is the interaction term 'sex times income.' As a whole, females viewed employment prospects as wider than did males. The interaction effect is easier to interpret if the equation is rewritten: NARROW-WIDE SCORE $= 2.15 + X_1 (-0.27 + 0.06X_2)$. If income ($X_2$) is high enough, i.e. £4.33 per week or more, which it was for all incomes recorded, males see employment prospects as significantly wider than do females, the difference increasing with income.

Career prospects for females with minimal academic qualifications are generally very poor, so this finding is not surprising.

There is no obvious explanation for higher female expectations amongst those still at school.

Apart from satisfying the usual assumptions of least squares regression, DVA is expensive of degrees of freedom, and requires large well-conditioned data sets, so that sample sizes do not fluctuate markedly from cell to cell. Careful inspection of the raw data is strongly recommended, and use of weighted regression and allied procedures often appropriate (Maxwell 1961, chap. VI; Goldberger 1964, 235–6; Cox 1970).

D.N.M. Starkie.

Bayliss, B.T., and S.L. Edwards, 1970 *Industrial Demand for Transport* (London).
Blaikie, P.M., J.A. Silk, and M.J. Moseley, 1970 *Predicting the Decision to Migrate*, unpubl. ms., U. Reading.
Cox, D.R., 1970 *Analysis of Binary Data* (London).
Draper, N.R., and H. Smith, 1966 *Applied Regression Analysis* (New York).
Goldberger, A.S., 1964 *Econometric Theory* (New York).
Gujarati, D., 1970 Use of dummy variables in testing for equality between sets of coefficients in linear regressions: a generalization, *Am. Stat.* Dec., 18–22.
Maxwell, A.E., 1961 *Analysing Qualitative Data* (London).
Orcutt, G.H., M. Greenberger, J. Korbel, and A.M. Rivlin *Microanalysis of Socioeconomic Systems* (New York).
Starkie, D.N.M., 1970 The treatment of curvilinearities in the calibration of trip-end models, *Urban Traffic Model Research* (London).
Suits, D.B., 1957 Use of dummy variables in regression equations, *J. Am. Stat. Assoc.* 52, 548–51.
Yeates, M.H., 1965 Some factors affecting the spatial distribution of Chicago land values, 1910–1960, *Econ. Geog.* 41 (1), 57–70.

P1222
Mathematical analysis of mobility matrices
LEROY O. STONE *Statistics Canada*

A central object of study in analysing regional mobility is the origin-by-destination flow matrix. The *ij*th cell of the initial flow matrix contains an integer, n_{ij}, showing the number of persons moving from origin i to destination j. Often the integers are transformed into proportions $p_{ij} = n_{ij}/(\Sigma_j n_{ij})$; so that p_{ij} gives the proportion of the persons (or other entities, e.g. firms) at origin i who move to destination j over a specified time interval.

In much of the existing literature the mathematical analysis of mobility matrices is undertaken within the context of the assumption that the proportions p_{ij} represent mobility probabilities and that the matrix containing these proportions is Markovian. This assumption prompts the analyst to place much emphasis on the derivation of the stable regional distribution of population that would be attained in most cases from continued 'operation' of the mobility matrix. As a by-product of this derivation a number of parameters are obtainable — such as mean time taken to move from region i to region j, and the mean time spent in region i.

Two major deficiencies may be noted in this approach to the analysis of a mobility matrix. Firstly, the analysts who have hastened to invoke and apply the Markov assumption tend to have their attentions diverted away from the social and economic factors that influence the regional mobility flows. Secondly, the derived latent parameters of the mobility matrix tend to be unduly inadequate because they fail to take into account the fact that a real regional population is subject to births, aging, and deaths.

A more adequate approach to the mathematical analysis of mobility matrices disposes of the assumption that movement among the regional locations may be characterized as a Markovian stochastic process, and instead anchors the generation of a sequence of mobility matrices (and their corresponding regional population distributions) to relevant explanatory social, economic, demographic, and political factors operating in the real world and subject to changes over time. We may make a mathematical analysis of the mobility parameters that is similar to that under the Markovian assumption; but the parameters thrown up by the analysis are treated as concise *transitory summaries* of certain features of the observed mobility regime, rather than as underlying properties of the system that somehow guide its evolution.

The purpose of this paper is to illustrate a more adequate approach to the mathematical analysis of mobility matrices than that which has been common. We shall define a general population change matrix operator that reflects regional mobility, births, and deaths. To demonstrate the flexibility of the approach being adopted, we shall simultaneously cover regional and occupational mobility, and shall include age and duration of residence (in a given region) in defining the 'dimensions' of the matrix operator. Then we will indicate the key to the derivation of several latent parameters of the matrix operator. From these parameters a host of synthesizing and characterizing measures of the mobility matrix properties can be defined. These measures include the stable occupational distributions in each region, the stable age distribution of each occupation, the stable region-by-occupation distribution of the population, the mean length of stay in each region, and stable in-, out-, and net migration rates for specific age and regional groups by occupation.

We shall develop the mobility matrix and the general population change matrix operator in terms of the following notation: Let: d refer to age, X refer to vector relevant to explanatory variables other than v, d refer to duration of residence in a given regional location, c refer to a vector of parameters, s be a given *point* in time, t be an *interval* of time, i and j be subscripts referring to one or another of a finite set of N geographic locations, a and b refer to one or another of a set of H occupation groups. The conditional probability, given duration of residence d in i and occupation a at time s, and age v at time s, of moving to jb over the interval t (and be found in jb at time $s+t$) is $p_{jb|ia,v_s,d}^{t,s}$. Then

(1) $\quad p_{jb|ia,v_s,d} = \lambda \cdot f(v, d, X, c) + u$,

where λ is a parameter and u is a stochastic disturbance term.

The mobility matrix is comprised by the set of values derived from (1) for all possible combinations of jb, ia, v_s and d (assuming categories are defined for all these variables). The terms indicated by (1) will reflect mortality (i.e., it is not to be assumed that all persons alive at time s are also alive at time $(s+1)$.

In order to approach reasonable realism in

forecasting or simulation, it will be necessary to embed the estimated mobility probability matrix into a larger population change matrix operator, which has additional terms to reflect the births in a given region between s and $s + 1$ who survive to be found in the region and in other regions (through migration of the parents) at time $s + 1$.

This general population change matrix operator for A age groups, $H \times N$ occupation-by-regional location categories, and D durations of residence, can be represented as G_s. Then if we let W_s be a column vector representing the *numerical* distribution of population by age, duration of residence in a given region, occupation, and region at time s (there is a category for no occupation), we generate an *expected* distribution at time $s + 1$ from:

(2) $G_s \cdot W_s = W_{s+1}$.

If λ represents its dominant characteristic root and W its corresponding characteristic vector,

(3) $G \cdot W = \lambda \cdot W$.

The scalar λ is the non-negative root with maximum absolute value of the determinantal equation

(4) $|G - \lambda I| = 0$.

Thus, solving (4) for this dominant root λ, we can (in principle) evaluate λ and substitute it into (3). With this substitution into (3), we may then solve (in principle) the system of linear equations $(G - \lambda I)W = 0$ for the non-negative vector mentioned above.

We may *define* this vector as the *latent* vector of system embodied by G. The formulas for calculating the various stable (latent) distributions of interest, are quite obvious and may be obtained as follows. Let a typical element of the latent vector W be W_{vaid}, the number of people aged v and located in the dth duration of residence in the aith location. Then the above-mentioned stable distribution is

(5) $y = \{W_{vaid}/(\Sigma_{v,a,i,d} W_{vaid})\}_{A \times M \times D, 1}$, where $v = 1, 2, \ldots, A$; $ai = 11, 12, \ldots (HN = M)$; $d = 1, 2, \ldots, D$.

From this column vector a host of stable age, duration of residence, occupational, and regional distributions, and migration rates, may be computed by appropriate summations. Due to shortage of space the reader will have to refer to Stone (1968) for several concrete examples. These stable distributions and rates would be the characterizing measures of the mobility regime embedded in G_s.

Rogers, Andrei, 1968 *Matrix Analysis of Interregional Population Growth and Distribution* (Berkeley).
Stone, Leroy O., 1968 Stable migration rates from the multiregional growth matrix operator, *Demography* 5, 439–42.
— 1971 *On the Analysis of the Structure of Metropolitan Area Migration Streams* (Toronto).

P1223
Spectral methods of statistical analysis and planning of observations of oceanographical fields
V.I. BELYAEV, G.A. MOISEYEV, and I.E. TIMCHENKO   *Ukrainian Academy of Sciences, USSR*

A common problem of processing and analyzing data on experimental observations exists in geography, geodesy, meteorology, oceanography, and other sciences. Usually the aim of such an analysis is to present information in a generalized form, e.g., as maps and atlases, permitting us to make a comparison between spatial co-ordinates and individual moments of time and certain numerical characteristics of objects and phenomena under study. The subject of our investigations will be fields expressing dependence of such values on co-ordinates.

The most general method of investigations of fields is their modelling using stationary random functions. Statistical analysis of data relating to fields is used successfully in meteorology in connection with problems of numerical weather forecasts (Gandin 1963). An attempt is made by us to apply this method to oceanography in calculations of some spatial fields, as well as in planning of oceanographical observations (Belyaev, et al. 1969, 121–4). Due to the above mentioned common character of the method basic results of the discussion remain valid in solving similar problems in geography.

Let us discuss the isotropic random field

$f(\mathbf{x})$ measured at the system of points $\{\mathbf{x}_i\}$. The value of the field at some arbitrary point $f_0(\mathbf{x})$ can be determined by interpolation. Let us call the interpolation optimal when the root-mean-square error of calculations of the field

(1) $E(\mathbf{x}) = M\{|f(\mathbf{x}) - f_0(\mathbf{x})|^2\}$

reaches a minimum. A linear approximation of the field values according to formula 2 will be an optimal interpolation under a Gaussian distribution of probabilities

(2) $f_0(\mathbf{x}) = \sum_{i=1}^{n} g(\mathbf{x}, \mathbf{x}_i) f(\mathbf{x}_i)$.

The weight coefficients of interpolation $g(\mathbf{x}, \mathbf{x}_i)$ are determined from the system of equations (Kholmogorov 1941, 3–14)

(3) $K(\mathbf{x}, \mathbf{x}_2) = \sum_{m=1}^{n} g(\mathbf{x}, \mathbf{x}_m) K(\mathbf{x}_2, \mathbf{x}_m)$.

in which $K(\mathbf{x})$ is the correlation function of the field.

Taking into account (2) and (3), the error of optimal interpolation takes the form

(4) $E(\mathbf{x}) = K(\mathbf{x}, \mathbf{x}) - \sum_{m=1}^{n} K(\mathbf{x}, \mathbf{x}_m) g(\mathbf{x}, \mathbf{x}_m)$.

The latter can be presented in a spectral form using a Fourier transformation (Petersen and Middleton 1962, 294)

(5) $E(\mathbf{x}) = [1/(2\pi)^N] \int_{(\omega)} \left[\phi(\omega) - \{G(\omega)/Q\} \right.$

$\left. \times \sum_{[l]} \exp(-i\mathbf{x}\mathbf{u}_{[l]}) \phi(\omega + \mathbf{u}_{[l]}) \right] d\omega$.

Here $\phi(\omega)$ and $G(\omega)$ are inverse Fourier transforms from the correlation and weight functions, $\mathbf{U}_{[l]}$ is a linear combination of basis vectors in wave number space (ω), forming a regular net, and N is the dimensionality of the field. Formula (5) is valid in cases when measurement of the field is performed on a regular lattice. Q is a constant equal to the volume of an elementary cell of the sampling lattice.

It follows from this formula that in order for the error of interpolation to approach zero, it is necessary that the following conditions be fulfilled: (*a*) the spectrum of field $\phi(\omega)$ must occupy a limited area in wave number space (ω); (*b*) the network of measurements must be chosen in such a way that 'secondary spectra' $\phi(\omega + \mathbf{u}_{[l]})$ do not overlap; (*c*) Fourier transformation of the weight function of the field must satisfy conditions

$G(\omega) = \begin{cases} Q & \text{everywhere where } \phi(\omega) \neq 0, \\ 0 & \text{everywhere where } \phi(\omega + \mathbf{u}_{[l]}) \neq 0. \end{cases}$

Fulfilment of all these conditions provides the optimal presentation of the field by discrete samples of its values.

The application of results presented, in practice, allows one to define rational means for measuring fields, which is of prime importance in the planning of observations. In addition, if a sampling lattice is chosen in accord with conditions (*a*), (*b*), (*c*), calculations of the field on an electronic computer can be performed using simple 'spectral' algorithms of interpolation not connected with the solution of the systems of equations (3) (Petersen and Middleton 1962, 308).

Assume, for example, that the spectrum of some two-dimensional isotropic field does not involve frequencies higher than $|\omega_c|$. Then, in order to satisfy conditions (*a*), (*b*), (*c*), it is enough to perform measurements on a 60° rhombic lattice having sides

(6) $a = 2\pi/\omega_c \sqrt{3}$.

In this case the spectral formula of interpolation is of the form (Petersen and Middleton 1962, 308)

(7) $f_0(\mathbf{x}) = \sum_{i=1}^{n} [J_1(\omega_c r_i)/\pi\sqrt{3}\omega_c r_i] f(\mathbf{x}_i)$,

where $r_i \equiv |\mathbf{x} - \mathbf{x}_i|$ and J_1 is the Bessel function.

The calculation of the field from formulae of type (7) proves to be a simpler operation than interpolation from formulas (2) and (3).

The present method has been used by the authors to evaluate the required number of measurements and their spacing for two oceanographical fields: the field of strontium-90 concentration on the Atlantic Ocean surface and the Black Sea surface temperature field. In each case we required that the average error of interpolation $\bar{E}$ should not exceed 20 per cent of the field's dispersion. Correlation functions of these fields are approximated by the general dependence

(8) $K(x_1, x_2) = \exp\{-\alpha(x_1^2 + x_2^2)\}$,

parameter α being taken equal to 3.2×10^{-4} miles^{-2} for the surface temperature field and 2.3×10^{-5} miles^{-2} for the field of strontium-90 concentrations. Under these assumptions it is easy to obtain the dependence of the distance between stations a on the average error of interpolation

(9) $a = (-3\alpha \ln \bar{E})^{-\frac{1}{2}}$.

Thus we find that the interval between stations for the field of surface temperature of the Black Sea must be of the order of 40 miles (64 km). For the fields of concentrations of strontium-90 this interval is about 300 miles (480 km). This means that calculation of the field of concentration with 20 per cent accuracy in the water area of the Atlantic Ocean of 2×10^7 miles^2 requires that about 300 measurements be made on a 60° rhombic lattice.

Similarly, estimation dependences of the number of measurements and their spacing for two-dimensional and three-dimensional fields, having standard correlation functions, have been constructed by us:

(10) $K(r) = e^{-\alpha r}$,
(11) $K(r) = e^{-\alpha r} \cos \beta r$,
(12) $K(r) = [(\alpha r)^\nu / 2^{\nu-1} \Gamma(\nu)] K_\nu(\alpha r)$,

where $r \equiv |\mathbf{x}|$, $K_\nu(\alpha r)$ is Macdonald's function of the order ν, and $\Gamma(\nu)$ is the Euler function. For example, for function (11) in the case of a three-dimensional field we have

$\bar{E}_a = 1 + (2/\pi)[\xi a/(1 + \xi^2)]$
$- (2/\pi) \arctan \xi a$, $\xi \equiv \alpha/\pi\sqrt{2}$.

The spectral method of calculation of the field is based on the assumption that the network of measurements is infinite, and the spectrum $\phi(\omega)$ is limited. It is evident that the first condition cannot be fulfilled in reality. Therefore it is of interest to test the spectral method in practice.

Since at present we have no regular network of stations in oceanography able to carry out the necessary measurements, we modelled a two-dimensional isotropic field with correlation function (8) on an electronic computer. The model of the field consists of 16,000 values of the 'field' on a 60° rhombic lattice. A rarer lattice of 'measurements' was chosen for interpolation so that it involved about 30 samples.

Numerical calculations indicate that the spectral method of calculation has a high accuracy. The average value of the interpolation error totals 5 per cent of the field's dispersion. On approaching boundaries of the field the calculation error increases insignificantly.

The results obtained show promise for the use of spectral methods in planning measurements of isotropic random fields and for the reconstruction of their maps on an electronic computer.

Belyaev, V.I., Doronin, I.F., Ermolenko, A.I., Ivanova, T.M., Nelepo, B.A., and Timchenko, I.E., 1969 Objective analysis of the field of strontium-90 concentration in the Atlantic Ocean, *Izvestiya AN SSSR, fizika atmosfery i okeana* t5 8.

Gandin, L.S., 1963 Objective analysis of meteorological fields, *Hydrometeoizdat* (Leningrad).

Kholmogorov, A.N., 1941 Interpolation and extrapolation of stationary random sequences, *Izvestiya AN SSSR*, ser. math. t5, N1, 3–14.

Petersen, D.P., and Middleton, D., 1962 Sampling and reconstruction of wave-number-limited functions in *N*-dimensional Euclidean spaces, *Information and Control* 5, 279–329.

P1224
Griffith Taylor and the beginnings of academic geography in Canada
G.S. TOMKINS *University of British Columbia, Canada*

The ideas and work of Griffith Taylor constitute a significant chapter in the progress of geography in Canada, covering the sixteen years (1935–51) when he worked here. This chapter also illustrates Taylor's philosophical stance and his view of the purpose of geography.

Philosophically, Taylor reflected a typical late nineteenth-century viewpoint, espousing the evolutionary geographic concepts of William Morris Davis, an early mentor. From Davis and others, Taylor acquired the notions of developmental stages, cyclical time scales of youth, maturity, and old age, and Spencerian ideas of evolutionary analogies (Tomkins 1967, 9–20). Like Davis, he applied these concepts to landform evolution but also to studies as diverse as racial origins and urban growth. For Harvey, Taylor's zones and strata theory is a common explanatory model, utilizing historical inevitability, i.e. the inevitability of man's adjustment to 'nature's plan,' a teleological approach embodying empirically untestable hypotheses (Harvey 1969, 78, 416, 423, 425). But in his *ex post facto* search for evidence, Taylor was a remarkable field geographer, possessing an 'eye for country' that resulted in enduring substantive works on marginal environments like Canada.

As a self-styled determinist, Taylor has been a convenient straw man in the environmentalist controversy. His early (1923) definition of geography as 'the study of the environment in relation to human values,' like his later 'stop-and-go determinism' stamp him actually as a possibilist (Taylor 1924, 438; Lewthwaite 1966, 11, 19). But Taylor, lacking the philosophic mind of the true theorist, exerted his influence in other ways. Anthropologists rejected his racial theories but were alerted to the importance of environmental factors and to geographical methods (Hooton 1940, 526; Penniman 1965, 249). He stimulated geographers to an interdisciplinary approach, and by insisting that religion, race, and language were proper geographic concerns, he anticipated cultural geography (Watson 1948, 58). His view of geography as a liaison discipline was reinforced at Chicago where colleagues espoused a social science approach to the discipline. In promoting geography as a correlative science, the subject of his inaugural lecture at Toronto, he anticipated the quantitative revolution (Taylor 1935). Ian Burton, in noting Rose's pioneering use of correlation analysis, might have also noted that Rose had been Taylor's doctoral student at Chicago and that Taylor introduced the technique at Toronto (Burton 1963, 153; Taylor 1938).

Warntz has deplored the loss of a macroscopic viewpoint that attended the eclipse of environmentalism. Significantly, Taylor is the geographer singled out for praise for maintaining this viewpoint after 1925 and for thus anticipating the macroscopic revival after 1955 (Warntz 1959, 447).

Taylor began to apply his theories of nation planning in Canada almost immediately upon his arrival in 1935. Before then, formal advanced geographical studies had begun in French Canada where the first professorship was established at Montreal in 1910. After 1920, eminent French geographers, notably Jean Brunhes and Raoul Blanchard, visited Canada. Except for the war years, Blanchard visited annually after 1927. He and Taylor were elected honorary presidents of the Canadian Association of Geographers in 1951. Young French-Canadian geographers began studying in France in the 1930s. Hamelin's claim that no other French discipline exerted such influence in Canada is thus borne out (Hamelin 1962–63, 6).

In English Canada, geography, though absent from the universities, maintained a precarious foothold in the schools. Resource studies, publication of a national atlas (1906), and the work of the Canadian Geological Survey, were notable achievements at the governmental level (Tomkins 1967, 164–9). Although generally regarded as a physical science, geography at Toronto was grouped with the social sciences when Taylor came to head the first autonomous department in Canada. It was so classified by Harold Innis, economic historian and future dean of Canadian social scientists, who had previously taught economic geography for some years. From the premises of his staples theory of Canadian development, Innis argued that Canada was defined by the territory of the fur trade, growing out of the east-west system of waterways that extended from the St Lawrence estuary via the Great Lakes, the Canadian Shield, and the Prairies to the Pacific. This environmentalist, anti-continentalist interpretation upset the conventional wisdom of the north-south grain of North American geography and influenced the Laurentian school of nationalist historians (Innis 1930). Taylor did not introduce environmentalist thinking to Canadian geography. It was already there, down to required readings from Ellsworth Huntington. Later, Taylor drew on Innis' theories for his own treatment of Canadian economic geography (Taylor 1947).

The work of Innis and others, and a sound geological tradition, comprised a *corpus* of knowledge that enabled Taylor to pursue his geographical task of description and synthesis. Beginning characteristically with broad strokes, he first produced basic papers that well illustrated his generalizing skill. Thus, using an existing scheme of physiographic divisions, he popularized a regional approach still widely used (Taylor 1936, 162). Later, more detailed studies resulted in monographs on most major regions of Canada. He had been hounded out of marginal Australia for effectively deflating irresponsible talk of a huge population potential and for demanding sensible 'nation planning' (Tomkins 1967, 71–82). Paradoxically, he exhorted the inhabitants of marginal Canada to plan for an ultimate population of 100 millions. Emphasizing indigenous resources as vital to national growth, Taylor foresaw a prairie industrial empire based on coal (Taylor 1936, 169). He set the 57° July and August isotherms as the limit of northern settlements where a Russian regime of hay-oats-rye-potatoes would nourish large populations (Taylor 1942, 24). His pioneering regional descriptions of northwestern Canada called national attention to these areas. Published widely in scholarly journals, by various public groups, in essay collections, and in the popular *Canadian Geographical Journal*, Taylor's many papers on Canada again proved his ability to communicate with scholar and layman and illustrated his view of the geographer's public role.

Taylor's 1947 text on Canada, combined with his earlier work on Antarctica and Australia, stamped him as the geographer of the marginal lands (Taylor 1947). Basically disdainful of theory in his explorer-scientist role, his unreflective philosophic stance was inappropriate to an empirically minded era. But his substantive geographic studies endure. His firm establishment of Canadian academic geography is witnessed by those of his ex-students now in the vanguard of the present remarkable flowering of the discipline in Canada. This suggests that his influence was more personal than intellectual. He did not found a school of Canadian geography. That task remains for the future.

Burton, Ian, 1963 The quantitative revolution and theoretical geography, *Can. Geog.* VII, 151–62.
Hamelin, L.-E., 1962–63 Petite histoire de géographie dans le Québec et à l'université Laval, *Cahiers géog. Qué.* VII, 1–16.
Harvey, David, 1969 *Explanation in Geography* (London).
Hooton, Ernest, 1940 Review of environment, race and migration by Griffith Taylor, *Am. Anthrop.* XLII, 525–6.
Innis, Harold, 1930 *The Fur Trade in Canada* (New Haven).
Lewthwaite, Gordon, 1966 Environmentalism and determinism: a search for clarification, *Ann. Assoc. Am. Geog.* LVI, 1–23.
Penniman, T.K., 1965 *A Hundred Years of Anthropology* (London).
Taylor, T. Griffith, 1924 Geography and Australian national problems, *Report of the Sixteenth Meeting of the Australasian Association for the Advancement of Science*, 433–87.
– 1935 Geography, the correlative science, *Can. J. Econ. Pol. Sci.* I, 535–50.
– 1936 Fundamental factors in Canadian geography, *Can. Geog. J.* XII, 161–71.
– 1938 Climate and crop isopleths for Southern Ontario, *Econ. Geog.* XIV, 89–97.
– 1942 *Canada's Role in Geopolitics – a Study in Situation and Status* (Toronto).
– 1945 Mackenzie doomsday, 1944, *Can. J. Econ. Pol. Sci.* XI, 189–233.
– 1945 A Yukon doomsday, 1944, *ibid.*, 432–49.
– 1947 *Canada: A Study of Cool Continental Environments and Their Effect on British and French Settlement* (London).
Tomkins, George S., 1967 *Griffith Taylor and Canadian Geography* (Ann Arbor).
Warntz, William, 1959 Geography at mid-twentieth century, *World Pol.* XI, 442–54.
Watson, J. Wreford, 1948 Geography and history: some observations based on 'our evolving civilization' by Griffith Taylor, *Can. Hist. Rev.* XXIX, 57–65.

P1225
The gamma function in geographic research: a technical note
BARRY S. WELLAR   *University of Kansas* and
GERALD J. LA CAVA   *Georgia State University, USA*

Some mathematical aspects of using the pointwise maximum of a finite family of gamma functions in geographic research are presented. It is noted here and developed in detail elsewhere (Wellar and LaCava 1972) that the gamma (Γ) function differs significantly from most probability density functions (PDF's) in that Γ has three parameters instead of two. The two parameters common to all PDF's are the location parameter, β, and a scale parameter, s. In addition to the β and s parameters, Γ has a third parameter, namely an intensity parameter, α.

The gamma family is modified in this discussion by the addition of a magnified indicator function (the magnified indicator function may be used in those cases where $\alpha \neq 1$ [no negative exponential member in the family of gammas] and it is desired that the left slope of the gamma closest to the origin not decrease to zero; for example, traffic congestion does not disappear in the CBD, so a magnified indicator function would seem to depict this), and some of the gammas are truncated at $x = \beta$ so that all are bounded (this is done when $\alpha = 1$ because the changing βs and ss could cause the value of the curve to be arbitrarily large). These modifications are included to achieve rigour and improved goodness-of-fit.

Define, for $\alpha > 1$,
$$\Gamma_{\alpha,\beta,s}(x) = \begin{cases} (x-\beta)^{\alpha-1} e^{-(x-\beta)/s}/\Gamma(\alpha)s^\alpha, & x \geq \beta, \\ 0, & 0 \leq x \leq \beta. \end{cases}$$
For $\alpha = 1$,
$$\Gamma_{1,\beta,s}(x) = \begin{cases} 1/s, & 0 \leq x < \beta, \\ (1/s)e^{-(x-\beta)/s}, & x \geq \beta, \end{cases}$$
where α is the intensity parameter, β the location parameter, and s the scale parameter.

Let $\mathscr{J}_K = \{(\alpha_i, \beta_i, s_i), i = 1, \ldots, K\}$ be a finite collection of ordered triples where the components satisfy the following conditions:
(1) $1 \leq \alpha_i, i = 1, \ldots, K$,
(2) $\beta_i \geq 0, i = 1, \ldots, K$,
(3) $s_i \geq 1, i = 1, \ldots, K$, with at least one $s_i = 1$.
Let
$$x_1 = s_1(\alpha_1 - 1) + \beta_1 = \min_i \{s_i(\alpha_i - 1) + \beta_i \mid \\ \Gamma_{\alpha_1,\beta_1,s_1}(x_1) > \Gamma_{\alpha_i,\beta_i,s_i}(x_1)\}$$
(the peak nearest the origin is labelled 1). Define
$$G_{\alpha_1,\beta_1}(x) = G I_{(0, \alpha_1 + \beta_1 - 1)}(x)$$
where
$$G = \max \Gamma_{\alpha_1,\beta_1,1}(x) \\ = (\alpha_1 - 1)^{\alpha_1 - 1} e^{-(\alpha_1 - 1)}/\Gamma(\alpha_1)$$
and $I_{(0, \alpha_1 + \beta_1 - 1)}(x)$ is the indicator function, which is 1 on $(0, \alpha_1 + \beta_1 - 1)$ and 0 elsewhere.

DEFINITION. $\mathscr{F}$ is called an *urban family of* Γ's if and only if $\mathscr{F}$ has the form
$$\mathscr{F} = \{G_{\alpha_1,\beta_1}\} \cup \{\Gamma_{\alpha_i,\beta_i,s_i}: (\alpha_i, \beta_i, s_i) \in \mathscr{J}_K,$$
for some positive integer $K\}$.
$\mathscr{J}_K$ is called the *index set* for the urban family.

DEFINITION. A *ray gamma*, Γ_*, is a function expressible as the pointwise maximum of an urban family of Γ's; that is,
$$\Gamma_*(x) = \max\{F(x): F \in \mathscr{F}, \text{ an urban family of } \Gamma\text{'s}\}.$$
In order to determine the ray gamma of the family of gammas it is necessary to first prescribe the urban family. This requires enunciation of the decision rules for the selection of the triples comprising the index set. Before that can be done, however, it is necessary to elaborate on the role of parameters (see also Greenwood and Durand 1960; Voorhees 1968).

As noted above the index set triples consist of the location, scale, and intensity parameters.

The location parameter, β, simply translates the Γ horizontally. Once the maximum value and scale have been determined, β can be used to shift the curve so that the maximum occurs at the desired point on the horizontal axis.

The intensity parameter, α, gives the shape of the Γ function. It also, together with s, determines the maximum value. The shape of each particular Γ is less important here because the model utilizes essentially the down-slope of the Γ, and s can be manipulated to such a degree that the influence of α on the shape is minimized for our purposes. The maximum value of $\Gamma_{\alpha,\beta,s}(x)$ is
$$(\alpha - 1)^{\alpha-1} e^{-(\alpha-1)}/\Gamma(\alpha) s$$
and is used to determine the intensity parameter.

The scale parameter, s, is a measure of the dispersion of the curve – the smaller values of s yielding steeper, more compact curves. The dispersions of two Γ's with β's and α's held constant are compared by computing the ratio of the two scale parameters. Thus, the Γ with the steepest slope in the family can have $s = 1$ and the others will have $s \geq 1$. For all curves in the family a shift of distance s in a positive direction from the maximum results in the same increase in the areas under the curve. The magnitude of this increase, ε, is determined using $s = 1$.

A new ordered triple of parameters results for each peak in the data (e.g., when graphed in two-space). If K is the number of Γ's in the family, then $K - 1$ is the number of times the scale changes. The following decision rules are used to determine K, α, β, and s:

1. K is equal to the number of peaks as one moves from, say, C_i to C_j (or from the CBD through other centres to the urban fringe, or any other comparable type of hierarchy).

2. One of the s_i's, say s_r, is set equal to 1. Let M_r be the maximum value of the gamma. Then,
$$M_r = (\alpha_r - 1)^{\alpha_r-1} e^{-(\alpha_r-1)}/\Gamma(\alpha_r).$$
For a given M_r, α_r can be determined. β_r is then chosen so that the gamma attains its maximum at $x = (\alpha_r - 1) + \beta_r$.

3. To determine the s_i, it is first necessary to set ε. Because of the nature of the s_i and since $s_r = 1$, ε satisfies the equation
$$\varepsilon = \int_{\alpha_r - 1}^{\alpha_r} \frac{x^{\alpha_r - 1} e^{-x}}{\Gamma(\alpha_r)} dx.$$
Hence, for a given α_r, ε can be found. s_i can now be determined as mentioned above, that is, s_i is the distance to the right of the maximum such that an increase of ε is achieved.

4. Choose α_i to satisfy
$M_i s_i = (\alpha_i - 1)^{\alpha_i - 1} e^{-(\alpha_i - 1)}/\Gamma(\alpha_i)$,
where M_i is the maximum value of the ith peak.
5. Choose β_i such that $s_i(\alpha_i - 1) + \beta_i = x_0$, where x_0 is the point at which the maximum occurs.

One test of the gammas that is now in progress is to determine that mix of parameters which best describes the distribution of certain commodities along selected links on the US Interstate Highway System. A second, and more complex task is to use the gammas for transportation planning. In brief, this task involves incorporating network characteristics (topology, capacity, density) and design/construction alternatives (add or delete lanes or links; make K lanes or links reversible, one-way, or two-way facilities; add, remove, and synchronize or de-synchronize traffic lights) into the parameters to meet specified performance levels.

Greenwood, J., and D. Durand, 1960 Aids for fitting the gamma distribution by maximum likelihood, *Technometrics*, 2 (1), 55–67.

Hoover, E.M., 1970 Transport costs and the spacing of central places, *Papers, Regional Sci. Assoc.* XXV, 255–74.

Voorhees, A.M., 1968 *Factors and Trends in Trip Lengths* (NCHRP Rept. 48, Highway Research Board, Washington, DC).

Wellar, B.S., and G.J. LaCava, 1972 Parametrization of the gamma function for description and planning of space use (prepared for IGU Commission Meeting, Ca. 19, Quantitative Methods, Queen's University, Kingston, Ontario).

P1226
Graph-theoretical analysis of ridge patterns
C. WERNER   *University of California, USA*

With the exception of areas without surface drainage the interlinked patterns of natural ridge lines form tree-like graphs which are topologically similar to natural channel networks. There are, however, a number of significant differences which prohibit the direct application of channel network research to networks of interlinked ridges. This paper will identify some of these differences; it will construct a mathematical framework which permits the determination of randomness in the topological lay-out of ridge networks, and it will test the hypothesis of randomness of selected sets of ridge network data sampled in eastern Kentucky.

Ridge networks do not possess a link which stands out against the others as does the outlet of a channel network. Thus, ridge networks do not have a 'natural' orientation by themselves, and any labelling of their topological layout must be arbitrary. Geometrically speaking, ridges usually possess saddlepoints (passes), which is in contrast to the strictly monotone decline of the elevation of stream channels. Disregarding these saddlepoints, the ridge patterns of humid areas form single trees of such a magnitude that for all practical purposes they can be considered to be infinite. If, however, ridge patterns are considered to be separated by saddlepoints, then they form a vast distribution of lower magnitude ridge trees (ridge networks). Each node in a ridge network interconnects at least three ridge lines (links), and, by close observation, no more than three ridge lines seem to be linked in one node. Since, in any real case, the location of a ridge node can never be identified with absolute precision, we will axiomatically assume that every node in a ridge network joins exactly three links. Hence, with the exception of the orientation problem mentioned before, ridge networks are topologically equivalent to channel networks. Their magnitude will be defined as the number n of their outer links (i.e. one more than for the corresponding channel networks), and it follows, that – for example – their total number of links is $2n - 3$; many other results from channel network research can likewise be translated to give equivalent statements about ridge networks.

A bowl of a ridge network is defined as a path connecting two 'neighbouring' outer links, i.e. a path which starts in an outer link and continues on the network by making only left (or right) turns until it reaches another outer link. Clearly, every dendritic network has as many bowls as it has outer links.

In addition to the concept of a bowl, the

notion of stream number and order in channel network topology will be used as ridge network parameters. We define all outer links of a ridge network as ridge streams of first order, and a ridge stream of the ith order is defined as that chain of links, which starts at a node produced by two ridge streams of $(i-1)$th order and ends in the first node in which it merges with a ridge stream produced in the same way. When a ridge network is reduced by stepwise elimination of the streams defined above, the remainder will be either a node in which three streams of the highest order merge or a chain of links which we define as two streams of the highest order.

Assume that all (labeled) ridge networks of magnitude n are equally likely. Consider a bowl of size k in a ridge network of magnitude n. It consists of k links and $k-1$ nodes. In these nodes the remaining network is linked to the bowl in the form of $k-1$ sub-networks which together have $n-2$ of the outer links of the network. Hence, the question of how often a specific bowl will be of size k considering all ridge networks of magnitude n is equivalent to the question, in how many ways one can form $k-1$ networks by merging $n-2$ outer links (not counting the links connecting the subnetworks with the bowl). This number is (Werner 1971, 10; Dacey 1971, 7)

(1) $Z(k-1, n-2) = \binom{2(n-2)-(k-1)}{n-2}$

$\times (k-1)/\{2(n-2)-(k-1)\}$

Since the total number of topologically different ridge networks with n (labeled) outer links is $Z(1, n-1)$ (Shreve 1966, 29), the probability $P(k, n)$ of a bowl in a ridge network of magnitude n to be of size k is

(2) $P(k, n) = \dfrac{Z(k-1, n-2)}{Z(1, n-1)} = \binom{n-1}{k}$

$\times (k-1) \Big/ \binom{2n-4}{k}$

$= (k-1) \prod_{i=1}^{k} \dfrac{n-i}{(2n-3)-i}$

so that for finite k and n approaching infinity we get

(3) $\lim_{n \to \infty} P(k, n) = (k-1)/2^k$.

Again we assume equal likelihood for the various topologically different layouts of (labeled) ridge networks of equal magnitude. The probabilities of networks when grouped according to their stream numbers can be calculated parallel to the corresponding derivation for channel networks (Shreve 1966, 29). If a ridge network has n_i streams of order i and n_{i+1} streams of order $i+1$, then $2n_{i+1}$ streams of ith order will merge pairwise to form those n_{i+1} streams of order $i+1$, whereas the remaining $n_i - 2n_{i+1}$ streams of ith order merge with the $2n_{i+1} - 3$ links of the higher order streams. According to Riordan (1958, 7) and Shreve (1966, 29) this can be done in

$\binom{n_i - 4}{n_i - 2n_{i+1}} 2^{n_i - 2n_{i+1}}$

ways. Additional topological configurations are obtained through n_i rotations of those constructed above. If the number of different sub-networks produced by the merger of the $(i+1)$th order streams – which includes n_{i+1} rotations – has already been taken care of, then the factor accounting for the necessary rotations is n_i/n_{i+1}. The total number of different ridge networks having n_i $(i = 1, 2, ..., m)$ streams of order i can be calculated through multiplication of the individual network frequencies for each pair of stream numbers n_i, n_{i+1}, and the probability distribution that a ridge network of magnitude $n = n_1$ has n_i $(i = 1, ..., m)$ streams of order i is

(4) $P(n_i, i = 1, ..., m)$

$= \left[\binom{2n-3}{n-1} \Big/ (2n-3)\right]^{-1}$

$\times \prod_{i=1}^{m-1} \dfrac{n_i}{n_{i+1}} \binom{n_i - 4}{n_i - 2n_{i+1}} 2^{n_i - 2n_{i+1}}$

$(n_i \geq 4$ for $i < m)$.

Both the distribution (3) of bowls in an infinite random ridge network (Table 1) and the distribution (4) of finite ridge networks by stream numbers under the condition of equal likelihood (Table 2) have been tested in an area of eastern Kentucky which is essentially free of geologic control, so that one might ask whether, under such conditions, the ridge pattern is topologically random. A hundred ridge network bowls as well as 212 individual ridge networks have been sampled from USGS 1:24,000 topographic maps; sampling of the first set was random; the second set constitutes all – and only – the ridge networks of three rectangular areas randomly selected. A comparison between expected and observed data shows a remarkable degree of correspondence, so that the hypothesis of topological randomness of

TABLE 1. Frequencies of network bowls

	2	3	4	5	6	7	8	9	10	>10	Total
Expected	25.00	25.00	18.75	12.50	7.81	4.69	2.74	1.56	0.88	1.07	100.00
Observed	23	31	20	13	3	6	1	1	2	0	100.00

$\chi^2 = 6.21$, $\alpha = .48$ and $\chi^2(0.95) = 14.1$, where DF = 7.

TABLE 2. Frequencies of networks, by stream numbers

	6/2	6/3	7/2	7/3	8/2	8/3 8/4/2	9/2	9/3 9/4/2	10/2	10/3 10/4/2
Expected	23.1	3.9	15.3	7.7	9.7	10.3	5.0	10.0	2.5	8.5
Observed	22	2	15	8	10	10	5	10	4	7

ridge networks in the absence of geologic controls is strongly supported.

National Science Foundation, grant GS-2989.

Dacey, M.F., 1971 Probability distribution of number of networks in topologically random drainage patterns, *Water Resources Research*, submitted.

Riordan, J., 1958 *An Introduction to Combinatorial Mathematics* (New York).

Shreve, R.L., 1966 Statistical Law of Stream Numbers, *Jour. of Geol.*, 74, 17–37.

Werner, C., 1971 *Patterns of Drainage Areas with Random Topology*, Research Report 5, NSF grant GS-2989, University of California, Irvine.

P1227
On metageography
JERZY ZABORSKI   *Sokrateion of Arizona, USA*

The problems of metascience have been looming on the scientific horizons for a long time. The name itself, meaning 'beyond science,' is derived from Aristotle's *Metaphysics*. Aristotelian 'metaphysics' is too broad a concept, however, to be helpful in the current discussion, John Dewey's feeling that all men are metaphysical when they preoccupy themselves with problems beyond experience, notwithstanding (Dewey 1938, 32).

The first definitely metascientific discipline, not only able to withstand rigorous criticism, but also to set the path for all metascience to follow, is the meta-algebra of classes created by Alfred Tarski (Tarski 1933, 1935, 1936). Tarski is also credited with the invention of metamathematics by virtue of creating the first metamathematical discipline, the meta-algebra of classes (Tarski 1956). With the publication of the original work in 1933 of this illustrious member of the Polish School of Analytical Philosophy (Skolimowski 1967), the scientific community was given a model for making other metasciences (Kalinowski 1971).

Accordingly, to paraphrase the original definition, the meta-algebra of classes is an axiomatised and formalised deductive science of structure of the algebra of classes, its components and properties. An example demonstrating the reasoning which led to the creation of the new discipline is called for at this time from *The Concept of Truth*, the fundamental work on the subject of the meta-algebra of classes (and easily available in the original Polish (Tarski 1933), German (Tarski 1935, 1936), English, as a part of a larger collection (Tarski 1956) and, the most up-to-date, French collection (Tarski, in press)):

'x is a true sentence,' is not sufficient; we must have:

'x is a true sentence in the algebra of classes'; or

'x is a true sentence in a given language.'
The name of the scientific discipline or a given language must always be stated, for

there are many disciplines, each with its own particular methodology and special criteria for the discovery of, so to say, 'the scientific truth.' Similarly, each language is guided by its own linguistic laws.

Such reasoning was already known to Aristotle: '... some recognise exclusively the language of mathematics; others demand only examples; still others want to fall back on the authority of some poet; and there are also those who expect exact evidence everywhere, that, in turn, is regarded by some as exaggerated, either because not having agile minds or not being able to follow the chain of reasoning, or from fear of drowning in the flood of detail. There is indeed something of this in an exaggerated demand for exactitude, which sometimes is looked at as unworthy of a free man, both in daily life as well as in a philosophical discourse. Because of this reason, one should, beforehand, decide upon requirements demanded in this respect of each kind of science; for it is absurd to search simultaneously for the method of a science and for the very science itself, especially that neither the former, nor the latter is easy. Particularly, one should not demand a mathematical exactitude in everything ...' (Aristotle)

The next milestone in the development of metascience came through the magnificent achievement of the Unity of Science Movement, its *Encyclopedia of Unified Science* (Neurath, Carnap, and Morris 1938). It is here, in a chapter on 'The Viewpoint of Scientific Empiricism,' one of his contributions to the *Encyclopedia*, that Charles Morris pleads for metascience (Morris 1938, 68-71). He devotes considerable space to the subject, defining metascience as 'the science of science' and as 'the science of the language of science.'

Recently, since Sputnik, interest in metascience has arisen among Soviet geographers. In their definition of the goals of metageography we find that it 'is concerned with the common basis of geographic regularities as well as the potential of geography as a science' (Gokhman, Gurevich, and Saushkin 1968, 1969a, 1969b). We shall return to some other ideas on the subject of metageography from the USSR after discussing my opinion of the discipline.

I would like to narrow down the band of scientific pursuits suggested both by Tarski and by Morris in their definitions of metascience: 'metascience is a science of symbols and concepts of science with behaviour and properties independent of the depicted actuality.' I feel that the above definition of metascience will suit its application to the new discipline of metageography better. Thus, 'metageography is a science of symbols and concepts of geography with behaviour and properties independent of the depicted spatial actuality.'

The above definition of metageography, deduced from that of metascience, brings into focus the contrast between actuality and the proper primary procedure for its investigation: observation, description, interpretation, and explanation, on one hand, and the secondary scientific procedure: the study of the creations of the above, the study of the tools, aids, and devices of the primary procedure. Thus metageography is twice removed from actuality. In this manner, the domain of metageography could be illustrated by a number of examples of those 'symbols and concepts' of geography: regions, landscapes, matrices, networks, lattices, linkages, nodes, compages, transitions, cores, cycles (both Darwinian and Davisian), vertical themes, hierarchies, nestings, central places, gravity points, thresholds, boundaries, densities, models, maps, globes, various diagrams illustrating spatial actuality, images (even those made by optical and other sensing devices), and, yes – the concept of 'surface' itself. All those are constructs of actuality, or reality, subjectively abstracted and selected, even though possibly based on logical criteria and procedure.

It appears from the foregoing that there is a direct evolution of metageography from the earlier models, with one exception: that of the Soviet definition of metageography. Actually, the discontinuity is more apparent than real. For, the Soviet 'geographic regularities' are reminiscent of Tarski's 'structure,' their 'potential' of Morris' 'science of science.' And, if we read further in the Gokhman, Gurevich, and Saushkin text, we shall find such concepts as stochastic analysis – our axiomatization and formalization; eigenspace – our demand for qualifying x; threshold scales – our primary and derived abstractions of reality; finally, classification and regionalization, where full accord is reached between our metageography, as discussed in the present article, and the one which the distinguished Soviet geographers are creating (Gokhman, Gurevich, and Saushkin 1968, 1969a, 1969b).

The present definition of metageography as the science of symbols and concepts of geography, endowed with autonomous behaviour and properties from the depicted-by-them actuality, and which can be objects of study in their own right, is here suggested as the unifying theme for the several regional schools of this lucid subject.

Aristotle, *Metaphysics*, 995 a 6–15.
Dewey, John, 1938 Unity of science as a social problem, in Neurath, Carnap, and Morris, *International Encyclopedia of Unified Science*, 29–38.
Gokhman, V.M., B.L. Gurevich, and Yu. G. Saushkin, 1968 Problemy metageografii (Problems of metageography), *Voprosy Geografii*, 77.
– 1969a Problems of metageography, *Sov. Geog.* (Sept.), 355–64.
– 1969b Problems of metageography (abstract), *Sov. Geog.* (Nov.), 563.
Kalinowski, Jerzy, 1971 O prawdzie (On truth), *Kultura* (La Culture), Paris, 1–2 (Jan.–Feb.), 191–8.
Morris, Charles, 1938 Scientific empiricism, in Neurath, Carnap, and Morris, 63–75.
Neurath, O., R. Carnap, and Charles Morris, eds., 1938 *International Encyclopedia of Unified Science* (Chicago). Reprinted in two volumes.
Skolimowski, Henryk, 1967 *Polish Analytical Philosophy* (London).
Tarski, Alfred, 1933 *Pojecie prawdy w jezykach nauk dedukcyjnych* (Warsaw).
– 1935 Pre-print of the 1936 position.
– 1936 Der Wahrheitsbegriff in den formalisierten Sprachen, *Studia Philosophica*, 1, 261–405.
– 1956 *Logic, Semantics, Metamathematics* (Oxford).
– (in press) *Logique, sémantique, métamathématique* (Paris).

P1228
A simple spatial diffusion model
D.L. ANDERSON   *McMaster University, Canada*

A simple model of the spatial diffusion of an innovation among a population located in several areal units is presented. The probabilities of an adoption in each areal unit are obtained.

The only individuals are farmers and change agents. Farmers are either adopters or potential adopters. Farmers are located in the various counties of a state. Each farmer is distinguishable by his county location and by whether he is an adopter or a potential adopter. Change agents are not associated with counties; they are located throughout the state.

Farmers who are potential adopters immediately become adopters when they communicate with either an adopter or a change agent.

Each individual is equally likely to initiate a communication. If the initiator of a communication is a farmer then he must communicate with another farmer located in the same county and he is equally likely to communicate with any other farmer in the county. If the initiator of a communication is a change agent, then he must communicate with a farmer who is a potential adopter and he is equally likely to communicate with any potential adopter in the state. Communications occur one at a time and the expected number of communications initiated by each individual in the state during a time interval of unit length is a constant.

An urn contains white balls and coloured balls, the balls representing individuals in the diffusion model. Each white ball represents a change agent. Each coloured ball represents a farmer and the colour of the ball identifies the county in which the farmer is located. Plain coloured balls represent farmers who are potential adopters and coloured balls with a white dot represent farmers who are adopters.

A sequence of draws of pairs of balls is made from the urn. Each pair of balls drawn represents two individuals participating in a communication. Each pair of balls is drawn from the urn in the following manner. The first ball in the pair is selected at random from all the balls in the urn, representing the randomly selected individual 'initiating' the communication. The procedure for selecting the second ball of the pair depends on whether the first ball is white or coloured.

If the first ball is white, representing a change agent, then the second ball is chosen at random from all the plain coloured balls in the urn, representing the potential adopters in the state. A white dot is placed upon this ball representing a potential adopter located in the county identified by the colour of the ball becoming an adopter upon communication with a change agent. The pair of balls is then returned to the urn.

If the first ball selected is a coloured ball, representing a farmer initiating a communication, then the second ball is selected at random from those remaining balls which are of the same colour as the first ball. This pair of balls represents two farmers in the same county who communicate 'at random.' If exactly one of these balls has a white dot, then the communication is between an adopter and a potential adopter. A white dot is placed on the plain ball representing the potential adopter becoming an adopter as a result of a communication with an adopter and the pair of balls is returned to the urn. If both balls have a white dot or if both balls are plain, then no new adopter results from the communication and the pair of balls is simply returned to the urn. If both balls have a white dot or if both balls are plain, then no new adopter results from the communication and the pair of balls is simply returned to the urn.

Denote by d the number of white balls and denote by b the total number of balls in the urn. Denote by a the number of different colours of (non-white) balls in the urn. Denote by b_j the number of balls of the jth colour, $j = 1, 2, ..., a$.

Denote by $R_j(n,m)$ the probability that in any draw of a pair of balls from the urn, a white ball is selected first *and* a plain ball of the jth colour is selected second, where n is the number of balls of the jth colour which have a white dot and m is the number of balls of all colours which have a white dot. Evidently, the assumptions of the urn scheme imply

(1) $R_j(n,m) = d(b_j - n)/b(b - d - m)$.

Denote by $S_j(n)$ the probability that in any draw of a pair of balls from the urn, *either* the first ball is a plain ball of the jth colour and the second ball is of the jth colour with a white dot, *or* the first ball is of the jth colour with a white dot and the second ball is a plain ball of the jth colour, where n is the number of balls of the jth colour which have a white dot. Evidently the assumptions of the urn scheme imply

(2) $S_j(n) = 2n(b_j - n)/b(b_j - 1)$.

Denote by $Q_j(n,m)$ the probability that any draw of a pair of balls from the urn results in a white dot being placed on a ball of the jth colour, where n is the number of balls of the jth colour which have a white dot and m is the number of balls of all colours which have a white dot. The (mutually exclusive) ways in which a white dot may be placed on a ball of the jth colour at any draw have probabilities given by $R_j(n,m)$ and $S_j(n)$, thus,

(3) $Q_j(n,m) = d(b_j - n)/b(b - d - m)$
$\qquad + 2n(b_j - n)/b(b_j - 1)$
$\quad = (b_j - n)[d(b_j - 1) + 2n(b - d - m)]/$
$\qquad b(b_j - 1)(b - d - m)$.

Evidently, the probability $Q_j(n,m)$ defined for the urn scheme is also the probability that any communication among individuals results in an adoption in the jth county, where n is the number of adopters in the jth county and m is the number of adopters in the state before the communication. This probability may be utilized for obtaining descriptions of model characteristics such as growth either by means of mathematical analysis or simulation.

P1229
A Geographia Generalis before Varenius
M. BUTTNER   *Ruhr-Universität Bochum, West Germany*

It has been generally accepted that general geography originates with Varenius. Recent investigation has, however, shown (Buttner, forthcoming) that it is Keckermann, the early seventeenth-century universal scholar, who may be credited with having written the first

Geographia generalis. He gave our discipline its theoretical foundations.

It is well worth asking why Keckermann's achievement fell into oblivion, and how Varenius came to be considered as the founder of our branch of knowledge. It should be remembered that, in the course of the seventeenth century, Keckermann was, in fact, held in higher reputation than Varenius. Alsted, another world famous scholar of that time, did not rank Varenius, but rather Keckermann, with the greatest geographers of all times (Alsted 1630).

Keckermann still subscribed to the Aristotelian notion of 'natural science,' whereas Varenius began to adopt in his work many reasonings of classical physics, which were a novelty then. This is the reason why Newton preferred Varenius' to Keckermann's work for publication, thus saving it for posterity. Though Varenius took freely from Keckermann and other scholars of his time he forebore to give references – in contrast to the common practice, which was by this time quite usual – and tried to create the impression that his *Geographia generalis* was exclusively based on his own personal erudition. It was only by intense investigation that Varenius' indebtedness to his sources could be traced out. The truth is that the basic structure was taken from Keckermann, whose name would have come down to us with the work of Varenius, if Varenius had given references.

The aim of my paper is to discuss two issues: (1) Which contemporary ideas may have inspired Keckermann to write a *Geographia generalis*? (2) According to which principles is his work organized?

It will not be claimed that Keckermann is a man *sui generis*. The growth of general geography takes place in the context of the ideological current of the early seventeenth century, which may be termed as pre-enlightenment. At the end of this epoch, for which the interrelation of geography and theology is characteristic, Keckermann takes an arresting position with his *Geography*, the first work of its kind that starts from a theologically neutral point of view and applies specifically geographic criteria. Before him, theological aspects used to prevail in geography. Out of the number of interrelations between these two disciplines, the three most important will be treated.

The first relation is directed from geography into theology. The developing Protestant doctrine of providence is displayed by adapting geographical terms. In this connection, it can be clearly worked out by a juxtaposition of Zwingli Melanchthon (1549) how, to back up one theology, only certain geographical notions are taken up as fitting. Zwingli explains his doctrine by means of geographical phenomena (rain, etc.) which make apparent the continuity from the past creation, by way of the flood, towards the present reign of God over nature. His geographical illustrations of providence doctrine aim at making evident the unity of the past and the present. Creation (the creation of water, for instance), flood, and present world government (note that it rains when God makes it rain!) are merely performances of the one plan of God that is prior to creation itself. Plan and performance form two sides of one and the same thing, that is, of providence.

Melanchthon, on the other hand, has practically no interest in the past. What matters to him is the present reign of God over the world, which is in turn illustrated by pointing to geographical processes. However, he picks up from the storehouse of geographical facts only those examples which make evident the present operating of nature under God's guidance. (He is, e.g., not concerned with the creation, or the rain at the flood, but with rain falling here and now, sent by God for the benefit of men, animals, and the vegetable world.)

Melanchthon's disinterest with regard to the past may be accounted for by pointing to the special purpose of Lutheran theology: for Luther it was only of secondary importance to ascertain what God had done or planned in former times. His attention was, above all, given to the merciful God. To find this merciful God here and now was his main endeavour.

The second relation is directed in reverse from theology into geography. The effect of providence doctrine linked up with certain geographical notions is that Reformed geographers always start from the past of creation, flood, etc. in order to account for the present state. The theologian and geographer, Sebastian Münster, is a case in point (Münster 1544). (Following this argument, it will become evident that Calvinistic countries provided from the start a much better basis for the development of geography than the Lutheran regions of Europe.)

Lutheran geographers (e.g., Melanchthon, Peucer, Neander, etc.) had only an eye for the present state. Neither with the aid of the Bible, nor by means of speculation (let alone on the basis of empirical research) did they try to recognize what things were like in the past.

The third relation leads to the liberation of geography from the grip of theology. It is Keckermann (a Reformist from Heidelberg), who has, by means of his new doctrine concerning sin and providence, created the theological basis for such a neutralization of our discipline, while nevertheless sticking to the theological relevance of geography. He states that our subjective relationship to God is upset by sin. We are no longer able to recognize God; however, in the objective sphere of the consideration of nature true understanding is possible. Here sin can exert no negative effect. Indeed, the more we explore nature in order to master it in the end, the more we regain the previous (primeval) image of God, for it consisted essentially in the complete knowledge and domination of nature.

Together with this doctrine of sin, Keckermann displayed his doctrine of providence, which provides a theological justification for the neutralization of geography from another angle. He states that in the teachings about providence we are no longer concerned with the explanation of God's operation in *nature*; rather providence merely occurs in the sphere of *man*. From this follows that the object of the geographer is no longer providence, that is, a theologically orientated geography; but he can pursue aims independently from theology.

With this, the emancipation of geography from theology is achieved. From now on, it is indisputably possible for geography to work along its own lines, apart from the influence of theological criteria. The geographer is no longer bound to show up God's providence in nature, but he can turn to geographical data without preoccupations. It remains to be asked what the first theologically neutral, 'un-preoccupied' *Geography* looks like.

Keckermann (1610) proceeds according to the analytical or distinctive methods respectively, which were in use in the seventeenth century and in the development of which he shared as one of the leading philosophers of his time. They were adopted by his successors up to Varenius.

This means that single geographical data are not left apart in an unconnected manner, but are associated with each other and hierarchized. Thus a certain purely geographic systematology is achieved for the first time (a step by which our discipline begins to come into its own). While the kind of 'systematology' that Münster or Mercator (1595) practised consisted in dealing with geographical facts by following the procedure of Genesis (we could nowadays treat this as a case of 'non-specific systematology'), Keckermann is conspicuous for his specific orientation (as also in his *Physics* and *Astronomie*). In his person geography begins to be established as an independent discipline after centuries of theological orientation.

Alsted, J.H., 1630 *Encyclopaedia septem tomis distincta* (Herborn).
Buttner, M., forthcoming *Habilitations-Schrift* (Wiesbaden).
Keckermann, B., 1610 *Systema geographicum* (Hanoviae). Also in *Systema Systematum* (Hanoviae 1613).
Melanchthon, Ph., 1549 *Initia doctrinae*.
Mercator, G., 1595 *Atlas sive Cosmographicae Meditationes* (Duisburg).
Münster, S., 1544 *Cosmographia* (Basel).
Neander, M., 1583 *Orbis terrae partium explication* (Islebii).
Peucer, C., 1556 *De Dimensione Terrae* (Wittenberg).
Zwingli, H., 1530 *De Providentia*.

P1230

Importance de la théorie géographique dans l'orientation de la recherche
MIHAI IANCO and OCTAVIAN BANCILA *Université de Bucarest, Roumanie*

En partant de l'importance accrue de l'élément rationnel et du rôle des recherches théoriques dans la science contemporaine, la communication souligne que la géographie physique est engagée dans une voie qui mène à la réalisation de nouvelles synthèses théoriques – la tâche de description de la terre étant, en grande partie, accomplie – synthèses destinées à établir l'essence, les causes et les lois de l'apparition, de l'évolution et de la diffusion dans l'espace des unités naturelles physico-géographiques. Ces synthèses théoriques tendent à unifier dans une ample perspective, à une échelle de plus en plus vaste, les données de l'observation, de l'expérience et de l'analyse.

La mise en valeur des recherches théoriques dues à des précurseurs tels que Al. von Humboldt, Vidal de la Blache, Emm. de Martonne, S. Mehedinți, G. Vâlsan, etc., est susceptible de contribuer avec succès à poser les fondements des catégories et des principes de la théorie de la géographie physique, ainsi qu'à élaborer des catégories et des principes nouveaux. Entre autres, on remarque la contribution de S. Mehedinți à l'identification et à la formulation de la définition des catégories géographiques, ainsi qu'à l'élucidation du spécifisme et du mécanisme d'action de la causalité physico-géographique, laquelle a trouvé son expression dans la loi de la subordination causale des enveloppes planétaires, loi à considérer dans la perspective des conquêtes de la géomorphologie depuis Davis jusqu'à nos jours.

La révélation du contenu de la théorie de la géographie physique et sa reconnaissance en tant que science indépendante sont d'autant plus actuelles que la diversification des disciplines physico-géographiques devient plus intense et favorise ainsi des réflexions sceptiques concernant une prétendue crise de la géographie.

L'importance de la théorie de la géographie physique, en ce qui concerne les directives à donner aux recherches effectuées dans le domaine des différentes disciplines géographiques spéciales, s'accroit d'autant plus que le nombre de celles-ci augmente. La théorie de la géographie physique, dont on sent de plus en plus l'amplification et la profondeur car elle contribue pleinement à la formation de la pensée géographique de tout chercheur, donne à ce dernier la nette conscience de l'originalité propre à son travail et accroit l'efficacité de son activité créatrice. Elle l'aide à aborder, dans une perspective spécifiquement géographique, l'objet soumis à la recherche, à la lumière de l'unité du tout et de la partie, du général et du particulier, du logique et de l'historique, de l'abstrait et du concret.

P1231
Growth of geography in India
S.C. CHATTERJI   *Kalimpong, India*

Recent excavations made in southern Baluchistan, at Mohenjodaro in Sindh, and Harappa in the Punjab show that in approximately 4000 BC a well-developed civilization existed in these regions. Archaeological materials found in these places indicate that during this prehistoric period there were trade relations between India and the western regions of Babylonia and Egypt. Recently discovered literature, found in the Mesopotamian cities, shows that there was a close commercial relationship. A cylindrical seal with engraved elephants on it, recently found in Mesopotamia, led some excavators to believe that at Ur and some other places in Babylonia Indian artisans had worked.

There is no consensus of opinion about the entry of the Aryans into India. It may, however, be safely assumed that they had established themselves in the western parts round about 2000 BC. There might have been a considerable gap between the disappearance of the Sind and the Indus valley civilization and the emergence of the Aryans as the chief inhabitants in northwestern India.

Various details about the life of the people in the early days are given in their sacred Vedic texts, which are the only source of the early geography of the Indian subcontinent. In those periods, this included territories far beyond the present limits of the region.

The literature of ancient India being chiefly of a religious character, her ancient geography, like her other early sciences, was chiefly dependent on religion. Consequently, all physical phenomena had a religious background, mountain peaks, crags, rivers, hoary trees, etc., being regarded as sacred. All the

Vedas and Dharmashastras discuss physical geography (atmosphere, land, and water) from a religious point of view.

The epics – the Ramayana and the Mahabharata – give valuable information about the natural and cultural resources of various regions, settlements, etc. Unlike the Vedas, the epics contained geographical knowledge based on observation and experience rather than on speculations and presupposed notions, knowledge which was used extensively for the composition of the later literature, particularly the Purans.

Like the epics, the Purans followed the same system of philosophy and religion. Although they accepted the general cosmology of the epics, they expanded and systematized the geographical notions of the earliest works. Almost all the Purans deal with the primary creation and subsequent secondary creation, which, as in Dante's cosmology, contains regions above or below the earth's surface called Bhuvana Kosa ('treasure of terrestrial mansions'). All the Purans, especially the Vayu Puran, deal with astronomical notions like the movement of the planetary system and the computation of time as well as the geography of the world with special reference to India.

Furthermore, the vast Mahatmya literature, an offshoot of the Purans, gives the origin, location, and site of places of pilgrimage which subsequently gave rise to towns and cities like Ayodhya and Varanashi, thus becoming a key to the urban geography of ancient cities as well as a gazetteer of old routes. Besides the sacred literature just mentioned, there are other miscellaneous works like Panini's grammar and Patanjali's Mahabhasya as well as some medical works which deal with geographical materials pertaining to India and her regions. A few centuries later, the Brihat Samita of Varahamihir and the Siddhanta of Bhaskaracharya brought about a reorientation of the cosmological and cosmographical ideas of the Puranic period; and Alberuni's treatise on India summarized the geographical ideas of the Indians of his time.

It may be mentioned here that geographical science as such began to take shape at a much later period everywhere in the world; and it was the advances in astronomy, mathematics, commerce, and pilgrimage that gave geography its definite shape. In India various names were given to the science from time to time, the word 'bhugola' for geography being of most recent origin.

In the course of time, men began to build up a general theory which embraced all phenomena regarding the nature of the earth: the shape and size of the terrestrial globe and the influence exerted by heavenly bodies in determining the geographical conditions on its surface.

During the Vedic and pre-Vedic period it was thought that a number of gods were responsible for the process of creation. They were great artists who applied their skill in the construction and completion of the universe. They used various building materials to shape it and complete its structure. In the post-Vedic period, various cosmogenic theories were formulated, depending on three main principles of existence – creation, preservation, and dissolution – relating to three main deities – Brahma, Vishnu, and Maheswara.

In referring to the origin of the earth, it is said that Prajapati, the great Creator, made water first and then the hirenyagarbha (golden egg) which contained an internal fire (it may be suggested that the golden egg gives the idea of a hot incandescent nebula, which is the starting point of modern cosmological theories). The blasting of this celestial egg-nucleus produced a nebulous condition resulting in the creation of our earth and other heavenly bodies. This Puranic cosmogeny thus suggests the presence of hot gaseous particles and the rotational force. On the dissipation of heat by the activity of particles, condensation took place, and the earth and the planets were created (this idea is very similar to Kant's hypothesis and later to Laplace's celebrated nebular theory).

Along with the notion of creation, came the idea of the dissolution of the world after a certain period (kalpa). Although the fundamental idea of destruction was the same among most of the civilized people of antiquity, some held that fire would destroy our earth, while others thought of recurrent inundation by water. Naturally, therefore, the concept of cosmic cycle entered into the minds of philosophers, and the cosmological theories about the main aspects – the creation, the preservation, and the dissolution – were formulated.

Obviously, therefore, the Puranic concept of four 'yuga' – Satya, Treta, Dwapar, and

Kali – had at least some similarity to the present theory of geological revolution, but it was based more on intuition and deductions than on scientific reasoning following upon experiments and observations.

According to Indian mythology Brahma created the earth and lives on it. The incarnations of Vishnu deal with the evolution of life in its various stages, namely matsya (fish), kurma (tortoise), varaha (boar), narasinha (half man and half animal), vamana (sub-man), rama-krishna (fully developed man). Another one or two more like kalki-avator and Buddha might have been subsequently added. This view of the evolution of life from fishes and reptiles to man is in consonance with modern views on the subject.

During the rational (Upanishadic) period various theories have been advanced about how life could achieve its present form. The Sankhya Patanjala system, accepting the duality of prakriti (matter) and purush (spirit), gives a scientific account of the process of cosmic evolution, whereas Nyaya Vaiseshika lays down a methodology of science insinuating that life has developed as a result of physico-chemical action. Materialists such as Charvaka have, however, argued that life is the outcome of a chemical combination of inorganic matter in organic forms; but no satisfactory explanation has been given as to how it is possible for a living organism to evolve from biologically dead substances. On the other extreme, the Vedanta regards 'maya' (illusion) as the material cause of the universe. Maya is both real and unreal, while Brahma (self) from whom life emanates and passes on from infinity to infinity, is absolute Reality.

The statement that the sun never rises or sets shows that the Vedic Rishis knew that the earth was spherical in shape, and that they thought that it was the sun and not the earth that moves. It was supposed that the earth was a globe hanging in the universe, and all the heavenly bodies were revolving round it. The ancients therefore referred only to the apparent motion of the celestial bodies with reference to the earth. Various estimates were made regarding the size of the earth, and it was thought that half the globe lay to the north of the mountain Meru. The Matsya Puran gives the size of the earth as 50 crore yojanas, and adds that the universe is divided into three portions – the earth, the atmosphere, and heaven. The atmosphere is again divided into three regions – the upper, the middle, and the lower – and was the realm of the gods Indra, Maruts, and Rudra. Maruts was a rain-bearing wind corresponding to the monsoon; and Rudra the storm and hurricane. Maruts were further divided into eastern Maruts and western Maruts, corresponding to NW and SE monsoons in northern India. They had a clear notion about rainfall, which they thought was due to the evaporation of water by heat and the condensation of the vapour. The clouds were formed by water vapour and destroyed by thunder and lightning. There is thus no waste of water in the universe, as it is in constant circulation. The Vishnu Purana says that the water remains the same at all times and never increases or diminishes; but the Matahya Purana gives a better estimate about the rise and fall of tides and connects them with the phases of the moon.

Both the epics and the Purans tell us of seven 'dwipas' (continents) and seven 'samudras' (oceans), The seven continents are Jambu, Saka, and Puskara, and they are surrounded by seven seas – the sea of salt water (lavana), of sugarcane juice (iksu), of wine (sura), of clarified butter (sarpi), of curd (dadhi), of milk (dugdha), and of fresh water (jalka). Jambu-dwipa is the centre of all these seas; and in the centre of this continental island stands the golden mountain, Meru.

The Mahabharat introduces the idea of concentric continents and oceans, stating that the central continent is Jambu, the centre of which had a table land, called Ilvarita, whose skeleton was Meru. From this mountain came the Ganga, which flowed to the south. To the east of Ilavrita is Bhadravarsa and to the west Ketumala. Towards the south are three ranges of mountains – Himachal, Himakuta, and Kimpurusha. To the north is the Nila (blue) mountain, farther north is the Sweta (white) range, and beyond it is the third ridge known as Sringavana. To the south of these mountains, which extend from one ocean to the other, lies Bharata which was probably confined entirely to India, especially in the beginning.

As the Aryans occupied the country south of the Himalayas, they designated it as Arya-varta, that is, the land of the Aryans; and it was so called during the days of Baudhayana and Manu. The Persians, who conquered a portion of Western India in the time of Darius, called it Indu or Hindu, as only the Indus valley was known to them. It was only about the fourth century BC that Katyayana and Megasthenes described the country south of the Vindhya mountain. The conquests of southern India by Ashoka and his successors gave Bharata a wider significance; and, subsequently, the spiritual conquests of central and southeastern Asia by Buddhist missionaries further extended the term Bharata from the river, Oxus, in Turkistan to the southeastern lands up to the islands of Java and Bali.

To sum up, the epics and the Purans give not only natural phenomena and the mode of living of the people of India, but also a fair amount of geographical knowledge of the whole world. No doubt other civilized countries of antiquity also gave a fair description of the configuration of the world; but India's contribution, as given in her sacred literature, was not less. Since the Indians did not care to write down their history, their conquerors are likely to look down on their achievements. It is therefore natural that historians have lost sight of many of the Indian works. It is only after some recent research, made by careful studies of ancient religious literature, that certain facts have been rediscovered and made known to the present world through other channels.

Basu, B.L. *The Aitreya Brahmana* (Allahabad).
Griffith, R.T.H. *The Hymns of the Rig Veda* (Benaras).
Kern, H. *Brihat Samita* (Calcutta).
Kingawadekar, R. *Harivansa* (Poona).
Pargiter, F.F. *Markandeya Purana* (Calcutta).
– *Ramayana* (Bombay).
Welson, H.H. *Vishnu Puran* (London).
– *Mahabharata* (Calcutta).

P1232
Modèles mathématiques en géographie
V.S. PREOBRAGENSKY   *L'Académie des Sciences, URSS*
Y.R. ARKHIPOV, N.I. BLAJKO, A.V. STOUPICHIN, and A.M. TROFIMOV   *Université de Kazan, URSS*

L'utilisation de modèles mathématiques en géographie est étroitement liée à la notion de 'systèmes' (géosystèmes). On s'imaginait que l'objet des recherches de la plupart des sciences géographiques (surtout la géographie économique, la géographie générale physique et la biogéographie) représente une certaine unité organisée et complexe qui se compose de parties variées, mais interdépendantes. Ce point de vue s'oppose à celui de l'unité simple (la substance physique), notion la plus courante dans les sciences physiques.

Presque toutes les sciences géographiques utilisent pour l'étude des ensembles, ou éléments de géosystèmes, les méthodes 'par branches' (monosystèmes) ainsi que 'territoriale' (polysystèmes). Chaque élément peut être considéré comme un sous-système, ou relativement indépendant du système, et est l'objet d'une étude spéciale. D'où la nécessité de deux types de recherches pour chaque géosystème: premièrement comme élément du système plus général (analyse des liens du système donné au super-système – aspect extérieur); deuxièmement comme un système autonome (étude de sa structure interne – aspect intérieur). L'image de n'importe quel géosystème ne peut être plus ou moins complète qu'en tenant compte des quatre aspects (par branches, territorial, extérieur, et intérieur).

L'existence des notions de chronologie territoriale (surtout nettement exprimée dans la théorie des divisions géographiques), de la stabilité des géosystèmes envers les influences aléatoires qui contrarient le milieu, d'éléments d'autorégulation des sites naturels, de l'aménagement de complexes industriels territoriaux et de la transformation de la nature en tant que processus dirigé prouve que la géographie considère ces objets non seulement comme des 'systèmes en général' mais comme des formations dynamiques complexes, correspondant précisément à de telles classes de systèmes complexes, dynamiques, et dirigés qui sont étudiées par la théorie générale des systèmes et la cybernétique.

En particulier, les modèles mathématiques

jouent un rôle de plus en plus important dans l'étude des formations complexes, celui d'un antidote contre l'élémentarité et la réductibilité qui sont des conceptions pour lesquelles toute l'activité des chercheurs est subordonnée à la division de l'unité en petits éléments. Le modèle se présente toujours comme un certain programme de synthèse, c'est-à-dire comme une réunion d'éléments extraits au début de la recherche.

Les modèles sont étroitement liés au langage scientifique. A son début, la géographie utilisait, pour la description de l'objet d'une étude, l'expression littérale et cartographique. Le développement de conceptions dynamiques a, peu à peu, réduit les possibilités du langage cartographique. Le langage par schémas-blocs est entré dans les études géographiques, en révélant non seulement la position des objets dans un espace bidimensionnel, mais aussi leurs rapports mutuels, et, ce qui est surtout important, leur situation dans le temps.

La pénétration active de la conception de système, la nécessité de poser clairement le problème et de sa stricte résolution, a conduit à l'élaboration d'un langage de symboles logiques mathématiques en géographie. Ce langage permet non seulement de montrer les rapports qui existent entre les éléments du problème, mais aussi d'effectuer (d'après les lois mathématiques) les transformations de ces rapports, sans lesquelles la construction de la théorie par pronostique est impossible. Au cours de ce processus, les moyens linguistiques, créés en géographie et empruntés aux autres sciences, deviennent un des éléments du langage géographique uni.

Par l'application des modèles au géosystème ou à certains de ses aspects, nous créons tout d'abord son équivalent. Les modèles complexes ne conviennent pas toujours pour résoudre les problèmes géographiques. Mais, simultanément, l'utilisation de modèles trop simplifiés peut conduire à des résultats erronés. Il est important que le modèle décrive les qualités principales de l'objet, qu'il puisse servir de programme à l'étude ainsi que de régulation au développement du géosystème lui-même.

Les modèles mathématiques géographiques représentent un processus de recherche en plusieurs étapes :

1) *L'étape préparatoire* au cours de laquelle est déterminée la nécessité d'appliquer la méthode des modèles et où est formulé le but des recherches.

2) *La systématisation de l'information existante dans le système* au cours de laquelle on découvre la composition de ses éléments, le caractère et la direction de leurs rapports.

3) *Le choix de la méthode de construction du modèle.* Le modèle peut être construit de deux façons : premièrement, en introduisant le plus possible d'éléments et en divisant ensuite le processus de l'étude en une série de buts particuliers qui seront résolus en ordre croissant, du problème particulier au problème plus général. Deuxièmement, et au contraire, en essayant de trouver les corrélations qui existent entre les grandes parties du modèle et en ne résolvant les problèmes particuliers qu'en cas de nécessité. En pratique la première méthode est plus fréquemment utilisée. On commence par diviser l'objectif général en objectifs particuliers. Envisageons d'une manière plus détaillée la résolution d'une tâche particulière d'un modèle de géosystème. La construction du modèle d'une telle tâche se compose toujours d'une série d'actions successives. L'objectif particulier détermine le choix du type de modèle (si toutefois existent des modèles mathématiques utilisables pour sa résolution), ou la modification d'un type de modèle déjà existant, ou la construction d'un nouveau modèle. Ainsi, l'objectif étant défini, l'information initiale et le type de modèle choisi prédéterminent la suite des opérations de recherche, en particulier la collecte d'informations complémentaires. Bien sûr, l'information doit correspondre aux possibilités et aux restrictions de l'appareil mathématique choisi lors de la construction du modèle. Puis, suivre la résolution du problème et l'analyse des résultats (examen de la stabilité de la solution, influence d'une modification des restrictions sur les résultats obtenus, etc.).

4) *Construction du modèle mathématique géographique général par la synthèse des modèles particuliers.* Le but des recherches défini lors de l'étape préliminaire est obtenu par la résolution de l'ensemble des problèmes, dont chacun correspond à l'étude de l'un des aspects du géosystème. Les résolutions particulières se réunissent en un modèle mathématique géographique général. Par la construction de ce modèle mathématique on résoud les problèmes qui garantissent la réunion des blocs.

5) L'analyse et l'évaluation des modèles dans leur ensemble s'effectuent: *a*) d'après le problème posé; *b*) d'après l'information initiale. Les résultats obtenus à l'aide du modèle peuvent être confrontés à ceux de l'activité du géosystème réel. Si la différence des caractéristiques ne dépasse pas une quantité donnée, on peut considérer le modèle comme satisfaisant au problème posé. Pour certains problèmes, par exemple la planification du développement d'un complexe industriel territorial, on utilise les évaluations optimales obtenues par le résultat de la résolution du double problème de la programmation linéaire.

Pour la construction du modèle par la méthode décrite, on utilise le plus souvent les statistiques, la géométrie élémentaire et l'algèbre linéaire, et les éléments de l'analyse mathématique.

La deuxième méthode consiste en une transposition progressive du modèle général abstrait du géosystème, en ne prenant en considération que ses caractéristiques de base principales, en un modèle plus concret qui dévoile les régularités principales de l'objet étudié. Cela est obtenu par l'introduction dans le modèle de facteurs particuliers et de statistiques plus simples qui dépendent des éléments composant le géosystème. Le modèle accumule progressivement les détails, et les opérations de recherche se ramifient. En construisant le modèle par la deuxième méthode, c'est-à-dire par la méthode de la transposition de l'abstrait au concret, on emploie les méthodes de l'analyse mathématique, de l'algèbre linéaire, de la programmation linéaire, non-linéaire, et dynamique, les méthodes stochastiques, la théorie des jeux, et autres.

Les deux méthodes sont pareillement légitimes. La première se fonde sur une matière empirique concrète et en même temps aboutit à la résolution de problèmes constructifs. La seconde joue un grand rôle dans le développement de la géographie théorique

P1233
Regional forecasting with an application to school-leaving patterns in the United Kingdom
ANDREW D. CLIFF and J. KEITH ORD   *University of Bristol, UK*

This paper continues earlier work by the authors (Cliff and Ord 1971) into the problems of regional forecasting. Three main aims of regional forecasting can be distinguished: (*a*) to account for regional variations in variate values at a given point in time (cross-sectional analysis); (*b*) to account for variations over time in variate values for a given region (longitudinal analysis); (*c*) to look for space/time relationships (interactions) between regions. For (*a*), we seek to isolate regional variations, and then attempt to analyse the underlying mechanisms which give rise to these differences. For example, the proportion of students staying on at school beyond the minimum leaving age may depend upon the socio-economic background of the region, employment prospects, and so on. Aim (*b*) is perhaps more genuinely described as forecasting. By modelling changes over time, we can try to forecast future shifts in the system. The third aim is potentially the most interesting and also the most difficult to achieve. However, even indications of the presence of such interactions may be a useful start.

In this paper, we outline a methodology for regional forecasting, omitting the mathematical details, and then apply the method to the analysis of school-leaving data for regions of the United Kingdom.

Suppose we have available, for the random variable Y, observations over N regions for each of T time periods. Then a standard analysis of variance could be carried out (King 1969, chap. 4) using the model

(1) $\quad y_{it} = \mu + \alpha_i + \beta_t + \gamma_{it} + \varepsilon_{it}$,
$\qquad\qquad i = 1, ..., N; \quad t = 1, ..., T$,

where μ = over-all mean; α_i = effect of ith region, $\alpha_1 + \alpha_2 + ... + \alpha_N = 0$; β_t = effect of tth time period, $\beta_1 + \beta_2 + ... + \beta_T = 0$; and γ_{it} is the interaction time-space effect, confounded with the random disturbance, ε_{it}, when only one observation is available for each (i, t) pair. The ε_{it} are assumed to be independently normally distributed with zero means and unknown variances, σ^2.

The analysis yields the estimators,

(2) $\quad \hat{\alpha}_i = \bar{y}_{i0} - \bar{y}_{00}, \quad \hat{\beta}_t = \bar{y}_{0t} - \bar{y}_{00}$,
$\qquad \hat{\gamma}_{it} = y_{it} - \bar{y}_{i0} - \bar{y}_{0t} + \bar{y}_{00}$,

where
$$T\bar{y}_{i0} = \sum_{t=1}^{T} y_{it},$$
$$N\bar{y}_{0t} = \sum_{i=1}^{N} y_{it},$$
$$NT\bar{y}_{00} = \sum_{i=1}^{N}\sum_{t=1}^{T} y_{it}.$$

The sets $\{\hat{\alpha}_i\}$, $\{\hat{\beta}_t\}$, and $\{\hat{\gamma}_{it}\}$ are uncorrelated (and hence independent) *between* sets, but not *within* sets. However, the following theorem may be shown to hold:

THEOREM *If the $\{\hat{\alpha}_i\}$, $\{\hat{\beta}_t\}$, and $\{\hat{\gamma}_{it}\}$, generated by (2) above, are examined separately using regression equations such as*
$$\hat{\alpha}_i = a(x_i - \bar{x}_0) + \varepsilon_i(1),$$
(3) $\quad \hat{\beta}_t = b(z_t - \bar{z}_0) + \varepsilon_t(2),$
$$\hat{\gamma}_{it} = c(u_{it} - \bar{u}_{i0} - \bar{u}_{0t} + \bar{u}_{00}) + \varepsilon_{it}(3),$$

then best linear unbiased estimators for a, b and c are given by the application of ordinary least squares to these equations directly.

Proof Follows by use of Cochran's Theorem and Aitken's Generalised Least Squares.

NOTES

1. The $\{\varepsilon(j)\}$, $j = 1, 2, 3$, denote error terms with a correlation structure given by the analysis of variance. Thus, the $\{\varepsilon_i(1)\}$, for example, are not mutually uncorrelated.
2. The x, z, and u variables are non-stochastic, with means defined as for y above.
3. The theorem extends directly to any number of explanatory variables, given sufficient degrees of freedom.
4. Any choice of explanatory variables is possible. However, if the variable depends solely on time, it will appear only in the $\hat{\beta}_t$ equation (assuming its regression parameter is constant). Polynomials in time are an obvious example. Similarly, region-dependent, time-independent, variables will only enter the $\hat{\alpha}_i$ equation, such as the polynomials used in trend-surface analysis. If the variables depend upon both space and time *or* if the variable depends on space [time] and the parameter on time [space], then the $\hat{\gamma}_{it}$ equation is relevant: e.g., $\hat{\gamma}_{it} = c_i(z_t - \bar{z}_0) + \varepsilon_{it}(3)$.

The data comprise the percentage of students who elected to stay on at school for at least one year beyond the statutory minimum leaving age of 15 years. To calibrate the model, data were used for the 5 years 1964–8 for 10 standard regions of the UK (Northern Ireland was excluded for several reasons). Since the random variable has a natural range, 0–100, the transformation
$$y_{it}^{\text{trans}} = \log_e [y_{it}^{\text{orig}}/(100 - y_{it}^{\text{orig}})]$$
was first carried out. The analysis of variance yielded the results given in Table 1. These effects were then analysed separately.

TABLE 1

Source of variation	Degrees of freedom	% variation*	F ratio
Between regions	9	60.2	112
Between times	4	37.5	157
Interaction/ residual	36	2.2	

*% variation = sum of squares expressed as % of total sum of squares.

(a) *Regional variations.* Average levels of income per head for adult males in each region were used as an approximate measure of socio-economic status. The ratio average income of region to national average income was used, as these quantities were very stable for the period under study. The correlation squared was only $r^2 = 0.362$, not significant at the 5 per cent level. However, a test for spatial autocorrelation among the $\hat{\alpha}_i$ yielded $I \simeq 0.1$, which is not significant, suggesting that the pattern depends on socio-economic, rather than spatial, variables. (See Cliff and Ord (1972), equation (1.44), for a definition of I. Binary weights were used with $w_{ij} = 1$ if the ith and jth standard regions had a common boundary and $w_{ij} = 0$ otherwise.)

(b) *Temporal variations.* The linear time trend, $\hat{\beta}_t = b(t - \bar{t})$, yielded $r^2 = 0.9988$, highly significant even with only three degrees of freedom for the residual. This trend is sustainable for the transformed data and was used for forecasting.

(c) *Interactions.* For each region i, the model, $\hat{\gamma}_{it} = c_i(t - \bar{t})$, was fitted to look for different trends between regions. A test of the hypothesis, H_0: all $c_i = 0$ against H_1: not all $c_i = 0$, rejected H_0 at the 1 per cent level. For details of the test see Cliff and Ord (1971, 52–8). A model of the form (3) using average incomes (as in (a)) gave no explanatory power ($r^2 = 0.016$).

On the basis of this analysis, the percentage of students staying on past the statutory school-leaving age in each region for 1969 and 1970 was predicted from
(4) $\quad y_{it}^{\text{trans}} = \hat{\alpha}_i + \hat{\hat{\beta}}_t + \hat{\hat{\gamma}}_{it},$
where $\hat{\hat{\beta}}_t = \hat{b}(t - \bar{t})$ and $\hat{\hat{\gamma}}_{it} = \hat{c}_i(t - \bar{t})$. The results were as given in Table 2.

Generally, the forecasts are optimistic,

TABLE 2

		North	Yorks & Humb.	East Mid.	East Anglia	South-east	South-west	Wales	West Mid.	North-west	Scotland
1969	Predicted	45.2	48.2	46.0	49.1	62.9	59.4	49.5	49.8	47.6	49.1
	Actual	44.7	47.5	45.7	47.0	62.5	58.6	49.9	49.7	48.1	50.0
1970	Predicted	49.9	52.1	49.5	53.8	65.9	63.1	51.8	53.2	50.7	51.5
	Actual	46.9	48.6	47.5	48.7	63.8	60.4	51.8	51.3	49.4	52.0

pointing to a slow-down in the rate of increase of those staying on. Clearly, more work is needed to explain this discrepancy. This is underlined by the fact that the present simple model reduces back to $y_{it}^{trans} = \alpha_i + d_i(t - \bar{t})$, $d_i = b + c_i$. Nevertheless, the analysis-of-components approach gives insight and provides the basis for further model-building.

Cliff, A.D., & J.K. Ord, 1971 A regression approach to univariate spatial forecasting, in *Regional Forecasting*, ed. by M.D.I. Chisholm et al. (London), 47–70.
– 1972 *Spatial Autocorrelation* (London).
King, L.J., 1969 *Statistical Analysis in Geography* (Englewood Cliffs, NJ).

P1234
Toward some dynamic concepts of spatial decision-making
H.P.M. HOMENUCK   *York University, Canada*

The spatial decision-making interest of geographers has been focused on industrial location and has been primarily concerned with point locations and movement along lines. The spatial decision-making concepts have been built upon the traditional idealizations of perfectly rational man, which resulted in the location models of Weber, Lösch, Hoover, and others. The assumptions underlying this body of theory – economic maximization, perfectly rational decision-making, and decision-making by a single individual – have been seriously brought into question over the past decade or so.

This reassessment of decision-making concepts has been sparked by immense changes in the economic structure of North America and, indeed, the world. The changes include the scale and complexity of organizations (approximately half of the economic activity in the United States is controlled by about 200 multilevel corporations), the rate of technological change, the impact of trade unionism, the influx of professionals into large-scale organizations, and shifts in the value system of the world community. Under such circumstances, the economic objectivity of the decision-making process in the large corporations can be easily exaggerated (Galbraith 1967). In these corporations, the executive behaves less as a profit maximizer than the classical theory assumes. It has thus become clear that the concepts of spatial decision-making must recognize and reflect the corporate complexity that dominates the economic scene.

Since actual location patterns and spatial behaviour are physical manifestations of decision-making, research has become concerned with the actual processes of decision-making. Some geographic research has indicated that social, psychological, and personal factors, though often unmeasurable, may influence decision-making (Pred 1967; Hamilton 1971). In fact, these factors may be the most important in explaining the spatial behaviour of some organizations. Considerable attention is now being given in geography to such factors and their spatial expressions (Smith 1971, 508). However, the greatest advances to date have come from some of the behavioural sciences which have uncovered many facts that are, explicitly and implicitiy, of interest to geographers.

Sociology and psychology have shaken the very roots of the classical theory in the understanding of firms and organizations. Considerable evidence exists to show that among other things man does not react solely on the basis of economic gain; man has a hierarchy of needs which change over time; interpersonal relationships are important and cannot be understood through conventional theory; these relationships can affect organizational effectiveness; communication gets distorted, particularly through a hierarchy (Bennis 1966, 185).

Various concepts have been outlined in regard to how organizations formulate goals; how goals are changed; how organizations are controlled; how decisions are reached (March and Simon 1958; Etzioni 1964). Considerable light has been shed on the interpersonal relationships of decision-makers and the use of the information and the communication network within complex organizations (Leavitt and Pondy 1964; Miller and Starr 1967).

The modern organization is much like a political organization established to govern a territory (Eells 1962). Both are political in two ways: first, in the sense that a group of individuals exists to promote a particular policy or to achieve certain goals and, secondly, in the internal politics of taking power, the control of power, or the compromise of power within the organizations. Power in any organization seems to play a significant role in how the organization operates and, in fact, in what spatial pattern will result (Homenuck 1969).

Every modern organization has a certain set of goals towards which it is working, and each organization has a set of parameters and rules which govern its functioning. These may include written rules such as job descriptions, constitutions, and the like or various unwritten rules with regard to status and certain rituals and roles (Packard 1962; Jay 1968). It is vital to understand the modern organization as such a political entity because many of its decisions, including spatial ones, are in fact the result of a process of assessing data within the context of certain goals, values, and parameters and sorting them out through a political process to reach a decision.

In addition to factors such as the above, there are many variables of a business management nature which can affect the efficiency and profitability of any spatial pattern. These include various marketing techniques, ownership of patents and copyrights, the total of investment decisions for the organization, product line expansion or termination. Such decisions can affect the profit picture of a firm or corporation over any time span. Thus the fortunes of an organization might fluctuate as might the relative merits of its spatial pattern.

Several case studies of how decisions are actually made within organizations have illustrated how organizations change over time (Chandler 1966; Homenuck 1969; Zald 1970). These studies shed light on the adaptiveness of organizations – how the goals change, how the internal power structure changes, and how the spatial pattern changes. Many organizations have learning processes continuously operating within them to identify fresh goals or to redefine existing ones as they grow and change. This can be seen in the changes and expansion in the product line of many firms. This ability of an organization to alter and change itself to meet the changing needs and desires of society is important to the understanding of the spatial behaviour of any organization.

These case studies, in highlighting the adaptability of a wide range of organizations, also illustrate the importance of the rapidly changing organizational environment. The number of variables and the pattern of relations between the leaders of organizations and the relevant areas of the environment is becoming more complex. For example, economic organizations are becoming increasingly enmeshed in legislation and regulation. In terms of adaptation and change, there is increasing reliance on research and development to achieve competitive advantage. Product and material substitutions are becoming increasingly important. Also, rather than competition between firms, there is a tendency towards maximizing cooperation. For example, the Canadian National Railway and the Canadian Pacific Railway operate together as much as they do in competition. Union Station in Toronto is jointly owned by both railway companies. There are industry leaders (such as in the steel and auto industries) which set wage rates for the industry as well as determining prices and price increases.

The studies referred to, and many others, have shown the dynamic and complex characteristics of organizations and their decision-making processes. Many have recognized the importance of adaptability and the changing organizational environment, but the conventional theory of the spatial behaviour of organizations has tended to neglect the dynamics of the decision-making process. It has not provided for assessment of the changing importance of locations and the spatial pattern, and it does not recognize that an organization has, in fact, a conscious responsibility for its own evolution. Indeed any organization, through the value system it adopts and the goals it sets, can grow or die; it can stagnate and regenerate.

The concepts of the spatial behaviour of organizations need to be elaborated to incorporate the dynamic variables – those that are accessible to control or to manipulations. Then geography will further the understanding of spatial decision-making and spatial behaviour.

Bennis, W.G., 1966 *Changing Organizations* (New York).
Chandler, A.D., 1966 *Strategy and Structure* (New York).
Eells, R., 1962 *The Government of Corporations* (New York).
Etzioni, A., 1964 *Modern Organizations* (Englewood Cliffs, NJ).
Galbraith, J.K., 1967 *The New Industrial State* (Boston).
Hamilton, F.E.I., 1971 Locating industry in East Europe: an approach to spatial decision making in planned economies, *Trans. & Papers, IBG*, 52.
Homenuck, H.P.M., 1969 Institutional spatial decision-making: a case study of the United Steelworkers of America 1936–1966, unpublished PH D dissertation, University of Cincinnati.
Jay, A., 1968 *Management and Machiavelli* (New York).
Leavitt, H.J., and L.R. Pondy (eds.), 1964 *Readings In Managerial Psychology* (Chicago).
March, J.G., and H.A. Simon, 1958 *Organizations* (New York).
Miller, D.W., and M.K. Starr, 1967 *The Structure of Human Decisions* (Englewood Cliffs, NJ).
Packard, V., 1962 *The Pyramid Climbers* (Greenwich, Conn.).
Pred, A., 1967 *Behavior and Location: Foundations for a Geographic and Dynamic Location Theory* Pt. 1 (Lund Series in Geography, Lund, Sweden).
Smith, D.M., 1971 *Industrial Location* (New York).
Zald, M.N., 1970 *Organizational Change* (Chicago).

P1235
Factoral-dynamic rows of elementary geosystems as a basis for modelling natural regions
A.A. KRAUKLIS   *Institute of Geography of Siberia and the Far East, USSR*

In the study of geosystems (Sochava) great significance is attributed to the investigation of the organization of the geographical sphere. In the solution of problems, the study of the lower steps in the hierarchy of geosystems ('topes') is of special interest, particularly elementary geosystems.

These structures, in view of their own small volumes and relatively short duration, provide the opportunity to use experimental methods of investigation in order to obtain the quantitative characteristics of the main parameters and their exact cartographical representation; these structures are also useful for formal analysis. An important application is to topes, which are initial and more active links in the system of geographical material-power revolution. These principal links functionally either are out of sight, or may be taken into account roughly when the natural environment is studied by means of planetary and regional categories.

The dimensions of elementary geosystems (EGS) may be judged by the following example. In one of the landscape regions (of area 4725km^2) that have been observed in detail in the south taiga of Central Siberia 30 EGS types (facies or elementary geomeres) have been distinguished. Every facie is represented by numerous plots (elementary geochores) of area 0.5–5.0ha. These systems spread upwards to 30–50m (including above-ground parts of vegetation and the layer of atmosphere with a specific microclimate) and to a depth of 3–10m (substratum with definite rhythm of hydrothermic and biogeochemical processes).

The minor volumes of topes are combined, as a rule, with major gradient parameters of environment in the horizontal and vertical directions. This means that EGS exist and function under conditions of intensive interaction with their nearest surroundings (environment) and are potentially subject to rapid and sharp transformations.

An EGS plot is not homogeneous, which reflects its functional differentiation, and its minimum area must evidently be enough to include some functionally necessary structures.

An EGS has various variable states, and certain periods of time are required for it to undergo the dynamic stages inherent to it.

The minimum volume and the minimum time depend upon all the other properties of the EGS, upon its interaction with the nearest surroundings and upon zonal-regional conformities. In the given example, the minimum area comprises 0.05–0.10ha and the minimum time 50–100 years.

Thus every EGS (1) includes subordinate structures, (2) has various variable states, (3) interacts with its surroundings. These properties are considered in more detail below.

1. Numerous subordinate structures within the plots of elementary geosystems (vegetation synusia, colonies of animals, pedons, minor roughness of surface, etc.) perform definite functions, 'consume' space, and organize particular systems around themselves. This heterogeneity ensures constant restoration of an EGS, continuity of its functioning, and 'optimum' space utilization.

It is important to define the degree of integration in these structures which 'utilize' the same plot of territory. The following categories may be distinguished: (a) weakly integrated (transient) particular systems, in which the indices of conjugation in different structures are low, usually showing the short duration of their existence; (b) moderately integrated (fragments of different stages of variants of one and the same facie) with rather close correlations between structures of some major components; (c) integrated (fragments of alien facies), remarkable for high indices of correlation between structures of all the major components (biota, soil, substratum) and regime parameters (hydrothermics, metabolism, and surface state).

From the standpoint of the subject under examination it is important that correlations of the categories mentioned are not equal in different facies. Three EGS groups are indicated, and, for these, the terms of G. Haase are used where the analogical classification is given: (a) monomorphic EGS consisting almost completely of transient particular systems; (b) semipolymorphic EGS, together with transient particular systems, fragments of stage variants, and (in reduced form) fragments of different facies distributed significantly; (c) polymorphic or particular systems of all categories.

2. Regime is a characteristic of regulation of dynamic phenomena connected with the interchanging of variable states of EGS. Seasonal rhythm is one of its major indices. It is also important to show variable states of an EGS (its dynamic stages) lasting from a few years to a few decades, and the sequence and character of the changes. If one takes into consideration the contrast between various variable states and the directions of the changes, the following subdivision of the EGS is possible: (a) EGS with cyclic repeated variable states with little contrast; (b) EGS with contrasting variable states, some of which are only partially reversible; (c) EGS with variable states of various degree of contrast and mainly irreversible changes.

3. Correlation of background (general, for the whole region, and local, where the influence is limited by local situations) factors is determined by the interaction of an EGS with its nearest surroundings. According to the 'localization' rate one may distinguish: (a) situations where the influence of the nearest surroundings is insignificant and an EGS corresponds in full measure to the background ('zonal-regional norms'); (b) 'weak-localized' situations where the influence of the surroundings is considerable but the EGS retains the main zonal-regional features; (c) 'strongly-localized' situations, when the zonal-regional peculiarities retire to the background under the strong stress of the surroundings.

Thus, in numerous continental south taiga regions of Central Siberia the zonal-regional standards are represented by facies of boreal dark-coniferous (mainly fir) forests with podzol soils. In strongly-localized situations they are substituted by reduced communities of normal appearance with peculiar dark-coloured soils, fragments of mountain taiga with skeleton soils of the sod-wood or podzol type, analogues of middle taiga fir forests with gley-podzol soils, and in some cases also impregnations of forest-steppe formations.

A very important geographical characteristic of an EGS is its belonging to one or another category. This must be taken into account when making regional comparisons of experimental results as otherwise incorrect conclusions can be made.

4. Data obtained from observations at field stations over many years, detailed thematic maps and plans, results of statistical processing (Sotchava 1969, 1970) show that relations exist between categories of EGS: (*a*) EGS of the zonal-regional type, as a rule, possess a monomorphic structure and are characterized by little-contrasted cyclically repeated dynamic stages; these are indigenous facies; (2) weakly-localized EGS belong to the semipolymorphic category and possess contrasting partially irreversible dynamic stages; these are quasi-indigenous facies; (3) strongly-localized EGS are polymorphic and usually with irreversible changes of dynamic stages; these are called serial facies.

The question of regulating ecological-geographical information about the numerous EGS and their representation in a compact form is of great importance. For these aims we suggest the principle of factoral-dynamic rows. This has been worked out for one of the taiga regions in Siberia (in Lower Preangarye) and used in compiling legends for landscape and geobotanical maps, territory classification, and regional comparison.

In the construction of factoral-dynamic rows the following EGS properties are taken into account: heterogeneity (structure), dynamic stages (regime), correlation of background and local factors (spatial-functional relations).

Factoral-dynamic rows unite facies possessing common structural-dynamic features and, as a rule, reflect the succession of facies which change in accordance with the intensification of a definite integral process of local importance. In particular, in many natural regions of Siberia the following rows may be distinguished: subhydromorphic (connected with the formation of initial links of hydrographic networks in watersheds and on slopes), sublytomorphic (corresponding to processes of drawing bedrock, exposed as a result of denudation, into an EGS), subcryomorphic (conditioned by periodic 'concentration' of cold), hydromorphic (complex phenomena, accompanying periodic floods and constant high levels of groundwater) and so on.

The system of factoral-dynamic rows of each region may be presented as a graphic model. The indigenous facie forms a central axis of the scheme. Quasi-indigenous facies are grouped around an indigenous one. Serial facies occupy extreme positions.

The regulating of facies as systems of factoral-dynamic rows makes it possible to model the topical properties of the natural region under study. Each natural region is characterized by its specific system of factoral-dynamic rows, and regional comparisons in this aspect will facilitate the cognition of the regularities of the structural-functional organization of large subdivisions of the geographical environment.

Thus, the suggested approach concerns some dynamic effects inherent in fractional subdivisions of the geographical environment which are often neglected when dealing directly with gradations of regional significance. Registration of these effects increases considerably the practical and theoretical value of geographical work.

In the same way as the notion of zonality became the basis for planetary geography, and the idea of landscape that for regional, the study of topes by geographers leads to the formation of a third aspect in geography, the topological.

Taiga méridionale de la Région d'Angara, redacteur en chef V. Sochava (Leningrad 1969; in Russian, abstracts in French).
Topologie des géosystèmes steppiques, redacteur en chef V. Sochava (Leningrad, 1970; in Russian, abstracts in French).
Topology of Geosystems (Irkutsk, 1971).

P13
Remote Sensing, Data Processing, and Cartographic Presentation
Télédétection, Traitement des données et Représentation cartographique

CONVOCATEUR/CONVENOR: Philip J. Howarth, *McMaster University, Hamilton*

This listing of short titles and first authors' surnames will assist in identifying articles, topics, and places of interest. A complete author and co-author index is located at the end of this volume, as well as a selected index relating papers to geographical locations. Note that the papers are *not* listed in alphabetical order. The organization of the volumes is described in full in the Preface.

Other papers of particular interest to readers of this section are to be found under c01 (Geographical Data Sensing and Processing), c08 (Geomorphological Survey and Mapping), c09 (Agricultural Typology), c15 (World Land Use Survey), and s02 (Water Resources).

Cette liste des titres abrégés et des noms des auteurs principaux permettra d'identifier les communications, les sujets et les lieux qui présentent un intérêt quelconque. Un index complet d'auteurs et de coauteurs se trouve à la fin de ce volume, ainsi qu'un index des lieux géographiques. Prière de noter que les textes ne sont pas classés par ordre alphabétique. On explique en détail le plan de ces volumes dans la Préface.

D'autres études qui intéresseront peut-être les lecteurs de cette section se trouvent dans c01 (Sélection et traitement de l'information géographique), c08 (Recherche et cartographie géomorphologiques), c09 (Typologie de l'agriculture), c15 (L'utilisation du sol dans le monde) et s02 (Ressources en eau).

P1301 Geometric aspects of remote sensing   DERENYI   965
P1302 Remote sensing in land-use mapping and urban change detection, San Francisco (USA)   ELLEFSEN   967
P1303 Developing remote sensing display modes for urban planning data input needs (Denver, USA)   HOWARD   968
P1304 Space photographs of the earth in the study of plate tectonics   KEDAR   970
P1305 Orthophotographic mapping – basis for information systems (USA)   KELLOM   971
P1306 Cover type identification capabilities of remote multispectral sensing techniques (Indiana, USA)   JOHNSON   973
P1307 Methods of medical cartography for geomedical research   JUSATZ   975
P1308 Cartographic presentation of three-dimensional urban information   LO   977
P1309 Maps of road travel speed and their evaluation (UK)   MORRISON   979
P1310 Data acquisition and presentation, the glacier inventory, Canada   OMMANNEY   980
P1311 Multilevel principle of generalization and problems of automatic map reading   SHIRYAEV   982
P1312 Satellite photography, a geographic tool for land use mapping (SW USA)   THROWER   984
P1313 Remote sensing system for detecting gross land use change in metropolitan areas (USA)   WRAY   986
P1314 Remote sensing of the environment with a Ka-band microwave radiometer   HOOPER   988
P1315 Le géocodage de Statistique Canada : système géographique automatisé au stade opérationnel   TERJANIAN   989
P1316 Thematic maps in physical planning (Poland)   WIECKOWSKI   991

P1317 La légende des cartes morphostructurales 1 : 50,000 et 1 : 500,000 (URSS) BACHENINA 993
P1318 Télédétection des eaux polluées POUQUET 995
P1319 Pour une plus grande diffusion et connaissance des 'panoramas' (Italie) SCARIN 996
P1320 Aggregate mapping of Ukrainian SSR agriculture ZOLOVSKY 998

P1301
Geometric aspects of remote sensing
EUGENE E. DERENYI *University of New Brunswick, Canada*

Sensors which operate outside the visible light region, such as infrared scanners and side-looking radar, are gaining an increased acceptance in the field of photo-interpretation. They are capable of operating by day and night, under adverse weather conditions, and are able to reveal information which otherwise would remain undetected.

Spectral characteristics of these sensors are relatively well known and intensive research is being performed at many agencies on this subject. However, little information is available on the geometric characteristics of these images, and investigation along these lines was begun only recently. It is important that one be familiar with the geometric properties in order to judge the potentials of the sensors properly.

Two aspects have to be considered in this connection: geometric or spatial resolution, and geometric fidelity.

Resolution is a rather complex phenomenon which involves not only the basic characteristics of an imaging system, but also other variable factors such as the shape, size, and contrast of the target.

In a conventional camera system there are two components to consider, the lens and the photographic emulsion. Electronic parts perform an important function in unconventional imaging devices, and the signal-to-noise ratio is an important criterion. One should note also that, in most cases, the resolution of the latter types is different in longitudinal and lateral directions with respect to the flight path.

For optical–mechanical scanners the theoretical limit of resolution is defined by the instantaneous field of view, usually specified in angular units. The best ground resolution is obtained when the scanner points vertically downward. Then it decreases as a function of $\sec \theta$ in longitudinal direction, and of $\sec^2 \theta$ in lateral direction, where θ is the scan angle (Holter et al. 1962).

For side-looking radar systems the ground resolution in the lateral direction is limited by the accuracy of time measurement, and as such it is proportional to the time duration of the pulse. In the longitudinal direction it is proportional to the slant range and to the beam width (A.S.P. 1966).

Table 1 presents a numerical comparison between the ground resolution of all three sensor types discussed. Results for the frame camera are based on the performance of a typical super wide-angle mapping camera. The instantaneous field of view of the scanner is taken as 1mrad by 1mrad, the pulse duration of the radar is assumed to be 0.1microsec, and the beamwidth is 0.1°, all of which are typical values. Furthermore, all values are given in metres and are valid for a flying height of 1000m.

It is apparent from these results that the resolution of unclassified scanners, and especially that of radar systems, is far inferior to the resolution of modern aerial camera systems.

The geometric fidelity of an image is influenced by the following factors:

1 Inherent distortions of the whole instrument package which produces the image (interior orientation). This factor can be ascertained only by a thorough calibration of the system. Little or no research has been done towards the calibration of unconventional imaging systems.

2 Distortion characteristics of the photographic material. This problem is rather

TABLE 1

Sensor	Direction	0°	30°	45°	60°
Camera	Longitud.	0.14	0.20	0.40	1.10
	Lateral	0.14	0.20	0.40	1.10
Scanner	Longitud.	1.0	1.2	1.4	2.0
	Lateral	1.0	1.4	2.0	4.0
Radar	Longitud.		2.0	2.5	3.5
	Lateral		30.0	21.2	17.3

general in nature and is well documented in the literature.

3 Distortions due to natural causes such as atmospheric refraction and earth curvature. Procedures followed in conventional aerial photography are readily available to deal with this problem.

4 Sensors which operate outside the visible light region employ scanning techniques to collect the reflected or emitted energy from objects with extended dimensions. In the direction of flight, the scanning is usually induced by the forward motion of the carrying vehicle, whereby the visual reconstruction of the sensed energy becomes a continuous strip image. Therefore, angular oscillations of the vehicle and variations in velocity and altitude have a direct distorting effect on the image, unless, of course, the sensor is gyro stabilized. The amount and direction of the displacement varies from image point to image point as a function of the changes in the position and orientation of the sensor (exterior orientation) (Derenyi 1971).

TABLE 2

	0°	30°	45°	60°	Av.
Unstabilized	18	37	42	78	44
Roll compensated	18	20	24	35	24

As an example, Table 2 lists the errors introduced into the ground position of points, situated at various scan angles, caused by a displacement of the corresponding image points (Derenyi and Konecny 1964). A change of 1° in the angular orientation and a flying height of 1000m above the terrain is assumed. A completely unstabilized and a roll-compensated scanner are considered separately. The values are given in metres but also represent the displacements in per mill (‰) of the flying height.

Although displacements of such magnitude may result from conventional frame photographs as well, nevertheless they can be easily corrected by established photogrammetric procedures. In the case of a continuous strip image, however, the amount and direction of displacements can fluctuate rather rapidly and can change from zero to a maximum within a section of the image a few centimetres in length. Rectification is a rather difficult task.

The geometric fidelity of imageries currently available is demonstrated by the following test. Identical points were selected on aerial photographs and two strips of infrared scanner imagery covering the same terrain. The co-ordinates of all image points were measured in a comparator and ground coordinates were deduced analytically. Co-ordinates computed from the two IR images were then compared with those originating from the aerial photographs, and with each other. Table 3 lists the maximum, average, and the root-mean-square (RMS) point errors obtained. All values are given first in metres and then in per mill (‰) of the flying height (1500m). The scanner in question was roll-compensated.

A comparison of these values with those presented in Table 2 indicates that discrepancies can mainly be attributed to the uncorrected effect of the changing exterior orientation of the sensor. A perfectly straight and level flight, without velocity and height variations, was assumed in the absence of any pertinent information.

The results of theoretical investigations and of tests performed on real imagery indicate that, from a geometric point of view, the performance of unconventional sensors is inferior to that of present-day camera systems. Furthermore, the sources of errors are rather complex and not yet thoroughly investigated. Unconventional aerial imageries must therefore be employed with these limitations kept in mind. At the same time one must strive for improved instrumentation and image utilization techniques.

TABLE 3

Point error	Photo-IR1		Photo-IR2		IR1–IR2	
	m	‰ h	m	‰ h	m	‰ h
Maximum	+66.1	+43.1	+41.7	+27.7	−33.2	−22.1
Average	+11.5	+7.7	+8.5	+5.5	−1.5	−1.0
RMS	±17.7	±11.8	±13.4	±8.9	±7.9	±5.3

National Research Council of Canada; Defence Research Board of Canada.

American Society of Photogrammetry, 1966 *Manual of Photogrammetry*, 3rd ed.

Derenyi, E., and G. Konecny, 1964 Geometry of infrared imagery, *Can. Surveyor* 18 (4), 279–90.

Derenyi, E., 1971 An Exploratory Investigation Concerning the Relative Orientation of Continuous Strip Imagery, Dept. Surveying Engineering, U. New Brunswick, Res. Rept. no. 8.

Holter, M.R., S. Nudelman, G.H. Snits, W.L. Wolfe, and G.J. Zissis, 1962 *Fundamentals of Infrared Technology* (New York).

P1302
An application of remote sensing in land-use mapping and urban change detection in the San Francisco Bay area
RICHARD A. ELLEFSEN and DUILIO PERUZZI   *San Jose State College, USA*

This paper reports on research in progress using remote-sensed imagery to map urban land-use in the San Francisco Bay area. We concentrate first on problems encountered in land-use mapping, and then give some preliminary indications in the detection of land-use changes. The experiment is being conducted by the US Geological Survey's Geographic Applications Program in cooperation with the Department of the Interior's Earth Resources Observation Systems (EROS) and NASA's Earth Resources Technology Satellite (ERTS) programs. Goals of the project are to measure and analyse land-use and to develop and test techniques for detecting and measuring land-use changes from 1970 to later points in time.

The urbanized area of the San Francisco Bay area test site, one of twenty-seven under study in the conterminous United States and Puerto Rico, was photographed with colour infrared film in the spring of 1970 coincident with the decennial population census. Photographs were taken from a reconnaissance aircraft at an altitude of 15,000 metres. Transparencies, each 230 by 230mm, were in two scales: (1) 1:100,000 (RC-8 camera, 15cm focal length lens) covering approximately 530 square kilometres, and (2) 1:50,000 (Zeiss camera, 30cm lens) covering just over 130 square kilometres. The entire site, including non-urbanized areas of the nine counties comprising the Bay area, was rephotographed (RC-8, 15cm lens from 18,000 metres) in the spring of 1971, to provide a comparison over a one-year time period.

Land-use identifications were made from colour infrared diapositives and then plotted on 1:62,500 black-and-white orthophoto mosaics gridded on the UTM rectangular coordinate system at one-kilometre intervals. In the projected user-oriented, looseleaf Atlas of Urban and Regional Change (scale 1:125,000) the following additional information will be presented on overlays: (1) boundaries of urbanized areas as delimited by remote-sensed criteria; (2) areas in discrete land-uses computed for each census tract as delimited for the 1970 Census; and (3) significant point-line identification features. A separate atlas is planned for each test site.

Detection of further change will be based on high-altitude photography and simulated colour, televised, orbital imagery from the Earth Resources Technology Satellite – both in 1972 – plus any subsequent imagery (Wray 1971, 2).

The classification scheme employed in the project attempts to portray the functional use of land (Table 1); it approximates the scheme proposed by Anderson (1971, 386) for use with orbital imagery supplemented with other information and at scales of 1:250,000 to 1:2,500,000. Use identification problems centred on the attempts to infer function from morphology. As, to paraphrase an expression, function does not always follow form, difficult identifications had to be made, if possible, from ground-truth observation. Where that was not feasible, identifications had to be made from clues suggested by observed associations with interrelated land uses or connecting linear links. Distinctions between industrial and commercial uses were especially troublesome.

Although the degree of generalization of the categories in the scheme is satisfactory

TABLE 1. Land use classification (Geographic Applications Program, US Geological Survey, 1971)

Categories	Subcategories	Code
URBAN		
Livelihood	Commercial	uLC
	Industrial	uLD
	Transportation	uLT
	Livelihood with non-agricultural residence	uLR
Residential	Single family	uRS
	Multi-family	uRM
Open-space	Improved	uOP
	Unimproved	uOU
	Water	uOW
NON-URBAN		
Livelihood and residential	Agricultural with residence	nLA
	Industry	nLX
	Livelihood with non-agricultural residence	nLR
Open-space	Improved	nOP
	Unimproved	nOU
	Vegetation	nOUv
	Wetland	nUOm
	Water	nOW

for the map scale employed, certain uses manifested themselves spatially more distinctly than did others. Single-family residential use, especially if in new, planned housing tracts, was strikingly obvious from the photographs. More difficult to distinguish was multiple-family residential use where it was mixed with commercial land-use, or where it was composed of converted single-family houses, duplexes, or other common-wall dwellings.

Although initial effort has been concentrated on the construction of a historical spatial data base, there are preliminary indications, from comparing 1970 with 1971 photographs, that recurring types of change patterns may be identified. For example, recently constructed, planned (controlled) shopping centres usually have tracts of adjacent land held in reserve for expansion; earlier centres are frequently confined by non-commercial uses. Similarly, major string streets in newly urbanizing areas feature vacant parcels large enough for free-standing stores or planned units. Industrial parks, with their vacant areas consigned for development, form another example.

Only general accuracy is possible in determining the land areas that the housing tracts at the city's expanding periphery will consume next: large amounts of space are still available and housing continues to be able to pre-empt agricultural land. A somewhat more predictable change is the filling-in of remaining open spaces with apartments and condominium town houses, especially along major arterials within the generally contiguously built-up sections of the city. Less conspicuous signatures are made by land set aside for new highway arterials. These swaths across the landscape frequently remain in agricultural use for many years prior to conversion, giving a misleading quantity of agricultural land within the city.

Patterns of urban land-use change have been identified. These appear to be replicable but further testing will be conducted to determine their universality. The principles so established will then be applied both to relatively large-scale air photography and to small-scale orbital imagery in an attempt to develop techniques for periodic monitoring, measuring, and analysing of land-use change.

Anderson, J.R., 1971   Land-use classification schemes, *Photogram. Eng.* 37 (4), 379–87.
Wray, J.R., 1970   Census cities project and atlas of urban and regional change, *Third Ann. Resources Program Rev., Geol. and Geog.* 1, sect. 2, NASA Manned Spacecraft Center, Houston, Texas, 2-1 to 2-16.
– 1971   Bay area in unique test to detect land-use change by satellite-borne remote sensors (unpublished report), Geographic Applications Program, US Geol. Survey, Washington, DC, 1–3.

P1303
Developing remote sensing display modes to satisfy urban planning data input needs
WILLIAM A. HOWARD   *University of Denver, USA*
and JAMES B. KRACHT   *Concordia College, USA*

The usefulness of satellite-derived remote sensing imagery to the urban planner depends on resolving some of the dichotomies now surrounding the very real problems of small-scale versus large-scale, and low resolution versus high resolution. This research investigated this problem area, with the principal question being whether planning information inputs can be changed to be compatible with current remote sensor output formats, or whether these formats are alterable enough to become more compatible with planning input needs. Primary emphasis was on land use information.

Small-scale photography at scales of 1:50,000 and 1:100,000 in conventional colour and colour infrared formed the source materials for the assessment of the feasibility of acquiring urban land use data. Throughout the research, land use data were derived based upon the master classification system adopted by the Denver Regional Council of Governments. This is an eight-digit code with digits assigned as follows: economic use, 0; economic activity, 0000; property class, 0; property sub-class, 00. This system has the advantage that the one- and two-digit levels of information are at comparable levels of detail for all use groups. Space does not permit a detailed breakdown of this code, and such a breakdown is really unnecessary for the purposes of this paper, as the research was concerned only with the first two digits and for the most part only with the first digit. At the one-digit level, categories of land use are as follows: residential, commercial, services, industrial, transportation, communications and utilities, public and quasi-public, parks and recreation, agricultural, and vacant (Inter-County Regional Planning Commission 1966).

Generally speaking, it was possible to identify most land uses at the one-digit level using the 1:50,000 colour infrared and conventional colour photography, with the former yielding better results than the latter. It was also possible to discriminate between land use at the two-digit level approximately 60 per cent of the time. The two-digit level discrimination was much more reliable in newer urban and suburban areas than in older areas. Residential, transportation, public and quasi-public, industrial, parks and recreation, and agricultural land were most readily identifiable at the two-digit level.

Some difficulties arose at the one-digit level in discriminating between commercial and service uses, but a person familiar with the city and knowledgeable in photographic interpretation could make fairly reliable discriminations between the two uses. In the older portions of the city, it was more difficult to identify land use types, even at the one-digit level, because of density of structures, heavier tree cover, strip developments containing commercial and service uses, and economic activities operating from buildings which were not built specifically for that purpose.

A lack of clear identification keys made valid and reliable interpretation of land use on less than a block basis nearly impossible. For these reasons, other sources of information are necessary where planning agencies are committed to a land use inventory on a parcel basis, as is the case with most municipal planning commissions. Where data are required on a census tract or a block basis and at the one-digit level, the 1:50,000 scale colour infrared photography is very satisfactory.

An attempt was also made to derive land use information from conventional colour and colour infrared photography at a scale of 1:100,000. It proved to be very difficult to determine any land use, except for residential, beyond the one-digit level. Certain categories of the classification system were often confusing or indistinguishable. Among these were public and quasi-public, commercial, service, and light industrial. Transportation rights-of-way were distinguishable, but often terminal facilities were not. Many utility installations were not identifiable. It was again apparent that land uses were more readily identifiable in the newer urban and suburban areas than in the older portions of the metropolitan area. Land uses of all types in the more densely built up and older areas of the city of Denver were difficult to identify with any degree of reliability or validity. Where detail is required in the discrimination of land use types, a great deal of supplementary information would be required in using the 1:100,000 photography. State and regional planning agencies could possibly find the 1:100,000 photography admirably suited to their needs.

Small-scale, high-quality remote sensing photography can be very useful in acquiring land use information. Changes in many metropolitan areas occur with such suddenness that the process of maintaining and updating a land use file is nearly impossible. The monitoring of change is, at present, often haphazard and unsystematic. Only with frequent and systematic coverage does monitoring become possible.

Small-scale photography, similar to the 1:50,000 scale colour infrared and 1:100,000 scale conventional colour and colour infrared used in this research, establishes a link between the large-scale photography which planners currently use and the imagery which will ultimately be available from an orbiting satellite. Through coverage acquired by means of an orbiting satellite, monitoring of change in cities can be effected. Frequency of coverage, quality of the imagery, and its resolution are most important considerations in the monitoring of urban areas.

Until imagery at a scale of 1:100,000 becomes available for widespread use by way of satellites, aircraft photography in a transparency form in conventional colour and colour infrared should be provided to major urban areas on a routine basis. Coverage twice a calendar year would be appropriate, with one overflight coinciding with the vegetative growth period and the other during the vegetative dormant period. The photography would facilitate a constant updating of the land use file.

The provision of small-scale aircraft photography to planning agencies, prior to satellite imagery, would establish such records as a routine data source. If satellite imagery is ultimately to become useful to planners, some of the traditional methods used in the planning process must be modified. Planning is fairly tradition-bound, and making small-scale imagery available on a routine basis and not on a 'catch-as-catch-can' basis will serve to modify traditional methodologies.

The central question of this research was whether planning information inputs can be changed to be compatible with current remote sensor output formats, or whether these formats are alterable enough to become more compatible with planning input needs. This question is a meaningless one in light of this study. Planners have been utilizing remote sensing techniques for a long period of time. Advances in remote sensing technology now make it possible for the planner to use the techniques in essentially the same format form but on a frequency never before possible. Planning commissions are generally very interested and receptive to the new possibilities that remote sensing technology offers, but availability is the essential requirement. If made available, classification systems commensurate with remote sensing imagery can be developed, and ultimately an effective interface can be established between alphanumeric and digital data and remote sensing imagery.

Inter-County Regional Planning Commission, 1966 *Land Use Classification Manual*, master plan report no. 26, 2nd ed. (Denver).

P1304
Space photographs of the earth in the study of plate tectonics
ERVIN Y. KEDAR *State University of New York, USA*

Photographs of the earth taken from manned orbiting spacecrafts make a unique contribution to plate tectonic research by generalizing, photographically, secondary and tertiary geomorphotectonic elements (geotectonic noise). The appearance of first-rank structural features of the earth is enhanced. In addition, a single photographic frame gives a synoptic view of an entire region.

National Aeronautics and Space Administration programs, such as Tiros, Nimbus, Gemini, and Apollo, have demonstrated that certain geologic features may be perceived more completely when studied from space than when studied on the surface of the earth. Earth phenomena that may be better understood from photographs taken in space are geotectonic movement, continental drift, plutonism, and rising and sinking crustal areas.

Sections of the surface of the earth between the latitudes of 35°N and 35°S are displayed in several series of photographs obtained from Gemini and Apollo missions. Many of these high-quality colour photographs may be viewed as stereopairs.

Two belts of similar configuration in the northwestern and southwestern parts of the United States have been recognized in space photographs. Interpretation of the photographs aided in the study of these belts, and many features of the belts have been identified and inferred from the photographs. The study was supported by the use of conventional geologic and geophysical sources. Landforms and surface morphology indicate similarities between the two belts. Both belts are described as consisting of crustal polygons of the crustal blocks, with boundaries caused by horizontal movements. These features are thought to be related to deep-seated structural disturbances such as deep-seated faults below the sedimentary veneer.

A deep-seated fault zone, which is a boundary between separate basins of sedimentation and rigid plates of the lithosphere, may be recognized by a regional gravity anomaly, a magnetic-field anomaly, a prominent terrain deformation, and localized volcanic and plutonic activities.

The inference from space photography is that both belts approach each other in the Mojave Desert, in an open-arched shape. Both belts, which have been described as left-lateral shear faults, trend into the Transverse Ranges and farther westward toward the Murray Fracture Zone. The evidence from space photography provokes speculation concerning the geotectonic forces and processes involved in the evolution of these two belts.

The hypothesis is advanced that both belts are master shear-fault zones that were caused by horizontal displacement in the crust of the earth. The inferences from space photography suggest that the belts may be the result of the structural movement, activated by the southeast Pacific Rise crest and the northeast Pacific Rise crest, generating two drifting platforms against one stationary platform, the Colorado Plateau complex. Thus, the belts are related to the East Pacific Rise drag phenomena that have been activated by a component of gravity in the substratum mantle flow.

Badgley, P.C., 1965 *Structural and Tectonic Principles* (New York).
Dibblee, T.W., Jr., 1966 *Geology of the Central Santa Ynez Mountains, Santa Barbara County, California*, Calif. Div. Mines and Geol. Bull. 186.
Eardley, A.J., 1968 Major structures of the Rocky Mountains of Colorado and Utah, in A Coast to Coast Tectonic Study of the United States, *UMR J.* 1, 79–99 (V.H. McNutt, Geol. Dept. Colloq. Ser. 1).
Gilluly, J., 1963 The tectonic evolution of the western United States, *Quart. J. Geol. Soc. London* 119, 133–74.
Isacks, B., J. Oliver, and L.R. Sykes, 1968 Seismology and the new global tectonics, *J. Geophys. Res.* 73, 5855–99.
Kedar, E.Y., 1970 Plate-margin morpho-tectonics case study – the Israel-Sinai section, in *Technical Letter* TF7-6, Manned Spacecraft Center, Houston.
– 1970 Texas lineament traced on earth photographs taken from space, *ibid.*, TF7-3. Manned Spacecraft Center, Houston.
– 1970 Transverse Range: tectonic inferences from earth photographs from space, *ibid.*, TF7-2.
King, L.C., 1967 *The Morphology of the Earth: A Study and Synthesis of World Scenery* (New York).
Larson, R.L., H.W. Menard, and S.M. Smith, 1968 Gulf of California: a result of ocean-floor spreading and transform faulting, *Sci.* 161, 781–4.
Le Pichon, X., 1968 Sea-floor spreading and continental drift, *J. Geophys. Res.* 73, 3661–97.
Molnar, P., and L. Sykes, 1969 Tectonics of the Caribbean and middle America regions from focal mechanisms and seismicity, *Bull. Geol. Soc. Am.* 80, 1639–84.
Moody, J.D., 1966 Crustal Shear patterns and orogenesis, *Tectonophysics* 3, 479–522.
Oliver, J., L. Sykes, and B. Isacks, 1969 Seismology and the new global tectonics, *Tectonophysics* 7, 527–41.
Zietz, I., P.C. Bateman, J.E. Case, M.D. Crittendon, A. Griscom, E.R. King, R.J. Roberts, and G.R. Lorentzen, 1969 Aeromagnetic investigation of crustal structure for a strip across the western United States, *Bull. Geol. Soc. Am.* 80, 1703–14.

P1305
Orthophotographic mapping – basis for information systems
JOHN B. KELLOM and M. FOLEY   *Raytheon Company, USA*

Many countries of the world are confronting the complex problem of how best to develop their physical and cultural resources. Administrators recognize that their programs often exert wide, unforeseen influences on the quality of life in areas where they are implemented. They also recognize the difficulty of controlling such influences through long-range planning because of the dynamic interactions which exist in any broad program between so many physical, social, economic, and political factors. Large volumes of information about the environment are required if these interactions are to be understood. Because of advances in data collection and reduction technologies, great quantities of data on the various disciplines have been accumulated. Unfortunately, much of the data are fragmented, recorded in varying levels of comprehensiveness, and stored at scattered locations. Usually, there is no designator to permit address, cross-correlation or interdisciplinary data synthesis; nor is there an optimum graphic generally available for displaying and updating the quantitative information dispersed in the various technical archives.

Many attempts have been made, therefore, to create a management information system which can collect, store, analyse, and output data in a common format to assist decision-makers in formulating plans. The design of any given information system depends on the user requirements, data sources, and the manner in which the information is to be used. However, a basic consideration of all systems is the need for an effective, uniform base for organizing, extracting, and portraying the available physical and cultural data and, since planning is heuristic in nature, to make use of new banks as they evolve. Any modern resources or environment-oriented data bank system can function only if there is an unambiguous identifier for address purposes. Of all the basic elements comprising man's environment, land is the only one that is static (i.e., non-moveable), and every parcel of land differs from all others with respect to its location upon the earth. Therefore, the use of land, expressed in geographic terms, provides a unique means of addressing data banks and permits all environmental data to be related to the ground unambiguously.

Any map with a co-ordinate grid permits the land to be used as a locator for spatial data. At the present time, the primary maps readily available to state, provincial, and local governments are of the conventional line and symbol type. A line and symbol map, however, is inherently limited by the small amount of information which it can portray and, hence, by the accuracy with which ground locations can be derived from it. In addition, the use of topographic maps as a reference base is often restricted because of scale considerations, a lack of up-to-date map sheets, and the inordinate time required for their production. Orthophotographic maps, on the other hand, are particularly suitable as a data base in that, while they meet the metric standards for class A topographic maps, they also retain about 90 per cent of the data recorded by the aerial photos used in their construction.

An orthophotograph is an aerial photo which has been fully scale corrected by electronically scanning stereoscopic models keyed to known geodetic control points. Differing from scale-limited photo mosaics and so-called photo maps, the orthophoto map exhibits unmatched positional accuracy since every point has been locationally adjusted from its inherent displacement on the original imagery. Orthophoto base maps are rapidly compiled by mosaicing individual orthophotographs to a prescribed map sheet format selected from an infinite range of scales. The resulting orthophoto map is overprinted with a grid from which the geographic position of discrete points, readily seen on the orthophoto map, can be expressed numerically. Thematic overlays can be keyed to the base map grid yielding a synoptic, land-related summary of the relevant data.

A few examples of information system-oriented projects using an orthophoto map as a data base are in order. In the State of Vermont, orthophoto maps, based on the state grid, were produced at a scale of 1:12,000 for general highway planning, land description, and regional planning efforts. Enlarged sheets at 1:6,000 were produced for more detailed urban analysis. A state-wide information system, using the state coordinates scaled from these base maps and a computer program, is being investigated. A pilot study was undertaken combining land use, geology, slope, transportation networks, water distribution, sanitary and storm sewer systems, and property map overlays keyed to an orthophoto base map. The land use and geology differentiations were derived, for the most part, from interpretation of the black and white aerial photography used to provide the base maps, while the slope zones were derived from contour data extracted coincidently with the map production sequence. Interrelationships of community facilities were observed by scribing lines of communication and various municipal utilities and services on the common map format. Of particular interest was the delineation of property parcels. Because of the high information content and metric fidelity of an orthophoto map, it provided an excellent base for checking the validity of existing tax maps. Variances and omissions from the property rolls could be readily detected, as well as notable property improvements such as swimming pools, garages, and house additions.

In Massachusetts, a basis for a cultural information system is being developed linked to orthophoto maps. A town was mapped using aerial photography which was transformed via stereoplotters, aerotriangulation controls, and a digital computer into orthophotos. Using a standard digitizer, streets were traced in the direction of increasing house numbers (which had been assigned in 50-foot increments from the beginning of each street) and all intersections and end points were recorded. Each house was then located in its proper spatial orientation with respect to the neighbourhood and in effect assigned its own digital address system. Computer software techniques were employed to display the houses on symbolized overlays keyed to the map. Any spatial data which had an address could be plotted. As an experiment, census data, referenced by street number, was tied into the system. From these data various distributions, such as selected age increments, unregistered voters, and new residents, were derived and displayed on a Gerber plotter. At the time this paper was written, a program was under way to assist school officials in determining the most efficient school bus routes by automatically plotting the location of all the school children on the base map, deleting those who lived within one mile walking distance of the school, and, still within the computer, defining proposed routes.

In summary, as a base upon which spatial data can be measured and related, the orthophoto map appears to have great potential. Orthophotos combine the detail of aerial photography with geodetic control points to produce the metric accuracy of a class A topographic map. Exhibiting a uniform scale throughout its area, distances can be scaled, areas computed, and azimuths determined on the orthophoto map as accurately as on a line map. Its geometric fidelity plus capacity to satisfy multiple map user requirements makes the orthophoto map an optimum pictorial data base for environment-oriented programs.

George Loelkes; Raytheon Company/ Autometric Operation.

American Society of Photogrammetry and American Congress on Surveying and Mapping, 1971 *Papers From the Orthophoto Workshop.*
Collins, S.H., 1968 Stereoscopic orthophoto maps, *Can. Surveyor* 22 (1), 167–76.
– 1969 The accuracy of optically projected orthophotos and stereo-orthophotos, *ibid.* 23 (5), 450–63.
Konecny, G., and D.H. Refoy, 1968 Maps from digitized stereomat data, *Photogrammetric Engineering* 34 (1), 83–90.
Raytheon/Autometric Company, 1968 *Technical Specifications for Orthophotomaps* (Wayland, Mass.).
Tomlinson, R.F., 1967 *An Introduction to the Geo-Information System of the Canada Land Inventory* (Ottawa).
US Department of Housing and Urban Development *Urban & Regional Information Systems: Support for Planning in Metropolitan Areas* (Washington).
Voss, F., 1968 'The Production of 1:5,000 Orthophoto Maps in North Rhine-Westphalia,' *NÖV-Nachrichten aus dem öffentlichen Vermessungsdienst Nordrhein-Westfalen.* No. 1, p. 3–13.

P1306
Cover type identification capabilities of remote multispectral sensing techniques
GARY E. JOHNSON *University of North Dakota, USA*

In order to evaluate cover type identification capabilities of remote multispectral sensing techniques, a remote sensing project was conducted at the Laboratory for Applications of Remote Sensing (LARS) at Purdue University. The project was conducted between February and May of 1971 using data collecter for LARS by the University of Michigan's scanner-equipped aircraft.

Remote multispectral sensing, as defined

relative to this project, refers to the sensing from a remote location of reflected and emitted electromagnetic radiation in thirteen discrete, relatively narrow spectral bands between 0.40 and 2.60 microns wavelength. This sensing technique is based on the premise that different surface features reflect and emit different amounts of detectable radiant energy.

The passive sensor of the remote multispectral sensing system is a multichannel optical-mechanical scanner whose use has been aptly described by Landgrebe and Phillips (1967, 3): 'The energy radiated by a specific ground resolution element at a given instant of time passes through the scanner optics and is divided according to its spectral wavelength and directed to an appropriate detector. The output of all such detectors are simultaneously recorded on a multiband instrumentation recorder. The transverse motion provided by the rotation mirror and the forward motion of the aircraft cause a continuous raster to be formed for each spectral band of the scanner output. An important feature of a system such as this is that by simultaneously sampling the output of all bands, one obtains a vector which contains all spectral information available about a given resolution element on the ground.'

Spectral responses thus collected on magnetic recording tape in analog form are converted to digital form and analysed using a series of computer programs developed by LARS researchers.

The steps leading to computer classification of remote multispectrally sensed data will not be detailed here because of space limitations. It will suffice to note that the resultant product of scanner collection and computer analysis based on pattern recognition techniques is a computer-generated map containing researcher-assigned symbolizations representative of those earth surface features which the computer has been trained to recognize.

Data selected for analysis consisted of five flight lines flown over Tippecanoe County, Indiana, on 5 August 1969 at an altitude of 1524 metres.

Flight-line dimensions averaged 32 kilometres by 1957 metres. Scanner coverage for the five flight lines represented a total area of 335.4 square kilometres or 26 per cent of the county area.

In computer printout form each printed symbol, referred to as a remote sensing unit, was representative of an average area of 12 metres by 12 metres. Because of scanner 'look angle' the representative width of each remote sensing unit varied from 9.5 metres at the nadir to 13.5 metres at the margin of the computer-generated map. The representative length of each remote-sensing unit remained constant at an average of 12 metres for the five flight lines.

Classes of cover type selected for performance evaluation were wheat, oats, corn, soybeans, pasture, hay, trees, water, and bare soil.

The method of evaluation was that of 'test field' performance evaluation. This method permits the researcher to select samples of a known cover type from photographs containing ground observations, to locate those samples on the computer-generated map, and to instruct the computer to calculate the percentage correct recognition of remote-sensing units contained in the sample.

Test fields representing 103,907 remote-sensing units; 4 per cent of the total available were selected at random from the ground observation photographs. Correct recognition was calculated at an average 85.0 per cent for the nine classes considered.

Wheat grown in Tippecanoe County is of the winter variety and is generally harvested by 30 July. The scanner therefore detected wheat stubble. Wheat was correctly classified, according to test field performance ratings, at an average of 73.8 per cent. This figure ranged from 67.8 per cent to 78.6 per cent for the five flight lines. Most wheat samples incorrectly identified were classified as oats (another stubble cover type at this date) or as pasture or hay. The latter is not surprising since it is a cultural practice of farmers in the study area to plant red clover with winter wheat. After wheat harvesting, the red clover is permitted to mature and is later cut for hay. It is encouraging to note that this mixture of wheat and red clover was detectable in individual fields.

A problem arises when the researcher attempts to classify such fields as either wheat or red clover. In this instance, all wheat stubble was considered as wheat whether red clover was present or not. Thus, test field methods tend to underestimate actual performance in this instance.

Oats was nearly all harvested by 5 August so the computer was trained to identify oats stubble. Correct recognition for oats averaged

71.2 per cent and ranged from 52.6 per cent to 83.6 per cent. Incorrectly identified samples were classified mainly as wheat, hay, and pasture.

Corn, the leading crop in acreage in the county, was correctly identified 93.7 per cent of the time. This evaluation ranged from a low of 87.7 per cent to a high of 98.1 per cent correct recognition.

Soybeans follow corn in amount of total acreage in the county; and these were the two most accurately identifiable crop cover types. Average performance for soybeans was 91.8 per cent, with a range from 83.6 to 94.0 per cent correct recognition.

Hay and pasture represented the two categories most difficult to classify accurately. Hay identification averaged 68.3 per cent correct whereas pasture was identified correctly 57.9 per cent of the time. The ranges of performance for both were large. As a result of the poor ratings, a test was conducted to determine whether performance could be improved by grouping hay and pasture as forage. The forage class performed at an average of 80.0 per cent correct recognition and ranged from 64.8 to 89.2 per cent correct recognition.

Water, as a class, yielded the highest performance ratings on test fields: 99.7 per cent correct. This rating is, however, somewhat misleading as only two water bodies were identifiable on the computer-generated maps, and these represented a large pond and the Wabash River. The fact that smaller water bodies were not identifiable as water may have resulted from waterways being dry, shallow, or weedy or may have been due to the dimensions of each remote-sensing unit.

Trees, as water, were classified with a high degree of accuracy: 98.0 per cent. This figure, like that for water, can be misleading since the average remote-sensing unit size does not permit the accurate identification of individual, scattered trees. As a result, only dense stands of trees could be used for training the computer and for testing performance.

Bare soil was identifiable at an average accuracy of 86.4 per cent. The range was from 72.4 to 93.9 per cent correct recognition. This category was representative of such features as fields lying fallow (few in number), bare spots in fields, and roads under construction.

Another category, that referred to as threshold, is composed of those remote-sensing units which the computer was not trained to recognize. Features found in the threshold class include roads, railroads, dry roughage, and urban areas. Such features were not assigned computer symbolization and were thus left blank on the completed computer map. It is the absence of symbolization which makes these features recognizable. The threshold class, being residual in nature, was not specifically tested for performance.

This study evaluated the performance capability of cover type identification using remote multispectral sensing techniques. The performance for some classes, specifically corn, soybeans, water, and trees is encouraging. The identification accuracy of other classes leaves much to be desired. Performance may be improved by using data from alternative dates, by obtaining repetitive coverage, or by altering scale characteristics. The geographer's concern is for a remote-sensing method which will permit him to identify accurately and analyse surface features of the earth. Remote multispectral sensing techniques potentially meet that concern.

Laboratory for Application of Remote Sensing, Purdue University; Indiana State University.

Landgrebe, D., and T.L. Phillips, 1967 A multichannel image data handling system for agricultural remote sensing. Paper presented at the Computerized Imaging Techniques Seminar, Washington, DC.

P1307
Methods of medical cartography for geomedical research
H.J. JUSATZ   *Heidelberg Academy of Sciences, West Germany*

The task of medical cartography is to record, by means of medical maps, observations concerning the occurrence and distribution of diseases and their changes with regard to time and location. A medical map not only depicts the topography, but also presents medical facts in a given region in a clearly arranged cartographical manner. This aim will be best achieved under the following basic conditions: (1) a minimum of cartographic representation concerning the projection of parallels and meridians, towns and rivers, international boundaries and landscape types, so that it is possible to locate the registration of the occurrence of a disease without difficulty and to compare with other maps; (2) a minimum of quantitative information about the occurrence of the disease in question in the portrayed area in a defined time; (3) a correct application of simple cartographic methods of representing locations or areas with a statement of the different intensity or temporary succession of the disease.

The following methods of applied cartography are used:
1. For limited amounts of information, it is possible to produce only a record map, on which the number of cases is registered according to the absolute method (dot method).
2. It should be noted whether the reproduction of the distribution of a disease in an area is based on a primary map of original references (e.g. based on exact observations and measurements in the area), or a derived one (e.g. with reproductions of revised material or mean values). The latter include density maps, as well as maps with isolines. However, the representation of density according to administration units (districts, etc.) is frequently a generalization of reality.
3. Only medical maps, in which areas with recording rates can be differentiated and marked off by occurrence and frequency of different diseases (the panorama of diseases) from other areas with equal conditions, give an objective picture as nosochoretic maps.

Diesfeld (1970) has pointed out a new method for an exact evaluation of bioclimatic and geographic-medical dependencies of the prevalence of diseases in a country: by the application of hospital returns. By means of 'hospital recording rates' which are referred to the 'hospital catchment area' and to the population living in that area, it is possible to set up comparable values for each disease which are correlated with the meteorological and geographical conditions observed in the same area.
4. The medical-cartographic methodology has to go beyond the simple registration of the territorial spread of a disease in a country. The medical map has – except the statement about occurrence – also to represent particulars about other factors of the physical, biological, and social environment of man which, according to the results of pathogenetic research, are also responsible for the occurrence of the disease in question in the area reproduced on the map. Outlining this correlation makes great demands upon the cartographer. Correlations with geofactors are of increasing importance in modern epidemiology. In order to be able to use an epidemic map for prognostic purposes, as well as for the study of the development of an epidemic, the geomedical maps have to give the reader a relatively uncomplicated survey of the ecological conditions or the so-called geofactors of climate, morphology, vegetation, etc. which may have an influence on the course of an epidemic. For instance, it is known that these influences can increase or limit the spread of animal vectors and animal reservoirs. Examples of this are climatic boundary lines, especially isotherms, isohyets, etc., which result from the ecology of animal vectors and which can easily be drawn in epidemic maps with a neutral colour.

The purpose of delineating the distribution of geofactors which influence the disease concerned, along with the disease, in a map is: (1) to give a better understanding of why this disease occurs especially intensively in the represented area, (2) to show that the spread of the disease is limited, and (3) to set down the potential area of a more extensive distribution of a disease in a map by representing the geofactors which favour it.

The task of medical cartography is to give in a map an outline of the ecological dependencies between the occurrence of a disease and the geographic and climatic conditions of an area as simply as possible.

The scientific target of all these investigations of medical geography has to be seen in the prognosis for a more extensive spread of a disease in location and time on account of a geo-ecological analysis represented by a medical map.

Diesfeld, H.J., 1970 The evaluation of hospital returns in developing countries, *Meth. Inf. Med.* 9, 27–34.

Jusatz, H.J., ed., 1967–71 Geomedical Monograph Series, Regional Studies in Geographical Medicine / Medizinische Länderkunde, Beiträge zur geographien Medizin (Springer-Verlag).
Vol. 1 Kanter, *Libya / Libyen*, 1967.
Vol. 2 Fischer, *Afghanistan*, 1968.
Vol. 3 Schaller/Kuls, *Ethiopia / Äthiopien*, 1971.
Vol. 4 Ffrench/Hill, *Kuwait*, 1971.
Rodenwaldt, E., and H.J. Jusatz, 1952–61 *World Atlas of Epidemic Diseases / Welt-Seuchen-Atlas*, vols. I–III (Hamburg).

P1308
Cartographic presentation of three-dimensional urban information
CHOR PANG LO *University of Hong Kong*

Cities today are increasingly characterized by prominent vertical growth so that the height of the building which constitutes the third dimension becomes a meaningful parameter in understanding the morphological and functional development of an urban environment. The cartographic presentation of the three-dimensional 'shell' of the city (i.e., the townscape) as well as the three-dimensional land-use patterns correlated with this 'shell' (i.e. the functions) should be invaluable as an aid in urban analysis. The following describes two methods whereby each of these problems can be dealt with.

It appears that the most effective means of depicting the three-dimensional townscape is still the age-old method of perspective which combines architectural and topographical details into a town plan (Lobel 1968; Hodgkiss 1969). The success of such an approach has been demonstrated by the recent work of a German graphic artist, Hermann Bollmann, who has been producing a series of *Bildkarten* or picture maps of towns with the perspective method ever since 1948 (Kinniburgh 1964; Hodgkiss 1969). It is particularly interesting to note his use of ground and aerial photography after 1960 to speed up his work and the application of his method to map such modern American cities as New York.

It has been found that aerial photography can be more properly used together with a photogrammetric machine and a special pantograph system to produce three-dimensional urban maps. A suitable machine for this purpose is the Galileo-Santoni Stereosimplex IIC, a topographic plotter, which can be coupled directly to a special pantograph system called the Perspektomat P-40 manufactured by F. Forster of Schauffhausen in Switzerland. The Perspektomat is a device for constructing block diagrams from contour maps using the parallel perspective principle.

This principle makes use of the normal axonometric projection in which the projection rays from the objects are perpendicular to the plane of projection (i.e. the drawing surface). As a result, the actual dimensions of the object are kept in their correct proportions so that direct measurement is possible.

The combination of the Stereosimplex IIC with the Perspektomat, therefore, forms an integrated photogrammetric-cartographic system which can be employed to produce three-dimensional urban maps directly from aerial photographs (Lo 1970).

In addition to presenting the morphology of the city, it is also necessary to present the land-use information at different levels of the city. Invariably, this involves mapping of correlations of relationships and patterns of uses. Previously, there have been attempts to map correlations by using superimposition of colour symbols as exemplified by the *Urban Atlas* (Passonneau and Wurman 1966). The same idea can be used in mapping the three-dimensional land-use patterns of the city. However, instead of employing the expensive and technically difficult overprinting of colours found in the *Urban Atlas*, the preprinted black-and-white patterns of lines, dots, and other geometrical shapes (such as the Letratone or Formatt) may be used. By superimposing one pattern on top of another, a new pattern can be produced characterizing a certain combination of uses, or an old pattern can be enhanced indicating the persistence of the same use at different levels. The superimposition of different families of curves gives rise to moiré patterns (Shulman

1970). Moiré patterns can be produced by the intersection of lines, or by overlapping two identical sets of dot pattern, two identical sets of rectangular patterns, or evenly spaced concentric circles, etc. In the case of intersection of lines, as the lines become more parallel, the moiré pattern line image becomes larger; similarly, in the case of overlapping dot patterns, as the dot pattern orientation tends to coincide, the moiré image pattern tends to enlarge. By using these properties of moiré patterns, the regular line, dot, or any other geometric patterns can present a graded series of intensity if desired, while the ungraded patterns serve as good qualitative symbols.

It is advisable to stick to a certain basic pattern for a certain type of use, and variations of this basic pattern will easily indicate subtypes of this use. Thus all offices may be shown by a dot pattern, but different types of offices can then be distinguished by different densities or arrangements of dots.

A great advantage of the moiré pattern maps lies in the fact that they can be easily revised to take into account the rapidly changing nature of the city. If the originals of these maps have been properly stored, the Letratone patterns on film can be easily peeled off again without causing too much damage to the surface of the drafting film, and a new pattern for a revised use can be stuck on again. Duplicates can then be made for further use.

A major limitation is that with more and more storeys of buildings involved, a larger number of sheets becomes necessary, and more material has to be used, more work has to be done, and the more difficult it becomes for superimposition and interpretation, for the light of the table will not be strong enough to penetrate so many layers of films to produce the desired moiré patterns. This is a case to be expected in an urban environment such as Hong Kong where multi-storey buildings abound. The answer to this is perhaps to combine two or more storeys into one sheet so that a reasonable number of sheets for superimposition or overlapping is always maintained; or a better solution is to map separately only those storeys at the lower levels, say, below the third floor, and combine the uses of all the higher storeys in one sheet. This appears to be a justifiable approach because the land-use pattern at higher levels tends to be more uniform and simpler than that at the lower levels.

An example of a set of moiré pattern maps of land use for a part of the city centre of Glasgow at the scale of 1:2500 has been done. This consists of four separate land-use maps by floors (i.e. ground, first, second, and third and higher) all produced on clear films.

In this paper, only two methods of presenting three-dimensional information cartographically have been suggested. There are other approaches which are equally suitable, such as the use of anaglyph maps (Adams 1969) and stereo-orthophotographs (Collins 1968). In the light of recent research into cartographic and photogrammetric digitizing (Petrie 1970), the production of three-dimensional urban maps and the moiré pattern maps of land use certainly stand a good chance of being automated and perfected in the near future.

J. Keates, A. Petrie.

Adams, J.M., 1969 Mapping with a third dimension, *Geog. Mag.* 42, 45–9.
Collins, S.H., 1968 Stereoscopic orthophoto maps, *Can. Surv.* 22, 167–76.
Hodgkiss, A.G., 1969 The depiction of towns, *Bull. Soc. Univ. Cart.* 3, 12–23.
Kinniburgh, I.A.G., 1964 Looking down town maps, *Cart. J.* 1, 52–3.
Lo, C.P., 1970 Determining and presenting the third dimension of a city centre: a photogrammetric approach, *Photogrammetric Record* 6, 625–39.
Lobel, M.D., 1968 The value of early maps as evidence for the topography of English towns, *Imago Mundi* 22, 50–61.
Passonneau, J.R., and R.S. Wurman, 1966 *Urban Atlas: 20 American Cities – A Communication Study Notating Selected Urban Data at a Scale of 1:48,000* (Cambridge, Mass.).
Petrie, G., 1970 Photogrammetric digitising: input for data processing, Paper presented at the Symposium of Commission IV, Int. Soc. for Photogrammetry, Delft (to be published *ITC Publications, Series A*).
Shulman, A.R., 1970 *Optical Data Processing* (New York).

P1309
Maps of road travel speed and their evaluation
A. MORRISON   *University of Glasgow, UK*

This paper is intended to explain a set of experimental road speed maps of part of Britain. The purpose of these maps is to show which is the minimum time route between any two points.

Journey speed information was obtained from unpublished records and from surveys on the road carried out by the author. The speeds were surveyed or adjusted to apply to a motor car driven with a desired speed of 96km/hr (60 miles/hr) between 0930 and 1630 hr, Monday to Friday, in spring or autumn 1968.

Owing to time limitations, this paper is confined to designs which resemble a conventional coloured road map, but I have also experimented with unconventional techniques in black and white. In each set of experimental maps only one design element has been varied. For reasons of space, illustrations could not be reproduced here.

The means of representation should preferably be able to show the conventional official road classification in addition to the speed classification (Morrison 1966, p. 21). The most satisfactory technique is to use a casing to represent the official classification, and to show the speed classification by the colour of the filling.

On a speed map, unlike a conventional road map, the pattern of the roads does not tell the reader which colour represents fast, which medium, and which slow roads, so the colours themselves must suggest this without ambiguity. There are several conflicting mental stereotypes which the reader may use. He may think of the colours as representing the range from high to low speeds, from good to bad conditions, from 'go' to 'stop,' or from urban to rural speeds, or he may think in terms of the colours used for direction signs on roads of different standards.

The six experimental colour schemes should also be compared according to which give a nearly correct impression of relative speeds; which are pleasing to the eye; which appear conventional; how many printing colours are needed; whether the map can be printed with the process colours; and how much of the fillings will appear in a screened colour. No scheme is preferable in all these respects, but a satisfactory compromise can be made.

The number of classes should be the minimum that will enable the reader to deduce correctly, in the majority of cases, which is the quicker of two alternative routes which differ significantly in travel time. Trial and error shows that at most scales this number is 5, but at 1:312,500 it is 6 or 7, and at 1:5,000,000 four are enough.

For most of the experimental maps, when the frequency distribution of speeds is plotted with respect to time rather than length, there is a symmetrical peak at 50 to 63km/hr, and a minimum at 90km/hr, which represents the distinction between motorways and ordinary roads. The class limits adopted were based on the mean ($\bar{x}$) and standard deviation (s) of this distribution, and are $\bar{x} - 2s$, $\bar{x} - 0.5s$, $\bar{x} + 0.5s$, and $\bar{x} + 2s$. For example, the limits used for the 1:625,000 map are 27, 49, 64, and 86km/hr. These yield five classes which contain respectively 1, 17, 46, 32 and 4 per cent of the total length of road shown, and represent respectively 3, 24, 45, 25, and 3 per cent of the travel time.

The maximum scale for a satisfactory speed map is about 1:300,000. One reason is that at larger scales the speed may be represented twice. At 1:300,000 one can estimate the delay due to bends from the plan shape; due to hills from the contours; and due to congestion from the extent of the built-up area. Another reason is that there can be differences in average travel time between the two directions of travel on the same section of road, and the time required to turn left at the end of a section may differ from the time required to turn right. These time differences, typically about 0.3min, introduce noticeable errors if the journey time for the section is less than about 3.0min. The expense of accurately measuring the average speed over short sections also suggests that they should be avoided. For example, to measure the average speed for a 2-minute rural section with 10 per cent accuracy, nine trial runs by car are needed, whereas for a 10-minute section the number is reduced to three (Dawson 1968, p. 20).

Thus there is no need for the map to have

a scale sufficient to represent sections less than 3 minutes in length. The shortest section should be represented on the map by a length of at least 5mm. On this basis a 3-minute section having the most usual average speed of 60km/hr would require a scale of only 1:600,000. A 3-minute section having a speed of 30km/hr typical of urban areas would require a scale of 1:300,000. Consequently, the experimental maps were prepared mainly on the scales of 1:625,000 and 1:1,250,000, while the larger scale of 1:312,500 was used only for the Clydeside conurbation.

In order to select the minimum time route from one of the experimental speed maps, the map reader must compare several alternative routes, each consisting of several sections of different lengths and colours. Examination of the maps suggests that, if the journey is between points more than 100mm apart on the map, then the map will show so many alternative routes (more than 10) made up of so many different sections (more than 30) that the reader will be unable to reach a definite conclusion. On the other hand, if the points are less than about 35mm apart, it is possible that the map may omit a route that is significantly faster than the routes which do appear. Thus a range of scales is required, each scale being optimum for planning journeys of a particular length, e.g. 1:625,000 for journeys of 20 to 60km, and 1:5,000,000 for 160–500km. Speed maps on scales smaller than 1:5,000,000 are practicable but perhaps unnecessary, because business journeys longer than 500km are likely to be made by air or rail rather than by road.

Some generalization of the basic data on speeds is obviously required to ensure that sections are long enough to be represented, say at least 2.5–5.0mm at map scale. But how much further generalization is needed to meet the user's two conflicting requirements?

These are: first, that alternative routes between distant points shall each consist of only a few sections of different colours, so that he can easily compare them; and second, that sections between every village and every road junction should be shown separately, so that routes can be compared between all pairs of places by every reasonable alternative route.

Consideration of five experimental maps leads to the conclusion that the best compromise is the map on which speeds have been averaged between all junctions appearing on the map, but on which there has been no further generalization. A link should only be divided at an intermediate village if the total trips generated by the village are several times the number of through trips. Anomalies arise on the more generalized maps, on which speeds on the higher classes of road are calculated for long sections, while those on the lower classes must still be calculated for short sections. A better way to assist route planning of long journeys is to omit some of the roads. This is equivalent to using a map at a smaller scale.

A computer simulation of the map-reading process, based on the trips between fifty random pairs of towns, indicates that by choosing his routes from a version of the speed map instead of from a conventional map, a motorist will on average reduce his travel time by 12.8 per cent. Tests on a random sample of real map users, to be carried out during the next six months, are intended to verify this result.

Dawson, R.F.F., 1968 *Economic assessment of road improvement schemes*, Road Research Technical Paper, no. 75 (HMSO, London).

Morrison, A., 1966 Principles of road classification for road maps, *Cart. J.* 3, 17–30.

P1310
Data acquisition and presentation for the glacier inventory of Canada
C.S.L. OMMANNEY and D. GAGNON *Environment Canada*

An inventory of Canadian glaciers was included in the Canadian program for the International Hydrological Decade (IHD) in December 1964. In 1970 formal guidelines for the completion of a glacier inventory, which are the basis of the Canadian program (Ommanney 1970), were published by UNESCO/IASH (1970).

All the information required for each glacier is contained on a standard data sheet.

The basic data are listed for key punching on to four cards (I–IV) in the following manner:
I: the region and basic classification (which is also repeated on subsequent cards), the glacier number, geographical and Universal Transverse Mercator (UTM) coordinates, orientation and elevations of different parts of the glacier;
II: date of the snowline, mean elevations, lengths, and width;
III: data on the surface area, accumulation area ratio, volume, and the glacier classification and description;
IV: general information on geomorphological features, a written description of the ice mass and its name.

The rest of the data sheet provides general information on maps and photos used and any other items of interest. It is estimated that there are about 100,000 glaciers in Canada covering just over 200,000km^2. Details of how the data on these glaciers are acquired and processed are given below.

Compilation of the basic glacier inventory data is based on standard air photo-interpretation techniques using mirror stereoscopes, parallax bars, proportional compasses, and a sketchmaster. On each photograph the areas of perennial ice and snow are identified, the ice divides and snowlines delineated, unusual surface features, moraines, glacier-dammed lakes are marked, and the glacier orientation is indicated with an arrow. Additional information is provided through a general description of the glacier and a glacier classification which identifies the major and minor forms in a 6 × 10 matrix, e.g. valley glacier – simple basin, mountain glacier – cirque; the snout characteristics, e.g. expanded foot; the nature of the longitudinal profile, e.g. cascading; the assumed main form of nourishment; and the estimated activity of the snout. The photography being used for this study is the vertical coverage available from the National Air Photo Library, Ottawa, mostly taken in the late 1950s and early 1960s from 30,000ft (9144m) above sea level at a scale of from 1:50,000 to 1:60,000 for most of the glacierized areas. Information from the photographs is then transferred to the best scale of maps available.

A D-MAC pencil follower interfaced with a PDP-8I computer is used for the accumulation of data from the glacier inventory work maps. Communication between the operator and the computer is through a teletype that requests the following input data:
1. The glacier number, region, and basin and UTM grid zone.
2. Four UTM fiducial points, usually the map corners which are then digitized.
3. The glacier identification point.
4. The appropriate code for the three different parts of the glacier to be digitized – accumulation area, ablation area, and debris-covered part.
5. The contour line elevations which are digitized with the outlines of 4.
6. The maximum length and width which are defined by the digitized line.

All digital information is written on magnetic tape for processing on a UNIVAC 1108.

Output from the UNIVAC 1108 is in the form required for the standard data sheet, excluding the volume estimates and the general information noted from the photographs. All figures are given in metric units.

The analysis of the glacier inventory data obtained from digitizing the work maps is carried out through a program developed by R. Goodman, of the Inland Waters Branch, which presupposes no prior knowledge of computer programming. A control card masking function in columns 1–7 permits the selection of one or a number of types of glacier from the first column of the 6 × 10 classification matrix. Subsorting is performed in an exclusive OR fashion on the subsequent 5-digit numbers (columns 8–51). On this same control card, the item to be analysed is identified by placing an alphabetic character (to indicate, for example, a range, an average, or a histogram) in the appropriate column of 52–62. Step sizes and range limitations are specified on a second control card. Column 80 is used for data printout functions and the remaining columns 63–79 are either not yet assigned or are designated for alphabetic data such as orientation (8-point compass). The program is written in Fortran V to run on a UNIVAC 1108. The program has been described (Ommanney et al. 1969).

Each glacier is identified by region and hydrologic basin codes and a glacier number. The glaciological information is shown on a series of map sheets at a scale of 1:500,000, photo-reduced from the compilation's base scale of 1:250,000, which will go to make up the Glacier Atlas of Canada. The format for the maps is standardized to reduce drafting and production costs, but this means that the orientation of north on the map sheets may vary by up to 180°. The sheet size is 11 by 15 inches.

The colours used for the maps are taken from the standard colour chart of the Map Compilation and Reproduction Division (MCR), Surveys and Mapping Branch, Department of Energy, Mines and Resources. A 30 per cent screen of Offset Transparent Map Brown (MCR 131) forms the border of each map; this contains the legend, a bar scale, and a small-scale index map of the region indicating the location of the specific map. The large-scale map shows the glaciers in a Cyan Blue (MCR 111) vignette with a black dot, located near the snout, and a number to identify the individual ice masses. Areas outside the main glacier basin are shaded with a 10 per cent screen of the outside border colour. Type and colour for the sea and lakes is the same as for the National Topographic Series 1:250,000 maps. Each basin is identified with a 10 per cent Map Drainage Blue (MCR 112) letter and the major area numbers are indicated in black. Every map sheet has a decimal number that will be used for indexing. The maps are being drafted by the Inland Waters Branch, Department of the Environment, and printed by the Surveys and Mapping Branch. For each major region, e.g. Baffin Island, small-scale general indices are being produced; these show the areas covered by the larger-scale maps outlined in red and identified by the appropriate decimal number.

Approximately 150 map sheets will be required to cover all the areas of perennial ice and snow in Canada. These maps will be (1) included in the detailed glacier inventory reports, (2) available as individual sheets, and (3) bound into an atlas when the project is completed.

To date, 25 maps covering Baffin and Bylot Islands and 5 covering Axel Heiberg Island have been produced; they are available through the Map Distribution Office of the Department of Energy, Mines and Resources.

R.H. Goodman, J.A. Gilliland, D. Mackenzie; Map Production Unit of the Department of Energy, Mines and Resources, Canada.

Ommanney, S., 1970 The Canadian glacier inventory, in *Glaciers, Proceedings of Workshop Seminar 1970* (Vancouver, Canadian National Committee for the International Hydrological Decade), 23–30.

Ommanney, C.S.L., R.H. Goodman, and F. Müller, 1969 Computer analysis of a glacier inventory of Axel Heiberg Island, Canadian Arctic Archipelago, *Int. Assoc. Sci. Hydrol. Bull.* 14 (1/3), 19–28.

UNESCO/IASH, 1970 Perennial ice and snow masses, *Tech. Papers in Hydrol.* 1, UNESCO no. A. 2486.

P1311

Multilevel principle of generalization and the problems of automatic map reading
E. SHIRYAEV   *Moscow State University, USSR*

Generalization, which we consider to be delineation of what is most typical and essential, implies along with its positive aspects the disadvantages which result in the loss of geometrical exactitude and the precision of information presented.

An attempt is being made to maintain the level of generalization of representation, so necessary for visual perception of the greatest detail and exactitude of representation on a map.

The second important task is an automatic selective reading of information from such maps and the feeding of it into a computer.

A technique of multilevel generalization is proposed. In essence, it consists of dividing the information given into levels of generalization depending on the amount of decrease of the object's geometrical size (i.e., according to how small the contour of the objects portrayed are, relative to the scale of cartography). This is done by the introduction of some quantitative values according to the area, brightness, raster parameters, etc.

To represent objects of smaller area, including those not depicted at the scale of the map, a technique is used to produce an areally different system of uniparallel linear rasters (i.e., a technique where every gradational stage is based on the thickness, or possibly on the frequency and colour of the raster lines, relative to the smallness of the contour). This appears to create a new scale, which is greater than that of the map.

Let us assume that we have three quantitative value gradations P_1, P_2, P_3, with corresponding raster thicknesses T_1, T_2, T_3. The quantitative value P_1 with the highest line thickness of T_1 is given to the large contours being depicted at the scale of a map; for the mean contour P_2 the mean thickness of T_2, and for the small ones P_3 with minimum thickness of T_3. We shall take as an elementary (discrete) length of a raster line a scanning pattern step d, or a division of the simplest measuring device. In this case, the formula for the areal quantitative value of the first gradation p_1 will be the product of the length of elementary line d, the interval between axial lines of the raster l, and the scale ratio of the map m_1 (i.e., the denominator of the scale of the map being made).

$$P_1 = d \cdot l \cdot m_1.$$

One can record the same values for the other quantitative gradations with ratios relating correspondingly to larger scales.

$$P_2 = d \cdot l \cdot m_2;$$
$$P_3 = d \cdot l \cdot m_3.$$

The area of a single contour $P_{(i)}$ is determined by measuring the general length of lines s and by adding together the lengths of elementary segments $d_1, d_2, \ldots, d_n$, taking into account the areal value p_i accepted for a given contour. In a sense it may be called the transformation of area into line lengths

$$P_{(i)} = p_i \sum_{k=1}^{n} d_k = S \cdot p_i.$$

It follows that the contour form can be arbitrarily changed, the important factor being the similarity of total length of lines with contour area. Taking this as a basis one can represent a small contour as a small segment of a line. Such techniques make it possible to show, on small-scale maps, objects of any size which retain their real area and location. The areas of these contours can be easily determined with a measuring device and a ruler.

It should be noticed that if the objects come too close together, contours can overlap. In this case the mutual penetration of raster lines becomes possible.

To represent linear objects both shown and not shown at the scale of a map, as for example river and ravine networks, the scale quantitative values are given only according to their width and relative to the brightness prescribed to one or some other group of objects. These are classified according to their size. Let us demonstrate this with the mapping of a river network characterizing water discharge. Most of the quantitative characteristics of rivers, including water discharge, can be given with two variable parameters: area and brightness of presentation. As the river width enlarges proportionally to its area, the change of this parameter will be taken as one of the river discharge characteristics. The scale of a river width is taken arbitrarily. Water discharge at some point is represented by the formula: $Q = S_i \cdot q$, where q is an arbitrarily chosen volume of water discharge per second for a unit of length with a river width S_i (e.g. 1.0mm = 10m^3 per second).

The second parameter characterizing water discharge is the brightness; the darker the colouring the greater the discharge. The necessity of introducing gradation of brightness is related to the fact that rivers with a large water discharge are depicted as unusually wide, while those with an insignificant discharge are not depicted at all.

Let us define the brightness ratio as $K_i = B_i/B_o$ where B_o is the brightness of the map's background and B_i is a brightness depicting a selected group of rivers. As B_o is larger than B_i, the introduction of the brightness ratio decreases the scale of representation for river width. This enables one to show water discharge of both large and small rivers on the same map. As a result the actual water discharge at the ith point will be characterized by the formula: $Q_i = (S_i/K_i)q$.

When producing a map, it is necessary to fix an optimal scale of brightness gradation. As experience has shown, three to four gradations will be quite enough. According to the chosen scale of brightness gradation B_1, B_2, B_3, ..., B_n, the river width is estimated only along straight sections of a river by the following formulae:

B_1 $S_1 = (K_1/q)Q_1h_1,$
      $S_2 = (K_1/q)Q_2h_1,$
      . . . .
      $S_t = (K_1/q)Q_th_1;$

B_2 $S_1' = (K_2/q)Q_1'h_2,$
      $S_2' = (K_2/q)Q_2'h_2,$
      . . . .
      $S_t' = (K_2/q)Q_t'h_2;$

B_l $S_1'' = (K_l/q)Q_1''h_l,$
      $S_2'' = (K_l/q)Q_2''h_l,$
      . . . .
      $S_t'' = (K_l/q)Q_t''h_l.$

A corrective ratio for generalization of river meanders, H, is found experimentally.

Thus the points recorded in this manner determine the conventional river width units.

It is advisable to carry out the work at a larger scale with a view to decreasing it further.

In both cases, the subjectivity of the process of generalization is reduced to a minimum. The visual estimation of information, along with singling out primary and secondary, or general and particular, is performed by the brightness grade or by the density of image, and in the case of rasters by the density or thickness of the raster's image.

The idea of multilevel generalization may be particularly effective by making use of luminescence phenomena when small objects are related to the unseen image.

The automatic reading of the above cartographic images can be performed with a photoelectron scanning device with both small and large scanning elements for reading. The computer recognition of information, together with differentiation by quantitative functions of areal objects in a raster image, is carried out by the form of impulses produced by the lines of various thickness and colour, and for linear objects by means of amplitude selection.

The amplitude selection can be carried out by the discrimination of amplitude signals formed by images of various brightness.

As the parameters are strictly determined and pre-known by the raster line thicknesses, and linear objects by their brightness, the process of reading is reduced to the mere identification of map image parameters by deciphering blocks with a reading device.

The combined computer and visual reading of maps, producing detailed and exact information, has great promise for research in geography.

P1312
Satellite photography as a geographic tool for land use mapping
NORMAN J.W. THROWER   *University of California, USA*

Research summarized in this paper concerns the application of satellite imagery from the United States Gemini and Apollo missions to land use mapping. The imagery provided the basic data source for the construction of land use maps of a very large region, the southwestern United States, at intermediate (1 : 250,000) and smaller (1 : 1,000,000) scales. In conjunction with this project, a land use classification system compatible with relevant observable phenomena on the photography was developed. The southwestern United States was selected as the area of study because it has a wide range of human activities with corresponding landscape expression. It also has nearly complete, and in many localities multiple, photographic coverage by Gemini IV and V and Apollo 6 and 9 missions. This imagery is largely unobscured by cloud cover.

The mapping project was undertaken by a team of researchers at the University of California, Los Angeles (UCLA), under the direction of the author, for the United States Geological Survey (USGS). This organization sponsored the research and provided imagery of a number of different kinds which had been generated from the 70mm colour transparencies brought back from space by US (NASA) astronauts. In addition to the colour transparencies themselves, the imagery included: colour and black-and-white prints (rectified and unrectified), some enlarged to the initial mapping scale of 1 : 250,000; controlled black-and-white mosaics of 1 : 1,000,000 and larger scales, annotated and unannotated; and USGS maps of various scales, including sheets of the 1 : 250,000 planning series.

Initial interpretations were made from the large colour prints using acetate overlays. This technique provided the basic input data on colour-textural patterns and, by extension, land use. These interpretations were then reinforced with analysis of transparencies using image projection, and magnification with an 8× tube magnifier on a light table. Black-and white prints and mosaics were used as sources supplemental to the colour prints and transparencies. All interpretation information was consolidated at 1 : 250,000 and plotted on USGS maps of this scale.

Among the various features interpreted from the satellite photos, agricultural land is very distinctive because of a dark colouring resulting from irrigation in generally water-deficient lowland areas, and the geometric shapes of fields related, typically, to the United States Public Land Survey. Transportation features were identifiable because of their linearity, primarily; however, it was frequently impossible to distinguish between roads and railroads from the images alone. Urban areas were discernible on the basis of a distinctive spectral signature. Hardrock mining sites, important in the US southwest, were characterized by a light-coloured, mottled-textured return, a function of disturbed land. Coniferous forests were associated with high mountains, while scrub woodlands occupied lower elevations in rural areas; these vegetation types could also be differentiated by colour, the former being dark blue and the latter a lighter, greenish-blue. Most of the area in the southwestern US is used for grazing, a land use not evident solely by inspection of the imagery.

A basic dichotomy existed among the phenomena observable on the photographs: (1) features that could be identified with a high degree of certainty; and (2) features delimited by colour-textural associations with no real certainty of identification. To serve as a verification, in the latter case particularly, several other sources were utilized, viz. published maps and reports, large-scale aerial photography, personal communication with knowledgeable individuals, and field check. To a greater or lesser extent, all of the above were employed and were especially valuable for establishing levels of interpretation confidence. In general, the drier the area the more satisfactory the interpretation; humid areas, towards the eastern margins of the region in this case, offered special problems to the researchers because of moisture in the air, as well as cloud cover itself.

An important goal of this project, as indicated earlier, consisted in the development and field testing of a land use classification scheme. Existing land classification systems were consulted (e.g. Stamp 1935; Marschner 1958; Christodoulou 1959). The final, modified land use classification system used for the interpretation of Gemini and Apollo imagery was as follows: (1) Transportation: (a) road and railroads; (b) airfields; (2) Settlements; (3) Cropland; (4) Arboreal Associations: (a) forest (coniferous), (b) woodland (scrub); (5) Extractive Activities: (a) mines and quarries, (b) oilfields; (6) Grazing Land (unimproved); (7) Water Bodies; (8) Unproductive Land; (9) Uninterpretable.

This classification is necessarily generalized owing to the scale of the mapping project and the quality of the satellite photography, with its limited but variable resolution. However, the system has been specifically constructed to permit expansion, which can be accomplished by adding major categories to the list above. New categories can be devised, as suggested by improved photography or as demanded by areas other than the ones considered here. Flexibility has been a dominant factor in the formulation of categories, since it was realized that undoubtedly better quality imagery will be forthcoming in the future. Actually the Apollo imagery, which was provided during the course of the research, made possible interpretations not feasible with the earlier Gemini coverage alone; for example, some pairs of Apollo photographs permitted limited stereoscopic examination possibilities. Moreover, other types of landscapes will be considered when satellite photography is put on an operational basis.

A working map scale of 1 : 250,000 was deemed the most suitable for displaying, cartographically, the observable land uses. A map possessing even greater potential value and importance is the 1 : 1,000,000 reduction (Thrower et al. 1970). This map shows the interpretable land uses in the southwestern US and also includes, as part of the map, a Gemini photo-mosaic base. Research leading to the production of the map revealed certain strengths and limitations of satellite imagery. The strengths include: geographical overview (without problems inherent in mosaicing many small frames); speed of interpretation including reduction of field work; and periodic coverage. Limitations, some of which can be overcome, relate to: lack of systematic overlapping coverage and therefore minimal stereoscopic viewing opportunities; and poor resolution as a function of scale and image quality. These problems can be resolved, or at least mitigated, by planning and improved technology.

The project demonstrates that satellite photography has definite utility for land use mapping at intermediate or smaller scales. Photography provided was of variable quality, but a significant number of solid land use categories were extracted from data generated from the imagery. Cartography resulting from this study provides a requisite first step in the analysis of the resources of an area and would be particularly useful for developing countries, where even rudimentary land use maps may not exist.

Christodoulou, D., 1959 *The Evolution of the Rural Land Use Pattern in Cyprus* (London).

Marschner, F.J., 1958 *Land Use and Its Patterns in the United States.* United States Department of Agriculture (Washington).

Stamp, L.D., 1935 *The Land Utilization Survey of Britain* (London).

Thrower, N.J.W., et al., 1970 Land use in the southwestern United States from Gemini and Apollo Imagery, Map Suppl. no. 12, *Ann. Assoc. Am. Geog.* 60(1).

P1313
A remote sensing system for detecting gross land use change in metropolitan areas
JAMES R. WRAY   *U.S. Geological Survey, USA*

This paper deals with developmental aspects of a system for detecting gross land use changes in a sample of United States metropolitan areas. One aim of the research is to identify for such application the operational role of remote sensors aboard aircraft and satellites.

Launched early in 1970, this work became known as the 'Census Cities Project.' It is an integral part of the Geographic Applications Program of the US Geological Survey. It has been supported through the Department of the Interior's EROS Program with funds and timely overflight data by NASA's Earth Observations Program. The Census Cities Project has been described in more detail in a paper ('The Census Cities Project and Atlas of Urban and Regional Change,' in *Proceedings of the International Workshop on Earth Resources Survey Systems*, University of Michigan, May 1971).

For twenty urban test sites, the USAF Air Weather Service and NASA's Manned Spacecraft Center acquired multispectral, high-altitude aerial photography at the time of the 1970 decennial census. The census returns serve as one form of 'ground truth' and one form of initial inventory for change detection. There are nine different combinations of film, spectral bands, and image scales. Simulation of imagery from the Earth Resources Technology Satellite (ERTS) is included. Photography is from the RB-57F aircraft flying at 50,000 feet or 15.2 kilometres above the terrain. This combination of census and sensor data is an unparalleled opportunity for comparative urban study!

The basic overflight data for urban change detection are provided by colour infrared photographs, which are recorded by an RC-8 metric camera at an image scale of 1:100,000. A controlled photomosaic of acceptable positional accuracy is being made from these photographs. We aim to procure similar aircraft photography over the same urban test sites when the ERTS satellite is launched in 1972. This will provide not only the basis for the detection of urban change but also for the interpretation and evaluation of the satellite imagery.

The mosaic is presented in sections, each a page in a looseleaf Atlas of Urban and Regional Change. It represents an early phase in environmental inventory and also one form of a user-oriented end-product. The sheet size is one standard computer printout page. The square image space accommodates modular maps or mosaic units at different scales. One such module will be a mosaic section, 20km by 20km, reproduced at 1:100,000. The one-kilometre grid is in the Universal Transverse Mercator (UTM) rectangular coordinate system. There is a panel for legends, including those to be added by users. The mosaic is used for locational control, area measurement, pagination in the atlas, and also as an underlay for computer-processed thematic maps. The mosaic is not intended to be the principal base for image interpretation, although it has some value for that purpose.

Another simulated atlas page shows boundaries and labels of census statistical areas. For cities, the basic unit is the census tract, an urban change detection device which has been in operation for several decades. A tract is a piece of real estate, usually bounded by streets, that remains essentially the same from one census to the next. The average tract has about 5000 persons. As pieces in a jigsaw puzzle, tracts can be combined to make up larger areas.

The land use classification scheme is based primarily on visible land use, not on knowledge of land ownership. This makes the system more applicable to observable change, but it is one more example of the classification dilemma confronting a prospective user who is already using a classification scheme based on data not readily detectable by remote sensing.

Land use is recorded by coloured pencil on a transparent film overlay fitted to the colour infrared photo at 1:100,000. Interpretation is done with the aid of a hand lens and light box; stereo examination also is possible. The smallest unit of observation is a 'use pattern' (I do not mean 'grid cell') that is not smaller than four hectares or ten acres. At photo scale, this is about the size of the blunt end of the coloured pencil. This unit is larger than the resolution capability of the aircraft imagery, but it may test the resolution limits of the ERTS imagery. Factors affecting usable informational detail include not only image resolution but also interpreter experience and available ground truth, the scale and minimum-area size of recording unit, the relative complexity of the areal phenomena being studied, and the range of uses of the information being acquired.

Land use interpretations are transferred to a rectified and gridded base (either map or mosaic) for area measurement and digitization of locations. Area measurement is done by eyeball method and dot planimeter at the present time. Land use interpretation and area measurement by semi-automated means are receiving some attention, but no immediate application is foreseen.

After land use is interpreted, the urban area is delimited. That is, the definition of 'urban' and 'non-urban' is actually done *after* the land use interpretation. The procedure for doing so seeks to define, for a given area, comparable urban real estate at different times, or for different urban areas at a given time. After the 1970 land use has been mapped and measured, the data are prepared for computer storage and retrieval.

'Change detection' was an earlier application of that part of remote sensing represented by conventional air-photo interpretation. Then, it was part of what was called 'damage assessment' in military operations, and it implies pre-event inventory as well as post-event analysis of change. There are similar applications in assessing sudden environmental changes resulting from fire, flood, earthquake, civil disorder, and civil defence. There are similar applications to the more gradual changes resulting from day-to-day constructive and destructive ecological processes, including pollution of land, water, and air resources.

There are two noteworthy existing systems for detecting change in urban and regional environments, only one of which uses remote sensing. The first of these is the census tract, or other statistical area, for which comparable census data are reported for successive censuses. The other is the modern 7½-minute topographic map published by the US Geological Survey, which shows, by magenta overprint, selected planimetric additions (mostly works of man) which were not present on the earlier edition where conventional classes of features are shown in traditional ink colours. The basis for the new map information is an interpretation of more recent aerial photography. Any comprehensive system of urban change detection ought to provide the basic 'what-ness,' 'where-ness,' and 'when-ness' of both the census tract and the magenta map overprint. This is what the Census Cities Project and Atlas of Urban and Regional Change attempt to do.

P1314
Remote sensing of the environment with a Ka-band microwave radiometer
JOHN O. HOOPER and ROBERT P. MOORE   *Naval Weapons Center, USA*

This paper describes recent advances in remote sensing of the environment with microwave radiometers. The advantages and disadvantages of microwave radiometry vis-à-vis other electromagnetic sensors are considered. Flight test results with a unique Ka-band mapping radiometer using a parametric amplifier are described.

Microwave radiometry (MICRAD) is the science of measuring naturally occurring microwave radiation. Radio astronomy is the best known example of MICRAD; terrestrial microwave radiometry or the mapping of the earth with a microwave radiometer is our interest here. The acronym MICRAD will refer to the terrestrial use of microwave radiometry in this paper.

All material bodies in the universe emit electromagnetic radiation as a function of the radiation frequency observed, the absolute temperature of the body, and the reflectivity of the body, among other parameters (Planck 1914, 168). In the microwave portion of the electromagentic spectrum the power received from a body is directly proportional to the absolute temperature of the body (Planck 1914, 170). Because of this relationship it is convenient to speak of the radiometric temperature in place of the microwave power emanating from a body. Consequently, the MICRAD map is less affected by changes in the terrain temperature than is an infrared map; hence, a MICRAD scene is more predictable if terrain reflectivities are known. In a given region on the earth's surface the absolute temperature will vary less than ten per cent (except for infrequent targets) while the reflectivity of the same region will vary approximately 900 per cent; therefore, the MICRAD signal is due primarily to changes in target reflectivity.

There are two components of the MICRAD signal for opaque targets: (1) The component of radiation emitted from a target is a function of the target temperature and emissivity. (2) The component of radiation reflected from the target is not a function of the target temperature, but of the target reflectivity. The usual source of reflected radiation is the sky, which radiates very little energy.

In addition, the microwave radiometer operates with incoherent radiation; that is, the MICRAD receiver operates on a wide band of temporally incoherent electromagnetic noise, and a spatially incoherent noise source, i.e., sky radiation. Since the sky radiation is a slowly changing function of angle of incidence at the frequencies of interest, the MICRAD map is very stable with respect to look angle. High resolution radar, however, uses both temporal and spatially coherent radiation, which produces the well known scintillation and clutter effects; a MICRAD system is essentially free of these harmful effects. Since MICRAD operates in the microwave region of the spectrum it is capable of operating in any weather conditions, with practically no degradation in performance if a proper choice of operating frequency is made.

A MICRAD mapping system has lower spatial resolution than visible spectrum photography, IR sensors, and high resolution radar. However, spatial resolution is only half of the resolution problem. Resolution (the ability to detect and separate close objects) is a function of spatial resolution and signal resolution. The high spatial resolution of optics is of no value if the items resolved are not bright enough to see or are obscured by clouds, fog, or smoke. A MICRAD sensor 'sees' some objects that other sensors do not see. For example, a Ka-band MICRAD map of cloud covered terrain 'sees' the land patterns almost as if there were no clouds, while the visible photograph of the same scene 'sees' only a white blanket of clouds (Hoover and Moore 1971, 31).

Research and development in MICRAD has been carried out for about a decade now (Porter and Parker 1960; Hooper 1966; Hoover 1971). Early work was primarily limited to the lower microwave frequencies for airborne applications, because of the lack of quality components at the higher microwave frequencies. Advances in microwave technology have allowed the Naval Weapons Center to construct a very sensitive Ka-band radiometer by using a parametric RF amplifier. The system maps the earth by scanning; three rotating parabolic antennas scan in the transverse direction, and the forward motion of the aircraft performs the scan in the flight path direction. The temperature sensitivity of the receiver is 0.08°K rms for a one second integration time but with losses this becomes 0.15°K. The spatial resolution is 1.15° between ½ power points of the antenna pattern. Recorded MICRAD signals are processed by an analog and/or digital processor to produce a photographic display of the data.

The Ka-band radiometric system has been mapping urban and rural land areas and the Pacific ocean during 1970–1. Most of the data have been collected from an altitude range of 300 to 3000m with some flights at 6000m. The data have been collected during the following weather conditions: very cold snow covered terrain, melting snow covered terrain, hot desert areas, with and without rain, clouds, and smoke. Some of these data will be described in the remainder of this paper.

Some comparisons of optical photographic intensity and relative radiometric temperatures for summer data have been accomplished. The comparisons were made between a panoramic aerial photograph of a rural farm land scene consisting of approximately 77.6km² and a Ka-band radiometric map, in a photographic format of the same area. The aerial photographs were taken about 6 weeks after the MICRAD maps were collected during the summer. The scene contained 247 farm fields, almost no trees (rows of trees for wind breaks), numerous sloughs, and small lakes. The terrain was very flat. Sixty-four of the farm fields were classed as radiometrically cold; the remainder as radiometrically hot. From the set of cold fields 63 of them appeared on the aerial photographs as fields of medium optical density with a unique mottled appearance. The other cold field appeared optically dark. The radiometrically hot fields appear on the aerial photograph in the following manner; 48 light (some mottled), 61 medium (not mottled, like cold fields), and 74 dark optical densities.

The MICRAD maps of urban areas show a great deal of detail with a photographic appearance. Most of the objects one finds in an urban area are visible in the MICRAD scene. Roads, buildings, and parks are some of the items observed in the MICRAD scene (Hoover and Moore 1971, 30).

The detection and location of forest fires through dense smoke has been accomplished. An example of this is the MICRAD maps obtained of the brush and forest fire near Santa Barbara, California during the autumn of 1971.

The uses of MICRAD as a remote sensor of the environment are yet to be exploited. Some uses are obvious to observers of good quality MICRAD maps such as forest fire detection and location through dense smoke, determination of agricultural area uses, and urban land uses.

Hooper, J.O., 1966 *Microwave Radiometry and the NOLC X-band Radiometer Program*, NOLC Report 641, Naval Ordnance Laboratory (Corona, CA).

Hoover, M.C., and R.P. Moore, 1971 *An Airborne Ka-band Microwave Radiometric Measurement Mapping System*, NWC (China Lake, CA).

Planck, M., 1914 *The Theory of Heat Radiation* (New York), Eng. trans. by M. Masius.

Porter, R.H., and M.S. Parker, 1960 *A Survey of Microwave Radiometers With Terrain-Mapping Applications*, AVCO Corp. Rept. No. RAD TR-9-60-20, AD243–29.

P1315

Le géocodage de Statistique Canada: système d'information statistique géographique automatisé au stade opérationnel
A.S.R. TERJANIAN and J.J. LEFEBVRE *Statistique Canada*

Durant les quelques années passées, les agences statistiques ont subit la foudre des critiques du monde académique et de celui de la recherche à cause de leur manière de traiter la dimension spatiale des données statistiques (Hägerstrand 1967).

Nous avons relevé le défi et avons mis au point ce système dont les objectifs premiers sont de: (1) permettre l'extraction des données du recensement pour les zones arbitrairement délimitées par l'usager, (2) créer une capacité d'exploitation très souple – susceptible de produire des tableaux statistiques variés sans nécessiter l'intervention d'un programmeur, (3) servir les usagers dans des délais minimums et à un coût raisonnable.

Ces objectifs sont maintenant atteints en même temps que quelques autres objectifs secondaires tel la capacité de géocoder automatiquement d'autres fichiers que ceux du recensement ainsi que la connection des fichiers du géocodage à des programmes tels que SYMAP dans le but de produire rapidement des cartes thématiques.

Ce système se distingue de la majorité des autres systèmes ayant les mêmes buts du fait qu'il est conçu à l'échelle nationale; ainsi le Canada est entièrement couvert par le système à deux niveaux de précision géographique: Le côté d'îlot dans 14 grands centres urbains (10 millions d'habitants, 250 mille côtés d'îlots) et le secteur de dénombrement du recensement (SD) dans le restant du pays (12 millions d'habitants, 30 mille SD). Le géocodage au niveau du côté d'îlot s'étend graduellement pour couvrir la majorité des centres urbains et des méthodes de géocodage plus souples et moins dépendantes du recensement sont à l'étude pour les parties rurales.

Le système se compose de trois phases principales: *Phase* I: Dans cette phase les tranches d'adresses correspondant aux côtés d'îlots sont codées et reliées à leurs coordonnées respectives; ces dernières sont relevées à partir de cartes urbaines au $1/10000^e$ à l'aide d'un coordinatoscope. Une opération similaire mais plus simple prend soin des SD. D'autres paramètres géographiques sont aussi ajoutés à ses fichiers et un nombre d'opérations de vérification (tracées par ordinateur) et de mise à jour viennent se greffer aux opérations normales (Boisvenue 1968).

Phase II: Dans cette phase, on extrait, à partir des fichiers géographiques de base, des fichiers correspondance (tranches d'adresses/coordonnées). Les données collectées comportant des adresses civiques sont mécanographiées, décodées et normalisées, puis l'adresse est comparée au fichier correspondance pour déterminer la tranche d'adresse qui la contient et ainsi les coordonnées du côté d'îlot correspondant sont rattachées aux données.

Phase III: (Podehl 1971) Dans cette phase le fichier est d'abord trié par ordre croissant des coordonnées. A ce stade, il contient, l'une après l'autre, toutes les variables (V) correspondant à chaque individu de la population (P); formant ainsi une matrice où il y a P enregistrements dont chacun contient V variables. Ce fichier est restructuré de façon à ce que chacune des variables V devienne un seul enregistrement de longueur P. En d'autres termes, au lieu d'avoir pour chacun des individus un enregistrement contenant les variables qui le caractérisent l'une après l'autre (par ex.: sexe, état civil, revenu, etc.), on aurait un enregistrement contenant par ordre géographique le sexe de chaque individu de la population, puis un autre enregistrement pour l'état civil et ainsi de suite. Une telle organisation du fichier en facilite l'accès étant donné que sur le nombre de variables qui peuvent être contenues dans une banque de données, on n'en analyse normalement pas plus qu'une demi-douzaine à la fois.

Cette opération accomplie, la banque de données est prête à servir les utilisateurs. Ceux-ci n'auraient qu'à soumettre sur carte les limites des régions qu'ils étudient et ils obtiendront les données désirées soit sous forme de tableaux statistiques ou sous forme de représentation cartographique.

Les opérations nécessaires pour arriver à ce résultat sont comme suit: (1) Les coordonnées des limites de la région sont établies. (2) Cette série de coordonnées définissant un ou plusieurs polygones se voit attelée à un programme de recherche spatiale pour s'attaquer au fichier géocodé et y identifier les points centroïdes contenus dans le(s) polygone(s) en question. (3) Seules les variables requises par l'utilisateur sont extraites, manipulées et présentées grâce au programme STATPAK qui est commandé par un language pseudo-anglais TARELA facilement utilisable par les intéressés sans l'intervention d'un programmeur.

Les usages possibles d'un système d'information à base géographique automatisé sont nombreux. Nous avons réussi jusqu'à présent à développer la possibilité d'extraire les données par zones arbitraires intéressant l'utilisateur, et la possibilité de cartographier automatiquement les résultats. Nos fichiers géographiques permettent aussi de produire des tracés urbains par ordinateur. Nous avons aussi développé des programmes permettant de produire de tels tracés avec une variété d'options quant à l'échelle, les caractères et les symboles désirés.

Nous avons aussi développé des programmes permettant d'identifier des régions selon certains critères spécifiés. Ces programmes ont été développés pour un usage interne et se basent sur la manipulation des coordonnées pour déterminer la proximité et la distance. Il n'y a pas de raison pour laquelle on ne pourrait pas de la même façon former par exemple les circonscriptions électorales du pays automatiquement selon des critères spécifiques.

Il existe aussi une variété d'autres applications possibles de ce système que nous n'avons pas encore exploré; tel que (1) le calcul de surfaces et de paramètres correspondants; par exemple, calcul de la surface à déblayer dans une rue donnée; (2) le calcul des trajets minimums à l'intérieur d'un réseau; (3) l'échantillonnage géographique; (4) la superimposition ou l'intégration de fichiers différents dans le but de comparaison ou dans le but d'étudier le changement des mêmes variables à travers le temps, etc.

En utilisant une ou plusieurs des facilités décrites plus haut, il est possible de faire rapidement des études de marché, de planification ou de rénovation urbaine ainsi qu'une variété d'applications de type administratif

tel la délimitation de territoires à patrouiller par chaque policier d'une ville ou par les vendeurs ambulants d'une compagnie de distribution, déterminer les meilleurs parcours d'autocar au service d'une commission scolaire, etc.

Dans une étape plus avancée, il serait possible d'envisager d'établir des systèmes de contrôle automatique à temps réel d'une variété de fonctions similaires à celles mentionnées plus haut.

Boisvenue, A., 1968 *Geocoding Technical and Clerical Manual.* Section de Géographie, Division du recensement, Ottawa (circulation interne).

Hägerstrand, T., 1967 *The Computer and the Geographer,* Transactions and Papers of the Institute of British Geographers.

Podehl, M., 1971 *An Introduction to the Generalized Tabulation Programs STATPAK and STATAPE and their Language TARELA,* Atelier sur l'Accès aux Données du Recensement, Ottawa, Oct.

P1316
Thematic maps in physical planning
MICHAŁ WIECKOWSKI *Ministry of Construction and Building Materials Industry, Poland*

The cartographic problems of spatial planning are closely connected wth map scales, both those used for analysing the existing information and those reflecting plans to be realized over longer or shorter periods of time, e.g., the trend up to the year 2000, a prospective period up to 1985 as well as the present period up to 1975.

The system of spatial planning in Poland is based on the regulations of the Act of 31 January 1961. This system is concerned with cartographic scales for a plan of spatial economic management for the whole country, general and detailed regional plans, plans of the complexes of settlement units, and general or detailed plans of towns and villages. All such plans are authorized by the Presidiums of the local People's Councils.

The plans concerning investment areas (defined in the Act dealing with spatial planning as realization plans) make another category of plans. Such plans, commissioned by investors, are produced by designing offices and confirmed by building supervision authorities together with preliminary project or technical designs of the investments.

In this paper, only problems connected with plans of the complexes of settlement units and general or detailed plans of towns and villages are discussed.

The concentration of our interest on scales from 1:10,000 or 1:25,000 to 1:1000 or 1:500, determine both the methods of cartographic approach to the problems described and their quantitative generalization. Of course, when applying the scales mentioned above, the *cartogrammes* (and especially *cartodiagrammes*) which may be useful in regional and national planning are not considered.

In physical planning, the method of areal presentation of facts and phenomena is used first of all. It is essential, when using this method, to apply qualitative generalization properly (i.e., adequate combining and analysing of certain information).

The range of items included in the documentation of a physical plan for drawing

conclusions concerning spatial planning is very wide. The following items are included:
1 Demographic information: population, sex, age, occupation – at the time of planning and by the end of the prospective period, for the area proposed in the plan.
2 Industrial information: distribution and range of investments, schemes of development for the industry and its effect on the surrounding area.
3 Agriculture and forestry information: when changes in the organizational and spatial structure of the agricultural and forest economy of a given area are planned.
4 Housing information: with due consideration of existing conditions and the foreseen demand for the future.
5 Social facilities and services (including marketing): the existing conditions and development schemes.
6 Recreation centres: their distribution and development schemes.
7 Transport facilities: the main problems and the needs of future development.

An assessment of the geographical environment and a cartographic survey are the basis for all the above listed analyses, since they reveal the existing relationships, problems, and the needs for the future.

It seems worth mentioning here that for the preparation of detailed plans, high accuracy and precision of cartographic surveys is necessary, which can be achieved through the use of large scales.

The subject matter of the maps made and used in the preparation of physical plans may be divided into the following sections: environmental information; social information; economic information; technical information.

To begin with, the question of environmental problems will be considered. Although we are in contact with the geographical environment in our everyday life, we are often not conscious of its existence until some irregularities strike us. We are likely to hear a traditional opinion that a survey of natural conditions should be limited to an assessment of land suitability for building purposes. In reality, an assessment of the land with regard to living conditions for its future inhabitants is no less important.

Such an assessment should not just be of the present state, but the effect of the new investments on future conditions should also be considered. This refers not only to residential settlements but also to industrial buildings, since their location may negatively influence local climatic conditions in the residential settlements.

It is obvious that various maps can answer only certain questions which in turn are fully readable only to specialists. Moreover, no single map can deal with all the characteristics of a given area. Only a complete assessment, however, of all the essential properties of the area makes possible an adequate consideration of the different ways of utilizing the area. Consequently, when analysing and estimating geographical environment conditions, no hasty conclusions should be drawn from superficial readings of various environmental components from maps, no matter whether these maps are drawn in either the small or the large scales used in physical planning.

Social information maps are commonly known about; in physical planning practice, they are demographic maps and maps of social services. As has been mentioned, *cartodiagrammes* and *cartogrammes* are generally used in small-scale maps. In physical planning, however, the dot method seems to be useful for drawing maps of population distribution in places of residence and work, which indicate the need for transport facilities. This method can also be used for indicating density of population in places of residence and work by the end of the prospective period (i.e., up to 1985). It also makes it easier to plan the distribution of services and transport networks.

Economic maps include three basic groups of information: industry, agriculture and its service, and transportation. It should be noted here that an areal method is used for the cartographic presentation of industry or agriculture; this method is widely used in physical planning, since one of the basic problems is differentiation of planned areas according to the actual and proposed ways in which they will be used. Agricultural problems can be presented, e.g., by the use of a method devised for maps of land utilization at the Department of Agricultural Geography of the Geographical Institute of the Polish Academy of Science.

For exceptional cases of cartographic presentation of transport, even at the scales used in physical planning, *band* diagrams can be used. Maps of technical information comprise: settlements, infrastructure, i.e., underground pipes and lines, and transport. These maps are normally made by geodetical or other design offices.

Systematic, although rather sketchy, discussion of the problems of thematic maps presented in this paper does not imply that in the course of designing, individual maps illustrating each problem should be done separately.

A complex of problems that will have to be solved when designing each settlement unit, will determine the basic and secondary topics of the analyses to be made – sometimes comprising a number of problems.

One thing should be kept in mind, that a complex map has to be the ultimate result of all this work, i.e., a basic sheet of the physical plan, a complete picture of actual and designed facts, a more general picture in the more general plan, or a more differentiated picture in a more detailed plan.

P1317
A propos de la légende des cartes morphostructurales aux échelles 1 : 50,000 et 1 : 500,000
N.V. BACHENINA et A.V. MIRNOVA   *Université d'Etat de Moscou, URSS*

Au XXI Congrès de l'UGI à New Delhi on discuta de la légende des cartes géomorphologiques à grandes échelles, établie par le groupe de travail des représentants de la Pologne, de l'URSS, de la France et de la RDA sous la direction du prof. M. Klimachevsky (1968). Les auteurs de cette communication en admettent les principes de base. Cependant actuellement, en plus des cartes géomorphologiques, se sont révélées nécessaires. La conception morphostructurale est née en URSS (Guérassimov 1959). Des équipes travaillent à la création d'une méthode d'analyse morphostructurale. Depuis quelques années, le Cabinet de cartographie géomorphologique, attaché à la Chaire de Géomorphologie de l'Université d'Etat de Moscou, prend également part à ce travail et a élaboré la légende. Ses auteurs ne prétendent pas à son 'universalité' et la proposent comme possible pour l'établissement des cartes morphostructurales.

La cartographie géomorphologique se développe avec succès au carrefour des sciences suivantes: géomorphologie, cartographie, géotectonique et géophysique. Une solution possible aux nombreux problèmes de géomorphologie peut être obtenue par l'analyse morphostructurale, principalement dans le but de cartographier les *morphostructures blocs*, les failles et accidents les délimitant. Par exemple, l'analyse morphostructurale permet de bien définir les morphostructures plissements et blocs et de différencier les morphostructures conditionnées par préparation ou par tectonique active.

Lors de la cartographie des morphostructures plissements, on constate assez souvent une dépendance du plan général du relief plissement de la base structurale bloc.

Les morphostructures blocs sont déterminées par une étude préliminaire du terrain, surtout dans les domaines de la jeune tectogenèse. D'après le dessin du réseau fluvial, des crêtes, versants, gradins de pente et la superposition de ces éléments du relief à des fissures, et des dessins linéaires du microrelief sur les photos aériennes et les cartes topographiques, on peut établir si la dislocation du relief est dûe à la tectonique ou à l'érosion et résoudre la question de la corrélation de la jeune tectonique blocs et des formes structurales résiduelles.

Le jeune âge des formes blocs est indiqué par: 1) une dissection accusée, évidemment tectonique, de grands blocs (présence de crêtes 'proéminentes,' de horsts séparés sans érosion, de sommets-horsts, etc; 2) une limitation des formes blocs par des accidents vivants, dont l'affaiblissement ou l'activité est confirmée par leur expression géomorphologique, ainsi que par les données géologiques et géophysiques; 3) une profondeur inégale des vallées, inclinaison et forme des versants en différents blocs; 4) présence de versants tectoniques et d'escarpements, accentuation de leur inclinaison vers la base et zones d'éboulis récents suivant des cassures rajeunies, versants tectoniques de blocs déversés, disloqués si faiblement (sans rapport avec leur inclinaison) que cela ne peut être expliqué que par la jeunesse des mouvements; 5) la grande quantité et la large ouverture de fissures béantes sur les blocs plus élevés;

6) le voisinage de massifs montagneux et crêtes avec dépressions d'entremonts, remplies de dépôts corrélatifs récents; 7) différence de la quantité des terrasses et de leurs hauteurs en divers blocs de la composition et de la puissance de leur alluvion; 8) cassures des profils longitudinaux des fleuves; 9) non correspondance des hauteurs et de l'accentuation des contours des crêtes et gradins du relief avec roches de dureté différente. Ainsi, les blocs les plus surélevés de l'Altai-Roudny sont formés des mêmes granites que ceux des chaînes situées 1000m plus bas. Ainsi c'est dans les Carpathes à flysch que se sont formés les blocs les plus surélevés et non dans les roches cristallines anciennes. Dans les cas complexes la *tectonique blocs* et la préparation se manifestent simultanément et en accord (par exemple, Oural-Sud).

Comme il est prévu dans la légende des cartes morphostructurales, le relief est représenté selon les trois mêmes caractéristiques que celles des cartes géomorphologiques: 1) aspect extérieur, 2) origine, et 3) âge. Cependant l'essentiel de ces caractéristiques est quelque peu différent, ce qui a modifié le style de la cartographie. Les limites naturelles sont des failles et des fissures, conditionnant le dessin plus 'rigide' du relief sur les cartes morphostructurales, parfaitement lu sur les photos cosmiques, aériennes et sur les cartes topographiques.

Aux échelles dont il s'agit, on montre les limites des zones de failles de fond et des fissures, les limites des formes structurales (en fonction de la nécessité), les morphostructures et leurs éléments, c'est-à-dire, par exemple, les surfaces des sommets et les crêtes des chaînes horsts de types divers (Bachenina et coll. 1971), les versants tectoniques ainsi que ceux modifiés par érosion, les fossés d'effondrement (graben), les vallées rift, etc. Tous les éléments morphostructuraux (tectoniques 'vivants' et volcaniques préparés par érosion) sont réunis dans la première division de la légende. On a aussi prévu dans la légende la représentation des formes et des éléments de formes d'origine exogène se développant inégalement dans les diverses morphostructures. Ces éléments du relief sont reportés à la deuxième division de la légende. Pour l'établissement de la deuxième division, on a également utilisé la Légende Internationale et les légendes des cartes géomorphologiques particulières (formes lits de cours d'eaux, géocryologiques, rives, etc.). De cette façon, on garde dans la légende le complexe ordinaire des cartes générales géomorphologiques, bien que sa représentation soit soumise à la figuration des morphostructures. En ceci consiste justement la spécificité des cartes morphostructurales.

Les morphostructures et leurs éléments sont représentés par un fond coloré qualitatif avec signes conventionnels en couleur. Cette combinaison permet de limiter l'assortiment des couleurs et des nuances. Pour l'édition des cartes une telle combinaison peut être élargie par l'emploi de trames typographiques en couleur.

La légende se compose de sept tableaux: I, le principal incluant environ 300 morphostructures et leurs éléments, formes du relief et éléments de formes; II, couleurs et nuances pour les principales subdivisions génétiques de la légende (35); III, couleurs pour signes conventionnels (6); IV, trames figuratives; V, signes conventionnels pour les dépôts corrélatifs; VI, signes conventionnels pour la lithologie des roches de fond; VII, index d'âge.

Etant donné que la légende est établie pour des cartes morphostructurales aux échelles du 1 : 50,000 au 1 : 500,000 la cartographie suppose une généralisation: 1) des lignes de failles et de fissures, formant le squelette de la carte; 2) des contours et limites des morphostructures et de leurs éléments en conservant les principaux traits du dessin naturel, par exemple de crêtes horsts, ramifiées de façon complexe, versants tectoniques, vallées rift, etc.; 3) des éléments des morphostructures par unification des différences génétiques (Bachenina et Zaroutskaya 1969). Par exemple, en généralisant certains terrains des versants, comme éléments des morphostructures, présentés sur la carte à l'échelle 1 : 50,000 selon les différences de genèse et les graduations de l'inclinaison, il est nécessaire de choisir pour la carte à l'échelle 1 : 200,000 et encore plus pour la carte au 1 : 500,000 le facteur déterminant de la formation du versant et une ou deux graduations prédominantes de l'inclinaison. La notion de 'versant' change lors de la diminution de l'échelle. Ainsi, on suppose une généralisation qualitative (d'après l'essence des objets) et non géométrique, dont le seul emploi peut mener à la perte du dessin naturel des morphostructures de leur plan.

Le principal tableau I est établi en rapport

avec les nécessités de la généralisation: la liste des éléments cartographiés est située au milieu de la page, à gauche les signes conventionnels de fond, à droite les signes conventionnels indices. Si l'échelle de la carte ne permet pas la représentation d'un élément quelconque par la couleur de fond, on emploie un signe correspondant.

La recherche morphostructurale, comprenant aussi l'établissement de cartes auxiliaires (Bachenina et Trechtchov 1971), permet d'établir rapidement la base des cartes générales des formes structurales et morphostructurales et d'obtenir une représentation claire de la tectonique contemporaine qu'il serait plus difficile et plus onéreux d'établir par des procédés géologiques: 1) intensité générale de la tectonique la plus récente; 2) degré de différenciation des déformations les plus récentes, leur signe et direction; 3) plan de déformations; 4) type des mouvements par zones de fissures: affaiblissement de l'activité de zones lors de dislocation par accident d'une autre direction, mouvements réguliers verticaux ou déversement des blocs etc.

La division des morphostructures voûtes-blocs par ordres a une importance génétique. Les grands blocs (de dizaines de kilomètres aux amplitudes de déformations de centaines et de milliers de mètres), ne peuvent probablement être expliqués que par des processus de fond, se passant dans le manteau. Les petits éléments du relief peuvent être conditionnés par des mouvements de petits blocs de l'écorce désagrégée en résultat d'une différenciation mécanique locale des tensions, apparaissant lors du mouvement de grands blocs.

Les investigations morphostructurales avec emploi de données sur la structure géologique du terrain et en contact avec la recherche géophysique aideront à l'éclaircissement graduel de la genèse des morphostructures voûtes-blocs. Il est très important que ces investigations concourent à la solution de nombreux problèmes pratiques. Ainsi, les vallées qui suivent les failles indiquent l'affaiblissement de ces dernières et conséquemment la possibilité de zones aquifères. Souvent l'affaiblissement le plus récent peut être hérité d'un ancien affaiblissement mésozoïque, et de plus les vallées suivent des fissures dans lesquels auparavant se formaient des zones métallifères.

Bachenina, N.V., & A.A. Trechtchov, 1971 Towards a method of morphostructural analysis for geomorphological surveys of mountain relief, *Geomorphology Publication 'Science,'* no. 3.

Bachenina, N.V., A.U. Mirnova, M.V. Piotrovsky, E.A. Rubina, N.N. Talskaya, & A.A. Trechtchov, 1971 About geomorphological cartography, at various scales, of the mountain relief with block structure, *Moscow U. Messenger*, no. 6.

Bachenina, N.V., & I.N. Zaroutskaya, 1969 Principles of generalization of medium- and large-scale maps, *Moscow U. Messenger, Geomorphology*, no. 2.

Guérassimov, I.P., 1959 Structural features of the relief of the earth's surface in the territory of the USSR and their origin, *Tr. Inst. Geog. USSR.*

— (ed.), 1970 Measurements of geomorphological methods and structural geological researches, sb. from NEDRO.

P1318
Télédétection des eaux polluées
JEAN POUQUET   *Université d'Aix en Provence, France*

La pollution des eaux, comme celle de l'air, peut être détectée dans l'infra rouge réfléchi et l'infra rouge émis depuis un satellite, un avion ou à partie du sol.

Le 19 avril 1969, par exemple, Nimbus 3 (radiomètre infra rouge haute résolution (HRIR), canal diurne, $0.7-1.3\mu m$; orbite n° 69), dans la bande spectrale $0.7-1.3\mu m$, dénonçait la turbidité de l'eau de mer de part et d'autre du delta du Nil, grâce à de plus hautes valeurs de réflectance, jusqu'à 8 pour cent, alors que l'eau non polluée demeurait au-dessous de 3 pour cent.

Le 11 mai 1970, Nimbus 4 (radiomètre infra rouge Température-Humidité (THIR), canal $10.5-12.5\mu m$, valeurs diurnes; orbite n° 444), dans la bande $10.5-12.5\mu m$, aux environs de midi, montrait un phénomène analogue dans la même région. Dans ce cas, les eaux polluées par une abondante charge solide en provenance du Nil avaient une température de radiance supérieure de 4 à 8°C à celle de l'eau non polluée.

Des travaux de terrain exécutés à l'aide d'un radiomètre portatif opérant dans la bande spectrale 10.5–12.5μm (radiomètre portatif BARNES PRT 5, 'spécial'; 10.5–12.5μm; ouverture d'angle 2°), ont prouvé que les polluants chimiques réduisaient l'émissivité alors que les particules mécaniques provoquaient une action inverse.

1) *Polluants chimiques.* Le barrage de Sherbrooke, Québec, Canada, est malheureusement le receptacle des détritus chimiques déversés par les usines de pâte à papier. Les eaux non polluées ont une émissivité de 0.984 environ, alors que celles contenant en solution une forte proportion de produits chimiques voient leur émissivité réduite à 0.807 (émissivité: quotient radiance de l'objet par radiance d'un corps noir à la même température réelle; en fait, la température réelle de l'objet a été prise comme la transcription thermique de la radiance du corps noir; le graphique, correspondance températures de radiance–radiance effective (mW/cm^2/ster), fourni par le constructeur, a été utilisé pour les calculs de l'émissivité).

A la NASA, Karl Sziekielda a démontré que l'addition d'huile de machine à coudre provoquait une baisse de température de radiance. Dans la Vallée de la Mort, Californie, les eaux de Bad Water, riches en micro-organismes, ont une émissivité anormalement basse, soit 0.906 pendant le jour et 0.891 pendant la nuit.

2) *Pollution d'origine mécanique.* Les remarques faites à propos de Nimbus 4 seraient suffisantes par elles-mêmes. Un contrôle réalisé de manière fort simple aboutit aux mêmes résultats. Au sud de Tucson, les eaux claires d'un canal d'irrigation ont été artificiellement polluées par addition de boues. L'émissivité de l'eau est alors passée de 0.956 à 0.971.

En Arizona, à quelque 20km au sud de Phoenix, l'air pollué provoquait une baisse des températures de radiance de 3 à 8°C. De nombreuses observations seraient faciles à réaliser en ce domaine par examen comparatif de cartes dressées à l'aide d'informations fournies par les satellites, cartes présentant une suite temporelle pour une région donnée.

La télédétection moderne peut contribuer de manière efficace aux progrès de la géographie, carrefour des disciplines liées aux sciences de la terre, *lato sensu*. Nos départements, instituts, et laboratoires de géographie doivent incorporer la télédétection dans leur enseignement, leurs techniques de recherche, seule façon de faire susceptible de conduire à l'exploitation rationnelle des documents relevant de la télédétection, que celle-ci soit réalisée au sol, depuis un aéroplane ou depuis un satellite.

Pouquet, J., 1971 *Les sciences de la terre à l'heure des satellites. Télédétection* (Paris).

P1319
Pour une plus grande diffusion et connaissance des 'panoramas'
E. SCARIN *Università di Genova, Italia*

Le problème de la représentation cartographique du 'panorama,' c'est-à-dire de l'aspect d'une partie de la surface terrestre vue d'un point de vue ou d'une série de points de vue de la surface même a déjà été étudié depuis plusieurs années. La vision du panorama a des buts variés, de l'examen comparatif des éléments qui le compose (à l'aide de l'interprétation de la carte topographique et du plan cadastral) jusqu'aux sensations que l'on peut éprouver sur le plan esthétique (cf. E. Scarin, *Il 'panorama,'* in 'Annali di ricerche e studi di geografia,' Genova, anno XXII, n° 3, 1966).

Le panorama varie logiquement par l'ampleur angulaire et par la profondeur, en relation avec l'altitude de l'observateur sur l'aire territoriale circostante. En outre, le panorama présente une simplicité monotone ou une grande complexité des phénomènes; une certaine harmonie ou disharmonie d'ensemble qui le caractérise une manière d'être considérée subjectivement digne d'être admirée, soit pour sa grandeur, soit pour ses caractères particuliers comme par exemple la variation de couleur et d'aspect suivant l'insolation et les saisons.

Les panoramas sont représentés cartogra-

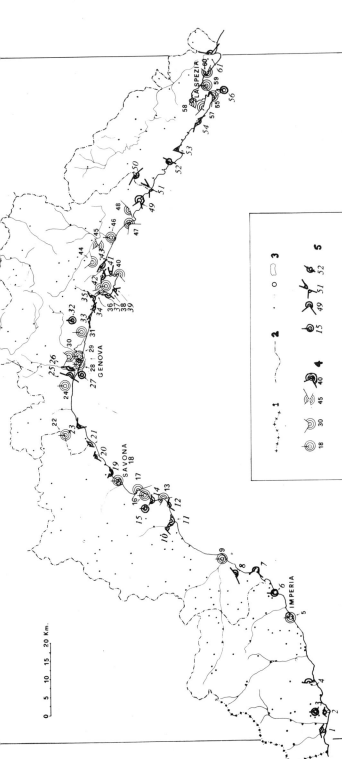

Una serie esemplificativa di panorami lungo le coste della Liguria. 1: Ventimiglia; 2: Bordighera; 3: Sasso; 4: San Remo; 5: Porto Maurizio; 6: Diano Marina; 7: Capo Mele; 8: Alassio; 9: Albenga; 10: Finalborgo; 11: Finalmarina; 12: Varigotti; 13: Capo Noli; 14: Noli; 15: Belvedere presso Voze; 16: Torre del Mare; 17: Vado; 18: Savona; 19: Celle; 20: Varazze; 21: Piani d'Invrea; 22: Monte Reisa; 23: Arenzano; 24: Brich del Gazzo; 25–26: Coronata e Belvedere; 27: Fanale di San Pier d'Arena; 28: Lanterna; 29: Grattacielo (Martini); 30: Righi; 31: Monte Moro; 32: Monte Fascie; 33: **S. Ilario**; 34: **Sant'Apollinare**; 35: **Megli**; 36: **Portofino Vetta**; 37: San Rocco; 38: Monte di Portofino; 39: S. Fruttuoso; 40: Portofino; 41: **Paraggi**; 42: S. Margherita; 43: **Rapallo**; 44: **Montallegro**; 45: **Leivi**; 46: Lavagna; 47: Sestri Levante; 48: **Macallè**; 49: Moneglia; 50: Passo del Bracco; 51: Deiva; 52 Levanto; 53: Monterosso; 54: Riomaggiore; 55: Portovenere; 56: Tinetto; 57–58: La Spezia; 59: Diga foranea; 60: Lerici; 61: Monte Marcello. *1:* confine politico; *2:* confini amministrativi; *3: Sedi comunali e città principali; 4: Tipi di segno convenzionale per panorami di grande interesse; 5: Tipi di segno convenzionale per panorami di interesse locale.*

phiquement dans les cartes d'intérêt touristique, avec un nombre limité de signes. Dans les recueils internationaux ayant trait à la cartographie, on distingue uniquement les panoramas circulaires et les points de vue partiels (cf. F. Joly et S. de Brommer, *Projet de normalisation des symboles de cartes thématiques*, in 'Internationales Jahrbuch für Kartographie,' Gütersloh, 1966). On considère en substance que la signalisation est secondaire si on la compare à d'autres phénomènes du milieu.

On devrait distinguer au moins les panoramas de grand intérêt régional, des panoramas locaux, en donnant des indications précises sur leur ampleur angulaire et sur leur profondeur tout en simplifiant au maximum le signe correspondant, afin de ne pas introduire de confusion dans l'ensemble des signes cartographiques traditionnels.

En considérant que la recherche et la diffusion des connaissances et des interprétations des panoramas constituent un enrichissement culturel, soit au point de vue technique, soit au point de vue esthétique, je pense qu'on devrait donner aux signes correspondants une plus grande variété, en les représentant en outre cartographiquement au moins dans toutes les cartes qui peuvent avoir un intérêt touristique et dans quelques-unes tout à fait spéciales construites dans ce but. Je présente à titre d'exemple une carte spéciale en couleur de la Ligurie (qui peut être reproduite facilement en blanc et noir) où ont été reproduits tous les panoramas principaux existants sur les côtes.

P1320
Problems of aggregate mapping of the Ukrainian SSR agriculture
ANDREY ZOLOVSKY   *Academy of Sciences of the Ukrainian SSR*

Phenomena and processes occurring in rural economies are interdependent and closely connected, not only with one another but also with conditions related to the development of production, which change from place to place. These circumstances in many cases require detailed study and analysis. The more problems have to be solved simultaneously, the more urgent is the necessity to study their interrelationships systematically and with a specific purpose in view. One must take into account complex and multiform relations between separate areas of the country, between branches of the national economy, including agriculture, between production and the natural environment, and between all the fields of social life. Thus, a complex, geographic approach is necessary to decide primary tasks related to the national economy.

Aggregate mapping, including the use of aggregate maps, occupies an important place among the numerous ways of studying problems connected with planning the development of agriculture and its branches. It is also important with regard to the rational use of environmental conditions, labour, material, and technical resources, and for the detection and analysis of interconnections and interrelations of phenomena and processes.

Aggregate thematic maps facilitate and simplify estimates of the location, level, and conditions for development of agriculture. They increase objectivity of characteristics and estimates of processes and phenomena, help to investigate many scientific problems and those of the national economy, and promote settlement of internal tasks in development and localization of agricultural production. Maps of this type are able to reflect existing and possibly future interconnections and interdependence of items that are mapped; they are of use in analysing and estimating economic potentialities and reserves, natural conditions and resources for the development of agriculture. The aggregate character of a map promotes enrichment of its contents in the various indices used. This extends the research, and broadens its range of users. It also increases the value of the map as a tool and a source of information; it essentially facilitates the possibilities of scientific generalizations; and simplifies a study of the interaction of environmental conditions and socio-economic factors.

The process of compiling aggregate maps has its peculiar problems due to the character of agriculture as a subject to be mapped; due to the wide range of research studies in this field of the national economy; insufficient development of methods for aggregate cartographic interpretation of agricultural production, etc.

These peculiarities, in their turn, raise the caliber and change the contents and significance of the details of the work, by producing a closer collaboration between the cartographer and the scientists engaged in economics and economic geography. This changes the extent and character of the work carried out at the first two stages of compiling the map, and it requires finding cartographic means and methods to map a complex of characteristics on one geographic basis. Compilation of these maps needs the application of process flow diagrams based on using source materials of different technical quality and completeness and varying in the degree of the author's participation in the complexities of map compilation and in the labour and time immediately spent in map compilation.

The uncommonness of agriculture as a topic to be mapped and that of aggregate maps as a particular means for mapping the development and location of objects, phenomena, and processes requires an additional theoretical development of special contents and the justification for combining specially selected detailed indices and characteristics in one map.

The latter may be carried out according to several principles, but it must be logical, not random, and must be determined by the purpose of the map. The main thing is that these indices be connected, and that they complement and develop each other and on the whole constitute a complex of economic and other characteristics to solve definite problems. The concrete requirements of scientific research in the field of agriculture, as well as the methods of guiding this production, are a principal criterion for the selection and determination of the aggregate map contents. Therefore, the higher the applied value of the map, the larger its scale, the smaller the area to be mapped, and the more versatile and extensive its contents will be.

The combination of indices in one map may be based on the following general principles: 1) there are several indices on the map which characterize the main object of mapping, e.g., the branch of agriculture seen from different points of view; 2) the map content involves characteristics connected by an index of one type which is transferred in some fields, either in various periods of development, or in two-three units of mapping, or in two-three periods and fields of economy; 3) one map conveys a complex of indices which are of different type but interconnected and characterizing agricultural production and the conditions of its development. They are compiled to solve some special problems within the definite limits of the research.

Extension of theoretical propositions and development of practical ways of generalization compatible with aggregate agricultural maps are important items of the given problem. Besides factors common for any map (purpose, scale, subject matter, peculiarities in development and location of objects and territory of mapping, methods of mapping, character of map delineation and nature of sources), some other factors should be taken into consideration which appear to influence the compilation of aggregate agricultural maps. The latter may involve items of national-economic importance, prospects for development and peculiarities of the objects and mapping units, extent of the research, and the extent to which special features are shown. To calculate the latter, formulae are suggested which indicate the amount of dependence on the map scale, the character of the location and the stage of development of the items being mapped.

Mathematical control of the map is an important consideration in aggregate mapping. This notion, besides the choice and control of the mathematical elements of the map, involves problems of controlling the choice or estimation of predetermined scale indicators, of calculating the dot weight and range of sizes, and of estimating the accuracy of representing economic characteristics and obtaining information.

When developing agriculture maps where the dot method is applied, a mathematical expression was found for the relationships between the dot weight and the map scale and the special circumstances of location of the items being mapped in the Ukrainian SSR. It is proved that these formulae may also be applied to indices which are not estimated by area.

In cases where scale indicators and dots are used, the scale of the map is mathematically dependent on the number of symbols or diagram figures, on the dot weight, and on the special circumstances of development and location of the items being mapped in the Ukrainian SSR.

Development of aggregate maps of the Ukrainian SSR agriculture is proceeding in two directions: *a*) improvement of the existing types of maps and those on traditional subjects, *b*) development of new types of maps. Successes in mapping the agriculture of the Ukraine are obvious. It is sufficient just to mention the agricultural atlases of the republic and western regions of the Ukrainian SSR, the aggregate atlases of the Ukrainian SSR and Moldavian SSR and the atlas of the Kiev region, and the numerous agricultural maps published in papers on geography, including the Ukrainian SSR and its regions. Agriculture of the republic is presented on a large scale in the Agriculture Atlas of the USSR and in the Atlas of Economy and Culture Development of the USSR. Maps of the national land reserves and their utilization, the second edition of the Agriculture Atlas of the Ukrainian SSR and the Atlas of Rational Nature Utilization and Protection are developed at the Academy of Sciences of the Ukrainian SSR.

14
Commissions

Commissions are working groups of the International Geographical Union which focus on a particular aspect of geography. They operate in various ways between and during the quadrennial Congresses. In 1972, certain Commissions solicited papers for meetings held in Canada, and some of the papers involved are included here.

The following list of Commissions, with the short titles of articles included and first authors' surnames, will assist in identifying articles, topics, and places of interest. A complete author and co-author index is located at the end of this volume, as well as a selected index relating papers to geographical locations. Note that the papers are *not* listed in alphabetical order. The organization of the volumes is described in full in the Preface.

Les Commissions sont des groupes de travail de l'Union géographique internationale qui s'occupent de certains aspects particuliers de la géographie. Elles sont actives d'une façon ou d'une autre au cours des congrès quadriennaux et entre les Congrès. En 1972, certaines Commissions ont sollicité des communications en vue des réunions qui se tiennent au Canada; on retrouve ici quelques-unes de ces communications.

La liste qui suit donne les titres abrégés des textes et les noms des auteurs principaux et permet d'identifier les articles, les sujets et les lieux géographiques. Un index complet d'auteurs et de coauteurs se trouve à la fin de ce volume ainsi qu'un index des endroits géographiques qui font l'objet d'études. Prière de noter que les textes ne sont pas classés par ordre alphabétique. On explique en détail le plan de ces volumes dans la Préface.

C01
Geographical Data Sensing and Processing
Sélection et traitement de l'information géographique
CONVOCATEURS/CONVENORS: R.F. Tomlinson, *Ottawa*; D. Steiner, *University of Waterloo*

c0101 Explaining and forecasting developed land in urban areas (Norway) AASE 1007
c0102 Automatic pattern recognition of Side-Looking Radar imagery statistical surface (Texas, USA) KEDAR 1008

C02
High Altitude Geoecology
Géoécologie des pays de haute altitude
CONVOCATEURS/CONVENORS: Stuart Harris, *University of Calgary*; J.D. Ives, *University of Colorado, Boulder, USA*

c0201 Pseudo-timberline: the southern Appalachian grassy balds (USA) GERSMEHL 1011
c0202 Geography of altitudinal belt ecosystems, the Caucasus (USSR) ZIMINA 1013

C03
Coastal Geomorphology
Géomorphologie côtière
CONVOCATEUR/CONVENOR: S. Brian McCann, *McMaster University, Hamilton*

c0301 Shoreline erosion, Barron delta, North Queensland (Australia) BIRD 1015
c0302 Zonal studies of the maritime coast, Canada OWENS 1017
c0303 Effect of typhoons on beaches, Hong Kong SO 1018
c0304 Accelerating crisis in beach erosion (USA, Florida; and Mexico) TANNER 1020
c0305 Classification of coastal classifications VALENTIN 1021

1002 / Commissions

c0306 Post-glacial emergence, Bay of Fundy coast (Canada) WELSTED 1023
c0307 Un banc corallien orienté : Reef Island aux Iles Banks, Nouvelles-Hébrides GUILCHER 1025
c0308 Caractéristiques des schorres des régions froides (St Lawrence R., Canada) DIONNE 1027
c0309 Aspects of importance of tidal range in coastal studies KIDSON 1027

C04
Present Day Geomorphological Processes
Les processus géomorphologiques actuels
CONVOCATEURS/CONVENORS: H.O. Slaymaker, *University of British Columbia, Vancouver*; K. Hewitt, *University of Toronto*; H.J. McPherson, *University of Alberta, Edmonton*

c0401 Relief and denudation: theoretical test of an empirical correlation AHNERT 1027
c0402 Role of catastrophic rainfalls in relief shaping, mountain areas (India, Lower Himalaya) STARKEL 1029
c0403 Soil erosion and sedimentation, six selected catchment basins, Tanzania RAPP 1031
c0404 Landslides, Mgeta area, western Uluguru mountains, Tanzania TEMPLE 1034

C05
Geography in Education
La géographie dans l'éducation
CONVOCATEURS/CONVENORS: M. Saint-Yves, F. Grenier, B. Robert, L. Trotier, *Université Laval, Québec*; B. Brouillette, *Université de Montréal*

c0501 Territorial decentration and geographic learning (USA, Georgia) STOLTMAN 1036
c0502 School broadcasting in a developing country (Kenya) BROWN 1038
c0503 Films in university geography teaching (UK) CLARK 1040
c0504 Regional geography in a school curriculum CLARK 1042
c0505 Geography in adult education, England and Wales COOPER 1044
c0506 Problem of hierarchy in objectives of pre-university geography teaching GRAVES 1045
c0507 Image mentale et connaissance réelle : l'exemple du Nord (Canada et URSS) HAMELIN 1048
c0508 The place of expert knowledge in curriculum development (UK) HANCOCK 1050
c0509 Application of central place theory to educational planning, Bath area, UK HONES 1053
c0510 Geography at the Open University (UK) LEARMONTH 1055
c0511 Influences of 'new geographies' on curricula and content in British schools NICHOLLS 1056
c0512 Geography in modern education (Brazil) OLIVEIRA 1057
c0513 Location as a factor in educational opportunity, England and Wales RAWSTRON 1058
c0514 Geography of education and educational planning RYBA 1060
c0515 Geography teaching and the rising level of expectations SCARFE 1063
c0516 Assessment of learning in geography in relation to Piaget and Bruner (USA, Iowa) SLATER 1065
c0517 Student-teacher involvement in creating new regions, US WARMAN 1067

C06
Humid Tropics
Les tropiques humides
CONVOCATEUR/CONVENOR: Theo. L. Hills, *McGill University, Montreal*

c0601 Adjustments of small-scale farmers to abandonment of estate agriculture, Nevis, West Indies WATTS 1069

C07
Man and Environment
L'homme et son milieu
CONVOCATEURS/CONVENORS: Ian Burton, *University of Toronto*; L. Hamill, *University of Calgary*; J.G. Nelson, *University of Western Ontario, London*

c0701   Air quality control legislation (USA)   BACH   1070
c0702   Human adjustment to agricultural drought, South Australia   HEATHCOTE   1073
c0703   Adaptation stages and man-induced vegetation change, interior Tamilnadu (India)   MURTON   1075
c0704   Human adjustment to volcanic hazard in Puna district, Hawaii (USA)   SHIMABUKURO   1076
c0705   New Zealand natural hazard insurance scheme: application to North America   O'RIORDAN   1076
c0706   The Indian desert of Thar, a man-made desert   RATHJENS   1079

C08
Geomorphological Survey and Mapping
Recherche et cartographie géomorphologiques
CONVOCATEUR/CONVENOR: Denis A. St-Onge, *Geological Survey of Canada, Ottawa*

c0801   IGU Commission on Geomorphic Survey and Mapping, activities (1968–72), program (1972–6)   DEMEK   1081
c0802   Recent trend in geomorphological mapping of alluvial plains in Japan, its application   OYA   1083
c0803   Geomorphological symbols for mass wasting (Colorado Plateau, Utah, USA)   SHRODER   1085

C09
Agricultural Typology
Typologie de l'agriculture
CONVOCATEUR/CONVENOR: Lloyd G. Reeds, *McMaster University, Hamilton*

c0901   Agricultural typology in Nigeria: problems and prospects   AGBOOLA   1087
c0902   Typology of agriculture in the American South   ANDERSON   1089
c0903   Indices for agricultural typology (Italy)   BONUZZI   1090
c0904   Agricultural space: experiment in classification (Brazil, São Paulo State)   CERON   1092
c0905   Systems approach to sugar cane type of agriculture   DINIZ   1093
c0906   Land-use classification in rural sections of metropolitan regions (Canada, Ontario, Toronto-centred region)   FOUND   1093
c0907   Terminology in typology – the problem of 'plantation'   GREGOR   1095
c0908   Preliminary attempt at typology of world agriculture   KOSTROWICKI   1097
c0909   Aspects of agricultural regionalization, South Korea   LEE   1100
c0910   Agricultural change in developing areas (West Indies, St Lucia)   MOMSEN   1101
c0911   Misuse of land in villages of Ganga Plain: a model in agricultural typology (India)   ROY   1103
c0912   Changing Japanese agriculture: typology notes   SOMA   1103
c0913   Land use and natural vegetation   VANZETTI   1105
c0914   Classification of agricultural land use for development planning (Latin America)   WOOD   1106
c0915   Aerial photographs: application to classification and analysis of agricultural land use (Canada, Southern Ontario)   RYERSON   1107
c0916   Tendencies in the development of Danish agriculture   KAMPP   1109
c0917   Towards a planning-oriented agricultural typology, Niagara-on-the-Lake, Ontario (Canada)   SUNDSTROM   1111
c0918   Regional differentiation in the productivity of field cultivation, Finland   VARJO   1113

C10
Regional Aspects of Economic Development
Aspects régionaux du développement économique
CONVOCATEURS/CONVENORS: Edward G. Pleva, *University of Western Ontario, London*;
R.S. Thoman, *Toronto*; W.B. Stohr, *McMaster University, Hamilton*

c1001 Problems of complex development of new areas (USSR, Komi ASSR)
        VITYAZEVA   1115

C11
Geography of Arid Lands
Géographie des pays arides
CONVOCATEUR/CONVENOR: Ronald Peel, *University of Bristol, England*

c1101 History of settlement, southeast Alberta (Canada), to 1920   FLOWER   1118

C12
Population Geography
Géographie de la population
CONVOCATEURS/CONVENORS: L.A. Kosinki, A.C.B. Allen, *University of Alberta, Edmonton*

c1201 Pragmatic approach to migration research (USA)   GALE   1120
c1202 Migration theory for an urban system   HUDSON   1121
c1203 Migration between labour market areas, England and Wales   JOHNSON   1121
c1204 Patterns of Maori migration (New Zealand)   JOHNSTON   1123
c1205 Industrialization, factor of recent internal migrations in Poland   LIJEWSKI   1125
c1206 Interregional migration, United States and Canada   LYCAN   1127
c1207 Migration patterns, Soviet republics   SHABAD   1128
c1208 Marriage migration, rural India   SOPHER   1130
c1209 Origins and composition of workers, Miferma iron ore mine, Mauritania
        SWINDELL   1131
c1210 Population concentration in South Africa, 1960–70: shift and share analysis
        BOARD   1134
c1211 Time-space and mobility among internal migrants, Ghana   ENGMANN   1135
c1212 Patterns of internal migration, India   GOSAL   1138
c1213 Internal migration, Mexico, 1940–70   GUTIERREZ DE MacGREGOR   1140
c1214 Internal migration, east-central Europe   KOSIŃSKI   1142
c1215 Internal migration, southeast Asian countries   NG   1144
c1216 Internal migration: review of data collection, southern Africa   PERRY   1145
c1217 Space and time dimensions in study of population mobility, Africa   PROTHERO   1147
c1218 Migration patterns in Uttar Pradesh: a geographical analysis (India)   ROY   1148
c1219 Aspects of internal migration to cities of Maharashtra, 1951–61 (India)
        TAMASKAR   1149
c1220 Some problems of internal population migration, northern Siberia (USSR)
        VOROBYEV   1151
c1221 Method and analysis in mobility study of non-literate populations (British Solomon Islands, Guadalcanal)   CHAPMAN   1153
c1222 Interregional migration, Latin America   GEIGER   1155
c1223 Migration and changing settlement patterns, Alberta (Canada)   PROUDFOOT   1156
c1224 La migration de retraite en France : étude de géographie sociale   CRIBIER   1158
c1225 Interregional migrations, tropical Africa: an overview   HARVEY   1159
c1226 Spatio-temporal model of mobility patterns in a multi-ethnic population, Hawaii
        MUKHERJI   1161
c1227 Recreational migration and central place hierarchy (Canada, Ontario)
        WOLFE   1162

C13
Applied Geography
Géographie appliquée
CONVOCATEURS/CONVENORS: Ralph R. Krueger, Peter Nash, *University of Waterloo*

c1301 Regional development policy, Hungary ENYEDI 1163
c1302 Application of climatological knowledge, particularly in future urban forms MCBOYLE 1165
c1303 La géographie appliquée au Québec (Canada) CERMAKIAN 1167
c1304 Development of navigation on the Danube, especially in Austria SCHEIDL 1169
c1305 Major accomplishments in applied geography in India since 1969 SHAFI 1171
c1306 University programme and requirements of applied geography STRASZEWICZ 1173
c1307 Geographer's role in clarifying environmental problems TIETZE 1175

C14
Quantitative Methods
Méthodes quantitatives
CONVOCATEURS/CONVENORS: Maurice H. Yeates, Robert H.T. Smith, *Queen's University, Kingston*

c1401 Model of ghetto growth (USA, Seattle) MORRILL 1177

C15
World Land Use Survey
L'utilisation du sol dans le monde
CONVOCATEURS/CONVENORS: A. Poulin, R. Paquette, *Université de Sherbrooke;* R.C. Hodges, *Department of the Environment, Ottawa*

c1501 Maps of soil use, Mexico ACEVES-GARCIA 1180
c1502 Land-use study by aircraft and satellite remote sensing (Alabama, USA) PALUDAN 1181
c1503 Land use maps for India, and recommendations of the world land use survey ROY 1184
c1504 Against land classification YOUNG 1184

C16
Geography of Transport
Géographie des transports
CONVOCATEUR/CONVENOR: Roy I. Wolfe, *York University, Toronto*

c1601 Recreational boat trips, Trent-Severn Waterway, Ontario, Canada HELLEINER 1186
c1602 Weekend tourism, an indicator of the process of urbanisation (Bavaria) MAIER 1187
c1603 Museum of transportation geography (UK) APPLETON 1189
c1604 Planification des transports et géographie, l'équipement maritime de la Basse-Loire, France VIGARIE 1191
c1605 Problèmes de sources dans l'étude régionale des transports au Canada CERMAKIAN 1193
c1606 Approach to a holistic conception of transportation geography HURST 1195
c1607 Intrinsic transportation and the Canadian north FRANCIS 1197
c1608 Highway network expansion, central Thailand, 1917–67 HAFNER 1200
c1609 Regional patterns of traffic in settlements HOTTES 1203
c1610 Commuter airlines: third level of air service, United States LANCASTER 1205
c1611 Rural road networks and land use change (Canada, Southern Ontario) MORLEY 1207

1006 / Commissions

c1612 Integration of industrial location and commodity flow studies (UK, Midlands)
WALLACE 1209

C17
Medical Geography
Géographie médicale
CONVOCATEUR/CONVENOR: J.L. Girt, *University of Guelph*

c1701 Agriculture, malnutrition, and deficiency diseases, rural Uttar Pradesh (India)
HUSAIN 1211
c1702 Un exemple d'étude des relations pollution-mortalité (Belgique, Anvers)
VERHASSELT-VAN WETTERE 1212
c1703 Training a medical geographer DEVER 1212
c1704 London and Glasgow: comparative study of mortality patterns HOWE 1214
c1705 Unusual multiplication and growth in man, animals, and vegetation KRÂL 1217
c1706 Les causes socio-économiques de l'alcoolisme-ouvrier (France, Italie et Pologne)
THOUVENOT 1218
c1707 Canonical correlation model relating mortality to socio-economic factors
MONMONIER 1221
c1708 Medical approach to the geography of oncochercosis, Mexico
MONTEIL-HERNANDEZ 1223
c1709 Academic training in Mexico of medical geography specialists
SAENZ DE LA CALZADA 1225
c1710 Copper, zinc, lead, and molybdenum in British and Canadian vegetables
WARREN 1226

Geographical Data Sensing and Processing
Sélection et traitement de l'information géographique

c0101
A model for explaining and forecasting developed land in urban areas
ASBJØRN AASE *Norwegian School of Economics and Business Administration*

A multiple regression model for explaining the pattern of developed land in urban areas is being tested for 6 towns with data from approximately the same year, and the various coefficients are compared (Chapin and Weiss 1962). The towns vary in size from 10,000 to 200,000 inhabitants and are very different topographically. For the biggest town, Bergen, the land use patterns for 1951 and 1970 are compared, and a model for explaining the increase in developed land is being tried out. On this basis, a model for forecasting future development under varying assumptions will be built (Malm, Olsson and Wärneryd 1966). As the computation techniques are still being improved the preliminary results will be given only in verbal terms.

The units of observation are small squares, 250m, giving about 800 units in small towns and 8000 in the biggest, allowing ample space for simulating expansion.

The dependent variable is 'developed land,' measured as proportions of each square on a 0 to 10 scale. The independent variables consist of distance variables, terrain barriers, and 'other variables.'

Among the distance variables only airline distance from CBD to each square has so far been computed. Allowance is made for the fact that the regression is heavily influenced by the big increase in observations as the radius from CBD increases, by basing the regression on a constant sample from each ring outwards from the centre. A program for calculating distances from CBD along the main road net and from all squares to the nearest point on the road net has been worked out. Distances from other points, such as local service and employment centres, may be included in the model after maps of residuals have been constructed.

Two variables describing topographical barriers have been found to be strongly correlated with developed area, without being excessively intercorrelated – elevation, i.e., the absolute difference between the CBD and the mean elevation of each square, and roughness of terrain, measured as the number of 5m contour lines crossed by the diagonals of each square. Other indices of topography have been found to be highly intercorrelated with one or the other of the variables mentioned, and do not improve the model. Areas of poor drainage do not seem to be important as barriers in the towns studied (Borchert 1961).

Other variables being computed are agricultural land, land in institutional ownership, airports, and planned open spaces.

Data are mostly collected from economic maps in 1:5000 or 1:10,000. It is a hopeless task to establish time series of urban development based on complete map editions, so a cheap and simple method of enlarging air photographs to the scale of the map and to transfer the grid net in an approximate way to the photographs has been developed.

The amount of variance accounted for by the 'macro variables' – airline distance from CBD and the topographic variables – is much higher in towns with a strong relief than in those on more featureless sites. This means that in the first towns, there is less room for 'micro variables' such as pattern of land ownership, planning decisions, or 'mere chance' in determining urban form.

The effect on the regression of increasing the size of the squares four times, to 500×500m, has been studied. The coefficient of determination for the 'macro variables' goes up (the variance of means being smaller than of individual observations), but the analysis of 'micro variables' becomes less rewarding.

As expected, the coefficients of regression for distance from CBD (related to proportion of developed land) decreases as town size increases, whether one compares a series of towns or the growth of one town over time. In small towns the slope of the curve seems to be linear until it flattens when reaching rural densities. As towns get bigger, the slope tends to take the form of an inverse logistic curve.

The coefficients of regression for elevation and roughness of terrain are rather homogeneous from town to town. This means that a given increase in barrier effect tends to affect the degree of development in more or less the same way in different towns, as long as the topographical features are strong enough to have any influence at all.

A standardized function of barrier effects is established, based on the average of the coefficients for the various towns.

A potential of land for development for a town, based on the two variables for barrier effects, is quantified in the following way: (1) A circle centred on the CBD and encompassing the area under consideration is constructed and the total area calculated. (2) All squares occupied by water are deducted. (3) From the total area of each of the other squares a deduction is made for the computed barrier value; no deduction if there is no barrier effect, big deduction if the barrier effect is great. (4) The potential can be expressed in absolute terms or relatively – as a percentage of the theoretical maximum, which is the area of the circle. The featureless plain would thus obtain the value of 100.

The following hypotheses are being tested for the 6 towns: (1) consumption of developed land per inhabitant increases with increasing relative land potential (i.e., with the natural supply of land); (2) consumption of developed land per inhabitant decreases with increasing town size. The preliminary computations suggest that the potential of land for development is an important factor in determining the overall density in a town.

A comparison of developed land in Greater Bergen in 1951 and 1970 shows an increase of 117 per cent, whereas population has gone up by only 26 per cent. A more detailed study of changes in land use is under way, by specifying various categories of land use and linking these to other data. For this purpose, data of an exceptional quality will be available, as the municipalities are getting their data on co-ordinates, with an accuracy of 10m (Hägerstrand 1967). This means that census data on population and housing and data on property value and income tax, etc. can be aggregated to cells 250 × 250m or any other geographical unit.

The mapping of the variables and of future land use projections is done by an IBM Calcomp Plotter. The map of Greater Bergen is made in 1:75,000, with symbols representing 11 classes for each variable.

Borchert, J.R., 1961 The twin cities urbanized area: past, present, future, *Geog. Rev.* 51, 47–71.
Chapin, Jr., F.S., and S.F. Weiss, 1962 *Factors Influencing Land Development*, Center for Urban and Regional Studies, University of North Carolina (Chapel Hill, NC).
Hägerstrand, T., 1967 The computer and the geographer, *Institute of British Geographers, Transactions*, no. 42, 1–19.
Malm, R., G. Olsson, and O. Wärneryd, 1966 Approaches to simulations of urban growth, *Geografiska Annaler* 48B.

c0102
An automatic pattern recognition system of the Side-Looking Radar (SLR) imagery statistical surface
ERVIN Y. KEDAR and SHIN-YI HSU *State University of New York, USA*

Possible applications of side-looking radar (SLR) imagery in earth science investigations have been suggested since SLR imagery became available ten years ago, and quite a few case studies have been made (Moore 1968). The imagery identification and interpretation techniques employed in these studies were basically confirmed in the framework of traditional photointerpretation. Recent development in SLR interpretative technology utilizing microdensitometer and false colour enhancement proved to be more sophisticated, but not efficient since the operations are essentially manual (Kedar 1971). This paper attempts to develop an automatic-computerized analytical system for SLR imagery interpretation by multivariate discriminatory analysis utilizing isodensitracer (microdensitometer) scanned density data.

In terms of sensor technology, side-looking

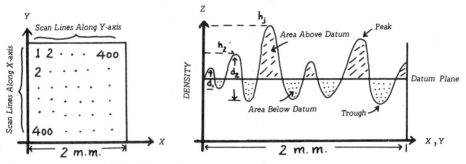

Fig. 1. Response variables of the SLR imagery statistical surface.

radar (SLR) uses an active microwave system to detect environmental features. Its imagery is the echo responding to the transmitted radar signals. It is, therefore, composed of thousands of individual resolution cells, representing the 'return' and 'no-return' of the radar signals striking at single and composite terrestrial features. Moreover, the 'return' components represent the terrain feature and the 'no-return' components indicate the shadow of the target. Hence, SLR imagery is an 'active' statistical surface representing through density variations and distributions the most essential, and almost the whole of, the characteristics of the target. It follows that if we are able to 'capture' these density surface characteristics by some parameters, target features can be successfully discriminated from each other by multivariate statistical analysis. The essential questions here become (1) how to effectively capture such surface characteristics and (2) how to design a method for surface discrimination by an automatic pattern recognition system.

The approach to the questions mentioned above involved two facets: (1) to select response variables of the statistical surface by analyzing the microcharacteristics of the statistical surface, and (2) to perform multivariate discriminatory analysis of these response variables attempting to see whether there are significant differences among the sampled (predetermined) targets, and finally, to classify unknown features into one of the predetermined categories.

In the analysis of the overall (or macro) characteristics of a statistical surface, such response variables as mean, variance, skewness, and kurtosis of the density surface are used. Although some of these variables can be used to differentiate a surface from the other by pairwise combinatorial comparisons, the analysis is not only tedious but also ineffective. This method, therefore, is not applicable to the analysis of SLR image surfaces. In order to solve this problem response variables are extracted from the micro-characteristics of the statistical surface. By microcharacteristics, we mean the surface roughness as depicted graphically by a three-dimensional block diagram. They are composed of undulating lines representing Z values of the data points along both X and Y axes, and can be captured by such response variables as density value, contrast value, and frequency value of the wavelines. More specifically, five response variables are selected along both X and Y axes as follows: (1) total area above the datum plane, (2) total area below the datum plane, (3) sum of contrast values or Σdi, (4) sum of the absolute high-density values or Σhj, and (5) sum of the number of peaks and troughs.

The purposes of multivariate discriminatory analysis are first to establish a discriminant function from several predetermined targets, called calibration samples, such as corn, wheat, barley, etc., and then test whether or not there are significant differences in the variables representing them; and secondly, to classify unknown features into one of the predetermined patterns.

In the analysis of SLR imagery data, ten crop types are selected as calibration samples. Each of the sampled areas covered four square millimetres (2×2) of the SLR firm, isodensitracer (IDT) scanning consisted of 400 scan lines along both X and Y axes and was performed to generate the statistical surface of each crop type. The data matrix for the ten groups can be summarized as follows:

GROUP 1   Response variables

$$\begin{pmatrix} & (1) & (2) & \ldots & (10) \\ \text{Scan 1)} & Y_{11}^{(1)} & Y_{11}^{(2)} & \ldots & Y_{11}^{(10)} \\ \text{Lines 2)} & Y_{12}^{(1)} & Y_{12}^{(2)} & \ldots & Y_{12}^{(10)} \\ \cdot & \cdot & \cdot & \ldots & \cdot \\ \cdot & \cdot & \cdot & \ldots & \cdot \\ \cdot & \cdot & \cdot & \ldots & \cdot \\ 400 & Y_{1400}^{(1)} & Y_{1400}^{(2)} & \ldots & Y_{1400}^{(10)} \\ \text{Mean} & \bar{Y}_1^{(1)} & \bar{Y}_1^{(2)} & \ldots & \bar{Y}_1^{(10)} \end{pmatrix}$$

GROUP 10   Response variables

$$\begin{pmatrix} & (1) & (2) & \ldots & (10) \\ \text{Scan 1)} & Y_{101}^{(1)} & Y_{101}^{(2)} & \ldots & Y_{101}^{(10)} \\ \text{Lines 2)} & Y_{102}^{(1)} & Y_{102}^{(2)} & \ldots & Y_{102}^{(10)} \\ \cdot & \cdot & \cdot & \ldots & \cdot \\ \cdot & \cdot & \cdot & \ldots & \cdot \\ \cdot & \cdot & \cdot & \ldots & \cdot \\ 400 & Y_{10400}^{(1)} & Y_{10400}^{(2)} & \ldots & Y_{10400}^{(10)} \\ \text{Mean} & \bar{Y}_{10}^{(1)} & \bar{Y}_{10}^{(2)} & \ldots & \bar{Y}_{10}^{(10)} \end{pmatrix}$$

Here the superscript stands for response variables, the first subscript stands for group identification, and the second subscript the scan line identification.

These data matrices can be simplified in vector form as:

GROUP 1   GROUP 10
$[Y_1^{(1)} \ldots Y_1^{(10)}]$ $[Y_{10}^{(1)} \ldots Y_{10}^{(10)}]$

The analysis first tests ten hypotheses *simultaneously* to see whether there are significant differences among the group-means, or

$H_0: \bar{Y}_{10}^{(1)} = \bar{Y}_{10}^{(1)} = \ldots = \bar{Y}_{10}^{(1)}$

$\qquad \cdot \qquad \cdot \qquad \cdot$
$\qquad \cdot \qquad \cdot \qquad \cdot$
$\qquad \cdot \qquad \cdot \qquad \cdot$

$\bar{Y}_1^{(10)} = \bar{Y}_2^{(10)} = \ldots = \bar{Y}_{10}^{(10)}$

or in vector form:

$H_0: \bar{Y}_1^{(1)} = \bar{Y}_2^{(2)} = \ldots = \bar{Y}_{10}^{(10)}$,

and the alternative when one of the equalities does not hold. (If it holds, we have to increase the response variables, thus the discrimination power, by taking the diagonal axis into consideration.) The approach to this test is to combine the ten response variables into a 'weighted total':

$Z = a_1 Y^{(1)} + a_2 Y^{(2)} + \ldots + a_{10} Y^{(10)}$

and choose $a_1, a_2, \ldots, a_{10}$ in such a way that the distance between the groups is maximized. The discriminant function is in fact the estimated values of the above function:

$\hat{Z} = \hat{a}_1 Y^{(1)} + \hat{a}_2 Y^{(2)} + \ldots + \hat{a}_{10} Y^{(10)}$

The significance test is done by Roy's Union-Intersection Method. Once we know that there exist some $a_1, a_2, \ldots, a_{10}$ (maximum values are used), the discriminant function is fully determined. We can then use it as a decision-making device to classify any unknown features into one of the calibration samples.

As mentioned above, we selected ten crop types in the Houston, Texas area as the calibration samples, and several 'unknown' plots for pattern recognition. The software to perform the above-outlined numerical analysis was developed at MSC, NASA, Houston, Texas.

The initial step was to display the SLR imagery statistical surface in a two-dimensional plot by the IDT. Then an automatic-computerized analytical system (GSI Program) took over the process of pattern recognition. It successfully classified the unknown plots into one of the predetermined patterns.

There might be some misclassification owing to variations in the size of the plants, the amount of moisture in the crop and soil, and even local micro-climatic conditions, which determine to some degree the dialectric characteristics of each field as a whole. In our case study, it was proved that the degree of local variations within individual plots on one single SLR frame was not significantly larger than the degree of variation existing between crop types, using ten response variables for discrimination analysis. This is the main reason for the success of our experiment.

Kedar, Ervin Y., 1971   Manual and automated-computerized systems in side-looking radar applications for land use survey, from Technology Utilization Ideas for the 70's and Beyond, 26, *Sci. and Technol.* (American Astronautical Society, Tarzana, Cal.).

Moore, Eric E., 1968   *Side-Looking Radar in Urban Research: A Case Study*, a report prepared under the terms of contract 14-08-0001-10654, Geographical Applications Program, US Geographical Survey, published by Remote Sensing Laboratory, Department of Geography, Northwestern University, Evanston, Illinois (June).

High Altitude Geoecology
Géoécologie des pays de haute altitude

c0201
Pseudo-timberline: the southern Appalachian grassy balds
PHIL GERSMEHL   *Concordia Teachers College, USA*

Near the summits of many southern Appalachian mountains are local treeless areas known as 'balds.' Although located in an otherwise forested region, these grassy anomalies have been described as natural prairies, yet decades of study have failed to produce a fully acceptable natural explanation for their treelessness. Early hypotheses of bald origin – cold, wind, exposure, fire, drought, insect attack, soil deficiencies – have been discounted. Theoretical thermal treeline is some 1200m higher than the tallest peak in the region. Wind velocities are greater in the forested gaps than on the summits. Many of the most exposed peaks are forested, while many balds seem well sheltered. Post-fire succession usually is dominated by woody shrubs such as *Prunus*, *Rubus*, or *Rhododendron*. Soil moisture tensions rarely exceed 1atm. Insect attacks are at best selective tree eliminators. Differences between bald and forest soils are minor and inconsistent (Gersmehl 1970b, appendix A).

Several workers have stressed the common location of balds near or 'just above the highest extent of the northern hardwood forest,' especially on peaks where, for some reason, the boreal conifer forest association is absent from the higher slopes (Davis 1929, 43). In such situations, the surrounding hardwood trees are low-growing and invade treeless areas very slowly. This has led to the development of a multiple-treeline or ecotonal hypothesis. According to this theory, a recent hypsithermal period induced an upslope migration of all forest types, reducing or perhaps eliminating the spruce-fir type from the higher elevations. With subsequent cooling, the hardwood forest has retreated downslope, often leaving a persistent grassy bald above it. Reforestation is delayed by the severe microclimatic conditions in deforested areas near the ecotone between the forest types (Mark 1958, 323–5).

Closer examination reveals inadequacies in this proposed vegetation history. Since a spruce-fir canopy profoundly alters its own micro-climate, it seems doubtful that raising average temperatures a few degrees for a few centuries could eliminate the boreal forest from a mountaintop. Such a 'fine-tuned' vegetative response to macroclimatic fluctuations is contradicted by the widespread persistence of spruce-fir stands as much as 600m below the average position of the present ecotone (Shields 1962, 139). Hardwood stands are likewise common some 400m above the ecotone. This suggests that the location of the ecotone is fixed primarily by non-climatic disturbances and competitive relationships between the two forest types, rather than by broad climatic gradients (Schofield 1960, 154). A complete tabulation of present and historical descriptions of balds includes many located within or above spruce-fir forests, some within the hardwood association, and a few on inverted ecotones, with spruce-fir forests below the bald and hardwoods above (Gersmehl 1970b, 309–419). Furthermore, there is a notable absence of corroborative evidence, such as pollen profiles or understory indicator species, to support the proposed vegetation history (Norris 1964, 140; Gersmehl 1970b, appendix E). Finally, the problem of tracing past vegetation changes is complicated greatly by the extensive grazing and logging during the two centuries of European occupance of the mountains.

The assertion that microclimatic severity retards tree invasion of deforested sites near the ecotone is likewise less convincing upon closer examination. During the past half century, grazing has ceased on most balds; this cessation is always accompanied by an intense, although areally limited, marginal encroachment by the surrounding forest. Direct measurement of seedling-height temperature extremes, soil temperatures, wind velocities, and crude water balance indicates that the microclimate in the zone of tree invasion closely resembles that on the treeless bald, while differing radically from conditions under the adjacent forest (Gersmehl 1971, 59).

The persistence of the balds throughout the nineteenth century is therefore best viewed as a consequence of heavy grazing. The present slow reforestation rate is due primarily to the limited viable-seed dissemination capabilities of the major trees found at the higher elevations. *Betula, Picea,* and *Abies* reproduction employs a winged seed with an effective dispersal distance seemingly limited to a few metres or tens of metres (Gersmehl 1970b, 193–207). *Quercus, Fagus,* and *Aesculus* reproduce by heavy nuts or direct root sprouting, with obviously limited range. Rodent-assisted seed dissemination onto a bald is also unlikely, since the common mountain rodents live in the forest and make only occasional forays onto the bald, usually to carry nuts back into the forest (Mark 1958, 323). It is therefore understandable that balds bordered only by nut-bearing trees are being reforested at a much slower rate than those surrounded by a more diverse forest. In either case, balds measured in hectares may require several tree generations for complete reforestation, though the grassy meadows may give way to woody shrubs in a far shorter time.

The imminent disappearance of the grassy balds raises serious questions concerning their pre-European existence. The handful of Cherokee Indian legends which mention the balds were recorded many decades after European settlement in the region; their variability and internal inconsistencies cast further doubt upon their validity as historical evidence (Gersmehl 1970b, 45–69).

More reliably documented Indian cultural practices, however, include deliberate burning of the Appalachian forests for the purpose of hunting and game habitat improvement (Gersmehl 1970b, 77–9). Moreover, as the early European settlers perceived the mountain environment, clearing of remote mountaintop pastures was a rational land use decision (Gersmehl 1970a). Burning or girdling mountaintop trees will not establish a grassy cover; like natural fires and windthrows, these merely initiate a woody succession back to the original forest. However, they can eliminate local seed sources for the canopy-dominant tree species. The ensuing shrubland is a more favourable grazing site, and at the elevation of the balds, heavy grazing and repeated groundfires suppress woody vegetation and favour grasses. This provides a plausible mechanism for the origin of the grassy balds (and one that is consistent with the historical descriptions, as compiled in Gersmehl 1970b, appendix B, 309–419).

In summary, natural disturbances at high elevations in this region yield shrublands that are persistent at least to the extent that local canopy-dominant tree seed sources are eliminated. The Appalachian grassy balds, however, are nineteenth-century cultural relics, originated and maintained by grazing and deliberate groundfiring. Their short-term persistence following cessation of grazing is a consequence of the limited seed-dispersal capabilities of the high-elevation tree species. Similar delayed responses to environmental changes may characterize other vegetation types, especially in the more severe environments.

Davis, J.H., Jr., 1929 Vegetation of the Black Mountains of North Carolina, PH D dissertation, U. Chicago.

Gersmehl, P., 1970a Factors leading to mountaintop grazing in the southern Appalachians, *Southeastern Geog.* 10, 67–72.

– 1970b A geographic approach to a vegetation problem: the case of the southern Appalachian grassy balds, PH D dissertation, U. Georgia.

– 1971 Factors involved in the persistence of southern Appalachian treeless balds, *Proceedings,* Ass. Am. Geog. 3, 56–61.

Mark, A.F., 1958 The ecology of the southern Appalachian grass balds, *Ecological Monographs* 28, 293–336.

Norris, D.H., 1964 Bryoecology of the Appalachian spruce-fir zone, PH D dissertation, U. Tennessee (Knoxville).

Schofield, W.B., 1960 The ecotone between spruce-fir and deciduous forests in the Great Smoky Mountains, PH D dissertation, Duke U.

Shields, A.R., 1962 The isolated spruce and spruce-fir forests of southwest Virginia, PH D dissertation, U. Tennessee.

Geography of altitudinal belt ecosystems in the Caucasus

R.P. ZIMINA, D.V. PANFILOV, and Y.A. ISAKOV   *Academy of Sciences, USSR*

The territory of the Caucasus represents a convenient model for a typological study of ecosystems in their dependence upon general physico-geographical conditions. We see here, on a relatively small scale, the development of cryophile, alpine meadows, mesothermophile, and thermophile forests, as well as ultra-thermophile desert ecosystems. Certain regions of the Caucasus differ essentially in their climatic regime determined, to a great extent, by the orographic peculiarities of this mountain country. Accordingly, different types of altitudinal zonality in the distribution of ecosystems are recorded in the Caucasus.

The main types of natural ecosystems are distinguished by us on the basis of a functional appraisal of edificators in the vegetational cover and animal population, as well as of their importance in the biological turnover. Such an approach makes it possible to establish direct and averaged relations between the structure of natural ecosystems and climatic features and economic activity of man. Nine main types of natural ecosystems are distinguished in the Caucasus characteristic for different altitudinal belts.

(1) Alpine meadows located at elevations of 2300–3500m asl under conditions of a cold alpine climate. The vegetation period is brief, 1–2 months only; the soils, predominantly peaty-humus. The total phytomass is not large (100–260cwt/ha abs. dry weight), 70–80 per cent being underground organs of the plants. The zoomass is also not large – from 40 to 45kg/ha (live weight). Animals use the green mass of plants rather fully, but only slightly their underground parts and the dead plant debris. This has contributed to the accumulation of poorly decomposed plant remains and a paludification of the soils. There is a considerable evacuation of substance by water into the underlying ecosystems. On the whole, the biological turnover is not complete, a substantial part of primary products being removed.

(2) Subalpine meadows (with elfin woodland and rhododendron shrublets) are developed on elevations of 2000–2300m asl. The vegetation period here lasts 2–3 months; the soils are mountain meadows, chernozem-like, humus, low-skeletal. The total phytomass is substantial (250–400cwt/ha), the yearly growth of green parts in plants coming to 20–25 per cent of its total reserves. The zoomass comes to 400kg/ha, 75 per cent of which represents the share of invertebrates – saprofagous forms. The animals consume quite intensely the green parts of the plants, the main part of the yearly growth being used by the detrivorous. The biological turnover in these ecosystems is intense and mostly closed. Because of the activities of soil-inhabiting forms and microorganisms, a considerable reserve of nutrients necessary for the plants is being created in the soils.

(3) Dark coniferous forests are best developed at elevations of 1200–1800m asl under conditions of a moderately warm climate (with a great amount of precipitation). The vegetation period lasts 4–4.5 months; the soils are similar to brown forest soils; under the spruce forest they are podsolic. The total phytomass comes to 5000cwt/ha, but its great bulk consists of perennial parts of the plants. The yearly increase, probably, does not exceed 200cwt/ha, the green mass forming only its minor part. The total zoomass is 450–500kg/ha, the animal population consisting mostly of saprophagous forms, numerous xylophagous, and some herbivores. The biological turnover is slow because of a conservation of a substantial part of organic substances in the perennial parts of the plants. It is not complete because there is a substantial substance removal by water into underlying ecosystems.

(4) Beech forests. Optimum conditions for the development of this type of ecosystem are formed at elevations of 1000–1500m asl, where the vegetation period lasts 4.5–5 months. Here brown forest, humus soils are formed. The total phytomass comes to 3500–5000cwt/ha with an annual increase in the green mass of 50–60cwt/ha. The total zoomass is about 750kg/ha, over 85 per cent of which is the share of invertebrates-detrivores. Here consumers of the green parts are nearly twice as numerous as in dark coniferous forests. The biological turnover is fundamentally similar to the ecosystems of dark coniferous forests, but is much more closed.

The activity of saprophagous animals here is much more effective, which contributes to the accumulation of nutrients in soils in an accessible form for the plants.

(5) Humid deciduous forests on lowlands and in the foothills of Transcaucasia in localities with a high mean annual temperature and a particularly abundant precipitation. The soils are low-humus zheltozems and red soils replaced in areas of low mountains by brown forest soils. The total phytomass is 3000–5000cwt/ha with a green mass increase of about 150cwt/ha. An intense activity by the animals in the use of live and dead parts of plants and a high activity of microorganisms contributes to a rapid biological cycle, whereas an abundance of precipitation and an intense run-off result in a decrease of nutrient reserves in the soils and, locally, to their swamping.

(6) Hornbeam-oak forests are developed in the northern Caucasus, usually, above 600m asl; in drier regions of Transcaucasia the limit is 1600m. Under them grey podzolized or brown forest soils are formed. The total phytomass comes to 2000–2500cwt/ha with the annual production of the green mass being about 100cwt/ha. The zoomass is 1000–1700kg/ha, invertebrate-detrivores coming to 600–650kg/ha. Both saprophagous animals and herbivores are taking an active part in the biological turnover, which here is intense and rather closed. It proceeds without any substantial loss of biogenic elements.

(7) Arid light forests and phryganas in the foothills and low-mountain areas with a dry, warm, and lengthy summer. The soils are brown. The total phytomass does not exceed 500cwt/ha with a production in the green mass of 40cwt/ha. The zoomass is relatively large – 500–600kg/ha, a substantial part consisting of herbivores (about 200 kg/ha). There are comparatively few consumers of plant litter – from 60 to 65kg/ha. The turnover in these ecosystems is mostly closed and rather rapid, because of an intense activity of the herbivorous and root-devouring animals. Part of the organic matter, however, escapes the rapid cycle as it is accumulated in the wood.

(8) Steppes are associated with plains and piedmont parts of the northern and eastern Caucasus, as well as with the Armenian highland. They are formed under conditions of a dry climate (not more than 400mm of precipitation per year). The total phytomass in these ecosystems is 150–250cwt/ha (about 30 per cent of which consists of above ground parts of the plants) and the increase of the green mass is 10–30cwt/ha. Abundant are herbivorous, root-devouring, and predatory animals, but saprophagous forms are also numerous. The biological turnover is intense and mostly closed, and results in a substantial accumulation of organic matter in the roots of plants and in the formation of chernozem soils.

(9) Deserts and semi-deserts occupy the driest regions in the east of Predkavkazie (Forecaucasus), in Azerbaidzhan and Armenia. The soils are grey-brown and grey. As a rule, the general phytomass does not exceed 110cwt/ha and the annual growth of the green mass is 40cwt/ha. The biomass of animals is comparatively large (380–400 kg/ha), and, consequently, their role in the phytomass consumption is quite substantial. Particularly great is the role of herbivores; of importance also are root-devouring and detrivorous forms. In general, the turnover of matter in these ecosystems is intense, but insufficient moisture frequently results in a salinization of the top soil horizons, which has a negative effect upon the vegetation cover.

Here are the main regularities in the distribution of natural ecosystems on the territory of the Caucasus. By their composition, structure, and elevation alpine meadow ecosystems are similar in all the regions of the Caucasus. Very similar also are forest ecosystems of medium-height mountain regions (dark coniferous and beech forests); they are developed, however, mostly in the west and are virtually absent in the east. Especially varied are ecosystems of the foothills, piedmont, and intermountain valleys. In the west (Kolchida) and in the extreme southeast (Talysh) humid forests are developed, whereas in the east of the Forecaucasus and Transcaucasia, as well as south of the Armenian highland, we find extremely xerophile ecosystems, predominantly of dry-steppe and semi-desert types. The reason for such a location lies in an interaction of humid air masses coming in from the west with the orography of the territory in the Caucasus.

At the high level of mountains the moisture and temperature regime are rather uniform everywhere. At lower levels, beginning with mountains of medium height, there is a great difference in the moisture of slopes facing the west and the east. Of particular effect upon the distribution of ecosystems on the territory of the Caucasus is the Main Caucasian mountain range. It protects Transcaucasia from the invasion of cold air from the north and intercepts the greater part of precipitation.

Because of these factors, the set of natural ecosystems that form altitudinal belts and the elevations at which they are located are different in various parts of the Caucasus. Most distinct are three types of vertical zonality that can be subdivided into six subtypes. The Kuban-Abkhazian type of altitudinal zonality includes the following series of ecosystems: steppes and shibliak (deciduous bush formation) – the latter along the sea, hornbeam-oak forests, beech forests, dark coniferous forests, subalpine meadows, and alpine meadows. This type of zonality is, predominantly, humid, with slightly expressed features of semi-aridity on piedmont plains and in the foothills. The biomass of all natural ecosystems here is especially great. The yield of the green mass and wood is substantial; as a rule, the role of animals in the biological cycle is considerable.

The Kolchida-Girkanian type of vertical zonality includes paludal forests on the plain, humid broadleaved forests in piedmonts, beech forests, and subalpine meadows. It is characterized by an extreme humidity (except that in Zuvant the humid forests of Talysh are replaced by steppe-like subalpine meadows and mountainous steppes). In the majority of ecosystems, the increase of the green mass is very large and the role of animals in the biological turnover is relatively small. Here we find a predomination of a matter migration by water that leads to its substantial evacuation into the lowlands and coastal parts of the Black Sea and the Caspian.

The Daghestan-Armenian type of zonality is characterized by semi-aridity and even aridity. Typical are ecosystems that form altitudinal belts: deserts and semi-deserts, piedmont steppes, phryganas and arid light forests, mountain steppes with dubravas (mesophilic high oak forests), steppized subalpine meadows, and alpine meadows. The biomass of these ecosystems is relatively small and the yield of phytomass not great, whereas the role of animals in the biological matter cycle is very substantial. The cycle, is to a great extent, closed.

Coastal Geomorphology
Géomorphologie côtière

---

c0301
Shoreline erosion on the Barron delta, North Queensland
ERIC C.F. BIRD   *University of Melbourne, Australia*

The geomorphological evolution of the Barron delta has been discussed elsewhere (Bird 1969). Quartzose sand, derived from deeply weathered granite outcrops in the headwater regions, has been carried down to the coast by the Barron River, notably during episodes of floodwater discharge following heavy rainfall in the coast ranges. Delivered to the coast, this sand has been moved northward and southward by wave action, to build a succession of beach ridges and spits which are incorporated in the prograded deltaic shoreline. The delta has a seaward fringe of sandy beach ridges, interrupted by swampy corridors which form the outlets for mangrove-fringed distributary creeks.

The dominant waves in coastal waters, generated by the prevailing southeasterly winds, produce a northward drift of sand along the shore. Ocean swell is largely excluded by the Great Barrier Reefs, but it does enter by way of the broad gap at Trinity Opening, and is then refracted in such a way as to approach the deltaic shoreline from the northeast. These waves, together with waves generated by occasional northerly winds, produce a southward drift of shore sand. In general the effects of southeasterly waves are much more prolonged and consistent, but the southernmost sector of the deltaic shoreline is less influenced by these dominant waves because it is in the lee of the Cape Grafton peninsula. In this sector the northerly waves are more effective. In consequence, there is a 'drift-parting' in the vicinity of the mouth of Redden Creek. Sand moves alternately northward and southward along the deltaic shoreline, but north of Redden Creek the predominant drifting is northward, as indicated by spit growth, creek mouth deflection, and sand accumulation against headlands, and south of Redden Creek the predominant drifting is southward to Casuarina Point.

The main outflow from the Barron River has been located at various points along the deltaic shoreline. At stages when it has opened north of Redden Creek, the bulk of the sand yield to the coast has been carried northwards. In recent decades the river mouth has lain south of Redden Creek, and there has been rapid sand accretion to build spits first at Ellie Point, and more recently at Casuarina Point. When Redden Creek itself was the main outflow from the Barron the sand yield was more evenly divided between northward and southward drifting, and under these conditions sufficient sand remained on this sector of shore to be built into the successive parallel sandy beach ridges which lie behind Machan's Beach. At an earlier stage, the Barron opened into Cairns Bay, farther south, and delivered sand which was built into the parallel beach ridges on which the city of Cairns now stands (Bird 1970).

The first settlers reached Cairns in 1876, and the main outflow from the Barron then lay just northwest of Ellie Point, its left bank consisting of a sandpit with an unbroken beach on its seaward side between the mouth of Redden Creek and Casuarina Point, then somewhat shorter than it is now. Under these conditions Ellie Point was an actively nourished sandspit.

In 1937 the Barron broke through a sandy isthmus to establish its present outlet, the old mouth soon becoming choked with mangrove swamp. Ellie Point is still growing slowly, but it is no longer receiving fresh sand supplies, and the shoreline has become muddy, with stands of mangroves developing in front of the sandy shore. Casuarina Point has been prograded and has extended southwards as the result of rapid accretion of sand since 1937. Machan's Beach, between Barr Creek and the present mouth of the Barron River, had long been a stable shoreline sector, but in recent years its shoreline has been cut back between 60 and 100 metres, and erosion continues despite the dumping of boulder walls. The erosion threatens the coastal township at Machan's Beach. Loss of sand from the foreshore was first noticed here in the late 1940s, but the phase of rapid erosion dates from 1963. Sand lost from this sector has moved away both northwards (to prograde Holloway's Beach and build the spit which deflects the mouth of Richter Creek) and southwards (to build a spit which constricts the mouth of the Barron until it is swept away to the southeast during episodes of floodwater discharge, and added to the growing spit at Casuarina Point).

The onset of erosion at Machan's Beach is due to two sets of factors. On the one hand the sand yield from the Barron River has diminished, first as the result of sand extraction from the river channel for use in building and road construction during the early 1940s, and then as the result of interception following the completion of Tinaroo Falls dam in 1958 and the Barron Falls dam in 1963. Since 1963, very little sand has come down to the mouth of the Barron River during floods, and the onset of serious erosion at Machan's Beach can be correlated with the trapping of sand in the reservoir behind the Barron Falls dam: a substantial sandy delta has developed at Kuranda, where the river enters the head of this reservoir. The supply of sand to the coast has thus diminished, and with continued drifting northward and southward along the shore, Machan's Beach has become a sector of sand deficit.

The second factor is the location of the main outlet from the Barron River. Since 1937 this has opened to the south of Machan's Beach, and in times of floodwater discharge it has swept away to the southeast sand that would otherwise have remained to move to and fro on Machan's Beach. Fluvial outflow has in effect increased the extent of southward drifting, and once it is south of the river mouth the sand cannot drift back to Machan's Beach, because of interception by fluvial outflow.

A possible solution would be to cut a main

outflow channel to the sea on the line of Redden Creek and seal off the existing mouth of the Barron. This would remove the fluvial contribution to southward drifting which now operates at the river mouth, and restore the beach sector which existed before 1937 by removing the interruption and interception due to existing fluvial outflow. Any sand brought down by the Barron River would then be delivered to Machan's Beach at the mouth of what is now Redden Creek, thus achieving optimum natural replenishment of this beach sector at the position of the drift parting. If this proved insufficient to conserve the beach, supplementary artificial nourishment could be achieved by dumping sand in the new cut, to be delivered to the shore by fluvial outflow.

Analysis of recent changes on the shoreline of the Barron delta has shown that, in an era of diminished sand supply and consequent shoreline erosion, location of the main river outlet may be critical in determining the extent of erosion. Under such conditions it may be wise to modify the natural distributary channels in such a way as to concentrate the diminished sediment yield to the shore at a place from which predominant longshore drifting (whether unidirectional or divergent) can deliver it to sectors of the shoreline where erosion control is required. If the diminished sediment supply is still too meagre to achieve this control, the selected outflow can be used as a feeder channel into which sediment is dumped for subsequent delivery to the coast by fluvial outflow and thence to eroding shore sectors by longshore drifting.

It is possible that engineering of distributary channels on the Nile delta could be used in this way to reduce the shoreline erosion that has developed since the completion of the sediment-trapping Aswan High Dam, and to aid beach replenishment on critical sectors of that deltaic shoreline.

ARGC Grant 67/16405.

Bird, E.C.F., 1969 The deltaic shoreline near Cairns, Queensland, *Australian Geog.* 11, 138–47.
– 1970 Coastal evolution in the Cairns district, *Australian Geog.* 11, 327–35.

c0302
Zonal studies of the coast of the maritime region of Canada
EDWARD H. OWENS *Bedford Institute, Canada*

Coastal geomorphological studies in Canada have been relatively few in number and usually have involved local investigations of particular process/response environments. Canada's shoreline is the longest in the world, being in excess of 500,000 miles, and covers a wide latitude of the northern hemisphere, 42°N to 83°N. Apart from Johnson's description of parts of the maritime region (1925) and some local arctic surveys, there have been no systematic projects designed to evaluate the nature or character of the coast. Probably less than 10 per cent of Canada's shoreline has been investigated at even the reconnaissance level.

The combination of a variety of geological constraints on coastal development and the diverse effects of the pleistocene ice sheets give a number of major coastal environments within Canada. The storm-wave dominated Atlantic coast and the ice-bound shores of the Arctic Ocean illustrate the range of climate and wave environments which modify and develop coastal forms. In addition the Great Lakes provide a land-locked system which may be more easily investigated than the ocean in terms of the air-sea interface.

A small project to describe one 400-mile section of coast was undertaken in Chedabucto Bay during 1970. This project resulted from the beach restoration program following the oil spill from the tanker ss *Arrow* in February 1970. The reconnaissance was carried out from a helicopter over a three-day period and provided a description and evaluation of coastal features and shoreline units. When assessed in the context of the geological and geomorphological history of this rapidly submerging region it was possible to provide a meaningful analysis of the character of the coast (Owens 1971).

In brief, the shoreline features of Chedabucto Bay are primarily related to the structural history of the region and the effects of the Pleistocene glaciation. Littoral processes have been successful in modifying the coast along the northern and western shores, whereas the south shore is a rocky fault line escarpment produced by differential erosion. The sedimentary Carboniferous rocks to the north of a major east-west oriented fault zone have been eroded to an undulating lowland while the resistant Ordovician and Devonian outcrops to the south remain as an uplifted tilted block. The rectilinear south coast is largely devoid of sediment and local variations in the coastal form result from the effects of wave erosion along secondary faults or lines of weakness. The character of the west and north shores of the bay is derived from the submergence of a low lying, glaciated region. Erosion of extensive till deposits and the redistribution of littoral sediments give a coast which is being constantly modified by wave processes.

The success of the reconnaissance survey in Chedabucto Bay has led to a project designed to study the coastal zones of the southern Gulf of St Lawrence. An air reconnaissance of 1200 miles of coast includes the identification of major shoreline units and this is supplemented by an investigation of the process environment and of the form of the coast within each unit. Using reconnaissance surveys and detailed zonal studies it is possible to describe and analyse the component features of a shoreline. Evaluation of air photographs dating back 40 years and a five-year survey of form and process will enable the correlation of shoreline changes and coastal geodynamics. Further projects of this nature should enable a better understanding of our coastal areas. Although reconnaissance coastal studies do not provide the data necessary for the understanding of particular phenomena or processes they are nevertheless necessary for the adequate development and protection of this valuable natural resource.

Johnson, D.W., 1925 *The New England–Acadian Shoreline* (New York).
Owens, E.H., 1971 *A reconnaissance of the coastline of Chedabucto Bay*, Marine Sciences Paper no. 4 (Ottawa).

c0303
The effect of typhoons on beaches in Hong Kong
CHAK LAM SO *University of Hong Kong*

The effect of typhoons on beaches in Hong Kong, promoted by the predominance of sand, makes itself felt on both a short-term and a long-term basis despite the existence of a 'mesotidal' and 'low-energy' coastal environment (Davies 1964, 36–8). The packing of wave energy into a short period imposes on the beaches patterns of 'cut and fill' not always emphasized in texts. The outcome of such changes, expressed in the loss and gain of beach materials of varied calibre and value, not only serves as a guide to any planned preservation of the beaches locally, but also provides experience which deserves wider attention.

The study emphasizes a field approach and incorporates relevant cartographic data. Field observations extended to the major sand beaches were synchronized to render comparative study more meaningful. To trace morphological changes in detail, some beaches, chosen on basis of varied exposure, offshore depth and lithology, were levelled at seasonal intervals and just before and after each typhoon or other tropical cyclone. Precision levelling is preferred as it leaves the sand surface undisturbed prior to the onset of a typhoon. Beach levels obtained were subjected to quantitative treatment and compiled into contour maps and profiles to depict beach changes in three dimensions.

Comparison of beach states before and after typhoons reveals considerable levelling of the beach and widening of the surf zone, even at the expense of pre-existing ridges. Both the beach and its submarine slope suffer a loss of sand which is hardly compensated for by gains elsewhere. The lower beach slope witnesses the encroachment of pebbles on sand which are more often scattered than thrown into ephemeral cusps. The flattened beach surface terminates abruptly on the backshore at a slope discontinuity that is concave upwards. Analysis of beach materials shows that after a typhoon coarser sand tends to occur, especially at the back of the beach where heavy minerals are also found to concentrate. Post-typhoonal beach changes indicate that materials lost from beaches in storms may be made good, and beach profiles steepened, in the following storm-free winter, but some beaches have not completely recovered from the effect of typhoon 'Wanda' experienced in 1962.

It would appear that beaches affected by a typhoon, by undergoing changes, tend to adjust themselves to the specialized wave dynamics and the intricate structures of the associated swash and backwash. In this the fetch becomes less significant as winds in a typhoon may raise a high sea within a short distance. The higher and steeper waves, locally reaching up to 6 metres in height when the typhoon is at some distance away and much more when the centre is near (Cuming 1967, 22), favour the development of plunging breakers and combing wave action. In such circumstances, the beaches are worked over and the submarine slopes are reorganized under conditions of a shortage of materials and considerable wave turbulence caused by gusts. The precise manner in which the process shapes the beach can be significantly modified by variation in the relative and absolute duration of each stage of the storm. The structure of a typhoon is such that swell development up to a day or more occurs with both the approach and the departure of the typhoon while the generation of forced waves with duration averaging 7 hours (Heywood 1950, 13) accompanies the passage of its centre. Thus the initial stage of the typhoon witnesses waves of rapidly growing strength levelling out the beach slope and increasingly throwing up pebbles on to the surf zone. The height of the typhoon is marked by maximum wave combing and displacement of the finer and lighter sand grains in suspension. The vertical limit of this effect is determined by the oversplash of storm surges which, varying in intensity according to coastal configuration (Cheng 1967, 3), have been recorded up to 8 metres above the predicted tide (Watts 1959, 28). With the passage of the centre of the typhoon, winds veer so that beaches little affected by swell at the approach of the typhoon would now experience its full effect. Beaches affected by typhoonal winds may therefore experience maximum wave intensity either early or late in the effective storm period.

The study reveals that the effect of typhoons on beaches in Hong Kong amounts to a rejuvenation of wave action and its extension to a higher level and a wider zone. To this process beaches of sand readily respond by appreciable transformation and by partial redistribution of their constituent materials. The resultant reduced beach gradient, through negative feedback, induces post-typhoonal deposition and restoration of the profile of equilibrium of the beach. Although the time it takes to complete the work ranges from a season to a decade or more, the general tendency is for the intensified, though short-lived, process responsible for the loss of sand in the storm to be followed by a long but slow process of recovery. Successive typhoons may terminate the recovery process or renew the pattern of 'cut and fill.' The state of the beach at any one time is determined by the progress of its recovery from the last typhoonal effect. On these beaches, short-term changes are often greater than the long-term, and the work performed by marine agents in the short-lived typhoon is many times that carried out over the long, storm-free period. Uniformitarianism based on this would appear to be taken too far. The enormous disparity between what happens at ordinary times and during periods of storms, which is felt in a unique way in Hong Kong by the very nature of the typhoons, ensures that the mode and the tempo of beach processes vary significantly in wave environments of contrasting energy.

University of Hong Kong.

Cheng, T.T., 1967 *Storm Surges in Hong Kong* (Hong Kong).
Cuming, M.J., 1967 *Wave Heights in the Southeast Approaches to Hong Kong Harbour* (Hong Kong).
Davies, J.L., 1964 A morphogenic approach to world shorelines, *Zeitschrift für Geomorphologie* 8, 127*–42*.
Heywood, G.S.P., 1950 *Hong Kong Typhoons* (Hong Kong).
Watts, I.E.M., 1959 *The Effect of Meteorological Conditions on Tide Height at Hong Kong* (Hong Kong).

c0304
Accelerating crisis in beach erosion
W.F. TANNER and F.W. STAPOR   *Florida State University, USA*

The beaches of the world are, by and large, undergoing erosion. Although the effects are spectacular on heavily populated or 'improved' coasts, rapid erosion is also taking place where man has not interfered in the operation of the littoral drift system. A good example of the latter can be seen in the state of Rio Grande do Sul, Brazil, from Torres to the southwest for at least 350km. Other examples could be cited, both from the literature and from our field experience.

This widespread erosion of beaches reflects a general shortage of sand. Not enough sand is being introduced to the beaches of the world to satisfy the requirements of the littoral drift system (Tanner 1958). The equilibrium beach form (as seen in plan view) is commonly maintained, but the sand deficit is made up by a systematic shift of the beach toward the land. The overall situation can be described as a littoral 'economy of scarcity.'

The present shortage of beach sand does *not* reflect, primarily, fluctuations of sea level. The shortage is a matter of the most recent decades or a few centuries, whereas sea level has occupied essentially its present position for several millenia. Extensive beach ridge plains which we have studied are (1) presently growing, (2) being maintained without visible gain or loss, or (3) undergoing erosion. Almost all of the examples which we know belong in class 3; most of the rest, in class 2.

A good example of the third class is found on St Vincent Island, off the panhandle coast of Florida. The dates available from this island show that it developed as a series of roughly 100 beach ridges between about 4000 years ago and perhaps 300 years ago. During that time sea level fluctuated slightly (Stapor 1972), but growth continued regardless of the fluctuations. The island developed in a littoral 'economy of abundance.' Each of the beach ridges, now preserved, exhibits the equilibrium outline in plan view, but it was a shifting equilibrium due to an excess of sand.

For this particular locality a coastal equilibrium was established near present msl about 4000 years ago, and the beach aggraded in an economy of abundance (but maintaining the equilibrium geometry) for nearly 4000 years. Two km^2 of new land were added each century. A few centuries ago the shift to an economy of scarcity was begun. The change can be seen in a sharp shift in the geometry of beach ridges, the dates of some of which are known from very old nautical charts. The geometry of equilibrium still remains, but there is no net growth now.

Other islands along the panhandle coast of Florida show essentially the same history.

Field work in the state of Tabasco, Mexico, showed that the three youngest beach ridges on the delta of the Grijalva-Usumacinta River system were built between 100 and 400 years ago. (They contain boards and nails, but large trees have grown on them.) It is assumed that the many older ridges, now well preserved, all date from late Holocene time. Up until a few centuries ago between 30 and 40km^2 of new land were added each century; the delta front is now undergoing extensive erosion. This area passed during the last few centuries from an economy of abundance to an economy of scarcity. At the time the change was made, *no* dams had been built in the drainage basin; hence man's actions do not appear to have been responsible for the onset of erosion.

In the state of Veracruz, Mexico, erosion is now general along the Holocene barrier island, after a prolonged period of beach ridge growth (Stapor 1972).

We have made similar observations in most of the places where we have been able to study well developed beach ridge plains. The history which we obtain can be summarized as follows:
1. Stabilization (more or less) of sea level, near its present position, 4000–5000 years ago (as is well known).
2. Construction of barrier islands, and seaward progradation of many of these islands into beach ridge plains; economy of abundance of sand.
3. Depletion of the reserve of sand which made step 2 possible.

4. Erosion of the barriers and their migration toward the mainland; economy of scarcity.

Step 3 took place, along many beaches, within the past few centuries, or is becoming obvious now. Because so much of the coast has already progressed to step 4, it is reasoned that many beach ridge plains, now in class 2, will soon be in class 3. That is, coastal erosion has been becoming more general in the last few centuries, and therefore is likely to become even more general in the near future. Sea level changes cannot enter, in any important way, into the development of the third and fourth steps.

One important question occurs: what significant event caused the shift from step 2 to step 3? In an equilibrium system the answer could be either an increase in wave energy levels, or a decrease in supply of sand. We much prefer the second of these two.

To explain the decrease in sand supply we suppose the following. Once sea level rise has ceased, an equilibrium geometry is quickly established along the new beach, and littoral drift processes then begin the task of rearranging the available sand (left over from a previous epoch of transport by non-marine agencies), until the geometry is such that the system will absorb the maximum amount of wave energy with a minimum of sediment transport. Until this has been done, no matter how long or short the time required, there is an economy of abundance. Once this task has been achieved, most of the available sand is 'out of reach,' in estuaries and lagoons, in sand bodies offshore, in dunes and beach ridges. Further operation of the system now requires the introduction of totally new sand: e.g. river sand. The volume of new sand, however, is not nearly as great as the former volume of 'available' sand. The system now passes into an economy of scarcity, and growth of shoals and barrier islands is replaced by erosion (except locally).

Such a shift cannot be made instantaneously, nor will it be made simultaneously at all points.

The implications are extremely important. Many of them can be summarized by saying that present beach erosion problems will be aggravated in the near future both in areal extent and in severity. It has been estimated recently (Tanner 1970) that maintenance of beaches in Florida, alone, by artificial nourishment, will shortly cost more than $10,000,000 per year, unless the people of the state choose to abandon their coasts. Our present assessment is that the problem is more severe than it was thought to be a little more than a year ago. The estimate made at that time was based on a level projection of current requirements; we now predict that the requirements will become larger with the passage of time.

In detail, the implications reach into economics, engineering, sociology, law, and government. It is not our present purpose to look into these matters, nor is it our purpose to be alarmist. We merely report a slow, subtle trend, which has become increasingly clear to us as we have extended our studies of beach growth and erosion. Geologically, the clear implication is that the barrier island is a chimera, having a lifetime of only some thousands of years, and that – barring a new large change in sea level – it will be pushed against the mainland coast and destroyed. The late John Hoyt (1967) developed a somewhat similar history, but without the present rationale.

Hoyt, J., 1967   Barrier island formation, *Bull. Geol. Soc. Am.* 78, 1125–36.
Stapor, F.W., 1972   Origin of the Cabo Rojo beach-ridge plain, Mexico, *Trans. Gulf Coast Asoc. Geol. Societies*, vol. 21.
Tanner, W.F., 1958   The equilibrium beach, *Trans. Amer. Geophys. Union* 39, 889–91.
Tanner, W.F., 1970   The thief who's stealing our beaches, *Florida County Govt.* 1, 27–9.

c0305
A classification of coastal classifications
HARTMUT VALENTIN   *Technical University of Berlin, West Germany*

Few geographers appear to realize that coasts can be classified from highly different points of view. Among the various possibilities are the following. (1) From the angle of cartography, coasts can be grouped according to the largest scale or the quality of the maps, charts, and air photos representing them. (2) The nature of coasts can be classified according to at least nine principles, viz., (21) the rocks, (22) geotectonics (Atlantic and Pacific types of Suess 1885, 6), (23) geomorphology, (24) hydrology (Davies 1964), (25) climatology (Bailey 1960), (26) phytogeography (Axelrod 1960), (27) zoogeography, (28) bioceanology, and (29) complex physical geography. (3) Intermediate between the physical and human coastal classifications are those relating to the potential use of the coast by man. (4) Classifications that refer to man and his works at the coast: their scope ranges from (41) divisions of the population to (49) complex human geographic classifications. (5) Finally, there may be synthetic classifications considering the entire geographical substance of the coast.

Until now, however, most geographers speaking of coastal classifications seem to think in terms of geomorphological classifications only. The following discussion is therefore restricted to that category.

(23.1) Some of the geomorphological classifications concentrate upon the relief of the land adjoining the seashore. For instance, the large world map by McGill (1958) shows primarily the 'major coastal landforms' of a strip extending inland about 8 to 16km, plus a few 'selected shore features.'

(23.2) The classifications of the coastal configuration proper occasionally consist of a simple juxtaposition of conspicuous coast types (23.21), but most authors arrange their types in classes which are then divided into sub-classes (23.22). According to the principles used, there are descriptive, semi-genetic, and genetic classifications.

(23.221) The descriptive classifications of coastal configuration often refer to (23.2211) the transverse profile of the coast. The old distinction between low coasts and precipitous coasts was refined by Richthofen (1886, 295) and various other German workers till Louis (1960, 248, and later), as well as by Ottmann (1963). (23.2212) On the other hand, there were also many attempts to classify coasts according to their outline. A brief survey of the numerical methods used by German writers from the early 19th to early 20th centuries was given by Johnson (1919, 170). (23.2213) Another kind of descriptive classification is concerned with the relation between the trend of the coastline and the strike of the land relief or of the rocks: longitudinal, diagonal, and transverse coasts of Richthofen (1886, 298) and other authors. (23.2214) A last group of descriptive classifications is characterized by a combination of the preceding criteria. One of them is Alexander's distinction of cliffed and non-cliffed shores which are then subdivided into regular and irregular shores of diverse shapes (1962).

(23.222) Richthofen also proposed a semi-genetic classification of coastal configuration (1886, 305). Here he distinguished types due to the ingression of valleys such as fjord or ria coasts, types due to deposition, etc. His idea was much elaborated by Schlüter (1924).

(23.223) The genetic classifications of coastal configuration either stress the rôle of vertical movements or of horizontal movements (exogenic processes) or try to consider both equally. (23.2231) Those giving priority to vertical movements originated with Davis (1898, 349). His two classes were shorelines 'produced by uplift or by depression of the land.' Johnson (1919, 172) renamed them 'shorelines of emergence' and 'shorelines of submergence' and added two further types, viz., 'neutral shorelines' and 'compound shorelines.' This well-known scheme was supplemented by other authors, particularly by Cotton (1952). (23.2232) The genetic classifications giving preponderance to horizontal movements date from Philippson (1893). He was the first to separate clearly primary or 'isohypse' coasts from those shaped by littoral agencies. His dichotomy was followed by Gulliver's division of initial and sequential forms (1899) and Shepard's partition into primary and secondary coasts (1937 and later). Six interesting classifications by Soviet scientists have attempted to get beyond this dichotomy. Only the system of Ionin, Kaplin, and Medvedev can be mentioned here. It was used on the geomorphological maps of the *Physical-Geographic Atlas of the World* (Moscow 1964) and translated into English in the journal *Soviet Geography* 6 (1965) no. 5–6, 21. (23.2233) A genetic classification of coastal configuration considering vertical and horizontal movements equally was submitted by Valentin (1952, 54). Its main types were coasts that have advanced (emerged or prograded coasts), and coasts that have retreated (submerged or retrograded coasts). The distribution of these types and various subtypes was shown on a world map.

(23.3) In contrast to the preceding classifications of coastal configuration, those dealing with the present coastal dynamics are rather rare.

(23.31) It is true that several maps of present vertical movements have been published. The latest world-wide attempt is the map by Newman (1968) showing rising, stable, and subsiding coasts.

(23.32) There is, however, no special world map of present horizontal movements of the coastline. Its major classes should be prograding, stationary, and retrograding coasts.

(23.33) A classification of present coastal dynamics comprising both vertical and horizontal movements was proposed by Valentin (1952, 50). Its main types were advancing (emerging or prograding) coasts and retreating (submerging or retrograding) coasts.

(23.4) Finally, a synthetic geomorphological classification of coasts is reached by combining coastal configuration and dynamics to the higher concept of *coastal condition* (Valentin 1952, 57).

For a fuller version of the present paper including a discussion of previous similar efforts and 125 references, see Valentin (1972).

Alexander, C.S., 1962 A descriptive classification of shore lines, *Cal. Geog.* 3, 131–6.
Axelrod, D.I., 1960 Coastal vegetation of the world, in W.C. Putnam, ed., *Natural coastal environments of the world* (Los Angeles), 43–58.
Bailey, H.P., 1960 Climates of coastal regions, *ibid.*, 59–77.
Cotton, C.A., 1952 Criteria for the Classification of Coasts, IGC 17 (Washington), 315–9.
Davies, J.L., 1964 A morphogenic approach to world shorelines, *Z. Geomorph.* 8, 127*–42*.
Davis, W.M., 1898 *Physical Geography* (Boston).
Gulliver, F.P., 1899 Shoreline Topography, *Proc. Am. Acad. Arts Sci.* 34, 149–258.
Johnson, D.W., 1919 *Shore Processes and Shoreline Development* (New York).
Louis, H., 1960 *Allgemeine Geomorphologie* (Berlin).
McGill, J.T., 1958 Map of Coastal Landforms of the World, *Geog. Rev.* 48, 402–5.
Newman, W.S., 1968 Coastal stability, in R.W. Fairbridge, ed., *Encyclopedia of Geomorphology*, 150–6 (New York).
Ottmann, F., 1963 Sur la classification des côtes, *Bull. Soc. géol. fr.*, sér. 7, 4, 620–3.
Philippson, A., 1893 Über die Typen der Küstenformen, insbesondere der Schwemmlandsküsten, in *Richthofen-Festschrift* (Berlin), pp. 1–40.
Richthofen, F.v., 1886 *Führer für Forschungsreisende* (Berlin).
Schlüter, O., 1924 Ein Beitrag zur Klassifikation der Küstentypen. *Z. Ges. Erdk. Berl.*, 288–317.
Shepard, F.P., 1937 Revised Classification of Marine Shorelines, *J. Geol.* 45, 602–24.
Suess, E., 1885 *Das Antlitz der Erde* 1 (Wien).
Valentin, H., 1952 Die Küsten der Erde, *Petermanns geogr. Mitt. Ergänzungsh*, 246.
– 1972 Eine Klassifikation der Küstenklassifikationen, *Göttinger geogr. Abh.* 60.

c0306
Post-glacial emergence of the Bay of Fundy coast
JOHN E. WELSTED *Brandon University, Canada*

Determination of sea level changes in the Bay of Fundy is complicated by one of the largest tidal ranges in the world; a range of over 16m having been recorded at Burntcoat Head. The great tidal range results from the relationship between the tide producing forces and the bay's length, depth, and shape. The Bay of Fundy averages 69m (225ft) in depth and its critical length is 258km (160 miles) which agrees closely with the measured length of 261km (162 miles). 'The natural period of oscillation of the bay is about 6.29 hours, which almost exactly fulfils the conditions for resonance to occur' (King 1962, 175). Deglaciation, isostatic readjustment, and eustatic variations of sea level caused changes in the length, depth, and shape of the Bay during late-glacial and post-glacial time; therefore, indicators of higher sea levels may have been formed under tidal conditions different from those of today. The following theoretical examples illustrate the errors that may occur if this fact is ignored.

Example 1: Along a contemporary cliff the sea at high tide reaches 1m above the base of the cliff, but the height of the shore platform/cliff contact is mistakenly taken as high-tide level. An error of −1m has been made. Inland a raised shore platform and cliff have a clearly defined contact 15m above the present contact. If it is assumed that tidal conditions during their formation were the same as they are now and that the sea reached 1m above the platform/cliff junction which is, however, taken as high-tide level at the time they were formed, the error is again −1m. Comparison of the height of the two platform/cliff junctions gives the relative change of high-tide level because the errors are the same in each case and 'like things' have been compared. However, the assumptions made with respect to the emerged platform and cliff are not valid. With relative sea level 15m higher than at present, the dimensions of the Bay would have been different and, consequently, the tidal range would not have been the same as it is today. Moreover, with change in sea level, coastal configuration would have been altered, affecting the fetch and the height reached by waves. For these reasons the errors made for the uplifted platform/cliff junction and the contemporary one would not be the same; the comparison would not be between 'like things'; and the result would not give a true indication of change of high-tide level.

Example 2: Greater errors occur if attempts are made to calculate and compare mean sea levels. For the sake of simplicity, it is assumed that the platform/cliff junction does represent high-tide level and that a contemporary platform/cliff junction and a fossil junction 25m above the present one are clearly defined and easily located. If the area in question has a tidal range of 10m, mean sea level is 5m below the platform/cliff junction. Assuming no change in tidal range, the old mean sea level was 5m below the uplifted junction. Mean sea level has, therefore, fallen 25m. However, if it is assumed that the tidal range was 2m when the uplifted features were formed, mean sea level would have been $25 - 1 = 24$m. Present mean sea level is −5m if the platform/cliff junction is taken as zero. Therefore, the change of mean sea level amounts to $24 + 5 = 29$m, not 25m as previously calculated. Because of the probability of the occurrence of this type of error the writer believes it is more meaningful to compare 'like things,' which have some morphological expression, than to attempt to compare calculated values.

In an attempt to summarize data about emergence of the Fundy coast, the writer produced a map and a table which together indicate the location and reliability of evidence of emergence of the Fundy coast (Welsted 1971). The information for the map and table was obtained in part from the literature (to June 1970) and in part from the writer's field observations. Since the compilation further information has been published for SW Nova Scotia (Grant 1971). Three guidelines were adhered to: (1) heights recorded were obtained from all types of indicators of emergence; (2) whenever possible, indication was given of the datum for the height of individual emerged features in order to avoid the confusion that might result from equating something which is, say, 15m above mean sea level with something which is 15m above high-tide level; (3) figures plotted on the map and recorded on the table were given as minima for the locations cited.

The reliability of the evidence for emergence at individual sites is suggested by the terms 'certain,' 'probable,' 'possible,' and 'doubtful.' 'Certain' indicates that shells have been found in sediments at, or extending up to, the altitude given. 'Doubtful' means that the height is usually out of line with figures for adjacent or nearby areas, or that there is reason for doubting the marine origin of the feature whose height is given. The distinction between 'probable' and 'possible' is subjective, taking into consideration the thoroughness with which the locality was studied in the field, impressions gained from map and air photo interpretation, and the opinions of previous workers in the area. An attempt was made to assess: (1) the detail and thoroughness with which previous studies were conducted, and (2) the validity of the reasoning used in labelling a feature 'marine.'

The map and table reveal the following

about the emergence of the Fundy coast: (1) the scarcity of post-glacial shells from the Nova Scotia coast is reflected in the existence of only one 'certain' indicator of emergence – at Middleton. (2) In contrast, shells are abundant on the New Brunswick coast south of Saint John, those that have been dated all being younger than 13,500 radio-carbon years. (GSC-795: 12,300 ± 160; GSC-882: 13,000 ± 240; GSC-886: 12,300 ± 160; GSC-965: 13,200 ± 200; GSC-1067: 12,600 ± 270; I(GSC)-7: 13,325 ± 500.) Generally, the values for emergence increase in a northwest direction. The highest for New Brunswick is Pennfield Plain (76m) and for Nova Scotia, Advocate Harbour (42m). It is significant that Advocate Harbour is nearer to the New Brunswick side of the Bay than any other place in Nova Scotia where there is evidence of emergence. (4) The figures may indicate the marine limit in some cases, but in others they clearly represent something less. Therefore, there is no basis for drawing isolines on the marine limit and, as there are few accurately heighted and dated sites, drawing of isobases is at this time impossible. (5) There are two negative areas as far as evidence of emergence is concerned: (*a*) Shepody Bay, Cumberland Basin, and the Nova Scotia side of Chignecto Bay; and (*b*) Cobequid Bay and Minas Basin east of Saints' Rest on the north shore, and east of the mouth of the Avon River on the south shore. (6) With few exceptions, emerged erosional marine features are absent from areas underlain by resistant pre-Triassic rocks and Triassic basalts. Where erosional landforms are preserved they are cut into non-resistant Triassic sandstones and shales or into surficial deposits. The conclusion is that, since glaciation, sea level has not remained stationary long enough for any significant erosion of the resistant rocks. Where there is a shore platform backed by steep cliffs cut into resistant rocks, as for example along the east coast of Chignecto Bay, it must have originated before the last glaciation. Glaciation and sea level changes probably played a part in steepening the cliffs.

Grant, D.R., 1971 Glacial deposits, sea level changes and pre-Wisconsin deposits in southwest Nova Scotia, *Geol. Surv. Canada*, paper 71-1, part B, 110–3.
King, C.A.M., 1962 *Oceanography for Geographers* (London).
Welsted, J.E., 1971 Morphology and evolution of the Bay of Fundy with emphasis on changes of sea level during the Quaternary. PhD thesis, U. Bristol.

c0307
Un banc corallien orienté: Reef Island aux Iles Banks, Nouvelles-Hébrides
ANDRÉ J. GUILCHER   *Faculté des Lettres, France*

Le récif corallien de Reef Island, ou Iles Rowa, situé dans les Iles Banks, Océan Pacifique, par 13°37′s et 167°32′E, est, avec le Récif de Cook (17°04′s–168°17′E), le seul de l'archipel des Nouvelles-Hébrides qui ne soit pas frangeant autour d'une île haute volcanique ou volcano-corallienne. Ce n'est pas non plus un atoll, mais un banc corallien en fer à cheval, dont la convexité fait face à l'ESE, alors qu'à l'WNW se dessine une baie concave. Il est donc adapté à l'alizé du Sud-Est, vent très prédominant dans cette aire marine, et à la houle qui en dérive.

Cependant, Reef Island n'est pas un récif actuel, mais, pour l'essentiel, un récif ancien. En ses différentes parties, en effet, le platier du récif est formé de corail mort, bien qu'en position de croissance comme le montrent de multiples observations, ce corail atteignant jusqu'au niveau actuel des hautes mers de grande marée (dénivellation de marée: 1,80m en vive eau). Plus haut, le récif ancien présente en général, nous semble-t-il, le caractère d'un conglomérat de récif très cimenté, mais non en position de croissance, atteignant 1,50m au-dessus des grandes hautes mers dans le NW, et jusqu'à 6,00m dans le NE à l'île Rowa où il constitue un ancien rempart du côté au vent. Le vieux récif a ainsi été édifié en fonction d'une houle d'ESE ou d'E analogue à l'actuelle. Son âge n'est pas jusqu'ici déterminé; des échantillons ont été prélevés par nous aux fins d'analyse. On peut cependant dire que l'âge est assez ancien pour que les parties les plus élevées aient subi une intense corrosion littorale (lapiés déchiquetés), et qu'il y ait même, dans le NW et dans l'ENE, des 'pipes' verticales résiduelles témoignant d'une évolution subaérienne relativement longue, comme celles de Point Peron (Australie Occidentale) et de certains dépôts marins anciens de la côte ouest de Madagascar.

Reef Island offre une répartition des formes de surface orientée en fonction de la houle d'ESSE comme l'est la forme générale du récif, à la fois du point de vue des aspects de bordure et des îles émergées.

1 En ce qui concerne la bordure, la face au vent battue par la houle a une morphologie de *spurs and grooves* immergés de quelques mètres, analogues à ceux décrits en divers autres récifs de l'Océan Pacifique (Iles de la Société, Iles Marshall, etc.). Dans la concavité abritée sous le vent, le récif se morcelle au contraire en nombreux pâtés de corail subcirculaires isolés, très vivants sur leur périphérie, s'élevant de quelques mètres au-dessus de fonds sableux qui descendent graduellement et assez doucement vers l'Ouest. Dans la zone du déferlement existe un certain revêtement de Lithothamniées roses, moins développé et moins beau cependant que la crête à Porolithon des Iles Marshall, des Iles de la Société et des Iles Tuamotu. Un petit lagon d'environ 800m sur 500 et de 7 à 8m de profondeur est englobé dans la partie sud du récif. Il est tapissé de farine corallienne.

2 En ce qui concerne les îles, on doit distinguer trois types.

a Sur le front battu au vent, se trouve une rangée d'îles séparées par plusieurs passes franchies par le *swash* à haute mer, lors des forts déferlements, passes analogues aux *hoa* des Tuamotu. Les îles sont formées de récif ancien consolidé; elles sont très basses en général, ne dépassant pas de 2m les hautes mers actuelles, sauf à Rowa où le rempart ancien atteint, comme on l'a dit plus haut, 6m. Les sédiments actuels y sont extrêmement peu importants.

b Dans la partie médiane, deux îles allongées NNE–SSW représentent une sédimentation actuelle de sable fin en arrière des déferlements. Ces îles ne portent que peu de *beach-rock* et sont soumises à certains remaniements (érosions et accumulations). Des bancs de sable intertidaux y sont associés.

c Dans le NW, un îlot consiste en une arête de sable fin meuble haute de 0,80m, convexe face au SE, avec des crochets multiples à ses deux extrémités NE et SW; en contrebas au NW se trouve un bas plateau de récif ancien saupoudré de sable, portant des salicornes et en son centre une mare avec palétuviers (Avicennia et Rhizophora) et prairie de phanérogames marines.

Les îles supportent dans leurs parties émergées la végétation habituelle aux îles basses coralliennes du Pacifique: Pemphis, Tournefortia, Pandanus, Cocotiers, etc., la zonation dépendant, comme d'habitude, de l'exposition. Le platier entre les îles des types *a* et *b* découvre entièrement à basse mer de grande marée, sauf en un endroit formant chenal bordé de vase corallienne à films d'algues. Sous le vent des îles de type *b* et jusqu'à l'île *c*, c'est un lagon assez vaste, large de 2km environ, peu profond (entre 0,50 et 4m à basse mer), couvert de sable avec végétation discontinue d'herbe à tortue (algues vertes filamenteuses). Habité jusqu'en 1939, Reef Island a été alors évacué à la suite de deux cyclones consécutifs qui ont inondé les îles sous les déferlements et fait fuir la population. Ces cyclones, ou d'autres, ont hissé en divers endroits de la crête externe des 'têtes de nègre' détritiques pouvant atteindre jusqu'à 3 ou 4m 3 pour les plus grosses.

Reef Island présente certaines analogies (rempart au vent, caye de sable sous le vent, faible profondeur du lagon intermédiaire) avec les récifs à caye et lagon peu profond définis par divers auteurs aux Antilles, dans les Océans Pacifique et Indien, et en Mer Rouge. L'existence d'une rangée d'îles médianes, comme la position en plein océan et non en des aires marines restreintes, sont cependant des originalités, de même qu'est intéressante la différence des bordures au vent et sous le vent qui reproduit, en un récif arqué mais non fermé, les contrastes observables sur les faces internes et externes du récif-barrière circulaire de Mayotte dans l'Océan Indien. Le Récif de Cook présente un intérêt moindre, quoique non négligeable. Sans aller jusqu'à faire de Reef Island un type spécial, il faut noter ses particularités qui sont des données à retenir dans la morphologie générale récifale.

Une description plus détaillée des formes et des sédiments, avec figures, paraîtra prochainement dans une revue française.

François Doumenge; le Centre National de la Recherche Scientifique français; M. Robert Langlois.

c0308
Caractéristiques des schorres des régions froides
J.-C. DIONNE *Ministère de l'Environnement, Canada*

Les schorres du Saint-Laurent, situés dans une zone climatique tempérée à hiver long et rigoureux, sont considérés comme appartenant au milieu froid. Ils diffèrent des schorres des régions tempérées à hiver court et doux (type européen) au point de vue pédologique et morphologique. D'une part les sédiments du schorre comprennent une forte proportion d'éléments grossiers (plus de 10mm) et de cailloux erratiques, de l'autre, la surface est découpée de marelles et plaquée de nombreux tertres gazonnés. Ces traits particuliers résultent de l'action des glaces flottantes qui tantôt arrachent des morceaux du tapis végétal qu'elles parsèment ici et là au hasard de la fonte des glaçons, tantôt abandonnent une charge plus ou moins substantielle de vase, de sable, et de cailloux qu'elles ont prélevée ailleurs le long des rivages. Le fauchage systématique du tapis de spartines ralentit la sédimentation vaseuse, alors que la présence du pied de glace réduit à huit mois par année la période des atterrissements de matériel fin. Cette double action se traduit par un rythme sédimentaire relativement faible au droit des schorres de l'estuaire maritime du Saint-Laurent.

c0309
Some aspects of the importance of tidal range in coastal studies
C. KIDSON *University College of Wales, UK*

The text is not included because of length.

Present Day Geomorphological Processes
Les processus géomorphologiques actuels

---

c0401
Relief and denudation: theoretical test of an empirical correlation
FRANK AHNERT *University of Maryland, USA*

Numerous authors have shown that under comparable climatic and lithologic conditions the mean rate of denudation correlates well with the available relief. This paper attempts to test the validity of this correlation for each of three different denudational processes (creep, viscous-plastic flow, and slope wash), and to trace the causal linkages that underlie it, by theoretical model experiments.

Mean denudation rates, derived usually from the measurement of stream sediment loads, represent the combined results of possible fluvial downcutting, of the denudation of slopes, and of the lowering of summits. For morphological purposes, identification of the separate effects of these three components is desirable and will be attempted.

The theoretical model is designed as a comprehensive computer program (Ahnert 1971) that permits the simulation of slope profile development under a great variety of conditions of fluvial downcutting, weathering, and downslope transport of waste.

Regression analysis of field data indicates an approximately linear correlation between mean denudation rate and relief (Schumm 1963; Ahnert 1970; Schumm's overall non-linear function is approximately linear for all but the highest relief values). In simplified form, the empirical regression equation for 20 large mid-latitude drainage basins is
(1) $d_m = 0.0001535\ H/1000$ yrs,
where d_m is the mean denudation rate and H is the mean relief in the drainage basin. When the probable effect of stream incision is eliminated, the equation becomes
(2) $d_m = 0.000106\ H/1000$ yrs.
For strictly geometrical reasons, one can postulate that in the absence of either fluvial downcutting or deposition, and with the val-

ley side slopes intersecting on the interfluves, the summit denudation rate d_s will be twice as large as the mean denudation rate d_m, i.e.,

(3) $\quad d_s = 2 d_m = 0.00021 \, H/1000$ yrs.

or, since the relief H itself decreases progressively,

(4) $\quad d_s = 0.189 \, H/1{,}000{,}000$ yrs.

Morphologically, d_s is more interesting than d_m because it denotes the rate at which the relief decreases when there is no fluvial downcutting. A graphic representation of $H = f$ (time), constructed from equation (4), would show that the relief decreases geometrically with time.

The foregoing empirical relationships can be simulated, and thus theoretically confirmed, by model experiments using creep by expansion-contraction cycles as the denudational process; the model program defines creep mathematically as

(5) $\quad R = h \tan \alpha_j$,

where R is the amount transferred per time unit (i.e., per program iteration) from a slope point to the next point downslope, h is a constant (expansion coefficient), and α_j is the local slope angle. Under conditions of fixed base level, the mean denudation rate d_m of this model decreases approximately linearly with decreasing relief, and the summit denudation rate trends towards $d_s = 1.5 \, d_m$; it is less than twice the mean rate because the slope profile of this model develops towards an overall convex shape, so that the slope angle at the summit is relatively small.

Viscous-plastic flow is defined after Souchez (1964) by

(6) $\quad R = b \, [V_j - (a/C_j)]$,

where a and b are coefficients of cohesion and fluidity, respectively, V_j is a function of the sine of the local slope, and C_j is the local waste cover thickness. With this process, the model's summit denudation rate decreases linearly as the relief decreases, but the mean denudation rate, computed from the entire profile, does not; instead, it stays very low, because the model profile becomes differentiated into a lower segment of net accumulation and an upper segment of net denudation. The accumulation has the effect of reducing the overall mean denudation rate. However, when the mean denudation rate is computed only from the net denudation segment, it decreases also linearly with decreasing relief, and has a value that is almost exactly half that of the corresponding summit denudation rate, thus confirming the postulate made earlier.

The third denudational process to be considered, slope wash, does not permit any accumulation on the model profile since it is programmed as a removal process without point-to-point transfer. It has been designed after Zingg (1940) as

(7) $\quad R = k \, V_j \, (1.0 + D_j^{0.6})$,

where R is the amount removed from a slope point per time unit, k is a coefficient, and D_j is the relative distance of the slope point from the summit ($0.0 \leqslant D_j \leqslant 1.0$). On the model slope that is being shaped by this process, the mean denudation rate decreases linearly with decreasing relief in an orderly fashion, but the summit denudation rate remains constant, and therefore independent of relief, until the slope has been almost entirely worn away. The slope profile developed with slope wash is concave, and the angle at the summit, where $D_j = 0.0$, stays approximately the same while the slope is worn down. This means that in equation (7) V_j at the summit, and with it the removal R and the summit denudation rate, also remain constant.

Just as for the field data, a graph $H = f$ (time) can be constructed for each of the three model slopes. It is rectilinear for the wash slope, but for the other two it descends in geometrical progression like the field-derived graph. This latter resemblance is useful for calibrating the process dimensions of the model to those of the real world; for example, if the field relief would be reduced to 50 per cent of its initial value in 3.5 million years, and the model slope is reduced to 50 per cent of its initial relief in 340 model time units (program iterations), then one unit of model time represents slightly less than 10,000 years of real time.

The model experiments indicate that the relationship between relief and denudation varies depending upon the type of denudational process. The mean denudation rate is proportional to the relief, but only when areas of net accumulation (as on the viscous-plastic flow model) are eliminated from the computation. The same would probably be true for a river basin that has areas of net accumulation upstream from the point where the sediment load is sampled.

Summit denudation is also proportional

to relief, and about twice as large as mean denudation, for slopes dominated by point-to-point waste transfers (creep or viscous-plastic flow) that have the tendency to reduce the slope angle in the summit area as the slope is worn down. It is not proportional to relief, however, in the case of wash because of the parallel slope retreat at the summit; the same conclusion probably holds true for all instances of parallel retreat of summit slopes, whether caused by process or by structure.

The denudational process equations (5), (6), and (7) show that relief itself does not enter as a factor into the downwearing of any point on the slope profile. The statistical-functional relationship between denudation rate and relief that is nevertheless observed on the models and in the field is due to the causal relationship between denudation rate and slope angle on one hand, and a statistically significant correlation between slope angle and relief on the other: since the spacing of valley bottoms and interfluve crests does not vary much from place to place (especially not under comparable climatic and lithologic conditions), higher relief is generally associated with a steeper mean slope and with greater probability of occurrence of steep local slope angles. The relationship between denudation rate and relief is thus associative rather than causal. Relief as a measure of potential energy is a cause of denudation only insofar as it affects the total amount of waste transported on the entire slope, but not the rate of lowering of any individual point on that slope.

Ahnert, F., 1970 Functional relationships between denudation, relief, and uplift in large mid-latitude drainage basins, *Am. J. Sci.* 268, 243–63.
– 1971 *A general and comprehensive theoretical model of slope profile development*, U. Maryland Occas. Papers in Geog. 1.
Schumm, S.A., 1963 *The disparity between present rates of denudation and orogeny*, US Geol. Survey Prof. Paper 454-H.
Souchez, R., 1964 Viscosité, plasticité et rupture dans l'évolution des versants, *Ciel et Terre*, 389–410.
Zingg, A.W., 1940 Degree and length of land slope as it affects soil loss in runoff, *Ag. Eng.* 21, 59–64.

c0402
The role of catastrophic rainfalls in the relief shaping of mountain areas
LESZEK STARKEL   *Polish Academy of Sciences*

Heavy rainfalls are the characteristic feature of mountain areas of the whole of Eurasia though the intensity and total amount of precipitation vary. It rises towards the southeast and reaches a peak in the ranges of the monsoon climate.

The author investigated the results of heavy rains and floods in some regions of India (Starkel, in press), especially in the Lower Himalaya near Darjeeling (Starkel 1970 and 1971). The area is at a height of 2500–3500m asl, and is dissected by v-shaped valleys to 2000m deep. It is built mostly of gneisses and crystalline schists and has heavy forest cover in those areas not affected by recent deforestation. During summer downpours (annual precipitation 3000–4000mm), water flows down in the permeable silty-sandy soils. It is only the catastrophic rainfalls of the order of 1000mm in 2–3 days (occurring 4 to 5 times in a century) that bring about a liquefaction of soil cover, mudflows, and, during deep infiltration, also big landslides. A single tearing down of cover from about 20–30 per cent of the deforested areas (only to 2 per cent of forest ones) leads to the retreat of slopes and to lowering of divides. The floods of a unit discharge exceeding $10m^3/sec/km^2$ lead to a deepening and widening of upper valley segments. In big valleys, with flood waves of 20m height overflowing the valley floor, there is a tendency to widen the valleys and carry away the material from the slopes to the mountain foreland. The floods remove the material accumulated as a result of the annual chemical weathering on slopes and poor fluvial transportation in the valley floors. The uplifting of mountains, relief energy, and the lithology are, as in Assam, the basic factors regulating the role of the mean and catastrophic precipitations. While the annual denudation lowers the mountains by 0.7–2.0mm (suspended load records), this value in October

1968 was 100mm. The mean annual denudation is 5mm/year. In other parts of the Himalaya and other uplifted mountains with monsoon rainfalls the value of denudation is similar and the mudflows play a most important role (Berry and Ruxton 1960; Starkel, in press; Wentworth 1943). Wentworth estimated the denudation at about 0.8mm/year in the Hawaiian Islands and 1.2–1.6mm/year in New Guinea (Wentworth 1943). Verstappen reported from Java 0.5–2.0mm/year (Verstappen 1955).

Catastrophic rainfalls leading to the development of new forms or to the rejuvenation of old ones are a characteristic trait of monsoon climate. During such rainfall there simultaneously occurs a modelling of slopes and valley floors. This occurs in regions of different relief energy and amount of precipitation. In the flattened plateaus of India the denudation rate sometimes exceeds 1mm/year (Sen 1968; Starkel, in press). In the humid variety of the monsoon climate the preparatory role in levelling the landforms is performed by the annual monsoon rains of a long duration causing a deep chemical weathering accumulation in valley floors which, in the spring drought, brings about deep soil cracks. In the dry variety, the same role is performed by wind and physical weathering.

Catastrophic downpours produce a different effect in a mountain area with considerable relief energy and uplifting, leading to a lowering of slopes with simultaneous deeping of valleys, and another effect in the poorly rejuvenated areas of plantations with inselbergs, where the retreat of slopes may be going on with a simultaneous accumulation of alluvia and deepening of river channels carrying away large quantities of suspended load (Starkel, in press). The intensive erosion during catastrophic rainfalls is also known in the other mountain areas of Eurasia.

In the mountains of central and west Asia, where in a semi-arid climate the vegetation is poor, the intensity of downpours amounts to 10mm/min (Ak. Nauk 1964). In Transcaucasia the flow rises to $8m^3/sec/km^2$ and the magnitude of denudation in the eastern part of the Caucasus in some catchment basins reaches 1.6mm/year. During a single downpour, masses of the order of $20,000m^3/km^2$ are sometimes carried away by a stream, which corresponds to a lowering of the basin by 20mm. The difference as compared with the region of Darjeeling is that while in the semi-arid areas heavy rainfall is the only morphogenetic agent bringing about every year a considerable intensity of processes (excepting the role of earthquakes), in the humid monsoon climate catastrophic rains constitute essential though unevenly distributed effects against the background of the yearly continuing denudation of less intensity and of the intensive chemical weathering.

In the mountains of Europe (Alps, Carpathians) the rate of denudation is smaller (in the Alps 0.3–0.6mm/year (Corbel 1964; Jäckli 1957), in Polish Flysch Carpathians 0.05–0.15mm/year (Starkel 1960)). But the winter season and the snow-melt period also play an important role, especially above the timber line. The material subject to movements was prepared by processes in the Pleistocene periglacial climate. In mountains of the temperate zone there occur also catastrophic downpours (in the French Alps in June 1957 (Tricart et al. 1962), in Italian Alps in November 1968 (Pellegrini 1969), in Carpathians in 1958, 1960 (Zietara 1968), 1970, etc.).

Here also catastrophic phenomena (24-hour rains of 200–400mm) are events setting the course for the evolution of the relief (rivers are carrying material from which can be inferred a denudation rate of ca. 1mm). In lower mountains, with mature slopes, the material from slopes is deposited mostly at the base of slopes, while rivers usually draw new material from undercuts and channels. Hence Tricart's thesis on the short-distance transportation and independence of the development of slopes and valley bottoms was developed (Tricart et al. 1962), a thesis which finds no support either in Himalaya or in many of the European ranges.

In the higher alpine belts of mountains and in the mountains of the arctic regions we observe the slow gravitational displacement on slopes and washing of weathering and gravitational covers by flowing water. The processes are concentrated during summer snow melt seasons (Rapp 1960; St Onge 1965). However they are not synchronized in time as in the case of downpours in monsoon climate. During arctic summer there is a sequence of processes: after a phase of meltwater erosion there is a period of congelifluction. That is why we observe in valleys an intertonguing of fluvial and slope sediments, and overloaded rivers have a very steep but levelled gradient. Even here can be observed that after very rare heavy rainfall there is a dissection of slopes, beginning a new cycle of evolution.

The review of catastrophic rainfalls in different climatic zones permits the drawing of a conclusion. Excluding arctic mountains, catastrophic rainfalls are the main factor in the transformation of mountain relief, leading to creation of new forms and to formation of relief contrasts, which are again slowly changed by other processes. The magnitude and intensity of changes depend not only on the amount of precipitation and relief energy but also on the resistance of substratum and on uplift movements. Therefore the degree of slope and valley transformation reflects the stage of the tectonic maturity of mountains. The present growing intensity of processes is usually in close relation with intensive deforestation of the mountain ranges.

Ak. Nauk., 1964 Sjeli w SSSR i meri bor'by s nimi. (Mud-flows in USSR and methods of wrestling with them), Inst. Geografii, Moskva.
Berry, L., and B.P. Ruxton, 1960 The evolution of Hong-Kong harbour basin, Zeitsch. f. Geomorph. 4 (2), 97–115.
Corbel, J., 1964 L'érosion terrestre, étude quantitative (Méthodes – Techniques – Résultats), Annales de Géographie 78, 386–412.
Jäckli, H., 1957 Gegenwartsgeologie des bunderischen Rheingebietes, Ein Beitrag zur exogenen Dynamik Alpiner Gebirgslandschaften, Beitrage zur Geologie der Schweiz, Geotechn. Serie, Liferung 36.
Pellegrini, G.B., 1969 Osservazioni geografiche sull'alluvione del Novembre 1966 nella valle del torrente, Mis. Atti e Mem. Ac. Patavina 81, 8.
Rapp, A., 1960 Recent development of mountain slopes in Kärkevagge and surroundings, northern Scandiavia, Geografiska Annaler 42, 2–3.
Sen, S., 1968 Bhagirathi-Hooghly Basin, in Mountains and rivers of India (Calcutta), 384–95.
St Onge, D.A., 1965 La géomorphologie de l'Ile Ellef Ringnes, Etude Géographique 38.
Starkel, L., 1969 Rozwój rzeźby Karpat fliszowych w kolocenie (Summ. Development of the Flysch Carpathians in the Holocene), Prace Geogr., IG PAN, 22 (Warsaw).
– 1970 Course and effects of a heavy rainfall in Darjeeling and in the Sikkim Himalayas (2–5 Oct. 1968), Jour. of the Bombay Nat. Hist. Soc. 67 (1), 45–50.
– 1971 The role of catastrophic rainfall in the shaping of the relief of the Lower Himalaya (Darjeeling Hills), Geographia Polonica 21.
– in press The modelling of monsoon areas of India as related to catastrophic rainfall, Geographia Polonica 23.
Tricart, J. et al., 1962 Mécanismes normaux et phénomènes catastrophiques dans l'évolution des versants du bassin du Guil (Hautes-Alpes, France), Zeitschr. f. Geomorph. 5 (2), 277–301.
Verstappen, M. Th., 1955 Geomorphologische Notizen aus Indonesien, Erdkunde 9 (2), 134–44.
Wentworth, C.K., 1943 Soil avalanches on Oahu, Hawaii, Bull. Geol. Soc. Am. 54, 53–64.
Ziętara, T., 1968 Rola gwałtownych ulew i powodzi w modelowaniu rzeźby Beskidów (Summ. Part played by torrential rains and floods on the relief of Beskid Mountains). Prace Georg., IG PAN, 60 (Warsaw).

c0403
Studies of soil erosion and sedimentation in six selected catchment basins in Tanzania
ANDERS RAPP *University of Uppsala, Sweden* and
L. BERRY *Clark University, USA*

In the period 1968–71 the writers in association with the University of Dar es Salaam (BRALUP, the Agricultural Faculty and the Department of Geography) have conducted a quantitative study of the erosion/sedimentation problem in some catchment basins in Tanzania. The studies concentrate on erosion by water and its effects. Wind erosion is not considered in this context because it deserves a study of its own.

The project was focused on contemporary soil erosion and sedimentation in order to obtain basic data on the types, extent, and rate of such processes in Tanzania and also their economic consequences. Examples of earlier studies of rates of erosion and sedimentation in small catchment basins are the published reports by Fawley (1963) and Pereira and Hosegood (1962).

The investigations were started in two widely contrasting environments: (*a*) deforested mountains with cultivation on steep slopes, and (*b*) interior plains with a long dry season.

The first-mentioned type of catchments are found in the Uluguru mountains, one near Morogoro, the other near Mgeta (Temple and Rapp 1972a and b). Three catchments are situated near Dodoma: Ikowa, Msalatu, and Matumbulu, in semi-arid inselberg terrain. Another catchment is Kisongo near Arusha (Murray-Rust 1971). The final reports of the project will be published in *Geografiska Annaler*, 3–4, 1972.

One of the relatively few gauged small mountain basins in Tanzania is the upper catchment of the Morogoro river. This site was chosen as one of the main areas of investigation. Like the other selected areas it was chosen because of the existing good background material and easy accessibility at all seasons. Good topographical and geological maps, air photo coverage, meteorological, and hydrological data existed.

The catchment covers an area of 19km^2 on the northern slopes of the Uluguru mountains. The uppermost slopes from about 1450m to the summits at 2100m altitude are under rain forest reserve, which occupies 30 per cent of the entire catchment area. The main part of the catchment is under cultivation or grass fallow on slopes as steep as 35°–43°. Some widely spaced, grass-covered contour ridges are the only conservation measures in practice. The lowermost slopes above the stream gauge at 550m altitude carry a secondary *miombo* woodland.

The bedrock is Archaean metamorphic rocks, mainly granulites. The soils on the slopes are sandy loams, derived from weathered bedrock and with a thin layer of topsoil, generally only 10–20cm thick.

The annual precipitation at 1450m altitude is 2400mm (1961–70 average), at 500m altitude only about 800mm. The rains in the October–May period of higher general precipitation are often storms of high intensity with up to 30–60mm of rainfall in 60 minutes. The runoff is highly fluctuating with flashflood peaks. They can rise suddenly at the stream gauge, in less than a minute, stay high as long as the rainfall in the catchment is intense, and then gradually sink to baseflow level during some hours time (cf. Rapp 1971, 15). Data on rainfall and streamflow have been provided by the Water Development & Irrigation Division of Tanzania (WD & ID) from long-term records.

In the present study, water sampling and analysis of suspended sediments were performed and correlated with data on rainfall and streamflow during three rainy seasons: March–May 1969, 1970, and 1971. The computations of these data are not yet finished. Some preliminary results can be mentioned.

The largest amount of transported sediments in the Morogoro River during peak flow measured in 1969 was around 400kg/sec. Maximum concentration of suspended sediment in the water was 10.6gram/lit. On the basis of the sediment sampling, a silt discharge rating curve was constructed for each year. The total sediment transport in 1970 was 7000 tons, calculated by means of the silt-discharge rating curve and the frequency and duration of flow peaks recorded by the automatic stream gauge. The year 1970 was below average in both frequency and duration of high flow peaks, so 7000 tons of sediments is probably a low annual figure. It corresponds to about 350 tons or 250m^3 of sediment yield per km^2 of drainage area.

Sheet wash from steep fields with no or only slight protection of vegetation supplies the flow of sediments, occasionally increased by small landslides and mudflows. Some few of these seem to occur in 'normal' years. Gullying is not observed at all in these valleys, rills only to a minor extent. One explanation of this remarkable observation is probably a high infiltration rate in the soils.

The main hazard of cultivating such steep slopes, besides the gradual exhaustion of the soils over a few years, is the danger of catastrophic soil erosion during extremely intensive rains, with recurrence intervals of, say, 10–20 years. In the Morogoro valley such rains have not occurred in the period of our

TABLE 1. Preliminary data on erosion and sedimentation in five catchment basins in Tanzania. Catchments no. 1–4 have lake reservoirs and are in semi-arid plains. Catchment no. 5 is in the Uluguru mountains and has no dam.

Reservoir	Orig. vol., thousands of m^3	Catchment area, km^2	Gradient, m/km	Sediment yield, m^3/km^2	Expected life, years
1 Ikowa	3600	640	4	160	30
2 Msalatu	388	8.5	15	600	40
3 Matumbulu	360	18.5	10	950	20
4 Kisongo	145	9.3	12	700	25
5 Morogoro stream	(no dam)	19	90	250	—

investigations, but data from the upper Mgeta valley nearby are indicative of effects of heavy and concentrated rains, triggering large numbers of small landslides from steep fields under cultivation or grass fallow (cf. Temple and Rapp 1972a and b).

Four catchment basins with man-made lake reservoirs have been chosen for study of contemporary erosion and sedimentation in the semi-arid inselberg plains. Three of these are situated in the Dodoma region on granitic rocks, while the other lies west of Arusha on young volcanic rocks. All these areas have an annual precipitation of about 600mm and a dry season of about 8 months' duration. The land use is a combination of stock grazing and cultivation.

In these areas streams are ephemeral and there are no long-term gauging records of small catchment streams available. Our investigation thus has to focus on the following two methods of approach: (a) Measurements of sediment accumulation in existing man-made lake reservoirs, by annual levelling of cross-profiles on the bottoms during the dry season. The data on annual sediment accumulation provide minimum figures of the rate of erosion in the catchment, as part of the eroded sediments are deposited upstream of the reservoir and another part of it passes through the spillway in times of flood. (b) Studies of erosion features in the catchment and reconstruction of their development. These inventories are made by air photo interpretation and field checking on the ground.

Some preliminary data on five catchments are given in Table 1. The gradient indicates the slope of the stream in metres per kilometre immediately above the reservoir or the weir. The annual sediment yield has been calculated on the basis of annual sediment accumulation in the reservoirs (1–4) and from data on sediment transport in the Morogoro river in 1970.

The sediment in the Ikowa reservoir is mainly clay and silt, carried in suspension and deposited as one layer from each individual flood peak. Continuing re-deposition of clay occurs during the dry season in the deeper part of the reservoir.

In the smaller reservoirs nos. 2–4, with steeper stream gradients, much of the sediment consists of sand, carried as bed-load and deposited in a delta. A considerable part of the bed-load is deposited above the full supply level and gradually raises the channel bed upstream of the reservoir. In the largest catchment, Ikowa, not only all sand but also much of the silt and clay eroded from the fields is deposited on gently sloping floodplain sections above the reservoir. Thus the figures of sediment yield for catchment 1–4 should actually be increased with the so far unmeasured quantities of sediments deposited above the full supply level. The results show high intensities of erosion/sedimentation, particularly in the semi-arid catchments and furthermore very short expected life-lengths of the reservoirs investigated.

H. Murray-Rust; P. Temple; C. Christiansson; A. Kesseba; Å. Sundborg; V. Axelsson; the staff of BRALUP, Dar es Salaam; the Faculty of Agriculture, Morogoro; the Department of Geography in Dar es Salaam; the Department of Physical Geography in Uppsala; the Tanzanian Water Development and Irrigation Division; Bank of Sweden Tercentenary Fund.

Fawley, A.P., 1953 Msalatu reservoir, Dodoma. Records of the Geol. Surv. of Tanganyika 3 (Dar es Salaam), 71–82.

Murray-Rust, D.H., 1971 Soil erosion and sedimentation in Kisongo catchment, Arusha region, Tanzania. BRALUP Res. Papers, 17, 1–69 (Dar es Salaam).

Pereira, H.C., and P.H. Hosegood, 1962 Suspended sediment and bed-load sampling in the Mbeya Range catchments, E. Afric. Agri. and Forestry Jour. 27, 123–5.

Rapp, A., 1971 Erosionen hinder för Afrikas gröna revolution, Forskning och Framsteg 2 (Stockholm), 13–18.

Temple, P.H., and A. Rapp, 1972a Landslides in the Mgeta area, western Uluguru mountains, Tanzania, this book.

– 1972b Landslides in the Mgeta area, western Uluguru mountains; morphological effects of sudden heavy rainfall, Geog. Ann. 54A (in press).

c0404
Landslides in the Mgeta area, western Uluguru mountains, Tanzania
P.H. TEMPLE   University of Dar es Salaam, Tanzania
and ANDERS RAPP   University of Uppsala, Sweden

The catchments of the upper Mgeta and upper Mbakana rivers (area 133km^2) are representative of the upper stream source regions of the western Uluguru mountains of Tanzania. Showing a considerable range of relief (over 1700m), an intricate dissection, rapidly weathering granulitic, and metaigneous rocks, its slopes are mainly steep and soil covered. Under natural conditions most of these slopes would be clothed with woodland or forest but much of this natural cover has been cleared to permit cultivation and to support a subsistence population with a density of 188/km^2. Cropped land proportions range between 39 and 60 per cent of the total area. Maize, millet, and beans are the main crops and there are few tree crops (Thomas 1970). Annual rainfall totals vary between 95 and 192cm over the area depending on position and altitude; the main rains fall between late February and early May.

On 23 February 1970 over 10cm of rainfall were recorded in 3 hours in this area. This storm caused slope failure and landslide damage over an area of approximately 75km^2 and flooding, fluvial erosion, and deposition both within the storm area and over a wide area beyond it. Almost 1000 landslides were counted, mostly small debris slides and mudflows. The purpose of this paper is to describe and explain this catastrophic event; a more detailed account is in press (Temple and Rapp, in press).

Landslide-affected slopes were surveyed by tape and clinometer from ridge top to valley bottom. Detailed recordings were made of form and nature of the slides. A selection of 34 representative slides were surveyed in this way soon after the event. Ground reconnaissance identified and recorded the gross features of a further 224 slides. Aerial reconnaissance was employed to delimit the area affected by landsliding and to count the slide numbers. Precipitation data from 4 stations within the area were analysed as were streamflow data available from an automatic gauge on the river Mgeta. An inventory of damage was made available by the area administration. Work was begun to study recolonisation rates of soil and vegetation on the scar surfaces.

The major effect of the storm was to cause damage to slopes through landsliding. Sheetslides originating from shallow, spoon-shaped scars were the most common mass movement type. Sheetslide scars were generally 1–1.5m deep and most were 5–10m wide and 2–4 times as long as they were wide. They narrowed downslope and, except close to streams, continued downslope as mudflows. Only steep slopes were thus affected (33–44°) and scars were generally located some way below ridge crests. 60 per cent originated on straight valley sides. 47 per cent of the slides affected cultivated land and 47 per cent affected grass and fallow: less than 1 per cent affected woodland and forest.

The mudflows which commonly spread downslope from the slide scars evidenced a very rapid transformation of displaced material into mudflow indicating a very high water content. This method of debris transfer had caused little erosion of the soil below scars but had wiped out growing crops. Mudflow depths of 1.2–1.5m were recorded from mudplaster on tree trunks and boulders.

Bottleslides were also observed. These slides were deep (up to 9m) with near-vertical sidewalls. These forms exhibited a spasmodic or continuous earthflow from the upper scar through a bottleneck outlet. The largest of these forms was 60m wide and 65m long, affecting a deeply weathered slope of 21°. These forms probably predate the 1970 rainstorm but experienced major flow accelerations because of it. The long profiles of the flows are convex both in section and length. From field evidence it appears that bottle slides develop from subsurface pipes which discharge water and regolith from deeply weathered slope localities. Once the pipe reaches a critical dimension collapse occurs; more drainage then reaches the scar and it progressively enlarges itself until the unstable regolith has been transferred downslope. Developmental stages in this process are described elsewhere (Temple and Rapp, in press). Descriptions of comparable forms have been published by Haldemann (1956).

It was remarkable how little evidence of gullying was observed in the area. This is probably explained by high infiltration rates and the dominance of subsurface flow and piping even on these steep slopes. Slope wash is active but does not appear to be the dominant degradational process even on cultivated land (Savile 1947).

Stream flooding, bed and bank erosion, and deposition were spectacular but not studied in detail. A large supplement to material excavated from stream channels was supplied by landsliding in this event.

It was concluded that the major mechanism which explained the simultaneous release of this large number of debris slides was a sudden rise in pore water pressure resulting from heavy rainfall. An alternative mechanism, namely earthquake shocks, was examined in detail but rejected.

Sudden rises in pore water pressure cause sudden reductions in the soil and regolith shearing strength. This results from shearing force due to seepage flow. This force creates an upward pressure on soil and debris below ridge crests, and this force mostly affects the friction material (regolith) in the profile. Shear failure then occurs along some discontinuity in the regolith despite the stabilizing pressure of wet surface soil. 'Such failures occur very suddenly and the whole mass appears to flow laterally as if it were a liquid' (Spangler 1951, 312).

Evidence in support of this hypothesis was as follows: the location of scars some distance below ridge crests, suggesting some critical hydraulic pressure was necessary to trigger off shear failure, the presence of pipes above and within slides, the rapid transformation of slides into mudflows together with oral evidence that the slides had 'burst forth' in the manner described by Hack and Goodlett (1960, 45). Following their terminology, the majority of the slides examined would be called 'water blowouts.'

Valley cutting and slope development in this area is much influenced by slides and mudflows. These probably occur in small numbers every rainy season but occur in large numbers as catastrophic events with a recurrence interval of less than ten years, related to exceptionally intense local precipitation.

Sheet slides and mudflows are probably the main mechanism involved in the widening of steep gradient tributary valleys. Bottle slides, showing headward recurrent slumping and episodic slow mudflow, create tributary valleys of low gradient. The fluted nature of the upper slopes of the area and the open basins of the lower slopes support this deduction.

Morphological activity of the type described is probably active under forest but is greatly accelerated after deforestation. Protection of land from landslide damage should therefore be by means of tree planting and conservation of existing forest and woodland cover.

Hack, J.T., and J.C. Goodlett, 1960 *Geomorphology and forest ecology of a mountain region in the central Appalachians*, US Geol. Surv. Prof. Pap. 347, 1–65.

Haldemann, E.G., 1956 Recent landslide phenomena in the Rungwe volcanic area, Tanganyika, *Tanganyika Notes Rec.* 45, 3–14.

Savile, A.H., 1947 Soil erosion in the Uluguru mountains, unpub. rept. Dept. of Agric., Tanganyika, Univ. microfilm, MF/1/8 (Dar es Salaam).

Spangler, M.G., 1951 *Soil Engineering* (Scranton, Ohio).

Temple, P.H., and A. Rapp., in press Landslides in the Mgeta area, western Uluguru mountains: morphological effects of sudden heavy rainfall, *Geog. Ann.* 54A.

Thomas, I.D., 1970 Some notes on population and land use in the more densely populated parts of the Uluguru mountains of Morogoro District, *BRALUP Res. Notes* 8, 1–51.

Geography in Education
La géographie dans l'éducation

Territorial decentration and geographic learning
J.P. STOLTMAN   *Western Michigan University, USA*

Territorial decentration is a question of major significance to geographic educators. The degree to which the geography curriculum complements the child's psychological development has wide-ranging implications for the learning process. Decentration is a developmental process in children and has been investigated by Jean Piaget in Geneva and Gustav Jahoda in Glasgow. The present paper reports the findings of a research project designed to determine the territorial decentration of an American sample of children (Stoltman 1971).

Piaget's (1928, 1951) theory of spatial stages postulates that territorial decentration in children occurs during the period from 6 to 12 years of age. During that time, children progress from prelogical verbal notions of political territories to a logical knowledge of those territories and their relationships to one another.

Piaget believes that decentration occurs through three stages. He observed that children of 6 and 7 years of age are usually in stage one. They know the name of their city or town and oftentimes their canton and nation. Most often, one of the territories is dominant in the child's thinking. When questioned regarding the territories, stage one children consider them to be mutually exclusive rather than related. Although the verbal notion that city x is within nation x emerges during this stage, there is little evidence that a logical understanding of the relationship of the two exists. For example, city and nation are consistently portrayed by stage one children as juxtaposed circles, rather than the nation being an inclusive circle.

Children in spatial stage two, 8 to 10 or 11 years of age, are verbally explicit that smaller territories are indeed within and part of larger territories; for example, city in nation. However, their reasoning is often inconsistent and reflects no clear understanding of logical territorial relationships. The major developmental change at stage two is that when asked to portray city and nation as circles, the children enclose the city within the nation.

Beginning at about 12 years of age, children provide consistent explanations of territorial relationships. At that age, children also complement verbal explanations with graphic representations enclosing smaller within larger territories. Piaget labels such attainment as stage three, the final stage of territorial decentration.

Jahoda (1963, 1964) observed that Scottish children also decentrate territorially with age. However, his findings are discrepant with Piaget's. Scottish children reportedly decentrate to the various stages at somewhat later ages than Swiss children.

The American study was designed, as nearly as possible, to facilitate the collection of data comparable to the prior studies. An interview-examination with standardized administration procedures was developed. Tasks from Piaget and Jahoda were pilot tested to determine their suitability. A sample of 204 children between the ages of 6 and 13 years was randomly selected and examined. All the children were residents of the State of Georgia, USA. The sample included black and white, male and female, and urban and rural children.

The primary purpose of the research was to determine if Piaget's territorial decentration stages are appropriate for American children. In order to test the theory, the researcher designed a decentration model based upon Piaget's reports. The American sample's observed decentration was then tested against the model by means of the Chi Square Test for Goodness of Fit.

Despite the testing of several age-decentration fits believed to reduce possible error variance in the model, the differences between the observed and theoretically expected distributions were significant ($p < .001$). Therefore, the researcher accepted the null hypothesis that Piaget's decentration stages are not appropriate for the American sample of children used in the study.

Inspection of the Piagetian decentration model and the American sample revealed near perfect correspondence for the age range 6 years to 8 years 6 months (8.6). The American children in that age range are usually aware of the name of their home town, but the state and nation are vague. No real awareness of larger territories is consistently expressed. American children at stage one of decentration, therefore, are somewhat older than expected, based on Piaget's reports. That observation is similar to the decentration lag revealed by Jahoda in his reports of Scottish children.

At age levels older than 8.6, the observed sample departed significantly from the Piagetian model. Of those children older than 8.6, 47 per cent remained in decentration stage one. Stages two and three each contained 26.5 per cent of the children older than 8.6. Therefore, 64 per cent of the American sample had not decentrated beyond their immediate community with regard to territorial perspective and relationships. It was expected that 36 per cent would be the maximum proportion at that stage.

Piaget (1951) attributed territorial decentration to the main effect of age. Jahoda (1963, 1964) attributed decentration to the combined effects of age and socioeconomic status. The author found a .71 ($p < .01$) multiple relationship between decentration and the independent variables, age, and socioeconomic status. Thus, the main effects of age and socioeconomic status explain 50 per cent of the variance in territorial decentration in the American sample. The remainder, or 50 per cent of the variance, is not explained by those variables. Identifying the remaining sources of variance at present requires speculation. The researcher believes that scholastic aptitude, home and school experiences, and travel are plausible explanations.

The decentration differential observed between Swiss and American children is probably attributable in part to curricular differences. The tradition of local studies in the European educational system is undoubtedly influential. The decentration effects of general social studies during the elementary school years of American children are questionable. Also, the absence of social studies and geography in American elementary grades is common. In other instances, only a superficial coverage of US history and descriptive geography is undertaken. The author has observed few instances in American elementary schools where the local community has been the subject for observational training and mapping of territorial relationships.

A second variable believed to affect decentration is the differing cultures of the US and Europe. Differences in language and other cultural patterns may facilitate the development of territorial differentiation in the European child. The combined cognitive and affective experiences associated with rather small geographical areas manifesting numerous cultural differences may result in earlier decentration. American children are not exposed to such overt differences.

It has been shown that age, as related to development in general, and socioeconomic status explain decentration to a degree. The effects of curricula have not been investigated. It is probable that children will decentrate at an earlier age when provided with certain types of school experiences. Perhaps the local studies tradition of European schools, a practice not found in the majority of US schools, accounts for the more rapid decentration of Swiss and middle-class Scottish children. Other types of curricula need to be compared for effect.

It appears that decentration involves logical thought processes in geography which are comparable to logic in the natural sciences and mathematics. Just as children use set theory and size relationships in mathematics, they seemingly need to investigate what it is that makes a town a town, a state a state, and a nation a nation, each involving sets of phenomena. To accomplish the latter, they must certainly determine the territorial relationships which underlie the nation.

Unquestionably, researchers in psychology have unveiled something which geographic educators have not yet resolved. A decentration curriculum is approximated in the expanding environment philosophy. However, that philosophy is disapproved of by many educators in the US. At present, little research evidence ties elementary school geography and social studies to the development of the child in other than generalities alluding to learning sequences. Despite the age differences in research findings, decentration appears to be a universal developmental sequence. Whether or not it should be left to chance is the remaining question. It is the responsibility of geographic educators to research the effects of geographic curricula on the developmental process of territorial decentration.

Jahoda, G., 1963 The development of children's ideas about country and nationality, *Brit. J. Educ. Psych.* 33, 47–60.

– 1964 Children's concept of nationality: a critical study of Piaget's stages, *Child Development* 35, 1081–95.

Piaget, J., 1928 *Judgement and Reasoning in the Child* (London).

– and Weil, A., 1951 The development in children of the idea of the homeland and of relations with other countries, *Int. Soc. Sci. Bull.* 3, 561–78.

Stoltman, J., 1971 Children's conception of territory: a study of Piaget's spatial stages. Doct. diss., U. Georgia.

c0502
The role of school broadcasting in a developing country
T.W. BROWN *Kenya Institute of Education*

One of the major problems which faces developing countries in the expansion of their educational systems is the provision of a sufficient number of adequately trained teachers from within. The use of expatriate teachers is at best only a partial solution, and only a temporary one, for the developed countries cannot afford to continue supplying aid personnel indefinitely. A good school broadcasting division is one method of filling the gap. All developing countries have a radio network, and some have television facilities. Ideally television is the most suitable medium for transmitting geography programs, but the ideal is rarely attainable. Even if the system is available its range is limited and the number of schools which are sufficiently wealthy to purchase a set is comparatively few. Besides, the problem of blackout in the steamy heat of the tropics looms large and probably there are many secondary schools without electricity. Ethiopia does in fact have a television program for schools, but it can be seen only in the environs of Addis Ababa and this is likely to be the area where the best teachers are located, who therefore have the least need of it. And so programs must be devised for sound radio, and techniques must be adopted which will assist both teacher and pupil in the presentation of material. In other words the broadcast should simulate a good classroom lesson, should direct the teacher into a variety of ways of presentation, and should provide a basis for further lessons on the same subject. It should stimulate the interest of the pupils and provide them with inspiration and material for further study.

One of the great drawbacks of radio is the absence of vision; since vision plays such an important role in the acquiring of knowledge this defect has to be diminished, even if it cannot be overcome in its entirety. Any technique employed must therefore have this as a first objective. Secondly, an unseen teacher on the radio has difficulty in impressing upon his listeners that portion of his personality which the classroom teacher usually finds so effective in controlling his class or getting across his material. Radio techniques must therefore replace this absence of personality with other methods of stimulating interest. Thirdly, if radio broadcasts in developing countries are to replace textbooks, rather than merely supplementing them, it will be necessary for material to be available for further study after the program is over, or to act as an introduction before it begins; for only a limited amount of knowledge can be imparted in the twenty minutes normally allowed for a broadcast.

Essentially therefore a series of programs will consist of three sections: the teacher's notes or guidelines; the broadcast program; and the pupils' handbook or workbook.

The teacher's notes should be issued well in advance of the beginning of term, so that the teacher can adequately prepare a work scheme for the series of programs. They will contain a synopsis of what is going to be covered, together with instructions and suggestions for pupils' work, both before and after the broadcast. In other words it shows the teacher how to conduct his introductory and follow up lessons, and advises him on the assistance he should give to pupils during the actual broadcast; this may consist of writing difficult words on the chalk board, pointing out names of places on a wall map, or listing points brought out by the broadcaster. It will also suggest work that the pupils can undertake during their homework, and may explain for the teacher technical terms which are used during the broadcast.

The pupils' handbook is a vital document for it is all that he will have in front of him during the broadcast, except possibly an atlas (a rare commodity in up country secondary schools in Kenya), and his notebook (which many pupils are loath to use unless they can copy down model notes off the chalk board). This book must therefore contain sufficient material to keep him interested and occupied during the broadcast, to provide a basis of work for further lessons or homework and to replace or supplement a textbook. It will therefore contain maps, diagrams, graphs, and, above all, pictures. Modern geography syllabuses centre round the sample study treatment of a topic or area; this necessitates a liberal use of pictures. They are more than ever desirable in developing countries, where pupils' horizons are limited and their travel experiences few.

Moreover it has been discovered that African pupils find picture interpretation extremely difficult, and they do not always see in a picture the points which it is intended to portray. The production of these pupils' handbooks can therefore be a difficult and costly process, and in order to ensure a clear artistically finished product, it may be necessary to have them produced abroad. Good picture reproduction requires expensive machinery and a picture which is not clear is valueless. In Kenya, the difficulties are largely overcome by entrusting the work to the Government Printer, who undertakes it for the cost of materials.

For the broadcast program a variety of techniques can be used and in general the greater the variety the greater the chance of holding the pupils' attention. The use of a single voice in a program should be used sparingly, unless the speaker is particularly skilled in holding his audience. The following are some of the methods which have been used by the Voice of Kenya and found to be effective:

(1) The broadcaster interviews an imaginary traveller on his return from the country being studied.
(2) The broadcaster takes an imaginary journey to the country and interviews different people.
(3) The broadcaster conducts a lesson in the studio, using one or two school pupils, who ask questions and make comments during the broadcast.
(4) The broadcaster asks questions based on the handbook and the listeners write down the answers.
(5) An imaginary traveller meets a student.
(6) Two or three teachers discuss a problem.

It is often advantageous to persuade nationals of the country being studied to appear in the program, as this adds authenticity, and their accent helps to get the atmosphere and feeling of the country.

However, even a first-class program can be spoiled by poor reception. Once again it is the rural schools, which are really the needy ones, that are going to suffer most. This can be overcome by making tape recordings of the programs and issuing the tapes to the schools. This has the added advantage that the geography lesson is not tied to the time of the broadcast and thus makes the timetable easier to construct. There are cheap battery operated tape recorders obtainable today and it might well be worth the expense for the Ministry of Education to supply each government secondary school with one. A tape recorder is cheaper than another teacher.

This leads to a refinement of radio broadcasting, which is getting near to television without the expense of that network. It has been called 'Radio Vision.' Essentially it consists of a broadcast program or tape, together with a set of colour slides which the teacher operates during the program. These are particularly convenient for geography teaching and the whole assembly is comparatively cheap to produce. They can be made even more effective by including teachers' and pupils' handbooks; the latter might contain black and white photographs of the colour slides, which the pupils can study in greater depth in their own time.

BBC, 1959   BBC *Handbook*, 109.
– 1962   BBC *Handbook*, 78.

- 1966 *Educational Television and Radio in Britain.*
Brown, T.W., 1971 (Term 3) *Notes for Teachers* (secondary ed.), 32–6. Voice of Kenya.
- 1971 (Term 3) *Pupils' Handbook. Australia and New Zealand.* Voice of Kenya.
Cassirer, H.R., 1960 *Television Teaching Today.* UNESCO.
Castle, E.B., 1966 *Growing up in East Africa* (London).
Maclean, R., 1968 *Television in Education* (London), 54–64.
Sherlock, G., 1971 *Exploration Earth. Teachers' Notes and Pupils' Handbook* (BBC).

c0503
Films in university geography teaching
MICHAEL J. CLARK *University of Southampton, UK*

The aims of this paper are threefold: to examine the university potential of films in their 'traditional' role as illustrative teaching aids; to consider how the function of the film can be extended beyond this traditional role; and to suggest ways in which some of the barriers to effective use of films in university teaching might be tackled at an international scale.

The essential quality that differentiates cine film from photographs or slides is its ability to depict movement or change in some object or process that is remote from the teaching situation in either space or time. Such movement may simply add to the reality of the perception, or it may be an integral and essential part of that perception. A second quality almost unique to the film medium is the ability to place a visual detail into its general context, either by panning (which eliminates the artificial boundaries established by the margins of the picture) or zooming (which changes scale between a local and a regional view), whilst retaining complete visual continuity. In this respect landscape shots from aircraft have considerable potential for helping the student with regional visualization. A further category of techniques manipulates time and space so as to give a visual experience that cannot be achieved directly; notably through time lapse (speeding up slow processes such as glacier flow or patterns of pedestrian movement) and slow motion (slowing down fast processes, such as blown sand movement), together with microphotography and global photography. These purely visual properties are welded into an audio-visual communication by the process of editing and by the intellectual direction that determines the order of concept presentation and permits the film to build up a prestructured argument. In this form the film transmits an educational message with an impact that can enhance student concentration and provide intellectual stimulation. It communicates a self-contained visual impression of or comment on any feature or process that has accessible visible manifestation – and indeed through the use of models and animation many abstract concepts can also usefully be depicted.

Such 'illustrative' films usually function as an adjunct to a formal teaching system in which the teacher constructs the argument or directs the enquiry, using the film mainly to reinforce his argument or give visual reality to his exposition. Increasingly, however, films are taking over part of this role of educational direction (Clark 1970). The film itself can follow an enquiry, suggest an hypothesis, and pose a series of questions for further consideration, thus transcending the role of a mere 'visual aid' to fulfil the more valuable function of an integral part of the educational process. Of course such a heuristic 'open-ended' film can still be used simply as a lecture insert by carrying on the theme of the lecture and preparing the student for subsequent lecture-based concepts, but it is more profitably used as a stimulus for student-based discussion and investigation. Clearly the intellectual structure of a film of this type is of critical importance. The obvious extension of this change from exclusive lecturer control to a lecturer/audio-visual partnership is the use of audio-visual packages for integrated self-teaching systems in which the teacher is not directly involved at all at the exposition stage. Although expensive in capital terms such systems can be extremely cost-effective in the service teaching of routine techniques and concepts, and in the revision or reinforcement of lecture-based concepts. Here the teacher is a guide to study and a source of authoritative discussion rather than simply a provider of straightforward exposition. Regardless of the chosen balance between teacher and technology it is axiomatic that satisfactory visual education demands greatly increased academic participation in the design and production of film material at all levels, together with enlightened and thorough integration of the audio-visual component into the learning process as a whole. The use of cine film as a research technique or as a medium for the communication of research findings is another field with immense but under-developed potential.

There are strong grounds for suggesting that films (and the related medium of recorded television material) could play an increasingly important part in university geography teaching (British Universities Film Council Limited 1971; Clark 1969). As well as providing interest, stimulus, and enhanced reality of perception they offer a particularly appropriate approach to the problems of escalating student numbers, an increasing component of routine technique teaching, and the need for greater cost-effectiveness in university education (Clark 1971). However, before the full potential of audio-visual geographical education can be realized, three groups of problems must be tackled.

Perhaps the main deterrent to increased film use is the impoverished supply of suitable material and the difficulty of finding, assessing, and obtaining such material as does exist. The most direct way in which the individual lecturer can supply his own needs is for him to make or at least supervise his own films, with or without the help of university or other professional audio-visual services (Everard 1969). This can be financially viable for short lecture inserts, though even here the subject matter is restricted to features in 'financial proximity' to the university concerned, but for longer projects the cost may limit technical sophistication if only one copy of the film is required. An alternative is to widen the market for both 'private' and commercial university level films, thereby increasing both the incentive for supply and the cost that can be borne by the production. It is in the nature of this academic level that only an international market (with the attendant need for foreign language versions or texts) can be really viable, and such an international market assumes an international information and demand evaluation structure that can best be provided through a body of the scale and authority of the IGU. A further need is for some centralization of film acquisition into national or inter-university film archives or libraries holding films of special academic merit or of non-replicable events that warrant an academic visual record, including perhaps some record of the work of major international personalities of the profession! Clearly, however, improved production is ineffective without improved distribution: the need for efficient and specifically geographical cataloguing and evaluation together with a comprehensive survey of the film needs of university geographers remains paramount, and again the IGU would seem to be the ideal body to co-ordinate such a task.

A second problem is academic conservatism in the face of the changing role of the teacher. This barrier can perhaps best be tackled through a combination of professional encouragement, a more thorough professional commitment to training in the use of audio-visual techniques, and the incentive that would follow an improvement in the supply of good university films.

Finally, the intending film user will have to overcome any institutional deficiencies of finance, equipment, and technical services. Whilst this is obviously a matter that can be handled only at the level of the individual university it cannot be too strongly stressed that the clear arguments of educational value and cost-effectiveness will be readily acceptable to individual departments only when they have been seen to be readily and actively acceptable to the policy-making bodies of the profession as a whole – amongst which the IGU Commission on Education in Geography must rank high.

British Universities Film Council Limited, 1971 *Visual Media in Geography and Geology: a Conference Report.*
Clark, M.J., 1969 'Problems in the use and supply of physical geography films at university level,' *University Vision*, no. 4, 5–12.
– 1970 'Physical geography on film,' *Geography*, 55, pt. 1, 16–26.
– 1971 'The use of film at university level,' *Times Educational Supplement*, London, 30 May 1971, 55–56.
Everard, C.E., 1969 'Film and geography,' *University Vision*, no. 4, 13–16.

c0504
A function for regional geography in a school curriculum
RICHARD CLARK *New University of Ulster, N. Ireland*

The analysis of curricula and the initiation of curricular change is currently based on comprehensive enquiry into such elements as aims and objectives, structure and relationships of content and concepts, design of learning experiences, etc. In the continuing attempt to clarify the role which regional geography might play in general education at school level it may be useful to reiterate some and propose other questions by which such a role could be examined. An initial working assumption is made that the position of regional geography in the school curriculum can be considered separately from those of other forms and methods of geography.

The semantic question of what are meant by 'regional' and 'geography' in this context may be postponed by the assumption that despite dispute about its merits there is fair consensus as to the prevailing character of school regional geography. The primary question becomes 'what educational function is regional geography intended and competent to meet?'; this can in no way be satisfied by assertions that the long history of regional study is adequate demonstration of its value. It is useful to have clear answers to questions of the type – what is a pupil better able to do, how better to operate – to conduct life – for studying regional geography? Even if this and the concomitant question of what knowledge encountered in regional geography is worthwhile are met, this form of study must sustain its claims for a place in the curriculum in a competitive context.

Certain attributes have been advanced as evidence of deserved inclusion; some of these are commented upon. The assertion is well known that through regional study geographers undertake that synthesis which is the core of their subject. It does not seem to be sufficiently investigated whether the processes of synthesis are attainable by a high proportion of school pupils, and whether the production of, or study of, syntheses is the more desirable educational activity depends upon criteria derived from educational objectives. The claim of regional synthesis to be a core of geography (though much regional writing is essentially analytical) can, in part, be judged by whether it is a means or an end, and if the former then to what end. Certainly from the viewpoint of the curriculum there is no self-evident position.

To view regional geography as the form whereby the geographer focuses on place rather than upon themes and systems establishes neither its *raison d'être* nor an educational validity. Furthermore recourse to the idea that regional study contains useful information loses credence for the lack of criteria of utility.

Whether the framework of concepts through which empirical data are organised and a structured relational understanding of the world attained is best developed through thematic or regional focuses does not appear to have been adequately investigated. There does not however seem to be any evidence that regional geography relies on any unique, unshared concepts. The man-land relationship which is to some at the heart of geography is a compound thing amenable both to systematic analysis and to located exemplification. It does not depend pedagogically upon association with a variety of other concepts and data in the confines of one area or more.

A way to capture the essential spirit of areas, or 'pays,' and to distinguish between them is advanced by some as a function of the regional approach. This is closely associated with the proposition that the description and explanation of landscapes is also a core geographical activity. Over and above the difficulties of determining landscape units and scales at which differences are meaningful, there is the need to articulate the purposes of this form of study in a school program.

Certainly topographic and regional writing continues in an unbroken tradition and the intrinsic interest in places continues to motivate both travel and the reading of well constructed topographic or chorographic description and interpretation; it may be worthy of examination how the non-academic so often succeeds in creating a lively picture of place while regional texts are so frequently formal, dull, and unable to stimulate such a picture. There does appear to exist a fairly general potential interest in places which might be harnessed to some identified objectives. It might be that a sensitivity to the identity and character of places and a balance to abstract, theory-based geography could be propagated through the study of regional literature as a 'humanity.'

Whether regional methodology owes more to analysis or to synthesis, the ever-present question for schools remains – what is worthy of study, by what criteria; how best may it be studied? Thus there is call both upon value judgments founded on contemporary interpretations of societal needs, and on pedagogic implications which depend upon whether there is sought a formal understanding, say, of the operation of a system, of the spatial relationships between phenomena, or of the ability to perceive consequences of the operation or relationships and to assume attitudes and undertake actions. The educational process will vary according to the emphasis placed on either the neutral understanding or the derived commitment.

The general trend engendered by the current interest in demographic and resource prognostication is towards a social commitment either to be achieved autonomously through an extrapolation from neutral evidence or by overt and directed indoctrination as purveyed by some conservation curricula. A relevant issue here is whether the identification and commitment is more soundly approached by analytical/synthetic study of particular regions, by the necessarily interdisciplinary study of relevant dynamic systems, or by a combination of both more planned than is the usual school relationship between systematic and regional geography. At least this broad and undifferentiated social goal affords a basis both for particularising objectives and for discrimination in selection of content. The past has seen school texts organised from geological structure to trade and communications. Both there and where alternative arrangements have been attempted concentration has been knowing 'what' and knowing 'how' but rarely upon considering 'should and shouldn't.'

The view that regional geography at school level should meet a societal function gives a basic criterion of relevance to the selection of themes and data. It does not in itself indicate a most effective way of organising learning and hence does not in itself wholly substantiate the regional approach. Three pragmatic propositions in support of its value are suggested. They are that the individual's associations with a home area provide an effective basis for examining the interplay of systems and the combination of perception with policy in regional management, a basis of reality and first-hand knowledge; that the understanding of earth dynamics (at whatever depth) remains an intellectual abstraction unless exemplified, analysed, and compared at regional level; that as the lives of all individuals are influenced by intra- and international policies the political unit is the most useful class of region.

These propositions (not new) give some lead to the selection and organisation of regions for study. Unit size will vary according to situations exemplified and comparisons being made; some units will demonstrate complex interactions, others perhaps one situation; formal attempts at comprehensive treatment will lose favour and the home region will be that most minutely studied assisted by the accessibility of data and the practicality of field study. Distinctions between systematic and regional geography may well be blurred in their integrated contribution to an understanding which calls on the concepts and perspectives of more than one discipline.

Some suggestions are made of themes for regional study which are relevant to society, of contemporary and unavoidable significance, have geographic parameters and are amenable to rigorous treatment: intra- and international conflict potential, e.g., urban sub-culture confrontations, socio-economic and political conflict, e.g., wealth and power distribution within nation states, conflicts of interest between states; the impact of forms of occupance upon ecosystems, the production and distribution of commodity and discommodity; the variations and distribution of states of public health, nutritional levels, disease vectors, and their relation to 'natural and social' environments; the location of various demographic types in relation to living standard and resources; the distribution of forms of economic activity in relation to concepts of 'under- and over-developed, and under- and over-populated'; the nature and distribution of resources in relation to regional interdependence and to concepts of exploitation and conservation; the regional variation in degrees and rates of urbanisation and the distribution of variations in their probable consequences.

Clearly physical and biotic systems are not given primacy as initiators of enquiry but are implicitly involved particularly when climate, space and form, soil, and ecosystem are treated as greater or lesser resources with the advantage that their nature is incorporated into a continuum which unites a range of disciplines.

In this sense the claim of a purposive regional geography to assist in reuniting two divergent cultures may have some validity at the expense of some loss of its own identity.

c0505
Geography in adult education in England and Wales
A.D. COOPER *Luton College of Technology, UK*

Adult education courses in geography in England and Wales may be obtained from three sources: (*a*) The polytechnics, colleges of technology, colleges of further education, and evening institutes providing higher and further education courses. (*b*) Formal courses from university extra mural studies boards, the Workers' Educational Association, and the educational branches of the armed services. (*c*) Informal courses provided by voluntary bodies such as the Geographical Association.

The greatest number of students and courses are found in the various colleges. The polytechnics provide high level and advanced courses of both academic and vocational nature. The colleges of technology have some university standard courses but a majority of intermediate standard, whilst the colleges of further education concentrate on introductory and lower level courses.

The students enter at a variety of ages and of education levels, attendance is voluntary, motivation being chiefly financial or social. The majority of students are between 18 and 25 years of age, with a substantial minority of younger students in full-time elementary courses and of older students in part-time courses. Attendance for advanced courses is usually full time for younger students and part time for the mature adult. Full-time attendance may be for the normal academic year or for 'block release' for one term – 13 weeks – or three months.

Geography is most important as a subject in school equivalent and post-school courses in the colleges of further education and the colleges of technology in terms of student numbers. It is also quite strongly represented in some of the polytechnics and colleges of technology offering university-level courses.

These latter colleges have played an important role during the last twelve years by offering courses for London University, External Degrees providing additional university level places for the increasing numbers of geography students qualified for them but who could not be accommodated in the universities. Management of these courses is now passing to the Council for National Academic Awards. Even so, in 1971 only one pure geography degree course was available under CNAA auspices. In other cases geography was offered in combination with economics, languages, or science, and similarly in vocational courses.

Since January 1971 geography has been available as part of the social science unit of the Open University. This utilizes sound radio, television, texts, and programmed learning systems with discussion groups, tutorials, and residential summer schools for 'second chance' home students who did not take a university course at 18 plus. Unfortunately, physical geography is in the earth science faculty and geographical specialization will be possible only in human geography.

The colleges of further education provide courses of the school type for the General Certificate of Education or of a vocational nature. Students taking GCE courses do so to provide themselves with qualifications for further academic study or for professional purposes. Many entering are school leavers, a few come to the colleges as an alternative from school. A smaller proportion are housewives, industrial workers, local government officers, sales representatives, and even teachers.

Vocational courses place geography in a somewhat equivocal position. Many lecturers justify it as a liberalising subject, and it is significantly regarded as this by lecturers teaching other subjects such as management in such courses. This undervalues the technological revolution which economic and other branches of geography are undergoing, and tends to hamper the introduction of new concepts and methods. As a liberal study, geography might be 'current affairs' or more strictly geographical studies of the home area based on concentric methods.

Vocational geography is usually economic geography. When further education was developing in the first decades of this century, bodies such as the Institute of Bankers and subsequently the Institutes of Export, Marketing, and Transport, included it in the syllabuses of their professional examinations. This dictated that it should be either commercial or economic geography. It is not included in the syllabuses of the Institute of Management.

Economic geography is also included in the National Awards scheme for business studies. The syllabuses for these are compiled by a joint committee in which the employers, trades unions, teachers, and the government are represented and the examinations are moderated by the Department of Education and Science. Teachers are free to operate their own schemes providing that they are approved by the committee whilst the syllabus itself allows considerable choice. In terms of student number, many more take the lower-level Ordinary National Certificate at Diploma courses than the higher awards courses. In the higher courses geography has to justify its place within greater competition and its inclusion reflects the personality and drive of the lecturer as much as the intrinsic value of the subject. Herein lies the weakness of emphasizing the liberal aspect of geography.

In the recently introduced National Awards scheme for surveying cartography and planning, geography is included to provide an awareness of landscape, landforms, and the natural environment. In the more advanced courses it develops into human geography and settlement studies particularly for the planners.

In the early days of development in the colleges, the teachers were employed part time and were frequently secondary school teachers who earned extra money by teaching in 'night school.' As the scope of geography in further education has extended, an increasing number of full-time lecturers has been appointed. Many of these are young graduates, some of whom have industrial or commercial experience. The majority of establishments have only one or two geographers on their staff. The largest geography sections are in those polytechnics and colleges of technology which have a concentration of advanced work where the team may be 15 strong and employ a proportional support staff of technicians and cartographers. Although styled lectures, the nature of instructional medium is usually a class.

By comparison with the colleges the other sources offering geography attract smaller student numbers. The students normally request the courses from the Workers' Educational Association or the University Extra Mural Boards. Courses tend to be more specific, frequently directed to particular regions or topics such as China or the Common Market (EEC). By comparison with courses in English literature or history, geographical sessions are few. Geography as a promotional subject is disappearing from the courses in the British armed forces, partially as the result of job-analysis which makes what is done more important than where.

Informal adult education is a still weaker field. The Geographical Association through its branches provides a national coverage but geography fails to have the amateur following that geology or history have. The Geographical Association has very few members who are not educationists in some field or another. Is the adult interest in geography stultified by the presentation of the subject, or is the rift between geography as it is and as it is popularly conceived unbridged? Certainly many television documentaries have a strong geographical content, yet the popular interest is not there.

c0506
The problem of hierarchy in the objectives of geography teaching at the pre-university level
N.J. GRAVES   *University of London, UK*

For the first 50 years of this century, discussion concerning the objectives of geography teaching was largely concerned with the general objectives and often took the form of a justification of the subject at school level. Evidence for this may be found in texts published on the teaching of geography (Geikie 1887; Fairgrieve 1926; Long and Roberson 1966) and in the suggestions and instructions issued by various ministries of education in those countries where geography was part of the school curriculum (HMSO 1960). These general objectives included training in observation, informing future citizens about world and national problems, promoting international understanding, and enabling students to bridge a putative gap between the sciences and the humanities. They amounted to articles of faith in what was a relatively new subject, but were seldom accompanied by evidence that such objectives were achievable.

In the past twenty years geographers and educationists through a process of cross-fertilization have become aware of each other's contribution to the continuing problem of defining objectives in education. As a result discussion has led to a realization that objectives may be given at various levels of generality or specificity (Bloom 1956; Kasperson 1967; Graves 1971). This has led to a tendency in recent literature to concentrate on behavioural objectives (Clegg 1970). The objectives are of help to the teacher requiring precise guidance, though they might be cramping the style of the more creative 'intuitive' teacher. Objectives which are specific without having the precision and circumscription of behavioural objectives might still be useful; for example: 'the learner shall be able to describe from 1/50,000 scale map evidence the relationship (if any) between relief and lines of communication.'

Such discussion on the specificity of educational objectives assumes that what is to be taught is valuable educationally. Some way of assessing the value of such objectives was therefore required and many have used Bloom's *Taxonomy of Educational Objectives* (*The Cognitive Domain*) to assess the intellectual worthwhileness of certain objectives, those objectives which could be classed as exhibiting characteristics of the abilities to evaluate, synthesize, and analyse, being considered superior to those which exhibited only the characteristics of knowledge or comprehension. Thus the use of Bloom's taxonomy may be seen as an attempt to structure the objectives of geography teaching into some sort of hierarchy from the point of view of the intellectual abilities required to achieve these objectives. Such a classification does not, however, dispose of the question as to whether what is taught, even if it may be classed as 'evaluation,' is educationally worthwhile. It is beyond the scope of this paper to discuss this philosophical question. Suffice it to indicate that if education is conceived of as a process of initiation into forms of knowledge (Peters 1965), then we may use the criterion that any objective which is superfluous to this process of initiation into the concepts and principles of geography should be eliminated.

Bloom's taxonomy of educational objectives has been widely interpreted in the hierarchical way suggested in the previous paragraph. Such a hierarchy is, however, incomplete if examined from the point of view of a teacher attempting to structure a course in geography. For example, under the category I 'Knowledge' one finds 'knowledge of terminology.' One could therefore place in this class both the terms 'westerly wind' and 'geostrophic wind'; or the terms 'coniferous forest' and 'ecosystem.' It would seem evident upon analysis that these terms are not on intellectually equivalent levels, that 'geostrophic wind' is a more difficult concept than 'westerly wind' and 'ecosystem' more complicated than 'coniferous forest.' This problem was examined by Gagné (1966) in relation to concepts in physics. He showed that simple concepts could be determined by observation; for example, students in geography may learn to distinguish between photographs representing an oxbow-lake and those not representing this feature. These he called 'concepts by observation.' Some concepts cannot be observed in this sense; for example, the relatively simple concept of density of population is not an observable concept. One cannot observe a series of landscapes and indicate which one has a 'density of population,' even though one might be able to make a judgment about whether or not a landscape was densely populated. The concept of 'density of population' has to be defined in terms of some more basic concepts in this case: population and area. Such concepts Gagné has called 'concepts by definition' or 'principles.' They involve relations between other concepts and usually with some operation implied. In the example given, 'density of population' involves the operation of division since population density is expressed in terms of numbers of people per unit area. To take another example, the concept of 'dynamic equilibrium' currently in use in geomorphology (Small 1969) involves the understanding of a whole series of other concepts, such as those of erosion, weathering, slope, mass movement, lithology, and so on. Such a concept could be termed an 'organizing concept' since it encompasses a wide range of related concepts concerned in the explanation of equilibrium in the processes of earth sculpture. Therefore, even in such a lowly category in the Bloom taxonomy as 'knowledge of terminology,' concepts of widely differing intellectual levels may be involved.

The above analysis yields the idea that, in some aspects of geography, there exists a logical structure in which more complex concepts are based on an infrastructure of simpler ones. A simple example would be that understanding the term 'drainage basin' implies some prior knowledge of such concepts as river, tributary, relief, and watershed (or water parting) probably in that order. A more elaborate example would be that an understanding of Hoyt's model of urban structure implies some understanding of urban functions, urban land use categories, the idea of models as ideal representations of certain aspects of reality, again probably in that order. There is, of course, no continuous linear structure in geography, but rather a series of parallel lines, some having cross-links. For example, a progressive development along conceptual lines of ideas connected with a vegetation ecosystem is bound to connect with a similar development of ideas connected with the concept of soil. It might therefore be thought that the development of a course in geography could depend on the close analysis of the logical structure of the concepts and principles inherent in the topics taught, and their careful arrangement as a series of hierarchical performance or behavioural objectives, in the same way as learning to solve simple equations in mathematics must precede the solving of simultaneous and quadratic equations.

Unfortunately there are certain difficulties which need to be resolved. When Gagné writes of 'concepts by observation,' he is referring to the ability to recognize certain things as belonging to a class or category of objects. Thus children may be able to distinguish oxbow-lakes, deltas, cuestas, cirques, arêtes, etc. from exemplars and non-exemplars. This, however, is a rather different ability from being able to explicate these concepts. Thus though a student may be able to recognize a delta, his 'expressed concept' of a delta may be rudimentary or sophisticated depending on the depth of his understanding of what a delta is. The concept delta may therefore be apprehended at many different levels of understanding. (It might be argued that at some levels, the concept was incomplete.) The same could be said of many 'concepts by observation' and the idea of using these as the bricks upon which the conceptual structure of geography could be built is probably an oversimplification.

Another difficulty lies in knowing whether the logical order which might be established by a close conceptual analysis of geography is mirrored in any way by students' psychological development. Apart from Piaget's and Inhelder's (1956) work on children's conception of space which revealed that topological relationships are acquired before Euclidean relationships, studies so far indicate that there is an increase in the quantity of geographic concepts acquired as children mature (Milbuon 1969; Lunnon 1969), but not how these concepts relate to one another in the temporal sequence of learning. We are still very much in the dark concerning this aspect of learning in geography.

The meaning of objectives in geographic education is now much clearer than it was. The task of structuring objectives in some hierarchical manner which would be valid both for the logic of the subject and for the mental development of students has hardly begun. Many studies on the logical structure of the discipline such as Harvey's (1969), on the conceptual understanding of geographic ideas and principles by children and on the way children perceive their environment will be needed before much progress can be made in this field.

Bloom, B.S., et al., 1956   *A Taxonomy of Educational Objectives*, 1 and 2 (David McKay Co. Inc.).

Clegg, A., 1970   *Developing and Using Behavioral Objectives in Geography in 'Focus on Geography' – 40th year-book*, National Council for the Social Studies.

Fairgrieve, J., 1926 *Geography in School* (ULP).

Gagné, R.M., 1966 The learning of principles, in H.J. Klausmeir and C.W. Harris, *Analysis of Concept Learning* (Academic Press).

Geikie, A., 1887 *The Teaching of Geography* (Macmillan).

Graves, N.J., 1971 *Geography in Secondary Education*. Geographical Association (UK).

Harvey, D., 1969 *Explanation in Geography* (Arnold).

Kasperson, R.E., 1967 On the process of curriculum reform, *J. Geog.* (Sept.).

Long, M., and B.S. Roberson, 1966 *Teaching Geography* (Heinemann).

– 1960 *Geography and Education* (HMSO).

Lunnon, A.J., 1969 The understanding of certain geographical concepts by primary school children. Unpublished M.Ed. thesis, University of Birmingham.

Milbuon, D., 1969 The development of vocabulary in geography by primary and secondary school children. Unpublished M.Phil. thesis, University of London.

Peters, R.S., 1965 *Ethics and Education* (George Allen and Unwin).

Piaget, J., and B. Inhelder, 1956 *The Child's Conception of Space* (Routledge and Kegan Paul).

Small, R.J., 1969 The new geomorphology and the sixth-former, *Geography* (July), 54, par. 3, 308–18.

c0507
Image mentale et connaissance réelle: l'exemple du Nord
L.-E. HAMELIN   *Université Laval, Canada*

'There have been as many geographies of the North as there have been illusions' (Watson 1969).

Cette recherche davantage basée sur des tests que sur des documents d'archives est passablement nouvelle tant sur le plan méthodologique que sur son champ d'application. Chez les géographes, l'utilisation des profils de polarité pour entrevoir les opinions des habitants ne semble guère remonter à plus de dix ans. Dans le Nord canadien, des anthropologues ont essayé de découvrir la mentalité de quelques types de résident: administrateur et missionnaire, mais ces portraits n'étaient pas établis dans un but pédagogique. La présente étude veut précisément s'intéresser aux mentalités et aux attitudes. Généralement, dans l'examen des facteurs de l'Arctique, les géographes considèrent des données physiques comme le froid, le pergélisol, et des données socio-économiques tels les Amérindiens et les mines; ils oublient l'aspect psychologique qui est très important dans les pays à faible élasticité. 'There are two kinds of arctic problems,' écrivait le Canadien W. Stefansson, 'the imaginary and the real. Of the two, the imaginary are the more real; for man finds it easier to change the face of nature than to change his own mind' (Stefansson 1945, 5). Il se crée donc un écart entre ce que l'on voit, ce que l'on perçoit et ce que l'on croit. Dans la coûteuse politique de la colonisation dirigée dans le pré-Nord abitibien, il a fallu deux décades et un quart de milliard de dollars pour que le réel l'emporte sur l'imaginaire. Nous nous intéressons aux opinions que l'on peut avoir du Nord et à l'écart de ces opinions suivant les sujets, les observateurs, et les régions. Pour mesurer cette différence, des questions du test devront se rapporter aux connaissances objectives du Nord ainsi qu'à l'image que l'on se fait de thèmes significatifs.

Il circule chez tout type de répondants: Indigènes étant très ou peu acculturés; Nordistes non-indigènes en résidence courte, prolongée, ou permanente dans le Nord; non-Nordistes comme le Canadien du sud ou les Soviétiques du sud-ouest mais ayant certaines connaissances du Nord; habitants de pays chauds n'entretenant aucun contact avec les régions polaires. Les observateurs occupent des fonctions différentes; ils sont administrateurs, ouvriers, étudiants, monsieur, tout le monde. Le répondant agit pour lui-même mais en certains cas, il lui est précisément demandé de répondre en épousant la cause d'un autre.

Les régions considérées sont d'une part le Nord canadien dans son ensemble et d'autre part le Nord soviétique. Nous nous intéressons non seulement à la façon dont les Canadiens considèrent leur propre Nord mais à celle suivant laquelle ils voient le Nord 'd'en face' dont ils ont des stéréotypes. Pour certaines questions, le Nord ne peut être apprécié que d'une façon comparative; dans le Nornam il est vu en rapport avec le sud de l'Ontario ou la plaine de Montréal; dans le Noruss, il l'est en relation avec la région de Moscou.

Il y a plusieurs façons de considérer le corpus des questions. La classification la plus importante tient aux formes mêmes du texte car celles-ci conditionnent directement les réponses. Nous utilisons un questionnaire à plusieurs sections, à remplir l'une après l'autre et sans revenir sur les précédentes. L'ensemble comprend (1) des questions d'information demandant une réponse brève ou absolue du genre, vrai ou faux, oui ou non, par exemple 'Au Canada, je préfère vivre plutôt dans le Nord que dans le Sud.' (2) des phrases types ou des statistiques à juger, à compléter, à corriger ou à dériver à partir de prémisses. C'est par un texte que le répondant exprime ici sa réaction. 'Commenter: les Esquimaux du Québec deviendront québécois.' (3) des assemblages appropriés de termes clés. 'Rassembler en trois groupes ces douze mots isolés: Rennes. Kayak. Indien. Isba. Esquimau. Lénine. Région sans arbre. Raquette. Iakoutes. Forêt. Castor. Phoque.' (4) des cartes muettes du Nord (ou des croquis) pour localiser des points d'intérêt. 'Quelle est la région nordique du Canada qui va le plus se développer au cours des 30 prochaines années? Ajouter quelques toponymes de référence.' (5) des cartes ou des croquis à bâtir. (6) des grilles d'opposition comportant des classes intermédiaires. 'Comment voyez-vous l'aventurier du Nord?':

Caractères	beaucoup	entre beaucoup et moyen	moyen	entre moyen et peu	peu	Caractères
sociable						insociable
buveur						sobre
travailleur						paresseux
gai						triste
célibataire						marié
batailleur						tranquille
sale						propre
pratiquant						non-pratiquant
né au Canada						né à l'étranger

Le test circule en deux versions principales, française et anglaise. Des adaptations sont prévues pour au moins deux autres langues.

En décembre 1970, un nombre restreint de questionnaires se rapportant alors à des informations géographiques ont circulé. Les premiers résultats ont surtout été d'ordre méthodologique et ils ont servi à la préparation d'une grille pour l'examen du cours sur le Monde nordique en avril 1971.

L'histoire relate les principales illusions dont le Nord canadien a été l'objet. Pour ne mentionner qu'un exemple donné par J.W. Watson, citons les variations dans les opinions que les Sudistes se sont faites du Nord; en 1947, les auteurs étaient beaucoup plus optimistes qu'ils ne le seront 20 ans plus tard. Les conceptions générales ont un profond impact sur les activités boursières, la mobilité de la main d'œuvre, les projets politiques, bref sur le développement même du Nord. Pour une part, un pays résulte du fruit d'une pensée.

Dans une communauté minière blanche du Moyen-Nord, un test a montré que le facteur le plus important n'était ni le froid ni la neige mais l'atmosphère amicale; par ailleurs, une diminution relative de l'isolement constituait le principal souhait des habitants. C'est donc avant tout par deux termes psychologiques que ces résidents définissaient leur propre situation.

Ces exemples montrent la nécessité des recherches sur les opinions pour ajuster au mieux les services d'enseignement. L'addition d'une aile psychologique modifie les objectifs généraux de l'enseignement; elle ajoute des thèmes à développer, partant, elle influence le programme des cours, la charge des travaux pratiques et la matière des examens.

Comité national canadien de la recherche géographique du Gouvernement fédéral; Maurice Saint-Yves; Henri Dorion.

Bloom, B.S., 1969 *Taxonomie des objectifs pédagogiques*. Montréal, tome 1, 232p, trad. de l'anglais par Marcel Lavallée.

Briggs, Jean, 1971 'Strategies of perception: the management of ethnic identity' in *Patrons and Brokers in the East Arctic*, édité par Robert Payne (Toronto), 11p.

James, P.E., 1967 'On the origin and persistence of error in geography,' *Annals of Association of American Geographers* (1967), no. 1, 1–24.

Rimbert, Sylvie, 1971 'Essai méthodologique sur des stéréotypes régionaux du Canada,' *Cahiers de géographie de Québec*, no. 36, 523–36.

Saarinen, T.F., 1969 *Perception of Environment*. Resources Paper, AAG no. 5, 37p.

Stefansson, W., 1945 *The Arctic in Fact and Fable* (New York), 96p.

Watson, J.W., 1969 'The role of illusion in North American geography,' *The Canadian Geographer – Le Géographe canadien*, 13, no. 1, 10–27.

Wonders, W., 1971 *Canada's Changing North* (Ottawa), 367p.

c0508
The place of expert knowledge in curriculum development
JOHN C. HANCOCK *University of Bristol, UK*

The paper examines a number of curriculum models currently being used for course design. Each of the models contains reference to the content of the course. The reference may be to the disciplines or areas of knowledge which are traditionally part of our school curriculum. At other times the model looks at concepts which are usually conveniently grouped by some common theme, so that for example those concepts which have a spatial element are grouped. A third alternative is to recognise certain needs of the pupils or society and then to search for contexts which may be used to examine those needs, so that, for example, an understanding of population growth is seen as one need. Then mathematicians, biologists, geographers, etc. show via theoretical constructs, experiments, and case studies the many aspects of population growth.

In each case, before classroom procedures and teaching styles are considered, the model requires some selection of content. It is at this point that one argues for an external input into the model. There is a need to present the teacher or group who will finally plan the course with a review of the content available to them. Failure to do this in the past has resulted in content which has become formalised and remote from current views on the subject. If this were available then many of the exciting developments in the subject of geography could be made readily available to schools.

There is, however, a thin path to be followed here between imposed content and freedom to opt out of change. Expert knowledge surely has its place, not as a threat to the professionalism of the teacher but as a service.

There are too those who state the place for expert knowledge. 'It has always been needed; it will always be needed. The question is how applicable is it to the local situation? The local group should answer this question' (Harnack 1968). The danger lies in the many interpretations there are of this process. A local group may screen expert information and produce a parochial course. There are examples of this type of syllabus available in Great Britain. However, assuming that a local group does wish to make use of expert knowledge, what are the problems it faces? Expert knowledge about geographical issues, instructional methods, and technological aids has been forthcoming for many years. But how can the local group know of its advantages and limitations, and decide how this knowledge can be used to improve the instructional program? There is, of course, today a growing volume of publications dealing with curriculum development but contributors to such publications tend not to be drawn from schools or university geography departments.

Usually those interested in the development of theory have naturally come to publish their results for an audience of university geographers rather than for an audience of teachers. This has sometimes meant that the practical relevance of their results has either not been examined at all, or has been missed because of difficulties of communication, or because of confusion about what are the problems of teaching and what are theoretical problems.

A good example of this can be found in a publication called *Progress in Geography* edited by Board, Chorley, Haggett, and Stoddart (1971). This occasional series aims to 'provide the means by which geographers, who wish to, may keep up with developments in fields outside their own, and in general forms a link between teaching and research in the discipline.' So far three volumes have produced fifteen excellent reviews. These include articles on urban gaming simulation (Taylor), environmental management (O'Riordan), the role and relations of physical geography (Chorley), in one volume. These are the areas which are interesting many of the best school teachers and college lecturers. Yet the impact of the reviews is as yet disappointing. The reviews are by experts; the teachers I believe want to know – yet the catalyst needed to create some interaction seems absent.

It may be that this theoretical interchange can be very confusing to the teacher who encounters it and who is asking what appear to be simple questions about how to handle change smoothly, what types of incentive are best, how to deal with conflicts that occur in the classroom, how to deal with the required specialist knowledge of the many aspects of his job. Of course, both teacher and theoretician know that the questions and answers are not as simple as they seem. The impasse occurs because different sets of difficulties appear to both of them.

To combat this situation there is a need for experts to put down, for school teachers, clear statements of the opportunities that their field offers. These statements should be intended to make us aware of these fields, of the possibilities and pleasures they offer and where it is needed the purpose, relevance, and value of their studies. It will then be for the local curriculum group to set these statements into their curriculum model and to decide on priorities for the inclusion of competing content. At present no one seems ready to bridge this gap in our system. The experts must in my mind come from the universities and colleges or be in contact with such institutions.

The final part of this paper contains an example of the type of supportive material that seems to be needed. It follows a year-long review of many recent publications. Urban society was taken as a central theme because of the increasingly urban world in which we live. Few would argue against some consideration of our urban areas. However, if one examines the syllabuses and textbooks mainly used in British schools one finds that the treatment of urban areas is very different from that suggested here. At present there is a concentration on the situation of a town, with physical factors called in to relate situation to site. The development of communications and more rarely trade is considered. Finally the major 'basic' products of a town are listed.

I would ask those reading the following list to remember it is intended for secondary school teachers in Great Britain, and that it will never be finished. As change takes place, as our knowledge improves so the opportunities available will change. Criticism, rewording, new questions, and added references will be most welcome.

The list is in three parts. Part I is an overview of the whole scheme and Part II fills in detail from Part I. (In this summary only a small part of Part II is included.) In this second section the content areas are presented as a series of questions or statements with a major explanatory reference for each. Part III contains explanations and summaries of the articles noted in Part II (not included in this paper).

PART I: A SCHEME FOR RECORDING CONTENT IN URBAN GEOGRAPHY (based on a scheme used by Morrill – Spatial Organization of Society)

A *Location factors*
 1 past developments
 2 natural environment
 3 comparative advantage
 4 social system
 5 economic system
 6 political influences
 7 time element
 8 distortion by lag

B *Towns as central places*
  1 existence of service centres
  2 central place theory
  3 spatial equilibrium and threshold
  4 central place activities
C *Towns as processing centres*
  1 industrial activities
  2 industrial location
  3 single plant location
  4 spatial behaviour of the firm
  5 specialisation
  6 basic/non basic concepts
D *Towns as places to live*
  1 rural/urban contrasts
  2 intra urban population change
  3 population gradients
  4 residential area characteristics
E *Movement of people and ideas*
  1 types of movement
  2 temporary movements
  3 transient movements
  4 permanent movements
  5 migration factors
  6 effects of migration
  7 transport systems
  8 population migration
  9 gravity models
  10 spread and diffusion
F *Systems of towns*
  1 location
  2 variety of cities
  3 urbanization
  4 urban life
  5 urban structure
  6 urban land use
  7 internal social structure
  8 growth of city
G *Networks*
  1 single route location
  2 route networks
  3 network geometry
  4 local density
  5 regional density
  6 network change models
H *Urban planning*
  1 history and control
  2 goals – identifying objectives
  3 plan formulation
  4 implementation and implication

EXAMPLES OF SECTION OF PART II OF SCHEME
(incomplete)

F *Systems of towns*
  5 *Theories of urban structure*
    (*a*) *Recognition of urban structure*
    (i) Do towns have an internal structure which can be expressed in terms of the differences in character and intensity of land use at various locations (Garner 1966)?
    (ii) Core, star-shaped, circular, and now axial – all forms of urban evolution. How has changing technology shaped this form (Nelson 1969)?
    (iii) Urban structure and land use represent a balance between mutual proximity and transportation (Mayer 1969).
    (iv) What forces are acting to change the spatial structures of retailing (Kelly 1955)? (Includes conceptual scheme for analysis of retail structure.)
    (*b*) *Functioning models*
    (i) *Spatial interaction models*
    Huff (1963), Lakshmanan and Hansen (1965), Haydock Study (Manchester University 1964), Lewisham Model (Rhodes and Whitaker 1967), Black (1966), West Midlands Shopping Model (West Midlands 1967), Lowry (1965).
    NB
    Framework for study using models
    (i) objectives, (ii) assumptions, (iii) hypotheses, (iv) data, (v) results, (vi) sensitivity (Nedo 1970).
    Divergence of explanatory and predictive models (Lowry 1965).
    What are the effects of changing the level of spatial and time integration? (Colenutt 1970)

Board, C., R.J. Chorley, P. Haggett, and D.R. Stoddart, 1971 *Progress in Geography*, 3 (Leeds).
Black, J., 1966 'Some retail sales models,' Paper for Urban Studies Conference, September 1966.
Colenutt, 1970 'Building models of urban growth and spatial structure,' *Progress in Geography* 2 (Leeds).
Garner, B.J., 1966 Aspects and trends of urban geography, *J. Geog.* (May).
Harnack, R.S., 1968 *Decision Maker and Curriculum Planner* (Scranton, Penn.).
Huff, D.L., 1963 A probabilistic analysis of shopping centre trade areas, *Land Economics* (Feb.).
Kelly, E.J., 1955 Retail structure and urban economy, *Traffic Quart.*, 9.
Lakshmanan, T.R., and W.G. Hansen, 1965

A retail market potential model, *J. Am. Inst. Planners* (May).
Lowry, I.S., 1965 A short course in model design, *J. Am. Inst. Planners*, 31.
Manchester University, 1964 Regional shopping centres in north west England.
Mayer, H.M., 1969 Cities and urban geography, *J. Geog.* (January).
Nedo, 1970 *Urban Models in Shopping Studies*. National Economic Development Office.
Nelson, H.J., 1969 The form and structure of cities, *J. Geog.* (May).
Rhodes, T., and R. Whitaker, 1967 Forecasting shopping demand, *J. Town Planning Inst.* (May).
West Midlands, 1967 Predicting shopping requirements, *Town Planning Inst.*

c0509
Some problems in the application of central place theory to educational planning – with specific reference to the Bath area of the UK
G.H. HONES *University of Bath, UK*

In a social climate where the demand is for more and equal educational opportunity, it is becoming increasingly evident that this can best be achieved through improved planning.

In this planning, the geographer surely has a contribution to make through a study of the spatial aspects of education systems – the location of individual establishments, their pattern of distribution, and the interacting relationships which comprise the network – as well as relating this analysis to the surface, physical and human, on which the system operates.

As a pre-requisite to future planning, the geographer may seek order or regularity in the everchanging patterns caused by present educational phenomena. In this context the application of an abstract 'model' of reality, as for example the theory of 'central-places,' could be valuable.

It is necessary at the outset, however, to recognize and evaluate the problems that will be met in the attempt to match theory with reality, whether of a general nature or stemming from the distinctive characteristics of education systems.

In a pilot study to test some aspects of the possible application of Christaller's theory of location of tertiary activity directly to education systems, rather than as one function in an over-all settlement pattern, the educational phenomena of the Bath area in the UK were examined and the resultant problems are discussed in this context.

The premise of a 'central place' providing a service for a surrounding tributary area seems relevant to education systems, with individual establishments functioning efficiently at the centre of minimum aggregate travel. However, most systems have evolved as a result of the needs of an existing settlement pattern and the resultant educational nodes are not often truly competing spatially. Instead of a pattern evolving through natural competition, such systems have been given some order by administration on a regional and/or national scale.

Another feature of Christaller's model is the hierarchy with places of greater centrality, thus of higher order, absorbing the lower order functions for that locality. In the educational 'hierarchy,' however, the status of an institution is not measured on a common index (using a separate criterion such as number of pupils, or a wider combination of indices) but derives from the age group served. Following this, the nodes normally remain discrete units, with those of the 'lower order' retaining their identity and separate functions even when adjacent to 'higher order' centres. As in the general model, these latter institutions catering for older students tend to be larger units, more widely spaced in larger catchment areas.

As a further complication, it should be noted that the colleges and universities of higher education are, in many countries, not truly connected with their local/regional framework but draw their students from a much wider, national catchment.

Although the principle of a 'threshold population' applies in general terms, the critical decisions concerning the number of students in an institution (minimum and maximum limits, or optimum sought) involve such issues as the distance a certain age group can be expected to travel daily, the variety of courses needed, and the age of transfer to other establishments.

It is this latter decision which then determines the precise position of the 'steps' in the hierarchy. With neighbouring administrative authorities free to decide on breaks at different stages in the educational process – even at times using different organizational structures for various areas within their control – there is little likelihood of a locational pattern emerging over a wide area with a generally applicable hierarchy of establishments.

The administrative boundary line thus often emerges as a major hindrance to the development of a logically ordered education system on a large scale. The effect on educational planning in areas such as that around Bath – where three counties and the city itself all operate as separate units – is unnecessarily complex. Even though some co-operative action is attempted – with some pupils attending the most suitable establishment irrespective of home location and boundary line – such rationalising is inadequate, perhaps imposed by financial pressure rather than stemming from a positive philosophy.

The whole problem of boundaries is further complicated in the UK by the fact that a reorganized system of local government is now being planned and the various proposals for the Bath area would all manifestly alter the existing pattern.

Another influence can be the effect of independent schools, often completely unrelated to the state system. Where serving a special national or regional catchment area there may be little direct local effect but those more locally oriented can affect the balance of the state pattern as, for example, when denominational schools are established. The magnitude of their effect naturally varies according to the country and region, its socio-economic structure, cultural, or religious bias.

Such norms and values of an area can be powerful influences in educational planning, not only stemming from the consumer, the local community, but also in terms of administrative behaviour. The possible effect of the attitude of the people involved, although difficult to define and impossible to quantify, must always be considered, for it can hinder any rational attempt to plan.

In fact, the effect of the environment, cultural and physical, on education systems is so great that the use of a theoretical model can be questioned. The irregularities of the existing settlement pattern, the problems of forecasting future population changes, the communication system as it affects educational journey patterns – these are all examples of real forces which modify any ideal network based on an imaginary homogeneous surface.

Nevertheless, in spite of this and the other issues discussed previously, the use of central-place theory in educational planning can be of real value. Although not likely to supply solutions to the problems that are met, the theory does provide a reference framework in which these problems are more readily understood.

Berry, B.J.L. and A. Pred, 1965 *Central Place Studies,* Regional Sci. Res. Inst. (Philadelphia).

Berry, B.J.L., 1967 *A Geography of Market Centres and Retail Distribution* (Englewood Cliffs, NJ).

Bracey, H.E., 1952 *Social Provision in Rural Wiltshire* (London).

– 1956 A rural component of centrality, applied to six southern counties of the United Kingdom, *Econ. Geog.,* 32, 38–50.

Brush, J.E. and H. E. Bracey, 1955 Rural service centres in south western Wisconsin and southern England, *Geog. Rev.,* 45, 559–69.

Bunge, W., 1966 *Theoretical Geography* (Lund).

Haggett, P., 1965 *Locational Analysis in Human Geography* (London).

Jacobs, R., 1964 The inter-disciplinary approach to educational planning, *Comp. Educ. Rev.,* 8, 17–23.

Krawetz, N., 1971 *Priorities in Educational Planning,* Report of 1970 Conference of Comparative Education Society of Europe, British Section.

Mitchell, L.S., 1968 An evaluation of central place theory in a recreation context, *Southeastern Geog.,* 8, 46–53.

Morrill, R.L., 1970 *The Spatial Organization of Society* (Wadsworth, Cal.), ch. 4, Towns as central places (p. 78).

Myers, C.L., 1969 Operational research in educational administration, in G. Baron and W. Taylor, eds., *Educational Administration and the Social Sciences* (London).

c0510
Geography at the Open University (UK)
A.T.A. LEARMONTH *The Open University, UK*

The Open University has been operating in a substantial way since September 1969, at first in preparatory work, and since January 1971 in course work for the first year's entry of 21,000 undergraduate students. The university's initial aim is to serve mature students, especially those who for some reason missed the opportunity of university education at the younger age which is normal in Britain. No formal entrance qualifications are required, but an advisory service offers advice on preliminary or alternative forms of education for those who appear unlikely to be ready for undergraduate work. However, if they wish to persist with an application to enter, this may go forward along with others. So far applications have exceeded the quota of places we can offer – the restriction is ultimately financial, and the source of money is entirely from government funds and fees (some paid by local authorities, etc.).

The courses are more interdisciplinary than is usual in Britain, except perhaps in one or two of the newer universities. Each student doing a full degree course for instance (i.e., *not* granted exemptions for some previous training) does *two* foundation courses put forward collectively by faculties of four or five disciplines each in humanities, social sciences, science, mathematics, technology, and education, and many other courses are inter-disciplinary.

The courses are also multi-media, using television and radio broadcasts, correspondence material, tutorial and counselling services offering one or two face-to-face contacts a month in one of 250 study centres mainly in borrowed premises, in the twelve regions in which much of the detailed work is organized. Education technologists are used for research on the project, but also to advise faculties and course teams on testing, computer marking, etc. The design of the foundation course material has received a good deal of thought.

Geography is situated in the faculty of social sciences, and is primarily human geography, since earth sciences, so far mainly geology and closely related disciplines like geochemistry and geophysics, are in the faculty of sciences. Staff is small on the campus (1 professor, 1 senior lecturer, 3 lecturers, and 1 research assistant), with 3 full-time staff tutors in 3 of the regions. Much of the marking, criticism, and face-to-face teaching are done by part-time staff.

Summer schools are held, at least for the foundation courses, using full-time and part-time staff. The geographical contribution to the teaching so far includes:

(1) Seven weeks of the students' work in the foundation course called Understanding Society; this comprises: (a) a 'unit' in which geography, and the other social science disciplines (economics, government, sociology, and psychology) each try to answer the question 'Why do men live in societies?'
(b) Four weeks of socio-economic geography and one of political in the main, middle part of the course, more in single disciplines, though establishing links between disciplines where possible. (c) A unit or week's work in the final series of programs in which each of the five disciplines offers an approach to the problems raised by the so-called population explosion.
(2) About five or six weeks' teaching in an interdisciplinary course called Decision-Making in Britain led by the political scientists.
(3) A single-discipline half-course called New Trends in Geography, in which some of the newer tools of geography are expounded as appropriate in sections dealing with continuity and change in geography, economic geography, social geography, the branches which have moved less rapidly into the new geography (regional, historical, and political geography), and a final section provisionally called Channels of Synthesis in which we hope the student may pull together the various strands of his course for himself in a fruitful way.
(4) Work is beginning on a full course on Urban Development, led by geography but with contributions from economics, sociology, psychology, and perhaps government.

c0511
The influences of the 'new geographies' on the curricula and content of geography in British schools
ALAN D. NICHOLLS   *St Clement Danes Grammar School, UK*

The 'use of the globes' and similar phrases which appeared in advertisements for Victorian private schools, can hardly be described as school geography, and the 'capes and bays' approach barely lasted until World War I. But in 1900 H.J. Mackinder gave a lecture on the 'new geography' which converted a young teacher (J.F. Unstead, who later became the first professor of geography at Birkbeck in 1921) and equally inspired others. Then in 1905 Herbertson read his paper on 'The Major Natural Regions' at the Royal Geographical Society and the then leaders of education recognized a possible development for the study of the environment (Herbertson 1905, 300). University schools of geography were few and professorial chairs even fewer.

The leaders in the new subject came from other disciplines, and were quick to see the value of synthesis and correlation. The English school derived much from the French determinists, possibly because of their proximity, though the German influence can also be traced and texts from both languages were translated into English for the use of students in universities. In Great Britain the unique nature of regions is early discerned, they are small, and striking variations often lie conveniently close to each other. So the natural method of study was based on regional geography which became the core of the curriculum. This was the first 'lamp' of Mackinder.

Of great significance was the group of early professors and teachers who formed the Geographical Association and produced a journal *Geographical Teacher* (later *Geography*). Herbertson was its first editor, while H.J. Fleure was the secretary. Thus, there was a close collaboration between the university departmental chiefs who were winnowing the content of the subject, and the teachers in the schools who were laying the foundations in the minds of future university geography students.

The provision of textbooks for the new subject was early seen to be of vital importance, and groups of school and university teachers together wrote them. The close collaboration is evident when Leonard Brooks' regions of increment, of difficulty, and the like are compared with Herbertson's regions, or when Fairgrieve's texts with their interspersed questions indicate the methods of enquiry to be followed. For the treatment of particular systematic topics, textbooks selected the continent which best exemplified them. Marion Newbigin in a manual on Ordnance Survey maps, said 'We are all drunk with writing,' and then suggested using a map 'upside down' to avoid using the print (Newbigin 1914, 10).

So by 1913, Dr Scott Keltie, the secretary of the RGS with O.J.R. Howarth an assistant secretary of the British Association, was able to write 'Geography as an educational subject of widely recognized value is coming by its rights, though the majority of the last generation may recall it as affording little else than superficial instruction in the position of countries, places, mountains and rivers' (Scott Keltie 1913, 145).

Textbook illustrations rapidly evolved from line drawings, often made by artists who had never seen their subjects, to be replaced by photographs of the actual example. More important probably was the careful, logical development of concepts (then called 'themes' or 'ideas') from the simplest to the more complex and abstract. During, and towards, the end of World War II, visual aids, other than lantern slides and pictures, notably films, film strips, and loops, brought more reality into the classroom. Indeed their misuse resulted in a surfeit.

Another 'new' geography developed with the affluent society – field work, both rural and urban. This had been fostered at an advanced level particularly by the Le Play Society. Now school children saw and examined for themselves physical features, experienced the unity of landscapes and learned Geography, as Fairgrieve suggested was proper, through the soles of their shoes.

World War II brought change. Its totality widened the interests of participants including teachers who had visited many lands, and a great social upsurge brought another 'new' geography which, under the name of social studies, endeavoured to bring a number of allied subjects together. This has been variously described, but rarely favourably. Whether the universities who have departments of combined studies will overcome the rejection by the schools, is doubtful. E.C. Marchant both criticizes and indicates the possibilities of such integration (Marchant 1968, 129). The over-riding problem is that of dilution or a lowering of standards in the separate subjects. My own experience was with VIth forms of boys of post school certificate standard. Whether the curriculum which was adapted to their locality in West London and included a land use survey or the broadening of their outlook as to future occupations was more attractive, I am unable to decide. Together we enjoyed the year's course which was dropped after a few years. The standards achieved overall in a number of subjects is less likely to be as high as in a single discipline. It has been suggested that depth of understanding is sacrificed for a butterfly sipping of the nectar of attractive blossoms. These possibilities led Professor S.W. Wooldridge to a spirited protest 'On Taking the Ge out of Geography' (Wooldridge 1949, 9). The team teaching method, associated with integrated studies, is based on a hierarchy of experts attended by less well qualified colleagues, which would seem to be contrary to 'the parity of esteem' which was a feature of British education.

The latest 'new' geography with its use of 'models' and quantitative methods has been introduced into schools with varying acclaim. Experiment and the use of new techniques on the frontier fringes of geography is right and proper as Professor K.C. Edwards indicated at a meeting of my Association with the Royal Geographical Society and the Institute of British Geographers in 1967 (Edwards 1967, 245). The questioning of the suitability of the methods and concepts for the immature minds of school children has been common not only among practising teachers, who

appreciate the standards of their pupils, but has been augmented by university professors. The apparent lack of reality in some of the simulation exercises begs the question – Why pretend that a fact is other than it is? Further, a number of leaders in our subject are clearly dubious of the exaggerated claims which are made for some systems (Spate 1966, 121).

A machine, no matter how complex, can only process material according to man directed instruction. As a US Information Service manual on Teaching and Learning aptly puts it – 'It is more important, rather than merely to teach facts, to teach people how to think, how to use facts, how to find facts, how to create new knowledge themselves.' 'If the primary function of the teacher is to issue an invitation to learning, and if the excellence of the teacher is a correlate of the persuasiveness and attractiveness of the invitation, then this function can never be superseded' (US Info. Service 1967, 15, 8).

Edwards, K.C., 1967   The broadening vista, *Geography* 52, 245–59.
Fisher, C.A., 1970   Whither regional geography, *Geography* 55, 373–89.
Herbertson, A.J., 1905   Major natural regions, *Geog. J.* 25, 300–12.
Marchant, E.C., 1968   Some responsibilities of the teacher of geography, *Geography* 53, 129–44.
Morris, J.A., 1966   Reality in geographical education, *Geography* 51, 87–98.
Newbigin, M., 1914   Ordnance Survey Maps (London).
Scott Keltie, J., 1913   *History of Geography* (London).
Spate, O.H.K., 1966   *Let Me Enjoy* (London), 121–30, 149–82.
US Information Service, 1967   *Teaching and Learning* (London).
Wooldridge, S.W., 1949   On taking the ge out of geography, *Geography* 34; 9–18.

c0512
Geography in modern education
L. OLIVEIRA   *Faculdade de Filosofia, Brazil*

When we think about the role of geography in modern education Piaget's name comes immediately to our mind. Because of the magnitude of Piaget's work no educator can ignore his conceptions. Although he is not a pedagogue and is not concerned with the application of his theory to educational problems, he has enriched teaching through his new conception of intelligence, stages of mental development, and the properties of logical thought. Piaget requires from us a reexamination of our own concepts about the learner, the learning process, and the nature of the subject matter. Thus, for the educator, it is important to know that cognitive structures are not fixed or given but developed through childhood and adolescence. Educators should use Piaget's system as a conceptual framework to reach after rationales and devise teaching strategies. In Piaget's system, knowledge derives from activity and not from perception. Therefore, activity is a significant requisite for mental development and opportunities to manipulate objects and ideas in active interaction situations with others should be provided for the student by the teacher. Of particular importance is the kind of material available and the way in which it is used.

On the basis of the previous considerations we should develop investigations in order to improve the quality of the teaching of geography. Devising experiments, through which to gain insight about how the child acquires the concept of geographic orientation, has great relevance for improving geographic education.

A series of experiments was carried out by the writer with children attending elementary school in a small town, in the State of São Paulo, Brazil, in 1971. The 180 subjects included in the sample were boys and girls between ages six and eleven from kindergarten through 4th grade.

The objective of the experiments was to measure the concept of relative location in terms of distance and orientation. The children in the experiment were exposed to a display of the devices constructed and asked to answer questions after the tester changed the relative locations of the devices. The display included three geographical situations: (1) a map of a town with houses, the church, the school, etc. (2) a map of the state of São Paulo with some cities, and (3) a map of Brazil with some states. The questions asked by the tester included, among others, the following terms: left and right; in front and behind; above and beneath; near and far. These terms define topological, affine, and metric spatial relationships.

The analysis of the children's responses suggests that the learning of relative location should be based on the concepts of spatial relationships. Thus, another study is being prepared trying to find an answer to the question of what are the mental operations needed by children to master the transition from spatial relations to relative location.

Aebli, Hans, 1956 *Didactique psychologique* (Paris).
Elkind, D. and J. Flavell, 1969 *Studies in Cognitive Development* (New York).
Flavell, John, 1965 *The Developmental Psychology of Jean Piaget* (New York).
Piaget, Jean and Barbel Inhelder, 1956 *The Child's Conception of Space* (New York).

c0513
Location as a factor in educational opportunity: some examples from England and Wales
ERIC M. RAWSTRON   *University of London, UK*

Rather more parents resident in Luton have stayed at school beyond the age of seventeen than in Bolton, yet the output of undergraduates from 1965 to 1969 inclusive was 68 per cent higher in Bolton than in Luton. At the 1961 census the East Riding and West Suffolk recorded 3 and 31 per cent respectively above the national mean for parental attendance at school beyond the age of seventeen, yet the East Riding sent 69 per cent more students with full value awards into higher education between 1965 and 1969 than West Suffolk. Indeed Bolton, which had substantially less than half of West Suffolk's quota of parents educated beyond seventeen recorded 42 per cent more full value awards per 1000 of its teenage population.

Locational differences in innate ability, in general environment, and in home background will each help account in some measure for these very wide discrepancies, but the major explanatory factor must be the comparative quality of the facilitative environment as between Bolton and the East Riding on the one hand and Luton and West Suffolk on the other. In short, the facilities for formal education in one pair of localities foster academic ability far more effectively than do the facilities in the other pair.

These contrasting but by no means unusual examples (Coates and Rawstron 1971, chap. 10; Taylor and Ayres 1969) starkly highlight a long neglected aspect of British educational planning, namely the geographical analysis of the provision and effectiveness of educational facilities. Without frequent geographical monitoring of the system it is impossible either to establish a sound base for planning or to check the results of past endeavours, and, without geographical analysis, it is quite impossible to ascertain whether the facilities provided by local education authorities, direct grant schools, and the private sector are reasonably close to creating equality of opportunity among local authorities. There is an enormous field for research here awaiting enquiry throughout the world and, to whet the appetites of potential researchers, it may therefore be helpful to devote the rest of this paper to a brief discussion of findings derived from a reconnaissance of education in England and Wales and to some suggestions for lines of research that may usefully be followed.

Opting out of the maintained sector shows marked locational variations as follows:
(a) Private boarding schools are mainly found in lowland England, their greatest incidence being south of the Midlands and especially in the Home Counties.
(b) Private day schools are heavily concentrated in certain of the Home Counties and especially in Surrey, which is residentially the wealthiest county in Britain.
(c) Direct grant schools are mainly to be found in Lancashire, the West Riding and Cheshire. This distribution is probably the result partly of general regional enthusiasm in the fairly recent past for education and partly of a strong regional Roman Catholic urge to stay out of the fully maintained sector. Wealth is not a significant factor. The locational origins of schools in these three categories would make an excellent research topic in historical geography.

The numbers staying on at school and in further education establishments (FEE) cannot be assessed wholly satisfactorily from published statistics because FEE data are lacking at local authority level. The locational conclusions that can confidently be drawn are: (a) the wealth of the southeast and the highly constructive parental attitudes in Wales promote proportionately large sixth-form populations in these areas; (b) southern England generally has higher proportions than the rest of England; (c) Lancashire and Yorkshire towns do rather better than midland towns; (d) Kesteven and the East Riding are examples of local authorities that appear unexpectedly high while East Anglian and certain midland counties appear unexpectedly low.

The locational analysis of those who remain in formal education from 16 to 18 inclusive is fraught with technical data problems. The scope for research is immense but the time taken in assembling really satisfactory data would be excessive unless official help were forthcoming. Since it is important that the variation from place to place in the proportion of each age-group remaining in formal education should be known and frequently monitored, the Commission might deem it worthwhile to press upon governments to take action to produce these and other educational data in forms designedly amenable to geographical analysis.

Higher education in Britain provides the most reliable data for geographical research. Division into male and female subheadings would be a very useful improvement, and a clear indication of the operation of the discretionary principle would permit a far more accurate assessment of local education authority attitudes than is now possible. Major findings are as follows:
(a) University catchment does not consistently reflect wealth, terminal age of education of parents, type of school, or efficiency of local education authority nor does it yield a smooth trend surface over England and Wales. Each major factor may be seen to operate more strongly in some areas than in others. Southeast England and Wales are, however, the leading areas in so far as undergraduate production is concerned.
(b) Entrants to colleges of education, when expressed as a percentage ratio of university entrants, yield a marked trend surface sloping outwards from London in favour of colleges of education. Perhaps one is getting a glimpse here of a trend surface of intelligence that future research might seek to verify and delineate.
(c) Awards to students for higher education

reveal, when mapped, a dim but sorry picture which must surely represent variable local attitudes – enlightened, fair, and generous administration in some localities contrasting with niggardly and backward administration in others. Unfortunately it is impossible from present data alone to be sure of more than a few of the most generous and a few of the most stingy authorities. Suffice it that the highly varied pattern is ample cause for concern, whether it is accurate in detail or not.

In addition to the suggestions already made for further research in Britain, research into the geography of education can be applied throughout the world and justified universally as a basis for effective planning and management. It should be emphasized in conclusion that the scope for research extends from the micro study of data for individual school catchment areas to macro analysis leading to the discernment of broad spatial trends.

Coates, B.E. and E.M. Rawstron, 1971 *Regional Variations in Britain* (London).
Taylor, George, and N. Ayres, 1969 *Born and Bred Unequal* (London).

c0514
The geography of education and educational planning
R.H. RYBA *University of Manchester, UK*

The geography of education is a generally neglected field of study to which increased attention, not least from geographers interested in educational planning, is long overdue. This paper attempts to elaborate the beginnings of a viable conceptual framework for such studies with particular reference to educational planning.

The relatively small role in educational planning so far played by geographers compares unfavourably with that already being played by experts in other social disciplines. The virtually non-existent contribution of a sub-discipline concerned with the geography of education, either within the field of education or within that of geography compares even more unfavourably with the already important contributions of educational sociology, the economics of education, and other similar areas of study. That the geographer's contribution to educational planning should be so limited is particularly paradoxical in view of the very high proportion of geographers in most countries who devote themselves on graduation either to education or to planning.

The development of the geography of education is particularly pertinent to the potential contribution of geography to educational planning. The dearth of interest revealed by academic geographers up to the present in the possibilities of the geography of education also contrasts unfavourably, as will emerge below, with the considerable but generally unsophisticated interest in its potential already evinced by experts in other related fields of educational enquiry (e.g., Bereday 1964, ch. 1).

Within the broad area of human geography, the scope of the geography of education overlaps, through its inextricable involvement in the fabric of society, with those of social and cultural geography. It is concerned with obvious questions related to the provision of colleges, schools, and other educational institutions. It should also be seen to extend to many other areas within the accepted purview of academic education, including the examination of geographical aspects of educational norms and attitudes, of the concept of educability, and of the related fields of educational opportunity and achievement.

In all these areas possibilities exist for studies throughout a range of scales, from macro-studies of world patterns, through studies at continental and national levels, to possibilities of regional, local, and even individual institutional levels. Moreover, educational phenomena are amenable to geographical studies not only of a static nature, whether related to present, past, prognosticated, or postulated theoretical areas, but also to themes related to the dynamics of developing geographical patterns.

These possibilities for the geography of education may usefully be classified within the framework of a threefold subdivision as follows: (1) the geographical examination of factors underlying education, (2) the examination of the spatial patterns of educational phenomena, and (3) the role of education as a factor influencing the geographical patterns of other social and cultural phenomena.

The remainder of this paper is concerned with some exploration of the possibilities in each of these categories, with particular reference to educational planning. It is illustrated as far as possible by pertinent studies already in existence and by suggestions for further research.

First, the geographical examination of factors affecting the development of education is of obvious relevance to educational planners concerned with the practical implementation of educational policies. Spatial variations inevitably exist in requirements for the provision of educational facilities and these have the same kind of differential relationships to aspects of the human and physical environment as is shown by other social phenomena which have long been subjected to detailed geographical examination. However, work in this field has so far been done mainly by non-geographers, notably comparative educationists and, while it demonstrates possibilities, it has been largely restricted to the application of concepts which most geographers would now consider to be outdated and crudely deterministic (e.g., Hans 1950, ch. 3). Thus the term 'geographical factors,' denoting essentially control by the physical environment, is not uncommon (Simons 1966). Yet fields of possible pertinent research are evident, concerned with geographical analysis of the relationship of educational phenomena to aspects of both the physical environment – e.g., climate and weather, relief and morphology, the availability of resources – and the human environment – e.g., patterns of population, social and economic organisation, and political and administrative structures. Thus possible studies to which geographers could make a valuable contribution include consideration of factors related to the development of distinctive national education systems in different parts of the world and to regional modifications made to these systems in the light of the different ecosystems of which they form a part. Similarly, as some existing studies show, local and regional problems related to the location of schools and colleges and to the delimitation of their catchment areas offer scope for further useful work (e.g., Philbrick 1949; Marklund 1969; Yeates 1963).

Second, problems related to the nature of spatial patterns created by educational phenomena themselves may be seen to offer a most fruitful and pertinent field for research. These patterns should be part of the basic social data upon which realistic educational planning is built and, although educational planners have largely neglected them until recently, they are becoming increasingly aware of their importance, for example, in France and the Netherlands (OECD 1969, vol. 2, ch. 3; Ruiter 1969). The first need in this area, the mapping and analysis of existing official statistical data related to education, has barely begun. Yet what has been attempted has already proved to be remarkably revealing, despite the limited nature of data used (e.g., Ferrez 1961; Gautier 1964; Goldblatt 1968; Coates and Rawstron 1971, ch. 10).

The collection and mapping of further data and their analysis and explanation offer an area of study of great importance as yet relatively untouched by geographers. Thus, educational values and attitudes vary spatially not only from country to country but also within countries. Sociological, psychological, and comparative studies already undertaken reveal some of the possibilities of this field of analysis (e.g., Husen 1967, vol. 1, ch. 6). The area of educability is a no less promising field for geographical work. Thus, for example, psychological enquiries have referred to intriguing spatial variations in the distribution of intelligence and personality characteristics (Vernon 1969; Price-Williams 1969). Similarly sociologists, uncovering variations related to social class, have noted considerable inter-urban, urban/rural, and regional disparities (e.g., Moser & Scott 1961; Wiseman 1964; Taylor & Ayres 1969). These and similar findings point the way to geographical studies of a more systematic nature which could be of considerable significance in the detailed planning of the location of educational provision.

Also, within this proposed second subdivision of the geography of education, are questions relating to the spatial patterns of educational opportunity and achievement. These are difficult to separate from each other since the latter depend to a large extent on the former. Some material is available with regard to spatial aspects of the provision of educational facilities, both institutional and curricular, and enormous, perplexing, and, from the point of view of educational planning, highly significant variations in space have already been revealed (Conant 1967; Geipel 1968; Coates & Rawstron 1971). Yet, once again in these areas, even preliminary description and mapping of data, let alone their analysis and explanation, have hardly begun.

Turning finally to the effects of education upon the geography of different parts of the world, these may be seen to be essentially concerned with education's part in forming individual and group perceptions of environments and, consequently, of the uses to which these can be put. Such a view, exemplified, for instance, by the role of Danish folk high schools in the development of Denmark since the mid-nineteenth century, lies at the heart of spatial differences of political, economic, social, and cultural phenomena over the earth's surface. Seen in this way, the educational process appears to be a shaper of intellectual capital or resource which, when acquired, is applied differentially through time and space on given environments. Economic aspects of education, looked at from this point of view, have been central to much work related to educational planning, especially in the last decade (stemming from Schultz 1961). Geographical aspects, already touched on in some of the work done by economists and educational planners, should also prove to be of interest and value (Harbison & Myers 1964).

Thus, in conclusion, within each part of the conceptual framework suggested in this paper for the geography of education, numerous possibilities appear to exist for research related directly or indirectly to matters of undoubted concern in educational planning. Some of the avenues suggested are already beginning to be explored but there would appear to be a need for much more emphasis on this field.

Bereday, G.Z.F., 1964 *Comparative Method in Education* (New York).
Coates, B.E., and E.M. Rawstron, 1971 *Regional Variations in Britain: Studies in Economic and Social Geography* (London).
Conant, J., 1967 *The Comprehensive High School* (New York).
Ferrez, J., 1961 Regional inequalities in educational opportunity, ch. 3 in A.H. Halsey, ed., *Ability and Educational Opportunity* (London).
Gautier, M., 1964 La répartition des effectifs scolaires en France, *Annales de Géographie*, 73, 42–66.
Geipel, R., 1968 *Bildungsplan und Raumordnung* (Frankfurt-am-Main).
Goldblatt, P., 1968 The geography of youth employment and school enrollment rates in Mexico, in A.M. Kazamias, and E.H. Epstein, eds., *Schools in Transition: Essays in Comparative Education* (Boston).
Hans, N., 1950 *Comparative Education* (London).
Harbison, F., and C.A. Myers, 1964 *Education, Manpower and Economic Growth: Strategies for Human Resource Development* (New York).
Husen, T., ed., 1967 *International Study of Achievement in Mathematics: A Comparison of Twelve Countries* (New York).
Marklund, S., 1969 School organisation, school location and student achievement, *Internat. Rev. of Educ.*, 15, no. 3, 295–320.
Moser, C.A. and Scott, W., 1961 *British Towns* (Edinburgh).
OECD, 1969 *Reviews of National Policies for Education; France.*
Philbrick, A.K., 1949 *The Geography of Education in the Winnetka and Bridgeport Communities of Metropolitan Chicago*, Res. Paper 8, U. of Chicago.
Price-Williams, D.R., ed., 1969 *Cross Cultural Studies*, Penguin Modern Psychology Readings (Harmondsworth).
Ruiter, R., 1969 Education and manpower forecasts, *Planning and Development in the Netherlands*, 3, no. 1/2, 66–183.
Schultz, T.W., 1961 Investment in human capital, *Am. Econ. Rev.* 51, 1–17.
Simons, M., 1966 What is a geographical factor?, *Geography*, 51, pt. 3, 210–17.
Taylor, G., and N. Ayres, 1969 *Born and Bred Unequal* (London).
Vernon, P.E., 1969 *Intelligence and Cultural Environment* (London).
Wiseman, S., ed., 1964 *Education and Environment* (Manchester).
Yeates, M.H., 1963 Hinterland delimitation: a distance minimizing approach, *The Professional Geog.*, 15, no. 6, 7–10.

c0515
Geography teaching and the rising level of expectations
NEVILLE V. SCARFE *University of British Columbia, Canada*

The purpose of this paper is to discuss the role of geography teachers in preparing future citizens to live in a world of rising expectations. How can a study of geography prepare young people to understand the political, economic, and social aspirations of the world's people?

Communication has made more people more acutely aware of the disparities between wealth and lack of it around the world. Young people in school will need to understand this unequal distribution of the world's resources. They will have to learn about those parts of the world which are overcrowded and those which are not. They will have to know where poverty and deprivation seem to be at a maximum and those where affluence is more prevalent.

It will be totally inadequate simply to describe the world's resources and its population distribution. It will be necessary to understand why things are as they are. Furthermore, children will want to know how greater equity and justice can be accomplished peacefully. In school, therefore, we have to deal with the problems of prejudice, ideological difference, misrepresentation of the truth, and the whole question of factual recording and the dangers of ignorant opinion.

If the world is to remain peaceful and there is to be a universal decrease in suffering then the necessary changes will certainly have to be carefully thought out, because every change has a tremendous number of unfortunate implications, ramifications, and side issues.

There is no way in which the majority of the people in the world will accept the present status quo. There has to be some raising of the level of the standards of living of all the poor. They have a militant and rising level of expectation.

On the other hand, there is no way in which we can get all the world's people to accept total equality. People are different. They have different expectations. Moreover, the world needs a great diversity of skills and abilities. The moment we encourage individuality so we encourage diversity and therefore inequality.

In our schools controversial issues must be raised in considerable detail. But this must be done, so far as possible, by impartial teachers who are prepared to see that all sides of a question are available to children. Young people must not be indoctrinated with biased views, but permitted to make up their own minds on the basis of full access to all the information and all the opinion on both sides of all questions. Unless there is freedom of access to all facts and all opinions true education ceases.

The teacher, therefore, in school is very much concerned with the development of humane people who have the milk of human kindness in their blood, who are prepared to try to understand all people, their motives, their attitudes, their prejudices, their needs, their aspirations, and their problems.

Geography deals with man's combat with nature. When man and nature come into conflict nature tends to win and human beings are forced to behave more sensibly. If they over-graze a hill slope then the soil is washed away. If they cultivate a dry land then the wind and the dust storms come. Because nature tends to win there are plenty of examples of human beings behaving in a very sensible fashion in adapting their lives to the laws of nature. They dress in warm clothing in Winnipeg in winter. They heat their houses and keep them insulated.

Geography teachers have to make sure that there is a balanced study of the world, even though they are dealing more and more with controversial issues, the problems of underdeveloped nations and deprived people with poverty and disease, with lack of resources and population pressures.

Controversial issues are not items in a program or curriculum. They are major ideas that grow slowly through the years as we accumulate data and information and start asking questions. A controversial issue is not something that is discussed this week and forgotten the next, not something that is fully analysed and discussed at any one time at all. It is a growing and developing complexity to which we add an increasing number of variables. Controversial issues cannot be discussed in any reasoned or objective way until one has accumulated a considerable amount of data of a descriptive nature. We have to know something of the thought processes, the methods of enquiry, and the points of view adopted by geographers, historians, sociologists, and psychologists long before we can effectively analyse the major political or economic issues of the world, especially if they involve a great deal of prejudice, emotion, or ideological bias.

Children come to school with all sorts of unconscious prejudices and poorly reasoned points of view. Until they start questioning their own acceptance of certain postulates or ideas they are not easily able to investigate more complicated problems in a rational fashion. The question, therefore, must always be asked, 'At what developmental level do we raise significant political issues for full and meaningful discussion?'

One of the objectives of education is to arrive at reasoned generalizations, usable principles, and apt guidelines about living on the earth. In doing this children are helped and encouraged if they devise models or invent games that will free their minds from the need to memorize masses of detail.

We may have to teach the skills of model construction, of map-making or gamesmanship just as we have to teach some of the skills of painting or drawing or sculpture but we always allow children to paint their own pictures, devise their own models, or invent their own games. Children tend to learn by doing, that is, by discovery, by planning, by expressing their ideas, and by creative and inventive performance. Inductive investigation tends to precede deduction.

Geography has sometimes been described as human ecology and, therefore, is very much concerned with the environment in which man lives and has his being. One of the most urgent challenges, in fact, now facing mankind is to halt and reverse the growing deterioration in the environment. The biosphere, that thin layer of air, water, and soil that surrounds our planet and on which all life on earth depends is in danger of destruction through abuse and neglect. Every country large and small is confronted with pollution hazards.

The problem now is how to alter man's activities so that the earth is not rendered uninhabitable. Pollution may be defined as being artificially induced deterioration of the natural environment – air, water, or soil. It is a health problem, an economic problem, it is a problem in conservation of natural resources, it is an aesthetic problem. It is also a geographic problem.

The central core of geography is an analysis of the idea that environmental conditions of the place where people live matter to those who live there. Furthermore, a study of these conditions always involves the primary consideration of air, water, and land at that place. If pollution means anything, it means destruction of the physical conditions necessary for human existence.

Although geographers cannot emulate the predictive ability of the physical scientist they can, nevertheless, imitate some of the technology used. We can, if we are careful, use operations research models and geographical games, provided that we allow children to construct their own models and devise their own games.

It is not likely that through the testing of explanatory hypotheses it will be possible to develop a controlled, consistent, and rational explanation of events. Even if such a thing were worthwhile it is not necessarily the most important task facing geographers. For we cannot make the assumption that man is predictable, rational, and behaves in a measurable manner in an objective world. Rising expectations are not necessarily rational or realistic.

The price of the employment of models is eternal vigilance since we must always try to look beyond the model to the reality that it is supposed to illuminate. Maps, models, and other abstract symbols can be very easily deceptive and lead us to a study of cartography or model construction rather than the real world that they are supposed to represent.

Models and games have a considerable quantitative flavour and content but geography must also be qualitative and lend itself to personal interpretation and comprehension. The quantitative and qualitative approaches to geography can and should illuminate and support each other. Geographers have always combined the study of particular places with the formulation of theoretical concepts. All geography students should receive regular training in quantitative methods and in model design but they should not be allowed to think that these are geography.

Geography as such does not necessarily produce humane, understanding people. Recent research has shown that the widely held belief that geography automatically assists in furthering citizenship upon an international scale has little justification. It does not, necessarily, generate goodwill by itself. The attitude of the person who teaches geography is much more significant than is the subject itself in developing an international outlook. Without a teacher who is himself humane in his understanding of human misery and who expresses through his own personal behaviour sympathy with the rising aspirations of people round the world we will not have an education directed toward that end.

If well taught, however, geography can encourage the idea that it is totally futile for men to be fighting men. Human beings must collaborate one with the other to win their 'daily bread' from the opportunities provided by nature. They may have to work hard to secure this advantage but the more conscientious the collaboration the more likely that world resources will meet the rising expectations of those who are now deprived.

Without a prior knowledge of the major geographical concepts, acquired by long, continuous, and diligent study, discussion of political issues is a futile exercise of prejudice exchange.

Burrows, H., 1923   Geography as human ecology, *Ann. Assoc. Am. Geog.*, 13, 1–14
Guelke, Leonard, 1971   Problems of scientific explanation in geography, *Can. Geog.*, 15, no. 1 (Spring), 38–53.
Harvey, David, 1969   *Explanation in Geography* (London).
Scarfe, N.V., 1970   Education for value judgement, *Education Canada*, 10, no. 2 (June), 12–14.
– 1971   Games, models, and reality in the teaching of geography in school, *Geography*, 56, pt. 3 (July), 191–205.

c0516
An assessment of learning in geography in relation to Piaget and Bruner
F.A. SLATER   *University of Otago, New Zealand*

The work of two modern psychologists, Jean Piaget and Jerome S. Bruner, provides theoretical bases for evaluating levels of learning in geography. Piaget's theory of intellectual development informs teachers of thinking and reasoning abilities students may be expected to attain; much of Bruner's work focusses attention on the value of presenting the underlying concepts of a discipline (Piaget 1926, 1952; Bruner 1960, 1966).

Piaget's investigations show that intellectual functioning is a special case of adaptation to environment and that intellectual growth develops in a series of stages. A sensory-motor aspect of thinking characterises the earliest years; a period of pre-conceptual thought exists between about two and four years; intuitive thinking dominates from four to seven years when the child sees the world from a single point of view only, his own, and his thinking is bound closely to his own experiences and movements. Piaget specifies a stage of concrete operations from about seven to eleven years when the child may achieve understandings of interrelationships through direct manipulation of the variables involved. Finally, onwards from eleven years, the capacity for formal thinking, the ability to reason logically and in abstract concepts is developed.

An important hypothesis presented by Bruner suggests that the structure of a discipline, the fundamental underlying ideas of a discipline, may be taught in some intellectually honest form to children at any grade level. Bruner is concerned that learners be given a sense of fundamental ideas in a discipline so that these can be transferred to new elements and problems in the subject as these are confronted.

As a child passes through stages of intellectual development and as he is exposed to curricula emphasizing the structure of a subject, one could expect enquiries and responses of a more abstract and ordered type. If this supposition were investigated in the subject of geography would such an intellectual development be revealed? Does geography, as it is taught, engage the formal and logical thinking abilities of high school children and impart an appreciation of an identifiable structure of basic concepts? (Slater 1969)

To a sample of 3000 children, aged 8–18 in grades 3–12 in city, town, and rural schools within 240 kilometres of Iowa City, Iowa, USA, the question was put: 'You want to know what this country is like (this country being an unnamed outline map in front of the classroom). What questions would you ask?'

The open-ended question used above was chosen in preference to a more structured schedule for two reasons. First, the open-ended question, pioneered in sociological research, has had a great deal of success: respondents have moved readily from specific to more general and abstract responses in the test situation (Goode & Hatt 1952). Second, an open-ended question ensured a lack of bias which would have been very difficult to eliminate from a structured test.

The responses were classified into three categories, based on the content nature of the responses, the level of concept development, and the structural framework in which the responses were placed. The variation shown through the grades was analysed in a series of statistical tests. The results of the tests were then evaluated with reference to the work of Piaget and Bruner.

Analysis of variance tests revealed statistically significant differences at the .05 probability level in the *same* type of responses between each grade and between *different* types of responses within each grade.

Variation and development in levels of learning in geography however, could be influenced by variables other than intellectual growth and concept attainment. Eight independent variables, socio-economic status; intellectual ability; level of attainment measured by IQ tests and the geography scores of the Iowa Tests of Basic Skills and the Iowa Tests of Educational Development; sex; rating of instruction in geography in terms of the amount of time devoted to geography; the textbooks used; the type of curriculum adopted by the school; and the qualifications of the teacher were selected and evaluated in order to account for the variation or to control for their effects so that any remaining variation could be examined in relation to Piaget's theory and Bruner's hypothesis.

Factor analysis was used to remove colinearity among these independent variables and three main factors were resolved from the analysis, an instructional rating factor, a socio-economic-intellectual achievement factor and a sex factor. The factor analysis output, also provided factor scores which were the control variables used in a series of covariance analyses. The covariance tests demonstrated a change in the relationships between the variables as the level changed. The variables were subject to different functional relationships within the groups.

The instructional factor showed the strongest relationship to the type of responses, which itself was most closely related to the curriculum component of the factor. The second factor combining socio-economic status and mental ability and attainment showed a significant relationship only one-sixth of the time. The implications become clear when the variation in the type of responses is assessed in Piagetian and Brunerian terms. If the design of the curricula being used do not take these theories into consideration then it is likely that learning remains at a concrete level and a relationship between ability and learning in geography is unlikely since the curricula have not provided experiences which lead the pupils to a higher level of thinking whether they are of high or low ability. The third factor, sex, was of no statistical significance in differentiating the level of learning being achieved.

The examination of the residual variation in the responses throughout grades and within grades did not exhibit a pattern which could be related to Piaget's theory of concept development. This suggests that the curricula and classroom techniques used did not provide students with the opportunities to use abilities which Piaget's theory suggests from eleven years onwards.

There was no evidence either to suggest that geographically accepted concepts such as spatial distribution, areal association, and spatial interaction were being taught, though these had been emphasized at the Iowa Geography Teachers' Spring Conference for more than a decade. Students were not receiving instruction which makes the principles of research and analysis available. Underlying concepts in geography are not taught as a template to be placed over occurrences of different phenomena.

The research findings revealed a lack of

differentiation and development in the thinking patterns of the test respondents when these were analysed in relation to Piaget's theory and Bruner's hypothesis.

Clyde F. Kohn.

Bruner, J.S., 1960 *The Process of Education* (New York).
– *Towards a Theory of Instruction* (Cambridge, Mass.).
Goode, W.J., and P.K. Hatt, 1952 *Methods in Social Research* (New York).
Piaget, J., 1926 *The Language and Thought of the Child* (London).
– 1952 *The Origins of Intelligence in Children* (New York).
Slater, F.A., 1969 An inquiry into the kind of geography being learned in grades 3–12, unpublished PH.D. dissertation, U. of Iowa

c0517
Student-teacher involvement in creating new regions of the United States
HENRY J. WARMAN *Clark University, USA*

This paper represents an account of college student-professor involvement in a regional course which had as one of its goals the acquisition of information about divisions and subdivisions of the United States. The major aim of the course was *to create regions and sub-regions* of the United States, not simply to read regional accounts of various authors. Training in regionalizing, solving problems of technique, introducing regional planning and developmental needs of selected parts of the United States were integers of the course.

The technique involved a 'trilogy' of geography – human factors, physical factors, and cultural factors. All three were to be used whenever possible in labelling each region and sub-region. Expressions such as the northeast, the northwest, central interior were outlawed, regarded as merely 'locating labels.' Terms like Lone Star State, the Middle Atlantic, and The Dakotas could not be used in characterizing the regions.

Physiographic regional compartments, one of the trilogy requirements, the physical one, were useful albeit they often stressed but one item. For example, the Atlantic and Gulf Coastal Plain province certainly had a single attribute 'uniform factor.' However, upon closer scrutiny, this plain has climate sub-regions ranging from dry to humid to wet and very wet. It also has crowded places (megalopoles) as well as sparsely inhabited sections. There is also a wide range of cultural differences – economic, industrial, agricultural, religious, social, and even behavioural. Add to these the varying population characteristics of age, sex, race, colour, stature, health, and creed.

One major requirement was to be met by each student: to produce on a large map of the United States as many regions and sub-regions as he (or she) could create. Three basic references were used: the newly published National Atlas of the United States, the Oxford Regional Atlas of the United States and Canada, and the atlas accompanying the Alpha set of Transparencies of Mid-Continental United States. An admonition at the start of the project was, 'If two students have final maps which are alike, even in a few parts, both students "flunk the course".'

Books were pored over but periodicals, papers, magazines, government reports, newspapers, and especially the large map resources of the Clark Geography School library were combed for information pertaining to human, physical, and cultural accounts of the United States. Some students criticized regional accounts which included no maps.

The ensuing exchange of ideas – defending, describing, explaining, and sometimes altering the 'trial regions' – opened the door to a freedom which at later times the instructor characterized as abandon. Caution coupled with integrity, not speed with carelessness, was suggested.

Halfway through the course each student wrote his justification for three regions including their respective sub-regions. Significantly, each student soon knew precisely where he stood in the evaluation of his own work and also in comparison with the application and progress made by others.

Films, field trip slides, and field studies were reviewed. Emphasis changed from securing information to analyzing and evaluating it, then compartmentalizing it. The composing of clear, concise regional titles proved to be an excellent device for highlighting the greater features and subordinating the lesser details.

The 'handouts' of created titles and map segments are illustrative of how detailed and diverse the regions became. They are lifts from the western and eastern United States regions of Mr Lee Markscheffel.

At the class session just preceding the final examination the instructor, in order to review the Trilogy Method, asked the students to select and list on separate sheets of paper those regions and sub-regions falling under one of these three headings: (a) regions and sub-regions exemplifying a trilogy – a balance of relationship of human, physical, and cultural factors; (b) regions and sub-regions showing by title a dualogy-relationship of human-cultural, or physical-cultural attributes; and (c) those showing by title but one dominant feature – human, physical, or cultural. Of the 15 major titles, 7 met the trilogy, 8 met the dualogy test. Only sub-regions appeared under the single factor heading. Drawing of relationships, was, for the most part, satisfying.

While there are quite a number of features or items which might be termed 'bad' about this regional involvement course, such as 'too map-oriented,' and a growing monotony especially in digging for subregions and sub-subregions bordering upon the inconsequential, the 'good' features seem to outweigh the 'bad.'

1 All students really were involved.
2 Involvement was independent, yet co-operatively experienced and shared.
3 Involvement was tangibly expressed; the students knew where they stood at all times.
4 No student failed.
5 Satisfactions with results (i.e., regions with titles that really seemed to fit) offset the frustrations.
6 Each student was continually self-evaluating, as well as comparing and ranking his performance and products with his peers.
7 The integrity of sources, mainly field studies, especially mapping, was realized and respect for excellence grew.
8 A solid acquaintance was established with the vast store of literature and the geography experts in regional work.
9 Incisive looks were directed at areas where new or more research is needed; where old research needs redoing or updating; where regional planning seems desirable, even imperative.
10 Desire to do field work was fostered. A common expression was 'I'd sure like to go take a look at ———.'
11 The social and behavioural actions and patterns of the human beings occupying the regions came into discussions frequently, indicating that this kind of geography is currently a student concern as well as that of the profession itself.
12 Several major concepts permeated the study; others grew in realization and scope as the course continued. Among these were: (a) the 'areal' concepts of distributions, similarities, differences, uniqueness, and especially relationships; (b) those of scale and methodology including map kinds and sizes, creating of such, and of regionalizing; (c) the relative location concept – or as Meredith Burrill puts it, 'the fifth dimension, that of context' also come before the regionalizers; (d) Human-cultural concepts, namely, culturally-defined resources, man – the ecological dominant, and that of comparative advantage laws being exercised, were inherent in the program; (e) Somewhat obscure, yet capable of being built into the study by some, were those dynamic concepts of change, choice, and spatial interaction.
13 One concomitant of the study, expressed in different ways by several participants was the knowledge gained about one's own country, its vast resources, the many problems present, and the great possibilities offered with a democratic frame of government. Students' opinions were expressed with a competence to do so; there was no 'discussion in a vacuum.'
14 Finally, one of the end-products, namely, the maps created to express knowledge now have become maps which may be used to learn from. This may be done, however, only if the professor can convince the students to give them up!

Humid Tropics
Les tropiques humides

c0601
Adjustments of small-scale farmers to the abandonment of tropical estate agriculture: the case of Nevis, West Indies
DAVID WATTS   *University of Hull, UK*

Nevis, in the Leeward group of the West Indies, presents an interesting case of recent agricultural adjustment following the demise of its once all-important plantation crop, cane-sugar. Formerly one of the richest islands in the Caribbean, it is now among the poorest, with a population increasingly dependent upon subsistent, rather than commercial agriculture. Although some traditional areas of small-scale crop planting for marketing have been maintained, the most recent trend in Nevisian agriculture has been towards developing herds of poor-quality small stock (sheep and goats), which are allowed to graze on old abandoned estate land.

Approximately one-third of this small, almost circular island (93km^2 or 36 mile^2 in area) is largely untouched by man, being rough mountain land covered by geologically-recent lava flows from its dominant and centrally-located relief feature, the dormant volcano Mt Nevis (1000m in altitude). Only a small proportion of the remainder (ca. 20 per cent) currently lies under cultivation, the most intensive area of cropping being in the south-central sector (Gingerland). Rising towards Mt Nevis from the coast over the whole of the rest of the island is former estate land, now covered to a greater or lesser extent by rapidly spreading *Acacia* scrub. Soils throughout the island are generally thin and stony, and sometimes poor in nutrients; frequently they are also extremely porous, and this, coupled with the generally low annual rainfall (75cm on coastlands, rising to 100 cm at the upward limits of cultivation) and a proneness to periodic climatic drought, makes for generally dry surface conditions. It is, accordingly, a difficult agricultural milieu. Nevertheless it supports an island-wide population of approximately 18,000, or just under 200 per km^2, most of whom reside in small rural villages established usually on the poorest quality land.

The origins of estate land in Nevis may be traced, as in so many West Indian islands, to the introduction in the 1650s of cane sugar as the prime commercial crop. The prosperity brought by this commodity was maintained throughout the seventeenth and eighteenth centuries (Merrill 1958). But subsequently, problems associated with the fierce competition from elsewhere, difficulties of trade and shipping, a decrease in the amount of capital reaching the island, absentee landlordism, and some soil erosion induced by poor planting methods, brought the beginnings of a decline in cane agriculture during the 1800s. Within the present century, increasing costs of production on small island units such as Nevis have made cane cultivation ever more uneconomic, so that by 1953 many estates no longer planted the crop. At this time only two sugar processing factories remained in operation on Nevis, and much of the cane cultivation had been taken over by small-scale farmers (i.e., those usually owning or working less than 4ha of crop land), whose labour costs were minimal. Both factories closed shortly afterwards, following which cane was sent for processing to the adjacent island of St Kitts, until the ferry costs of transportation across the short stretch of water separating the two islands rose so greatly as to make this also an uneconomic proposition. The consequence is that almost no cane is produced in Nevis today. Meanwhile on estate land, Sea-island cotton had replaced cane to some extent (there were 1600ha of cotton in 1958, cf. 320ha of cane), but this was never a healthy crop, and its cultivation also ceased during the 1960s, when most of the estate land was effectively abandoned, to become ripe for the colonization of weeds such as *Acacia*.

Bereft of their former income from cane, and unwilling to take a gamble on cotton, small-scale farmers have recently begun to seek other means of financial support to supplement that obtained from the sale of their traditional tree and ground crops, such as papayas, limes, bananas, plantains, yams, and sweet potatoes. This produce has always been marketed both in Charlestown, the capital of Nevis, and (in greater quantities) on the quayside at Basseterre, St Kitts. Usually the women of the farming families, termed hucksters, are responsible for marketing these crops, and in so doing have come to play an important role in the social and economic structure of the island.

Within the last five years the main search for additional prosperity has centred around the acquisition of small stock, which is grazed on abandoned estate land. The latter has indeed to a large extent come to be regarded as common grazing land. Not all of this is however available as such, for some has been bought or retained by local entrepreneurs (e.g., Prospect estate in the southwest), or by the island government (e.g. Indian Castle estate in the southeast), in attempts to establish high-quality dairy farms, using Zebu and Holstein crosses. But even so, approximately 95 per cent of former estate land is used by the small-scale farmers. Herds of small stock have been acquired either through direct purchase, or through the 'caring' system, in which those who do not initially have enough money to buy stock will look after the stock of other farmers, on the understanding that they will receive a share of the profits when the stock is sold. In this way they may build up cash reserves of their own. Usually these reserves, and also the profits accrued from the sale of stock, will be put back immediately into the purchase of more stock, except where immediate debts need to be paid. Slaughtering is usually a haphazard business, for the farmer usually regards the small stock as representing solely a source of capital, which is to be converted to cash only when it is absolutely necessary to do so, though it is the case that some more progressive owners do sell regularly to markets in Nevis, St Kitts, Antigua, or Montserrat. These developments have meant the very rapid build-up in numbers of small stock throughout the settled part of the island, particularly since 1968.

Little is known concerning the numbers of small stock currently present in the island. The most recent governmental agricultural census indicates that 9399 sheep and 5661 goats were present in 1961, but it is likely that these figures have now been greatly exceeded. Although a full analysis has not yet been completed, the preliminary results of a small-stock census undertaken by the author in 1970 support this latter proposition. This census records some remarkable local grazing densities, particularly along the southern and eastern coasts, where individual herds of over 50 sheep and 35 goats are not unusual, and the small stock density may reach over $175/500km^2$. Although some of this stock is herded into *Acacia* pens each night as a protection from ravaging dogs, most represent a highly mobile population, which makes its way back towards individual village folds at night, returning unguided to the grazing areas the following morning. Much smaller numbers of cattle, donkeys, horses, and some pigs are also present on the 'common' land.

It is difficult to forecast what the ultimate consequences of these changes in land use might be. But already there are some signs that overgrazing, soil erosion, and a rapid spread of *Acacia* will result from such a major increase in small-stock numbers. Indeed the changes may in time add to, rather than diminish, the problems of the small-scale farmer in this island.

The Caribbean Project, McGill University.

Merrill, G.C., 1958 *The historical geography of St Kitts-Nevis* (Mexico City).

Man and Environment
L'Homme et son milieu

---

c0701
Air quality control legislation
WILFRID BACH *University of Hawaii, USA*

The major objective of any air pollution legislation should be to maintain air that is as clean as possible but not dirtier than a specified quality (Wohlers et al. 1969, 58). Legal goals, more specifically, should then be (1) to preserve the health and welfare of man; (2) to protect plant and animal life; (3) to prevent any damage to physical property; (4) to ensure visibility for safe air and ground transportation (Edelman 1968, 592); and (5) to guarantee an esthetically pleasing environment.

Legal action to control air pollution should be guided by the following axioms: (1) Control actions should be initiated before an air pollution disaster has occurred. (2) Prevention of air pollution emissions rather than establishment of costly monitoring and warning systems should be the major control criterion. (3) State and local control agencies should seek the pollutors' and the pollutees' co-operation in implementing air quality standards (Schueneman 1968, 720).

Legal authority to control air pollution is usually vested in municipal, county, and state agencies. The federal government, however, has often found itself compelled to take action when either a local agency failed to act because of conflicting interests, or when local authority was non-existent. It is thus clear that control legislation for air pollution is necessary. Constant public awareness, supervision, and pressure are essential in the enforcement of such legislation.

In 1306, during the reign of Edward I of England, the first smoke abstainment law prohibited the use of 'sea coal' because of its deleterious effects on health (Air Conservation 1965, 212). In 1307, a violator of this law was executed. In later years prosecutions became apparently more humane, but the inhumane use of the air as a giant sewer continued unabated. In 1863, when the Alkali Act was passed in Britain, the world had its first comprehensive clean air act controlling emissions of offensive gases, smoke, grit, and dust.

In the US the first federal air pollution control program was passed in 1955 (Environmental Quality 1970, 73). The major components of this act stipulated that the federal government would provide research and technical assistance, but that states and local governments would be responsible for local air pollution control. Federal technical assistance and training has been provided under the auspices of the Department of Health, Education and Welfare (HEW) through the Division of Air Pollution, later renamed National Air Pollution Control Administration (NAPCA). In 1970 the agency was renamed Air Pollution Control Office (APCO), administratively operating now within the newly established Environmental Protection Agency (EPA).

The Clean Air Act of 1963 promulgated that federal enforcement authority should supervise control programs of inter-state air pollution problems. The amended Clean Air Act of 1965 regulated nationally air pollution from new motor vehicles starting with 1968 models.

It was not until the passage of the Air Quality Act of 1967 that a major breakthrough occurred with the acceptance of responsibility by the Federal Government if local control agencies failed. The major objectives of the 1967 Air Quality Act were 'to protect and enhance the quality of the Nation's air resources so as to promote the public health and welfare and the productive capacity of its population' (Guidelines for the Development of Air Quality Standards and Implementation Plans 1969, 2); and 'to insure that air pollution problems will, in the future, be controlled in a systematic way' (Air Quality Act of 1967, 1967, 13).

The major steps for action included (1) the establishment of 91 Air Quality Control Regions (as of August 1970); (2) the promulgation by HEW of Air Quality Criteria for the five pollutants suspended particulates, sulfur oxides, carbon monoxide, hydrocarbons, and photochemical oxidants, and of control techniques; (3) the establishment of air quality standards after public hearings; and (4) the submission of implementation plans to HEW for approval.

The nation was in the midst of establishing Air Quality Standards, when on 30 December 1970, the President of the US signed into law the Clean Air Amendments, which will have a considerable impact on the nation's air quality at least throughout the 1970s.

As one of the major deviations from the 1967 Air Quality Act, under the new law the Environmental Protection Agency (EPA) has now proposed national primary (health related) and secondary (welfare related) ambient air quality standards (AQS) plus recommended measurement techniques by 1 February 1971 (Federal Register 1971, no. 21). By 1 April 1971, EPA completed the designation of any new air quality control regions. Under the new law no public hearings are required for setting AQS as long as the national AQS are met. This may be beneficial for highly polluted areas, but it is unfavourable for relatively clean areas, such as Hawaii, whose pollutant levels fall far below the national standards. The law gives states and political subdivisions the right, however, to adopt and enforce standards more stringent than the national standards.

Public hearings on the adoption of implementation plans are mandatory for the five previously listed standards, and for nitrogen oxides. EPA has just published regulations (Federal Register 1971, no. 67) according to which implementation plans have to be prepared and submitted to EPA by 1 February 1972. If EPA approves, primary AQS have to be attained by 1 June 1975, or, on a governor's extension request, by 1 June 1977. If EPA disapproves or a state fails to submit implementation plans, EPA prescribes implementation plans for attainment by 1 August 1975. Secondary AQS are to be attained within a 'reasonable' period of time, a rubber clause, which is one of the major weaknesses of the present law.

In addition EPA will propose performance or emission standards for each category of stationary sources by 1 October 1971; and it will prescribe hazardous emission standards by 1 May 1972. This most logical approach to air pollution control will minimize pollutant emissions and help prevent hazardous levels under episodal conditions.

Any person has the right to bring civil suit against any person or governmental agency that violates any emission standard under the Act. Any person knowingly exceeding the AQS is subject to fines up to $25,000 per day of violation and/or imprisonment up to one year or double the fines for second convictions. Any person knowingly falsifying information required under the Act will be subject to fines up to $10,000 and/or imprisonment up to 6 months. The word 'knowingly' provides a loophole which may lead to acquittals under the established injunctive court action. Under injunctive enforcement, ignorance of the law has to be considered and intent of the violator has to be proven. Misdemeanour enforcement would be preferable because it need not consider ignorance or intent, but only the fact of violation. District courts have jurisdiction to enforce all standards.

Under the new Act EPA prescribes emission standards for any class of new motor vehicles for air pollutants endangering public health and welfare. EPA will prescribe by 1 July 1971, the following emission standards for light-duty motor vehicles: for carbon monoxide (CO) 37 g/mile, for hydrocarbons (HC) 2.9 g/mile, and for nitrogen oxides (NO_x) 3.0 g/mile. This would mean a 90 per cent reduction of CO and HC from 1970 vehicle emission by the 1975 model year, and a 90 per cent reduction of NO_x from 1971 emissions by the 1976 model year (Clean Air Amendments of 1970, 1971, 16).

Effective with 1970 model year vehicles and engines manufactured after 1 March 1971, manufacturers must warrant that vehicles conform with applicable emission standards, are free of defects, and would not fail to conform to its useful life. Manufacturers or any other persons after the effective date of applicable standards of 1 July 1971 are prohibited from selling any new vehicles unless they comply with the standards, knowingly removing control devices installed, or selling or leasing any vehicle which does not comply with the standards. Any person in violation of the above is subject to a civil penalty up to $10,000.

No state (*except* California) may adopt or enforce standards applicable to new motor vehicles. States may, however, establish their own emission standards for older and used cars. States may regulate motor vehicle fuels and fuel additives only for attainment of federal air quality standards.

Regarding aircraft emission standards, EPA will hold public hearings on its own proposed standards, and it will issue regulations by 1 October 1971. In contrast to vehicle emission standards, aircraft emission standards will be enforced by the Department of Transport. No state or political subdivision may adopt or enforce any aircraft emission standard unless it is identical to federal standards.

The Clean Air Amendments of 1970 constitute a healthy start on the long road toward maintaining breathable air. For polluted areas the setting of national pollution limits may well result in a temporary reduction of air pollutant levels. For relatively clean areas recent promulgations of local air quality standards reveal an inclination of health departments to permit locally cleaner air to deteriorate to federally prescribed standards. Because of conflicting interests, state and local legislatures do not seem to be suitable vehicles through which legislation that protects the public health and welfare will pass. It is clear that only strong vigilance on the part of the citizenry and a federal supervising agency such as EPA can guarantee progress toward improving the ambient air. Real progress can be expected through federally sponsored economic incentives for industry to install control devices; for oil companies to develop new fuels; for automobile manufacturers to develop new propulsion systems, e.g. electricity, gas, steam, or hybrid; for urban developers to construct rapid transit systems, etc. Economic incentives by themselves, however, are often not sufficient. A legal catalyst is still required to trigger control action.

National Science Foundation.

Air Conservation, 1965   Report Air Conservation Commission of the AAAS, Pub. no. 80, Washington, DC, p. 212–33.
Air Quality Act of 1967, 1967   Report of the Committee on Interstate and Foreign Commerce, House of Representatives, Washington, DC.
Clean Air Amendments of 1970, 1970   Public Law 91-1783.
Edelman, S., 1968   Air Pollution Control Legislation; in A.C. Stern, ed., *Air Pollution*, 2nd ed., 3 (New York), p. 553–7.
Environmental Quality, 1970   *The First Annual Report of the Council on Environmental Quality*, chap. 4, 'Air pollution,' US DHEW, PHS, p. 61–91.
Federal Register, 1971   36 (21), pt. 2, Environmental Protection Agency, Washington, DC.
Federal Register, 1971   36 (67), pt. 2, Environmental Protection Agency, Washington, DC.
Guidelines for the Development of Air Quality Standards and Implementation Plans, 1969   US DHEW, PHS, Washington, DC, 53p.
Schueneman, J.J., 1968   Air Pollution Control Administration; in A.C. Stern, ed., *Air Pollution*, 2nd ed., 3 (New York), p. 719–96.
Wohlers, H.C., et al., 1969   Can air pollution be controlled by legislation? *Scientia* 104 (681–2), ser. VIII, p. 58–64.

c0702
Human adjustment to agricultural drought in South Australia
R.L. HEATHCOTE   *Flinders University of South Australia*

This paper presents the preliminary results of research undertaken as part of the Commission on Man and Environment – Human Adjustment to Natural Hazard Research Program and forms part of a comparative study of human adjustment to agricultural drought in Australia and Tanzania.

Prior work on resource appraisal and land settlement in semi-arid Australia (Heathcote 1965) and more recent surveys of the national significance of drought in Australia (Heathcote 1967 and 1969) were the background to this more specific study of human adjustment to agricultural drought in South Australia. Agricultural drought is defined as a shortage of water harmful to man's agricultural activities. It occurs as an interaction between agricultural activity (i.e., the demand) and natural events (i.e., the supply) which results in a water volume inadequate for plant and animal needs.

The aims of this study were to test hypotheses and techniques derived from previous work by the Natural Hazards Research Group of the universities of Clark, Colorado, and Toronto (Natural Hazards Research Group 1970). Briefly these hypotheses were:

(1) Occupance of hazardous areas is rational according to the inhabitants.

(2) Responses to hazards comprise 3 types: (*a*) folk (pre-industrial) – more in harmony with than implying control over nature. (*b*) technological (industrial) – more attempting control over nature. (*c*) comprehensive (post-industrial) – comprising elements of both above types.

(3) Variation in perception of hazard may be accounted for by the magnitude and frequency of its occurrence, personal experience, the importance of the hazard to income or local interests, and personality factors (risk-taking propensity, etc.). Such variations do not seem to be related to socio-economic indicators.

(4) For the individual, choice of adjustments is a function of individual perception of the hazard, perception of the choices open, command of technology, the economic efficiency of alternative adjustments, and perceived linkages with other people.

(5) Individually, economic efficiency may be estimated by the time horizon chosen, relationship between losses forecast and reserves, and personal choice. Communally, economic efficiency may be a function of perception of hazard, and choice of adjustments as influenced by government action.

With these hypotheses in mind the specific aims of the study are to assess the extent of the hazard zone and its occupation; to identify the full range of possible adjustments; to study the perception of the hazard from farm level through local officials and community leaders to state officials and leaders; to study the process of adaptation of damage reducing adjustments in their social context and finally to attempt to identify optimal sets of adjustments and their anticipated social consequences.

Two basic methods are employed. The first comprises archival search of private and official documents – to assess the historical sequence of drought occurrences over the study period (1946–70), and the impacts (material and mental) of these occurrences, as indicated in a range of sources from private diaries, newspapers, and production statistics, to legislation. From these, chronologies of drought occurrences, intensities, and impacts can be drawn up and the results from different sources compared both with each other and with evidence from the second (field) sources. The second method involves field application of a questionnaire, based upon one developed by the Natural Hazards Research Group. This questionnaire is currently being applied to four study sites (of basically similar terrain) within the Mallee-Wimmera Wheat-Sheep Zone of Southern Australia. Two sites of high drought risk, one in South Australia and one in Victoria, are balanced by two sites of medium to low drought risk similarly chosen. Interviewees will be stratified according to educational background and family responsibilities, and originally 120 interviews per site were hoped for. Preliminary work has shown that 60–80 per site will be a more realistic return.

Prior to the drawing up and field testing of the final questionnaire, a much simpler open-ended enquiry was made in an area straddling the irrigated and dry farming areas of the Murray River. A survey in 1970 provided 470 interviews from both urban and rural population and a follow-up 25 per cent sample survey of urban and rural households provided 230 interviews to a shorter open-ended questionnaire. Basically, both these questionnaires asked for definitions of drought and whether respondents felt that their locality was in any way threatened. The complexity of definitions of drought suggested a range of perceived impacts, from environmental (dust and soil erosion) to economic (reduced retail business) to social (population migration and communal agitation for relief measures).

The field survey of the 4 sites will be completed late in 1971, but preliminary results have already suggested some interesting comments:

(1) For the two completed (South Australian) sites, farmers in the high as well as medium risk sites tend to deny drought as a major problem. Its impact is generally seen as serious but controllable (even of potentially mobile sand dunes), by careful farm management. The current world crisis in agricultural prices is a much more serious 'drought' as many farmers observed.

(2) This ability to cope now is generally seen as the result of successful adaptations of new farm management techniques following on from the disastrous droughts of 1944–5, when soil erosion, crop losses, and farm abandonment were widespread.

(3) The bad publicity given to this area at that time lingers on and locals claim that unfair reporting – painting pictures of erosion and abandoned farms – is still put out by 'big city' (i.e., Adelaide) newspapers unwilling to recognize the new, less dramatic, image of relatively successful farming. Search of newspaper articles tends to substantiate this dichotomy between 'local' and 'outside' images.

(4) Despite modern farming technology, the decision when to seed the crop seems to be only partly technical, and often almost mystical in origin. Nearly all farmers keep rain gauge records assiduously, but use the records mainly to compare notes with neighbours after the event, rather than as a guide

to soil moisture conditions.

(5) In 1865 the South Australian Surveyor General Goyder drew a line through the state to delimit the edge of a major drought. This line was extended from the Murray River over country which Goyder did not see, probably by a Lands Department clerk, and separates the two South Australian study sites. In both sites and even across the border into Victoria (to which the line was never applied) virtually two-thirds of all farmers thought that their locations with regard to that line affected the frequency of drought years!

In the presented paper final results of the surveys will be provided and an attempt made to assess the validity of the hypotheses tested.

Heathcote, R.L., 1965 *Back of Bourke: A Study of Land Appraisal and Settlement in Semi-arid Australia* (Melbourne).
- 1967 The effects of past droughts on the national economy, in *Report of the ANZAAS Symposium on Drought* (Melbourne), 27–45.
- 1969 Drought in Australia: a problem of perception, *Geog. Rev.* 59, 175–94.

Natural Hazards Research Group, 1970 *Suggestions for comparative field observations on natural hazards*, Working Paper 16, Natural Hazard Research (Toronto), rev. ed.

c0703
Adaptation stages and man-induced vegetation change in interior Tamilnadu
B.J. MURTON   *University of Hawaii, USA*

The text is not included because of length.

Allchin, F.R., 1963 *Neolithic cattle-keepers of south India. A study of the Deccan Ashmounds* (Cambridge, Eng.).
Arokiaswami, M., 1956 *The Kongu Country* (Madras).
Brookfield, H.C., 1968 New directions in the study of agricultural systems in tropical areas, 413–39 in E.T. Drake, ed., *Evolution and Environment* (New Haven, Conn.).
Carol, H., 1964 Stages of technology and their impact upon the physical environment: A basic problem in cultural geography, *Can. Geog.* 8, 1–7.
Clarke, W.C., 1971 *Place and People: An Ecology of A New Guinean Community* (Berkeley).
Cohen, Y.A., ed., 1968 *Man in Adaptation: The Cultural Present* (Chicago).
Darby, H.C., 1956 The clearing of the woodland in Europe, 183–216, in W.L. Thomas, ed., *Man's Role in Changing the Face of the Earth* (Chicago).
Deutsch, K.W., 1966 *The Nerves of Government: Models of Political Communication and Control* (New York).
Gaussen, H., P. Legris, and M. Viart, 1962 *Notes on the Cape Comorin Sheet*. International Map of the Vegetation and of Environmental Conditions at 1/1,000,000. Indian Council of Agricultural Research (Delhi).
Gordon, D.H., 1958 *The Pre-historic Background of Indian Culture* (Bombay).
Hamilton, D., 1862 *Report on the Shevaroy Hills* (Madras).
Leighly, J., 1963 *Land and Life: A Selection From the Writings of Carl Ortwin Sauer* (Berkeley).
London, India Office Library. MacKenzie Collection. Translations of Vernacular Tracts. Sketch of the history, of the rajahs of Anagoondy, Mysore, Baramahal, etc., 10, no. 3; History of Salem, 40, no. 9.
Madras Records Office. Proceedings of the Board of Revenue.
Madras Records Office. *The Baramahal Records*, sect. v, Property, 1915; sect. XXI, Miscellany, 1925.
Maloney, C., 1970 Archaeology in south India. Accomplishments and prospects, presented at meeting of Soc. for South Indian Studies, U. Wisconsin, Madison, April.
Maruyama, M., 1963 The second cybernetics deviation-amplifying mutual causal processes, *Am. Sci.* 51, 164–79.
Mikesell, M.W., 1967 Geographic Perceptives in Anthropology, *Ann. Assoc. Am. Geog.* 57, 617–34.
Murton, B.J., 1970a Agricultural change

in south India before 1800. Historical and spatial perspectives, paper presented at meeting of Ass. for Asian Studies, San Francisco, April.

Murton, B.J., 1970b   Man, mind, and land. A peasant production system in late eighteenth century south India. PH D thesis, U. Minnesota.

Odum, E.P., 1969   The strategy of ecosystem development, *Science* 164, 262–70.

Raghunatha Rao, S., 1936   Forests and forest products of the Salem district, *J. Madras Geog. Ass.* 11, 130–40.

Ramaswami, A., 1967   *Salem*, Madras District Gazetteers (Madras).

Richards, F.J., 1918   *Salem*, Madras District Gazetteers (Madras).

Sampath, C.R., 1963   A pioneer iron industry in Madras, *Hindu Weekly Magazine*, 12 May (Madras).

Sauer, C.O., 1956   The agency of man on the earth, 49–69 in W.L. Thomas, ed., *Man's role in changing the face of the earth* (Chicago).

Shortt, J., ed., 1870   *The Hill Ranges of Southern India* (Madras).

Stein, B., 1967–8   Brahman and peasant in early south Indian history, *Adyar Library Bulletin* (Madras), 31–2, 229–69.

Stoddard, D.R., 1966   Darwin's Impact on Geography, *Ann. Assoc. Am. Geog.* 56, 683–98.

Stoddard, D.R., 1967   Organism and Ecosystem as Geographical Models, in R.J. Chorley and P. Haggett, eds., *Models in Geography* (London), 511–48.

Waddell, E., 1970   Methodology and explanation in cultural geography, or the quest for an interface in man-milieu relationships, paper presented at meetings of *Assoc. Am. Geog.*, San Francisco.

Zeuner, F.E., 1950   *Stone Age and Pleistocene Chronology in Gujarat.* Deccan College Monograph Series 6 (Poona).

c0704
Human adjustment to volcanic hazard in Puna district, Hawaii
SHINZO SHIMABUKURO and BRIAN J. MURTON   *University of Hawaii, USA*

The text is not included because of length.

County of Hawaii Civil Defense Agency, 1969   *Natural Disaster Instructions* (Hilo, Hawaii).

Kates, R.W., 1971   Natural hazard in human ecological perspective: Hypotheses and model, *Econ. Geog.* 47, 438–51.

Macdonald, G.A., 1962   The 1959 and 1960 eruptions of Kilauea Volcano, Hawaii, and the construction of walls to restrict the spread of the lava flows, *Bull. Volcanologique* 24, 248–94.

– and A.T. Abbott, 1970   *Volcanoes in the Sea: The Geology of Hawaii* (Honolulu).

– and J.P. Eaton, 1964   Hawaiian volcanoes during 1955, *Geol. Survey Bull.* (Washington), 1171.

Natural Hazard Research, 1970   Suggestions for comparative field observations on natural hazards, Working Paper no. 16.

Report of the Puna Volcanic Fact Finding Committee, April 1955   Mimeo.

Swanson, D.A., D.B. Jackson, W.A. Duffield, and D.W. Peterson, 1971   Mauna Ulu eruption, Kilauea Volcano, *Geotimes* 1 (May), 12–16.

c0705
The New Zealand natural hazard insurance scheme: application to North America
TIMOTHY O'RIORDAN   *Simon Fraser University, Canada*

New Zealand offers a unique case study of a national natural hazard insurance policy which covers all classes of natural hazard and which, in theory at least, comes very close to the optimum in hazard management. In practice, however, the scheme is running into numerous difficulties which tend increasingly to divert implementative policy from legislative policy to the ultimate detriment of the program. This paper will review the New Zealand scheme, account for this unfortunate divergence of policy, and offer suggestions for incorporating the lessons learnt from this analysis into hazard management practice in North America.

The New Zealand policy is essentially a national insurance scheme to cover damage from all classes of natural hazard. Any building covered by a fire insurance policy is automatically subject to an additional levy of $0.05 per $100 of coverage. This revenue (which amounts to $13 million annually) is paid to an administrative body called the Earthquake and War Damage Commission, and is diverted into two accounts. Nine-tenths goes into the Earthquake and War Damage Fund. This fund now stands at about $120 million and from it the commission pays out on any damages (up to the indemnity value of the property involved) caused by earthquake or, in the event of war, enemy action. Ten per cent of the annual revenue is diverted into the Extraordinary Damage Fund which now stands at about $600,000. This fund is used to repair or replace (to its indemnity value) property damaged by events considered to be 'abnormal and unforeseen and of extraordinary effect.' Such events include volcanic eruption, flood, storm, tempest, landslide, or tsunami – in fact any natural disaster is covered.

In theory the New Zealand scheme offers a valuable example of near-optimum hazard management in a natural context.

1 The scheme is based upon the philosophy that the 'true' hazard is that event which exceeds the normal adjusting and buffering capabilities of the human system to absorb it (Kates 1970). In New Zealand such events may occur anywhere, affect any property owner, and are essentially unpredictable. Potential damage (particularly catastrophic earthquake) may be considerable (a severe earthquake in Wellington could result in over $600 million in damages [Power 1968, 25]), and since private insurance companies are unable to provide the necessary cover (because of lack of statistical knowledge of the event, unwillingness to cover only the risky clients who are the only ones to apply, and inability to build up large reserves or collect the necessary reinsurance cover [Steinbrugge 1968]), a communal fund is the only feasible means of providing disaster relief. On the assumption that the 'true' hazard is random in incidence and unpredictable in occurrence, the communal fund is also theoretically as well as administratively appropriate.

2 The 'normal' or 'reasonably expected' event is supposed to be safeguarded against by the property owner. The commission can enforce this by ensuring that all properties meet building code requirements designed to protect each class of building from 'foreseeable' damage. Should a property not meet these requirements, the commission can classify it into a category of higher risk and set deductible clauses and premiums accordingly. Where a property is in an area of very high risk (e.g. a flood plain or unstable slope), the commission can refuse to provide cover. In addition all property owners must seek private insurance cover as protection against the 'normal' event. The commission is supposed to cover damage only in the case of 'extraordinary and unforeseen' disasters.

3 The commission can initiate updated building codes and can, in theory, advise against any new development in perilous areas. Failure on the part of local authorities to meet either requirement should the commission wish it could result in reclassification of any property involved into higher-hazard risk or even rejection of its eligibility for cover.

4 Claim files provide an invaluable source of information on the extent and nature of damage for every natural disaster in the country. This is a vital record to enable managers to determine more precisely the characteristics (location, frequency, extent, patterning) of the natural events system. Armed with this information, the commission is in a better position to assess hazard risk and set appropriate rates and conditions.

In practice, however, the New Zealand scheme is running into difficulties. This is due largely to public misinterpretation of the nature of the commission's work and reflects how the public evaluates the incidence of natural hazard in their lives. The commission is seen as a backroom, claim-processing administrative body, sitting on millions of dollars of public money which should be used, not saved. (The concept of hedging against a long-term catastrophic damage is not widely understood, yet a severe earthquake in Wellington could set back the economy by ten years [Power 1968, 26].) As a result the commission has been forced to expand its coverage to events which are not

TABLE 1. Cost of claims on the Extraordinary Damage Fund 1966

	Domestic	Commercial/ industrial	Farm outbuildings
Percentage of total claims	7.7	7.7	84.6
Amount paid	$2,850	$76,178	$52,204

SOURCE: Earthquake and War Damage Commission

in any sense 'unforeseen,' such as landslip damage on slopes of known instability or flood damage in areas with a previous record of flood. Because its role is seen at best as advisory and at worst as merely paper-pushing, the commission has no powers to enforce building codes nor to restrict development in areas of 'foreseeable' hazard. To reclassify property would require knowledge of the hazard risk which the commission has on file but does not have the technical personnel to enforce. In any case such an action would affect local property values and would be strongly resisted locally. As a result local authorities fail to enforce earthquake building codes (which anyway are designed more to protect loss of life than to control structural damage [Steinbrugge 1968]), and blissfully ignore potential hazards in their development proposals.

Another problem is political. When the commission is faced with a number of damage claims from a particular area, no matter how poorly maintained each or all of the buildings and no matter how foreseen the disaster, it must pay up. To discriminate between claims would be political suicide in view of the fact that the average New Zealander feels he has paid his share, the funds are vast, and he suffered from a 'natural' disaster. Thus spurious claims are frequent – cracking of walls due to clay shrinkage attributed to earthquake, poorly constructed sheds blown over by a strong breeze, slump-damaged houses sitting on clay-lubricated subsoils. Consequently claim payments are rising annually though the incidence of hazard does not appear to be increasing.

Private insurance companies are loath to cover high-risk high-value property and whenever possible leave such coverage to the commission. Many properties do not seek private insurance coverage as the rates may be 10 to 100 times higher than the subsidized scheme provided by the commission. As a result, though about 7 per cent of all claims on the Extraordinary Disaster Fund are from commercial properties, these claims account for about 70 per cent of the annual payments (see Table 1). Damage to poorly constructed barns makes up much of the remaining claims.

Thus, unless public clarification of the theory of the scheme is made, the commission appears locked into a system from which it has little escape. The larger the funds grow, the greater the pressures after every disaster to expand the coverage; the wider the coverage, the greater the amount of claims; increasing claims deplete the funds; depleted funds encourage the commission to raise the premium; raising the premium enlarges the funds; and so on.

The New Zealand scheme would appear to be applicable on a state- or province-wide basis in those areas subject to random and unpredictable natural disasters such as earthquakes in California and Alaska, tornadoes in the midwest, hurricanes in the south and southeast, hail storms in the prairies. It could be designed to provide coverage where private insurance interests cannot or are unwilling to protect property at reasonable rates. To be successful the appropriate hazard management agency would have to provide constant public information on the extent of the policy to offset the inevitable selfish desire to make use of a communal fund. The agency would have to be armed with the necessary teeth, with the necessary technical personnel from the appropriate agency of government to ensure that building and zoning regulations were being met. Part of the revenue could be diverted to research aimed at improving hazard forecasting and upgrading techniques of building construction and risk estimation. The agency would then be in a position to devise a suitable combination of hazard avoidance and hazard modification behaviour depending upon the characteristics of both the human use and the natural events systems. (This would be the 'post industrial multiple means' stage of Kates' model [Kates 1970].) The agency could initiate this policy with all new development, then ensure that high-value property was suitably protected, and finally upgrade existing lower-value property. As a result, damage from 'normal' events should be drastically reduced while damage from truly catastrophic events would be covered by legitimate relief.

Natural Advisory Council for Water Resources Research.

Kates, R.W., 1970 Natural hazard in human ecological perspective: hypotheses and models, Natural Hazard Research Working Paper no. 14, Department of Geography, U. Toronto; also in *Econ. Geog.* 47 (1971), 438–51.

Power, C.A., 1968 Earthquake insurance in New Zealand and the problems of reconstruction, *NZ Eng.* 34, 23–7.

Steinbrugge, K.V., 1968 Earthquake hazard in the San Francisco Bay area: a continuing problem in public policy, U. California Institute of Governmental Studies (Berkeley).

c0706
The Indian desert of Thar as a man-made desert
CARL RATHJENS   *University of Saarbrücken, West Germany*

The desert of Thar in the arid region of northwestern India is an example of a locale where changes in the natural environment introduced by man, and the dangers for man and his economy resulting from these changes, are well known and have been explored to a relatively great extent. I want to acknowledge especially the research work of some Indian geographers (e.g. Mishra 1968), the work of the members of the Central Arid Zone Research Institute at Jodhpur and the work of the members of some Indian universities in Rajastham and in the Punjab. Their findings have been discussed intensively at the Symposium on Arid Zone at Jodhpur (1968), held in connection with the last International Geographical Congress in India. The reports of some European geographers like Dresch (1965), Nitz (since 1966), and of myself (since 1957) have made substantial contributions in this area. In publications about the Thar and also in papers read at the Congress of New Delhi in 1968 and in the European Regional Conference of Budapest in 1971, I have tried to draw attention to some of the questions of geoecology and anthropogenic geomorphology which emerged from a study of the Thar. Here we shall attempt to investigate some problems of the geographical environment which may be seen particularly in our example of a man-made desert.

The questions if and how far man depends on his natural environment, and in what way his environment is influenced and affected by human settlement and economic activity, are among the oldest problems which have been treated by geographical science. Modern environmental research, within the framework of geography, therefore should not fall into the error of developing new positions of geographical determinism. On the contrary, we should direct our efforts especially to those cases where interference by man and his economy has disturbed the natural equilibrium of the surface of the earth: where natural processes and formations of the geomorphological features, of the soil, and of the vegetation, are changed and destroyed in such a way that there arise dangers for the productivity or even for the future existence of human settlement, economy, and life itself. On this question very little scientific work has been done up to now; here lie important tasks for geographical research. We should bear in mind that, of the hazards threatening human life on earth, man himself is by far the greatest.

Physiognomically the Thar offers all the features of a normal desert caused by climate. Large parts of the country are covered by sand which has formed dunes. In addition, there are wide areas of rocky desert. Near the frontier between India and western Pakistan the groundwater table sinks to a depth of more than 100m. The water flow in the rivers is either periodical during the summer monsoon, or only episodic after single monsoonal showers of rain in the plains or in the bordering Aravalli mountains. Efflorescences of mineral salts are very widespread. In the western part of the Thar there are a number of endorheic basins without outlet, filled with water only during the monsoon and for a short time after it. The vegetation is sparse and distinctly xeromorphic. Human settlement and agricultural land use are dependent on the artificial production, transaction, and distribution of subterranean water.

These characteristic signs of a desert, however, do not correspond with the climatic conditions of the area and cannot be explained by the natural structure of the landscape alone. Only the most western parts of the Thar are arid; the centre of the arid region is situated in the middle of the Indus basin in western Pakistan. To the northeast in the direction of the Gangetic plains, to the east as far as the Aravalli mountains, and to the south in the direction of Saurashtra, precipitation rises to 600–800mm per year, although it is concentrated in the monsoon period of the summer. The natural vegetation, a thorny savannah with different acacias, grades to the east into a relatively dense and exuberant prosopis savannah, and into a deciduous dry forest along the border of the Aravallis (Meher-Homji 1970). Moreover, the climatic conditions allow agriculture on unirrigated rain fields in a large part of the eastern Thar.

Therefore we have to look elsewhere for reasons why all the other factors of the natural structure do not correspond to the climatic conditions: and the answer is the interference of man. In particular, the following facts must be taken into consideration:

1 The protecting cover of the forests and of the arboreal formations has been destroyed by man: partly on purpose, to create areas for settlement and agriculture; partly unintentionally, for instance, by the lowering of the groundwater table, by fire, and so on.
2 Overstocking of the grazing areas hinders the growth of young forest trees: it has reduced and cleared the remaining vegetation and driven back the grasses, but favoured the thorny plants.
3 From the great demand for fodder for the herd animals there has developed a special method of gaining forage, namely, lopping off the prosopis trees; as a result there has been a loss of the wood stocks.
4 The great demand for burning material for the purposes of cooking and heating has forced the people to collect all vegetation consisting of wood and further diminished the vegetation as a protective cover.
5 Agriculture dependent on rainfall, with the main crops of millet (bajra), sorghum (jowar), and pulses, together called kharif, is carried on in wide areas of the Thar, in favourable years far beyond the isohyet of 200mm of average annual precipitation.
6 The demand for water for man and cattle, which has to be drawn from tanks and wells, is reducing the storage of groundwater, especially in the vicinity of settlements.
7 Irrigation of fields, carried on without special technical knowledge, with an insufficient quantity of water, and without drainage measures, very often brings about an accumulation of mineral salts in the upper horizons of the soil.

As a result, there occur the following changes and disturbances of the ecological structure of nature, or ecosystems, which are disadvantageous and even dangerous for man and his economic activity:

1 There is extensive deterioration of the dunes and movement of the sand; this activity affects not only the older dunes originating in the Pleistocene period, which are renovated, but also material essential to soil formation, and results in weathering and decomposition of the bedrock. The arrangement of the recent dunes shows clearly that wind transport of the sand is related to the areas of greatest human activity. The continuing extension of the sand fields to the northeast, i.e. in the prevailing wind direction, can be interpreted as the most significant phenomenon in the increasing of the Thar desert, and must be considered a threat to the cultivated land along the Jumna and Ganges rivers.
2 Soil erosion is common in the entire Thar. Its action works on all surfaces, in many cases denuding the deeper soil horizon of calcareous concretions, the so-called kunkur, but also creating extended systems of gullies. Both effects make the soil useless for further agricultural utilization.
3 Change in the water balance can be observed generally in two ways: the groundwater table sinks so far that it can no longer be reached by the primitive technical installations of the agricultural population for raising water; also, the runoff after precipitation becomes more irregular and is more loaded with sediments, so that the use of water for agricultural purposes is made more difficult.
4 Change in the vegetation and in the soil

humidity are also causing change in the microclimates and the local climates, generally in the direction of the more unfavourable conditions of desiccation.

5 Formations of crusts and of salt horizons near the surface of the soil make agriculture more and more unproductive and finally quite impossible.

It seems apparent, therefore, that the arid conditions of the Thar have deteriorated as the result of human land use, and that possibilities for human use are, in turn, diminished by them. An equilibrium between the natural ecosystem and the challenge of man and his economy may be established, but with the increasing density of the population, and its increasing demand for goods, the situation will become more and more unfavourable for man. The moment will soon come when large areas of the Thar will not be serviceable at all for human settlement and land use, if radical steps are not undertaken quickly. New technological methods of agricultural land use and other measures for protecting the water storage, the natural vegetation, and the soils must be introduced.

Dresch, M.J., 1965 Le 'désert' de Thar, *Bull. de l'Ass. de Géogr. Franc.*, Nr. 332–3, 36–47.

Meher-Homji, V.M., 1970 Some phytogeographic aspects of Rajasthan, India, *Vegetatio* 21, 299–320.

Mishra, V.C., 1968 The Marusthali, *India: Regional Studies* (Calcutta), 245–65.

Nitz, H.-J., 1966 Bericht über die Ergebnisse einer Forschungsreise nach Nordwest-Indien, *Geog. Zeitschrift* 54, 144–56.

Rathjens, C., 1957 Physisch-geographische Beobachtungen im nordwestindischen Trockengebiet, *Erdkunde* 11, 49–58.

– 1959 Menschliche Einflüsse auf die Gestalt und Entwicklunge der Thar, *Arb. a. d. Geogr. Inst. d. Univ. des Saarlandes* (Saarbrücken), 4, 1–36.

– 1968 Some important principles of anthropogenic geomorphology in arid climates, XXI Int. Geog. Cong. (Delhi), 1968, 97.

– 1970 Die Wüste Thar. Beispiel einer vom Menschen geschaffenen Wüste, *Dt. Geogr. Forschung in der Welt von heute* (Kiel), 61–7.

– 1971 Desiccation by human influences, Europ. Reg. Conference, Budapest, 1971, 22–3.

Symposium on Arid Zone, 1968 Scientific papers, XXI Int. Geog. Cong. (Delhi), 1968 (Central Arid Zone Research Institute, Jodhpur).

Geomorphological Survey and Mapping
Recherche et cartographie géomorphologiques

c0801
IGU Commission on Geomorphic Survey and Mapping: its activities 1968–72 and program for 1972–6
JAROMÍR DEMEK   *Czechoslovak Academy of Sciences*

Approval for the foundation of the Commission on Geomorphic Survey and Mapping was given by the 12th General Assembly of the IGU on the occasion of the 21st International Geographical Congress in New Delhi, 6 December 1968. The commission is a direct continuation of the IGU Commission on Applied Geomorphology (President: J. Tricart, France) and its Sub-Commission on Geomorphological Mapping (Chairman: M. Klimaszewski, Poland).

The commission was established at its first meeting, held in the Institute of Geography, Czechoslovak Academy of Sciences, Brno, 19–21 March 1969. Its composition is as follows: President, J. Demek (Czechoslovakia), Secretary H.T. Verstappen (the Netherlands), and full members N.V. Bashenina (USSR), J.F. Gellert (GDR), F. Gullentops (Belgium), F. Joly (France). Simultaneously numerous corresponding commission members were elected from various countries. For the activities of the commission in the period 1968–72 the following program was accepted at the constituent meeting:

1 elaboration of the project of the Interna-

tional Geomorphological Map of Europe (1:2,500,000) and organization of work connected with the compilation and publication of the map;
2. elaboration of the Manual of Detailed Geomorphological Mapping (for geomorphological maps, 1:25,000 to 1:100,000); and
3. tackling of problems of geomorphological regionalization.

The commission held further meetings as follows:

2nd meeting in Paris, 1 September 1969, on the occasion of the International INQUA Congress (organizer F. Joly);
3rd meeting in Tupadly, 1–7 June 1970, at the invitation of the Czechoslovak Academy of Sciences (organizer J. Demek);
4th meeting in Abano Terme, 21–28 February 1971, at the invitation of the University of Padova (organizer G.B. Castiglioni);
5th meeting in Budapest, 12 August 1971, on the occasion of the IGU Regional Conference.

In September 1970 the compilation of the International Geomorphological Map of Europe was included by the General Assembly of UNESCO in its program of activities, and the commission was given financial support.

The commission activities were oriented mainly towards the elaboration of the project and legend of the International Geomorphological Map of Europe (1:2.5 mill.). The proposal to compile the map was discussed for the first time at the meeting of the IGU Sub-Commission on Geomorphological Mapping in Brno and Bratislava, Czechoslovakia, in 1965. At the 21st International Geographical Congress in New Delhi, J. Demek submitted the reasons for the need to compile and publish the geomorphological map of Europe. After World War II geomorphological mapping became, particularly in Europe and the USSR, the principal research method of geomorphology. The geomorphological map makes it possible to represent the appearance of the relief and to establish the distribution of the different relief elements, their relationships, origins, and ages. The compilation of the geomorphological map requires more exact geomorphological research methods, thus leading to further progress in the development of geomorphology as a science. In recent decades several geomorphological maps of different countries with diverse content and on different scales have appeared. In spite of this, geomorphological maps have not been compiled so far for considerable parts of Europe. The compilation of the geomorphological map of the whole of Europe on the basis of a unified key on 1:2,500,000, on the unified topographical base of the world map on 1:2.5 mill., can contribute to:

– the progress in geomorphological research in Europe, mainly in regions little investigated so far, and in regions where geomorphological maps have not yet been compiled;
– the establishment of new or less familiar laws governing the development of relief, mainly of the relationships between morphostructure and exogenous processes;
– the establishment of the laws controlling the development of the natural environment in Europe by comparison with other maps of natural elements of Europe on 1:2.5 mill. compiled under UNESCO's aegis;
– obtaining the knowledge for the compilation of geomorphological maps of other continents, especially of regions not known from the geomorphological point of view. In such regions even the general (synoptic) geomorphological map can be of immediate practical significance.

The content of the geomorphological map of Europe has been vigorously discussed at the meetings. Commission members were of three different opinions as to the map content. The first group thought that the map should, above all, express the appearance of the relief – the distribution of large orographical units, the degree of relief dissection – and then their origin and age. The second group believed that the map should represent the basic morphostructural units of Europe which are the dynamic base for the development of concrete landforms (morphosculptures). The third group, represented mainly by geomorphologists from Leningrad, submitted a proposal for a legend which preferred the representation of concrete landforms on the map. In the course of discussion five variants of the legend were elaborated. Commission members arrived at the conclusion that the geomorphological map of Europe should express the basic morphostructural elements of the continent which are divided according to dissection into lowlands, hilly lands, uplands, mountains, and high mountains. These relief types, delimited on the basis of morphostructures, are completed with concrete landforms classified according to origin and age.

The topographical base of the geomorphological map is the new map of Europe 1:2,500,000, compiled and published by cartographical institutions of European socialist countries. The map involves contour lines, a detailed hydrographical network, and other elements necessary for orientation (towns, communications).

Recently the first sheet was sent to the press – the Warsaw sheet, involving the territory of southern Sweden (compiled by S. Rudberg), the GDR (J.F. Gellert), Poland (Galon, Starkel), Czechoslovakia (J. Demek), a part of western USSR (team of geomorphologists from VSEGEI, Leningrad), and a small part of the FGR, Austria (Fink), and Hungary.

The commission paid special attention to the elaboration of the Manual of Detailed Geomorphological Mapping. The manual should unify the key and the methods of compilation of detailed geomorphological maps at scales of 1:25,000 to 1:100,000. The manual is expected to appear in several languages, including English, French, German, Russian, and Spanish. In the compilation of the manual many commission members took part. In January 1971 the preliminary edition of the English version was printed by the Institute of Geography, Czechoslovak Academy of Sciences, Brno, and sent for comments and remarks to all commission members. On the basis of comments and supplements the English version of the manual with colour maps will be published by Academia, the publishing house of the Czechoslovak Academy of Sciences, in 1972.

The commission is expected to continue its activities in 1972–6. Special attention will be paid to the completion and publication of other sheets of the International Geomorphological Map of Europe at 1:2,500,000, with financial support by UNESCO. Explanatory notes to the different map sheets will appear. At the conclusion the knowledge obtained in the map compilation will probably be treated in a book tentatively entitled 'Geomorphology of Europe.' The possibilities for preparing geomorphological maps of other continents will be examined. Further versions of the Manual of Detailed Geomorphological Mapping will also be prepared. On UNESCO's initiative the possibility of inserting into the program of the commission the investigations into questions of the standardization of methods of geomorphological research is being considered.

c0802

A recent trend in the geomorphological mapping of the alluvial plains in Japan and its application
MASAHIKO OYA   *Waseda University, Japan*

In Japan the study of geomorphologic land classification has been advanced by studies of two types: one is the basic study of topography; the other is applied geomorphology for land development, conservation, and use. The latter has been especially significant in advancing the discipline.

Work on the study of basic topography has been done by T. Okayama. According to him there are four landform classifications: (1) classification by agents or formation process, (2) classification by historical geomorphology or period of formation, (3) classification by geological structure, (4) classification by morphology. At present almost all landform classifications belong to classes (1) and (2). Class (3) is a geologic classification and not a landform classification. Class (4) is the most important classification in geomorphology or geography, but there are few studies using this method.

In Okayama's work, as in preceding studies, time, agents, structure, and process of formation are adopted as criteria in geomorphology. However, geographers use technical terms for landforms unsystematically, basing them sometimes on the morphology, sometimes on the agents or formation. Because of this some landforms are not even given technical terms.

In order to correct this defect in the landform classification mentioned above Nakano has proposed a landform classification utilizing the method used in pedology. He has classified the surface of the ground into landform units and given the name 'landform type' to these units.

A landform type includes surfaces formed in about the same age, by similar agents, and displaying similar morphology. For example, fans and natural levées belong to the same landform type.

A river terrace, fan, natural levée, and delta which were formed by the same river are grouped as a landform series. Collected landform types are named a landform area. The landform area is divided into three groups: mountain and hill, terrace and upland, and lowland.

In 1958, in order to provide fundamental data for better land use and conservation, a land classification survey was made by the government of Japan. It consists of a soil survey, a lithology survey, and a landform survey.

The landform survey in this land classification is not a basic geomorphologic land classification but a classification for land conservation and land use. The method of the classification is as follows: The area is divided into three primary groups: mountain and hill, terrace, and plains. Second, by its morphology and state of geologic structure, the mountain group is divided into two parts: gentle slopes and steep slopes. Gentle slopes are further classified as gentle slope on mountain ridge, gentle slope on mountain side, or piedmont gentle slope. Terraces are secondarily classed as gravelly terrace, rocky terrace, karst upland, or lava plateau. Furthermore, each terrace is classed by its altitude as higher terrace, middle terrace, or lower terrace. Plains are secondarily classed as valley plain, fan, delta, tide land, single bed and beach, or rocky ridge.

As has been mentioned above, the classification is not a basic geomorphologic land classification, but a classification for land use and land conservation. Therefore some classification has been based on morphology and some has been based on agents or formation processes.

Recently H. Kadomura (1967) has established a geomorphologic land classification based on the agents or formation process of the landforms. A student of Nakano, his study is similar to Nakano's. He states in his paper, 'Systematic aerial photo analysis of soft ground conditions,' the following: The first step in inferring ground conditions is the delineation of micro-landform units. These are the smallest land units and are composed of similar geologic materials having similar physical properties. The landform unit is correlated with the landform type described by Nakano. But the landform unit described by Kadomura is a more detailed division than the landform type described by Nakano. A landform area is a complex land form which consists of various kinds of micro-landform units, and occupies a wider area. According to him a fan or a delta is not a landform unit but a landform area.

As has been mentioned above, there have been studies of landform classification by agent or formation process but few based on the morphologic viewpoint. Therefore the writer has classified the plain in terms of morphology.

In the case of Japan a lot of sand and gravel is transported from the upper reaches in flood time, and deposited at the foot of the mountains. The material of the area is sand and gravel, and bank erosion occurs very easily in flood time, because the viscosity of the material is low. Because of the bank erosion in flood time the river channel becomes wide and the river displays the braided river course pattern. Shifting of the river courses has occurred frequently, because of the variation of flood discharge, velocity, and duration of flooding. This process has formed the fan as a rivet at the valley mouth.

In Japan large fans were formed in the Würm Ice Age. After the formation of the fans transgression occurred. The lower part of the fans was covered by sea water, and deltas were formed on them. Following formation of the deltas, regression occurred. The river channels prolonged their courses to the deltas and formed natural levées. Of course deposition continued on the fans too. Thus the main parts of the fans were formed in the Pleistocene, and the upper fans were thinly formed in the Holocene. Natural levées have been heightened by the deposition of sand which has been transported by the flood water. Back swamps occupy the spaces between two natural levées.

Deltas have developed between the natural

levée areas and the seashore. They are formed by the deposition of sand, silt, clay, etc. when the velocity of the current of a river becomes slower and is almost stopped by the sea or lake water. This type of deposition of the delta is similar to that of the fan i.e. the fan is formed at the mouth of the valley as a rivet, and the delta is formed at the mouth of the river as a rivet.

In Japan the boundary between the natural levée area and the delta area is clear, partly because of variations in the velocity of regression and partly because of variations in the volume of sand which has been transported from the upper reaches.

Typically in Japan the alluvial plain is divided into three parts: alluvial fan, natural levée area, and delta area.

The topographic elements are formed by the repetition of floods. The types and distribution of these topographic elements show the history of the floods. Therefore, if we classify the alluvial plain into the above mentioned topographic elements, we can foretell the type of future flooding in each plain, as well as read the past. From this point of view the writer has prepared geomorphological land classification maps of the river basins, indicating areas subject to flooding in the main river basins in Japan and some river basins in Southeast Asia.

These maps have been used to predict not only flooding from river courses, but also the results of high tides and inundation caused by local rainfall.

c0803
Geomorphological symbols for mass wasting
JOHN FORD SHRODER, JR. and WILLIAM C. PUTNAM   *University of Nebraska at Omaha, USA*

For the past several years Shroder has been mapping the geomorphology and Quaternary history of the Aquarius Plateau, located in the High Plateaus section of the Colorado Plateau in southern Utah. This plateau, which is capped by several hundred metres of strongly jointed basalt and underlain by weak limestones, marls, and tuffaceous fluvial clastics, has great relief and is highly susceptible to a wide variety of mass-wasting phenomena. During the Pleistocene and early Holocene the dominant process at work, judging from the preponderance of landforms, was the movement of massive landslips and rock glaciers.

Most of the area is now US National Forest and various parts of it are being developed, in a minor way, for camping, fishing, and hunting. Occasionally this building of roads, dams, and campgrounds has resulted in slope instability. For example, the construction of the Barker reservoir dam and others nearby in the early 1960s was apparently sufficient to raise the water table in the area enough to initiate or strongly contribute to the movement of the large Holby Bottom earth-flow. This flow has destroyed the road to Barker reservoir on numerous occasions.

Detailed geomorphologic mapping of the mass-movement activity and deposits of part of the Aquarius Plateau was initiated by Putnam, partly in order to develop a useful mapping technique for this multi-varied phenomenon, and partly to serve as an aid in the intelligent development of a highly unstable area.

In our opinion geomorphologic map symbols should be useful both in field mapping and in the final cartographic presentation. Furthermore, the symbols should be so designed as to be easy to use and to remember without constant reference to a legend. This thus implies that a separate symbol for each type of mass wasting would produce too great a proliferation of immemorable new symbology. We have accordingly chosen to use combinations of simple symbols whenever possible. In addition, the most memorable sort of symbology is that which relates directly to an existing classification of a particular phenomenon.

We believe that the best classifications of mass wasting are those based on the type of movement and the type of material. Rate of movement and amount and phase of water may also be included. Perhaps the most popular and comprehensive classification of this type is that devised by Sharpe (1938) for the entire spectrum of mass wasting. This classification was modified by Varnes

(1958) for mass-movement (landslip) phenomena and excluded rock glaciers, solifluction, creep, and the like. Shroder's (1971) classification is based upon that of Varnes but uses the terms *rocks*, *debris*, and *earth* for type of material rather than the somewhat nebulous term *soil* used by Varnes. The three main types of landslip movement are *falls*, *slides*, and *flows*. Creep and the *isolated unit falls* and *slides* of talus and colluvial accumulation complete mass-wasting movement types. Combining the types of material with the types of movement produces the classification. Thus there are rock-, debris-, and earth-falls. Slides are divided into two main types; the block-slides of few units, and the rock-, debris-, and earth-slides of many units. Earth-slides have been referred to as *failure by lateral spreading* (Varnes 1958). Flows include the rock fragment-flow (avalanche), the debris-flow, and the earth-flow (slow or rapid). Mud-flows are included as a more liquid type of earth-flow. Creep includes the movement of rock glaciers, solifluction (earth or debris), and soil creep (earth or debris), all of which may be a variety of flow.

Arrow symbols for the four main types of gravity movement are as follows:

fall (─────────→)
slide (─ ─ ─ ─→)
flow (∼∼∼∼∼→)
creep (ⅠⅠⅠⅠⅠⅠⅠⅠⅠⅠ▷)

Each of these arrow symbols should be oriented in the dominant direction of the movement. Several of the same symbols may be used in the case of multi-directional or multi-lobate features. Those symbols are then combined with any of the three types of material which constitute the main body of the mass-wasting feature. These symbols are as follows:

rock (▲▲▲▲▲▲)
debris (▪▪▪▪▪▪)
earth (∴∴∴∴∴∴)

Each of these latter symbols is contained within lines delimiting the mapped extent of the deposit. Talus and colluvial accumulation is indicated by the appropriate material symbol with no arrow symbol of movement.

Additional symbols include those for block-slides; the differentiation among debris-flows, mud-flows, and earth-flows; the differentiation between solifluction and soil creep; and the various slope stability categories.

Block-slides include four main types of slide blocks (Shroder 1968): slump, tilt, glide, and ridge blocks. Each of these may be differentiated by inserting the first letter of the name of the appropriate block into the dashed arrow symbol of sliding movement; thus ─ ─ R ─ → for ridge block. Slump blocks, with backward rotation, tilt blocks, with forward rotation, and glide blocks, with planar motion along bedding planes, have ridge crests which we map as a simple line. It is important to distinguish the relative antiquity of these features; thus small diamonds (─◇─◇─) indicate a crest with an angular surface, and small circles (─○─○─) indicate an older crest with a rounded surface. Ridge blocks, with non-rotational downward movement, may have no crest, only a flat summit. No symbol is used unless the block is narrow enough to appear at the appropriate scale as a ridge crest.

Mud-flows may be distinguished from the drier earth-flows by the addition of a wavy water symbol (∽∽∽∽∽) to the mud-flow. The same symbol may be used to differentiate the wetter debris-flows from the drier variety. Solifluction may also be differentiated from soil creep by the use of this symbol.

Three main types of slope stability occur: slopes which are actively moving (slide, flow, or creep); slopes which are potentially active (any of the four types of movement); and slopes which are stable. These stability classes are partially related to the Quaternary chronology of movement in the area. Thus the older landslips tend to be more stable. St-Onge (1968, 402) has suggested the use of a genetico-chronological classification in order to distinguish active from inactive forms. We are, however, at the present time unable to date adequately much of the topography and so must use other, more precise, means. Colour, preferably an obvious or distinctive hue to aid planners, may be laid over the appropriate material and movement symbols. Solid red for active slopes, red stripes for potentially active slopes, and no colour for stable slopes may be used.

This classification and related symbology represents an attempt to order our knowledge of mass-wasting into a coherent and easily used mapping system. As St-Onge (1968, 403) has noted, geomorphological mapping must reflect not only a theoretical and scientific point of view, but also a practical economic viewpoint for use in planning.

University of Nebraska at Omaha Senate; Glen Graves.

St-Onge, D.A., 1968 Geomorphic maps, in *Encyclopedia of Geomorphology*, ed. R.W. Fairbridge (Reinhold), 388–403.
Sharpe, C.F.S., 1938 *Landslides and Related Phenomena* (Columbia U. Press).
Shroder, J.F., Jr, 1968 Landslide landforms and concept of geomorphic age, 21st Int. Geog. Cong., *Abstracts of Papers* (New Delhi), 75.
– 1972 *Landslides of Utah* (Salt Lake City).
Varnes, D.J., 1958 Landslide types and processes, in *Landslides and Engineering Practice*, ed. E.B. Eckel, NAS-NCR 544 (Washington), 20–47.

Agricultural Typology
Typologie de l'agriculture

c0901
Agricultural typology in Nigeria: problems and prospects
S.A. AGBOOLA   *University of Ife, Nigeria*

The Commission on Agricultural Typology of the IGU has suggested criteria for identifying agricultural types on a worldwide basis. Although there has been considerable controversy about the details of what indices would be suitable to make such a delimitation as objective as possible, yet there is some agreement on broad criteria that could be used. These relate to social and ownership features, organizational and technical properties, and production characteristics of agriculture. Suggested indices have been tested in many parts of the developed world, but as yet only in a few developing countries. The aim of this paper is to examine some of the problems which have made response from African countries like Nigeria very limited so far, and to assess the prospects of their contributing towards the achievement of the work of the Commission.

The dearth of agricultural data in many tropical African countries is now well known, and Nigeria illustrates the seriousness of the problem rather well. In the first place, owing to lack of funds and personnel and the widespread illiteracy among farmers, no full agricultural census has taken place in the country earlier. Second, the first Sample Census of Agriculture took place only in 1950, and this and subsequent ones have been criticized for the smallness of their samples and the limited accuracy achieved. Third, the Sample Censuses to date have covered only a limited range of agricultural features, initially only acreage and production data, and later aspects of land tenure, agricultural population, farm sizes and dispersion, cropping patterns and fields, and livestock numbers.

However, the data collected to date should not be dismissed as completely unsuitable for use in agricultural typological studies. While it can be admitted that, in accuracy and details, they fail to compare with similar data in developed countries, yet it has been demonstrated that they are more useful than their theoretical degree of accuracy would lead one to expect (Coppock 1964; Agboola 1968). With his relevant experience at the FAO, Smit has also expressed the view that 'for many purposes, reasonable estimates based on sample surveys are just as useful as fully accurate figures' (Kostrowicki 1967). The small number of items covered in the Nigerian agricultural surveys is a prevalent characteristic in tropical Africa and perhaps other parts of the developing world. This does not necessarily make the identification of agricultural types impossible; rather it underlies the necessity for limiting the criteria used to a few basic, simple, and manageable features which can be universally applied.

The three broad criteria suggested so far by the Commission can at this point be examined against the background of available data and other information in Nigeria. Perhaps one of the most dominant features of Nigerian agriculture relates to the social aspect, especially land and labour. The relationship between the farmer and the land has undergone considerable changes in the recent past, and instead of the traditionally prevalent communal ownership of land, there are now some variants which aid differentiation of agriculture. This is particularly so as the variations tend to be related to differences in cropping patterns; e.g. while communal ownership is still common in areas which emphasize food crop production, other tenurial patterns like tenancy and individual ownership have emerged in the cocoa-growing areas. Similarly, the use of hired or family labour is related to the profitability of the crops grown, with export crops having an edge over food crops. Although it is only on land tenure patterns that data are available with regard to the social characteristics of farming, yet it is possible to base other deductions on cropping patterns.

With regard to the functional characteristics of agriculture, the country's ecological environments have to be emphasized, owing to their considerable influence upon aspects of cultivation, land, and crop-rotation systems. Available data on farm sizes, dispersion, and crop-combination systems can then be used as supplements leading to a realistic, although somewhat subjective, agricultural classification. Owing to the absence of appropriate data, such measures as labour and capital intensity in farming cannot at present be used in identifying agricultural types.

Only yield and production figures are available to throw light on the production characteristics of different crops. Owing to frequent fluctuation in the prices of agricultural products and the wide variations from place to place, it is impracticable under present conditions in Nigeria to convert the production figures into monetary units as a means of rationalizing among different crops. Neither are other indices like labour inputs and grain equivalents any easier to use owing to the absence of suitable data. No figures are available on notions like commercialization and specialization, although crop categories like export and food crops production give some subjective indication.

In the circumstances a classification of agriculture into types in Nigeria has to depend on approaches which recognize the data situation and the ecological conditions in the country. In the attempt made here, western Nigeria has been used as an example. As a result of the low level of technology used in agricultural production and the limited application of scientific knowledge, agriculture is still very much influenced by ecological conditions. As a first step in the classification of agriculture in western Nigeria, therefore, the significance of ecological differences between forest and savannah must be recognized, for, against such a background, the identification of areas in which certain crops or crop-combinations are dominant becomes more meaningful. Acreage rather than production data are used to identify the dominant cropping patterns. This is because production data may distort the patterns of agricultural production in favour of the more bulky but less valuable crops. However, the use of acreage data is supplemented by further consideration of indices like the proportion of farmers producing different crops, the land tenure and cropping patterns, and farm dispersion. Through the analysis and mapping of each of the above indices and the super-imposition of the maps on one another as well as on the ecological background, broad agricultural types have been identified.

The following provisional agricultural types are suggested:
(1) Plantation agriculture emphasizing tree-crop cultivation;
(2) Peasant export-crop production (cocoa): (a) mixed export-crop/food-crop economy, and (b) predominantly export-crop economy;
(3) Other tree-crop cultivation for internal trade (Kola);
(4) Food crop production for internal use: (a) cassava-producing areas, (b) mixed roots (yams, cassava), and (c) mixed roots/grains (yams/maize/guinea-corn).

It is hoped that, as the data situation improves, sophisticated analytical techniques can be used to classify a more complex system of agricultural types in Nigeria. Such an approach is hardly justifiable at present.

Agboola, S.A., 1965 The collection of agricultural statistics in Nigeria: the example of the agricultural censuses 1950–65, *Nig. Agric. J.* (2), 53–60.
Coppock, J.T., 1964 Agricultural geography in Nigeria, *Nig. Geog. J.* 7, 67–90.
Kostrowicki, J., 1967 Methods and techniques of agricultural typology; discussion on the Commission questionnaire no. 2, mimeo. (Boulder, Col.).

c0902
Towards a typology of agriculture in the American South
JAMES R. ANDERSON   *University of Florida, USA*

The American South has been selected for testing the applicability of the recommendations made by the Commission on Agricultural Typology for typological study at a regional or meso-scale (Kostrowicki and Tyszkiewicz 1970, 48).

The main purpose of this paper is to appraise and justify the selection of characteristics which was made in order to give a balanced and accurate representation of southern agriculture as it exists today. From the list of agricultural characteristics applicable to the South which were proposed by the Commission on Agricultural Typology for use in the delineation of types of agriculture, seven characteristics were chosen for initial study. Data on a county unit basis were available for the analysis of several other characteristics; however, some of these were excluded because of the extensive amount of analytical work needed to provide a meaningful typological analysis. Others were omitted because they were considered to be of secondary value for a comprehensive typological study of present-day agriculture in the American South. The characteristics which were chosen are:

1 The mosaic of major land uses expressed in terms of dominance and combinations of uses.
2 Proportion of all land in agricultural land, which is expressed by using area of land in farms as a percentage of total land area.
3 Land tenure expressed as a percentage of land in farms operated by owners, part-owners, tenants, and managers. A more common approach has been to make a proportional analysis of the number of operators belonging to the different tenure categories.
4 Average size of farm. Although derived by dividing the number of farms into land in farms, the grouping by number of farms by size of farm was also explored as a possible analytical approach.
5 Source of income for farm families. The gross income from the sale of farm products was related to off-farm income.
6 Agricultural sales were expressed by showing the percentage of total sales derived from crop and livestock products.
7 Crop and livestock production patterns. A generalized characterization of the distribution of production of individual agricultural commodities was made by classifying production patterns into four main categories with several subcategories. The main categories used were: (*a*) production widely dispersed throughout the region; (*b*) production spatially concentrated with one to three dominant concentrations; (*c*) production localized but occurring over a larger area than a clustered pattern; (*d*) production clustered with a high degree of concentration within a limited area.

Among the characteristics which were carefully considered for inclusion in a typological analysis but which were eventually excluded because of complex problems of handling available Census data are the following: (1) land management practices (drainage, irrigation, terracing, etc.); (2) use of inorganic fertilizers; (3) degree and kinds of mechanization; (4) use of family and hired labour; (5) average yields of principal crops and average productivity of livestock.

In justifying the selection of characteristics which was made for the typological study of agriculture in the South, it must be emphasized that the dynamics and transformation of southern agriculture constituted a major focus of the study being made. Therefore, it was essential to examine particularly the present patterns of those agricultural characteristics which had undergone considerable change.

The marked degree of mixture or combination of much forest land with cropland, along with the importance of non-agricultural land in many areas, has not been generally stressed in the description of southern agriculture. The long-standing dominance of such labour intensive crops as cotton and tobacco, which were being grown on a relatively small part of the total land base, has tended to obscure the extensive importance of forestry, not only as a part of farm operations but on tracts having no agricultural operations as well. The significance of highly improved grassland pastures in some areas as well as the presence of an extensive use of rangeland or rough grazing land in parts of the South has also never been widely recognized.

For many years following the abolition of

slavery, the use of the sharecropper system in the production of cotton and tobacco led to a widespread notion that the tenant farmer was the dominant tenure type in the South. Although sharecroppers were numerous in some areas from the 1870s to the 1950s, the total area of farmland operated under this system of tenure was relatively small. Following the demise of the sharecropper system, the importance of the tenant who owns no land has been greatly reduced. In order to improve the efficiency of their operations, many farmers now both own and rent land. Therefore, it seemed more fitting to express land tenure as a percentage of land in farms operated by owners, part-owner tenants, and managers rather than using the more common proportional analysis of the number of operators belonging to the different tenure categories.

Although the plantation, as a type of organization of agricultural production in the United States, has been widely recognized for a long time as unique to the South, the typological analysis of size of farm clearly indicates that the plantation as a unit of agricultural production today is not as large as it is often thought to be. Furthermore, plantations are now intermixed with farms that do not have the characteristics of plantation agriculture. Another interesting feature of the size-of-farm analysis is the clearcut importance of large farms in some parts of the South where plantation agriculture was never much significance.

In a study of present-day agriculture, it seemed essential to stress the high degree of intermixture of agriculture with other economic activities in many parts of the South. The South is still remembered as a predominantly agricultural region with relatively little industrial activity. In many parts of the South today there is much income earned by members of farm families from employment in non-farm occupations. While the part-time farmer in some areas does not contribute significantly to the total agricultural production, he may be a relatively significant producer of crops and livestock products in other areas. Thus the emphasis of the significance of off-farm income seemed warranted in the study of southern agriculture as it exists today.

Showing the sale of crop and livestock products as a percentage of total farm sales seemed an appropriate means of generalizing and complementing the presentation of the distribution patterns of individual agricultural commodities. By using three categories for crop sales dominant, livestock sales dominant, and mixed crop and livestock sales, an effective generalization was made.

In summary, the selection of a limited number of characteristics for use in making an initial first or lower-order typological analysis of southern agriculture indicates to the satisfaction of this researcher that the individual characteristics of agriculture can be typologically studied quite effectively. The much more complex development of a systematic hierarchy of agricultural types in which 'individual types of lower order are grouped together into the types of higher order irrespective of their distribution over the earth's surface' awaits my further study (Kostrowicki and Tyszkiewicz 1970, 15).

Kostrowicki, J., and W. Tyszkiewicz, 1970 *Agricultural Typology: Selected Methodological Materials* (Warsaw).

c0903
Indices for agricultural typology
V. BONUZZI   *Universita di Padova, Italy*

Agricultural typology interests a large field of research which aims at classification of types of farming. In this respect, carefully chosen series of indices or variables are all-important.

In order to get to such a classification, some problems arise in regard both to the size of the area under study and the choice of the variables or indices which might be the most suitable for a meaningful classification based on the data available for that area and the methods to be used in elaborating the data.

Choosing the area in which one wants to operate is the first factor which conditions the research. It is possible to study rural typology at different scales according to the territory which has been chosen, for example, a region, a country, a continent, or the whole world. This first choice strongly influences the choice of the variables to be selected. If the study is carried out in a small territory where there is pronounced agricultural activity and many statistical data are available, the farm is, no doubt, the basic unit to be referred to and all the variables adopted must be related to the farm. If the territory under study is larger but in the same national context, it is necessary to examine the statistical data which are available before any choice of the variables is made. In this case, the choice must abandon any specific reference to the farm and choose other variables capable of characterizing a wider area and related to administrative units such as the county or the region, which are taken as the basic units.

When the study is carried out on an international basis, the difficulty of gathering the data and comparing them is really remarkable. The World Atlas of Agriculture has, in this regard, contributed a great deal to emphasizing the various degrees at which the data are to be found and the different methods of comparing the statistical data available for agriculture in the different countries all over the world. In this sector of research, however, there are still many problems to be solved. The core of the problem lies in the comparability of economic data on an international basis. If these data undergo the official monetary exchanges, they either become less meaningful or result sometimes in something completely wrong as far as some types of comparisons are concerned. To rely on indices such as the ratio of the economic value expressed in a certain currency to the cost of living (which is ascertained according to standard quantities of commodities) expressed in that same currency, might remove these inconsistencies to some extent at least. An example of such an approach is the following: *Gross domestic product expressed in currency/Standard cost of living* expressed in the same currency. The difficulty lies in defining a standard which might be applied to different economic systems in different physical and environmental conditions and to different stages of development.

Other solutions which have often been adopted are the ones proposed by Kostrowicki who calculates the unit values by relating all the values to wheat market prices. But can wheat constitute a valid means of comparison?

The choice of the methods of elaborating the data constitutes the third problem and it is related to the previous ones. Several methods may be applied to the elaboration of the data from production function to Kostrowicki's graphic methods, from the method of clustering variables in index numbers to the method of factor analysis. It seems that Kostrowicki's method which is essentially graphic, and the factor analysis method which is essentially mathematical, are the most likely to obtain the most reliable results.

Recently Kostrowicki's method has been used by Kostrowicki himself in the classification of rural typology in Poland, by Bonnamour from France, and by myself in Italy at the Agriculture Economics Department of the University of Verona (Italy) to determine the typology of a very small mountainous region. These researches have shown that the number of variables to be selected must be very small and that the variables must represent the most characteristic and prominent aspects of the environment to be studied. A carefully chosen variable can, in fact, directly or indirectly cover several aspects which are useful to the research. Thus great care is to be applied in the choice of the variables since all subsequent procedures are influenced by it.

The cluster analysis method which was used for that small mountainous region has shown itself as more easily applicable as it requires a small number of variables but it has to be based on a rather large number of basic units – the farms – in order to get meaningful results.

The factor analysis method has very often been applied to the study of rural typology; some reports based on such analysis were offered at the last meeting of the Commission for Rural Typology which was held in Verona (Italy) in September–October 1970. Some of these research works, those done in the United Kingdom for example, are mainly carried out on a mathematical basis and use a large number of variables whose interaction and interrelationship are studied in order to define various levels of homogeneity. The knowledge of localization of the variables makes it possible to draw homogeneous areas as far as some variables are concerned and to obtain a really effective representation of rural typology. This methodological approach is, at present, used at the Department of Agricultural Geography and Agriculture Economics in the University of Verona to which I belong.

These approaches, however, suggest that both graphic and mathematical methods are progressively improving and becoming more widely used, but the results which are attainable depend mostly on the choice of the variables which is made by the research worker and hence on the very ability of the research worker.

c0904
Agricultural space: an experiment in classification
A.O. CERON   FFCL de Rio Claro, Brazil

In the classification of agricultural types applied in the northwest region of the State of São Paulo, Brazil, we have used 30 different variables which express the internal characteristics of agriculture – social and ownership properties (10 variables), organizational and technical properties (11 variables), and production properties (9 variables). These variables classified 93 municipalities corresponding to 43,000km^2 or 17 per cent of the State of São Paulo.

The classifying method used was factor analysis that reduced the 30 variables into 8 factors or agricultural types. The explained percentage variation represents 73 per cent of the total variation. Each type was analysed and defined according to the factor loadings. The classification of the space *sensu stricto* was obtained by cluster analysis.

The crucial problem in the development of this experience relates to the complex of variables considered as significant, selection, and measuring, according to the theory of agricultural types (or pre-theory) recommended by the IGU Agricultural Typology Commission. To solve this problem, it was necessary to adopt an exhausting work of preclassification and to find techniques which permitted the transformation of variables into measurable units.

The results obtained were not entirely satisfactory, but this does not mean that factor analysis is not a very useful classifying method. Many of the types identified showed a very fragile structure whereas in reality they are much more complicated. We believe that to find a useful classifying space type form which can be generalized and reapplicable, it would be necessary to define clearly the main structure supporting a certain model of space type.

It is improbable that, based on local classifying experiences, we can reach a large classifying system with world-wide applicability. It seems that social agricultural characteristics should be considered as supplementary to a space definition and classification characterized by agricultural attribute type. The political organization and economic orientation of a certain country have strong influences on the social properties of agriculture. When we strongly emphasize these variables, or when we give them the same emphasis as others, such as functional and production ones, the results lose their universal validity.

We decided to do an experiment in the northwest area of the State of São Paulo, classifying spaces basically according to the property organization type – crops or grasslands – and according to economic orientation. The experiment was performed by the application of factor analysis upon two matrices containing the data on the internal characteristics. We hypothesized that spaces vary from a maximum crop predominance to its total inexistence and from a maximum livestock predominance to its total inexistence.

All the variables related to crops were placed in a separate matrix, entirely independent from the one containing the variables related to livestock. At a certain level, the space types would be exclusively defined by the existence of crops or livestock or both. On the other hand, there is no social agricultural property exclusively specific to either of these dimensions.

The space type defined by coffee crops, as an example, may show any agricultural social attribute at any level of differentiation.

Finally we consider the notion of agricultural space type, as it was defined, as a notion of agricultural region. The size of these spaces depends on the scale of analysis used by the researcher.

c0905
Systems approach to a sugar cane type of agriculture
J.A.F. DINIZ *Faculdade de Filosofia, Brazil*

In 1968, we tried to arrive at a classification of types of agriculture according to the IGU Commission on Agricultural Typology and its first questionnaire, using land holdings as the basic unit and qualitative information. This classification did not allow a level of generalization adequate to build models.

In 1970, using the IGU's second questionnaire, in 'Depressão Periférica' of São Paulo State, Brazil, we carried out a further study using the county as the basic unit and applied factor and cluster analyses. From internal properties of agriculture, 34 of the more representative variables were selected. A similarity coefficient of 13.50 was used to establish the final classification, and sugar cane represented three types of agriculture. Other typological works have revealed that sugar cane is an important factor in characterizing agricultural space, at least in the form of plantations.

The findings of both studies have provided knowledge for thinking about the problem, in terms not of individual cases, but of taxonomic models that present a high degree of selectivity. We are building a probabilistic model of sugar cane distribution on the basis of the recommendations of the IGU Commission on Agricultural Typology. For us, the type of agriculture is viewed as a system, and the sugar cane type is an example. The components of this system were analysed, including the demand factors (understood as the needs for sugar and alcohol), production factors (the chief output), and the energy and level of entropy.

It is expected that this deductive approach to a classification problem will contribute to the establishment of a world agricultural typology.

c0906
Land-use classification in rural sections of metropolitan regions
W.C. FOUND and C.D. MORLEY *York University, Canada*

The movement into the countryside of urbanites is one of the major forces affecting the physical, social, and economic life of rural areas surrounding contemporary cities. The effects of this process include the tendency for rural properties to become the permanent or temporary homes of those employed in cities, the generation of weekend flows of traffic as people visit rural retreats or view the countryside from automobiles, the decline of the existing social structure in former agrarian communities, and the temporary abandonment of land purchased by speculators anticipating urban growth. The various forces operating in this situation create a special problem for the scientist attempting to classify land use in the rural area within the urban 'shadow,' particularly since the purposes underlying the use of land are many and must be measured along a number of dimensions. This is in contrast to the use of land in the solely agricultural area, where the generation of farm income is the single or dominant objective of land use.

This paper is a brief introduction to research undertaken by the writers in the rural area surrounding Toronto, Ontario (the study area corresponds roughly with the provincial government's planning area called 'the Toronto-Centred Region'), in which an attempt is made to classify rural land types in a way which differs considerably from normal methods, and which, hopefully, will lead to a more accurate picture of land-use dynamics in this type of area. A basic premise is that a classification based solely on activities reflected in the physical appearance of land is not very useful, since similar landscapes may reflect any one of a number of management objectives. The writers also feel that models which might be used to forecast or predict land uses according to such a classification would fail to incorporate the active elements of the process of change. The key element in determining the appearance of land is the land owner; and the writers have concluded that the best classification of land for purposes of planning, prediction, or theoretical 'understanding' is a classification of the land's owner. For example, it seems to be far more useful to classify a property as owned by an urbanite who uses it as a weekend retreat than to note that it is largely forested and includes a trout pond – even if the objective of classification is solely to present a picture which permits the casual observer to understand current trends in resource use. Furthermore, a classification based on the user permits the calculation of some very important derived statistics (such as the seasonal demand for local services, expected traffic flows on local roads, or probable political behaviour of local residents). These could only be inferred indirectly and quite imperfectly from more traditional typologies. As a final point, it is felt that models can be constructed which permit 'prediction' (based on hypothesized situations) of probable land categories for a given area far more realistically for categories based directly on the actors in the land-use process – i.e. the land user – rather than on an indirect phenomenon such as physical appearance. This is not to deny, of course, that the appearance of the land has a relationship with ownership which requires investigation.

The writers have hypothesized that properties in the study area represent ten basic types, where the typological categories refer primarily to ownership:
1. Prestige rural estate (resident, wealthy, urban-based owner)
2. Short-term rural retreat (urban owner)
3. Social and/or ethnic group property (urban or resident owners)
4. Rural residence (where owner commutes to city or to local employment centres)
5. Retirement property (intended or actual retirement)
6. Property of semiretired farmer
7. Rural recreational enterprise
8. Commercial farm property (includes part-time farming)
9. Speculative land holding (non-resident owner)
10. Public conservation or recreation area.

To test the classification scheme, four sample areas, each approximately 256km^2 in size, have been selected for detailed study. Selection of the areas was carried out so as to include the full spectrum of suspected land ownership types. The selection procedure took account of distance from metropolitan area, transportation networks, physical characteristics, agricultural productivity, and existing proportion of urban-based owners. Within each area, 60 land owners (30 resident and 30 non-resident) have been selected on a random basis for detailed interviewing. Each interview has yielded standardized questionnaire data which have been recorded on computer cards for analysis. The data obtained include nature of land tenure, physical property characteristics (size, land uses, etc.), proportion of time spent working the property (if a farm), proportion of time spent on property, location and characteristics of main residence and employment, routes used in commuting, local and non-local social contacts, detailed classification of leisure activities on and off property, shopping habits, family socio-economic characteristics, and an interviewer's detailed evaluation of property features critical to recreation and farming. The paper presented at the symposium will report on the extent to which the analysis of these data confirm or reject the hypothesized ten-part classification scheme.

The writers feel that the verification of a preliminary classification of land owners like that described above is vital for further research in rural areas surrounding major towns. Certainly, for planning purposes methods of evaluating and predicting changes in rural resource uses must be developed, and the current set of tools and models seem to be inadequate for this purpose. To begin with, highly aggregated models tend to ignore one of the basic components of change, which is decision making at the *level of the individual*. Unless one can become familiar with common classes of individual land owners and potential land owners, and with the behavioural environment within which they operate, one cannot make safe predictions about the response of such individuals to changed conditions in factors such as urban population, land price, transportation networks, leisure time, social structure, or public attitude towards differing life styles. Existing normative and empirical models which emphasize accessibility and distance-decay functions are not considered by the authors to offer reliable powers of explanation or description except, perhaps, at very general levels of abstraction. Inventories of land quality and land use which do not relate to the perceptions and leisure-time behaviour patterns of non-farming populations are similarly suspect. To overcome these deficiencies, research must be undertaken into the informational and decision-making mechanisms which really influence land owners in rural areas.

Of course, not all relevant decision making occurs at the level of the individual. Research allied to that described above is being undertaken by the authors in an attempt to identify the important 'actors' at the group level (for example, real estate companies, local and provincial governments, and citizen action groups). A report on this endeavour will also be included in the symposium paper.

To summarize, changes in the rural sections of metropolitan regions require, initially, new typologies of land ownership and use before meaningful models with descriptive or predictive powers can be devised. Specifically, typologies at the level of the individual land owner which take into account his behavioural tendencies must be developed. This paper reports on such research being undertaken by the authors in the Toronto-Centred Region of Ontario.

c0907
Terminology in typology – the problem of 'plantation'
HOWARD F. GREGOR   *University of California, Davis, USA*

Terminology is inseparable from typology, for if typology is needed – through classification – to give names to things, names are vital to the handling of concepts in typologizing. Unfortunately, concepts are conceived, if not phrased, in diverse languages and are influenced by personal experiences and associations. Quite rightly, then, the first major activity of the Commission for Agricultural Typology was the clarification of 'type' and 'system' with respect to the 'supreme notion in agricultural typology' (IGU 1966). The need for such action can be naturally expected to increase as the numerous agricultural types and their variants are derived. Plantation farming will certainly be one of those most eligible for review because of the increasing approximation of the modern plantation by several agricultural types and in the face of the disagreement over the meaning of 'plantation.' A solution, therefore, can provide a major organizational link in the classification system that hopefully will be developed by the Typology Commission.

The term plantation, although accepted in many languages and used for many centuries, has undergone changes in meaning. Moreover, the definition is still complicated because the views on what characteristic of the plantation is the dominant one are so various. 'Plantation' in its original sense referred to a 'planting' of people in a new area (i.e. a colony). In the seventeenth century, the meaning changed to a planting of crops, and more specifically a special crop for export. Since this kind of planting was associated with the establishment by Europeans of large-scale farming operations in the tropics, the plantation definition was automatically extended to the farm unit. The coincidence of the expansion of the plantation definition with the areal expansion of the operating unit in low latitudes thus gave the basis for the still heavy emphasis on the areal, and particularly tropical, link to the plantation. Later, as more expert attention focused on the plantation, various and conflicting recommendations were made for restrictions to the use of the term even within the tropics. Many of these additional qualifications were based on what was considered to be that aspect of a crop that presumably made it most amenable to plantation operations. A perennial nature, the need to be planted, and high perishability were some of the more important indicators recommended. Other opinions emphasized such social and economic criteria as cheap labour and land and dependence on foreign markets, and although these definitions of the limits to the plantation term were more liberal than those based on the nature of the crop, they also reinforced the view that the plantation was a peculiarly tropical institution.

In the last few decades, however, rapidly advancing agricultural technology has fostered an expansion of extratropical farms that closely resemble the plantation in many of its most important characteristics: crop specialization, use of advanced farming techniques, large size, centralized management, labour specialization, and massive capital investment. Traditional plantations in turn have been undergoing some changes that are narrowing the small remaining gap. Yet the spread of what appears to be a modern version of the plantation has been met by a mixed reaction terminologically. A few researchers have apparently been willing simply to extend the term plantation to these changing units, but more have preferred an additional, qualifying word or prefix in order to emphasize a particular aspect of this expanding farm, thus 'modern,' 'new,' or 'neo-' for its relative recency; 'usina' and 'industrielle' for its economic organization; 'temperate' for its areas of very rapid growth; and 'vegetable' or other crop terms for its particular specialization, in contrast to the more traditional plantation crops. Further, many of the combination terms have been the results of individual regional studies, so that it is an open question of how willing many of these writers would be to extend their terminology to similar farms in the rest of the world.

Far more national and regional terms do not even include the word plantation, although they are applied to both traditional and more rationalized types of large commercial farms and are used where the latter version is widespread. The two best illustrations are the Soviet Union and the United States. 'Kolkhoz' is supreme in the Union, despite a vast array of large farms that come very close to the more modern plantations. In the United States, tradition appears to be the main reason for the persistence of 'plantation' in Hawaii and the south, and even on the margins of this last region the term is giving way to the nondescript 'farm.' However, 'farm' is an integral part of those terms used most to typify the large, intensive commercial farm in its over-all national distribution, with the most popular being 'corporate,' 'industrial,' or 'factory' farm. In several Commonwealth countries, where the modernization of the plantation is proceeding in full swing in many sections, the English word estate dominates. In Latin America, 'hacienda' and 'fazenda' are the traditional and inclusive terms.

Finally, there are also scholars who reject the expansion of 'plantation' to cover this highly rationalized farm. For some, the more recent modifications of the plantation portend its end, its nature being regarded as more political and social rather than economic. For others, it is a change that can only be acknowledged terminologically for plantations in the tropics, the argument resting on the assumption that only tropical climates allow a year-round industrial type of operation.

The possibility, therefore, of obtaining general agreement on applying the term plantation to this rapidly growing and intensive farm type seems only fair at best. The long association of the term with the tropics may well be the main reason why those writers who suggest its expansion are still not followed by the majority of their colleagues and more critically, by the public, in their usage. But no less serious is the resistance to the term in those areas where the pace of farm rationalization is at its greatest, particularly the Soviet Union and the United States. Ideology strongly militates against the use of 'plantation' in both countries, not to mention the situation in those newly independent countries where plantations previously played on exploitative role. The solution nevertheless could very well lie in selecting a word well typifying the advanced state of agriculture in these two nations and the level of advancement eagerly sought by the less developed ones: 'industrial.' 'Industrial' has old linguistic roots and 'industrial farm' is an eminently neutral and not uncommon term. The term also provides a basis for a scale that can indicate the level of 'industrialization' of a farm, thus giving planners an index by which they can judge how much still remains to be done in improving the farming situation. As to how many of the terms already noted could be properly applied to the various 'industrial' subtypes, that would have to await the intensive work and cooperation of the typology and terminology Commissions of the IGU.

IGU Commission for Agricultural Typology, 1966 *Principles, Basic Notions and Criteria of Agricultural Typology, discussion on the commission questionnaire no. 1*, Warsaw (mimeo.).

A preliminary attempt at a typology of world agriculture
J. KOSTROWICKI   *Polish Academy of Sciences*

The present attempt at a typology of world agriculture is based on the principles and methods developed by the IGU Commission on Agricultural Typology during seven years of its activity.

The main methodological problems that underline the whole problem are the proper selection of variables characterizing types of agriculture and then the adequate and objective method of comparing individual cases characterized by sets of selected indices.

As a result of discussion the following 20 variables representing various agricultural properties have been adopted to characterize world types of agriculture.

I SOCIAL AND OWNERSHIP CHARACTERISTICS: (1) system of land tenure C; (2) average size of farms J.

II ORGANIZATIONAL AND TECHNICAL CHARACTERISTICS: (3) inputs of labour J; (4) inputs of animal power J; (5) inputs of mechanical power J; (6) organic manuring J; (7) chemical fertilizing J; (8) extent of irrigation J; (9) system of irrigation C; (10) system of land use C; (11) system of crop (or land) rotation C; (12) intensity of cropland use J; (13) cropping system C; (14) system of livestock breeding C.

III PRODUCTION CHARACTERISTICS: (15) land productivity J; (16) labour productivity J; (17) level of commercialization J; (18) degree of commercialization J; (19) ratio of animal to total production within gross production J; (20) ratio of animal to total production within commercial production J.

Fourteen of these variables are expressed in indices (J), six others of non-measurable character (C) by symbols representing various categories within the proposed classifications. All of them were normalized by reduction either to five thresholds based on world range of a given phenomenon (J) or five classes by simplifying their classifications.

These indices or symbols have been ascribed to 33 model types of world agriculture selected on the basis of the previous classification, statistical yearbooks, and vast literature concerning areal differentiation of world agriculture.

For the reasons that have been explained elsewhere, more refined quantitative methods used in many research works to distinguish homogeneous units could not be used in typology of world agriculture at least at this stage of the Commission work. Therefore the combination of the graphic method of typograms (star diagrams) and the deviation from the model type method both tested already by several regional studies have been adopted in the present proposal.

Using variables enumerated above, typograms have been constructed for each of the assumed 33 model types of agriculture in a way presented on a model typogram (Fig. 1).

MODEL TYPOGRAM

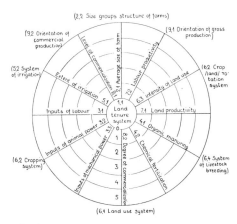

Fig. 1

As the indices represent in fact certain classes each of the typograms constructed consists actually of two typograms showing – for each model type – maximal and minimal range of indices within which individual cases representing given type could be located.

The use of the deviation method implies the formalized presentation of variables. The following formula has thus been applied:

$T = SO/P$

in which T means type of agriculture, S social characteristics, O organizational and technical characteristics, and P production characteristics.

In such a formula variables can be arranged in the following way:

TABLE 1. Model types of agriculture

Type	Variables																			
	1.1	2.1	3.1	3.2	3.3	4.1	4.2	5.1	5.2	6.1	6.2	6.3	6.4	6.5	7.1	7.2	8.1	8.2	9.1	9.2
1	A	0	1	1	0	0	0	0	0	C/D	A	1	A	O/D	1	1	1	1	1	1
2	A	0	1/2	1	0	1	1	0	0	C/D	A	1	A	O/D	1	1/2	1	1	1	1
3	A	0	1	1	0	0	0	0	0	A	0	0	0	A	1	1	1	1/2	1	5
4	B	1/3	1/2	2/3	0/1	1/2	1	0/1	0/A/B	C	B	2	B	B/D	1	1/2	1	1	1	1/2
5	A/B	1/3	1	2/3	0/1	1/2	1	0/1	0/A/B	A/C	B	2	B	B	1	1/3	1	1/2	2/4	3/5
6	B/C	1/3	2/3	3/5	1	1/3	1	0/1	0/A/B	C	C	3	B/C	D	2/3	1/2	2	1/2	1/2	2/3
7	B/C	1/2	3/5	1/3	1	1/3	1	0/1/2	0/B/C	C	C	3/4	B/C	D	3/4	1/2	1/2	1	1	1
8	B/C	1/2	3/5	1/2	1	1/3	2/3	3/5	B/C	C/D	C/E	4/5	A/B/C	D	3/5	2	1	2/3	1	1
9	C	1/2	3/5	1	1/2	1	2/3	0/1/2	D	C/D	C/E	4/5	B/C/D	C/D	3/4	2	2/3	2/3	1	1
10	C	1/2	3/4	2	1/2	1	1/2	0/1	0/D	C/E	C	3/5	B/C/D	D	3	2	2	2/4	1	1
11	C	1/3	2/3	1	1	1	1	0/1	0/A/B	C/E	A/C/E	3	A/B/C	0/D	2	2	3	3	1	1/3
12	C	4/5	2/3	2/3	1/3	1/2	1/2	0/1	0/A/D	A/C	B/C	2/3	B/C/D/E	B/C/D	1/2	2/3	1/2	3/4	1/2	1/3
13	C	2/3	3	1/2	3/4	3/5	1/2	0/1	0/E	C	C	3	D/E	D/E	3/4	3/4	3/5	4/5	2/3	3/4
14	C	2/3	2/4	1/2	3/3	3/5	3/5	0/1	0/D/E	B/C	C/E	3	D/E	D/E	3/4	3/4	3/5	4/5	3/4	5
15	C	1/3	3/4	1/2	1/3	3/5	3/5	0/3	0/D/E	C/D/E	C/E	3/4	A/E	0/D	4/5	3/4	4/5	4/5	1	4
16	C	4/5	2/4	1/2	2/4	3	2/4	0/2	0/D/E	B/C	C/D	2/3	D/E	D	2/3	3/4	2/3	4/5	4	5
17	C/D	4/5	2/4	1	2/4	1	3/5	0/4	D/E	E	E	3/4	E	0	3/4	2/4	2/4	4/5	1	1
18	C/D	2/4	2/3	1	3/4	1/2	3/5	3/5	0/D/E	B/C/E	C/D/E	3/5	E	0/D	4/5	2/4	4/5	4/5	1	1
19	C/D	4/5	1	1	4/5	1	3/4	0/2	0/D/E	C	C/E	2/3	0	0/D	2/3	4/5	5	5	1	1
20	C/D	5	1	1	1/2	1	1	0	0/D/E	A	0	3	0	C	1/2	3/4	5	5	5	5
21	E	5	2/3	1/2	2/4	2/4	3/4	0/2	0/D/E	B/C	B/C	3	E	D/E	2/3	2/4	4/5	4/5	2/4	2/5
22	E	4/5	3/4	1	2/3	3/4	4/5	0/4	0/D/E	C/E	C/E	3/4	D/E	0/D/E	3/4	3	5	5	1	1
23	E	4/5	2/3	1	2/4	1	2/4	0/4	0/C/D/E	C/E	C/E	3/4	0/E	0/D	2/4	2/4	4/5	4/5	1	1
24	E	5	1/2	1	4/5	1	2/3	0/1	0/E	C	B/C	2/3	E	D	2	3/4	5	5	1	1
25	E	5	0	1	1	1	1	0/1	0/E	A	0	2	0	C	1/2	3	5	5	1	5
26	E	5	3/4	1	1/2	1	1/2	0/2	0/B/D	C	C	3/5	A/B/E	0/D	3/4	1/2	1/2	1/2	4/5	1
27	E	5	3/5	1	1/2	1	1/2	4/5	B/C/D	C/E	C/E	4/5	D/E	0/D	3/5	1/2	1/3	1/3	1	1

$T = 1, 2 (3, 4, 5, 6, 7, 8, 9, 10, 11, 12, 13, 14)/(15, 16, 17, 18, 19, 20)$.

Each individual case can thus be compared with the formula representing the most similar model types. Up to certain number of deviations a given case could still be considered as being of the same type. The cases with the deviations going in the same direction could then be grouped into subtypes.

The comparison of the typograms and formulas made for the preliminary 33 model types of world agriculture (for examples see Fig. 2) revealed that some of them are very similar to each other and might be considered as subtypes rather than types of the first order. In result the number of types has been reduced to the following 27, each characterized by a particular set of variables (Table 1) that are only partly reflected in their names. These types were then assembled into 5 groups of types:

I PRIMITIVE AGRICULTURE: (1) shifting agriculture with long (forest) fallow; (2) shifting agriculture with short (bush) fallow; (3) nomadic herding.

II PEASANT AGRICULTURE: (4) current fallow agriculture; (5) current fallow agriculture with migratory herding; (6) continuing extensive, mixed agriculture; (7) labour intensive non-irrigated crop agriculture; (8) labour intensive irrigated crop agriculture; (9) labour intensive irrigated semi-commercial crop agriculture; (10) labour intensive non-irrigated semi-commercial crop agriculture; (11) low intensive semi-commercial crop agriculture.

III LATIFUNDIUM AGRICULTURE: (12) large-scale, low intensive, semi-commercial agriculture.

IV MARKET-ORIENTED AGRICULTURE: (13) intensive mixed agriculture; (14) intensive agriculture with livestock breeding dominant; (15) intensive agriculture with fruit growing and/or market gardening dominant; (16) specialized large-scale agriculture with livestock breeding dominant; (17) plantation agriculture; (18) specialized irrigated agriculture; (19) specialized large-scale grain crop agriculture; (20) specialized large-scale grazing (ranching).

V SOCIALIZED AGRICULTURE: (21) mixed agriculture; (22) specialized fruit and vegetable agriculture; (23) specialized industrial crop agriculture; (24) specialized grain crop agriculture; (25) specialized grazing; (26) intensive non-irrigated crop agriculture; (27) intensive irrigated crop agriculture.

The typology of world agriculture as presented in an abridged form above has to be considered as a preliminary step to be discussed by correspondence by the Commission regular and corresponding members. An improved version of this typology will be presented for discussion at the fifth meeting of the Commission. Only when it is accepted could it be presented on a map and recommended as a framework for more detailed regional studies.

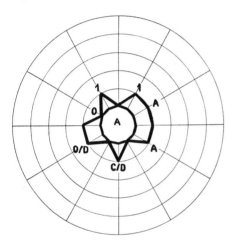

Fig. 2, type 1

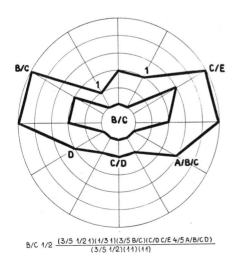

Fig. 2, type 8

Certainly those regional studies going deeper into areal differentiation of world agriculture will modify again the framework typology of world agriculture both as to the number and as to characterization of individual types of agriculture.

Kostrowicki, J., 1964 Geographical typology of agriculture, principles and methods: an invitation to discussion, *Geog. Polonica* 2, 159–67.
- 1966 Tipologia geografica de la agricultura mundial, principos y metodos, IGU, *Conf. Reg. Latino-americana* 5 (2), 793–807.
- 1968 Agricultural typology, agricultural regionalization, agricultural development, *Geog. Polonica* 14, 265–74.
- 1971 Commission on Agricultural Typology, Verona, Italy, 28 Sept.–2 Oct., 1970, *IGU Bulletin* 22.
- and N. Helburn, 1967 *Agricultural Typology, Principles and Methods: Preliminary Conclusions* (Boulder, Col.).
- and W. Tyszkiewicz, eds., 1970 Agricultural typology, selected methodological materials, and Essays on agricultural typology and land utilization, *Geog. Polonica* 19.

c0909
Some aspects of agricultural regionalization in South Korea
CHUNG-MYUN LEE   *Fresno State College, USA*

In any country the character of agriculture differs from one place to another. As such the impact of various regions differs in its effect upon the national economy, and it is desirable to identify those areas which are growing the fastest or contributing the most to the development of the nation. Such agricultural regionalization needs to be analysed.

The approach to agricultural regionalization is essentially through establishing a process of land classification. Individuals are grouped into classes on the basis of similarity through some objective quantitative measures. For this, a number of statistical procedures are available.

Heretofore, the method of agricultural regionalization has varied with each researcher and the practical aim of the research project; as well, the studies of large areas have employed loose regional schemes whereas studies of small areas have lent themselves to more rigorous breakdowns in classification systems. The latest prospect for enriching research methods in this field comes through efforts at standardizing the criteria (typology) in order that the pattern of agricultural activity may be made internationally comparable.

In South Korea today, about 60 per cent of the people are engaged in agriculture. Not only is agriculture the main form of employment, but it also contributes a significant proportion of the national income. Since 1960 the author has been extensively studying the agricultural regions in South Korea in order to shed more light on regional differences in land use and also on agricultural typology.

A study of agricultural management relating to the distribution and specialization of agriculture is now under way in South Korea. Unfortunately, too little attention is still being given to the relationship between types of agricultural production and the physical environment or the classification system. Indeed most of the systems of agricultural regionalization now in use in Korea do not give sufficient attention to classifying agricultural data or explaining the factors which determine regional differences in agriculture.

In an article titled 'A study of agricultural regions in South Korea' based upon the 1960 agricultural census, the author attempted to set up new agricultural regions. First the article deals with general problems related to the selection of sundry standards and indices for demarcating agricultural regions in South Korea. Secondly, the article provides a statistical analysis of the agricultural census data along with an interpretation and clarification of the problems relating to the agricultural regionalization schemes devised. In this analysis, the author mapped out crop combinations, sources of incomes, labour intensity, type of tenure, size of farm management, and land ownership systems. A system of comparable and uniform quantitative indices was used, which made it possible to achieve the necessary objectivity in classifying and delimiting the agricultural regions.

In the first study, the indices for classification were largely those of other scholars; physical factors were sometimes paramount, and in other cases they were used in equal association with human factors. When both physical and human factors are used, the relationship between the two is often very obscure. Until recently, most research on agricultural regions has continued to employ administrative units as the basic areal unit for classification in order to simplify the work, with the result that physical regions are not infrequently treated as agricultural regions. In reality however, differing climatic conditions give rise to quite different crop combinations and, more importantly, differing agricultural techniques and organization.

In the present study therefore, focus has been on the interrelationship between institutional and organization features and farm size and land tenure systems. Furthermore, the more traditional approaches have been broadened to encompass the consequences of changing marketing conditions and other socio-economic factors. Thus, in order to emphasize the functional nature of farm management, the author has incorporated more human factors into the classification scheme which follows.

INDICES

(1) Land use and farm management. (a) Ratios of paddies against total cultivated acreage; (b) superior crops; (c) farming pattern.
(2) Size of management. (a) Cultivated acreage per farm population; (b) cultivated acreage per farming household; (c) farming acreage for total population.
(3) Extent of intensive management. (a) Input of fertilizers per tanbo (1 tanbo = 0.099 ha); (b) labour hours per tanbo for a year; (c) rates of double crops for paddies; (d) rates of double crops for fields.
(4) Pattern of land ownership.
(5) Basic source of farm income.

What is apparent in this new study of agricultural regions based upon the 1970 agricultural census is that the scale of the agricultural enterprise becomes increasingly important in the differentiation of regions as the traditional rice farmer moves from the traditional subsistence economy to a commercial economy and from a pre-industrial agricultural system to an industrial agricultural system around the urban and urbanizing areas in South Korea.

Young-Hyun Keel; student of geography at Kyung Hee University.

Berry, B.J.L., 1958 A note concerning methods of classification, *Ann. Assoc. Am. Geog.* 48, 300–3.
Grigg, David, 1969 The agricultural regions of the world: review and reflections, *Econ. Geog.* 45 (2), 95–132.
Kostrowicki, Jerzy, 1964 Geographical typology of agriculture, principles and methods, *Geog. Polonica* 1, 111–46.
Lee, Chung-Myun, 1970 A study of agricultural regions in South Korea, IGU Commission on Agricultural Typology, *Geog. Polonica* 19, 147–70.
Rakitnidov, A.N., and V.G. Kryuchkov, 1966 Agricultural regionalization, *Soviet Geography – Review and Translation* 7 (5), 48–58.
Whittlesey, D., 1936 Major agricultural regions of the earth, *Ann. Assoc. Am. Geog.* 26, 199–240.

c0910
A model for agricultural change in developing areas
JANET H. MOMSEN *University of Calgary, Canada*

A comparative factorial analysis of small-scale agriculture in three Caribbean territories suggested that the different stages of overall economic development were associated with certain types of agriculture in a regular sequence (Momsen 1970, 9). Thus the historical process observed in the most highly developed area provided a predictive model for other similar but less-developed areas. In order to test this hypothesis, a survey of the decision-making processes of farmers in one territory which was undergoing particularly rapid economic growth was undertaken.

On the basis of previous patterns in the region, it was hypothesized that growth of the tourist industry with the associated increased demand for fresh foodstuffs was the catalyst precipitating the development of farm enterprises concentrating on production for the domestic market. In the area studied, the eastern Caribbean island of St Lucia, tourism grew 78 per cent between 1967 and 1969 and in the period 1970-1 the number of hotel beds doubled (British Development Division 1970, iii; and St Lucia Tourist Board 1971, 4). The spread effects of the tourist industry led to a rapid rise in the general level of per capita incomes creating an additional demand for an improved diet high in fresh fruit, vegetables, and animal products. The increased demand was largely met through imports. The cost of food items imported rose from $4 million Canadian in 1968 to $5.5 million in 1970, and their proportion of the total value of imports increased from 20 to 30 per cent (*Voice of St Lucia* 1971, 4). The increasing trade deficit of the state made import substitution through local food production imperative. The St Lucian government set up a marketing organization in May 1968 and gave encouragement and advice to farmers wishing to produce foodstuffs for domestic consumption. The preconditions for a new stage in the agricultural development of the state, introducing domestically oriented polyculture into the existing system of export-oriented monoculture, were established, as they had been in Barbados a decade earlier.

In order to examine the relationship between these preconditions and the actual development of a particular type of agriculture, a questionnaire survey of 35 per cent of the commercial vegetable producers of St Lucia, as listed by the Ministry of Agriculture, was carried out in 1971. Three-quarters of these farms, both smallholdings and plantations, had increased their production for the local market during the previous twelve months but the decision-making process involved in this land-use change varied considerably. The most commonly cited reason for the change, as mentioned by one-third of the interviewees, was the decline in the price of bananas, the major export crop, although nearly half the farmers still obtained most of their income from export agriculture. One-fifth of the farmers were influenced by the higher prices being offered for local produce, and a similar proportion had been encouraged to change their production pattern by their local agricultural extension agent. There was no spatial concentration evident. None of the farmers admitted to being influenced by their neighbours and only 4 per cent of the smallholders, as against 50 per cent of the plantations, had responded to direct demand from the many new hotels.

It has been said that farmers are reluctant to go into vegetable production because of its time-consuming labour demands in comparison to those of the traditional export crops of sugar cane and bananas. In fact, the labour relationship is far from simple. The decreased profitability of bananas was due not entirely to lower prices for the fruit but also to the increased cost of labour resulting from competition with the tourist industry and a recently introduced minimum wage law. Many farmers stated that they found it easier to supply the daily input of light labour needed by vegetables than the full-day once-a-week heading-out of 13kg stems of bananas from steep, hillside fields located far from paved roads. Children, a plentiful commodity in the region, could be used to help with vegetable production. Women, who constituted one-fifth of the farmers interviewed, could combine vegetable growing with housework and childcare, and the old and infirm, making up nearly one-third of the interviewees, could obtain a small income from the relatively light work of raising vegetables. Market gardening in St Lucia is, in many cases, the occupation of the weak.

Yet despite the high proportion of family labour used in vegetable production, labour shortage was the major problem mentioned by two-thirds of the farmers visited. Almost half the vegetable producers also had serious difficulties with poor roads and poor marketing facilities. This would suggest that the official policy of building feeder roads and setting up a marketing organization is not yet fully effective.

However, domestic food production in St Lucia has increased in parallel progression with the growth of the tourist industry. In particular, the area in commercial vegetables increased fourfold between 1968 and 1970 (Ministry of Agriculture, St Lucia, 1968, 1969, and 1970). Thus the original hypothesis is upheld. In fact the strength of the factor analysis of Caribbean agriculture as a dynamic model is strongly supported. Experience in St Lucia has followed that of Barbados very closely in many unexpected ways, although with a time lag of almost a decade (Ingersent et al. 1969). In both cases, a high proportion of the farmers changing to vegetable production were in the over-45 age group rather than being the supposedly innovative younger farmers; the official marketing organization set up to stimulate local food production had little effect (Brathwaite and Nurse 1971, 9–15); and many of the more efficient plantations that moved into vegetable production did so in order to diversify their exports rather than to sell on the domestic market.

The analysis emphasizes the credibility gap between the 'planner's prescriptive reality,' the 'cultivator's pragmatic reality,' and the 'cultivator's ideal reality' which inhibits change (Clarkson 1970, 715). Yet the model was successful in both predicting and explaining the introduction and spread of a new type of agriculture.

Brathwaite, A.H., and J.O.J. Nurse, 1971 Marketing of food crops in the agricultural diversification programme in Barbados, Paper presented at the Sixth West Indian Agricultural Economics Conference, Guyana.

British Development Division in the Caribbean, Ministry of Overseas Development, 1970 *Economic Survey and Projections. St Lucia.* September (Bridgetown).

Clarkson, J.D., 1970 Ecology and spatial analysis, *Ann. Assoc. Amer. Geog.* 60 (4), 700–16.

Ingersent, K.A., A.H. Brathwaite, and J.O.J. Nurse, 1969 *Vegetable Production in Barbados: An Economic Survey*, Ministry of Agriculture, Science and Technology (Bridgetown).

Momsen, J.H., 1970 Classification of agriculture: A case study from the Caribbean, paper presented to the Fourth Meeting of the IGU Commission on Agricultural Typology (Italy).

St Lucia Tourist Board, 1971 *Report, 1970* (Castries).

Voice of St Lucia, 1971 8 May 1971.

c0911
A method to determine the misuse of land in some selected villages of Ganga Plain: a model in agricultural typology
B.K. ROY   *Office of the Registrar General, India*

Determination of misuse of agricultural landscape is difficult, especially when the use of land is very diversified. Through field work, it has been observed that the Indian peasants are very conscious about the use of their holdings and exercise definite planning. With this in mind, the author has carried out intensive fieldwork in selected villages (chosen mainly from geographical areas) of Uttar Pradesh to determine the use and misuse of agricultural land in specified holdings. Three types of cultivators namely owners, sharerenters, and cash-renters, numbering about 30, have been selected for this study in order to develop the concept of the determination of misuse of land. This shows that the cultivators plan to grow food crops, vegetables, and cash crops in more than 12 categories on their holdings, ranging from below one acre (0.45ha) to above 2.5 acres (1.01ha), with definite planning based on weather and climate, type of land, market and storage facilities. In addition, this study shows how the good intentions and planning considerations of the selected cultivators are vitiated by unfavourable elements which cause deviations among various categories of plots (holdings) marked for specific use by the cultivators because of the physiographical and economic conditions of the areas. The study has been supplemented with maps which have been drafted in accordance with empirical surveys of the selected villages.

c0912
Changing Japanese agriculture: typological notes
MASATANE SOMA   *Ehime University, Japan*

Japanese agriculture as a whole has commonly been characterized as 'peasant semi-commercial agriculture.' But remarkable changes in it have been occurring as a result of the great growth of Japan's whole economy together with rapid technological developments and advancing foreign trade. It is not enough now to see it only in terms of the traditional types of agriculture. A new characterization of Japanese agriculture will require a new approach or a new point of view: it is necessary that we should try to grasp it in the totality of its changing features and not merely relative to each changing feature separately. Two case studies recently carried out by the writer are reported here; they seem to support his observation in its essential points.

Yoshida-cho, which is on the shore of the Uwa Sea at the western end of the Shikoku Mountainous Region, is developing as a citrus-growing centre utilizing terraced fields. According to the 1970 statistics, it has a population of 16,166, 1884 farm households, and 2468ha of cultivated land. In the size of the farmland they operate, the farmers in Yoshida-cho rank relatively high among those in southwestern Japan as well as in Ehime-Ken, 56 per cent of them each operating 0.5–2.0ha. Sixty-one per cent of the farmers are full-timers – the prefectural highest. The remarkable progress and high productivity of citrus-growing in Yoshida-cho are mainly due to the fact that the middle-class and upward farmers made 'joint' investments in fixed capital in order to have better management and new or improved means of transportation – ropeways, agricultural roads, etc. The development was made possible through the setting up of a communal cooperative corporation with governmental subsidies. To this extent, the farming in the town assumed the form of cooperative agriculture, although this was a temporary phase, the final outcome being the break-up of the corporation leaving a co-operative management just sufficient to compete with success in a capitalistic system. It cannot rightly be said that the form of farming has reverted to the former type of 'peasant semicommercial agriculture.'

Honkawa-mura in Kochi-ken is northeast of Yoshida-cho and in the uppermost reaches of the Yoshino River. According to the 1970 statistics, it has a population of 1766, 199 farm households, 177 of which are part-timers, that is, those engaged in some side business or other as well. Ninety-five per cent of its area is wooded land, of which 11,873ha is owned by the state and about 2000ha is owned by private corporations. It has only 40.09ha of cultivated land. Up to 10 years ago the *mura* had been one of the typical burning-cultivation areas in this country, but almost all of the burning fields have been afforested. Although *mitsumata* (*Edgeworthia papyrifera* Sieb. et Zucc.), which used to be the main crop in these fields, is now grown by a special new method in ordinary fields near where the people live in the *mura*, the development of forestry has recently been a chief feature of the productive activity of the *mura*. This has been brought about through the policies of the central and local authorities (building motorable roads, etc.) and the efforts of the *mura* forestry association. The latter has introduced useful machinery and organized labour squads of local people. These mechanized squads take care of forests outside the *mura* as well as their own forests within the *mura*. Thus the association is coordinating local economic activities on a cooperative and commercial basis.

A few tentative observations may be made on the basis of the foregoing case studies. (1) There has been little change in the form of land ownership. (2) The present great-sphere capitalistic system has necessitated various improvements in the form of farm or forest management, in the means of production or transportation, or in the organization of labour. Some form of local cooperation, extending beyond the limits of peasant farming, is needed, if only for a time, to bring these improvements into being. (3) Specialization in a particular form of agriculture cannot be avoided if a large scale of operation is necessary to survive economic competition.

If these observations are correct, the factor-analysis type of approach is not adequate in considering the type of agriculture. An integrated point of view is also needed.

Kostrowicki, J., and W. Tyszkiewicz, eds., 1970 *Agricultural Typology – Selected Methodological Materials* (IGU), 7–60.
Shirahama, H., 1969 The characteristics of Japanese agricultural region, some angles of analysis in agricultural typology, *Bull. Faculty Education, Chiba U.*, 136–48.
Soma, M., 1970 A geographical study of the Nanyo region in Ehime Prefecture, *ibid.*, 74–82.
– 1971 The evolution of land utilization in the villages of the Shikoku Mountains, *Geog. Rev. Japan*, 44 (4), 301–18.

c0913
Land use and natural vegetation
C. VANZETTI   Università di Padova, Italy

Let us hope that the methodological researches which are carried on in farm analysis will lead not only to the individualization of environmental characteristics through concise and yet meaningful notions, but that the results of this research may be used to promote economic development, and to make comparisons among the different stages of development of both territories which are in the same national context and in different states.

In farm analysis also, the knowledge of land use is most effective for detecting the general characteristics of the environment as the knowledge of land tenure and types of farming is effective, but in this case one should detect some elements which can be matched with the ones which are essential to farm analysis. There exists also a regional or territorial analysis, as you may call it, which without neglecting the farm, cannot study so small a unit in detail properly; fundamentally it is based on statistical data which are already available or have been especially gathered for the total area under study. In this case, the general data of land use assume a relevant importance.

The concept of land use has already been discussed in this Commission and elsewhere and it has been pointed out that land use aims not so much at representing what exists at a precise moment but rather at representing what is, at a certain period of time, characterized by such a rotation as to be easily represented by averages. If the idea of rotation is fairly clear when it is continuous, namely when one cultivation is followed by another one, it becomes less and less easy to be defined when the cultivation is followed by either a short, or long, or very long fallow.

The solution adopted by the *World Atlas of Agriculture*, whereon the co-operators on this work have agreed, gives some examples of the way the so-called shifting cultivation and bush fallow may be represented, namely by showing the areas cultivated yearly by conventional areas spread out on a background of vegetation.

Natural vegetation is not often so easily defined and represented for those areas where man has deeply changed the environments. So the Mediterranean maquis is of little relevance where agriculture is developed, since this type of vegetation is represented by small patches and will not reconstitute any longer during the fallow which is generally too short to allow the most characteristic species to develop.

Natural vegetation becomes a less abstract concept when it corresponds to a concrete reality, and what has to be represented really exists on the ground: the classifications 'woods and forests' and 'savannah' correspond to a concrete reality and 'woods and forests' are represented such as they are even if they are dotted by temporary shifting cultivation and bush fallow.

The problem of representing 'woods and forests' from an economic point of view is the same problem as detecting whether they are exploited or not, the way they are exploited, whether the exploitation is a mere abstract possibility or is inconceivable because of technical and economic reasons. The problem might be solved through subclassification which, at the present stage of research, cannot be realized on a world scale and sometimes on a regional scale too.

The *World Atlas of Agriculture* has limited itself to indicating as non-exploited those vast areas of forests which are now excluded from exploitation, but may somewhere else be 'woods and forests,' even in the most developed countries, and which cannot be exploited owing to their location, to poor communications, and for economic reasons.

At present we can be but vague and imprecise because research giving these details and data is lacking.

Another big problem is the classification of permanent vegetation which is utilized for fodder or rich grazing, and is called 'meadows and permanent grassland,' and the vegetation which constitutes 'rough grazing land,' that is to say poor vegetation which could be exploited by the livestock which grazes it.

The problem is complex in itself, but it may be that in those countries where agriculture has been developed for a long time and economic life has a high standard, it is probable that the utilization of such natural vegetation should be extended over most of the area; but exodus to the town, which has been a worldwide phenomenon these last years, raises some doubts as to the large scale exploitation of natural vegetation.

The problem of 'savannah,' 'steppe,' and 'rough grazing land' is still more complex, since we do not know if these natural vegetations are utilized; as for 'savannah' and 'steppe,' utilization is often precarious and uneven, as it is exclusively dependent upon rainfall.

There exist some semi-desert or desert areas which are utilized for grazing at long intervals of years, since the rainfall is rather uneven and alternates with very long dry periods.

Elsewhere the problem may be different, and the 'savannah' may be utilized for grazing according to existing trypanosomiasis and to whether there is livestock available for grazing.

The same problem rises in regard to 'steppe' and 'savannah' which are mixed with 'non-agricultural land.'

The solution adopted by *World Atlas of Agriculture* seems to be careful and correct: the classification adopted for all the world is 'rough grazing land' and 'rough grazing land and non-agricultural land,' but it has always been pointed out whether it was 'savannah' and 'steppe' in order to make it clear to the reader that their exploitation is uneven, unpredictable, and dependent upon special, environmental conditions.

c0914
A classification of agricultural land use for development planning
HAROLD A. WOOD   *McMaster University, Canada*

Before considering the classification of any class of phenomena, the class must first be defined precisely. In this paper, 'land use' may be defined as *that group of associated human activities by which the land is made to yield products of value to man*. In other words, it is asserted that 'land use' is not simply another name for 'vegetative cover,' as is commonly assumed. This definition is basic to the ensuing discussion.

The characteristics of any classification will depend upon its principal purpose. In the field of economic and social planning, land use information is obviously relevant. However, such information must be provided in a form that can be related not only to the natural resource base but also to fundamental human problems, such as poverty, unemployment, etc.

These two requirements cannot easily be met at any single cartographic scale, since variations in the resource base involve ecological changes which commonly occur within even a single field, while variations in living conditions are often regional in character and, in any case, involve generalizations over at least an entire farm or other unit of production.

In the past, most land use classifications have had an ecological bias, and are suitable for use in planning only if the planning is also essentially ecologically oriented. Where the purposes of planning are more specifically social or economic, a land use classification should be based on production units. Under these conditions, two characteristics of such units are of fundamental importance: the identity of the product or products and the intensity of production. Both characteristics, however, may be highly complex and a classifier might proceed with almost limitless subdivision unless some other criteria can provide guidance on where to stop and on the validity of any groupings which are accepted.

Such criteria may be found in two basic parameters of any given land production unit, namely the labour requirements and the value of production per hectare (or other unit area) per year. These parameters will be useful in land use classification if it can be shown that they are relatively uniform where given items are produced at given intensities of production.

The validity of this assumption has been examined in rural areas of six countries of Latin America. It has been established that acceptable levels of homogeneity, with respect to these two parameters, may be achieved where major commercial crops are identified precisely, subsistence crops are placed in three groups, the ratio of used to unused land is noted, the seasonality of production is defined, and the level of technology is indicated. Nevertheless, fairly substantial errors arise when results from one country or geographic region are extrapolated to another, because of the influences of soil fertility, local price structures, and certain cultural features. Full applicability of the principles involved in the classification will thus require the use of 'regional coefficients.' The precise construction of such coefficients has not yet been investigated. Apart from the theoretical desirability of including certain elements in a land use classification, it is important, for practical reasons, that much of the data be obtainable through airphoto interpretation. In the case of the proposed classification, the bulk of the needed information can, in fact, be secured by this means, especially for areas which are also covered by an agricultural census and for which reliable climatic statistics are available, but some field work will normally be required, especially to determine precisely the level of technology employed.

Strauss, E., 1969 *Metodología de Evaluación de los Recursos Naturales,* Cuadernos del Instituto Latinoamericano de Planificación Económica y Social, Ser. II/Anticipos de Investigación no. 4 (Santiago, Chile).

Polish Academy of Sciences, Institute of Geography, 1962 *Land utilization, methods and problems of research,* Geog. Studies no. 31 (Warsaw).

c0915
Aerial photographs: their application to the classification and analysis of agricultural land use
ROBERT A. RYERSON   *University of Waterloo, Canada*

Historically air photo interpretation (API) (American Society of Photogrammetry 1960) parallels the development of many other techniques in that one moves ever from the general to the specific. This move has been facilitated by those familiar with both aerial photographs and some other area of academic interest. A number of these people have been, and are, geographers.

Before discussing recent basic research in API, one should be familiar with the interpretative criteria used by photo interpreters. These are tones, colours, relative size, texture, shadow, and stereoscopic height. These criteria are in turn related to crop phenology, sun angle, degradation of tone over the film, scale, etc. Thus, for a specific mission, one must consider the best time of year (based on crop phenology), film-filter combinations, and scale.

Using such considerations, Table 1 has been drawn up relating these variables to API in southern Ontario. These analyses have been successfully applied and are no longer considered to be in the realm of 'basic research' in API.

The development of API techniques rests on one premise: the researcher must possess a thorough understanding of the phenomena being interpreted. A more general requisite is a systematic approach to the posing and solution of the photo interpretation problem.

Over a period of several years Wood noted that, in southern Ontario, one could determine crop types on 1:12,000 photography flown in mid-summer. The next logical step in the analysis of agriculture, moving ever from the general to the specific, was believed to be the identification of farm type, to be followed by determination of livestock numbers (see Ryerson and Wood 1971).

It was found that each type of animal (beef steer, dairy cow, beef calf, etc.) requires specific crops (namely corn, small grains, hay, and pasture) rationed by weight of feed per day dependent on the age, use, and size of animal. One could then determine the number of hectares of each crop (given local yields) required to feed each type of animal. For each type of animal several such areal equivalents involving these crops exist. These equivalents, when changed to percentages and graphed on a triangle graph (1ha of pasture taken as 0.6ha of hay) yield several areas for beef and two for dairy.

To determine farm type, one need only determine the percentage of each farm in each crop and compare them to the graph. Of 269 farms so tested, 88 per cent fall into the predicted location. Of 83 farms for

TABLE 1. Analysis of temperate agriculture using aerial photographs

Purpose	Film type	Filter	Scale giving high accuracy	Number of coverages	Approximate date
Separating agric. from non. agric.	panchromatic	minus blue	1:60000	1	July
Cultivated from pasture	"	"	1:30000	1	July 1
Crop identification (all grains as one)	"	"	1:12000	2	mid July mid August
Separation of grains	colour false colour	none yellow	1:8400	1	"
Detection of diseased areas	"	"	1:8400	1	middle to height of growth
Detection of diseased spots	"	"	1:3600	1	"
Ranking levels of infection	"	"	1:800	1	"
General farm management practices	panchromatic	minus blue	1:12000	1	mid to end of August
Specific practices, farm boundaries, silo capacity, etc.	colour	none	1:8400	2	July 1 mid to end of August

which more detail existed, 63 fell into the proper locations. Of the anomalies, it was found that some were part-time farmers, some had mixed operations, and others were cow-calf beef operations where beef cows were fed as dairy. Many of the anomalies were explicable in terms of building type. Like crops, buildings too can provide an insight into the farm type. Each building design, as seen from the air, can be associated with a particular use.

The means of determining cattle numbers follows directly from the above. In the rigorous climate of southern Ontario, cattle must be sheltered in winter and thus require some form of housing. Further, all cattle need a certain amount of feed to live, reproduce, and provide meat and/or milk. Thus a two-pronged approach may be developed to determine number of cattle.

In the case of feed, all that remains, once type of operation is determined, is to measure hay and corn silage available and divide by the weight of feed required per animal per year for the given type of operation. The accuracy of prediction in a regional context (i.e., per county sample of 20–30) is high, with a mean accuracy per farm of −3.6 per cent, but with a standard error of 14.3 per cent.

Using building capacity, one takes into account that barns of given dimensions and design can accommodate only so many animals. These capacities were predetermined and then applied to the barns on the sample farms. The accuracy of the building analysis is quite similar to the crop analysis, mean error per farm of −3.3 per cent with a standard error of 16.6.

These large errors at the individual level are a result of unused space, inefficient use of cropland, or yields lower than expected. It was then decided to accept (essentially) the lower of these two estimates.

Using this procedure the results are considerably more accurate (see table below).

	Number of farms	Mean error of prediction per farm in %	Standard error per farm	Number of cattle	
				Actual	Predicted
Beef	52	−1.9	5.9	3816	3765
Dairy	29	−0.5	7.3	1335	1330

Teaching students to carry out this task proved inexpensive and took only several hours of lab work.

With reference to land classification, the foremost limitation is here seen to involve the level of classification. Ideally each level of a classification scheme should be equally easy to interpret.

Other problems are associated with the particular research areas, type of vegetation, and cropping practices (e.g. multiple cropping in the tropics) or the similar appearance of crops (e.g. cabbage and tobacco in early summer in southern Ontario).

Other limitations include constant cloud cover (now radar is used); improper coordination of aerial surveys and ground data collection; undirected interpretation resulting in confusing maps; the plethora of classifications, some tested, some not; and lastly the lack of communication between potential users of techniques and those developing them.

Some of the above are simply questions to be resolved; others are indeed limitations. The list, not intended to be complete, shows some of the wide range of problems encountered by users of API for agriculture.

Itemizing potentials of API must be tentative, ill-defined, and largely in the realm of conjecture. In this light then, the author suggests that 1:60000 colour IR imagery may prove useful in agriculture in place of 1:12000 black and white (panchromatic) imagery for crop identification, farm type recognition, and regional livestock inventories (Ryerson 1971a). Accuracy is high, with a dollar saving related to time to interpret, number of prints handled (1/25) and additional feature of a larger overview.

Basic research now under way is aimed at obtaining a more complete economic survey of agricultural producing units utilizing API and ground sampling (Ryerson, 1971b).

Closing on an optimistic note, 'photo surveying and mapping can provide the preliminary cartographic exploration needed to alleviate world-wide under-nourishment' (Brandenberger 1969). It is largely up to informed agricultural geographers to apply their knowledge and abilities to aid in answering this call.

American Society of Photogrammetry, 1960 *Manual of Photo Interpretation* (Falls Church, Va.).

Brandenberger, A.J., 1969 'Economic impact of world wide mapping,' *Photo. Eng.*, 341.

Haefner, H., 1967 'API of rural land use in western Europe,' *Photogrammetria*, 143.

Ryerson, R.A., 1971a A Comparative Analysis of Several Aerial Imagery Types for Agricultural Studies, paper for Integrated Aerial Surveys, University of Toronto, McMaster U., U. Guelph, and U. Waterloo, mimeo.

– 1971b An Investigation of Agricultural Data Collection from Aerial Photographs: The Prediction of Land Use Change, PH D dissertation proposal, U. Waterloo, Geog. Dept. Division of Environmental Studies, mimeo., 37 pp.

– and H.A. Wood, 1971 'Air photo analysis of beef and dairy farming,' *Photo. Eng.*, 157.

Steiner, D., and H. Haefner, 1965 'Tone distortion for automated interpretation,' *Photo. Eng.*, 269.

– 1967 *Index to the Use of Aerial Photos for Rural Land Use Studies.* Bad Godesberg, Selbstverlog der Bundesstalt für Landeskunde und Raumforschung (in English – over 1000 annotated entries).

– and H. Maurer, 1969. The use of stereo height as a discriminating variable for crop classification on aerial photographs, *Photogrammetria*, 223.

Wood, H.A., and R.A. Ryerson (in press) *Air Photo Interpretation of Agricultural Land Use: Southern Ontario, a Case Study* (working title) (Toronto).

c0916
Tendencies in the development of Danish agriculture
AAGE H. KAMPP   *Royal Danish School of Educational Studies*

Until about the year 1900, agriculture was the main industry of Denmark, as regards both the national income and employment, and even after secondary, tertiary, and quaternary trades became the principal employers of manpower, agriculture still continued to earn the greatest amount of foreign currency. Today, Danish agriculture as employer and money maker has been surpassed by the manufacturing industries.

Danish agriculture is an industry which till now has been dependent on a rather one-sided production of animal goods. A failing market will influence the farmer's standard of living, the agricultural manufacturing industries, marketing organizations, carrying trade, processing industries, foreign trade, and the budgets of the state and the municipalities.

The number of farms is being reduced with accelerating speed. Two hundred years ago, the enclosure movement started to bring together the allotments of the individual farms, and Danish farms have been rounded off better than in other European countries, involving advantages as to labour, transportation, and crop rotation, thus promoting effectiveness and competitiveness.

During the first half of the 20th century, 30,000 state subsidized holdings were established, but the parcelling out has now been replaced by a reversely directed accelerating movement; since 1960, the number of agricultural holdings has been diminished by a third, though in Jutland by only a quarter.

Owner-operated farms have dominated in Danish agriculture for more than 150 years, and today more than 66 per cent of the Danish farmers are freeholders. Tenant farmers, however, enjoy a great independence as to making decisions, and several farmers have begun to realize that the right of use is more important than the right of ownership, and it has become common for a landowner to rent the land of a neighbour when the latter uses the buildings only for residence. Joint operation in 1967 was already twice as common in the Island as in Jutland.

Labour is now being drawn towards the more permanent and better paid jobs in the manufacturing industries. In this way, those who remain farmers have the possibility of a larger profit. At the same time, state subsidies render possible modernization of the agricultural production, amalgamation, or abandonment of unprofitable holdings, and investments in new buildings and implements in step with the technological development. The annual production per male worker in agriculture is now 3 times as large as in 1950; at the same time, there is an increase of the number of people employed (1) in factories supplying goods for agriculture, (2) in the processing industries, and (3) in connected branches of industry such as transportation, commerce, financing, advertising, administration, science, education, etc. The tractor made available 250,000ha which instead of horse forage can produce pig feed or vegetables.

The reduction of farmland to a large extent takes place at the expense of the most fertile parts of the rotation area. The largest decrease has taken place in the Islands and in Eastern Jutland, but a tendency to a parallel development is already noticeable towards the west. The cereal areas decreased from 1938 to 1950, but the reduced working staff and effective pesticides and herbicides have since then brought about a strong increase.

Add to this, that cereal growing is state subsidized, that the possibilities of growing other cash crops is very limited, and that the low prices on products of animal husbandry have encouraged a reduction of the feed crops.

Spring barley amounts to 51 per cent of the rotation area. All other cereals have decreased. As of 1968, winter barley was prohibited for 5 years because it is the winter host of mildew, yellow rust, and brown rust, thus diminishing the yield of spring barley.

The beet root areas have increased to a certain extent, but the areas of swedes have decreased more because they yield less solids. New types of beetroot seed combined with herbicides spraying have reduced the manual work in the root fields.

The areas with grass have decreased by 1/4 mill.ha, but the reduction of the total yield is smaller than the reduction of livestock; accordingly, grass makes up a larger percentage of the feed than earlier. For a grassfield, only 1/3 of the amount of fertilizer demanded by a beet crop is used. On the other hand it yields only half of the caloric value but a greater yield of protein per ha. When expenses are taken into consideration, it yields 5 times as much per 100dkr. Silage yields are quite reliable, thus compensating for fluctuations in the production of roughage.

The above-mentioned variations in area reflect that several holdings have reached a considerable simplification. A rising number of farms approach the fully specialized production with a single or a few crops and none or only one single form of animal production. The rotation of crops has been replaced by deep tilling, pesticides and herbicides, and special machinery.

Livestock has been a very sensitive branch of production, and is, for the moment, severely affected by economic conditions. The number of horned cattle has been only slightly decreasing during the last 30 years, but the structure of the herds has changed. Herds of 15–30 dairy cows are now dominant. The average percentage of dairy cows has decreased from 50 per cent to 40 per cent today; the yield of milk per cow has increased from 3000 to 3900kg. The number of calves has increased from 26 per cent to 34 per cent because they are slaughtered at a later period of life. When a cow disappears, ½ha of land becomes vacant for cereal growing. The favourable market for beef has increased the difficulties for dairy products in that the result has been larger production of these; more than 6000 holdings out of 135,000 have, however, now exclusively beef cattle.

In spite of the decrease in quantity and quality of the farming land, there has been a production increase of 2 per cent yearly because of plant breeding, extended use of fertilizers and chemicals against weed and plant diseases, mechanization, rationalization, soil improvement, and a close co-operation between science and practical agriculture. The yield per ha has increased more than 20 per cent since 1950. The increase was largest towards the west because of larger use of fertilizer.

The yield of animal husbandry increased even more because of an increased import of forage, the above-mentioned increase in plant production, and the continuous improvement of the stock and of the feeding science.

The above-mentioned tendency to simplification of production has among other things a connection with regionalizing, i.e., specialized regions with common agricultural adviser, common cultivation experiments, and use of the waste products. Such crops are sugar beets, potatoes, carrots, green peas, and alfalfa for bolus. Likewise there is a tendency to establish industrial pig-breeding centres and 'egg-factories' as well as to extend the grass growing in marginal regions and the cereal areas towards the east. The transport problem is no longer decisively significant.

In the years following 1939, Denmark was divided into regions according to the same quantitative principles as later used by Shafi with the modification that he has put each figure in relation to the national Indian average.

At present, there is the extra uncertain factor for Danish agricultural export that there is as yet no definite decision as to the membership in the EEC of Denmark and UK.

c0917
Towards a planning-oriented agricultural typology for the municipality of Niagara-on-the-Lake, Ontario
MARVIN T. SUNDSTROM   *McMaster University, Canada*

The role that agriculture will play in regional official plans for long-range development in Ontario constitutes a vital problem. In the past, agriculture has typically been viewed as a residual land-use category which provided vast land reserves for future urban development. If a strong case is to be made for an agricultural role in regional planning, some assessment of the past trends, the current situation, and the future prospects for agriculture is essential. The purpose of this paper is to develop a number of variables which can be used to assess the future of agriculture with the possibility of incorporating the variables into a planning-oriented agricultural typology.

Emphasis will be placed on evaluating the future prospects and agricultural adjustments in one municipality in Regional Niagara. Niagara-on-the-Lake, the focus of this study, is one of the important agricultural areas of Regional Niagara. Agricultural assessment accounts for some 25 per cent of the total assessment for Niagara-on-the-Lake. Approximately one-third of the tree fruit and grape area in Regional Niagara is found in Niagara-on-the-Lake. Although fruit production dominates, a number of intensive animal operations give it a much more diversified agricultural base than the 'fruitbelt' designation might suggest. Increasing land values and zoning reflect the pressures for non-farm development. Because of the importance of agriculture in Niagara-on-the-Lake and the mandatory requirement of an official plan, it has been selected as a suitable area and scale for which to assess agricultural adjustments and prospects.

Variables representing physical, economic, and social characteristics important to a continued high level of production have been selected. In addition, factors reflecting the process of adjustment within the agricultural sector have been included. Randomly selected blocks used in a study by Reeds have been employed to assess agricultural change on a more detailed level. Pending changes involving the loss of agricultural land for the new Welland Canal route and rural non-farm residential development will further reduce the agricultural acreage.

The procedure used in the study consisted of four steps, which are similar to those used by Conklin in his county level evaluation of income expectancy in New York State. In stage one, background data, contained in the ARDA agricultural classification, the soil survey, together with information on land values and current zoning for the municipality, are assembled. Agricultural land use was mapped from air photographs. Stage two consists of field checking and field observations supplemented by information from farm management studies. Stage three includes data obtained from interviewing residents in the sample blocks and other key people who have a comprehensive knowledge of the local industry. Market prospects and regional agricultural significance are examined in the final stage. Supplementing this will be an overview of agriculture in Niagara-on-the-Lake along with a detailed look at agriculture in selected 40.5ha (100 acre) sample blocks.

Because of the small area involved, one can assume that climatic differences will not constitute a significant variable. Difficulty was encountered in attempting to assess the physical base solely on the ARDA agricultural land capability classification. Although the specialized nature of agriculture in Regional Niagara is discussed in the text, mapping is based on the suitability of the soil for general farming. Consequently, it was necessary to consider a supplementary source in this area of specialized production. This source was the Lincoln County Soil Survey Map, which has a rating of the suitability of the various soil types for a range of crops.

Completion of an updated version of a general agricultural land-use map and comparison with a map based on 1961 mapping indicated little change in the area devoted to fruit production. Although there is a gradual reduction in the amount of agricultural land, it might be hypothesized that these reductions are more than offset by increasing intensity of production on the remaining land. The pattern of land sale values for the municipality provided an indication of where the major pressure has been exerted in the area over the past ten years. Analysis of the sale values on a per-acre basis and the size of holdings reveals a significant inverse relationship. That is, the larger the holding, the less the per-acre sales price. This analysis suggested that the greatest pressure for land sales was on the smaller tracts of land. These findings suggested that more detailed information on farm size on a sample block basis would be a desirable input. Finally, under economic considerations, one should look at the type of commercial operation, i.e., whether it is part-time or full-time, and the nature of the long-term demand for products of the fruit industry.

One of the factors relating to the future of the individual operation is the presence or absence of an heir to the farm, the age of the operator, along with recent changes and pending changes in tax legislation. These variables will be assessed on a sample block basis and related to the farmer's plans for the future of the farm.

The ability of agriculture to maintain itself in an area under pressure depends in part on how successful it has been in intensifying production over the years in the face of competing land uses. When combined with the changes in land use on a sample basis, some understanding of the process of agricultural adjustment can be gained.

In addition to the changes occurring within the agricultural sector, continual pressure is being exerted on the agricultural land resources by rural non-farm development. These pressures are represented by demands for rural non-farm residential development, the new Welland Canal route, and demands for land for part-time farming.

In conclusion, an understanding of the adjustment process and future prospects of agriculture can be gained by the use of selected variables representing physical, economic, and social factors in conjunction with measures of agricultural change and external pressures. An over-all assessment for the municipality as well as an assessment of selected 40.5ha (100 acre) sample blocks can be made for Niagara-on-the-Lake. These sample blocks can then be used as a basis for monitoring future change. Furthermore, selected variables can be incorporated into an agricultural typology relating to the ongoing process of agricultural adjustment. Included in such a typology at a sample block level would be such factors as: (1) physical base rating, (2) land values, (3) farm size, (4) nature of commercial operation: full time or part time, (5) future plans for the farmer, (6) value of production for various types of farm enterprises, (7) pending changes that will reduce the agricultural resource base.

A typology which would take some or all of the above variables into consideration provides a sounder basis for assessing and evaluating the future prospects of agriculture.

Conklin, H.E., 1959   The Cornell system of economic land classification, *J. Farm Econ.* 41 (3), 548–57.
Gregor, H.F., 1963   Urbanization of Southern California agriculture, *Tijd. Econ. Soc. Geog.* 54, 273–8.
– 1964   A map of agricultural adjustment, *Professional Geog.* 16, 16–19.
Griffin, P.F., and R.L. Chatham, 1958   Urban impact on agriculture in Santa Clara County, California, *Ann. Assoc. Am. Geog.* 48 (3), 195–208.
Krueger, R., 1970   The disappearing Niagara Fruit Belt, from *Regional and Resource Planning in Canada*, rev. ed. (Toronto), 134–49.
Reeds, L.G., 1969   *Niagara Agricultural Research Report*, Part 2 Fruitbelt (Regional Development Branch, Ontario Department of Treasury and Economics).

c0918
Regional differentiation in the productivity of field cultivation in Finland
UUNO VARJO   *Oulu, Finland*

There are many factors which contribute to agricultural productivity. These are examined here within the context of field cultivation in Finland.

In order to appreciate fully the factors which affect the productivity of arable farming, one should first refer to Fig. 1, in which the various factors involved in the productivity of wheat cultivation are presented in diagrammatic form. The starting-point for this diagram is the gross return, the total value of the products of wheat cultivation. By subtracting the production costs from this gross return, we obtain the net profit. Net profit, the interest on capital invested in agriculture, and the farmer's estimated wage

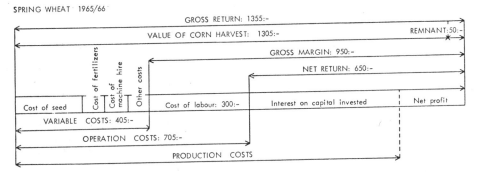

Fig. 1.

TABLE 1. Labour costs, variable costs, and gross margin per hectare for certain of the principal agricultural crops in 1965–6 on farms of less than 10 arable hectares in Finland (MKJ 518; Westermarck 1964a, 49; MTTJ 16, 1968)

Crop	Labour costs (Fmk)	Variable costs (Fmk)	Gross margin (Fmk)
Autumn wheat	360–	430–	1200–
Autumn rye	360–	290–	1020–
Autumn rape	110–	280–	1320–
Spring wheat	300–	410–	950–
Barley	300–	290–	590–
Oats	300–	290–	560–
Potatoes	1160–	780–	1560–
Sugarbeet	1400–	840–	3600–
Kale	1300–	540–	540–
Hay	100–	220–	300–
Silage	200–	290–	290–

constitute the gross margin, which conversely represents the gross return less the variable costs, i.e., expenditure of materials, taxes, employees' wages, etc.

A central position in agricultural planning has been accorded to gross margin as a measure of commercial achievement because it is obtained as an absolute number and may be calculated separately for the production of each branch of agriculture (Westermarck, 1964b, 7). Thus the productivities of different branches of agriculture, different regions, and even different years may easily be compared one with another. Similarly, gross margins from different branches of agriculture may be combined, the gross margin from field cultivation being considered as the sum of the gross margins from the various crops raised on the farm.

The gross margin per arable hectare may be calculated for each crop from the proportion of the field area and variable costs devoted to it and from the size of the harvest (SVT III: 63, 1968; 64, 1969; 65, 1970; Table 1). The resulting tabulation for the period 1967–9 is presented in Fig. 2 (not shown). Here the columns denote the gross margins from the crops cultivated, and the shaded portions for each area the combined gross margins per hectare.

Among the many crops cultivated in southern Finland, sugarbeet, potatoes, autumn rape, and autumn wheat yield a relatively large gross margin per hectare, while that obtained from oats, hay, and silage is in general small. In Lapland only oats, potatoes, hay, and silage remain, and the gross margin from these is very small. The largest gross margin from arable farming is obtained in Varsinais-Suomi, over 600Fmk/ha. This decreases northwards and eastwards, so that north of the Oulu region, it is below 100 Fmk/ha – scarcely one-sixth of the value in Varsinais-Suomi.

Since a large proportion of the gross margin is taken up by wages, that part of the gross margin from arable farming remaining after these have been deducted is shown in Fig. 3 (not included). Here, too, the best results per hectare are obtained from sugarbeet, rape, and autumn wheat in southern and southwestern Finland. In Lapland, no crop gave a gross margin sufficient to repay labour costs, production being uneconomic in every case.

If we now combine the proportions of the gross margins from all the crops cultivated which remain in excess of labour costs, we see that the largest net profits (ignoring debts) for small farms are obtained in southern and southwestern Finland, where these may in places exceed 250Fmk/ha. The situation is similar in South Bothnia and in the Lake Region (excluding the southwestern corner), where a net profit of 50–100Fmk/ ha is found. The remaining parts of Bothnia and Lapland are in a weaker position, for here the net income from arable farming does not meet more than a part of the labour costs. Thus the farmers here must either accept a lower return in wages, or else, in Lapland especially, concentrate almost exclusively on fodder crops, which it is not worth while bringing in from outside but which may be used to the advantage of the farm as a whole in the more profitable exercise of cattle-rearing.

MKJ 518   Maatalousseurojen keskusliiton julkaisuja 518.
MTTJ 16, 1968   Tutkimuksia Suomen maatalouden kannattavuudesta, 1966 (Publications of the Agricultural Economics Research Institute, Finland 18).

SVT III, 63–65, 1968–70   The official statistics in Finland III, Agriculture 62 and 63.
Westermarck, Nils, 1964a   Viljelijän suunnitelmaopas, Maatalousseurojen keskusliiton julkaisuja 513.
– 1964b   Ohjeita katetuotto-menetelmän mukaisen taloussuunnitelman laatimiseksi, Maatalousseurojen keskusliiton julkaisuja 514.

Regional Aspects of Economic Development
Aspects régionaux du développement économique

c1001
Problems of complex development of new areas
V.A. VITYAZEVA   *Academy of Sciences, Komi ASSR, USSR*

The European northeast of the USSR, including the territories of the Komi Autonomous Soviet Socialist Republic (Komi ASSR) and of the Nenetski National Okrug of the Arkhangel'sk Oblast', were in the past undeveloped outlying districts of tsarist Russia. Nowadays the development of this vast territory shows to the whole world the enormous transforming and constructive powers of the socialist system and of Lenin's national policy. As a result of the purposeful complex work of scientists and planning and construction organizations, the northeastern area, formerly known as 'land of unpassed paths,' has changed during the short time of a single generation into a rapidly developing region of intensive industry.

The up-to-date European northeast of the USSR no longer presents a severe, poorly inhabited, and undeveloped area, as it had been described (Encyclopaedia Britannica 1958, 475). The writers there report that industrial enterprises and railways are lacking and mineral resources poorly exploited. Today the Komi ASSR is a large-scale industrial district, well provided with all modern means of transportation.

During the years of Soviet power many important deposits of coal, oil, natural gas, titanium ores, bauxites, rock salt, and other mineral resources have been discovered and carefully investigated in this territory. Furthermore, enormous wood stocks and considerable water-power resources are concentrated here and rich natural resources of raw materials are combined with favourable conditions of exploitation. This fact is a major precondition for a rapid drawing of complex natural resources of the northeastern region into the national economy of the state.

The economic efficiency of the complex utilization of all these natural resources is considerably higher than an ordinary sum total of the resources of fuel, energy, and mineral raw materials of the region when exploited separately. The specific character of the territorial distribution of these resources and the high level of their areal concentration make it possible to achieve an essential saving of materials and of labour force by means of a common internal structure of leading industries, and as a result of combining and territorial rapprochement of technologically combined productions.

A relatively favourable geographical situation of the region favours its rapid economic development under the conditions of modern scientific and technical progress. Mineral deposits as well as other natural resources of the region are distributed considerably nearer to the major industrial centres of the European part of the USSR, as compared with the distribution of the natural resources in many other regions of the country. Besides, they are concentrated here mostly in populated areas in the zone of the main railway line.

During the years of Soviet power fundamental changes in the life of the European northeast peoples have taken place. The population has increased more than 5 times and exceeds 1 million at present. Gross industrial output of 1970 has increased 220 times in the Komi ASSR and 80 times in the Nenetski National Okrug since 1913.

The following main branches of industry were newly established: coal, oil, gas, soot, cement, sawing, veneer, woodworking, pulp and paper. Additional necessary industrial branches such as power, metal working, and production of building materials have been created. Industrial branches connected with material needs of the population (light industry and food industry) as well as agriculture have been developed. Agricultural zones were created in the vicinity of industrial centres. The general industrial development of the region was connected with railway building. The territory of the region is now crossed from southwest to northeast by the main railway line Kotlas-Vorkuta with railway branches to Khal'mer-Yu, Labytnangi, Syktyvkar, and Koslan.

More than 30 new towns and settlements of urban type with 598,000 inhabitants (62 per cent of the total population) have been built in the Komi ASSR (Komi Branch Acad. of Sci. 1969). In the pre-revolutionary time there was only a single small town, Ust'-Sysol'sk (now Syktyvkar) with 5000 inhabitants.

Before the October Revolution the Komi and the Nenets, who inhabited the territory of the region, had no national written language. Nowadays the whole population is literate. In the Komi ASSR 14 scientific-research institutions are established with a staff of more than 1500 scientists, including more than 200 doctors and candidates of sciences. The major complex scientific institution is represented by the Komi Branch of the USSR Academy of Sciences, which is in charge of fundamental problems of the development of the European northeast of the state.

The scientific investigations of the Komi Branch of the USSR Academy of Sciences show that the ways of the economical development of new areas in the European northeast of the USSR are distinct as compared with the further development of older inhabited regions of the country (Geog. Soc. 1952–70; *Science* 1967; Problems of the economy 1970; Acad. of Sci. Pubs. 1953, 1954).

Specific features of the industrial development of the Komi ASSR were always revealed during the whole process of industrialization of this region. Industrial enterprises were established mostly on the basis of extractive industry and of wood stocks. In older inhabited regions of the country many industrial branches have been created on the basis of the existing agricultural backgrounds and have been 'gemmated' from it. Under the environmental conditions of the European northeast of the USSR agriculture could not constitute a base for the major industrial branches. Moreover it could not satisfy the increasing food needs of the growing population engaged in industry. In accordance with that situation it was necessary to deliver most of the agricultural products from southern regions and at the same time to create and to develop the local agricultural production in the vicinity of industrial centres.

The local conditions of industrial development have a considerable influence on the geographical distribution of the industrial enterprises and on the types of industrial centres. The 'spot' or 'patch' distribution of industry is characteristic of areas of the European northeast. Industrial enterprises are established here in the 'green wilderness' in the form of separated, isolated, and primarily not connected industrial 'spots' which arose on exploited deposits of mineral resources. Thus, a process of gradual penetration of industrial objects in wild taiga and tundra took place here. As a result of this process new towns such as Vorkuta, Inta, Ukhta, and dozens of urban settlements now exist.

However, in the process of the development of the productive forces the character of economical occupation of the territory is getting more and more diversified. Various new natural resources are beginning to be used in compliance with the modern level of technical development. Neighbouring, less-developed territories are affected by new interrelationships between the productive forces. Thus, new industrial centres and groups are appearing (for example, the towns of Pechora, Sosnogorsk, Mikun', and the settlements Khal'mer-Yu, Vuktyl, Vozhayel', and others).

Our investigations show a high economic efficiency of the development of the productive forces in the north in general and especially in the European northeast of the USSR. Development of new northern regions proves that technical and economic efficiency may be considerably raised when many kinds of works are produced by means of a new technique and by a new technology specially adapted to low temperatures of air, permafrost conditions, and extensively distributed swamps and bogs. Maximal industrialization of building works and use of special building materials approved for the north are also necessary.

Further progress in the efficient development of new northern areas requires the creation of territorial industrial complexes with the most rational use of natural resources and labour force. The principle of proportionality between main and attending industrial branches as well as a rational structure of the economic connections between regions are to be secured. The establishment of high productivity objects results in a considerable saving of capital investments.

In our country a complex development of productive forces exists. For example, we can mention major industrial complexes of Khibinogorsk, Noril'sk-Igarka, Bratsk, Sayany, and the biggest territorial complex of western Siberia, now being established. The Directives of the 24th Congress of the CPSU provide for the organizing here of the most powerful fuel base of the country (Directives ... 1971).

Nowadays the main problem of the European northeast of the USSR is the exploitation of natural resources in the Timan-Pechora region. Potential resources of oil and gas are estimated here as many billions of tons of conventional fuel; geological resources of the Pechora coal basin exceed 214 bill. tons; wood stocks are more than 700 million m³. Rich deposits of titanium ores, bauxites, rock salt, potassium, and of various building materials are concentrated on a comparatively small territory which is well connected by transportation routes with central districts of the country. Because of all these favourable factors the Timan-Pechora region may be considered as an area of the first and foremost occupation, where a high rate of development may be combined with a high economic efficiency (Increasing the efficiency ... 1971; Problems of increasing ... 1969).

Many capital investments have been made at present in that area: oil and gas deposits are being intensively exploited; the main gas pipeline 'Northern Lights' is in operation; the gas and oil pipelines Vuktyl-Ukhta-Torzhok, Nadym-Ukhta, Usinsk-Ukhta-Yaroslavl' are under construction. In the near future the construction of the Pechora fuel power station, of the Ukhta gas-oil factory, and of many woodworking enterprises and mills of building materials will be started. The ninth Five Year Plan (1971–5) envisages a new upsurge of the development of the productive forces in this area.

Thus we can say that on the basis of scientific planning which provides for the specialization and complex development of the national economy, the new Timan-Pechora industrial complex is now being established. It will be one of the most important and highly efficient territorial industrial complexes in the Soviet north.

About the influence of the transference of the flow of northern rivers into the basin of the Caspian on the national economy of the Komi ASSR, 1967 Len. section, Science.
Directives of the 24th Congress of the CPSU for the development of the national economy of the USSR in the Five Year Plan 1971–5, 1971 Edition of Polit Literature.
Encyclopaedia Britannica, 1958, vol. 13 (London).
Increasing the efficiency of social production in northern conditions. Syktyvkar, 1971 Komi Books.
Komi Branch of the Academy of Sciences of the USSR. Syktyvkar, 1969 Komi Books.
News of the Komi branch of the Geographical Society of the USSR, nos. 1–13 1952–70.
Problems of the economy of Komi ASSR, 1970 Komi Books.
Problems of increasing the efficiency of production and capital investment in the industry of the Komi ASSR. Syktyvkar, 1969 Rotaprint Komi branch AN, USSR.
The productive strength of the Komi ASSR, 1953, 1954 Vols. 1, 2, 3, publications of the Academy of Sciences of the USSR.

Geography of Arid Lands
Géographie des pays arides

c1101
A history of settlement in southeast Alberta to 1920
D.J. FLOWER *Medicine Hat College, Canada*

Southeast Alberta is a marginal agricultural region in the semi-arid part of the prairie provinces. Its precipitation averages 33cm annually and, following Thornthwaite, the area has a moisture deficiency of over 12.8cm during the normal year (Sanderson 1948, 298). Settlement has always reflected this moisture situation: early settlers flowed in and out of the region, attracted by good years and repelled by droughts. In this century the farming collapse of the 'dirty '30s' was the most spectacular example of this dependency. The economic potential of the area has always been subject to a variety of assessments. In 1857 Palliser described it as 'the central arid desert,' a view upheld by Hind in 1858 but completely reversed by Macoun in 1879 (Roe 1946, 131). The latter claimed 'the apparent aridity vanishes before the first efforts of husbandry' (Macoun 1882, 144), and this opinion provided one reason for relocating the CPR line through Calgary rather than Jasper. Even then the controversy was not ended, for the CPR did not consider this southern land 'fit for settlement' and, therefore, would not accept it as part of their contract (Hedges 1934, 50). Macoun's optimism encouraged subsequent homesteading and has been blamed for later economic difficulties. Land classification and soil surveys carried out more recently have supported Palliser's views rather than those of Macoun. For example, in 1942, 88 per cent of the land lying between Medicine Hat and the Red Deer River was classed as submarginal and unsuitable for wheat production (Stewart and Porter 1942, 33). It is against this controversial background that the history of settlement in southeast Alberta must be seen.

By 1920 the bulk of the agricultural settlement of the area had taken place: since then the farm population has fluctuated considerably though with a constant downward tendency. The era under discussion can be divided into three broad periods: (1) up to 1883 – pre-CPR construction; (2) 1883–1910 – ranching and some homesteading; (3) 1910–20 – 'boom' homesteading.

The first ranch was established in the Medicine Hat region in 1880–1 by James F. Sanderson (Gershaw 1967, 13), some 10 years after the first settler arrived in the Alberta foothills (MacEwan 1960, 42). By this time the federal government was leasing grazing land to settlers at 2.4 cents per hectare and permitting 40,500 hectares to be held for up to 21 years. Such terms attracted many ranchers, mostly of English and Scottish background. Some brought their cattle north from the USA, others from the east; all wanted to benefit from the vast rangelands. Most of the early ranches lay south of Medicine Hat. The poorer grasses in the east were not occupied until homesteaders took over the older western ranching districts.

The arrival of the railway line led to a great influx of settlers. Ranching thrived, and 'between 1885 and 1892 the number of cattle on the Alberta ranges trebled' (Burton 1941, 45). Well known local families such as Hargrave, Ross, and Cavan became, and still are, synonymous with the region's cattle industry. Alongside this expansion of ranching were the first attempts to establish arable farming. In 1884, the CPR established ten small experimental farms, three of which, at Tilley, Stair, and Dunmore, lay in southeast Alberta. These farms were to demonstrate the fertility of the land and thereby attract homesteaders. This *Financial Times* report of 17 February 1888, typifies attempts to persuade farmers to migrate to the then North West Territories: 'The excellent harvest of last year [on CPR lands], gives abundant encouragement to settlers and the pressure upon the poverty stricken farmers and crofters in this country [British Isles] ought to lead to a very active immigration into the North West of Canada in the coming spring and summer, if it be properly encouraged and assisted with money ...' At the same time, 1884, Sir John Lister-Kaye and his Canadian Agricultural Coal and Colonization Company acquired ten 4000 hectare parcels along the railroad, four of which lay in the southeast. Each farm was elaborately planned in the hope that large-scale cultivation and stock rearing would stimulate settlement (MacDonald 1966, 245). Unfortunately by 1893 the whole scheme had collapsed through drought and lack of capital. The only other attempt at organized settlement occurred in 1888 when two colonies of German settlers were established at Rosenthal and Joseph-burg; by 1891 both had failed because of drought and the people moved to Manitoba. Settlers slowed to a trickle because of these failures. However, there was some homesteading, and the federal government's 1906 Disposition of Lands Map shows a belt of occupied even-numbered sections, 40km on either side of the railroad. Equally significant, however, are the large unoccupied areas north and south of this belt. It was obvious that standard European farming methods would not work in this semi-arid region and so in 1903 several private companies were established to promote irrigation around Medicine Hat. By 1910 several thousand hectares were irrigated.

A dramatic change took place around 1910, with a huge influx of settlers. Most of the suitable land elsewhere had been filled, and the continuing demand for land, plus several years of good precipitation, led to rapid settlement in this arid zone. The settlers were optimistic that the droughts would not recur. New varieties of wheat, new farming methods and rising wheat prices attracted settlers, and they came from the USA, from Europe, primarily Scandinavia and eastern Europe, and from eastern Canada. 'The land was easily and inexpensively cultivated, the wheat produced was of extremely high quality, prices were good, and the bumper crops of 1915 and 1916 made this area seem a veritable Eldorado' (Alberta government 1936, 25). Some stayed, some moved on, some went back, but the land was settled. Wheat farming prospered and pushed the ranching south and yet even by 1920 there were signs that crop growing was extremely precarious, that without adequate water, the 'desert' conditions could rapidly return. Irrigation proved to be a partial answer and the real impetus to its large-scale development came in 1914 with the start of the CPR irrigation project at Bassano.

This arid southeast corner of Alberta was one of the last areas of the southern prairies to be settled and it was this same area which was to cause major relief and rehabilitation problems in the droughts of the 1930s. Its unreliable and inadequate precipitation eventually proved too great a problem for many homesteaders, resulting in vast tracts of land gradually being returned to grass.

Alberta government, 1936 *Report on the rehabilitation of the dry areas of Alberta and crop insurance 1935–36* (Edmonton).
Burton, G.L., 1941 The early development of cattle ranching in Alberta, *Econ. Annalist* 11, 41–6.
Gershaw, F.W., 1967 *Saamis* (Medicine Hat).
Hedges, J.B., 1934 *The federal railway land subsidy policy of Canada*, Harvard Historical Monographs (Cambridge, MA).
MacDonald, N., 1966 *Canada: immigration and colonization, 1841–1903* (Aberdeen).
MacEwan, G., 1960 *John Ware's cow country* (Edmonton).
Macoun, J., 1882 *Manitoba and the great north west* (Montreal).
Roe, F.G., 1946 Early opinions of the 'fertile belt' of western Canada, *Can. Hist. Rev.* 27, 131–49.
Sanderson, M., 1948 Drought in the Canadian North West, *Geog. Rev.* 38, 289–99.
Stewart, A., and W.D. Porter, 1942 *Land use classification in the special areas of Alberta*, pub. no. 731, Canada Department of Agriculture (Ottawa).

Population Geography
Géographie de la population

c1201
Some notes on a pragmatic approach to migration research
s. GALE   *Northwestern University, USA*

This paper presents some methodological notes on the systematic examination of patterns of short-distance residential movements. Disaggregated, individual level information on the structure, distribution, and processes of residential change forms the empirical basis of the program. The study focuses on the analysis of the contingent relationships of various social and economic characteristics and their effect on the propensity to change residences.

The methodology is thus viewed as providing a foundation for the investigation of the determinants of the demand aspects of intra-regional residential location processes. The distribution and structure of the population at several time periods is depicted and, using methods similar to those employed in the analysis of demographic change and social mobility patterns, the mechanisms (or processes) of transformation are identified. Moreover, since these data consist of observations of individual movements and their associated socioeconomic conditions, inferences concerning the behavioural antecedents are also made. The intention here is therefore to detail a methodology which employs a rich descriptive foundation as the basis for explanatory inferences and hypothesis formulation in the investigation of residential mobility; in effect, it may be regarded as a means for using individual-level data to describe and explain macro-level process.

Three issues have been of particular concern in the development of the methodology: (*a*) description; (*b*) explanation; and (*c*) hypothesis formulation.

The empirical foundation of the proposed program is the use of disaggregated historical information on individual residence locations and socioeconomic status in the detailed description of the intra-urban migration patterns in a small community in the United States. A time-dependent contingency table or K^N model is employed as the basis of the description; the description itself is composed of the distribution of individuals in the classificatory states of the contingency table (say, by residence and social class categories) and a transition matrix which expresses the (observed) probability of changing states in a given time period. Additional descriptions of those individuals who enter the system and those who leave are obtained. This part of the program is, of course, closely linked with data collection and classification.

Explanation may be defined as 'discerning patterns from descriptive information.' Thus, given a descriptive characterization in the form of a multi-way contingency table, four classes of explanatory hypotheses are examined: (1) temporal and spatial stationarities; (2) temporal order properties of the processes of change; (3) degree of association (or correlation) among attributes; and (4) differences among tables. Methods for evaluating each of these classes of hypotheses are outlined. The methodology is therefore concerned not as much with deductions from known laws, but rather with the organization and unification of observations and low-level generalizations; it may thus be regarded as a means of 'structured induction' or 'pattern recognition and specification' rather than a hypothetical-deductive process of theory confirmation.

Perhaps the most important part of the present research program is its capacity to provide a rational basis for the formulation of general hypotheses. The specific questions which are examined are answered only in terms of information relating to the data of one specific community. The over-all concern is, however, with residential mobility and neighbourhood development as a more comprehensive issue. Thus the extension of these questions to general hypotheses which can be tested on a wide variety of communities is a major aim of the study. It should be noted that the use of disaggregated data and an inductive view of analytical methodology is particularly important here. The data, for example, can be collected (say, from the available published sources) on a uniform (disaggregated) level. Classification procedures can also be clearly specified which are functionally oriented rather than locationally or temporally specific. The contingency tables and the associated non-parametric statistical tests then provide a representational model from which additional explanatory insights can be formulated.

In the short run then, the aim of the proposed program is to present a foundation for giving a detailed, historical account of the demand aspects of the process of residential mobility and neighbourhood development in one community. In the long run, however, by systematically testing hypotheses on a wider range of communities, it is expected that a comprehensive picture of the antecedents, processes, and consequences of residential change and neighbourhood growth can also be developed.

The discussion of the program is developed as follows: after outlining some of the background material on formal models of social and residential mobility, a detailed discussion of the methodology is then presented. To illustrate the argument, several examples from a recent study are given. The final section summarizes the main points of the discussion and offers some tentative conclusions regarding the efficacy of the proposed research strategy.

c1202
Some observations on migration theory for an urban system
JOHN C. HUDSON   *Northwestern University, USA*

ABSTRACT
Certain questions in human migration theory are of the type that require simultaneous consideration of growth rates of populations in the various subregions of a system. Growth and migration processes for a system of spatially distributed places may be studied by applying the theory of linear, ordinary differential equations and by studying properties of the solutions of these equations. The differential equations themselves are nothing more than the usual formulations of population growth involving natural increase and migration rates. By solving the equations the time path of population distribution is obtained. The application of this theory in population geography is especially interesting in the case of an urban system in which migration flows are hierarchical, some regional populations are growing rapidly, others are naturally declining, and so on. Various hypothetical migration and growth processes are illustrated.

c1203
Migration between labour market areas in England and Wales
JAMES H. JOHNSON, JOHN SALT, and PETER A. WOOD   *University College, London, UK*

This paper discusses a research project concerned with the links between housing characteristics and the mobility of labour in England and Wales. It attempts to outline the background to this project and to introduce an analysis of population movement between urban areas in England and Wales being undertaken as part of the broader study.

The 1966 census recorded over five million residential moves in Britain during the year before the enumeration; and a Ministry of Labour estimate, made in the same year, has suggested that over eight million job changes are made every year. Many of these moves of homes and jobs were independent of one another, since they took place over a limited distance, so that a move of job did not necessarily demand a change of home or vice versa. Nevertheless, the absolute number of people moving longer distances is large: between 1961 and 1966, for example, gross interregional movement was over two million people. Many of the longer moves, presumably, were stimulated by job reasons. In Britain it is often assumed that housing acts as a major barrier to population mobility and that this barrier restricts economic growth, but despite unanimity of opinion on the existence of connections between housing and labour mobility, the actual relationship is far from clear.

A search of the literature shows that the overlap between labour market and housing market has been relatively neglected. It is possible, however, to weave together some of the relevant strands that appear in the separate literatures to form some kind of coherent pattern. In this project we conceive of labour migration (the simultaneous change of employment and home, assumed for simplicity to be largely motivated by a move to a new job) as part of the labour market mechanism, which matches the supply and demand for particular skill groups, whose remuneration is related to their short-term elasticity of supply. Elasticity is low for occupational categories which involve long periods of training and high for the unskilled and semiskilled. Thus the occupational structure of the work force is determined by 'barriers' to movement between different occupational categories, which are relatively fixed, since early education, training, and skill strongly determine the subsequent role of an individual in the labour force.

A second barrier to the ready attraction of suitable numbers of workers to a point of employment demand is the geographical location of this demand in relation to sources of supply. Jobs and homes are mutually linked in space by the commuting 'tolerance' of workers, since the area over which housing opportunities will be sought by a household depends on the amount of commuting it is prepared to accept. Similarly, given the location of the family home, the workers in a household seek employment opportunities over an area defined in terms of commuting tolerance. These commuting tolerances may be thought of as varying between individual workers or between different income and skill groups; or they may be considered as exerting a net effect, delimiting the general size of the labour force from which an employer may draw workers. Commuting hinterlands around centres of employment provide the basis for the definition of local labour market areas, within which workers may normally change jobs without necessarily moving their homes. On aggregate, the movement of a worker's place of employment from one of these labour markets to another would be expected to involve a change of residence, and thus cause labour migration.

The locational substitution of jobs between local labour market areas probably takes place rather more easily than between skill-groups and results in labour migration; but the close interdependence of occupational and spatial barriers to the free allocation of workers to jobs needs emphasis, since one cannot be adequately considered without reference to the other. Because of this there are probably strong contrasts in the motivation and behaviour of labour migrants from different occupational groups and from different places.

To complement the relative scarcity and inelasticity of supply of these skilled occupational groups, a higher level of geographical mobility is usually found among them. Unskilled and lower-paid workers, on the other hand, are usually recruited locally and scan much more restricted horizons for job opportunities. In Britain they depend on council housing or on cheap accommodation with controlled rents, to which access would be difficult in a new area. Thus their housing, their sources of information, and the pattern of location of suitable jobs all combine to reduce their propensity to migrate.

Less is known about intermediate groups, which are of particular interest because their behaviour is more open to influence by government decisions. Skilled manual, technical, 'middle' managerial, and trained clerical workers are enormously varied in their characteristics; and the effective general constraints on their migration behaviour as well as their individual patterns of motivation have been little studied in the sociological literature. It may be suggested that in general these workers are occupationally 'fixed' by their training and that thus their geographical mobility might be expected to be relatively high. On the other hand, in Britain these intermediate groups occupy either council housing or relatively low-cost privately owned houses. Thus, in changing jobs over long distances, housing availability might be expected to be particularly critical for them, compared with the mobile, wealthy, professional, and higher managerial groups, or with the immobile, unskilled, and semiskilled groups.

The pilot survey operates within this general frame of reference and is attempting to investigate some of the likely relationships between housing and labour mobility in two major ways. One approach involves the study of a sample of labour migrants by means of a questionnaire survey. Here the aim is to study in some depth the total pattern of motivation involved in the decision-making process of a migrant household, for whom housing characteristics are only one thread in a complex design.

A second approach involves the analysis of migration data from the 1966 census in a manner that throws some light on the aggregate character of labour migration and hence forms a basis for organizing the questionnaire survey. A matrix of moves between every local labour market area in England and Wales during the year before the 1966 census has been produced. The assumption underlying this method of handling the statistics is that most of the movement between labour market areas will have been for job purposes.

One use of the information derived for the 100 labour market areas has been to produce a classification to indicate the degree of similarity in the general range of migration and housing characteristics found in different parts of the country, thus providing a basis for sample design. A second use has been to provide data for an analysis of migration flows, which provide a surrogate for labour migration. In all, almost 600,000 people moved between the various local labour market areas of England and Wales during the year studied.

In addition to information on migration flows, data have been collected on the distances between labour market areas and on the characteristics of the housing market, labour market, and population of these areas. In its crude form this information has proved useful in the organization of the questionnaire survey, but we have also analysed the matrix of flows further to test some of our ideas about labour migration and to see if any further hypotheses are suggested which can be examined more effectively by field survey. Although there are difficulties in applying the most significant techniques of analysis to this data, it is useful because it provides information which reflects some of the geographical realities of population movement which are sometimes masked when larger, more artificial areas like economic regions are used as a basis for study.

Working papers are being issued as research proceeds by the Housing and Labour Mobility Study: 1, Housing and labour mobility: an introduction study; 2, The multivariate classification of local labour markets; 3, Housing and Labour Mobility Study, Questionnaire design for the survey (all 1971).

c1204
Patterns of Maori migration
R.J. JOHNSTON   *University of Canterbury, New Zealand,*
M.F. POULSEN   *University of Waikato, New Zealand,*
D.T. ROWLAND   *Monash University, Australia*

The Maori population has been in a state of flux since the coming of Europeans to New Zealand. During the twentieth century, following a long period of decline in numbers, recovery has taken place as a result of the group's persistently high birth rate and a pronounced fall in the death rate. Accompanying this rapid growth has been a large-scale rural to urban redistribution, in which the percentage of Maoris living in administratively constituted urban districts rose from 9.8 in 1926 to 45.4 in 1966. The major part of the change has come since the 1950s. The aim of the present paper is to chart the salient features of this redistribution since 1951, and the consequent patterns within one urban area. Since no data are available on migration, net migration estimates have been computed for the three periods 1951–6, 1956–61, and 1961–6, using age and sex data derived from the censuses in the end years of each, and estimates of gross migration and migration flows have been compiled, for male adults only, from electoral rolls.

Using unpublished single-year tabulations of age and sex data for five-year categories, the following average survival rate formula has been applied:

$\text{NM}_c = (1 + \text{LTSR}_c)[P_{2c} - (P_{1c} \times \text{LTSR}_c)]/2\text{LTSR}_c$, where NM_c is the net migration estimate for the cohort, P_{1c} and P_{2c} are the cohort populations at dates 1 and 2, and LTSR_c is the life table survival rate for the cohort.

Applying this method of estimation to the constituent local government areas of New Zealand for the period 1951–66, the following major conclusions emerge:

1. During the 15 years there was a net redistribution of Maoris, across local government area boundaries, of some 63,000 persons, or 39 per cent of the 1966 total Maori population. The redistribution rates increased over the three intercensal periods.

2. Over the period there was a marked increase in the amount of family migration, as indexed by both the sex ratio (percentage of females) and by a dependency ratio (children aged 5–14/males aged 20–59) among the estimated net migrants. This suggests that migration is becoming a much more permanent affair than it was in the earlier 'pioneer' period when young males predominated in the migration stream (Rowland 1971b).

3. The highest rates of mobility (net migrations/total population) were in two age groups: young adults and elderly persons.

4. Although for both sexes the greatest number of migrants was in the 15–19 age group, female migrants were more highly clustered in this group and a large proportion of males (especially in the later intercensal periods) were aged 20–24.

Maps of these data indicate a growing concentration of the net migration of Maoris in a few places. Net migration has been dominated by outflows from the rural areas of the north, central, and eastern parts of the North Island, and by inflows towards the major urban centres of these areas, plus Wellington-Hutt: over time the inflows have become more concentrated in certain places. There has also been a net flow into the South Island, apparently mainly of young single males, in contrast to the more usual 'family-type migration' to the North Island towns.

A basic weakness of the net migration estimates discussed above is that they give no indication either of the destinations of migrants from areas of net loss, or of the origins of those who come to live in areas of net gain. In addition, they conceal a probably much larger amount of gross migration into and out of every constituent area. To compensate for these disadvantages, data have been abstracted from the electoral rolls to give some indication of the dimensions of the migration streams within a seven-region division of New Zealand.

Established migration theory states that most moves are over only a very short distance, and are directed towards larger settlements than the place of origin (Lee 1966). With New Zealand Maoris, therefore, one would expect most migration to be directed towards Auckland, perhaps in a number of steps for those originating from the remote rural fastnesses.

Analysis of the direct and indirect connections suggests such a national migration system, with Auckland as the major destination and Hamilton and the rural central North Island as important intermediate stops. When only the direct connections were considered, it was found that most movement was over relatively short distances, between town and hinterland. These separate local regions are linked into a national system of three cells, one based on Auckland, one on the central North Island, and one on Wellington, each cell of which is considerably self-contained and dominated by short-distance rural-rural, rural-urban, and urban-rural migration. Connection between the cells is mainly at the interurban level, and is directed mainly towards Auckland.

In New Zealand urban areas, the clustering of low-status immigrant groups is not a feature of the inner districts only, though it was until 15–20 years ago (Whitelaw 1971). The change results from the provision of large estates of subsidized government housing for Maoris in the outer suburbs. Thus, the spatial patterns of migration of Maoris into and within New Zealand's largest urban centre (Auckland) are more diffuse than is the case with minorities in many other countries.

Once in the city, Maoris are a highly mobile group, indirect evidence suggesting that they move twice as often as non-Maoris (Rowland 1972), no doubt partly because fewer of them have home-ownership ties. In general, the spatial lineaments of their mobility serve to emphasize their bipolar distribution. Of 509 adult intraurban migrations, 47 per cent of the moves originated in the inner city. One-fifth of these remained within the inner city, while of the remainder almost four-fifths went to the outer suburbs and just under a fifth went to the adjacent inner suburbs. There was also considerable migration (30 per cent of the total) within the outer suburbs, but very little back towards the central suburbs and inner city.

Within Auckland, therefore, the main role of the inner city is as a source for centrifugal movements, mainly to the outer suburbs. It is the major destination for Maoris unused to urban life and those whose family situation makes living in inner suburbs unattractive. Households consisting of parents and young children and possessing either or both greater knowledge of urban life and friends or relatives in the outer suburbs are much more likely to move directly there. Because of overcrowding, however, they will seek their own home, leading to considerable migration within the outer suburbs.

Lee, E.S., 1966 A theory of migration, *Demography* 3, 47–57.
Metge, A.J., 1964 *A New Maori Migration* (London).
Pool, D.I., 1964 The Maori population of New Zealand, PHD thesis, Australian National U.
Poulsen, M.F., 1970 Internal Maori migration, 1951–1969, MA thesis, U. Canterbury.
Rowland, D.T., 1971a Maori migration to Auckland, *New Zealand Geog.* 27, 21–37.
– 1971b Age structures of Maoris in Auckland, *Proc. Sixth New Zealand Geog. Conf.* 111–17.
– 1972 Patterns of Maori urbanisation, *New Zealand Geog.* 28, 1972.
Whitelaw, J.S., 1971 Migration patterns and residential selection in Auckland, New Zealand, *Australian Geog. Studies* 9, 61–76.

c1205
Industrialization as the main factor of recent internal migrations in Poland
TEOFIL LIJEWSKI   *Polish Academy of Sciences*

Major population migrations in Poland were due, in the first instance, to such political events as World War II and the associated Nazi occupation, the destruction of a majority of the towns, and changes of frontiers in 1945.

About 1949, these politically inspired population movements stopped, once the territories recovered by Poland were resettled and the towns essentially reconstructed. In the following year, 1950, Poland's first economic development plan was started. Its principal objective was intensive industrialization of the country. New industrial plants were located partly in existing big cities (one of the aims was to increase the proportion of industrial workers in their social structures), and partly in economically underdeveloped small and medium-sized towns; some plants were constructed beyond the towns also, which led to the creation of new urban settlements. Over the period of the plan's operation, 1950–5, national industrial production increased more than twofold, and employment in industry by 35 per cent.

After 1955 this policy was changed. Emphasis was put on the maximization of capacities in existing plants; the construction of new plants was curtailed; and the growth of industrial employment slowed down. As a result, national industrial production in the period from 1955 to 1960 increased by 60 per cent, while related employment rose by only 11 per cent. There were no major shifts in the distribution of industry.

In the 1960s, outlays for industrial investment were increased. However, over this period the development of industry was restricted to about 70 'developing centres' and not spread throughout the country. Areas with high natural environmental value were excluded from the industrialization program. The development of industry was also curtailed in the biggest cities where negative consequences of the overconcentration of industry had already been noticed. Overall, in the period from 1960 to 1970 the production of industry grew by 123 per cent, industrial employment by 37 per cent, and major changes occurred in the spatial distribution of industry. The part played by underdeveloped regions increased, while that of big cities and old industrial districts declined, though they have retained the bulk of the country's industrial capacity.

The most important social phenomena resulting from this development of industry have been: (1) an increase of industrial employment from 2 million in 1950 to 4.1 million persons in 1970; (2) an increase of employment in construction, transportation, and tertiary activities; (3) an increase of population in many towns and the construction of new industrial centres; (4) migrations of people from agricultural areas, especially those with high agrarian population densities, to towns and industrial districts.

Between 1950 and 1970, about 23 million people in Poland changed their place of residence (people counted every time they changed their residence). Of these moves about 5.6 million were between towns and 6 million were from rural to urban locations. The net migration gain of the towns amounted to about 2.2 million people, i.e. almost the same number as the increase in industrial employment.

The present distribution of population results from the historical development of particular regions, and especially from the partition of Poland by the three powers in the nineteenth century. A high density of agrarian population is characteristic of the southeastern area and to a lesser degree of central Poland. Most of the migrants originate from these parts of the country. A shortage of workers, on the other hand, exists in big cities, in industrial agglomerations, and in the developing mining districts which are located mostly in the western half of the country.

The following forms of economic migration associated with industrial employment are the most common in Poland:

1 The inflow to big cities which offer a wide choice of jobs and various other attractions. This inflow was initially uncontrolled and exceeded the capacity of new housing construction. Consequently several cities gradually introduced restrictions against in-migration, and now special permission of the city council concerned is required to settle in these cities.
2 The inflow to medium-sized and smaller towns, in which large new industrial plants were previously located; their labour demands far exceeded available labour supplies. Such migrations are not restricted and the recruited workers obtain new apartments in a relatively short time.
3 The inflow to suburban zones of large cities. In-migrants, to whom permission for settlement was denied, often try to overcome this restriction by settling in the suburban zone of the given city and commuting to work. After a lengthy period of employment they may be granted permission to settle in the city, if their applications are supported by their employers.
4 Temporary migrations chiefly connected with the construction of new industrial plants. The workers come alone, without their families, and stay in workers' hostels or rented rooms. When the plant is ready, they either go home, are engaged to work elsewhere, or start work in the newly built plant. In the last case they usually get accommodation in a short time and bring their families to join them.
5 Daily movements (commuting). This form of mobility is the most frequent, as it involves daily about 2 million persons (one million of whom are employed in industry). Where the distance between the places of work and residence is not long and the existing transport facilities enable convenient travel, commuting replaces the above-mentioned forms of migration. It should be added here that the prices of the workers' monthly railway tickets are very low, and that many industrial plants drive their commuter workers home by special buses. Commuting to work is also popular because many commuters possess small farms which bring them some additional income. These so-called peasant-workers are in most cases not interested in migration to town, as this would mean giving up their land.

Individual types of migrations do not exclude one another, and often one type develops within the other. In particular, daily commuting and temporary migrations sometimes develop into permanent changes of residence; similarly migrations to suburban zones are frequently complemented by a final migration from the suburban zone to the city itself.

Migrations have been the chief element of population increase in rapidly developing towns, while in the remaining big and medium-size towns they have made a considerable contribution to any increase. Comparing the distribution of population increases with that of industrial employment, one can observe a striking concordance between the two variables. This confirms the hypothesis that industrialization is the main location factor determining the contemporary settlement pattern. It channels the direction of migration of those people who are not satisfied with their existing living conditions.

When the principal cities are ranked according to their population and industrial employment increases for the years 1950–70, a high positive correlation is revealed. The majority of new towns located after 1945 owe their origin to the establishment of large industrial plants. Among them are Nowa Huta, an administrative district of the city of Krakow (150,000), and Tychy, a dormitory town for the Upper Silesian Industrial District (70,000).

A positive correlation was also found to exist between the percentage shares of the national urban population increase and the industrial employment increase obtained by the various voivodships. Industrialization is the main cause of urbanization and, in turn, gives rise to other economic and social processes which collectively are transforming a previously agrarian community, with a rather uniform population distribution, into an urban-industrial society with distinct concentrations of population in a number of areas providing the best conditions for the development of industrial activity.

c1206
Interregional migration in the United States and Canada
D. RICHARD LYCAN   *Portland State University, USA*

Migration in North America is characterized by a great variety of types of moves ranging from shifts between adjacent labour market areas such as Baltimore and Washington, DC to transcontinental moves between points such as Halifax and Victoria. The rate of movement between various regions is inversely related to intervening distance, but distance has important qualitative effects on the composition of the migrant streams as well. Long-range streams have a much larger proportion of high socioeconomic status persons than do short ones.

As Canada and the US become increasingly urbanized, the importance of interurban migration increases while the movement from rural to urban areas declines. Excluding the movement that takes place within labour market areas, two-thirds of all migration in the US is between metropolitan areas. Generally, migration tends toward areas of economic expansion but the movement from economically stagnant areas does not occur consistently.

At a broad regional scale there are several specific regional trends. In Canada the cultural separateness of Quebec results in a lowered rate of exchange of population with the remainder of the nation. In Canada and the US, the most geographically central areas – Ontario and the north central states of the US – seem to have a lower rate of interchange of population with surrounding areas than their central location would suggest. In the US the east south central states are in a special sense the 'deep south' in that they have a particularly low rate of exchange with the rest of the nation. The most prominent regional pattern has been the westward drift toward the Rocky Mountains and Pacific coast. Some view this as induced by the natural amenities of these regions, but this movement is also consistent with a job opportunity hypothesis.

There are several broad questions that have come to my mind in the process of writing this paper.

(1) The gravity model formulations of interregional migration have not resulted in very complete explanations, particularly for net migration. Can this technique be improved to the point where results useful for population forecasts can be made? If not, what alternative approaches might be more fruitful?

(2) The subdivision of space into data units and the aggregation of these data units seem to have important effects on the results of subsequent analyses. For the US the problem is compounded by a need to limit the size of the data set for economy reasons. Is there some optimal set of regions for the

study of interregional migration? Can we say what are the effects of aggregating and disaggregating our data?

(3) The results of the 1970 US and 1971 Canadian Census are gradually becoming available. What approaches should be used in the analysis of these data? Would a more co-ordinated approach with several major foci be more productive than a larger number of individual efforts?

(4) How can we operationalize the concept of amenities as it bears on migration patterns? It seems to me that aggregative approaches will not shed much light on this factor and that carefully designed research on the decision-making process of movers is required.

Adams, Russel B., 1969 U.S. Metropolitan migration: dimensions and predictability, *Proc. Assoc. Am. Geog.* 1, 1–6.
Gallaway, Lowell E., 1969 *Geographic labor mobility in the United States, 1957–1960*, US Social Security Administration Research Report no. 28 (Washington).
Gould, Peter, 1969 Problems of space preference measures and relationships, *Geog. Anal.* 1 (1), 31–44.
Hansen, Niles M., 1971 *Rural Poverty and the Urban Crisis: A Strategy for Regional Development* (Bloomington, Ind.).
Heide, H. ter, 1963 Migration models and their significance for population forecasts, *Milbank Memorial Fund Quarterly* 7 (1), 55–76.
Lansing, John B., and Eva Mueller, 1967 *The Geographic Mobility of Labor* (Ann Arbor).
Lowry, Ira S., 1966 *Migration and Metropolitan Growth: Two Analytical Models* (San Francisco).
Lycan, Richard, 1969 Interprovincial Migration in Canada: The Role of Spatial and Economic Factors, *Can. Geog.* 8 (3), 237–54.
– 1970 An Analysis of Migrants' Distance and Preference Spaces in Canada, 1949–1968, a paper presented to the Western Regional Science Association.
Portland Metropolitan Planning Commission, 1965 *Population Mobility, Portland Metropolitan Area* (Portland).
Rogers, Andrei, 1968 *Matrix Analysis of Interregional Population Growth and Distribution* (Berkeley).
Schwind, Paul J., 1971 *Migration and Regional Development in the United States, 1950–1960* (Chicago).
Stone, Leroy O., 1969 *Migration in Canada: Some Regional Aspects*, Canada, Dominion Bureau of Statistics census monograph (Ottawa).
Toffler, Alvin, 1970 *Future Shock* (New York).
Vanderkamp, John, 1968 Interregional mobility in Canada: a study of the time pattern of migration, *Can. J. Econ.* 1 (3), 595–608.
Walpert, Julian, 1965 Behavioral aspects of the decision to migrate, *Papers of the Regional Science Association* 15, 159–69.
White, Harrison C., 1970 *Chains of Opportunity: System Models of Mobility in Organizations* (Cambridge, Mass.).

c1207
Migration patterns in the Soviet republics
THEODORE SHABAD   *The New York Times, USSR*

Preliminary analyses of the 1970 census of the Soviet Union have already noted a distinctive southward trend in internal migration (Harris 1971, 110; Kovalev 1970, 15; Perevedentsev 1970). If the USSR is broken down roughly into three macroregions – west, east, and south – the south is found to have increased in population during the intercensal period 1959–70 at a significantly higher rate than either the west (European Russia) or the east (Siberia).

Further publication of census results by republics yields further refinement of the general migration pattern, although a more comprehensive analysis will have to await the appearance of more detailed data. The present paper uses the results published in the local newspapers of the Soviet Union's 15 constituent republics to categorize migration patterns, including both rural-urban migration and interregional net migration, at the republic level.

Four types of migration patterns are distinguished: (1) net in-migration associated with an increase in rural population; (2) net in-migration associated with rural depopulation; (3) net out-migration associated with a rural increase; and (4) net out-migration associated with rural depopulation.

Type 1, in which net in-migration during the intercensal period 1959–70 has been accompanied by rural increase, includes the four central Asian republics (the Uzbek, Kirghiz, Tadzhik, Turkmen SSRS) and Kazakhstan. The pattern is also found in Moldavia, a southern European republic, and in Armenia, one of the Transcaucasian states. In these republics natural increase accounts for most of the population increase, and rural population growth is adequate both to supply migrants to the cities and to bolster the rural sector. However, ethnic cultural peculiarities, particularly in the central Asian republics, inhibit rural-urban migration and thus require in-migration from other republics for economic development of the cities. In such republics as Kirghizia and Tadzhikistan as little as 10 per cent of the natural increase of population in rural areas moves to the cities. Net in-migration is particularly marked in Kazakhstan and Uzbekistan, two republics in which a high rate of industrial development exceeds the potentialities of internal rural-urban migration. The migration to Kazakhstan and central Asia during the intercensal period amounted to a net influx of 1.2 million from other republics.

Type 2, in which net in-migration is associated with rural depopulation, includes the three Baltic republics – Estonia, Latvia, and Lithuania – as well as the Ukraine. The pattern is found in its most extreme form in Estonia and Latvia, where a low rate of population growth, particularly in rural areas, has made net in-migration the main source of population increase. In Estonia the rural population sector is virtually stagnant, with a natural increase of 1,100 in a rural population of roughly half a million during the intercensal period 1959–70. In Lithuania, a republic with a predominantly Roman Catholic population, a higher rate of natural increase provides for a greater share of manpower needs than in the Lutheran republics of Estonia and Latvia. In the Ukraine internal rural-urban migration accounts for most of the migration to cities, and the relatively small net in-migration represents less than 10 per cent of total population growth.

Type 3, distinguished by net out-migration and rural increase, includes the two Transcaucasian states of Georgia and Azerbaidzhan. These republics, especially Azerbaidzhan, have high rates of natural increase that provide most of the manpower required by the cities and generate a substantial rural out-migration. However, a limited rate of industrial development evidently cannot absorb the rural migrants, thus accounting for a net out-migration from the two republics.

Type 4, made up of the two republics of Russia and Belorussia, is characterized by net out-migration and rural depopulation. The Russian Federated Republic, by far the largest state of the Soviet Union, recorded a net out-migration of 1.7 million during the intercensal period and represents the largest source of migrant manpower for the other republics. There are, of course, also significant variations in migration patterns within the Russian Republic, such as net out-migration from parts of central Russia and Siberia and net in-migration into southern regions, such as the North Caucasus. But analysis of migration within the republic will have to await publication of more detailed data. A more coherent picture is offered by the smaller republic of Belorussia, in which rural overpopulation generates a large rural-urban migration that cannot be entirely absorbed by Belorussia's industrialization program. In the intercensal period 1959–70 roughly four out of five rural migrants were absorbed by Belorussian cities.

Two basic long-range tendencies are suggested by the migration patterns of the Soviet republics. In the Baltic region, particularly in Estonia and Latvia, the needs of industrial development combined with the low rates of natural increase may be expected to result in a continuing influx of population from other republics, thus reducing the percentage share of the indigenous population component. The share of ethnic Estonians in Estonia's population declined from 74.6 per cent in 1959 to 68.2 per cent in 1970; in Latvia, the corresponding decrease was from 62 to 56.8 per cent of the total population. In the other distinctive region of net in-migration, constituted by the republics of central Asia, a high rate of natural increase among the indigenous population overshadows the rate of in-migration. But the relative immobility of the rural component of the population is producing a growing stratification of Russian-dominant urban population and indigenous rural population. It remains to be seen whether the continuing industrialization of central Asia will be accompanied by measures to stimulate the use of indigenous manpower from overpopulated rural areas.

Harris, C.D., 1971   Urbanization and population growth in the Soviet Union, 1959–1970, *Geog. Rev.* 61, 102–24.

Kovalev, S.A., 1970   Changes in numbers and distribution of the population of the USSR during the period between the censuses of 1959 and 1970, *Geografiya v Shkole*, no. 6, 13–20.

Perevedentsev, V., 1970   All of us, taken together, *Literaturnaya Gazeta*, 6 May.

Shabad, T., 1970–1   News notes, *Soviet Geography: Review and Translation*, Oct., Nov., Dec. 1970; Sept., Oct., Nov. 1971.

c1208
Marriage migration in rural India
DAVID E. SOPHER and M.J. LIBBEE   *Syracuse University, USA*

This paper undertakes to identify variation in the size of marriage fields in rural India by utilizing census data dealing with population movements. In India, the village marriage field comprises many discrete fields of different endogamous groups ('castes'), further modified in varying degree by local exogamy. Where the latter is rigorously adhered to, a crater effect appears in the curve of distance-decay, which has a gentle slope for tens of kilometres beyond the village. Detailed data on marriage distances from two villages in the western Gangetic Plain show these features clearly. Since an Indian village like these may be, through its combined affinal links, in contact with a potential information field of a quarter of a million persons, variation in marriage field size can have important implications for the spatial structure of rural Indian society.

Village studies providing reliable estimates of marriage field size are rare and impressions of regional variation have hitherto been attempted only at a gross scale. Until 1961, census data could not be used effectively to identify variation in marriage migration. Place of birth data in the 1961 census differentiated rural males and females born in the village of enumeration from those born elsewhere in the district and those born outside the district, showing for the first time the full extent of female migration between villages. The proportion of females enumerated in the village of birth shows no correlation with comparable male figures and positive correlation with percentage of females never married. The proportion of female immigrants into villages who were born within the district is positively correlated with district area.

Holding district area and shape constant, this proportion should provide an index of relative size of the average district marriage field.

Preliminary investigation showed that substantial rural colonization created large anomalies in a few districts, and a correction based on the number of male migrants was applied to the female numbers. Because of severe distortion from the Pakistan refugee component of the rural population in some border districts, persons born outside the present national territory were omitted from the computations.

Libbee (1971) derived a formula to predict in effect the probability that any straight line of fixed length, D, taken at random and having at least one end within a circle of radius R, will lie completely within that circle. The apparent simplicity of the problem is deceiving and in the end an approximate formulation was obtained for the probability value, G. Only a graphic solution could be found for the next step, the calculation of D, given G and R. G was then replaced by the ratio of females born outside the village but within the district to all females born outside the village (the number of females being corrected as above), and R by the value (A/π), A being the district area. The corresponding value, D, read from the derived probability tables, was taken as the estimated median marriage distance.

This procedure requires the following additional assumptions: (1) marriage migration is random in direction; (2) district boundaries have no effect on marriage migration; (3) the population is evenly distributed; (4) every female who has migrated for marriage has moved the median marriage distance;

(5) a district is generalized as a circle of area equal to the district area. The median marriage distance does not include the zero travel distances of intra-village marriages.

Estimated marriage distances for all districts were divided into five classes of equal size and mapped. Validation of the model was sought in a number of ways. Comparison with a small number of field study data showed strong positive correlation and provided a rough calibration. A full inter-district rural migration matrix available for Bihar was found to be of limited value because of large inter-district distances and small numbers of long-distance migrants. In the expectation that it should vary inversely with estimated marriage distance, a measure of local endogamy, ER, was calculated from census data and mapped. (ER = ((FBIV/(RF − (.75 × MBOV))) − (RFUW/RF)) / (1 − (RFUW/RF)), where FBIV = females born in village of enumeration; RF = rural females; MBOV = rural males born outside village of enumeration; RFUW = rural females never married.) A moderate negative correlation appears: the marriage-distance 'ceiling' decreases regularly with the estimated level of local exogamy, but high levels of exogamy do occur with short marriage distances. The distribution of large residuals is strongly clustered and appears to be associated with differences in 'surface friction.'

Using the calculated endogamy ratios, 1931 place of birth data were used to generate crude marriage distance estimates for that year for comparison with similarly generated 1961 estimates for 68 districts that could be compared. (As before, D, the estimated marriage distance, is extrapolated from the computed table giving G, the ratio of female marriage migrants from within the district to all female marriage migrants, for all necessary combinations of D and A, the district area. G is now estimated thus: $G = ((FP − FUW) (1 − ER) − (FBOD − (.75 × MBOD))) / ((FP − FUW) (1 − ER))$, where FP = total female population; FUW = females never married; ER = district endogamy rate as computed for 1961; FBOD = females born outside the district; MBOD = males born outside the district. Considerable variation in ER, such as might result from using a standard endogamy rate for each state, would not change G by much, and would have hardly any effect on the correlation of the 1931 and 1961 values.) A high positive correlation is found, and the stability in the pattern of areal variation so demonstrated lends support to the interpretation of the statistic as a measure of marriage distance. The regression equation predicts an increase of 50 per cent in the estimated distance over the 30-year period and this is interpreted as a consequence of changes in local transportation.

The mapped patterns show departures from conventional regionalization of Indian kinship structures. Critical features of the so-called 'northern kinship' structure are confined to the northwestern plains and plateau margins, with a conspicuous southward extension through Malwa to the central Deccan, while extension of the region eastward through eastern Uttar Pradesh and Bihar is called into question, so far as these criteria are concerned. The contrasting 'southern' kinship type is well developed in Kerala and some adjoining districts, but is not predominant throughout the southern, Dravidian-speaking zone.

Local exogamy and the fetching of brides from some distance seem to be more widely practised throughout India than is generally acknowledged in the literature, although by no means as universally in the north as is supposed. The size of the average Indian field is well above what has been indicated about the rural European range.

Libbee, Michael J., 1971 A model estimating marriage distances in India. MA thesis, Syracuse U.

c1209
Some observations on the origins and composition of workers at the Miferma iron ore mine, Mauritania
K. SWINDELL *University of Birmingham, England*

The MIFERMA (Société Anonyme des Mines de Fer de Mauritania) iron ore mine was opened in 1963, and constitutes the only large-scale industrial operation in Mauritania. With an annual production of over 8 million tons of ore it now ranks as the second largest

TABLE 1. Expected and observed origins of the Mauritanian labour force, MIFERMA, 1969

Cercle	Expected no. of workers	Rank order	Observed no. of workers	Rank order	Rank difference between observed and expected
Adrar	493	1	568	1	0
Inchiri	55	6	44	7	−1
Baie du Levrier	65	3	25	8	−5
Tiris-Zemour	342	2	302	2	0
Trarza	65	3	61	4	−1
Brakna	35	9	65	3	+6
Gorgol	47	7	57	5	+2
Guidimaka	8	12	22	9	+3
Hodh Occidental	24	10	22	9	+1
Hodh Oriental	18	11	13	12	−1
Assaba	39	8	14	11	−3
Tagant	57	5	55	6	−1
	1248		1248		

iron ore mine in West Africa. Like several other West African mines, it is a highly capitalized international venture, operating on a large scale, with fixed forward contracts. Its location, however, is perhaps more remarkable than its counterparts elsewhere. Situated in one of the most arid parts of the western Sahara, close by Fort Gouraud (F'Derik), its development presented a challenge to mining engineers and operators. When this enquiry took place, there was virtually no other demand for labour for the potential migrants at Fort Gouraud (F'Derik). These factors, together with the mine's location in an arid environment, largely peopled by nomadic Moors, prompted this appraisal of the origins and composition of the daily paid labour force.

The workers' record cards revealed the home town or village to which they professed allegiance, what kind of job they held at the mine, and their marital status. Details of the general stability of the labour force were supplied by the company's personnel office. The data were collected in August 1969, when there were 1338 manual workers employed at Fort Gouraud (F'Derik), including 1248 Mauritanian, 82 expatriate Africans, and 8 Europeans. In view of the almost perfect labour market which obtained, an expected supply situation, for Mauritania, in terms of the distribution of workers' origins by administrative regions (cercles) was derived by use of a modification of the gravity model:

$M_{im} = Ed(P_i/im^2)/\Sigma_d(P_i/im^2)$,

where M is the expected number of migrants, i = cercles, m = mine, P = urban population, d = distance, E = number of job opportunities (1248). The model predicted the number of workers per region as a proportion of the 1248 jobs occupied by Mauritanians at the mine (Table 1).

In terms of the general distribution of the population relative to the mine, which is the basis of the supply situation, it needs to be emphasized that the mine is located in a region of extreme aridity and low population density. The most densely populated areas lie along the southern margins of the country, at greatest distance from the mine. Thus the introduction of a large mine into this particular spatial context, and the likely origins of its labour force, posed an interesting situation. The expected values seem to be of the right order for each administrative unit. The bulk of the labour force is recruited within 300 kilometres of the mine, with a marked southward orientation towards the towns of Atar, Chinguetti, Ouadane, and Akjoujt. However, although a large part of the explanation of the migration is accounted for within the 300km zone, it is interesting to look at the more distant areas.

In the case of Baie du Levrier, the shortfall of observed workers can be accounted for by the presence of alternative employment at Port Etienne (Nouadhibou). The underestimation of the number of workers from the cercles farthest away from the mine (some 800km) focuses attention on an interesting component in the migration field. The southern cercles (Brakna, Guidimaka, and Gorgol) lie along the north bank of the Senegal and supply twice, a quarter, and three times as many workers, respectively, as the expected values. The reasons for this lie in the ethnic and socio-economic differences found within Mauritania.

Although 80 per cent of the population occupying northern, central, and lower Mauritania may be classified as nomadic Moors, the Senegal valley is characterized by sedentary semisubsistence villages inhabited by predominantly sudanic negro peoples (Toucoleurs, Peuls, Sarakoles, and Ouolofs). This region is one of complete contrast to any other in Mauritania, and is marked by a relatively high density of population. The people are principally rice and millet cultivators.

Informed opinion at the mine seemed convinced that southern Mauritania is an important supply area for workers, because of the negro's inclination to accept the conditions attaching to regular wage-earning employment. Also the southerners provide an important element in the skilled labour force (Table 2). These remarks equally apply, if not more so, to those expatriate workers from Senegal, who have not been included so far in the discussion.

Of the 1338 workers investigated at Fort Gouraud, 510 could be classified as semiskilled, and of this total 114 were highly skilled. The company has a rigorous system of grading its workers, and by means of its training school promotes workers to higher grades, with a view to building up a skilled Mauritanian personnel. The bulk of unskilled workers come from the immediate vicinity of the mine (up to 70 per cent), with a slight increase shown when the riverine area is reached at 800km distant. By comparison skilled workers show a more even spread of origins relative to the mine. The number of opportunities for skilled employment are so limited and widely spread in Mauritania that long-distance migration is necessary to find employment in such jobs.

The numbers employed at MIFERMA and other mines may be relatively small, but their impact is considerable, especially since over the years there has been a constant turnover of labour. Initially unskilled workers in particular had a very high turnover rate (70 per cent) but this has steadily declined and by 1968 the unskilled and skilled turnover rates were approximately the same (15 and 12 per cent respectively).

The emergence of a stable labour force is

TABLE 2. Highly skilled workers by region of origin, MIFERMA, 1969

Region	No. of workers
Lower, central, and northern Mauritania	51
Senegal valley (Mauritania)	22
Senegal	35
Other African countries	6
Total	114

a compound of numerous factors. Workers must accept at least a semipermanent domicile near the mine, and this may have been encouraged by the provision of housing and other amenities.

The small township of Zouerate, occupied by over 5000 people, has grown up alongside the MIFERMA mine, in addition to housing and labour lines provided by the company. It is often accepted that the African male migrant is young and unmarried. No age data were available. Three-quarters of the labour force, in 1969, was married. However employment cards showed that only 18 per cent of the married men in 1969 were already married on arrival. Thus approximately 80 per cent of the men seeking job opportunities over the period 1960–9 were single on arrival.

In conclusion, although the number of employment opportunities is relatively small, they are nonetheless significant in a small country such as Mauritania. Turnover of workers and their dependants swells the number of those who make a response to the opening of a mine such as MIFERMA. In addition there are those who are attracted by the opportunities of trade, and of providing services in mining settlements. Within west Africa as a whole, mines produce enclaves of modern industrialization, which activate social and economic change, as well as having a recognizable impact on the population geography of several areas.

Centre for West African Studies, University of Birmingham, England.

Hilling, D., 1969 Saharan iron ore oasis, *Geog. Mag.* 41 (12), 909–17.
Rose, A.M., 1970 Distance of migration and socio-economic status, *Readings in the*

Sociology of Migration, ed. C.J. Jansen (London), 85–91.

Swindell, K., 1967 Iron ore mining in West Africa: some recent developments in Guinea, Sierra Leone and Liberia, *Econ. Geog.* 43 (4), 333–46.

c1210

Population concentration in the republic of South Africa, 1960–70: a shift and share analysis

CHRISTOPHER BOARD   *University of London, UK*

Preliminary results of the 1970 census of population have disclosed a 34 per cent increase to nearly 21.5 million in the last decade. Although there has been a 38 per cent increase in the African population of the Republic as a whole, the increase in the White areas was only 18 per cent. However, the increase in the Bantu homelands amounted to 70 per cent, with the result that these areas now hold 46 per cent of the African population instead of 38 per cent as in 1960. In the same period the White population increased by only 21 per cent. A previous study of population change (Fair 1965) observed that between 1911 and 1936 South Africa's major core regions (accounting in 1960 for 55 per cent of her total Gross Domestic Product) attracted an unexpectedly high proportion of the population increase. Although this concentration continued between 1936 and 1960 there was also evidence of 'spilling out' by population increase into areas immediately adjacent to the Witwatersrand, Cape Town, and Durban.

At the time of writing the only detailed regional statistics on population change between 1960 and 1970 are available only for White areas and nine Bantu homelands or national units. The following account is based on population changes in the White areas *only*. Nevertheless, because they are widely spread throughout most parts of the Republic it is still possible to reach general conclusions about changes in the core and peripheral regions studied by Fair (1965).

Losses in total population were recorded in many districts of the drier interior of the western Cape Province, northern Natal, and the northern Transvaal. Increases were greatest in suburban districts near major cities in the Southern Transvaal, Western Cape, and Natal. However, when analysed by ethnic groups different patterns emerge. Many more districts are losing White population, chiefly agricultural areas in the Cape Province and Orange Free State. Such changes gave cause for alarm in the 1950's resulting in the appointment of the Commission of Inquiry into European Occupancy of Rural Areas which reported in 1960 on the importance of maintaining a White rural population. On the other hand many fewer areas have suffered a loss of non-White population and again these are chiefly in the northern Transvaal and wholly urban districts such as Cape Town, Johannesburg, and Durban. Such losses appear to be directly related to efforts to resettle non-Whites, particularly Africans in new townships, at some distance from old and more cramped quarters near city centres (Fair 1969, 345–6).

There are no statistics on migration as such within South Africa so that the attractiveness of particular regions as destinations of population movement has to be assessed indirectly. An attempt to isolate that component of change *not* attributable to 'national' population growth *nor* to the differential growth rates of the four ethnic groups was made by using shift and share analysis.

Regional shifts of each ethnic group were calculated and mapped by computer. For Whites the areas of upward shift are extremely localized. Large positive shifts characterize newly urbanized areas near major cities and also occur in important overbounded city districts such as Pretoria and Port Elizabeth but not in East London. A fairly limited zone of upward shifts lies east and south of the Witwatersrand associated with non-agricultural economic development. However, underbounded cities such as Johannesburg, Cape Town, and Durban all present downward shifts compensated by neighbouring suburban upward shifts. The other main types of area experiencing upward shifts are new mining areas, developing agricultural areas and the area affected by construction for the Orange River Development Project. The Letaba district in the Transvaal

Lowveld had 9569 Whites in 1960 and 17,497 in 1970: an 83 per cent increase. It includes the new township of Phalaborwa where copper and phosphate mining are important and where forestry and irrigation projects concentrating on fruit and vegetable production have recently been established. Elsewhere, spectacular upward shifts occur in the Natal Coastal belt with its sugar plantations and growing industrial developments supported by improved road and rail communications. Downward shifts were recorded by most other agricultural areas whether cropping regions in the Southwestern Cape or the High Veld or pastoral farming regions in the western or eastern Cape.

Upward shifts in the African population are more widely distributed, showing some concentration in major urban areas or core regions. More significant is a horseshoe-shaped arc of upward shifts stretching from the Swaziland border, running along the Vaal river and back across the middle of the Orange Free State between Kimberley and Bethlehem. This suggests that the spilling out of economic activity noted by Fair (1965, 69) has extended with the progressive integration of this area with the Southern Transvaal core region. Industrial development supported by much improved communications and a now quite dense electricity power grid as well as new mining developments on the eastern High Veld underpin this change. An isolated area of upward shift is again associated with the Orange River Development Project some 150km south of Bloemfontein. Negative shifts characterize the northern Transvaal in general and, locally, some districts near the Witwatersrand. These are undoubtedly white areas from which Africans have been resettled onto nearby Bantu homelands. The large negative shifts recorded in East London and Durban are also the result of short distance movements rather than an indication of economic decline. Very few upward shifts are found in the western Cape Province suggesting that the Government has successfully limited African population expansion there. Shifts in the Cape Coloured and Asian groups are mostly small, suggesting a general lack of mobility except for local reorganization of their residential areas in conurbations under the provisions of the Group Areas Act. The development of new Indian townships near Durban has limited Indian population increase in Durban itself. In Cape Town new growth of Coloured townships on the Cape Flats has taken most of the increase in the Coloured population in that urban area.

The pattern of upward shifts in the African population extends considerably beyond the original core regions and may prove, as Fair (1970) has suggested, to be a better indicator of economic growth than White population change. Nevertheless local population movements picked out by high negative shifts cannot be regarded as the result of economic forces. The complete picture of the spatial pattern of population shifts cannot be established until detailed statistics for Bantu homelands become available.

L.L.H. Baker; R.J. Johnston; University of Canterbury.

Fair, T.J.D., 1965 The core-periphery concept and population growth in South Africa, 1911–1960, *S. African Geog. J.* 47, 59–71.
- 1969 Southern Africa: bonds and barriers in a multi-racial region, in R.M. Prothero, ed., *A Geography of Africa* (London), 325–79.
- 1970 Population indicators and regional economic development, *S. African Geog. J.* 52, 124–28.

c1211
Time-space and mobility among the internal migrants in Ghana
E.V.T. ENGMANN *University of Cape Coast, Ghana*

Migration, as a force which attempts to redress the balance between the distribution of population and resources, tends to be selective in its impact on both the source and receiving areas, depending on the age, sex, place of birth, residence, and aspirations of the migrant; its effects vary according not only to the volume and composition of the migrant population, but also to the time during which it has operated. Two subdivisions of time

TABLE 1. Percentage of migrants by duration of residence

Duration of residence	Percentage of urban population			Percentage of rural population			Males per 1000 females	
	M	F	Total	M	F	Total	Urban	Rural
Migrants								
Less than 2 years	24.0	18.5	21.5	17.8	15.7	15.5	1,587	1,361
2 years to 5 years	15.5	11.8	14.0	9.0	10.4	8.8	1,536	1,042
5 years or longer	35.0	32.9	34.0	25.8	47.9	33.2	1,258	720
Non-migrants since birth	25.5	36.8	30.5	44.2	49.3	41.8	806	1,076
All periods	100.0	100.0	100.0	100.0	100.0	100.0	1,179	1,202

Source: Based on data derived from correspondence No. 510/18/60, Vol. 2/453, 2 February 1968 Census Office, Accra.

were identified, viz.: (1) the length of time it took the migrants to travel from their home areas to their host residence, and (2) the length of time during which the migrants have been staying in their present residence. Similarly two segments of space were distinguished; (1) the distance over which the migrants travelled from their home areas to present residences, and (2) type of areas where the migrants have their new residences. The effects of the distance travelled with the influences which supervene were indicated. This paper examines the effects of duration of stay on the migrants with regard to areas of residence, age, sex, and occupational structure, and assesses their mobility rates.

Data used were derived mainly from census reports and other official sources. Results of the 1970 population census were not yet available for analysis.

The population was grouped according to birthplace, and persons who were enumerated in places other than their birthplaces were considered as migrants; the duration of residence of migrants in their present areas was divided into periods of (1) less than 2 years, (2) from 2 to 5 years, and (3) 5 years or more.

In the urban areas of Ghana, newly arrived migrants represent 24.0 per cent of the males and 18.5 per cent of the females (Table 1). The minimum period of residence forms the basis on which the mobility rates were computed. On the basis of these figures it is shown that it will take a little over 4 (i.e. 100/24) years for the male migrant population to be completely replaced or depleted; and considerably more than 5 (i.e. 100/18.5) years to deplete the female population. Thus the mobility rate of the male migrants in the urban areas is about 25 per cent higher than that of the females; conversely in the rural areas the mobility rate is higher among the females than the males. Also the urban population as a whole is about 50 per cent more mobile than the rural population. The table suggests that among the urban population the male ratio declines with the length of residence, and among the residual urban population females are numerically predominant. The low male ratios may be accounted for partly by the outmigration of urban-born males, and partly by the relative resistance to mobility of the urban-born females. The analysis suggests that, given similar circumstances, girls born in the urban areas are less likely to move out of the urban area than are urban-born boys. In contrast the high male ratio of the residual population in the rural areas may be explained partly by the relative outmigration of adult females to the adjoining towns and cities. Other factors may be locally important. But the inference is that given similar circumstances rural-born males are less likely to move out of the rural environment than are rural-born girls.

The structure of the migrant population is sharply differentiated from that of the overall population (Table 2). The average age of the migrants rises with the length of residence; and the median age of the newly arrived population is about 10 years lower than that of the non-migrant population. Also new migrants in the towns form a higher proportion of the female migrants than is the case with the male migrants. Thus it may be deduced that, for the females, migration to the towns is a relatively new experience. With each successive age group the diminution of numbers is more rapid among the female mi-

TABLE 2. Age and sex structure of internal migrants by duration and place of residence (percentage within each sex)

Age groups	Urban								Rural								Ghana total (5)	
	Less than 2 years (1)		2 to 5 years (2)		5 years and over (3)		Since birth (4)		Less than 2 years (1)		2 to 5 years (2)		5 years and over (3)		Since birth (4)			
	M	F	M	F	M	F	M	F	M	F	M	F	M	F	M	F	M	F
15–24	47.3	54.4	39.5	47.0	18.9	25.8	39.5	35.0	37.1	50.5	29.0	44.1	16.2	18.5	28.8	32.1	24.2	26.2
25–34	34.8	27.8	36.0	37.2	31.2	31.6	23.2	26.2	34.3	28.8	34.6	35.0	21.9	32.3	24.2	27.3	39.3	40.6
35–44	11.4	11.2	14.5	9.8	23.4	19.5	16.3	15.6	19.0	11.6	20.9	13.3	23.1	22.6	18.9	17.0	16.1	14.9
45–54	4.0	3.7	4.7	3.9	14.5	10.8	9.8	10.0	6.1	5.0	9.1	5.0	16.7	12.2	13.0	10.4	9.8	8.1
55–64	1.6	2.0	4.7	1.1	6.7	6.4	5.8	5.6	2.3	2.2	4.0	1.6	10.6	7.3	7.2	6.3	5.4	4.7
65+	0.9	1.0	0.6	1.0	5.3	5.9	5.4	7.6	1.2	1.9	2.4	1.0	15.0	7.1	7.9	6.9	5.2	4.8

Source: see Table 1.

TABLE 3. Persons employed in each occupational group by duration of residence, and sex (percentage within each sex group)

Type of occupation	Duration of residence					
	Less than 6 months (1)		6 months to 2 years (2)		2 years or longer (3)	
	Males	Females	Males	Females	Males	Females
Professional workers	24.6	12.9	22.1	23.8	53.3	63.3
Administrative workers	14.5	14.5	19.4	12.2	66.1	73.5
Clerical workers	11.4	6.7	15.7	16.9	72.9	78.4
Sales workers	10.8	7.7	11.1	6.9	78.9	85.4
Farmers etc.	7.6	5.2	3.3	4.3	87.1	90.1
Miners etc.	17.8	17.7	20.8	17.9	81.3	70.4
Workers in transport etc.	14.3	17.5	9.9	12.2	75.8	70.3
Craftsmen	12.8	10.3	13.2	8.4	73.9	81.3
Service workers	13.4	14.2	19.7	16.5	66.9	89.3

Source: See Table 1.

grants, again suggesting partly that migration among the females tends to cease at an earlier age than for the males. The Chi Square (χ^2) statistic is applied to test the meaningfulness of these observed differences, and it appears that the age structure of the male and female migrants is significantly different from that of the overall population. In the rural areas the age structure of the female migrant is different from the overall population, but not so the structure of the male migrants. The inference is that, for females, age is a more important factor in migration in both the urban and rural areas; while among male migrants in the rural areas age is not so important.

The distribution of various categories of workers according to the duration of residence shows substantial differences (Table 3). Application of the Chi Square (χ^2) statistic suggests that the occupational structure of the migrant workers is significantly different from that of the non-migrant section of the population.

Analysis of the occupational structure of migrants by duration of residence suggests that 9.7 per cent and 45.5 per cent of the newly arrived migrants were occupied respectively in professional jobs and farming. If these values are compared with the expected ones, the degree of mobility of the group in question can be shown. Among newly arrived migrants there are 3.13 (i.e. 9.7/3.1) times as many professional and allied workers as expected; similarly, farmers account for less than 75 (45.5/62.8) per cent among the newly arrived migrants. Thus farmers are among the least represented of the new migrants. Further analysis suggests that professional and allied workers are about 4.33 times more mobile than farmers. When reduced to indices it is shown, for example, that the mobility rates of professional and allied workers and of executive, managerial, and clerical workers are respectively, 433, and 208. Accumulation of more data is needed for useful conclusions to be drawn.

c1212
Patterns of internal migration in India
GURDEV SINGH GOSAL and GOPAL KRISHAN   *Panjab University, India*

An overwhelming majority of India's massive population spends its life cycle in or near the place of birth. At the 1961 census, 88 per cent of the population was enumerated within the district of birth, and almost 95 per cent within the native state. The continuing dependence of most of the people on agri-

culture, the inadequacy of employment opportunities in the non-agricultural sector, the high incidence of illiteracy, and strong family ties are among the factors accounting for this phenomenon. The increasing supply of local labour (arising from an improving rate of natural growth) and a recent acceleration in commuting, especially to big cities, have also been responsible for confining some of the prospective migrants to their native places.

Nevertheless, mobility has been increasing in recent decades, particularly since independence. In 1921, 24.7 million persons (9.8 per cent of total population) were enumerated outside the district of their birth, the figure rising to 38.5 million (10.8 per cent) in 1951, and to 52.9 million (12.1 per cent) in 1961. The partition of the country in 1947, the post-independence emergence of a new zeal for improving standards of living, socio-economic development under the five-year plans, and intensifying pressure of population on agricultural land were the chief factors stimulating this change. It is remarkable that of the total migrants almost half had moved during 1951-61 alone.

As defined by the Indian Census code, a migrant is one who is enumerated at a place other than that of birth. Accordingly, almost one-third (144.1 million out of 439.2 million) of India's population was migrant in 1961. Of these, 67.6 per cent were females. The preponderance of female migration is associated with marriage, for under the prevailing system of patrilocal matrimonial residence it is the wife who moves to the husband's place and in the process becomes a migrant. Nonetheless, 10.7 per cent of the male population which migrated outside its native district, and another 10.1 per cent which moved within it, together make no less than 47 million.

It is noteworthy that as high a proportion as 74 per cent of the migrants moved within rural areas. Another 15 per cent were involved in rural to urban migration. Urban to urban migrants accounted for 8 per cent of the total, and the remaining few were the urban to rural migrants. Thus, as far as magnitude goes, rural-rural migration has been of overwhelming importance though a sizable segment of it was due to matrimonial mobility among females. By contrast, the rural-urban flow was distinctly male-dominated and was directed chiefly towards the large industrial cities. In comparison, the urban to urban migration channelled itself mainly to principal administrative centres (especially the capitals of states), educational centres (particularly those with universities), and manufacturing centres with specialized industries.

In-migration has been characteristic of the following types of areas:

(1) Urban-industrial concentrations, such as the Calcutta conurbation; the Bombay-Thana-Poona industrial complex; the Gujarat plain; western Tamil Nadu; and scattered districts and territories such as Delhi, Kanpur, and Bangalore.

(2) Plantations (mostly of tea) in the foothills zone of the Brahmaputra and Surma valleys in Assam; the Darjeeling-Jalpaiguri tract in West Bengal; and the Southern Ghats in parts of Kerala, Mysore, and Tamil Nadu.

(3) Newly developed agricultural lands in the Assam valley; the area newly irrigated by the Bhakra Canal system in Panjab, Haryana, and Rajasthan; the Dandakaranya in Madhya Pradesh and Orissa; and the Terai in Uttar Pradesh.

(4) Mining areas, like the Damodar valley; the north Orissa plateau and its contiguous tract in Bihar; and southeastern Rajasthan along with its adjoining areas in Gujarat and Madhya Pradesh.

(5) Construction sites of various multipurpose projects, such as Bhakra-Nangal, Hirakud and Tungbhadra.

On the other hand, the heaviest outflows of migrants emanated from those rural areas where the per capita agricultural productivity was low due mainly to a high nutritional density of population, the small size of agricultural landholdings, and overdependence on agriculture. These areas include (1) the Ganga Plain in Uttar Pradesh and North Bihar, (2) coastal parts of Tamil Nadu, Orissa, and Konkan, (3) northeastern Rajasthan, and (4) the North Panjab Plain. Minor flows originated in the Bundelkhand plateau (Uttar Pradesh and Madhya Pradesh), Ryalaseema tract (Andhra Pradesh), Rarh plain (West Bengal), and parts of the Western Himalayas.

In the context of the predominantly agricultural economy and consistently fast-growing population of India, the small and declining size of agricultural landholdings (national average about 3 hectares in 1961) has been basic to the process of out-migration, notwithstanding considerable out-migration from areas like Panjab, where agricultural landholdings are comparatively large. The zeal to improve one's standard of living has been another vital consideration in increasing migration from small towns to big cities, and movement of landless agricultural labourers (mostly belonging to scheduled castes) from rural areas to urban places. Proximity to urban-industrial concentrations and major construction sites works as a strong pull factor. During the pre-independence period, mobility within and between the areas under British rule was strikingly more than in those under the princely rule. Apart from economic stresses and strains, however, the enterprising spirit of certain communities, such as those of Panjab, Gujarat, and Tamil Nadu, played its own role in migration, not only within India but also abroad.

The history of the existing patterns of economically induced migration in India can be traced back to the mid-nineteenth century, when the process of industrialization was initiated, railways were laid out, and plantation agriculture introduced. A second spurt to the economy and to migration was experienced after the First World War when industries started growing under protection, mining activities intensified, plantations expanded, and transport networks thickened. Developments during and after the Second World War provided an additional stimulus to migration. Lastly, independence in 1947 was a landmark in the recent history of migration in the country. It resulted not only in displacement of millions of persons but was followed by a period of planned development oriented towards a socialistic pattern of society.

What of the future? With consistently staggering increases in population and limited scope for expansion of the area under cultivation, a shift of some agricultural population from the rural areas to other activities is inevitable. Prevailing socio-economic forces will also accelerate the tempo of migration of non-agricultural population. On the other hand, many of the in-migration areas are becoming saturated, the supply of local labour is increasing everywhere, and regional feelings against inflow from outside the state are intensifying. Commuting is putting some desirable restraint on the influx into cities. Inter-state migration on any considerable scale is not apparent. It is likely, therefore, that the rate of interstate migration may not increase, while that of intra-state migration promises to rise. The multi-regional complex of migration to metropolitan cities is likely to continue.

c1213
Internal migration in Mexico, 1940–70
MARÍA TERESA GUTIERREZ DE MACGREGOR   *Ciudad Universitaria, Mexico*

The purpose of this paper is to present the evolution of the internal migration in Mexico from 1940 to 1970, and to analyse the inter-regional movements of population which are responsible for regional population changes. Correlation analyses between regional migration movement and such factors as economic, social and cultural characteristics have been carried out. This information will cast light on the strong regional contrasts that are observed in Mexico among the different regions, not only from the demographic point of view but from the economic, social, and cultural as well. Special attention has been given to internal migration towards the three largest cities: Mexico, the capital city, Guadalajara, and Monterrey.

In order to understand how internal migration has developed in Mexico from the sharp contrasts observed between cities and regions, it is perhaps necessary to present in summary form some facts about the phenomenon.

(*a*) in 1900 only 1,706,131 inhabitants (12.5 per cent of the total population) were living in urban centres (towns having 10,000 or more inhabitants). By 1970 this percentage had increased to 48.6 per cent; that is to say, 23,526,556 inhabitants were living in urban centres.

(*b*) Although annual natural increase remains high, the rapid urban growth is due, very largely, to extensive internal migration from rural towns toward larger urban centres, and in lesser degree to migration from small urban centres to large ones (70.9 per cent of the whole urban population is located in 34 cities with more than 100,000 inhabitants).

(c) The concentration of the urban population of the three main cities has been intensified by the heavy and rapid outflow of rural population to cities. In 1970 more than 47 per cent of the urban population of the country was concentrated in three cities: Greater Mexico 8.5 millions (35.5 per cent), Guadalajara 1.4 millions (6.2 per cent), and Monterrey 1.2 millions (5.1 per cent).

(d) During the period 1960 to 1970 Mexico's total population increased by approximately 38.5 per cent, but this increase was unevenly spread between the rural and urban sectors. The rural sector grew by only 14.7 per cent, while the urban population increased by 77.4 per cent.

(e) The high rate of population growth tends to equal or outpace economic growth rates.

(f) During recent years there has been a rapid development of the communications network and the transportation system.

(g) The low rate of productivity gains, together with the mechanization of agriculture, have brought a reduction in the number of agricultural labourers in some regions of the country. The proportion of the economically active population engaged in agriculture in 1960 was 54.9 per cent. By 1970 this figure decreased to 39.5 per cent of the people working in agriculture or related activities. This represents a decrease of 15.4 percentage points in 10 years.

The total number of persons changing the state of origin amounted to 2 millions in 1940 and 7 millions in 1970. No doubt more people move from their locality of origin but censuses register changes only at state level. We may assume that migration displacements will remain strong in Mexico in spite of the inability of the larger cities to provide adequate facilities such as housing, services, and employment for the floods of migrants.

An examination of the curve showing the ratio of immigrants in relation to the total population for the same period reveals a reduction in the last ten years. A possible explanation may be that the Mexican population in 1970 was probably better distributed in terms of economic opportunities than was the case ten years earlier, attributable perhaps to government policy during the last decade.

Efforts have been directed towards decentralizing the economy in industry, commerce, education, etc., bringing some benefits to provincial towns away from the capital.

In 1970 people changing their state of origin were concentrated in the three main cities: Greater Mexico, 39 per cent; Guadalajara 4 per cent; and Monterrey 6 per cent.

In order to study the spatial movements of migrants the country has been divided into 8 geo-economic regions: Region I Northwestern; Region II Northern high Plateau; Region III Northeastern; Region IV South Pacific; Region V Central-West; Region VI Central; Region VII Gulf of Mexico; Region VIII Peninsula of Yucatan.

Region VI receives the greatest number of immigrants; it shows an increase from 49 per cent in 1940 to 53 per cent in 1970. This region is located in the central part of the country and had approximately half of the nation's immigrants during the whole period 1940–70. It is also important to note that in 1970 43.4 per cent of the immigrants to this region were concentrated in two of the seven states that comprise this region: the Federal District and the State of Mexico.

In Mexico we may consider different causes for migrating. Apart from the historical reasons prevailing since colonial times, in which the economies were very centralized and transport systems were structured accordingly, there are other reasons in modern times:

(1) Migrations towards the Central Region VI, caused by the attraction of the capital city. There, the people find more opportunities for employment, higher salaries, and more social and educational benefits, which have a decisive influence upon migration.

(2) Migrations caused by the rapid development of Monterrey, the second most important industrial city of the country, situated in the Northern High Plateau, Region II.

(3) Migrations caused by improvements in methods of cultivation, as in the Northwestern Region I, in which the irrigation system of the valleys of Mexicali, Yaqui, and Fuerte y Mayo, attracts a great number of migrants.

(4) Migrations caused by the attraction of high salaries offered by the border cities of the southern United States.

c1214
Internal migration in east-central Europe
LESZEK ANTONI KOSIŃSKI   *University of Alberta, Canada*

Three types of data and consequently three methods of study can be used in an enquiry into internal migration in east-central Europe. (1) From regular censuses, estimates of population, and vital statistics, the residual method can be used provided changes in administrative boundaries are not too frequent and/or appropriate adjustments can be made. One such study of the whole area was based on data of 1961–5 (Kosiński 1968) and more detailed studies for Poland and Czechoslovakia are also available (Kosiński 1970; Carter 1971). (2) Questions concerning mobility have been included in the Polish, Yugoslav, and Bulgarian censuses: they concerned previous residence in Poland and place of birth in Yugoslavia and Bulgaria. Such data provide valuable material for national studies (Latuch 1970; Sentić 1968) but cannot be used for international comparisons. (3) Finally, in almost all countries of this region (except Yugoslavia and Albania), registration data on directions, size, and frequency of population flows have been published since the early 1950s. Several authors have used them for national or international analyses (Thirring 1967; Pivovarov 1970).

Two periods of post-war internal migrations in east-central Europe can clearly be distinguished. In the late 1940s internal movements were still influenced by the recent war, territorial changes, and transfers of population. Internal movements represented a continuation of external migrations, since repatriation of nationals who found themselves outside of changed or reduced territory of a country was followed by their internal transfer and settlement. East Germany, Poland, and Czechoslovakia all experienced this type of move. In Germany new arrivals created an additional burden for the war-shaken society. In Poland and Czechoslovakia the vacuum created by departing Germans was partially filled by the repatriates; internal settlers also participated in this movement to the vacated areas. In Yugoslavia migration from the less developed, mountainous areas to the more fertile lowlands was accelerated by the departure of Germans from the plains of Vojvodina and by the land reform.

Beginning in the 1950s a more 'normal' pattern of migration developed. Rural-urban flows reflected the rapid drive towards industrialization, rural-rural migrations were partially caused by changes in the pattern of rural settlements, and urban-urban migrations followed the accelerated process of urbanization.

In the present paper basic trends in mobility in six countries are examined. An analysis based on current registration indicates that the mobility of population decreased in all countries. The decrease was more substantial in more developed countries – East Germany, Poland, and Czechoslovakia – which may suggest that a new equilibrium has been approached (Pivovarov 1970). Another explanation for the apparent decline of mobility is that the three northern countries were more affected by the post-war transfers and the higher proportion of uprooted people tended to be more mobile. Labour deficits in rural areas of Czechoslovakia and Germany also adversely influence the mobility of population.

Rural-urban flows have by far exceeded the urban-rural counterstreams resulting in consistent population gains of urban at the expense of rural communities to the extent of over 1 million per year in 1951–60 (Pivovarov 1970). In an area where the average level of urbanization is not very high and the process of urbanization is fairly rapid, internal migration accounts for a high and an increasing share of the urban growth: 43 per cent in Poland in 1960–7; in Czechoslovakia and Romania about 50 per cent in 1956–60 and 1956–61; in Hungary 92 per cent in 1961–5.

The urban-urban migration has been less important until now but it will undoubtedly increase. In Poland it accounted for 22 per cent of the total flows in 1960–7, and in Bulgaria 14 per cent in 1961–5. Rural-rural migration continues to play a rather important role, representing 30 per cent of the flow in Bulgaria (1961–5) and 36 per cent in Poland (1960–6).

The distance of moves is decreasing since the proportion of migrations between the regions of the first order declines and more people change address within them (Pivovarov 1970). In Poland migration within voivodships represented 61 per cent in 1960 and 66 per cent in 1967 (Poland 1969); in Romania migrations within regiunea accounted for 58 per cent in 1966 as compared to 48 per cent in 1955 (Enache 1967). A similar trend has been observed in East Germany (Bose 1970). Nevertheless, interregional migrations do take place. In all countries the areas attracting migrants are relatively restricted to large cities and industrialized regions. On the other hand extensive rural regions are losing population. In a sense inter-regional migration represents rural-urban transfer. The correlation between industrial employment and gross migration in Czechoslovakia, in 1960–4, was fairly high (+0.754) (Kuba 1968). In Yugoslavia the coefficient of correlation between net migration in 1958–61 and percentage of urban population was 0.894 (Sentić 1968).

In three countries migrants are asked the reasons for moving. Housing conditions are the most important causes in Czechoslovakia, exceeding economic reasons (44–5 per cent and 30–3 per cent respectively in 1966–7); but in Bulgaria (1964) and Hungary (1967) economic reasons clearly dominate. As might be expected, migrations are highly selective with respect to age.

Internal migrations play an important role in the growth of urban population. However, since a large proportion of them are restricted spatially they have not changed substantially the distribution of population in east-central Europe during the last two decades.

Bose, G., 1970 Entwicklungstendenzen der Binnenwanderung in der DDR im Zeitraum 1953 bis 1965, *Pet. Geogr. Mit.*, 114, 117–31.

Carter, F.W., 1971 The natural and migrational components of population change in Czechoslovakia, 1950–1970, *Inst. Brit. Geogr., Symp. on population change in eastern Europe* (mimeo.).

Compton, P.A., 1969 Internal migration and population change in Hungary between 1959 and 1965, *Trans. Inst. Brit. Geogr.*, 47, 111–30.

Cucu, V. et al., 1966 Contribuţii la studiul geografic al migraţiei interne a populaţiei din România, *Anal. Univ. Bucureşti, Geologie-Geografie*, 15, 137–46.

Enache, C.M., 1967 Industrializarea socialistă – baza utilizării eficiente a forţei de munca, *Probleme economice*, 20, 119–34.

Korčák, J., 1959 Vnitřní migrace v Československu. In: *Sbornik v chest Akad. A.S. Beshkova* (Sofia), 55–64.

Kosiński, L., 1968 Population growth in east-central Europe in the years 1961–1965, *Geographia Polonica*, 14, 297–304.

– 1970 The internal migration of population in Poland, 1961–1965, *Geographia Polonica*, 18, 75–84.

Kuba, R., 1968 *Racionálny pohyb obyvateľstva*, Vydavateľstvo politickej literatúry, Bratislava.

Latuch, M., 1970 *Migracje wewnętrzne w Polsce na tle industrializacji (1950–1960)*, Państwowe Wydawnictwo Ekonomiczne, Warszawa.

Pivovarov, J., 1970 *Naselenie sotsialisticheskikh stran zarubezhnoi Evropy*, Nauka, Moskva.

Poland, Główny Urząd Statystyczny, 1969 *Ruch wędrówkowy ludności w Polsce w latach 1960–1967*, Studia i Prace Statystyczne, no. 18.

Sárfalvi, B., 1965 *A mezőgazdasági népesség csökkenése Magyarországon*, Akadémiai Kiadó, Budapest.

Sentić, M., 1968 Some aspects of migration movements in the Yugoslav population. In *World Views of Population Problems*, E. Szabady, ed., Akadémiai Kiadó, Budapest, p. 321–8.

Thirring, L.L., 1967 Internal migration in Hungary and some Central and East European countries, In *U.N. World population conference, Belgrade 1965* (New York), 4, p. 527–31.

Todorov, V., 1968 *Migratsionnite protsesi i niakoi sotsialno-ekonomicheski i kulturno-bitovi problemi na seloto*, Bulgarska Akademia Nauk, Sofia.

Tufescu, V., 1969 Modifications dans la structure de la population en Roumanie, après 1948, *Revue Roumaine de Géologie, Géophysique et Géographie, Ser. Géographie*, 13, 111–27.

c1215
Themes on internal migration in southeast Asian countries
RONALD C.Y. NG *University of London, UK*

Southeast Asia is a major geographical region with continental dimensions and a total population exceeding 220 million. A common feature of the eleven countries in the area is their rapid rate of natural population increase. Although most of these countries have now adopted family planning programs as state policy, the trend of fast growth is not likely to decline substantially in the near future in spite of the strenuous efforts being devoted to this problem. As the result of rapid population growth in the recent past, large numbers of job-seeking youths are entering the labour market every year.

With the exception of the Republic of Singapore, these countries are predominantly rural and have extremely uneven population distributions. Population pressure is considered to be high on the alluvial deltas and coastal plains, but outside these 'core areas' each country has its rather extensive 'frontier zones' with high agricultural potentials. Strenuous, but often disappointing, efforts have been devoted by governments to encourage the development of these less densely populated areas. At present, local pockets of more densely populated areas away from the major concentrations are associated with centres of plantation development and mineral extraction. The basic disparity of population densities between the 'core' and 'frontier' zones is one of the primary motivating forces in generating internal migration. Of Malaysia, for example, it has been said that 'the contrast between close and sparse settlement will be far greater (in the near future), so great indeed that some movement of population to lessen it seems inevitable' (Jones 1965, 50). A similar comment has been made for Thailand (Sternstein 1965, 20-2).

While population pressure is building up in individual countries, there is little prospect of international migration within the region itself. Nations conscious of the experience of massive Chinese and Indian immigration during pre-independence days are reluctant to accept further migrants even from neighbouring countries. With international boundaries becoming almost entirely impermeable, the rapidly increasing populations are redistributing themselves within their respective countries.

Most countries in the region are in the course of executing their economic development plans, which, among other objectives, are calling for accelerated industrialization of a labour-intensive type. The interaction between the 'pull' of the growth points and the 'push' of population pressure in the surrounding countryside has generated substantial streams of urbanward migrants with the attendant results of high levels of unemployment and rapidly expanding urban slums.

Inter-urban migration is of relatively little significance in southeast Asia because of the primate nature of its central cities. Most of the non-agricultural activities are heavily concentrated in the capitals and their satellites, while upcountry 'urban' areas tend to have predominantly rural characteristics. The primary functions of these secondary centres are marketing and the provision of services. The special skills required in the successful operation of these trades are not commonly possessed by the rural people. The very growth of these country towns is dependent on the prosperity of their agricultural hinterland. There is therefore little rural-urban migration directed towards these lesser towns away from the capital.

Of much greater significance are the inter-rural migration streams, as the populations of southeast Asia are predominantly rural. The characteristic uneven distribution of settlement demands redistribution of population as the best possible solution to the dual problem of extreme population pressure in the core areas and a lack of labour for effective economic development of the traditionally sparsely populated frontier zones that exist in every country of the region. In the Philippines, for example, the number of settlers moving to Mindanao is estimated to be much larger than that of rural-urban migrants.

Elsewhere, although there is in many cases administrative provision for peasants in crowded areas to obtain land grants in more sparsely populated parts, spontaneous pioneering efforts are often stifled by the great difficulties and high costs of land clearance and the economic problem of subsistence during the initial period of re-establishment.

Apart from the spontaneous inter-rural migrants, who display a definite and stable pattern of movement under normal circumstances, there are large numbers of migrants sponsored by the state to settle in specially designated areas. Earlier settlement schemes were conceived primarily as measures for achieving the desired goal of population redistribution, but the results were generally more significant in terms of the development of the sparsely settled areas than actually relieving the problematic population pressure of the congested areas.

Java is a case in point. During the 1930s, a transmigration policy aimed at massive transfers of population from congested Java to the outer islands was adopted by the Dutch East Indies government. Although the density of the province of South Sumatra, which received most of the Javanese migrants, rose from 13 persons per km^2 in 1930 to 17.5 in 1961 (Withington 1963, 205), the transfer of 46,096 individuals in 1960, the best year on record, is hardly significant to the problem of population pressure on Java where annual increase is in the order of a million. Malaysia has had similar schemes for population redistribution. These have on the whole proved more successful, perhaps because of their greater support for the settlers and because of the proximity of the resettlement projects to the communities whence the settlers originated. In the Philippines the focus has been on the frontier island of Mindanao, which received no fewer than 1.5 million new settlers between 1948 and 1960. Nevertheless, only a third of the estimated 6,000,000ha of potential agricultural land has been occupied (Antonio 1958, 18), though with further expansion of roads in the interior, migration to Mindanao will continue.

Rural population pressure might be relieved by measures such as the provision of vital irrigation facilities, but these often entail the displacement of large numbers of people who would either have to migrate spontaneously or be resettled by the state. When the mammoth projects of the Mekong Basin Development Scheme get under way, movement thus stimulated may be the most significant type of internal migration in the country concerned. Displacement caused by the construction of infrastructural facilities has already occurred, most noticeably in the case of the irrigation projects of northeastern Thailand.

Over the past few decades southeast Asia has experienced much internal and external conflict which has caused substantial population drifts. The ravages of the Pacific war created thousands of refugees throughout the region, and further disturbances were created in the post-war period as different political factions vied for control of the nascent independent governments (McGee 1960, 49–58). The latter were frequently complicated by the high degree of plurality of southeast Asian society. Migration trends in the affected areas will continue to be complicated by a large number of interacting factors of both transient and permanent character.

Antonio, G.E., 1958   Economic development of Mindanao, *Philippine Econ. Rev.* 4, 15–18.

Jones, L.W., 1965   Malaysia's future population, *Pacific Viewpoint* 6, no. 1, 39–51.

McGee, T.G., 1960   Aspects of the political geography of southeast Asia: a study of a period of nation-building, *Pacific Viewpoint* 1, no. 1, 39–58.

Sternstein, L., 1965   A critique of Thai population data, *Pacific Viewpoint* 6, no. 1, 15–35.

Withington, W.A., 1963   The distribution of population in Sumatra, Indonesia 1961, *J. Tropical Geog.* 17, 203–12.

c1216

Internal migration: a review of data collection, with particular reference to southern Africa

J.W.B. PERRY   *University of Botswana, Lesotho and Swaziland*

Internal migration is one of the vital processes which alter the size and distribution of a population, yet the data available are often limited. Measurement and analysis of migrations is complicated by the fact that they vary both in space and in time.

To facilitate data collection it is useful to identify migration types. Three principal criteria for classification of migration have been identified on the basis of place, purpose, and time.

The most widespread movements in the first category in southern Africa are rural-urban movements and rural-rural movements. Urban-rural movements are significant among the African population of South Africa.

The most important purposes of movements within the countries of southern Africa are labour migration and political movements.

On the basis of time, movements may be daily, seasonal, short-term, long-term, temporary, and permanent. In southern Africa the principal daily movements, apart from those of the cultivator from his village to his fields, are urban commuter movements. Short-term and seasonal movements are mainly agricultural. Long-term temporary and permanent migrations are directed mainly to the urban, industrial, and mining areas.

The study of migration is concerned with the following variables: numbers and characteristics of the migrants; characteristics of the source region; distance between the source and the destination and the characteristics of the region in between; characteristics of the destination area; differences between the community of origin and the community of destination.

Normally the only choice for the demographer or population geographer collecting data on internal migration is to select from among the existing censal, administrative, and/or political boundaries, which therefore limits the kind of movements that can be studied. In most southern African countries, apart from South Africa, the only available data boundary is the census enumeration area.

Two main data sources are recognized: direct sources which include continuous national registration of population, direct census questions, and sample field surveys; indirect methods such as the 'vital statistics' method, the 'survival-ratio' method, the 'place of birth' method, and measurements of ethnic imbalance. Records of state organizations such as national health services or employment registers, which can be indirect data sources, do not exist in southern Africa. South Africa is the only country in southern Africa which has a national population register.

The indirect methods all have limitations, one of the most important of which is their inability to identify migration flows. More important, however, is the inability of either of these principal data sources to provide information, except on a very general level, of the qualitative characteristics of movement; they reveal only quantitative change. Qualitative information is normally obtained only from intensive field studies which are usually limited in size and scope. The principal limitation of such surveys is that they are usually time-specific.

A considerable number of studies of internal migration, based on both direct and indirect data sources, exists. The industrialized countries usually have more access to direct data, but a number of studies of internal migration in east and west Africa have utilized such indirect census-derived data as 'place of birth' statistics, age-sex differentials, and the measurement of ethnic imbalance.

Not only is there generally a lack of data, but illiteracy, vagueness of knowledge of time periods, difficulties of communication, and resistance to questioning for political, economic, or cultural reasons may impede data acquisition. Data comparisons are further complicated by continually changing statistical units, a problem not shared to the same degree by most industrialized countries. The 1966 census of Lesotho, for example, was based on different enumeration areas from the 1956 census.

Southern Africa as a whole is characterized by the paucity of data available. The Republic of South Africa is exceptional in that population movements are well documented as part of a political policy. Labour migration is the principal type of population movement. Unfortunately most countries use a general category 'absentees' in their censuses, which does not distinguish between internal and external movements. Lesotho, for example, geographically a part of South Africa, is characterized by very large recurrent movements of labour migrants to South Africa which are technically external because of political boundaries, but nevertheless of the same type as occur within South Africa.

An analysis of the available data sources by country, including short-comings and possibilities for improvement, will be attempted. In view of the limited and often crude data at present available, there is a need for more particular information via precise census questions and detailed field studies.

c1217
Space and time dimensions in the study of population mobility in tropical Africa
R.M. PROTHERO, A.D. GODDARD, and W.T.S. GOULD   *University of Liverpool, England*

There is need for greater attention to be paid to spatial and temporal dimensions in the study of population mobility. Some years ago one of the authors of this paper outlined an as yet unpublished tentative typology of population mobility in tropical Africa, directing attention to these dimensions. It considered types of mobility within two major categories associated respectively with economic and non-economic factors. Since that time a three-year research project, 'A Survey and Analysis of Population Mobility in Tropical Africa,' has been set up. The project has been working with a further typology which is outlined here.

The formulation of a typology is an important element in the project's aims to 'undertake a systematic appraisal of what is already known of mobility phenomena' and 'to contribute to theory and methodology.' While an inductive rather than a deductive approach is clearly implicit, nevertheless it has been essential to make some preliminary attempt to formulate a typology, if only to facilitate bibliographic work with a system for the automatic retrieval of information. The basic framework for the typology is shown in the accompanying chart. It incorporates the two basic geographical dimensions of mobility: space expressed in the four categories of rural/urban relationships, and contemporary time in the relative sense of the span of each movement, generally but not invariably increasing in length from left to right. A further dimension of historical time might be introduced. For the present this perspective is omitted and only contemporary movements are considered, though it is important to recognise that even within recent time the relative importance of various types of mobility has changed.

Fundamental to the typology are distinctions made between the terms mobility, migration, and circulation. Mobility is a sufficiently broad term to include all types of movement. Migration is sometimes used in this broad sense; others recognise that migration is more limited in scope than mobility.

Migration may be defined in a strictly operational sense as movements revealed in official statistics. This sort of definition does not have any rigorous conceptual basis, but rather is one of convenience. Most definitions of migration include reference to permanent change of residence; movements which are not within this category therefore are those which do not involve any permanent change but are of a rhythmic or oscillatory nature. These movements, where there is no permanent change, could be designated as circulation.

The principal difference between migration and circulation lies in the permanence of the former and the non-permanence of the latter, but permanent has been and may be defined in different ways, so that the dividing line between migration and circulation may be blurred. Hence the definition of the United Nations of migration as those movements lasting more than one year has been

Population mobility in Africa

Space	Time					
	Circulation				Migration	
	Daily	Periodic	Seasonal	Long term	Irregular	Permanent
Rural-rural						
Rural-urban						
Urban-rural						
Urban-urban						

generally ignored in practice in African studies involving field investigation, though it has been followed by statisticians and formal demographers relying on census data. Other discussions of permanence have considered the economic and social commitments of a mover to destination as compared with those to his home area, but there has not been any universally accepted definition of permanence. It is suggested that if there is a specific desire on the part of the individual or group of individuals who are moving to return to their place of origin, and when before leaving in the first place this intention is clear, then the movement may be considered as circulation rather than migration. However, some movers know only the timing or direction of future movements, or neither, and their movements may be considered as migration. This distinction between circulation and migration is not directly related to the duration of each movement, for some circulatory movements may last longer than migratory ones, but to the long-term changes in distribution that result from the movements. In circulation changes in the distribution of population in the long term are not significantly different from those in the short term; in migration changes in the long term are very different from changes in the short term.

Circulatory movements have been subdivided according to the length of their cycle into four main groups – daily, periodic (from one night to one year), seasonal and long-term (more than one year).

The presence or absence of 'permanent' elements in movement is the basis for distinguishing between migration and circulation. However, different interpretations of 'permanent' give rise to two types of migration.

Permanent migration in the conventional usage of the term is relatively uncommon in tropical Africa, though less so now than was the case in the past. Irregular migration is not wholly permanent in that further movement is likely in the future, but neither time nor direction of such movement is known.

All African population movements may be accommodated in one of the six major categories outlined and in addition within the framework of the typology each has a spatial component to further identify it.

The typology outlined has been formulated for working purposes. It has already been suggested that the typology should incorporate in the spatial dimension some expression of distance (e.g. local, regional, national, international, or short, medium, long). These and other refinements will be incorporated to provide a more sophisticated insight into the variety and nature of population mobility in tropical Africa.

c1218
Migration patterns in Uttar Pradesh: a geographical analysis
B.K. ROY   *Office of the Registrar General, India*

The area selected for this study is confined to Uttar Pradesh. In the present paper certain patterns of migration in Uttar Pradesh have been delineated on the basis of census data as well as field study investigations. This study has been carried out with a twofold objective: (1) to evaluate the migration data of the 1911 and 1921 censuses and (2) to identify distinctiveness in the pattern of migration by carrying out field investigations during 1968–9 of certain classes of families from four selected districts of the state. The analysis of the data from the censuses of 1911 and 1921 forms a historical background in establishing levels of migration mainly on the basis of birthplace considerations. To highlight some of the chief characteristics of family considerations, four districts of eastern Uttar Pradesh (Jaunpur, Ballia, Varanasi and Pratapgarh) have been selected for detailed study of migration patterns and to show the flow of migrants according to place of birth and place of occupation. This also elucidates certain outstanding family socio-economic features which cause shifts in population.

Some forty families have been selected from these four districts of Uttar Pradesh according to financial characteristics (financially poor families, medium-income, and rich families), to identify migration patterns in qualitative and quantitative terms. Data have been collected and processed and on this basis the origins of migrants in relation to sex, age, and employment have been discussed. The patterns of migration have also been mapped to show the effectiveness of rural-rural and rural-urban migration in Uttar Pradesh as a model study.

c1219
Some aspects of internal migration to cities of Maharashtra, 1951–61
B.G. TAMASKAR   *Amravati Camp, India*

The objective of this paper is to examine the patterns and problems of migration from other states to the cities of Maharashtra (excluding Bombay, about which a separate work has been published). For this purpose, immigrants to each city with a population of over 100,000 in 1961 are classified according to industrial groups.

The total volume of migrational flow of both rural and urban population into Maharashtra from, (1) Uttar-Pradesh, (2) Mysore, (3) Gujarat, (4) Andhra-Pradesh, (5) Tamil Nadu (Madras), (6) Kerala, (7) Rajasthan, (8) East Punjab, and (9) other states of India, rank in decreasing order during the decade under study. On the other hand, 66,000 persons left Maharashtra for Madhya-Pradesh as compared to 918,000 in-migrants into Maharashtra. In-migrants represented 12.7 per cent of the state's population in 1961. The cumulative figures of migration reveal that net migration from Madhya-Pradesh to Maharashtra in the previous decades, preceding 1951, has been reversed. Gujarat, which occupied the first place in earlier decades, has dropped to third, Uttar-Pradesh has risen from third to first, Tamil Nadu (Madras) has outstripped Rajasthan, and the remaining states have retained their respective ranks.

A large proportion of in-migrants from other states (35 per cent of the total and 17 per cent of the working group) were concentrated in eleven cities (Nagpur, Poona, Sholapur, Kolhapur, Amravati, Nasik, Malegaon, Akola, Ulhasnagar, and Thana).

The sizes of the first four cities bear some direct correspondence with the magnitude of migration from within and outside the state, though Sholapur records a higher percentage of in-migrants from other states than that to Poona and Kolhapur. Ulhasnagar is a peculiar case of an industrial city of recent origin which came into existence after the rehabilitation of displaced persons from West Pakistan on the eve of Indian independence, forming a large colony made up largely of Sindhis (Kulkarni 1963, 156). Nasik and Amravati have relatively higher percentages of in-migrants from within Maharashtra, possibly because their functional diversity attracts a larger number of in-migrants from nearby areas. Generally, centres in which manufacturing, household services, or a combination of both provide the employment base, seem to exercise a greater pull on in-migrants from other states than cities of other classes. Ahmadnagar, with its large military establishment, is the solitary exception.

From an examination of the volume of in-migration from different states the following conclusions emerge: (1) The number of in-migrants coming from other states of India varies inversely with the distance of migration (Gosal 1961, 119). (2) At intermediate cities, situated on busy routeways such as Nagpur and Poona, large numbers of in-migrants have concentrated. Many of these originate from the distant state of Tamil Nadu. (3) A relatively large number of in-migrants were attracted into cities of Maharashtra from urban centres rather than from rural areas of other states. (4) Invariably males exceed females among the working in-migrants from other states. (5) The working in-migrants to cities of Maharashtra from other states represent between 25 and 75 per cent of the total number of in-migrants in the respective cities.

The patterns of selective migration from various states to cities of Maharashtra are summarized in Table 1.

Streams of in-migrants from other states throw into bold relief the increasing cultural diversity in the cities of Maharashtra. This results in the isolation of migrants due to the language barrier and differing customs and

TABLE 1. Pattern of migration from various states to cities of Maharashtra

Source state	Industrial categories		
	1st preference	2nd preference	Remaining preferences
Gujarat	Commerce & trade at Nagpur, Poona, Amravati, Akola, & Thana	Manufacturing at Poona and Nasik	Other services at remaining cities
Madhya-Pradesh	Other services at Nagpur, Poona, & Amravati	Construction at Nagpur	Transport, storage, and communication at Nagpur
Andhra-Pradesh	Construction at Nagpur	Other services at Nagpur, Poona, & Malegaon	Manufacturing at Sholapur
Mysore	Other services at Nagpur	Manufacturing, transport, storage, and communication, trade and commerce, household industry at Poona & Sholapur	
Kerala	Other services at Nagpur, Poona, & Ahmadnagar	Manufacturing at Thana	
Rajasthan	Other services at Nagpur & Poona	Manufacturing at Thana	
Uttar-Pradesh	Other services (as labourers) at Nagpur, Poona, Ahmadnagar, & Akola	Household industry and manufacturing at Nagpur, Poona, & Malegaon	Trade & commerce at Nagpur & Akola
Punjab	Manufacturing at Nagpur & Poona	Other services at Ahmadnagar, Nagpur, & Poona	Trade & commerce at Nagpur & Poona
Tamil Nadu (Madras)	Other services at Nagpur & Poona	Transport, storage, & communication at Nagpur & Poona	

food habits. The sex disparity among the in-migrants poses another social problem of no mean degree. The problem of residential accommodation for unanticipated numbers of in-migrants, resulting in congestion of people in parts of the cities, remains still to be solved. The problem of providing primary and secondary education in the migrants' own languages is aggravated by the preference for the regional language as the medium of higher education at the university level. Without attaining proficiency in the regional language (Marathi), the in-migrants are likely to be disadvantaged in public and private employment. Their separate schools, congregation halls, places of worship, and recreational centres isolate them from the mainstream of city life. In a nutshell, the growing heterogeneity of population in cities, stemming from immigration of peoples from other states, generates social tensions, strains, and stresses which may find vent on occasions of elections in municipal and state electoral constituencies.

Most of these problems call for planning of physical facilities and social amenities in cities. It is imperative to assess the volume of migration periodically if adequate housing, domestic and industrial water-supply, and transportation facilities are to be provided.

M.G. Marathey.

Gosal, G.S., 1961 Internal migration in India – a regional analysis, *Indian Geog. J.*, 36, no. 3 (July–Sept.), p. 106–21.

Kulkarni, B.A., 1963 *Census of India, 1961*, vol. 10, Maharashtra Part II-A, General Population Tables, Delhi.

c1220
Some problems of internal migration of the population of the northern part of Siberia
V.V. VOROBYEV and V.I. CHUDNOVA *Institute of Geography of Siberia and the Far East, USSR*

The northern part of Siberia, defined here as the area situated north of 60°N (and, to the east of Lake Baikal, the area north of 55°N), is a region of great perspectives where the greater part of the country's raw material resources are concentrated: coal, mica, non-ferrous and rare metals, large deposits of oil and gas, hydropower, and timber. It is one of the most sparsely populated parts of the country. The local population, concentrated mainly in separate industrial-transportation centres scattered about the immense territory, is engaged in hunting, fur-trapping, and reindeer breeding, and have none of the industrial skills necessary for the development of mineral and fuel resources. Industrialization here is made possible only by attracting manpower from the more densely populated areas of the country and by the redistribution of population within the territory.

In the Siberian north two different types of areas can be defined. In one the aboriginal population (or Russian old-timers) predominates; this group is characterized by comparative stability and low participation in migrations in and out of the area. The aborigines migrate mainly within the limits of their own ethnic territories, which cover large areas but have comparatively small populations. The second type – newly-developed regions – are connected with developing industries and cover rather small territories, but the bulk of the population of the north is concentrated here. The population is formed chiefly by migration and the role of natural increase is very small initially, though it increases as development progresses. In the areas of the development of the oil industry in the Tumen Region migration provided 75 per cent of the general increase of population (Vorobyev et al. 1966, 10). A similar situation existed in other rapidly developing areas of the North (Mirniy, Chukotka, Talnakh).

A high volume of migration is characteristic of the northern parts of Siberia, particularly of the newly developed areas whose populations possess great mobility. In some such areas about 20 to 25 per cent of the population is renewed annually. In the past decade in the Magadan region and Yakutia, 1.8 million persons arrived and departed, which represents twice the population of this area (Yanovskiy 1969, 34). In the Tumen region in 1962–6 more than a million persons arrived and departed (Chudnova 1969, 142). The most intensive migration occurs in the Ob River North, in some industrial districts of Yakutia and the Magadan region, that is, in districts with widescale development of mineral resources but no conditions for the stabilization of population.

Three-quarters of the population arriving in the northern regions of Siberia belongs to the age group 20–39, and males predominate (Orlovskaya 1964, 23; Chudnova 1969, 146). As a result, the sex and age structure of the population is much affected. The greater part of the population is of working age with a low percentage of children and aged people. Throughout the north the proportion of men in the population is greater than in the USSR as a whole. In the comparatively well-developed regions the sex ratio approaches the USSR average, but in regions of pioneer development (Chukotka and others), there are few women. In this respect the rural areas with their permanent old populations differ greatly from the recently-built towns and workers' settlements.

The balance of migration in the northern areas in the past 10 to 15 years is mainly positive, but it represents only 8–10 per cent of the total volume of migration during these years. Simultaneously with mass arrivals there is high emigration of population, as only a small proportion of the arrivals settle permanently. Investigations in the Magadan region showed that the greater part of the arrivals from other parts of the country stay here not more than 1 or 2 years (Orlovskaya 1964, 39).

The structure of the economy in the northern areas makes it difficult to form permanent stable populations with a balance between sexes. The mining, timber, fishery, and other industries requiring male labour prevail. About three-quarters of the population are engaged here, whereas the service establishments where women may be employed require only a quarter.

People from all parts of the Soviet Union are represented in the present population of northern Siberia. The major source areas are the southern regions of West and East Siberia and the Far East, whose people are adapted to the severe climatic conditions and have experience in those economic fields which predominate in the north. However, these same areas experience a shortage of labour resources and cannot relinquish large contingents of manpower for the development and settlement of the north without detriment to their own economic development. This situation makes it absolutely essential to create a system of 'stepped' population migrations in which the southern regions of Siberia and the Far East, while giving up part of their population to the north, will get replacements from the more densely populated western regions of the country.

A considerable role in population formation is played by inter-regional migrations within the north. Approximately 25 per cent of those leaving the north do not return to their native places (or last residences) but to neighbouring regions of the north. This is a result of the fact that the northern regions of Siberia do not have similar natural, transportational, economic, or living conditions. The population, migrating individually, takes these differences into account and seeks regions with complexes of more favourable conditions.

It has been noted that regions of variable and permanent populations can be defined in the north. In the extremely severe regions of Siberia, close to the coasts of the northern seas, in the tundra and in the mountains, there are no conditions (and no necessity) for creating a permanent population. For the industrialization of these regions it is thus expedient to focus on manpower of variable composition, enrolled for definite periods. This will naturally result in intense migration movement in these regions and will influence population composition and reflect on the complex of enterprises, services, and cultural establishments which arise. It will require a new approach to the solution of problems connected with the planning and development of such localities.

At the same time, the Siberian north has substantial permanent aboriginal and newcomer populations, and all the conditions for further growth of permanent populations exist there. Such populations will be formed in key places such as transportation-organization centres of large regions and areas where new construction is concentrated. In general, at the present stage of development they will be concentrated in the southern part of the zone under observation, but will also, undoubtedly, be built in the northern part.

The intensity and character of migration in regions of the second type will approach that observed in the old industrial centres in the southern part of Siberia. The area of the region where it is expedient to form a permanent population will develop because of the technical progress and further development of the economy. The fulfilment of the XXIV Party Congress resolutions in creating conditions for the influx of personnel and retaining them in Siberia will further the formation of permanent populations in many northern industrial regions.

Chudnova, V.I., 1969 'Osobennosti formirovaniya naseleniya raionov promyshlennogo osvoyeniya Obskogo severa', in *Geograficheskiye osobennosti osvoyeniya tayezhnykh rayonov Zapadno-Sibirskoy nizmennosti* (Irkutsk), 135–67.

Orlovskaya, L.I., 1964 'Osobennosti formirovaniya kadrov narodnogo khozyaystva Magadanskoy oblasti,' in *Problemi razvitiya proizvoditelnykh sil Severo-Vostoka SSSR* (Magadan), 21–41.

Vorobyev, V.V. et al., 1966 'Geograficheskoye obosnovanye meropriyatiy po osvoyeniyu severa Zapadno-Sibirskoy ravniny,' in *Geograficheskiye aspekti osvoyeniya tayezhnykh territoriy Sibiri* (Irkutsk), 6–54.

Yanovskiy, V.V., 1969 *Chelovek i sever* (Magadan).

c1221
Method and analysis in the mobility study of non-literate populations: case study and implications
MURRAY CHAPMAN *University of Hawaii, USA*

This paper focuses upon data collected in the British Solomon Islands Protectorate between October 1965 and February 1967 for a study of population movement from the standpoint of the village. It outlines the technical and methodological traps embedded in the detailed investigation of a non-literate society, as well as the conceptual rewards that derive from the collection of fine-grained data.

A detailed investigation was made of a coast and a peripheral bush village, Duidui and Pichahila respectively, on the Weather (south) Coast of Guadalcanal. On 1 March 1966, the former community had a *de jure* population of 221 residents in 40 households; on 1 October 1966, the latter contained 110 persons in 18 households. Both cross-sectional (field census) and longitudinal data (wage-labour records; population register) were collected to test the applicability of the concept of 'internal migration' when defined as persons resident in the place of enumeration for the previous year or whose absence was not expected to exceed more than one year. This exercise reveals that, for these two communities, the low incidence of 'internal migration' is a definitional artifact that camouflages a high degree of population movement defined as physical transfers involving an absence of 24 hours or more. A second finding is that the village is, in conventional terms, both the place of origin and of destination since the patterns of mobility are almost entirely circular.

At census date, those who have been absent for 24 hours or more accounted for thirty per cent of their village populations (Duidui 67 out of 221; Pichahila 34 out of 110), were predominantly male, and left mainly for schooling or wage labour. The distinct seasonality of this movement is revealed by the continuous record developed for all persons away for one or more months during the years 1965–6. For the majority of children, going to and from school follows 10–11 month cycles while absence for wage labour averages three (Pichahila) to six (Duidui) months. The latter generally occurs between February and August, since the yams and taro have to be planted in August–September and January–February, and any leaf replaced on houses during the sunniest months September to March.

The daily, weekly, and monthly flow of people across a village's boundaries is indicated by the population register maintained for five months for each study community (Duidui: 1 December 1965 to 31 May 1966 but excluding the month of April; Pichahila: 1 July 1966 to 30 November 1966). From this, the ratio of initial to completed moves is found to be virtually identical so that mobility, when viewed through village lenses, involves a continual transfer of individuals both singly or in groups.

With the kinds of mobility established, the conventional wisdom in census analyses is to correlate with such variables as sex, age, marital status, education, and income both to identify who moves and to infer for what reasons. There are two glaring weaknesses in this strategy. First, research which attempts to deduce reasons by indirect methods reflects, at least in part, the way in which census data not meant to reveal such information were originally collected, and second, 'reasons' become an omnibus category into which various levels of information are fitted. To describe such 'reasons' as 'causes' of mobility is therefore a transition rather too comfortably and frequently, if erroneously, made. One solution to these problems is to invert customary procedures so that the dependent variable becomes not the type of movement – for example, inward or outward – but rather statements about its occurrence made by the participants.

To establish the objectives of such mobility, statements made by the participants must be disaggregated in terms of the destination, the objectives (activity pursued or intended to be pursued), and the underlying reasons why a physical transfer happened at the particular moment in question. Based upon field experience, a list of 162 objectives was abstracted from the population register and grouped into nine clusters: relatives/kinsmen, return home, gardening, general visiting, wage labour/church, schooling, administrative, medical, and other. When this intuitive procedure was tested through the use of a contingency coefficient (C), these objective-clusters were found to be statistically significant in all but three minor cases.

In Duidui, residents and aliens made 324 moves in pursuit of 541 objectives during the two months December 1965 and February 1966; over a similar period, July and October 1966, the bush community of Pichahila experienced 316 moves undertaken for 519 objectives. For both settlements, the number of objectives per move ranged from one to eight, with averages of 1.7 (Duidui) and 1.9 (Pichahila). A single activity is therefore normally not sufficient to lure a villager away from his household and future analysis of the underlying reasons embedded in the movers' statements would even further emphasize this fact.

Objectives that manifest the needs of the family or the social relationships amongst tribesmen far outweigh those that derive from the still peripheral impact of a colonial administration or metropolitan commercial interests. Journeys undertaken for activities associated with gardening, relatives and kinsmen, and local visiting comprise 71.9 per cent (Duidui) and 90.3 per cent (Pichahila) of 1,135 recorded objectives while those concerned with wage earning, Christian religion, schooling, medical services, and the district administration account for the remainder. The preeminence of the undifferentiated category 'return home,' first out of nine clusters in Pichahila and second in Duidui, underscores previous indicators of a conspicuously mobile people; it also conclusively demonstrates the validity of viewing the village as both the place of origin and ultimate place of destination of completed moves.

If the undifferentiated clusters 'return home' and 'other' are set aside, the enduring importance of customary activities becomes even more apparent and the dominance of 'pre-contact' over 'post-contact' objectives established (Duidui: 72.2 to 27.8 per cent; Pichahila 87.8 to 12.2 per cent). This result contradicts the impression derived from the field census that residents apparently are responding to post-contact and predominantly economic goals. That difference, illuminating rather than spurious, tends to confirm a growing suspicion about analyses of the reasons why people move. Given the definitional constraints associated with census data that have been collected across rather than through time and given the analytical strategy of explanation by inference, generalizations that 'economic' forces are primarily responsible for internal migration seem unwittingly tautological.

The strategy of the micro-study reported here reflects no desire to pursue the unique or the exotic. On the contrary, it was adopted purposely to sharpen the cutting edge of technique and analysis, to focus upon mobility from the standpoint of the village in all its ongoing complexity, to yield definitions that are locally relevant rather than predefined from external conventions and, finally, to generate inductive models based upon a specified nucleate population.

There are larger questions that derive from this research to which the Symposium on Internal Migration hopefully will address itself. First, the cultural specificity of the conventions of population movement, for example, 'migration,' and the need to elucidate ones that are locally relevant but reported in ways that still permit cross-cultural and international comparison. Second, our ignorance about the range of mobility experienced by a particular society and the refinement of available techniques necessary to identify them, one remedy for which might be the maintenance of a population register by semi-literate villagers when the academic is thousands of kilometres away! And third, the need for more imagination in data collection, analysis, and subsequent reporting to foster (a) identification of the logical links in the decision to move; (b) a distinction between the different kinds of information contained in movers' statements and fieldworkers' observations; and (c) field investigations that proceed at three levels: the objective, the normative, and the psycho-social.

Chapman, Murray, 1969 A population study in South Guadalcanal: some results and implications, *Oceania* 40, 119–47.
– 1970 Population movement in tribal society: the case of Duidui and Pichahila, British Solomon Islands (unpub. PH D thesis, U. of Washington, Seattle).
– 1971 Population research in the Pacific Islands: a case study and some reflections, *East-West Population Institute Working Paper 17* (Honolulu).
McArthur, Norma, 1961 *Report on the*

Population Census of 1959 [British Solomon Islands] (Honiara).

Pirie, Peter, and Ward Barrett, 1962 Western Samoa: population, production and wealth, *Pacific Viewpoint* 3, 63–96.

Reissman, Leonard, 1967 The metrics of migration, *Pacific Viewpoint* 8, 211–12.

Shryock, Jr., H.S., and Elizabeth A. Larmon, 1965 Some longitudinal data on internal migration, *Demography* 2, 579–92.

c1222
Interregional migration in Latin America
PEDRO PINCHAS GEIGER   *Instituto Brasileiro de Geografia, Brazil*

In the modern literature of the social sciences, the expression 'interregional migrations' is seldom used. Even considering regions as 'self-organizing systems' and migration as one of the forms of interaction between regional units, they are generally complex enough to be sources of common variables external to the individual that motivate migrations.

Consequently, migrations are more frequently defined by such notations as interurban, rural-urban, stepwise, etc. Little work has been done as yet in Latin America on studies of perception and decision-making among migrants, with the application of multiple dimensional scaling techniques.

In geographic space, under some circumstances certain points stand out in the migratory process as major attractive nodes, e.g., Brasilia, Monterrey, and Guayana City. But under other conditions, whole areas with common characteristics linked to a determinate aspect of the regional structure, which is relevant to the migratory movement, are affected. An example is to be found in the out-migration from semiarid areas in the Brazilian northeast. The prosperity of northern Paraná, Brazil, and its identification with the state of São Paulo are attributed, among other reasons, to the massive presence of 'Paulistas' in settlement and in-migration.

Process continuity may cause a migration-attracting point to develop in a regional structure. Hyperurbanization around a national metropolis or primate city is a phenomenon present in nearly all the countries of Latin America, though on a varying scale and at different stages, and involves a high percentage of the total migrants. Around the city of São Paulo, however, there stretches more than a simple metropolitan agglomeration; rather it is an urbanized, developed region with tentacular linkages, representing the extension of a hyperurbanization process. Migrations, as elements of social and economic processes, therefore, enter into the structuring of regions either as in-migration sharing in the build-up of regions or as out-migration influencing their evolution or involution.

The regional approach has its problems of hierarchy and scale. In countries like Argentina, Brazil, or Mexico, even the meso-regions may be more extensive than an entire country of Central America. This makes it interesting to identify, in a system of regions where those of a lower order nest within those of a higher order, to what level or levels the various expressive aspects of migration are related.

A series of aspects of migrations are common to Latin America, viz., a large volume of movement towards the urbanized areas, even without technological evolution in agriculture; a recent accentuation of movements, promoted by a greater accessibility reaching to the most distant corners of the country; and sequences in the constitution of new regions by settlement of empty areas. Similar divisions of opinion occur in various countries: some critics complain that migration to an agricultural frontier is relatively weak, while the flow to the big cities is excessive; others believe that the cost of occupying new areas is too high, and that it would be more beneficial to renovate the older and denser regions.

The settlement of new regions often involves urban colonization with populations originating in the larger cities of the rearguard. In Maranhão, Brazil, for instance, traditional practices in trade relations tend to make the migrants from Ceará into a factor in the development of relations with the latter state.

Governments attract occupants to empty regions by infrastructural works, notably roadbuilding, and even by guided settlement. In the private sector the spontaneous and disorderly settlement of deliberate 'colonization' is significant.

In areas such as the Peruvian selva or the tropical forest of Central America, the new agricultural regions grow products for export besides supplying the domestic market. In larger countries such as Brazil, which besides more highly developed areas, has far-spreading agricultural ones, the distant fringe is more likely to be turned over to stockraising and resembles an outer ring of the Von Thünen model.

Some authors would like to see an even greater exodus from traditional, stagnating regions, such as certain semiarid areas in the Brazilian northeast. The cultural patterns of Indian populations have been stressed as a factor of lesser mobility in the Andean regions.

Nevertheless, as a general rule, planning for peripheral regions is characterized by measures designed to turn their major cities into poles of growth. Channelling an increasing flow of urban-urban and rural-rural migration from their areas of influence, these cities tend to promote a relative slackening of interregional migrations, as has been observed in the state capitals of the Brazilian northeast or in the northern coastal cities of Peru. Expansion of university education in those cities may contribute, on the one hand, to their role as steps in migration, but a greater number of students remain in them after graduation and the rise in the cultural level of the regional centres is instrumental in reactivating regional development and attracting a return flow.

As the large metropolises continue to grow, they tend to receive only an urban-urban flow of migrants. In Buenos Aires, 50 per cent of the migrants were born in cities of more than 20,000 inhabitants. In their immediate areas of influence, where farming is a flourishing occupation, a process may be observed in which the most highly qualified local elements and others involved in agricultural pursuits move on to the cities and are replaced by migrants from poorer regions. This is what has happened around São Paulo, and within the area of influence of Buenos Aires with the immigration of Paraguayans.

Finally, emphasis must be laid on the role of interregional migrations in the shaping of national self-realization or awareness through the contact between populations originating in different geographical units, hitherto isolated.

c1223
Migration and changing settlement patterns in Alberta
BRUCE PROUDFOOT   *University of Alberta, Canada* and
GLENDA LAMONT   *Government of Alberta, Canada*

While urbanization in Canada has received considerable academic and political attention, study of the processes leading to rural depopulation and to the decline of small rural service centres has been neglected. The present paper investigates some aspects of changing settlement patterns in Alberta and of migration into smaller centres with population in 1970 between 500 and 12,000. Sample centres were selected on the basis of their recent population history, and thus present apparent viability.

There has been little uniformity in the spatial patterns of population change in Alberta, although there has been a gradual northward shift of the centre of gravity of the population. The rural southeast has tended to lose population, while most of the small towns north of Edmonton, those along the Edmonton–Calgary axis, and those within a 32km radius of these two cities have tended to grow more quickly than the provincial average. There is no clear relationship between population change and population total, apart from the fact that no centres larger than 3500 in 1966 had grown less quickly than the provincial average in the previous five years. Similarly, retail trade figures by themselves do not provide a useful classification of the apparent viability of small towns.

To achieve a classification which could serve as a basis for studying migration into centres of different types, twenty-seven variables, considered to bear some relationship to centre viability, were used in a principal components analysis. Factor scores for the first three components were calculated for each town and plotted each against the other.

Five of the 27 components explained more than their share of the total variance. Two of these, however, each accounted for less than 4 per cent of the variance and were discarded (King 1969, 174). Together, the other three components accounted for 56.4 per cent of the variance. The first component, accounting for 40.2 per cent, is an index of the size of centre. The second component characterized the centres primarily by the age structure of their population and subsequently by percentage population growth, and distance to a competing centre. Major loadings on the third principal component were the number of government offices and the number of hospitals in each town. Distance to a competing centre again contributed. This third factor can usefully be described as an index of centrality, and, as such, high factor scores would be positively associated with viability. When factor scores were plotted, five clusters of points exhibiting a reasonable degree of internal consistency of membership could be distinguished. These clusters could be regarded as varying in viability. Resource and new towns, dormitory suburbs, and institutional centres lay outside these clusters and they were omitted from the total population for purposes of sample selection. Centres in the Peace River region were also omitted since a similar study of migration had been done for those centres in that area in 1968–9 (Lamont 1970). From the five sets of centres grouped according to their similarities, fourteen were selected for study. The number chosen from each group was approximately proportionate to the total membership of that group, and took into account the distribution of members of that group throughout the province. A random sample of in-migrants to each of the fourteen centres between 1965 and 1970 was chosen and a questionnaire mailed to them in late 1970. The questionnaire was designed to obtain data concerning the migrants themselves, the nature of their move, and their reactions to their destinations.

Age and occupation suggest significant differentials in mobility. Forty-three per cent of the migrants fall into four distinctive age-occupation groups (Table 1). Some 23 per cent of all migrants had been farmers before moving, and most of these had moved when they retired from active farming. Generally, they had moved relatively short distances,

TABLE 1

Occupation	Age	Percentage of all migrants
Retired	Over 60	16.2
Professional and technical workers	Under 35	13.4
Craftsmen and foremen	Under 30	8.8
Clerical workers	20–24	4.7

and one-half still owned all their land. Apart from the farmers, more highly skilled people were more mobile than those with fewer skills, and many young professionals seem to be going to small centres as a 'stepping stone' in their careers.

Responses to an open-ended question regarding the motives for moving were coded using nine empirically derived categories. Most motives for moving were economic and associated directly with employment. About 43 per cent of all migrants moved to take advantage of a job opportunity while an additional 17 per cent moved because of a job transfer. Retirement accounted for a further 18 per cent, most of whom were farmers. About 8 per cent gave the desire for better services and facilities as their primary motive for moving, and 6 per cent cited nearness to family and friends. Some of these were migrants who had retired. Secondary reasons for migration included such additional factors as costs of living, housing, social reasons, improved services and facilities, and dislike of large cities. Job opportunities are relatively more important as reasons for moving among younger than among older migrants. Better services and facilities and nearness to families and friends tend to be more important reasons for moving among the older and the less-skilled migrants. Those who moved for better services and facilities came predominantly from Albertan centres smaller than those they were living in at the time of study. Those moving because of their dislike of the big city tended to be young and more highly educated, although most of the migrants from Edmonton and Calgary had moved primarily for job opportunities. When reasons for moving were examined in relation to the viability of the clusters of centres it was found that more migrants moved for job opportunities to more viable clusters than to less viable clusters although there was considerable variation within clusters.

Some 58 per cent of the migrants intended to move to another centre, but there was considerable variation between different age groups. Of those over 60 years of age about 90 per cent intended to stay compared with only 15 per cent of those under 50. Migrants who were 55 years or older at the date of moving and intend to remain at their destination represent exactly 20 per cent of all migrants questioned, and clearly form a fairly large, older, stable group in the centres to which they have moved. Intentions also vary with skills and occupations. In-migrants with professional training exhibited the least propensity to stay and many had already made plans to leave in 1970 when the survey was carried out. Those technically trained had proportionately the greatest numbers whose intentions for moving were uncertain (43 per cent), and potentially this technical group is the most flexible regarding its future. Clerical workers, craftsmen-foremen, and operatives have a low propensity to remain, as do those who have migrated from outside Canada or have moved from Edmonton and Calgary to smaller centres. Those who moved shorter distances within Alberta are more likely to remain than those who moved longer distances.

The high propensity, especially of the more skilled migrants, to move from smaller centres is surely symptomatic not only of the high mobility of North American society in general, but more particularly, of the inability of many small central places in Alberta to offer the variety of economic and social opportunities that characterize viable larger centres.

Alberta Human Resources Research Council.

King, L.J., 1969   *Statistical Analysis in Geography* (Englewood Cliffs, NJ).
Lamont, G., 1970   Migrants and migration in part of the South Peace River Region, Alberta, unpub. MA thesis, U. of Alberta.

c1224
La migration de retraite en France: étude de géographie sociale
FRANÇOISE CRIBIER   *Centre National de la Recherche Scientifique, France*

La distribution de la population âgée, en France, comme dans tous les pays très développés, est de moins en moins régulière, et varie de plus en plus selon le type de commune, selon les régions. Les différences dans la structure par âge ne dépendent plus seulement de différences dans la natalité, dans l'exode ou l'afflux des jeunes, mais, de plus en plus souvent, du comportement propre des gens âgés: la plupart restent sur place, mais de fortes minorités se dirigent, à l'âge de la retraite, vers un autre type de commune, vers une autre région. Cette mobilité, en rapide croissance, joue déjà un rôle dans la structure du peuplement des différents types de communes, et plus encore des diverses régions.

Elle a pourtant très peu été étudiée, et pour deux raisons: parce qu'elle affecte une partie inactive de la population et parce qu'elle affecte une population migrante âgée, doublement minoritaire: minoritaire parmi les migrants, dont les 9/10e ont moins de 60 ans, minoritaire parmi les gens âgés, puisque, au cours d'une période censitaire de 6 ans, six sur sept restent dans la même commune, et que dans l'ensemble, les 2/3 des gens vieillissent sur place.

Les migrations de gens âgés ont affecté, en France, entre 1962 et 1968, 900,000 personnes de 65 ans et plus, soit presque une sur sept et, au total, 1,600,000 personnes de plus de 55 ans, dont 300,000 seulement étaient encore actives en 1968. Mais surtout la migration des gens âgés prend une ampleur rapidement croissante: de 1954–62 à 1962–8, le volume annuel de la migration de gens âgés s'est accrû de 35 pour cent et pour les seuls Parisiens quittant définitivement la capitale et ses environs proches, de plus de 50 pour cent.

On doit distinguer plusieurs catégories dans les migrations de gens âgés. Deux d'entre elles, les plus importantes, sont spécifiques de ce groupe d'âge: celle qui accompagne, suit, ou parfois précède, la fin de la vie active, et lui est liée: nous l'appelerons 'migration de retraite'; et celle qui accompagne ou suit le passage des gens les plus âgés au stade de la dépendance (entrée en hospice ou maison de retraite, refuge chez les enfants).

Ces deux migrations ont une importance inégale, obéissant à des règles propres; la plus importante, par le nombre de gens qu'elle concerne, et par les changements de la répartition régionale qu'elle entraîne, est la *migration de retraite*. Qui y participe, quelles régions, quelles communes sont quittées, et par qui, et pourquoi? Quelles régions, quelles communes attirent les retraités, et lesquels, et pourquoi? C'est tout le mécanisme qu'on voudrait comprendre, en observant les relations entre les caractéristiques des gens et celles des lieux qu'ils quittent ou qu'ils choisissent.

Pour l'étudier, nous avons eu recours à la fois aux renseignements massifs des recensements de population (1962, 1968), et aux renseignements beaucoup plus fins de documents individuels des fichiers de retraite. On a cherché à constituer des catégories pertinentes, qu'il s'agisse de groupes de migrants, ou des ensembles régionaux ou communaux dans lesquels on a fait les observations. Les résultats ont été systématiquement cartographiés.

Le concept le plus utile à notre analyse a été la distinction en trois grands groupes de personnes arrivées à l'âge de la retraite: (A) ceux qui restent sur place; (B) ceux qui retournent dans leur région ou commune d'origine: on parlera de *migrations de retour*; (C) ceux qui se dirigent vers une région, une commune, dont ils ne sont pas originaires. On parlera cette fois d'une *autre migration*: elle a ses caractères propres, et se développe beaucoup plus rapidement depuis 10 ans.

Pour chacun de ces groupes, et à partir d'échantillons différents, on a analysé certaines caractéristiques socio-économiques et géographiques, et on les a mises en relation avec le comportement, puis, pour les migrants, avec le type de commune ou de région où ils s'installaient. Les résultats ont montré la pertinence de la distinction entre les deux types de migrations: qu'il s'agisse, en effet, des caractéristiques socio-économiques des migrants, des différences de comportements selon les régions de départ, selon les régions d'arrivée, selon le type de commune délaissée ou choisie, les deux migrations obéissent à des règles différentes et n'évoluent pas de la même manière. On a réussi à dégager quelques-unes de ces règles, à comprendre le sens de certaines évolutions, et les résultats les plus intéressants concernent:

(1) Le rôle fondamental du niveau socio-économique (revenus, ancien métier, instruction), le rôle très important de l'origine géographique, et l'existence, autour de chaque ville de départ, d'une aire de retraite plus ou moins étendue, parfois discontinue, fort comparable à l'aire de vacances, et dans laquelle les personnes des différents groupes sociaux se répartissent inégalement.

(2) Le fait que les régions et communes de retraite ne sont pas nécessairement celles qui offrent aux retraités le meilleur milieu de vie; la redistribution de la population âgée qui s'opère ainsi ne correspond pas toujours à un optimum social, ni à un optimum économique.

(3) L'inégale capacité des régions françaises à faire revenir les retraités originaires, et la très inégale capacité d'attraction sur les non-originaires.

(4) La différenciation croissante qui en découle, puisque les différents types de commune, de région, voient leur population âgée se diversifier en nombre et en qualité (selon le niveau socio-économique et l'âge). On observe bien souvent, dans les quartiers et banlieues des grandes villes, dans ces villes de retraite que sont tant de villes touristiques, dans les campagnes, un renforcement de la structure sociale antérieure.

c1225
Interregional migrations in tropical Africa: an overview
MILTON E. HARVEY   *Kent State University, USA*

Historically, the reasons for population relocation in Africa have changed as man's outlook on resource utilization, technology, and socio-cultural ties has been modified. Concomitant to these changes has been the modification of the patterns and intensity of mobility in the African continent. This review paper attempts to isolate the motives for interregional migration in the continent, analyse the type of data that have been used by researchers. and identify areas of future research.

In terms of processes which have induced interregional migrations in Africa during historic times, three patterns can be identified: the tribal-based, the traditional state-based, and the colonial. Each of these epochs is characterized by distinctive input variables which aggregatively induced interregional population relocations. As most independent African countries have largely continued the colonial pattern of development with excessive socio-economic concentrations in a few nodes, a post-colonial migration type has not yet evolved. Such a pattern is justifiable only after massive inputs have restructured the space economy and instituted a new locational matrix through which growth is canalized.

During the tribal based period, migration essentially involved either a tribe or a group of lineages. Although it sometimes reflected a gradual diffusion process, warfare and coercion were, however, often an essential prelude. Such *en masse* migrations are exemplified by that of the Fulani from northern Nigeria westwards into the headwaters of the Niger and the Senegal and the dispersal of the Bantu from an original heartland.

The introduction of new technology, notably the plough and the art of extracting minerals, during the traditional state-based epoch, were among the factors that generated long-distance trading in Africa, increased population densities at advantageous locations, and ultimately caused the growth of towns. For example, the rise of the Axumite Kingdom of the African Horn was in response to trade on the Red Sea, whereas in west Africa, trade in gold and forest products like kola nuts, on the one hand, and salt and manufactured goods, on the other, was paramount in the evolution of the savanna states. Within the framework of these centralized and non-centralized states, migration in post-eighth century Africa was induced by trade, population pressure and regional economic disparities, the continuous fission and fusion of states, and the Islamization process. Population shifts during this period include those of the Somali southwards, and the Ashanti southwards into the forested belt.

The ultimate decline of the savanna states coincided with the development of forest kingdoms, a disruption of trans-Saharan trade, and the inception of coastal-based trade with Europe and Asia in tropical products and slaves. This new trade orientation and the general change in the nature, intensity, and directionality of modernizing forces was finally institutionalized at the 1884–5 Berlin Conference. The colonial period introduced factors like schools, hospitals, Christianity, and modern communications. Conjointly, these factors, reinforced by physical and climatic variations, created marked unconformities in the existing socio-economic surface. The resultant push-pull forces resulted in marked interregional migration to mines, plantations, and towns; especially the capitals and ports.

In Africa, studies of the processes discussed above have largely focused on labour migration. These researches have concentrated on seasonal labour migrants, short-term migrants, and long-term or permanent migrants. Generally, seasonal migrants are labourers, mainly men, who migrate and stay for four to six months. Unlike seasonal migrants, short-term migrants leave home for a specified period, usually under two years. Permanent or long-term migration naturally entails longer time spans. If a migrant were a bachelor at the time of migration, he may get married either to a girl he met in his new abode, or more frequently, to a 'ready made' wife sent by his family.

Although literature on African interregional migration is rather limited, there is however a consensus that migration reflects the individual's assessment of comparative place utility. In this process, prior information about the existing opportunity surface has played a very important role. The frictional effect of distance on African mobility has also been stressed. Analysis of the relative importance of information flow, distance, and economic motives in migration have shown that the information factor was most important in smaller ethnic groups. For the large groups, however, the single most important factor was economic differential. In all instances, except for the largest groups, the friction of distance was significant.

Besides purely economic motives, certain politico-sociological factors have also induced migration. For example, the delimitation of national boundaries cut across the historical migration routes of nomads. Consequent readjustment did sometimes involve the creation of new migration circuits. As in other parts of the world, African migration has been selective as regards sex, education, and age. Studies have also shown that certain ethnic groups and lineages have a greater propensity to migrate.

For Africa south of the Sahara, migration data have largely come from three main sources: samples collected in the field, employment and unemployment records, and national censuses. Sample surveys are very expensive and do require technical skill and time; therefore they have been either localized or very limited in scope. As regards the second data source, two basic problems have tended to limit their use: they are unreliable and very limited in information content. The former stems from the fact that people who register under such circumstances usually give wrong answers about ethnicity and place of birth; the latter reflects the types of questions asked by the data-collecting organization. Invariably, the type of information useful to an employer is different from that required by a researcher interested in the complex interrelationships between processes.

Although most African countries had their first real census only in the last fifteen years, migration studies based on these data have mushroomed. In both west and east Africa, the basic mobility information is birth-place data.

In spite of the data problems, many researchers have attempted, in the last decade, to apply statistical and mathematical models to African migration studies. For example, empirical studies employing regression models and/or factor analytical models have been done for Ghana and Sierra Leone and Tanzania. In addition, analytical models designed to explain African interregional and rural/urban migration in terms of cost benefit concepts have been developed.

Because of the size of the area, this review paper does not pretend to have been comprehensive. However, it shows that more methodological research designed to test hypotheses and explain processes is required. Process-oriented research designed to understand how individuals and groups migrate over time and space is a *sine qua non* for any regional planning aimed at reducing rural depopulation and accelerated rural/urban migration. In the long run, because of the present data scarcity in Africa, more detailed field work by scholars from various social sciences is essential if the future development of man and resources in Africa is to be efficiently planned. Interdisciplinary research may be the best approach to the future study of interregional migration in the African continent.

c1226
A spatio-temporal model of the mobility patterns in a multi-ethnic population of Hawaii
SHEKHAR MUKHERJI *University of Hawaii, USA*

The text is not included because of length.

National Science Foundation; Murray Chapman; Robert J. Earickson; Brian J. Murton; Paul J. Schwind.

Chapman, M.T., 1970 Population movement in tribal society: The case of Duidui and Pichahila, British Solomon Islands, PH D dissertation, U. Washington.
Hågerstrand, T., 1957 Migration and area: Survey of a sample of Swedish migration fields and hypothetical considerations on their genesis, in *Migration in Sweden*, ed.,
D. Hannerberg, T. Hågerstrand, and B. Odeving, *Lund St. in Geog.* 13, 27–158.
Hågerstrand, T., 1962 Geographic measurements of migration, human displacements, in *Measurement and methodological aspects*, ed. Jean Sutter, Entretiens de Monaco en Sciences Humaines, 61–84.
Ishwaran, K., 1965 Kinship and distance in rural India, in *Kinship and geographical mobility*, ed. Ralph Piddington (Leiden), 81–94.
Kohl, S., and J.W. Bennett, 1965 Kinship, succession, and the migration of young people in a Canadian agricultural com-

munity, in *Kinship and geographical mobility*, ed. Ralph Piddington (Leiden), 95–116.
Prothero, R.M., 1968 Migration in tropical Africa, in *The population of tropical Africa*, eds. J.C. Caldwell and C. Okonjo (London), 250–63.
Rogers, E.M., 1969 *Modernization among Peasants* (New York).

c1227
Recreational migration and the central place hierarchy
ROY I. WOLFE *York University, Canada*

It is fitting that the father of central place theory should have spent the last years of his life examining recreational migration, for the two subjects are more intimately related than even he recognized. Walter Christaller thought of tourism in terms of the periphery, with people moving from central places to beaches, forests, and mountains occupying outlying spaces – but he may not have been aware that the very hierarchy of central places that he himself adumbrated is subtly reflected in the patterns of movement between central and peripheral places.

In broadly generalized terms, automobile traffic leaving and entering towns and cities during the summer months, in those parts of the industrial world that experience peak traffic conditions during those months, may be classified into two contrasting trip purposes and two contrasting times of the week. These are recreational and non-recreational trips (the latter encompassing trips made to and from work, shopping and commercial trips, and so on), and trips made in midweek and during the weekend. In neither case are boundaries clear-cut and immutable – one can, after all, mix business with pleasure, and the weekend, though it always includes all of Saturday and Sunday, also includes parts of Friday and Monday, and on long weekends even the early hours of Tuesday morning.

In terms of recreational trips made in midweek, all central places except resort towns occupy a broad boundary zone that divides two antithetical types of rural areas: recreational and agricultural. The further away one moves from the concentration of towns and cities into resort country, the greater becomes the proportion of recreational trips among the total number of midweek trips. The deeper one moves into the agricultural hinterland, the less important do recreational trips become.

In the metaphorical buffer zone occupied by central places between the two types of rural zones, there is a mild progression from level to level in the hierarchy in the relative importance of recreational trips made in midweek. The higher the position of an urban area in the central place hierarchy, the lower the proportion of recreational traffic in the weekday traffic mix.

In small towns total daily traffic may be half again as heavy during the weekend as during the middle of the week, whereas in metropolitan areas there is little to choose between one part of the week and another. The main reason for the discrepancy is that recreational traffic looms so much larger in small towns than in metropolitan areas. In Ontario, for example, it has been found that, whereas about two out of three vehicles leaving and entering metropolitan areas are making recreational trips, the proportion for smaller cities is more than four out of five, and for still smaller towns it is well over nine out of ten. Note that weekend travel is everywhere largely devoted to recreation: even in the larger areas, recreational trips outnumber non-recreational trips by a factor of two to one, and by the time we give attention to small towns, we find that the proportion of non-recreational trips is negligible – on the order of one to eleven or twelve.

Increasing attention is being given to what is commonly called the 'problem' of urban recreation. The patterns here under discussion are not, strictly speaking, part of the 'problem.' They are exurban rather than urban; that is, they appear on highways on the outskirts of towns and cities, rather than on roads within them. On most intra-urban roads, at most times of the year, non-recreational trips outnumber recreational ones, and traffic on weekdays is as heavy as or heavier than on weekends. Hierarchic considerations do not apply; or, if they do, they do so with much less force than with respect to exurban highways.

Nonetheless, the character of an urban place at a given level in the central place hierarchy does help explain the exurban patterns here described. We need not labour the fact that central places at higher levels of the hierarchy generate and attract greater volumes of all types of exurban traffic than do places at lower levels, and do so at all times of the week. Their hinterlands are greater, the varieties of activities performed and opportunities available are greater. On the working days of the week there are great numbers of commuters moving into and out of the larger cities, numerous trucks and commercial vehicles, and so on and so on. In a word, however great the absolute volume of recreational traffic generated by large cities may be, it has to contend with even greater volumes of non-recreational traffic.

The above comments apply in reverse to smaller centres. As one moves progressively down the various levels of the central place hierarchy, hinterlands contract, activities and opportunities are fewer, and general travel into and out of the community becomes correspondingly less; but the recreational travel patterns of individual people in small places are virtually indistinguishable from those of their opposite numbers in large ones. They take vacations, they leave town on weekends, they move to summer homes and camp in parks. A similar volume of recreational traffic will therefore loom much larger in a small place than in a large one, and, as has been pointed out, it will loom largest of all on the weekend. Work in a big city never stops; in a small one, we may presume, it virtually does so on weekends. Here, then, may be the reason that recreational traffic is only twice as great as non-recreational on a weekend on the outskirts of a large city, whereas it is well over ten times as great near small towns.

Wolfe, Roy I., 1967 *A theory of recreational highway traffic*, DHO Report RR 128, Department of Highways, Ontario (Toronto).
- 1969 A tentative procedure for estimating recreational highway traffic, *Traffic Quart.* 23, 105–21. (Reprints are available in French (*Transports*, 150 (1970), 104–16); Portuguese (*Urbanização*, 5 (1970), 163–77); and Japanese (*Kosoku Doro To Jiosha*, 12 (1969), 85–91).)

Applied Geography
Géographie appliquée

---

c1301
Regional development policy in Hungary
GYORGY ENYEDI *Hungarian Academy of Sciences*

A significant problem of economic growth in industrial countries is the unequal development of certain regional units. Regional differences might hinder economic growth (e.g., through the settlement costs of populations migrating from the underdeveloped areas) and cause political tensions, too. Hence, governments of both western and socialist countries strive to assert a regional equalization in their regional policies.

According to certain opinions, the rise in economic standards leads automatically to the decrease in regional differences. Evidence of this should be that the differences have in fact decreased in the most developed industrial countries during recent decades. This, however, is not due to automatic processes of economy. On the contrary, modern economy is of a concentrating character. Both industry and agriculture, and the service sectors as well, and especially the scientific-technical capacities playing an increasingly important role, show a concentration of factories or institutions of ever growing sizes, and, geographically, in metropolitan areas. It is only due to regional development policies of many decades that the regional differences have been moderated in the developed industrial countries.

One of the most important fields of Hungarian applied geography is to provide a scientific basis to regional development plans, especially long-term ones. There are not very high regional differences within the small country of Hungary, but the importance of spatial problems is indicated by the very fact of having 24 per cent of the total population within the area of the Budapest agglomeration which gives 47 per cent of the net production value of the industry.

From the beginning it has been a target of planned economy, started more than 20 years ago, to decrease regional differences. 20 years ago more than 50 per cent of industrial workers were occupied in Budapest. As well, the industry was represented by the mining regions of the Hungarian Middle Mountains and the energy or metallurgic works, with a few exceptions. The industrialization of the less urbanized agricultural areas, especially those of the Great Hungarian Plain, has always been an objective, but hardly anything could have been done in this respect in the first half of the period of planned economy. After the reconstruction of war damaged areas, planned economy has developed heavy industries and energy production, both of which were inherently bound to the sites of coal and ores (i.e., the line of the Hungarian Middle Mountains). The next planning period set the objective of technical reconstruction of industry, which, of course, could be performed only where there was an industry developed before. Only in the second half of the 1960s was there a more marked structural transformation of the economy (i.e., a priority development of modern industrial sectors – chemical industry, electronics – and a more rapid growth in the non-productive sectors). It was this period that made it possible to create a higher number of non-agricultural working places in the underdeveloped agricultural areas.

In the 1970–85 period of economic development the economy of the country will pass into a stage of higher development (the $1000 GNP per inhabitant value will be attained by ca. 1975). This period will provide an opportunity to promote the regional development of the country according to a definite conception. In the following short summary the research objectives and conclusions will be described which have been attained in the course of developing a long-term development policy.

The basic objective of regional development policy is to promote regional equalization with a concurrent and efficient development of the economy. The objective represents, thus, a certain compromise, economic development tending to ever higher concentration, on the one hand, and the conceptions of local policies urging for decentralization on the other. To have a more exact definition of the objective, the concept of measuring regional development and the concept of equalization were to be defined more exactly.

The most important definitions are as follows:

(1) The meaning of economic underdevelopment is of relative value. It is represented by a deviation from the county averages rather than by absolute terms (i.e., per capita GNP).

(2) Regional underdevelopment has been expressed in Hungary mostly by the living conditions of the population. The regional differences in per capita GNP differ by a factor of only 1.3 between the highest and lowest county value. The per capita income in the counties varies by about 30 per cent only, since the average income of industrial employees and farm workers are in fact equalized. In this way, the regional policy of the development of infrastructure might be a very important measure of industrialization.

(3) 'Regional equalization' means the gradual approach of backward areas to the average; a marked equalization process in living conditions, a less marked equalization in per capital GNP: and an even lesser one in the degree of industrialization.

The main results of research (and optimum calculations) in the long-term regional structure of the economy are:

(1) The decentralization policy should not touch the most important branches of Budapest industry. It might slow down the economic growth of the whole country.

(2) The efficiency of industry is rather low in the old (traditional) industrial regions because of extensive mining and heavy industry. The costly modernization of the industrial structure seems to be necessary for fully utilizing the high-level infrastructure and to avoid growing political tension (e.g., miners' underemployment, etc.).

(3) The industrialization of rural areas has to be concentrated into urban centres, and should not be dispersed. As a consequence, in many areas agriculture should have a leading economic activity in the future, too. The

large-scale, vertically integrated agriculture could fulfil this task in the fertile chernozem zones. The critical zones are the sandy and eroded hilly territories, where outmigration of the population and a diminution of the cultivated area can be expected.

(4) The development of the non-productive sectors and of infrastructure represent important factors of regional equalization. A certain concentration is necessary for an efficient development even in this field. The significant industrial and large-scale infrastructural development is planned to be concentrated into 50–70 settlements; the basic service institutions in about 1000 settlements (of the total 3000 settlements).

'The concentrated decentralization' represents the basic concept of our regional policy, including a careful selection of the means of development, acording to the geographical and economic conditions of the various regions.

c1302
The application of climatological knowledge, particularly in future urban forms
G.R. MCBOYLE   *University of Waterloo, Canada*

In its simplest sense, applied geography means the application of relevant geographical concepts, skills, and expertise to problem-solving. In the solution of environmental issues most geographical knowledge is relevant and necessary even if not sufficient. With regard to climatology it is the writer's contention that it forms a very viable input to any complete appraisal of the ecosystem and that urban climatology is becoming increasingly so.

Just as our cities have reached a certain stage in their functional evolution, so also has the study of urban climatology as a discipline. Much has been done to discover, quantify, and analyse the effects of the built environment on meteorological elements; but, at a time when the minds of urban geographers, planners, sociologists, etc. are being increasingly cast forward toward speculation of future urban environments, the urban climatologist's interest in this aspect seems lagging.

The present and future hazards of air pollution, it is true, receive considerable thought and experimentation, but questions of the possible use or abuse of urbanized features of natural climate have not yet been asked. Knowledge of climate in general has been used for centuries, e.g., for agricultural practices such as the sowing of rice immediately after the beginning of the monsoon rains, and for military strategy, viz., the use throughout history of foggy conditions for surprise troop movements. Today awareness of incoming weather conditions allows the prediction of disaster warnings such as those for hurricanes and pollution warnings based on knowledge of imminent anticyclonic conditions. The stage seems to have been reached, however, when such passive use of weather is giving way to acquired modification of undesirable elements, for example, the alteration of hurricane intensity and direction by 'seeding,' the 'seeding' of potential rainclouds to provide additional water supply, or perhaps hail-suppression.

Neiburger discusses the viability of various means of smog removal for Los Angeles including some alterations of weather conditions vast in scope and fascinating to the imagination, and rejected them, it seems, solely on economic grounds. The next step, then, is a predetermined use of weather for some human purpose; and active misuse might be the succession to passive use.

Science fiction has toyed with the idea of climatic control as a means of unprecedented power. Several popular films and books have included such fiendish plots as melting the polar ice-caps and altering the normal precipitation rates to make temperate lands instant deserts! Professor Joseph Samuels of the University of Western Ontario, pointing out the potential conflict in hurricane rerouting, claims that 'in perhaps 10 years weather modification could be a military weapon.'

Such power is not far from man's grasp, with all the ensuing potential for yet another series of ecological disasters to add to the already impressive list. It would be nice to be, for once, able to plan before the event. What is needed then, is a broad, comprehensive, unified approach to the environment present and future. Into this environmental policy framework, or system, climatology would provide one input along with all the other considerations of geology, ecology, demography, etc., but the formation of any policy, and particularly one so all-embracing, raises questions not only of organization and implementation but the most basic dilemma of all – the moral dilemma of who determines the 'good.' Since climate knows no borders, *international*, if not global, agreement will become essential.

At the national level, in a democratic framework, the application of a policy decision should require the sanction of the public; and for the public to provide informed contributions to decision-making it must first have a broad but accurate knowledge not only of all the problems, but also of all alternative solutions available and their probable implications. Today this means a greater willingness on the part of professionals to impart their specialized knowledge to the public, and greater but qualitative use of mass-media and public education programs; tomorrow, probably use of what Nigel Calder (1970) refers to as 'The World Box.'

Living in a period of environmental crises has conditioned most minds into a realization of a need to plan our future environment, and plan wisely. Most of us view the creeping expansion of urban agglomerations with some apprehension and not a little distaste. The Chipitts, San Sans, and Boswashs of 'The Year 2000' (Kahn and Wiener 1967) conjure up images frightening and depressing, rather than stimulating and exciting. The question that is begged is, then, *ought* they to become reality?

What does the urban climatologist predict for these urban conditions of tomorrow? Are we to live in giant cities, as arid as the desert, subject to sudden torrential downpours engendered by vast upwellings of warm air over the city heart, and battered on the outskirts by severe pulsing winds which never reach the dry, windless centre? Or a giant hothouse enclosed by a dome of substance sufficiently translucent to allow diffusion of sunlight, and in which the climate is totally controlled by a series of buttons which can produce pure, filtered air along with any degree of warmth, moistness, and air movement which our whims might demand? If so, do we want this and all the implications which accompany it? We don't know what conditions might evolve, but we need to.

The most pressing need for urban climatology is surely to include future considerations within its scope and to attempt some sort of forecasts of possible conditions (perhaps by use of scenario or Delphi techniques), such that the implications deriving from these be made available to society to allow at least a well-grounded basis for choice of future environments.

The conception of a vast cost-benefit analysis of urban life along the lines Perloff (1969) has suggested has considerable value in this respect, especially if the elements of desirability and personal choice can be stressed, i.e., factors pre-eminent in nurturing man's sense of well-being but regarded for their 'untidiness' and random nature as inconsequential in computer technology.

But it is only axiomatic to state that future urban climates depend on future urban forms! The Lithwick Report on *Urban Canada – Problems and Prospects* (1970) envisages a Canada almost 95 per cent urbanized by the year 2001 if present rates continue, with about 73 per cent of the people living in 12 major cities. Toronto, Montreal, and Vancouver alone will account for about 15 million Canadians. The trend is undoubtedly in evidence and one has only to monitor developments in the USA to detect the potential for problems.

However, another choice seems available. Although decentralization processes employing concepts of new communities (or new towns) have had until now only partial success as a means of positive reorganization of regional form, better results might be achieved in the future. Probably the most far-reaching innovation to influence urban form will be the anticipated extension of the coaxial cable as a means toward the 'Wired City' (Department of Communications, 1971), the implications of which are enormous. Other than for, say, water supply and waste disposal, all urban services could be provided by a combination of telephone, television, and computer link-up on both a one-way, and a two-way system. The effects this will have on urban form are only speculation, but the options for a more dispersed settlement pattern appear to be at least technologically feasible in the next decades.

This paper has attempted only to raise

some problems and encourage some speculation among climatologists. Climate can be viewed as a resource; we are on the brink of active modification of it. Whether this be exploitation or careful management depends on a number of factors, including the quality of climatologists' knowledge and the integrity of political decision-making.

Urban climate has been overlooked both as a resource and as a future headache. Assuming urban agglomerations continue to develop according to principles exhibited in the past two decades, what type of urban climate will develop? Do we want what is anticipated? Can it be considered a potential resource to be utilized productively, or will the distaste for future urban climates be a significant input in hastening a decentralized and dispersed settlement pattern which technological innovations seem about to permit?

Calder, N., 1970 *Technopolis: The Social Control of the Uses of Science*, Panther Science.
Department of Communications, 1971 Multiservice cable telecommunication systems – the wired city, *Department of Communications Study 8 (a)*.
Kahn, H., and A.J. Wiener, 1967 *The Year 2000*, Macmillan.
Lithwick, N.H., 1970 *Urban Canada: Problems and Prospects* (Ottawa).
Perloff, H.S., 1969 A framework for dealing with the urban environment: introductory statement, in *The quality of the urban environment*, H.S. Perloff, ed., 3–34.

c1303
La géographie appliquée au Québec
JEAN CERMAKIAN   *Université du Québec à Trois-Rivières, Canada*

Depuis quelques années, nous assistons à un développement considérable des recherches en géographie appliquée dans la province de Québec. Ces recherches couvrent un grand nombre de domaines qu'il serait difficile de tous discuter dans ce rapide tour d'horizon. Quatre thèmes retiendront ici notre attention: les recherches sur le nord québécois, les travaux d'aménagement régional, les études sur les zones urbaines et péri-urbaines, et les inventaires et autres études économiques.

Un certain nombre de géographes travaillent en collaboration avec des spécialistes d'autres disciplines (géologues, archéologues, sociologues, économistes) à l'étude du milieu naturel et humain et aux problèmes du nord québécois. On les trouve notamment à la Direction Générale du Nouveau-Québec au Ministère des Richesses Naturelles, sous la direction de Benoît Robitaille, et au Centre d'Etudes Nordiques de l'Université Laval. Dans ce dernier, les recherches dirigées par les Professeurs L.-E. Hamelin, A. Cailleux et feu le Professeur Jacques Rousseau, sont multi-disciplinaires et s'intéressent aussi bien au grand nord québécois qu'à la côte nord du golfe du Saint-Laurent entre Sept-Iles et Blanc-Sablon. Les recherches ont donc porté jusqu'à présent sur une grande variété de thèmes relatifs à ces régions: géomorphologie, climat, ethno-biologie, problèmes socio-économiques, archéologie amérindienne et esquimaude, linguistique, problèmes de transports aériens, problèmes de frontières entre le Québec et Terre-Neuve au Labrador, etc.: un grand nombre de ces recherches ont été publiées par le Centre, d'autres le seront bientôt. Certaines de ces études ne sont que des inventaires de la réalité existante, d'autres (comme l'étude des transports aériens au Nouveau-Québec, de M. Pouliot) suggèrent des solutions ou des améliorations aux problèmes actuels, après en avoir fait l'analyse détaillée. Un domaine qui mériterait d'être approfondi dans ces recherches nordiques serait celui des ressources minérales et forestières, de leur exploitation et de leur conservation. Il reste encore beaucoup de travail à faire dans ces domaines et sur d'autres problèmes du nord québécois, tant pour les géographes que pour les autres chercheurs.

Lorsqu'il est question d'aménagement régional au Québec, le Plan du BAEQ (Bureau d'Aménagement de l'Est du Québec) vient tout de suite à l'esprit. Il s'agit là d'un inventaire détaillé de toutes les ressources naturelles et humaines du 'Bas du Fleuve,' de la Gaspésie et des Iles-de-la-Madeleine, et d'un schéma d'aménagement dont le but était le relèvement du niveau de vie des habitants de cette région. Le Plan a fait l'objet d'une publication considérable: vingt volumes et un atlas de 1964 à 1966. Si cette entreprise en fut une de grande envergure, la participation des géographes y fut nettement minoritaire.

D'autres projets d'aménagement régional ont sans doute pu compter sur une collaboration plus intense de la part des géographes: ainsi, la mission d'aménagement au Saguenay–Lac Saint-Jean entreprise par l'OPDQ (Office de Planification et de Développement du Québec) avec la collaboration de plusieurs ministères provinciaux a-t-elle compté dans ses rangs plusieurs géographes. Son rapport a paru en 1970. L'année précédente, le Ministère fédéral de l'Expansion Economique Régionale publiait *Le Royaume du Saguenay en 1968*, autre étude d'aménagement fort intéressante dirigée par le géographe Pierre-Yves Pépin.

De grandes lacunes subsistent dans le domaine de l'aménagement régional, tant en ce qui concerne les régions périphériques du Québec (Abitibi, Côte Nord) que dans les régions plus anciennement peuplées et industrialisées (Mauricie, Cantons de l'Est). Ici, les universités à vocation régionale (Trois-Rivières, Sherbrooke, Chicoutimi, Rimouski) ont un rôle très important à jouer en collaboration avec les autorités gouvernementales. D'ores et déjà, certaines de ces universités travaillent activement sur des projets régionaux d'inventaire et d'aménagement.

L'aménagement urbain et péri-urbain fait depuis quelques années l'objet de nombreuses recherches au Québec. Un certain nombre de géographes travaillent pour les services d'urbanisme municipaux tels que ceux de Montréal, Québec et Ville-de-Laval. Leurs travaux portent sur la rénovation urbaine, les aménagements récréatifs, les problèmes de circulation et les problèmes de fusions et regroupements municipaux. D'autres géographes travaillent pour le Ministère des Affaires Municipales ou la Société d'Habitation du Québec. Enfin, il y a quelques années, des économistes et géographes du Ministère de l'Industrie et du Commerce ont élaboré une hiérarchie des villes du Québec et de leurs zones d'influence: la carte de ces zones a paru dans l'*Annuaire du Québec 1966–67*.

Plus récemment, la décision du gouvernement fédéral de construire le nouvel aéroport international de Montréal à Sainte-Scholastique, au nord-ouest de la ville, a poussé le gouvernement provincial à mettre sur pied un bureau d'aménagement des environs de ce nouvel aéroport: ce bureau comptait plusieurs géographes. De plus, les recherches sur le même sujet ont également été confiées au Centre de Recherches Urbaines et Régionales (CRUR) de l'Institut National de la Recherche Scientifique, organisme affilié à l'Université du Québec. Ce centre, créé en 1970, a pour objectif principal la recherche multi-disciplinaire sur le tissu et les fonctions urbaines, et compte des géographes, urbanistes, économistes et sociologues parmi son personnel permanent. Le CRUR, en liaison avec les différentes universités, fera bientôt un travail très utile dans le domaine des recherches urbaines et péri-urbaines.

La division des Etudes régionales du Ministère de l'Industrie et du Commerce est composée entièrement de géographes, dont le travail consiste à faire des inventaires et des recherches économiques régionales. Le directeur de la division, Jacques Girard, est l'auteur d'une *Géographie de l'industrie manufacturière au Québec*, publiée par le gouvernement provincial en 1970. La même année paraissaient des inventaires économiques des régions de Québec, Baie Comeau-Hauterive et Trois-Rivières. La même année également était publiée l'étude fort intéressante de Gilberte Leclerc et Jean-Guy Blouin intitulée *Le port de Québec: facteur de localisation industrielle*, résultat d'une recherche portant sur le trafic et les possibilités d'implantation industrielle du port de Québec. Pour 1972, la division des Etudes régionales prépare une série d'inventaires économiques de chaque région administrative du Québec. La plupart de ces études existent en éditions séparées française et anglaise. Elles constituent de précieux outils de recherche et de documentation à l'échelle régionale et même parfois municipale.

Il resterait à mentionner bien d'autres travaux de géographie appliquée concernant des domaines aussi divers que l'utilisation du sol, l'hydrologie, la cartographie géomorphologique et les études de population. Dans ces domaines, comme dans ceux qui ont été détaillés ci-dessus, il reste encore beaucoup à faire même si le travail accompli est déjà considérable.

c1304
The development of navigation on the Danube, especially in Austria
LEOPOLD G. SCHEIDL *Vienna School of Economics, Austria*

Before 1918 the Austro-Hungarian monarchy controlled the greater part of the catchment area (about 500,000km^2 out of 817,000km^2) and of the course of the Danube River (1340 km out of 2860km). The Austrian Danube Steamship Navigation Company (Donau-Dampfschiffahrts-Gesellschaft, DDSG), founded in 1829, was about 1850 the biggest of its kind in the world. In 1913 it owned 142 steamships and 868 barges and transported 2.3 million tons of goods and 2.2 million persons (Scheidl 1970, 43).

After the First World War the Federal Republic of Austria retained only a small part of the population and of the territory of the monarchy. The length of the Danube within the country amounts to only 351km but its drainage area in Austria covers about 77,000km^2, and the bigger share of the population (7.44 mill. in 1971) and of the economic activity of the state concentrate on the regions along the Danube (Scheidl 1968, 29–31 ff.).

After the First World War the Danube Steamship Navigation company had to cede less than half of its fleet, while during and after the Second World War the company lost two-thirds of its fleet and most of its other property. In 1969 it again had 246 boats at its disposal, but shipped merely 2.3 mill. tons of commodities or little more than it did 80 years ago (Haeseler 1969, 451–3).

In the meantime the aggregate fleet on the Danube and the volume of freight have grown considerably, but the Austrian share has become smaller and smaller. Nowadays she has to compete in river transportation with the other seven riparian states connected by the Danube.

The total fleet on the Danube had a tonnage of 1,609,000t and a propelling power of 186,000 HP in 1914 and of 2,699,000t and 589,000 HP in 1969. In 1967 it consisted of 775 towboats, 143 self-propelling boats, and 3165 barges, altogether 4083 units with a tonnage of 2,452,000t and a propelling power of 543,000 HP. Austrian companies possessed 39 towboats, 21 self-propelling boats, and 259 barges, and occupied fifth place in the number of boats and fourth place in tonnage (Korompai 1971, 66–7).

The aggregate commodity flow on the Danube amounted to 13 mill.t in 1913, 18.3 mill.t (3.5 bill. t/km) in 1950, 25.6 mill.t (9.3 bill. t/km) in 1962, 44.5 mill.t (15.5 bill. t/km) in 1967, and 48 mill.t. (15.1 bill. t/km) in 1969. It is still increasing corresponding to the industrialization of the riparian states. Among these Hungary has been leading for years as regards the volume of transport (8.4 mill.t in 1967), whereas the Soviet Union has the first position in terms of ton-kilometres. Inland traffic is predominant (Korompai 1971, 68). The commodity flow upstream exceeds the flow downstream by far, except on the section between Linz and Regensburg (Schauer 1968, 72–4).

Austrian navigation companies shipped 2.9 mill.t (nearly 2.36 bill. t/km) in 1970 (Österr. Güterverkehr, 1971, 830). It was estimated that boats contributed only 8.19 per cent to the aggregate freight transport in Austria in 1966 (17,265 mill. t/km), while railways contributed 50.77 per cent, and trucks 41.04 per cent. The passenger transport is negligible (Ebner 1968, 4; Scheidl 1970, 45).

In 1969 almost 13,000 incoming and outgoing boats with a carrying capacity of nearly 10.3 mill.t, belonging to different nations, navigated on the Austrian Danube. They transported 7,238,000t of goods (7,609,000t in 1970), of which 5,718,000t (5,692,000t in 1970) were shipped in the international, 836,000t (1,050,000t in 1970) in the transit and 6,884,000t (866,000t in 1970) in the inland traffic (Stat. Hb. 1970, 202; Wirtsch.-Soz.-stat.Tb. 1971, 251). The prognosis expects a slight increase in the international and transit traffic and stagnation in the inland traffic (Bundesmin. f. Verk. 1968, 51).

Before the World Wars the navigation on the Danube carried chiefly agricultural goods upstream and industrial products downstream. Today mainly oil products from Hungary, Czechoslovakia, and Austria, coal and coke from Czechoslovakia and Poland, and sheet iron from Austria are shipped upstream, whereas coal and coke, iron and scrap, and oil products from Germany and from overseas to Austria, chemical fertilizer, pipes, and machines from Austria to the Soviet Union, and wood are freighted downstream (Stat. Hb. 1970, 203). Transportation of crude mineral oil has practically stopped since the completion of the Adria-Wien-Pipeline in 1971. Imports of iron ore from South America via Bakar to Osijek on the Drava (Yugoslavia) and then up the Danube to Linz (Austria) started in the middle of 1971 (DDSG 1971, 1174–5).

Although the Danube has a much longer course (2860km, of which 2379km are navigable) and a much larger catchment area (817,000km^2) than the Rhine (1320km, of which 886km are navigable, and 224,400km^2 (Scheidl 1970, 45)), the freight traffic on the Danube (44.5 mill.t in 1967) is small compared to that on the Rhine (225 mill.t in 1966). Some reasons for this fact lie in the relatively high gradient of the Danube, the higher velocity of the current, the more irregular discharge, the longer duration of the ice blockade, the drainage into a remote sea, the setback caused by the World Wars, the lower degree of industrialization in the eastern parts of Europe, the still insufficient organization and technical development of the river traffic, and its poor or lacking connections with other waterways (Schauer 1967a, 14; Korompai 1971, 66).

In future, navigation on the Danube needs, therefore, to be better organized and more rationalized, means of traffic have to be modernized and the waterway must be improved and connected by canals to other rivers.

More and more diesel-engine boats will be used instead of steamships. The number of self-propelling boats, including tankers, which can be directed easily and with a minimum of personnel, will increase, mainly on the upper course of the Danube. The present smaller types of boats will be replaced by the 'Europe-boat' of 1350t. Push boat convoys, which need less personnel, will gradually replace towboat convoys, especially on the middle and lower course of the river (Schauer 1967b, 10–11). Container boats will be used increasingly in the future. Structures for river improvement and control started more than a century ago, dams of river power plants, secure a certain depth of water. The program of the Danube Commission calls for a minimum depth of navigable water of 270cm between Regensburg and Vienna and a depth of at least 350cm between Vienna and Braila (Scheidl 1970, 46). The Rhine-Main-Danube or Europe-Canal and the Danube-Oder-Elbe-Canal will link the Danube basin with the networks of waterways in other parts of central and in western Europe.

The Rhine-Main-Danube-Canal (168km) will be completed in 1981, and the canalization of the Danube in 1985 (Scheidl 1970, 46; 50 Jahre 1971,1099). Then, the canal will be a part of a waterway approximately 3500km long and suitable for carrying Europe-boats, leading from the north to the Black Sea, and connecting highly industrialized regions with areas still under industrialization. It is expected that the commodity flow on the canal will amount to 16–20 mill.t (of which 3–4.7 mill.t will come from and go to Austria) after a few years, and 30–5 mill.t (10–11 mill.t from and to Austria) later on (Ifo-Inst. 1969, 100; 50 Jahre 1971, 1100).

The Danube-Oder-Elbe-Canal (488km) is planned to connect the industrial and mining areas of Czechoslovakia, Poland, the Democratic Republic of Germany, and Austria with Hamburg and Szczecin. Construction work has not yet started and will last 8–10 years. The freight volume is estimated to comprise 32–5 mill.t a few years after the completion of the canal (Bundesmin. f. Verk. 1968, 56; Baláž 1970, 30–3; Scheidl 1970, 46–7).

Baláž, V., 1970 Der Donau-Oder-Elbe-Kanal, *Der Donaurau* 15, 27–33.
Bundesministerium für Verkehr und verstaatlichte Unternehmungen, 1968 Gesamtverkehrskonzept der österreichischen Bundesregierung (Wien).
DDSG Wieder auf der Donau, 1971 *Verkehr* 27, 1174–5.
Ebner, J., 1968 Probleme der Verkehrspolitik und ihre wirtschaftlichen Aspekte, *Österr. Wirtschaft* 122, 3 ff.
50 Jahre Rhein-Main-Donau-Ausbau, 1971 *Verkehr* 27, 1099–1100.
Haeseler, P., 1969 Das 140-jährige Bestandsjubiläum der DDSG, *Verkehrsannalen* 16, 451–6.
Ifo-Institut für Wirtschaftsforschung, 1969 Die internationale Bedeutung der fertigge-

stellten Rhein-Main-Donau-Grosschiffahrtsstrasse, Bayer. Staatsministerium f. Wirtschaft u. Verkehr (München).

Korompai, G., 1971 Changes in the structure and direction of expanding commodity transport on the Danube, *Geoforum*, no. 6, 63–74.

Österreichs Güterverkehr im Jahre 1970, 1971 *Verkehr* 27, 829–30.

Schauer, R., 1967a Hohe Kosten der Schiffahrtsbetriebe auf der Donau, *Berichte u. Informationen*, no. 1094/95, 14–15.

Schauer, R., 1967b Technische Entwicklung in der Donauschiffahrt, *Berichte u. Informationen*, no. 1106, 10–11.

Schauer, R., 1968 Grundfragen der Güterschiffahrt auf der Donau, *Verkehrsannalen*, 15, 68–77.

Scheidl, L., 1968 Die Verkehrslage Österreichs und Wiens im europäischen Grossraum, *Verkehrsannalen* 15, 23–38.

Scheidl, L., 1970 Österreichs Verkehrslage, Verkehrseignung und Verkehrsentwicklung, in *Geographie und Wirtschaftsentwicklung* 1 (Hirt, Wien), 9–61.

Statistisches Handbuch für die Republik Österreich, 1970 Österr. Statist. Zentralamt (Wien).

Wirtschafts- und Sozialstatistisches Taschenbuch, 1971 Arbeiterkammer (Wien).

c1305
Major accomplishments in the field of applied geography in India since 1969
MOHAMMAD SHAFI   *Aligarh Muslim University, India*

Since the International Geographical Congress held in Delhi in December 1968, applied geography has made appreciable progress in India. Work in the field of applied geography has progressed at a brisk speed in the university geography departments while some of the government departments have also sought the assistance of geographers.

At the university level, research in applied geography is focussed mainly on the following aspects: (a) agricultural land use, (b) urban land use, (c) location of industries, and (d) transport and tourism. The application of geographical techniques and methods towards the solution of some of the agricultural problems is one of the most important aspects of geographical research. The university geography departments of Aligarh, Calcutta, Patna, Gauhati, Sagar, and Udaipur have been actively engaged in this aspect. One of the techniques that has been developed is to study the agricultural land use at the village level and how the food crops that are produced are converted into calories and other nutrients according to the international scale. The nutrition thus available is related to the population dependent upon it. This gives a measure to the planners to assess the nutritional deficiency of each region from which the villages have been selected. Further at the current level of technology the authors have attempted to determine the potential production of each type of agricultural land which may serve as guidelines for the preservation of good agricultural lands (Husain 1969; Mohammad 1969).

Another serious attempt that has been made recently is towards the delimitation of relative productivity regions of India for planning purposes. With the help of quantitative methods, the country has been divided into food crop productivity regions in a paper entitled *Measurement of Food Crop Productivity in India* presented at the annual meeting of the Indian Council of Geographers in January 1970 (Shafi 1970). In this paper the attention of the planners has been drawn to the specific regions which have low and very low productivity compared to the national scale. The precise increase in food production that may result from increasing the level of production of the deficient regions has also been calculated. There is still considerable scope for applied geographers in refining the techniques for the delimitation of such regions. The government of India set up a high-powered National Commission on Agriculture in 1971 to examine the current progress of agriculture in India and to make recommendations for its improvements and modernization. One of the tasks entrusted to the commission requires classification of the country into suitable agro-climatic regions and determining their production potential for different time perspectives. The Commission also aims to assess the types of crops best suited for each region along with their yield potential and to estimate the possible levels of production of various crops that can be achieved by adoption of suitable cropping pattern and the agronomic technology during the different time perspectives, say 1975, 1985, and 2000 AD. The commission also aims to identify and enumerate various geographical factors, aids, and methods by which agriculture can be improved in dry and rainfed areas and to suggest the possibilities of diversification of agriculture in different regions. The commission has appointed a working group of nineteen scientists including one geographer to examine the problems. It is perhaps the first time that the importance of the application of geographical techniques has been realized at the highest national level. The task accomplished at the national level may help the absorption of applied geographers at the state and district levels (Abel and Easter 1971).

Another important aspect on which applied geographers are working in the universities is the delimitation of land resources regions and areas of India. The land resource region attempts to synthesize the information about soils, water, climate, topography, vegetation, and land use. The regions have been divided into areas and for purposes of detailed planning they may be subdivided into resource units. This is a task which the geographers have to accomplish.

In the realm of urban geography, the emphasis is mainly on the delimitation of urban fields based on relevant indices, delimitation of slums, and a geographical analysis of the problems associated with them, delimitation of central place hierarchy employing quantitative methods.

At Ranchi a doctoral thesis on 'Regional Pattern of Cities in Bihar' attempts to project the population in each of the towns in 2001. It has pointed out that two cities in Bihar (Jamshedpur and Ranchi) have been increasing their population at a rate of 150 per cent while in the remaining cities, it is not more than 30 per cent. The study further points out that three out of the present seven cities would become one million cities in 2001 and by that year the total urban population of the state would be 60 per cent as against the present 33 per cent. Such urban projection for each town would help to plan the future requirements within the limits of existing and potential resources (Sinha 1970).

A number of useful dissertations have been completed by geographers in the realm of industrial location. A thesis on the 'Location of Industries in the Son Valley' (Azam 1969) completed recently makes recommendations for the specific location of future industries in the valley in Bihar, while another thesis has considered in detail the 'Location of Small Scale and Cottage Industries in the Delhi State' and, after pointing out the disadvantages in the location of some of the industries, makes recommendations for the future location of these industries. The recommendations are likely to be considered at the highest official level (Singh 1971).

Applied geographers in India have turned their attention to the geographical aspects of tourism. Dissertations at the MA level have been completed in several universities on geographical aspects of tourism, e.g., *Udaipur as a tourist centre*; *Recreational Use of Land in Udaipur City* (Bhattacharya 1969-70), which particularly examine the trends in the growth of recreational areas and their future planning both in terms of local population and tourists. There are indeed significant gaps in the study of recreational geography and in view of the scenic potential of the country, it may be recommended that the study be taken up on a micro-regional basis and the factors which retard the healthy development of the tourist industry be mapped and indicated by the geographers.

Abel, Martin E., and K. William Easter, 1971 Division of the country into homogenous crop regions, identifying agroclimatic regions of India, papers presented before the National Commission on Indian Agriculture New Delhi.

Azam, S.S., 1969 Location of industries in the Son Valley, Aligarh, PhD thesis.

Bhattacharya, A.N., 1969-70 Udaipur as a tourist centre, recreational use of land in Udaipur City (Udaipur), unpub. M Phil thesis.

Husain, S. Sajid, 1969 Land utilization in Budaun and Shahjahanpur districts, PhD thesis, Aligarh Muslim U.

Mohammad, Noor, 1969 Geographical bases of agricultural land use in Ghaghra-Rapti Doab, PhD thesis, Aligarh Muslim U.

Shafi, Mohammad, 1970, 1972 Measurement of crop productivity in India, presidential address, Indian Council of Geographers, Kharagpur, 1970, Transaction, 1, no. 3, in press.
— Measurement of agricultural productivity of the great plains of India, paper presented at the European Regional Conference of the IGU Budapest.
Singh, Amer, 1971 Location of small-scale and cottage industries in the Delhi state, Aligarh, PhD thesis.
Sinha, S.P., 1970 Regional patterns of cities in Bihar, Ranchi, PhD thesis.

c1306
University programme and requirements of applied geography
LUDWIK STRASZEWICZ *Lodz, Poland*

Until recently geography was a clearly and unequivocally defined branch of science. It was a separate discipline of university training, subordinated to the domain of natural or classical sciences, and its pupils were predestined to exercise the teacher's profession. Thus, geographers held positions in higher and secondary education as university professors or lecturers, or as high-school teachers. Following the First World War geographers began to take employment in special scientific institutions involving research, but at that time their number was limited. It was principally university professors who took up strictly scientific research work; at times even high-school teachers did the same.

In conformity to its etymology 'geography' deals with the description of the earth, of particular continents, countries, and regions. For several generations, geographers have attended to informing the population about environment, social and economic problems occurring in the world, about foreign countries and one's own country. However, writing books is by no means the geographer's only job and cannot serve as the means of his sustenance. Geographers are in no way professional writers; as a rule they consider their literary production a mere supplement to their scientific and educational activities.

Every group of humanity which has gained some definite stage of education is characterized by a professional activity conformable to its ability – meaning the scope of professional tendencies aspired to. Hence one speaks of a narrow professional activity of a graduate when he exercises exclusively, or for the most part, but one profession or, in other words, when a given scientific discipline makes its graduates pursue only one or very few trends in their occupation. To give an example: all physicians are engaged in the medical services. On the contrary, many specialities taught at universities lack specific counterparts in definite professions, and in cases like this we speak of a wide potential range of occupational activities. As an example may be cited law graduates who either work in independent professions or are engaged in jurisdiction, in offices of public administration, the diplomatic service, etc.

The professional calling of geographers is passing through a remarkable transformation. In some countries geographers are teachers and nothing else. In contradistinction to this, the career of geographers in Great Britain, the United States, Poland, the Scandinavian, and a number of other countries has expanded over a variety of further professional work besides schooling. Thus, in all these countries we observe what might be called a wide-reaching professional activity of geographers. From the resolutions arrived at by the Commission on Applied Geography of IGU it may be said that in countries with a wide-reaching activity, geographers pursue an enormous variety of employment and appointments.

Since the end of the Second World War a steadily growing number of graduated geographers are being directed to work in a variety of jobs outside scientific institutions and besides teaching in colleges and high-schools. Gradually but incessantly, in all countries the percentage of college-trained geographers employed outside of teaching is on the increase. In the 1940s, in Poland the vast majority of geographers worked in this kind of teaching; in the 1960s more than 60 per cent of all graduated geographers have found employment in practical professions (Straszewicz 1967).

Investigating the present-day career of geographers employed outside teaching, we may divide all the professions they pursue and the positions they hold into two principal categories, depending on the degree to which they apply their geographic training, and the capacities acquired since then, to the professional work they are actually performing.

The first group comprises professions and jobs in no way connected with the professional training and the abilities of a geographer. Among these is employment by railways, in trade, in police work, in journalism, etc., for which no particular trend of previous geography studies is required. Here geographic schooling rather takes the place of a general higher education.

The second group, many times larger than the first, occupies positions in what on the whole may be called applied geography. This sort of work demands both a geographic education and ability of applying this knowledge. Geographers who occupy this type of position are frequently called 'professional geographers' or even 'geographers-engineers.' It should be stressed that the lack of a proper nomenclature and the inappropriateness of many of these terms lead to many a misunderstanding, apt to be of serious consequence for the evolution of geographic branches.

There are thousands of the geographers of the world who occupy positions of this kind. Actually there are six principal fields in which what might be called 'practising geographers' are working: (1) the field of geomorphology and geology represented by a variety of offices and institutions engaged in studies of cover deposits – such as geological institute, research laboratories of building materials, and others; (2) the field of hydrological and meteorological services; (3) the field of cartography, attended to by departments of geodesy and cartography, and of the editing of maps; (4) the field of physiography, covering all sorts of work aiming at describing and evaluating land for purposes of planned investment; (5) the field of spatial planning, performed by departments of planning offices, town-building authorities, investment departments of communities, etc.; (6) the field of tourism, which comprises agencies financing undertakings providing tourist services, and develops the country economically for purposes of tourism and recreation, including editing offices for maps and propaganda literature.

With regard to the three first-mentioned items, geographers are face to face with strong competition: geologists, geophysicists, geodesists. Here it often happens that geographers are exposed to reproach because of the assertion that they are insufficiently prepared to cover the topics involved.

When it comes to fields 4, 5, and 6, these indeed are exclusively or nearly so, the domain of action for geographers who are perfectly capable of undertaking this sort of work. For this reason these three fields may be called those of principal interest to geographers. In particular it is spatial planning which is the obvious task for which geographers are qualified, and a majority of them are actually employed in attending to this kind of work in their daily routine.

Worth emphasizing is that in some countries a considerable number of geographers have found employment in politics, in diplomacy, or in the military services. In Poland many geographers work in the agricultural sector, classifying soil types, etc. Apart from this, a steadily growing number of geographers are occupied in matters of public transportation.

There can be no doubt that geographic science finds a wide field of application in many provinces of our everyday life. Even those geographers who question the concept of applied geography cannot deny that geographers are enlisted in a wide variety of social activities (George 1961). As said before, practising geographers occupy many positions. Besides teaching, the most propitious career of a geographer is spatial planning. Every geographer working in this capacity should be considered a professional planning expert. The same refers to one working in climatology, cartography, etc. However, this is not merely the matter of a name for the position of a geographer, because any work in this capacity implies essential responsibilities: (1) In order to attain a high degree of specialized ability, the geographer has to acquire qualifications which are not transmitted to him in standard university education. (2) In most positions occupied by geographers, this pursuit is only one of the basic disciplines which prepare him for his new calling. On the other hand, graduates from other branches of sciences must possess a comprehensive knowledge of geography in order to become, in time, prominent specialists. (3) Even when attaining the rank of specialists in given professions like planning, physiography, or problems of tourism, geographers must be certain to remain geographers. Their value in the occupation assigned to them rests to a large degree upon their geographic training.

Unfortunately it is a fact that in the majority of countries the present-day schedule of geographic education fails to meet the demands imposed by modern life. The reason is that up to now this education is based on traditional rules and regulations which do not fit true requirements and rather adhere to matters important, in the first place, to the teaching profession of geographers, whereas each of the above enumerated fields of practical work presupposes specific abilities.

The principal conflict between what universities teach and what life demands, lies first and foremost in the province of specialization which by now has been included in many countries in the curriculum of geographic studies. These studies can be made conformable to the requirements of applied geography by introducing a suitable schedule of special subjects taught either throughout the period of normal studies or during the last term only. There would have to be drawn up a list of professions in which geography graduates are either occupied now, or might or should be occupied after attending special schedules of studies towards those trends for which specialists would have to be specially trained. Thus, apart from specializing in training geography teachers, professional specialization would prepare particular specialists for work in those occupations in which geographers are now employed, making them experts equal to those who have graduated in other branches outside of geography. A program of university studies arranged in this way would take into account the changes which are taking place in the discipline of geography, and would prepare students for the tasks awaiting them in consequence of the rapid progress made by applied geography in its application to what modern humanity demands.

George, P., 1961 Existe-t-il une géographie appliquée, *Annales de Géog.*, no. 380.
Straszewicz, L., 1967 L'utilité de la profestion de géographie. *Bull. de la Soc. Géog. de Liège*, 3.

c1307
Geographer's role in clarifying environmental problems
WOLF TIETZE *Geoforum, West Germany*

It is a well known fact that generations of geographers all over the world have applied considerable effort in studying the development of localities, areas, and regions. Some have focused their investigations on either physical *or* human matters. Only a few managed to recognize or even to command the interdependence of physical *and* human matters, which both make up the character of a locality, area, or region. Many got stuck half way, captured by methodological and technical difficulties, and further remaining tied up by long lasting and tiring endeavours to solve such problems. We owe the deepest gratitude to this group among our fellows, although we may in some cases have reasonably become reluctant to see them still as geographers, because they rather often have lost the final goal: full understanding of a locality's, area's, or region's character. This thorough understanding is the fundamental purpose of any useful application of the results of geographical research.

It is very tough to achieve this proper understanding. Since the oldest days of our science this has more or less successfully been done by accumulating experience of observation and subsequent interpretation. As long as this procedure is favoured by a good intelligence, awareness of prejudices, and some fortune, we may still yield satisfying and stimulating results. This approach occasionally proves to be the least time consuming one. It is, however, rarely reachable and scarcely repeatable by anybody, nor are the results easily checkable. This method – if it is one at all – therefore increasingly meets mistrust. Again and again geography is even questioned as a science, and, honestly, we must admit that a considerable part of our literature proves certain religious or political background attitudes of the respective authors or any myths and wishful thinking introduced by succeeding interpreters. However, such behaviour is human, and geographers are also – please believe me – human beings. Furthermore, they have to account for the impact of human beings on their natural environment and vice versa. They even have to deal with the variety within the human society and with the interdependencies of the many social groups. This does sometimes affect them personally, a stimulating and therefore welcome process.

In turn we have good reasons to welcome new quite sober research techniques as they have developed during the past two decades with respect to processing numerically transformed data. These steps are indisputably a real progress *in* geography – not necessarily *of* geography, a distinction that appears to be essential. A considerable amount of methodological work still needs to be done with respect to acquisition of the basic material, its homogenizing and transformation. Well balanced generalizing is indispensable, but usually it is not evident by itself at what stage of the whole process of investigation and to what degree this generalization should best be done. Mathematical accuracy is the means, not the purpose, of the procedure. Occasional confusion about this principle has already led to quite a pile of nonsense literature, and some progress certainly has only been pseudo-progress.

So far we find ourselves confronted by a strange dilemma: traditional ways of research – those on the base of experience – need admittedly to be supplemented, not replaced, by modern mathematical techniques. These, however, all too frequently have proven a bluff. Apparently applicants of these techniques are so fascinated by the elegance and speed of their processing that they have become blind to their feeding in of geographical garbage; and because of the accuracy of the mathematical process garbage in means garbage out.

Consequently, an improved comprehensive methodological approach of data acquisition, analysis, synthesis, and presentation is necessary. This approach must be guided by one overall goal. Otherwise it will get lost. This goal is the regional geographical aspect, in German understood by *Landschaftskunde*. It is the only platform from which geographers can provide any valuable advice for any kind of environmental planning. As environment is an overall ecological system (that *is* Landschaft), not the tiniest fraction of it can be changed without producing a respective impact on the entire rest of the system. Within the global ecumene even 'natural' or 'physical' ecology cannot be separated from 'social' or 'human' ecology without serious disadvantage.

Here stands geography, the nucleus of the wide range of geo-sciences, having branched off, and by this time having grown up to become more or less independent. Most of the political decision makers and a vast, steadily growing part of the general public do realize today the power of our environment and the meaning of our dependency thereupon. This is the right moment to apply geographical recognition.

It is understood that human settlements are the foremost tools of mankind to produce material as well as cultural values. Furthermore settlements are the places where human abilities are created, partly restored, but also, settlements are the places where a considerable human and natural potential is being spoiled and destroyed. The intensity of this kind of function of settlements is different of course according to the wide variety of settlement types, of social, political, and economic circumstances. All factors are continuously changing. The amplitude of these alterations and hence their consequences are biggest in and around the big cities. Consequently the values – positive as well as negative – turned over in the big cities are remarkably huge. Careful consideration should therefore be given to all measures directing the functions of the cities. Their productivity or rather, their efficiency, should become a main research topic of urban geography, considering every city dweller as a shareholder. In contrast to a factory, a city is a social organism, too, which should guarantee an equally fair chance to everybody to participate in the benefits of the urban functions. Strictly, this would, for instance, morally exclude any kind of privileges. Privileges are still maintained and newly established by legally protected processes, which do not move congruently with the development of the majority of other urban functions, but which, produce a profit, frequently a very substantial profit, to the advantage of an extremely small minority at the expense of everybody. Such incongruency is usually found with immobile values – first of all with real estate values – which alter their status only by means of a huge transfer of speculative money. As the necessary amounts cannot always and in time be made available, urban renewal is commonly lagging behind planning, if this stands for current and future public needs, thus increasing the urban problems. If the project can be financed, it goes in favour of somebody, who usually has not adequately contributed to this urban development. This contrasts with the fact of a city being a social organism.

As most urban planning measures are still confined to construction, the wide field of urban function development comprising city and umland is still mostly vacant, due not the least to an outdated administrative structure. There are but very rare exceptions where political decision makers are supplied by sophisticated economic-geographical studies of an urban area and its future perspectives. Instead, the very primitive point of view of calculating the voting power or – likewise bad (or worse?) – any political doctrine commands the same. This, however, raises the standard of living as badly as speculation does.

On a statewide level it may be ascertained that a nation with a high average per capita income resulting from a huge low income substratum with a small (i.e., under 30 per cent) middle class and a very narrow peak of high income people appears to be as underdeveloped as any other nation with a much lower per capita average. It is very significant, that a socially poor environmental situation is regularly accompanied by a depressing lack of responsibility for the natural environment, which consequently is exploited and polluted without any scruple. In other words, pollution of the physical environment is characteristic of an underdeveloped social environment.

Briefly concluding, a number of important tasks of geography may be formulated: (1) Geography has to acknowledge the overall significance of Landschaft as object of research. Landschaft–environment should by no means be confined to the physical aspect, but include the social one as well. (2) All partial investigations – and, accordingly, all methods and techniques applied in those contexts – must be planned from and afterwards their results be incorporated into the overall Landschaft- (-environment) concept. (3) The evaluation of the globe's surface in favour of mankind is different. The evaluation is in particular characteristic of the ecumene, escalating sharply in settlements, especially in urban settlements. These are crossroads of human energies and abilities where nature's potentials are hard pressed. (4) No efforts should be spared to investigate the man–land relationship and to develop models of optimal ecological interdependencies, to develop methods of continuous control of this manifold interlinkage – and to teach the vital importance of careful observation of the man–land-interdependencies. Mankind must acknowledge that environment is their strongest teacher, geography being the mediator.

Quantitative Methods
Méthodes quantitatives

c1401
A model of ghetto growth
RICHARD L. MORRILL   *University of Washington, USA*

Ghettoes are the homes of some fifteen million black Americans – spatially separate colonies developed mainly during the last 60 years. The ghettoes grew as blacks left the farm and plantation for the hope of better opportunities in the cities. Blacks were confined to separate ghetto spaces, owing to both external social, economic, and political discrimination and to internal pressures of group solidarity and cultural preference. The black occupancy of the ghetto has proved to be permanent, having grown continuously larger to accommodate both natural increase and further migration out of the rural south. The main reason for this permanence is the simple fact of blackness. Blacks are social prisoners of the ghetto, whether or not they achieve middle class educations and incomes.

The typical ghetto originated in the mixed residential-industrial-wholesale slums near the city centre. At times growth was accommodated by overcrowding, but physical expansion was usually necessary. Sometimes this growth was concentrically into areas of similar lower class housing, but more often expansion was sectorally, into areas being abandoned by lower middle and middle class ethnic groups. The growth has been channelled by real estate and financial institutions, which have supported the popular aim of minimizing social contact between races. The extension is often accomplished by a block-by-block transition, so that the border of the ghetto is often surprisingly sharp.

In 1965 a simple model of ghetto growth was published. Criticism of that model and new knowledge impel me to offer now a more realistic and behavioural model, which will better predict such geographic outcomes as (a) intensification or reduction in the proportion of blacks; (b) areal expansion or stagnation or contraction of the ghetto; and (c) deterioration, no change, or upgrading in the residential quality of the ghetto. The principal 'actors' appear to be (1) black families, (2) black organizations, (3) white families, (4) white organizations, (5) the real estate industry, (6) financial institutions, and (7) the government. But each group contains people with conflicting goals: as black and white individuals and organizations who favour integration or segregation or who are indifferent.

Some blacks may desire integration as a means of improved social and educational opportunities and of obtaining better housing; others may prefer to reduce social friction and maintain social relations by staying within the black community. Many whites prefer the existing community character or may fear blacks. Encroachment of the ghetto may lead to either resistance or flight. Some whites and blacks believe in integration as a means of achieving equality and reducing racial conflict. The extent to which these goals are realized and the likelihood of various geographic outcomes depends as well on such factors as the absolute size, rate of growth, income, and educational levels of the black and white communities; the nature of housing stock, the history of race relations locally, and long-term governmental bias.

Intensification without expansion may result from increasing identification of blacks with the community or from very strong white and institutional resistance or both. If the population growth and immigration are high, intensification and areal expansion are likely; racial conflict is possible. Areal expansion, with dilution of the proportion of blacks, might occur where the black community is fairly small, or where the whites tend to be integrationist and stay, or fearful and flee. Deterioration of housing quality may occur with high growth of lower status blacks; an upgrading is possible where status is improving and where whites are able to flee to the suburbs, allowing fairly peaceful expansion into better housing.

Most individuals are not totally committed to an inflexible position, but rather their decision to move results from a weighing of alternative benefits and costs. The black may choose between newer and better housing and the solidarity of the black community. The white family may fear blacks becoming too numerous but be very attached to their home.

In order to incorporate these behavioural complexities, a stochastic simulation is proposed. It is deterministic in that only one outcome is possible from a given set of initial conditions and exogenous forces, but behaviour is controlled by a rather complex set of probabilities or proportions which are sufficiently diverse to permit a wide range of outcomes.

Operation of the model (flow diagram, not shown here)

Initial conditions: By small areas, we need to know the numbers of homes occupied by whites and blacks, and which are vacant, by at least three classes of dwelling. We need also to estimate the rate of net new household formation by class.

Mobility: For blacks we need to estimate the proportions who are integrationist, separationist, or indifferent. Next we need to estimate the proportions of people with a motivation to move at all during a given time period, and whether they want very much to move or only somewhat want to move.

Availability of housing: For those who wish to move, there is a problem of the objective possibility and capability of a move. For any pair of origins and destinations, it is necessary to match potential movers of a given class and housing in that class. A 'first come, first serve' rule is used to deal with housing shortages.

Allocation of searching by distance: Since information about housing and the desire to live there may be expected to decline with distance from present location, moves will be biased in favour of closer opportunities. We must estimate then the probability of people in various classes having information and desiring to live various distances beyond the ghetto. If insufficient housing is available, remaining potential movers wait for housing to open up. This step yields a distribution of potential moves.

White reaction to black searching: The very act of blacks searching for housing leads to white reaction, for example, some initial white flight, which is proportional to the intensity of the black search. This in turn opens some vacancies, permitting further moves.

Real estate barriers: The prospective mover must overcome institutional resistance. We then estimate the probability of an initial rebuff from real estate agencies, a function of both the class of area and the present proportion of blacks. The weakly motivated will probably give up at this rebuff, but the strongly motivated will try again, perhaps several times.

Financial barriers: Even if shown a home, the prospective buyer must arrange finances. We may hypothesize a probability of resistance, largely a function of the income class or the mover. Those who have searched and found housing, and passed through real estate and financial barriers are temporarily successful movers.

White reaction to black entry: (1) Resistance. Whites may react by organizing resistance. Probably resistance will be higher in lower class areas and where present proportions of blacks are rather low.

White reaction to black entry: (2) Flight.

The flight of whites governs the creation of vacancies and is of fundamental importance. The probability that lower and upper classes will move is less than for the middle classes, and for all groups the probability of flight increases with the proportion of blacks.

At the end of these steps, a new pattern of black and white population will exist. New vacancies will have been created, which sets the stage for repetition of the entire process in an ensuing time period, and which changes the probabilities or resistance and the areas of preferences for those who would move from the ghetto.

The model was applied to the Seattle ghetto ('Central Area') for the periods 1950 to 1970 in five-year increments. The replication of the pattern of expansion for 1950–60 was quite acceptable. The results for 1960–70, which depended on poorer information, were not so successful. For both decades upward social mobility was underestimated, more severely for the 1960s. The main failure was the inability to predict a radical shift in the direction of growth. In hindsight the model could have predicted this, and the shortfall was largely due to my placing unnecessary restrictions on the free operation of the model.

Beauchamp, A., 1966 Processual indices of segregation: some preliminary comments, *Behav. Sci.* 11, 190–2.

McEntire, Davis, 1960 *Residence and Race*, Berkeley.

Morrill, Richard L., 1965 The negro ghetto: problems and alternatives, *Geog. Rev.* 55, 339–61.

Rose, Harold M., 1969 Social processes in the city: race and urban residential choice, Commission on College Geography Resource Paper No. 6, Washington, DC, Assoc. of Am. Geog.

Rose, Harold M., 1970 Development of an urban subsystem: the case of the negro ghetto, *Ann. Assoc. Am. Geog.*, 60, 1–17.

Wolf, E.P., 1963 The tipping point in racially changing neighborhoods, *J. Am. Inst. of Planners* 29.

Zelder, Raymond, 1970 Racial segregation in urban housing markets, *J. Reg. Sci.* 10, 96–101.

World Land Use Survey
L'utilisation du sol dans le monde

c1501
Maps of soil use in Mexico
M. ACEVES-GARCIA *National University of Mexico*

Studies of the soil use, undoubtedly related to the development of Mexico, have become important during the last four years since research has been carried out in specific zones of the country by geographers and agronomists dedicated to the problems of soil use.

The country is in need of a map of the soil use since, in the first place, the map on a scale of 1:4,000,000, which is being made by the Institute of Geography of the National Autonomous University of Mexico, gives only a view as a whole, presenting all the country in one sheet.

In the second place, the map of 1:1,000,000, which, following the outlines of the Commission on Land Use of the International Geographical Union, is being prepared by the College of Geography of the National Autonomous University of Mexico, in fact, only attempts to give, in a more detailed manner, the soil use in Mexico; it is composed of approximately 18 sheets which will enable a deeper study of the different regions of the country. This map has, furthermore, the advantage of being comparable to a world map on the same scale. It is hoped that in approximately one year this map will be finished.

In the third place, there is a map, 1:50,000, where apart from the soil use there appear the planimetric details obtained from the basic topographic map, as well as the infrastructure works in existence, the toponomy, and services which are established in the localities classified on the map.

The elaboration of the general maps has not required complicated or detailed procedures, since its distribution will be wide, and the data they contain can be used for general studies.

For this purpose, the existing data have been compiled in different institutions of the country, and mention has been made of the experience of the research fellows who have carried out studies of this nature, in small regions, which are considerable, both in number and in quality. The main sources of research belong to work carried out on the basis of aerial photointerpretation.

The map, 1:50,000, is being made by the Office of Soil Use of the Commission of Studies of the National Territory, which is associated with the Direction of Economic Studies of the Mexican Ministry of the Presidency. The photographic interpretation of the office is done by the agronomists and geographers, who carry out the corresponding photointerpretation.

Said map has a format of 65 × 90cm, which corresponds to 15 minutes in latitude and 20 in longitude, respectively. The legend, the diagrams for localization, the name and the code of the sheet, and other supplementary information are on each sheet so that they give the person consulting them ample information on the region embraced.

In July 1971, 216 sheets were in the process of re-interpretation and another 39 sheets were ready for printing; the average monthly map production is 15 sheets; it is considered that in an eight-year period the soil use map will cover all the national territory. Furthermore, the above-mentioned institution is producing, besides the soil use map, maps of soil and geology, and with the information obtained in such a way is produced the map of potential soil use.

Ramón Solís Vega.

Kedar, Yehuda, 1966 *Un nuevo enfoque para la clasificación de las regiones agrícolas y del uso del suelo en el mundo*, Unión Geográfica Internacional, Conferencia Regional Latinoamericana (México).

c1502
Land-use study in Alabama by aircraft and satellite remote sensing
CHARLES T.N. PALUDAN   *National Aeronautics & Space Administration, USA*

Land-use studies have never had their importance under-estimated in the United States but difficulty in obtaining or updating has severely limited data availability. Some of the history and difficulties were reported in a report by the Resources for the Future, Inc. On the nation-wide level, the 1:7,500,000 scale map in the National Atlas of the United States, by Anderson, is based primarily on Marschner's study of the 1940s (National Atlas 1970, 157–9). Of the 50 states, only New York state has an up-to-date comprehensive land-use data base (Shelton and Hardy 1971, 1571–5).

Several factors focus attention on larger scale land-use studies at the present moment in US history. First, there is the national public concern over the condition of the environment – to a large degree touched off by the rearward observations of the 1968 Apollo lunar flight; second, federal legislation has forced each of the 50 states into immediate action on land-use planning; third, federal judicial rulings have forced some states to revise and equalize property tax assessment which may result in taxation according to land use; and fourth, the remote sensing technology and the National Aeronautics and Space Administration's (NASA) Earth Observations Program has made the needed data available comprehensively and cheaply for the first time.

The state of Alabama has been particularly affected by all the factors mentioned. NASA's largest field centre, the George C. Marshall Space Flight Center (MSFC), is located there, giving the entire state an unusually receptive attitude toward space-acquired observations. Alabama might well be termed a 'developing state,' because of its many untapped resources and, paradoxically, its general substandard per capita income and other indicators of wealth. For that reason, the state government has been very active in attempts to encourage new industry. It is recognized, however, that Alabama's relatively unspoiled environment (Birmingham excepted) is fragile, and state and local officials charged with development are anxious to develop a rational approach to land use. There has been some historical experience in the state with this approach. The great land development project of the 1930s, the creation of the Tennessee Valley Authority, was greatly aided by land use studies conducted using aerial photography (Hudson 1936).

In the spring of 1971, in response to the Announcement of Flight Opportunities, the University of Alabama, supported by a number of state agencies including, especially, the Geological Survey of Alabama, the Alabama Development Office, and the governor, submitted a proposal on the use of data acquired by the Earth Resources Technology Satellite toward natural resource planning within the state. Considerable state resources were pledged to support of the proposal, which was accepted by NASA and is at present an on-going project. MSFC is also a contributor to the overall proposal, especially in the field of computer-aided data analysis. A statewide land-use study is an important part of the overall proposal.

In an effort to acquaint state and local officials with the advanced technology techniques NASA has learned to apply, MSFC has been working with these officials since early 1971 in a number of demonstration projects, one of the most important of which is a land-use study. Prior to the flight of the Earth Resources Technology Satellite, data acquired by aircraft have been used in this demonstration project.

In northeast Alabama, a five-county, 10,000km^2 region known as the 'Top of Alabama' constituted MSFC's initial demonstration area.

Resources were not available for extensive use of traditional field study. In fact, resources were judged to be inadequate for manual photo interpretation of conventional aerial photography. Advanced remote sensing techniques were clearly indicated. But MSFC was only partially equipped for this; i.e., no in-house experience with side looking airborne radar was available. Some experience with multispectral photography and a considerable experience with automated data analysis dictated our approach (Paludan 1970, 373–82).

A methodology was developed which permitted rapid translation of data from four-band photography into land-use information. The goal of this demonstration project was to develop techniques that could be applied to data returning from the Earth Resources Technology Satellite.

Of the many small-scale land-use studies conducted within the United States it is very difficult to find any two that use the same classification system. Certainly it is difficult to find any that adhere to the classification system proposed by the International Geographical Union in 1949 for the World Land-Use Survey. The most comprehensive standard is the Standard Land-Use Coding Manual published by the Urban Renewal Administration and Bureau of Public Roads in 1965. In recognition of the peculiar problems involved in use of a classification system with the data primarily coming from orbital altitudes the National Aeronautics and Space Administration and the Department of Interior jointly sponsored a conference in the summer of 1971 to consider what a new national classification system should entail (Gerlach 1972). Several well-known systems were put forward as straw men. These included the Canada Land Inventory, the Land Use and National Resources Inventory of the State of New York, and a system developed by the Commission on Geographic Applications of Remote Sensing of the Association of American Geographers (Department of Regional Economic Expansion 1970; Center for Aerial Photographic Studies 1970; Anderson 1971, 379–87). Potential users of the data from a broad cross section of the United States were given the opportunity to contribute to the final recommendations which are still being developed by the steering committee. The preliminary indications are that the classification scheme will be as shown in Table 1. The system is a multi-level hierarchy with the nine first level categories being those that can primarily be determined solely by remote sensing with little need for ground truth. The second level categories would require some ground truth in many cases. The third and fourth level categories would contain information not available by remote sensing techniques, which are the concern of this paper. It is this classification system that MSFC has applied to its preliminary studies in the state of Alabama.

Another general problem considered early in the demonstration project was that of a locational grid. Several systems were considered but in final analysis the Universal Transverse Mercator (UTM) grid was chosen. One reason for its selection is that NASA in general has made a choice that the Earth Resources Technology Satellite data will have the UTM grid super-imposed upon the data that are being sent out to the various investigators. A metric grid was readily acceptable in the state of Alabama where already some speed limit signs appear in metric units.

The earlier study of land-use in the Top of Alabama Region was based solely upon four-band multispectral photography – a relatively sophisticated, fast, but non-automated technique was used for data analysis. Resulting maps are being published by the Tennessee Valley Authority at the scale of 1:24,000. Several approaches to truly automated data analysis have been pursued at MSFC.

Data management activities at MSFC run the gamut from the remote sensing of data through data processing and data analysis to the design of a computerized multi-purpose information system.

For the past few years, MSFC has developed a repertory of computer aided data analysis techniques. Automatic pattern recognition techniques using both electronic (analog and digital) and optical methods were developed to analyze the geometric properties of images. Automatic signature recognition techniques were developed to perform feature classification or species separation, using spectral characteristics.

Contrast enhancement and background removal algorithms using classification decision logic based solely on contrast differences were successful in several cases where the distinctive categories were partially obscured by background. However, the limitations incurred by this technique necessitated the development of decision logic based on pattern recognition techniques as a means for accounting for the correlation between different features in the classification algorithms. The vector representation of spectral signatures in Hilbert space is particularly useful since the highly sophisticated algorithms of feature classification can be applied directly (Bond and Dasarathy 1970). This technique has been applied to the interpretation of multispectral photography with spectacular results. For example, in one case involving a Chandler Mountain farm in Ala-

TABLE 1. Land-use classification

	Symbol	Colour
Urban & built-up	U	Red
residential	Ur	
commercial & services	Uc	
primarily industry	Ui	
extractive	Ue	
major transport routes & areas	Ut	
public & institutional	Up	
open land	Uo	
Agricultural	A	Light brown
cropland/pasture	Ac	
orchards, vineyards, groves, bushfruit, including horticultural areas	Ah	
feeding operations	Af	
other agricultural land	Ao	
Rangeland	R	Orange
rangeland	R	
Forestland	F	Green
forest (solid crown)	Fc	
aforesting brushland (intermittent crown)	Fb	
arid woodland	Fa	
tree plantations	Fp	
Water	W	Light blue
streams	Ws	
lake	Wl	
reservoir	Wr	
coastal & estuary waters	Wc	
Non-forested wetland	M	Dark blue
vegetated	Mv	
bare	Mb	
Tundra	T	Grey
tundra	T	
Barren land	B	Yellow
beaches	Bb	
exposed rock	Br	
tailings, abandoned pits, & quarries	Bt	
salt flats	Bh	
lava flows	Bl	
sand (other than beach)	Bs	
other	Bo	
Permanent snow & ice fields	S	White
permanent snow & ice fields	S	

bama, ground tomatoes were successfully distinguished from staked tomatoes.

In addition to maps, an essential element of the demonstration project has been adaptation of an existing, space-related, data retrieval system to meet user needs in southeastern US. The system has the capability to accept and manipulate analytic data based on lab and field research and archival data such as census information and courthouse records.

Anderson, James R., 1971 A tentative land use classification scheme for use with remote sensing complemented by other information sources, *Photogrammetric Engineering* 37, no. 4, 379–87.

Bond, A.D., and B.V. Dasarathy, 1970 *Feature classification techniques*, Computer Science Corp. Report M00030 (Huntsville, Ala.).

Center for Aerial Photographic Studies, 1970 A review of the New York state land use and natural resources inventory, Cornell U. (Ithaca).

Department of Interior, 1970 *The National Atlas of the United States of America* (Washington, DC), 157–9.

Dept. of Regional Economic Expansion, 1970 A guide to the classification of land use, *The Canada Land Inventory*, report no. 1, 2nd ed. (Ottawa).

Gerlach, Arch C., ed., 1972 *Proc. of the Conference on Land Use Information and Classification*, Dept. of Interior.

Hudson, G.D., 1936 The unit area method of land classification, *Annals AAG* 26, 99–112.

Paludan, C.T.N., 1970 Potential contributions of orbital earth resources data to urban planning, *Planning Challenges of the 70's in Space*, G.W. Morgenthaler, and R. Morra, eds. (Tarzana, Cal.), Am. Astronautical Soc., 373–82.

Shelton, R.L., and E.E. Hardy, 1971 The New York state land use and natural resources inventory, *Proc. 7th Int. Symposium on Remote Sensing of Environment*, U. Michigan (Ann Arbor), 1571–5.

c1503
Land use maps for India: a consideration for coverage and classification in view of the recommendations of the world land use survey
B.K. ROY   *Office of the Registrar General, India*

It is more than two decades since the establishment of the World Land Use Survey Commission at the Lisbon Congress in 1949 of the International Geographical Union, which proposed a scheme for delineation of land use classification on a uniform scale for all the countries of the world. Since then, the mapping of land use has been adopted in many countries to evolve a sound approach to resource mapping. This paper outlines in brief the developments which have occurred so far in various countries in this direction. A detailed discussion in this paper has been presented regarding several operational problems of land use mapping and the problem of agricultural statistics and their correlation with land use maps. Further, the problem of land use classification as recommended by the World Land Use Survey Commission into the categories as available in India's agricultural/non-agricultural landscape has been tested with a view to presenting a sound compromise for international appreciation and measurement of the land use categories in India.

Various aspects of the problem of mapping have been presented and finally a proposal has been put forward to demarcate six main and thirty-six sub-categories of land use for coding and classification on large scale maps. Considering the present need of the country in the planning programs, a proposal for coverage of land use maps has been outlined which may satisfy all organizations connected with agricultural administration, land classification, productivity surveys, crop cutting surveys, etc. Besides, the need by metropolitan areas for maps has been assessed and a coverage has also been designed so that a complete picture of land for agricultural and non-agricultural uses may be obtained in India.

c1504
Against land classification
A. YOUNG   *University of East Anglia, UK*

Most natural resource surveys and land development plans include a land classification map. This is intended as a link between the natural scientists (e.g., soil surveyors, agroclimatologists) who survey the resources of the physical environment, and the social scientists (e.g., economists, planners) who seek to organize the utilization of these resources.

I need not repeat existing reviews of classification systems (e.g., Mabbutt 1968; Robertson et al. 1968), but shall only distinguish two main approaches. The first, land classification *sensu stricto*, is simply a regional division of the physical environment. The land systems method (Brink et al. 1966) is the most widely used of this type. The mapping units are defined in terms of the environment only, e.g., 'Gently undulating plain, Basement Complex rocks, ferallitic soils, savanna woodland.' The second approach is land capability classification, in which the suitability of the land for one or more types of utilization is assessed. This may either take the form of a single ordered grading, e.g., the Classes I–VIII of the USDA system; or the land may be rated according to its suitability for various uses, producing mapping units described as, e.g., 'Potential for arable crops, high; tree crops, moderate; improved pasture, moderate; irrigated padi, low.'

Three sources of information are used in compiling a land capability map. First, environmental conditions, either obtained from land classification maps of the first type or derived directly from data on climate, landforms, soils, etc. Secondly, technological factors, derived for example from agronomy or engineering. In the case of agricultural land classification, the requirements are commonly stated in the form of 'limitations,' e.g., of soil depth, acidity, liability to waterlogging. Thirdly there are considerations of economics. The extent to which these are taken into account is variable, a matter of controversy – and usually ill-defined. They certainly cannot be ignored entirely; to adapt a well-known geographical example, it is on both technical and economic, as well as environmental, grounds that the banana-growing potential of the North Pole is classified as 'low.'

What are the supposed purposes of land classification, and why is there such a demand for it? First, to simplify the complexity, and categorize the continuous variability, of the environment. This is a justifiable aim, and serves a useful descriptive purpose. Problems arise because of the desirability that planning units should be spatially contiguous, and the fact that areas of uniform environment frequently are not (Young 1969). A second purpose is to extract from the multiplicity of environmental information those parts which are relevant to the user, e.g., slope angle for soil conservation, bearing ratio for road construction. One can envisage two groups of data: 'all that we can find out' as supplied by the scientists, and 'all that we would like to know' as specified by the planners. Land capability classification is based on the overlap between the two groups. In a perfect system this overlap would be 100 per cent, but it is easy to demonstrate that this is not at present attainable: the economist wants to know crop yields – and the soil surveyor cannot tell him.

But I believe there is a third 'purpose,' and that is as an attempted link between otherwise non-communicating groups of (to use the FAO term) 'experts.' To the natural scientist, the land capability map is the end of the road: an uninteresting, but not time-consuming, addition to the field survey. The economist finds in the map a means of extracting the information he needs (or, rather, the best he can get), uncluttered by discussions of wilting point, dystrochrepts, and the Gondwana surface.

Objections to this approach are first, that there is inevitably a loss of information; the scientist does not know precisely what the users want, nor can the latter specify their requirements without knowing the nature of the information available and the terms by which it is described. Secondly, it leads to a lack of flexibility. Economists and planners experiment with alternative development plans, involving different land use patterns; but they are faced with one, unvarying, land capability map. Indeed, the personnel who constructed the map may have already moved on to another project. It has been convincingly demonstrated (Gibbons 1961) that different land uses have different suitability criteria; this applies not only between different types of use (e.g., a good agricultural soil may make a poor road foundation), but also between two crops of similar type (e.g., rubber and oil-palm, or pine and spruce). Technological, economic, and even political changes can rapidly alter the desirability of different land uses. Moreover, the relation between environmental and economic factors is not a one-way flow, but involves mutual interaction (Young 1968). In such circumstances it is better to have the basic environmental information clearly presented than to rely on the monolithic, and partly obsolete, basis of a land capability map.

It will be apparent that my opposition is not so much to land classification as to the attitude of mind that it engenders. Both land classification and land capability classification have their uses, the former as a first step in regionalization of the environment, the latter in linking environmental information with technological and economic considerations. What I am against is using such classifications as a cover for mutual ignorance. The soil surveyor who knows nothing of cost-benefit analysis, or even agronomy, cannot direct his observations in such a way as to supply the information sought from him. The economist unaware of the, admittedly confusing, nomenclature of tropical soil types cannot form a valid assessment of the reliability of the land use recommendations with which he is presented. An extreme case of confusion arises when an area is classified as 'marginal' (or even 'sub-marginal'). This term has a technical meaning in economics; but it is doubtful the compilers of land classification who employ it intend this meaning – if, indeed, they know what they intend at all.

Brink, A.B., et al., 1966   *Report of the working group on land classification and data storage*, MEXE Report no. 940 (Christchurch, Hants, UK).
Gibbons, F.R., 1961   Some misconceptions about what soil surveys can do, *J. Soil Sci.*, 12, 96–100.
Mabbutt, J.A., 1968   Review of concepts of land classification, in *Land evaluation*, ed. G.A. Stewart (Macmillan of Australia), 11–28.
Robertson, V.C., et al., 1968   The assessment of land quality for primary production, in *Land evaluation*, ed. G. A. Stewart (Macmillan of Australia), 88–103.
Young, A., 1968   Natural resource survey in Malawi: some considerations of the regional method in environmental description, in *Environment and land use in Africa*, eds. M.F. Thomas and G.W. Whittington (London), 355–84.
Young, A., 1969   Natural resource surveys for land development in the tropics, *Geog.* 53, 229–48.

Geography of Transport
Géographie des transports

c1601
Recreational boat trips on the Trent-Severn Waterway, Ontario, Canada
F.M. HELLEINER   *Trent University, Canada*

The Trent–Severn Waterway is a series of lakes and rivers extending from Georgian Bay to Lake Ontario. Navigation along its entire length, by boats with a maximum draught of 1.22 metres, has been made possible through a series of locks and a marine railway. Almost 100 per cent of the boats using these navigational facilities are classed as recreational boats.

At each of 42 points on the Waterway, where navigation is interrupted by a lock or other facility, the Ministry of Transport of the Government of Canada collects data concerning every boat that passes through that point. The data collection points are not uniformly spaced along the Waterway, but most of the major lakes and river segments are indeed separated by such points. Since each boat is identified in the data by its registration number, and since boat movements are differentiated into upbound and downbound traffic, and since the date and time of passage are also recorded, it is possible to trace the course of any given boat through all of the recording stations at which it appears. The author acquired these data for the 1969 boating season (16 May to 15 October) and, after correcting the noticeable errors (about 10 per cent), he extracted a 25 per cent stratified sample of the boats. He then extrapolated from this sample a total of 56,675 boat trips.

A 'trip' is defined in this study as a chronological progression through a sequence of adjoining lakes involving no reversal of direction. Where a reversal of direction occurs, a second trip is said to begin. Thus, a round trip from a given point to another point and back consists of two trips rather than one. The origins and destinations of trips are generalized rather than specific. From the available data, it is impossible to pinpoint these locations any more precisely than at a particular lake or river segment.

A study of the lengths of boat trips reveals some interesting findings. Less than half of one per cent of the recorded trips (256) travel the entire length of the Waterway between Lake Ontario and Georgian Bay, in either direction. Unlike traffic on a commercial waterway, where economies are achieved only on the longer trips, recreational boat traffic consists predominantly of short trips. Over 60 per cent of the trips in this study (35,714) involve passage through only one lock station. The mean trip length involves passage through 2.84 recording stations. For large boats (over 12.19m in length) the mean is 5.47; for medium-sized boats (over 5.49m and up to 12.19m in length) the mean is 4.58; and for small boats (up to 5.49m in length) the mean is 2.03. Inasmuch as many trips are known to occur without passing any of the recording stations, the actual trip lengths are really even shorter, on average, than those mentioned here.

Boat trips on the Trent-Severn Waterway exhibit no strong directional bias. Many of them are, in fact, return trips, i.e., pairs of reciprocal trips between pairs of points. At any given recording point, there may be far more trips bound in one direction than in the other direction, but the number of incoming trips from a given direction is in most cases roughly equivalent to the number of outgoing trips in the same direction. This feature, which the author calls 'directional balance,' contrasts markedly with the patterns of cargo flow on many of the world's commercial waterways, where a pronounced directional bias is often the most obvious characteristic.

There are relatively few lakes on the Trent-Severn Waterway which serve as major terminal points for trips. Sturgeon Lake and the adjoining Pigeon Lake/Buckhorn Lake/Chemung Lake group together account for 20 per cent of the destinations. Lake Simcoe, Balsam Lake, Cameron Lake, Sparrow Lake, Georgian Bay, Gloucester Pool, and Stony/Clear Lakes each account for between 5 per cent and 10 per cent of the total. This concentration of destinations reflects a similar concentration of origins which, when considered in conjunction with the short trip lengths, results in a highly localized pattern of boat traffic.

Nystuen and Dacey have shown how a graph theoretic analysis of flow phenomena may be applied to the problem of establishing functional regions (Nystuen and Dacey 1961, 29–42). The writer has applied their methodology to the analysis of boat trips in the Trent-Severn Waterway. The result is the identification of seven sets of lakes and river segments having strong dynamic links with neighbouring lakes and river segments. These seven sets of lakes are spatially discrete in most cases, and are classic cases of functional or nodal regions. Between them are parts of the Waterway where traffic flows are intermediate in orientation between the neighbouring regions. These areas are termed 'cols,' following the terminology used by Wallace in a different context (Wallace 1958, 360).

In summary, the peculiar properties of recreational boat trips on the Trent-Severn Waterway (short trips and directional balance) give rise to a set of functional regions separated by 'cols' of lower traffic density. It is suggested that these features are typical of waterways where the recreational function is the dominant one.

Ministry of Transport, Canada; Trent University.

Canada Department of Transport, 1951 *Trent Canal System, Improved Natural Waterway Connecting Lake Ontario and Georgian Bay* (Ottawa).

Canada-Ontario Rideau-Trent-Severn Study Committee, May 1971 *The Rideau Trent Severn Yesterday Today Tomorrow*, Report of the Canada-Ontario Rideau-Trent-Severn Study Committee (Toronto).

Nystuen, J.D., and M.F. Dacey, 1961 A graph theory interpretation of nodal regions, *Papers and Proceedings of the Regional Sci. Ass.* 7, 29–42.

Wallace, W.H., 1958 Railroad traffic densities and patterns, *Ann. Assoc. Am. Geog.* 48, no. 4, 352–74.

c1602
Weekend tourism as an indicator of the process of urbanisation
JÖRG MAIER *University of Munich, West Germany*

In order to place the following explanations into their scientific context, it is necessary to mention some aspects of the methodological scope in which our investigations are laid out: (1) According to the conception of social geography we try to explain spatial phenomena of our society from the combination of various basic functions of human existence (Ruppert and Schaffer 1969). (2) The importance of leisure-time and its spatial effects is seen in close relation with the respective social situation (Ruppert and Maier 1970). (3) By urbanization we understand both the state of high intensity as well as the process of distribution of urban patterns of behaviour into rural regions (Paesler 1970).

Weekend tourism as one aspect of the tendencies of urbanization brings about a change both in the appearance of places and settlements and in the social and economic structure of rural areas. By weekend tourism those kinds of short-time recreation are meant which from a time point of view extend over the weekend, i.e., recreation lasting half a day or a day. As a determinant of decision we are not so much making use of the metric distance, which in the course of proceeding transport development and growing motorization has lost some of its significance. We rather apply the temporal limitation of leisure time (Ruppert and Maier 1969).

The few studies on hand so far do not yet allow the formation of theories, but by means of several theses initial points for a model for consideration may be found.

As far as the extent of weekend tourism is concerned, a census among owners of individual cars in Bavaria carried out in 1969 showed that 25 per cent of the miles driven per person fell to this share of travel. This figure was 44 per cent with regard to civil servants, which was well above average compared with only 10 per cent for independent professions. Besides these differentiations by specific groups, we know from the analysis of vacation transport that the intensity of recreation in urban areas is above average, whereas in rural districts it rates below average.

Does this pattern of behaviour also apply to weekend tourism? Our investigations in the surroundings of Munich proved that weekend tourism is not as distinct in suburbs as it is in cities, where between 25 and 35 per cent of the population are concerned. The different patterns of structure became even more evident, however, from various studies in rural areas. 27 per cent of the population of Garmisch-Partenkirchen, being an increasingly urbanized centre of tourism, participated regularly in weekend tourism, whereas in very agricultural areas like Mainburg (centre of hop-culture) or Griesbach (central place on inferior level in Lower Bavaria) this figure reached only 10 per cent. Even if weekend tourism does, in the individual case, depend on leisure-time at the disposal of persons seeking recreation, financial possibilities for realization of their leisure-time demands, means of transportation that may be used, weekend recreation areas with their infra-structural establishments which are already existing and may be chosen alternatively, as well as on the preferences of social groups, the following general statement may be made:

Participation in weekend tourism is more strongly developed in cities than it is in suburbs, and it is greater in urban than in rural areas. Since the development of a professionally specialized division of labour and since a diversification of professions is one of the characteristics of the urbanization process, weekend tourism is also being influenced by the patterns of behaviour of the various social groups. In this connection it shows that with passing from the basic social group to the middle group there is an increase both in the frequency and in the distance of weekend tourism. Whereas in metropolitan areas the upper social group often prefers short-time recreation in urban surroundings, this group represents in rural areas one of the most essential participants in weekend tourism.

Different leisure-time behaviour of the social groups is having an influence on spatial patterns of weekend tourism in urban and rural areas. Leisure-time habits are being influenced not only by the social economic structure of the population but also by a series of factors related to behaviour. Thus, for example, we were able to state for Munich that those groups of persons that have moved to Munich only in recent years show a different behaviour in weekend tourism from those that have been residents for a longer period of time.

Groups of persons that newly move to a place prefer recreation areas of inter-regional image value. Moreover, it becomes clear that the regional scattering of weekend tourism depends decisively on differentiations of the natural and cultural geographic potential of landscape as well as on leisure-time habits of the demanding persons. They contribute to the formation of seasonal areas of attraction. Because of the diversification of determinants it is not admissible to fix the distance of journey of weekend tourism to a certain radius. The linear concentration to few roads or highways respectively gives weekend tourism a radiate extension along these lines of travel.

In addition to the potential of landscape offered, the distance of journey is most of all being defined and changed by the influence of urban patterns of behaviour. In accordance therewith three zones of destination are to be found in urban areas: at the edge of urban areas, in the closer surroundings (20–50km), and in the more distant range (80–150km). Compared therewith there are only two zones in rural areas: up to 20km and above 100km.

The structure of the means of transportation used is playing an important part in the different spatial effects. Although, according to all recent investigations, individual cars are ranking far on top, their share in metropolitan areas is lower than in rural areas. Whereas, for example, in Munich 50 per cent of all persons seeking weekend recreation are using individual cars, the share was 72 per cent in a central place of inferior level like Griesbach and 95 per cent in agrarian places in the district of Mainburg.

Because of the better transport service and the efficiency of public means of transportation in urban areas the share of individual cars is more distinct in rural areas. By means of these theses an attempt was made to discuss some aspects of inductive formation of models, whereby the main problem is to be found in the heterogeneity of phenomena of weekend tourism as well as in the diversification of preconditions and parameters, respectively.

Bearing all that in mind, it must however not be forgotten that because of the process of urbanization the rural areas, especially the recreation areas, are subject to increasing changes. This not only transfers the dwelling function from urban to recreation areas in a constantly growing way. The change in the consciousness of the population also contributes to the reduction of differences between urban and rural areas by accepting urban ways of life and economy. For this reason, weekend tourism, being one of the indicators of this urbanization process, must accomplish the task to draw attention to regional differences of intensity in the degree of urbanization.

Paesler, R., 1970   Die Urbanisierung als raumgestaltender und raumbeeinflussender Prozess, unpub. essay (München).

Ruppert, K., and J. Maier, 1969   'Der Naherholungsverkehr einer Grossstadtbevölkerung, dargestellt am Beispiel Münchens,' in *Information, Bonn–Bad Godesberg*, 2 (19 Jg.), 23–46.

– 1970   'Zur Geographie des Freizeitverhaltens,' 6 of *Münchner Studien zur Sozial- und Wirtschaftsgeographie* (Kallmünz/Regensburg).

Ruppert, K., and F. Schaffer, 1969   'Zur Konzeption der Sozialgeographie,' in *Geographische Rundschau* 6 (21. Jg.), 205–14.

c1603
A museum of transportation geography
J.H. APPLETON   *University of Hull, UK*

The term 'Museum of Transport History' suggests a repository for the storage, study, and display of the paraphernalia of transport. Such museums have proved invaluable for transport historians as the complement of their literary and documentary sources, but there is a limit to the kind of relics they can house, and transport geographers may regret that relics of most interest to them – often the handiwork of civil rather than mechanical engineers – tend to be too large for housing in orthodox museums.

The recent recognition of 'industrial archaeology' has prompted various attempts to solve this problem, such as the setting up of special sites in which large structures like bridges, furnaces, and buildings can be reassembled for exhibition (Atkinson 1964). In England, for instance, such sites are being developed in County Durham and Shropshire. Such 'open-air museums' have three advantages. First, they may permit the preservation of structures which would otherwise be lost; secondly, they may enable several such exhibits to be seen more conveniently and economically than if they remained on widely separated sites; and thirdly they may facilitate management.

Against these advantages, however, must be set one serious objection, the destruction of context. While a locomotive may be studied as satisfactorily in a museum as in the works where it was built, a structure of civil engineering may lose intelligibility when moved from its original environment. For example, at Blist's Hill, Shropshire, the Shropshire Canal passed immediately above the ironworks. In developing this site as a museum the proposal is to construct to the derelict ironworks a short canal branch into which can be fitted Telford's Longden aqueduct – one of the earliest iron aqueducts in the world – and a lock of unusual if not unique design, both from the neighbouring derelict Shrewsbury Canal. That these structures should be saved and exhibited is admirable. But the great problem in constructing the Shropshire Canal in 1788 was that it traversed a small upland with no water supply adequate to operate locks, and only the use of inclined planes (of which the best surviving example, at Hay, is also within the museum site) enabled the canal to be built at all. The introduction of the lock, therefore, makes nonsense of the geographical arrangement and invites misapprehension of the engineering structures within the hydrological context.

When exhibits are too large to move, or when there are special grounds for retaining them in their proper context – and both these conditions frequently apply to the remains of transport systems – preservation *in situ* is the only solution (Rix 1964), and our museum of transportation geography can be nothing less than the entire area within which the exhibits were originally located. Although my examples come from Britain, they illustrate general principles.

Three major questions arise: (1) What does preservation involve? (2) By what criteria should selection be made? (3) What organizational measures are required?

(1) The first question demands a simple classification of exhibits by their functions and conditions:

(*a*) Routes still in use for their original purpose or its modern equivalent. Preservation of the line of route is assured with no additional expense, but its visual appearance may well have changed with technological advance. The English road system includes many kilometres of Roman road, while many early railways are incorporated in the network of British railways. Stephenson's London & Birmingham Railway was one of the first trunk lines in the world. Electrification is the last of many innovations which have altered its appearance, but the tunnels, cuttings, and embankments manifest a relationship to the land surface which still reflects the civil engineering practices of 1833.

(*b*) Routes still in use but not for their original purposes. Canals converted to railways over a century ago reflect physical characteristics, such as gradients, typical of canal rather than railway engineering. Railways in turn have been converted to other uses, such as roads, footpaths, or cycle-tracks (Appleton 1970).

(*c*) Remains of routes which are unused for any regular purpose. In Scotland much of General Wade's military road system of the seventeen-twenties and thirties can still be seen in fields, moors, and forests. Routes in this condition often have the advantage of displaying more accurately the true character of the original. Roman road surfaces, for instance, cannot be examined under a modern highway, whereas they may be excavated and studied in a field. But they are more vulnerable than routes which have been preserved by usage, and often only a positive policy involving financial expense can save them for posterity.

(2) To preserve everything is as absurd as to preserve nothing. There must be selection, but on what grounds should selection be made? Mere antiquity is not enough. (The nineteenth-century 'atmospheric' railway was far rarer than the Roman road.) There must be clear identification of objectives and rational assessment of the purpose which could justify preservation in each case. I suggest five criteria:

(*a*) Historical association. Connection with an event, process, or person may bestow value beyond the intrinsic properties of the place concerned. The railway at Rainhill, Lancashire, is not visually exceptional, but it was the site of the 1829 trials which did much to establish locomotive traction on railways.

(*b*) Singularity. This differs from (*a*) in that

it indicates an unusual or unique property inherent in the object itself. Telford's aqueduct completed in 1805 at Pontcysyllte – a cast-iron trough 307m long and 38m above the River Dee – is not only a remarkable structure for its period; in its geographical context it has much to teach about the relationship of canals to watersheds and drainage systems. This leads on to

(c) Explanation. The clarification of some idea or process which is otherwise difficult to communicate. The inclined planes of the Shropshire Canal, already mentioned, were simple in principle but more complex in detail. The Hay plane, studied in conjunction with Telford's diagrams, clearly demonstrates the operation of the device.

(d) Utility. Preservation of railway routes by conversion to other uses has already been mentioned. Mineral railways and canals have been reprieved as tourist attractions; harbours as marinas. A special kind of utilization is

(e) Amenity. Disused routes are often scenically attractive even in dereliction. Railway embankments properly planted or canal fragments converted to water gardens can escape total elimination and contribute to the quality of contemporary life.

Exceptional exhibits may qualify by more than one criterion. The Iron Bridge, Shropshire, could claim distinction under all five. (a) The first iron bridge, (b) built according to principles of timber rather than metal construction, (c) which are lucidly explained in the structure itself, (d) still provides a useful river crossing, though now only for pedestrians, and (e) in its picturesque gorge environment makes an aesthetically satisfying composition.

(3) My third major question, that of organizational measures, raises problems likely to vary in different countries. In Britain responsibility is shared by individuals, voluntary societies, industry, and local and central government; but any policy involves planning and finance – whatever the source – and priorities of competing projects can be properly determined only if acceptable criteria are established.

Appleton, J.H., 1970 *Disused railways in the countryside of England and Wales* (London).
Atkinson, F., 1964 An open-air museum for the north-east, *Industrial archaeology*, 1, 3–8.
Rix, M.M., 1964 A proposal to establish national parks of industrial archaeology, *Industrial Archaeology*, 1, 184–92.

c1604
Planification des transports et géographie: l'exemple de l'équipement maritime de la Basse-Loire, France
ANDRÉ C. VIGARIE *Université de Nantes, France*

Depuis 1966, la politique française d'Aménagement du Territoire repose sur la constitution ou le développement des métropoles d'équilibre: ce sont des villes maîtresses que l'on dote des fonctions et des équipements grâce auxquels elles pourront freiner l'exode provincial vers Paris, et devenir les centres d'une armature urbaine régionale réorganisée. Sur la façade Atlantique, et pour une partie essentielle de l'ouest de la France, Nantes et St-Nazaire, sur l'estuaire de la Loire, ont été conjointement choisies pour jouer ce rôle; un Organisme d'Etude pour l'Aménagement de l'Aire Métropolitaine (OREAM) a été constitué pour préparer la planification de la future métropole. L'auteur de la présente communication en a été le conseiller technique plus spécialement chargé des questions portuaires. Ce sont les résultats de sa participation de géographe à une équipe pluridisciplinaire pour l'organisation de l'espace régional qui sont donnés ci-dessous, du seul point de vue de la géographie maritime.

Les buts recherchés par la création de la nouvelle métropole sont multiples. Elle doit posséder un poids démographique, économique, fonctionnel suffisant pour contrebalancer au moins partiellement l'attraction parisienne; il lui faut étendre son aire de rayonnement à un ample secteur de l'ouest français, l'ouest armoricain pour l'essentiel, c'est-à-dire constituer une large région d'influence urbaine, qui est aujourd'hui partagée entre plusieurs capitales provinciales séculaires; elle doit tenir compte des perspectives générales de l'aménagement du territoire français, dans lesquelles sept autres métropoles sont en cours de constitution, qui ne peuvent avoir toutes les mêmes équipements de niveau supérieur, par crainte de double emploi et de dispersion, et parce que certains ne peuvent être regroupés qu'à Paris; enfin elle doit pouvoir s'intégrer dans le contexte d'une Europe du Marché Commun sous peine de priver l'ouest Atlantique des bénéfices du développement économique vigoureux que suscite l'européanisation.

La promotion de Nantes–St-Nazaire au rang de ville-maîtresse est donc un problème difficile; cette difficulté se trouve accrue d'abord par le fait que les diverses capitales historiques – Rennes, Angers, Le Mans etc. – pourraient, en divers domaines, prétendre jouer un rôle directeur, et sont par conséquent des concurrents permanents; ensuite parce que la nouvelle métropole doit, par décision gouvernementale, être bicéphale et s'appuyer sur les deux agglomérations inégales de Nantes (394,000 hab.) et de St-Nazaire (111,000 hab.) qui délimitent à l'amont et à l'aval, l'estuaire de la Loire, lequel les sépare sur 50km.

Cependant, c'est précisément cet estuaire qui va constituer l'une des principales bases du programme de planification: en dehors d'autres avantages économiques (développement industriel et tertiaire plus poussé que dans les cités rivales), il donne à la future métropole de l'ouest à la fois une unité fonctionnelle et une activité spécifique: celle de la vie portuaire et océanique.

Actuellement, la Basse-Loire a un trafic maritime de 12 millions de tonnes, réparti sur les trois ports de Nantes (industries et *general cargo*), St-Nazaire (trafic industriel, mais encore très limité), et Donges (port pétrolier: 8.5 millions de tonnes). La circulation des tankers dont le tonnage maximum atteint à peine 70,000 Tdw. (navires spécialement conçus pour un tirant d'eau de 11.50m), fournit donc le plus gros volume de marchandises (83 pour cent); le trafic de Nantes est davantage générateur de richesses, mais le chenal y est limité à 8.25m (9m dans quelques années). L'ensemble ne dispose pas d'arrière-pays véritable: ce sont les industries locales qui provoquent, dans la proportion de plus de 90 pour cent, l'activité de ces ports.

A partir de cette situation qui révèle une puissance actuellement très moyenne, le programme d'aménagement a retenu le projet de constitution d'un grand port capable d'exercer une vigoureuse incitation de développement local et régional.

Bien évidemment, le géographe n'intervient pas seul: il soumet ses recherches et ses conclusions à la critique très sévère de l'équipe pluridisciplinaire, qui peut les adapter et les compléter.

Quelle orientation de pensée a marqué ces recherches et conclusions? Au point de départ, se situe une interprétation géographique large des faits d'économie portuaire, c'est-à-dire une analyse, sous l'angle à la fois actuel et prospectif, de la politique française, peu propice aux ports atlantiques, mais aussi de la conjoncture européenne – et alors, les villes maritimes ici étudiées sont évidemment écrasées par l'énormité des compétiteurs tels que Rotterdam, Londres, Anvers – pour déceler les domaines dans lesquels une promotion de la Basse-Loire était possible. Il a fallu le faire en tenant compte de la révolution des transports océaniques, marquée par le gigantisme des transporteurs de vracs et l'étroite spécialisation des navires de marchandises générales (cf. containérisation). Corrélativement, une minutieuse étude des caractères nautiques de l'estuaire et de ses abords, menée conjointement avec les services techniques du Port Autonome qui gère les ports locaux, a permis de déceler les limites matérielles ou économiques que l'on devait accepter ou assigner au recreusement des accès, c'est-à-dire au tonnage des plus grands navires que l'on voulait recevoir; cette étude a été orientée en particulier vers l'hydrologie et la sédimentologie estuairiennes, et vers la recherche du paléolit würmien de la Loire, afin de l'utiliser au mieux par des dragages. Ces analyses techniques et économiques ont abouti à l'élaboration du projet suivant: création d'un port poly-industriel accessible à des minéraliers de 120,000 Tdw. (environ 15m de calaison), dont les équipements sont répartis dans le secteur aval de l'estuaire, autour de Donges (4000 hectares disponibles selon une programmation prévue en 5 étapes). La pièce maîtresse, la première réalisée, en est la Zone Industrielle Portuaire de Montoir (5000m de front d'accostage direct en Loire, 1000 hectares de terrains pour usines à l'intérieur, avec une darse initialement prévue pour la desservir de façon optimale, et creusée à la cote −15). Là, 16,000 emplois directs sont possibles, provoquant un afflux de population de quelque 50,000 personnes vers St-Nazaire; avec les emplois indirects, c'est le développement démographique et tertiaire de cette ville qui est obtenu jusqu'à un niveau compatible avec la fonction métropolitaine qui lui est conjointement assignée; et des virtualités de développement diversifié sont offertes par ce nouveau port à toute la région, ce qui permet de concevoir la constitution d'un hinterland véritable.

Entre ce projet et la réalisation se place l'austère cheminement du souhaitable au possible. Les contacts avec le Port Autonome de Nantes–St-Nazaire, qui doit rester le maître d'œuvre pour les nouveaux équipements portuaires, ont été essentiels pour confronter son point de vue de gestionnaire avec celui, plus large, de la prospective et de l'aménagement régional. Il a fallu faire accepter les diverses modalités des installations prévues par les autorités locales, telles les Chambre de Commerce, par l'opinion publique, et par les autorités nationales: Délégation à l'Aménagement du Territoire (DATAR), ministères, Conseil des Ministres. Bien des divergences d'intérêts ou de conceptions ont dû être éliminées, ce qui a été souvent l'œuvre de l'équipe de recherche dans son ensemble, avant que, par décision gouvernementale de septembre 1970, le projet soit enfin officiellement retenu: seule la darse de Montoir, sans être rejetée, était momentanément laissée dans l'ombre malgré sa grande importance. Les premiers quais à grande profondeur sont déjà construits: la réalisation, lentement, est en cours.

Dans cette expérience, quel fut le rôle du géographe? Au sein de l'équipe d'aménagement dont il est difficile de démêler l'apport, c'est d'abord un rôle de conception: au point de départ, il y a une appréhension géographique prospective de la vie portuaire internationale et de ses impacts locaux; puis intervient l'étude technique des problèmes: morphologie dynamique de l'estuaire, conditions de navigation, etc.; cela sous-entend une spécialisation dans le domaine de la géographie des transports océaniques. Mais à côté de ce travail, de la minutie et de la sûreté duquel dépendait l'efficacité des arguments présentés en discussions, les efforts pour convaincre les autorités et administrations responsables apparaissent aussi importants que le rôle de conception du projet présenté; il a fallu modifier les opinions courantes, parfois démolir des convictions locales mal fondées, s'appuyer sur les milieux locaux dont l'expérience pratique a permis de perfectionner le projet, pour faire accepter les idées essentielles par les milieux responsables nationaux.

C'est là toute une action psychologique et sociologique pour laquelle le géographe – quant à la part qu'il y a prise – n'est pas toujours préparé, et qui lui donne profondément, et de multiples façons, le sens de ses responsabilités, à la fois comme conseiller technique et comme membre d'une collectivité dont il contribue à modeler l'existence à court et à long terme.

c1605
Problèmes de sources dans l'étude régionale des transports au Canada
JEAN CERMAKIAN   *Université du Québec à Trois-Rivières, Canada*

La recherche dans le domaine de la géographie régionale des transports au Canada se heurte à des obstacles sérieux sinon insurmontables. Les difficultés dans ce sens sont sans doute moins grandes pour certains moyens de transport que pour d'autres, et la qualité de la documentation statistique peut varier d'une province à l'autre. Mais il reste que dans un pays connu pour la qualité et la variété de ses statistiques démographiques, agricoles, et manufacturières, les données concernant les transports font figure de parent pauvre.

Les raisons de cette situation sont multiples. Il y a tout d'abord le fait que Statistique Canada (l'ancien Bureau Fédéral de la Statistique) compile et publie une somme énorme de données statistiques de toutes sortes, mais le plus souvent à l'échelle du pays tout entier ou parfois à l'échelle de chaque province. Le mandat de Statistique Canada est d'ailleurs de tenir constamment à jour les séries statistiques qui reflètent l'ensemble de la réalité socio-économique canadienne et de procéder périodiquement à des recensements, et non pas d'entreprendre des études détaillées au niveau d'une région provinciale, d'un comté, ou d'une municipalité. D'autre part, il y a le problème du partage des juridictions entre les gouvernements fédéral et provinciaux en matière de transports: la circulation aérienne, ferroviaire, fluviale, lacustre et maritime dépend d'un certain nombre d'agences fédérales (Ministère des Transports, Commission Canadienne des Transports, Conseil des Ports Nationaux, Administration de la Voie Maritime du Saint-Laurent), alors que la circulation routière (construction et entretien, législation et règlementation de la circulation des automobiles et camions, code de la route) est entièrement sous la responsabilité des gouvernements provinciaux individuels (Ministères de la Voirie et des Transports de chaque province). Ce partage des responsabilités entraîne une multiplication des sources de données statistiques et souvent un manque de comparabilité entre les données présentées par ces sources. Enfin, pour diverses raisons, certaines données sont confidentielles ou soumises à la loi du 'secret statistique' pour protéger les intérêts de certaines compagnies de transport.

Les données publiées par Statistique Canada et susceptibles de nous intéresser proviennent de la Division des Transports et Services d'utilité publique d'une part, du Centre des Statistiques de l'Aviation d'autre part. La Division a pour responsabilité de compiler et de publier les statistiques nationales du transport par eau, ferroviaire, et routier. Pour chacun de ces modes de transport, la Division publie un 'bulletin de service' qui contient des données fort intéressantes, le plus souvent à l'échelle nationale mais, pour la navigation, par port. D'autre part, les statistiques mensuelles et annuelles du transport ferroviaire et routier sont encore une fois à l'échelle du Canada tout entier ou dans certains cas par province ou par compagnie ferroviaire. Pour des statistiques plus détaillées du transport ferroviaire, le responsable de la Division nous a renvoyé aux services de recherche des deux principales compagnies (Canadien National et Canadien Pacifique). Pour le transport routier, la Division a lancé, au printemps 1971, une enquête sur les transports commerciaux par camion entre plusieurs grandes villes du Canada. Le succès d'une telle opération dépendra cependant de la bonne volonté des transporteurs et des gouvernements provinciaux, souverains en matière de transport routier. Enfin, c'est dans le transport par eau que la Division publie les statistiques les plus détaillées: depuis 1961, la publication *Shipping Report/Transport maritime* donne chaque année le trafic d'entrée et de sortie de tous les ports intérieurs et maritimes du Canada, les origines et destinations précises des huit produits principaux du commerce maritime, et le trafic détaillé, par origines et destinations, des ports canadiens les plus importants. Depuis 1968, le recours à l'ordinateur permet d'étendre cette statistique à presque tous les ports du Canada. Voilà donc un précieux outil de travail. De même, le Centre des Statistiques de l'Aviation publie les données par origine et destination du trafic voyageurs entre les différents aéroports canadiens, et entre ces derniers et ceux des Etats-Unis, à côté d'autres données moins utiles.

Les statistiques des transports publiées par les gouvernements provinciaux sont bien moins abondantes et détaillées que celles de Statistique Canada. Pour prendre l'exemple du Québec, le Bureau de la Statistique du Québec ne publie aucune donnée sur les transports, sauf dans son *Annuaire du Québec* dont les séries statistiques proviennent toujours d'autres sources. Par contre, le Ministère des Transports compile le nombre des véhicules, par catégorie, pour chaque comté et municipalité, d'après les immatriculations. D'autre part, le Ministère de la Voirie publie un recensement annuel de la circulation pour toutes les routes numérotées et autoroutes de la province. Dans son rapport annuel, ce même ministère publie un inventaire complet de la longueur des routes et des ponts qu'il doit entretenir. Ces renseignements sont cependant publiés avec un certain retard et restent souvent inédits au niveau de la municipalité ou même parfois du comté.

Si les statistiques de la voirie sont parfois fragmentaires, les données régionales sur le transport par rail sont à peu près inexistantes. Les deux principaux réseaux ne publient guère ces données qu'elles estiment confidentielles. Tout au plus est-il possible, grâce à une publication annuelle de la Commission

Canadienne des Transports, d'avoir une idée du trafic de marchandises par wagons complets entre les trois grandes 'régions' ferroviaires du pays: la région de l'Atlantique, qui comprend les quatre provinces atlantiques et le Québec à l'est de la ville de Lévis, la région de l'est, entre Lévis et Armstrong (Ontario) (près de la tête du lac Supérieur), et la région de l'ouest, à l'ouest d'Armstrong. L'étendue de ces régions est bien entendu trop vaste pour permettre une analyse tant soit peu sérieuse des flux ferroviaires intra- ou interrégionaux. Le même problème de confidentialité des données existe dans le cas de certaines compagnies aériennes possédant un droit de desserte exclusif dans le nord du pays.

Que signifient ces limitations au niveau des transports régionaux? Dans le centre du Québec méridional (Mauricie et Bois-Francs), nous possédons d'excellentes statistiques origine-destination pour le port de Trois-Rivières de 1961 à 1970, et pour l'aéroport de Trois-Rivières les deux ou trois dernières années. Les données du trafic routier sont disponibles pour les axes principaux (numérotés), ainsi que le kilométrage des routes par comté, mais il nous manque la différenciation entre voitures particulières et véhicules commerciaux. Enfin, pour le trafic ferroviaire, nous ne disposons que de quelques données, à savoir le nombre de wagons chargés et déchargés dans les différentes gares du Canadien Pacifique en 1970, mais sans détail aucun sur le nombre de voyageurs ou le tonnage des marchandises. Dans une étape ultérieure, il sera sans doute possible d'obtenir certaines de ces données, sous forme manuscrite, aux sièges des réseaux à Montréal.

c1606
A tentative approach to a holistic conception of transportation geography
M.E. ELIOT HURST   *Simon Fraser University, Canada*

In many senses, transportation is a measure of the relations between areas, and is essentially therefore geographical, in that it is involved with concepts like spatial interaction and areal association. The relations and connections between areas are frequently reflected in the character of transportation facilities and in the flow of traffic. This should not however be taken as a *carte blanche* to think of transportation as merely a network of conduits through which inanimate objects stochastically flow, nor as simply a matrix of origins and destinations representing particular points of input and output. As long ago as 1894 we were warned 'there can be no adequate theory of transportation which has regard only to some one aspect of its social function, as the economic aspect. That is not the only aspect, nor can one truly say that it is more important than the others. All are coordinate, equally indispensable to social progress' (Cooley 1894, 263). An early plea for systems analysis! 'The character of transportation as a whole and in detail, at any particular time, and throughout its history, is altogether determined by its interrelations with physical and social forces and conditions. To understand transportation means simply to analyze these interrelations' (Cooley 1894, 262). As one step towards a more realistic framework for transportation geography, the simplistic systems schema of figure 1 (not printed here) is suggested.

The *first stage* in any transportation system is the need or desire for interaction to take place; interaction arises because of the spatial separation of the means of satisfying those needs or desires. However, interaction, movement, transportation, are not ends to themselves; they are but part of a broader set of utilities influenced by individual and group life styles, aspirations, political motivations, and myriad other causes and counter causes. A more realistic thesis therefore is to assume that transportation is governed by the ways in which utilities are assigned, and by the ways in which the dimensions of space are conceptualised; these parameters can be summed up by the term *'movement space'* (Eliot Hurst 1969).

The basic structural components remain the modes, stocks, and networks (*stage two*). Movement is constrained by the channels of the network, the characteristics of particular modes, and the facilities of the fixed and mobile stocks (Garrison and Marble 1965). The mobile stocks, such as vehicles, provide the interface between the items being transported and the fixed stock of the network, such as the roadway or rail track. With some modes – pipelines, television – while there are no mobile stocks, there is still some interface between the 'goods' and the fixed stocks. The networks consist of nodes and links; each node or activity site corresponds to the demand or supply, each link to specific transportation channels. The links may be tangible and well defined (railroad) or may be relatively diffuse and intangible (radio, face-to-face contacts). Some nodes may be interchange points between links of the same mode (railyards, highway interchanges), whilst other nodes may be interchange points between links of different modes (the seaport, ferry, rail, or air terminal), where modes are complementary. Often, however, modes may compete for traffic over similar link paths.

This second point establishes the concern with movement of objects, people, or ideas; that networks exist to channel the movement; that modes and stocks interact over space and time, competing or complementing, but flowing through the channels in a variety of interacting paths.

The decision maker has open to him a range and variety of movement options (*stage three*). The individual, whether potential traveller, broadcaster, or shipper sees the transportation systems as essentially fixed; on the whole, he can choose his own particular routing and time schedule only through the system. The operators and policy makers, however, can establish the routes and schedules, pricing and reliability, and other factors which determine the 'level of service' (*stage four*). In addition such decision makers govern the types, number, and availability of stocks in the system; add new links and abandon old ones; improve operating characteristics – dredge rivers, widen highways; and at higher levels, regulate competition, assign operating rights and rates. This full set of systems operations is usually open to one agency or organisation only in socialist planned economies; in other types of economic organisation many individuals and agencies may be involved. To all, however, the types of options will differ from time to time – a specific trip decision can be implemented rapidly, but network or technological changes may take a considerable time to take place.

Before movement finally occurs a decision maker considers several characteristics, making up the level of service. Such consideration may include not just cost and travel time, but safety, comfort, reliability, indirect costs (e.g., total transfer costs), and frequency of service, covering a wide range of variables, from total trip and schedule times, direct transport charges, loading and warehousing costs, to the various accident probabilities, physical comfort of passengers, their privacy, general amenities, and aesthetic experiences. The level of service at which transportation is provided varies spatially and over time; it is therefore a function of the spatial characteristics of demand and supply, the characteristics of modes, stocks, and networks, and their operational milieux.

In response to demand, routes, systems operations, and level of service, a flow of persons, objects, ideas, moves through this system (*stage five*). Flows are a volumetric measure of the degree of interaction and the successful interrelationship of the other components. Flows vary in actual composition and in their time-space variations, as for example in the flow of traffic through a city during a weekday. There is a feedback effect to level of service, operating characteristics, routes, and ultimately to demand itself.

Finally, transport or movement decisions are influenced by a wide variety of directly related (*stage six*) and indirectly related factors (*stage seven*). The direct factors include general investment decisions to achieve the overall socio-economic development of an area or region; in urban areas they might include decisions to plan an integrated multi-mode system of public transit and freeways which would obviously have feedback effects throughout a region and its transport network; general investment decisions which affect supply and demand, and the general welfare of the people in a region; and there may be military decisions based on strategic mobility, to best achieve national military objectives through the rapid deployment of forces (e.g., the Interstate and Defence Highways System in the United States). Other transport-oriented developments include the introduction of basically new transport technologies, such as new modes or vehicles, new networks, new operating systems. All of these options can shift the workings of the previous five factors.

However, besides direct influences, there are a number of other external and often operatively non-transport factors which must also be considered, since they can affect the demands for, and use of, the transport system. National economic policies can affect the distribution of demand over space and time, and hence the demand for freight services; other effects will result from land use controls, the provision of public utilities, industrial logistic systems, differential regional growth policies, as well as the impact and the reaction of non-users of the system.

The common threads presented in this heuristic device underlie the great variety of interactions present in the socio-economic landscape. It is not an exhaustive scheme, but the principles are equally applicable to communications in general, urban transportation, intercity movements, regional and national interactions, and world trade flows. Hence the system reflects the relations between areas and underlies the geographical concepts of spatial interaction and areal association; at the same time such a schema brings together disaggregated components – modes, networks, stocks, flows, the reasons for interaction, and shows how they interact and operate within a holistic framework.

Cooley, C.H., 1894   The theory of transportation, Publications of the *Am. Econ. Assoc.* 9 (3), 225–370.
Eliot Hurst, M.E., 1969   The structure of movement and household travel behaviour, *Urban Studies* 6, no. 1, 70–82.
– 1972   *Transportation Geography: Comments and Readings* (New York).
Garrison, W.L., and D.F. Marble, 1965   A prolegomenon to the forecasting of transportation development, Northwestern U., Transportation Center, *Research Report.*

c1607
Intrinsic transportation and the Canadian north
KARL E. FRANCIS   *University of Toronto, Canada*

In terms of function we may consider transport systems as either intrinsic or extrinsic, depending on whether the system is primarily designed to serve the resident population or some outside interest. Thus the Roman roads, assuming their fundamental purpose was to serve Rome and especially to facilitate the movement of the Roman army, formed an extrinsic transport system. Virginia's Skyline Drive, since it serves mostly people from outside the Blue Ridge and is so designed, is also an extrinsic system. On the other hand, the Los Angeles freeways are distinctly intrinsic since they are designed for and serve the population of Los Angeles and, for a variety of reasons inherent in their design, are rather difficult for outsiders to use.

It follows that the relative proportion of extrinsic to intrinsic transport systems in any given area would reflect the degree to which that area is dominated by and used to the advantage of outsiders. Conceivably one might develop indices of 'intrinsivity' based, for example, on parameters of road patterns as they relate to population centres, indices which could be sensitive measures of local political power. The ratio of local to outside users of airports might similarly reflect local political or economic circumstances. Whatever profit may be had from this concept, this paper uses it only qualitatively to draw attention to one of the peculiar features of northern Canada, a predominance of official and scholarly interest in extrinsic transportation.

In the rather considerable literature on northern transportation and in the certainly not infrequent conferences dealing either wholly or partially with northern transportation there is an almost exclusive concern with extrinsic systems. Intrinsic transportation and indigenous requirements for transportation have either been ignored or presumably considered trivial or non-existent. While vast sums have been spent on efforts to develop vehicles and procedures for seismic operations such that they do not offend the aesthetic consciences of outside conservationists, the mechanized toboggan has become ubiquitous in the north (Francis 1971b) and the first modest effort has yet to be made even to recommend the most pressing modifications necessary to make these machines suitable and safe for northern use.

At Herschel Island for the benefit of extrinsic craft plying the Beaufort Sea there is an elaborate navigational aid installation operated through the entire shipping season. For the hundreds of small craft running up and down the lower Mackenzie River, all across the Delta and along a most extraordinarily hazardous coast, carrying men, women, and children about their ordinary affairs, there is neither navigational aid, suitable meteorological services, nor even the most primitive of back-up systems.

Perhaps the most spectacular demonstration of overbearing concern with extrinsic transportation in the North was the passages of the Manhattan; and this event will soon be overshadowed by the building of petroleum pipelines.

There are more subtle examples of the extrinsic bias in northern transportation. Winter roads might appear to be serving local needs. Yet at a conference last year I recall a winter road advocate citing the use of the roads by local hunters as one of the nuisance factors with which he has to contend. It was certainly made clear that serving local needs was not the primary function of the road.

There is frequent reference to the importance of air service to local communities in the north. Yet if one were to sort out the great mix of demand, potential demand, and service, clearly the best service would fall to outside demand or to transients from the south. If you want to take your annual leave 'back home' in Vancouver, there are two flights a day out of Inuvik, both jets. You can be home in a few hours. If you want to visit some of your family in Barrow, Alaska, good luck. It will take about the same time by scheduled aircraft as it would take to go by dogteam, and it will cost a small fortune. If you want to bring your furs into Inuvik from Old Crow, you may as well charter a plane.

One has only to look at a map of scheduled air routes in northern Canada to see dramatic evidence of the overwhelming extrinsic function of major air transportation. The routes tie all of the north firmly into packages anchored to one or another major southern city. Edmonton controls the Mackenzie River drainage, Vancouver gets the Yukon, Montreal has the eastern arctic, and so forth. There is no scheduled service between Whitehorse, the capital of the Yukon and Yellowknife, the capital of Northwest Territories. In fact, there is virtually no east-west routing.

It would be both futile and trivial to argue that expenditures for extrinsic transportation be reallocated to intrinsic needs. In fact, as is so frequently argued in support of even the most locally irrelevant extrinsic work (pipelines?) there is some 'trickle down.' The 'trickle down' effect characterizes intrinsic northern transportation. It would seem reasonable, however, to argue that some more attention be given to intrinsic transportation *per se*. One of the great needs is simply to expose the existence of intrinsic transportation and the role transportation plays in the ordinary affairs of northern people.

At this writing there are very few firm data on intrinsic transportation in northern Canada. The travel and transportation demand of northern people remains very inconspicuous to outsiders. In part, this is because there is no great fanfare when northern people take a trip. They rarely say goodbye, let alone file an itinerary or flight plan. I am aware of some trips, which had they been taken by an outsider, would have made international news. As it was, they were not even reported in the bush news of the northern press. Yet even the observant visitor to the north will surely be impressed by the travel of the northern people. Contrary to the southern notion that the north is an empty place, there are really very few places in the north where one can sit for long without seeing someone go by. One seems constantly to be bumping into someone from someplace else and often places very far away.

Among other things, of course, this movement of people has some bearing on land claims. If it were possible to believe that northern people simply sat around their dwellings all the time, one might rationalize a rejection of any claims to land occupance. But they surely do not. Far more than southern people, northern people live on the land. The land is their home more than any particular dwelling. This perspective is radically different from the mid-latitude urban perspective which views the dwelling itself as home and castle. It could also account, in part, for some seriously faulty northern housing policy (not to mention the incidence of cabin fever among southern transients in the north).

Today the outboard motored boat and the mechanized toboggan (or snowmobile) are two of the more important intrinsic transportation modes. While there has been a number of recent observations and comments on the advent of the mechanized toboggan (Francis 1969c; Hall 1970?; Usher 1970; Pelto and Müller-Wille 1971?; Francis 1971b), it seems necessary to admit that the implications of the event are far from understood. Curiously, there is no significant literature on the role of the outboard motor. Also conspicuously missing is any detail of information on the locale, extent, and intensity of travel for any surface form.

One may hazard that a very large proportion of the resources of northern people is invested in travel; but we have only estimates and personal experiences and impressions to support this assertion. A current survey of the extent and effect of the advent of the mechanized toboggan in settlements of northern Canada suggests that in some cases over half of the cash income of the owners of these machines goes into their purchase and operation (Francis 1971b). In the lower Mackenzie River area there is now a co-operative study of resident travel set up to record for several co-operating settlements the total resident travel for a full year, specific to route, mode, purpose, cost, return, and particular operational problems. By the time this paper is read (almost a year after its writing) I would hope to be able to comment more fully on the returns from that work. The project is staffed and operated by local northern residents, a scheme designed, among other reasons, to bring to bear on the research local perspectives and knowledge.

It has been observed previously (Francis 1969b) that the strength of mid-latitude bias has produced a variety of intellectual and operational discrepancies and inefficiencies in northern affairs. Even the design and execution of topographic maps of the north illustrate a remarkable foreign bias, a bias which makes them of the north (as viewed from the south) but not by the north or for the north. They are seriously faulty for most intrinsic functions and suspect for almost any function (Francis 1967).

We might consider briefly possible explanations for the neglect of intrinsic transportation and intrinsic travel demand in the Canadian north. One could contend that the self-interest of the economically and politically more powerful parts of the country (and the world) provide all the explanation needed; but this is both simplistic and questionable. Were that the case, the north and northern people, not to mention Maritimers and farmers, would likely be far worse off than they are. There is surely another factor, which I would suggest to be the invisibility of northern life and northern perspectives amongst outsiders.

The notion of immersible places occupying the same space has been noted previously (Francis 1969a, Francis 1971a), and the concept is useful here. The north could be thought of as two immersible worlds, the north of the outsider and the north which is home. There is surely a vast difference between the 'north which is barren and cold and, incidentally, where the Eskimos live' and the 'land which is my home, the land of my fathers and my children.' It is hardly necessary to observe that official Canadian northern policy shares the former percept. The latter percept is both strange and bothersome. This discrepancy in perspective would go far towards accounting for the lack of interest in or awareness of northern intrinsic transportation.

Today there are many encouraging signs that we are in process of recovering from what was surely a great and tragic error in judgement: the assumption that the people of the north should accept southern values and disappear into some variety of southern life. We are coming, I trust, to recognize that there are values in northern life and that, in any event, the people of the north have every right to continue to operate within their own value system and to build their own evolving life systems. If we accept that notion, it follows that we must be prepared to think in the context of northern life if we are to do meaningful work in the north. In transportation research this surely demands that we give some thought to intrinsic transportation. In geography it could be the key to a whole new perspective on northern diversity and landscape dynamics.

Francis, Karl E., 1967 A consideration of operationally oriented terrain parameters in arctic Alaska, Ass. of Pacific Coast Geographers, Annual Meeting, Chico, Cal.
- 1969a Black America, California Council for Geographical Education, 23rd Annual Meeting, San Diego, Cal.
- 1969b Environmental adaptation and Alaskan development, 20th Alaska Sci. Conf., College, Alaska.

- 1969c Decline of the dogsled in villages of arctic Alaska: a preliminary discussion, *Yearbook of the Ass. of Pac. Coast Geog.*, 69–78.
- 1971a Beyond four dimensions, Note to editor, *Prof. Geog.*, Jan.
- 1971b Expanded perspective on the decline of the dogsled, Ass. of Pac. Coast Geog., Annual Meeting.

Hall, Edwin S., Jr., 1970? The 'Iron Dog' in northern Alaska, ms.
Pelto, Pertti J., and Ludger Müller-Wille, 1971? Snowmobiles: technological revolution in the Arctic, ms.
Usher, Peter, 1970 The use of snowmobiles for trapping on Banks Island, NWT, Am. Anthrop. Ass., 69th Annual Meeting, San Diego, Cal.

c1608
Highway network expansion in central Thailand, 1917 to 1967
JAMES A. HAFNER   *University of Massachusetts, USA*

That economic development requires adequate and effective transport services is axiomatic. However, there exists no consensus on the role of these services in the development process. Although transport development may exert a positive, permissive, or negative effect on economic development there is a general absence of studies documenting the spatial and temporal variations in highway network expansion and impact within the context of the development process (Wilson et al. 1966). In Thailand, where increasing emphasis has been placed only recently on developing the road network, a melange of variables have conditioned the impact of the expanding highway network. The following discussion constitutes an exploratory analysis of changes in the structure of the highway network in central Thailand and considers several examples of the impact of highway network expansion on the rural economy.

Up until the beginning of the twentieth century land transportation in Thailand as a medium for trade, communications, and regional interaction was essentially non-existent. The only region of the country which contained any type of functionally homogeneous transport system was the central plain where an extensive network of waterways spread across the inundated wet-rice plain of the lower Chaophraya river delta. This waterway network linked the capital at Bangkok with 'upcountry' provinces and facilitated the relatively higher levels of economic and administrative integration which have characterized this region for several centuries. By comparison the highway network in this region had no first- or second-class roads outside of those in a few municipal areas and less than 100km of dirt cart tracks which at best could be used for only limited periods in the dry season. The complete dominance of water transport is reflected in Van der Heide's estimate that of the total rice exports of 14 million piculs in 1905/6, no more than 2 per cent originated outside of the central plain (Van der Heide 1903, 57).

The existence of an extensive inland waterway network, functionally compatible with the physical and agricultural environments of the central plain, historically precluded the development of alternative forms of land transport in this region. The development of rail lines to the north and northeast after 1896 was never conceived as a stimulant to the expansion of rice cultivation which took place in the central plain between 1860 and 1930, although they did effectively stimulate agricultural production and commodity exports from the north and northeast (Ingram 1971, 47). The characteristic emphasis on water transport in the central plain which lasted well into the twentieth century was a response to (1) the desire to expand areas under cultivation to rice, (2) the necessity to increase the flow of income from the provinces to the capital where bureaucratic growth was straining the kingdom's income, and (3) the need for greater administrative control over sub-regions within the central plain (Bunnag 1968, 29).

The expansion of the highway network in central Thailand beginning in 1917 can be represented sequentially in a series of discrete stages (Haggett and Chorley 1970, 261–8). The initial stage covered the period from 1917 to 1927 and appears to have represented a policy of building low-cost feeder roads to rail lines and major waterways rather than to penetrate areas where roads would serve as stimulants to agricultural production. The second stage (local node differentiation) is characterized by developments in 1937. The Taaffe model represents this stage as one in which a few major ports with inland lines of penetration begin to emerge and inland trading centres develop at these ports (Taaffe, Morrill, and Gould 1963, 520). The incipient road networks emanating from Bangkok, Ratchaburi, and Chonburi-Chachoengsao paralleled portions of the inland waterway network, which may have accentuated their impact as lines of inland penetration. Some of the most marked changes in the connectivity and structure of the highway network in the central plain took place between 1937 and 1947. The third stage (node interconnection) in 1947 reflects the linking of emergent coastal ports through lateral interconnections, the rapid linking of inland provincial centres, and vertical and lateral network expansion across the plain. Lateral interconnections in this stage correspond to similar developments reported for highway expansion in Ghana and Nigeria (Taaffe et al. 1963, 511–14). The fourth and final stage (regional node differentiation) began in the 1950s and is characterized by increased interconnection, linkage, and concentration featuring the emergence of 'high-priority' linkages. In central Thailand all of these linkages centre on Bangkok and connect the capital with major provincial centres to the north, northeast, and south. Feeder-road development has also expanded in this final stage in contrast to its development in an earlier stage in the Taaffe model, and appears to have been a result of the continued reliance on the inland waterway network to provide local feeder services while capital and resources were devoted to developing a minimal priority network.

During the entire fifty-year period of highway-network expansion considered here the overall effect of structural changes in the network on the rural economy has at best been permissive. Only in certain areas and in concert with other factors has growth in the highway network elicited any measurable changes in the productivity and structure of the rural economy. Furthermore, the failure of the rural economy to respond positively to the stimulus of transport development suggests that the concept of 'change clusters' and exogenous factors condition the extent to which transport network growth and regional development are positively correlated (Wilson et al. 1966, 3). This generally permissive effect of highway development can be related to a number of ecological, economic, and spatial factors (Hafner 1970, 239–49).

The inland-waterway network has in the past conditioned, and continues to condition, the impact of road-network development. Trade, social activities, economic patterns, and even life styles continue to be oriented toward the canal and river system. Even during the more rapid extension of the highway network in recent years it has not provided the majority of transport users with cost, convenience, or service advantages over those offered by the waterway system. A comparison of the area planted to rice, rice yields, and highway kilometerage in the central plain from 1917 to 1967 indicates that (1) rice yields have steadily declined, (2) area planted to rice has expanded at somewhat less than 3 per cent per year, and (3) highway kilometerage has increased at an average annual rate of over 13 per cent. If improved transport services are assumed to stimulate productivity through the distribution of new factor inputs and streamlining of marketing systems, then yields should have risen rather than declined. Recent increases in rice yields can be attributed to the impact of irrigation rather than distribution and adoption of new factor inputs (Soothipan and Ruttan 1965, 26). Silcock (1970, 38–50) also argues that indebtedness and tenancy conditions inhibited farmers from improving yields rather than yields having declined as a result of the expansion of rice cultivation into areas of lower productivity per rai as Ingram has argued (1971, 43–50). In fact, the expanding highway network, at least until 1950, may have actually had a negative effect on yields and productivity as oppressive tenancy and credit systems have followed entrepreneurs into new areas along expanding transport routes. Environmental considerations as factors influencing the impact of transport network expansion are often only marginally acknowledged, although Taaffe has shown that 'hostile' environment is an important variable in determining network density. In central Thailand the annual inundation of the lower delta has long been recognized as a deterrent to road construction (Siam 1919, 12). It has an equally deleterious effect on the ability of wet-rice farmers to benefit from the real and potential advantages of road development. In contrast to this situation road network expansion in the transitional uplands bordering the lower central plain to the north and northeast has had a more dramatic effect.

Corn as a primary export crop in Thailand has undergone marked expansion in area cultivated and production within the past fifteen years, most expansion having occurred in a group of contiguous provinces in the upper central plain. Highway kilometerage for these provinces increased significantly during the decade 1947–57. However, this was primarily in the form of a radial 'penetration-capture' sequence emanating from provincial centres and there was no evidence of significant increases in corn area harvested or yield. Highway-network expansion after 1955, however, resulted in an increased interconnectivity of provincial centres and coincided with an almost phenomenal increase in corn production and, more recently, a significant rise in vehicle registrations. This pattern is suggestive of sequential developmental change through 'change clusters' in which road expansion in combination with improved marketing systems, better transport services, increased market demand, and availability of hybrid seeds has stimulated a rapid expansion in corn production. This simplified interpretation of what is undoubtedly a more complex process does, however, emphasize the importance of highway network expansion in relation to changes in the rural economy. Certainly the existence of a highway does not in and of itself explain the growth of corn production. Yet without concurrent expansion of highway facilities it is doubtful if this area could have developed so quickly into the primary centre of export corn production in Thailand.

In assessing the implications of these examples it is important to emphasize the differentials in both ecologic conditions and impact of highway-network expansion. The permissive effect of highway expansion during the first three stages of network growth (1917–47) reflects the nature of planning and investment priorities, environmental constraints on changing traditional agricultural land use patterns, the economic limitations imposed by high transport costs, seasonality, and trafficability limitations on road use, and technological restrictions inhibiting the adoption of new forms of agrotechnology. In recent years the evidence of more positive effects of highway development in the lower central plain has increased although the rate at which traditional constraints on network impact are being overcome is slow and subject to distinct spatial limitations. In contrast, the more dynamic and positive impact characteristic of highway-network expansion in the drier uplands reflects the absence of traditional restraints associated with the wet-rice environment of the lower plain.

Appropriate planning and investment in transport facilities is imperative if the required rates and goals of economic development programs are to be attained. The success and efficiency of these investments must be based, however, on a realistic appreciation of the purposes for which transport services are designed and their position in an integrated transport system. If the variability of highway-network growth as a factor in regional development programs is considered more judiciously, the ultimate benefits from these programs should be more extensive and appropriate to the needs of the entire country.

Ford Foundation.

Bunnag, Tej, 1968 The provincial administration of Siam from 1892 to 1915: a study of the creation, the growth, the achievements, and the implications for modern Siam of the Ministry of Interior under Prince Damrong Rachanuphap. PhD thesis, St Anthony's College, Oxford U.

Hafner, J.A., 1970 The impact of road development in the central plain of Thailand, Ann Arbor, Mich., University Microfilms.

Haggett, P., and Richard J. Chorley, 1970 *Network Analysis in Geography* (New York).

Ingram, J.C., 1971 *Economic Change in Thailand, 1850–1970* (Stanford).

Siam, Department of Royal State Railways, 1919 *Department of Railways first annual Report B.E. 2461* (1918) (Bangkok).

Silcock, T.H., 1970 *The Economic Development of Thai Agriculture* (Ithaca, NY).

Soothipan, A., and V. Ruttan, 1965 An

analysis of changes in rice production, area, and yield in Thailand, *Proceedings*, Fourth Conference on Agricultural Economics, Kasetsart University (Bangkok), 1–43.

Taaffe, E.J., R.L. Morrill, and P.R. Gould, 1963 Transport expansion in underdeveloped countries: a comparative analysis, *Geog. Rev.* 53 (4), 503–29.

Van der Heide, J.H., 1903 *General report on irrigation and drainage in the Lower Menam Valley*, Bangkok, Ministry of Agriculture.

Wilson, G.W. et al., 1966 *The impact of highway investment on development*, Washington, DC.

c1609
Regional patterns of traffic in settlements
KARL H. HOTTES *Ruhr-Universität Bochum, West Germany*

Geographers have established divisions, which are based on physio-regional, economic, and central-place aspects. As far as regions have been subdivided on the basis of traffic, these attempts were based on (*a*) a differentiation of predominant ways and means of transportation on a world-wide scale, mostly in view of physical suitability, (*b*) different densities and networks of communication, (*c*) regional units of technical and administrational delimitation, (*d*) a complex view of input and output areas, and (*e*) an identification of integrated economic regions and traffic regions referring to both passenger and goods traffic.

Regional subdivisions apparently suffer from the fact that in the case of wide-meshed networks and/or low frequencies, traffic seems to be only a subordinate determining factor for a given region. If, however, transportation movements in agriculture are considered as essential for internal rural requirements, and if we accept Andreae's statement that 'agriculture is a transport-trade against its own will,' we find that rural regions may be institutionalized as transport-regions. These concepts, based on transport-regions, will have even more validity when applied to settlements as high-density traffic units, which have so far only been classified after social patterns, after the regional distribution of economic functions, after architectural periods and statistical units, and very rarely on an integrated view of these factors. In our modern environment, especially in urban areas, traffic requires extensive areas, creates environmental problems, and determines locations, and thus becomes more and more relevant for the geographical shape and function of settlements. There is no doubt about the fact that urban traffic research is a main object of regional traffic research. Traffic outside urban areas is arranged along lineaments which are characterized by a number of concentrations and crossings; in settlements, however, we find such a heavy concentration that a disentanglement by means of one-way traffic, diversions, and multi-storeyed traffic sometimes meets with only limited success. Settlements, however, consist of zones of highest frequencies in contrast to those of quietly flowing traffic; and there are streets of high transparency and predominant seclusion. Attractiveness includes a casual slowing down, a stop here and there, a strolling along. The actual traffic cells are not made up of traffic itself, of traffic installations, of facilities, of passengers, but – as local functional elements – the flats, the production-and service-units, which are both sources and destinations of transportation movements. The street itself may adopt additional secondary functions, e.g. it may partly become a market street, a 'play street,' a 'discussion street.'

In front of the above-mentioned cells the actual transportation movements take place – and if we try to examine integration patterns of traffic in settlements, the geographer finds the integrational phenomenon to be concentrated on the 'way.' Streets may be classified by their functions; types of streets, on the other hand, link up with the adjoining cells into traffic-regional, i.e. eventually cultural-geographic integration units. So we regard the settlement, which is a complex cultural region, as a system of structural traffic-regions, which consist of the way or street, where traffic happens, and the adjoining traffic cells, which in the case of business streets will be in close functional connection, in the case of thoroughfares in a loose one.

Such a regional subdivision, based on aspects of traffic, is, of course, possible only through the application of traffic-specific parameters. They are primarily (a) traffic installations as a form of immobile investment, (b) the means of transport as a mobile form of investment and in addition the pedestrians as participants of traffic, and (c) the density of traffic. The individual decision for a specific means of traffic or specific traffic installations depends on the possible investments, the readiness to take over costs, and the general human attitude. We will find that in different regions to some extent different 'styles' of traffic have developed, which also give an impression of the respective region's specific transport installations and facilities, as for example certain draught or riding animals, rickshaws, etc. In our table (not printed here) we tried to classify under these regional aspects by pointing out the general difference of urban traffic in developing countries and in industrialized nations. (We consider this to be more geographical and also more neutral than Voigt's (1965) classification, who makes a difference between regions with traffic-systems of high and low 'value.')

Traffic installations may be defined by their construction, their cross-section, and safety equipment, i.e. their general technical equipment. The construction is closely connected with the kind of road-surface. Streets may be completely without a surface (mud roads) or covered by very light to very heavy surfaces. (These may consist of many different sorts of material, which is of no particular importance in this connection.) The question will be more important if rails are built in the road level or if they have their own separate embankment. The level of equipment is attached – if traffic planning has been undertaken at all – to the real or expected volume of traffic: pedestrian streets require a paving different from those which are used by heavy lorry traffic.

The geographic appearance and efficiency, however, will be determined more by the actual cross-section of a street. By means of multi-lane and storeyed traffic a possible chaos of vehicles of different speeds may be avoided, road-traffic may speed up, room may be left for short stops (also for pedestrians before shop windows), and safety may be generally improved. Splitting up the traffic through distribution over separate lanes requires extremely wide roads: there may be pavements, cycle-paths, parking-strips, up to 8 lanes for car traffic, and rail embankments. Solutions to this problem may be a partial accommodation of high-speed traffic on elevated lands or underground, an underground installation of supply and drainage systems as well as communication systems. A limited one-way traffic plus installations for crossing traffic such as bridges, flyovers, tunnels, etc. will be favourable for the process of safe acceleration and direction finding.

Through the various vehicles, a street receives its real function, i.e. it becomes a 'car-street,' a 'cycle-road,' a 'pedestrian-street.' Vehicles therefore represent most important criteria. Vehicles, apart from pedestrians and bearers (coolies), represent mobile investments. Vehicles may run along planned routes (public conveyances). Generally, however, they run on the basis of individual decisions, not on a fixed plan.

The third determinant, traffic-density, is a measure for the actual efficiency of a street, resulting from the frequencies of all regular public transport and the volume of all other vehicles and other participants. The density may change periodically, episodically, and seasonally (city in the summer). Its rate may be so high that the traffic slows down or sometimes becomes impossible. Density also reflects the function of a particular street, conforming with the equipment of the street with traffic cells. In front of those cells, which are both sources and destinations of transport, we also find both barrages and quiet zones of flow of traffic. The number of cells differs according to the type of street which is shown in the last column in our table.

The table is made up on the basis of mapping in many European, North American, and Asian settlements and on the basis of our own and foreign traffic sample studies. This table, which must not be regarded as complete, tries to find a classification of streets, seen as integrated regional traffic units in settlements. With regard to applied geography they may be seen as 'planning-units.' It will be possible to computerize the respective determinants without difficulty by using punch-cards. On the other hand the specific character of a street will not only be seen by the geographer as a result of added elements, but it will be regarded, described, and explained as an integrated unit. The general result is that the functional differentiation of transport is at a much more developed level in industrialized countries: such a differentiation has to be one of the ideal planning objects of traffic planning. If therefore a theatre street shows also aspects of a thoroughfare, which could be seen as variations of the typical determination groups, the planners should try to find separating solutions to the problem. Medium-sized towns, small towns, and villages in industrialized countries mostly show a lack of differentiation in their service functions, so that we are used to finding more mixed types.

Finally I should like to say that this approach of finding specific traffic units as integration patterns of traffic in settlements should result in a new traffic-regional division of settlements and lead to a better understanding of the urban cultural landscape; but also attention should be drawn to this main task of geography, i.e. to establish a modern concept of applied traffic-regional division.

c1610
Commuter airlines: a third level of air service in the United States
JANE LANCASTER *Boston University, USA*

Commuter airlines offer scheduled, short-haul transportation between points generating sufficient daily traffic to justify the use of light, mostly twin-engine aircraft. Their passenger capacity ranges from 3 to 19; some are larger; mail and cargo are also carried. Although in existence since 1948, these scheduled third-level air carriers have expanded operations markedly over the past decade.

First-level (trunk) and second-level (feeder) airlines require the Certificate of Public Necessity and Convenience to operate. By contrast, commuter airline operation requires no such certificate, hence is not subject to federal regulation of entry restrictions, compulsory extension of existing services, and service abandonment. However, all third-level carriers operate under the aegis of the Civil Aeronautics Board (CAB) and the Federal Aviation Administration (FAA). In 1952 the CAB exempted all aircraft not over 5670kg from rate and route regulation, an act which helped both scheduled and on-demand air taxis.

Four categories of operating techniques and route structures stand out: (1) operations under contract to trunk and feeder lines: Apache Airlines with American Airlines in Arizona; (2) non-competitive shuttle services: TAG Airlines linking downtown airports in Cleveland, Detroit, and Chicago; (3) independent, short-haul, high density operations, some competitive with certificated carriers: Executive Airlines in New England and Florida; (4) scheduled night airmail: Buker Airways in the northeast.

Many communities abandoned by certificated airlines and/or railroads have turned to third-level air service, thus restoring connections with hubs. Some commuter lines link one community with several hubs: Hub Airlines, Ft Wayne, Ind. Others link one hub with several communities: Altair Airlines, Philadelphia; or link a regional set of communities: Key Airlines, Ogden, Utah. Special situations include connecting resort or residential areas with hub terminals: Provincetown-Boston Airline.

Unlike many feeder and trunk airlines which perforce started out with inefficient aircraft (some designed for other purposes) commuter lines pioneered a new concept of service initially utilizing small twins and some 15- to 21-place light transports designed solely for third-level service. Over 80 per cent of the equipment is multi-engine. Two types of STOL aircraft are currently in use: Canada's de Havilland Twin Otter and Germany's Dornier Skyservant. One key to effective commuter aircraft design is substitution of payload for fuel; the average commuter stage length is only 120km. Also emphasized is rapid conversion between passenger and freight configurations. Like the certificated carriers, these aircraft use sophisticated flight and navigation instrumentation, thus facilitating 'all weather' capability.

The FAA listed 12 commuter airlines operating in 1964. For 1969 one source reported over 200, another listed 116, and another 145. These figures reveal rapid growth but also point up statistical inequities. The comparative lack of federal and state regulation of these carriers is reflected in loose statistics compared to the voluminous data available on certificated carriers. Five official sources each compiled a list of commuter airlines operating in 1969; the totals ranged from 60 to 215; only 31 carriers appeared on all five lists.

Yearly maps and tables are helpful; regrettably, space precludes their inclusion here. Substantial commuter growth occurred between 1964 and 1970 as measured by increases in number of airlines, aircraft, employees, points served, departures, revenue plane miles, and passengers. It was possible in 1968 to fly from the Florida Keys to the state of Washington, solely on commuter airlines, if the passenger had two weeks to spare!

Yearly maps show marked expansion, but not the numerous starts, stops, bankruptcies, and mergers which characterize third-level airline development. Nor is seasonality of routes shown; e.g., Provincetown-Boston Airline shifts some employees and aircraft to its Florida division in winter. Likewise, Executive Airlines of Boston serves Tampa and Sarasota, Florida, to balance its seasonal income flow in New England.

Over 10,000 landing facilities exist in the United States; 4000 are public airports. Some 500 are served by certificated airlines but 70 per cent of their passengers board at the 22 large hubs. Commuter airlines serve over 400 airports; only 34 per cent board at those hubs.

The curtailing effects on short-haul transportation wrought by railroad abandonment and highway congestion are somewhat offset by commuter air service. The CAB's 'Use-It-Or-Lose-It' rule requiring a minimum of 1800 passenger boardings per year has led to isolation of over 100 communities from feeder and trunk airline service. Their larger aircraft and jet power led to fewer landings per day, inconvenient scheduling, hence fewer boardings. Moreover, some airports could not accommodate the new aircraft for reasons of topography, altitude, or space limitations. Just as subsidized feeders once accepted marginal routes dropped by trunk carriers, the commuter carriers are now taking over, but with quite different route structures, scheduling, and equipment, and no subsidy.

Co-operation between certificated airlines and scheduled third-level carriers has been notable. Among the most successful commuter airlines have been those contracted to certificated carriers who underwrite the commuter system, cover losses, and supply terminal facilities, expertise, and much traffic generation; e.g., Allegheny Airlines has developed a substantial Allegheny Commuter system in six northeastern states.

Another positive factor is favourably-inclined airport management. This can result in promotional policies such as low or no landing fees, counter fees, gate charges, or fuel taxation.

The CAB authorized commuter airlines to carry mail in 1965. This greatly benefitted the post office overnight delivery program, and airmail has added significantly to the payload, status, and economic stability of third-level carriers.

In contrast with the above are many factors of attrition. Poor original planning (with little research on establishing routes), undercapitalization, and inept management are three such factors. Public attitude, too, can be negative.

The attrition rate also reflects mergers; one example is Golden West Airlines – a merger of five California commuters in 1969. This pattern is consistent with that of second and first-level mergers, and will probably continue.

CAB regulation of third-level air carriers has been minimal when compared with certificated carriers. Concomitant with lack of regulation is lack of regulatory protection.

In some states no certification or registration is required of third-level carriers; in many more no route protection is afforded. However, many states are considering certification to provide route protection, and insurance and safety standards.

The carriers asked the CAB in 1968 to amend its air taxi regulations. The CAB complied; now all 'commuter air carriers' must file an annual registration statement, a certificate of passenger liability insurance, and quarterly traffic reports and flight schedules. Also in effect are new procedures which upgrade operating rules for safety and reliability.

Government attitude has changed. Commuter air carriers are no longer *merely* a branch of general aviation; they are now officially recognized as scheduled, although uncertificated, airlines – significantly because airlines as such are granted priority use of airways and airports at hubs where traffic is deemed too heavy for both airline and general aviation operations. Over half of commuter passenger traffic connects with certificated carriers.

Third-level operations will continue to burgeon although the rate of feeder suspension/commuter replacement will possibly decline. Longer stage lengths and larger aircraft are foreseen. The FAA forecasts 100 new public airports built each year until 1981, many in small towns, where the linkage functions of commuter air carriers are especially promising. Passenger traffic is forecast to increase from 149 million in 1970 to 669 million domestic enplaned passengers in 1985.

The full potential of commuter airlines is yet unrealized. Doubtless, they will continue functioning – a flexible and promising unit of the local/regional communications structure.

c1611
Rural road networks and land use change
C.D. MORLEY and W.C. FOUND   *York University, Canada*

In its broadest terms, this paper deals with the future of rural space within areas extending up to 160km from the centres of major metropolitan cities. It presents the transportation component of a wider project, the aim of which is to develop a behavioural process model of land use and land user change in metropolitan-centred regions. In particular, this paper focuses on the role played by the evolving country road network in the process of land transfer in rural metropolitan regions.

Three elements of this relationship will be considered: (1) the direct impact of improvements in country road networks on traffic flow; (2) the indirect effect of network changes on the propensity of existing land users and potential land purchasers to sell and buy rural land; and (3) the impacts that the travel demands of a changing rural resident population have on traffic flow in local road networks. These interdependent impact elements have potential planning and policy-making significance, for, with a fuller understanding of the implied relationships, changes in the country road network may be used to provide a mechanism for assisting in the achievement of regional planning goals.

The study reported here focuses on user and potential user responses to changes in the network, and the nature of their decisions regarding the origins, destinations, routes, and trip frequencies associated with use of the network. Viewed in this way, the transportation impact problem appears in a different form from that usually defined. It becomes necessary to distinguish between the structure of the transport system (that is, network topology) and the functioning of that system. The mathematical analogues which underlie the existing transportation impact models derive their stability from deductive laws and use self-regulating, internal transformations to explain changes over time. However, quite different mechanisms are implied if the models used focus on the responses of the different classes of decision-maker involved. Changes in road networks are seen as elements in the wider behavioural environments of users. The structure of such models focuses on model operation rather than results. Their functioning relates to a process involving exchanges with the external environment, and their internal mechanisms act as experimental settings, rather than reductions to normative rules.

The paper will be illustrated by reference to a detailed case study of the impact of country road improvement in Ontario. The case in point is the 'Airport Road,' a combination of three county roads which stretches for 112km to the northwest of Toronto International Airport, and on which a variety of major improvements have been carried out. This activity has produced an alternative direct road route to the Georgian Bay cottage country and beaches and the Blue Mountain skiing district. It passes through an area of morainic ridge country in which considerable numbers of rural retreats and country estates owned by Toronto residents have already been established. The Airport Road, therefore, provides a useful setting for the examination of the effects of country road improvement in terms of each of the three impact elements defined above. On-going work on the road network influenced by these changes includes the application of an interview survey to existing land owners, contacts with real estate and commercial operators, an aerial photographic survey, and the analysis of existing traffic flow data.

There is a clear link between this empirical work and the decision process which leads to the allocation of funds for the improvement of particular sections of the country road network. The funding of major changes in rural road networks is typically out of the hands of local authorities and under the control of regional, state, or provincial governments. In Ontario, even local township road improvements are subsidized to the extent of between 50 and 80 per cent of costs by the provincial government. Allocation procedures in Ontario demand that detailed information relating to existing physical structure, annual average daily traffic flows, design hour volumes, and commercial enterprises and public buildings served by the road be taken into account. Future traffic estimates over a ten-year period are made; while it is suggested that the generating effects of road improvement be taken into account, no formal evaluation is demanded. Typical traffic forecasting is based on future population forecasts, which are determined by the use of least squares projections of existing trends. Under the present procedures applied in the province, the long-term impact of road improvement is not directly considered.

Traffic flows through the rural road networks of metropolitan regions are dominated by the inhabitants of the main urban population centres. There are likely to be significant changes in the nature of local traffic flows as the permanent resident population within any section of the system becomes urban-oriented. The most obvious tendency is towards traffic peaking, particularly in upper tier roads within the network. Peaking will operate on a daily cycle if connected with commuter traffic, and it may be possible to distinguish separate movements through the network relating to travel to the major metropolitan centre as opposed to local employment centres. Weekly and seasonal cycles will be common where recreational travel is the primary source of traffic. Extreme peaking is likely where the road network intervenes between the metropolitan city and primary recreation areas (e.g., weekend cottage areas, camping associated with national parks, beaches, or major ski resorts). A final traffic flow component relates to vehicles associated with industrial or natural resource sites. A combination of peaked traffic flow and concentrated route selection puts severe demands on existing networks and generates the attempts to disperse traffic by improving networks.

It has been asserted earlier that the amount and the distribution of future traffic in an improved rural road network is dependent on the mix of the different classes of user operating within the current social and economic environment. The distinctive use patterns of the different groups demand that an effective impact model be disaggregated to that level. At this stage in the model's development, a breakdown on the basis of primary automobile trip purposes is suggested. In these terms the following classes of user trips are suggested: (1) trip to/from permanent home; (2) trip to/from second home; (3) trip to/from particular recreation site; (4) sightseeing trip; (5) search for rural property; (6) business trip; (7) tertiary service trip to and from retail service locations; and (8) social trips to and from personal contacts.

The user groups associated with these trips may be expected to vary in their use of the network on the basis of a number of elements: trip frequency and repetition cycles; trip structure in terms of use of the road hierarchy; degree of commitment to routes; and the likelihood of making stops en route. For each group a distinct decision setting may be defined, and therefore differential reactions to improvements in the road network may be expected. Separate runs of the model will be required to define the flow patterns of each trip type.

The key to determining the behavioural responses of user groups is the operation of a series of indices based on information levels, image structure, attitudes, and comparative evaluation. These define the choice environment in terms of which the existing users respond and new traffic is generated. Actual components at the decision (i.e., selection-allocation) stage are: (1) expected changes in time-distance ratios – the traditional route minimization assumption; (2) a social differential which relates to the social networks with which the different classes are linked and the levels of information associated with such networks; (3) a perceived environment differential concerned with the relative importance of the quality of environment on the route travelled and likely to be of importance when a significant proportion of traffic is related to discretionary recreation trips; and (4) a safety and comfort differential.

On this basis, changes in the use of the new network relating to frequency and timing of trips may be hypothesized for the various existing users. It is more difficult to determine the response of transfer users (those whose origin and destination remain the same but who switch to the system following network changes), and of the new users who begin moving to the area after network changes. The model's primary concern regarding new traffic flow is its allocation according to user classes and the nature of the trips they generate.

The classes of existing land owners in the rural sections of metropolitan regions have been discussed elsewhere in relation to this project (Found and Morley 1972). It may be assumed that the location of properties within the network will have a direct bearing on the nature of the response to the changing network of traffic flow. Owners of the properties adjacent to the upper tier roads within the local road system may be expected to respond in different ways from the owners of more remote properties. First there will be the effect on trip movements made by local residents; internal service trips, external service trips, and linkage with the major metropolitan city will all be influenced by the improved local and external access; but with the improvement is likely to come additional traffic throughout the network, and invasion of privacy with the potentially increased interest in the potential of land for private recreational use and for residential subdivision development. Responses to these pressures are of primary concern in this study.

It is evident that much empirical work needs to be carried out before such disaggregated behavioural models can be operationally defined. In many cases new types of research instruments must be designed and detailed hypotheses listed. It is the contention of the authors that despite the problems implicit in such work, the potential value to the decision-maker of experimental rather than predictive modelling in this field, justifies the attempt.

Found, W.C., and C.D. Morley, 1972
Land-use classification in rural sections of metropolitan regions, IGU Symposium on Agricultural Typology.

c1612
Toward the integration of industrial location and commodity flow studies
IAIN WALLACE   *Carleton University, Canada*

The correspondence between studies of industrial location and those of commodity flows, while logically clear-cut, is complex in detail. Each area of interest boasts a body of theory by geographical standards quite substantial, but the nature of these theories does not make for their easy marriage. Moreover, it is being claimed that in both spheres the development of theory along traditional lines has begun to stagnate, suggesting that new points of departure are called for. This paper defines a context in which industrial location and freight traffic studies can be united to provide a new analytical perspective relevant to both. In so doing, it gives due weight to empirical findings.

The intellectual achievements of classical location theory, epitomized by Isard (1956), are not in question, even if it is now fashionable to call for a more behavioural, 'realistic' theory (Pred 1967). Our concern must be not with what is 'wrong' with classical theory, but with defining an analytical framework which will help answer the most pressing *unresolved* questions about industrial location. Chisholm (1971) advocates both a new economic framework (welfare economics) and a change of scale (from the single plant to the economic region).

Neither is lack of achievement the problem with spatial interaction theory, including commodity flow analysis. Smith's (1970) review of work in this field confirms the power of distance and population substantially to explain the volume of traffic moving between centres of economic activity. But this 'apparently unbreakable contract with the concept of gravitation' (Pred 1967, 19) is something of a theoretical liability, especially with respect to freight flows (Heggie 1969). Treatment of aggregate data is required by the derivation of the model, yet detailed analysis calls for disaggregation, destroying the model's rationale. The gravity model cannot be paired at the micro level with the prescriptive determinism of classical location theory. Rather, it is when regional industrial distributions are paired with aggregate flow data that we see the most complete analytical conjunction of the two phenomena, in a linear programming/transportation model (e.g., O'Sullivan 1970).

This still leaves an undesirable theoretical vacuum for microscale study; and, despite obvious objections, the single plant appears the most appropriate basis for a new analytical framework. It can no longer be claimed as the seat of independent entrepreneurial decision-making; but the case for adopting it rests more on technical than organizational criteria.

Despite geographers' interest in industrial linkage (Wood 1969), there has been an unwillingness to view the production process as a *functional* whole. At whatever scale, input-output data focus on the *formal*, inter-sectoral characteristics of commodity movements/linkages. While complex linkage patterns do reflect intense *product* specialization, they are also amenable to a *process*-oriented analysis. An initial threefold categorization of manufacturing plants is proposed – 'processing,' 'fabricative,' and 'integrative' – related to the way in which material inputs are handled. This division recognizes: (*a*) different relationships between the contributions of capital and labour inputs to value-added by manufacturer; (*b*) allied differentials in the time-budget of material handling; respectively high volume throughput/low mean passage time (mpt), lower volume/higher mpt and high volume/low mpt (the essential contrast here between processing and integrative plants is the degree of differentiation of products handled); (*c*) a production sequence in which a unit product becomes increasingly endued with a specific identity (e.g. iron ore to auto component). As a result of (*c*), one can define a *product* time-budget, which becomes progressively more stringent. The climax is reached on the assembly line of an integrative plant, where production involves the continuous and co-ordinated supply of hundreds of components, which itself demands, notwithstanding the buffer role of inventory management, a high degree of reliability in the transport system maintaining industrial linkages.

This conclusion is supported by empirical studies of the factors which influence manufacturers in decisions relating to the transport of their products (Bayliss and Edwards 1970). 'Quality' variables (speed, reliability, freedom from damage, etc.) are a prime concern, more so than an explicit commitment to the lowest-cost form of transport, narrowly defined. By detailed attention to the service provided, it is possible to rank formal divisions of the transport industry in terms of these qualities. The manufacturer's own fleet of trucks appears to offer most: those modes involving greatest delay and rehandling en route are the least favoured. If it can be shown that the manufacturer's choice of transport is significantly related to the time-budget of his plant and the goods passing through it, some claim can be made to have begun to integrate theories of industrial production and commodity flow at the micro level.

The related spatial hypotheses and their operational definition are more complex than might appear, in view of the multivariate nature of the choice of transport provision. Also, it remains to be shown that relationships which hold good in Britain are equally valid in North America, where distances are much greater; but there is evidence that they may be. Black (1971) has confirmed the theoretical expectation that the more sophisticated the product, the flatter the distance-decay curve will be. This implies that it is the goods with the most stringent transport demands which travel furthest, and the significance of this is that distance of shipment appears to be the major discriminator, if not determinant, of modal choice (Bayliss and Edwards 1970). Correlation of commodity time-budgets with a service-quality categorization of modes is thus complicated by the distance factor, the influence of which is related to economic and statutory (length of drivers' shift) controls, as well as the adequacy of the road network. Indeed, given manufacturers' revealed modal preferences, one questions whether surveys of industrial traffic flows reflect an *ex post* or *ex ante* situation. If the most reliable form of transport has a known range of viable operation, this range may be a determinant, rather than a function, of observed patterns of industrial linkage.

The author's study of industrial traffic in the English Midlands (Wallace 1971), using data categorizations outlined above, confirmed the link between the transport mode and the position of the consignment in the overall production cycle. Spatial conclusions were much less well-defined, but revealed the influence of factors so far little discussed. It cannot be claimed that a complete, synthetic theory of location and flow characteristics has been forged; but fruitful avenues for further research have been outlined.

Bayliss, B.T., and S.L. Edwards, 1970 *Industrial demand for transport* (London).
Black, W.R., 1971 The utility of the gravity model and estimates of its parameters in commodity flow studies, *Proc. Assoc Am. Geogr.*, 3, 28–32.
Chisholm, M., 1971 In search of a basis for location theory, in C. Board et al., eds., *Progress in geography*, no. 3 (London), 111–33.
Heggie, I.G., 1969 Are gravity and interactance models a valid technique for planning regional transport facilities?, *Opl. Res. Q.*, 20, 93–110.
Isard, W., 1956 *Location and space-economy* (Cambridge, Mass.).
O'Sullivan, P.M., 1970 *The spatial structure of the Irish economy* (London).
Pred, A., 1967 Behaviour and Location, Part 1, *Lund Studies in Geography*, Ser. B, no. 27.
Smith, R.H.T., 1970 Concepts and methods in commodity flow analysis, *Econ. Geog.* 46 (supplement), 404–16.
Wallace, I., 1971 Freight traffic of industrial firms in the north-west Midlands: a study in spatial and commercial interaction, D Phil thesis, Oxford U.
Wood, P.A., 1969 Industrial location and linkage, *Area*, no. 2, 32–9.

Medical Geography
Géographie médicale

c1701
Agriculture, malnutrition, and the deficiency diseases in rural Uttar Pradesh (India)
S. SAJID HUSAIN *National Council of Education Research and Training, India*

ABSTRACT
Uttar Pradesh is 294,364km^2 in area and has 88.3 million people (86 per cent rural). The main sources of calories are foodgrains and sugarcane. The usual diet of villagers consists of cereals, pulses, *gur*, sugar-juice, and starchy roots. Low productivity, particularly in medium and poor quality lands, is a major cause of poverty – both of land and the people. Fallowing, most common in the medium and low quality lands, is mainly due to want of good manuring and irrigation. Their shortage results in low production. Only 26.8 per cent of total cultivable area is cropped twice a year. For a long period of time the rate of manuring could not keep pace with the rate of exhaustion of the soil.

The caloric sufficiency of food cannot guarantee a well-balanced diet. Deficiency of fat, vitamin A, vitamin C, and calcium are found in almost all types of villages, and protein deficiency is found in many. The villages with low calorie intake suffer more from malnutrition. There is absence of good nutritive diets. In UP, there are a large number of cases of six major diseases in one year. The daily diet mostly contains no vitamin C, and is generally deficient in calcium. Hence anaemia, tuberculosis, rheumatism, scurvy, scabies, poor growth of bones and teeth, dysentery, diarrhea, respiratory diseases, rickets, night blindness, dental caries, and stunted growth are the result. In UP, 26,000 cases were ricketsial in 1970. *Tarai* areas of Uttar

Pradesh have a high incidence of goitre. Infant mortality in Uttar Pradesh is 87 per 1000.

Malnutrition, unscientific nourishment, and unhygienic conditions have promoted various deficiency diseases. Developments in agriculture, manuring, irrigation, rotation of crops, yields, and hygienic improvements will improve the health, nutritional, and living standard of the rural population.

c1702
Un exemple d'étude des relations pollution-mortalité
Y. VERHASSELT-VAN WETTERE   *Vrije Universiteit de Brussel, Belgique*

Nous avons tenté une première approche de recherche des relations entre le degré et la répartition de la pollution et la distribution spatiale des causes de décès par catégorie. L'étude porte sur l'agglomération anversoise.

L'on constate des répartitions différentielles nettes des causes de décès en ce qui concerne les maladies respiratoires. Une relation semble exister avec la distribution de certaines formes de pollution. D'autres facteurs furent également envisagés tels que la population (densité, composition par âge, activité professionnelle), les logements (qualité, taux d'occupation), le trafic (intensité, fréquence).

c1703
Training a medical geographer
G.E. ALAN DEVER   *Georgia State University, USA*

This paper is one of the first to relate specifically to the training of a medical geographer (McGlashan 1965 and 1969; Armstrong 1965). The type of training a student should receive, no matter in what discipline, is a debatable issue. The development of a curriculum in a new or even existing discipline is always a major concern and represents continued discussion in many departments. For this reason, it is essential to provide a framework for the training of students. Of course, there may be skeptics who will ask why we should train medical geographers if employment is not readily available. Frankly, this position warrants considerable merit and I will briefly comment on this at the conclusion.

Most medical geographers do not have direct training in medicine. There are, however, exceptions. Jacques May and Howard Hopps are physicians who have been influenced by geography to an extent that it became appropriate to develop a parallel interest. Notably, May has explicitly stated his parallel interest is 'medical geography' (May 1950) while Howard Hopps has termed his parallel attraction to geography as 'geographic pathology' (Hopps 1968). Besides medically trained researchers there are also geographers who have been influenced by medicine. Either approach is of tremendous value but I do not believe it to be of prime importance that a medical geographer have a degree in both medicine and geography. For instance, I was graduated as a medical technologist before entering the field of geography. Another yet somewhat different example of allied health training of which I am aware is one, R.W. Armstrong, who received a master's degree in public health which was beyond his formal education in geography. Thus, there are those who have medical degrees; those who have backgrounds in allied medical fields and lastly those who are self-made having acquired their knowledge of medicine through diligent reading. It is, therefore, correct to assume that most medical geographers have widely divergent backgrounds.

Now, the emphasis in training the new medical geographer will depend on the following features which must be incorporated into the forthcoming framework. Such features are: (1) the student's association with an active member in the field, (2) didactic training, (3) reading specific textbook material, and (4) perusing periodical literature (Dever 1970). Accordingly, these work-lines should be pursued before conferring of degrees but also it is important to continue this to an even greater extent after graduation. Actually, these features should be pertinent to the training of students no matter what type of discipline is involved. Recognizing that the above approaches and features are important to the training program let us consider this relative to medical geography.

A few articles have attempted to shed light and comment upon the specialty. Such works as 'The Scope of Medical Geography' (McGlashan 1965), 'Medical Geography – an Emerging Specialty' (Armstrong 1965), and 'The Nature of Medical Geography' (McGlashan 1969) have been written essentially as position papers; but more, I believe, they suggest the role a medical geographer must play and therefore create an implicit statement about training. On the basis of the information stated in the above papers, it is crucial that the prospective medical geographer be aware of the many problems encountered in the field. It is not my intent to enumerate these problems but to suggest a method of training that will be equal to the efficient handling of these problems (Banta and Fonaroff 1969). Let me stress, first of all, that a program for training a medical geographer should not be completely structured but rather it should be tailored to fit the students' interests and to comply with existing curriculum.

Let us take an aspiring medical geographer and assume that he or she has had little or no training in medicine or allied health disciplines. Their training, ideally, should begin in undergraduate school, even though most students are magnificently indecisive at this stage. Realizing this, then, some course work should be in the area of biology and physiology or more generally the natural sciences. Also, he or she should take ecological course offerings and at least a beginning mathematics or elementary statistics course. Beyond these stated courses are those encountered in the major – hopefully geography. This rather flexible structure would be basically sound for many divergent professions and not hamper the student should he change his mind. It might be well to remember that many majors in college are rigorously structured and the flexibility of this particular program would be a definite advantage to the perplexed undergraduate.

At the beginning graduate level it is still rather premature to entrench the mind with medical phenomena. However, the master's thesis should be on a medical problem and the students' program should include a solid geography core along with advanced statistics. Also, an awareness for existing medical problems must be developed and these problems are several. They range from disease ecology and health planning to location of health facilities. The latter is representative of location theory.

At the PHD level the student must be explicitly directed. This points to a slight difference in the training of a medical geographer as opposed to other PHD's. It is true, at least in American and Canadian schools, that taking the PHD degree becomes one of the most flexible periods in a student's life, but a PHD degree emphasizing medical geography should be slightly structured. This again is assuming no prior medical affiliation. If the work is not or cannot be structured, then the alternative is to take a degree in public health beyond the PHD in geography.

In the PHD program many of the courses will be taken outside the department of geography. Those courses are (1) demography; (2) epidemiology; (3) advanced statistics; (4) chronic and infectious diseases; and (5) genetics. This again should be tailored to the student. Clearly, specialization occurs in the area of the student's dissertation topic.

Thus, the theme being suggested is that the majority of training should be reserved for the advanced degree. The reason is purely pragmatic. The field is still new and developing, although with more rigor in recent years, and to get down to hard facts, places of employment for those trained are not available. The majority of the employment exists only in the teaching field where this specialty becomes their prime area of research. However, future programs, and I mean 10 years or even 5 years hence, if possible, should be geared to produce medical geographers at an earlier level in their education. This advancement must be co-ordinated with our endeavours to make medical geography a more recognized and – more importantly – necessary field! This, of course, means that those of us who are active in the field must strive to produce substantive results that relate to unravelling disease association or causation. We must make our influence count! Otherwise medical geography will always remain an obscure, nefarious field; strictly a research oriented discipline open only to PH D's. I firmly believe that it can be a geographer who can provide a positive causal link to a heretofore mystifying disease!

Armstrong, R.W., 1965   Medical geography – an emerging specialty? *Internat. Pathol.*, 6, 61–3.
Banta, J.E., and L.S. Fonaroff, 1969   Some considerations in 'The Study of Geographic Distribution of Disease,' *Professional Geog.*, 21, 87–92.
Dever, A.G., 1970   Medical geography – a selected bibliography, *Internat. Pathol.*, 11, 13–16.
Hopps, H.C., 1968   Environmental Geochemistry and Geographic Pathology, editorial, *Internat. Pathol.*, 9, 64–5.
McGlashan, N.D., 1965   The scope of medical geography, *South African Geog. Jour.*, 47, 35–40.
– 1969   The nature of medical geography, *Pacific Viewpoint*, 10, 60–4.
May, J.M., 1950   Medical geography: its methods and objectives, *Geog. Rev.*, 40, 9–41.

c1704
London and Glasgow: a comparative study of mortality patterns
G. MELVYN HOWE   *University of Strathclyde, Scotland*

Unmistakable differences in the level of mortality rates and of life expectancy may be identified between continents, between countries, between cities, and even within cities. *The National Atlas of Disease Mortality in the United Kingdom* (Howe 1970) reveals a range of contrasting distributional patterns for the major causes of death and this despite the fact that the National Health Service applies to all classes of society. This paper presents an examination of the mortality experience of males in London and Glasgow respectively and of the contrasts in levels of mortality between and within these large cities of Britain.

Arteriosclerotic heart disease, a major public health problem in all urbanized, industrialized countries, is the main cause of untimely death for men in Britain, certified in almost a third of the deaths of those in middle-age. Why, when death rates are adjusted to take into account differences in age structure of the male populations of London and Glasgow, should the standardized death ratio for Glasgow be 25 per cent above the United Kingdom average yet much the same as the national average for London? (Fig. 1.) (On this and the next map, standardized mortality ratios have been represented on both demographic and geographical base maps.) In London the most unfavourable mortality experience is in the boroughs of Hampstead and St Marylebone, 18 and 16 per cent respectively above that for the country as a whole; in Glasgow the experience is 21 per cent above the United Kingdom average in the Langside ward and in nine other wards (Pollokshaws, Cathcart, Parkhead, Shettleston. Tollcross, Ruchill, Whiteinch, Yoker, Knightswood) it equates with that in the two unfavourable London boroughs mentioned. In contrast there are eight wards in Glasgow (Cowlairs, Townhead, Exchange, Calton, Dalmarnock, Hutchesontown, Gorbals, and Pollokshields) where the mortality experience is 15 per cent or more below the national average and only one borough in London, the City, in this same favourable category. Contrasts such as these demonstrate the dangers inherent in generalization and in assuming homogeneity in environments and life styles within and between cities.

On a national basis mortality experience for chronic bronchitis is much worse in the urbanized, industrialized areas of Britain than in the rural areas. Despite the overriding importance of cigarette smoking in the aetiology of the disease, there appears to be a correlation between presumptive pollution levels and mortality from bronchitis in middle life. The level of mortality from bronchitis is higher in London than in Glasgow but there are certain boroughs in London

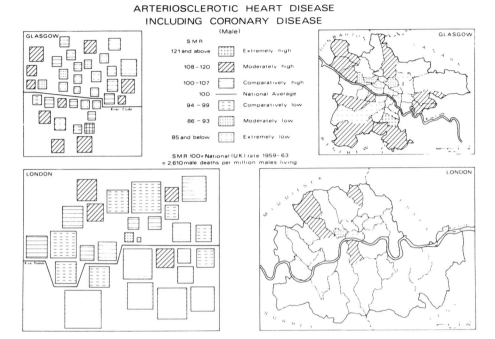

Fig. 1.

and wards in Glasgow which are far worse than others. (Map of standardized mortality ratios for bronchitis omitted because of space limitations.) The mortality ratio in the metropolitan borough of Shoreditch, for instance, is two and a half times the national average and a little over twice that in Stepney and Bethnal Green. In Southwark and Poplar the ratio is 80 per cent above the United Kingdom average. For Glasgow as a whole the standardized mortality ratio (SMR) is 145 (UK average = 100). This ratio reflects in particular the high ratios in the four wards of Ruchill, Mile End, Dalmarnock, and Hutchesontown.

Both London and Glasgow have unfavourable mortality experience from *Lung-Bronchus Cancer* (map of standardized mortality ratios omitted because of space limitation). The SMR's in London range from 185 in Shoreditch to 95 in Battersea. The ratio for Shoreditch, the highest in the United Kingdom, is followed closely by that for the metropolitan boroughs of Finsbury (SMR 173), Islington (SMR 170), Southwark (SMR 164), and Stepney (SMR 162). The severity of the mortality experience of Glasgow may be judged from its SMR which, overall, is 161, and yet, different from London where the SMR's for all boroughs are in excess of the national average, there are several wards in Glasgow which have relatively favourable experience, particularly Kelvinside (SMR 66) and Langside (SMR 87).

London and Glasgow demonstrate quite remarkable geographical variations in levels of mortality from *Stomach Cancer* (map of SMR's omitted because of space limitation). Partick West and Govan wards in Glasgow and Poplar, Finsbury, and Southwark boroughs in London record very high mortality levels from the disease and the experience of Provan, Parkhead, Dalmarnock, Cowcaddens, Maryhill, Pollokshaws, and Cathcart wards in Glasgow, and St Pancras, Islington, Stoke Newington, Stepney, and Holborn boroughs in London is only slightly better. In contrast, Kelvinside, Anderston, Exchange, Cowlairs, Pollokshields, Camphill, Langside, and Govanhill in Glasgow and Hampstead, Chelsea, and Westminster in London have standardized mortality ratios which are 25 per cent or more below the national average.

1216 / *Medical Geography*

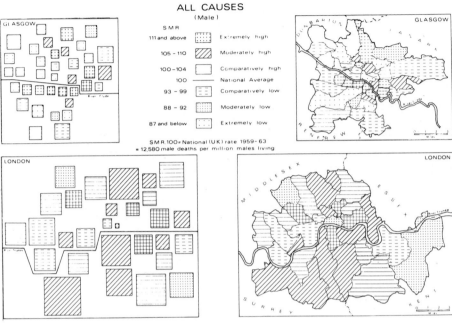

Fig. 2.

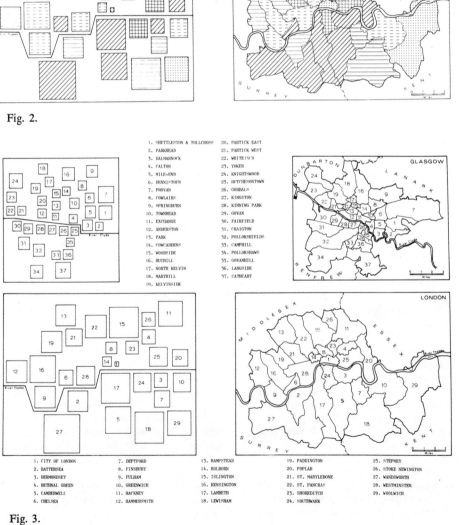

Fig. 3.

The stark contrast between *Infant Mortality* rates in Glasgow and Greater London is obvious (map omitted because of space limitation). Rates in all London boroughs are appreciably less than the United Kingdom average (19.1 per 1000 live births), but in Glasgow several wards have rates which are well above the national average. In Dalmarnock and Exchange wards, for instance, the rates are 38 and 39 per 1000 live births respectively.

Mortality from all causes, the sum total of human tragedies, reflects human response to the complex of environmental hazards and genetic susceptibilities within the two cities (Fig. 2). Some parts of both London and Glasgow have better than average mortality experience, indicating scope for increased longevity. Other areas may be identified where, in contrast, mortality is high. In the favourable category come the Hampstead and Westminster boroughs of London and Kelvinside, Exchange, Pollokshaws, and Langside wards of Glasgow. In the unfavourable category are the London boroughs of Southwark, Stepney, Shoreditch, and St Marylebone and the Glasgow wards of Ruchill, Mile End, Dalmarnock, Hutchesontown, Gorbals, Kingston, Kinning Park, and Govan.

Areal differences of disease mortality, such as have been demonstrated for London and Glasgow, reflect the deleterious influence of local environmental factors, present and past, on human life. The study of such differences not only contributes to the elucidation of aetiological relationships of chronic and degenerative disease but assists in the formulation of plans for a better quality of life.

Registrars-General, London and Scotland.

Howe, G. Melvyn, 1970 *National Atlas of Disease Mortality in the United Kingdom*, rev. and extended ed. (London).

c1705
The problem of unusual multiplication and growth in man, animals, and vegetation in various parts of the world
JIŘI KRÁL *Prague, Czechoslovakia*

The causes of the unusual multiplication and growth of man, animals, and plants have to be sought in the local geographical environment, where a main factor is radioactivity, which governs the life of the whole universe and thus also of our planet. This is why the problem can be explained in the first place by radiogeography investigating the influence of radiation on nature, and by radioanthropogeography investigating the influence of radiation on man, animals, and vegetation. The two sciences were developed by the author many years ago. Recently natural radiation has been intensified by artificial radiation produced by man. Its effects can be traced especially where there is an appearance of rich radioactive minerals and radioactive springs or excessive amounts of fallout from radioactive aerosols, mainly from atomic tests. Naturally the investigations are still in an initial stage, so that it is so far impossible to draw final conclusions, all the more as the effects of absorbed radiation are manifested only after a long time. Thus, it still remains an open question why, for example, in Liberia earthworms attain the length of up to one metre, spiders are as large as plates and snakes are up to ten metres long, or why there exists in Australia excessive multiplication of rabbits, sheep, foxes, mice, dingo dogs, wild horses and donkeys, camels, wild goats, pigs, buffaloes, and toads; or why there are plants, for example, prickly cactuses – the so-called prickly pears in Australia – depreciating the value of the soil, and which cannot be eradicated. The problem of abnormal human and locust proliferation in India has not so far been solved. The same applies to fish multiplication in some rivers in Vietnam, rat and bear multiplication in Japan, multiplication of vipers in Italy, and of marine fish in some seas. The question of abnormal human growth up to 234cm still remains unsolved. Obviously the above problems can be solved only by co-operation of numerous specialists, especially geographers, and among these especially radiogeographers and radioanthropogeographers.

c1706
Les causes socio-économiques de l'alcoolisme-ouvrier
C. THOUVENOT   *CNRS, France*

La consommation des boissons enivrantes a été de tous temps mais c'est au 19ème siècle qu'on a forgé le terme d'alcoolisme et que le fléau a pris un aspect profondément social en touchant essentiellement la masse des nouveaux ouvriers au service de la grande industrie triomphante.

On a été long à admettre que l'alcoolisme puisse être un signe de pathologie sociale grave et que lorsqu'une société boit c'est pour des raisons très sérieuses et respectables (Tremolières 1961).

Nous avons cherché à dégager quelques unes de ces raisons dans trois milieux ouvriers: celui du textile lillois entre 1840 et 1870; du textile vosgien entre 1850 et 1890; des mineurs et ouvriers du fer italiens entre 1890 et 1910 et polonais après 1920.

Les nouveaux ouvriers qui affluent vers les mines, les fabriques et les usines à feu sont tous des prolétaires campagnards qui viennent spontanément pour gagner de l'argent. Ils perdent brusquement toutes les traditions et les règles et sont privés de points d'appui. La rupture avec leur milieu d'origine est pour eux un événement capital, généralement douloureux, même lorsque des compatriotes, voire des gens ou des parents du même pays, se chargent de l'accueil.

Les fabricants utilisent cette nostalgie, chez les Polonais par exemple, en décorant les bouteilles d'alcool qu'ils produisent, d'une image de la vierge miraculeuse de Czestochowa! Les débitants polonais recréent, eux aussi le pays: chez eux le déraciné contemple l'aigle blanc de Pologne, les portraits des héros polonais, écoute les mélopées nostalgiques et les chants guerriers de sa lointaine patrie. L'œil humide, il boit jusqu'à l'ivresse sont eau de vie préférée (Lyzinski 1932).

Dans le textile lillois, les femmes travaillent comme les hommes dès 8 ans. Elles n'ont aucune formation ménagère, épuisées de le travail, elles n'éprouvent guère l'envie de faire face à des tâches familiales. Dès 5 heures du matin maris, femmes et enfants partent à la fabrique – les petits étant gardés par une aïeule ou une voisine – dînent dans les cabarets, épiceries ou gargotes qui offrent de modestes portions largement arrosées de vin, de bière, ou d'alcool (Pierrard 1965).

Les Italiens du pays de Briey sont jeunes et sans femmes. Ils prennent pension, passent leur journée de repos et leurs soirées dans des cantines, où ils jouent, dansent, et boivent (Ruben Dorf 1913).

Même si les Polonais arrivent plus nombreux mariés, ils prennent, vers 1920, des 'pensionnaires' célibataires jeunes et vigoureux qui convoitent rapidement la femme de l'hôte et l'obtiennent généralement à l'issue de formidables ripailles où hommes, femmes et même enfants roulent sous la table (Lyzinski 1932).

Personne ne s'intéresse aux problèmes nouveaux qui se posent dans les agglomérations, personne ne tente d'atténuer les chocs psychologiques de la transplantation et de l'insertion dans la société existante. Pour les nantis, l'ouvrier est un pêcheur qui n'a ni morale, ni religion. Nulle part sauf chez le logeur, le revendeur, et le cabaretier l'ouvrier ne trouve d'accueil. Le cabaret est plus pour lui un centre d'accueil, un autre chez soi où il recherche la chaleur humaine autant que celle de l'alcool, qu'une 'école de vice où se détendent les ressorts du bien' (Pierrard 1965).

Les activités sont totalement différentes de celles du milieu campagnard, très dures certes, mais rythmées par le jour et la nuit et les saisons.

Les enfants resteront très longtemps selon le mot de Villermé des 'forçats cloués au pied d'un métier,' les horaires de travail subis sont démentiels (12 à 15h. parfois 16 heures par jour) dans des conditions d'hygiène déplorables.

Ainsi les nouveaux ouvriers arrachés de leur milieu social et familial, transformés en machine à produire ne trouvent qu'une seule forme d'accueil, le débit de boisson, le seul endroit où ils peuvent chercher à oublier.

Le choc du salaire est décisif chez des ruraux jusque là insérés dans une économie quasi autarcique.

La paie, si insuffisante qu'elle puisse paraître si on la compare aux prix des bonnes denrées consommées par les nantis, est un événement qui transforme la vie d'hommes n'ayant connu jusqua là qu'exceptionnellement la 'couleur de l'argent.'

Les ouvriers se comportent avec leur argent en grands enfants. N'ayant pas eu dans leur existence passée l'occasion d'économiser, ils vivent au jour le jour et dépensent facilement les jours de paie, sans songer à faire des réserves pour leurs vieux jours ou pour les moments de crise.

L'argent leur permet ainsi d'assouvir des besoins simples, les seuls à leur portée. Il devient ainsi un facteur d'alcoolisation chez ces êtres humains travaillant dans des conditions très difficiles et attirés par un ensemble de satisfactions à assouvir et considérés par eux comme l'image de leur bonheur: l'amour physique, le manger, et surtout le boire. Ce comportement justifie d'une certaine manière l'opinion des patrons du 19ème siècle: augmenter les salaires, c'est envoyer les ouvriers au cabaret!

Dans ces conditions les ouvriers sont généralement gros buveurs d'alcool. La 'ninique' chez le Lillois (alcool de pomme de terre, de grain et de betterave à 85°), 'le pétrole' ou 'le chien' chez le Vosgien, l'eau de vie de 'poivre,' l'alcool à 90° chez le Polonais. Si l'Italien ne dédaigne pas l'alcool il préfère plus encore le 'champagne' le 'spumante' que, les jours de paie, le cantinier lui revend bien cher.

Une multitude de distributeurs, graisseurs, épiciers, marchands de vin, limonadiers, fruitiers, voire crémiers s'entend à récupérer l'argent gagné par les ouvriers.

En Lorraine industrielle chaque maison est un cabaret, tout le monde 'vend à boire.' Les grandes brasseries, les gros distillateurs, les marchands de vin et d'alcool favorisent les changements de profession.

Toute cette faune commerciale, issue le plus souvent des rangs ouvriers, s'attaque à des milieux vulnérables où l'isolement moral, la fatigue physique, l'augmentation et la régularisation des ressources sans naissance de besoins concurrentiels au manger et au boire, peut faire naître toutes sortes de névroses favorables à une alcoolisation plus ou moins poussée selon les facultés d'adaptation des individus aux situations nouvelles et selon les tendances d'intempérance préexistantes.

Selon le Dr Romme (1904) l'avènement de l'alcoolisme coïncide avec la fabrication de l'alcool autrement que par la distillation des boissons fermentées, avec des pommes de terre, des betteraves, des grains, des mélasses. La première distillerie industrielle de la betterave est fondée en 1854. Depuis cette date la production passe de 300,000hl de moyenne annuelle à 550,000hl pour la période 1880-4, 806,000hl pour 1895-99, 1,345,000hl pour 1900-13. En Lorraine annexée le 'Sprit' fabriqué en quantité industrielle par les grandes distilleries du Nord de la Confédération Allemande inonde le marché. En France la production industrielle d'alcool qui atteint à peine 70,000hl de moyenne annuelle entre 1840 et 1850, passe à 486,000hl vers 1855, 780,000hl vers 1865, plus d'un million d'hl vers 1876, 2m d'hl en 1892, 3m d'hl en 1912.

Les alcools 'dits de moyen goût' théoriquement impropres à la fabrication des spiritueux sont employés une fois sur six. Selon l'administration des Contributions Indirectes en 1895, plus du 1/3 des alcools impropres à la consommation et soumis à la dénaturation retourne en fraude à la confection de boissons alcoolisées très aromatisé (Romme 1904).

Le prix très bas de l'alcool industriel (28F. l'hl) assure jusqu'à la première guerre mondiale le triomphe des gros distillateurs betteraviers. A Lille la 'ninique' vaut 5 à 6 fois moins que l'alcool fin (3.15F. le litre). En Alsace l'alcool Allemand vaut 50 à 60 centimes le litre. Le 'tord boyaux' le pétrole est à la portée des bourses les plus modestes.

Les 'vins d'Espagne' fabriqués avec les excédents d'alcool de betterave allemand et introduit en France jusqu'en 1888, les vins artificiels fabriqués à partir du 3/6 de contrebande auquel on ajoute de l'eau, du sirop de glucose, de l'acide sulfurique, des colorants végétaux, et surtout chimiques (ex. la fuchsine) sont produit sans trop de risques sinon fiscaux. En Lorraine annexée les hygiénistes dénoncent le Kistlewin 'le vin qu'on fabrique au grenier' et A. Vigneron cite telle ville de Lorraine française d'où il est sorti deux fois plus de vin qu'il n'en est entré sans tenir compte de la consommation des habitants!

En dépit des protestations d'hygiénistes, d'économistes, ou d'hommes d'œuvres aucun gouvernement n'inquiète les grands fabricants d'alcool qui savent s'allier au pouvoir et obtenir distinctions honorifiques et tranquillité. Vers 1900 l'alcool procure à l'Etat Français 6 à 700 millions sur un budget de 3 milliards. La question des bouilleurs de cru qui agite la Lorraine et la France après 1870 n'est qu'un combat fiscal entre les défenseurs des intérêts campagnards et ceux des intérêts des gros marchands de vin et d'alcool.

Aussi la protection de la santé publique n'est-elle qu'illusoire. Les lois sont faites de façon à donner satisfaction à la fois à l'opinion publique qui les réclame et aux producteurs d'alcool ainsi qu'à leur armée de débitants qui empoisonnent la foule. En 1900 Triboulet et Mathieu constatent la faillite des moyens de protection employés. Personne ne veut ou ne peut s'attaquer aux racines du mal et les plus hardis parmi ceux qu'on écoute proposent de frapper tout alcool de droits très élevés et d'affecter les sommes ainsi perçues aux œuvres d'assistance qui répareront et préviendront les méfaits de celui-ci (Triboulet et Mathieu 1900).

L'alphabétisation des masses semble être l'espoir de tous les réformateurs.

Toutes les conditions sont réunies pour une alcoolisation de masse.

– Un grand nombre de nouveaux ouvriers affluant sans cesse des campagnes, coupé de son milieu d'origine, soumis à de nouvelles conditions de vie et qui se trouve brusquement confronté à une pléiade de spéculateurs: revendeurs, épiciers, cabaretiers débitants de toutes sortes, fabricants ou distributeurs d'alcool et de vin à bon marché produits à une échelle industrielle par une minorité de très honorables fournisseurs.

– Des pouvoirs publics soumis à des pressions de classe, peu désireux de porter remède à une situation qui leur rapporte en dépit des cris d'alarme des médecins, fonctionnaires ou hommes de cœur.

– Une majorité de nantis que l'alcoolisation des ouvriers ne concerne pas, n'inquiète guère sinon pour déplorer l'immoralité de ces fauteurs de trouble et la mauvaise santé des éventuels défenseurs de la patrie.

Si on en croit les moralistes c'est toute la nouvelle classe ouvrière qui est atteinte. En fait les avis diffèrent, Jules Simon estime qu'à Lille vers 1850 25 pour cent des hommes et 12 pour cent des femmes s'adonnent à la boisson (Pierrard 1965, 286). En Alsace certains médecins concluent à une intoxication alcoolique générale, d'autres estiment qu'on n'observe que rarement des cas d'intoxication alcoolique proprement dite.

Dans le pays de Briey, vers 1900 75 pour cent des ouvriers adultes sont réputés alcooliques ou alcoolisables (Rubendorf 1913).

Ainsi l'alcoolisme ouvrier semble être une conséquence du grand mouvement industriel du 19ème siècle qui a apporté une plus grande abondance de richesses, un essor rapide de la science et de la culture, une disparition des famines et une augmentation générale du pouvoir de consommation.

Plus ou moins acteurs et victimes de cet essor nombre d'ouvriers ont cru trouver en l'alcool un secours et un soulagement dans la surexcitation que procure celui-ci. Alcooliques de la pénurie et de l'isolement, les nouveaux ouvriers sont-ils moins à plaindre que les drogués de l'abondance et du désenchantement?

André, R., 1909 L'alcoolisme dans le Bassin de Briey – Hygiène de l'industrie du fer (Paris).

Bresard, Marcel, 1969 Alcoolisme et conscience collective, *Revue de l'alcoolisme*, no. 2 (avril, juin).

Claudian, Jean, 1970 Comportement de l'homme vis-à-vis du liquide, *Cahiers de Nutrition et de Diététique*, 5, fasc. 2.

Duveau, Georges, 1942 *La vie ouvrière en France sous le Second Empire.*

Faurobert, Louis, 1962 *Les degrés de l'alcoolisme.*

Jellinek, Pr., 1951 Colloque européen sur l'alcoolisme (Copenhague).

Ledermann, Sully, 1956–64 *Alcool, Alcoolisme, alcoolisation,* 1 et 2.

Lyzinski, Henri, 1931–2 Aperçu sur l'alcoolisme dans le Bassin de Briey, thèse médecine (Nancy).

Perrin, P., 1950 Les travaux du Centre Américain de Yale, no. 13, 14, 16. Etudes antialcooliques.

Pierrard, Pierre, 1965 *La vie ouvrière à Lille sous le Second Empire* (Paris).

Romme, R., 1904 L'alcoolisme et la lutte contre l'alcool en France, *Encyclopédie Scientifique des aide-mémoires* (Paris).

Rubendorf, J., 1913 La syphilis dans les agglomérations ouvrières du Bassin de Briey, thèse médecine (Nancy).

Sauvy, Alfred, 1969 Sociologie de l'alcoolisme. *Actes du 3ème Congrès National contre l'alcoolisme* (Versailles), octobre.

— 1963 Alcool ou santé. *Actes du Congrès de Rouen*, n° 3 et 4.

Tremulières, Jean, 1961 L'alcool erreur alimentaire. La santé de l'homme, no. 131 (mars, avril).

Triboulet, H., et F. Mathieu, 1900 *L'alcool et l'alcoolisme: notions générales, toxicologie, physiologie, pathologie, thérapeutique*, Bibliothèque générale des sciences (Paris).

c1707
A canonical correlation model relating mortality to socioeconomic factors
MARK S. MONMONIER   *State University of New York at Albany, USA*

Medical geography attempts to answer not merely the query 'Where?' but also the question 'Why?' Since disease systems are often quite complex – involving a multitude of interacting physical, social, and genetic factors – it is, at times, desirable to reduce the problem to an abstraction of reality in which only the principal elements are retained. To do this is to model reality.

Models can, in general, be assigned to one of three types: iconic, analog, and symbolic (Ackoff, Gupta, and Minas 1962, 108–10). Mathematical models are an important subset of symbolic models, a necessary subdivision since the larger set also includes cartographic symbols. Further partitioning segregates statistical models from other subclasses such as simulation and linear programming models. Statistical analysis reaches its maximum utility in those techniques whose primary goal is explanation and prediction rather than description. Canonical correlation is an explanatory statistical model with great potential for medical geography. This paper outlines the principles of canonical correlation and illustrates its use in medical geography by relating causes of death to socioeconomic conditions for the United States.

Given two sets of n and m variables, X_i and Y_j, canonical correlation finds two linear transformations

$$U = \sum_{i=1}^{n} u_i X_i = u_1 X_1 + u_2 X_2 + \ldots + u_n X_n$$

and

$$V = \sum_{j=1}^{m} v_j Y_j = v_1 Y_1 + v_2 Y_2 + \ldots + v_m Y_m,$$

such that the correlation between U and V is maximized (Hotelling 1936; Phillip and Gibson 1970). This correlation is referred to as the *canonical correlation* or *canonical root*. The coefficients u_i and v_j are the *canonical variates* or *weights* used to indicate the relative importance of the original variables in the linear expressions; they are similar to multiple regression Beta-weights (standardized Beta coefficients). As in multiple regression, it is possible to obtain expected values for U and V by evaluating these linear equations for each observation, using Z-scores from the original data. Thus, we can compute the *canonical scores* for observation k as

$$U_k = \sum_{i=1}^{n} u_i X'_{ki}$$

and

$$V_k = \sum_{j=1}^{m} v_j Y'_{kj},$$

where X' and Y' are the variables standardized to a mean of zero and a standard deviation of one. In essence, canonical correlation is akin to a multiple regression where there is no restriction on the number of dependent variables.

These canonical scores are, however, similar to principal component scores in that they represent the values for each observation scaled along a vector in a space defined by the original variables. This analogy is strengthened when one extracts successive pairs of canonical vectors; a separate pair can be obtained for as many variables as there are in the smaller of the two sets of original variables. For the second pair, the canonical variates are selected to yield the highest possible correlation between these two new vectors while keeping each of them orthogonal to the one previously extracted from the same set of variables. Subsequent pairs are chosen in a similar fashion so that all canonical vectors within a set are orthogonal.

A word of caution is needed in interpreting the results. While canonical variates are like factor loadings in the sense that they represent the relative contributions of the original variables to the canonical correlation between a pair of canonical vectors, they are *not* – as are loadings on principal compo-

TABLE 1. Correlations of canonical vectors with original variables[a]

Criterion variables[b]	Canonical vectors				
	I	II	III	IV	V
Diseases of the heart (400–2, 410–43)	−96				
Malignant neoplasms (140–205)	−99				
Vascular lesions of central nervous system (330–4)	−60	−70	31		
Accidents (E800–E962)	60	−50			44
Diseases of early infancy (760–76)	45		62	−32	38
Influenza and pneumonia (480–93)	−40		45	−32	
General arteriosclerosis (450)	−66	−37			
Diabetes mellitus (260)	−76				
Other circulatory diseases (451–68)	−72				53
Cirrhosis of the liver (581)	−39	70			50
Predictor variables					
Population increase, %, 1950–60	45	41	−36	−30	
Median age, 1960	−90				
Urban population, %, 1960	−33	83			
Negro population, %, 1960			88		
Aged (65+) population, %, 1960	−88	−39			
Families with income over $10,000, %, 1960		80			
Sound housing units, %, 1960	−34	74	−43		
Median income, per capita, 1967	−33	71			
Median no. of school years completed, 1960		38	−63		54
Population completed high school, %, 1960		35	−62		55
White-collar employment, %, 1960		69	−33	−32	36
One-unit housing structures, %, 1960	45	−67			
Population in an SMSA, %, 1960	−34	78			
Canonical correlation	98	96	93	82	80

[a]Correlations are shown as 1/100ths; those with absolute values less than .30 have been eliminated for clarity.
[b]Crude mortality rates; data are for 1962. Numbers in parentheses are classification numbers from *International Lists of Diseases and Causes of Death.*

nents – correlations between the vectors and the original variables. These correlations can, however, be obtained by correlating the canonical scores with the raw data; this should always be done to facilitate interpretation.

To illustrate the applicability of the model to medical geography the relationships between 10 leading causes of death (*Statistical Abstract of the United States* 1962, 61) and 13 socioeconomic traits were subjected to canonical correlation. Although the mortality data are only for a single year, 1962, and are in the form of crude death rates, the use of the 50 states and the District of Columbia as enumeration areas minimizes the effects of random fluctuations that would have arisen from the use of smaller sampling units; the effects of differences in age-structure can be considered in interpretation. The following table shows the canonical correlations between the first five pairs of vectors and the original variables. Their canonical roots are high and, if the data were normally distributed, would be statistically significant at the one-percent level, based on a chi-square test (Bartlett 1941).

The elimination of low vector-variable correlations from the table aids interpretation. Since similarity of sign and absolute value are all that need be considered at this stage, Vector I can be interpreted as showing a strong association between the age of a state's population, on one hand, and heart disease, cancer, arteriosclerosis, diabetes, and circulatory ailments on the other. Other factors have moderate loadings for this pair of vectors, but the above are dominant. The association between urbanization and affluence, on the predictor side, and vascular lesions and cirrhosis, as indicated by the second vector-pair, is also consistent with medical opinion. Vector III points out the correlation between significant numbers of Negroes and higher than average childhood mortality, influenza, and pneumonia. This correlation, however, should not be interpreted largely as a racial effect, since the limited access of Negroes to health care is also important. Vectors IV and V show the less significant influences of educational level, white-collar employment, and population growth.

Canonical correlation is one of the most useful statistical models for medical geographers since it provides an efficient means of relating sets of criterion and predictor variables. The criterion set may include any disease incidence data of interest, while the predictor set can measure physical, social, and – when better information is available – genetic traits of the sample areas. The method's utility lies in its finding the highest correlations across the two sets of variables under an orthogonality constraint. Computation of the correlations between the original data and the canonical vectors increases interpretability.

Ackoff, R.L., S.K. Gupta, and J.S. Minas, 1962 *Scientific method: optimizing applied research decisions* (New York).
Bartlett, M.S., 1941 The statistical significance of canonical correlations, *Biometrika*, 32, 29–38.
Hotelling, H., 1936 Relations between two sets of variates, *Biometrika*, 28, 321–77.
Phillip, P.J., and S. Gibson, 1970 Simple, multiple and canonical correlation, *Inquiry*, 8, no. 2, 55–9.
Statistical Abstract of the United States, 1962 (Washington).

c1708
Medical approach to the geography of oncochercosis in Mexico
A. MONTIEL-HERNANDEZ   *Ministry of Education, Mexico*

The study of the geography of this disease in our country is very important because of the large area which it embraces and the importance of the population affected by it. Oncochercosis is a parasitic disease caused by the filaria *Oncocherca volvulus* and communicated by various species of mosquitoes of the *Simulidae* family. The filaria can be found on the skin or in the subcutaneous cellular tissues, where it can create fibrous nodules. The gravest consequence of this disease is blindness.

Oncochercosis was discovered by Leuckart, a German missionary doctor, who found, in two natives of the Republic of Ghana, subcutaneous tumours filled with worms, which afterwards were called *Filaria volvulus*. A most recent hypothesis on the appearance of oncochercosis in Mexico is the one sustained by Dr Javier Torroella, who established that oncochercosis was brought to Mexico by the Sudanese soldiers who came with the French troops during the second half of the nineteenth century, and remained in Villa Alta, State of Oaxaca.

The staff of the medical school of Guatemala was the first to establish a diagnosis and treat oncochercosis.

The surveys made by our health authorities on epidemic diseases from late in the 1940s to the present have revealed the existence of only 3 zones affected by oncochercosis, located in the southeastern part of Mexico; two in the state of Chiapas and the third one in the state of Oaxaca.

We find two areas of oncochercosis in the state of Chiapas. The first one, located in the northern part of the state is named 'Chamula area.' It takes its name after a Tzotzil group of natives who live in this zone. The population of this region is 225,000 inhabitants; 4000 are infected. The second one, known as 'Soconusco,' covers part of the region of that name. Here we find a population of 93,000 inhabitants; 30,000 are infected.

The oncochercosis area of Oaxaca is located in the northern and central part of the state, with a population of 45,000 inhabitants; 5800 are infected. The territorial extent of the three areas covers a total of 8900km^2.

The various factors in the geographical environment that have a decisive influence on this disease are:

(1) *Latitude*. The endemic areas are located in tropical zones between 18° and 15° north latitudes. The most distant limits of the three centres of oncochercosis are: (a) Soconusco area; 15°4' and 15°57' north latitude; 92°5' and 93°7' w. longitude, (b) Chamula area; 16°22' and 17°7' north latitude; 92°29' and 92°40' w. longitude, (c) Oaxaca area; 17°25' and 17°48' north latitude; 96°12' and 96°40' w. longitude.

(2) *Altitude*. This is one of the most important factors in the proliferation of the *Simulidae*. In the oncochercosis areas of Mexico these insects are found at an altitude ranging from 600 to 1300m asl.

(3) *Climatology*. (a) *Temperature*. This factor affects the growth of the *Simulidae*. The average temperature in endemic zones is between 18° and 20° C. (b) *Light*. This factor affects the activity of the *Simulidae*, as observations to this effect have concluded that there is a decrease in the activity of these insects as light increases. This explains the low rate of bites between eleven o'clock and one o'clock noon. (c) *Wind*. The direction and speed of the wind extends or shortens the area of activity and the flight capacity of the *Simulidae*. (d) *Humidity*. In the endemic zones the average humidity ranges from 70 per cent to 90 per cent. Increases in humidity are most favourable for the insects and their power for infection reaches its highest point when the average humidity is from 70 per cent to 80 per cent. (e) *Rainy season*. The areas infected by oncochercosis are closely influenced by the rainy season, as during this period the adult mosquitoes do not find any adequate resting places; this affects their flight as they have to be constantly flying and, as a consequence, they do not have enough strength to search for their food. Furthermore, the larvae are washed down by water streams running at very high speeds and the number of mosquitoes is considerably reduced; but, during the dry season these insects quickly recover and multiply as the speed of the runoff waters diminishes.

(4) *Hydrology*. This factor contributes greatly to the survival of the *Simulidae*. The rivers, with high speed streams, serve as breeding places for these insects and have the following traits: (a) *Volume*. The breeding places of the *Simulidae* are located in streams of medium volume and high banks; usually young rivers, without beach. (b) *Turbulence*. The slope angle and stream course as well as the presence of obstacles both on the bed and banks of the rivers, have a great influence on the biological conditions relative to feeding and breathing. The presence of microorganisms is necessary for the *Simulidae* breeding places, as it nourishes the larvae. During the rainy season the increase in speed of the streams and the rivers washes down the microorganisms and the larvae, causing their death.

(5) *Orography*. The slopes in the ground determine the angle of the rivers where the *Simulidae* breeding places are found and here in Mexico the slope of the ground in the zones infected by oncochercosis is very sharp.

(6) *Flora*. This factor acts on oncochercosis by developing ecological conditions that either help or hinder the dissemination of this disease. The areas infected by oncochercosis are located in places having a luxuriant vegetation that keeps the air humid and lessens the amount of light; this circumstance favours the multiplication of the *Simulidae*.

(7) *Fauna*. Animals also play an important role in the proliferation of the *Simulidae*. Some of the fish eat the larvae, while others eat only the food the larvae need for their nourishment.

(8) *Human factors*. Oncochercosis appears mainly within the rural population, as the low standard of living of these people favours contagion. The endemic areas are mainly located within coffee plantations that require a great amount of man labour in the open country, thus leaving the workers exposed to mosquito bites.

Furthermore, as temporary labour is very common in the coffee plantations, this compels the workers to frequently travel from infected to non-infected zones, spreading the disease in this manner.

The economy in the endemic zones is also affected by oncochercosis, since the people who are left blind by this disease become non-productive citizens.

Based on the data obtained in Mexico by Drs Samuel Villalobos and José Larumbe, an oncochercosis program was established in 1930. Their principal objectives were to give medical and particularly surgical, attention to the patients and to keep this disease under control.

This institution has the following basic activities: (1) census in the area of the population affected by oncochercosis; (2) diagnosis of the sickness; (3) medical and/or surgical attention; (4) health education.

Oncochercosis is confronted with two problems: pathological and geographical; and in order to be able to eradicate this disease it is necessary not only to be acquainted with it from a medical point of view but to have a wide knowledge of the geographical environment which has a decisive influence on the population of the endemic zones. In this respect the work performed by medical geography is essential and, therefore, we highly recommend the inclusion of medical geographers in the staff of the oncochercosis program. These geographers will be in a position to make a careful analysis of each and every one of the geographical factors that favour or hinder the development of this disease.

We sincerely hope that a close co-operation and mutual assistance among physicians and geographers will contribute to the eradication of oncochercosis.

Baez, Manuel Martínez, 1971 *Manual de parasitología médica*. Prensa Médica Mexicana (México).

Casahonda, Enoch Cancino, 1967 Algunos aspectos de la oncochercosis en Chiapas. *Revista del Instituto de Ciencias y Artes de Chiapas*, no. 18, Tuxtla Gutiérrez (Chiapas).

Mackie, Hunter, and Worth, 1946 *Manual de medicina tropical*, Science Service (Washington, DC).

Secretaría de Salubridad y Assistencia, Dirección de Salud Pública, 1962 *El Problema de la Oncochercosis en México*. Revista de Salud Pública, 4, no. 6 (México, DF).

c1709
Academic training in Mexico of medical geography specialists
C. SAENZ DE LA CALZADA *National University of Mexico*

A remarkable event for the future of medical geography in Mexico was the special meeting of the Commission of Medical Geography, which, sponsored by the International Geographical Union of which it is a part, was held in the capital of Mexico during the Latin American Regional Conference, in August 1966. Present at the meeting, presided over by the British geographer, A.T.A. Learmonth, were specialists in nosoctonology from Great Britain, the United States, Australia, New Zealand, Canada, and Mexico (the latter submitted papers by Fernando Latapi, Yolanda Ortiz, Somolinos D'Ardois, and Saenz de la Calzada); all the papers considered were compiled in volume 6 of the report of that conference, which was attended by more than 700 geographers from all parts of the world.

This event immediately affected the structure of the program of studies for the degree in geography at the National Autonomous University of Mexico, which included, in 1967, for the first time in the world, medical geography as a regular subject (and we also believe that the subject of nutritional geography, which we founded a year later at the School for Nutritionists of the Ministry of Health and Assistance, has been the first in the international field). Later, in 1971, the university council approved a new five-year study program for a bachelor's degree in geography, establishing the specialization in applied geography, and within it, a specific option for medical geography, in accordance with the following plan which covers the last year of this career.

FIRST SEMESTER: seminar on medical geography, seminar on Mexico's climatology, geomedical statistics, didactics of geography and practices, geographical research methods.
SECOND SEMESTER: seminar on medical geography II, seminar on Mexico's climatology II, medical climatology, soil and water chemistry, nutritional geography, fundamental medicine for geographers.

There is, furthermore, during the fourth year of the curriculum, a course on general geographical medicine, and for the doctor's degree, a seminar on specialized studies on medical geography, with research work in the field, devoted exclusively to the Mexican Republic.

The seriousness of the geography studies, and the specific specialty of medical geography, have been duly assessed by the health authorities in Mexico, who accept our graduates in their courses for a master's degree in public health, which had previously been reserved for physicians and engineers, and this opens up new prospects for nosoctonological work.

Nevertheless, we believe that the plan of studies is insufficient, if it is not supplemented by three additional subjects: epidemiology, preventive medicine, and ambient hygiene, and if the college of geography is not provided with the necessary didactic materials; because, even though we do have a laboratory for edaphology and meteorology, we do not have any anatomic models nor proper physiological laboratory work to adequately assess the relations between the meteorological and homeostatic variables of the individuals (in their most prominent projections).

Apart from this, all the personal research work which we have successfully begun in the rural and urban environments, especially in the 'misery belt,' by means of carefully prepared research, should be increased in extent and improved in depth.

The other aspect, contact with the great research fellows of the world in the field of medical geography: Learmonth, from England; Jusatz, from the German Federal Republic; Ignatiev, from the USSR; May from the United States, and many others, as well as the participation in International Geographical Congresses (let us remember the extraordinary importance of the 21st International Geographical Congress, held in 1968 in New Delhi, and which was attended by thirty-six Mexican geographers) keep our specialists permanently informed on the world development of geography and the subtle, but definite, changes which this *mater* discipline is experiencing at present, and which are so obvious in the nosoctonological field.

Therefore, the circumstances are favourable to the future of this specialty in Mexico, since the geographers are in agreement to sponsor this new discipline. For instance, at the Institute of Geography of the National Autonomous University of Mexico, there are already two research fellows efficiently and enthusiastically working in this new specialty: Zaida Falcon and Luis Fuentes Aguilar. The interest of the future geographers in Nosoctonology, both at the National University and at the Higher Normal School, is rapidly increasing, making it possible to foresee in the immediate future, an enthusiastic group of specialists in medical geography who will promote this specialty which has such a deep historical tradition in Mexico.

If, as explained by Schlegel, the historian is a prophet in reverse, we, who have been circumstantial historians of the brilliant development which medical geography has experienced in our country since the pre-Hispanic era, and are at present modest collaborators in its actual 'status,' set our courses towards the future, in a prophesy filled with faith in Mexican nosoctonology.

Universidad Nacional Autónoma de México,
1971 *Guía de Carreras, 1971* (México).

c1710

The copper, zinc, lead, and molybdenum contents of some British and Canadian vegetables: a reconnaissance
HARRY V. WARREN *University of British Columbia, Canada*

During the past two years numerous studies, largely resulting from an increased public interest in contamination of air, water, and food, have provided many data on the trace element content of food. Most of these studies have been based on what are considered to be 'average' diets. This paper must be considered merely as a contribution to show that any one individual's trace element intake through eating vegetables may vary widely from whatever may be considered normal.

Local geological and geographical factors as well as urban and industrial contamination are responsible for greater variations in the trace element concentrations found in vegetables than is generally appreciated. The vegetables reported in this paper were selected largely because they were most commonly

TABLE 1

Vegetable	A Dry weight as percentage of wet weight		B Ash as percentage of dry weight	
	Between 75% and 85% fall between	Working average	Between 75% and 90% fall between	Working average
Lettuce	4–10	6.5	15–20	19.0
Cabbage	4–10	7.0	8–11	9.3
Potato	15–20	19.5	4–6	4.7
Bean (except broad)	7–14	11.0	5–9	7.3
Carrot	8–14	11.0	5–8	6.8
Beet	8–14	11.5	7–10	8.5

grown and could be collected from a minimum of seventy-five separate localities. Because different gardens provided different assemblages of vegetables it is obvious that it is not desirable to draw too precise conclusions concerning the ability of different vegetables to take up various trace elements from a soil.

Representative samples of each vegetable growing in a garden were selected. Unfortunately different vegetables mature at different times and furthermore their trace element concentrations vary during the growing season. Unfortunately our resources usually permitted only one visit to each locality during a growing season. Thus at different localities varying collections of vegetables were made. Obviously the resulting data must be accepted merely as a reconnaissance. Nevertheless the results should serve to emphasize the importance not only of geological and geographical factors but also those of urban and industrial contaminations, factors which have not always received the attention they deserve in those more sophisticated studies to which reference has been made.

Each vegetable sample was prepared as it would be by a 'prudent housewife.' This meant that young potatoes or carrots were scraped and older ones peeled. Lettuces and cabbages were stripped of their outer leaves and carefully washed with tap water – we were interested in what people were going to eat.

After the vegetables were prepared, as they would be for consumption, they were oven dried at approximately 60° C. until they were dry enough to be milled in a small Wiley Mill. From one to five grams of the milled material were then ashed overnight at 550° C.

If practical, the 'wet' or natural vegetable was weighed and also the resulting oven-dried material, and likewise the ash obtained therefrom. In this way it was possible to make crude comparisons between trace element content of ashed, oven-dried, or of 'wet' vegetables. Various writers report their results differently – in parts per million (ppm) – of ash, of oven-dried, or 'wet' material.

Table 1 lists the ranges and 'working averages' for relating 'wet,' oven-dry, and ash contents of vegetables. It was not always practical to determine the relationship between 'wet' and oven-dried weight so that these relationships have actually been tentatively determined on from twenty to forty samples and may be expected to be revised when more data become available. It is expected that these figures will be modified as more data are made available. Tables 2 and 3 give, for sundry localities, the 'commonly encountered' or tentatively 'normal' copper, zinc, lead, and molybdenum contents of lettuce, cabbage, potato, bean (except broad), carrot, and beet in terms of ash, oven-dried, and fresh or 'wet' sample, and deviations that have actually been encountered in one or more localities.

From an examination of the data presented above the following observations appear pertinent: (1) The trace element contents of vegetables vary much more than is generally appreciated. (2) In so far as copper, zinc, lead, and molybdenum are concerned, their relative order of concentration in vegetables tends to parallel that found in normal soils in the earth's crust as a whole, i.e. zinc is most abundant, then copper, then lead, and lastly molybdenum. (3) The greater the normal concentration of an element in a vegetable the smaller the deviations from the normal are likely to be. Conversely the smaller the normal concentrations of an element the larger may be the deviations from

TABLE 2. Summary of copper, zinc, lead, and molybdenum contents commonly found in some vegetables and deviations that may be encountered under anomalous circumstances (in ppm)

		Lettuce	Cabbage	Potato	Beans (except broad)	Carrot	Beet
A	*Copper*						
	1 In ash	60	40	100	100	70	80
	2 In oven-dried	11	4	5	7	5	7
	3 In fresh or 'wet'	.7	.25	.9	.8	.5	.8
B	*Zinc*						
	1 In ash	400	300	320	450	450	420
	2 In oven-dried	75	30	15	35	30	35
	3 In fresh or 'wet'	5	2	3	4	3	4
C	*Lead*						
	1 In ash	20	16	40	30	30	20
	2 In oven-dried	4	1.5	2	2	2	2
	3 In fresh or 'wet'	.25	.1	.4	.25	.2	.2
D	*Molybdenum*						
	1 In ash	5	20	16	60	30	4
	2 In oven-dried	.95	3	.75	4	2	.35
	3 In fresh or 'wet'	.06	.2	.15	.5	.2	.04

TABLE 3. Summary showing the range and order of magnitude of deviations in the copper, zinc, lead, and molybdenum concentrations that may be found in vegetables grown under anomalous conditions

		Lettuce	Cabbage	Potato	Beans (except broad)	Carrot	Beet
A	*Copper*						
	1 Lowest contents found[a]	1/15	1/6	1/9	2/5	1/9	1/8
	2 Highest contents found[b]	8	2½	4	2½	2½	2½
	3 Extreme range of concentrations found[c]	120	15	36	6	22	20
B	*Zinc*						
	1 Lowest contents found[a]	1/6	1/2	1/2	1/2	1/2	1/4
	2 Highest contents found[b]	16	6	5	2	8	12
	3 Extreme range of concentrations found[c]	90	12	10	4	16	48
C	*Lead*						
	1 Lowest contents found[a]	1/10	1/8	1/10	1/5	1/3	1/6
	2 Highest contents found[b]	30	2½	15	4	9	11
	3 Extreme range of concentrations found[c]	300	20	150	20	27	66
D	*Molybdenum*						
	1 Lowest contents found[a]	1/8	1/30	1/16	1/30	1/4	1/30
	2 Highest contents found[b]	12	8	7½	7	3½	10
	3 Extreme range of concentrations found[c]	96	240	120	210	14	300

[a]Expressed as a fraction of those commonly encountered.
[b]Expressed as a multiple of those commonly encountered.
[c]Expressed as multiples of one.

the normal, e.g., zinc anomalies are usually an order of magnitude lower than are those of molybdenum. (4) Although not demonstrated in the data presented above, it may be stated that in many urban and industrial areas contamination adds greatly to the trace element concentrations found in vegetables. (5) Some vegetables appear to have a greater affinity for one or more elements than do others: beans and cabbages have a much greater affinity for molybdenum than do beets and lettuce.

Because so many people eat vegetables that come from many localities and because vegetables constitute only a modest proportion of our intake of food it is obvious that only under unusual circumstances are people likely to suffer from a surplus or deficiency of trace elements. Nevertheless epidemiological studies do suggest that unusual prevalences of some diseases can be related to specific geographical areas: Itai-Itai disease, which is caused by too great an intake of cadmium, was first found to be related to smelting operations in Japan. Minamata disease, which is caused by too much mercury being ingested, was eventually related to an area where sea food had unexpectedly high concentrations of mercury.

In other studies made in our laboratories unusual concentrations of one or more elements, including zinc, lead, and/or molybdenum have been found in vegetables grown in areas where medical men have reported unusual prevalences or mortality rates for such diseases as multiple sclerosis and stomach cancer.

If epidemiologists and environmentalists, and this might include medical geographers, pooled their knowledge, it seems probable that some correlations of significance may emerge.

Canada Council; Donner Canadian Foundation; The Geological Survey of Canada; the National Research Council of Canada; Defence Research Board of Canada; several mining companies; R.J.F.H. Pinsent; Dr and Mrs L.E. Lefevre; A. Bradshaw; R.E. Delavault; K. Fletcher; A.S. Dhillon; Anne Baxter.

Cannon, Helen L., and Howard C. Hopps, eds., 1971 *Environmental Geochemistry in Health and Disease* (Am. Assoc. for Advancement of Science Symposium, Dallas, Texas, December 1968). Geol. Soc. of Am. Memoir 123.

Furst, Arthur, 1963 *Chemistry of chelation in cancer* (Springfield, Ill.).

McAlpine, Douglas, Charles E. Lumsden, and E.D. Acheson, 1965 *Multiple Sclerosis: A Reappraisal* (Edinburgh & London).

Underwood, E.J., 1956 *Trace Elements in Human and Animal Nutrition* (New York).

Voisin, Andre, 1959 *Soil, Grass and Cancer*. Trans. from the French by Catherine T.M. Herriot and Dr Henry Kennedy (London).

Warren, Harry V., Robert E. Delavault, and Christine H. Cross, 1967 Possible correlations between geology and some disease patterns, *Annals of the New York Acad. of Sci.* 136, art. 22, 657–710.

Warren, H.V., R.E. Delavault, K. Fletcher, and E. Wilks, 1971 A study in lead pollution, *Western Miner* 44, no. 2, 22–6.

Warren, H.V., R.E. Delavault, and K.W. Fletcher, 1971 Metal pollution – a growing problem in industrial and urban areas, *The Can. Mining and Metall. Bull.* (July), 34–35.

15
Symposia
Symposiums

Symposia on selected topics were held in various Canadian universities before and after the International Geographical Congress in Montreal. Some of the papers solicited for these symposia are included here.

The following list of Symposia, with the short titles included and first authors' surnames, will assist in identifying articles, topics, and places of interest. A complete author and co-author index is located at the end of this volume, as well as a selected index relating papers to geographical locations. Note that the papers are *not* listed in alphabetical order. The organization of the volumes is described in full in the Preface.

Certaines réunions de symposiums traitant de thèmes spécifiques se sont tenues dans diverses universités du Canada avant et après le Congrès international de géographie à Montréal. On retrouve ici quelques-unes des textes de ces symposiums.

La liste qui suit donne les titres abrégés des textes et les noms des auteurs principaux et permet d'identifier les articles, les sujets et les lieux géographiques. Un index complet d'auteurs et de coauteurs se trouve à la fin de ce volume ainsi qu'un index des endroits géographiques qui font l'objet d'études. Prière de noter que les textes ne sont pas classés par ordre alphabétique. On explique en détail le plan de ces volumes dans la Préface.

S01
Cultural Discord in the Modern World
Les discordes d'ordre culturel dans le monde moderne
CONVOCATEURS/CONVENORS: L.J. Evenden, F.F. Cunningham, E.M. Gibson, R.B. Sagar, *Simon Fraser University, Burnaby*; R. Leigh, J.V. Minghi, *University of British Columbia, Vancouver*

s0101	Race and culture: urban conflict and reconciliation (USA) BROWN	1235
s0102	Modernization trends in Gabon CHAUSSADE 1236	
s0103	Voting behaviour in a Belfast parliamentary constituency (UK) DOUGLAS	1237
s0104	Tourist as counter-agent in cultural diffusion (Denmark and Turkey) MILLER	1240
s0105	Territorial implementation of black power, United States MORRILL	1242

S02
Water Resources
Ressources en eau
CONVOCATEURS/CONVENORS: F.M. Leversedge, W.R.D. Sewell, *University of Victoria*; T. O'Riordan, *Simon Fraser University, Burnaby*

s0201	Clear water erosion: geographical and international significance BECKINSALE	1244
s0202	Dam hindsight in Texas (USA) COOK 1246	
s0203	Public participation in decision-making DAVID 1248	
s0204	Benefit-cost analysis of multiple-purpose reservoir, Deer Creek, Ontario (Canada) DAY 1248	
s0205	Environmental impacts of water resource development FISCHER	1250
s0206	Neophilicism, environmental stress, and water resource management: the price of democracy FOSTER 1252	
s0207	Residential water demand: an econometric analysis and implications for management LINO GRIMA 1254	
s0208	Pricing and efficiency in water resource management HANKE	1256

s0209 Perception of phosphate pollution in an urban area (Canada, Windsor) LAVALLE 1258
s0210 The decision to control eutrophication (North America, Great Lakes) LEE 1260
s0211 Water-landuse policy on the amphibious Piedmont Corridor, Colorado (USA) LOEFFLER 1262
s0212 Problems and issues of implementing national water legislation at subnational levels (USA) MUCKLESTON 1264
s0213 Canadian glacier inventory applied to static water balance studies: Vancouver Island (Canada) OMMANNEY 1266
s0214 Approach to evaluation in multiple objective basin planning (North America) O'RIORDAN 1268
s0215 Attitudes towards water quality and water-based recreation, Qu'Appelle Valley, Saskatchewan (Canada) PARKES 1270
s0216 Drainage and politics of western water protection: defining area of origin (North America) QUINN 1273
s0217 Interbasin transfer of water, England and Wales RAUP 1275
s0218 International Joint Commission: a model for international pollution control organizations? (North America) ROSS 1277
s0219 Role of hindsight evaluations (USA, California) THOMAS 1279
s0220 Competition among water users within the framework of statute law (Australia, New South Wales, Hunter Valley) TWEEDIE 1282
s0221 Water and regional development, Murray River Basin, Australia WOOLMINGTON 1283
s0222 Nature of attitudes toward renovated wastewater (USA) BAUMANN 1286
s0223 Hindsight evaluation: Tryweryn reservoir in Wales (UK) MITCHELL 1288
s0224 Methodological problems in water resources management SOLODZUK 1290
s0225 Water use map of Strait of Georgia and Puget Sound: preparation and application (Western North America) ROBERTSON 1292

S03
Role of the City in the Modernization of Developing Countries
Les villes dans les pays en voie de développement
CONVOCATEURS/CONVENORS: J. Spelt, D.P. Kerr, J.B.R. Whitney, *University of Toronto*

s0301 Formation of central place systems in developing countries (Mexico) GORMSEN 1293
s0302 Market centres in western India KULKARNI 1295
s0303 Cities of the third world: industrialization and spatial repercussions (South America) SANTOS 1297
s0304 Middle East urbanization with special reference to Syria, Lebanon, Iraq, Iran WIRTH 1300
s0305 Modernization of inland southwestern Nigeria: the role of Ile-Ife IBIODUN 1303
s0306 Spatial structure of cities in India: pre-industrial or post-colonial? BRUSH 1303

S04
Frontier Settlement on the Forest/Grassland Fringe
L'aménagement du territoire en régions frontalières prairies/forêts
CONVOCATEURS/CONVENORS: R.G. Ironside, *University of Alberta, Edmonton*; C.J. Tracie, J.H. Richards, E.N. Shannon, *University of Saskatchewan, Saskatoon*

s0401 Generation waves and spread of settlement (Northern Sweden) BYLUND 1306
s0402 Government land policies, southern Africa CHRISTOPHER 1309
s0403 Evolution of Australian pastoral land tenures: challenge and response in resource development HEATHCOTE 1311

s0404   Unit size as a factor in land disposal policy (USA)   PATERSON   1313
s0405   Edaphoclimatic frontiers (Canada)   WILLIAMS   1315
s0406   Railroads and urban settlement in Nebraska: a case study (USA)   DAVIS   1317
s0407   Organized land opening, a facet of settlement geography, Venezuela   EIDT   1319
s0408   Colonization and retreat in cut-over lands, Lake Superior region (USA)   JAATINEN   1320

S05
Karst Geomorphology
La géomorphologie karstique
CONVOCATEURS/CONVENORS: D.C. Ford, *McMaster University, Hamilton*; M.F. Goodchild, *University of Western Ontario, London*; M.C. Brown, *University of Alberta, Edmonton*

s0501   Development of limestone caverns (Europe and North America)   FORD   1322
s0502   Comparison of littoral with other karren (Puerto Rico)   MIOTKE   1324
s0503   Alternate solutional and nonsolutional effects in cavern genesis   OZORAY   1327
s0504   Geological environment of cave development   WALTHAM   1328
s0505   High altitude karst in the Himalaya (India and Nepal)   WALTHAM   1330
s0506   Classification of cave sediments   WOLFE   1332
s0507   Karst et préhistoire (Sud Europe)   FENELON   1333
s0508   Cuban coastal plains and lowlands as a specific type of tropical karst   PANOŠ   1333

S06
Developing the Subarctic
La mise en valeur des régions sub-arctiques
CONVOCATEURS/CONVENORS: J.R. Rogge, A.K.W. Catchpole, *University of Manitoba, Winnipeg*

s0601   Development in the fringe settlement zone, northern Sweden   PORENIUS   1335
s0602   Problems of the subarctic USSR   SOCHAVA   1337

S07
Minority Group Settlement
Installation de groupes minoritaires
CONVOCATEURS/CONVENORS: T.R. Weir, H.L. Sawatzky, *University of Manitoba, Winnipeg*

s0701   Landscape perception, minority group viability, and cultural group extinction (USA, Indiana, Yellow Creek)   LANDING   1342
s0702   Albanese communities in Pollino, Italy, faced by new consumer mass civilization   CATAUDELLA   1343
s0703   L'intégration des Canadiens-Polonais de la région de Barry's Bay, Ontario   JAROCHOWSKA   1347
s0704   Landscape of early group settlements, western Manitoba (Canada)   TYMAN   1348

Cultural Discord in the Modern World
Les discordes d'ordre culturel dans le monde moderne

s0101
Race and culture: urban conflict and reconciliation
WILLIAM H. BROWN, JR. *University of California, USA*

The expansion of European powers during the past six centuries has, as viewed by Frazier, resulted in the creation of worldwide 'frontiers of race and culture' (Frazier 1957, 3–36). In the United States the major racial frontier was created through the importation of African slaves whose descendants have had little exposure to other than western models of society and culture. Central to ascendant American economic and political power have been the growth of its cities and the urbanization of its people. The urban revolution of the twentieth century has seen blacks outstrip the white urban population so that more than 70 per cent of all blacks are now classified as urban dwellers while approximately 67 per cent of all whites are so classified. The major problem in this conversion from an agricultural to an urban-industrial society is not to be found in the small difference in degree of urbanization between blacks and whites, but in the vast differences in the physical, social, and economic amenities and opportunities ordinarily offered by urban life to the two races. The frontier of race and culture in America has therefore assumed a sharp urban focus with easily measured differentials in terms of education opportunities, employment, housing, family life and organization, and mortality.

Social ecologists of the 1920s envisioned urban cultural groups as existing in a more or less harmonious spatial arrangement determined partly by choice and partly by the complex but naturally occurring functions of city life of which these groups were a part. However, to the extent that the essential principle of a viable ecosystem is one of an intrinsically harmonious interaction of all of its parts, a closer analysis, particularly in terms of energy flow, would have revealed an inordinate consumption of social energy in perpetuating the existence of some of these groups, particularly blacks, at a given social level and in a given space.

Why has America been willing to spend so much of its energy in 'keeping the black man in his place,' as it were? And what does it mean in terms of reaching a satisfactory racial/cultural accord?

Tocqueville despaired 'of seeing an aristocracy disappear which is founded upon visible and invisible signs' (Tocqueville 1957, 373). Erik Erikson notes 'the unpleasant fact that our god-given identities often live off the degradation of others (Erikson 1965, 236). In the case of white Americans this 'pseudo-speciation,' as Erikson calls it, is the mythology of race, supported by the 'pseudologia' of racism. The spatial and social isolation of blacks from the mainstream of American life which has followed has, to be sure, conferred a general sense of community and culture upon blacks, something often romanticized by blacks and whites alike. But as a speaker at a black power rally once pointed out, 'the black man is no more native to the ghetto than the white man is to America.' Certainly in providing a culture area for the development of a people's unique potential, something far better can be offered than our urban ghettos. 'For,' as Alaine Locke said with respect to Harlem in 1936, 'no cultural advance is safe without some sound economic underpinning, the foundation of a decent and reasonably secure average standard of living; and no emerging elite – artistic, professional or mercantile – can suspend itself in thin air over the abyss of a mass of unemployed stranded in an overexpensive and crime-ridden slum. It is easier to dally over black Bohemia or revel in the hardy survivals of Negro art and culture than to contemplate this dark Harlem of semi-starvation, mass exploitation and seething unrest. But turn we must. For there is no cure or saving magic in poetry and art, an emerging generation of talent, or in international prestige and interracial recognition, for unemployment or precarious marginal employment, for high rents, high mortality rates, civic neglect, capitalistic exploitation on one hand and radical exploitation on the other' (Locke 1936, 457).

But ghetto life and culture do not entirely define the black experience. What does is the racism which every black American encounters in one form or another, at one time or another. It is this which has kept blacks from jumping into the great American 'melting pot' with the hopeful abandon of European culture groups.

'One of the grim dangers facing *Homo sapiens* is the continuing homogenization of cultures, the erosion of man's spectacular array of cultural differences. ... These priceless resources are in short supply, they are dwindling rapidly, and they are non-renewable' (Ehrlich and Ehrlich 1970, 303). The important point in the foregoing statement is the need for diversity in the ways of life offered to people; but diversity need not be gained solely through the perpetuation of old culture forms, which may in fact be outmoded and counterproductive in terms of a society's survival. Diversity can also be obtained through the cultivation of *new forms* which emerge naturally in the course of a society's development.

The resurgence of black nationalism during the recent decade has had a contagious effect in emphasizing the persistence of yet other self-conscious unassimilated culture groups in American life, each proclaiming its own version of the black power slogan. In addition, there has emerged an active and articulate youth culture now made fully potent politically in America by granting the vote to eighteen year olds. What this portends is not entirely clear as yet, but this much is apparent: the ideological death of integration, and the emergence of a white radical youth culture now represent the cornerstones of a polycentric cultural development in American life. As a political force an alliance between the two has not been easy to forge, but it has been carried to its furthest development yet in Berkeley, California, where in the most recent city council elections (April 1971) such an alliance was responsible for giving the city its first black mayor. Of even greater significance are the kinds of questions which were raised during the election, such as a proposal (which was defeated) to give official recognition to three distinct sub-communities, the hill community (white), the university community, and the flatlands (black), each with its own police force.

Positive recognition and cultivation of diverse cultural groups in American life, particularly in its large cities, runs counter to the accepted notion of harmony by amalgamation or acquiescence. Hopefully America is destined for constructive conflict in this respect. The Berkeley experience is possibly a straw in the wind.

Ehrlich, Paul R., and Anne H. Ehrlich, 1970 *Population, Resources, Environment: Issues in Human Ecology* (San Francisco).
Erikson, Erik, 1965 The concept of identity, in Talcott Parson and Kenneth B. Clark, *The Negro American* (Boston).
Frazier, E. Franklin, 1957 *Race and Culture Contacts in the Modern World* (Boston).
Locke, Alaine, 1936 Harlem: dark weather vane, *Survey Graphic* (August).
Tocqueville, Alexis de, 1957 *Democracy in America*, 1 (New York).

s0102
On modernization trends in Gabon
J. CHAUSSADE   *Université de Moncton, Canada*

In the countries of Africa, as in many other places no doubt, the city still constitutes the *milieu par excellence* where the traditional patterns of living break down the most rapidly and completely. This paper attempts a brief description of the effects which the city, as a place where habits are weakened, has on its neighbouring villages, and the example chosen is the encroachment of Libreville, in Gabon, on the village communities of the equatorial forests, situated less than 60km from the capital.

Traditionally villages of 100 to 200 inhabitants were located near a river. The men hunted and fished while the women and children tended manioc 'plantations' and banana groves situated in previously cleared forest areas. The village community thus corresponded to a family, this term being applied in its broadest sense, since it comprised the father (the village chief), his sons and the sons' wives who had always been chosen from another clan in accordance with the practice of exogamy. When the family became too large, that is, when the number of clearings surrounding the village could not be increased without the soil becoming exhausted too rapidly and the survival of the entire community being thus endangered, one of the sons with his family would leave to settle elsewhere near some other village which had just been created by the same process. The traditional Fang village always had the same plan: two parallel rows of rectangular huts on each side of a wide passageway guarded at the entrance and exit by small forts perforated with loop-holes. In the middle of the passageway there was always a guard-house or 'abègne' where meetings were held to apportion the work, and in a more general way, to palaver.

The dominance of the capital has had, as principal effect, the breakdown of the village and clan family. The sons no longer leave to found new villages, but go instead to Libreville to pursue their studies in hopes of becoming civil servants in some government agency. As in many other places, the rural exodus results in a progressive aging of the rural population, but this is somewhat attenuated by the fact that aging men who remain in the village can have three or four wives and consequently an astonishing number of young children. From grouped habitats composed totally of habitations constructed with bark and covered with raffia palms, the change has led to a scattering of heteroclite huts which are preferably strung along the track. This dispersion solidifies geographically the splintering of families of the same village and the integration of families already splintered off from other clans.

The African patriarchal family is thus on the way to extinction, with all the consequences this implies for the countryside and rural life.

This disaggregation of clan society here brings into question the role of women. In continuing, as in the past, to undertake the heavy work in the fields as well as the upkeep of the home, women acquire more and more freedom; selling bananas or manioc loaves along the track or in Libreville has become for them an extraordinary means of emancipation. Through contact with the city, they adopt the new fashions and keep abreast of the news which they report in the evening to the family around the fire. In this way they acquire an authority they certainly did not have a few years ago. It is reasonable to assume that in many village families the wife, because she is the only dynamic element of production, directs the family unit. By the same token, the husband has lost much of his past prestige, and, most of the time, merely plays an honorific role as family representative in the village council and as host to any stranger that comes to his home.

However, it should be noted that this disaggregation of clan society is not accompanied by the abandonment of community practices. Confronted with a thousand problems occasioned by the forest (that 'green hell' as it has been called), individualism is quite out of place and mutual aid among villagers is retained, come what may. For the same reasons, nothing is changed in cultivation methods, namely, itinerant agriculture in the burnt-off forest with its problems of following fields. The few attempts by the government to develop, for example, the cultivation of rice in this equatorial region have had no success so far, even though the population suffers from malnutrition.

Thus, it is evident once again that the influence of the city on the surrounding countryside is totally negative. It monopolizes the energies, progressively erodes the traditions, and breaks down tribal structures. In short, it imposes its own set of values. In this aspect its action can be called imperialistic.

s0103
Voting behaviour in a Belfast parliamentary constituency
J.N.H. DOUGLAS and J.M. ROBINSON   *Queen's University, Northern Ireland*

The political choice made by each individual in an election is the end product of a complex behavioural process involving the perception, values, attitudes, and opinions of the voter at one point in time in a given milieu. Analysis of the voting decision and of the

TABLE 1. Political choice and occupation (percentages)

Occupational group	Unionist	Labour	Protestant Unionist	Sample size
Non-manual	59	22	19	37
Skilled manual	54	21	25	48
Semi-skilled and unskilled manual	43	40	17	53
Retired and non-employed	67	17	16	71

pre-decision stage can thus provide insights into the causes and consequences of cultural conflict and into constraints on voting behaviour in conflict situations.

In Northern Ireland the present conflict results from conditions which have existed in political terms since the partition of Ireland in 1921 and in general cultural terms since the Protestant Scots were settled amid a Catholic Irish population in north-east Ireland during the seventeenth century. Analysis of voting behaviour as a guide to the nature of this conflict is hampered by the method of conducting elections which leaves the researcher with aggregate data of very limited value. Studies in depth have to be based on social surveys and, until recently, few such surveys existed. In their absence it has been generally assumed that political choice, which can be expressed at local government, Northern Ireland government (Stormont), and national government (Westminster) levels, closely reflects the religious dichotomy in the community. The Catholic population is thus believed to vote for the political parties which are anti-partition (the Nationalist and the Republican Labour parties) while the Protestant population votes for the Unionist Party. The lack of success of the N. Ireland Labour party and the Liberal party, which base their platforms on non-sectarian socio-economic issues, gives support to this general assumption. Such rigid voting behaviour reflects a polarized community and means that the Unionist party, representing the larger Protestant population (60 per cent of total population), has uninterrupted political power.

The results of the survey set out below show the characteristics of voting behaviour in the overwhelmingly Protestant (97 per cent) Belfast constituency of Bloomfield at the February 1969 N. Ireland general election. The constituency includes the full range of socio-economic groups and the sample survey, comprising 230 interviews, was taken during August 1969. The election had been brought about largely by increasing criticism of the Unionist administration led by Prime Minister O'Neill. The administration had been developing, albeit in a slow and halting fashion, closer relations with the Eire government and making conciliatory gestures towards the Catholic population. This policy sent reverberations throughout the Unionist structure, created dissent among Unionist MP's and caused the Reverend Paisley to form the Protestant Unionist Party to act as a focal point for right-wing anti-O'Neill voters.

The result of the election in the constituency was: Unionist (pro-O'Neill) candidate, 9084 votes; Protestant Unionist candidate, 3569 votes; N. Ireland Labour party candidate, 2196 votes. Percentage poll: 70.43.

Evidence of class voting preference is seen in Table 1 in the stronger support for the Labour party from among the semi-skilled and unskilled workers. The significant conclusions, however, relate to the Unionist and Protestant Unionist parties which gain support from all occupation groups. The belief, widely held in N. Ireland, that support for the Rev. Paisley comes mostly from the 'working class' Protestant is clearly untrue. Concerning the level of support, it is noteworthy that the election was held at a time when open conflict and civil disorder were still germinating.

TABLE 2. Voting changes 1965–9

| 1969 | 1965 voters | |
	Unionist	Labour
Unionist	72	16
Labour	—	70
Prot. Unionist	22	4
Did not vote	6	10
Sample size	92	50

Shows 1969 political choice of 1965 voters in percentage figures.

TABLE 3. Perception of election issues

	Unionist	Protestant Unionist	Labour
Leadership of the Unionist party	33	13	—
Maintenance of Protestant religion	51	81	43
Maintenance of NI as part of UK	36	67	19
One man one vote	13	—	14
Improving housing situation	27	13	50
Better conditions for working man	33	20	60
More workers' control in industry	7	5	7

Percentage values based on issues ranked 1st and 2nd in importance. Total 200%.

TABLE 4. The N. Ireland boundary issue and party support (percentage figures)

Attitude to the boundary issue	Unionist	Protestant Unionist	Labour
Stay as it is	91	80	85
Join the Irish republic	—	—	4
United Ireland linked to UK	7	2.5	11
Go it alone as an independent nation	2	17.5	—
Sample size	118	40	53

TABLE 5. Party choice and attitudes towards social issues

	% Who answered 'yes' to each question		
Question	Unionist	Protestant Unionist	Labour
Do you think RC's are discriminated against in jobs and housing?	23	10	45
Do you approve of mixing RC and Protestant children in schools?	79	5	91
Do you approve of the RC church's influence in NI?	58	17.5	50
Do you think inequalities in wealth and privilege are wrong?	18	25	64
Sample size	118	40	53

Comparing the voting behaviour in Table 2 of the interviewees in 1965 and in 1969 makes it clear that there has been a swing away from the Labour party and a general closing of Protestant ranks around Unionism. The entry of the Protestant Unionist party establishes the presence among the Protestant population of a significant anti-moderate, anti-change group. The shift towards Unionism reflects a concern for maintenance of traditional values. Perception of the major issues believed to be at stake in the election shows the nature of this concern.

To the Unionist voters, religious and constitutional issues were important though they were also caught up to some extent in a leadership struggle. Socio-economic issues were perceived as most important by Labour supporters, though even among this group the religious issue was significant. To the Protestant Unionists religion and constitutional issues were of overwhelming significance.

In order to clarify the attitudes of the interviewees towards the constitutional position of N. Ireland, they were asked to select from a number of alternatives the line of policy which the government should follow with regard to partition.

The support for maintenance of the status quo is overwhelming yet the results do show that a section of Protestant Unionist supporters are willing to go to the extreme lengths of attempting to set up N. Ireland as an independent nation. The single-mindedness of Protestant Unionists was also highlighted by another aspect of the survey. Most of them refuse to accept that discrimination exists, they are against integrated education, are anti-Catholic church, and relatively disinterested in socio-economic inequality.

In conclusion, the survey emphasises the importance of constitutional and religious issues among all occupation groups in the Protestant section of the community in Northern Ireland. This means that any change of policy which can be perceived as threatening the political status quo pushes the Protestant population towards the right and, in particular, towards the Protestant Unionist party. The decline in the Labour party vote indicates that even voters who place real importance on socio-economic issues rally in support of the constitutional position in time of stress. The Unionist party clearly brings together under a large umbrella supporters with a very wide range of opinions and attitudes and consequently its policy direction suffers in its attempts to satisfy the demands put upon it. The recent emergence of the Protestant Unionist Party has already struck at the foundation of the traditional Unionist structure. In the 1969 election no Protestant Unionist candidate was successful, though the Rev. Paisley contributed significantly to the final downfall of the Prime Minister by polling 6331 votes against O'Neill's 7745 in the Bannside constituency. Since 1969 the party has won two Stormont seats and one Westminster seat at by-elections and its success reflects the further hardening of attitudes among the Protestant population as conflict and civil disorder have developed. It is safe to surmise that a similar shift to the extreme viewpoint has taken place among the Catholic population so emphasizing the intractable nature of this conflict situation.

Barritt, D.P., and C.F. Carter, 1962 *The Northern Ireland Problem* (Oxford).
Boal, F.W., and R.H. Buchanan, 1969 The 1969 Northern Ireland election, *Irish Geog.*, 6, 78–84.
Campbell, A. et al., 1960 *The American Voter* (New York).
Heslinga, M.W., 1962 *The Irish Border as a Cultural Divide* (Assen).
Kasperson, R.E., 1969 On suburbia and voting behaviour, *Ann. Assoc. Am. Geog.* 59, 405–11.
– and J.V. Minghi, 1970 *The Structure of Political Geography* (London).

s0104
The tourist as the counter-agent in cultural diffusion
JOHN J.B. MILLER *International Travel Agency, USA*

The purpose of this paper is to call attention to the problem of ever-increasing temporary migration of persons for tourist or holiday purposes, because it is contended that this results in the erosion of the native cultures of the host areas. The problem is manifested in its greatest extent when tourists from the affluent or 'developed' countries visit countries or areas considered to be at a lower level of 'development.'

It will be observed that the author of this paper is a travel agent. It is the ostensible purpose of travel agents to facilitate the movement of travellers to unfamiliar destinations, and to encourage this movement. The profit motive in the travel industry leads to certain economies such as group movement and standardization of accommodations.

As travel increases, the host areas must provide amenities to serve the real or assumed needs of their paying guests. In the less developed areas, it may be debated whether the limited funds available for economic growth should be expanded in the construction of hotels and the like. More significant is the erosion of the local cultures as a result of the tourist influx.

Denmark, for example, is considered one of the most developed countries of Europe in terms of economic standard of living. Values in the Danish culture, however, are not the same as in the American culture, if such a term may be used, or were not until acculturation took place. Several visits to Copenhagen over a period of years have shown changes in the life-style. One item is the bed in the hotel room! Gone is the hard wedge-shaped bolster under the pillows. The hotel maids tired of picking them up off the floor where they had been heaved by the discomfitted Americans. Likewise, sheets and blankets are replacing the Scandinavian down comforter as a bed covering. The proprietor of one of the hotels has instituted a sidewalk cafe, certainly not a Danish fixture, but it has become a popular afternoon watering-spot for tourists, and is being patronized by an increasing number of Danes. One building in the famous Tivoli amusement park is now filled with slot machines. True, they do not dispense actual cash, but only tokens to be used at facilities within Tivoli, but a new, and decidedly un-Danish, facility has been introduced, supposedly to gratify the desires of the tourists. The same hotel that introduced the sidewalk cafe has just opened a bar with the same name as one in Rome, which has long been a favourite with Americans because of its resemblance to a New York counterpart. The proprietor hopes for a similar success, and probably will achieve it. Of course, the pornography craze in Denmark would have run its course long ago were it not for the continual flow of tourists keeping it alive.

The level of development evident in Turkey falls far below that of Denmark, but the travel agent can assure his clients that they need not fear: Istanbul boasts an hotel where the standards of amenity that prevail 'at home in America' may be found. For fifteen years the Istanbul Hilton, recently enlarged, has been an exclave of United States material culture overlooking the Bosporus. Not that the phenomenon is new to that city: in the 1880s Thomas Cook's was responsible for the erection of the Pera Palas to provide the accustomed comforts for the then-dominant British tourist industry. The venerable edifice still functions, reminiscent more of Queen Victoria than of Abdul-Hamid. The Hilton was constructed with funds of a workers' endowment trust, which owns the hotel (and another in Bursa), but has it managed under contract by Hilton International. The trust seems to have made a wise choice in the selection of management because of the reassurance of the familiar name to the wary excursionist, who takes his meals in the hotel dining room lest he should be exposed to some kind of poisoning elsewhere. The hotel obliges with simulated Turkish food, at double the prices charged by the city's better restaurants, but accompanied by butter for the bread, and unlimited ice-water. The bathrooms, however, satisfy the guests with the standard Hilton fixtures rather than the plethora of gigantic towels that might be found at a more typically Turkish establishment such as the same trust's hotel in Bursa, which is locally managed. The success of the Hilton has been such that a contract has been signed with the Sheraton chain to operate a similar hotel in Istanbul.

As geographers, perhaps we should feel some dismay at the diminution of the cultural experience by these people who have spent so much money to have their hamburger and french fries (oh, yes!) on the shores of the Bosporus, and perhaps some alarm at the allocation of funds that makes this happen. Still, without these facilities, many of the tourists would not come, bringing their hard currencies with them.

Perhaps the most extreme example of a cultural exclave created by tourism is Acapulco, Mexico. The glittering forest of modern hotels, linked to the United States by direct jet service, is removed by only a few kilometers from a pre-industrial society.

Are these tourist amenities of benefit to the societies upon which they are superimposed, are they simply self-nurturing, or do they inflict actual harm, tearing down traditional institutions in the process of their operation? Further study is needed to determine this. Until now, most studies of tourism have been in the realm of Economic Geography (Fiabane 1971, 234–6). Economic studies will indicate the magnitude of the problem, but the cultural impact has been alluded to only rarely (Jackson 1963, 27).

Therefore, this travel agent urges the participants in this symposium to consider further the problem of the effect of tourists upon the cultures they visit. His experience is supposed to enrich his culture rather than to litter others'. Let me conclude with a quotation from one of the most peripatetic of geographers, Freya Stark (1933, xiv): 'To awaken quite alone in a strange town is one of the pleasant sensations in the world. You are surrounded by adventure. You have no idea of what is in store for you, but you will, if you are wise and know the art of travel, let yourself go on the stream of the unknown and accept whatever comes in the spirit in which the gods may offer it. For this reason your customary thoughts, all except the rarest of your friends, even most of your luggage – everything, in fact, which belongs to your everyday life, is merely a hindrance. The tourist travels in his own atmosphere like a snail in his shell and stands, as it were, on his own perambulating doorstep to look at the continents of the world. But if you discard all this, and sally forth with a leisurely and blank mind, there is no knowing what may not happen to you!'

Fiabane, Dino, 1971 Information sources on international travel and tourism, *Prof. Geog.* 23, 232–4.
Jackson, J.B., 1963 Tourism: more give and less take, *Landscape*, 13, 27–8.
Stark, Freya, 1933 *Baghdad Sketches* (London).

s0105
The territorial implementation of black power in the United States
RICHARD L. MORRILL *University of Washington, USA*

During its early phase, 1950–64, the civil rights movement was dominated by the liberal belief in the moral and constitutional necessity of integration – geographically as well as socially and economically. Progress was so little, so costly, and met with such violent resistance that the younger black leadership began to wonder whether joining the white society was worth it. Consciousness of a black cultural identity and pride in its values grew. Many blacks came to fear that assimilation, including spatial integration, given the ingrained racism in society, would involve a high risk of even easier and more permanent economic and social suppression. Concepts of 'black power' and 'black nationalism' developed (Carmichael and Hamilton 1968). More blacks chose to enhance the quality of life within the ghetto, rather than escape it, but the viability of the ghetto was seen to depend on acquiring political and economic control over these territories.

Economic and political control can take several forms. At the extreme are advocates of a totally separate black state or nation. Since probably rather a small minority of the nation's 23 million blacks would want to come together into such a state, but mainly since such a grant of land and power to blacks would place large numbers of whites and their economic activities under black control, a separate state is utterly inconceivable.

More realistically, the problem is to achieve some equality of political and economic power within the areas where blacks in fact live. The achievement of economic equality and proportional economic power is fundamental and extremely difficult. For over three hundred years black people have been used as a source of slave and then cheap labour for agriculture and industry. The chronic labour surplus in the American economy has been perhaps the single most important factor behind racial discrimination. Equal access to the full spectrum of occupations, industries, professions, and income is the key to economic equality, and I do not believe this can be achieved without a more radical public intervention in the private economy than we have seen up to now. However, I will not develop this complex theme in this paper, but discuss mainly the problem of political power.

Some major attributes of true political equality include: (1) representation in proportion to the black (voting?) population in local, state, and national political (and educational) elective and appointive bodies; (2) when a minority, the power of veto over private and public plans which would very adversely affect vital community interests; and (3) community control of critical local functions in areas of black majority. The first two of these characteristics have not been seriously considered, let alone remotely approached. Indeed blacks have not even been able to wrest securely the minimal right of voting. Certainly these notions are very radical, quite probably not in conformance with the Federal and state constitutions, and very contrary to American political tradition.

Proportional representation for blacks, because of colour, would shatter all precedent in its grant of unique privilege to one group, and in its formal admission of an unbridgeable gap between white and black. The veto power would be an unprecedented violation of the tradition of majority rule, even if limited to a power of delay. The third characteristic, community control, is both rare and extremely limited in scope thus far, but is possible, because many white segments of the population perceive utility in it for themselves.

I frankly believe that the ultimate survival of American society will depend on that degree of integration which approaches the creation of a new joint race, but I am as frankly convinced that this will be generations off. I would like to believe that political equality can be achieved by the traditional route of the electorate 'voting for the best man,' but such an ideal must be set against the basic reality of a black and a white America that cannot be wished away, a division that is maintained both by ingrained white racism and by black distrust of whites and preferences for its own community. One hundred years after emancipation the inequality of political power remains so extreme that only the most naïve could believe that whites will freely and normally grant equality. In 1971, only .3 per cent of officeholders are black, although 10 per cent of voting age Americans are black. For example, in Chicago, with over one million blacks who are unusually concentrated and educated – that is where conditions for gaining power are peculiarly good – only 29 of 364, or 8 per cent, of elected officials are black (Baron 1968; Lewis 1971).

The argument may be reasonably raised that if blacks are given proportional representation, then why not Chicanos, Puerto Ricans, the elderly, the poor, and countless other cross-sections of the population? Any honest observer of the American scene, any student of American history must see that since 1619, the unequal relation of black and white has been and will continue to be our overriding social problem. Because of the simple fact of slavery and because of the sheer large numbers and the concurrent history of black and white, the situation of blacks is unique. The only comparably suppressed group is the American Indian, which also deserves special consideration.

In the next section of the paper, I examine the geography of black power as it might emerge if the radical proposals outlined above were to be implemented. For states and major county and city concentrations of black people, the proportionally determined representation is calculated and compared with current representation. The more important areas that would be affected by full implementation of community control are identified. Even given these radical changes, representation would rarely reach the actual black proportion of the population, because of the high degree of dispersion, especially in the south. However, greater segregation of blacks from whites and concentration in fewer larger ghettoes must be considered socially undesirable, since overcoming discrimination and raising economic status depend on increased friendly contact. These proposals have the advantage of guaranteeing adequate representation without requiring excessive physical separation, currently the only means of increasing political representation and power.

Distribution of the black population: 1970. Figure 1 (not printed) shows the relative proportion of the population that is black, for counties and major cities over 25 per cent black. Relatively few areas have a black majority, only 2½ million of 23 million blacks living in these areas. The most important, including Washington, DC, Atlanta, Newark, and Gary, are listed in table 1 (not printed). The map does illustrate these wide rural and small town areas of the South for which proportional representation of congressional seats and legislative positions would be especially important. Figure 2 (not printed) depicts the *absolute* numbers of black people in counties with over 5000 blacks. This map is especially useful in visualizing the probable concentration of black representation and political power. Although blacks are relatively numerous throughout the south, urban and rural, they are absolutely almost as numerous in the large metropolitan areas of the north and west.

The numbers of Congressmen and legislators that would be expected according to proportional representation in the states, in comparison with current numbers elected, are given in table 2 (not printed). Data for major metropolitan areas are included. The glaring inequality, north as well as south, is apparent and needs little elaboration. Underrepresentation is more extreme in the south, but is inexcusably severe in all areas. The greatest increase in US congressional representation would occur in Texas (3) and in Alabama, Florida, Georgia, Louisiana, Mississippi, New Jersey, North Carolina, South Carolina, and Virginia (2 each), while Maryland, Michigan, and Missouri have the representation expected. The greatest proportional increases in state legislatures would occur in Arkansas (the only southern state with no representation), Mississippi, South Carolina, Louisiana, Virginia, Alabama, and Texas. The states with the highest ratio of actual to potential legislators are Arizona, Colorado, Ohio, Washington, and Michigan.

The distribution of potential black power would be regionally concentrated in the south and in relatively few large northern and western metropolises, but at the same time would be very discontiguous. Except for a few rural counties and central cities, the dispersion of the black population is such that even proportional representation would bestow majority control in very few areas.

For the above reason, community control over at least critical local functions – most notably education and police – is important to black people. Central city communities over 50,000 contain about 10.4 million or 46 per cent of the black population in 1970.

Finally, in which areas would the power of veto be important? If we assume that all areas (counties and cities) with over 25 per cent black but less than a majority were granted such power, then wide portions of the south would be affected, as well as several large northern cities and counties, in all about 6.4 million black people. Only five states still have more than 25 per cent of the population black: Alabama, Georgia, Louisiana, Mississippi, and South Carolina. If we further assume that within counties and cities, large black communities (over 50,000 but less than 25 per cent black) in their political units would exercise such power, then 55 such communities with 9.2 million black people would qualify. Together, approximately 70 per cent of the black population lives in such areas.

Having conducted this investigation, I am not convinced of the wisdom of such changes. Nevertheless, it is of value for all the people to know just what 'black power' might really mean, if it were implemented in a meaningful way.

Baron, Harold, 1968 Black powerlessness in Chicago, *Transaction* 6.
Carmichael, Stokely, and Charles Hamilton, 1968 *Black Power, The Politics of Liberation in America* (New York).
Detwiler, Bruce, 1966 A time to be black, *New Republic*, 17 Sept. 1966, 18–22.
Lewis, Denise, 1971 Victory and defeat of black candidates, *Black Politician*, 2 (4), 34ff.

Water Resources
Ressources en eau

---

s0201
Clear water erosion: its geographical and international significance
ROBERT P. BECKINSALE   *University of Oxford, England*

The sudden abstraction or a large reduction in the amount of sediment in a river can be achieved naturally or artificially. The commonest natural way is by lakes impounded by natural barrages or accumulated in eroded hollows. In these, lake-head deltaic deposition has attracted much attention but its corollary, the erosive effect of a clearwater overflow, has been ignored by geomorphologists. The commonest artificial way of abstracting river load is the construction of a dam and reservoir. These man-made lakes have become common in the last three thousand years and most of them are of the solid overspill type.

In the drier Mediterranean lands solid dams were popular especially in Roman times whereas in the wetter parts of Europe they did not proliferate until the common introduction of the overshot wheel in the later Middle Ages. Some of these millpond barrages have been functioning for several centuries and the water channels downstream from them show distinct signs of incision and quickened erosion. Geomorphologists who found a slight nickpoint or break of gradient in the thalweg near a mill assumed that the mill had been placed there to take advantage of the nickpoint. None suggested that the mill had created the nickpoint by means of clearwater erosion downstream of the dam. The explanation came after 1930 from observations at large barrages in Europe, southern Asia, and the American west.

Large lakes, natural or artificial, lessen the peak discharge of floods, regularize the seasonal regime and decrease the mean annual discharge of rivers issuing from them. These changes must be correlated with the abstraction of the load, which may involve all the bed load and all but the finest particles of the suspended load. The absence of a new supply of bed load would tend to increase the cross-sectional velocity of the river, while the absence of coarser suspended matter would increase turbulence and in some streams might tend to lessen the velocity of a clearwater discharge. The combined effect of the removal of bed traction load and coarser load must be considered in relation to changes in competence, in regime, and in the channel morphometry.

Recently Komura and Simons (1967–9) have supplied a differential equation for the calculation of river-bed degradation downstream of dams if the channel width, particle size distribution of bed material, and the bed and water surfaces are known. They have also supplied a numerical solution for the 'final equilibrium profile' which takes account of the armouring of the channel beds. The many participants in the discussion showed the desirability and the difficulty of coming up with a complete answer. Among the difficulties not mentioned were the effects on buried valley infill, in some areas, of changes from pluvial to nival or glacial climates in the Pleistocene. The hydrologist must know the detailed stratification of the valley infill, which involves problems of palaeohydrology and often also of archaeology (Schumm 1969).

The clearwater discharge below dams attempts to pick up much of its former load. In doing so, it degrades its bed and usually creates a slightly narrower incised channel. Recent rates of incision and amounts of material removed from alluvial channels are already given in standard texts (Ven te Chow [ed.] 1964; Leopold et al. 1964; Kuiper 1965). The morphometric effect resembles that of rejuvenation because of relative uplift of base level, but the direction of the process works downstream not upstream. Tributaries entering a degraded reach will also be rejuvenated.

Clearwater erosion below dams may cause serious undermining of artificial river structures and increased damage to the channel sides, which are liable to cave in. In countries with a widespread irrigation network, as in West Pakistan, the lowering of the river-bed can disastrously decrease the head of the diversion structures of the downstream canals. Similarly the lowering of the water table of former flood-plains by a few metres may be harmful to the local water supply. On the other hand, where clearwater flow reaches deltas presumably it will improve the drainage of the damper lowlands and be beneficial to navigation channels. Here, however, the clearwater itself may bring great economic difficulties.

A reservoir scheme that resulted in relatively clear water reaching deltaic plains would no doubt improve their drainage but it would also deprive their floodplains of most of their former supply of silt. Moreover, the dominance of fine particles and of colloidal clays in the sediment that was still carried might well in time be unfavourable to agriculture. Similarly the coastal fisheries may become less productive through the absence of a mixed debris supply. In the rules on the uses of the water of international rivers adopted in 1966 at Helsinki (UN 1970), Article IX states 'As used in this Chapter (3), the term water pollution refers to any detrimental change resulting from human conduct in the natural composition, content, or quality of the waters of an international drainage basin.' Had anyone then considered that for deltaic nations clear water might be an extreme form of pollution?

Bogardi, J.L., 1968 *Sediment Transportation in Alluvial Streams* (Budapest, Res. Inst. Water Resources Development, ed. M. Erdélyi).

Borland, W.M., and C.R. Miller, 1960 Sediment problems of the Lower Colorado River, *Proc. Am. Soc. Civil Engrs. J. Hydr. Div.* 86, p. 61–87.

Chow, V.T. (ed.), 1964 *Handbook of Applied Hydrology.*

Einstein, H.A., and N. Ching, 1952 Second

approximation of the suspended load theory, *Univ. of California Series 47*, issue 2, Berkeley.

Hathaway, G.A., 1948 Observations on channel changes, degradation and scour below dams, *Intern. Ass. Hydr. Res. Report*, p. 287–307.

Komura, S., and D.B. Simons, 1967–9 River-bed degradation below dams, *Proc. Am. Soc. Civil Engrs. J. Hydr. Div.*, 93, p. 1–14; 95, p. 1042–8; discussion in 94, p. 336–40, 589–98.

Kuiper, E., 1965 *Water Resources Development*.

Lane, E.W., 1934 Retrogression of levels in river beds below dams, *Engng. News Rec.*, 112, p. 836–8.

Leopold, L.B., M.G. Wolman, and J.P. Miller, 1964 *Fluvial Processes in Geomorphology*.

Schumm, S.A., 1969 River metamorphosis, *Proc. Amer. Soc. Civil Engrs. J. Hydr. Div.*, 95, p. 255–73.

Tinney, E.R., 1962 Process of channel degradation, *J. Geophysical Research*, 67, p. 1475–80.

United Nations, 1970 *Integrated River Basin Development* (rev. ed.).

Vanoni, V.A., 1946 Transportation of suspended sediment by water, *Trans. Am. Soc. Civil Engrs.* 111, p. 67–133.

Vanoni, V.A., and G.N. Nomicos, 1959 Resistance properties of sediment laden streams, *Proc. Am. Soc. Civil Engrs. J. Hydr. Div.*, 85, p. 77–107.

s0202
Dam hindsight in Texas
EARL COOK   *Texas A & M University, USA*

In the first 70 years of this century, 152 large water impoundments have been made in Texas, at a cost of several billion dollars. All Texas has been affected in one way or another by these impoundments, but no comprehensive study of the total impact of any of them has been made. The proposed Texas Water Plan would cost more than $10 billion. It would seem helpful in assessing the future impacts of implementation of that plan to know what the effects of large completed multipurpose water resources projects in Texas have been.

In searching the literature, one finds that the idea of trying to measure the total impact of a large water-resources project appears surprisingly novel. North American post-construction studies address themselves mainly to the decision process and to the economic impacts, but they neglect environmental, social, land-use, and even geohydrological impacts.

The reasons for such neglect appear psychological, political, institutional, and social. Large dams are perceived as permanent additions to the crafted environment, and to a considerable extent each large impoundment seems to be regarded as unique in its setting. At least in the arid west, the use of water, especially for agriculture, is virtuous, regardless of engineering efficiency or opportunity costs; a dam then becomes a perceived good in a way a mine never could be. What is the point of a post-construction audit of an irrevocable, unique good?

The politician who birled his log to a favoured position in the legislative millpond, and by artful alchemy nursed its transmutation into a dam for his grateful constituents, will hardly encourage a critical review of the total costs and benefits of that dam, especially if it has his name or bust on it.

The basic *institutional* deterrent to hindsight stems from a simple fact: very few of us have yet made a living by *not* doing something, and consequently, we have not been prepared to institutionalize the non-development alternative. Moreover, those who have built and are building dams have a tendency to wish to continue building dams. Their built-in obsolescence factor, siltation, not having been programmed by General Electric or General Motors, affords them little assurance of a viable replacement market. Consequently, dam builders face a shrinking primary market in which the real money has already been made, where all the good corners are taken, and where survival depends on projects with lower and shakier benefit-cost ratios and ever greater environmental impacts. The boat's already low in the water; why shoot a hole in its bottom by studies that may show that some existing dams should never have been built?

As long as social pricing of water dominates cost or efficiency pricing, we shall have also a *social* deterrent to hindsight. A society which demands that water be made available to any of its communities at a price that the community can afford to pay, regardless of the real cost to the society, will not look kindly on hindsight studies that might cast the troubling light of doubt on the basic premise, which is that the society can be permanently unconstrained by the caprices of resource distribution, that it will continue to be able to subsidize people where they want to live rather than where it may be efficient for them to live.

If, despite these deterrents, we can find support for a hindsight or impact study, we shall encounter several difficulties.

First, there may be serious gaps in available baseline information needed to reconstruct the situation in the impact area at the time of authorization of the project.

Second, it may prove difficult to distinguish the changes attributable to the water project from those produced by other factors.

Third, there is the problem of calculating the changes that would have occurred in the impact area had the dam not been built. Comparing what exists now with what existed at the time of authorization will usually give an inflated picture of a dam's impact, even when the influence of other developments can be separated out.

Fourth, there is the question of defining the area of impact of the project. Even the most isolated project has some impact far beyond its immediate vicinity.

Fifth, there may be some difficulty in getting accurate measurements of the present situation. In most areas, adequate ecological and natural-resource basic data do not yet exist, stream-gauging may still be poor, land-use maps inadequate, and good cultural information hard to come by.

Sixth, the question of quantification of non-market benefits and costs remains a vexing one. Not, of course, that everyone is yet agreed on the best ways of quantifying *market* benefits and costs! But the non-market case is more thorny, because it involves not only numbers that reflect more or less accurately the changes over time attributable to the impoundment but numbers that can be compared with numbers similarly derived in another resource context, with some assurance that the comparison will be valid.

Now, if we shall have overcome these difficulties, and turn to relating the impacts measured to the decision process, we shall find ourselves in another maze, the heart of which is the difficulty of reconstructing, from records available, the planning process and decision rationale. Traditional benefit-cost analyses rarely if ever include all the significant decision elements. Perception of secondary economic benefits and of non-market political benefits, or a demand for local or regional shares of public resources or redistribution of income can sway decisions greatly; a hindsight study should seek to weigh the results of the project against those expectations, inchoately recorded as they may be, and to relate the expectations to human values.

The promise of a good hindsight study is twofold: it should provide information and methodology vital to improvement of the contemporary planning process, at least for similar projects; and it will yield a baseline survey for later impact studies of the same project area as it matures.

Criteria for choice of a project for hindsight study would include, but not be limited to, the following: (1) The impoundment should be large enough to have produced a significant impact; (2) The project should be relatively isolated from similar projects built at about the same time, as well as from the confusing impacts of dissimilar projects or developments; (3) Construction should have been completed long enough ago, say 10 years, that the impacts of the project can have reached a certain maturity, yet not so long ago, say 40 years, that its impacts have become overlaid thickly by the effects of other developments, some of which may be nationwide; (4) Adequate baseline information must be available; (5) The planning process and decision history should be available; (6) There should exist similar projects in differing stages of development, so that after methodology has been worked out in a simple case, it can be tested on a more complex project in a similar setting.

Several impoundments in Texas come close to meeting these criteria. Because the perfect example was not found, a two-stage approach will be taken. Methodology will be worked out on a suitably isolated, but immature project, and then tested on a more mature project not so isolated.

s0203
Public participation in decision-making
E.L. DAVID   *University of Michigan, USA*

The text is not included because of length.

Borton, Thomas, and Katharine P. Warner, 1971   Involving citizens in water resources planning, *Environment and Behavior* (Sept.), p. 284ff.
Borton, Thomas, Katharine P. Warner, and J. William Wenrich, 1970   *The Susquehanna Communication-Participation Study*, US Army Corps of Engineers, Institute for Water Resources, IWR Report 70-6. Available from the Clearinghouse for Federal Scientific and Technical Information, Springfield, Va. 22151.
Cartwright, Dorwin, and Alvin Zander, 1958   *Group Dynamics: Research and Theory*.
David, E.L., forthcoming monograph on outdoor recreation.
– 1971   Public perceptions of water quality, *Water Resources Research*, 7, no. 3 (June).
Feldt, Allan, 1971   Water and land resource utilization simulation, Environmental Simulation Laboratory, U. Michigan, Ann Arbor. Mimeo.
Fox, Irving K., 1971   *Institutional Design for Water Quality Management: A Case Study of the Wisconsin River Basin, Volume I, Section A, Summary*. US Office of Water Resources, Technical Report OWRR C-1228. Department of the Interior.
Kelnhofer, Guy J., Jr, 1968   *Metropolitan Planning and River Basin Planning: Some Interrelationships*, Water Resources Center, Georgia Institute of Technology, Atlanta (July).
Lansing, John B., and James N. Morgan, 1971   *Economic Survey Methods*, Institute for Social Research, U. Michigan, Ann Arbor.
Nelson, Paul A., 1970   *The Impact of Accurate One Week Weather Forecasts on the Demand for Outdoor Recreation*, SSRI Workshop Series, U. Wisconsin, Madison.
Ostrom, Vincent, and Elinor Ostrom, 1971   Public choice: a different approach to the study of public administration, *Pub. Admin. Rev.* (Mar./Apr.), 31, no. 2.
Strong, Ann Louise, 1969   *The Plan and the Program for the Brandywine*, Institute for Environmental Studies, U. Pennsylvania, Philadelphia.
Thompson, Peter, 14 Mar. 1969   Brandywine Basin: defeat of an almost perfect plan, *Science*, 163, p. 1180–2.

s0204
Benefit-cost analysis of a multiple-purpose reservoir, Deer Creek watershed, Ontario: a reassessment
J.C. DAY   *University of Western Ontario, Canada*

Economic feasibility studies have been advocated in Canada to improve water and land management efficiency. This study is an assessment of the Deer Creek reservoir and road financed jointly by the Long Point Region Conservation Authority, Norfolk County, and the Ontario Conservation Authorities Branch and the Department of Highways (Vance et al. 1967). It presents a comparison of benefits and costs anticipated during project planning and a reassessment two years following construction.

Deer Creek enters the Big Creek 19.2km above Long Point Bay in Lake Erie. Deer Creek is incised from 7.77m to more than 15m in the Norfolk Sand Plain. Because of steep walls and a narrow floor, the valley is used for woodlots, cattle grazing, or left idle. The dam intercepts drainage from 17.92km^2 forming a .31km^2 water surface with

0.00185km³ (1500 acre-feet).

Only a restricted acreage of the watershed has high agricultural capability. Most land has severe limitations favouring the production of perennial forage crops. Excessive soil moisture, susceptibility to erosion, and low fertility are common problems. Forests occupy 1/4 of the watershed concentrated mainly along the valley.

The project is intended to satisfy many needs. Increased agricultural productivity because of improved irrigation water availability and a new road are the primary benefits: increased recreation, livestock watering, firefighting, domestic water supply, downstream summer flow, groundwater tables, and flood control opportunities are intangible benefits. Project capital cost was estimated at $597,000; repayment obligation was shared among Ontario, Norfolk County, North Walsingham Township, and the Big Creek Conservation Authority in the ratio of 87:4:6: and 2 respectively.

The project is justified on the basis of anticipated increases in farm productivity due to irrigation. The analysis is based on the assumptions that provision of water would influence 6.2km² contiguous to the reservoir and 1.6km² downstream; farmers in this area would irrigate a greater variety of crops, and more frequently, if water were available; and all tobacco, field corn, vegetables, and strawberry acreage as well as 1/3 of the hay and pasture land in the area would be irrigated.

Interviews in 1971 with all farmers on the 7.8km² to be benefitted by the reservoir indicated that the anticipated cropping pattern is unlikely to develop because of faulty assumptions made in the feasibility study. Most serious was the adoption of a set of crops typical of the cropping pattern throughout the Deer Creek watershed to estimate irrigation benefits rather than the crops on the land to be benefitted. A gross overestimate of tobacco area resulted. Instead of 1.4km² as anticipated on the 6.2km², only .52km² of tobacco exist, and instead of experiencing an areal decrease to no more than 75 per cent of total tobacco rights, these diminished to 50 per cent by 1971. A likelihood exists of further reductions. Moreover, irrigation of .34km² of tobacco anticipated on the 1.6km² downstream allegedly affected by the project was counted a benefit although at least .18km² were irrigated previously from the creek. Irrigation of this area should not be considered a benefit because there is no history of water shortage. As a consequence, two years after project completion only 35 per cent of the minimum area to be watered from the reservoir was irrigated (.64km² of tobacco instead of 1.8).

Other factors also question the validity of project benefits. None of the farmers to be assisted previously experienced detrimental water shortages, none at present irrigates crops other than tobacco, and none intends to adopt vegetables or strawberries as cash crops. The report further assumes that without a dam, 18 per cent of the tobacco crop would not be irrigated, although there is no precedent for such a cropping practice during periods of aridity.

Correction of these faulty assumptions reduces the benefit-cost ratio from 2.2:1 to 1.38:1. However, until 1.8km² of tobacco, or other crops equivalent in water use, are irrigated, no agricultural benefits should be claimed as the dependable watershed yield prior to reservoir construction would not be exceeded.

The report identifies specifically neither the area, nor the farmers, to be benefitted. This is an important omission since virtually all primary project benefits identified accrue to 4 farms which irrigated previously from the reservoir area. This is a direct subsidy exceeding $70,000 per family for which the beneficiaries repay no more than other residents of North Walsingham Township. Moreover, should the project's intangible benefits attributed to land-value appreciation and raised groundwater tables materialize, the same families encircling the project area would also benefit most.

A road crossing the Deer Creek Dam site was planned in 1965 by Norfolk County at a cost of $375,000, half to be paid by Ontario. Subsequently, a section including the dam was declared a development road meaning that 100 per cent of road and 80 per cent of bridge costs were paid by the province.

More than 57 per cent of the total project expenses were assigned to transportation improvements. Two routes were considered in evaluating the new road. The cost of the more expensive alternative was subtracted from the cost of the combined dam and highway plan adopted, which follows the least expensive route.

This accounting stance enhances project feasibility from a local perspective because a large proportion of conservation expenses are transferred to transportation accounts. Since the road was subsidized almost completely by the Ontario Department of Highways, the project appears more desirable to the local municipality repaying part of the conservation costs.

When the adjusted benefit-cost ratio discussed above is further modified by subtracting the expense of the least-cost highway route from the road and reservoir combined cost, project feasibility becomes highly questionable. The benefit-cost ratio decreases to 1.24:1 at 4 per cent interest (0.99:1 at 5 per cent and 0.80:1 at 6 per cent). However, the social desirability of providing virtually complete provincial subsidy for such a local service road is not explained nor is the road value reflected in the economic feasibility calculations.

The feasibility study presents insufficient information for determining the social desirability of the Deer Creek project. There is little likelihood of realizing agricultural benefits claimed as project justification; insufficient information is presented to permit determination of the desirability of the associated road. Moreover, many intangible benefits appear illusory: a pre-project livestock water shortage was not experienced and there are few livestock in the reservoir area; Langton has no history of municipal water shortage and no intentions of building a pipeline to the reservoir; the reservoir surface is too far below the highway for rapid firefighting use; and partial control of $.1km^2$ in a $745km^2$ watershed offers little flood-control advantage. Therefore, extremely large benefits must accrue to recreation and transportation improvements if the project is to become a good investment.

The social desirability of this multiple-purpose water and land management project was inaccurately and incompletely justified. The problem is mainly due to inadequate review procedures by, and ineffective interagency co-ordination between, the responsible Ontario governmental agencies.

National Advisory Committee on Water Resources Research; Environment Canada.

Vance, Needles, Bergendoff, and Smith, 1967 *Preliminary Engineering Study and Report, Deer Creek Dam, Deer Creek* (Woodstock, Ont.).

s0205
Environmental impacts of water resource development
D.W. FISCHER and GORDON S. DAVIES *University of Waterloo, Canada*

Water development programs and associated water development projects are undertaken to yield an array of benefits in the form of increased goods and services to various groups of users. Water development may be seen as an activity whereby an existing environment is purposefully changed into a 'new' environment deemed to be more useful in quite significant ways. Traditionally, environmental quality has not been a recognized factor in the formulation of water policies and programs and its introduction poses a number of problems other than how to achieve direct economic benefits to specific groups of users.

The concept of 'environment' is very broad and goes well beyond pollution in the kinds of concerns it embraces. 'Environment' refers to the whole array of diverse surroundings in which people perceive, experience, and react to things and events. The immediate and long-term resultant changes brought about by some intervention, such as the activities associated with the management of water resources, not only alter some related ecological system, but they also affect the perceptions of people living in the locality affected. People may well respond more to their somewhat diffused and perhaps aesthetic perceptions of changes in their surroundings than to the potentially measurable but more restricted and less visible changes in an ecological system.

The environmental impacts of water development decisions are becoming increasingly important. At the present stage in most countries no method exists for recognizing, quantifying, and incorporating environmental impacts into the water development decision process. The purpose of this paper is to present a preliminary approach which will aid in the recognition of environmental impacts and to indicate how these impacts can be incorporated into the decision-making process.

A three-step approach is suggested as a method of developing an environmental analysis. First, the constituent 'environmental units' which compose the river basin or other regions to be affected are identified. Second, an environmental base study for each of these units is carried out and third, the probable effects of changes resulting from water development are assessed. These latter two steps involve a listing of the major elements within environmental units, the preparation of judgmental indices of present conditions, and an assessment from diverse points of view of probable impacts from proposed changes. This process can also help identify key environmental factors for which more quantifiable indices can be monitored over time in order to assess the actual impact of actions taken.

An environmental unit or community, both biological and social, can be equated with the concept of the firm in economics where the firm is an independent internalised unit. Judgement is necessary to identify these units and their bounds. Local citizens will be invaluable in helping to identify such units.

After the appropriate environmental units have been identified it will be necessary to collect base-line data. At present no satisfactory indices to express the base condition or quality of environmental units have been developed. In the absence of adequate indices there is an obvious need to use an interdisciplinary environmental evaluation team in order to define and assess the characteristics of the base condition of the environmental units.

The third step is to carry out an environmental impact study. The objective of an environmental impact study is to assess the probable effect of a series of alternative water projects or proposals that could alter the environmental factors of the environmental units in a river basin. The impact study uses the environmental base study as the foundation on which to compare changes resulting from projects in a with and without approach. A method of describing such an impact is found in the National Environmental Policy Act of the United States (Section 102c) which requires a statement including five separate subheadings: the probable impact, the unavoidable adverse effects, the alternatives, the short-versus long-term impact, and the irreversible commitments of resources. In addition to these descriptions, the incidence of environmental changes and possible compensation in the form of replacement and indemnification costs should be included. Finally, the impact study should incorporate the costs of maintaining the environmental units in their present condition versus the costs necessary to sustain such units after the project is completed and operational. Such a with and without comparison of environmental costs will aid in making the water development decision. Again, an inter-disciplinary team will be important in assessing the above impacts and determining the inherent costs.

Matrices (Leopold et al. 1971, 1–3) are a convenient method of tying the above three steps together. For example, the environmental units could comprise the column headings and the base and impact variables comprise the row headings. In addition, this format would be useful for identifying gaps in information, setting the stage for environmental input-output analysis and providing a device for the basis of a permit or licensing system. Indeed, many countries use permit systems in some form or another. Consideration should be given to modifying these permit systems to include any agency or applicant who may modify any significant part of the existing environment of a river basin or coastal zone. Requiring a detailed permit application from all agencies desiring environmental changes will shift the burden of proof onto the developer. The information found in the permit application can also become an integral part of the information required to accomplish an environmental impact analysis.

In conclusion, rather than listing the many adverse impacts that have resulted from previous water development projects a framework for studying and mitigating these impacts has been suggested. Whether or not the suggested approach is used by water development agencies, the environment as a force in water development decisions has become a major factor.

Leopold, L.B., F.E. Clarke, B.B. Hanshaw, and James R. Balsley, 1971 A procedure for evaluating environmental impact, USA Geological Survey Circular 645, p. 1–3.

s0206
Neophilicism, environmental stress, and water resource management: the price of democracy
HAROLD D. FOSTER *University of Victoria, Canada*

Established institutions are facing mounting public criticism. Such dissatisfaction is to a large degree a consequence of the increased polarization of western society. Protagonists are vociferously supporting two, largely incompatible, value systems; one of which has economic growth and the other environmental stability as its major objective (Sewell and Foster 1971, 123). In consequence, in attempting to satisfy the demands of one 'public,' resource managers almost inevitably offend against the perceived rights of another. Criticism has, therefore, become automatically associated with action. Perhaps nowhere is this dilemma more obvious than in the field of water resource management. This paper illustrates how increased public participation has undermined normal water resource decision-making procedures, demonstrates institutional responses to the resulting flux, and suggests certain measures which, if adopted, might improve the situation.

Water management systems have not traditionally functioned in a social vacuum but have normally responded to preference signals, from society as a whole, pressure groups, and other managers, when establishing priorities for action. Such stimuli have been collectively termed 'environmental stress' by Kasperson, who has demonstrated their necessity in overcoming institutional inertia and precipitating problem solving responses (Kasperson 1969, 481). Water managers have, for example, frequently relied upon public hearings, opinion polls, letters to elected officials, and the news media when assessing 'what the majority of the people want.' The views of other concerned managerial groups have been determined by either ad hoc informal consultation or permanent institutionalized contact. In this way opinion has been assessed, problems identified, and alternative ways of dealing with them ranked. Strategies adopted to meet perceived problems have depended to a large degree on the attitudes of water managers to the limits of their responsibilities and on their grasp of the range of alternative solutions open to them. The effectiveness of such decisions has normally been judged by the degree to which the environmental stress which precipitated them has diminished. Such reductions in social pressures have usually been assessed by manifestations of outside satisfaction which have included the continued growth of the water resource decision-making unit, its funding trends, and public acclaim.

Recently, however, water resource managers have been experiencing increasing difficulty in reducing the level of environmental stress under which they must function. Public and managerial criticism has followed both decision and indecision, activity and inactivity. The reasons for this discontent are closely related to society's complex and constantly changing relationship to space, time, energy, and matter. As these parameters alter, so too do social goals and associated problems and so too must the institutions established to satisfy public demands.

Technological change has, for example, reduced the friction of distance, allowing water transfer over ever increasing distances. Nation states appear increasingly less constrained by national boundaries and plan to solve their water shortages on an international scale. Both the NAWAPA scheme and the Central Africa Water Plan illustrate this tendency (Sewell 1967, 9; Raney 1970, 175). Similarly, urban centres, exploding in both size and per capita consumptive demands, have resorted to transporting water from increasingly far afield. New York and Southern California provide examples of this process (Quinn 1968, 108).

As the scale of water transfer has increased, so too have the social and environmental problems associated with it. International engineering projects may cause major climatic changes, extensively damage floral and faunal resources, reduce national sovereignty, and displace numerous communities at great social cost. As the range of alternative solutions to problems increases, so too does the tempo of criticism to which water resource administrators are subjected. Its source is both that segment of the public that requires water and which can correctly claim that a technological solution to their problem is available and could easily be implemented by farsighted administrators, and from those who fear the implications, both social and environmental, of such solutions (Quinn 1968, 108).

The significance of time depends upon the speed of process. Technological innovations have increased both physical mobility and the speed of communication, so intensifying the environmental stress to which water managers are subjected. At a time when the interrelatedness of systems has never been more clearly understood, information concerning any individual component has never been greater. Because of this, the information explosion, water managers are faced with the impossibility of achieving competence in all aspects of their field or correctly assessing a plethora of conflicting advice from a wide variety of experts. Such confusion frequently leads to error, which when it occurs, or when social goals have altered so that an initial success appears later as a failure, is immediately exposed by the media to widespread criticism. For example, the Athabasca Delta, the Skagit Valley, and the Aswan Dam have received extensive publicity. Unlike some professionals, surgeons for example, water managers are unable to bury their mistakes.

The environmental stress facing decision makers in water management is also increasing because of society's apparently insatiable energy demands, and the associated increase in newly discovered compounds and industrial processes. These trends towards increasing use of power and material variety, together stimulate economic growth, intensifying the conflict with those dedicated to environmental preservation and conservation.

In an effort to meet mounting criticism, water managers have adopted a variety of strategies. A common response, seen for example in the establishment of the Canadian Department of the Environment, has been the administrative shuffle. This normally involves the renaming of agencies, their realignment, and the internal reorganization of responsibilities. A cynic might suggest its major aim is to divert attention from the perceived problems to the administrative framework.

A second device, the stall, is also being used by water resource managers. A posture of indecision in the public interest is adopted and a plea for more research into all aspects of the problem is issued. Typically, little effort is then made to ensure that such information is collected. When such data are already available, their validity is seriously questioned.

Perhaps the most successful reaction to the dilemma has been the concerned but fettered approach. Decision makers have requested more sweeping legislation and greater financial support. Where such additional power has been granted there has frequently been little attempt to apply it in an effort to solve problems effectively.

If criticism of water managers is to be significantly reduced, several major adjustments must take place in the administrative framework of this resource. Institutional changes are required to allow greater public participation in both the planning and decision making phases of water resource management. A recent innovation in this field was the circulation of proposed water legislation prior to its enactment, with a request for constructive criticism, by British Columbia's provincial government. Similarly, institutional flexibility must be increased, allowing the consideration of constructional and non-constructional alternatives in solving perceived water problems. Although the implementation of these and similar measures might reduce the level of criticism, it is unlikely to subside to earlier levels. Water managers must, therefore, learn to accept increased environmental stress, which is, after all, the price of greater public participation in the democratic process.

Kasperson, R.E., 1969 Environmental stress and the municipal political system, in Kasperson and Minghi, *The Structure of Political Geography* (New York).

Quinn, F.J., 1968 Water transfers: must the American west be won again? *Geog. Rev.* 58, 108–32.

Raney, F.C., 1970 A west Africa water and power plan, *Western Geog. Ser.* 3, 175–82.

Sewell, W.R.D., 1967 Pipedream or practical possibility? *Bull. of Atomic Scientists* 23, 9–27.

Sewell, W.R.D., and H.D. Foster, 1971 Environmental revival promise and performance, *Environment and Behavior* 3, 123–34.

s0207
Residential water demand: an econometric analysis and its implications for management
A.P. LINO GRIMA   *Royal University of Malta*

Although water is a ubiquitous commodity, the development of a water resource for municipal purposes requires considerable investment which is characterized by discontinuities and which is almost wholly irreversible. In the United States investment in water utilities, according to Bogue (1968 267), is greater than the combined investment in iron and steel, more than twice the investment in gas utilities, and much more than the investment in railroads. Total investments in urban water supply systems will be substantial in the next few decades, partly because of the backlog of deficiencies, partly because of rapid urban growth. Investment costs per capita as well as per unit of capacity are likely to increase due to higher per capita rates of use, more marked peaks in demand, the need to develop less accessible sources of supply as the present ones prove to be inadequate to meet increasing demand, and the tendency for urban areas to 'sprawl.'

The cost of providing additional urban water supplies may be gauged by the vast scale of urban growth in the world as a whole. In 1962 there were about 1200 million people living in cities; this number is expected to increase by 42 per cent to 1700 million by 1980, and for the year 2000 the estimate is 2500 milion. The investment required to close the gap between existing levels of supply and needs in urban areas in 75 developing countries with a total urban population of 336 million has been estimated at 5834 million US dollars over the period 1961–75 (Dieterich and Henderson 1963, 13, 56). In the United States total new investment during 1956–65 was 10,200 million dollars (AWWA Staff Report 1966, 772). In addition, the expenditures on sewage collection and treatment are often greater than the costs of urban water supply.

The problem of urban water supply should be tackled with urgency: one in every five human beings (500 million in 1960), mainly but not exclusively in the developing countries, suffers from waterborne diseases (Logan 1960, 469). For example, according to a US Department of Health, Education and Welfare task force report (1967, 13), fifty million Americans drink water that does not meet the standards of the Public Health Service.

In general the emphasis in water utility management has been to develop new sources of municipal water in order to meet projected requirements for municipal water. This emphasis on the 'supply fix' is based on the premise that residential water is a single 'good.' Like land, water is complementary to many activities and it may be said to have a composite demand (e.g. water for drinking and washing and water for lawn watering). Therefore there are priorities in the total requirements for water, and these may be articulated by building a model of water residential demand, i.e. by identifying variables that affect the level of residential water use and the level of the related investment in water supply. Some of these variables are less amenable to manipulation by municipal water supply management (e.g. level of evapotranspiration, size of family, income level, density of housing). Other variables may be varied by management (e.g. metering, minimum charge, price) with a view to reducing investment requirements. This approach requires the formulation of hypothesized structural relationships among the component variables and then verification of the proposed model against a sample of observations. The results would apply in varying degrees to differing milieux but the experiment can be repeated and evaluated as often as required.

The level of residential water use is a response to the environmental conditions, i.e. the physical conditions (e.g. climate, size of residence) and the managerial decisions (e.g. price, quality of service). Attempts to model residential/municipal water use, reviewed by Grima (1972) and Wong (1970), indicate a wide variety of approaches and of the type of sample data collected.

If data for individual residences are available it may be said that the amount of water used by a water-using appliance i in a given time period is

$$d_i \equiv n_i p_i w_i \quad (i = 1, 2, ..., m),$$

where n_i stands for the number of appliances of type i, p_i stands for the probability of the appliance being turned on, and w_i stands for the rate of water use by appliance i.

By summing d_i for m types of appliances one could calculate the amount of water used by a household:

$$(1) \quad D \equiv \sum_{i=1}^{m} d_i \equiv \sum_{i=1}^{m} n_i p_i w_i.$$

The hypothesized structural relationship should identify the important variables that affect the magnitude of m, p_i, w_i, and n_i in the identity (1) above. Since residential water use is complementary to other household activities, income level may be expected to be an important explanatory variable. Second, residential water demand is a composite demand; therefore (a) there is a hierarchy of water-complementary activities, and (b) various measures of D are available depending on the seasonal and diurnal variations in the use of water-using appliances. Third, the level of water use by a household is likely to change gradually over time since the purchase of durable and semidurable water-complementary equipment (such as lawn-watering equipment, washing machines, air conditioning units) does not adjust instantaneously to changes in price or income or technology. Fourth, since there are innumerable variables that affect m, p_i, w_i, and n_i, one should expect large errors from the estimating equations, and therefore it is advantageous to concentrate on the significance - or otherwise - of variables that are relevant to policy making (e.g. metering, price).

The choice of variables and of the functional form of the equation are discussed fully elsewhere (Grima 1970, 93ff). The hypothesized equation

$$D = f(X_1, X_2, ..., X_n),$$

where X is an independent variable, was fitted to a sample of 91 households in the Toronto region using 1967 meter readings of water use. A second sample of groups of townhouses whose water use was not metered was also taken. Over a score of models were fitted to the data, using average water use during the year (wu_a), average water use during summer (wu_s), average water use during winter (wu_w) as the dependent variables. Statistically significant regression coefficients were obtained for four explanatory variables: value of residence (a surrogate for income level), size of household, the charge for water beyond the minimum bill, and the size of the fixed bill per billing period.

Satisfactory results were obtained in the linear equations. However the *a priori* expectations called for a curvilinear equation. In addition, water use data are expected to have a distribution of residuals which is positively skewed and homoscedastic in the linear form. Therefore the regression coefficients would be unbiased but would not have minimum variance. Since the standard errors of the regression coefficients are of primary importance in the application of the model by management, a transformation of the data to logarithmic form was required.

The main findings from the fitted equations were as follows:

1. Metering *per se* does not influence the level of water use; a substantial marginal (or variable) charge for water makes the difference by giving the consumer an incentive to lower his water bill.

2. The efficacy of pricing as a policy variable is amply demonstrated. The price elasticity for water use during summer was −1.07 (i.e. a 10 per cent change in price would be followed by a similar percentage change in quantity demanded). This compares well with Howe's reported price elasticity of −0.9 for sprinkling use on an average summer day in the eastern United States (Howe 1968, 70).

3. Water use during winter is less responsive to price (elasticity is −0.75). This is compatible with the identity (1) above: as the range of water uses increases (from winter to summer), the impact of price becomes more marked. In fact, Howe and Linaweaver (1967, 24–5) found that the price elasticity is least (−0.23) for indoor uses of water in winter (e.g. washing, cooking).

4. The curvilinear (convex to the point of origin) form of the model with respect to price bears emphasis. Empirical evidence confirms *a priori* expectations that a (small) increase in the marginal charge for water has a greater impact on water use when the marginal price of water is low.

5. Through the application of appropriate moderate pricing policies to a hypothetical city of 200,000 people, the water use on a summer day could be cut by 25 to 65 per cent and investment requirements by 15 to 40 per cent (or $4.5 million to $12.3 milion out of a total $30.0 million).

The formulation of an investment policy based on tested models of demand rather than requirements obtained by rules of thumb would increase the efficiency of managing a resource that is likely to become increasingly important – and expensive – in an age of increasing urban growth.

AWWA Staff Report, 1966   The water utility industry in the U.S., *J. Amer. Waterworks Assoc.* (AWWA) 58, 767–85.

Bogue, S.H., 1968   Financial management of a water utility, *J. Amer. Waterworks Assoc.* (AWWA) 60, 267–72.

Dieterich, B.H., and J.M. Henderson, 1963   *Urban Water Supply Conditions and Needs in 75 Developing Countries*, Public Health Paper no 23, World Health Organization (Geneva).

Grima, A.P., 1972   Residential Water Demand: Alternative Choices for Management, Ph.D. thesis, U. of Toronto.

Howe, C.W., 1968   Municipal water demands, in W.R.D. Sewell and B.T. Bower (eds.), *Forecasting the Demand for Water* (Ottawa).

– and F.P. Linaweaver, 1967   The impact of price on residential water demand and its relation to system design and price structure, *Water Resources Research* 3, 13–32.

Logan, J., 1960   The international municipal water supply program: a health and economic appraisal, *Amer. J. Tropical Medicine and Hygiene* 9, 469–76.

US Department of Health, Education and Welfare, 1967   Task force on environmental health and related problems, in *A Strategy for a Livable Environment* (Washington).

Wong, S.T., 1970   *An Econometric Analysis of Urban Municipal Water Demands*, mimeo.

s0208
Pricing and efficiency in water resource management
S.H. HANKE   *Johns Hopkins University, USA*, and
R.K. DAVIS   *George Washington University, USA*

The general function of prices in an economy is to assert proper checks and balances on production and consumption. The amount that beneficiaries pay for goods and services influences efficiency, since the assessment of charges affects the rate at which services are used. The absence of charges may induce waste, whereas excessive charges may result in a failure to meet project potentials. When a facility is put into operation, output should be set at a level so that the incremental value per unit output (price) equals the increment to costs (marginal costs); i.e., the price set for the output should equal the marginal cost of providing that output. If the prices for public output are not properly conceived, programs will not coincide with their original objectives. The result will be that services will not be provided by the least costly alternative or the scale of provision will be either too large or too small.

The history of water resource development in the United States has been marked by a separation between the evaluation of alternatives and the pricing of services rendered by these alternatives. The result has been little reliance on pricing or user charges, outside the municipal field, to recoup costs of water resource development. With a few exceptions to federal water resource programs today, there is virtually no effort being made to recover costs from the beneficiaries of water resource programs.

The separation of evaluation and pricing cannot be made if there is to be a proper economic analysis of water resource programs (Krutilla 1966, 60–75). If evaluation procedures are to have consistency, the following pricing principle should, in general, be followed by policy-makers in the present national water resource economy: the beneficiaries of water resource projects should be charged the full cost of goods and services provided them unless either the cost of imposing and collecting such a charge is prohibitive or a specific equity criterion is violated (Cicchetti et al. 1972, 6).

A further consideration regarding the implementation of this principle is how to give an incentive to public agencies which lack the profit motive. It is suggested that an institutional structure must be introduced which requires that an agency's budget depend at least in part on the revenues which it collects from pricing.

Based on the criteria stated above, we reviewed water resource services and pricing policies in the United States (Davis and Hanke 1971). This review has led us to conclude that there are some areas in which there is both high potential for improvement in performance and some promise that progress in the use of prices can be rather easily made. These areas include municipal water services, industrial use of municipal sewer systems, navigation, and flooding and related natural disasters.

Municipal water pricing practices already contain the elements for efficient pricing; the major change needed is to base prices on discounted future rather than past historical costs. Municipal water rates should reflect more accurately existing cost differentials; most notably, the spatial and temporal cost dimensions. Industrial and commercial use of municipal sewers should entail more general and systematic application of pricing policies sensitive to the costs of these wastes imposed on municipal systems. A major improvement in response to natural disasters has been the initiation of a compulsory flood insurance program; this could be extended to hurricane damage. Toll-free navigation on inland waterways has produced congestion losses and inefficient investments; a system of segment tolls and congestion fees is needed to cope with this problem.

It is always difficult to classify matters of policy, but in our survey of pricing practices and the possibilities for significant improvements, we have encountered certain areas in which the obstacles to improvement seem a bit more than routine. These areas include outdoor recreation, fish and wildlife habitat, hydroelectric power, and agricultural irrigation.

We recommend that the agencies collecting fees be allowed to keep all or a portion of those fees; in addition, we recommend that a general pass as well as special service fees for federal recreation areas be required. Wetlands valuable for fish and wildlife habitat pose a general problem of being non-market goods whose beneficiaries may be widespread and elusive. Instead of basing hopes on direct pricing systems, we suggest a more realistic rationale to counter the destructive effects of agricultural subsidies and to correct tax incentives and federal channelization and drainage programs which are threatening large areas of this habitat. Hydroelectric power is an area in which marginalist pricing policies are called for by federal laws, but are not always followed in practice. Agricultural irrigation is a large and important problem area in which nothing short of a change from current ability-to-pay pricing to a practice of cost-based pricing will succeed in correcting misallocations.

Pricing is not a panacea for all problems of allocation and equity in federal water resource programs. However, sensitive pricing can contribute to more rational public policy in the field of water resources.

Cicchetti, C.J., R.K. Davis, S.H. Hanke, R.H. Haveman, and J.L. Knetsch, 1972 Benefits or costs? (An assessment of the Water Resources Council's proposed principles and standards), Dept. Geography and Environmental Engineering of the Johns Hopkins U., Baltimore, Md., March.

Davis, R.K., and S.H. Hanke, 1971 *Pricing and Efficiency in Water Resource Management*, Natural Resources Policy Center of the George Washington University, Washington, D.C., December.

Krutilla, J.V., 1966 Is public intervention in water resources development conducive to economic efficiency?, *Natural Res. J.* Jan.

s0209
Perception of phosphate pollution in an urban area
PLACIDO LAVALLE   *University of Windsor, Canada*

Within the last three years a strong public outcry has been raised concerning the pollution of local waterways by various phosphate compounds derived from industrial, agricultural, and residential activities. This concern is based on observations which indicate that the addition of excessive quantities of phosphate compounds into fresh water ecosystems may be the catalyst that stimulates the accelerated eutrophication of local lakes and rivers (Beeton 1965; Hasler 1947; Task Group 2610P 1967; US Department of the Interior 1968). In the light of these research findings numerous governmental and private agencies have promoted programs leading to more stringent control of the addition of phosphate compounds into Canadian waterways (Canada Water Act 1970). These events have been closely monitored by the local press, and several editorials have been written urging more careful use of our precious waterways, in addition to the adoption of phosphate-pollution abatement programs (*The Windsor Star* 1971a; 1971b). Now the big question is how does the general public perceive the phosphate-pollution problem, and what are people prepared to do about it? It is the purpose of this investigation to study certain aspects of the perception of the phosphate problem by residents of Windsor, Ontario.

In this investigation an attempt will be made to assess certain attitudes towards phosphate pollution relative to some of the socio-economic characteristics of the population and the willingness of the population to change its phosphate-use patterns.

In order to achieve these goals a survey of Windsor households was made in order to assess public attitudes towards phosphate pollution in relation to socio-economic traits and household detergent-use behaviour patterns. First, an areally stratified random sample of 1530 households was selected, and then each householder was interviewed regarding his attitudes towards phosphate pollution, use of phosphate detergents, and socio-economic status. Included in this questionnaire were questions concerning family income, the occupation of the principal breadwinner, and the estimated monthly use of phosphates in detergents. Next, the respondents were asked questions on the following topics: (1) to what extent do they believe phosphates cause pollution; (2) which activity is the prime source of phosphate pollution; (3) to what extent does phosphate pollution affect them; (4) the need for remedial action to abate phosphate pollution; (5) their willingness to pay higher taxes for improved sanitary facilities to alleviate phosphate pollution; and (6) their willingness to change to low-phosphate detergents.

Once these interviews were collected, coded, and tabulated, certain observations were made. The respondents were asked whether they agreed or disagreed with the following statements: (1) 'Phosphates cause pollution'; (2) 'The phosphate problem must be dealt with as soon as possible'; (3) 'The spoiling of our rivers and lakes will continue unless the amount of phosphates used is substantially reduced.' To each of these questions the response was one of general agreement, since over 93 per cent of the respondents supported all three statements. Because of the fact that there was very little variance in the response pattern, these results seem to indicate only that a vast majority of the people acknowledge that there is a phosphate-pollution problem but not much else.

However, when the respondents were asked whether they agreed with the statement, 'Phosphate pollution concerns me personally,' or 'I would be willing to spend extra tax money on improved sewage disposal treatment of phosphates,' the response pattern was much more varied. While 80 per cent of the respondents agreed with the statement that phosphate pollution does concern them, 20 per cent of the respondents were not personally concerned with phosphate pollution: these results indicate that a significant amount of apathy exists. While everyone seems willing to admit that a phosphate-pollution problem exists, a significant minority of the respondents in this study were willing to admit that they were unconcerned, and it would be interesting to see how this response pattern is related to various socio-economic traits. When the respondents were asked whether or not they would agree to paying higher taxes to combat phosphate pollution, the response pattern was even more revealing. Only 24 per cent of the respondents were in agreement with paying higher taxes to combat phosphate pollution, and 19 per cent strongly opposed paying higher taxes for phosphate-pollution abatement. It seems that when it comes to footing the bill for pollution control, the pious opposition to phosphate pollution evaporates!

Perhaps a partial answer to the respondents' unwillingness to pay extra taxes for pollution abatement stems from the fact that 62 per cent of the respondents felt that industry was the primary source of phosphate pollution, while only 26 per cent of the respondents felt that householders were the primary source of phosphate pollution: thus this study suggests that the public feels that improved waste disposal will not get at the primary source of phosphate pollution. Another manifestation of this collective attitude is the fact that only 41 per cent of the respondents were willing to change to a low-phosphate detergent. However, because 43 per cent of the respondents were not sure whether or not they would change detergents, a change in detergent-use patterns may be influenced by effective programs of education and legislation.

When the data pertaining to the respondents' personal concern with phosphate pollution were compared to the householders' income, occupation, and average weight of phosphate used per month, several significant relationships were noted. Generally speaking, groups with high family income tended to show less concern over phosphate pollution than low income groups. In addition, a significant relationship between occupation and concern over phosphate pollution was noted ($\chi^2 = 22.37$). A higher proportion of labourers and skilled outdoor workers tended to show concern over phosphate pollution than clerical and professional personnel, indicating a tendency towards apathy relative to phosphate pollution on the part of high-income white collar workers. Finally, a significant relationship between average monthly use of phosphate in detergents and degree of concern with phosphate pollution was detected in an analysis of variance, but these results are not quite what one would expect. It seems that those respondents who were either very concerned or not at all concerned with phosphate pollution tended to use less phosphates than those respondents who tended to be moderate in their response. Perhaps those who showed the greatest concern over pollution have already switched to low-phosphate detergents, while those who are least concerned may use small amounts of detergents and feel that their use of phosphates does not contribute to phosphate pollution.

When data on the respondents' relative willingness to pay more taxes were compared to the householder's occupation, income, and phosphate use, two statistically significant relationships were observed. First, as family income tended to increase, the willingness to pay for pollution abatement declined. Next, it was found that a slightly smaller proportion of professional and clerical workers supported increased taxation for phosphate control, which indicates that a slightly smaller proportion of high-income professional and clerical workers were sufficiently concerned about phosphate pollution to pay for abatement. Finally, there was no significant relationship between phosphate use and willingness to pay higher taxes for pollution abatement.

Several conclusions may be drawn from this survey. Over 90 per cent of the public feels that a phosphate problem exists. Many people place the blame for phosphate pollution on industry rather than on household phosphate use. While many people acknowledge the existence of a phosphate problem, 20 per cent of the population express a lack of concern over phosphate pollution and only 24 per cent are willing to pay for pollution abatement. These findings, combined with the fact that only 41 per cent express a willingness to change detergents, indicate a significant level of public apathy towards the phosphate problem; also it seems that the higher-income white collar workers tend to be less concerned with phosphate pollution and less willing to pay higher taxes for pollution abatement. While further analysis of the data is planned for the future, these findings indicate several significant trends, which should be considered prior to the formulation of phosphate-control strategies. The first step in phosphate-pollution abatement programs should be the development of a more effective educational program, which effectively portrays the implications and sources of phosphate pollution.

Beeton, A.M., 1965 Eutrophication of the St. Lawrence Great Lakes, *Limnology and Oceanography* 10 (Apr.), 240–54.

Canada Water Act, 1970 *Statutes of Canada*, 1969–70, chap. 52, part III.

Hasler, A.D., 1947 Eutrophication of lakes by domestic drainage, *Ecology* 28 (Oct.), 383–95.

Task Group 2610P Report, 1967 Sources of nitrogen and phosphorus in water supplies, *J. Amer. Water Works Assoc.* 59 (Mar.), 344–66.

US Department of the Interior, Federal Water Pollution Control Administration, Great Lakes Region, 1968 *Lake Erie Report: A Plan for Water Pollution Control* (Chicago), p. 107.

Windsor Star, 1971a Clean lakes far off, Mar. 30.

– 1971b Canada's phosphate ban explained, Mar. 31.

s0210
The decision to control eutrophication
T.R. LEE *Canada Centre for Inland Waters*

In September 1969, following four years of investigation, advisory boards submitted a report to the International Joint Commission which identified eutrophication as the major pollution problem of the lower Great Lakes and the rate of the addition of phosphorus as the factor which led to the development of this advanced state of eutrophication (International Lake Erie Water Pollution Board 1969, 7). In the period following the submission of that report the problem of the control of phosphorus loadings became a major issue of international concern and fundamentally affected the policies of a major industry. There is no evidence that, when the investigation of the pollution problems of the lower Great Lakes was being made, either the US or the Canadian governments were aware of the implications of the findings. The events following the submission of the report provide, therefore, a valuable insight into decision-making behaviour in an international setting in which the response to the problem of the agencies involved developed empirically as the situation evolved. It serves also to illustrate the real restraints that exist on the authority of international institutions.

A eutrophic lake is one with a very high biomass productivity. Eutrophication is a natural stage in the aging of lakes. The problem of concern is that of cultural or accelerated eutrophication caused by the addition of nutrients by man (Vollenweider 1969, 13–15). These conditions have become increasingly common in lakes surrounded by high levels of human activity. It was this condition that was found to exist in the lower Great Lakes and traced to the high level of nutrients, especially phosphorus, added over the last forty years. The phosphorus loading was found to come largely from municipal sewage, with a smaller contribution from land run-off and industry. In 1967 some two-thirds of the phosphorus load to Lake Erie was from municipal sewage (International Lake Erie Water Pollution Board 1969, 74). The sources of the phosphorus in municipal sewage are human waste and heavy-duty laundry detergents. The contribution of the latter has been growing constantly since 1930 at a rate more than proportional to the growth of population (Gilbertson, Dobson, and Lee 1972). The report to the International Joint Commission recommended that sewered wastes be treated for the removal of phosphorus and other nutrients and that phosphorus, in the form of sodium triphosphate, be replaced in detergents.

Two months after receipt of the report from the advisory boards the International Joint Commission initiated a series of public hearings on the recommendations contained in the report. The hearings focused on the proposals for the control of phosphorus. At the hearings the detergent companies attempted both directly and indirectly to attack the central hypothesis that phosphorus was the limiting nutrient. This argument spilled over from the hearings to continue through presentations before committees of both Congress and Parliament and to be debated in the academic and technical press. The controversy centred on the relative roles of carbon, nitrogen, and phosphorus in limiting the eutrophication process (Canada, Department of Energy, Mines and Resources, Inland Waters Branch 1970, 1).

This debate concentrated attention on the detergent regulation problem. In the public view the issue became one of industry versus ecology, although the establishment of the role of phosphorus in eutrophication did not necessarily require that it be removed from detergents. During 1970 the International Joint Commission itself recommended to the two governments that a nutrient-control program was required, involving both waste treatment and phosphate replacement in detergents. The Government of Canada moved to reduce the level of phosphates in detergents by means of an amendment to the Canada Water Act. A Committee of the House of Representatives recommended that regulation be undertaken in the United States. Most of the Great Lakes states did regulate the levels of phosphorus in detergents through 1970 and 1971.

There was, therefore, an apparent acquiescence on the part of all parties to the recommendations of the International Joint Commission. This situation was changed, however, when the US Surgeon-General announced reservations over the advisability of using the most probable replacement for sodium triphosphate in detergents, nitrilotriacetic acid (NTA) (US Surgeon-General 1971). This action produced a gap in the gradually developing consensus on the required reaction to the eutrophication problem.

The two countries continued negotiations towards a general agreement on the control of the pollution problems of the Great Lakes, including adherence to the reduction in the nutrient loadings recommended by the International Joint Commission, but the differing policies on the use of NTA raised serious questions as to how to construct an international agency effectively if there was not international agreement on the means as well as the goals.

This somewhat cursory review of events is intended to show that the recommendation of the advisory boards was passed through the decision-making process with very little hindrance. The whole advice of the scientific investigators rapidly became the policy of the international agency and one of the national governments involved. All bodies accepted the recommendations up to the point of regulating the level of phosphates in detergents.

The unanimity of the water-quality agencies was broken only by the action of the US Surgeon-General. It was this action that revealed, however, the very difficult administrative conflicts which existed in the two countries. In Canada the members of the advisory boards were drawn from the senior levels of management in Ottawa and from the two leading water research institutes, the Canada Centre for Inland Waters and the Fresh Water Institute. The control of phosphate rapidly became a national policy; partly, no doubt, this is a reflection of the significance of lakes in Canada, but it is also an event traceable to the central position of the advisory board in the federal regulatory authorities. The rapidity of action was facilitated by the coincidental passage of the Canada Water Act through Parliament.

In contrast, the US members of the advisory boards were largely drawn from regional offices. Consequently, in the subsequent escalation of the magnitude of the decision that had to be made, the Great Lakes were a peripheral issue. Despite congressional support it was not possible to make the quality of the lakes a national issue in the United States. The potential problems of the disruption of the soap and detergent industry and the possible health implications of the substitutes for phosphates were national issues. It did not make sense, therefore, to control detergents for a minority interest. The ostensibly more rational federal solution to the problem of phosphate control became waste treatment alone rather than the dual policy recommended by the International Joint Commission and accepted by Canada. Most of the Great Lakes states had adopted phosphate-content regulations.

There have been a number of suggestions in recent years for a new comprehensive management institution on the Great Lakes to replace the International Joint Commission (Ostrom, Ostrom, and Whitman 1970; Jordan 1969; Dworsky 1967, 71–8). It is the contention here that the history of the phosphate decision does not substantiate any claim for a comprehensive management agency. Such an agency would be forced to limit itself to issues that were of regional concern, although many issues are not. A regional agency is not appropriate beyond the management of waste treatment (Kneese and Bower 1968, 281–92; Freeman and Haverman 1971, 70). Industrial material use and health standards decisions are national issues and beyond the competence of a regional pollution control authority. The International Joint Commission was established to perform functions that were beyond national solution. It would be a serious error to move back to replacing it with a limited regional agency for the Great Lakes alone. Any modification to the present decision-making system must take account of the unequal position of the Great Lakes in the two countries.

Canada, Department of Energy, Mines and Resources, Inland Waters Branch, 1970 The control of eutrophication, *Technical Bulletin* no 26.

Dworsky, L.B., 1967 Political environment of water resources, in *The Fresh Water of New York State: Its Conservation and Use* (Dubuque).

Freeman, A.M., and R.H. Haverman, 1971 Water pollution control, river basin authorities, and economic incentives, *Public Policy* 19, 53–74.

Gilbertson, M., H. Dobson, and T.R. Lee, 1972 *A History of Phosphorus Loadings and Hypolimnial Dissolved Oxygen in Lake Erie*, in press.

International Lake Erie Water Pollution Board and the International Lake Ontario–St Lawrence River Water Pollution Board, 1969 *Pollution of Lake Erie, Lake Ontario and the International Section of the St. Lawrence River*, report to the International Joint Commission.

Jordan, F.J.E., 1969 Recent developments in international pollution control, *McGill Law J.* 15, 279–301.

Kneese, A.V., and B.T. Bower, 1968 *Managing Water Quality: Economics, Technology, Institutions* (Baltimore).

Ostrom, V.A., E. Ostrom, and I.L. Whitman, 1970 Problems for institutional analysis of the Great Lakes Basin, *Proceedings 13th Conference, Great Lakes Research*, 156–67.

US Surgeon-General, 1971 Statement before the Federal Trade Commission, April 26, 1971.

Vollenweider, R.A., 1969 *Scientific Fundamentals of the Eutrophication of Lakes and Flowing Waters*, report to the Organization for Economic Co-operation and Development.

s0211
Water-landuse policy on the amphibious Piedmont Corridor, Colorado
M. JOHN LOEFFLER *University of Colorado, USA*

This paper is concerned with the water-landuse policies of the privately owned storage reservoirs on the northern Colorado Piedmont.

Centrally located at the east base of the Southern Rocky Mountain front, the northern Piedmont occupies the South Platte drainage. Its mixed nature of land and water makes it unique among world amphibious landscapes. In December 1971, there were 461 privately owned storage reservoirs on the land, nearly all of them man-made, reflecting the cultural and human responses to an erratic water-deficit environment.

Precipitation on the northern Piedmont is marginal for any kind of landuse. Of the 35.56 centimeters total, 74 per cent falls in the spring and early summer months and contributes to excess water only when there is occasional flooding. As a consequence, the cultural amphibious landscape is an anachorism and the result of reservoirs built to retain excess water from the spring melt and from floods that supply the needed water to accommodate intensive Piedmont cropping practices.

Water requirements for early maturing crops are often met from direct stream diversions, although additional water is requisite to Piedmont agriculture. From 1872, the building of the first reservoir, to the present, the primary concept of reservoirs on the Piedmont has been holdover storage for late irrigation. Three crops, sugar beets, alfalfa, and corn, all late-season water hogs, are the core of the agricultural economy and storage water is essential for their successful production. Since 1950, however, a subset of different needs has evolved from a rapidly increasing and dominantly urban population of four to one ratio that forces the consideration of new concepts and policies of reservoir and landuse by the irrigation farmer vis-à-vis the urban dweller.

In 1971, 110 reservoirs were randomly selected for a water-landuse study. Two selection criteria were invoked: (1) *Size*. Only reservoirs of four hectares or more in surface were included. Lakes under four hectares noticeably narrow the range of use schemes. (2) *Location*. All reservoirs are in Weld County, the current growth locus.

Whether for storage, recreation, or homesite use, reservoir size and storage water sources are significant aspects of the economy and the landscape. Of the 110 lakes in the model, 51 range from four to 12 hectares in size, 26 from 12 to 20 hectares, 14 from 20 to 40 hectares, 9 from 40 to 200 hectares in area, 4 from 200 to 400 hectares and 3 exceed 400 hectares. Water sources for these pondages determine reservoir fluctuation levels and are the key to reservoir use. Seventy-six reservoirs are supplied with diverted water from permanent streams assuring a relatively constant supply whereas 34 reservoirs gather seep and run-off water, mainly flood catchment.

Reservoir ownership and administration are essential aspects of current and future use policies. Eighty of the 110 reservoirs belong to individual or small unincorporated groups. Mutual Ditch Companies own 12 reservoirs as part of incorporated canal systems for water storage and delivery and 13 are the property of reservoir companies supplying water to stockholders via irrigation canals. From this total of 105 reservoirs, 90 are used solely for agricultural storage whereas 15 have limited public or leased private recreational rights as well as farm storage. Two reservoirs in this group have sizable residential lake frontage development. Of the remaining five reservoirs, three are for domestic water supply and two for public recreation.

From growth mania pressures and from the use and ownership inventory there evolves the question of whether increasing demands for water-based recreation and reservoir frontage for homesites are compatible with agriculture's preeminent position. It is already clear that the position of agriculture is not jeopardized by human activities on or near the reservoir sites. It is equally clear that the continuously increasing scale of involvement of the land developer does threaten agriculture. The developer buys agriculture tracts for housing development. With the agricultural tracts are the water rights which he sells to amortize some of the land cost or which he may hold for future speculation.

The Constitution of the State of Colorado leaves not a scintilla of doubt that the water resources of the state are public property: 'Every natural stream not heretofore appropriated within the State of Colorado is hereby declared to be the property of the public and the same is dedicated to the use of the people of the state.' Further, no water appropriation is complete until the water that is diverted has been applied to the beneficial use for which it was intended.

Arguments pro and con are beginning to crescendo concerning storage water for agriculture. The question is whether water, which is public property, stored in pondages on private property on land with no access rights, denies public ownership and beneficial use. Inasmuch as ninety of 110 reservoirs store water for farm use exclusively, it is urgent and prudent to examine the entire philosophy of water use and ownership on the Piedmont landscape.

In the seven-county area that embraces most of the amphibious landscape there is as well the population growth locus, the Piedmont Corridor. In the 1950 and 1960 decennia, 61 and 74 per cent, respectively, of the total state population increases occurred in these counties. From this growth there is a cancerous engorgement of the land and a substantial alteration of its surface from bulk building, houses, freeways, and factories that affect the capacity of the land to provide enough quality water for all. These conditions lead inevitably to concern for adequate water supplies and to a rethinking of the role of stored water on the Piedmont landscape.

Of first order concern must be the role of evaporation as the main dissipator of Piedmont water supplies. In the reservoir model, 6872 hectares of water are exposed to a dry atmosphere with frequent and extended wind conditions. Whatever the method of calculation, water losses are staggering. It is prudent, therefore, to rearrange the Piedmont priorities to focus on these losses.

The first order of planning and work must, therefore, concern reservoir consolidation. Close examination of the amphibious landscape indicates that at least twenty reservoirs, 40 hectares or larger, can be eliminated. Their water can be stored in existing pondages with minimal increases in surface exposure. Several reservoirs in the Colorado-Big Thompson transmountain diversion, a federal system, have been overbuilt. High elevation storage from the point of view of evaporation losses is preferred and available space should be utilized. With reservoir consolidation and elimination there is attendant ditch and canal reduction and additional saving of water from seepage, evaporation, and phreatophytic incursion.

If attended to, this initial priority will refocus the entire beneficial use concept of publicly owned water. It is clear that a Federal agency and dollars for an undertaking of this scale are needed. If such be the case, then the 1965 Federal Water Project Recreation Act, a mandate to recreation, will be invoked: 'full consideration shall be given to the opportunities, if any, which the project affords for outdoor recreation and for fish and wildlife enhancement and that, wherever any such project can reasonably serve either or both of these purposes consistently with the provision of this Act, it shall be constructed, operated and maintained accordingly.'

With this kind of reordering of the Piedmont landscape, it is important that 'take lines' for each pondage be fully and adequately established. This will provide space for the needed recreation facilities, fish and wildlife habitat, and for adequate disposal facilities. Water-based recreation for a large population will become a reality by making Piedmont water accessible and usable. Reducing canalage will enhance the aesthetic quality of the Piedmont landscape even more by removing the existing bulwark of raw, excavated land now occupied by a network of canals and ditches. High elevation storage and fewer Piedmont reservoirs will noticeably stabilize agricultural water and reduce the pollution potential already rampant. The remaining canals, small rivers, and the basal pondages, must, in keeping with the qualities of Piedmont location, be appropriately landscaped to enhance the aesthetic qualities already imparted to the land by the mountain front.

s0212
The problems and issues of implementing national water legislation at subnational levels
KEITH W. MUCKLESTON   Oregon State University, USA

Because of the physical complexities of water and the characteristic disconformity of units of supply, jurisdiction, and demand, implementation of national water legislation is often difficult. This difficulty is particularly challenging in large countries, which usually encompass a great diversity of water resources that are managed by a wide variety of administrative-political entities. Commonly, these entities represent two or three levels of government and are organized either by function or by territory. Changing social priorities and increased pressure on water-derived services exacerbate what already may have become dysfunctional water management systems. This paper seeks to illustrate the general problem of fitting national water legislation to subnational levels by focusing on the implementation of the Federal Water Project Recreation Act (hereinafter the Act) in the United States of America.

The Act, passed in 1965, reflects new national concerns in several respects. First, it gives recognition to the idea that recreation and fish and wildlife must share equal consideration with the traditional and more utilitarian benefits from water development. Second, it recognizes that man has the technological and political means not only to alleviate environmental damage caused by water projects, but also to improve selected desirable aspects of his biophysical surroundings (Cooley and Wandesforde-Smith [eds.] 1970, 121–2). Third, by further encouraging user fees, it seeks to assess direct beneficiaries for project derived services. Fourth, it attempts to broaden the decision making base by devolving upon nonfederal entities increased planning and financial responsibilities for two water derived services stemming from federal projects, thus interposing another set of interests between the federal agencies and their clientele interest groups that have traditionally benefitted the most from federal water projects (Hammond, Beard, and Muckleston 1970, 238–47).

The Act contains several provisions that render its implementation difficult. First and foremost, non-federal entities are required to pay for part of recreational enhancement, which was often a free byproduct of federal water projects. Moreover, participation in the Act usually requires assumption of long-term debt, a situation against which most non-federal entities have safeguards. Other factors which can impede implementation of the Act are: a frequent lack of planning expertise among non-federal entities; the distance of proposed federal water projects from centres of population; and priorities of non-federal entities that do not mesh with the federal schedule of project authorization, appropriation, and construction.

The extent of its implementation was examined through questionnaires, interviews, and correspondence six years after passage of the Act. Questionnaires, the principal source of information, were sent to all Corps of Engineers District Offices and Bureau of Reclamation Regional Offices; to state agencies responsible for recreation and fish and wildlife; and to local entities that had been identified as participating in the Act. Response from federal, state, and local levels was excellent, adequate, and poor, respectively.

Analysis of the response revealed a notable pattern of discrepancies between federal and state respondents, the latter being markedly less positive about the Act. In addition, perception of the problems by state fish and game departments was often markedly different from that of state agencies responsible for recreation and/or combined functions. Major problems of implementation are now considered (Muckleston 1971, 13–29).

The most formidable obstacle is the high cost of participation. Non-federal entities are not infrequently either unable or unwilling to cost-share. Federal agencies placed more emphasis on this point than the states, which blamed unsuitable project location and/or poor timing as frequently as high cost. State agencies responsible for fish and game are, however, not nearly as concerned with the questions of location and of priority as are the state agencies responsible for outdoor recreation. When recreation is the principal consideration, ready access to reservoirs by large numbers of urbanites may well be one of the major factors affecting the decision to participate in the Act.

Desire not to assume long-term debt would appear to be another prominent obstacle to implementation. The states place considerable emphasis on this point; and most particularly on the significance of the states' constitutional restrictions against incurring long-term debt. The federal agencies, in contrast, infrequently perceived state constitutions as impeding implementation. The appraisal by federal agencies probably reflects more accurately the reality of the situation. Evidence suggests that non-federal entities circumvent constitutional restrictions when desiring to participate in the Act, and cite the restrictions when participation in the Act is deemed undesirable.

In at least several cases, however, state laws have effectively impeded implementation. In Florida eminent domain cannot be used to acquire recreational land, which successful implementation of the Act requires; in North Dakota reimbursement is required for tax loss when lands are developed for recreation, which results in higher costs to non-federal entities cost-sharing under the Act; and in Oregon constitutional restrictions against incurring debt have shifted practically all non-federal participation to the counties, only some of which have adequate staff and funds for participation.

A third noteworthy impediment to implementation is the dissatisfaction of some state agencies with their input in planning. Under terms of the Act, a non-federal entity that agreed to cost-share the enhancement of recreation and/or fish and wildlife would plan these facilities. Although the planning of enhancement facilities such as boat-launching ramps and picnic tables has been largely assumed by the non-federal participants, reservoir operations remain under the jurisdiction of the federal agencies and are apparently managed to maximize traditional water derived services. This has prompted some non-federal entities to charge that despite the Act, interests favouring recreation and fish and wildlife remain residual legatees, and, that the federal agencies are only using these benefits to justify project authorization.

There are additional obstacles to implementation: counties that do not wish to provide recreation for non-county residents, as they ostensibly bring more problems than benefits; intervening recreational opportunities, particularly in sparsely populated parts of the country well endowed with lakes, mountains, and forests, most of which are under federal management and therefore offer free access; fish and game departments that will not consider cost-sharing under the Act because other federal acts require them to share a much smaller proportion of the costs of enhancement and, because they desire mitigation in kind, not for example, replacement of fur bearers with trout.

It is submitted that the problems of implementing the Federal Water Project Recreation Act illustrate many of the general complexities of fitting national water legislation to the myriad patterns of biophysical and human phenomena present in large countries.

Office of Water Resources Research through the Water Resources Research Institute at Oregon State University. Grant agreement number 14-31-0001-3237. OWRR project number A-004-Ore.

Cooley, R.A., and G. Wandesforde-Smith (eds.), 1970 *Congress and the Environment* (Seattle).

Hammond, K.A., D.P. Beard, and K.W. Muckleston, 1970 *The Impact of Federal Water Legislation at State and Local Level* (Completion Report for OWRR Water Research Project B-019-Wash), Ellensburg, Washington.

Muckleston, K.W., 1971 Problems and issues of implementing the Federal Water Project Recreation Act (PL 89-72) in the Pacific northwest (Phase 1) (OWRR project number, A-004-Ore.), Corvallis, Oregon.

s0213
Application of the Canadian glacier inventory to studies of the static water balance. I. The glaciers of Vancouver Island
C. SIMON L. OMMANNEY   *Environment Canada*

Perennial ice and snow masses constitute some 75–80 per cent of the fresh water in the world and cover $16 \times 10^6 km^2$. Although their contribution to the dynamic water balance is quite small, in arid areas of low summer stream flows the contribution of glacier-melt can be as much as 80 per cent of the discharge in July and August. However, glaciers do form the major part of the fresh water component in the world's static water balance and are also of great importance to the earth's energy balance. Estimates of the glacierized area in Canada are 151,000km^2 for the Arctic and 50,000km^2 for the mainland with a total volume of approximately 40,000km^3; for comparison, the area of the Great Lakes is 242,455km^2 with a volume probably about the same as that of the glaciers. Recent evaluations indicate that currently accepted figures may be quite incorrect. Hydrologists in the United States have found about 20 per cent more glaciers in the Cascades than had previously been noted (US Dept. of the Interior, 1970, 39), whereas an inventory of Axel Heiberg Island showed 25 per cent less ice than previously measured (Ommanney 1969). These figures do not indicate any substantial fluctuation of the glaciers but are more a reflection of the very poor data we have for the glacierized area. To assess the true quantitative role of glacier ice in the static water balance a glacier inventory program has been undertaken as part of the Canadian government's water resource inventory and the International Hydrological Decade (IHD). The inventory will provide the datum from which the role of glaciers in the dynamic water balance can be assessed on a regional basis.

This paper discusses some aspects of the Canadian Glacier Inventory Project using the inventory of Vancouver Island as an example. Although the importance of some 27.6km^2 of perennial ice and snow to the water balance of Vancouver Island is obviously not great, this brief discussion of the data obtained will indicate the type of information that will be available for all the glaciers in Canada and which, in areas such as those of the North Saskatchewan, Columbia, and Kootenay river systems, will provide much needed data for the assessment of the contribution of glacier-melt by the extension of existing runoff models and thus the role of glaciers in the dynamic water balance. Techniques are already being developed for increasing or decreasing glacier-melt and thus long term planning of Canada's water resources can only properly be carried out when such information is available.

In 1968 a Glacier Inventory Section was formed in the Glaciology Subdivision of the Inland Waters Branch.

The standardized guide to the worldwide inventory of glaciers, formulated by UNESCO (UNESCO/IASH 1970), is being followed in the Canadian program with the addition of some information on literature, photographs, moraines, glacier-dammed lakes, and special glacier features. Under the international proposals a glacierized area is divided into hydrological basins, the ice mass subdivided into individual glaciers and each then numbered and the constituent parts measured.

At present the only completed glacier inventory report is that for Axel Heiberg Island in the Queen Elizabeth Islands (Ommanney 1969). This island, of 37,185km^2, has 11,735km^2 of ice with an estimated volume of 3222km^3. Almost half of the 1121 glaciers found there are in the main ice cap which covers 7222km^2 of the island. 11,364 glaciers on Baffin and Bylot islands and 1835 glaciers on Devon Island have been mapped, as have 1616 glaciers in the Nelson River drainage basin and 219 glaciers on Vancouver Island.

Complete data on the glaciers of Vancouver Island have recently been compiled. Geographical and Universal Transverse Mercator co-ordinates were measured for each glacier, as well as the elevation of the upper and lower parts of the glacier, the snowline, and the mean elevations of the accumulation and ablation areas. Lengths, the width, and the surface area of the constituent parts of each glacier were obtained and estimates of the depth and volume made. Based on a preliminary analysis of these data some tentative conclusions can be drawn.

Of the 219 glaciers covering 27.6km^2 of Vancouver Island almost half are located in Strathcona Provincial Park. 72 per cent of the glaciers are classified as glacierets and all but one of the remainder are small mountain glaciers of the niche, cirque, or ice apron type. The orientation is predominantly in the NW to NE quadrant, 84 per cent of the ablation areas facing this direction and 76 per cent of the accumulation areas. The elevation range is from 670m to 2073m above sea-level with a mean of the highest elevations of 1466m and of the lowest of 1256m. Unfortunately the photographic coverage of the island, on which the inventory was based, varies from July 1953 through to September 1969, with the bulk of data from 1957 (151 glaciers), so that it is difficult to draw any conclusions about the transient snowline. The weighted mean accumulation area elevation for 1957, using glacier area as a weighting factor to eliminate undue bias from the large number of small glacierets, is 1500m and that for the ablation area 1175m above sea-level. For 1953 it seems to be about 25m higher and for 1956 about 50m lower but the figures for both these years are mainly from glaciers in the north western part of the island where the differences might be attributable to different orographic and precipitation patterns. Only three glaciers had measurable snowlines, the rest falling either above or below the transient snowline at the time of photography.

An assessment of snow survey data from four stations, with an elevation range 975–1280m, in the vicinity of Strathcona Park, Forbidden Plateau – 76, Burman Lake – 94, Upper Thelwood Lake – 95 and Memory Lake – 96, based on data from 1954 to 1969 (BC Dept. of Lands, Forests and Water Resources 1965) has been made to determine the probable activity of the Vancouver Island glaciers. The data show that in 1956 the maximum winter accumulation was 53.8cm above the mean, whereas in 1957 it dropped to 58.7cm below the mean. Glacier data from the May 1957 photography are too early in the season for snowline assessment but taking data from 77 glaciers inventoried from the July to October 1957 photography, covering 7.44km^2, an accumulation area ratio (AAR) of about 80 per cent is obtained which indicates a fairly strong tendency towards a positive balance for that year; a 70 per cent AAR is generally considered indicative of a balanced glacier. This apparently positive year may be an anomalous reflection of residual snow from the heavy falls in 1956 or it could be in balance with the much lower snowfalls of 1957, which would indicate a very healthy state for the small glaciers of Vancouver Island. Five-year running means of the snow survey data show a steady drop from a value of almost 178cm in 1956 to 149cm in 1963 and then a steady rise to 198cm in 1967. Thus, if 1957 were considered a positive year then the movement of the five-year means, showing winter accumulation increasing from 1965 to 1967 above the 1957 values, would indicate that the glaciers of Vancouver Island are healthy and could expand if present climatic conditions persist. A more sophisticated analysis of the climate and precipitation patterns will be carried out to see whether these tentative conclusions can be further substantiated.

BC Department of Lands, Forests, and Water Resources, 1965 *A Summary of Snow Survey Measurements, 1935–1965*. Water Investigations Branch, Water Resources Service, BC Department of Lands, Forests, and Water Resources, 170 pp.

Ommanney, C.S.L., 1969 Glacier inventory of Canada – Axel Heiberg Island, Northwest Territories, *Inland Waters Branch Technical Bulletin No. 37*, Dept. of Energy, Mines and Resources (Ottawa), 97 pp.

UNESCO/IASH, 1970 Perennial ice and snow masses, *Technical Papers in Hydrology No. 1*, UNESCO no. A.2486, 59 pp.

US Department of the Interior, 1970 River of life: water, the environmental challenge, *Conservation Yearbook Series Vol. 6*, Govt. Printing Office (Washington), 96 pp.

s0214
An approach to evaluation in multiple objective basin planning
J. O'RIORDAN   *Department of the Environment, Canada*

The recent introduction of multiple objective planning in water resource management has initiated a reassessment of the traditional approaches to the planning process (US Water Resources Council 1970). This paper explores the implications of this new strategy on the evaluation process and following a brief critique of traditional concepts of evaluation develops a new empirical approach within the context of comprehensive river basin planning.

The main concept of multiple objective planning is the recognition that society's welfare is an integration of many values represented by economic, environmental, and social goals. Thus, water resource management alternatives should be evaluated in terms of all the values contained in a chosen set of multiple objectives and not just in terms of the economic growth objective as has traditionally been the case. Because other major objectives such as preservation or enhancement of environmental quality, equitable distribution of income and opportunity, and public health and safety are now explicitly included in comprehensive river basin planning (Pacific Northwest River Basins Commission 1970), benefit-cost analysis which is at present developed to examine efficiency criteria associated with the economic growth objective can be viewed as only a partial approach to evaluation.

Two techniques for extending benefit-cost analysis to evaluate multiple objective water resource plans have been devised in recent years. The first involves the attempt to measure all contributions of water management alternatives to multiple objectives in monetary terms so that they are commensurate and then apply efficiency criteria. Although great advances have recently been made in developing measures for evaluating such values as recreation amenity and wild river aesthetics (Water Resources Engineering, Inc. 1971), many of the elements associated with environmental and social goals simply defy monetary evaluation. The second approach is to treat certain non-monetary consequences as constraints and apply economic efficiency criteria within these defined limits. Unfortunately, the implications of these constraints, such as specified water quality standards, are seldom analysed in the context of other objectives such as economic growth or equitable distribution of income and thus there is a real danger that society's total welfare may not necessarily be improved. An approach that systematically takes into consideration the impact of each alternative plan on all relevant multiple objectives is required and is discussed below.

Within the context of water resource planning, the first step of this evaluation approach is to identify economic, environmental, and social values associated with both consumptive uses (irrigation, industrial, domestic) and non-consumptive uses (recreation, aesthetics, ecological system needs) of water and to determine the potential contribution of each to all major study goals. Just as few elements of the environmental quality goal can be readily expressed in monetary terms, it is also true that money does not measure all the values associated with irrigation or municipal water supplies. Where possible, water quantity and water quality parameters associated with a range of desired water uses should be translatable into criteria associated with the multiple goals such as dollars (economic growth objective) or employment opportunities (social well-being). Where such transformations are not feasible, water resource parameters must be used as surrogates for social values; for example, surrogate measures for aesthetics could include decreases in turbidity or incidence of algae in streams or lakes.

The next step is to define instrumental or planning objectives which relate aspects of water management (water supply, flood control, water quality improvement) to each identified water use. To indicate the degree to which each planning objective (and thus related major goals) can be achieved by alternative courses of action, each such objective must be clearly specified in time, place, and magnitude for each of the water quantity and water quality parameters chosen to represent social values.

Once these planning objectives have been specified, a set of alternative projects, plans, or policies can be developed to achieve them. It is important that included in this set of alternatives is an analysis of present conditions so that other plans can be directly compared with the status quo. Furthermore, all alternatives including the status quo should be modelled so that their output is specified in the same water quantity and water quality parameters as are contained in the planning objectives. In this way, the degree to which planning objectives are achieved can be properly assessed and evaluated.

Analysis of alternative water resource plans in terms of their effect on the complete set of major goals is achieved by the development of an evaluation matrix. Four steps are involved in the construction of such a matrix. Firstly, the consequences of achieving and of failing to achieve the water quality and water quantity parameters specified in the planning objectives must be assessed for all alternatives including the present or null alternative. Secondly, consequences associated with all development alternatives should be compared with those associated with the null alternative (model of present conditions), positive increments being considered as benefits and negative increments considered as costs in terms of each major goal. Thirdly, those incremental consequences that directly contribute to the economic growth objective can be evaluated in dollar terms and then analysed according to established economic efficiency criteria.

Incremental impacts, for which only surrogate values are available, accruing to planning objectives related to the environmental and social goals can be ranked according to a scoring procedure. For each such planning objective, the extreme values of impacts (for example, increases in dissolved oxygen) with the set of alternatives should be noted and the difference between the value for the null alternative and these extreme values (above and below the zero alternative) calculated. The larger difference, whether positive (benefit) or negative (cost) can be scored as 100. Differences in impacts for all other alternatives and the zero alternative can then be scored as a percentage of this absolute maximum difference, incremental gains being scored positively and incremental losses being scored negatively. Quantitatively, the scoring process can be stated as follows:

(1) $S_n = (X_i - X_0)_n / |(X_i - X_0)_n|_{max} \times 100$,

where S_n is the score for planning objective n, X_i is the impact of alternative i on planning objective n, and X_0 is the impact of the null alternative on planning objective n.

The fourth step involves weighting the scores to reflect the relative value of these gains and losses to society. Because of the mechanics of the scoring procedure outlined in equation 1, at least one alternative will always have a maximum score of plus or minus 100 in terms of achieving (positively or negatively) each planning objective. Unless the achievement of planning objectives is equally valued by society, scores should be adjusted by a weighting factor which reflects society's preferences.

In practice, it is not feasible to develop an acceptable weighting system a priori as society's preferences are generally unknown at the start of the evaluation process. Once the monetary benefits and costs associated with the economic objective have been calculated, an initial set of weights can be devised by comparing incremental benefits and costs accruing to the non-economic objectives with economic net benefits. Iteration using different weighting systems and public feedback can then proceed to test whether plan selection is sensitive to small changes in weights.

The evaluation approach described above does not possess the analytical elegance of traditional benefit–cost techniques but it does represent a systematic way of evaluating all impacts of water resource plans in terms of a full range of study goals. Obviously, all planning objectives cannot be analysed in equal depth, otherwise the planning costs would quite likely exceed the marginal benefit of the improved results. Early in the planning process, planners must approach the public to obtain information on their value preferences and then select the major planning objectives to be included in the quantitative analysis.

It is difficult, if not impossible, for any person to express an absolute measure of satisfaction for some amenity values unless these values are put into the context of what has to be sacrificed in terms of other major goals to achieve such values. The main advantage of the matrix approach to evaluation is that all values can be analysed in this context and that it provides the basis for establishing incremental bargaining procedures so that consensus decision making can be achieved.

Pacific Northwest River Basins Commission, 1970 Puget Sound and adjacent waters, App. xv, Plan Formulation (Vancouver, Wash.).

us Water Resources Council, 1970 Standards for planning water and related land resources, Special Task Force Report to the us Water Resource Council (Washington, DC).

Water Resources Engineering, Inc., 1971 The Wild rivers project, Report to Office of Water Resources Research (Washington, DC).

s0215
Attitudes towards water quality and water-based recreation in the Qu'Appelle Valley, Saskatchewan
J.G.M. PARKES   *Environment Canada*

The problem of providing public involvement in the planning process is not new. Devices such as meetings, hearings, and mass education programs have been employed in the past to ensure that 'the general public' will be involved in the final choice of planning alternatives. Too often, however, these efforts are post facto exercises in tokenism, designed to soothe consciences.

One promising new vehicle which has appeared to facilitate public participation is the perception and attitude study. The general aim of such studies is to discover, for a certain segment of the population, just what individuals 'see' in their environment, how they feel about it, what it means to them, and how they would be disposed to act towards it (White 1966, 118). By using certain sampling procedures, a researcher can obtain a larger, less emotional, and more representative set of public views than by conventional methods. The perception and attitude study can complement public hearings, in that information is gathered from those non-vocal, non-organized segments of society that are hesitant or unable to come forth with their comments (Schiff 1970, 1). In addition, responses are comparable from one individual to another. A study of public responses can help planners estimate, rather than assume, the social costs that may be incurred by changes in the environment.

A number of federal-provincial water resource planning agreements were signed in anticipation of the assent of the Canada Water Act on 26 June 1970. One such agreement involved the governments of Saskatchewan, Manitoba, and Canada, who jointly have set about the task of producing a plan for the comprehensive administration of water resources in the Qu'Appelle River Basin. The agreement calls for a final plan to be produced in 1972. Under the Qu'Appelle Agreement, Section 3(e) provides for public participation in the planning process, so that the final plan will 'reflect the needs and desires of the population for whom it is intended.'

After discussions between officials of the

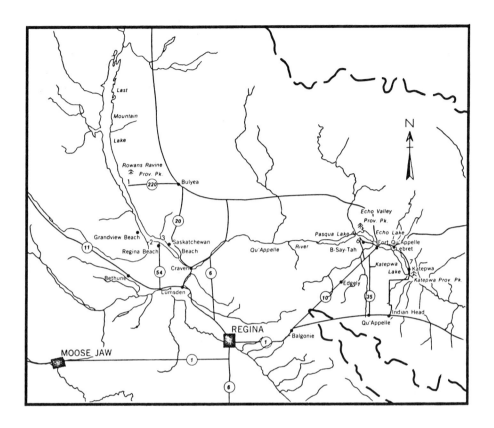

LEGEND

WATERSHED BOUNDARY — — — — ⌒ ⌒ ⌒

SAMPLE POINTS — — — — — — — —

MAJOR ROADS — — — — — — — —⟨6⟩—

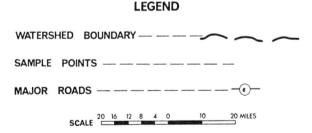

Fig. 1. Last Mountain Lake and fishing lakes, Qu'Appelle Valley (20 miles = 32km).

federal government and Saskatchewan regarding a study of public user attitudes towards the quality of water in certain lakes within the Qu'Appelle chain, a proposal for a perception and attitude study was drafted and accepted by the Qu'Appelle Study Board. Briefly, the objectives of the study were: (1) to examine how the present level of water quality influences water-based recreation in certain Qu'Appelle Valley lakes, and (2) to determine the extent of people's willingness to pay for improvements in water quality.

Four lakes and their adjacent recreational areas were chosen for study. Water quality sampling and testing was carried out three times a week during the period of the study at the provincial park beach areas at Echo, Pasqua, Last Mountain and Katepwa lakes (Fig. 1). This program enabled a comparison to be made over time of the physical con-

TABLE 1. Recreation characteristics by lake

	Last Mountain	Pasqua	Echo	Katepwa
Average number of days spent in water-oriented recreation	19.5	13.6	16.7	17.2

TABLE 2. Perception of algae problem by lake

	Last Mountain	Pasqua	Echo	Katepwa
Per cent of users mentioning algae as a major problem	31.0	60.0	78.0	71.2
Per cent of user reduction due to algae	4.6	30.0	29.0	26.5

TABLE 3. Willingness to pay by lake

	Last Mountain	Pasqua	Echo	Katepwa
Per cent willing to pay for improved water quality	48.0	79.0	80.0	98.0
Average amount willing to pay (per user-day per season)	$0.29	$0.76	$0.62	$0.66

dition of the water with the public's perception of the resource.

A stratified random sample of 560 recreation users (campers, cottagers, and day users 18 years of age and over) was interviewed (Berry and Baker 1968, 93). The sample sites included provincial park and private beaches at Katepwa, Echo, Pasqua, and Last Mountain lakes. In addition, a sample of 240 shoreline cottage residents located around these lakes was taken. Interviews were taken at each lake daily, averaging 6 per day during the week and 18 per day on the weekends from 4 July to 1 September 1970.

The results of testing indicated that the surface waters in the recreational areas studied met accepted water quality criteria during most of the study period. Algae growths visibly increased as the study progressed and the increased algae levels were reflected in higher values of pH, colour, turbidity, and BOD during August.

A comparison was made of the water quality trends over the summer for each sample lake, and each lake was ranked with the others to determine a crude scale of ranking. Last Mountain Lake had the best water quality; Katepwa the poorest. Pasqua and Echo fell between the two.

In theory, the lake with the poorest water quality should have different effects on the recreation population than the lake with the best water quality. An analysis of the interviews taken at each lake confirmed this hypothesis. Results are summarized in Tables 1–3. The users at Last Mountain Lake (ranked as the lake with best water quality): (a) recorded the highest average amount of use in terms of primary water contact sports per user per season (Table 1); (b) recorded the least amount of activity reduction due to water quality problems (Table 2); (c) did not spontaneously mention algae as a major pollution problem (Table 2); (d) were least willing to pay for improved water quality (Table 3); (e) recorded the lowest average amount indicated to improve water quality (Table 3).

On the other hand, the users at Echo and Katepwa Lakes (which were ranked as the lakes with the worst water quality): (a) spontaneously mentioned algae as the major water quality problem with the highest frequency (Table 2); (b) were overwhelmingly willing to pay for improved water quality (Table 3); (c) were willing to pay a significantly larger amount for cleaner water than those users at Last Mountain Lake (Table 3); (d) registered the highest degree of concern over pollution.

In addition, with the aid of a multiple regression analysis, it was found that the most significant independent variables which positively affected willingness to pay for improved water quality were: (*a*) income levels, (*b*) time of the season, and (*c*) amount of participation in water-oriented activity. As these variables increased, so did willingness to pay.

From the results of this study it is evident that water quality in general and algae levels in particular are severe enough to cause a significant reduction in recreational use of water in three of the lakes surveyed. Users in the study area are aware of the mounting concern in society over pollution and seem to be becoming more sensitive to environmental quality in their immediate milieu. Also, the users at the lakes with the poorest water quality are willing to pay a significant amount per user day per season over and above the additional costs that are normally incurred in the recreational experience.

Each recreation experience has a quality dimension. However, a definition of levels of quality is not easy to complete. The lakes in the Qu'Appelle Valley provide water-based recreation for about a third of the population of Saskatchewan (based on attendance figures). As such, the quality of their water is of prime importance for the users. However, recreation is not the sole use made of the water resource. The question arises as to what level of water quality is acceptable. Competing uses of the water such as supply for domestic consumption, irrigation, and sewage disposal exist. Somehow these uses must be weighed and optimum levels determined. This type of study should aid the decision-makers in this process.

G.C. Mitchell; J. Glenn; D. Silliphant; C. Burton; D. Carter.

Berry, Brian J.L., and D. Marble, 1968   *Spatial Analysis* (Englewood Cliffs, NJ).
Saarinen, Thomas F., 1966   *Perception of the Drought Hazard on the Great Plains*, Department of Geography Research Paper no. 106 (Chicago).
Schiff, Myra R., April 1970   The Concept of Perception, paper submitted at a symposium on Attitudes and Perceptions, University of Victoria, Victoria, BC.
White, Gilbert F., 1966   Formation and role of public attitudes, in *Environmental Quality in a Growing Economy*, Henry Jarrett (ed.) (Baltimore).

s0216
The drainage and politics of western water protection: defining an area of origin
F.J. QUINN   *Environment Canada*

Public interest has focused so completely on proposals for interregional and international water diversion that few have noticed the effective opposition to them put forth by those areas with the most to lose. No area of streamflow origin in the North American west has ever parted willingly with its water. The inhabitants of Owens Valley, the upper Missouri, the Pacific northwest, and Canada have that much in common.

In the several controversies surrounding western water diversion since the turn of the century, how have these areas organized spatially to protect their interests? It will be instructive to identify the area of origin in two stages: one, an approximation in which the drainage pattern assumes paramount importance, the other essentially a refinement of the outline in that those who feel their rights or their welfare threatened by external control of the basin, or a part thereof, are likely to group along more traditional jurisdictional lines. The resolution of basin and jurisdictional units toward defining the area of origin is the purpose of this paper. In the process it should become clear that the basin is not an isolable unit for defining water development interests.

A cursory map examination will reveal that the configuration of drainage basins had very little influence on the evolution of political regions in the North American west. The reverse is not true. Once the west was organized politically, these jurisdictions had a great deal to do with the pattern of subsequent water development, as will become clear below.

No matter how suitable a drainage basin might otherwise be as an areal unit for water management, the absence of political institutions which can adequately represent the interests of basin inhabitants limits its effectiveness. There is no legislature in the basin. Governmental agencies make decisions outside its context. Integrated development has proven illusory everywhere that its proponents have pictured the basin as a self-contained unit, free of centrifugal forces and loyalties (Wengert 1957, 271).

At a subnational level, the perspective may be less clear than on an international river but the outcome is not much different. Conceptually, the states of an interstate basin should try to develop the most efficient or appropriate combination of projects in the basin, modified by considerations of equity in sharing benefits. But in fact, the real business of a state like Wyoming has been in establishing how much of the Colorado and Missouri rivers it owns and in promoting development in those parts of the rivers above all for the good of Wyoming, rather than for the good of the basins, the West, or the United States (Upper Colorado Commission 1948, I, 27).

The states, like the provinces in Canada, have proprietary rights over the resources within their borders. It is the basin states, and not the basin inhabitants otherwise constituted, which pursue their claims to an interstate river. The experience of past litigation and compact negotiation has been that each state will try to maximize its rights of ownership, thereby emphasizing the differences rather than the common interests in the basin. On the other hand, once effected, a compact may facilitate the basin states' common interest in gaining federal funds for development or in gaining water from other basins and states.

The bias of state over basin in the matter of interbasin transfers is made explicit in the Colorado River Compact (Widmer 1956, 39). The several states were the signatories and it was to them that allocations of water were eventually made, *not* to the basin portions of the states. More water is diverted out of the Colorado Basin than any other river basin on the continent, but in each case diversion consists of redistribution *within* the state. The raison d'être for heavy extrabasin diversion is evident on a map of water and people: the important centres of water consumption among the basin states almost all lie outside the basin itself (see Figure 1). A state like Wyoming, therefore, has a claim to Colorado River water because a portion of that basin lies within the state; Wyoming's use of that water, on the other hand, is not limited to the

Fig. 1. Basin and basin states of the Colorado.

basin portion of the state. Associated power generation likewise serves the greater numbers outside the basin.

It is more than coincidence that all transfers of water that have been effected thus far on the continent fall within state and provincial borders (Tinney 1969, 411). Although the effects are certainly felt downstream across these borders, it is still fair to say that the present pattern of interbasin diversions strongly reflects the political regionalization of Canada and the United States.

The literature speaks of the protection of rivers like the Columbia, the Mackenzie, or the upper Colorado against outside or downstream appropriation, and certainly area-of-origin considerations have no meaning except with reference to the river whose future is at stake. The water in the river is the object of contention.

Yet cries for protection come not only from the basin or part thereof; there is no effective representation along strictly basin lines. The term 'area' is preferred to 'basin' of origin because it is the encompassing political area which speaks for basin interests. Nor is this simply a matter of political authority or administrative convenience. That part of the political area outside but close to the basin may identify closely with the basin, as does Seattle with the Columbia, both as a consumer of the basin's power or water or recreation and through other commercial or social relationships. The object of political action in this case is to extend protection on the basin of proximity to, and jurisdiction over, the source of supply rather than on a narrow hydrologic interpretation. Whether the streamflow remains inside the basin is of less concern than whether the appropriate jurisdiction guards its rights of origin against its neighbours, who may or may not own a share of the basin.

There is no one area of origin in the west. Rather, basin and jurisdictional units interplay at various levels to form what might be called a hierarchy of areas of origin. At the lowest level is found the local area of origin, whose protection against adjacent areas depends on a higher authority like the state; the counties-of-origin legislation enacted by California is an early example (California Stats. 1931, 720). At the next level in the hierarchy are the states or provinces which by adjudication or compact protect their rights to interjurisdictional waters, as does New Mexico, for example, on the Pecos and Wyoming on the Colorado. At a third level might be included a group of states which have a common stake in a river basin and move to reserve it against the importation ambitions of other states; the four states sharing most of the Columbia River succeeded in this effort against the southwest (Quinn 1970, 155). Finally, Canada as a whole might be considered an area of origin protecting any of its lakes or rivers from those who would divert them south of the border.

There is, of course, a paramount difference between what residents of a county in California and what the parliament of Canada can do at either end of the hierarchy. Sovereignty, it became apparent with the diversion of northern California waters, counts for little at the lower end.

California Statutes, 1931 Sacramento.
Quinn, Frank J., 1970 Area-of-origin protectionism in western waters, unpub. doctoral thesis, U. of Washington, Seattle.
Tinney, E. Roy, and Frank J. Quinn, 1969 Canadian Waters and Arid Lands, in *Arid Lands in Perspective*, W.G. McGinnies and B.J. Goldman (eds.), American Association for the Advancement of Science and University of Arizona Press, Tucson.
Upper Colorado River Basin Compact Commission, 1948 *Official Record*, 3 vols.
Wengert, Norman, 1957 The politics of river basin development, *Law and Contemporary Problems*, 22, 258–75.
Widmer, Richard T. (ed.), 1956 *Documents on the Use and Control of Waters of Interstate and International Streams*, us Dept. of the Interior, Washington.

s0217
Interbasin transfer of water in England and Wales
HENRY A. RAUP   *Western Michigan University, USA*

In England and Wales, as in many regions of the world, there is a growing disparity between the natural distribution of water supplies and the location of population and industry which are dependent upon those supplies. Interbasin transfer of water is being used increasingly to overcome water shortages, but it is commonly accompanied by sharp controversy. However, interbasin transfers have been made in England and Wales for more than a century, and, unlike the situation in North America, these transfers have been accomplished with minimal controversy. It is the purpose of this paper to establish the magnitude of interbasin transfers in England and Wales, and to suggest those conditions which have contributed to the freedom from controversy. Data on interbasin transfers (daily averages for 1967 and 1968) have been obtained from water supply agencies, from appropriate river authorities, and from published sources – especially Wilkinson and Squire (1968).

The basic pattern of existing water transfers in England and Wales has developed as a matter of expediency, to meet specific local water deficiencies from the nearest, cheapest source of clean water. Some 24 major interbasin transfer schemes (those in excess of 22,730.45 cubic metres per day, or 5 million imperial gallons per day) have been identified, ranging in scale from 26,367m^3/d to 447,335m^3/d. These 24 schemes account for a total transfer between drainage basins of 2,779,024.82m^3/d. Since England and Wales have an estimated *net* water consumption of approximately 15,456,706m^3/d for public water supply and industry (Rowntree 1968, 506), the major transfer schemes represent about 18 per cent of all water supplied in the two countries. There is, thus, a significant dependence, by cities and industries, upon imported supplies of water.

Recipients of imported water are clustered in five distinct regional groupings: (1) the Manchester–Merseyside area, supplied with an average of 1,006,504m^3/d from the Lake District and from north Wales; (2) the West Midlands (particularly Birmingham) supplied with 464,610m^3/d from central Wales; (3) the London–South Essex area, supplied with 427,332m^3/d from the Thames and from several Essex rivers; (4) the West Riding of Yorkshire, supplied with 382,781m^3/d from the eastern flanks of the Pennines and from the Yorkshire Ouse; and (5) a number of dispersed coastal cities, usually tapping small nearby rivers for a total of 497,797m^3/d. Most of the major English conurbations are included within these recipient groups which are supplied wholly or in part with imported water.

Although most British transfer schemes generated opposition at the time of their initial installation, the objections have not been directed towards the export of the water, but rather towards the inundation of the reservoir site. Fears are expressed over the inundation of communities, of arable land, or of particularly distinctive flora, but little concern is expressed over the loss of the exported water itself. The relative absence of controversy surrounding British interbasin water transfers may be traced to a combination of conditions which facilitate transfers – the generally humid environment, the absence of strong topographic divides, the use to which imported water is put, and the absence of legal restraints on interbasin transfers.

First, for England and Wales as a whole there is no threat of an absolute water shortage. The mean annual precipitation is about 890mm, ranging from more than 5000mm in Wales and the Lake District to less than 508mm in the Great Ouse basin. Annual runoff is estimated at 181,843,600m^3/d, of which only 8.5 per cent is at present being abstracted for public and industrial uses (Rowntree 1968, 507). The abundance of surplus water, especially in the least populated areas, has reduced fears of inadequate water supply and the concomitant proprietary interests in water which develop where surplus water is in demand by arid regions.

Second, the early use (beginning in 1849 with Liverpool's importation from Rivington Reservoir) and increasing dependence upon imported water is the result, in large part, of the ease of establishing transfer schemes. The character of the topography, in combination with the proximity of sources to recipients, presents no unusual or serious engineering problems. Divides are low and generally may be breached by cut-and-cover techniques. Furthermore, the supply reservoirs of some schemes (such as Vyrnwy and Elan) are sufficiently elevated that gravity-fed siphons are employed to lift water over intervening divides. Because other supplies have been available, no transfers have yet been developed across the Pennines or the Welsh crest, although both such transfers are now being considered.

Third, unlike North America where large-scale interbasin transfers are made for a variety of purposes – including irrigation, hydroelectric generation, low flow augmentation, sewage disposal, and others – interbasin transfers in England and Wales are exclusively for public water supply and industry. Two schemes, at Fawley and at Grimsby, transfer degraded water for industrial use; all of the remaining schemes supply water to the general distribution systems of the individual recipients, in which industrial and domestic consumers share an interest. No significant transfers between basins in England and Wales are made for purposes other than industrial and public water supply. With imported water being utilized exclusively for public water supply – the highest priority of usage – controversy over transfer schemes is diminished and complaints which arise in North America, that transferred water is misused or misallocated when applied to lower priorities, are unknown in England and Wales.

Finally, and of greatest importance in dampening controversy over interbasin transfers, is the absence of political 'divides' in England and Wales. The unitary system of government in England and Wales gives Parliament ultimate control, directly or through its designated agencies, over any specific interbasin transfer. While there may be considerable debate over a proposed transfer scheme, once Parliament has approved a transfer project, there is no legal means by which any lesser political unit can prevent its development, in contrast to the United

States where individual states may prohibit the export of water from their respective territories. There is no equivalent to 'states' rights' among the counties of England and Wales. Furthermore, the Water Resources Act (Great Britain 1963, Section 1) charges the national Water Resources Board and the 29 river authorities with the duty to redistribute water among the various drainage basins. This is the only alternative to water shortages specifically mentioned in the act, and interbasin transfers are now a matter of official policy.

Given a perceived need for more water, then, British engineers regularly turn to importation to meet their requirements, and interbasin transfers now constitute a major component of British water management. With most major English and Welsh cities importing at least part of their water supplies, the dependence of industrial and population centres upon imports may well be expected to increase in the future, as increasing demands for water are being matched by new proposals for interbasin transfer schemes. With neither physical nor institutional barriers to hamper them, interbasin transfers have been accomplished with relative ease, which in turn has contributed significantly to the absence of strong controversy surrounding new transfers. The British public, as well as the engineers involved in providing adequate water supplies, have come to accept interbasin transfer as a normal and customary water management device.

Great Britain, 1963   *Water Resources Act*, HMSO (London).
Rowntree, N.A.F., 1968   The problem of future water supplies, *Water and Water Engineering* 72, 505–10.
Wilkinson, D., and N. Squire, 1968   *Water Engineer's Handbook* (London).

s0218
The International Joint Commission (IJC): a model for international pollution control organizations?
WILLIAM M. ROSS   *University of Washington, USA*

Politicians, academics, and laymen have grown increasingly apprehensive over the increasing degradation of the world environment. The United Nations attested to this concern by organizing the 1972 Conference on the Human Environment.

One particular area of concern has been the water quality of many inland lakes, especially the Great Lakes. Several attempts have been made to clean up and prevent pollution in these waters. Most of the programs have emanated from the International Joint Commission (IJC) and some individuals have now suggested that other organizations designed to control international pollution could be modelled after the IJC (Ablett 1971a; 1971b). This paper evaluates the capability of the IJC to meet the threat of international pollution, its ability to enforce pollution regulations in the boundary area between Canada and the United States, and its appropriateness to control international pollution on a global scale.

The IJC is but one of several bilateral commissions in Canada and the United States.

The organization was established by the Boundary Waters Treaty of 1909. Article IV of that treaty specifically prohibits the pollution of boundary waters to the detriment of property or personal health. Boundary waters have been defined as fresh water; thus pollution transmitted by air or salt water is not specifically covered by the treaty. The treaty does provide, however, a mechanism whereby the commission can investigate such problems, but only upon a joint reference from Canada and the United States (International Joint Commission 1965). The recommendations are not binding, however, on either government.

The commission consists of six members, three appointed by the United States and three by Canada. Commissioners, while appointed by national governments, are pledged to view problems from a continental perspective and to act as a single body pursuing the joint interest.

When problems arise, a letter is sent by each government to its own national chairman. While the IJC does not have a large technical staff, it selects experienced and competent people from the civil service of both countries who conduct independent technical investigations. The IJC in turn holds hearings, studies the technical reports, and submits recommendations to national governments.

The record of the commission in dealing with a limited number of pollution problems does not lead to an optimistic conclusion about the ability of the IJC to deal with international pollution. In 1912 the commission was asked to undertake a broad study of pollution problems from the Lake of the Woods to the Saint John River. The IJC found that most of the lakes were not polluted but that areas near major urban centres were seriously contaminated. Recommendations for alleviating the problem were made but subsequently ignored by the national governments.

In 1928 an air pollution reference was given to the commission. Farmers in the state of Washington had complained that fumes from the smelter at Trail, BC had ruined their orchards and crops. Investigations by the IJC resulted in the Consolidated Mining and Smelting Company paying compensation of $350,000 (Bloomfield and Fitzgerald 1958, 20).

Water pollution references in 1946, 1955, 1959, and 1964 for differing water bodies along the boundary resulted in Objectives for Boundary Waters Quality Control, as the minimal criteria to be used by enforcement agencies. In all these cases the commission was given the task of continuing supervision, but these recommendations and supervision have failed to arrest the decline in water quality (Bourne 1971, 133). Funding of national programs has been small, subnational authorities have resisted commission recommendations, and the commission is powerless to enforce its own objectives. As a result, the Great Lakes and many other boundary waters are in a greater state of deterioration than in 1912.

References in 1964 and 1966 considerably broadened the scope of the commission's investigations. The 1964 reference asked the commission to investigate water pollution in the Great Lakes while the 1966 reference asked the IJC to monitor pollution in the Detroit River and to investigate all sources of air pollution along the Canada–United States boundary (Rempe 1968, 142). Recommendations from the 1964 reference resulted in a June 1971 meeting during which Canada and the United States agreed on a plan to clean up the Great Lakes by 1975.

Given the limited responsibilities that the commission possesses, its record of accomplishment is commendable. It has identified many of the critical problems, undertaken exhaustive research studies, and recommended remedial actions. These successes may be attributed to the structure of the IJC, to the thoroughness of its investigations, and to the non-binding nature of its recommendations. Other international organizations seeking to control pollution would do well to emulate these characteristics. The commission has several weaknesses, however. It has no power to deal unilaterally with pollution problems and no experience outside the southern boundary.

The IJC, by the very nature of its mandate, derives its power from, and reflects the will of, the signatory nations (Wilson 1971, 508). The reluctance of either nation to take forceful action under the treaty demonstrates the fundamental problem with international pollution control. Problems along the international border are numerous and complex and the work of the IJC cannot be considered independently of attitudes to pollution control, the respective economic strength of the contracting states, foreign policy, and social ties. Much of the criticism that can be directed at the IJC, therefore, is not a condemnation of the commission, but rather of the framework within which it must operate.

Given the existing attitude of national governments, it is not likely that the IJC will become a supranational institution with commensurate power. This does not mean, however, that pollution control efforts of the commission cannot be improved. Two avenues are open. One would involve increasing the power of the IJC by giving it authority to deal with all pollution problems and the right to report violations to the appropriate Attorney-General for prosecution. The other would entail, with IJC concurrence, the establishment of regional committees to deal with a pollution problem in a specific area. Both of these suggestions could be incorporated within the scope of the present treaty.

These suggestions are made with the assumption that national action can reduce international pollution to tolerable levels. If such an approach fails, more radical changes will be necessary, giving international organizations the power to determine potential problems, undertake studies, and enforce recommendations (Ross 1971, 326–7). Such proposals would require negotiation of a neoteric treaty.

The IJC is, therefore, a useful model for those who believe environmental problems are not critical and can be solved through international as opposed to supranational approaches.

The Stockholm Conference met earlier this year to consider international environmental problems. In a sense it was the first concrete recognition of a global environmental crisis and it represented a determined, if perhaps futile, effort to reduce environmental pollution by co-operative nation-state action. The attempt to clean up the Great Lakes by 1975 will be a specific and perhaps final test of the IJC and its approach to pollution control. If the lakes are not cleaned up or if significant improvements cannot be made by 1975, then we would do well to bury the 'international approach' in favour of some experimental 'supranational' programs.

Ablett, Dave, 1971a US-Canada near pact on cleaning up the Great Lakes, *Vancouver Sun*, June 9, p. 16.
– 1971b Canada-US approve plan to clean up Great Lakes, *Vancouver Sun*, June 11, p. 21.
Bloomfield, L.M., and Gerald F. Fitzgerald, 1958 *Boundary Water Problems of Canada and the United States* (Toronto).
Bourne, C.B., 1971 International law and pollution of international rivers and lakes, *U. British Columbia Law Rev.* 6, 113–36.
International Joint Commission, 1965 *Rules of Procedure and Text of Treaty* (Washington).
Rempe, George A., III, 1968 International air pollution – United States and Canada – a joint approach, *Arizona Law Rev.* 10, 138–47.
Ross, William M., 1971 The management of international common property resources, *Geographical Rev.* 61, 325–38.
Wilson, Thomas W., Jr, 1971 International environmental management: some preliminary thoughts, *Natural Resources J.* 11, 507–12.

s0219
The role of hindsight evaluations
H.E. THOMAS   *US Geological Survey, USA*

The traditional role of hindsight evaluations has been summed up by the National Academy of Science (NAS) Committee on Water (1966, 15): 'There has been surprisingly little analysis of the effects of water-use decisions of the past ... no major water project in the United States has been studied with sufficient care and precision to determine its full effects on the systems of water, soil, plants, and human activity which it has altered.'

My first involvement in hindsight evaluation was as participant in a survey of the first large artificial reservoir – Lake Mead – undertaken in 1949, fourteen years after Hoover Dam first interrupted the flow of the Colorado River (Thomas 1954). The survey was concerned with the sediments, dissolved salts, and heat, as well as the water impounded in the reservoir (Smith et al. 1960), and was a multidisciplinary landmark for its day, but our studies were chiefly of the reservoir and not of the effects upon ecology or upon human activities.

Today let us consider the California Water Plan, its implementation, and hindsight evaluation. This plan (CDWR 1957, 37) 'is a master plan for the control, conservation, protection, and distribution of the waters of California, to meet present and future needs for all beneficial uses and purposes in all areas of the State to the maximum feasible extent. It is a comprehensive plan which would reach from border to border both in its constructed works and in its effects. The Plan is a flexible pattern susceptible of orderly development by logical progressive stages, the choice of each successive incremental project to be made with due consideration to the economic and other pertinent factors governing at the particular time.'

The implementation of the plan was summarized about a decade later (CDWR 1968, 3): 'California is currently in the midst of constructing an unprecedented water project for one essential reason – the State had no alternative. Nature has not provided the right amount of water in the right places at the right times. Eighty per cent of the people in California live in metropolitan areas from Sacramento to the Mexican border; however, 70 per cent of the State's water supply originates north of the latitude of San Francisco Bay. Throughout the State, the bulk of rainfall occurs in a few winter months, while the summers, when water needs are greatest, are long and dry. The solution to California's maldistribution of water resources has been one of conserving the sporadic stream runoff in surface storage reservoirs and transporting the regulated supplies to areas of use. Progressively larger storage works and longer conveyance systems have been required to meet the continuing growth in demands for water ...'

Currently the 13 largest of the 'progressively larger storage works' have capacities ranging from 1 to 5.5km^3 (1km^3 = 811,000 acre-feet), and aggregating more than 30km^3. An additional 20km^3 of storage is provided by more than a hundred reservoirs of 50 million cubic metres or more. The conveyance system includes 4,000km of canals and tunnels with capacities greater than 10m^3/s. These developments have been constructed by the federal Bureau of Reclamation and Corps of Engineers, the state Department of Water Resources, county water agencies, municipalities, local irrigation districts, and private and public utilities. For several decades the State of California has held in trust the water rights to practically all undeveloped surface waters, and has released or assigned those rights for projects that are 'not in conflict with a comprehensive and co-ordinated plan of development of the water resources of the State.' Thus the California Water Plan is well established as a master plan, comprehensive and co-ordinated, and with ample flexibility particularly with respect to the sources of funds for development.

Hindsight evaluations are available in large volume and variety, ranging from sweeping generalizations to detailed appraisals of specific items. These are concerned with implementations and operations to date as well as portents for the future, for the California Plan generally contemplates 'so full a stage of development of the water resources of the State that almost no drop of fresh water would reach the sea without having been first put to some commercial use' (Bain 1966, 104). Many hindsight evaluations lead to recommendations similar to the NAS Committee on Water (1966, 48) concerning water planning in general: 'All organizations, public and private, engaged in planning for the use of water, should give increased attention to alternative approaches and courses of action, to the appraisal of social costs and benefits, and to the use of research as one of the means by which new effective solutions could be reached.' The Sierra Club (Watkins 1971) claims to chronicle not a shortage of water but a prodigious shortage of human foresight, and points to the book on *California Water* edited by David Seckler (1971) as the single most useful source in determining the present and future prospects of the California Water Project. Seckler's book declares that some very bad mistakes may be made unless old ways of thinking are adapted to new and changed conditions and potentialities, but notes also (p. xii) that in the most recent report on the California Plan (CDWR 1970) the Department of Water Resources is 'much more in accord with the thinking in this book.' Thus hindsight is evidently leading to a consensus that advances in the knowledge of water and its possible uses have changed the character of water problems, and have made it possible to deal with these problems in a greater variety of ways and more effectively than in the past.

However, plans for the future call chiefly for additional storage and conveyance works – more 'big dams and ditches' – and it looks as if water development programs have a sort of inertia of uniform motion or 'progress' that is demanded of, by, and for government bureaucracies everywhere. As pointed out by the Committee on Water (1966, 13): 'the consideration of valid alternatives often has been prevented by the rigidity of the mission of the agency proposing the project, by imperfect apportionment of costs, by neglect of nonmarket benefits and social costs, and by a simple lack of awareness of available alternatives.' The space remaining to me, I hope, is just enough for status-or-progress reports on several of the alternatives that have been proposed, considered, implemented, or committed to further study, by the major water-development agencies.

Ground-water storage is the alternative to surface-water storage, and the only way mother nature can keep fresh water indefinitely in arid regions. Man can do no better, and surface storage is a losing proposition in any area of perennial water deficiency, where evaporation exceeds precipitation. In California two-thirds of the surface storage capacity is in water-deficient areas; this is comparable to the storage in Lake Mead, but the water-surface area is more than twice as great, and annual losses by evaporation may well exceed the annual loss of about a cubic kilometre from Lake Mead.

The total ground-water storage capacity in California is far greater than the surface storage capacity, and the annual pumpage from underground is about $20km^3$, but the federal and state agencies that are prime movers in the California Water Plan are only indirectly and slightly involved in subsurface storage, and then chiefly in research. Partly this is because man is far more adept at withdrawing than at putting water underground, and artificial recharge is deserving of more research and demonstration projects. Partly it is because the federal and state agencies have become chiefly wholesalers of water who can encourage ground-water recharge by their pricing structure, but leave management largely to local agencies; some of these are exceedingly competent and foresighted, others nearsighted or dominated by landowners who stand on their rights to the water and the storage space beneath their lands.

Of more than $40km^3$ of water withdrawn annually in California, about half is used for washing, flushing, processing, and cooling – non-consumptive, pollutive uses. Traditionally these waste waters have been dumped into the ocean or estuaries or rivers or on the ground. Standard sewage treatment produces waters that are reusable but generally about 300mg/l more mineralized; tertiary treatment removes much of the dissolved mineral matter, but is more costly. In southern California more than $0.5km^3$ of used water is reclaimed annually and reused (*waste water reclamation*), chiefly for ground-water recharge.

Desalination of sea water is more costly than the conventional storage and delivery of continental water to the point of use, but advancing technology is decreasing the cost, and projections of current trends indicate that by 1980 desalted water may be competitive with waters transported great distances by conventional means. Particularly will this be true if the energy required for the desalination can come from nuclear fuel or geothermal sources at sufficiently low cost.

The unthinkable alternative is that people might change their wasteful ways in time to do some good. For disposing of the wastes of our existence we have developed a flush-toilet syndrome – get them out of sight and out of mind, dilution is the solution to pollution, and spray frequently with aerosols. Water is extremely important to cleanliness and we have argued long and loudly over rights to use water; but without assigning any responsibility to the water user for cleaning up his messes. Effective water management would require that all those who would put water to beneficial but nonconsumptive use return it to the resource essentially unchanged.

Bain, J.S., R.E. Caves, and Julius Margolis, 1966 *Northern California's Water Industry* (Baltimore).

California Department of Water Resources (CDWR), 1967 The California water plan, *Bull. 3.*

– 1968 The California State water project in 1968, *Bull. 132-68*, Appendix C, p. 3.

– 1970 Water for California – the California Water Plan outlook in 1970, *Bull. 160-60.*

Committee on Water, 1966 Alternatives in water management, Natl. Acad. Sci. – Natl. Research Council, Pub. 1408.

Cooper, Erwin, 1968 *Aqueduct Empire* (Glendale, Calif.).

Seckler, David, ed., 1971 *California Water* (U. California Press).

Smith, W.O., C.P. Vetter, G.B. Cummings, et al., 1970 *Comprehensive Survey of Sedimentation in Lake Mead, 1948–49*, US Geol. Survey Prof. Paper 295.

Thomas, H.E., 1954 *First Fourteen Years of Lake Mead*, US Geol. Survey Circ. 346.

Watkins, T.H., 1971 California – the New Romans, in R.H. Boyle, John Graves, and T.H. Watkins, eds., *The Water Hustlers* (Sierra Club).

Competition among water users within the framework of statute law
A.D. TWEEDIE   University of Newcastle, Australia

Throughout the world the current trend in water resource use and management runs strongly against the development and operation of individual choice, and statute law is being more widely accepted as the institutional basis of water resource allocation (Thomas 1969). Rapidly changing situations in technology and demand are being met by the supposedly increased flexibility which statute law makes possible. State authorities concerned with the operation of such law, however, can themselves introduce an element of inflexibility as traditional thinking colours decision making.

In Australia the need for change in the institutional basis of water management became obvious in the late nineteenth century, and in the colony of Victoria statute law replaced common law as early as 1886 (East 1950). Following federation in 1901 a framework of statute law of broadly comparable aspect was adopted in each of the Australian states in which ownership of surface and, more recently, of sub-surface waters is vested in the crown (the states), and water use by individuals is by licence issued by the appropriate state authority.

Within this framework applications for licences are advertised and objections from existing users along the stream are considered by a land court which subsequently advises the state authority on the terms of licence to be issued and the period for which it is valid. A system of equitable distribution for irrigation agriculture, with restrictions in periods of low supply being uniformly applied, is the result.

There is one significant exception to this seemingly complete control by a single state authority. Rights to the organisation and use of water are also lawfully vested in the statutory body charged with water supply to each of the state capitals and a few other major urban centres, and allow the statutory authority in this case 'from time to time [to] divert and impound water from any stream it may seem fit.' Intended initially to organise and guarantee the water supplies on which the health of these urban communities depended, urban water authorities today are responsible for the needs of almost all of Australian manufacturing industry.

Statute law in each Australian state has thus created two independent authorities charged with separate tasks but not necessarily regionally separated. Initially designed on the one hand to provide an equitable distribution of irrigation water in an agrarian society, and on the other to protect the health of urban populations, these statutory authorities now face the need to resolve problems of competition of much greater complexity. Traditional approaches do not provide more than a temporary solution. Two recent examples of competition, in the Hunter Valley of New South Wales, illustrate the nature of the problem (Dickson 1970).

Electricity generation in New South Wales is heavily dependent upon thermal power production, and a $200 million plant to generate 2000 megawatts (40 per cent of the foreseen energy demands of the state) is currently under construction on the coalfields of the Middle Hunter. The design includes a cooling pond, and cooling losses of 0.037km^3 (30,000 acre-feet) annually are predicted. Operating as the law requires, the State Electricity Commission applied for a licence to make good these losses by pumping from the river. Farming interests, whose total irrigation requirements from the river were no more than treble the predicted requirements of the generating plant, strongly opposed the application, and the land court's subsequent recommendation was for a pumping licence to replenish the cooling pond only when the flow of the river was more than adequate to meet irrigation requirements. Industrial users, in effect, were granted junior appropriator status. A long history of resolving the rivalry of irrigation farmers gave the licensing authority no other basis for considering the relative merits of agriculture and of industry as competitors for water. No declaration of principle concerning this competition was made, no serious consideration was given to other possible sources, and at no time during the debate was any appeal made to the market mechanism. The farmers won their case on prior right of beneficial use. Whether, under drought conditions, 40 per cent of the state's power supply can be endangered by upholding this right remains to be seen.

A second competitive situation developed

in the 1960s in the Williams Valley (tributary to the Hunter) when the Hunter District Water Board, to meet rising domestic and industrial demands in Newcastle, augmented its established headwater storage in the catchment with a pumping and storage unit downstream close to tidewater. The total catchment of the Williams thus became the board's immediate concern, and, in keeping with its act of authority, it opposed the issue of further irrigation licences in the valley. Pressures by farming interests, however, forced a compromise. A joint committee of the two statutory authorities agreed to comparable restrictions in periods of low supply, and the principle of equitable distribution was upheld.

An institutional base which fails to recognise the multiplicity of demands for services from water in modern society, and the need to integrate these in some regional scheme, must result in decisions which proceed in this *ad hoc* fashion. There are some who argue that total administrative control of water can proceed in no other than this arbitrary way, and that only by private decisions and a market mechanism can an efficient, flexible, and rational use of water be effected (Hirschliefer et al. 1960). There is, of course, no reason why, within a statutory framework, economic arguments cannot be used to influence decisions, but to date in Australia the decisions have been reached by political process rather than by economic argument (Davidson 1969). The Federal States Grants (Water Resources) Act of 1964 is the only statement to date that requires some attention to an economic analysis of proposed water resources projects.

Within the institutional framework that operates at the state level of statute law, the competition of agriculture and of industry for water use in Australia remains unresolved. The prospective competition of recreational use is not even contemplated, although storage reservoirs not directly concerned with urban supplies usually make some attempt to accommodate these needs as a secondary benefit. The Australian states, though early in the field of statute law, or rather because of this, have yet to find a satisfactory answer to the problem of competing water uses in a time of growing complexity. The statutes, in their present form, may allow the prospect of an equitable distribution of resources that appeals to the egalitarian attitudes which are supposedly a distinctive feature of Australian life, but the increasing complexities of competition for water uses noted in developed countries calls for something more than the two separate and now competing statutory authorities the present laws have created.

A single authority, with some regional rather than a state sphere of operation, might be a more efficient mechanism to meet the present situation. The United Kingdom, more recently moving into the field of statute law, has been able to approach the problem in this fashion. In Australia such a solution, adopted typically in an *ad hoc* fashion, has temporarily resolved the conflicts in the Williams River near Newcastle and the even larger conflict between the interests of Metropolitan Melbourne and the irrigation interests of the Murray Valley, but existing statute law derived from a non-urban and agrarian society is in itself inadequate to resolve the complexities of competition for uses in the modern industrial and urban state.

Davidson, B., 1969   *Australia, Wet or Dry?* (Melbourne).
Dickson, J.A., 1970   The Organisation of Hunter Valley Water Resources 1885–1969, MA thesis, U. Newcastle (NSW).
East, L.R., 1950   *Victorian Water Law: Riparian Rights* (Melbourne).
Hirschliefer, J. et al., 1960   *Water Supply, Economies, Technology and Policy* (Chicago).
Thomas, H.E., 1969   Water Laws and Concepts, *Trans. Am. Geophys. Union* 50.

s0221
Water and regional development in the Murray River Basin, Australia
E. WOOLMINGTON   *University of New England, Australia*

The ambition to develop the Australian continent has provided a continuing motive for government policies at both state and federal levels – irrigation projects, rural closer settlement schemes, and, more recently, a groping towards effective policies of decentralisation and regional integrated development. The most notable single project was initiated after World War II. This was the Snowy Mountains project, involving the diversion of eastward-flowing streams of the Australian alps westward into the large but low-discharge river systems of the Murray and Murrumbidgee. The scheme had the two-fold objective of hydroelectric power generation and the further provision of irrigation water for the inland. The diversion required massive Commonwealth government investment and by 1970 the project was substantially complete.

However, during the period of its construction, the trend of other, socio-economic events has undermined the 'development utility' of the project. Hydroelectric power is now regarded as at best marginally economic, and irrigation, in an era of agricultural overproduction in the 'developed world,' as downright uneconomic (Davidson 1965, 1969). Thus, concern for the more effective use of the expensively augmented Murray waters has led to consideration of them in the context of urban decentralisation.

In sparsely populated Australia the problem of urban drift is acute; meagre populations in rural areas are diminishing rapidly as younger people move out to take up residence in the handful of metropolitan centres. This drift is a result of rural 'push' and urban 'pull' factors. Rural push derives from agricultural overproduction and diminishing financial returns, resulting from the application of accelerating science/technology to rural production without attendant expansion of markets (Woolmington 1970). This produces not only rural depopulation but also creeping redundancy among rural service towns. Urban pull is a compound of employment opportunities, deriving from the well known snowballing process of the industry/population growth cycle that is productive of urban centralization, plus the generally perceived greater 'facility' of urban life.

Rural depopulation has produced something of a social crisis in Australia's spatially vast rural hinterland and, more importantly, attendant incipient political crisis, in which one of the major parties in the coalition that has held federal office since 1949 (the Country Party), being rurally based, is losing both votes and voters – and faces the danger of extinction! Urban decentralization is perceived as one way of halting the drift to the cities, and additionally in recent times as a method of mitigating the alleged twin problems of environmental degradation and social pathology attending modern metropolitan life. Thus some policy guidelines on decentralization are emerging (Development Corporation of New South Wales 1969) which have suggested government involvement to stimulate the growth to 'mini-metropolitan' dimensions at a few selected rural locations. This is known as the 'selective decentralization' policy.

It is in this context of planning for 'selective decentralization' that the research work reported in the rest of this paper was conducted. Notwithstanding the fact that the key questions in urban decentralization have not been solved, so that nothing is known with any confidence about either the psycho-social desirability of urban decentralization or its economic feasibility, the author sought to establish, in *a priori* fashion, what might be regarded as the most suitable locations for decentralized urban entities in terms of certain critical locationally biased parameters.

Water remains one of the natural resources without which urban entities cannot function. Moreover, urban water consumption in Australia is escalating in almost a quasi-exponential fashion (Tweedie 1967). Thus, as water resources over the greater part of the Australian continent are meagre, water availability must impose a constraint on possible locations for urban decentralization. In this context two macro appraisals of urban water availability were carried out in 1966 and 1967 (Hobbs and Woolmington 1972).

The first of these was an attempt to map 'water availability potential,' based on available stream discharge data, and on the assumption that most urban water systems are supplied from stream discharge. Water availability potential was assessed on the basis of two interacting parameters, these being (1) volume and (2) reliability of discharge. From this exercise, conducted for the state of New South Wales (the most populated of the three southern states which among them contain almost 80 per cent of the national population), two general areas emerged as providing the most favourable water availability potential situations. These were: (1) the far north coast of the state, with its array of unregulated, high-discharge and flood-prone streams, and (2) the upstream areas of the Murray/Murrumbidgee river systems, to which the augmentation and regulation of water flow has made an important contribution.

Although the investigation did not extend into Victoria or South Australia, general knowledge of discharge situations suggests that, apart from possibly the southeastern corner of Victoria, no area in either state might be expected to equal the two identified within New South Wales.

The second exercise represented an attempt to construct an urban water opportunity model which identified *urban water opportunity* as a function of the capability of water availability situations of meeting urban water need situations. Water consumption at Newcastle, New South Wales, was used as the model for water need (Tweedie 1967) and was calibrated against consumption in a series of larger country centres. The model was located in proximity to ten country centres (Bathurst, Bourke, Casino, Cowra, Dubbo, Grafton, Gunnedah, Tamworth, Wagga Wagga, and Albury), providing a full sample cover of the various hydraulic environments identified in the previous exercise. Two criteria were used to assess the favourability of locations: (1) total sustainable population, and (2) sustainable population per square mile of catchment. Albury, on the Murray river, emerged in first and second place respectively for these criteria, suggesting it as the most favourable location in water terms for urban decentralization planning.

Equally important along with water availability as a test of urban viability would appear to be the accessibility of possible urban locations to markets for goods which might be manufactured in the decentralized entity. In classical location terms transport cost to market is the critical parameter involved. On the basis of these assumptions the author conducted a further exercise (Geissman and Woolmington 1971) aimed at assessing the favourability of all locations in southeastern Australia in terms of access to all the people of southeastern Australia (i.e. the market). This exercise involved the adaptation of gravity models, widely used for the study of urban interaction (Carrothers 1956; Reilly 1929; Huff 1964; Woolmington 1965; McPhail and Woolmington 1966), to the search for a new central place. From this exercise the insignificant rural centre of Crookwell emerged as the place having the greatest potential, but Albury and Wagga, on the Murray and Murrumbidgee respectively, were not far behind. Moreover, population shifts through time were suggestive of future enhancement of the favourability of locations south of Crookwell, this circumstance further favouring the Murray system locations.

On the basis of the criteria investigated the upper Murray would appear to be the place to initiate a policy of selective urban decentralization, and such locations are favoured by the further and all-important criterion of political feasibility. The Murray is the border between Victoria and New South Wales, and the Murray waters are used by these states plus South Australia. Thus, under the Australian constitution, the Commonwealth government has a right and obligation to intervene, and it alone of Australian government agencies has the financial resources to implement urban decentralization. Further impetus derives from the need to find a 'new use' for the Snowy project. There are complicating issues, such as questions of pollution, recreation, and preservation, but at the present point in time the area concerned would appear to be a front-line candidate for national development planning.

Carrothers, C.P., 1965 An historical review of the gravity and potential concepts of human interaction, *J. Amer. Inst. Planners* 22.

Davidson, B.R., 1965 *The Northern Myth* (Melbourne).

– 1969 *Australia: Wet or Dry?* (Melbourne).

– Development Corporation of New South Wales, 1969 *Report on Selective Decentralisation* (Sydney).

Hobbs, J.E., and E.R. Woolmington, 1972 Water and urban decentralisation in New South Wales, *A.A.A.G.* (Mar.)

Huff, D.L., 1964 Defining and estimating a trade area, *J. Marketing* 28 (July).

Geissman, J., and E.R. Woolmington, 1971 A theoretical location concept for decentralisation in southeastern Australia, *N.Z. Geogr.* (Mar.).

McPhail, I.R., and E.R. Woolmington, 1966 Changing functions of the state border as a barrier to Brisbane influence in northern New South Wales, *Aust. Geog. Studies* (Oct.).

Reilly, W.J., 1929 *Methods for the study of retail relationships.* Monograph no. 2944 (U. Texas).

Tweedie, A.D., 1967 Water and the city: prospects and problems, *Aust. Geog. Studies* (Apr.).

Woolmington, E.R., 1965 Metropolitan gravitation in northern New South Wales, *Aust. Geog.* (Sept.).

– 1970 Theoretical implications of the Malthusian Inversion, *Proc. 6th N.Z. Geog. Conf.*, Christchurch.

s0222
The nature of attitudes toward renovated wastewater
DUANE D. BAUMANN   *Southern Illinois University, USA* and
ROGER E. KASPERSON   *Clark University, USA*

The manner in which an individual perceives the potential of recycled renovated wastewater appears to be a function of his perceptions of the numbers and quality of alternative sources of supply, his education, and past experience. James Johnson discovered that the individual who perceives sufficient alternatives for municipal water supply did not consider re-use as part of his range of choice (Johnson 1971). It may be, however, that individuals actually perceive re-use as a lower quality among several alternatives of water supply provision. This does not lead to the conclusion that the public would find re-use unacceptable.

A most significant finding in several studies as well as in our own research, is the relationship of a person's attitude toward re-use and his formal education and knowledge of re-use. Johnson found that the more formal education a person has had, the higher the probability of his receptivity to consuming renovated wastewater. A telephone survey of 321 respondents conducted at Johns Hopkins also found education important both for low and high order uses of reclaimed water. Similarly, Ackerman and our pilot project both found a relationship between a respondent's knowledge concerning re-use and his acceptance (Ackerman 1971). In fact, during our interviews, as a person learned more about renovated water and the types of re-use systems his initial negative reactions were frequently modified. Our present research effort will further test the strength of relationship between the amount of a respondent's formal education and his attitude toward re-use; however, another dimension may be a more influential one in shaping a person's assessment of re-use, that is, his faith in technology. The more formal education may actually represent a greater acculturation and faith in technology as the alternative to solve most of our problems. Although Johnson found no relationship between confidence in technology and his attitude toward water re-use, we consider the dimension sufficiently important to warrant continued analysis. In any event, if education or faith in technology are important dimensions to a person's acceptability of water re-use, a properly directed educational campaign should contribute to allaying reservations toward drinking renovated water. However, the content of an educational campaign will necessarily be directed to the dimensions which are important in shaping a person's attitude; if not, then the program may have the reverse effect of causing the existing weak negative attitudes to solidify.

In any effort of formulating a public education program, the message would not necessarily be the same as in the standard issuance of a tornado watch. Instead, there may be regional variation in the components of attitudes by water managers and the general public toward water re-use. Johnson speculated that in the United States, the southwest, midwest, and great plains are most favourable to implementation of recycling renovated wastewater, whereas the northwest and southeast find the notion unacceptable. This regionalization was based upon his finding that compared to those in the humid environments, persons living in arid environments were more willing to accept water re-use, especially those who have experienced a shortage. Furthermore, in our survey of 300 water managers, we always find striking regional variations in attitudes toward the acceptance of water re-use. For example, in California, 85 per cent of the managers considered re-use acceptable; however, over 80 per cent of these same managers believed that the public would not accept it; but, among the Massachusetts respondents, 78 per cent of the managers who responded were opposed to re-use of renovated wastewater.

Even when we look at the individual layman's attitudes, as we did earlier, we find evidence of regional variation; but, the reasons for such variation are not clear. Although the amount of formal education may be important, we also find marked attitudinal differences among groups with similar amounts of formal education and similar experiences in having endured drought. Another dimension was found to be significant in both our earlier survey of 300 water managers and Johnson's work.

The manner in which an individual perceives the quality of his existing *source* of water supply profoundly influences his attitude toward re-use. If he perceives his present supply as of poor quality or polluted, then he tends to be more willing to accept renovated wastewater. Our survey indicates that such a relationship may indeed exist. As shown in Table 1, water managers, whose source of water supply was from ground water, are more reluctant to consider renovated wastewater as safe, potable water than were managers whose source of supply is a lake or a river. Individuals who see their supply to be polluted also gain the satisfaction of knowing that a water reclamation system will also contribute to pollution abatement.

In Johnson's study and in our reconnaissance study during the summer of 1970, a respondent's age, sex, and religion were unrelated to his attitude toward re-use of renovated wastewater. The Johns Hopkins study showed a weak tendency for men to be more receptive to drinking renovated water. In addition, Johnson found that a person's evaluative attitude toward local government was not influential in determining his attitude toward recycled wastewater.

Finally, the effects of costs of water on an individual's assessment of renovated wastewater have been inadequately measured. Some preliminary evidence suggests that by increasing the price of water, public acceptability rises proportionately. In a study conducted in a Worcester suburb, three groups were given the choice of accepting reservoir

TABLE 1. Managers' attitude toward recycled renovated wastewater and existing source of supply

Source of supply	Favourable		Unfavourable	
	No.	%	No.	%
Well	94	66	49	34
Impoundment	19	68	19	32
Lake	41	80	10	20
River	87	87	14	13
TOTAL	241	75	82	25

water over recycled water but at a 25 per cent, 50 per cent, and 100 per cent increase in the water rates. Another group was merely given the choice of reservoir water or recycled renovated wastewater. Clearly the responses were as expected in that the group faced with no price change preferred reservoir water; but with each increase in price, more individuals began to prefer renovated wastewater over reservoir water as a source of supply. The Baltimore study, however, found no significant differences in acceptance related to price of water. As one might not expect, however, those willing to pay more to retain a traditional supply (not re-use) are the lower income portions of the sample. (This may, in fact, be measuring education.)

In conclusion, then, we should note that some success in public acceptance is clearly being registered at the recreational lakes of Santee and that produced near Lake Tahoe. Acceptance at Windhoek has been so great that managers are now increasing the proportion of the renovated effluent in supply to 50 per cent. The state of the art now suggests that the reality of public acceptance is that with proper education and preparation of the public, support from appropriate authorities, with perhaps a sequential process of adoption, and with retention of some natural buffers, acceptance *can be gained* from the public.

Ackerman, Neil, 1971 *Attitudes toward water reuse*, Master's thesis, Dept. of Geog., Southern Illinois U.

Bruvold, William, and Paul C. Ward, 1970 Public attitudes toward uses of reclaimed water, *Water and Sewage Works* 117, 120–2.

Johnson, James, 1971 *Renovated Waste Water*, Chicago, Dept. of Geog. Paper 135, U. Chicago.

s0223
Hindsight evaluation: the Tryweryn reservoir in Wales
B. MITCHELL   *University of Waterloo, Canada*

In principle, the notion of hindsight evaluation is one with which few would disagree. Yet, in practice, such evaluation seems infrequent. Reasons for the infrequency of assessment after the event by resource managers may range from lack of awareness, lack of time, to political constraints. Despite these difficulties, however, hindsight evaluation offers at least two advantages. First, such studies may lead to projects being brought closer to design objectives through modification and adjustment. The major problem in this context is achieving adjustment for capital intensive resource projects such as hydroelectric dams or irrigation projects. Second, hindsight evaluations should improve future projects by minimizing errors discovered in earlier schemes. Operationally, a constraint exists here in that political considerations may eliminate evaluations if politicians fear resurrection of old skeletons which could return to haunt them.

With the above difficulties of, and opportunities for, hindsight evaluations in perspective, this study assesses one reservoir project in Wales (Mitchell 1969). The project, known as the Tryweryn reservoir, proposed in 1955 and officially inaugurated in 1965, will be discussed in terms of the perceived flood control, fishing, and resettlement benefits and costs visualized prior to and subsequent to project completion. This evaluation seems appropriate since to the summer of 1969 no comprehensive evaluation had been attempted. Furthermore, at the height of discussion on the project in 1957, Tryweryn was such an emotional issue that it was being debated in the British parliament (House of Lords 1957a; House of Commons 1957).

Tryweryn was proposed in November 1956 by Liverpool as a water supply project to provide 291 million litres per day to the industry and people located in its statutory area of supply. The reservoir, located in northwestern Wales on a tributary of the River Dee, was perceived by Liverpool as a large-scale project, providing savings from scale economies and incentive for economic growth and stability throughout the northwest by provision of a stable water supply. In addition to water supply, multiple benefits would be conferred by flood control, recreation, and hydroelectricity functions. The project was opposed on the grounds that a valley with traditional Welsh cultural attributes would be inundated to further development in an English city, and that water should not be transferred from Wales without Welsh consent. The opponents desired a smaller project at a different location. Thus, Liverpool perceived Tryweryn as a large-scale, comprehensive, and multiple-purpose project bringing benefits to the northwest while its opponents viewed it as a shortsighted, parochially conceived scheme conferring benefits to a small area at the expense of those living in Wales. Within this context, some of the claims made prior to and after completion of the reservoir concerning several of the multiple-purpose aspects are considered.

Liverpool Corporation argued that benefits would accrue to downstream farmers as a result of flood control (House of Lords 1957b, Day 9, 15). The reach of the Afon Tryweryn and River Dee below the reservoir site was susceptible to flooding with resultant disruption to activity upon the floodplain. After the project began to function, downstream farmers complained that their livelihood had been affected adversely by the flood control function of the reservoir. First, they stated that the fluvial deposits which enriched fields during spring freshets were no longer received, and thus that it was more expensive to maintain soil fertility. Second, farmers argued that prior to the reservoir, peaks in the spring were high but of short duration, disrupting farming for not more than several weeks (Braine 1959). With flood control, peaks were reduced but their replacement, medium-high flows of longer duration, appeared to be water-logging much of the farmland adjacent to the River Dee. The farmers argued that their equipment could not work in such fields, and that eventually land would have to be abandoned. On this evidence, there would seem to be some doubt as to the accuracy of the pre-project estimation of flood control benefits.

Commercial fisheries are unimportant on the River Dee. However, over 1000 licences are issued annually for sport fishing (Dee and Clwyd River Authority 1967, 39). The advocates of Tryweryn saw two benefits for fisheries. Raising of minimum flow during the dry season was expected to improve salmon spawning in the autumn. Conversely, reduction of high flows would decrease riverbottom scouring and associated disruption of spawning beds. While the Fisheries Officer for the Dee and Clwyd River Authority estimated a definite but immeasurable improvement to fisheries following flow regulation, a number of difficulties have arisen.

In 1965, with the reservoir operational, it was noted that salmon were experiencing difficulty in negotiating fish weirs at Bala Lake, downstream from the reservoir. A conventional fish pass was recommended if fish were to continue spawning in the headwaters. Following this recommendation, Liverpool provided £40,000 to construct a fish trap in which salmon were to be stripped of eggs, or removed to be kept until ripe for stripping. After hatching the eggs, the River Authority would distribute fry into the Dee and Tryweryn Rivers. With available data to 1967, results had been very disappointing at the trap. In the 1966 season, for example, only 64 males and 39 females were trapped (Dee and Clwyd River Authority 1967, 29).

In 1966, after further observation, it was noted that although peak flows in the Dee were well controlled, prolonged duration of moderately high flows resulted in salmon not settling in middle reaches of the river between Erbistock and Llangollen. As a result of this behaviour, many fishing beats no longer produced as many fish as they had done prior to control. Deterioration was also noted below Bangor-on-Dee, where salmon appeared reluctant to settle at all. Balancing this decline, however, was some improvement in fishing above Llangollen. The mixed results in regard to sport fishing thus suggest that ecological and social benefits and costs were not perceived adequately prior to the project. Of particular importance in this regard was the change in incidence of benefits and costs. While the total benefits and costs may not have shifted, considerable redistribution appears to have occurred as a result of the new fish patterns. This outcome is as serious to the resource users as an absolute loss.

Opponents of the scheme argued that the residents of the upland valley in which the reservoir was to be located would be forced to leave the area, thus breaking up a rural community strong in Welsh culture and tradition. They maintained that there was more to life than economic considerations, and that if the traditional Welsh way of life were lost through such disturbances, mooted in the name of progress, then all of Britain would be the poorer.

Liverpool countered with the argument that it accepted the fact that disturbance would occur, but that the disturbance at Tryweryn was minimal for a project of such magnitude. The corporation maintained that wherever such a project was built some people would suffer, and in fact the selected site resulted in less disruption than would occur at alternative locations. Furthermore, it was stressed that depopulation resulting from lack of employment opportunities would inevitably scatter the valley inhabitants over time.

Sixty-nine people were affected by the project, thirty-three of whom had to leave their homes which were inundated by the reservoir. The others were able to keep their homes but lost land. To accommodate those who lost homes, Liverpool provided houses several miles downstream at the village of Frongoch. As a result, the corporation argued that the people would not be scattered and could stay together as a unit. Calculations show that the average distance travelled by the thirty-three people to resettle was 6.4 kilometers, with the longest and shortest distances being 22.5 and .40 kilometers respectively. It would thus seem that disruption to settlement was in fact minimized and that effort was taken to aid those valley residents directly affected by the project. In this manner, it might be suggested that the benefits and costs were perceived accurately. On the other hand, the resettlement issue emphasizes that the benefits and costs may vary sharply with different spatial and interest perspectives. As one Tryweryn resident commented at an inquiry, 'Nothing will compensate me – nothing on earth – for losing my home and sacred community' (House of Lords 1957b Day 8, 30).

This paper has attempted to illustrate that the benefits and costs perceived prior to the construction of the Tryweryn reservoir were not necessarily those which occurred when the project was completed. Particularly, problems of the reservoir arising from flood control and fisheries management were at least partially underestimated while costs perceived for resettlement were over-estimated.

Such hindsight evaluation is unlikely to have much impact upon the design and operation of Tryweryn itself. On the other hand, the findings should emphasize the need to consider ecological, social, and economic aspects of future projects. For Tryweryn, initial planning emphasis seems to have been placed upon economic aspects, with less attention to social aspects and virtually none to those of an ecological nature. At a different level, the project suggests that a clear distinction should be made between long-term planning and long-term or large-scale projects. The proponents of Tryweryn perceived it as representing a long-term, comprehensive plan because it was large-scale and would satisfy requirements for some 30 years into the future. On the other hand, objectives of a long-term plan might just as easily have been realized by a series of short-term projects which would have provided a greater degree of management flexibility. Mixing of time horizons and magnitudes of programs, plans and projects has occurred in subsequent reservoir projects in Britain. Examples are Manchester's proposals for the Lake District in the mid 1960s and the Tees Valley and Cleveland Water Board's reservoir scheme in the northeast of England during the late 1960s. A comprehensive evaluation of Tryweryn might have illuminated this issue and smoothed the managerial path for later projects.

In conclusion, it is fitting to note that at least three considerations require attention regarding hindsight evaluations. First, it is essential that more hindsight evaluations be done if ecological, social, and economic benefits and costs are to be identified and evaluated in future planning. Second, if such evaluations are to be done it would appear that a body of theory is sadly lacking to guide such investigations. And third, along with theory is a need for improved methodologies which will allow the multiple dimensions of resource management to be assessed. In short, it is not enough to call for more hindsight studies without urging a search for, and development of, theory and methodology.

Brain, C.D.C., and Partners, 1959 *Report of flooding of the River Dee*, prepared for the Three Counties Committee, National Farmers Union, April.
Dee and Clwyd River Authority, 1967 *Annual report, 31 March 1967* (Cheshire, Chester).
House of Commons, 1957 Liverpool Corporation Bill, *Parliamentary Debates*, July 3, cols. 1170–1124.
House of Lords, 1957 Liverpool Corporation Bill, *Parliamentary Debates*, 20 Feb., cols. 1067–80, 1091–1123.
House of Lords, 1957 *Minutes of evidence taken before the Select Committee of the House of Lords on the Liverpool Corporation Bill*, 1–13 May.
Mitchell, B., 1969 *Decision-making for water supply in England and Wales*, PhD dissertation, U. Liverpool, 59–152.

s0224
Methodological problems in water resources management
W. SOLODZUK and T.V. MUSSIVAND   *Alberta Department of the Environment, Canada*

One of the basic problems that has always existed in the field of water resources management is the wide gulf between research carried out by institutions and the actual problems encountered when planning the development of water. For example, the usual method of problem-solving taught and practised in universities, research agencies, and scientific institutions, is that of extracting one specific problem from its surrounding fabric and attacking it in isolation. This piecemeal approach is artificial and theoretical, and the results are of little practical value, for it is extremely difficult to combine one such solution with other related problems. Unfortunately, the majority of research in the field of water resources management has been of this piecemeal nature, and directed towards solutions of isolated and sometimes trifling problems, while the basic and funda-

TABLE 1

Topic	No. of papers published 1967–71	Approx. no. pages
Aquifers and groundwater	218	3,600
Water cycle and hydrology	1,742	13,000
Groundwater conservation	2,150	23,000
Design	919	14,000
Pollution	600	5,000
Planning techniques	30	200
Objective of water resources management	Perhaps 1 or 2	Perhaps 10
Local, regional, and national priorities for water use and their effect on water resource management	0	0
Co-ordination among institutions involved in water resources	0	0

mental problems have been left entirely unanswered.

For example, an examination of the many papers published on all aspects of water management in the last five years yields the breakdown shown in Table 1. It can be seen that many vital topics have been completely ignored. Above all, there has been no research on that most basic problem whose solution must precede everything else: What is the objective of water resources management? This must be clearly defined, for it could be one of several things: maximum economic return, conservation of the resource, utilization of the resource, improvement in the quality of human life, ecological and social benefits, or some combination of these.

Even after the final objective has been determined, there are many other thorny problems. What are the techniques or mechanisms necessary to achieve the defined objective? There is no simple answer to this and other questions, because there has been very little thought or research devoted to them. In fact, the following crucial topics have been generally untouched: (1) objective of water resources management, (2) local and regional priorities and their effects on water resources management, (3) priority of water use, if any, (4) criteria for identifying priorities, (5) local, regional, and international water pollution problems and solutions, (6) criteria for evaluation of water resources undertakings, (7) techniques of planning, (8) cost-sharing and pricing arrangements, (9) co-ordination among institutions involved in water resources research, (10) allocation of water between competing uses, (11) determining the effects of man's activity on water, (12) ecologic impact of water resources projects, (13) criteria for preparing alternative proposals, (14) methods of making economic and ecological evaluations, (15) methods of communicating with the public at large.

As well, in academic institutions the emphasis is generally on theoretical concepts. Theory and pure research have an important role, but they should not be divorced entirely from actual problems.

There is also a tendency to deal only with classic examples of problems, and to assume that there is only one unknown factor in each problem. In real problems there are far more unknowns than knowns. For example, it is good to postulate that ideally the cost of a project should be shared by the beneficiaries in the same proportion as benefits received, but how does one actually determine who the beneficiaries are and what proportion of the benefits they receive?

Another stumbling block is the fact that the people engaged in actual water resources planning and management have few mechanisms for communicating to research institutions the problems which trouble them and thus should be studied. Lack of communication also hampers effectiveness of research in another way. Because there is little exchange of information among research institutions, similar studies are needlessly duplicated all over the world. For example, a study on the input-output of a river basin was carried out simultaneously by nine different agencies known to the author in various parts of the world.

The lack of practical techniques is evident

when one tries to formulate a sound approach to water management. Very few items in this area have been adequately researched, and consequently few applicable methods have been evolved; only a basic framework exists as a general guideline.

For example, in a logical planning process, the first activity should be identification of the problem or demand. This is a step of crucial importance, for if the problem is not accurately identified, it can never be solved. Unfortunately, there are few guidelines to follow. Can problems be quantified in units, as demands are? To take another instance, it is at present very difficult to predict future problems or needs. Projections can be made from past trends, but as yet there is no accurate, reliable method for anticipating future problems. Can research devise a practicable way for doing this?

Assuming that the problem has been identified, the second logical step is to formulate all the possible solutions to the problem. However, this is only the ideal; in practice, no search for solutions can be complete because a technique does not yet exist. Usually one proposal is fastened upon, other ideas are not explored, and the one alternative is implemented without further thought.

Once all the alternatives have been formulated, one must evaluate each one in order to select the optimum solution. On what basis should the alternatives be assessed – ecological, sociological, economic, or regional? Until now the economic consideration has usually been paramount, but assuming that a combination of all these factors could serve as a desirable criterion, what simple and practical technique can be used to evaluate proposals in terms of this criterion?

Once an alternative has been selected, the next activity would likely be implementation. At this step one could ask several questions. How should the project be financed? What type of cost-sharing arrangements should be instituted? How can one go about getting local, regional, and overall approval of the project? Should one try as a rule to satisfy the majority of the populace, or is the approval of a minority sufficient? What other problems might affect implementation of a project, and have they been resolved?

It is generally thought that the implementation of a project completes the planning process and solves the problem, but this is not necessarily true. Review of the project is necessary to determine whether the project has met the need, and whether it has itself created any new problems. However, what methods are available for evaluating a project's performance, and what factors should be considered? These questions, as many previous ones, do not yet have an answer.

These problems have been pointed out at such length in the hope that they may be brought to the attention of this international gathering and some headway may be made on their resolution. Since the solution of most of the problems raised demands a form of co-operation and communication which is at present lacking, suggestions to remedy the situation are: (1) the formation of an international organization to direct and co-ordinate the research in water resources planning and management; (2) as the first step of such an organization, an inventory of existing research; (3) next, definition of those areas which need more attention; (4) allocation of the topics to various institutions; (5) communication of the study results to all other agencies. This procedure would ensure that topics of common and vital concern were thoroughly studied, with little overlapping of effort. Such an organization would be an invaluable aid in shifting the whole direction of water resources research away from less important areas of work, and aiming it toward the solution of more fundamental and pervasive problems.

s0225
Water use map of the Strait of Georgia and Puget Sound: preparation and application
JOHN C. ROBERTSON   *University of British Columbia, Canada*

The text is not included because of length.

Role of the City in the Modernization of Developing Countries
Les villes dans les pays en voie de développement

s0301
Considerations on the formation of central place systems in developing countries
ERDMANN GORMSEN   *Universität Mainz, FDR*

According to my knowledge of central place studies, two important aspects have hardly been taken into account: (1) The hierarchy within the Christaller model, so far considered as a static principle, should be looked at rather as a dynamic system where the importance of central places changes with the evolution of the structure of society. (2) Even within the same phase of development, different social groups of an area avail themselves differently of the existing central places of lower, medium, and higher order.

Both aspects can be found in many countries and of course also in southern Germany, where Christaller worked out his theory forty years ago, but about the second aspect scarcely any investigations have been made. Regarding point (1), however, a few recent studies have proved that only since the beginning of the industrial revolution has the highly evolved hierarchy of central places developed from a simple two-level-system; that is to say, a narrow network of weekly markets with a few larger towns as higher centres (Faber 1967; Hellwig 1970). Moreover, nowadays in densely populated and industrialized areas a tendency towards disintegration of the lower stages of centrality can be observed, because their utilities and services are to be found in all places.

During my investigations in the region of Puebla-Tlaxcala, east of Mexico City in the Mexican meseta, I collected some relevant data on both aspects, which allow one to draw a certain parallel between the existing stages of the central place system and the historical phases of the model in central Europe. Furthermore, following some of my own observations in the middle east as well as a few references in the literature, it seems to me as if the second point mentioned above would apply to many developing countries (Mikesell 1958; Hausherr 1968; Grohmann-Kerouach 1971; Gormsen 1971a).

The Puebla region, surrounded by the highest Mexican volcanoes, proved to be very convenient for a study of this problem because it includes three areas differing substantially in natural environment and social structure as well as in central place relations. The central highland basin of Puebla (with the cities of Puebla and Tlaxcala, the prehispanic religious and commercial centre of Cholula, and a high density of agrarian and industrial population) slopes with various faults towards the south from some 2000m to about 1200m above sea level. Here we reach the Mixteca Poblana – an arid, scarcely populated area with higher concentrations only in a few better irrigated valleys. Towards the northeast of Puebla, beyond the old volcano of the Maliche, we find some wide highland basins of more than 2400m leading to the Sierra Norte de Puebla. Here the density of population is low because relatively large farms, producing wheat, barley, and pulque, are dominant in spite of the landreform. Contrasting with this, the extremely abrupt descent of the Sierra towards the coastal lowlands is densely populated by Indian subsistence farmers.

In the whole region, the original system of markets has changed only little since the colonial period, when every main parish had its weekly *día de plaza* (market day), the size of which varied merely with the extent and the population of the municipio. The selection of goods in such small markets corresponds still to the modest needs of peasants who are producing most of their subsistence themselves. A certain surplus of agricultural or handicraft production is also taken to the market, which is also a place to exchange news, thus having also a social function (cf. Gormsen 1971b). This type of semi-autarchic economic organization is still very common in remote areas, e.g., the Sierra Norte de Puebla. During the last years a few branch roads have been constructed, but the steep landforms do not allow any truck- or bus-traffic on the unimproved tracks. Therefore goods are transported largely on human backs. In this area only a very few inhabitants of the market-towns themselves have higher demands which can be satisfied only in better equipped centres.

More significant alterations have taken place in the Mixteca Poblana, in the south, for two reasons. On the one hand the great Mexican revolution of 1910 caused the destruction and partition of most of the large haciendas and monasteries, and consequently brought about the emigration of the well-to-do population; on the other hand the advent of the automobile made it possible to travel over longer distances in a day, even on tracks – at least during the dry season. As a result of both events, many villages have lost their former importance whereas others moved up to a higher grade because of their favourable location on main roads. This stage of transition might be compared with the period of railroad construction in Europe during the middle of the last century when the hierarchy of central places was developing.

In contrast with the semi-autarchic district in the north and the transitional one in the south, we can point out urban-rural relations in the central area of Puebla-Tlaxcala which can be compared with those of central Europe. The starting point was the textile industry which dates back to 1834. It not only remoulded the city of Puebla itself, but also the surrounding area for about 30 to 40km. The factories attracted more and more workers from the vicinity of the cities, thus changing their way of life. They became commuters who, besides their industrial job, continued to plough their fields like many peasants in Germany (*Arbeiterbauern*).

Even in this central area, however, an important part of the population does not participate in the urban standard of living. Only a few kilometres to the north of Puebla, on the slopes of the Malinche, some conservative Indian villages exist where many people do not speak Spanish, and the percentage of illiteracy is far above average. Furthermore, the traditional market network is surviving in the whole area. Many small places, however, have died, while others have grown tremendously, e.g., Texmelucan where I counted more than 1850 market stalls on one day.

As a first conclusion we can demonstrate three different stages of development of the market network in three neighbouring areas: in the north, most of the markets are of medium size and their tributary zones are of similar dimensions, taking into account the landform problems. In the south, many peasants tend to visit not only the local market, but also more distant places like Tepeaca, a small town of some 5000 inhabitants (1960) which reckons with much more than 10,000 visitors every Friday, when buses commute uninterruptedly between the large market place and villages up to 100km away.

If we now classify the towns of our region into a central place system of lower, medium, and higher order according to criteria normally used in industrialized countries, we find certain discrepancies compared to the traditional market network. Important market towns like Tepeaca have almost no centralizing function besides the weekly market, whereas others show all the characteristics of a centre of medium order with its clearly defined zone of influence, although their market is relatively small. Obviously, the traditional exchange system which is not static itself is overlain by a modern structure of rural-urban interrelations, whereby only a part of their respective tributary zones are congruent (Gormsen 1971a, map IIIb).

A closer interpretation leads furthermore to the conclusion that not all groups of the population use the different systems in the same way. Most of the needs of small farmers and farm-hands are provided for by the weekly markets except a few special services like district-court, notary's office, and dispensary which can be found in centres of lower order. These centres meet the common necessities of the more or less urbanized working class as mentioned above; sometimes these people visit the centres of medium order with wider selections, including, besides a few simple shops of durable goods, banking, hotel, and cinema facilities, an automobile repair shop, a secondary school, a small hospital, and a dentist, as well as a few members of liberal professions. These services meet the needs of the recently developing middle classes including teachers, employees, foremen, *rancheros* (small farmers mainly producing for the urban market), etc. Finally, the few upper-class people who live in the countryside (*hacendados*) or in small towns, taking advantage of modern roads, pass all smaller centres formerly important for them and go directly to the big city of Puebla or even to Mexico City where they can satisfy their wants. Moreover, Mexico City is not only the centre of highest specialization for the whole country but also for the upper classes of the other central American republics. (Similar observations can be made in the middle east, where Beirut, with its liberalized commerce as well as with its universities, hospitals, and amusement installations, is the supra-national centre for all eastern Arab countries.)

This example suggests the general conclusion that, in developing countries with great class differences, these classes can be correlated with the main levels of central place systems coexisting in neighbouring areas and partly superimposed upon each other. To a certain extent, these different stages can be compared with historical phases of central place systems in Europe. Finally, it should be stated that the actual situation in developing countries should not be taken as a static phase but as one in process which eventually will reach the stage already existing in industrialized countries.

Faber, K.G., 1967 *Neuzeitlicher Wandel der Stadt-Land-Beziehungen in der Pfalz,* Institut für Landeskunde, 25 Jahre Amtliche Landeskunde, Beiträge der Mitarbeiter, Bad Godesberg, 226–50.

Gormsen, E., 1971a Zur Ausbildung Zentralörtlicher Systeme beim Übergang von der semiautarken zur arbeitsteiligen Gesellschaft, *Erdkunde,* xxv, 108–18.

– 1971b Wochenmärkte im Bereich von Puebla/Mexico; Struktur und Entwicklung eines traditionellen Austauschsystems, *Jahrbuch für Geschichte von Staat, Wirtschaft und Gesellschaft Lateinamerikas,* 8.

Grohmann-Kerouach, B., 1971 Der Siedlungsraum der Ait Ouriaghel im östlichen Rif, Kulturgeographie eines Rückzugsgebietes, *Heidelberger Geographische Arbeiten,* H35.

Hausherr, K., 1968 Die Entwicklung der Kulturlandschaft in den Lanao-Provinzen auf Mindanao unter Berücksichtigung des Kulturkontaktes zwischen Islam und Christentum. Diss. Bonn. (maschinenschriftl.).

Hellwig, H., 1970 Der Raum um Heilbronn – sein zentralörtliches Bereichsgefüge aufgrund der Stadt-Land-Beziehungen unter besonderer berücksichtigung der Entwicklung seit dem ausgehenden 18. Jahrhundert. (Veröff. des Archives der Stadt Heilbronn, 16).

Mikesell, M.W., 1958 The Role of Tribal Markets in Morocco, *Geog. Rev.* 48, 494–511.

s0302
Market centres in western India
GOPAL S. KULKARNI *Indiana University of Pennsylvania, USA*

An intricate network of small market centres dotted over the countryside forms the necessary foundation for the spatial organization of economy of a predominantly rural country like India. These market centres form a vital link between the villager and the national economy by serving the basic need of assembly and distribution of the required goods for the surrounding areas. In this process some of these centres also serve as foci, of varying importance, for the social, administrative, recreational, and political activities of their respective tributary regions. Such an areal functional organization consists of a hierarchy of centres with increasing functional complexity and population size of tributary areas (hinterlands) served by them. Meaningful aggregates of such contiguous areas form a viable spatial unit of the national economy. These spatial subsystems may display areal functional specialization having internal balances and external flows (Friedmann and Alonso 1964, 4).

The Indian Statistical Institute, which developed the basic structure of Indian national plans, has displayed an active interest in a regional planning approach. Consequently, the spatial dimensions of economic development are being increasingly emphasized in India's Five Year Plans (Planning Commission, Government of India 1969, 316). For a realistic formulation of a plan of regional economic development an adequate understanding of the physical and activity patterns of functional regions is very essential (Friedmann and Alonso 1964, 9). The main concern of this study, which is a part of a larger investigation, is to examine, in the context of central place theory, some aspects of the physical and activity patterns, namely, the spatial arrangement of market centres and certain aspects of the structure of their hinterlands in Kolhapur – a district in western India.

Central place theory, postulated by Walter Christaller in 1933, deals with the location, size, nature, and spacing of settlements. Empirical verification of this deductive theory has been undertaken in various parts of the world but mainly in the areas of western culture (Berry and Pred 1961). The present study area, representing one of the most ancient cultures in a relatively low stage of economic development, provides a different empirical base for verification.

In this context, the following hypotheses will be tested:
1. There is a positive relationship between the population size and the number of functional establishments of a market centre.
2. There is a positive relationship between the population of a market centre and the population of its hinterland.
3. There is a functional basis for the hierarchy of centres in Kolhapur district.

These hypotheses will be tested in Kolhapur district (about 8,260km² in area) which is located in the southern part of Maharashtra, a state in western India. The city of Kolhapur, approximately 400km south-southeast of the city of Bombay, is the administrative headquarters of this district. Some of the main characteristics of this district are:
1. The district is located sufficiently far away from the direct influences of such large urban centres as Poona and Bangalore. The city of Kolhapur thus represents a type of lower-order regional centre where many new facilities are being located during the post-independence era of national planning.
2. Historically and politically the district is one of the active parts of Maharashtra and displays a regional 'personality.'
3. The district is productive and relatively prosperous both agriculturally and industrially. Consequently, its population density has always been higher than the state average.
4. The district located on the Deccan Plateau consists of narrow river valleys separated by the spurs of the Western Ghats in its western parts and a relatively level area in its eastern parts. These physiographic differences are further accentuated by the differences in the amount of rainfall which decreases rather abruptly eastwards. Fairly well developed irrigation facilities and responsive lava-derived soils make the district's agricultural yields among the highest in western India. Industrialization is necessarily limited to urban centres and, excepting the six sugar mills, it is mainly of the small scale type (Maharashtra State Gazetteers 1960; Census of India 1964).

In 1961 there were 1,086 villages and 11 towns having a population of approximately 1.6 million in Kolhapur district. Of these, 729 settlements with the Panchayat form of local self-government have been used to gather data on 115 functional types. These settlements together had 1.2 million people in 1961. The 115 functional types belong to the following main categories:
1. Retail
2. Wholesale
3. Commercial
4. Financial
5. Professional
6. Medical facilities
7. Entertainment
8. Religious
9. Education
10. Transportation and communication
11. Industrial
12. Amenities

A tentative list of functions normally found in market centres in western India was first prepared. Copies of this list were then sent to the District Officer at Kolhapur for eventual distribution to the local self-government officers throughout the district. These lists were collected personally at the Taluka headquarters and a random sample of them was field checked. One of the many other questions on this list pertained to the identification of the nearest market centres and the alternate market centres which the majority of villagers normally visit.

A simple correlation matrix (116 × 116), i.e. population and 115 variables, was computed, using data on all 729 settlements. With such a large number of observations very few r values were rejected at 0.01 level. The following fourteen variables showed a coefficient of correlation (r) value of over 0.50:

FUNCTION	r
Provisions	.76
Tailoring	.71
Firewood and fuel	.67
Hotel	.67
Cycle shop	.65
Hindu temple	.64
Cloth	.59
High school	.55
Co-operative society	.54
Muslim mosque	.53
Tobacco and betel nut	.53
Flour mill	.52
Cinema	.52
Post office	.51

Hypothesis no 1 was therefore accepted.

The questionnaire was used in delimiting the hinterlands of 108 market centres. Almost all adjoining hinterlands were found to be overlapping. Their populations were computed without any adjustments however. The coefficient of simple correlation (+0.50) for population of market centres and population of their hinterlands was significant at 0.01 level. Hypothesis no 2 was therefore accepted.

A principal components analysis was then performed. Factor scores on the first two principal components, after Varimax rotation, were used in identifying an hierarchy of 729 settlements (Abiodun 1967, 350). Hypothesis no 3 was accepted. A linkage analysis was carried out for a more specific combination of functions and of settlements at various levels.

Spatial distribution of settlements of various levels of hierarchy is, however, far from even (Johnston 1966, 541). The eastern part of the district contains settlements of all levels but has the majority of the higher-order centres, whereas the western areas have only the lower-order settlements. Such incomplete spatial hierarchy may be attributed to the large spatial variations in topography, rainfall, land productivity, transportation facility, and population and purchasing power distribution in Kolhapur district.

Abiodun, Josephene Olu, 1967 Urban hierarchy in a developing country, *Economic Geog.* 43, 347–67.
Berry, Brian J.L., and Allen Pred, 1961 *Central Place Studies: A Bibliography of Theory and Applications*, Bibliography Series no 1 (Philadelphia).
Census of India 1961, 1964 *District Census Handbook: Kolhapur* (Bombay).
Friedmann, John, and William Alonso, eds., 1964 *Regional Development and Planning: A Reader* (Cambridge, Mass.).
Johnston, R.J., 1966 Central places and the settlement pattern, *Annals Assoc. Am. Geog.* 56, 541–9.
Maharashtra State Gazetteers, 1960 *Kolhapur*, rev. 2nd ed. (Bombay).
Planning Commission, Government of India, 1969 *Fourth Five Year Plan 1969–74: Draft* (Delhi).

s0303
The cities of the third world: industrialization and spatial repercussions
MILTON SANTOS *Massachusetts Institute of Technology, USA*

If we are to believe the plentiful literature brought out by many specialists, economic development can be equated with an increase in overall quantities, measured in terms of national product or per capita income. Similarly, modernization can be defined as the adoption of new forms of production, consumption, and organization, imported from more developed countries, in order to promote growth in certain sectors of the economy, or in other words to encourage development.

In spite of its macroeconomic pretensions, such an interpretation cannot but remain microeconomic, because it tends to balance individual successes and failures, whether or not there are any functional links between them.

The geographer, however, must approach the problem from an entirely different angle. He is concerned with the spatial and social productivity of investments, in that these must have both a stimulating effect on the country or region affected by modernization and results which are fairly distributed among the residents. This is indeed one of the conditions for the maximization of spatial productivity as we understand it.

There is another ambiguity which we feel might confuse the issue as well, and this concerns the types of cities involved. In this paper we will consider only those types of urban agglomeration established in developing countries as a result of modernization according to the definition given above; we will also examine the aims and problems of development, as defined by numerous economists, international agencies, and governments. We will be concerned here mainly with forms of urbanization attributable to certain methods of industrialization. Though we will not give an exhaustive description of a phenomenon which may be observed in countries on all underdeveloped continents, we will ask some questions about the problems raised by the establishment of such urban centres, which nevertheless contribute to the punctual modernization of the countries involved and to an undeniable increase in their overall quantities from an accounting point of view.

The geographer may make a substantial contribution in this area to the formulation of theories by economists and planners. With this thought in mind, it would be interesting to study the theories of growth poles developed in France mainly by François Perroux (1969) and Jacques Boudeville (1966), and also the parallel points of view presented somewhat later in the United States by Lloyd Rodwin (1961) (concentrated decentralization) and by John Friedmann (1968) (deliberate urbanization).

Industrial cities and *incomplete metropolises* also lend themselves to the type of discussion which we are suggesting. We may pose the question in the following terms: is industrialization in itself a decisive factor in transmitting the force thus acquired to the surrounding region or country via the city?

This question is important from the point of view of both methodology and theory. Just as geographical literature is saturated with studies on relations between commercial cultures and urban growth, so it is penetrated by the idea that the very presence of industry is enough to strengthen the ties between the city and its surrounding region, while at the same time multiplying jobs and wealth there and in the rest of the region or country — and it could well have been imagined during the sixties that such a trend would be irreversible. In our discussion we will consider this question from a new angle.

Industrial cities are agglomerations set up in the majority of cases from scratch to accommodate the new industries considered essential for the development of the country, generally basic or heavy industries. Due to the fact, however, that the implantation of such industries is dependent not only upon greater economies of scale but also on decisions made outside the country by multinational firms, these industries do not have functional links with the rest of the economy and their ability to act upon it is therefore limited. The problem is further aggravated in cases where the location of the city is determined by sources of raw material, and is far from the economic and demographic centres which may already exist in the country. Paradoxically, the possibilities for exchanges with such a region only become apparent when a population not directly involved in the activities of modern factories but, drawn by the hope of finding work, finally comes to settle in the city which offers job opportunities. It is not the modern activity itself, however, which acts directly as a magnet. In order to operate, the modern firm must, because of its high level of technology and organization, internalize its externals (*sic*), and the fact of so doing is both a result and a cause of minimized relations with the host environment.

We do not wish to enter the moral debate on the value of modernization, a debate which nevertheless is very necessary; we are, however, entitled to ask ourselves to what extent the modern activities of these industrial cities contribute in themselves to the development of the region in which they have been implanted.

The *incomplete metropolis* represents a different case. We have already defined it as an agglomeration which can achieve macro-organization of its territory, but which itself cannot produce all the goods which form the object of its trade, even though it can provide for all the basic requirements of its territory. This type of city is represented by large cities and capitals in underdeveloped countries which are not yet adequately industrialized (industrialized underdeveloped countries such as Brazil, Mexico, Argentina, India, and Egypt have both complete and incomplete metropolises). Thus some incomplete metropolises find their complete metropolis within their own country, while others have to look abroad (Santos 1971).

In countries lacking complete metropolises industrial activity in the incomplete metropolises tends to be monopolistic or quasi-monopolistic. By way of explanation, we would point out that the existence of such metropolises, parallelling that of industrialization, is a modern phenomenon occurring at the same time as the revolutions in consumption and technology.

Monopolistic activity tends to generate considerable problems as regards the organization of space. The specialists, however, do not seem to have concerned themselves with this basic problem either from the standpoint of developing a theory of space in underdeveloped countries or from that of determining rules for their planning.

Monopolistic activity, which is already concentrated in areas which are privileged because of the presence of certain favourable factors, encourages a growing concentration of manufacturing activity by product type. At the same time similar activities tend to disappear from other cities in the national system. It is for this reason that in a country like Venezuela, which is characterized by this type of development, urbanization is tertiary outside the industrial centre, even though there are cities which are important from the demographic and commercial point of view. In the sub-metropolises of the interior we find only those industries which meet the needs of the local and regional population, and of the traffic (Fr. 'circulation'). These are mainly small manufacturers, for medium-sized industries can only survive here in activities in which distance makes it easier for them to compete with the monopolies or in which the monopoly is not involved.

Since the monopoly practices a policy of deliberately restricting consumption, we can easily predict the economic consequences for the country as a whole which finds its growth slowed down in this way. Under such conditions it becomes difficult to apply such theories as development poles or concentrated decentralization.

Industrial monopolies may also be created as a result of favourable local conditions, such as the accumulation of capital from regional agricultural activity. In such a case, however, no matter what its size, the city containing the industry will appear to be in a transition stage between incomplete metropolises, as described above, and the industrial city. Capital is accumulated by commercial speculation on agricultural exports and brings no benefits to the rural population. This was the case in Medellin, Columbia (Santos 1969), where a conjunction of favourable circumstances made it possible to establish a textile industry in the city. Because of the restricted nature of the market, however, aggravated by the low salary policy in effect in the city and country, by modern marketing methods, and by production crises brought about by the aging of the plantations, the factories, which had very great capacity for production, were easily able to capture distant and even foreign markets, thanks to the low cost of manpower (*sic*). These commercial successes encouraged the industries involved to increase their production capacity by obtaining even more modern machinery. The number of jobs decreased while an ever-growing rural exodus helped to disrupt the labour market even more.

Since these industries are highly integrated, their behaviour is similar to that of steel complexes and of industrial cities. They internalize their externals, thus discouraging the development of other industries in the city. The disappearance of these industries and the corresponding price increases reduce the opportunities of the poorer people in the city and surrounding countryside to become modern consumers. As a result, a large part of the local population still turns to the older economic circuit when purchasing many goods.

We are not confronted by a paradox here, but rather by one form of dialectic between modernization and permanence in a rural-urban continuum where modern activities, overly oriented toward the outside world, are not strong enough to propagate other forms of modernization outside their own forms of production (McGee 1971).

Is it possible to build a model from the information which we have presented here? Once again we must place ourselves in the stream of history to observe the conditions of the impact of international temporal systems on the underdeveloped countries. Incomplete metropolises and industrial cities of every kind are a contemporary geographic phenomenon which, though of more or less recent origin, is only now showing its full dimensions.

Is this to say that the operating conditions of the present world system are such as to constitute an obstacle preventing the cities which are basically representative of the present period from fulfilling their roles as development poles?

There are no specific studies on this subject, but we may state that, as far as development – not to mention modernization – is concerned, the forms of organization of national space and the types of relations between cities and the national and regional space depend upon the existence of complete metropolises. At a lower level, it is possible to imagine relations of reciprocal growth between city and region when neither one is overly specialized. The very fact, however, that agglomerations lower on the scale are dependent on a metropolis shows that the problem of national urbanization remains to be solved.

If such proves to be the case, the general problem of the types of city would lose its importance, because the organization of the territory would be determined by the date and level of industrialization of the dynamic centre, by the degree of modernization of the country, and by the forms taken by such industrialization.

Would it be possible to interpret the problem entirely in terms of historical data, in that incomplete metropolises and industrial cities are for the most part recent phenomena postdating World War II, and are therefore the direct result of the present temporal system controlled by the large industrial units, technology, and advertising? We should add that countries which were industrialized earlier on were able to avoid falling under the sway of the monopolies, whereas those where industrialization was introduced at a later date were unable to do so.

We can nevertheless isolate the variables of the model, namely: the date modernization was first introduced into the country or in other words the date the country was first integrated into the world market; the date of the beginning of industrialization; the time of the demographic revolution; the time, in relation to the first three variables, when territorial integration through transportation took place; the area of mobility of these factors; acceptance of modernization and the corresponding political framework; the total population of the country and its area of concentration at a given moment, as well as the area integrated by modern transportation.

Such information might well provide us with the instrument needed first of all to classify underdeveloped countries by category and then, for each country, to determine the possibilities offered by the world system, in conjunction with the internal variables of the country and their dynamics, which will enable the cities to involve the rest of the country.

Boudeville, Jacques, 1966 *Problems of Regional Economic Planning* (Edinburgh).
Friedmann, John, 1968 The strategy of deliberate urbanization, *AIP J.* (Nov.).
McGee, T.G., 1971 *The Urbanization Process in Third World* (London) esp. chap. 2.
Perroux, François, 1969 *L'Economie du XX⁰ siècle*, 3ème éd. especially part 2 (Paris).
Rodwin, Lloyd, 1961 Metropolitan policy for developing areas, in Walter Isard and John H. Cumberland eds., *Regional Economic Planning*, OEEC (Paris).
Santos, Milton, 1969 Un Essai d'Interprétation de l'Evolution Economique de Medellin et de sa Région, Centre National de la Recherche Scientifique et Institut de Hautes Etudes de l'Amérique Latine de l'Université de Paris, mimeo.
– 1971 *Les Villes du Tiers Monde* (Paris).

s0304
Urbanization in the Middle East with special reference to Syria, Lebanon, Iraq, Iran
EUGEN WIRTH   Erlangen–Nürnberg University, West Germany

1. Most of the current definitions of the term 'urbanization' can also be applied in the countries of the Middle East. But here 'urbanization' comprises two fundamentally different aspects:

(a) On the one hand 'urbanization' includes the form of life of the oriental town, as it evolved over thousands of years. It is still deeply rooted in old traditions and stands out clearly against the other two traditional ways of life represented by the fellaheen and the nomads. Cultural elements of the ancient Orient, the Hellenistic–Roman antiquity, and the Islam come together in this traditional form of urbanization to create a unique synthesis which we find only in the towns of the Middle East. An essential element of this traditional urban form of life is the so-called 'Rentenkapitalismus' or rent capitalism (Bobek 1959).

(b) On the other hand in the countries of the Middle East 'urbanization' implies the increasing influence of a modern western form of urban life. At first this process does not seem to differ considerably from similar processes in many other developing countries. Yet because, in the Middle East, the modern western way of life is being confronted not with a comparatively primitive or agrarian culture, but with a highly civilized and traditional urban one, special problems arise.

If we discuss the major problems involved in the measurement of urbanization in the Middle East, we should not ask how to measure extent or proportion of urbanization compared with non-urban forms of life. Far more important seems to be the question of how far modern western forms of urban life have already influenced, changed, or replaced the traditional forms of life of the oriental town. It is only natural that it will be difficult to measure such a process quantitatively.

2. There are many indices showing that already in the centuries before 1800 AD European influences contributed to the development of the traditional oriental town and its way of life. Probably the high points and the times of prosperity of the oriental urban culture between 1400 and 1800 AD can only be understood if the contact with western culture, commerce, and relations with Europe are considered. The greatest and most impressive examples of the oriental bazaar have been shaped partly by European influences. But only since about 1800 AD has the modern western urban culture begun to replace the traditional culture and way of life of the oriental town.

This influence now appears to be a competitor of greater prestige. Thus the process of urbanization in modern times in the countries of the Middle East means above all westernization. The military, political, and economic superiority of the European states during the nineteenth century, together with the growing prestige of western forms of life, may have been the decisive factor which, during the past hundred and fifty years, has brought about the gradual supplanting of the traditional way of life of the oriental town by the modern western form of urbanization.

This process cannot reasonably be measured quantitatively but can only be shown by means of indices. Such indices of an advance of western forms of life into the oriental town are, for instance, housing and consuming habits; organization of leisure time; transformation of the old city and growth of new residential districts; functional changes of traditional town quarters.

3. The current forms and patterns of urbanization in the Middle East are the result of the causes mentioned above. They are phenomena resulting from the westernization of urban forms of life. Consequently most near-eastern cities today appear to be a very interesting mixture of ancient, traditionally oriental elements and modern, purely western elements. As a result of these influences the larger cities show quite a number of the highly characteristic features of the traditional trading centre (the bazaar and the neighbouring khans), the modern commercial quarters of western style, the medieval part of the city, the more recent residential quarters outside the medieval city, the long-distance road traffic districts (Wirth 1968).

4. One of the major problems created by the modern western urbanization in the Middle East stems from the fact that the traditional forms of life of the fellaheen and the nomads were dependent on the forms of life and economic activity of the traditional oriental town. Thus its westernization necessitates a far-reaching adaptation and structural change in their worlds.

Essentially the position of the fellaheen profits by this development, for modern western ideas in the towns mean the replacement of the rent-based capitalistic system and its exploitation; they mean land reform, better education, medical welfare, and higher income.

The world of the nomads, however, will presumably not be capable of surviving the impact of western ideas and western institutions. Urbanization, in so far as this means westernization, could lead to the disappearance of the nomadic way of life during this century.

Many of the more difficult problems of the countries of the Middle East are shared by other developing countries. Among these is the growing migration of the rural population to the towns and, in close connection with this, the formation of slums or bidonvilles at the outskirts of towns and the creation of an unemployed white collar proletariat. The western concept of social prestige joins here with the old Islamic traditions of education with catastrophic results. Although in the countries of the Middle East today people are particularly needed in practical and technical professions, those careers are largely avoided, since they mean leaving the town, doing field work, and getting one's hands dirty. Preference is given to white collar work, although the number of those trained for such jobs already considerably exceeds the number of available positions.

5. The most impressive present and late twentieth century trends in urbanization in the Middle East are:

(a) As in most other developing countries there is an influx of the rural population to the town. As a result of this process the capitals of some of the Middle Eastern countries threaten to get the character of an oversized agglomeration, almost a 'hydrocephalus,' within their territory. Attempts are being made by means of a forced industrialization to absorb the new working force. In medium-sized and smaller towns, however, the process of immigration is already a selective one. Those towns which persist in their old traditions stagnate; the influx of rural population is compensated for by the departure of urban population towards more dynamic centres. Medium-sized and small towns which are sympathetic to modern tendencies and have a sufficient number of free jobs in modern industries and in the westernized tertiary occupations have often got growth rates as great as the largest towns.

(b) Qualitatively, urbanization in the decades to come will mean an ever growing advance of modern western urban forms of life, as opposed to the traditional world of the Islamic-oriental town. It is uncertain whether the large towns of the Middle East in the year 2000 will still be different in essential elements from the large towns of Europe and North America. The political, economic, and social upper classes of the Middle Eastern countries mostly strive for a quick and more complete adoption of western concepts, Nevertheless, the power of inertia and the resistance of traditional forms of life, economics, social orders, and institutions in oriental towns must not be underestimated. They often show an astonishing flexibility and seem to be able to adapt themselves to present conditions as well as the western innovations. Thus they may have a real chance of survival in the decades to come.

(c) As in almost any other country of the world urbanization in the states of the Middle East will mean an ever growing impact of urban forms of life, patterns of behaviour, and standards on the countryside. In the Orient too, in the year 2000, there will probably be hardly any agrarian or rural regions which would not appear to be urbanized to a larger or smaller extent.

Bobek, H., 1959 Die Hauptstufen der Gesellschafts- und Wirtschaftsentfaltung in geographischer Sicht, *Die Erde* 90, 259–98.

Dettmann, K., 1969 *Damaskus: Eine orientalische Stadt zwischen Tradition und Moderne*, Erlanger Geogr. Arb. 26 (Erlangen).

Hourani, A.H., and S.M. Stern (eds.), 1970 *The Islamic City: A Colloquium*, Papers on Islamic History 1 (Oxford).

Ruppert, H., 1969 *Beirut: Eine westlich geprägte Stadt des Orients*, Erlanger Geogr. Arb. 27 (Erlangen).

Wirth, E., 1966 Die soziale Stellung und Gliederung der Stadt im Osmanischen Reich, in *Konstanzer Vorträge und Forschungen* 9, 403–27 (Konstanz).

– 1968a Strukturwandlungen und Entwicklungstendenzen der orientalischen Stadt: Versuch eines Überblicks, *Erdkunde* 22, 101–28.

– 1968b Tradition, Westernization and Re-Orientalization in the Modern Near-Eastern City, paper presented at the 21st Intern. Geogr. Congr., New Delhi.

s0305
The modernization of inland southwestern Nigeria: the role of Ile-Ife
J. OLU ABIODUN *University of Ife, Nigeria*

The text is not included because of length.

Abiodun, J.O., 1971 Service centres and consumer behaviour within the Nigerian cocoa area, *Geografiska Annaler* 53B, 2.
- 1970 Central place functions and consumer behaviour in Ife Region, paper presented at the 15th Annual Conference of the Nigerian Geog. Assoc.

Forde, Daryll, 1962 *The Yoruba Speaking Peoples of South-western Nigeria* (London).

Johnson, Samuel, 1921 *The History of the Yorubas* (London).

s0306
The spatial structure of cities in India: pre-industrial or post-colonial?
JOHN E. BRUSH *Rutgers University, USA*

The typical spatial structure of cities in India may be identified as 'pre-industrial,' using Sjoberg's descriptive criteria of demography and social ecology (Sjoberg 1960). In the vast majority of cases a compact indigenous nucleus is readily identifiable as the past or present focus of political, religious, or economic activity (Brush 1962). In a previous analysis of the urban patterns of population distribution in India (Brush 1968), I was able to show that much of the growth in recent decades has been absorbed by continuing concentration at the old city centre, where mixed land use and strong localization of castes and other ethnic groups prevails. Berry and his associates have shown also that high-status neighbourhoods are to be found in the central areas of large Indian cities, while low-status neighbourhoods are typically located on the periphery (Berry and Rees 1969; Berry and Spodek 1971). There are two variant morphological patterns arising from British colonial rule: first, the Indo-British seaports of Calcutta, Bombay, and Madras, which have the highest population densities outside of the old waterfront business districts, and second, the twinned cities of the interior, such as Hyderabad, Bangalore, and Lucknow, where centres of indigenous origin co-exist with satellite centres. A fourth spatial pattern is found in planned industrial and administrative cities, which have been built during the twentieth century. Here low population density and separation of residential land use from the locations of employment is suggestive of a new phase of India's urbanization.

In order to gain further understanding of the spatial structure of traditional and modern cities in India, I analyze wardwise census data from 1961 from three unplanned indigenous cities to compare with two new planned cities. Nine socio-economic variables, based on the meagre data available in the District Census Handbooks, are selected to measure differences in terms of gradients between the urban centre and periphery. Correlation of the variables with distance from the centre involves regression analysis, subjected to tests of significance, following procedures used in my previous study of population density. (In the previous analysis [Brush 1968, 371], the data for density were treated in terms of logarithms.)

The five cities under examination provide a regional and chronological representation of urbanism in the subcontinent. Varanasi (Banaras) on the River Ganga in the eastern Uttar Pradesh plain is the epitome of traditional India, its origins shrouded in prehistory (Singh 1955). The city has grown during the last 100 years around the ancient religious town and now extends its boundaries from the central bazaar close to the sacred riverbank over a radius of five to eight km. The population of greater Varanasi in 1961 was 489,000 in some 79km^2, including Varanasi Cantonment, Banaras Hindu University and the Government of India's Diesel Locomotive Works, all outside the municipal boundaries. Sholapur, a regional trading and manufacturing centre in the lava plateau of southern Maharashtra State, had its origin about the twelfth century AD, as a religious site, an annual fair and a military stronghold (Gadgil 1965). The present city with a population of 338,000 (1961) covers some 24.3km^2, including the former walled city and fort, the district administrative and housing tracts which were developed under British rule, together with new industrial and residential tracts towards the periphery. Coimbatore with a population of 286,000 in 1961, comprises 22.68km^2 (Chettiar 1939). Located in the interior lowlands of Tamilnadu in the southern peninsula, Coimbatore had its start in the ninth century AD at the locus of three temples and a fort. After being occupied by the British in the nineteenth century, its expansion was related at first to its role as a district administrative centre and later to its advantageous situation for transport, trade, and manufacturing. Jamshedpur is the creation of the Tata Iron and Steel Company, which was organized in 1907 (Misra 1959; Dutt 1961). Shortly thereafter a tract of rural land was chosen in the hills of southern Bihar for establishment of India's first large iron and steel works and a company town for the employees. By 1961 a largely planned city with 269,000 inhabitants had grown up around the Tata works and the plants of associated industries, covering an area of 64km^2 under the direct or indirect control of the parent company, while an additional 64,000 persons lived in the adjacent South-Eastern Railway colony and unplanned town and village areas. Chandigarh is a completely planned governmental city, situated in the piedmont plain of northern India, created after the partition of India and Pakistan in 1947 (Evenson 1966; D'Souza 1968). It serves as the capital for the Indian Punjab and jointly (since 1968) for the new state of Haryana. In 1961 the city, then in an early stage of development, comprised about 31km^2 inhabited by only 89,000 persons.

Results of regression analysis of all five cities under examination consistently show certain spatial patterns. Negative relationships are found for population density, households per house, per cent of population literate, and per cent of employed population engaged in trade and commerce, i.e., higher values occur in the centre than on the periphery. Household size is negatively correlated in four of the five cities. Positive relationships are observed for per cent of population belonging to scheduled castes, that is to say, low socio-economic status, who are relatively more concentrated on the periphery. Per cent of population engaged in manufacturing is also positively correlated in four of the five cities. Relationships for sex ratio and employment in trade and services are less consistent. The constancy of these tendencies in both unplanned and planned cities suggests the degree to which cities in India may be alike in socio-economic structure.

Closer scrutiny, however, reveals that it is the indigenous cities which exhibit most of the statistically significant correlations and highest gradients. Correlation coefficients for population density and employment in trade and commerce in Varanasi, Sholapur, and Coimbatore range from .57 to .90, in all cases exceeding the one per cent level of probability. Significant coefficients are obtained for households per house and household size only in Varanasi and Coimbatore, while those for scheduled castes are significant in all three cities. Typically, the density of population is over 100 persons per acre (0.405 hectare) within a mile from the city centre, ranging to 350 or 370 in certain central wards and up to 490 in Sholapur, while it declines towards the periphery to less than 50 at about 2.4 to 3.2km. There are commonly 1.5 to 2.5 households per house in the centre of the city and 5 or more persons per household. Varanasi is exceptional in that the mean number of persons per household is 4.5 to 5.0 in the centre and increases to 5.5 on the periphery. In Sholapur and Coimbatore the sex ratio is commonly between 900 females per 1000 males in the centre, rising to 950:1000 on the periphery. Varanasi is also exceptional in having a less balanced sex ratio on the periphery (650–750:1000) as compared to the centre (850–900:1000). Employment in trade and commerce varies from 10 to 15 per cent on the periphery to 20 or 30 per cent in the centre, reaching a maximum of 40 per cent in Sholapur. Scheduled castes are almost completely excluded from the central areas of all three cities, while their relative percentage is 10 to 15 on the periphery with maximum of 20 to 30 in some wards. This evidence tends to reinforce the view that recent growth in India's indigenous cities has strengthened the preëminence of the centre and maintained the essential spatial features of 'pre-industrial' urbanism.

On the other hand, in Jamshedpur and Chandigarh differences between the centre and the periphery diminish and are statistically significant in both only in the case of literacy. Population density in the central areas of these new cities rarely exceeds 40 persons per acre (0.405 hectare), diminishing gradually to less than 10 on the periphery. In the case of Chandigarh, population density declines sharply in the central commercial area and the correlation is not significant. In both cities the level of literacy among the inhabitants of the central area is between 60 and 80 per cent, as compared to about 40 per cent on the periphery. While employment in trade and commerce is significantly correlated with distance from the centre in Chandigarh, this kind of activity supports less than 15 per cent of the population residing in the central area. In Jamshedpur, there is a significant correlation of employment in miscellaneous services, which supports as much as 30 per cent of the employed population in the central area. There is also evidence showing that the relative concentration of low-status population, which in this city is composed largely of aboriginal tribes, is significantly higher on the margins of company-controlled land and in unplanned areas outside the company's jurisdiction. The observed relationships suggest that in India's planned cities preference for central location has been maintained by the educated people and those engaged in trade and services, while the less well educated and socially-disadvantaged classes of the population continue to be relegated to the periphery as in the traditional cities.

Thus, it appears that the fundamental structure of India's traditional cities has not yet been altered and that India's modern, planned cities retain central site preference for residence, although the demographic and social gradients are diminished. It has been found that education and income level, rather than caste or ascriptive status, determine social rank and residential location in Chandigarh as well as in newly-built Chittaranjan, an industrial city in West Bengal. The author agrees with Fox, who calls for recognition of an intermediate stage in India's urban development, not strictly 'pre-industrial' in the sense of Sjoberg, yet not exhibiting the characteristics of modern commercial and industrial ecology. I interpret these phenomena to be caused by the general scarcity of public or private resources for intra-urban transport and areal expansion of urban settlement. It is this intermediate developmental stage, which I describe as 'post-colonial.'

American Institute of Indian Studies, Rutgers Research Council; governmental officers and town planners.

Berry, Brian J.L., and P.H. Rees, 1969   The factorial ecology of Calcutta, *Am. Jour. of Soc.* 74, 445–91.

Berry, Brian J.L., and H. Spodek, 1971   Comparative ecologies of large Indian cities, *Econ. Geog.* 47, no. 2 (supplement), 266–85.

Brush, John E., 1962   The morphology of Indian cities, in *India's Urban Future*, ed. Roy Turner (Berkeley and Los Angeles), 57–70.

Brush, John E., 1968   Spatial patterns of population in Indian cities, *Geog. Rev.* 58, no. 3, 362–91.

Chettiar, C.M. Ramchandra, 1939   Growth of modern Coimbatore, *Jour. of the Madras Geog. Assoc.* 14, 101–66.

D'Souza, Victor S., 1968   *Social Structure of a Planned City: Chandigarh* (Bombay).

Dutt, Ashok K., 1961   Jamshedpur and its Umland, PH D dissertation, Patna U., Dept. of Geog.

Evenson, Norma, 1966   *Chandigarh* (Berkeley and Los Angeles).

Fox, Richard G., 1969   *From Zamindar to Ballot Box: Community Change in a North Indian Market Town* (Ithaca, NY).

Gadgil, D.R., 1965   *Sholapur City: Socio-Economic Studies* (Poona).

Misra, B.R., 1959   *Report on socio-economic survey of Jamshedpur City* (Patna), Patna U., Dept. of Applied Economics and Commerce.

Mohsin, Mohammad, 1964   *Chittaranjan: A Study in Urban Sociology* (Bombay).

Singh, R.L., 1955   *Banaras: A Study in Urban Geography* (Banaras).

Sjoberg, Gideon, 1960   *The Preindustrial City: Past and Present* (New York).

Frontier Settlement on the Forest/Grassland Fringe
L'aménagement du territoire en régions frontalières prairies/forêts

s0401
Generation waves and spread of settlement
ERIK BYLUND   *Umea University, Sweden*

When studying a regional spread of population or colonization, variations in frequency between different periods are often found. One period shows for example a much larger number of new settlements than the next period. Sometimes it seems possible to trace a regular periodicity in the intensity of colonization. Often an explanation for these ups and downs by reference to the economic situation and its variations is attempted.

This paper has been written to draw attention to another factor whose importance should possibly be considered before opening the difficult discussion on the role of general economic fluctuations. Let us examine the fluctuations in the spread of colonization as demographic generation effects and at the same time point out the influence of different divisions of time in shorter or longer periods on the number of new settlements accounted for periodically.

It appears that the colonization of the inland district of 'Pite lappmark' in northern Sweden was only to a minor extent the result of people moving into the district. Instead, after a first stage of immigration, the development of colonization mainly came about by an interior colonization; that is, it was the sons of settlers of several generations who fulfilled the colonization work. Nor was there any question of a very large first immigration wave. No more than a handful of men with their families started the colonization, which by the effort of the following generations filled the interior of Norrland with settlements and people to the limits of the agricultural resources of the country. The parish of Arjeplog is a fairly pure example, where, from the start of the colonization in the 1720s until the year 1867, some 80 per cent of the 82 settlement units can be traced back to only six progenitors.

The question which now can be asked is: 'Can periodic variations in the colonization activity be explained by this klon-like growth of settlements?' (To characterize the course of events I use the word 'kloncolonization' from the Greek word 'klon' = branch, sprout.) Let us assume that during an initial period of ten years of colonization one settlement per year is established, that is, ten settlements in all; no continuous immigration takes place but the proceeding settling activity is carried out merely by the fact that male descendants of the first colonizers establish new settlements when they become 19 or 20 years of age.

If we now analyse the development of every ten-year period, which is the conventional procedure, we should obtain, as a consequence of these assumptions, a regular periodicity where every second ten-year period would show ten new settlements and every other one none at all.

The assumption that new settlements are established by 19-year-old men is, of course, not at all realistic. The age of 25 would be more adequate. This has been supposed to be the case in the simple time-schedule for the development of colonization in Fig. 1, which is based on the same assumption as above. The figures of the diagram, as well as those of the following diagrams of the same kind, indicate year of establishment during a period of colonization, in this case of 110 years (1–110), divided into 11 ten-year periods. Naturally it is because of the short period of ten years that the generation effect shows in the periodic fluctuation of the figures.

Let us complicate the time model somewhat more by supposing that immigrants from outside set up new settlements every two years after the first ten-year period. Not counting the generation effect, an addition of five settlements per period is thus brought about; of course, these will have a generation effect too, 25 years later. Fig. 2 displays the development. We notice also now fluctuations in the growth of colonization as generation effects. Per ten-year period during a hundred years, we find the following figure series: 10 – 5 – 10 – 12 – 10 – 20 – 15 – 20 – 22 – 20.

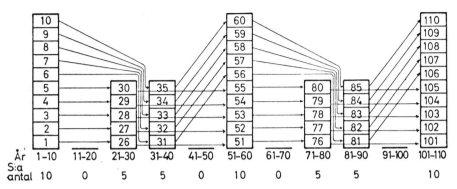

Fig. 1. Generation table of the colonization development. Rectangle = settlement; figure in rectangle = year of establishment; arrows link maternal and filial settlements.

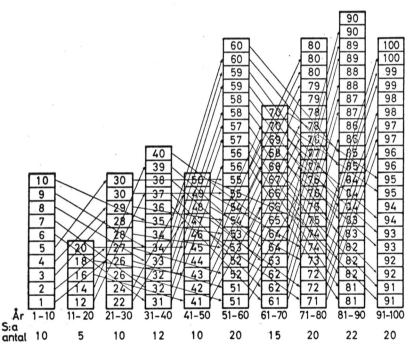

Fig. 2. Settlement rectangles lacking arrow from maternal settlement are settlements by immigration.

In Fig. 3 the time model has been changed in such a way that the periods have been adjusted to cover 15 years, and the sons of settlers who are assumed to set up new colonies are born the year after the foundation of the maternal settlement. This means, for example, that the first son of a settler is 25 years old and builds his settlement in year 27. No immigration increase is included.

However, compared with the factual course of events in 'Pite lappmark,' the time models presented above cannot be considered realistic with respect to the time interval between maternal and filial settlement. In most cases the generation gap proved to be 35–45 years. An interval of 40 years has been cal-

1308 / *Frontier Settlement*

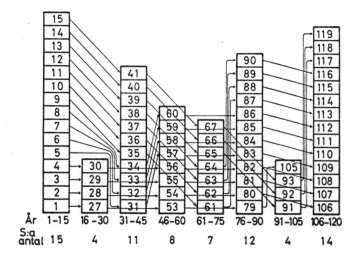

Fig. 3.

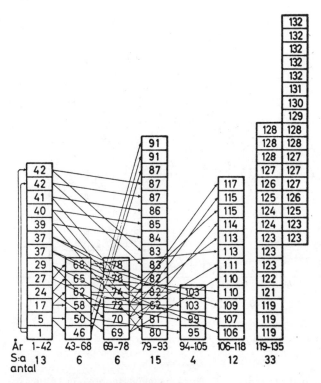

Fig. 4. Arrows omitted, for technical reasons, after year-of-establishment 118.

culated into Fig. 4, which furthermore has been adjusted to the division of periods which for various reasons (among others, historic-administrative ones) has been used in my thesis (–1775, 1776–1800, 1801–1810, 1811–1825, 1826–1837, 1838–1850, 1850–1867). Thus year 42 in the model represents the year 1775 in reality. The number of settlements in the first period of the model, covering 42 years, is equivalent to the real number of

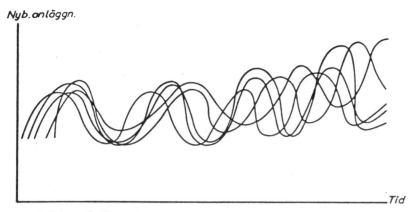

Fig. 5. Schematic illustration of the generation effect on the frequency of settlements.

settlements before 1775 in the parish of Arjeplog. This parish has been chosen because it is characterized, more than all parishes examined, by an interior colonization, relatively 'undisturbed' by immigrations.

Thus no 'immigration settlements' have been included in the model. On the other hand, the rather realistic assumption has been made that every maternal settlement renders two filial settlers every four years. The correspondence between the periodical figures of the model and the real evolution is good with respect to the first four periods; after that the correspondence is less good. The number of new settlements in the model per period is 13 – 6 – 6 – 15 – 4 – 12 – 33 = 89. The same in reality is 13 – 4 – 7 – 10 – 15 – 16 – 17 = 82.

After some time the colonization activity of several generations can be interlaced simultaneously, which means that the generation effect will appear only as small ripples on the evolution graph and no longer as big waves (see Fig. 5). There is always an appreciable generation effect, however, and the more marked the 'kloncolonization' has been, the more clearly perceptible it is. Also there is every reason to study this simple demographic factor before opening other more circumstantial, but maybe, in the end, still necessary discussions on the importance of general economic situations in order to find an explanation for a temporal fluctuation in a course of expansion such as, for example, the spread of the population and the colonization of a region. As I said before, the object of this essay has been only to point out the generation effect in a complex like the one mentioned, as well as the effect which the division of time into longer and shorter periods may have on the variations in the statistical data and accordingly on the time-series analysis too.

Bylund, Erik, 1956 Koloniseringen av Pite lappmark till och med år 1867 / The spread of settlement in Piteå Lappmark up to 1867, *Geographica* 30 (Uppsala).

Bylund, Erik, 1960 Theoretical considerations regarding the discussion of settlement in inner north Sweden, *Geografiska Annaler* 42 (Stockholm).

Rudberg, Sten, 1957 Ödemarkerna och den perifera bebyggelsen i inre Nordsverige. En diskussion av vissa orsakssamband bakom fördelningen bygd – obygd / Unsettled areas and frontier settlement areas in inner northern Sweden – A discussion of some causes underlying this division, *Geographica* 33 (Uppsala).

s0402
Government land policies in southern Africa
A.J. CHRISTOPHER   *University of Port Elizabeth, South Africa*

In the nineteenth and early twentieth centuries two major trends in land policy for the extension of settlement in southern Africa may be discerned. The first was related to the concepts of the Cape-Dutch peoples, descended from the seventeenth-century settlers at the Cape of Good Hope. The second was evolved by the British government in London for general use throughout the British Empire.

The Cape-Dutch land policy had evolved slowly as settlers had penetrated the dry interior of the Cape from the area around Cape Town. The semi-arid lands of the Karoo, with average rainfalls of less than 500mm per annum meant that the European concepts of farming had to be abandoned and extensive cattle runs were evolved. The government of the Dutch East India Company largely followed the colonial practice, so that by 1730 a well-defined system had taken shape. For a small annual rental a settler was allowed to appropriate an area of land within half an hour's ride from a selected middle point, provided it did not overlap another holding. The area involved was approximately 2500ha. Such a size was accepted by the British administration which superseded the Dutch in 1806. However, regular survey was insisted upon and instead of circular farms, rectangular farms covering a standard area of 2500ha were allowed (Christopher 1970).

The Cape-Dutch system of alienating land was clearly incompatible with the ideas of the colonial reformers, who framed the British colonial land policies of the 1830s. Concentrated settlement on small farms (20–100ha) was the desired colonial model, where crop growing could take place and control could be exercised. The extent of the settlement was to be controlled by fixing a suitable price for the land – often a high one. Thus in the 1830s and early 1840s an attempt was made to bring south African land regulations for extending the settlement frontier into line with those in operation in other colonies (Duly 1968).

The introduction of British land policies to South Africa brought legal extension of the colonial settlement to a halt. The price demanded for land (£0.25–£0.62 per ha) was clearly unattractive to settlers where land had previously been free. Undoubtedly land policy played a part in the decision of a substantial group of the Cape-Dutch population to leave the Cape Colony and open up new lands beyond the colonial boundary. The Great Trek of 1836–40 led to the extension of settlement to large areas of the Orange Free State, the Transvaal, and Natal. Understandably the Cape-Dutch took their traditional ideas on land settlement with them. In 1841 their government laid down the law allowing grants of 2500ha, with two grants for those who had taken part in the Great Trek.

Clearly, throughout the newly settled areas regular survey was impossible at first and farms had to be surveyed as best they could by inspection. The inconveniences of circular farms were recognized and square farms were allowed, with each side one hour's ride in extent, giving an area of 3200ha to the standard unit. Many farms had increased to twice that size when survey was finally effected, but the owner was left in undisturbed possession of his supposedly square 3200ha-farm (Baumann and Bright 1940). Such policies could be adopted only where the population was sparse and where there was unlimited land.

Land was not inexhaustible. The British government was aware of this and attempted in the Cape Colony and Natal to allow land sales only at prices in agreement with colonial policy. Such laws were frequently evaded, or settlers simply squatted on their land. Extension of settlement was therefore slow in Natal and non-existent in the Cape Colony until the mid 1850s. Self-government for the Cape allowed the colony to revert to the previous extensive grants, but with significant differences. Land in any sized block could be sold, but the purchaser still had to pay an annual rent. Thus within the Cape Colony the idea of a fixed area for a farm was abandoned. In Natal, after free grants were temporarily reintroduced (1856–8), sales were maintained, with slight extension of settlement until the introduction of highly generous credit facilities in the 1880s.

Purchases were therefore introduced permanently to the two British colonies for varying sized portions of land according to the ability of the purchaser to pay. In the area controlled by the Cape-Dutch Republic the standard sized farm concept was adhered to until the South African War (1899–1902). The various ephemeral republics established on the borders of the Transvaal between 1845 and 1890 all attempted to provide every citizen with an equal share of land and prevent any one person obtaining too much.

Wherever possible the standard Transvaal farm of 3200ha was granted.

The settlement of Rhodesia in the 1890s is therefore of particular interest as, although it was organized by Cape Colonial interests, the Cape land laws were not introduced. The old idea of a standard sized farm as preserved in the Transvaal was introduced, with the important exception that only 1250ha was allowed per pioneer in Mashonaland in 1890. This was largely because most of the pioneers viewed Mashonaland as a gold field and were more interested in the gold claims they received than in farming. However, with the opening up of Matabeleland and Gazaland in 1893, standard sized 2500–3200ha farms were offered, to attract pioneer farmers.

The nineteenth-century assumption of inexhaustible land reserves came to an end in the period 1902–14. The frontier of settlement had been pushed into modern Zambia and large intervening tracts remained unoccupied. Intensification of settlement now became the main concern of the various governments, and closer-settlement programs were adopted throughout southern Africa, in an attempt to make better use of the available land resources. Irrigation was clearly the key to this and a highly flexible policy of leases and sales was introduced to allow for both intensification and extension of settlement. Increasing assessment of the land's potential led to the abandonment of both the fixed price in Natal and the fixed area in the Transvaal, and the adoption of Cape practice throughout southern Africa by 1914. This policy continues to the present day.

Unlike the contemporary settlements in the Americas and Australasia the land policies adopted in southern Africa were not responsible for the introduction of large numbers of rural colonists. Undoubtedly much of the extension of settlement was the result of pressures from an expanding local population rather than massive immigration. Control of land policy, with the exception of the Cape from 1831 to 1856 and Natal from 1848 to about 1875, was in local hands which framed the land laws according to the desires of the settlers. Their requirements were for cheap or free land, preferably in large blocks, but such a policy could last only until the available suitable land had been exhausted. Furthermore such a policy was not really designed to develop the land to its full capabilities. Development on any scale came later because greater use had to be made of the land as settlement was intensified.

Baumann, G., and E. Bright, 1940 *The Lost Republic* (London).
Christopher, A.J., 1970 The European concept of a farm in southern Africa, *Historia* 15, 93–9.
Duly, L.C., 1968 *British Land Policy at the Cape 1795–1844* (Durham, NC).

s0403
The evolution of Australian pastoral land tenures: an example of challenge and response in resource development
R.L. HEATHCOTE   *Flinders University of South Australia*

Pastoral land use in Australia began as a profitable private enterprise but became recognized by all Australian state governments in the latter half of the nineteenth century as a useful and often necessary, although temporary, precursor to the ultimate goal of land settlement, i.e. small-unit, freehold family agriculture or 'yeoman farming.' In fact, livestock grazing has remained unchallenged as the dominant continental land use and, in the three-quarters of the continent which is arid, is mainly carried on still in extensive units owned by partnerships and companies (often non-Australian). The official policies of settlement intensification through time, or 'closer settlement,' have had to be modified in the arid environment and in this and the evolution of the pastoral tenure systems can be seen evidence of challenge and response in the sequence of resource development (Heathcote 1965, 1969, chap. 12; Powell 1970).

Pastoral land use in Australia initially copied British practices and took place on land held under tenure systems imported from Britain. In the late eighteenth century British horses, cattle, and sheep were grazing on a variety of tenures, from the freehold estates of absentee landlords in Scotland, through grazing leases for up to 25 years, to 21-year agricultural leases (as part of heath reclamation in Norfolk) and on the traditional village 'common' lands. This variety was imported into Australia, and grazing during the first forty years of settlement took place on all forms of tenures, from freehold grants such as Macarthur's at 'Camden Park' (New South Wales) through 'Depasturing Licences' to town 'commons.'

From mid-nineteenth century, however, evidence began to appear of a recognition of the need for modifications of pastoral land tenures to cope with specific problems of the arid interior. The pastoralists faced four challenges in the interior plains. First was the sparse nature of the natural feed and consequent low productivity per unit area. Second was the challenge of competing users, whether indigenous fauna and the Aboriginal population or alternative introduced resource uses such as mining. Third was the challenge of resource variability over the arid areas in time and space, both feed and water supplies, and fourth was the challenge of specific catastrophic variations in the availability of resources.

Before the Imperial Waste Lands Act of 1846 the initial reaction to this challenge had been to propose increases in the sizes of the free land grants, but these had failed because of fears of land monopoly. The sales of land, after the abolition of free grants in 1831, did not meet the challenge either, because of limitations on the maximum size of blocks and relatively high reserve prices for the land. Despite sales on credit and free selection from the 1860s in some states, arid lands were not purchased or acquired for freehold as *bona fide* pastoral production units but usually only as strategic locations (waterholes and stockyards) from which adjacent crown lands could be controlled and grazed at will. The exceptions were stud properties where high capital investment was matched by high unit area productivity from improved pastures and high value livestock.

As a result of the 1846 Act tenures were diversified according to the estimated quality of the land. By 1850, New South Wales had been divided into Settled, Intermediate, and Unsettled (mainly arid) lands, with pastoral leases varying from 1, through 8, to 14 years respectively, but with no limits on the total lands so held; South Australia had lands within the 'hundreds' (proclaimed local government areas) which were to be sold, and lands outside the hundreds (mainly arid areas) which were to be leased for grazing for up to 14 years; and Western Australia had a parallel division with up to 8-year leases in the interior. The basic division between leasehold pastoral (arid) tenures and freehold and other agricultural (humid land) tenures had been established and was to be re-emphasized throughout the following century of land settlement.

Subdivision of large leaseholds to intensify settlement from the 1880s onwards appears to have been at most only half as successful as intended and, indeed, policies to provide minimum economic living areas for settlers have in fact reduced numbers of settlers eventually, as rising living standards and declining stock carrying-capacities in overgrazed ranges required property amalgamations to provide viable economic units.

European livestock faced competition from indigenous herbivores as well as imported game animals and were victims of Aboriginal and dingo attacks. Controls on 'vermin' (arbitrarily defined from imported rabbits to indigenous eagle hawks) have been written into most leases, and native fauna has been protected only relatively recently and over very limited areas. The acquisition of Aboriginal hunting grounds has only partly been offset by Aboriginal Reserves, created in the most unpromising pastoral country, but it is interesting to note that Aboriginal hunting rights have been preserved in the pastoral leases current in South Australia and Northern Territory. Here, as with the separate allocation of mining licences, complementary use of grazing lands is envisaged.

To meet the challenge of scarce and variable resources, the initial official reaction was to preserve such resources for general use. Public water reserves were set up around strategic waterholes and spaced along river frontages from the 1840s, and from the 1880s governments were drilling and reserving for public use the wells which tapped the newly discovered artesian basins. Land grants to successful explorers and special 'developmental' leases encouraged private well digging and land clearance for pasture improvement. Large leases, initially seen as but a passing phase of settlement, came to be recognized as permanent features needed to increase the chances of catching the random rainfalls and areally diverse fodder responses. Although the leases normally allowed use of timber for domestic fuel and building needs only, recognition of edible browse among many *Acacia* species has resulted in special in-drought provision for tree lopping or 'scrub pushing' for stock feed. Finally, the need to move livestock to market on the hoof has been met by reserves for travelling stock routes, which incidentally have often offered unscrupulous 'grass pirates' free commonage in droughts.

At the extreme level of resource variability, when droughts have destroyed pastoral resources, the official relief policies have included tenure modifications. One early and continuing method of relief has been the remission of rents until the drought ended; another, the extension of the term of lease to allow drought losses to be recouped by profits in the good seasons which are assumed to follow. More recently, the limitations of such policies have been recognized and official insurance in the form of 'drought bonds' (redeemable for cash in drought periods) has been made available to pastoralists (Heathcote 1969b).

The short-term effects of droughts, however, are somewhat overshadowed by the long-term effects of misuse of pastoral resources. Since the stimulus of massive erosion in eastern Australia in the 1940s, in the extreme cases of misuse, leases have not been renewed and land has reverted to the Crown, while increasing control of land management has been written into most current pastoral leases. Control of noxious weeds and limits upon the number of stock to be grazed have replaced the old provisions of a minimum number of stock to be grazed for the lease to remain valid. From initial encouragement for any use of the land, the official policies have swung back to controls upon that use.

The sequence of pastoral land use and the associated land tenures in arid Australia have reflected the modification of introduced British techniques and tenure systems to cope with the stresses of the arid environment. Official policies aimed at intensive family unit land settlement have been modified to allow extensive company operations; attempts to dispose of land as freehold have been frustrated and leaseholds still dominate the tenure patterns.

At a time when world economic trends threaten to remove the incentive for pastoral land use, one bright spot is the system of leasehold tenures which will allow flexible and multiple alternative use of resources with minimum legal disruption, for the general public, through the various Federal and State governments, is still the ultimate landlord.

Heathcote, R.L., 1965 *Back of Bourke* (Melbourne).
– 1969a Land tenure systems: past and present, in R.O. Slatyer and R.A. Perry, eds., *Arid Lands of Australia* (Canberra).
– 1969b Droughts in Australia: a problem of perception, *Geog. Rev.* 59, 175–94.
Powell, J.W., 1970 *The Public Lands of Australia Felix* (Melbourne).

s0404
Unit size as a factor in land disposal policy
JOHN H. PATERSON *University of St. Andrews, UK*

In the great migratory movements which represented Europe's response to the opening of new lands in the nineteenth and early twentieth centuries, one of the factors most influential in determining their strength and direction was the question: how much land could an individual obtain? In most cases the answer depended on the policy of the government in effective control; that is, on its method of subdividing the new lands and the conditions of entry it imposed.

Since these policies were diverse, there are a number of areas of broadly similar relief and climate, settled contemporaneously and under much the same conditions of technology, in which quite different patterns of landscape and land use have emerged. In particular, the possibilities open to the settler have been limited by the size of unit to which he could obtain title. This paper considers the familiar case of land disposal policy in the USA. Other areas' policies are much less well known. The Australian settlements have received some attention but, perhaps owing to language problems, little has appeared in English about, for example, the Pampas or the Steppes. There were certainly homesteads in South America – 25 hectares (62.5 acres) on the Pampas and up to 400 (1000 acres) in Colombia. Contact is sought with any colleagues who may be pursuing parallel studies in these other regions.

In the USA during these years one particular unit size achieved such common usage that it came eventually to be regarded as sacrosanct – the unit of 160 acres (65 hectares) or the quarter-section. It first appeared in national legislation in 1804, and in certain important contexts it remains on the statute books to the present day. Its influence on the landscapes and settlement pattern of the North American interior is well known (for it was, of course, adopted in Canada too). Equally well established has been its unsuitability in the dry west, where 160-acre units proved too small to be viable. It therefore remains to ask why this unit size achieved such universal acceptance, even when it was being applied in a region to which, on climatic grounds, it was manifestly unsuited.

One thing which can be stated quite unequivocally is that the choice was not governed by the realities of agricultural settlement. Later rationalizations that the quarter-section was chosen as the size of holding which would support a family find no confirmation in the debates of the period. There was nothing equivalent to the estimates of 'living area' or 'home maintenance area' which were defined (and defined, moreover, in the courts) in Australia in the period 1880–1905. The quarter-section units bore no relation to the average farm size in the USA at any time during the first century of the republic's life. The reasons for the choice of this size of unit were financial. The reasons for its perpetuation were political.

During the earliest period of United States land-policy making, financial considerations were paramount. The land was a potential source of revenue, while the principle of prior survey which had been adopted involved the government in costs of survey to be set against this revenue. The object was to maximize one and minimize the other.

To achieve this, the unit must not be too small, for the smaller the unit the higher the cost of survey. The rectangular system of survey which was adopted was the cheapest and simplest that could have been employed, but even so there was a shortage of surveyors, the terrain was difficult (the national survey began in southeastern Ohio), and for these reasons the original instructions were for the survey simply of townships 6 miles (9.5km) square. Not until 1796 were the 36 sections in each township to be surveyed, and not until 1804, in the kindlier terrain of Indiana, did the survey include quarter-sections.

On the other hand, however, the unit could not be too big. There were few men in North America who could afford to buy a whole township, even as an investment, at the price the government proposed to charge ($2 an acre or $5 per hectare approximately), and experience in the colonies before the Revolution had shown that where land was sold in small lots the price realized per acre was considerably higher. That the quarter-section eventually became available was simply due to the fact that the larger units – township, section, half-section – were proving very slow sellers and revenue was the foremost consideration.

It was for equally non-geographical reasons that the quarter-section figured in the long legislative battles of the mid-century period, leading up to the Homestead Act of 1862 (which had been under debate for 17 years). During this period, while Congress was deeply divided on the principle of the free homestead, the size of the unit under discussion was never seriously challenged. Whenever it was suggested that the figure should be changed from 160 acres, the Congressional Record simply reports 'Negatived without debate.' Why was this?

The figure of 160 was adopted partly because it already appeared in earlier legislation, especially in the Pre-Emption Act of 1841, to which the Homestead Act was seen as a logical successor. But it was also adopted because, very significantly, this was a proposal for homesteads by anti-slavery interests. It was therefore important not to make the homestead unit too large, lest it be capable of supporting slavery. This was why, in fact, the South opposed the homestead for so long – the act was, of course, only passed after the Southern states had seceded. The South was not against homesteading as such: Virginia and North Carolina had pioneered the technique on their own public lands, with the ideals of Jefferson to guide them. It was against homesteading on a 160-acre unit. And as settlement spread into the drier west, the doctrinaire Northern commitment to the 160-acre unit returned to plague its inventors.

It would be interesting to know whether, on any other of the century's land frontiers, settlement policy was determined by factors bearing so little relationship to the geographical realities.

Ford, A.C., 1910 *Colonial Precedents of Our National Land System* (Madison, Wisc.).

Heathcote, R.L., 1965 *Back of Bourke* (Melbourne).

Johnson, H.B., 1957 Rational and ecological aspects of the quarter section, *Geog. Rev.* 47, 330–48.

Pattison, W.D., 1957 *Beginnings of the American Rectangular Land Survey System, 1784–1800* (Chicago).

Sanborn, J.B., 1900 Some political aspects of homestead legislation, *Am. Hist. Rev.* 6, 19–37.

Stewart, L.O., 1935 *Public Land Surveys: History, Instructions and Methods* (Ames, Iowa).

Thrower, N.J.W., 1966 *Original Survey and Land Subdivision* (Chicago).

Truesdell, W.A., 1908 The rectangular system of surveying, *J. Assoc. Eng. Socs.* 41, 207–30.

s0405
Edaphoclimatic frontiers
G.D.V. WILLIAMS   *Canada Department of Agriculture*

Edaphoclimatology may be defined as the science that deals with the influence of soils and climate on living things, particularly plants, including man's use of land for plant growth. This definition was derived by adding the words 'and climate' to a textbook definition of edaphology (Buckman and Brady 1970, 607). The edaphoclimatic frontier of a crop is the limit beyond which the climatic or soil resources, or both, are inadequate for growing that crop.

In Canada frost risk, severe winters, summer droughts, and relatively infertile soils typify agricultural frontier areas (Ehlers 1968). The importance of physical limitations in such situations is emphasized by the similarity between the present position of the climatic limit for grain maize, which is around the 2300 corn heat unit isoline (Chapman and Brown 1966, 10), and the northern frontier of maize growing, prior to the settlement of North America by Europeans, which included southern Ontario and a small part of southern Manitoba (Moodie and Kaye 1969). Perhaps a major part of the reason for the northwestward expansion of our agricultural ecumene since then has been the introduction of grains such as barley and wheat, which, unlike maize, inherited the ability to adapt readily to climates as severe as those of the mountain valleys of Asia.

In some cases it may be adequate to define climatic limits using simple climatic variables. Examples are the $-40°c$ and $-10°c$ annual mean minimum temperature isolines at the northern and southern limits of the sugar maple region (Dansereau 1957, 62–4), the $21°c$ mean June to August temperature at the southern limit of spring barley production, and $-1°c$ mean December to February temperature at the northern edge of the winter barley zone in eastern North America (Weaver 1950, 55–8).

That more complex bioclimatic expressions are usually needed has long been recognized. In 1912 wheat's climatic limits were mapped using accumulated temperatures and day length (Unstead 1912). Switzerland was recently mapped for maize, using temperature, sunshine duration, and precipitation (Primault 1968).

The edaphoclimatic frontier for barley in Canada has now been mapped. For the climatic aspects the techniques were mainly similar to those used for wheat on the Canadian great plains (Williams 1969). A biophotothermal time scale was developed using Olli, an early maturing barley variety. With this the date of reaching any of five phenological stages, the last being the ripe stage, can be estimated if the planting date and the subsequent daily temperatures and photoperiods (day lengths) are known.

As in the wheat study, temperature normals estimated by regression equations from the latitudes, longitudes, and elevations were obtainable for nearly 1200 survey control points on the Canadian great plains.

These, plus published normals for 900 climatological stations across Canada, were used in obtaining interpolated normal maximum and minimum temperatures for each day of the year. These daily normal temperatures were used with daily photoperiods, which were computed readily from the latitudes, in the biophotothermal time scale equations to determine the normal dates of reaching various stages. It was assumed that planting took place when the smoothed normal mean daily temperature rose to 10c in the spring.

Using the data obtained from this equation the line where barley would just reach maturity, i.e. the climatic frontier beyond which it would not normally ripen, was drawn. Within the barley maturing zone the suitability for barley of the soils as mapped on a preliminary draft of the Soil Map of Canada were rated on a scale from 0 to 100. Stony, rocky, permafrost, or mountainous areas were considered non-arable and rated zero. The barley edaphoclimatic frontier can be taken as the limit beyond which either barley would not normally mature or the soil is non-arable.

The climatically defined barley maturing zone includes not only the present agricultural ecumene, but also as much area again to the north. Much of this additional area is edaphically unsuitable, but among the substantial exceptions is an area, as large as the entire Great Lakes–St Lawrence lowlands, in northern Ontario and Quebec between the 48th and 52nd parallels.

The soils of such areas beyond the present ecumene, while not generally rated very highly, might conceivably be used for barley production if the need arose. These may be the agricultural pioneering areas of tomorrow. In considering such resource evaluations it should be kept in mind that in these areas the climatic and soil resource data available were generally much less adequate than they were in the ecumene, and that the mapping will therefore be less reliable for these fringe areas.

In the great plains the main northern climatic frontier for barley is around 63°N,
or about a degree farther north than for wheat; the northernmost extremity is near 66°N, or about 3° farther north than for wheat; and the non-maturing upland areas farther south are much smaller for barley than for wheat. These differences emphasize the need for separate assessments for each likely crop. In the meantime, since barley is the fastest maturing cereal, the edaphoclimatic frontier derived can probably be taken as the limit for all cereals, at least for present technology and varieties.

Information on edaphoclimatic limits should be very useful in development along the agricultural frontier. In view of the lack of awareness of Unstead's findings which seems to have been evident over the past 50 years, however, the task of informing potential users, from pioneer farmers to senior land resource use planning authorities, may be just as difficult as that of deriving the information in the first place.

Buckman, Harry O., and Nyle C. Brady, 1970 *The Nature and Properties of Soils* (New York).

Chapman, L.J., and D.M. Brown, 1966 *The Climates of Canada for Agriculture* (Ottawa).

Dansereau, P., 1957 *Biogeography, an Ecological Perspective* (New York).

Ehlers, Eckart, 1966 The expansion of settlement in Canada: a contribution to the discussion of the American frontier, *Geographische Rundschau* 18, 327–37.

Moodie, D.W., and Barry Kaye, 1969 The northern limit of Indian agriculture in North America, *Geog. Rev.* 59, 513–29.

Primault, B., 1969 D'une application pratique des indices biométéorologiques, *Agri. Meteorol.* 6, 71–96.

Unstead, J.F., 1912 The climatic limits of wheat cultivation with special reference to North America, *Geog. J.* 39, 347–66, 421–46.

Weaver, John C., 1950 *American Barley Production, A Study in Agricultural Geography* (Minneapolis).

Williams, G.D.V., 1969 Applying estimated temperature normals to the zonation of the Canadian Great Plains for wheat, *Can. J. Soil Sci.* 49, 263–76.

s0406
Railroads and urban settlement in Nebraska: a case study
J.F. DAVIS   *University of London, UK*

The latter part of the nineteenth century saw a considerable departure from the earlier development of the railroad network. Previously railroads had been built mainly to link existing towns, to serve existing farming areas, and to reach and span the major rivers and so connect with and compete for river traffic. Now in the last half of the century the railroad was turning from a passive to an active role as it moved from serving existing settlements to creating its own.

The need to foster economic development near to the railroad tracts was not new; already the railroads prospering most were those which served the richer and faster developing areas anywhere in the USA but now the need arose for a far more positive approach to this development of economic activity. Already companies pushing out into the great plains were faced with an area largely empty except for buffalo, Indians, a few traders, trappers, and miners. Thus the fostering of settlement and agriculture became vital for economic reasons. Apart from the very small amount of land needed by the railroads themselves for switching yards, depots, etc., most was of no economic value unless it could be sold, ideally to farmers and manufacturers who would thus bring money to the railroads both by land purchase and by generative traffic. (Federal land grants to the railroads in Nebraska amounted to 3.076 m.ha [Union Pacific 1.943m and Burlington and Missouri .972m] and state grants to .202m.ha; a combined total of 3.306 m.ha.)

This paper focuses on the development of settlements prior to 1900 in the eastern half of Nebraska where there is a transition from grassland to woodland.

The task is complicated. The prior settlement of some land makes it difficult to gauge the railroads' full impact on settlement and farming especially as US census figures are available only at ten-year intervals and sporadically in between from Nebraska's own census. Although the area was relatively unsettled it is often difficult to unravel the sequence of events, whether railroad followed settlement or vice versa. Sometimes speculators would buy land along an anticipated railroad route. Towns were sometimes established and then the railroads 'invited to link them to their system.'

Railroads offered certain advantages to settlers in their disposal policy of land. Although railroad land often was more expensive than Homestead Act land it could be bought over a period of years; it could also be purchased in a wide range of sizes and could be larger than 65 hectares if desired; railroads offered more fringe benefits too, for example rebates on fares and freight of personal and farm belongings for the new settler; they also often granted land, materials, or money or carried materials free for churches, schools, etc., in new settlements. The railroad companies sponsored agricultural improvements, and encouraged innovations by sponsoring agricultural shows and reducing interest payments on money outstanding on farm purchase for farmers who carried out improvements or by lengthening the time over which capital had to be repaid if a farmer was a good farmer.

Perhaps the greatest advantage was that many railroads through their land offices and overseas agents were very willing to help establish group settlements, which were not possible through the Homestead Act. Exact figures for the number of groups helped are very difficult to obtain, but one estimate is that 4850 families participated in Union Pacific group colonizations, of which there were 96; of these, 20 groups involved more than 50 families each (Spencer 1950). The Burlington and Missouri built an emigrants' home in Lincoln to provide temporary accommodation. By the end of 1871 the company had five agents and 150 sub-agents in the United Kingdom and took advertising space in some 150 newspapers mainly in farming areas and also in some agricultural and religious journals, e.g., *The Methodist Recorder*. Thus the railroads were often instrumental in the establishment of communities which often rapidly became rural service centres and small towns. Certainly in many areas the railroad settlements were more nucleated than those of the Homestead Act.

In 1867 the Union Pacific crossed Nebraska, and by 1882 the Burlington and Missouri had built right through the Republican

Valley in southern Nebraska to Benkelman (Dundy Co.). Railroads pushed northwest and west to Valentine (1883) and on to Wyoming (1885). By 1900 all but ten of Nebraska's counties had a railroad and of these, six still have had no railway, and the State's railroad patterns had been established.

A close look at available population data shows that most counties had their greatest absolute population increase in the ten year period following the establishment of a rail link within that county; exceptions were normally either in parts of the extreme east where settlement was already well established before the railroad, e.g., Richardson, Nemaha, and Cass Counties; or in the west where, although in some cases the railroad came early (Union Pacific), maximum population growth was not until 1890–1900 because of the less attractive terms for farmers in the area under the original 65 hectares Homestead rule, greater distances and therefore transport costs, and the Indian menace, together with the important fact that this was cattle country initially and cattlemen did not take kindly to agricultural homesteaders.

The colonizing work of the railroads is particularly interesting as in this they were, in the main, working along an avenue untouched by the federal government. With better resources, more land, more agents, and greater need for settlers it was the big railroad companies who undertook most of this work. 'The Burlington and Missouri lands offer an inducement to the immigrants in a financial point of view, superior to the government for the reason that when a purchase is made from the government, the business is transacted on a cash basis entirely, while a purchase from the Burlington and Missouri allows the immigrant to use his surplus funds to improve and stock his farm and is virtually a loan to him of so much money at the extraordinary low rate of 6% interest per annum' (*Seward County Reporter* 1872).

The colonies came both from other parts of the USA and from Europe; some were of religious groups, others linguistic, others national, some large, some small. Depending on the size of the individual group, their proximity to others, and the date of their establishment relative to other settlements in the same regions, many of these groups played an appreciable role in the settlement history of Nebraska, as indeed similar groups did in other states and provinces across central North America. In 1870 the foreign born population constituted about 32 per cent of the Nebraska total, in 1880 25 per cent and in 1890 44 per cent. Numerous colonies were established by both the Union Pacific and the Burlington and Missouri railroads on the lands granted to them by the Federal and state governments. Examples of settlements established through a church were especially numerous among Catholics, e.g., the Irish Catholic Colonisation Company established a settlement in Greeley County, though Mennonite and others were also founded, e.g., Russian Mennonites around Jansen (Jefferson Co.). Group settlement was furthered by individuals, e.g., the Soldier's Free Homestead Colony of Ohio which settled Gibbon. The Burlington and Missouri railroad in particular through its subsidiary – the Lincoln Land Company – platted and established a large number of towns, especially in Republican Valley. Between 1885 and 1891 the Lincoln Land Company platted over 18 towns and others were platted by the Republican Land Company.

Railroads established towns in other parts of Nebraska too and were certainly instrumental in fostering the growth of very small groups of farms into towns in some instances as well as providing the magnet, once built, by which settlements were attracted to them. Another Burlington company was the Eastern Land Associates, organized in 1870. This group was particularly concerned with the lands along the S. Platte and established the 'alphabet' towns to the west of Lincoln: Crete, Dorchester, Exeter, Fairmont, Grafton, Harvard, Inland, Juniata, Kenesaw and Lowell. At Blair, Washington Co., the Sioux City and Pacific Railroad sold the first town lots in 1869 and by 1872 there were 'substantial buildings' including six hotels, three agricultural machinery makers, a bank, and 20 stores (*Newberry Library Collections*).

The Homestead Act played a significant role in the establishment of farming in Nebraska. It was the federal government too, which indirectly, it could be argued, was really responsible for the settling of land through the sale of the railroad lands. However, it was the railroad companies themselves which had the necessary economic drive and indeed necessity to see their lands, especially the better ones, sold off to established farmers, manufacturers, and others who would serve the growing farm communities. The railroads stood to gain twofold from these sales; first from the actual sale of acreage and, secondly, from the increased business they would hope for from the farmed land and growing towns. Consequently the railroads were, as has been seen, anxious to promote settlement and did so in various ways. The considerable number of group settlements from both Europe and other parts of the USA in Nebraska bear witness to at least the partial success of the railroads' efforts. The fact that many of the settlements were group settlements and that a considerable number of these were established with embryonic urban nuclei was an important contribution by the railroads to the settlement pattern of the great plains belt in general and of Nebraska in particular.

Newberry Library Collections (Chicago), 769.8.
Seward County Reporter, 9 March 1872.
Spencer, M.N., 1950 'The Union Pacific's utilisation of its land grant with emphasis on its colonisation programme,' unpub. PH D thesis, U. of Nebraska.

s0407
Organized land opening as a facet of the settlement geography of Venezuela
ROBERT C. EIDT *University of Wisconsin, USA*

During the last two decades existing urban settlements in Latin America have expanded at rapid rates primarily because of the attention given to industrial development. However, the impact of the subsequent rural-urban movement has brought into question the ability of these urban settlements to survive as initially intended. The emergence of *cinturones de miseria* or *tugurios* (slum quarters), which now completely ring many of the urban centres, and the inability to provide employment for rural migrants after the fashion of more advanced nations have signaled a breakdown in the function of older Latin American cities. This failure has gradually compelled nations which formerly gave top priority to expansion of industrial programs to undertake far-reaching agrarian reform with the hope of stemming the tide of movement to existing cities by establishing productive new settlements in unused or underused areas. With the exception of Uruguay, where there has been little emphasis on agrarian reform, all South American countries now have long-term projects involving various new types of rural settlement, short-term efforts to speed the transfer of modern technologies to existing farm settlements, and/or the development of new urban facilities where they are lacking.

One of the most successful countries in adopting major reforms to date is Venezuela, where the position, form and structure, and distribution of planned agrarian settlements now contrast strikingly with those of its immediate neighbours, whose progress has not been as marked. Access to special reports, both governmental and private, field work carried out in many settlements, both old and new, and the lack of an integrated picture of the settlement results of Venezuelan agrarian programs would seem at this time to justify a general investigation of these settlements.

Unlike its neighbours, Venezuela has long been considered as a nation with a settlement problem based on a low population density. Therefore, even before oil exports began to provide funds for large-scale national projects in the 1930s, a series of foreign agricultural settlements was begun. Most of these settlements failed, but experience was gained which was applied later. During the 1950s food imports became too expensive even with large oil revenues so that new programs for establishing farm colonies were begun. An agrarian reform law was passed in 1961 which did not stop at a simple division of property but was an integrated concept of change based on consideration of large-scale land availability, a small, backward, and scattered farm population, urgent need for technical assistance of all types, availability of credit, and access to markets. The Instituto Agrario Nacional was established as an autonomous agency with the responsibility of implementing the new law. It was given a separate budget, authority to purchase or expropriate lands if necessary, and to found new *centros agrarios*, or rural settlements, throughout the nation.

The agency began to implement this program during the early 1960s by offering reasonable compensation for arable land – a feature which accounted for an unusual voluntary break-up of surviving latifundium holdings as owners sometimes flocked to IAN authorities with offers to sell idle property. Whenever land was acquired, the existing rural inhabitants, the so-called *conuqueros* (isolated subsistence farmers), were gradually gathered into group settlements. These new entities are popularly called *asentamientos rurales, asentamientos campesinos,* or *centros agrarios,* i.e., rural settlements. Their populations vary in size from 15–20 families to over 50, and they have now been established in every state in the country. By 1970 there were over one thousand and their uniform and new appearance could be easily noted by the most casual observer travelling through Venezuela. Co-ordination of each *asentamiento rural* and an associated field system was developed under the early guidance of technical experts from Israel. The result has been a series of adaptations derived from Israeli kibbutzim and moshavim settlement types.

Accomplishments of the program are measurable in terms of statistics from individual segments. For example, approximately ten per cent of the land available in 1960 had been subdivided by 1970. About half the subdivided land is now in use. Approximately 30 per cent of the rice produced in Venezuela is from the new rural settlements and the country has recently become self sufficient in production of that staple crop. The story with potatoes is similar. Statistical estimates like these become even more significant when it is realized that only about 25 per cent of the *asentamientos rurales* are mechanized and that, to date, they affect only about one-third of the farm population.

Much may be learned from the Venezuelan experience that both corroborates and contradicts various aspects of current thinking about new land opening processes, and at the same time adds an original element to the fund of knowledge about successful modern colonization in the Americas.

s0408
Colonization and retreat in the cut-over lands of the Lake Superior region
STIG T. JAATINEN   *University of Helsinki, Finland*

The upper midwest region, particularly the part lying round Lake Superior which is denoted as the cut-over lands, was the last pioneering fringe in the USA. Colonization was active here into the early 1930s. Rural settlement in the region partly derived from urban-industrial nuclei developed in connection with the utilization of mineral resources (copper and iron) and forest resources. Direct immigration played a lesser role.

The general conditions for rural colonization were not favourable because of climate and soil, while in addition the distances to the big urban markets were long and communications poor. Since agricultural colonization took place from the 1880s onward, regardless of these adverse factors, it can be viewed as a result on one hand of exceptional land hunger among the immigrant and urban populations of the region, and on the other hand of the land policy carried out by landowners of that time, who were the railroad companies and forest owners.

The role of the settler in this colonization process is well illustrated by the Finnish immigrants. The supply and sale of land has to be seen in the context of both larger-scale projects and a general policy of attracting people to the land.

The region under consideration lies well north and east of the prairie boundary and was originally coniferous forest, although in some areas deciduous forest was dominant, especially in Wisconsin. Climatic and edaphic conditions corresponded to the boreal lands of Europe. Locally, clay and loam have created better opportunities for agriculture, for example on the southern coastland of Lake Superior, and in western Wisconsin.

The following points provide briefly the human background to the relatively intensive colonization of this rather inhospitable and remote region. (1) In general, the immigrants to this region, for instance the Finns, had an exceptional urge to own a piece of their own land in the new country. This was a heritage from the situation in their home country. The Finns were mainly from the landless rural population, whose view of prosperity was to a major extent derived from the image of economic and social security through land ownership. (2) The Finns were also well accustomed to forest work and the combination of agriculture and forestry. This disposition was well suited to the prevailing conditions in the region around Lake Superior, where the big forest companies were looking for manpower of exactly this type, who represented at the same time eventual buyers of the cut-over lands. (3) Most of the Finnish immigrants were originally attached to the mining industry around Lake Superior (to which they were introduced through contact with Norwegian miners). The opportunity to change to a rural way of life, with agriculture and forestry instead of mining, was thus close at hand. Social and economic difficulties (unfamiliarity with mining, strikes, and blacklisting) in the mining towns strengthened their desire to move to the countryside. (4) These immigrants were late in arriving in the USA, and thus alternative areas of agricultural colonization were few or non-existent. The cultural barrier created by difficulties in understanding English, poverty, and poor opportunities for saving in mining-work contributed to limit their choice to the area adjacent to the mining districts. (5) The perception of a similar and familiar environment compared with their home country might have played a certain role, but can not be said to have been the decisive factor.

Those who owned the land – the railroad and forest companies – tried energetically to sell their land either directly or usually indirectly through various land agencies. The motivation was quite straightforward: to ensure future traffic and to get the remaining value out of the land after the timber had been removed. Eventually, the local and state authorities began to sponsor the colonization of the cut-over lands in order to keep population from declining, and because they feared they might lose the basis for economic prosperity.

Most of the land bought by the settlers was in blocks of 16.25ha. When land was acquired under the Homestead Law, the settlers sold part of their property as soon as possible under the law, usually keeping only 16.25ha. A considerable degree of cultural cohesion was thus displayed during the colonization stage, resulting in rather homogeneous ethnic areas. These still retain much of their original character and have thus provided suitable cases for the study of geographic adaption and change.

The agriculture of the new settlements was to begin with characterized by a high degree of subsistence. The newly cut-over lands provided natural and comparatively fertile pastures for dairy cattle, and thus the sale of milk for local consumption soon became the main source of cash. At this time, forestry still played an important role both by giving employment during the winter and by providing a demand for dairy products on a local scale.

Eventually, forestry declined and dairying was directed towards the growing urban centres. This change in the economy was supported by the vigorous development of co-operative processing and marketing organizations. Fresh milk became the main product of this period until the depression of the 1920s. As time progressed, dairying on these small farms, of which usually only 4–6ha was cleared, could not compete with the larger farms situated farther south and on better land. Rationalization and specialization nevertheless took place, resulting for instance in the growing of such crops as peas and strawberries and the keeping of beef cattle. Dairying was further restricted on small farms through the introduction after World War II of new hygienic regulations which were beyond the capability of these units to attain.

At the same time, demographic changes took place. During the depression of the 1920s, a certain urban exodus affected these areas by temporarily raising the number of rural inhabitants, but depopulation resumed and accelerated after World War II. Part-time farming and complete farm abandonment became pronounced features and have now resulted in total dereliction of former agricultural settlements. Depopulation has been highly age selective, leaving a disproportion of elderly people. Social-security schemes and the soil-bank program contributed substantially to retirement from farming.

The net result of these trends has been both economic and social backwardness. Thus this region is one of the depressed areas of the USA. Different schemes to stimulate activity have been proposed, but so far with poor results. Recreational land use of former agricultural areas has to a certain degree alleviated the problems, but usually recreation and farming have such different requirements that they are impossible to combine. Although recreational activity has undeniably led to economic improvement for the region as a whole, most of the former agricultural settlements have been bypassed by such development. The difficulties met by the mining industry around Lake Superior during recent years have contributed to the problems of the rural settlements, especially as part-time farming and working in the towns became common after the war.

From the viewpoint of regional differentiation, a number of types of regions with different backgrounds and sets of problems can be discerned. (1) Marginal areas which were originally predominantly supported by mining. Also rural settlement and agriculture developed from sheer necessity due to the conditions at certain times in the towns. These areas have now suffered almost complete abandonment or at least the virtual disappearance of agriculture (for example in northeast Minnesota, the Upper Peninsula of Michigan, and adjacent parts of Wisconsin). (2) Areas where the marginal situation has been alleviated by a number of factors such as soil conditions, climate, and proximity to prosperous urban markets. Here the settlements have usually survived, although greatly changed both in their economy (declining agriculture, but a selection of new industries mainly connected with services and recreation) and their social and demographic structure (retired people, commuters, generally fairly marked depopulation and concentration of population at a few points with good communications). Examples of this type are the coastal strip east of Duluth-Superior south of Lake Superior, and similar areas west of Duluth, in north-central Wisconsin, and in parts of the Upper Peninsula of Michigan. (3) The last type is found along the southern border of the cut-over lands for instance Barron and Price County in Wisconsin. They still have a fairly vigorous agriculture as their main support. Population has declined, as has the number of farms, but the size of the latter is much bigger than originally and they have been able to keep level with development elsewhere.

Borchert, J.R., and D.P. Yaeger, 1968 *Atlas of Minnesota, resources and settlement* (Minneapolis).

Garland, John H., ed., 1955 *The North American Midwest: A Regional Geography* (New York and London).

Helgeson, Arlan, 1962 *Farms in the Cut-over: Agricultural Settlement in Northern Wisconsin* (Madison).

Hoglund, A. William, 1960 *Finnish Immigrants in America 1880–1920* (Madison).

Jesness, O.B., and R.I. Nowell, 1955 *A Program for Land Use in Northern Minnesota* (Minneapolis).

Kolehmaninen, John I., and George W. Hill, 1965 *Haven in the Woods: The Story of the Finns in Wisconsin* (Madison).

Martin, Lawrance, 1932 *The Physical Geography of Wisconsin* (Madison).

Michigan's Agriculture (K.T. Wright and D.A. Caul), 1964 Michigan State U.

Nelson, Lowry, 1960 *The Minnesota Community: Country and Town in Transition* (Minneapolis).

Pincus, Howard J., ed., 1962 Great Lakes Basin, Am. Ass. for the Advancement of Sci. (Washington).

Senninger, Earl J., ed., 1970 *Atlas of Michigan*, 3rd ed. (Flint, Mich.).

Sobotka, Stephen P., *Profile of Michigan* (New York).

Wisconsin's changing population, 1942 U. of Wisconsin Bull. (Madison).

Karst Geomorphology
La géomorphologie karstique

---

s0501
Development of limestone caverns
D.C. FORD   *McMaster University, Canada*

For a long while a central problem of the origin of solution caves in carbonate rock has been the nature of the long profile, i.e., the development of cave systems in the dimensions of length and depth. Early writers contended that the bulk of development would occur in the vadose water zone *above* a pre-established watertable or piezometric surface (Warwick 1962). In 1930, W.M. Davis argued that most cave systems develop at random depth *beneath* a watertable; Swinnerton (1932) and Rhoades and Sinacori (1941) advanced the view that major caves develop *along* a pre-established watertable or parallel to it and at shallow depth below.

The three arguments are mutually contradictory but each was presented as an explanation of the general case; i.e. the authors presumed that a majority of limestone caves have developed in the particular manner that they outlined. Many recent regional studies have supported the 'watertable' argument (Sweeting 1950; Davies 1960), but no conclusive evidence has been adduced.

The present author has investigated more than 500 caves in limestone cavern regions of Europe and North America. The morphology of many caves has been mapped and studied in detail (Ford 1964, 1965, 1971). The principal finding is that there is no one general case of limestone cavern development. Rather, there are three common cases – the predominantly vadose cave, the deep phreatic cave and the watertable cave; and one special case – the cave in a true artesian setting. Between the sinkpoints and a spring, an accessible cave may be wholly of the vadose type, the deep phreatic type, etc., or it may be a combination of vadose type in the higher parts with deep phreatic or watertable type below. In exceptional cases, vadose, deep phreatic, and watertable cave elements may all be found in a single system although all will not have been created at the same time, e.g. Swildon's Hole, sw England (Ford 1965).

Three sets of factors interact to determine which of the common cases, or which combination of them, will apply in a given situation. The factors are: the frequency of fissures significantly penetrated by groundwater, the structural altitude and lithology of the host rock, and the local relief between sinkpoints and springs.

In most cavernous limestones only particular parting planes, joints, and faults possess significant permeability. The frequency of these features and of their interconnection within the host rock determines whether a deep phreatic or a watertable cave will develop. It is difficult to assign absolute dimensions to the frequency but states of 'low' or 'high' frequency can be defined. Where fissure frequency is low, the probability that groundwater streams will be able to follow a route of continuous shallow gradient to exit springs is also low. Much of a cave will be composed of deep loops ('phreatic loops'), below local piezometric surfaces. Where fissure frequency is high, shallow 'watertable' exit passages predominate.

Initial fissure frequency is a product of the bedding characteristics, diagenetic and tectonic history of limestones. It varies within and between regions and often varies from zone to zone in a given formation. Younger limestones, particularly tertiary ones, appear to have higher fissure frequencies, ceteris paribus. Despite a characteristic high vug porosity, many reef limestones have low effective fissure frequency.

Once karsting is initiated in a region, it is important to appreciate that groundwater continues to circulate in other fissures about and below cavern passages of accessible dimensions whilst the latter are developing. Thus fissure frequency increases with the passage of time in a karst region, increasing the probability that the watertable type of cave will be found in the younger parts of a multi-phase system.

The structural attitude or gradient in well-bedded limestones is an important secondary determinant of the type of cave which will develop. Bedding parting planes in a limestone mass are continuous entities, extending to the boundaries of the mass. Joints are smaller features with discrete terminations. Where strata are flat-lying, groundwater intake is via the discrete joints and discharge via the continuous parting planes. It is probable that there will be no great measure of descent via the discrete joints before a continuous exit line in a parting plane is discovered and utilized.

The converse is true where stratal dip is steep. Groundwater intake is guided to great depth by continuous parting planes. The likelihood of shallow discharge is reduced by the discrete character of joints that must guide the discharge route. Thus, where fissure frequency is constant, the watertable-type cave is more probable in flat-lying strata and the deep phreatic cave in steeply dipping strata. A dip as gentle as 5° constitutes 'steep' dip in one instance, Castleguard Cave, Canadian Rockies (Ford 1971).

Bands of shale and chert and beds of dolomite are common in many limestone formations. Where strata are flat-lying, groundwaters are often perched upon them (Waltham 1970). They increase the probability of a watertable-type cave developing. Where stratal dip is steep, such obstructive layers often assume the role of an artesian trap, compelling groundwater to pass to great depth before a breach is found.

Very tightly folded strata (the scale of folding is smaller than the scale of accessible caves), are associated with high fissure frequencies. The watertable-type cave is probable (Davies 1960).

The concept that a watertable at depth in limestone precedes significant cave development is wrong in most instances. It occurs only where fissure frequency is exceptionally high. Normally, the watertable is at the surface when cavern development commences. It falls as conduits expand to take all available water. A substantial fall (tens or hundreds of metres) may occur whilst conduits are still very small. Their enlargement to accessible dimensions will then be of a vadose character although the system skeleton is phreatic, e.g. G.B. Cave, sw England, where 95 per cent of the void space is a product of vadose erosion (Ford 1964).

The extent of fall of the watertable to a quasi-permanent rest level in a developing cave system is a function of the local relief between sinkpoints and springs, combined with fissure frequency. The greater the relief or the frequency, the greater is the fall and thus the proportion of vadose cavern in the entire system.

Davies, W.E., 1960 Origin of caves in folded limestone, *Nat. Speleol. Soc. Bull.* 22, 5-18.

Davies, W.M., 1930 Origin of limestone caverns, *Geol. Soc. Am. Bull.* 41, 475-648.

Ford, D.C., 1964 On the geomorphic history of G.B. Cave, *Univ. Bristol Speleol. Soc., Proc.* 10, 149-88.

– 1965 The origin of limestone caverns: a model from the central Mendip Hills, England, *Nat. Speleol. Soc. Bull.* 27, 109-32.

– 1971 Geologic structure and a new explanation of limestone cavern genesis, *Cave Res. Grp. Gt. Brit., Trans.* 13, 81-94.

Rhoades, R.F., and M.N. Sinacori, 1941 Patterns of groundwater flow and solution, *J. Geol.* 49, 785-94.

Sweeting, M.M., 1950 Erosion cycles and limestone caverns in the Ingleborough District, *Geog. J.* 115, 63-78.

Swinnerton, A.C., 1932 Origins of limestone caverns, *Geol. Soc. Am. Bull.* 43, 663-93.

Waltham, A.C., 1970 Cave development in the limestone of the Ingleborough District, *Geog. J.* 136, 574-85.

Warwick, G.T., 1962 The origin of limestone caves, in C.D. Cullingford, ed., *British Caving* (London), 55-82.

s0502
Comparison of littoral karren with other karren
FRANZ-DIETER MIOTKE   *Technische Universität, Hannover, West Germany*

Carbonate-cemented Pleistocene dune sands occur between Holocene and contemporary dunes along the north coast of Puerto Rico near Arecibo. The sand grains consist chiefly of limestone fragments; quartz and volcanic rock fragments derived from the interior of the island are rare. The cemented dunes are up to 25m thick (Briggs 1968; Kaye 1959).

The solution of the aragonite or calcite cement of these aeolianites and the outwashing of loose sand grains as well as the solution of the limestone grains themselves produces littoral karren forms at Pta. Morilla de Puerto de Arecibo ranging from a few millimetres to a few metres wide. Although solution of limestone is the primary cause for the development of these karst landforms – their geometry is due to the flow dynamics of the corroding water.

For the sake of brevity, the complex chemistry of solution by salt water, rain water, or biological processes is not discussed. Emphasis is instead on the flow dynamics of such waters as the causal agent for the karren. It will be shown that flowing water may erode and/or corrode rock or sediment under various conditions to produce astonishingly similar microforms.

In order to explain the genesis of the littoral karren, the flow dynamics within the surf must be analysed: every breaker throws seawater onto the coast. As the water returns to the sea it causes both currents and swirling eddys that, because they have longer contact with the rock and because they are often confined to near-parallel runnels, are much more effective in dissolving the rock. The areas between the aligned runnels are corroded only by the incoming surf, and for a shorter period of time. The consequent relief synergistically promotes further turbulence in the surf and thus a greater rate of solution. The water returns to the sea in thin films and rivulets. Depending upon the local surf conditions and the pre-existing relief of the coast a great variety of scallops, kamenitzas, trittkarren and rinnenkarren will be developed on the rock by the numerous water movements. All these forms are partly the result of the direction of landward movement of the surf – a function of the prevailing wind direction, but primarily the result of water flowing down slope and returning to the sea. Both directions of flow are determined by sea swell and by tide and because each of the two directions are quite different in their solution activity, the developing karren have a very complex morphology. During their course of development they necessarily dissect each other and leave bizarre remnants.

During low tide, when the waves do not reach the karren, the seawater in local ponds or kamenitzas, corrodes the rock until saturation is reached. Also, rainwater flowing along the surface of the littoral karren is capable of corrosionally modifying them. Guided by, and indeed actually controlled by the relief of the existing karren, rainwater can only modify these forms.

If the limestones are very coarse-grained and porous as certain coral reefs are coarse-structured, a continuous aligned flow is possible only for short distances because the water penetrates the rock as if it were a sieve and corrosionally enlarges the interconnected pores so they become small conduits. I have witnessed such flow also on marine terraces of Nassau in the Bahamas and in the Dominican Republic.

Above the direct activity of the breakers the limestone surface can only be reached by sea water spray. Every big breaker throws a cloud of seawater drops on the rock. Smaller water drops are drifted by wind into the surrounding vicinity, so that surface of the limestones is nearly always wet. Smaller kamenitzas are water-filled. Within this spray zone a very intensive solution activity occurs. The small droplets of water wash CO_2 out of the air and become even more aggressive than non-dispersed seawater. Presumably, seawater that has been diluted by rainwater is very aggressive.

Littoral karren can be classified according to their height above the sea level into four gradational and overlapping zones of occurrence:

1. *Zone of breakers* – surf usually or even rarely covers the limestone. All kinds of littoral karren may develop; mechanical erosion may occur.
2. *Spray zone* – gradational between the breaker zone and dry land; corrosion is by sea water spray and rain water; sometimes the surface is dry; water slowly returns seaward or is impounded in kamenitzas, some of which are aligned and some of which are filled with soil. The development of such soil-filled pits may be aided by reactions with humic acid and soil CO_2.
3. *Zones of fossil littoral karren* – because of a change in sea level the once-actively-forming littoral karren may be: A. drowned; B. elevated above the spray zone.

Non-littoral karren are those that are too high above the sea or too far from it to have their development influenced by sea water; they are formed solely by meteoric water and/or soil water.

In contrast to littoral karren, non-littoral karren are formed by water that flows only in one direction, following the gradient; but in both types the water quantity and flow velocity have greatly varied from time to time. The origin of non-littoral karren is less complex because only one flow direction is involved.

The kamenitzas that are formed exclusively by standing water are classified separately from kamenitzas and similar depressions that are formed mostly by moving water. All these depressions are developed not only in limestone but also in sandstone, granite, and other relatively insoluble rocks.

All types of karren are subject to the influences of lithological variance (homogeneity and purity of limestone, etc.) and structural variance (joint pattern, thickness of beds, etc.). Internal rock differences become apparent because of locally greater solution or the result of other weathering processes. Pure limestone generally shows sharper, acute karren. Relatively impure limestone shows more rounded forms.

If the limestone surface is steeply inclined rinnenkarren develop; if a low gradient surface is smooth, meandering rinnenkarren develop. Again it is clearly seen that solution is necessary but is not what gives the karren its form! Rinnenkarren similar to those in limestone may develop also in recent marine muds, granite residuum, clay, and shale. The dimensions may range from a millimetre to a metre. Microkarren (rillenstein) may develop in dry climates where small quantities of water condense as dew on relatively soluble rocks. The films of dew water spread along the rock surface; surface tension plays an important role there and solution of the rock forms a close net of micro-meanders on the formerly smooth pebbles. Steeper zones on pebble margins have deeper and less sinuous meander karren because the waterfilm moved more; flat zones have extremely meandering rillenstein. Similar relationships between slope angle and karren morphology can also be seen on other non-littoral karren and on littoral karren.

Since corrosion, as discussed by Miotke (1968, 119–24), is not sufficient to form the karren morphology, one must also pay special attention to the flow dynamics of the corroding waters. It is known that the shear stress caused by water flowing past a rock surface increases with higher flow velocity and thus favours increasing erosion. The conditions which locally lead to quicker, and therefore greater, limestone corrosion where there is a high flow velocity are more complex.

When a turbulently flowing stream passes over a rough, irregular surface, there is increased turbulence on the downstream side of a slight projection or within a slight depression that is on this surface. At such points the flow velocity increases, so the boundary layer – that portion of a fluid flowing past a surface which, in the immediate vicinity of that surface, has a reduced or zero flow due to the forces of adhesion and viscosity – is locally thinner and the solution rate is locally greater than it is on the upstream side. Downstream from, for example, a projection, vortices occur and a cavity tends to develop behind the projection; it continues to develop even after the complete destruction of the projection. Goodchild and Ford (1971) have experimentally confirmed the theoretical analysis by Curl (1966) of the clear relationship of flow velocity to the form and size of scallops.

Why is the rate of solution so intimately related to the flow velocity of the water? The thickness of the boundary layer has a dominant influence on the rate of diffusion of the dissolved calcium. The rate of diffusion (i.e. the rate of solution) is inversely proportional to the thickness of this boundary layer. Because of the shear stress transmitted by moving water, the boundary layer becomes thinner where a higher flow velocity produces a higher shear stress. If at a certain place on a rock surface the flow velocity is higher, solution will be more rapid than at nearby places with a lower flow velocity. The flow conditions determine the morphology of the karren forms. But it must be remembered that natural flow conditions are highly variable; thus karren forms are the complex result of summation and averaging of all the different flows that had operated for a certain time. It is fully realized that local structure and lithologic differences are also important controls on the morphology of karren.

Because the fluid dynamics of flowing water – be it flowing on limestone or other kinds of rock – must be alike, it is not astonishing that erosion and corrosion forms sometimes look so alike.

James Quinlan.

Briggs, R.P., 1968 Geologic map of the Arecibo Quad., Puerto Rico: *US Geol. Survey, Geol. Quad. Map* I-551.

Curl, R., 1966 Scallops and flutes, *Cave Res. Grp. Gt. Brit. Trans.* 7, 121–60.

Goodchild, M.F., and D.C. Ford, 1971 Analysis of scallop patterns by simulation under controlled conditions, *J. Geol.* 79, 52–62.

Kaye, C.A., 1959 Shoreline features and quaternary shoreline changes, Puerto Rico, *US Geol. Survey. Prof. Paper* 317-B, 49–140.

Miotke, F.-D., 1968 Karstmorphologische Studien in der Glazial-überformten Höhenstufe der 'Picos de Europa,' Nordspanien, *Geographischen Gesellschaft zu Hannover, Jahrbuch, Sonderheft* 4.

s0503
Alternate solutional and nonsolutional effects in cavern genesis
GEORGE F. OZORAY *Research Council of Alberta, Canada*

The characteristic and decisive process in karst development is solution. Caverns are characteristic karst phenomena. However, the giant caves of underground river systems are shaped by underground corrasion. Corrasion in itself is not a karstic process, but underground corrasion is deeply connected with karst, because such corrasion: apart from special cases, acts inside soluble rocks; is often supported by solution; is a genetic consequence of solutional processes (before it can assault the inside of the limestone masses it is in need of pre-existing underground channels; these channels are initially a result of solution).

However, solution itself attacks along pre-existing paths. Let us keep in mind that most of the controlling factors and genetic processes of cavern-development act both in karstic and non-karstic rocks (Ozoray 1962). Non-karstic caverns can be found in limestone or dolomite (Ozoray 1963). Most numerous of them are tectonic crevices. Their unaltered presence tells us about a genetic void: about the absence of moving, corrosive water. Otherwise solution modifies or wipes out the original features. Tectonic crevices in limestone are embryonic karst-caverns, but may remain mere abortive possibilities.

Solution by moving cold water has been discussed in many theories of cavern genesis, putting the emphasis on the water table, on the vadose or the phreatic phase or on their alternation. A clever theory even projected the effect of deep groundwater flow to the water table (Rhoades and Sinacori 1941). Really, modern hydrogeology theoretically explains (Hubbert 1940; Tóth 1963) and practically proves (Geraghty 1960) that groundwater flows from surface to surface through considerable depth.

Hot-water solution, often under high pressure and helped by H_2SO_4 and by abundant CO_2, results in a rich diversity of cavern forms (Ozoray 1961, 153).

Many of the antagonistic genetic theories may be valid under relevant circumstances. There is no master theory to explain the genesis of all caverns. Monographs, aiming to grasp general understanding, usually agree up to that point (Thornbury 1969, 336; Scheidegger 1970, 413). Solution caverns can develop below, at and above the groundwater table, in different kinds of rocks, by cold or hot waters or even by vapours, affected by many processes and controlled by faults, joints, bedding, orographic, hydrographic and climatic conditions. Some general remarks, however, can be hazarded: (1) there are caverns, even in limestone, in which solution has not played an important role; (2) for most caverns solution is a crucial genetic factor; (3) caverns formed entirely by solution are rare – usually other genetic processes have expanded the size of the caverns; (4) fractures, joints and faults are important controlling factors.

The developmental phases of some caverns are: (1) preformative (tectonic) phase: the formation of the controlling joint-system or of tectonic crevice-caverns; (2) solutional phase: solution by cold or hot water; (3) corrasional phase. A fourth phase of rock-fall and collapse might be added, but cave-volume increases in this way only as long as the removal of debris occurs.

It is notable that: (*a*) there is a solutional phase between two non-solutional phases; (*b*) the later non-solutional phase is fundamentally different from the earlier one; (*c*) the solutional phase is a genetic necessity for a well-developed underground river-system type cave.

During each phase one process predominates but the others may be active. The dynamic balance between the counteracting processes can be very delicate. For instance, there may be simultaneously occurring precipitation by dripping water and solution by flowing water, lateral and vertical erosion and sedimentation, rock falls and active faulting. The balance is further complicated by factors such as seasonality and by the mixing of surface water, entering by swallow holes, and groundwater in various ratios (Brown, Wigley and Ford 1969).

Each cavern and each karst-area has a unique history, rooted in the history of the geologic framework. There are no pure cycles. Any phase may be interrupted and may start again under entirely changed circumstances. Theories of cavern genesis can, therefore, promote understanding of actual case-histories, not as general models, but as tools, selected from the kit according to the special need.

R. Bibby, R. Green, O. Tokarsky and J. Tóth, Contribution no. 553, Research Council of Alberta, Edmonton, Alberta, Canada.

Brown, M.C., T.L. Wigley, and D.C. Ford, 1969 Water budget studies in karst aquifers, *J. Hydrol.* 9, 113–6.
Geraghty, J.J., 1960 Movement of contaminants through geologic formations, *Nat. Water Well Assoc., Techn. Div. Acts*, 33–42.
Hubbert, M.K., 1940 The theory of groundwater motion, *J. Geol.* 48, 785–944.
Ozoray, G., 1961 The mineral filling of the thermal spring caves of Budapest, *Rassegna Speleologica Italiana, Mem.* 5, 2, 152–70.
– 1962 The genesis of non-karstic natural cavities as elucidated by Hungarian examples, *Karszt- és Barlangkutatás* 2, 127–36.
– 1963 Einige genetische Probleme der Höhlen in Karstgesteinen an Hand von Beispielen aus Ungarn, Internat. Congr. of Speleol., 3rd, Vienna 1961, Proc., 2, 95–8.
Rhoades, M., and M.N. Sinacori, 1941 Pattern of ground-water flow and solution, *J. Geol.* 49, 785–94.
Scheidegger, A.E., 1970 *Theoretical Geomorphology*, 2nd ed. (Berlin).
Thornbury, W.D., 1969 *Principles of Geomorphology*, 2nd ed. (New York).
Tóth, J., 1963 A theoretical analysis of groundwater flow in small drainage basins, *J. Geophys. Res.* 68, 4795–812.

s0504
Geological environment of cave development
A.C. WALTHAM   *Trent Polytechnic, UK*

In any area of limestone karst, the regional topography, climate, hydrology and limestone solubility will all influence the possibilities of the very existence of underground drainage and cave systems. However the detailed pattern and form of cave development is disproportionately controlled by various geological parameters.

Almost invariably the details of cave passage morphology reveal that there has been some form of stratigraphic control over patterns of development. Non-calcareous horizons such as shale or sandstone are very important even when forming only thin partings in a limestone succession. In northwest England nearly all the sub-horizontal vadose caves have roofs formed by single, almost smooth, bedding planes, immediately above the shale horizons. The soft impervious shale beds guided the drainage along their tops in the initial pattern-forming stages of slow downward percolation; the cave streams, subsequently developed on them, rapidly abraded them, and large cave passages then developed in the limestone just below the shales. Sandstone horizons in the same area of northwest England had a similar effect in the initial stages, but their mechanical strength resisted abrasion so that the vadose streams still flow on top of the sandstones, but in enlarged caves.

Initial hydrological penetration of these thin shale beds was most commonly via open fractures, but also through areas of non-deposition. Dye testing has proved that streams also penetrate the sandstone beds of up to 5m thickness. Faults, common in the area, would offer a favourable locus of such penetration, but the one observed example is via a local facies variation.

In the Pokhara Valley of Nepal, essentially vadose caves have formed in stratabound limestone beds immediately below bands of conglomerate that now constitute the passage roofs. A very high primary porosity of the conglomerate meant that the vadose water initially flowed along the base of the clastic beds; only subsequently did the caves enlarge into the limestone. Caves may also be preferentially located near the tops of limestone beds where an impervious caprock has defined initial flowpaths of dominantly upward-moving percolation in a deep phreas.

Many cave passages were clearly initiated along bedding planes. The reason for this concentration of development is not always apparent – it may be due to very thin shale partings, open bedding-place joints or differential solubility of adjacent beds.

Lithological variation within the limestone succession must influence solubility rates and hence loci of cave development. Rauch & White (1970) have demonstrated that high dolomite and non-carbonate contents, and large grain size, restrict cave development in some Appalachian limestones. The influence of dolomite content seems universally applicable, though examination of limestones from Great Britain indicate that the impurity content and grain size do not bear a constant relationship to solution rate. Primary porosity and permeability are generally low in cavernous limestones but where high may control cave development to the disregard of all other lithological parameters. This is well demonstrated in Kashmir where some secondarily dolomitised, very porous beds are preferentially dissolved from a limestone sequence.

Joints are the most important initial openings along which karst water may penetrate and consequently the influence of their pattern remains impressed on subsequent cave development. Vertical, or steeply inclined, joints commonly characterize avens, waterfalls and sharp descents in cave passages; in northwest England almost every vertical feature, connecting the sub-horizontal cave passages, has formed on, or receded from, a joint. Also, numerous cave surveys show how joints control the plan-form of passage development when the levels are restricted by horizontal features.

Faults exercise similar controls to joints, but faults are more thinly dispersed and individually on a larger scale: many of the world's largest cavern features owe their origins at least in part to concentrated erosion in fault zones. There exist examples where mineralised faults have acted as hydrologic barriers in karst development, and yet in the English Pennines large caves are common on dolomitised, essentially less soluble, fault zones. The directional influence of breccia zones, fault-opened cavities and even slickensides are difficult to assess and warrant further research.

Largely due to the major influence exercised by stratigraphic parameters, the amount and direction of the dip of the limestone is most important. On a worldwide scale, there is more extensive cave development in limestones of low dip than in those approaching a vertical inclination. This is at least partly due to the increased vertical transmissibility down the steeply inclined bedding, which carries the percolation water efficiently down to a complex phreas and precludes the adequate drainage concentration which may form cave systems; a contributory factor to the situation is that vertical bedding almost excludes the possibility of the existence of an overlying impervious catchment area supplying corrosive water to the limestone outcrops.

Particularly in gently dipping limestones, dip direction is most important. Vadose stratabound caves must flow downdip, and consequently concentrate their drainage down the plunge of synclines. Furthermore the tendency for fractures to open along fold axes facilitates axial drainage of this type.

Almost any dip angle is significant in the pattern of phreatic cave development. The majority of phreatic drainage routes have sawtooth profiles, and in many caves the water flows alternately up and down bedding planes and joints; the relative scale of these phreatic lifts will then be a function of the relationship between hydrologic and geologic gradients.

Undoubtedly, the groundwater rest-level, whether or not it resembles a classical water-table, has influenced cave development in a number of areas around the world. However, detailed examination of many cavernous regions formed in sub-horizontal limestones shows that the horizontal influences on cave development exercised by the water-table are commonly subordinate to the stratigraphic controls. The caves of NW England, once thought to be of water-table origins, are in fact mainly stratimorphic. In contrast, steeply dipping limestones, where stratigraphic parameters can have little influence in the vertical plane, seem particularly prone to containing caves formed at, or near, a water-table.

As controlling factors of cave development, various geological parameters are of widely different importance. Major structures are significant, and the concentration of karst drainage in synclines is a general tendency. However, the presence of joints is probably an even more basic influence. The lack of cave development in some Himalayan limestones is most reasonably ascribed to a deficiency of significant joints in the plasticly deformed rock.

Furthermore, contrasting geological features influence the patterns of cave development within the different hydrological environments. In parts of the northwest England karst, vadose caves are dominantly stratimorphic, while phreatic caves, within similar geological terrains, are most commonly oriented along fault lines (Waltham 1971). Consequently a change of cave pattern is commonly found where the formative cave streams enter, or entered, the phreas.

The distribution of 'water-table cave systems' indicates the importance of geological detail – they appear to be common only where horizontal geological features are minimal. Undoubtedly the geological environment exercises considerable control over cave development, and it can only be unreal to theorise on the complex problems of cave genesis without consideration of geological detail.

Rauch, H.W., and W.B. White, 1970 Lithologic controls on the development of solution porosity in carbonate aquifers, *Water Resources Research* 6, 1175–92.

Waltham, A.C., 1971 Controlling factors in the development of caves, *Cave Res. Grp. Gt. Brit., Trans.* 13, 73–80.

s0505
High altitude karst in the Himalaya
A.C. WALTHAM   *Trent Polytechnic, UK*

Though dominantly formed of pelitic rocks, the Himalayan Mountains include minor, but topographically important, outcrops of massive limestone. The best developed karst areas are in the foothills of Assam and Northern India, at altitudes less than 2500m, but this paper concerns the geomorphology of two more elevated limestone regions in Indian Kashmir and Central Nepal.

In Kashmir there are considerable thicknesses of strongly folded Triassic limestones, unfortunately with important local lithological variations, some being almost pure dolomite. Altitudes of the outcrops range from below 2000m to around 4000m, in the Vale and along the Sind Valley just to the north. There is a general restriction of karst development due largely to the dolomitic nature, and locally to the vertical attitude, of the thin-bedded limestone, but also related to the extent of frost shattering in the higher regions.

Karren forms are rare at levels above 3000m except for restricted occurrences of various types. Below this level rillenkarren are fairly common and the dominant solutional form – there is noticeably less mechanical weathering.

Dolines and closed depressions are almost non-existent in the region – there is just a single group of boulder-floored depressions at an altitude of 3700m. The extent of karst drainage varies directly with altitude. In the areas over 3000m only a few small karst springs are known, though extensive drainage takes place through and under the numerous alluvial fans, composed largely of limestone fragments. However, at the eastern end of the Vale of Kashmir, at about 1700m asl, are five large karst risings (mean flow of each about 1–2 cumecs). They are fed largely by percolation water, and the only known sink is in the bank of a river bed where it locally swings to the margin of its wide alluvial plain. This feeds one spring 13km away and 200m lower, and the flow characteristics may suggest significant large cave development. Some small fossil phreatic caves exist in the Vale, though at higher altitudes the only known cave is at Amarnath, at 3900m; this is a single 20m diameter frost pocket, not a solution cave.

The Kali Gandaki River cuts right through the main Himalayan chain in what is probably the deepest valley in the world, and exposes outcrops of two different types of limestone, both with considerable vertical ranges.

At Jomsom, in the almost arid rain-shadow north of the main ranges, are outcrops of Jurassic dolomitic limestones at altitudes of 2800–3800m. In the cold dry 'Tibetan' climate, solution of the limestone is limited, and its effects are further minimised by the great extent of frost shattering. Rillenkarren are dominant, but become steadily rarer with increase in altitude, and small irregular pitting is the only other common solutional form. Other, larger, karst forms, including some small caves, occur only in isolated cases and are mainly fossil features dating back to a more pluvial climate. A group of springs at Jomsom are at least partly fed by juvenile water, and the importance of karst groundwater circulation is difficult to estimate.

Fifteen kilometers further down the Kali Gandaki Valley, the Nilgiri Limestone forms a belt of outcrops continuous from the valley floor at 2600m almost to the summits of Dhaulagiri and Annapurna, each at over 8000m. The limestone is Ordovician in age, about 2500m thick, and occurs in a huge series of recumbent folds; it is almost completely non-dolomitic, but very impure, with insoluble content ranging from 15 to 35 per cent, and this has undoubtedly restricted its solutional erosion. Furthermore, it has been plastically folded and very little jointed; these characteristics have been responsible for an almost complete lack of underground karst development.

Karren are only patchily distributed, occurring almost entirely on a few of the more massive, and purer, beds. Rundkarren are almost the exclusive form and are always associated with the thicker areas of glacial drift and vegetated soils; wandkarren are developed on cliffs on the same massive beds. None are present above altitudes of 4000m except for a few scattered outcrops with rillenkarren up to 4500m, and minor amounts of solutional etching only continue to about 4800m, above which mechanical shattering is the only visible effect of weathering.

Closed depressions are rare, only one group of deep shakeholes being observed at 4250m. There are numerous small springs along the margins of the alluvial plain flooring the Kali Gandaki Valley, though they are quantitatively insignificant compared to the surface runoff from the limestone. Much higher are two isolated groups of large karst springs – the White Peak Risings at 3700m on the flanks of Dhaulagiri, and the Hum Khola Risings at 4100m on the slopes of Nilgiri. In both cases the water rises from narrow fissures with no associated cave development, and the large catchment areas are devoid of active sinkholes; the rising water appears to be of karstic percolation origin.

Caves are almost non-existent in the Nilgiri Limestone. The only caves in the area are formed in the superficial deposits; a locally famous tufa cave occurs on Dhaulagiri at an elevation of 3000m, and similar, smaller, rock-shelters range up to 4400m. The main rivers draining off the mountains cross the limestone outcrops in spectacular gorges, hundreds of metres deep and only a few metres wide, perhaps giving some indication of the size of vadose trenches which would have been formed underground had the limestone been sufficiently fractured to initially engulf these streams.

Undoubtedly, geological factors have largely been responsible for the limited extent of karst development in both described areas in Kashmir and Nepal. However, the ranges of altitude with uniformity of lithology of the limestones do make some conclusions valuable.

The extent of solution in all areas decreases with increases in altitude. This was best revealed by numerous analyses of surface, spring and seepage waters on the Nigiri Limestone outcrops in Nepal, where the solute concentration shows a remarkably linear inverse relationship to elevation (Waltham 1971); total hardness of the water ranged from about 200ppm (expressed as $CaCO_3$) at 2700m, to less than 50ppm at around 5000m. Exceptions to the general uniform trend are provided by anomalously low glacial waters, and two samples of spring waters, with unusually high hardnesses, which however, were probably so due to prolonged circulation in an atypically large phreas.

There is little evidence to support a theory of vertical zonation of karst types. Nowhere is there a zone of doline karst below a zone of karren karst, though the outcrops do descend into the temperate climatic altitude zones. Nearly all the observed karst forms range in elevation up to the permanent snowline at about 4600m, above which solutional erosion appears to be almost negligible. In Kashmir the largest karst springs are restricted to the lower levels, but in Nepal the two largest known underground drainage systems are the two at the highest altitudes.

Similarly, elevation has little influence on the distribution of the karren forms. On the Nilgiri Limestone rundkarren is dominant and there is scarcely enough rillenkarren at higher levels to designate a more elevated rillenkarren zone. Accepting the lithological control, the distribution of rundkarren is related to soil abundance and type; there is thick till and soil cover on much of the Nilgiri Limestone. Conversely, rillenkarren is commoner on the barer limestone outcrops of Kashmir and on the arid exposures of the Jomsom limestones.

Clearly, geomorphological examination of further Himalayan limestone is needed, but the present study does suggest that altitude has little control over the karst forms, except to almost entirely preclude their development above a level near the permanent snowline.

British Karst Research Expedition to the Himalaya, 1970.

Waltham, A.C., 1971 Full report of the British karst research expedition to the Himalaya, 1970, Full report, Geology Dept., Trent Polytechnic, Nottingham, England.

s0506
A classification of cave sediments
THOMAS E. WOLFE   *State University of New York at Buffalo, USA*

There is no generally accepted classification of cave sediments. Various authors have grouped cave sediments within their personal interests of study. Kukla and Ložek (1958) classified cave sediments on the basis of source of material found in the caves. They used the terms 'autochthonous' (derived from within the cave) and 'allochthonous' (derived from outside the cave). They also divided the cave sediments into 'entrance facies' and 'interior facies.' White (1964) proposed 'clastic' (including autochthonous and allochthonous fills) and 'chemical' deposits. Frank (1965) refers to 'tardigenic' (slowly deposited by infilling through small cracks) and 'torrigenic' (rapidly deposited through larger openings). Link (1966) proposed a textural classification using a three end-member triangle of clay, silt and sand.

In the broadest sense, all material found in caves other than the host rock in situ can be considered to be 'cave sediment.' In this study of fluvial sediments, process rather than source, texture or position of the deposits is the basis of their classification. The following classification utilizes the terminology of previous investigators with an emphasis on geomorphic processes responsible for deposition in the caves.

I. CLASTICS
A. Gravitional fills: products of weathering and gravitational transfer (without a transporting agent). (1) Infiltrates: material derived from outside the cave. This includes entrance detritus, soil, sinkhole fills, material which enters the cave via vertical shafts, joints or faults which does not show evidence of transport (see erosional fills). (2) Breakdown: material derived from within the cave. This includes fine weathering detritus, large blocks and whole sections of ceiling collapse. ('Autochthonous' fills of Kukla and Ložek 1958). B. Transported fills: products of erosion which show evidence of transport by water, wind or ice. (1) Fluvial sediment: Material which shows evidence of stream transport. This includes all material derived from within or outside the cave which shows rounding, graded bedding, a systematic vertical variation in grain size, or any particle or structural features characteristic of fluvial transport. (2) Glacial sediment, (3) Aeolian sediment, (4) Marine sediment, (5) Lacustrine sediment: these are rarely found in most humid temperate, well-drained, inland karst regions. They are also difficult to identify where other processes have altered them.

II. CHEMICAL DEPOSITS (after White 1964)
A. Carbonates. B. Evaporites. C. Manganese and iron hydrates. D. Ice. E. Phosphates.

III. ORGANIC DEPOSITS (not showing evidence of transport)
A. Floral remains. B. Faunal remains.

IV. ARCHEOLOGICAL DEPOSITS
A. Evidence of human presence or activity.

'Genetic' classifications have been criticized for their closeminded 'pigeon-hole' attitude of fixing a label on a given species and attempting to explain how deposits were laid down in the caves. The assumed explanation given may be at fault, i.e., a coarse grained poorly sorted 'fluvial deposit' may later prove to have been deposited by gophers (organic deposit). However, by attempting to explain the origin of the processes responsible for the deposit; the classification is more than a list of ordering of names and features. Such a classification is more than a list or ordering of names and features. Such a classification encourages scientific inquiry and investigation. The genetic approach is not finite and is subject to change. The use of modern field and laboratory techniques enable the geomorphologist today to make 'educated guesses' about process to the point where 'genetic' classifications are useful in furthering the study of splean deposits and an understanding of the processes responsible for their transport, deposition and reworking.

The following modifications were suggested as a result of the discussion following the presentation of this classification to the Cave Research Associates on 14 October 1971 at McMaster University, Canada.

(1) Where the work of several erosional agents is evidenced in clastic deposits, the one which appears to have the most recent effect upon the deposit is used to describe the deposit. As an example: stream transported and deposited glacial till found at the entrance of a cave in New York State would simply be referred to as – Transported: fluvial sediment. A more detailed description of provenance particle and mass properties could then follow.

(2) Organic deposits: A. floral remains or products of plants (living or dead); B. faunal remains or products of animals (living or dead).

(3) 'Organic deposits' refers to *in situ* remains. For example, a log jam in a syphon would be classified as – Transported: fluvial sediment.

(4) The classification of archeological deposits should be expanded.

This classification is a proposal; it is not meant to be final. The author welcomes comments and criticisms from interested readers.

Brain, C.K., 1962 Procedures and some results in the study of Quaternary cave fillings.

Collier, R.C., and R.F. Flint, 1964 Fluvial sedimentation in Mammoth Cave, Kentucky, *USGS Prof. Paper 475/D*, D141–D143.

Davies, W.E., and E.C.T. Chao, 1959 Report on sediments in Mammoth Cave, Kentucky, *Admin. Report, USGS*.

Ford, D.C., 1966 The sedimentary record at the Cheddar Caves, Somerset, England, *Bull. Nat. Speleo. Soc.*, 28 (2), 93.

Frank, R.M., 1965 Petrologic study of sediments from selected central Texas caves, MA thesis, U. Texas.

Kukla, J., and V. Ložek, 1958 On the problems of investigation of cave deposits (in Czech and English), *Czechoslovensky karas*, 11, 19–83.

Link, A.G., 1966 Textural classification of sediments, *Sedimentology*, 7, 249–54.

White, W.B., 1964 Sedimentation in caves – a review, *Bull. Nat. Speleo. Soc.*, 26 (2), 77–8.

Wolfe, T.E., 1964 Cavern development in the Greenbriar Series, W. Virginia, *Bull. Nat. Speleo. Soc.*, 26 (2).

s0507
Karst et préhistoire
PAUL FENELON *Institut de Géographie de Tours, France*

Durant une période assez courte, celle de la glaciation würmienne, et dans une zone qui s'étend de la péninsule ibérique á l'Anatolie, les tribus du Paléolithique supérieur trouvèrent dans les régions calcaires des conditions favorables à leur existence: les abris sous roche les protégeaient des intempéries, les balmes leur servaient de refuge; les grottes et les cavernes furent les sanctuaires où ils représentèrent scènes de chasse et symboles magiques. Leur civilisation disparut avec le retour d'un climat tempéré.

s0508
Cuban coastal plains and lowlands as a specific type of tropical karst
VLADIMÍR PANOŠ *Palackeho University, Czechoslovakia*

The Cuban insular platform consists chiefly of Laramide structures that include infolded relics of probably Acadian (Carboniferous) tectonic units that are mantled and fringed by post-Laramide rock complexes. These youngest Cuban tectonic units consist of marine, terrestrial, and mixed deposits that are mostly calcareous and characterized by offlap and many facies changes. They range in age from Late Eocene through Quaternary and are moderately warped but locally intensely faulted.

Because of recurrent Neocene tectonic activity in the Caribbean area the post-Laramide structures were irregularly subjected to temporary emergence and submersion. At the end of the Tertiary some marginal zones were considerably uplifted. These uplifted portions are characterized by large bare surfaces with ferruginous or allitic duricrusts and calcareous evaporitic coatings alternating with patches of ancient sedimentary and weathered mantles. Also during the Pleistocene, certain post-Laramide areas were alternately flooded and exposed in response to weak crustal movements and considerable glacioeustatic sea level fluctuations. These regions are now covered, usually by thick and continuous, marine, terrestrial, and mixed sedimentary mantles. Even during the Late Pleistocene and Holocene some regions emerged slowly above the sea level and are noted for neritic and littoral deposits as well as for dead coral barriers, fringing reefs, and peripheral cemented bars, all of which camouflage relics of the pre-transgressional exogenous landscape. However, other regions with the same morphology are still covered by salt and fresh-water swamps or invaded by a shallow sea. An exception is the low terrains that are the base of numerous small isles (cayos or keys) of the Cuban Archipelago or narrow peninsulas of the main islands, specifically of the Isle of Cuba and the Isle of Pines.

The post-Laramide structures represent initial constructional macroforms of present Cuban coastal plains and lowlands. As their mostly calcareous bedrock was exposed at least temporarily to morphogenetic processes, there developed in the warm and humid tropical climate a destructional landscape consisting of various surface and subsurface sets of corrosional, corrosional-erosional and corrosional-suffosional mesoforms. The variability of the individual morphological sets depends on a number of geologic and physiographic factors that modified the effects of the uniform climatic conditions.

The most important characteristic of these coastal plains and lowlands is an extensive level surface with absolute prevalence of negative (concave) karst forms and numerous cave systems controlled by bedding planes.

The surface of coastal plains and lowlands is due mostly to its origin by corrosion and erosion and partly by accumulation. It is flat or slightly rolling. The flats dip gently seaward or they are horizontal. The rolling areas consist of shallow basins in downwarped zones and of low domes or elongated ridges in uplifted regions. Only elevations of mendips, very numerous in certain areas, rise abruptly above the regional surface.

Individual portions of this surface now appear in various stages of development. The most advanced regions were transformed by corrosional-erosional processes to a subdued cuesta-landscape, on which subsequent lowlands display the characteristics of shallow poljes, valley-poljes, and wide blind or semi-blind valleys. The less advanced stage of geomorphic development may be recognized in regions that emerged a relatively shorter time ago and have the morphology of fossil coral barriers, fringing reefs, and consolidated calcarenitic bars – all of which may determine the distribution of various karst forms.

Bare flats, interstream uplands, and back slopes of cuestas are intensely corroded and characterized by extensive fields of deep clints with resistant duricrust or evaporitic coatings and isolated relics of former allogenic lateritic mantles. The lower surfaces are also mantled by a thick and continuous similar regolith or by peat, and they are well jointed. All parts of coastal plains and lowlands are dotted by numerous cylindrical or pot-shaped, flat-bottomed depressions that are identical with the *cenotes* characteristic of Yucatan and that were formed by vertically percolating waters and late stage perforation of subjacent caves. Mantled surfaces also display large but shallow, dish-like corrosional-suffosional depressions that developed in response to subsidence of disintegrated bedrock and downwashing of supra-jacent soil and other sediment.

There are two types of caves, both of which are controlled by bedding planes. One is characterized by dendritic sub-horizontal fluvial passages that occupy two or three levels. The systems on the uppermost level have streams usually just below the topographic surface, where those having lower levels proceed deep below the water-table as well as below sea level. The second type of cave is characterized by dome-like cavities, the flat floor of which is covered by fallen blocks, downwashed supra-jacent sediments, or occupied by lakes. These caves with the perforated or collapsed ceiling are the cenotes mentioned above. Usually they do not connect to horizontal fluvial passages but are drained by the extremely porous calcareous bedrock that is a highly integrated aquifer.

The basic hydrological characteristics of karst in the Cuban coastal plains and lowlands are the prevalence of subsurface drainage and the unusual composition of its subterranean water.

Only thickly mantled surfaces have discernible valleys with ephemeral or perennial surface streams. Such streams are very rare or entirely lacking on the naked bedrock surfaces. The valleys of allogenic streams disappear, usually in subsequent lowlands or in swampy synclinal basins. Some of the rivers flow in broad valleys that cross cuestas and other hills. Only a few valleys reach the coast; there they disappear in tidal swamps or flow directly into the sea. In higher inland belts of coastal plains the valleys display typical broad cross-sections with a wide channel and long, smooth, and moderately inclined slopes that resemble surfaces of outwash pediments.

Subterranean water occurs in the karst in more or less independent bodies determined by folding of the bedrock. Because of the intense primary porosity and secondary (karstic) porosity of calcareous beds, the ground water has a uniform and continuous water-table as well as the hydrodynamic characteristics of water percolating through loose and very permeable sediments. Since the karst water-table is situated usually just below the topographic surface, the coastal plains and lowlands have many lakes, conduit-flow and diffuse-flow springs, and extensive swamps. Diffuse-flow springs often create a pronounced spring-line along the inland margins of interior or coastal swamps. Numerous submarine springs occur at various distances from the coastline.

Moreover, the porosity of the karsted calcareous bedrock favours salt water intrusion far beyond the theoretical limit. Consequently the karst water bodies consist of two different layers, the fresh water layer and the subjacent salt water mass. Subsurface permeability is very high, and hydraulic connection and tidal fluctuations are detected far inland by diurnal fluctuations and subsurface karst water bodies. Considerable annual fluctuations of water levels in individual karst basins are caused by the seasonal distribution of precipitation.

Consequently, there can be no doubt that the sets of karst forms in Cuban coastal plains and lowlands should be classified as a distinct type of tropical karst that can be differentiated into various sub-types. This karst landscape and subsurface differs substantially from other types of karst landscapes that are also developed in the same seasonally humid tropical morphoclimatic zone.

Developing the Subarctic
La mise en valeur des régions sub-arctiques

s0601
Development in the fringe settlement zone of northern Sweden
PER PORENIUS *The Agricultural Board, Sweden*

The fringe settlement zone of northern Sweden is big. It extends from the northern tip of Sweden, 340km north of the Arctic Circle, to the county of Kopparberg. The length is about 950km, and it is 20–30km wide for a total area of some 200,000 square km. It is an interesting region because depopulation goes on at a rate uncommon in Europe and in strong contrast to the development in similar, nearby regions of Norway and Finland.

In 1960 the population was 240,000 and in 1970, 200,000. Only the mining centre of Kiruna shows an increase. However, the population balance is delicate and easily influenced as the total number of in- and out-migrants is four times bigger than the loss of population. Today's situation is better understood when some data from the colonization history are known.

The last glaciation ended about six thousand years ago. We know very little about the early millenia, but we know that 4000–3000 years ago a climate warmer than now prevailed and offered fine living conditions. During that time a population that might have been ancestors to the modern Lapps appeared in the region. About 1000 years ago the Lapps domesticated the reindeer but they still led a nomadic life. Colonization by a resident population within the fringe zone did not start until the 18th century.

As far back as can be recorded Lapps have been organized in clans – 'Lappbyar.' That organization is still maintained and since 1 July 1971 it has received solid official status through the new Reindeer Farming Legislation. Reindeer farming has always been an exclusive Lapp industry. Their rights were confirmed by law in 1886 and again in this new legislation.

Each 'Lappby' claims a right to reindeer pastures within a vast district. There are many families within a 'Lappby'. Traditionally each family follows the migratory movements of its herd from spring land to summer land to autumn land to winter land. The reindeer farming was run along oldfashioned but well established lines.

After the confrontation during the last few decades with industrialism, modern communications, and tourism, the Lapps have adopted a new system for reindeer farming – an extensive system. At the same time they adopted new ways of living and an economy on a monetary basis. Today the net Swedish reindeer stock numbers about 250,000 head. Pasture capacity is estimated at a maximum of about 300,000 head. Of the 10,000 Swedish Lapps, only about 2500 depend on reindeer farming today and only 800 are active males.

The modern reindeer farming has not been a success. From 1 July 1971 reindeer farming therefore is furnished with the rationalization support and the vocational and advisory facilities available to Swedish farming in general and besides that with special financial and technical support. Among other things rationalization will mean that the number of reindeer farming Lapps must be reduced by 50 per cent. An interesting detail is the occupational program launched in 1971 meant to encourage Lapps leaving reindeer farming to stay within the region and to provide them with income sources. This is contrary to Swedish occupational policy, but the scheme is, above all, supposed to help preserve the Lapp culture.

The dominant group in the zone represents an extension of the main Swedish population although a group in the northeastern parts of the zone has Finnish origin. The colonization and settling within the zone have always been controversial because of the confrontation with reindeer farming. During the 1750s the Swedish government established a 'Lapp Region Boundary' very nearly coinciding with the eastern boundary of the actual fringe settlement zone. Further settling to the west of that boundary should not take place. However, the urge to colonize was strong and supervision was weak and settlements to the west of the boundary became numerous.

In 1867 the government therefore set a new boundary further west – the cultivation boundary – and secured by legislation the rights for the Lapps and for the settlers. Settling to the west of the new line became very restricted and the reindeer herds were not allowed to appear to the east of it after May 15th and during the summer. In that way a vast region was opened for colonization. The old settlements were established singly or in small groups far apart. When more settlers now moved in, the net of settlements grew denser until a saturation of the region was reached. The colonization went on right up to the 1950s. The settlers were farmers and they ran their farms on lines of self-sufficiency. Hunting and fishing were important to them as well as to the Lapps.

During the late 19th century and onwards forestry provided a complementary occupation, but during the last ten years the rationalization within forestry has meant a decreasing need for labour. Furthermore big forest areas are now classified as zero-zones: zones where there is no profit in forestry.

In 1916 the colonization according to the Mountain Farm Scheme started to the west of the cultivation boundary. The idea was to encourage settlement in very isolated places where the settlers should help to protect against forest fires, provide emergency aid, sustain people working in the forests, and aid military intelligence. The scheme has now stopped but contrary to what might be expected there are always applicants for existing mountain farms when they are free.

On the whole there seems to be no future for the farming settlers within the zone. Like the Lapps they have given up self-sufficient farming during the last decades. The launching of new agricultural and occupational policies during the 1960s has made it clear that agricultural products should be produced in the good agricultural districts where farming will be run on rational lines and engage as little manpower and capital as possible. Surplus farmers should be re-educated and then transferred to industrialized districts.

Of today's 200,000 inhabitants, about 40,000 rely on farming and forestry, and another 30,000 to 40,000 are indirectly depending on these industries. If the current trend holds, 80,000 to 100,000 will move out of the zone within little more than a decade. However, depopulation has been too fast. A special agricultural and occupational policy has come into operation since 1 July 1971 making it possible for settlers to stay on their holdings for the rest of their active life.

The mining industry with connected service industries will provide a living for about 50,000. The industry is flourishing and has a good future. The hydroelectrical development works culminated during the 1960s and will end altogether during the 1970s. They were extensive and provided good sources for occupation and income. Tourism is of little importance today, but may eventually grow. The situation in the Northern settlement fringe of Sweden is at present characterized by encouraged depopulation. And yet there is no program for an organized retreat where the continuous availability of social services and modern facilities is secured. There has been no limit set for depopulation. The demographic situation will react rapidly to depopulation. These factors should provide a challenge to administrators.

s0602
Problems of the subarctic on the territory of the USSR
V.B. SOCHAVA, G.V. BACHURIN, V.V. VOROBYEV, U.P. MICHAILOV, B.B. PROKHOROV, and V.P. SHOTSKIY *Institute of Geography of Siberia and the Far East, USSR*

In Soviet geographical literature there is no generally accepted opinion about the boundaries of the Subarctic (s). Soviet physico-geographers (Grigoriev 1946, and others) usually consider it as not extending southward beyond the limits of the forest-tundra. This corresponds to the opinion adhered to by most of the speakers at the Symposium in Helsinki (Ecologie des régions subarctiques, UNESCO 1970). In the present paper the Subarctic is regarded as territory covered by tundra and forest-tundra (excepting arctic spaces proper) and by northern taiga regions. Within these boundaries the Subarctic corresponds to the concept accepted in Canadian literature (Court 1969; Hamelin 1968).

Figure 1 (not published here) represents the boundaries of the s which are adhered to by the authors of the present paper. The lines, separating the tundra and forest-tundra from the taiga correspond to the southern limits accepted by many physico-geographers. The boundaries of the 'North,' to which special laws extend in the USSR are also marked on the map.

The total area of the Subarctic in the USSR is about 7 million sq km. It stretches in a narrow strip to northern Europe and covers vast territories in northern Asia, where there are massive mountain chains and a severe continental climate.

Its location in relation to the oceans leaves its imprint on the nature of territory. In this connection the s of Euro-Asia is divided into pre-Atlantic Ocean, continental, and pre-Pacific Ocean territories. On the plains and plateaus the s is represented by: (1) geomes of tundra and forest-tundra; (2) geomes of northern taiga; (3) middle taiga geomes. In the mountains the vertical belts are in the following sequence: taiga geomes of optimal development, taiga geomes of limited development, reduced taiga geomes, mountain-tundra geomes. The mountainous s, which is located only in Siberia has an area of about 4 million sq km.

Tables 1 and 2 give the data characterizing the radiation balance, thermal regime and humidity. Many hydro-climatic parameters vary considerably according to regions. The pre-Pacific Ocean s stands out especially in the hydro-climatic respect. The climatic variations correspond to the peculiarities of soils, vegetation cover, and animal life.

Tundra and forest-tundra provide pastures for northern deer and provide the sphere for

TABLE 1. Radiation balance and heat in the Subarctic of the USSR

	Total sun radiation (kcal/cm²)	Radiation balance (kcal/cm²)	Average monthly temperature of air in January (°C)	Average monthly temperature of air in July (°C)	Total temperature of air during periods with temperature over +10°	Number of days in year with average day temperature of air over +10°	Absolute minimum temperature of air (°C)	Absolute maximum temperature of air (°C)	Annual amplitude of temperature of air (°C)	Average duration of non-frosty period in days
Territories of Atlantic Ocean influence										
Western:										
Tundra and forest tundra	70	20–25	−7 − 20	+7 + 13	0–600	20–50	−35 − 50	+25 + 30	18–32	90–105
Northern taiga	75	25	−10 − 20	+11 + 15	450–1200	40–90	−40 − 50	+30 + 35	22–34	75–105
Middle taiga	80	25–30	−11 − 19	+15 + 17	1000–1400	90–100	−45 − 50	+30 + 35	26–34	90–120
Eastern:										
Tundra and forest tundra	70–75	15–20	−19 − 29	+4 + 14	0–800	10–60	−50 − 60	+25 + 30	30–42	40–75
Northern taiga	75–85	20–25	−23 − 28	+15 + 16	800–1300	60–90	−55 − 60	+30 + 35	36–44	75–100
Middle taiga	85–90	25–28	−20 − 25	+16 + 18	1200–1600	90–110	−50 − 60	+ 35	34–42	100–105
Continental territories										
Tundra and forest tundra	65–70	10–15	−30 − 40	+2 + 12	0–800	0–50	−55 − 65	+25 + 35	34–52	45–65
Northern taiga	75–85	15–23	−29 − 40	+12 + 16	400–1000	50–70	−60 − 65	+30 + 35	44–56	60–80
Middle taiga	85–100	23–28	−24 − 42	+14 + 18	800–1600	60–100	−60 − 65	+30 + 35	40–62	60–90
Baikal-Djugjur Mountainous region	90–100	20–30	−26 − 38	+10 + 18	400–1200	30–80	−55 − 65	+30 + 35	38–60	60–90
Eastern part of the Yano-Kolym mountain region	70–90	15–20	−22 − 48	+4 + 14	0–1000	10–60	−45 − 70	+20 + 30	30–60	60–90
Territories of Pacific Ocean influence										
North pre-Pacific Ocean region										
(a) Anadar and Penjinsk tundra	70–80	15–20	−20 − 28	+8 + 12	0–800	10–60	−30 − 65	+25 + 30	30–44	60–120
(b) Koriyak highland, Kamchatka, Kurils	65–70	15	−8 − 24	+8 + 14	0–1200	30–100	−50 − 60	+25 + 30	22–36	60–90

TABLE 2. Precipitation, outflow and evaporation in the Subarctic of the USSR

	Elements of water balance (mm)				Number of days with snow cover	Maximum height of snow cover (cm)	Supply of water in snow (mm)	Duration of ice cover on rivers in days
	Precipitation	Outflow	Underground runoff	Evaporation				
Territories of Atlantic Ocean influence								
Western:								
Tundra and forest tundra	550–650	300–450	30–80	220–250	200–220	60–70	120–140	190–220
Northern taiga	600–700	300–350	30–80	250–350	180–220	70–80	140–160	180–200
Middle taiga	700–750	300–350	65–95	350–400	180–200	70–90	140–180	160–180
Eastern:								
Tundra and forest tundra	400–600	200–250	15–65	200–350	230–260	50–80	100–160	220–240
Northern taiga	600–700	250	65–95	350–450	200–220	70–90	140–180	190–220
Middle taiga	650–700	200–250	65–95	450–500	180–200	60–80	120–160	180–190
Continental territories								
Tundra and forest tundra	300–600	200–400	15–30	100–200	240–270	30–80	60–160	230–240
Northern taiga	300–600	150–250	15–30	200–300	220–250	40–80	80–160	210–230
Middle taiga Baikal-Djugjur mountainous region	300–700	20–250	10–95	250–350	200–240	40–90	80–180	180–210
Eastern part of the Yano-Kolym mountain region	400–800	200–300	15–65	100–300	220–260	40–70	80–140	200–240
Territories of Pacific Ocean influence								
North pre-Pacific Ocean region								
(a) Anadar and Penjinsk tundra	500–600	300	30–65	250	200–240	70–80	140–160	200–220
(b) Koriyak highland, Kamchatka, Kurils	500–1000	300–800	160–300	200–300	200–260	80–200	160–400	100–200

TABLE 3. Timber resources of the Soviet Subarctic

Territory	Forest area		Forest covered		Reserves		
	mln ha	% of total	mln ha	% of total	mlrd cu.m	% of total	m³ to 1 ha
Subarctic USSR	510.6	12.2	373.6	9.9	37.3	20.3–23.1	100
Total USSR	915.8	21.9	746.2	19.7	80.1	44.4–50.0	107
Total world	4184.0	100	3792.0	100	160–180	100	42.3–47.6

hunting. Very large areas in the mountainous tundra (1.25 million sq km) are practically untouched by man, but they are important as areas of snow accumulation. In the future mountainous tundra of the s may be utilized more effectively, but at the present time this remains problematic.

A great part of the s is primordial taiga. It is covered with coniferous forests, predominantly Daurian larch (*Larix dahurica*) in the pre-Pacific Ocean and the continental sectors, while in the pre-Atlantic Ocean sector the Siberian fir (*Picca sibirica*) is predominant. In the West Siberian Plains there are huge massifs of sphagnum bogs and large peat-fields. Half of the timber resources of the USSR and a quarter of the world's supply of timber is to be found in the forests of the s (see Table 3), including almost half of the coniferous forests of the earth.

Only the central and part of the northern taiga forests in the Ural and the European regions of the USSR are being intensively utilized. Exploitation of the forests in the Asian s is insignificant so far. The Subarctic is rich in many mineral resources that have not yet been completely explored (non-ferrous and rare metals, gold deposits, platinum, lead, wolfram, nickel, copper, titanium, and other metals). About two-thirds of the USSR coal resources, mica, unique deposits of apatites and nephelines are to be found here.

In connection with the latest geological discoveries in West Siberia and Yakutia, the s has become a large oil and gas-bearing region.

The regions of the Subarctic possess a large hydropower potential (more than 50 per cent of the total supply in the USSR). The taiga and tundra are rich in fur-bearing animals. Over 2 million head of deer are bred in the s. The northern seas, as well as the inner reservoirs (lakes and rivers) abound in fish.

The whole territory of the s in the USSR is inhabited by only 5 million persons. Although this population is insufficient, it provides a density of 8.5 times more than in the Subarctic regions of North America. In the s there are two types of regions of differing population characteristics. In one type the aboriginal population predominates. This type covers large territories, has sparse populations which are comparatively stable and do not participate in out-migrations. The second type of region is connected with developing industry, situated on comparatively small territories. The greater part of the population of the s lives here. This population is extremely mobile and is only in the primary stage of development. In many northern areas about 20–25 per cent of the population is renewed annually. The most intensive migrations can be observed in the Ob River north and the Magadan Region where conditions for the stabilization of population have not yet been created.

A series of ecological problems and questions concerning the optimization of the human environment arise in connection with the development of population in the s. Among them are such problems: (1) protection from cold (choice of micro-climatic conditions for settlements, suitable planning of inhabited localities, the building of large hangars for isolating people from low temperatures, rational choice of clothing, etc.); (2) the creation of optimal sanitary and hygienic conditions (often causing great difficulties in the s); (3) carrying out measures to protect people from blood-sucking two-winged insects; (4) taking preventive measures against diseases specific to the North (rabbit-fever, toxoplasmosis, tick-encephalitis, and so on); (5) conducting regular medical examinations to hasten the adaptation of people arriving in the s.

An important factor in the expansion of the population in the s is the development of agriculture which must be regarded as a subsidiary field to the development of industry and forestry. Deer-breeding and farms breeding fur-bearing animals use up the wastes from the fishing industry and are important commodity producers.

One of the possible centres of the Subarctic agriculture is the West Siberian Plain where potatoes can be cultivated up to 65°N. The largest historically occupied agricultural region in the s is Central Yakutia. Middle Siberia and Zabaikalye are very suitable for stock-breeding.

In the future agriculture will not occupy large areas and will not affect the taiga aspect of the Subarctic plains, in spite of rather severe changes and thermal amelioration.

One of the chief scientific problems to be solved with the help of geographers is forecasting of the future of the s. At the present time the basic trends in the development of the s up to the year 2000 are being outlined. The scale of hydropower development is clearly defined. The supplies of timber will increase considerably. Annual output will reach 25 million m³ by 2000. Great changes in the geography of the s are to be expected in connection with the construction of hydropower stations in the lower reaches of large rivers (Yenesei, Lena, and others). Regulating the flow of the Ob and Irtysh rivers will allow establishment of forage bases for commercial stock-breeding in the flood-plains of these rivers.

There will be serious changes in the natural environment and economic conditions of the s as a result of transferring the water of the rivers which flow into the Polar basin (Pethora, Ob), to the south. The prognosis of changes which may occur as a result of this is one of the chief tasks of geographers.

In the future we see the s as a great power base (oil, gas, hydropower, coal, peat) supplying local developing industries and exporting power resources.

The growth of the economic potential will increase continuously because of the discovery of new deposits of raw materials. This is a very important problem for the next two decades.

The s is one of the richest sources of pure water for utilization in industry and other spheres. Taking into account all geographical factors, there will be further changes of the water balance and effective solution of questions dealing with the expanding demand for water resources.

Intensified productivity of forests in many Subarctic regions may give greater amounts of raw materials for the wood-chemical industry. However the taiga landscapes should be preserved on considerable areas of the northern taiga and in the mountains in a natural condition as an environment-forming factor (humidity turnover, hunting, berry-picking, and so on).

Drainage of bogs and the simultaneous utilization of peat increases land resources considerably, but this must be carried out without altering the optimal humidity turnover in the subarctic regions.

The most important constituent in the development of the s is the solution of problems connected with the reclamation of the taiga (Sochava 1964), the southern regions of which do not belong to the s, but represent important industrial, agricultural, and demographic outposts of its development.

The s possesses great opportunities for the growth of population provided medical measures are improved, and stock-breeding and vegetable farming are developed. However, the s's importance is still great as a source of vital resources for other densely populated regions which are more favourable to man's life and work.

Court, A., 1969 Definition and characteristics of the Subarctic, *Geog. Rev.* 59 (3), 444–5.

Grigoriev, A.A., 1946 *Subarctic: experience of characterizing the main types of physico-geographical environment* (Moscow).

Hamelin, L.E., 1968 Un indice circumpolaire, *Ann. géog.* 77 (juillet-août), 414–30.

Sochava, V.B., 1964 Geographical problems of development of the Taiga, *Soviet Geog.: Rev. and Translation* (Sept.), 40–52.

Minority Group Settlement
Installation de groupes minoritaires

Landscape perception and minority group viability: some insight into cultural group extinction

JAMES E. LANDING *University of Illinois at Chicago Circle, USA*

In a broad sense the various governmental units of the United States and Canada, federal, state and provincial, and local, have been dealing with minority group problems since the earliest days of both countries, and have faced problems quite similar involving, frequently, the same groups. In a very specific sense both countries have had to deal, at some time and in some way, with the groups referred to as the Old Order Amish, the Mennonites, and the German Baptist Brethren. These three sectarian cultural groups form the basis of this inquiry.

Amish and Mennonite history and settlement are intimately linked, since the Amish emerged as a schism within the Mennonite movement in Europe. The German Baptist Brethren, however, have had an independent history, being a later development and never functionally linked with either the Amish or the Mennonites. Despite this independent historical origin their life style as expressed in North America is quite similar to that of the Amish and the more conservative Mennonite groups.

All three groups have had their external troubles with the larger society which surrounds them, and all three have had innumerable internal divisions and schisms, largely in North America. As a result, Mennonitism, Amishism, and German Baptist Brethrenism are now expressed in countless sectarian and denominational groups which, in some cases, have departed so greatly from the traditional ways that the linkages can be established only through historical analysis. In other cases the various groups are so similar that the outside observer is frequently puzzled as to why there is any ceremonial division at all.

The latter problem is best exemplified in North America in only a single geographical locality, the Yellow Creek area of western Elkhart County, Indiana, between the cities of Goshen and Wakarusa. The regional name is derived from a sluggish stream which has its headwaters in a small lake and winds its way tortuously some 12 miles northward from Wakarusa where it empties into the much larger Elkhart River. This is the only locality remaining where all three groups, Amish, Mennonites, and German Baptist Brethren, live in an overlapping residential pattern and still maintain traditional sanctions against automobile ownership and electricity utilization (Landing, in press). The three groups, the Old Order Amish, the Old Order Wisler Mennonites, and the Old German Baptist Brethren, are buggy driving people (Landing 1970, 3–4).

Despite the fact that the three groups live in a common geographical area there is tremendous variation in the viability of the groups. Based on population and membership trends the Old Order Amish are thriving, the Old Order Wisler Mennonites are more than holding their own, but the Old German Baptist Brethren are becoming extinct. The latter group is now represented in the Yellow Creek area by only five families and two widows. Only one of the five families includes children still young enough to become members of the group. If they decide not to do so this particular group will pass out of existence within a few years. Why do these three groups, living so similarly in the same environment, have such widely differing survival probabilities?

One looks in vain for an answer in such variables as the climate, vegetative and edaphic patterns, and even in basic contemporary social interaction patterns. Analysis of the latter simply indicates that there has been change, that the changes can be identified and ordered chronologically and spatially, but that the recognition of such changes does not answer the basic question.

There does, however, seem to be a glimmer of insight into the question when approached from the perspective of the basic core values of these societies, those values which immediately set them apart, not only from each other, but from all other groups. The five basic core values were derived from analysis of the Old Order Amish, but the values are shared in common by the other two groups (Huntington 1956). The five basic values are: separation from the world; adult responsibility initiated by baptism, church-community rules and discipline; excommunication and shunning; and, closeness to the soil (Hostetler 1969, 20–25).

All three groups are socially nonconformist and maintain strong social isolation from those not of their group. All accept the basics of adult confession of faith and initiation into the church through baptism at an advanced age, although the ages have differed through time; and each group has its own set of church rules by which the group traditions are maintained and ordered. Excommunication and shunning are still practised, although at different levels. The Old Order Amish do not consider themselves to be in fellowship with any other group, or with any other Amish group which has a life style they can not accept. The same is true for the Old Order Wisler Mennonites and the Old German Baptist Brethren. It appears, at this stage of investigation, that the differential viability rate of these groups can be explained, in part, by a quite different in-group perception of the actual meaning of the last of the five core values, closeness to the soil.

Traditionally, closeness to the soil has been interpreted as owning and operating a farm and high esteem was granted to those that did this successfully. But the actual meaning and its application, in practice, has changed and differs among the three groups. Although to all three, closeness to the soil implies a rural, as opposed to an urban, setting, from this point their perspectives diverge.

Closeness to the soil can be verbalized, but it can also be statistically analysed.

Among the Old Order Amish of Yellow Creek, who are an integral part of the Nappanee, Indiana, settlement, at least 65 per cent of the employed heads of households are engaged in non-agricultural occupations (Rechlin 1970, 79). Although the figures for the Old Order Wisler Mennonites are not yet clear, the percentage is less, but a considerable number are engaged in full-time non-agricultural forms of employment. Among the Old Brethren German Baptists, however, none of the surviving families in Yellow Creek have ever been anything but farmers. They apply the basic core value of closeness to the soil most literally. They have, apparently, not been successful in raising their children to find fulfilment in this type of society.

Thus, the Old German Baptist Brethren of Yellow Creek are becoming extinct, not through assimilation, but through a lack of accommodation to a changing economic and social milieu. It is paradoxical, however, that their death is the vindication of their belief.

Hostetler, J.A., 1969 *Educational achievement and life styles in a traditional society, the old order Amish*, us Dept. of Health, Education, and Welfare, Office of Education (Sept.), 20–5.
Huntington, A.G.E., 1956 Dove at the window: a study of an old order Amish community in Ohio. PhD dissertation, Yale U.
Landing, J.E., 1970 The buggy cultures, *Mennonite Hist. Bull.* 31 (4), 3–4.
– in press A morphology of cultural disintegration among German-American minorities in Elkhart county, Indiana, *Indiana Magazine of History*.
Rechlin, A.T.M., 1970 The utilization of space by the Nappanee, Indiana old order Amish: a minority group study. PhD dissertation, U. Michigan, 79.

s0702

The Albanese communities in Pollino (southern Italy) faced by new consumer mass civilization
MARIO CATAUDELLA   *University of Salerno, Italy*

An examination of the profound changes which developments in modern methods of information have brought about nowadays reveals that one of the principal results has been that of the levelling out of the cultural differences existing among the various human groups living in the world today.

Two characteristic phenomena of so-called

mass civilization contribute towards this levelling. The first is the growing possibility of communication between the various social groups and classes, which has brought about a kind of cultural osmosis between various populations, resulting in the diffusion of a common system of symbols. The second characteristic phenomenon has been the endeavour on the part of those centres which decide for the mass consumer civilization, to choose and standardize the symbols to be transmitted, with the intention of arriving at the best possible conditions for both production and sale.

As we already know, these conditions are often arrived at by conditioning the consumer, and by limiting his choice, taking care that the mechanics of mass production avoid stagnation, or any kind of cultural obstruction, in the course of expansion. In this way the world is destined to become more and more uniform, thanks to those regions or those countries which find themselves technologically ahead of the others.

It is, in fact, obvious, that in this process of cultural assimilation, or rather of absorption, the countries bound to suffer most are those which we may call underdeveloped. Unless their traditions and customs can be altered and adapted to suit the system, they are destined to fall into disuse, and disappear. But if, in theory, the whole of mankind runs the risk of having an alien culture imposed upon its real, legitimate culture, it is the minor ethnic groups which run the greatest risk, especially if they are limited by out-of-date economic systems, incapable of autonomous modernization, for these minor ethnic groups undergo a two-fold cultural attack: to that which comes from the world outside them is added a second cultural attack which takes the form of artificially created desires imposed upon them. The results of these external pressures are invariably grave, and often lead to the social disintegration and alienation, or even to the complete destruction of the ethnic group itself.

On the basis of these considerations it seems opportune to turn our attention to the actual situation in which a number of isolated ethnic groups find themselves today and examine the changes which technological progress is forcing them to undergo. We shall turn our attention to the eight Albanese communities living in Pollino whose particular geographical position in southern Italy offers the student a rich field of research (Zangari 1941; Nasse 1964). Here, in fact, the typical problems of the minority ethnic group are mixed with the problems of an area which has itself all the characteristics of an underdeveloped zone: mass emigration, the decline of centres, the inadequacy or total absence of agricultural renewal and only intermittent interventions on the part of industry.

Both in the past and today – even if on a lesser scale – the economy of the Albanese communities in Pollino has been founded on agricultural resources, and on the rearing of sheep and goats in isolated mountainous districts, cut off from the principal routes of communication and commerce. Up until the end of the Second World War this isolated way of life enabled these Albanese communities to preserve not only their traditions and customs, but also their native language.

Based on agriculture, their way of life seemed destined to continue indefinitely, but from 1955 onwards unexpected changes resulted in an upheaval which profoundly altered things. The reasons for this upheaval may be found in the disintegration of the large estates which existed previously, in the barrenness of the land, and in the irregular spacing of properties, thanks to continual division through inheritance. (In an area of 23399ha of agricultural land, there are 8250 farms, averaging 2.7ha; 5952 of them are smaller than 1 ha.)

Built on foundations as weak as these, the economy of these communities could not hope to resist in a country undergoing rapid expansion in the field of industry, and in fact they suffered their first losses as a result of mass emigration of the young people living in

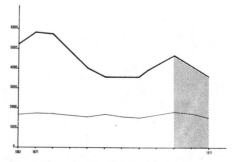

Fig. 1. Diagram plotting the fluctuation in population in two communities.

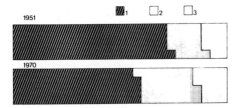

Fig. 2. Division of the working population:
1 Agriculture; 2 Industry; 3 Other activities.

the area, who left home in order to find work in the industrial zones of north Italy. This crisis in agriculture, which up until then had represented the only source of life, inevitably led to crises in other fields. Although in the eight communities under examination the working population has apparently increased by 1 per cent in the last twenty years, from 43 per cent in 1951 to 44 per cent in 1971 – there has in fact been a decrease of almost 1000 workers during the same period, which may be explained by an overall decrease in the number of inhabitants.

In the last twenty years there has been a decrease of 35 per cent in the number of agricultural workers in these Albanese communities, while the number of workers in other camps (industry, commerce, and so on) has increased by 43 per cent. From the point of view of percentage the greatest increases in the number of workers have taken place in the building industry, while the smallest percentage increases have been seen in the various manufacturing industries.

In relation to the population as a whole, the decrease which has taken place in the number of agricultural workers has been only partly compensated for by the increase which has taken place in other activities (49 per cent), while many other workers have emigrated.

The fact that the land above a certain altitude is no longer cultivated, while less fertile land has been abandoned altogether, shows that the most dominant note is that of inexorable decline in the field of agriculture. (Up until 1950 the grain producing zones went high up into mountains, limiting the pasture land available: at Acquaformosa they reached 1430m in the plain of Ferrocinto; today they scarcely reach 800m.) Moreover, an alteration in the existing situation would seem to be impossible, as the wild, mountainous condition of the land does not facilitate the modernization of farming techniques. Nor is the economic picture at all promising when taken on the whole, if we consider the fact that the present boom in the building industry will fall off within a few years when various public works reach completion, while the rise in the number of workers involved in commerce has been determined both by a rise in state expenditure and by an increase in the amount of ready cash in circulation, thanks to emigration.

Thus the only solution to these problems for a large number of people has become emigration, and its gradual spiral escalation has served to worsen an already precarious situation. Indeed, emigration is one of the factors which most threaten the existence of this particular ethnic group, for it both brings about the break-up of the group itself and exposes the individual members of the group to the influences of other cultures.

Official statistics show that during the period between 1951 and the present day, the number of inhabitants in the areas under consideration went from 19,164 in 1951 to 16,390 in 1971.

If, however, we base our calculations on the real population, and not on the statistics presented by the registry office, we find that the drop in number has been much greater. Basing the calculations on a limited poll, one may state with a fair measure of certainty that in the last five years 78 to 90 inhabitants

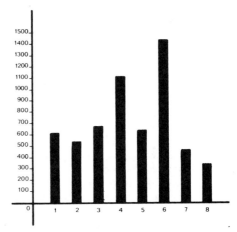

Fig. 3. Emigration within the last ten years.
1 Acquaformosa; 2 Castroregio; 3 Civita; 4 Firmo; 5 Frascineto; 6 Lungro; 7 Plataci; 8 S. Basile.

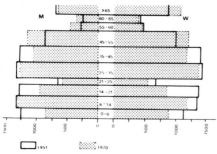

Fig. 4. Comparative age structures, 1951 and 1970.

for every 1000 have emigrated.

Because the Federal Republic of Germany is the country preferred by many emigrants, it is often called the 'New America,' and both the way in which the country is described by people already living there, and the monthly wage, regularly paid, greatly tempt those of the Albanese communities who have remained faithful to the land, or encourage some modest trade. Even if the mass exodus often attenuates the situation, it is nevertheless no more promising for those who leave the areas than for those who remain in the ethnic oases of Pollino, where an increased number of old and infirm workers, and female workers, has helped make agricultural labour uneconomic.

This ethnic group, which has found itself under pressure economically, and which has been virtually decimated numerically, has been placed at a further disadvantage thanks to mass communication. Although a certain number of periodicals, such as *Sheizat, Sgjmi,* and *Vetra Jonë* are published in the hope of keeping the cultural traditions of the Acbëreshë alive, their circulation is extremely limited (50 copies for 1000 inhabitants) and there is a circulation of magazines and papers published in other parts of Italy.

The fact that daily newspapers are not widespread may be explained by the fact that the sale of a newspaper depends on the speed with which it reaches its destination. In the eight Albanese communities under examination, the lack of communications and of adequate means of transport often means that the paper arrives either in the afternoon, or worse still, in the evening when the information it contains is already stale. Consequently weekly magazines are more widespread because the news they contain is less immediate.

However, the number of the reading public when compared to the number of publications entering the area is much higher than might be supposed, because a printed page decorated with some suggestive image of the consumer civilization may pass from hand to hand within the entire community.

The importance of television and radio as methods of information is not to be doubted, and it is for this reason that they have been qualified as services in the public interest by the authorities of state. Television is undoubtedly the more desirable commodity as far as the consumer is concerned, and in this, too, the Albanese communities are behind Calabria taken as a whole, and even more behind the rest of Italy, for they have by far the lowest number of subscribers. Once again, however, statistics tend to falsify the picture, for although the number of subscribers in these communities may be small, the number of people who watch television is extremely high. The evenings which were once spent by the whole family grouped around the fire, telling stories or rhymes, or talking of work, are now spent with the family united to watch television (Scura 1962).

That television may well play an important part in encouraging the mass rural exodus which is taking place has not yet been proven, but is an hypothesis which could have some foundation in truth. As the high rate of illiteracy which exists amongst the inhabitants of these communities often precludes the possibility of written contacts, radio and television frequently represent their only link with the outside world, and consequently they are of fundamental importance in influencing the way of life of these Albanese. As far as the cinema is concerned, there are only five small halls where films are projected in the whole of the eight communities, and of these two do not function during the winter. Nor are the cinemas well attended, and the films themselves are of a low intellectual level, calculated more to satisfy the demands of a public which is not at all well versed in film language, than to increase its critical abilities. Thus the only means of cultural formation with any real possibility of penetration in such an environment are radio and television, this latter having the possibility of really giving the inhabitants a push in the direction of a balanced process of evolution. Unfortunately, however, and thanks to the Albanese communities' complete lack of the kind of defence which a culture derived from other sources would represent, television has preferred to become an efficient and economic means of publicising those products which the more economically and technologically advanced areas of Italy are anxious to impose upon the national market.

Television has also preferred to act as a means of capturing the existing culture, rather than to act as an instrument of cultural formation. This method of mass communications has radically altered a great many customs, and has enabled a large number of people who had previously had no opportunity of doing so, to hear the Italian language spoken in their own homes. The traditional internal structure of the Albanese communities, based on the family as a unit, has disintegrated.

With the disappearance of a cultural world which sprang from a particular economic situation, a desperate search for a new system of values has begun, identified all too often with the possession of those objects which modern, mass, consumer-based civilization has rendered attractive through publicity and propaganda.

One is induced to think that in those areas a whole world of traditions, a whole way of life is coming to an end. In fact, the young people of Pollino, encouraged by the messages of the new mass media, want a different kind of social system and have rejected a tradition which they no longer feel to be theirs. To a certain extent, even if unjust, this is at least natural, for it is a direct result of the loss of those economic resources which helped both to generate and preserve that cultural world.

Recently, however, new interest has been taken in these areas. On the one hand there has been an increase in the number of conferences summoned for the purpose of studying the problems of these Albanese communities, while on the other there has been a notable increase in the plans to develop the amount of tourism in the area, based not only on the natural attractions which the area has to offer, but also on the artificial interest which has been created around the local festivals and esoteric costumes of the population.

While a serious study of the problems is being undertaken by the Italian section of the 'World Wildlife Fund' with a view to creating a national park which would constitute a means of help in the future as far as southern Italy is concerned, and would indicate a new policy in the setting up of tourist zones, there has also come into being a 'plan for mountain reclamation' in the area of Pollino. This is one of the usual exploitations of the land which result in the ruin of the countryside, and which, in the space of a few years will bring about the existence of five thousand beds, distributed throughout three tourist villages, and numerous slopes for winter sports. In this lies the hope of the Albanese communities of reforming their economy; in this lies the renewed interest in Albanese folklore.

The 'Plan for mountain reclamation in the area of Pollino,' in fact, intends to preserve an ethnic group on its way to dignified extinction, under glass, as it were, for the delight of the tourist.

In the place of a dignified disappearance, there will now be an embalming in extremely bad taste, which will merely keep alive the shadow of what was once a cultural reality, and which publicity will undoubtedly represent as an object indispensable to our happiness as consumers.

Napoli, Casella, and G.N. Nasse, 1964 *The Italo-Albanian Villages of southern Italy*, National Acad. of Sciences, NRC.

Scura, A., 1962 *Gli Albanesi in italia e i loro canti tradizional,* Cosenta, Casa del libro.

Zangari, D., 1941 *Le colonie italo-albanesi di Calabria: storia e demografia* (sec. XV–XVI).

s0703
L'intégration polonaise des Canadiens-Polonais de la région de Barry's Bay, Ontario
M.A. JAROCHOWSKA   *Université du Québec, Canada*

Dans les sociétés modernes et développées, l'intégration totale à l'intérieur d'une nationalité ne touche qu'une partie de la population. Le reste est composé de groupes, qui sont venus des autres sociétés ou vivaient pendant un certain temps dans les autres pays et s'intégraient partiellement à eux. Ce phénomène, résultant de la mobilité humaine et de l'augmentation des liens parmi les pays du monde, place les sociétés face au fait que le concept d'intégration uninational est insuffisant.

En réponse à cette nouvelle situation, et pour en découvrir la véritable signification, un intérêt considérable est né, stimulé par des recherches sur divers types d'intégration (Iwańska 1971).

Dans les études géographiques, cette approche amène entre autres vers la question suivante: quelles sont les intra-relations des types d'intégration tels qu'ils sont définis par M. Werner S. Landecker (1967). Cette question ne peut être répondue que par les recherches opérationnelles où le chercheur doit établir l'étalon individuel pour chaque groupe analysé (Lazarsfeld 1967).

Il me fallait donc, construire un schéma de la structure créée par la participation de types particuliers d'intégration et tel que je considère représentatif du groupe polonais du village de Barry's Bay et des environs. (Le village a aussi la plus grande (plus que 95 pour cent) concentration des Canadiens-Polonais.)

Je l'ai construit comme un triangle placé à l'intérieur d'un carré. Chaque côté du carré représente un type d'intégration, et sa surface libre correspond à l'intégration anglo-saxonne. Le triangle par contre représente

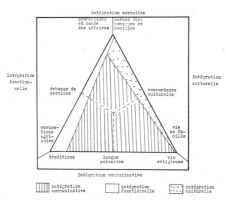

Fig. 1. Intrarelations des 4 types de l'intégration polonaise à la région de Barry's Bay.

l'intégration polonaise. La participation des 3 types d'intégration est transférée sur les côtés du triangle et ses 3 médiatrices expriment les phénomènes suivants: l'emploi de la langue polonaise, la concordance culturelle avec les traditions et coutumes polonaises, et les échanges de services comme appui aux membres du même groupe minoritaire. Le côté d'intégration polonaise normative est réduit au minimum, parce que là domine surtout l'intégration anglo-saxonne. Le triangle est de plus partagé entre les 3 surfaces hachurées différemment. Leurs tailles correspondent aux rôles joués par chaque type d'intégration dans la vie du groupe polonais. Elles sont calculées sur la base des entrevues réalisées par l'auteur.

L'utilisation du schéma présenté a permis de tirer certaines conclusions, ainsi: (1) C'est l'intégration communicative qui maintient l'intégration polonaise la plus profonde. Elle polarise dans l'intégration culturelle par tous les faits culturels en utilisant les signes communicatifs autres que la langue. (2) L'intégration polonaise est vivante parce qu'elle donne aux membres du groupe minoritaire polonais les éléments nécessaires pour fonctionner normalement, c'est-à-dire protection contre la pression du monde à l'extérieur; appui dans le choix des nouvelles décisions et actions; sens de sa propre dignité humaine comme un individu ou comme un groupe ethnique. (3) L'intégration polonaise cesse d'exister quand elle n'a plus d'éléments productifs dans la réalité dans laquelle vivent les membres du groupe. Donc puisque l'éducation fournit le plus grand nombre d'appuis psychiques, ce sont les gens instruits qui peuvent se désintégrer du groupe minoritaire.

Iwańska, A., 1971 O grupach etnicznych. *Kultura*, no. 10/289 (Paris).

Landecker, W.S., 1967 *Les types d'intégration et leurs mesures: le Vocabulaire des sciences sociales* (Mouton).

Lazarsfeld, P., 1967 *Le vocabulaire des sciences sociales* (Mouton).

---

s0704
The landscape of early group settlements in western Manitoba
J.L. TYMAN *Brandon University, Canada*

The mosaic of cultural groups in western Manitoba is rather less complex than that characteristic of the Red River valley, and its impact upon the landscape is less obvious as a result. Southwestern Manitoba in particular was solidly Anglo-Saxon to begin with, and to this day bears the imprint of Upper Canada more clearly than any other part of the prairie provinces. The Red River Colony extended no further west than Portage la Prairie; and the formal reserves for Mennonites and Icelanders, which were so conspicuous a feature of the early days in eastern Manitoba, have no real counterpart here.

Significant concentrations of minority groups were, however, established in the area during the 1880s and 1890s. In some cases the initiative was provided by colonization railroads and related companies, in other cases it rested with individuals. Colonists from Sweden, from Norway, and from Denmark were settled at Scandinavia (north of Minnedosa) under the auspices of the Scandinavian Colonization Society of Manitoba, with additional help provided by the Manitoba and North Western Railway. Destitute crofters and cotters from the west coast of Scotland were settled near Pelican Lake (south of Brandon) by the Imperial Colonization Board with the assistance of the Canada North West Land Company, and a sizable Icelandic settlement was formed close by in the rural municipality of Argyle. Further north, in the forest belt, Ukrainian communities were established north and south of Riding Mountain.

Flying over these areas today, however, it is impossible to distinguish any of the abovementioned settlements from those surrounding them. The whole of western Manitoba was subdivided according to a rectangular grid of some sort and none of the ethnic groups settling there were excused the residence requirements prescribed for homesteading (as were the early Mennonite and Icelandic colonists of eastern Manitoba). Nevertheless, there are two exceptions to the general sameness of the region – two group settlements which, though they were fitted to the rectangular grid, were noticeably different from their surroundings, and in both cases their original distinctiveness is recognizable even today. The settlements to which I refer were those of Hun's Valley and Turtle River.

The nucleus for a Hungarian colony at Hun's Valley was provided by a group from Pennsylvania selected by Count Esterhazy. The Manitoba and North Western Railway took charge of the party and located them on Stony Creek in Township 16, Range 16 W in August of 1885. Seventeen families had entered for homesteads by the end of the year and seven houses had been built. The houses, which were shared initially, were sited close to the river, which trends N/S; and the farms were laid out in strips running E/W across the valley (Stephenson 1889). The sections here instead of being divided into NE, NW, SE, and SW quarters were split into four strips of 65ha (160 acres), each 1.6km (1 mile) in length. These were designated as the N 1/2 of the N 1/2, the S 1/2 of the N 1/2, the N 1/2 of the S 1/2, and the S 1/2 of the S 1/2.

Several parcels of land here had already been taken up by 'Canadians,' but the railroad released three of their sections to allow the colonists to settle close together (Smith 1886). Later entries by Hungarians here and in adjacent townships conformed to the regular quarter section format, but lands purchased from the Manitoba and North Western Railway (and later from the provincial government) were frequently split into lots of 32ha or less (Patent Register, Range 16). No lands were purchased from the Dominion, on instructions from Esterhazy (Immigration Report 1885, 118).

At the time of Stephenson's inspection in 1889 the colony numbered 26 settlers with 45ha (112½ acres) broken altogether. In 1893 there were 29 Hungarian families here, numbering 122 persons, with 120ha (300 acres) cultivated (Interior Report 1893, 6). In later years the colony was overrun by Ukrainian and Polish settlers. It was a marginal farming area at best and few of the original colonists chose to remain there.

Today few people in Manitoba have even heard of Hun's Valley: the post office was renamed Polonia. Yet the marks of this early phase of settlement with its elongated land grants can easily be traced in the landscape of today. Since the field boundaries run E/W between section lines instead of diagonally as in the Red River Colony the 'long lot' pattern is less obvious, as one would expect; but it is discernible nevertheless.

The Turtle River settlement of Ste Rose du Lac is another interesting example of a departure from the supposed norm of prairie settlement. In 1890 some 60 families of French-speaking half-breeds from eastern Manitoba squatted on lands alongside of Turtle River in what was subsequently designated Townships 23 and 24 of Range 15 W. When A.F. Martin of the Dominion Lands Survey subdivided these townships in the following year he found settlers in occupation of farm lands on either side of the stream which, like Stony Creek, ran more or less from north to south. The squatters, as usual, were forced to adjust their claims to the regular grid, but took land in strips with a frontage of 0.4km (¼ mile) to give as many people as possible access to the river (Field Notes, T.24, R. 15w). As a result, most of the homesteads here were defined originally in terms of river banks and legal subdivisions and where necessary were supplemented by the grant of additional parcels of land nearby. The Hudson's Bay Company accepted lands elsewhere in lieu of quarters occupied by squatters, and exchanges of school lands were also arranged.

With the initiation of a program of active colonization sponsored by the Roman Catholic Church in the early years of this century, the linear concentration of farms along the river gave way to a more dispersed pattern of homesteads spread over a wider area. The arrival of a more aggressive group of agriculturalists led to the displacement of the early Métis colonists by French-speaking settlers from points further afield (Dawson 1936, 344). The original pattern of land grants, once again, can readily be distinguished amid the landscape features of today. The narrower blocks of land alongside the river, with E/W axes, are noticeably different from the pattern prevailing elsewhere.

Neither Hun's Valley nor the Turtle River Settlement rank as major elements in the landscape of western Manitoba. The character of both areas today, however, offers vestigial evidence of the manner in which the otherwise uncompromising quarter section format could on occasion be modified to meet the felt needs of minority groups, and of how such persons could in consequence have a real impact on the landscape.

Canada, Department of Agriculture, Immigration Branch, 1885 *Annual report.*

Canada, Department of the Interior, 1893 *Annual report.*

Canada, Department of the Interior Field notes of the Dominion Lands Survey, in Archives of Surveys Branch, Winnipeg.

Canada, Department of the Interior Patent Registers, in Archives of Lands Branch, Winnipeg.

Dawson, C.A., 1936 *Group settlement* (Toronto).

Smith, H.H., 1886 Letter to Minister of Interior, 1 April 1886, Public Archives of Canada, Record Group 15, Series B1a, File 90895(1).

Stephenson, R., 1889 Hun's Valley, Manitoba, report dated 20 March 1889, Public Archives of Canada, Record Group 15, Series B1a, File 90895(2).

Addenda

P0257
An application of Lettau's climatonomy to an alpine valley
DAVID GREENLAND *University of Canterbury, New Zealand*

Previously published applications of Lettau's theory of climatonomy (Lettau and Lettau 1969; Lettau 1969) have dealt principally with macro-scale considerations. However, some parts of his unpublished work may be applied on a micro-scale. Such a procedure was used in a study of the radiation and energy balances of the Chilton Valley (780 masl) (Soons and Rayner 1968) in the New Zealand Southern Alps for the period 15 August 1969 to 14 August 1970.

In the part of climatonomy used here, net short-wave radiation is regarded as a forcing function, F, given by
(1) $F = (1 - \alpha)\text{sw}\downarrow$,
where α is albedo and sw$\downarrow$ is the total incoming short-wave radiation. For a standard terrestrial case F gives rise to a response function F' which is
(2) $F' = \varepsilon_\sigma T_0^4 + \text{LW}\downarrow + A + P + LE$,
where ε is surface emissivity, σ the Stéfan Boltzmann constant, T_0 surface temperature, LW$\downarrow$ incoming longwave radiation, A soil heat, P sensible heat, and LE evaporative heat flows. Equations 1 and 2 are instantaneously valid and can be applied to a time series of values formed by F and F'. With Fourier expansion F becomes
(3) $F(t) = \bar{F} + \Sigma_i \Delta_i F \cos(n_i t - \delta_i)$,
where Δ_i is the amplitude, $n_i t$ the harmonic number, and δ_i the phase angle. The response function reformulated in terms of T_0 becomes
(4) $T_0(t) = \bar{T}_0 + \Sigma_i \Delta_i T_0 \cos(n_i t - \delta_i^*)$,
where δ_i^* is the phase angle of the response function. Two important parameters of the climatonomy are the *total climatic impedance* Z_i defined as $\Delta_i F/\Delta_i T_0$ and the *total phase lag* ζ_i defined as $\delta_i^* - \delta_i$. Equation 4 may also be written as
(5) $T_0(t) = \bar{T}_0 + \Sigma_i (\Delta_i F/Z_i) \cos(n_i t - \delta_i - \zeta_i)$.
After defining i sets of partial impedances (Γ_i, β_i, ψ_i, Φ_i, χ_i) and i corresponding sets of phase constants (α_i, b_i, ψ_i, ϕ_i, χ_i) for all n_i, where both sets refer respectively to upward and downward longwave radiation, soil heat, sensible and evaporative heat flows, it is possible to derive equations giving values of Z_i and ζ_i (Greenland 1971). Z_i and ζ_i, and $T_0(t)$ (from equation 5), can be computed, and the time series of the energy balance components can be obtained following the use of a series of parameterizations (Greenland 1971).

Values of the radiation and energy balance components were estimated or measured by standard methods, and in so doing input data for climatonomic theory became available. The theory was applied to these data with the results shown in Table 1. All of the series of actual and climatonomic values show generally reasonable agreement, but important differences lie in the amplitudes of the series and in certain individual values. The average discrepancy revealed, when the energy balance equation is closed for each time point of the climatonomic estimates, is 37 ly day^{-1}. These discrepancies, and differences from the actual values, may be attributed to at least two groups of factors. Firstly, the climatonomic analysis in which only four harmonics were used tends to give

TABLE 1. Climatonomic estimates of heat balance data, for the study year. Heat balance and radiation values in ly day^{-1}. Flow towards surface is positive.

	15 Aug.	1 Oct.	15 Nov.	1 Jan.	14 Feb.	1 Apr.	15 May	1 July
Net LW	−95	−110	−118	−124	−115	−109	−96	−96
A	+3	−2	−2	−2	+3	+5	+7	+6
P	−1	−77	−108	−133	−92	−54	+3	−1
LE	+11	−114	−164	−206	−138	−76	+18	+12
T_0 (°C)	0.8	9.1	13.9	17.2	12.6	8.7	1.5	0.9
T_{200} (°C)	0.8	8.2	12.7	15.7	11.5	8.1	1.5	0.9
Net SW	142	279	449	430	380	209	104	58
Rn	47	169	331	306	265	100	8	−38

climatically 'normal' values and is unable to detect important variations in a particular year. In the present case, for example, an unusually dry November gave rise to high P values and low LE values. Secondly, several of the parameterizations employed could be changed allowing a more sensitive application than was used in obtaining the above results. For example, an adjusted Bowen ratio (which is advocated by Lettau) could be applied, prohibiting large, physically unlikely downward flows of LE. Furthermore, it has been shown elsewhere (Greenland 1971) that the level of accuracy of the input data is critical, as might be expected. More attention should therefore be given to the above aspects. However, the overall, general agreement between the two sets of results suggests that if this were done, clima-tonomic theory might be powerful in estimating values of the radiation and energy balance on this scale, at least in the present location.

Greenland, D.E., 1971 Heat balance studies at the Chilton Valley, Cass in the New Zealand Southern Alps. PhD thesis, U. Canterbury.

Lettau, H., 1969 Evapotranspiration climatonomy ..., *Monthly Weather Review*, 97, 691–9.

Lettau, H., and K. Lettau, 1969 Shortwave radiation climatonomy, *Tellus* 21, 208–22.

Soons, J.M., and J.N. Rayner, 1968 Microclimate and erosion processes in the Southern Alps, New Zealand, *Geografisker Annaler*, 50, ser. A, 1–15.

P0258
Classification des types de circulation et des types de temps pour le Québec (période 1963–9)
J. LITYNSKI   *Université du Québec à Trois-Rivières, Canada*

La présente classification est conçue de façon telle qu'on puisse déterminer séparément les types de circulation et les types de temps et, ensuite, trouver la dépendance entre eux. Le but principal de ce travail est de constater jusqu'à quel point le type de circulation dans le Canada oriental détermine le type de temps local à Québec.

Pour déterminer les types de circulation, on a choisi la classification numérique basée sur l'indice zonal de circulation (I_z), l'indice méridional de circulation (I_m) et la pression (P). Les indices sont calculés dans la zone délimitée par les parallèles et méridiens suivants: 40° et 60°N; 60° et 85°W. La pression est prise à Québec, qui se trouve à peu près au milieu de cette zone. Chaque facteur est divisé entre trois classes de probabilité égale, ce qui donne 27 types de circulation (Tableau 1). Si tous les composants étaient indépendants, la probabilité de chaque type de circulation serait de 3.7 pour cent (Litynski 1970).

La classification des types de temps faite par l'auteur en Pologne était basée sur deux éléments: la température et les précipitations. Cette méthode est assez simple et commode, mais elle laisse de côté d'autres facteurs de temps qui peuvent être très importants. Pour élargir la base de classification et en même temps conserver le nombre des types de temps (neuf), on a introduit l'indice basé sur trois

TABLEAU 1

Classes de I_z	Classes de I_m		
	N	O	S
E	NEc NEo NEa	Ec Eo Ea	SEc SEo SEa
O	Nc No Na	Oc Oo Oa	Sc So Sa
W	NWc NWo NWa	Wc Wo Wa	SWc SWo SWa

Remarque: les petits indices c, o et a désignent des classes de pression.

éléments: les précipitations, l'insolation et la vitesse du vent. Cet indice, qui est appelé l'indice d'aggravation de temps, est calculé de la façon suivante: on classifie chacun de ces trois éléments en trois classes marquées 1, 2 3, et on fait la somme: $I_a: C_p - C_i - C_v$, où I_a = indice d'aggravation, C_p = classe de précipitations, C_i = classe d'insolation inversée, C_v = classe de vitesse du vent. L'indice d'aggravation peut varier de 3 à 9.

Ensuite on procède à la classification de I_a en trois classes de probabilité égales. De même façon on classifie la température, ce qui donne en somme 9 types de temps (Tableau 2).

Pour le moment la classification est faite pour

TABLEAU 2

Classes de I_A	Classes des temp.		
	1	2	3
1	11	21	31
2	12	22	32
3	13	23	33

une période de 7 ans (1963–9). Cette période est assez courte et il faut être prudent avant de tirer les conclusions statistiques. Néanmoins la dépendance entre les types de circulation et les types de temps est bien visible. On va discuter séparément la période froide de l'année (décembre, janvier, février et mars) et la période chaude (juin, juillet, août et septembre).

Pour cette analyse, on se base surtout sur l'indice de dépendance (D) calculé d'après la dispersion des types de temps dans chaque type de circulation. Cet indice peut prendre les valeurs 0, 1, 2 et 3 dépendemment des écarts moyens de la température et de l'indice d'aggravation de temps: $D = 3$ quand $\Delta \leq 0.50$ – très bonne dépendance; $D = 2$ quand $0.51 \leq \Delta \leq 1$ – assez bonne dépendance; $D = 1$ quand $\Delta > 1$ – faible dépendance; mais il existe le type de temps caractéristique $D = 0$ – pas de dépendance; il n'existe pas le type de temps caractéristique:

$\Delta = \Delta T + \Delta I_A,$

ou ΔT = écart moyen de la température; ΔI_A = écart moyen de l'indice d'aggravation de temps.

Durant la saison froide (Tableau 3), les types de circulation qui ont très bonne dépendance avec le type de temps ($D = 3$) comprennent 18 pour cent des cas, dans 75 pour cent des cas la dépendance est assez bonne et seulement dans 7 pour cent des cas elle est faible ou n'existe pas du tout (Tableau 3).

On peut désigner ici quelques groupes caractéristiques des types de circulation. (*a*) Types anticycloniques des secteurs w et N; fréquence totale 21 pour cent, type de temps caractéristique 11 – beau et froid; (*b*) types cycloniques du secteur N (NWc, Nc, NEc); fréquence totale 16 pour cent, type de temps 13 (ou 23) – tempêtes de neige froides; (*c*) types cycloniques des secteurs E et S; fréquence totale 13 pour cent, type de temps 33 – tempêtes de neige chaudes, parfois avec de la pluie; (*d*) types anticycloniques du secteur SE (SEa, Sa); fréquence totale 7 pour cent, types de temps 31 et 21 – beau et assez chaud. Ces groupes comprennent en tout 57 pour cent des cas. Le type le plus fréquent est NEc avec fréquence 8.4 pour cent et type de temps 23 ou 12 (tempête de neige froide).

On voit bien que la dépendance entre types de circulation et types de temps est bien meilleure pour l'été que pour l'hiver. Durant la période chaude (Tableau 3) de l'année, 44 pour cent des cas ont une très bonne dépendance entre le type de circulation et type de temps ($D = 3$), ce qui dépasse presque 3 fois le nombre de cas analogiques en hiver; 47 pour cent des cas ont $D = 2$ (assez bonne dépendance) et 8 pour cent des cas ont $D = 1$ ou $D = 0$. Le type pour lequel $D = 0$ est la même qu'en hiver (O_0), ce qui n'est pas étonnant, car dans le type O_0 on peut trouver des distributions de pression assez différentes.

Pour la saison chaude on peut désigner les groupes caractéristiques suivants. (*a*) Types cycloniques des secteurs N et E (NW$_c$, N$_c$, NE$_c$, E$_c$, SE$_c$); fréquence totale de ce groupe 16 pour cent, type de temps caractéristique 13 – froid, mauvais temps; (*b*) types anticycloniques des secteurs N et E (NW$_a$, N$_a$, NE$_a$, E$_a$, SE$_a$); fréquence totale 13 pour cent, type de temps caractéristique 11 – assez froid et beau; (*c*) types cycloniques du secteur sw (S$_c$, SW$_c$, W$_c$); fréquence totale 12 pour cent, type de temps caractéristique 33 – chaud et humide; (*d*) types anticycloniques du secteur sw (S$_a$, SW$_a$, W$_a$); fréquence totale 17 pour cent, type de temps caractéristique 31 (ou 21) – chaud et beau temps.

Ces quatre groupes comprennent 58 pour cent des cas. Le type le plus fréquent est SW$_a$ avec fréquence 11.1 pour cent et type de temps caractéristique 31 (plus rarement 21).

D'après le matériel classifié on peut constater qu'il existe une dépendance bien prononcée entre le type de circulation dans le Canada oriental et le type de temps à Québec. Cette dépendance est plus forte en été qu'en hiver. On peut constater aussi qu'elle est meilleure pour les types anticycloniques que pour la circulation cyclonique. Les types de circulation qui définissent assez mal le type de temps sont les types intermédiaires (classe de pression '0').

La durée moyenne des types de circulation est de 2.1 jours. Il faut souligner que cette durée est presque 3 fois moins longue qu'en Europe centrale (5.7 jours). On peut expliquer ce fait par l'absence de phénomène de blocage sur le continent nord-américain.

Même si plusieurs conclusions peuvent être tirées du matériel déjà classifié, une plus longue période de classification permettra de mieux analyser certains aspects de la circulation. Pour cette raison on pense prolonger la période de classification sur une base de 30 ans.

TABLEAU 3. Les types de circulation et de temps pour Québec, 1963-9

Type	Types de circulation Fréquence (%)	Durée (jours) moyenne	max.	min.	Types de temps Principaux	Secondaires	Dépendance Ecart moyen	Indice
L'HIVER (mois XII, I, II, III)								
Nc	2.6	2.4	4	1	13	13	0.62	2
No	2.3	2.1	4	1	21 12	21	0.80	2
Na	4.1	2.1	5	1	11	11 21	0.58	2
NEc	8.4	2.8	6	1	23 12	23 13	0.99	2
NEo	3.0	2.2	5	1	21 12	21	0.98	2
Nea	3.4	2.0	3	1	11	—	0.49	3
Ec	3.5	1.5	3	1	23 33	23	0.60	2
Eo	4.0	2.9	7	1	32 33	32	0.78	2
Ea	3.5	3.3	5	1	32 21	32	0.97	2
SEc	3.5	1.6	3	1	33	—	0.47	3
SEo	2.0	2.2	4	1	32 33	32	0.72	2
Sea	2.9	3.2	7	1	31 21	31	0.85	2
Sc	2.1	2.0	4	1	33	—	9.49	3
So	4.7	2.2	5	1	32 33	32	0.73	2
Sa	4.0	1.6	3	1	31 22	31	1.00	2
SWc	3.5	1.9	3	1	33 32	33	0.71	2
SWo	3.1	2.2	5	1	32 22	32 23	1.03	1
SWa	4.9	1.9	4	1	11 12	11	0.77	2
Wc	3.0	1.9	3	1	23 33	23	0.99	2
Wo	0.9	2.0	3	1	11	11 21	0.54	2
Wa	3.5	1.5	2	1	11	—	0.44	3
NWc	4.0	2.1	6	1	13 22	13 23	0.84	2
NWo	4.5	1.9	5	1	12 11	12 21	0.89	2
NWa	5.0	2.1	5	1	11	—	0.42	3
Oc	2.2	1.8	4	1	32 33	32 22	0.96	2
Oo	3.5	2.1	3	1	—	—	1.25	0
Oa	3.2	1.8	4	1	21 11	21	0.58	2
X	2.8							
LA SAISON CHAUDE (mois VI, VII, VIII, IX)								
N_c	4.5	1.9	3	1	13	—	0.48	3
N_o	3.6	1.8	4	1	12	22 23	0.66	2
N_a	2.1	1.8	3	1	11	—	0.25	3
NE_c	1.4	2.3	3	2	13	12 22	0.75	2
NE_o	1.4	1.8	3	1	13	12 23	1.25	1
NE_a	2.6	1.9	4	1	11	—	0.28	3
E_c	2.9	2.4	4	1	13 12	—	0.45	3
E_o	3.0	2.2	7	1	22	13 12	0.79	2
E_a	2.3	1.7	4	1	11 21	—	0.28	3
SE_c	1.8	2.0	6	1	13	—	0.42	3
SE_o	1.8	1.7	3	1	13	23	0.74	2
SE_a	0.5	1.5	2	1	21 11	—	0.36	3
S_c	3.6	1.6	4	1	33	23	0.58	2
S_o	2.0	1.9	3	1	22	21 13	1.08	1
S_a	1.5	2.0	4	1	31	—	0.12	3
SW_c	4.0	2.6	5	1	33	32	0.63	2
SW_o	5.7	1.9	4	1	32	31 23	0.83	2
SW_a	11.1	2.5	6	1	31 21	—	0.47	3
W_c	4.1	1.8	5	1	33	32 23	0.70	2
W_o	6.0	2.0	4	1	22	33 32	0.96	2
W_a	4.7	2.0	4	1	21	—	0.39	3
NW_c	5.6	2.0	4	1	13 23	12	0.71	2
NW_o	3.8	1.6	3	1	22	12 13	0.96	2
NW_a	5.6	2.1	4	1	11 21	—	0.48	3
O_c	3.8	1.9	4	1	23	13 22	0.81	2
O_o	3.9	2.0	3	1	—	—	1.35	0
O_a	3.8	2.3	5	1	31 21	—	0.30	3
X	2.9	—	—	—	—	—	—	—

Index of Authors and Co-authors
Index des auteurs et coauteurs

This index lists alphabetically the authors and co-authors of all papers in Volumes 1 and 2. The entries are by paper number. The organization of papers within the volumes is described in the Preface.

Cet index est le répertoire alphabétique des auteurs et coauteurs de toutes les communications du Volumes 1 et 2. Les communications sont identifiées par leur numéro d'ordre. Le plan des volumes est décrit dans la Préface.

Aase, Asbjørn　c0101
Abiodun, J. Olu　s0305
Aceves-Garcia, M.　p0924, c1501
Adams, W.P.　p0244
Adejuyigbe, O.　p0710
Agarwal, P.C.　p1140
Agboola, S.A.　c0901
Ahnert, Frank　p0101, c0401
Alaev, Enrid B.　p0401
Alberti, Maria Paola Pagnini　p1141
Alexander, Charles S.　p0315
Aliev, H.A.　p0301
Al-Khalaf, Jassim M.　p0402
Allen, J.L.　p0501
Amiran, David H.K.　p1003
Anderson, D.L.　p1228
Anderson, James R.　p1004, c0902
Andrews, J.T.　p0102
Andryushchenko, A.A.　p1201
Appleton, J.H.　c1603
Are, Felix　p0142
Arkhipov, Y.R.　p1232
Aschmann, Homer　p0901
Auliciems, Andris　p0906
Aussin, C. Murray　p1101
Auwerter, John　p0403

Babonaux, Yves　p1143
Bach, Wilfrid　p0201, c0701
Bachenina, N.V.　p1317
Bachurin, G.V.　s0602
Bagdassarian, A.B.　p0915
Bailey, Harry P.　p0202
Bancila, Octavian　p1230
Bandyopadhyay, M.K.　p0103
Banerjee, Gouri　p0103
Bariss, Nicholas　p0156
Barrett, Frank A.　p0601
Bastian, Robert W.　p0502
Batalla, A. Bassols　p0404
Batchelder, Robert B.　p0242
Bateman, M.　p0602

Baumann, Duane D.　s0222
Beaubien, J.　p0307
Beaujeu-Garnier, Jacqueline　p1143
Beckinsale, Robert P.　s0201
Beguin, Hubert　p0927
Belair, J.L.　p0307
Bell, E.L.　p0316
Belyaev, V.I.　p1201, p1202, p1223
Ben-Arieh, Y.　p0503
Bernát, Tivadar　p1102
Berry, L.　p0918, c0403
Bhardwaj, Surinder M.　p0801
Billingsley, Douglas　p0903
Bird, Eric C.F.　c0301
Blackbourn, A.　p0802
Blacksell, Mark　p0903
Blagovolin, N.S.　p0104
Blajko, N.I.　p1232
Blazek, Miroslav　p1142
Board, Christopher　c1210
Bockheim, J.G.　p0329
Bohland, James R.　p1005
Bohra, D.M.　p0405
Bonuzzi, V.　c0903
Bora, Gyula　p0430
Box, Elgene　p0214
Brandt, W.　p0511
Braun, Juliusz　p0916
Brazel, Anthony J.　p0203
Breton, Roland J.-L.　p0610
Brown, Eric H.　p0902
Brown, Robert Wylie　p0804
Brown, T.W.　c0502
Brown, William H. Jr.　s0101
Browning, Clyde E.　p0805
Brunet, Roger　p1143
Brunger, A.G.　p0244, p0504
Brush, John E.　s0306
Bruyelle, Pierre　p1143
Bugmann, E.　p1006
Bunting, B.T.　p0302
Burghardt, A.F.　p0701

Burtenshaw, D.   p0602
Burton, Ian   p0903, p0906, p0914
Butler, J.H.   p0904
Butlin, R.A.   p0505
Buttner, M.   p1229
Bylund, Erik   s0401

Cailleux, André   p0157
Caldo, C.   p0603
Cameri, Giovanni   p0838
Cameron, James M.   p0506
Candida, L.   p0926
Carr, Alan P.   p0105
Casetti, E.   p1203
Cataudella, Mario   s0702
Caviedes, C.N.   p0106, p0511
Cermakian, Jean   p0406, c1303, c1605
Ceron, A.O.   c0904
Chang, Jen-Hu   p0245
Chapman, Murray   c1221
Chappell, John E. Jr.   p0204
Chatterji, S.C.   p1231
Chauhan, V.S.   p1007
Chaussade, J.   s0102
Christopher, A.J.   p0507, s0402
Chudnova, V.I.   c1220
Clark, Michael J.   c0503
Clark, Richard   p1103, c0504
Clark, W.A.V.   p1104
Clarke, Colin G.   p1105
Clarke, John   p0508
Clement, Pierre   p0158
Cliff, Andrew D.   p1233
Cole, Alan L.   p0246
Cook, Earl   p0202
Cooper, A.D.   c0505
Coque, Roger   p0159
Correa-Perez, G.   p0921
Costa Barbosa, Ignez   p1122
Cowen, David J.   p1008
Cribier, Françoise   c1224
Crozier, M.J.   p0148
Curry, Leslie   p1204

Das, Nikhilesh   p0240
David, E.L.   s0203
Davies, Gordon S.   s0205
Davies, R.J.   p1106
Davies, Wayne K.D.   p1107
Davis, J.F.   s0406
Davis, R.K.   s0208
Davitaya, F.   p0241
Day, J.C.   s0204
De Koninck, R.   p1009
Demangeot, Jean   p0160

Demek, Jaromír   p0107, c0801
Demko, Donald   p1131
Denis, Robert   p0143
Denman, D.R.   p0928
Derenyi, Eugene E.   p1301
Dever, G.E. Alan   c1703
De Vorsey, Louis Jr.   p0509
Dickenson, John Philip   p0806
Dickinson, Joshua C.   p1010
Dickson, K.B.   p0510
Diniz, J.A.F.   c0905
Dionne, J.-C.   p0307, c0308
Dobrzański, B.   p0320
Doddridge, K. Pat   p0807
Dojcsak, G.V.   p0149
Domanski, Ryszard   p0808
Dorion, H.   p0709
Douglas, J.N.H.   s0103
Dresch, Jean   p0161
Dugrand, Raymond   p1143
Dutt, Ashok K.   p0809
Dutt, G.K.   p1011
Dynowska, Irena   p0205
Dzus, Roman   p0910

Edgar, D.E.   p0132
Edgell, Michael C.R.   p0902
Eidt, Robert C.   p0511, s0407
Eiumnoh, A.   p0325
Eliot Hurst, M.E.   p0810, c1606
Ellefsen, Richard A.   p1302
Elliott, Laurence Hugh   p0407
Endriss, Gerhard   p0537, p1012
Engmann, E.V.T.   c1211
Enjalbert, Henri   p1143
Enyedi, Gyorgy   c1301
Ermolenko, A.I.   p1201
Estall, Robert C.   p0811
Evans, Ian S.   p0108

Falconer, A.   p0109
Fedina, Aleksandra   p0408
Fedoroff, Nicolas   p0330
Fenelon, Paul   p0150, s0507
Fenwick, I.M.   p0331
Figueroa-Alcocer, E.   p0922
Findlay, B.F.   p0206, p0243
Fischer, D.W.   p0205
Fletcher, Roy Jackson   p0207
Flower, D.J.   p1013, c1101
Foley, M.   p1305
Ford, D.C.   p0110, s0501
Foster, Harold D.   s0206
Found, W.C.   c0906, c1611
Francis, Karl E.   c1607

Fränzle, Otto   p0303
Fraser, D.A.   p0332, p1043
French, Hugh M.   p0144
Fuentes Aguilar, Luis   p1014
Fuggle, R.F.   p0224

Gabert, Pierre   p1143
Gabrielian, H.K.   p0162
Gade, Daniel W.   p0323
Gaertner, Erika   p0247
Gagnon, D.   p1310
Gale, S.   p1205, c1201
Garcia-Leon, F.   p0812
Garnier, B.J.   p0208
Gassaway, A.R.   p0409
Geiger, Pedro Pinchas   c1222
Gensiruk, Stepan   p0917
Gentilcore, R. Louis   p0512
Gentileschi, M.L.   p0604
Geradin, V.   p0307
Gersmehl, Phil   c0201
Gill, Don   p0304
Giovinetto, Mario B.   p0248
Goddard, A.D.   c1217
Goldenberg, L.A.   p0513
Goossens, M.   p0846
Gormsen, Erdmann   s0301
Gosal, Gurdev Singh   c1212
Gould, W.T.S.   c1217
Goward, Samuel N.   p0242
Gradus, Y.   p1108
Graniel-Graniel, H.   p0111
Graves, N.J.   c0506
Gray, James T.   p0112
Greenland, David E.   p0257
Greer-Wootten, B.   p1146
Gregor, Howard F.   p1015, c0907
Grelou, G.   p0410
Groves, Paul A.   p0514
Guilcher, André J.   c0307
Gutierrez de MacGregor, M.T.   p1109, c1213
Gutierrez de Velasco, A.   p0847
Guzman-Villanueva, R.   p0813

Hafner, James A.   c1608
Hails, John R.   p0105
Hamelin, L.-E.   p0431, c0507
Hamilton, Patrick   p0426
Hancock, John C.   c0508
Hanke, S.H.   s0208
Hannell, F.G.   p0209
Hare, F.K.   p0210
Harley, John Brian   p0515
Harris, S.A.   p0326

Hartshorn, T.A.   p0834
Harvey, Milton E.   p0801, c1225
Hathout, Salah A.   p0302
Haupert, John S.   p1016
Head, C. Grant   p0516
Heathcote, R.L.   c0702, s0403
Heinritz, Günter   p0517
Helburn, N.   p0905
Helle, Reijo K.   p1017
Helleiner, F.M.   c1601
Henoch, W.E.S.   p0249
Hess, M.   p0250
Hesselink, H.   p0518
Hewes, Leslie   p0519
Hewings, John   p0906
Hill, Alan R.   p0305
Hirt, M.S.   p0243
Hodgin, Jean   p0411
Hoffman, J.   p0911
Hoffman, Wayne L.   p1110
Hofmeister, B.   p1111
Höhl, Gudrun   p0412
Holmes, R.M.   p0238
Holtzman, Betty   p1039
Homenuck, H.P.M.   p1234
Hones, G.H.   c0509
Hooper, John O.   p1314
Hottes, Karl H.   c1609
Howard, William A.   p1303
Howe, G. Melvyn   p0929, c1704
Hsu, Shin-Yi   p1206, c0102
Hudson, John C.   c1202
Hufty, André   p0251
Hulbert, François   p0925
Hung, Frederick   p1018
Husain, S. Sajid   c1701

Ianco, Mihai   p1230
Ichikawa, Masami   p0113
Ilesic, Svetozar   p0432
Innes, F.C.   p0520
Innis, Donald Quayle   p1019
Inyang, Paul E.B.   p0211
Isakov, Y.A.   c0202
Ishida, Hiroshi   p0605
Ivanička, Koloman   p0839

Jaatinen, Stig T.   p0408
Jackson, James C.   p1020
Jackson, William B.   p0212
Jacob, Günter   p0433
Jahn, Alfred   p0114
Jarochowska, M.A.   s0703
Jay, Leslie Joseph   p0413
John, B.S.   p0533

Johnson, Claude p0202
Johnson, Gary E. p1306
Johnson, H.B. p0521
Johnson, James H. c1203
Johnson, P.G. p0115
Johnston, R.J. p1146, c1204
Jolliffe, Ivan Phillips p0907
Joshi, Tulasi R. p1207
Journaux, André p0163
Joynt, Marsha I. p0306
Juilliard, Etienne p1143
Jurdant, M. Michel p0307
Jusatz, H.J. p1307
Juvik, James O. p0245, p0308
Juvik, S.P. p0308

Kaikow, J. p0930
Kampp, Aage H. p1021, c0916
Kar, Nisith R. p1112
Karrasch, H. p0116
Kasperson, Roger E. s0222
Kasumov, Rufat M. p0840
Katasonov, E. p0117
Kates, Robert W. p0918
Kaushic, S.D. p0168, p1002
Kavrichvili, K. p1208
Kayser, Bernard p1143
Kedar, Ervin Y. p1304, c0102
Keeble, David E. p0841
Kellman, Martyn C. p0309
Kellom, John B. p1305
Kenntner, Georg p0333
Kerr, William Sterling III p0908
Kidson, C. c0309
Kielczewska-Zaleska, Maria p1113
Kimber, Clarissa p0310
King, L. p1203
King, Roger Hatton p0334
Kirchherr, E.C. p1114
Klingeman, P.C. p0229
Knight, David B. p0702
Kohl, H. p1134
Kolbert, C.F. p0909
Komar, Igor V. p0931
Kondo, Naruo p0311
Konstantinov, Oleg p1145
Kosiński, Leszek Antoni c1214
Kostrowicki, J. c0908
Kracht, James B. p1303
Král, Jiří c1705
Krauklis, A.A. p1235
Kress, W.D. p0414
Krishan, Gopal c1212
Kriz, Hubert p0213
Krumme, Gunter p0814

Kularatnam, K. p1144
Kulkarni, Gopal S. s0302

Lacasse, J.-P. p0711
Lacava, Gerald J. p1225
Lamont, Glenda c1223
Lancaster, Jane c1610
Landing, James E. p0522, s0701
Lanegran, David A. p0415
Lantis, David W. p0416
LaValle, Placido p0910, s0209
Learmonth, A.T.A. c0510
Lee, Chung-Myun p0815, c0909
Lee, T.R. s0210
Lefebvre, J.J. p1315
Le Heron, Richard B. p0816
Libbee, M.J. c1208
Lieth, Helmut p0214
Lijewski, Teofil p0848, c1205
Lino Grima, A.P. s0207
Litovka, Oleg p1135
Litynski, J. p0258
Liu, Hung Hsi p0417
Lo, Chor Pang p1308
Loeffler, M. John s0211
Longley, Richmond W. p0239
Lonsdale, Richard E. p0817
Lopez-Santoyo, A. p0118
Lougeay, Ray p0215
Lovingood, Paul E. Jr. p1008
Lu, Jonathan J. p0818
Luckman, B.H. p0119
Luedemann, Heinz p0434
Lundgren, Jan O.J. p1022
Luyot, L. p0253
Lycan, D. Richard c1206

Macar, Paul p0164
MacFarlane, M.A. p0120
MacGregor-Loaeza, R. p0335
Maderey-Rascón, Laura p0216
Maier, Jörg p0932, c1602
Mainguet-Michel, Monique p0121
Malek, Richard B. p0315
Manzi, Elio p0418
Margulis, Harry L. p0819
Massam, B.H. p0703
Matsui, Takeshi p0311
Maxfield, Donald W. p1209
McBoyle, G.R. p0217, c1302
McCann, L.D. p1115
McCann, S.B. p0122
McCune, Shannon p0704
McCutcheon, John G. p0230

McIlwraith, Thomas F.   p0523
McKay, G.A.   p0206
McNulty, M.L.   p1116
Mednis, Roberts J.   p0312
Megee, Mary   p0820
Melamid, Alexander   p0821
Meland, Nils   p0911
Merlin, Pierre   p0419
Meynen, E.   p1210
Michailov, U.P.   s0602
Michie, George H.   p1023
Mielke, Howard W.   p0336
Miller, David H.   p0218
Miller, E. Willard   p1117
Miller, G.H.   p0102
Miller, John J.B.   s0104
Mints, A.A.   p0420
Miotke, Franz-Dieter   s0502
Mirnova, A.V.   p1317
Miron, John Robert   p1211
Mistardis, G.   p0421
Misztal, Stanislaw   p0822
Mitchell, B.   s0223
Mitchell, Robert D.   p1024
Mitsui, Kazuo   p0219
Moiseyev, G.A.   p1223
Momsen, Janet H.   c0910
Monmonier, Mark S.   c1707
Montiel-Hernandez, A.   c1708
Mookerjee, S.   p1212
Mookherjee, Debnath   p1118
Moore, Robert P.   p1314
Morain, Stanley A.   p0313
Morissonneau, Christian   p0534
Morley, C.D.   c0906, c1611
Morozova, T.D.   p0327
Morrill, Richard L.   c1401, s0105
Morrison, A.   p1309
Mosley, M.P.   p0132
Mottershead, D.N.   p0123
Muckleston, Keith W.   p0212
Mukerjee, Sudershan   p0124, p0606
Mukherji, S.   p1119, c1226
Müller, Paul   p0337
Muller, Peter O.   p0823
Muller, Robert A.   p0220
Munn, R.E.   p0243
Murton, Brian J.   p0607, c0703, c0704
Muscara', Calogero   p0435
Mussivand, T.V.   p0224

Nader, G.A.   p1120
Naidoo, Munsamy B.   p0145
Newcomb, Robert M.   p0524
Ng, Ronald C.Y.   c1215

Nicholls, Alan D.   c0511
Nigam, Raj Kumar   p0422, p1042
Nitz, H.J.   p1025
Nömmik, Salme   p0842
Nonn, Henri   p1143
Nunley, Robert E.   p1213
Nystuen, John D.   p0609

Oberlander, Theodore M.   p0125
Occhietti, S.   p0165
Ofomata, Godfrey Ezediaso Kingsley   p0126
Ogundana, B.   p0824
Oguntoyinbo, Julius Sunday   p0221
Ohmura, Atsumu   p0223, p1214
Ojo, G.J. Afolabi   p0825
Ojo, Oyediran   p0222
Oke, T.R.   p0224
Oliveira, L.   c0512
Ommanney, C. Simon L.   p1310, s0213
Ongley, Edwin David   p0166
Ord, J. Keith   p1233
O'Riordan, J.   p0214
O'Riordan, Timothy   c0705
Owens, Edward H.   c0302
Oya, Masahiko   p0127, c0802
Ozoray, George F.   p0128, s0503

Pal, Indra   p1121
Palamarchuk, Maxim   p0843
Paludan, Charles T.N.   c1502
Panda, B.P.   p1030
Pande, Shyam Narain   p0525
Pandey, J.N.   p1037
Panfilov, D.V.   c0202
Panoš, Vladimír   p0151, s0508
Papageorgiou, George J.   p1215
Parker, M.L.   p0249
Parker, R.S.   p0132
Parkes, J.G.M.   s0215
Paskoff, R.P.   p0152
Paszyński, Janusz   p0225
Paterson, John H.   s0404
Patmore, John Allan   p0912
Paviani, Aldo   p1122
Payne, G. Frederick   p1026
Pecora, Aldo   p1027
Pecsi, Marton   p0129
Peeters, Leo   p0153
Péguy, Ch.-P.   p0252
Perry, J.W.B.   c1216
Peruzzi, Duilio   p1302
Phillips, David W.   p0226
Pike, Richard J.   p0423
Pissart, Albert J.G.   p0169

Place, John Louis   p0913
Plesnik, P.   p0338
Popp, N.   p0167
Porenius, Per   p1028, s0601
Poulsen, M.F.   c1204
Pouquet, Jean   p1318
Preobragensky, V.S.   p1216, p1232
Price, Larry W.   p0130
Probald, Ferenc   p0227
Prokhorov, B.B.   s0602
Prothero, R.M.   c1217
Proudfoot, Bruce   c1223
Putnam, William C.   c0803
Pyle, G.F.   p0809

Quinlan, James F.   p0131
Quinn, F.J.   s0216

Racine, J.-B.   p1217
Rahman, Mushtaq-ur   p1044
Rakitnikov, André   p1029
Ramdas, L.A.   p0228
Randall, Howard Adrian   p0526
Rao, D.N.   p0324
Rapp, Anders   c0403, c0404
Rathjens, Carl   c0706
Raup, Henry A.   s0217
Rawstron, Eric M.   p0826, c0513
Rees, John   p0827
Rengert, A.C.   p0608
Rengert, G.F.   p0608
Riddell, J. Barry   p0828
Roberts, M.C.   p0229
Robertson, John C.   s0225
Robinson, J.M.   s0103
Roglic, J.   p1218
Rognon, Pierre   p0146
Romsa, Gerald   p1110
Roper, Geoffrey W.   p1213
Rosenfeld, Charles L.   p0154
Ross, William M.   s0218
Roth, Irvin J.   p0844
Rouse, Wayne R.   p0230
Rowland, D.T.   c1204
Roy, B.K.   c0911, c1218, c1503
Rumley, Dennis   p0705
Rutherford, G.K.   p0314
Ryba, R.H.   c0514
Ryder, Roy H.   p0339
Ryerson, Robert A.   c0915

Sada, Pius O.   p1116
Saenz de la Calzada, C.   c1709
Saint-Laurent, Gilbert   p0424
Salt, John   c1203

Samano-Pineda, C.   p0923
Sandru, Ioan   p0425
Sant, Morgan E.C.   p0829
Santos, Milton   s0303
Sargent, Charles S. Jr.   p1123
Sari, Djilali   p0147
Saxena, J.P.   p1030, p1031
Scarfe, Neville V.   c0515
Scarin, E.   p1319
Scarlett, M.J.   p1219
Scheidl, Leopold G.   p0830, c1304
Schiel, Joseph B. Jr.   p0315
Schiff, Myra   p0906, p0914
Schmidt, Charles G.   p0831
Schulz, Peter   p0706
Schumm, S.A.   p0132
Schwarcz, H.P.   p0110
Schwartzberg, Joseph E.   p0527
Sdasyuk, Galina V.   p1124
Seguin, M.B.   p0253
Semevskiy, B.N.   p1220
Shabad, Theodore   p0849, c1207
Shafi, Mohammad   p1032, c1305
Shannon, Gary W.   p0609
Sharma, H.S.   p0133
Shear, James A.   p0231
Shepherd, R.G.   p0132
Shimabukuro, Shinzo   c0704
Shiryaev, E.   p1311
Shlemon, Roy J.   p0316
Shotskiy, V.P.   p1045, s0602
Shroder, John Ford   p0134, c0803
Silk, John A.   p1221
Silva, W.P.T.   p1033
Singh, Basant   p0707, p1034
Singh, R.L.   p1035
Singh, Ram Bali   p1035
Singh, Surendra   p1007
Singh, Ujagir   p1036, p1037, p1125
Singh, Vijaya Ram   p1036
Slater, F.A.   c0516
Smith, Peter J.   p1115
Smith, Robert H.T.   p0832
Smotkine, Henri   p1136
Smyth, Anthony John   p0317
Snytko, V.A.   p0318
So, Chak Lam   c0303
Sochava, V.B.   p0328, s0602
Sokolovsky, Igor   p0135
Solodzuk, W.   s0224
Soma, Masatane   c0912
Sopher, David E.   c1208
Spence, E.S.   p0232
Sporck, José A.L.   p1132
Sprincova, Stanislava   p0919

Squires, Roderick Hugh p0319
Stafford, Howard A. p0833
Stapor, F.W. c0304
Starkel, Leszek c0402
Stauffer, Truman P. Sr. p0920
Stebelsky, I. p0528
Steed, Guy P.F. p0803
Stepanov, V.N. p0233
Stephenson, R.A. p0834
Stoltman, J.P. c0501
Stone, Jeffrey C. p0529
Stone, Kirk H. p1038
Stone, Leroy O. p1222
Stoupichin, A.V. p1232
Stouse, Pierre A.D. Jr. p1039
Straszewicz, Ludwik c1306
Stutz, F.P. p1126
Suddhiprakarn, A. p0325
Sugden, David E. p0426
Sunamura, Tsuguo p0136
Sundstrom, Marvin T. c0917
Suzuki, Takasuke p0136
Swindell, K. c1209
Syers, J.K. p0511

Takahashi, Ken'ichi p0136
Tamaskar, B.G. c1219
Tanner, W.F. c0304
Tarmisto, V. p0850
Taylor, Chris p0906
Temple, P.H. p0137, c0404
Terjanian, A.S.R. p1315
Thomas, H.E. s0219
Thompson, P.L. p0110
Thouvenot, C. c1706
Thrower, Norman J.W. p1312
Tietze, Wolf c1307
Timchenko, I.E. p1201, p1223
Tomkins, G.S. p1224
Trimble, Stanley W. p0530
Trofimov, A.M. p1232
Tschinkel, Tim p0905
Tsvetkov, D.G. p0104
Tweedie, A.D. s0220
Tyagi, B.S. p1036
Tyman, J.L. p0531, s0704
Tyson, P.D. p0234

Udo, R.K. p1040
Ugolini, F.C. p0329
Urlanis, Boris p0536
Utenkov, N.A. p0427
Uziak, Stanislaw p0320

Valentin, Hartmut c0305

Valessyan, L.A. p0845
Valussi, G. p0535
Valverde, Carmen p1109
Van Hylckama, T.E.A. p0235
Vanzetti, C. c0913
Varep, E. p0538
Varjo, Uuno p1046, c0918
Varma, O.P. p0428
Verhasselt-Van Wettere, Y. c1702
Verma, V.K. p0138
Veyret, Germaine p1143
Veyret-Verner, Germaine p1137
Vigarie, André C. p1138, c1604
Villeneuve, Paul Y. p1127
Villmow, Jack p0254
Vishwakarma, Y.B. p1121
Vityazeva, V.A. c1001
Volkowitsch, Maurice p1143
Vorobyev, V.V. c1220, s0602

Walczak, Wojciech p0139
Walker, Gerald p1041
Wall, Geoffrey p0903
Wallace, Iain c1612
Waller, Peter P. p1128
Waltham, A.C. s0504, s0505
Wang, I-Shou p1129
Warman, Henry J. c0517
Warren, Harry V. c1710
Watts, David c0601
Waxmonsky, Raymond W. p0835
Webster, Robert Arthur p0532
Weir, T.R. p1047
Wellar, Barry S. p1225
Welsted, John E. c0306
Wendland, Wayne M. p0255
Werner, C. p1226
Wieckowski, Michał p1316
Wiek, Klaus D. p0836
Wilkinson, Paul p0914
Williams, F. p1203
Williams, G.D.V. p0236, p0256, p0306, s0405
Williams, Owens p0837
Wirth, Eugen s0304
Witayarut, Prasert p1139
Witkowski, Stefan p1133
Wolfe, Roy I. c1227
Wolfe, Thomas E. p0140, s0506
Wood, Harold A. c0914
Wood, Peter A. c1203
Woodruff, James F. p0141
Woolmington, E. s0221
Wray, James R. p1313
Wright, J.L. p0238

Yano, Yoshiji p0311
Yap, D. p0224
Yatsu, Eiju p0109
Yeates, Maurice H. p1130
Yonekura, Jiro p0429
Yoshino, Masatoshi M. p0237
Young, A. p0321, c1504

Zaborski, Bogdan p1001
Zaborski, Jerzy p1227
Zaidi, Iqtidar H. p0708
Zimina, R.P. c0202
Zimm, Alfred p1134
Zimmerman, Carol p0403
Zobler, Leonard p0322
Zolovsky, Andrey p1320
Zonneveld, J.I.S. p0155

Selected Index of Locations
Index des noms de lieux

This index lists alphabetically the area or areas of the world to which the papers in Volumes 1 and 2 relate, using, wherever possible, the geographic areas adopted by the Association of American Geographers. In addition, many papers are identified with a locality within the broad geographic areas, where a major portion of the paper is explicitly related to a location. The entries are by paper number. The organization of papers within the volumes is described in the Preface.

Cet index est un répertoire alphabétique des noms des lieux sur lesquels portent les communications des Volumes 1 et 2. Autant que possible, on a classé chaque communication dans une des zones géographiques définies dans la classification de l'Association des géographes américains. De plus, pour plusieurs communications, on a pu identifier une localité à l'intérieur d'une grande région géographique. L'index ne comprend cependant pas tous les noms de lieux mentionnés dans les communications. Les communications sont identifiées par leur numéro d'ordre. Le plan des volumes est décrit dans la Préface.

Aberdeen P0217
Aconcagua Valley P0106
Africa P0121, P0126, P0137, P0145, P0146, P0147, P0211, P0221, P0222, P0234, P0303, P0507, P0702, P0710, P0804, P0824, P0825, P0828, P0837, P0838, P0918, P1040, P1106, P1114, P1116, C0403, C0404, C0502, C0901, C1209, C1210, C1211, C1216, C1217, C1225, S0102, S0305, S0402
Africa, East P0137, P0918, C0403, C0404, C0502
Africa, North P0121, P0146, P0147, P0804
Africa, south of Congo P0145, P0234, P0303, P0507, P0702, P0837, P1106, C1210, C1216, S0402
Africa, tropical C1217, C1225
Africa, West P0126, P0211, P0221, P0222, P0710, P0824, P0825, P0828, P1040, P1114, P1116, C0901, C1209, C1211, S0102, S0305
Alabama C1502
Alaska P0203
Alberta P0110, P0119, P0249, P1013, P1115, C1101, C1223
Algeria P0147
Andes Mountains P0339
Anglo-America See North America
Antarctica P0329
Antwerp C1702
Appalachians, Southern C0201
Aquarius Plateau P0134, C0803
Arasaki coast P0136
Arctic Ocean P0117
Argentina P1123
Arizona P0235
Asia P0103, P0113, P0133, P0136, P0138, P0142, P0159, P0168, P0219, P0228, P0237, P0240, P0301, P0311, P0325, P0327, P0402, P0405, P0417, P0418, P0420, P0427, P0429, P0503, P0607, P0801, P0815, P0840, P0844, P1002, P1007, P1009, P1016, P1018, P1020, P1025, P1030, P1031, P1033, P1034, P1035, P1036, P1037, P1042, P1044, P1108, P1119, P1121, P1125, P1129, P1139, P1140, P1141, P1144, P1207, P1231, P1235, C0303, C0402, C0703, C0706, C0909, C0911, C0912, C1215, C1218, C1219, C1220, C1608, C1701, S0302, S0304, S0505, S0602
Asia, South P0103, P0133, P0138, P0168, P0228, P0237, P0405, P0607, P0801, P0844, P1002, P1007, P1025, P1030, P1031, P1033, P1034, P1035, P1036, P1037, P1042, P1119, P1121, P1125, P1140, P1141, P1207, P1231, C0402, C0703, C0706, C0911, C1218, C1219, C1701, S0302, S0505
Asia, Southeast P0127, P0309, P0325, P0418, P0815, P1009, P1018, P1020, P1139, C0909, C1215, C1608
Asia, Southwest P0159, P0402, P0503, P1016, P1108, S0104, S0304
Atsumi Peninsula P0113
Australia P0313, P0821, P0902, P1047, C0301, C0702, S0220, S0221, S0403
Austria P0225, P0830, C1304
Avignon/Montfavet P0253
Axel Heiberg Island P0223, P1214
Azerbaijan P0301, P0840

Baffin Island P0102
Baghdad P0402
Baja California P0901, P0924

Baltimore s0222
Baluchistan, Southern P1231
Banks Island P0115, P0144
Banwaripur P1007
Barrie P0302
Barron Delta c0301
Barry's Bay s0703
Barvaux P0164
Bath c0509
Bavaria P0932, c1602
Beacon Valley P0329
Belfast s0103
Belgium P0164, P0846, c1702
Belmont Township P0518
Berlin P1134
Black Sea P1201
Borneo P0418
Boston P0242
Botswana P0702
Brant County P0243
Brasilia P1122
Brazil P0806, P1122, c0512, c0904, c0905
British Isles P0123, P0217, P0319, P0331, P0407, P0413, P0505, P0526, P0532, P0533, P0602, P0826, P0829, P0841, P0903, P0907, P0909, P0912, P0929, P1043, P1103, P1107, P1221, P1233, P1309, c0503, c0505, c0508, c0509, c0510, c0511, c0513, c1203, c1603, c1612, c1704, c1710, s0103, s0217, s0223
British Solomon Islands Protectorate c1221
Broken Hill P0902
Buckinghamshire c0510
Budapest P0227, P1102
Buenos Aires P1123
Bundelkhand P1031

Caen P0163
Calabria P0604
California P0125, P0316, P0416, P0913, P1302, s0219
Canada P0102, P0108, P0109, P0110, P0112, P0115, P0119, P0120, P0122, P0128, P0130, P0143, P0144, P0149, P0157, P0158, P0165, P0166, P0206, P0209, P0215, P0223, P0224, P0226, P0230, P0232, P0239, P0243, P0244, P0247, P0249, P0256, P0258, P0302, P0304, P0306, P0307, P0312, P0332, P0334, P0406, P0424, P0431, P0504, P0506, P0508, P0512, P0515, P0516, P0518, P0531, P0534, P0601, P0602, P0701, P0703, P0711, P0802, P0803, P0836, P0906, P0910, P0914, P0925, P1001, P1013, P1023, P1043, P1115, P1130, P1131, P1214, P1217, P1224, P1310, P1315, c0302, c0306, c0308, c0507, c0906, c0915, c0917,
c1101, c1223, c1227, c1303, c1601, c1605, c1607, c1611, c1710, s0204, s0209, s0213, s0215, s0405, s0703, s0704
Canada, British Columbia P0108, P0602, P0803, s0213
Canada, Maritime Provinces P0312, P0516, c0302, c0306
Canada, Ontario P0109, P0166, P0226, P0230, P0243, P0247, P0302, P0504, P0506, P0508, P0512, P0518, P0701, P0703, P0802, P0906, P0910, P0914, P1023, P1131, c0507, c0906, c0915, c0917, c1227, c1601, c1611, s0204, s0209, s0703
Canada, Prairie Provinces P0110, P0119, P0128, P0232, P0239, P0249, P0256, P0531, P1013, P1115, c1101, c1223, s0215, s0704
Canada, Quebec P0120, P0143, P0158, P0165, P0224, P0244, P0258, P0307, P0406, P0424, P0534, P0711, P0925, P1217, c1303
Canada, Territories P0102, P0112, P0115, P0122, P0130, P0144, P0209, P0215, P0223, P0304, P0334, P0431, P1214, c1607
Caribbean P0310
Caucasus P0162, c0202, c1001
Ceylon P1033
Chalk River P0247
Chambal Valley P0133
Chedabucto Bay c0302
Chhattisgarh Basin P1030
Chile P0106, P0152, P0333
China P0429, P1129, c0303
Chitistone Pass P0203
Christchurch P1146
Cincinnati P0827, P0833
Colorado P0519, P0905, P1303, s0211
Colorado Piedmont s0211
Colorado Plateau c0803
Corsica P0410
Crimean Mountains P0104
Crowsnest Pass P0110
Cuba P0151, s0508
Czechoslovakia P0213, P0839

Danube c1304
Darjeeling P1119
Deer Creek s0204
Denmark P0524, P1021, c0916, s0104
Denver P1303
Detroit P0609
Devon Island P0209, P0334
Durban P0145, P1106

East Germany P0433, P0434, P1134, P1136
Ecuador P0339, P1027

Index des noms de lieux / xxiii

Edmonton P1115
Elkhart County S0701
England *See* British Isles
England, Northern P0319
England, Southeast P0841
England, Southwest P0123
Erie, Lake S0210
Essex County P0910
Estonian SSR P0538, P0850, P0851
Europe P0104, P0123, P0135, P0139, P0150, P0154, P0161, P0163, P0164, P0167, P0205, P0213, P0225, P0227, P0252, P0253, P0314, P0319, P0320, P0330, P0331, P0338, P0407, P0409, P0410, P0412, P0413, P0419, P0421, P0425, P0430, P0432, P0433, P0434, P0435, P0505, P0517, P0524, P0532, P0533, P0535, P0538, P0603, P0604, P0814, P0822, P0830, P0841, P0844, P0846, P0847, P0850, P0851, P0917, P0919, P0929, P0932, P1001, P1006, P1012, P1017, P1021, P1028, P1038, P1043, P1046, P1102, P1107, P1113, P1134, P1136, P1137, P1138, P1142, P1143, P1201, P1203, P1221, P1316, P1319, P1320, C0101, C0202, C0503, C0509, C0510, C0903, C0916, C0918, C1001, C1203, C1205, C1214, C1224, C1301, C1304, C1306, C1602, C1604, C1702, C1704, C1706, S0103, S0104, S0217, S0223, S0401, S0501, S0601, S0702
Europe, Central P0213, P0227, P0430, P0433, P0434, P0517, P0537, P0932, P1102, P1134, C1214, C1301, C1304, C1602
Europe, Eastern P0139, P0167, P0205, P0225, P0320, P0425, P0822, P1113, P1142, P1316, C1205, C1214, C1306
Europe, Scandinavia P0314, P0409, P0524, P1012, P1017, P1021, P1028, P1046, C0101, C0916, C0918, S0104, S0401, S0601
Europe, Southern P0161, P0253, P0330, P0338, P0410, P0421, P0432, P0435, P0535, P0603, P0604, P1319, C0903, S0507, S0702
Europe, Western P0150, P0154, P0163, P0164, P0225, P0252, P0419, P0517, P0814, P0830, P0844, P0846, P0847, P1006, P1132, P1136, P1137, P1138, P1143, C1224, C1604, C1702, C1706

Finland P1012, P1017, P1046, C0918
Finnmark P0409
Florida P0207, P0509, C0304
Fogo Island P0312
France P0150, P0154, P0163, P0252, P0253, P0330, P0410, P0419, P0517, P0844, P1137, P1138, P1143, C1224, C1604, C1706
France, Southern P0330, P0410

France, west coast C1604
Franconia P0517
Fundy, Bay of C0306

Gabon S0102
Ganga Plain C0911
Gangapar Plain P1036
Gaza-Sinai coast P1016
Georgia, USA P1005, P1026, C0501
Georgia, Strait of S0225
Germany *See* East Germany *and* West Germany
Ghana C1211
Glasgow P0929, C1704
Gondwana P0138
Great Lakes P0226, P0246, P0502, S0210, S0408
Great Whale P0120
Greece P0421
Guadalcanal C1221
Guadalupe Island P0922
Guyanas P0155

Haifa P1108
Haldimand County P0243
Hamilton, Canada P0230, P0701
Hawaii C0704, C1226
Himalayas P0133, P0168, P1002, P1119, P1207, C0402, S0505
Hong Kong C0303
Hungary P0227, P0430, P1102, C1301
Hunter Valley S0220
Huron, Lake P0246

Ibadan P0221, P0222
Illinois P0254
India P0103, P0133, P0138, P0228, P0240, P0324, P0405, P0422, P0429, P0525, P0527, P0605, P0606, P0607, P0707, P0801, P0809, P0844, P1002, P1007, P1011, P1025, P1030, P1031, P1032, P1034, P1035, P1036, P1037, P1042, P1118, P1119, P1121, P1124, P1125, P1140, P1141, P1231, C0402, C0703, C0706, C0911, C1208, C1212, C1218, C1219, C1305, C1503, C1701, S0302, S0306, S0505
Indiana P1306, S0701
Iowa C0516
Iran P0159, P1141, S0304
Iraq P0402, S0304
Ireland *See* British Isles
Israel P0503, P1108
Italy P0435, P0535, P0603, P0604, P1319, C0903, C1706, S0702

Jamaica P1022
Jamshedpur P1121
Japan P0113, P0136, P0219, P0311, P0704, C0802, C0912
Japan, Southern Alps P0311

Kansas P0519
Kansas City P0920
Kanto District P0219
Kathmandu P1207
Kentucky P1226
Kenya C0502
Knob Lake P0244
Komi ASSR C1001
Korea *See* South Korea
Kumaon P1025

Labrador P0244
Lagos P1116
Lake District P0532
La Mauricie P0406
Lebanon S0304
Legend Lake, Wisconsin P0911
Leicester P1107
Leith Hill P0331
Lesotho C1216
Libreville S0102
Libya P0804
Liguria P1319
Liverpool P0912
London, Canada P0504, P0914
London, England P1221, C1704
Los Angeles County P0913
Lut Desert P0159

Mackenzie River Delta P0304
Madhya Pradesh P1140
Maharashtra C1219, S0302
Mälar Valley P1028
Malaysia P0815
Malehra P0131
Manchester P0912
Manchuria P1129
Manitoba, Western S0704
Margarita Island P0924
Maria Magdalena Island P0923
Mars P0141
Mauretania C1209
Medicine Hat P1013
Mediterranean P0161
Megalopolis P1117
Mexico P0111, P0118, P0216, P0335, P0404, P0414, P0608, P0812, P0813, P0901, P0921, P0922, P0923, P0924, P1014, P1109, C0304, C1213, C1501, C1708, C1709, S0301

Mexico, arid zones P1109
Mexico, Central P0813
Mexico City S0301
Michigan City P0522, P1104
Michigan, Lake P0246
Middle (Caribbean) America C0304
Mildura P0902
Milwaukee P1104
Minas Gerais P0806
Mindanao P0309
Miura Peninsula P0136
Mojave Desert P0125
Montreal P0224, P1217
Moon P0124
Munich C1602
Murray River Basin S0221

Natal P0507
Nebraska S0406
Nepal P1207, S0505
Nevis C0601
New Hebrides C0307
New Jersey, Eastern shore P0212
New South Wales S0220
New Zealand P0148, P0257, P1146, C0705, C1204
Newark P0819
Newfoundland P0312, P0516
Niagara-on-the-Lake C0917
Nice P0252
Nigeria P0126, P0211, P0221, P0222, P0710, P0824, P0825, P1040, P1114, P1116, C0901, S0305
Norfolk County P0243
Normandy P0154
North America P0210, P0236, P0238, P0335, P0336, P0411, P0414, P0501, P0515, P0523, P1038, P1127, P1234, P1318, C0705, C1206, S0203, S0207, S0210, S0214, S0216, S0218, S0501
North America, semi-arid areas P0336
North Polar Region P0102, P0115, P0117, P0122, P0144, P0209, P0223, P0334, P0431, P1214
Northern Hemisphere P0255, P1001, P1038, S0501
Northern Ireland *See* British Isles
Norway P0314, P0409, C0101
Nsukka Plateau P0126

Ogilvie Mountains P0112
Ohio P0827, P0833
Ojuelos P0608
Ontario, Lake P0226
Oregon P0229

Index des noms de lieux / xxv

Pakistan P0844, P1044
Pamir Mountains P0250
Papaloapan Basin P0111
Pembroke P0533
Peru P1128
Peyto Lake P0249
Philippine Islands P0309
Pisque Valley P0339
Pite Lappmark s0401
Poland P0139, P0205, P0225, P0320, P0822, P1113, P1316, c1205, c1306, c1706
Polar Regions *See* North Polar Region *and* South Polar Region
Pollino s0702
Portsmouth P0602
Puerto Rico P0322, s0502
Puget Sound s0225
Puna District c0704
Pyrenees P0150

Qu'Appelle Valley s0215
Queen Elizabeth Islands P0122
Queensland P0313, P1047, c0301

Rajasthan P0405, P0801, P1034
Revillagigedo Archipelago P0921
Romania P0167, P0425
Rome P0603
Romerike P0314
Rovaniemi P1012
Ruby Range P0130
Ryukyu Islands P0704

Sacramento Valley P0316
Saguenay P0307
Sahara P0146
Saint-Gabriel-de-Brandon P0143
Saint-Jean, Lac P0307
Saint-Narcisse P0165
Salem District c0703
Salerno s0702
San Fernando P1105
San Francisco Bay P1302
São Paulo c0512, c0904, c0905
Saryupar Plain P1037
Saskatchewan s0215
Scotland *See* British Isles
Seattle c1401
Sherbrooke P0158
Shillong P0103
Siberia P0142, P1235, c1220
Siberia, northern c1220
Sierra Leone P0828
Sind P1044
Singapore P1009

Sorocco Island P0921
South Africa, Republic of P0145, P0507, P1106, c1210, s0402
South America P0106, P0152, P0153, P0155, P0160, P0333, P0337, P0339, P0806, P1027, P1122, P1123, P1128, c0512, c0904, c0905, c0914, c1222, s0303, s0407
South America, Northern South America and West Coast P0106, P0152, P0153, P0155, P0333, P0339, P1027, P1128, s0407
South Australia c0702
South Carolina P1008
South Dakota P0415
South Devon P0123
South Korea c0909
South Polar Region P0329
Southern Hemisphere P1112
Spain P0847
St Elias Mountains P0115, P0215
St Lawrence River c0308
St Lucia c0910
Sudetes P0139
Superior, Lake P0246, P0502, s0408
Surprise Valley P0119
Sweden P1028, s0401, s0601
Switzerland P1006
Syria s0304

Taiwan P0417
Tamilnadu P0607, c0703
Tanzania P0137, c0403, c0404
Tchad P0121
Texas c0102, s0202
Thailand P0325, P1139, c1608
Thar Desert c0706
Tippecanoe County P1306
Tizar River Valley P0216
Toronto-centred region c0906, c1611
Transvaal Lowveld P0303, P0507
Trent–Severn Canal c1601
Trieste P0535
Trinidad P1105
Tropics P1010, P1019
Tryweryn Reservoir, Wales s0223
Turkey s0104

Ukrainian SSR P0135, P0917, P1320
Uluguru Mountains c0404
Union of Soviet Socialist Republics (Asia) P0104, P0135, P0250, P0528, P0538, P0850, P0851, P0917, P1135, P1145, P1201, P1317, P1320, c0202, c0507, c1001, c1207, c1220
Union of Soviet Socialist Republics (European) P0142, P0301, P0318, P0327, P0401, P0408, P0420, P0427, P0840, P0843,

xxvi / *Selected Index of Locations*

P0849, P1235, C1220, S0602
Union of Soviet Socialist Republics (Subarctic)   S0602
United Kingdom   *See* British Isles
United States of America   P0125, P0134, P0140, P0156, P0201, P0202, P0203, P0207, P0212, P0220, P0229, P0231, P0235, P0242, P0246, P0254, P0308, P0315, P0316, P0322, P0403, P0414, P0415, P0416, P0423, P0502, P0509, P0511, P0514, P0519, P0522, P0530, P0609, P0805, P0811, P0818, P0819, P0823, P0827, P0833, P0834, P0835, P0904, P0905, P0908, P0911, P0913, P0920, P1004, P1005, P1008, P1015, P1024, P1026, P1104, P1110, P1117, P1225, P1226, P1302, P1303, P1305, P1306, P1312, P1313, C0102, C0201, C0304, C0501, C0516, C0517, C0701, C0704, C0803, C0902, C1201, C1226, C1401, C1502, C1610, C1707, S0101, S0105, S0202, S0208, S0210, S0211, S0212, S0219, S0222, S0404, S0406, S0408, S0502, S0701
United States of America, East Lakes   P0140, P0609, P0827, P0833, S0210
United States of America, Great Plains–Rocky Mountains   P0134, P0315, P0415, P0519, P0835, P0905, P0920, P1303, C0516, C0803, S0211, S0216, S0406
United States of America, New England–St Lawrence Valley   P0242, P0423
United States of America, New York–New Jersey   P0212, P0819, P1117
United States of America, Pacific Coast   P0125, P0229, P0316, P0416, P0913, P1302, C1401, S0219, S0225
United States of America, Southeastern   P0134, P0207, P0509, P0514, P0530, P0834, P1005, P1008, P1024, P1026, P1110, P1226, C0201, C0304, C0501, C0902, C1502, S0219, S0222
United States of America, Southern New England   P0423
United States of America, Southwestern   P0125, P0235, P1312, C0102, S0202
United States of America, West Lakes   P0156, P0246, P0254, P0502, P0511, P0522, P0911, P1024, P1104, P1306, S0408, S0701
Upper Teasdale   P0319
Utah   P0134, C0803
Uttar Pradesh   P1007, P1036, P1125, C0911, C1218, C1701

Valais   P1006
Vancouver area   P0803
Vancouver Island   S0213
Varanasi District   P1035
Venezuela   P0153, S0407
Virgin Islands   P0322

Walapane Division   P1033
Wales   *See* British Isles
Washington, D.C.   P0514, S0219
Wernecke Mountains   P0112
West Bengal   P0240
West Germany   P0537, P0814, P0932, C1602
West Indies   P0151, P0310, P0322, P0520, P1022, P1105, C0601, C0910, S0502
West Virginia   P0140
Wick   P1043
Windsor   P0802, S0209
Wisconsin   P0156, P0511, P0911

Yoshida-Cho   C0912
Yugoslavia   P0432
Yukon   P0112, P0115, P0130, P0215